ORGANOBORON COMPOUNDS IN ORGANIC SYNTHESIS

ORGANOBORON COMPOUNDS IN ORGANIC SYNTHESIS

B. M. Mikhailov
and
Yu. N. Bubnov

*N.D. Zelinsky Institute of Organic Chemistry,
USSR Academy of Sciences*

Revised English Edition
Translated from the Russian by
Leo Wiener

Harwood Academic Publishers GmbH
Poststrasse 22
7000 Chur
Switzerland

42 William IV Street
London WC2N 4DE
England

58, rue Lhomond
75005 Paris
France

P.O. Box 15053
3501 BB Utrecht
The Netherlands

P.O. Box 786
Cooper Station
New York, NY 10276
United States of America

Library of Congress Cataloging in Publication Data

Mikhaĭlov, B. M. (Boris Mikhaĭlovich)
Organoboron compounds in organic synthesis.

Translation of: Bororganicheskie soedineniia v organicheskom sinteze.
1. Organoboran compounds. 2. Chemistry, Organic—Synthesis. I. Bubnov, Yu. N. (Yuriĭ Nikolaevich) II. Title.
QD412.BIM5413 1983 547′.05671 83-6094
ISBN 3-7186-0113-3

Printed in Great Britain by Bell & Bain Ltd., Glasgow

Table of Contents

PART II

Boracyclanes

PART III

Alkenylboranes

PART IV

Allylboranes

PART V

Alkynyl- And Allenylboranes

PART VI

Esters Of Thioboric And Organothioboron Acids

Preface

The investigation of organic boron compounds, which were discovered as early as the 1860s, was for a long time episodic in nature and gained in scope only during the last 25–30 years. During this period, the chemistry of organoboron compounds has developed into a large branch of organic chemistry that is of great theoretical and applied significance.

The achievements in the development of methods of synthesis and related theoretical problems of the structure and reactivity of organoboranes as electron-deficient systems have not only made possible the evolution of the chemistry of organoboron compounds but have also served as a prerequisite for the development of new methods of preparative organic chemistry.

In connection with the great achievements in the area of the synthesis and use of boron compounds in synthetic organic chemistry, it has become necessary to familiarize broad groups of chemists with these new, interesting, and promising directions.

This monograph presents a systematic treatment of a vast amount of material on the synthesis, reactivity, and use of different classes of boron derivatives in many different areas of preparative organic chemistry. Organoboron compounds are used as synthetic reagents in the chemistry of acyclic and cyclic saturated, olefinic, acetylenic, allenic, dienic, and polyenic hydrocarbons and of oxygen, sulfur, nitrogen, and other functional derivatives of different types of hydrocarbons, various heterocyclic compounds, and also in the chemistry of terpenes, amino acids, steroids, sugars, and other natural substances.

This book considers the characteristics and areas of application of hydroboration, allylboration, and thioboration; stereospecific methods for the introduction of functional groups into simple and complex natural substances; and methods of alkylation, arylation, homologization, and cyclization of many classes of substances of practical importance with the use of boron derivatives. It also describes original methods of synthesis of cyclic and cage structures from simple unsaturated compounds. The book covers questions of the use of organoboron compounds in asymmetric synthesis and also deals with new methods for the protection of functional groups and the generation of radicals.

It also considers some general questions of the chemistry of organoboron compounds that are necessary for an understanding of the specific features of their chemical behavior and also the reaction mechanisms, which are in many cases of general theoretical value.

In the preparation of this monograph, difficulties were encountered in the systematization of the diverse and interrelated problems of the synthesis and reactivity of different types of organoboron compounds. The authors considered it to be most rational to present the material by classes of organoboranes, which are the starting materials in the synthesis of different types of organic compounds.

The monograph is intended for a broad group of scientific workers, instructors, postgraduate students, and students at chemical institutes working in the area of organic and organometallic chemistry.

The authors consider it their duty to express their deep appreciation to V. A. Kasparov, V. G. Kiselev, N. A. Nikolaeva, T. V. Potapova, O. D. Smirnova. K. L. Cherkasova, and E. A. Shagova for their great help in the preparation of the manuscript for publication.

Introduction

The use of organoboron compounds as reagents has enriched organic chemistry with original synthetic routes that successfully complement the methods of synthesis based on organic compounds of magnesium and of alkali metals.

The chemistry of organic boron compounds, because of its original nature, occupies a special place among the other branches of organic chemistry. Its peculiar features result from the electronic structure of the boron atom, which contains three valence electrons and four orbitals, of which three are occupied by electrons while the fourth is vacant. As a rule, in the compounds BX_3 (*1*), the boron atom is in a trigonal valence state and is located in the same plane as the substituents, while the vacant orbital is perpendicular to this plane. An exception to this rule is represented by the electronic structure of 1-boraadamantane (*2*), in which, because of the molecular geometry, the boron atom is present in a tetrahedral valence state, and the vacant orbital represents an sp^3 orbital, rather than a p orbital. The most important feature of the compounds BX_3 consists of their ability to be transformed by means of the vacant orbital into coordinate compounds of different types with a boron atom in a tetrahedral valence state. These include neutral complexes (*3*), ionic compounds with a tetravalent, negatively charged boron atom (*4*), and ionic compounds with boron as the central atom of a complex cation (*5*).

X_3B

1

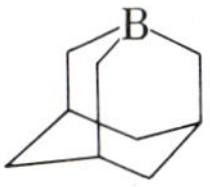

2

$X_3B \leftarrow L$

3

$[X_4B]^- M^+$

4

$[X_2BL_2]^+ Y^-$

5

However, this does not exhaust the variety of types of boron compounds.

The partially substituted borane derivatives R_2BH and RBH_2 form dimers with three-center two-electron bonds between the boron and hydrogen atoms (*6*). Different derivatives of borane containing nitrogen, phosphorus, or sulfur exist in the form of dimers, in which a special type of coordination interaction takes place with the participation of electrons from both the donor atom and the boron atom ("semicoordinate bonds") (7).

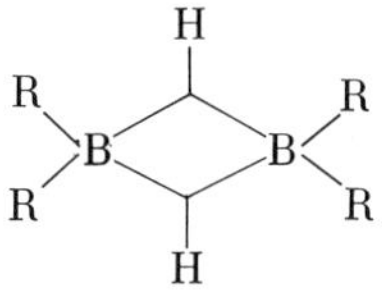

6

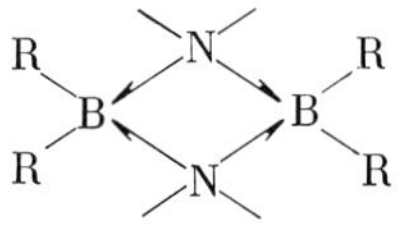

7

The ability to form coordination compounds with different characteristics, which is inherent in boron, is responsible for the unique nature of the chemistry of this element, which differs fundamentally from the chemistry of the organic compounds of magnesium and alkali metals.

The reactions of organoboron compounds can be divided into several general types:

1. Electrophilic deboration (protolysis, reaction with some salts, halogenation of arylboranes, etc.)
2. Electrophilic addition to some compounds with multiple bonds
3. Radical reactions, in which the primary act is a homolytic cleavage of the B—C bond (autoxidation; 1,2-, 1,4-, and 1,6-addition to compounds with multiple bonds; coupling of radicals under the action of silver salts, aqueous H_2O_2, etc.)
4. Substitution reactions at the α-carbon atom (radical halogenation of alkylboranes, abstraction of α-hydride hydrogen)
5. Reactions with the participation of a β-hydrogen atom (transalkylation of trialkylboranes by higher olefins, dienes, and acetylenes; reduction of carbonyl compounds, nitriles, etc.)
6. Transformation of an organic radical (hydrogenation, halogenation, nitration, and other conversions).

The initial act in most reactions of organoboranes is the formation of adducts of the type $R_3B \cdot AX$ or $R_3\overline{B}$-AX. The further behavior of the adducts is determined by the nature of the base, the organoborane, and the conditions of the process. Some rearrangements leading to the formation of new C—C, C—O, C—N, and C—S bonds are of fundamental importance. One of the most important types of conversion is the intramolecular 1,2-anionotropic rearrangement of organoboron compounds containing electronegative groups on the α-carbon atom:

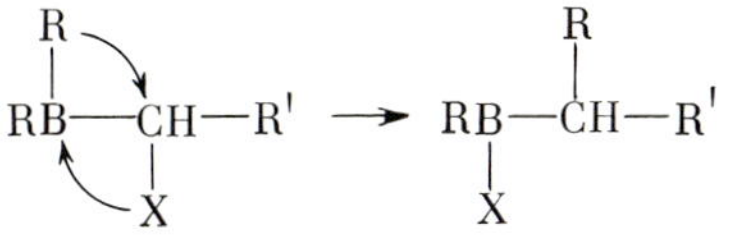

$$R_3\bar{B}\text{—}\underset{\displaystyle X}{\underset{|}{CHR'}} \longrightarrow R_2\bar{B}\text{—}\underset{\displaystyle R}{\underset{|}{CHR'}} \quad (\text{with X on B})$$

(X=Hal, OR, SR).

In compounds of trivalent boron, a rearrangement of this type takes place under the action of nucleophiles (attack on the boron) or of electrophiles (attack on the X); in complex anions, migration usually proceeds spontaneously.

The oxidation of organoboranes with amine N-oxides, rearrangements of peroxyboranes, and reactions with chloramine, sulfenyl chlorides, ylides, diazo compounds, and a number of other conversions can be considered to be of the same type:

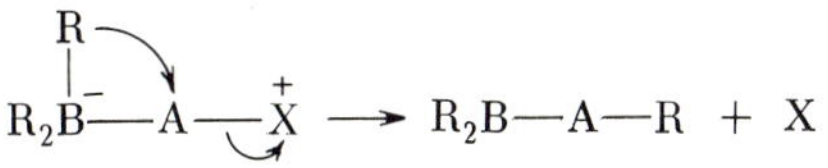

($AX = R_3NO$, NH_2Cl, RSCl, $\bar{C}H_2\overset{+}{S}R_2$, N_2CHR).

A similar rearrangement takes place in adducts of organoboranes with isonitriles and carbon monoxide:

$$R_3\bar{B}\text{—}\overset{+}{A}{=}X \rightarrow R_2B\text{—}\overset{\displaystyle R}{\overset{|}{A}}{=}X$$

(AX = CO, RNC).

In the presence of bases (water, alcohols), two or even three radicals migrate from the boron to the α-carbon atom.

Many adducts $R_3B{\cdot}AX$ or salts $M^+[R_3\bar{B}\text{-}AX]$ are stable under ordinary conditions and rearrange only under the action of electrophilic reagents:

$$R_3B + \bar{C}{\equiv}X \rightarrow R_3\bar{B}\text{—}C{\equiv}X \xrightarrow{E^+} R_2B\text{—}\overset{\displaystyle R}{\overset{|}{C}}{=}X\text{—}E \qquad (C{\equiv}X = CN,\ C{\equiv}CR'),$$

$$R_3B + \bar{A}{=}X \rightarrow R_3\bar{B}\text{—}A{=}X \xrightarrow{E^+} R_2B\text{—}\overset{\displaystyle R}{\overset{|}{A}}\text{—}X\text{—}E \qquad (AX = CH{=}CR_2,\ Ar),$$

$$R_3B + \bar{A}\text{—}X \rightarrow R_3\bar{B}\text{—}A\text{—}X \xrightarrow{E^+} R_2B\text{—}A\text{—}R + E\text{—}X \qquad (AX = HO_2^-).$$

In some conversions, the electrophilic reagent abstracts an α-hydride hydrogen from the tetraalkylborate anion (oxidation with hydrogen peroxide, reactions with benzyl halides):

$$R_3\bar{B}CH_2R' \xrightarrow{E^+} R_3\overset{-+}{B}CHR' \rightarrow R_2B\overset{\overset{\displaystyle R}{|}}{C}HR'$$

* * *

The methods of preparations, properties, and uses in organic synthesis of alkyl- and arylboranes are presented in Part I.

Methods for the preparation of trialkyl- and triarylboranes based on reactions of boron halides or of borate esters with organic compounds of magnesium, lithium, aluminum, and other metals are characterized and the most important methods of preparation and the properties of esters of organoboron acids are covered (Chapter 1).

Methods of synthesis of organoboranes by the hydroboration of hydrocarbons with a C=C bond are considered (Chapter 2). The organoboranes obtained by this method, as a rule, undergo further conversions without isolation in the free state. Their oxidation with hydrogen peroxide leads to alcohols. This method of olefin hydration has found broad application, including its use for different classes of natural compounds. The method of synthesis of the simplest trialkylboranes by the hydroboration of olefins with diborane has the disadvantage that individual compounds can be obtained only from olefins of the type $R_2C{=}CH_2$ and $RCH{=}CHR$, while the hydroboration of α-olefins ($RCH{=}CH_2$) and also that of unsymmetrically substituted non-terminal olefins ($RCH{=}CHR'$) leads to a mixture of isomeric trialkylboranes in a ratio of 7:93 to 49:51.

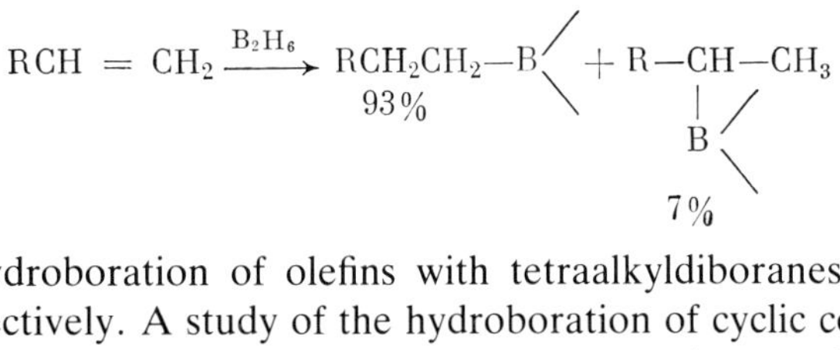

The hydroboration of olefins with tetraalkyldiboranes proceeds more selectively. A study of the hydroboration of cyclic compounds with an exomethylene group and of cyclenes has made a great contribution to the stereochemistry of addition reactions at a multiple carbon-carbon bond.

Hydroboration with tetraisopinocampheyldiborane, one of the

most common and most readily available chiral reagents, has found application in the synthesis of optically active compounds and also for a kinetic separation of alkenes, allenes, and cyclenes.

Functionally substituted organoboranes obtained by the hydroboration of functional derivatives of olefins are of great preparative importance (Chapter 3). The hydroboration of enamines, followed by acidolysis, leads to olefins. In this way, it is possible to achieve a smooth transition from ketones to unsaturated compounds, which is particularly important in steroid chemistry, and also to obtain unsymmetrically substituted compounds, particularly those of the terpene series, from symmetrically substituted alicyclic ketones.

The hydroboration of ω-aminoalkenes, followed by oxidation of the organoboranes formed, makes it possible to synthesize α,ω-amino alcohols; from the vinyl esters, alkoxy alcohols of the aliphatic and alicyclic series can be obtained in this way, and trialkylsilyl enol esters of alicyclic ketones can also be converted to 2-trialkylsilyloxycyclanols. Hydroboration-oxidation, as a method for the hydration of allyl esters, is finding preparative use for a number of acyclic and alicyclic compounds and also in carbohydrate and steroid chemistry. Esters of unsaturated acids are converted to alcohols and diols by the action of diborane and subsequent oxidation, and protolysis of the intermediately formed organoboranes produces olefinic compounds. These reactions are being successfully used in terpene and steroid chemistry.

Hydroboration-oxidation is a valuable preparative method for the hydration of unsaturated alcohols, mercaptans, aldehydes, and ketones, making it possible to carry out a transition from these classes of compounds to bifunctional derivatives of acyclic and alicyclic hydrocarbons. This method is finding use for the hydration of unsaturated carboxylic acids and their derivatives.

Methods for the synthesis of cyclopropane compounds, based on the hydroboration of allyl halides, have been developed. With the use of diborane, it is possible to carry out the hydration and conversion to unsaturated aliphatic and aromatic alcohols of different oxygen-containing heterocyclic compounds: Δ^2-dihydropyran, dihydrofurans, chromenes, flavenes, and coumarins. Hydroboration-oxidation is used for the hydration of different derivatives of tetrahydropyridine, tropidine, and other nitrogenous heterocyclic compounds.

Trialkylboranes with secondary and tertiary hydrocarbon radicals are capable of isomerization to organoboranes with primary radicals upon heating. On the basis of this property of trialkylboranes, olefins with a non-terminal double bond can be converted to primary alcohols by hydroboration-oxidation (Chapter 4).

Different methods of oxidation of organoboranes are considered in Chapter 5. Autoxidation, proceeding by a radical mechanism, is used for the preparation of alcohols and hydroperoxides and for initiating the polymerization and radical-addition reactions of organoboranes to functional olefins and α,β-unsaturated carbonyl compounds. Oxidation with hydrogen peroxide in alkaline, acid, or buffer solution and oxidation with amine N-oxides in hydrocarbons is used for the transition from organoboranes to alcohols, phenols, aldehydes, ketones, and thiolenones. The oxidation of *sec*-alkyl- and of cycloalkylboranes with chromic acid or manganese compounds leads to the corresponding ketones.

The protolytic cleavage of B—C bonds in organoboranes by different reagents is used for the synthesis of hydrocarbons and their functional derivatives. A combination of protonolysis and hydroboration is used as a method of cis-hydrogenation of olefins and acetylenes and also for the monohydrogenation of allenes (Chapter 6).

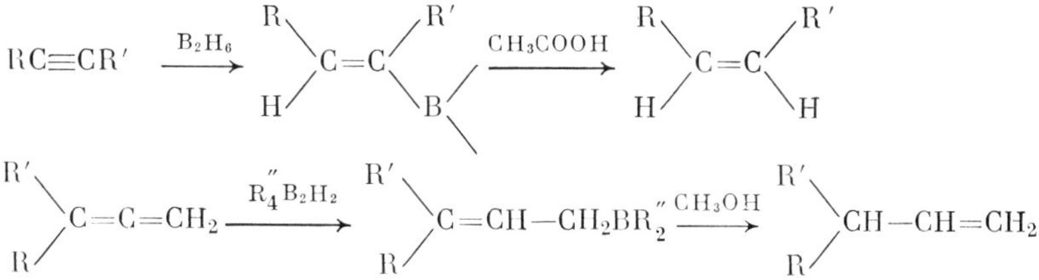

The monodealkylation of trialkylboranes, catalyzed by mercaptans and pivalic acid, is used in the quantitative determination of water or of hydroxy groups in inorganic and organic compounds and also for the protection of hydroxy groups in alcohols, phenols, polyols (sugars), and amino acids.

The boryl group in organoboranes may be easily replaced by an amino group by the action of chloramine, hydroxylamine derivatives, or copper phthalimide:

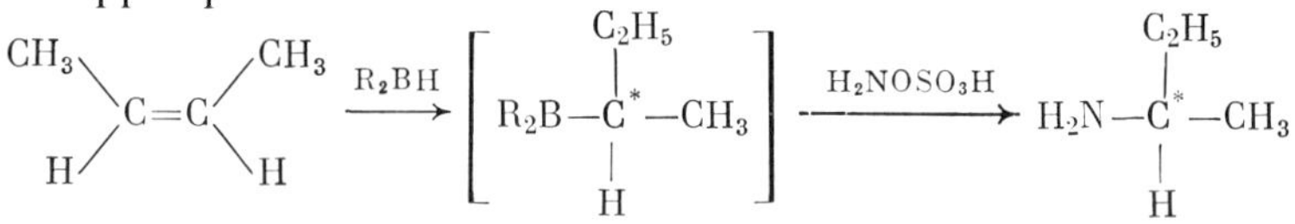

Secondary and tertiary amines, aziridines, amides, and enamines are also prepared from organoboranes (Chapter 7):

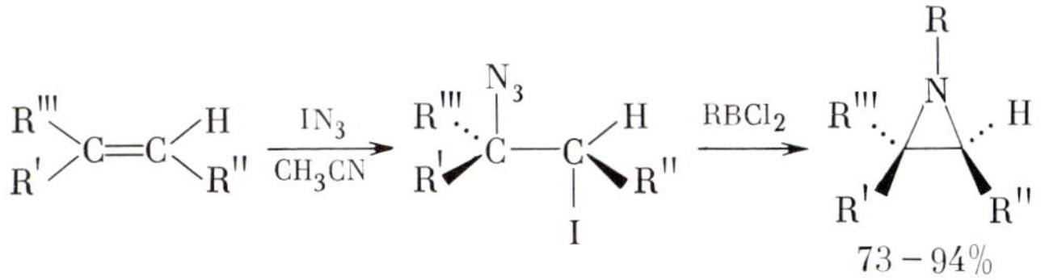

The halogenation of alkyl-, cycloalkyl-, and arylboranes (Chapter 8) is used for the preparation of alkyl and cycloalkyl halides, branched alcohols, ketones, α-olefins, 1,5-dienes, dibenzyl compounds, and heterocyclic compounds and also for the hydrohalogenation of olefins. In this case, the intramolecular 1,2-anionotropic rearrangements of the α-haloalkylboranes proceeding under the action of electrophilic and nucleophilic reagents are of great importance:

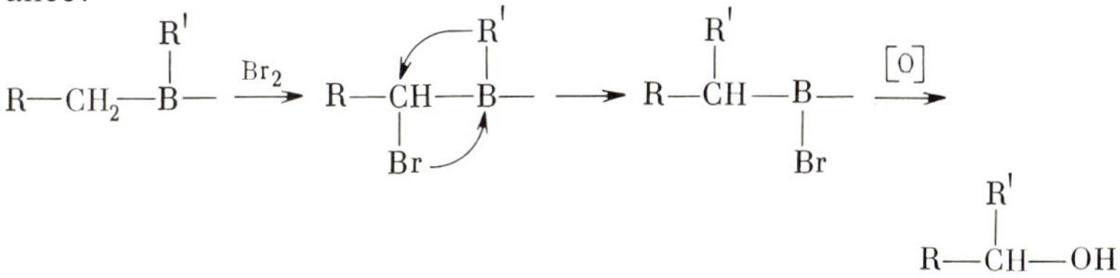

Reactions of this type proceed by the coupling of organoborane radicals (with the simultaneous introduction of a functional group). The same chapter also presents data on the electrophilic halodeboration of arylboranes and on the halogenation of these compounds in the nucleus and in the side chain.

Trialkyl- and tricycloalkylboranes, in the presence of radical initiators (O_2), alkylate α,β-unsaturated carbonyl compounds, including those of the steroid series, Mannich bases, and quinones (Chapter 9).

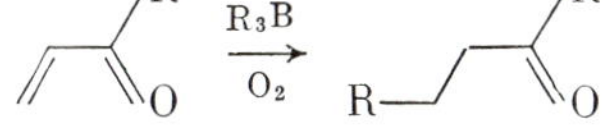

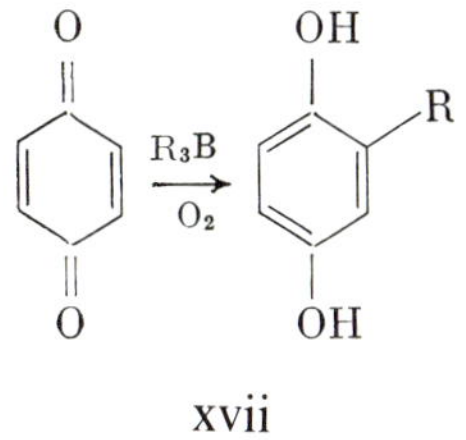

Reactions with carbonyl compounds are used for the preparation of higher alcohols, saturated and ω-unsaturated ketones, 2-alkylquinones, olefins, and polyenes and also for the counterthermodynamic isomerization of internal olefins to terminal olefins and of methylcyclenes to methylenecyclanes. The same chapter includes a consideration of the properties (hydrolysis, deuterolysis, halogenation, alkylation, condensation with aldehydes, ketones, and nitriles) of vinyloxyboranes–boron ethers of enols.

Depending on the conditions, the carbonylation of organoboranes proceeds through the migration of one, two, or three radicals from boron to the carbon atom of the CO molecule. The reactions of trialkyl- and triarylboranes with isonitriles proceed analogously to the carbonylation. Both of these reactions are used as the basis for simple methods of synthesis of alcohols, dialkyl- and trialkylcarbinols, aldehydes, ketones, olefins, heterocyclic compounds, and a number of other types of organic compounds (Chapter 10).

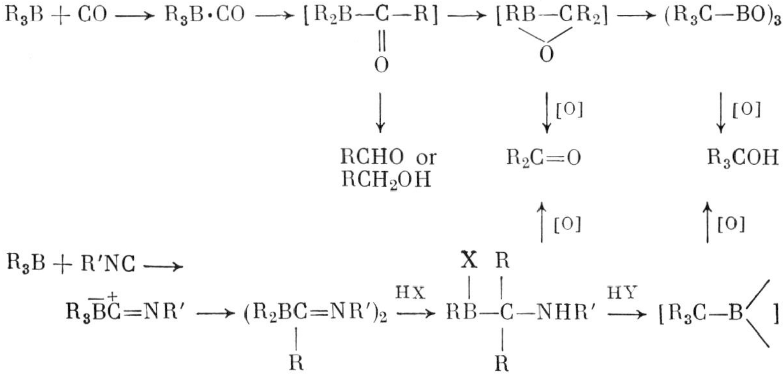

Symmetrical and unsymmetrical ketones, cycloalkanones, trialkylcarbinols, and amines can also be obtained from alkali metal cyanotrialkylborates:

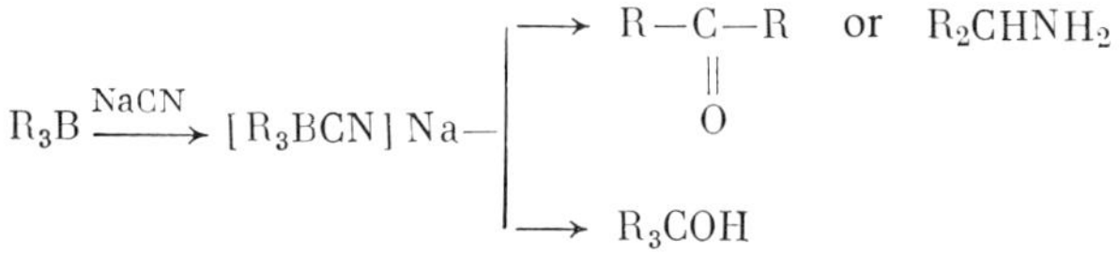

Organoboron compounds are capable of alkylating and arylating diazo compounds, ylides (Chapter 11), halomethanes, α-halocarbonyl compounds, α-halonitriles, thioacetals, and sulfonyl compounds (Chapter 12). These conversions are used as the basis for simple methods of synthesis of primary, secondary, and tertiary alcohols, aldehydes, higher ketones, derivatives of carboxylic acids, their nitriles, dinitriles, cyano esters, and sulfonyl compounds:

$$R_3B + N_2CHCHO \longrightarrow RCH_2CHO,$$

$$R_3B + (CH_3)_2\overset{+}{S}\overset{-}{C}HCOOR \longrightarrow RCH_2COOR,$$

$$R_3B + X_3CH \longrightarrow \xrightarrow{[O]} R_2CO,$$

$$R_3B + BrCH_2COX \longrightarrow RCH_2COX.$$

The reactions of organoboranes with unsaturated, aromatic, and cyclopropane compounds are described in Chapter 13. Photochemical and oxygen-initiated addition reactions of trialkylboranes with cyclohexene compounds, 1,3-butadiene monoxide and vinylacetylene oxide, acrylonitrile, α-nitroolefins, α,β-unsaturated sulfones, and sulfoxides are used in the synthesis of *cis*-2-alkylhexanols, allylic and allenic alcohols, nitriles, nitroalkanes, and *trans*-olefins.

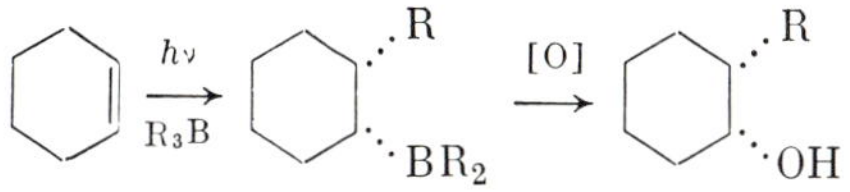

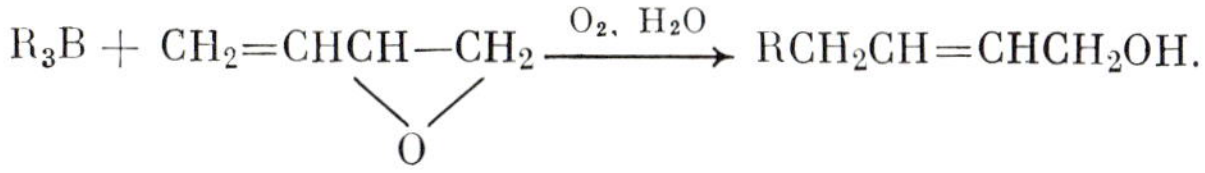

The ability of aryl- and benzylboranes to add to acetylene derivatives represents the basis for a number of methods for the preparation of derivatives of styrene and acetophenone. The reaction of trialkylboranes with 1-methylcyclopropene is the basis of a simple method for increasing the length of the carbon chain of olefins and alkyl halides by a methallyl group and, in the reaction with isopropenyllithium, by an isopropenyl group.

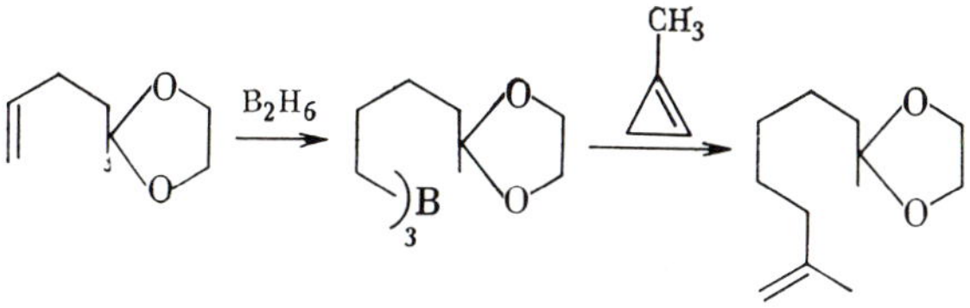

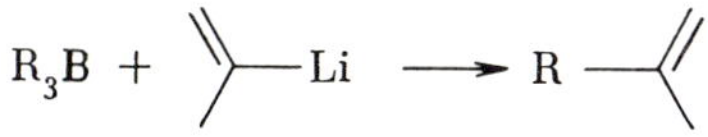

Trialkylboranes can be used for the alkylation of acridines to 9-alkylacridanes in a high yield and for the dialkylation and reductive dialkylation of aromatic compounds.

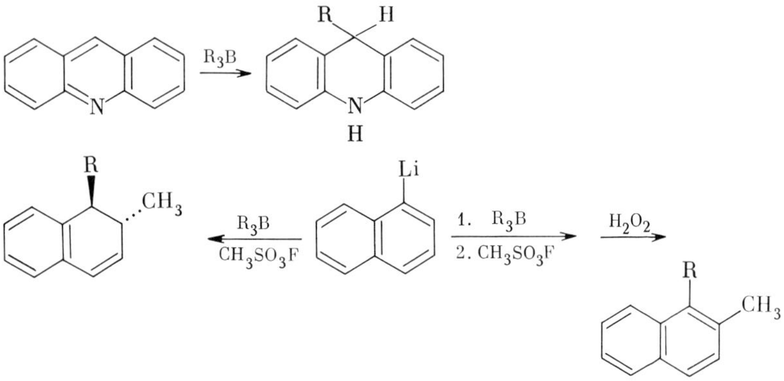

The reactions of organoboranes with metal salts (Chapter 14) are used for the preparation of several types of organometallic (Ag, Hg, Cu, Pb, Tl, etc.) and metallocene compounds and also for the transition from boron compounds to higher hydrocarbons, organic halides, thiocyanates, phenols, and esters. Under the action of silver salts, alkylboranes generate radicals; a combination of hydroboration and mercuration is used for the anti-Markownikoff hydroacylation of olefins.

Part II is devoted to a consideration of the methods of synthesis and chemical conversions of monocyclic, bicyclic, and tricyclic compounds containing boron in the ring. Boracyclanes are obtained with the use of dimetalloorganic compounds, by the thermal conversions of organoboranes, and by the hydroboration of acyclic and alicyclic dienes and polyenes. The oxidation of boracyclanes produces α,ω-bifunctionally substituted aliphatic hydrocarbons, while carbonylation leads to cyclic alcohols and ketones:

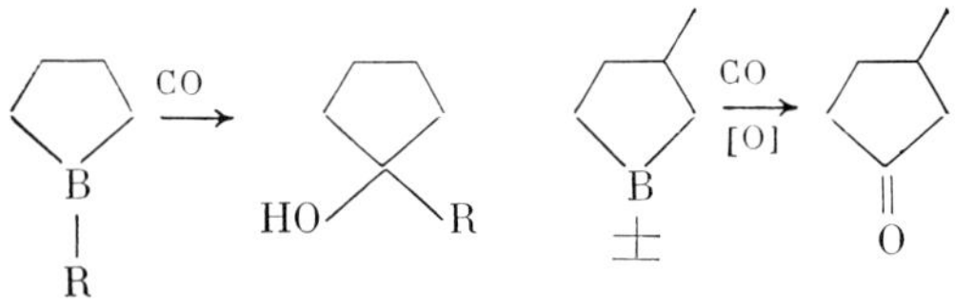

The bromination of boracyclanes, followed by oxidation, can form ω-bromoalkanols:

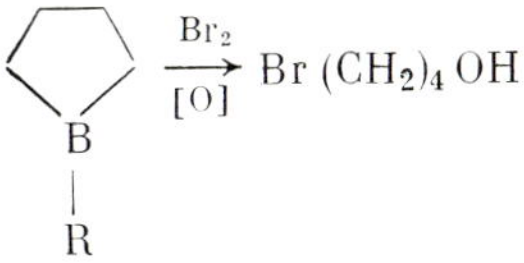

and cyclic alcohols and ketones:

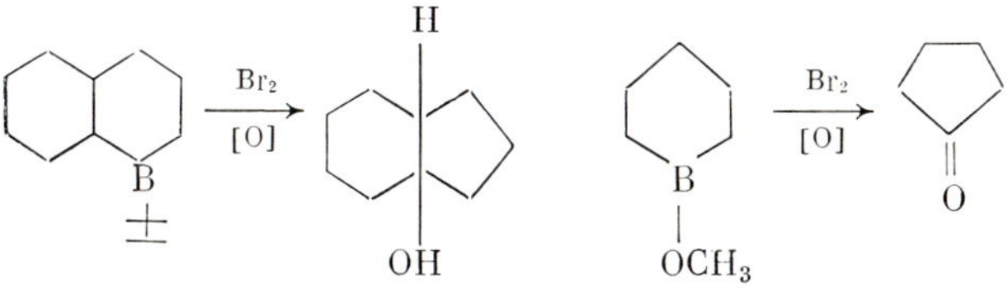

1-Alkenylboranes, obtained from alkenyl compounds of metals, 1-alkynyltrialkylborates, or by the hydroboration of acetylenic hydrocarbons and their functional derivatives are finding different synthetic applications (Part III).

The protolytic cleavage of 1-alkenylboranes leads to olefinic hydrocarbons and their deuterated derivatives:

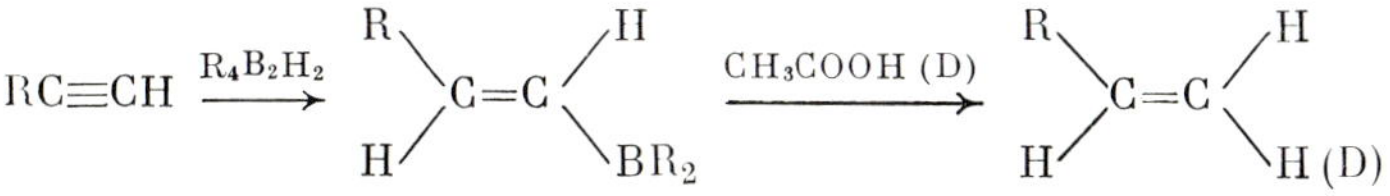

Starting from the corresponding functionally substituted acetylenic hydrocarbons, this method produces *cis*-haloolefins and *cis*-α,β-unsaturated acids, while the hydroboration-protolysis of vinylacetylenes forms dienes and alcohols. The oxidation of 1-alkenyldialkylboranes leads to aldehydes and esters of α-keto acids. The hydroboration-oxidation of diacetylenic hydrocarbons produces α,β-acetylenic ketones.

Bromoolefins (*cis* or *trans*) are synthesized by the bromination of *trans*-1-alkenylboranes:

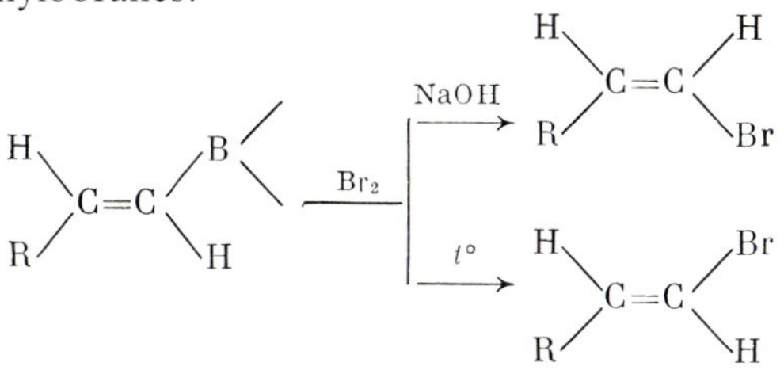

Under the action of iodine in the presence of alkali, *trans*-1-alkenylboranes are converted to *trans*-olefins, *trans*-alkenylboronic acids to *trans*-iodoolefins, and tris(1-alkenyl)boranes to *cis*,*trans*-1,3,-dienes:

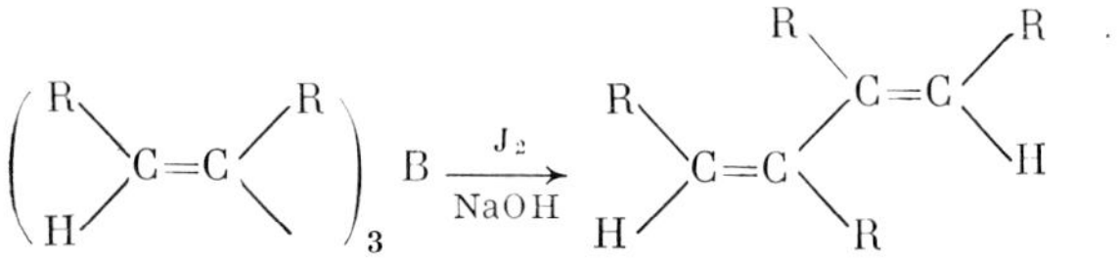

A stereoselective general method for the synthesis of conjugated *trans*,*trans*-dienes and of *trans*-alkenyl ketones based on the hydroboration of haloacetylenes has been developed.

trans-1,4-Dialkyl-1,2,3,-butatrienes are formed from bis(α-iodo-1-alkenyl)alkylboranes under the action of sodium alkoxide. Carbethoxy-substituted vinyldialkylboranes serve as a starting material for the synthesis of β-hydroxycarboxylic acids.

Chloro-substituted 1-alkenylboranes, formed by the action of tetraalkylboranes on propargyl chloride and its homologs, undergo decomposition, with the formation of allene hydrocarbons, under the action of an anion:

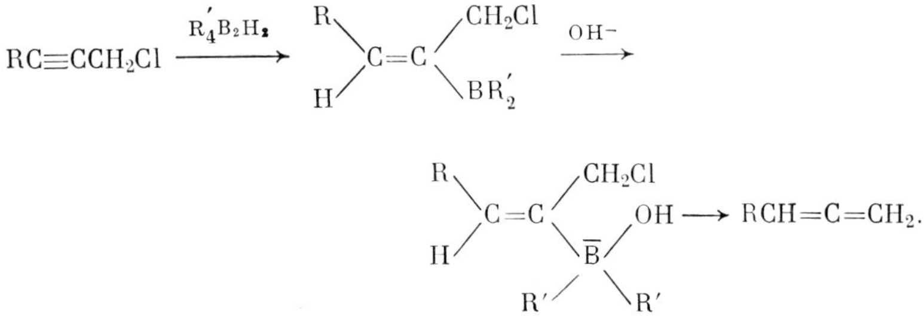

1,1-Diboron compounds, obtained by the hydroboration of 1-alkenylboranes, serve as starting materials in the synthesis of malonic acids, olefins, and alcohols and also of derivatives of cyclopropane and cyclobutane.

2-Arylglutaric acids and unsaturated organomercury and organothallium compounds can be obtained from 1-alkenylboranes.

A number of methods for the synthesis of organic compounds have been developed on the basis of metal 1-alkynyltrialkylborates which, under the action of electrophilic reagents, undergo anionotropic rearrangement with the migration of an alkyl group from the boron atom to an α-carbon, as a result of which 1-alkenylboranes are formed:

$$R_3B + R'C{\equiv}CNa \longrightarrow [R_3BC{\equiv}CR']^- Na^+ \xrightarrow{R''J}$$

$$\longrightarrow R_2B{-}\underset{\displaystyle R}{\underset{|}{C}}{=}CR'R'' \xrightarrow{H^+} RCH{=}CR'R''.$$

This method has been used for the preparation of olefinic hydrocarbons, saturated and α,β-unsaturated ketones, unsaturated alcohols, allyl methyl ethers, 1-alkenylsilanes, allylamines, allenic hydrocarbons, and other compounds.

Allylboranes occupy a special place among the organic boron derivatives (Part IV).

In reactions with different types of carbonyl compounds, allylboranes behave as organometallic reagents, and their use for synthetic purposes frequently has a number of advantages over allylic Grignard reagents. A specific feature of allylboranes is their ability to react with compounds containing multiple carbon-carbon bonds. The reaction of allylboranes with acyclic and cyclic vinyl ethers is a convenient preparative method for the synthesis of 1,4-dienes and their functional derivatives:

$$(CH_2{=}CH{-}CH_2{-})_3B + CH_2{=}CHOR \longrightarrow (C_3H_5)_2BCH_2\underset{\displaystyle OR}{\underset{|}{C}H}CH_2CH{=}CH_2 \longrightarrow$$

$$\longrightarrow CH_2{=}CHCH_2CH{=}CH_2 + (C_3H_5)_2BOR.$$

Allylcyclopropanes and 2-substituted derivatives of 1,5-hexadiene are obtained by the action of allylboranes on cyclopropene hydrocarbons.

Particularly valuable from the preparative aspect are the reactions of allylboranes with acetylenic compounds, which proceed in three consecutive stages, with the formation of 2-substituted derivatives of diallyl(1,4-pentadien-1-yl)borane, 3-substituted derivatives of 1,5-diallyl-1-boracyclohex-2-ene, and 7-substituted derivatives of 3-allyl-

3-borabicyclo[3.3.1]non-6-ene, respectively:

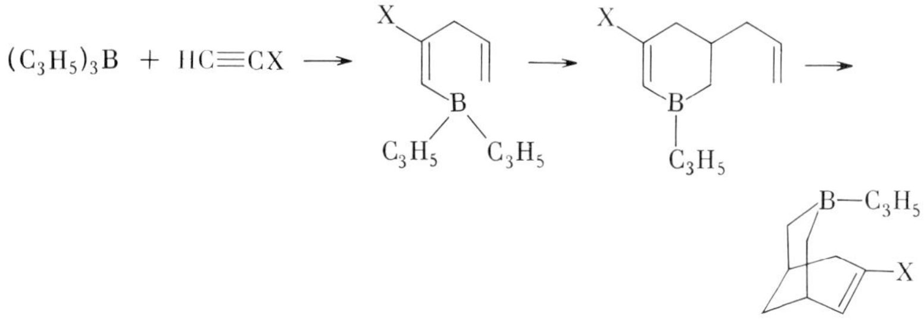

Each of these types of organoboranes is used successfully for the synthesis of different classes of organic compounds: 1,4-pentadiene and 1,6-heptadiene hydrocarbons, 2-alkoxy-1,4-pentadienes, dienic alcohols, unsaturated aldehydes and ketones, unsaturated keto alcohols, γ-keto acids, trienic hydrocarbons and alcohols, and derivatives of cyclohexanone, biphenyl, and other cyclic systems. The allylboron-acetylene condensation is a stereospecific method for the synthesis of different cyclohexene and cyclohexane derivatives.

Allylboranes react with allenic hydrocarbons with the formation of derivatives of 7-methylene-3-borabicylo[3.3.1]nonane, the hydroboration of which leads to 1-boraadamantane compounds:

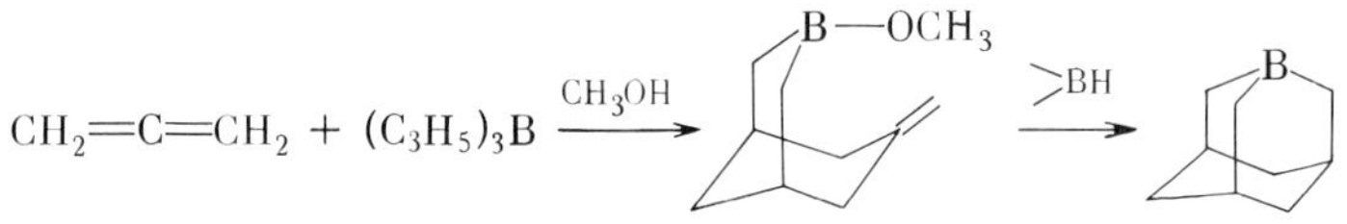

1-Boraadamantane and its 4-functional derivatives can also be obtained from triallylborane and propargyl derivatives.

The reaction of 1-boraadamantane and its derivatives with carbon monoxide gives rise to adamantane compounds.

The synthesis and some applications of alkynyl and allenyl boron derivatives are described in Part V. However, little attention has been paid to chemical conversions of these compounds up to now, despite their ready availability. Synthetic use is made of esters of allenylboronic acids, whose reactions with carbonyl compounds lead to alcohols of the propargyl and allenyl series. The use of 1-alkynyltrialkylborates for the synthesis of alkenylboranes is described in Part III.

The methods of synthesis and chemical conversions of esters of thioboric and organothioboronic acids are considered in Part VI. Thioboranes are used for the preparation of thioketals and thioacetals and for the synthesis of vinyl sulfides and 1-alkoxyvinyl(alkyl) sulfides, disulfides, thiophene derivatives, substituted thio acids, β-imino acids, sulfur-containing ketones and keto alcohols, analogs of coenzyme A, and alkyl dithiocarbamates. Thioboranes open the oxirane ring mainly or completely with inversion. The reaction of trialkylboranes with sulfur can be used for the synthesis of mercaptans and disulfides.

The development of methods of organic synthesis using boron compounds is receiving ever increasing attention and new advances in this rapidly developing branch of chemistry should be expected.

The Nomenclature Of Boron Compounds

The nomenclature of boron compounds used in synthetic organic chemistry and considered in the present monograph has been mainly established and is generally accepted. However, some types of boron derivatives are still named differently; this frequently causes difficulties, particularly when no structural formulas are given in the text.

For the naming of organic boron compounds in the monograph, we have mainly used the rules and recommendations of the International Union of Pure and Applied Chemistry (IUPAC) [1–8]. It should be noted, however, that these rules are not always simple to apply in organoboron chemistry, particularly in the case of boracyclanes.

Acyclic compounds of trivalent boron. Acyclic compounds of trivalent boron are considered as derivatives of borane, BH_3. This nomenclature is accepted for all compounds containing one or two B—H bonds.

BH_3	borane
RBH_2	alkylborane (arylborane)
R_2BH	dialkylborane (diarylborane)
H_2BNH_2	aminoborane
H_2BNHR	alkylaminoborane
$HB(NR_2)_2$	bis(dialkylamino)borane
$RHBNR_2$	dialkylamino(alkyl)borane
H_2BPH_2	phosphinoborane
H_2BOR	alkoxyborane
$HB(OR)_2$	dialkoxyborane
H_2BOCOR	acyloxyborane
$HB(OCOR)_2$	diacyloxyborane
H_2BSR	alkylthioborane (alkylmercaptoborane)
$HB(SR)_2$	dialkylthioborane (dialkylmercaptoborane)
$HB(NR_2)SR$	alkylthiodialkylaminoborane (alkylmercaptodialkylaminoborane)
H_2BCl	chloroborane

$HBCl_2$	dichloroborane
RHBCl	chloro(alkyl)borane

Trialkyl and triaryl boron compounds, R_3B, have been called "trialkylboron" and "triallylboron" since the time of their discovery. At the present time, the "borane" nomenclature is accepted for these compounds and for most other derivatives of trivalent boron with a wide variety of substituents. Alkoxy and alkylthio dervatives of boron are also considered as derivatives of borane, BH_3, but these, in accordance with their chemical nature, can also be named as esters of boric, alkyl- and arylboronic, dialkyl-, and diarylborinic acids.

R_3B	trialkylborane, triarylborane
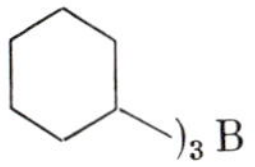	tricyclohexylborane
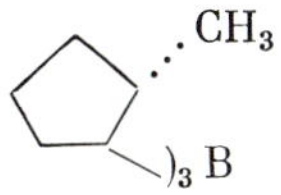	tris(*trans*-2-methylcyclopentyl)borane
$(RO)_3B$	trialkoxyborane, trialkyl borate, trialkyl orthoborate, ester of boric acid
$RB(OR)_2$	dialkoxyalkylborane, ester of alkylboronic acid
R_2BOR	alkoxydialkylborane, ester of dialkylborinic acid
$B(SR)_3$	trialkylthioborane
$RB(SR)_2$	dialkylthio(alkyl)borane, ester of alkylthioboronic acid
R_2BSR	alkylthiodialkylborane, ester of dialkylthioborinic acid
$RBCl_2$	dichloro(alkyl)borane
R_2BCl	chlorodialkylborane
$RB(OH)_2$	dihydroxy(alkyl)borane, alkylboronic acid
R_2BOH	hydroxydialkylborane, dialkylborinic acid
R_2BSH	mercaptodialkylborane, dialkylthioborinic acid
$RB(SH)_2$	dimercapto(alkyl)borane, alkylthioboronic acid

Substituents on boron, as in the case of all organic compounds, are listed by order of preference in the following sequence [7]:

1. Hydrogen
2. Methyl
3. Ethyl
4. *n*-Propyl
5. *n*-Butyl
6. *n*-Pentyl
7. *n*-Hexyl
8. Isopentyl
9. Isobutyl
10. Allyl
11. Neopentyl
12. 2-Propynyl
13. Benzyl
14. Isopropyl
15. Vinyl
16. *sec*-Butyl
17. Cyclohexyl
18. 1-Propenyl
19. *tert*-Butyl
20. Isopropenyl
21. Ethynyl (acetylenyl)
22. Phenyl
23. *p*-Tolyl
24. *p*-Nitrophenyl
25. *m*-Tolyl
26. 3,5-Xylyl
27. *m*-Nitrophenyl
28. 3,5-Dinitrophenyl
29. 1-Propynyl
30. *o*-Tolyl
31. 2,6-Xylyl
32. Trityl
33. *o*-Nitrophenyl
34. 2,4-Dinitrophenyl
35. Formyl
36. Acetyl
37. Benzoyl
38. Carboxy
39. Methoxycarbonyl
40. Ethoxycarbonyl
41. Benzyloxycarbonyl
42. *tert*-Butoxycarbonyl
43. Amino
44. Ammonium $\overset{+}{N}H_4$
45. Methylamino
46. Ethylamino
47. Phenylamino
48. Acetylamino
49. Benzoylamino
50. Benzoyloxycarbonylamino
51. Dimethylamino
52. Diethylamino
53. Trimethylammonium
54. Phenylazo
55. Nitroso
56. Nitro
57. Hydroxy HO—
58. Methoxy
59. Ethoxy
60. Benzyloxy
61. Phenoxy
62. Glycosyloxy
63. Formyloxy
64. Acetoxy
65. Benzoyloxy
66. Methylsulfinyloxy
67. Methylsulfonyloxy
68. Fluoro
69. Mercapto HS—
70. Methylthio CH_3S—
71. Methylsulfinyl
72. Methylsulfonyl
73. Sulfo HO_3S—
74. Chloro
75. Bromo
76. Iodo

The radical with the highest number is listed first.

The naming of boron-containing radicals is derived from borane, BH_3, and its derivatives.

$H_2B—$ boryl

$Cl_2B—$ dichloroboryl

$OB—$ oxoboryl

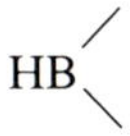 borylene

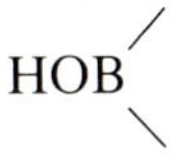 hydroxyborylene

$(CH_3)_2B—$ dimethylboryl

$(HO)_2B—$ dihydroxyboryl

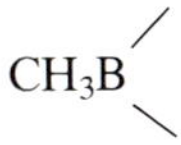 methylborylene

Di- and polyboron compounds are named in the usual manner:

$(CH_3O)_2BCH_2CH_2CH_2B(OCH_3)_2$ 1,3-bis(dimethoxyboryl)propane or tetramethyl ester of 1,3-diboronic acid.

Radicals formed by the removal of one or two outer hydrogens from the diborane molecule are named as follows:

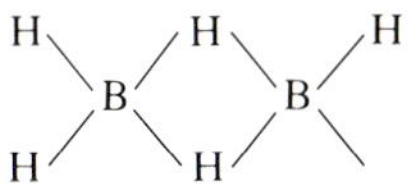

diboranyl or diboran(6)yl

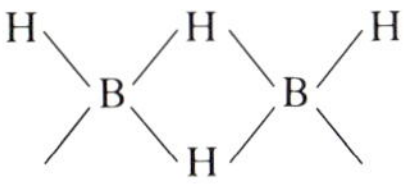

1,2-diborandiyl

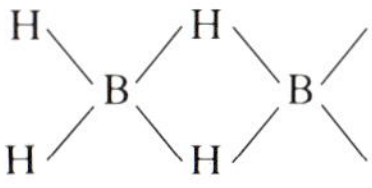

1,1-diborandiyl

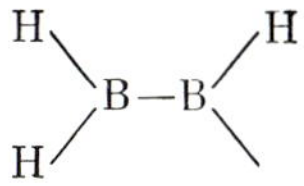

diboran(4)yl

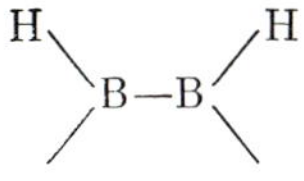

1,2-diboran(4)diyl

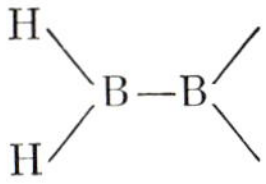

1,1-diboran(4)diyl

The use of the names of these radicals for the naming of organoboron compounds can be illustrated by means of the following examples:

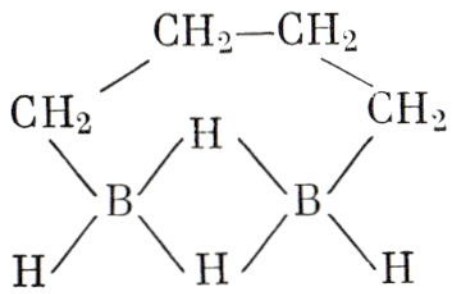

1,4-(1,2-diborandiyl)butane

1,2-tetramethylenediborane

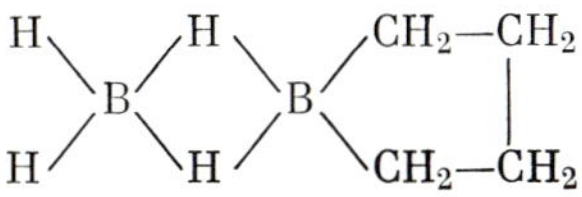

1,4-(1,1-diborandiyl)butane or

1,1-tetramethylenediborane

Substituted diboranes. Many boron derivatives containing one or two B—H bonds are dimeric compounds (as a result of two-electron, three-center bridging bonds) and are considered as derivatives of diborane.

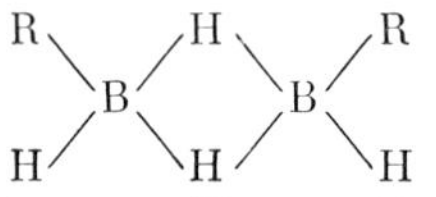

or $(RBH_2)_2$, 1,2-dialkyldiborane

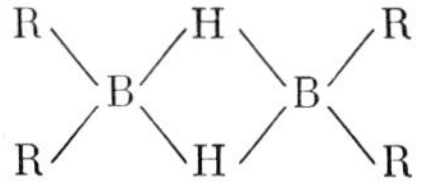

or $(R_2BH)_2$, tetraalkyldiborane

Cyclic compounds. According to the IUPAC nomenclature rules for the naming of heterocyclic compounds, the Hantzsch-Widmann system or a substitutent "a" nomenclature is used, with the two systems having equal validity [5, 6].

According to the Hantzch-Widmann system, saturated boron-containing heterocyclic compounds are named as follows: three-membered ring: borirane, four-membered ring: boretane, five-membered ring: borolane, six-membered ring: borinane, seven-membered ring: borepane, eight-membered ring: borocane, nine-membered ring: boronane, and ten-membered ring: borecane.

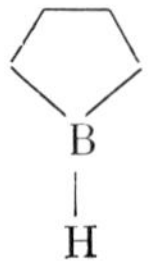

borolane
(boracyclopentane)

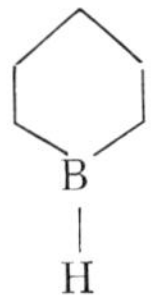

borinane
(boracyclohexane)

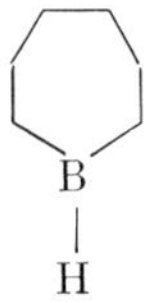

borepane
(boracycloheptane)

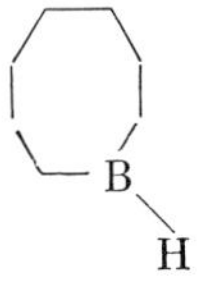

borocane
(boracyclooctane)

In the case of more complex systems, the substituent "a" nomenclature is used:

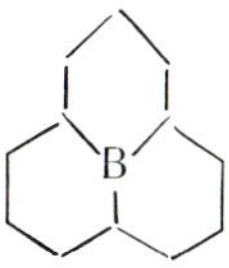

perhydro-9b-boraphenalene

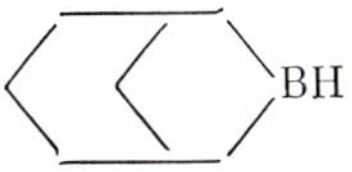

3-borabicyclo[3.3.1]nonane

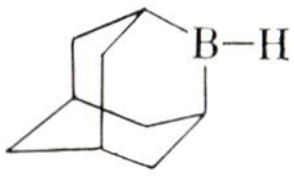

2-boraadamantane

The names of boracycloalkyl radicals are formed in the usual manner:

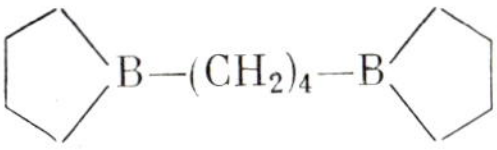

1,4-bis(1-borolyl)butane

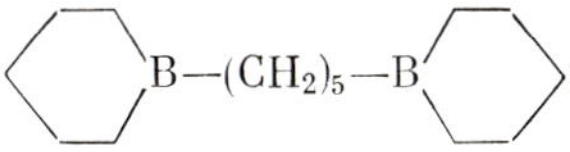

1,5-bis(1-borinyl)pentane

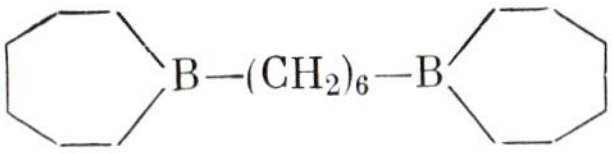

1,6-bis(1-borepyl)hexane

In this monograph, we have given preference to the "a" nomenclature in the naming of partially unsaturated heterocyclic compounds, since the use of the Hantzsch-Widmann system for these compounds is difficult in many cases.

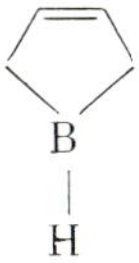

boracyclopent-3-ene (3-borolene)

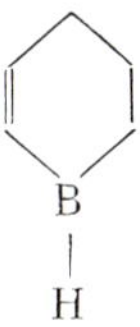

boracyclohex-2-ene (2-borinene)

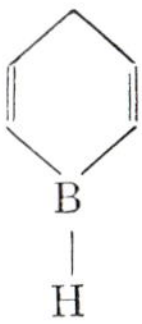

boracyclohexa-2,5-diene (2,5-borinadiene)

The names of heterocyclic compounds containing several heteroatoms in the ring are formed by adding to the name of the ring according to the Hantzsch-Widmann system the appropriate prefixes and numbers indicating the position of the heteroatoms; in this process, the heteroatoms are arranged in the following order: O (oxa), S (thia), Se (selena), Te (tellura), N (aza), P (phospha), As (arsa), Sb (stiba), Bi (bisma), Si (sila), Ge (germa), Sn (stanna), Pb (plumba), B (bora), and Hg (mercura).

All saturated five-membered rings containing several heteroatoms are referred to as borolanes, with the exception of the nitrogen-containing heterocyclic compounds, for which the name "azaborolidines" has taken root. The names of the six-membered rings are derived from borinane, those of the seven-membered rings from borepane, etc.

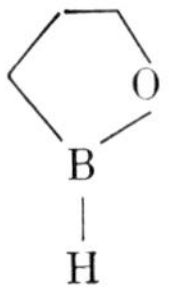

1,2-oxaborolane

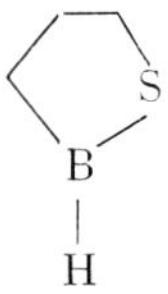

1,2-thiaborolane

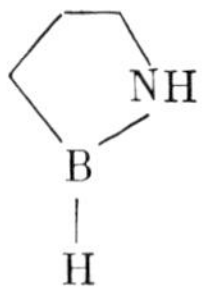

1,2-azaborolidine

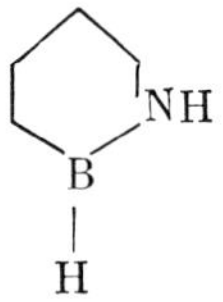

1,2-azaborinane

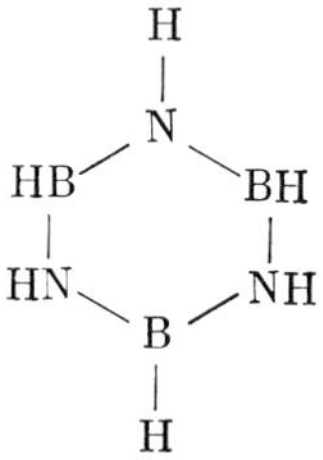

borazole

For six-membered cyclic compounds with alternating B—N bonds, the name "borazole", rather than "borazine" has taken root in the Russian and German nomenclature.

Unsaturated boron-containing heterocyclic compounds are named according to the "a" nomenclature or according to the Hantzsch-Widmann system:

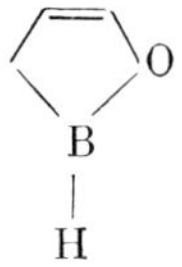

1,2-oxaboracyclopent-4-ene
(1,2-oxaborol-4-ene)

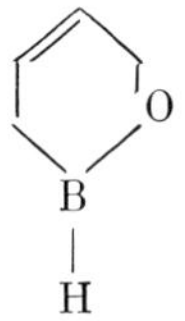

1,2-oxaboracyclohex-4-ene
(1,2-oxaborin-4-ene)

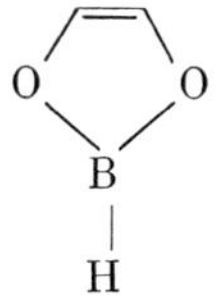

1,3,2-dioxaboracyclopent-4-ene
(1,3,2-dioxaborol-4-ene)

Neutral complexes. A large number of compounds formed by means of an electron pair of the donor and a vacant boron orbital are known. They are named as follows:

$(CH_3)_3N \cdot BR_3$	trimethylamine-borane
$CH_3NH_2 \cdot BR_3$	methylamine-trialkylborane

The method of addition of the ligand to the boron is indicated if an alternative complexing is possible:

$OC \cdot BH_3$	carbonyl(C—B)borane
$(CH_3)_3N \cdot BH_2NH_2$	trimethylamine(N—B)aminoborane
$CH_3ONH_2 \cdot BH_3$	methoxyamine(N—B)borane

A number of unsymmetrical triorganoboranes and aryl(alkyl)boranes, containing an amino group in the δ or γ position, are chelated compounds. In naming such compounds, the prefix "cyclo" is added and the type of coordination is indicated in parentheses†:

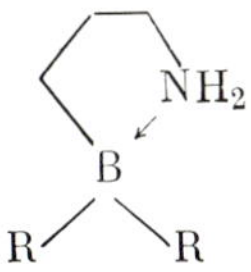

Cyclo(N—B)-3-aminopropyldialkylborane

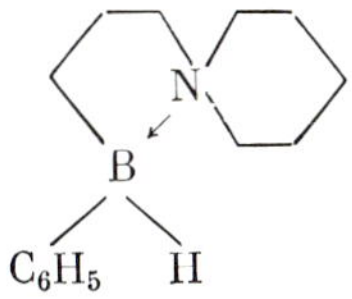

Cyclo(N—B)-3-piperidinopropyl(phenyl)borane

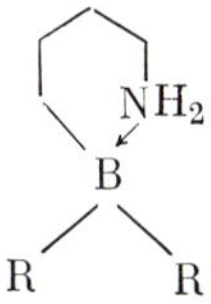

Cyclo(N—B)-4-aminobutyldialkylborane

Complex salts with tetravalent negative boron. These salts are called borates in this monograph:

$Li[B(C_6H_5)_4]$	lithium tetraphenylborate
$K[R_3BCN]$	potassium cyanotrialkylborate

† The use of the prefix "cyclo" has been recommended by Dr. U. W. Gerwarth (Mainz, Federal Republic of Germany), in a private communication.

REFERENCES

1. H. Steinberg: "Organoboron Chemistry", vol. 1, New York, John Wiley (1964).
2. B. M. Mikhailov: "The Chemistry of Borohydrides," Moscow, Nauka (1967).
3. "The Nomenclature of Boron Compounds", *Inorg. Chem.* **7**, 1945 (1968).
4. "Nomenclature of Inorganic Boron Compounds. Information Bulletin of IUPAC, Appendixes on Tentative Nomenclature; Symbols, Units and Standards," No. 8, Oxford (1970).
5. "Chemist's Handbook. Supplementary Volume. Second Edition. Nomenclature of Organic Compounds. Safety Techniques. Cumulative Subject Index." Leningrad, Khimiya (1968).
6. "Nomenclature of Organic Chemistry, A, B, and C." Third Edition. London, Butterworths (1971).
7. *Pure Appl. Chem.* **45,** 11 (1976).
8. *Steroids* **13,** 277 (1969).

PART I

ALKYL- AND ARYLBORANES

Alkyl and aryl derivatives of boron show a high reactivity and are widely used in organic synthesis.

In this part, we shall consider methods of preparation of aliphatic and aromatic organoboranes with the use of organometallic compounds, by the reaction of boron halides with aromatic hydrocarbons, and also by the hydroboration of different types of olefins.

The boryl group in organoboranes may be practically replaced by any functional group, hydrogen, deuterium or a metal; the deboration reaction, as a rule, proceeds with a retention of the structure and configuration of the radical.

$$R(CH_2)_nX \leftarrow R{-}B\langle \rightarrow RX$$ (X = H, D, Hal, OH, SH, SR, NH_2, NR_2, CHO, COOH, CN, Hg, Ag, Cu, etc., $n \geqslant 1$).

Organoboranes are used for the introduction and protection of functional groups, for the alkylation and arylation of organic compounds, for lengthening carbon chains, and for other types of homologization. Such specific properties of organoboranes as the ability to generate radicals under the action of oxygen and of some metal salts or to isomerize upon heating have found important practical application.

The numerous reactions of alkyl-, cycloalkyl-, and arylboranes considered below are divided into several general types, the most basic of which were described in the Introduction.

PART I

CHAPTER 1

Preparation of alkyl- and arylboranes with the use of organometallic compounds and of the Friedel-Crafts reaction

The general method of synthesis of alkyl- and arylboranes consists of the reaction of organometallic compounds with boron derivatives, BX_3:

$$RM + BX_3 \xrightarrow[-MX]{} RBX_2 \xrightarrow[-MX]{RM} R_2BX \xrightarrow[-MX]{RM} R_3B.$$

Organic derivatives of magnesium, lithium, or aluminum and boron halides or alkoxy compounds of boron are mainly used in the reaction; boron compounds containing nitrogen or sulfur and boranes may also be used.

Reactions of this type may be of the equilibrium type [1], but in those cases where the metal (M) is more electropositive than boron, the equilibrium is completely shifted towards the organoborane and MX [2].

The reactivity of alkyl [3] and aryl [1] organometallic compounds with respect to $(C_4H_9O)_3B$ and $[(CH_3)_2N]_3B$ decreases with decreasing electropositivity of the metal: Li, Mg, Al > Zn, Cd > Pb, Hg, Sn. The ease of replacement of the X group under the action of RMgX or RLi decreases in the order: $>$B—Cl > $>$B—OR > $>$B—NR_2 [4]; with respect to triethylaluminum, the highest reactivity is shown by boron fluorides: $>$B—F > $>$B—Cl > $>$B—Br > $>$B—I [5]. Although it is assumed [6] that the activity of boranes toward RM decreases in the order: $BX_3 > RBX_2 > R_2BX$, this is not always true. Thus, the reaction of n-$C_6H_{13}MgBr$ with BF_3 leads to trihexylborane, even when a sixfold excess of BF_3 is used [7].

Methods of synthesis of organoboranes using organometallic compounds have been discussed in reviews [8–11] and in books [6, 12, 13].

1.1 TRIALKYL- AND TRIARYLBORANES

Trialkyl- and triarylboranes are among the most readily available boron derivatives. They are prepared by the reaction of boron halides (BX_3), their complexes or salts of the type KBF_4, borates, metaborates, and a number of other compounds with organic derivatives of lithium, magnesium, zinc, and aluminum. Lithium and magnesium compounds are used for the synthesis of both aliphatic and aromatic organoboranes, while zinc and aluminum compounds, as a rule, are used only for the synthesis of aliphatic compounds.

In most cases, the preparation of triorganoboranes is a simple operation and can be carried out with the use of simple laboratory techniques. The only requirement is the use of an inert atmosphere (nitrogen or argon), since many organoboranes are easily oxidized in air. It is also necessary to make some comments on the methods of carrying out the experiments.

1. After the addition of BX_3 to Grignard reagent, many authors treated the reaction mass, which had separated into layers, with a solution of sulfuric or hydrochloric acid until the salts had completely dissolved, and then separated the ether layer, dried it, and isolated R_3B by distillation or crystallization. Although most trialkylboranes are stable toward water and solutions of acids and alkalis, it is difficult to obtain the pure R_3B compounds by this method, because of their easy oxidizability, with the porous drying agents ($CaCl_2$, $MgSO_4$, etc.) being able to serve as oxygen sources, for example. It is impossible to use an acid treatment in the preparation of tribenzyl- and triallylboranes, and also of some triarylboranes, because water easily cleaves one or two B—C bonds in these compounds.

2. It is well known that the higher trialkylboranes, starting with $(C_3H_7)_3B$, undergo isomerization upon heating above 150°C, with migration of the boron along the carbon chain, or a different type of

cyclization (Chapter 4). Meanwhile, some authors have carried out the reaction or distillation of the product at 150–230°C, for example, distilling R_3B from solid magnesium or aluminum salts.

In this connection, critical reference should be made to some early work on the synthesis of compounds of the type R_3B, particularly trialkylboranes.

1.1.1 Trialkylboranes

Trialkylboranes are most frequently prepared by the action of Grignard reagent on halides or oxygen compounds of boron:

$$BX_3 + 3RMgX \longrightarrow R_3B + 3MgX_2.$$
$$80\%$$

All possible isomers of lower trialkylboranes (C_1—C_5), trineopentyl-, trihexyl- and tricyclohexyl-, *tris*(ω-alkoxyalkyl)-, *tris*(ω-alkenyl)-borane and many other trialkylboranes have been synthesized from boron halides and organomagnesium compounds [8, 9,12].

The most convenient from the preparative aspect are the reactions using boron trifluoride etherate, $(C_2H_5)_2O \cdot BF_3$.

The following method of synthesis of trialkylboranes with a high degree of purity can be recommended [14]: one mole of boron trifluoride etherate, in the form of a 50% solution in ether, is added dropwise to three moles of Grignard reagent prepared from magnesium and RX in ether. After this, the reaction mass, which has separated into two layers, is refluxed for one hour and is then left for 5–12 hours. After this time, the lower layer solidifies as a rule. The ether layer, containing the R_3B, is decanted and the solvent is distilled off. The trialkylborane is distilled under appropriate vacuum at a temperature below 100°C in order to prevent isomerization of the radicals on the boron. For example, tributylborane for thermo-chemical investigations has been prepared in this way [15].

Tribenzylborane is prepared from benzylmagnesium chloride and boron trifluoride etherate with subsequent removal of the ether and extraction of the organoborane with benzene [16]. The synthesis of this compound described by Krause [17], including decomposition of the magnesium salts with water, actually leads to the anhydride of dibenzylborinic acid, $(C_6H_5CH_2)_2BO—B(CH_2C_6H_5)_2$ [16, 18]. In this connection, the data of [19], which presents a detailed investigation of the UV spectra of a number of substituted benzylboranes, $(p\text{-}X—C_6H_4CH_2)_3B$, ($X = H$, F, CH_3, CH_3O) and also $[2,4,6\text{-}(CH_3)_3C_6H_2CH_2]_3B$, obtained by the Krause method, should be revised.

The tribenzylborane used in [20, 21] was synthesized in a yield of over 60% from benzylmagnesium bromide and BF_3 etherate, taken in a ratio of 1.5:1, with subsequent distillation of the ether and extraction with hexane.

Instead of BF_3 etherate in the preparation of trialkylboranes using a Grignard reagent, esters of metaboric acid, $(ROBO)_3$, can be used. The reaction is carried out in ether [22, 23], toluene [24], or ethylbenzene [25]; the yield of trialkyl-, tricyclohexyl-, and tribenzylborane by this method is 60–90%.

Many trialkylboranes are prepared from readily available aluminum trialkyls ($R = C_2H_5$, *iso*-C_4H_9) or from sesquialkylaluminum halides, $R_3Al_2X_3$, prepared by the reaction of the corresponding alkyl halides with aluminum. The boron component used is BF_3, BF_3 etherate, BCl_3, esters of boric and metaboric acids, boric anhydride, sodium tetraborate, KBF_4, and other compounds [10, 12, 26]:

$$R_3Al + (C_2H_5)_2O \cdot BF_3 \longrightarrow \underset{85-95\%}{R_3B} + (C_2H_5)_2O + AlF_3 \text{ (or } RAlF_2),$$

$$R_3Al_2X_3 + BX_3 \longrightarrow \underset{90\%}{R_3B},$$

$$R_3Al + (R'O)_3B \longrightarrow \underset{90-95\%}{R_3B}.$$

These reactions, which proceed with vigorous evolution of heat, are carried out by careful mixing of the reagents, with subsequent distillation of R_3B at atmospheric pressure or under vacuum. Usually, no difficulties are encountered in the isolation of the trialkyboranes. An exception is represented by the reaction of tri-*tert*-butylaluminum with BF_3, which leads to *tert*-butyldifluoroborane, $(CH_3)_3CBF_2$, and $[(CH_3)_3C]_2AlF$, as a result of steric effects [27].

Organoaluminum compounds are intermediates in the synthesis of trialkylboranes from aluminum, alkyl halides, and BCl_3 or trialkyl borates in a flow or stationary system [10, 12].

In a number of cases, exchange processes between trialkyl- or triarylboranes and aluminum trialkyls are of preparative significance [16, 28]:

$$(C_6H_5CH_2)_3B + (C_2H_5)_3Al \xrightarrow{95^\circ C} (C_2H_5)_3B + \underset{96\%}{(C_6H_5CH_2)_3Al} \xrightarrow{H_2O} C_6H_5CH_3.$$

The exchange of radicals takes place at room temperature; in this process, the equilibrium may be shifted to the right by distilling off the low-boiling triethylborane. These reactions, which proceed with the intermediate formation of associates with three-center two-electron bonds, are used for the preparation of triphenyl-, tri-α-naphthyl-, and tribenzylaluminum [16] and of heterocyclic aluminum compounds [29]. Optically active (+)-tri-[(*R*)-2-methylbutyl]borane, (+)-tri-[(*S*)-4-methylhexyl]borane, (+)-tri-[(*S*)-2-phenylbutyl]borane and (+)-tri-[(*S*)-2-phenyl-3-methylbutyl]borane form chiral aluminium trialkyls of the type $[RR'\overset{*}{C}H(CH_2)_n]_3Al$ ($n = 1, 2, 3$) upon heating with triethylaluminum [30].

Exchange reactions are used for the conversion of organoboranes to hydrocarbons under mild conditions, since the Al—C bond is easily cleaved by water, whereas trialkylboranes do not react with water at 20°C. Examples of compounds prepared by this method are deuterocyclopropane [31] and 2,2'-dideuterodiphenyl [29]:

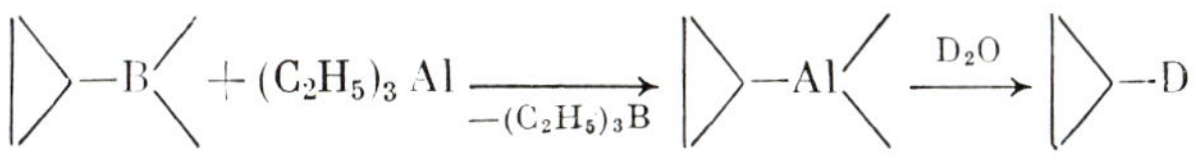

The reaction of butyllithium with BF_3 leads to $(C_4H_9)_4BLi$ even when an excess of boron trifluoride is used [32]. The action of cyclopropyllithium on trimethyl borate produces tricyclopropylborane [33]. Tri-*tert*-butylborane and its complexes with ammonia and pyridine have been synthesized from *tert*-butyllithium and $(CH_3S)_3B$ or $(CH_3O)_3B$ [162]. Lithium triethylborate reacts with CH_3Cu to form triethylborane [163].

1.1.2 Triarylboranes

Symmetrical triarylboranes are usually prepared from boron halides and a Grignard reagent. An excess of ArMgX should not be used in the reaction, since salts of the type $[Ar_4\overset{-}{B}][\overset{+}{M}gBr]$ are obtained in this case. In early work on the synthesis of triarylboranes from BF_3, the decomposition of magnesium salts and complexes of the type $[R_4B][MgX]$ by water or by solutions of inorganic acids was used [12]. It was later found [34] that many triarylboranes react rapidly with water, particularly in the presence of traces of acid, splitting off one or two aryl radicals:

$$Ar_3B + H_2O \xrightarrow[-ArH]{} Ar_2BOH \xrightarrow{H_2O} ArB(OH)_2 + ArH.$$

In this connection, two methods of isolation of the organoboranes are recommended.

After the addition of BX_3 to the Grignard reagent, ether is distilled off from the reaction mixture and the triarylborane is extracted from the salts with benzene or another hydrocarbon [35].

In other cases, the ether layer (sometimes after the addition of water) is decanted from the magnesium salts and saturated with ammonia. In this process, the complex $Ar_3B\cdot NH_3$, which is stable in air and can be stored for a long time, is precipitated. The triarylborane is isolated from the complex by treatment with an ether solution of HCl [34]; in many cases, solutions of inorganic acids can not be used [34].

A convenient method for the preparation of triphenylborane with a yield of up to 90% has been developed [36]. The reaction is carried out by adding Grignard reagent to a solution of BF_3 etherate (in a ratio of exactly 3:1) in xylene, with subsequent complete distillation of the ether and separation of the hot xylene solution of the organoborane from the solid salts.

Phenyllithium reacts with BF_3 etherate to form triphenylborane [37]. Tri(*o*-diphenyl)borane and tri(*p*-dimethylaminophenyl)borane have been synthesized in an analogous manner [35]; tris(pentafluorophenyl)borane has been prepared from BCl_3 and C_6F_5Li [38] or Grignard reagent [39].

One of the simplest methods of synthesis of triphenylborane consists of the reaction of sodium tetraphenylborate with BF_3 etherate [16]:

$$3Na[B(C_6H_5)_4] + BF_3\cdot O(C_2H_5)_2 \xrightarrow{20-135^\circ C} \underset{75\%}{4(C_6H_5)_3B} + 3NaF + (C_2H_5)_2O.$$

The same compound is conveniently obtained by the pyrolysis of trimethylammonium tetraphenylborate [40]. Some α-thienylboranes can be prepared without the use of organometallic compounds, for example, by the reaction of iodides with BI_3 [41]:

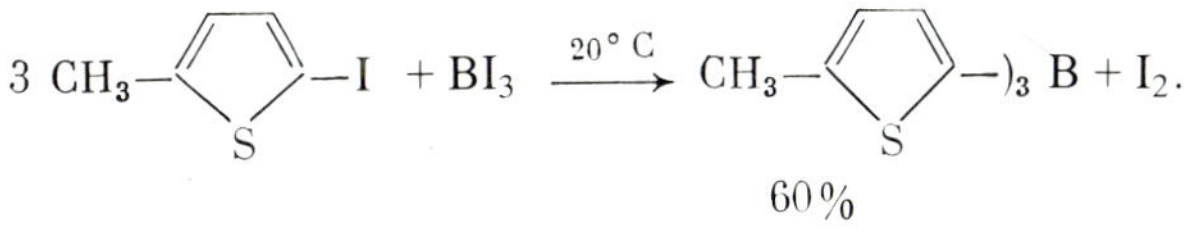

The UV spectra of triphenylborane [42] and of tri-*p*-chloro-, *p*-methoxy-, and *p*-dimethylaminophenylboranes have been investigated [43]. In an investigation of the crystal structure of triphenylborane, it was found that there is no intermolecular interaction in this compound [44]. Trimesitylborane and its analogs have a "propeller" configuration both in the solid state and in solutions at low temperatures [45, 46].

1.1.3 Unsymmetrical compounds

Mixed trialkylboranes, triarylboranes, dialkylarylboranes, and alkyldiarylboranes are relatively stable substances. They symmetrize in the presence of compounds with $>B—H$ bonds [16, 47–49], RMgX[10], R_3Al[16, 49], and probably some other RM compounds, and also upon heating.

When specific conditions were maintained (low isolation temperatures [below 100°C], use of complexing agents), a large number of different unsymmetrical triorganoboranes were obtained and isolated. Some methods of synthesis of boron compounds of this type are presented below.

Dibutyl(ethyl)borane and dibutyl(propyl)borane have been prepared by the action of lithium alkyls on esters of dibutylborinic acid [50]:

$$(C_4H_9)_2BOR' + RLi \longrightarrow Li\left[(C_4H_9)_2B\begin{matrix} R \\ OR' \end{matrix}\right] \xrightarrow{HCl} (C_4H_9)_2BR.$$

Such mixed trialkylboranes as methyl-, isopropyl-, and *tert*-butyldicyclohexylborane, isopropyl-, *tert*-butyldibutylborane, and a number of other compounds have been synthesized by this method from the methyl esters of the corresponding dialkylborinic acids in hydrocarbons [51].

1-Adamantyldimethylborane has been synthesized by the action of CH_3MgI on 2-(1-adamantyl)-1,3,2-dioxaborinane [163a].

The reactions of anhydrides, $(R_2B)_2O$, and halides, RBX_2, with Grignard reagent, aluminum trialkyls, or sesquialkylaluminum halides are used for the preparation of some unsymmetrical trialkylboranes [10, 12], while the aryl(alkyl)boranes Ar_2BR and $ArBR_2$ are prepared by the action of RMgX on 2-alkyl- or 2-phenyl-1,3,2-dioxaborolanes [52]. The pyridinates of diaryl(alkyl)boranes of the type $Ar(C_6H_5)BCH_2CH_2R$ have been synthesized in 60% yield from β-aminoethyl esters of diarylborinic acids and RCH_2CH_2MgX or by the hydrogenation of pyridinates of unsymmetrical acetylenic boron derivatives, $Ar(C_6H_5)BC{\equiv}CR$ over platinum [53].

Unsymmetrical triarylboranes with two different radicals and their amine complexes have been prepared in a yield of 45–90% by the action of lithium aryls [34] or ArMgX [54] on esters of diarylborinic acids. Triarylboranes are easily obtained from $Ar_2Ar'B \cdot NH_3$ complexes by treatment with dry hydrogen chloride in ether.

A large number of substituted phenyldimesitylboranes (*1*), including the compound (*2*), have been synthesized by the reaction of dimesityl borofluoride with the corresponding phenyllithium compounds and their fluorescent properties have been investigated [55].

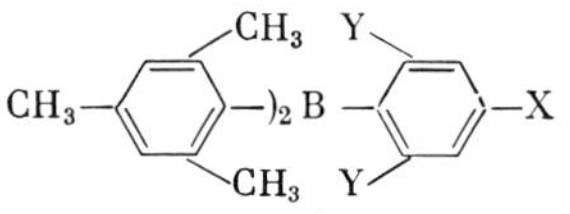

1

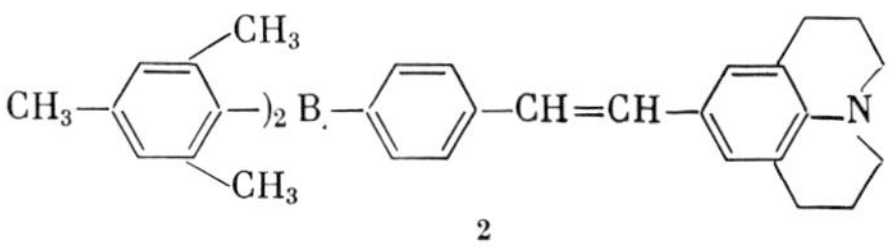

2

A method of preparation of triarylboranes with three different radicals on the boron — phenyl-*o*-tolyl-*p*-tolylborane and phenyl-*o*-tolyl-*p*-chlorophenylborane — which have been used for the preparation of lithium borates with an asymmetrical boron atom has been developed [54].

In some cases, an isomerization of the radicals takes place during the synthesis of symmetrical triorganoboranes, and unsymmetrical compounds are obtained. Thus, the reaction of *tert*-butylmagnesium chloride with BF_3 etherate or $(CH_3O)_3B$ leads to *tert*-butyldiisobutylborane [56, 57]. Isomerization of the radicals (to an extent of 26–87%) is observed in the alkylation of 9-chloro-9-borabicyclo[3.3.1]nonane by means of RLi [51].

A trialkylborane with three different radicals on the boron — *tert*-butylisobutyl-*n*-amylborane—has been prepared by the action of *tert*-butylmagnesium chloride on $n\text{-}C_5H_{11}BF_2$ [58]. The reaction of dimethylzinc with the diboron compound $Cl_2BCH_2CH_2BCl_2$ leads to 1,2-bis(diethylboryl)ethane in 88% yield [59], while cyclopropyldimethylborane has been synthesized by treating dicyclopropylmercury with chlorodimethylborane [33].

In the reaction of trialkylboranes with aluminum trialkyls, an exchange of the radicals takes place and mixed trialkylboranes are formed [16]:

$$R_3B + R_3Al \longrightarrow R_2BR' + RBR'_2 + \underbrace{RAlR'_2 + R_2AlR'}_{\downarrow H_2O}$$

$$RH + R'H + Al(OH)_3$$

Organoaluminum compounds hydrolyze upon treatment with water, whereas organoboranes remain unchanged.

Unsymmetrical organoboranes have been prepared by the action of an alkyl- or aryllithium on tetraalkyldiboranes, with subsequent cleavage of the lithium borate according to the following scheme [51, 60]:

$$(R_2BH)_2 + R'Li \longrightarrow [R_2BHR']\,Li - \begin{cases} \xrightarrow{CH_3I} R_2BR' + LiI + CH_4 \\ \xrightarrow{CH_3SO_3H} R_2BR' + H_2 + LiSO_3CH_3 \end{cases}$$

Diphenyl(1′-phenylferrocenyl)borane pyridinate is formed by the oxidation of triphenylferrocenylborates by air [61]. Diphenylboron bromide reacts with $Hg[C_6H_5Cr(CO)_3]_2$, with the formation of tricarbonyl(diphenylborylbenzene)chromium(O), $(C_6H_5)_2BC_6H_5\text{-}Cr(CO)_3$ [62].

Trialkylboranes with two or three different radicals are also prepared by the stepwise hydroboration of olefins (Chapter 2 and 3).

Esters of dialkylborinic acids form mixed trialkylboranes upon reduction with lithium aluminum hydride in the presence of olefins [63, 64]:

$$R_2BOCH_3 + LiAlH_4 \xrightarrow{\text{olefin}} R_2BR'.$$

Alkyldiphenylboranes are obtained in an analogous manner [164].

The thioesters R_2BSR', obtained from trialkylboranes and mercaptans [23, 65, 66], react analogously [67]:

$$R_3B + R'SH \xrightarrow{O_2} R_2BSR' \xrightarrow[65-84\%]{LiAlH_4,\ \text{olefin}} R_2BR'.$$

Mixed trialkylboranes, R_2BR', are also formed by the reduction of R_2BBr by NaH in the presence of an olefin [68, 165, 166].

1.2 ESTERS OF ORGANOBORON ACIDS

Organoboron acids and esters of the type $RB(OR')_2$ or R_2BOR' are widely used in organic chemistry. They are used for the synthesis of many classes of organic compounds (Chapters 5–14), for the production of carbocyclic and heterocyclic structures (Part II), for the protection of some functional groups, and for the separation of mixtures of substances and isomers. Organoboron acids have played, and continue to play, an important role in the chemistry of ferrocene and other sandwich structures.

By this time, simple methods of synthesis of organoboron acids and their esters have been developed, and convenient methods for going from these to the corresponding halo-, amino-, and thioboranes and vice versa have been found.

The methods of preparation and properties of organoboron acids have been partially considered in reviews [8, 9] and books [6, 12, 13].

1.2.1 Esters of dialkyl- and diarylborinic acids

a) Preparation

Dialkyl-, diaryl-, and alkylarylborinic acids and their esters and other boron derivatives with two identical (R_2BX) or different

(RR′BX) organic radicals are prepared by the following methods: 1) by the action of organometallic reagents on boron derivatives of the type BX_3 or RBX_2; 2) by the monodealkylation or monodearylation of triorganoboranes; this method is used mainly for the preparation of aliphatic compounds (Chapter 6); 3) by exchange reactions of the type $R_3B + B(OR)_3$; 4) by the hydroboration of olefins by means of H_2BX (Chapter 2); 5) by the hydrolysis or alcoholysis of compounds of the type R_2BX (X = H, Hal, NR_2, SR, OR).

At this point, we shall consider methods of preparation of the esters R_2BOR' using organic derivatives of magnesium and lithium and also by means of exchange reactions.

The action of two moles of Grignard reagent or of an organolithium compound on trialkyl borates at −70°C leads to the replacement of two alkoxy groups by organic radicals [9, 12, 13]:

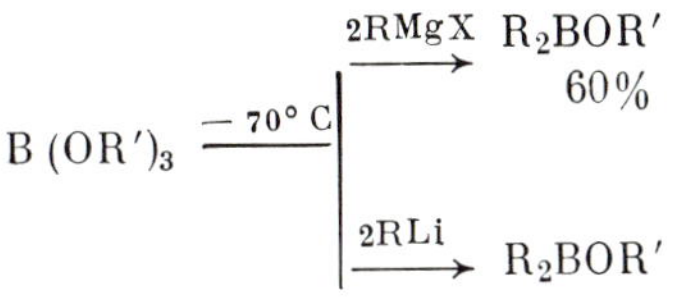

The esters formed can be hydrolyzed to the corresponding acids or converted to the stable β-aminoethyl esters. For the preparation of esters of dialkylborinic acids, the use of glycol esters $ROB(OCH_2)_2$ and RMgX is recommended [69]; however, this method has not found broad use, since the esters of symmetrical dialkylborinic acids are prepared more conveniently from trialkylboranes and alcohols (Chapter 6). The esters $R_2BOSiR'_3$ can be obtained by the reaction of R_2BOH and R'_3SiH in the presence of H_2PtCl_6 [167].

It is better to use the reaction of arylmagnesium halides with metaborates, $(ROBO)_3$, for the synthesis of diarylborinic acids and their esters [9, 70]. This method is more universal and does not require the use of low temperatures.

Dimesityl- [71] and di-9-anthrylborinic acids [72] are prepared from mesityl- or 9-anthrylmagnesium bromide and BF_3 etherate. The replacement of the third fluorine atom by the organic radical in these two cases proceeds slowly, because of steric hindrance.

The action of two moles of butyllithium on one mole of boron triiodide in hexane produces a 77% yield of $(C_4H_9)_2BI$ [73], the hydrolysis of which leads to the anhydride of dibutylborinic acid [8, 9].

Organolithium compounds are used for the synthesis of esters of unsymmetrical dialkylborinic acids, RR′BOR″ [74, 75] and alkylarylborinic acids, ArRBOR′ [76]. The alkyl esters of mixed diarylborinic acids are prepared by the treatment of esters of arylboronic [77] or chloroarylboronic [54] acids with Grignard reagent.

A number of intramolecular esters of mixed diborinic acids [78], such as R′O(R)BR″B(R)OR′, have been prepared by the action of alkyl-, cyclohexyl-, phenyl-, and thienylmagnesium halides on cyclic 2,2-dimethyltrimethylene esters of methylenediboronic ($n = 4$, 10), *p*-phenylenediboronic, *p*-diphenyldiboronic, and 2,5-thienyldiboronic acids with subsequent esterification with ethanolamine or 8-hydroxyquinoline.

The action of diethylzinc on dialkyl- or diarylborinic acids produces the esters $R_2BOZnC_2H_5$ [168].

In the preparation of derivatives of the benzyl and aromatic series, the ease of protolytic cleavage of the B—C bonds in these compounds should be taken into consideration (Chapter 6). Thus, lithium diisobutoxydiphenylborate (*3*), formed by the action of phenyllithium on $C_6H_5B(OC_4H_9)_2$ forms isobutoxydiphenylborane in a yield of 82% when treated with dry hydrogen chloride [79]:

$$C_6H_5B(OC_4H_9)_2 + C_6H_5Li \longrightarrow \underset{3}{[(C_6H_5)_2B(OC_4H_9)_2]Li} \rightarrow$$

$$\xrightarrow{HCl} (C_6H_5)_2BOC_4H_9 + C_4H_9OH$$

$$\xrightarrow{H_2O,\ HCl} C_6H_5B(OC_4H_9)_2 + C_6H_6$$

At the same time, treatment of the borate *3* with 20% hydrochloric acid leads to a mixture of esters of diphenylborinic (28%) and phenylboronic (62%) acids.

Esters of dialkylborinic acids (together with esters of alkylboronic acids) are formed by the prolonged heating of trialkylboranes with borates at 200°C [80]. Exchange reactions of this type are catalyzed by boranes, which act as carriers of alkyl and alkoxy radicals (by means of bridging bonds) [48, 81–85].

$$2R_3B + B(OR')_3 \xrightarrow[60-70\%]{>B-H} 3R_2BOR'.$$

In the presence of 1–2% of tetraalkyldiborane at room temperature, the reaction is completed within 90–100 hours; at 100°C, only 2–3 hours is required. It should be noted, however, that the yield of the esters R_2BOR' is 70% at 20°C and approximately 50% upon heating [83]. The exchange reactions of trialkyl- and tricycloalkylboranes with trimethyl or triaryl borates in the presence of the complex $H_3B \cdot THF$ lead to the methyl or aryl esters of dialkyl- and dicycloalkylborinic acids [86].

b) Some properties. Use for the protection of functional groups

Organoboron acids R_2BOH and their anhydrides and esters are easily oxidized in air; the alkyl esters also hydrolyze. At the same time, the β-aminoethyl esters (*4*), many of their pyrrolidine, piperidine, and morpholine analogs, and also the α-pyridylmethyl and *o*-aminophenyl esters are stable toward oxygen and atmospheric moisture as a result of internal coordination [13]. Conversion to the stable 8-quinoline esters is used for the identification of dialkyl- [86] and diarylborinic acids [13].

The methyl, ethyl, and glycol esters of dialkyl- and diarylborinic acids form the corresponding acids under the action of water and of solutions of alkalies and acids [9]. Butyl and other higher esters react with alkalies with the formation of salts $[R_2B(OR)OH]Na$. Upon acidification of solutions of these salts, the starting ester is regenerated almost quantitatively [9, 75]. Esters of diarylborinic acids are hydrolyzed by aqueous solutions of ammonia or barium hydroxide [9] and those of dialkylborinic acids by heating with alkali solutions [75]. The β-aminoethyl esters saponify under the action of acids [87].

When heated with higher alcohols, the esters R_2BOR' undergo transesterification [13]. Methanol is capable of displacing higher alcohols from esters of dialkylborinic acids [88].

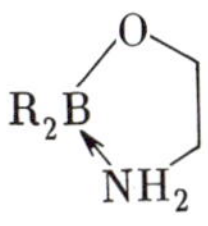

4

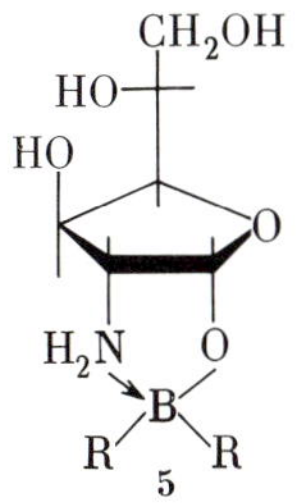

5

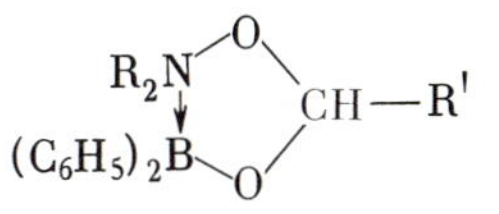

6

The ability of diphenylborinic acid to form chelates of type *4* is the basis for its use for the protection of functional groups in γ-morpholino- and γ-piperidino-β-hydroxybutyronitriles [89], D,L-serine, D,L-carnitine [90], and glucosamine (5) [91]. The diphenylboryl group is removed by heating with 1,3-propanediol at 90°C or by oxidation with dilute hydrogen peroxide or periodate.

An interesting method for protecting the carbonyl group in aldehydes consists of their conversion to cyclic acetals of the type *6* under the action of *N,N*-dialkylhydroxylamines and diphenylborinic acid [92].

When the butyl ester of diphenylborinic acid is heated in pyridine with guanosine, adenosine, inosine, or *N*-benzoylguanosine, a cleavage of one B—C bond takes place and a quantitative yield of 2′,3′-cyclic esters of phenylboronic acid (7) is obtained, the subsequent phosphorylation of which leads (after removal of the protective group with 1,3-propanediol) to the corresponding 5′-phosphates [93, 94].

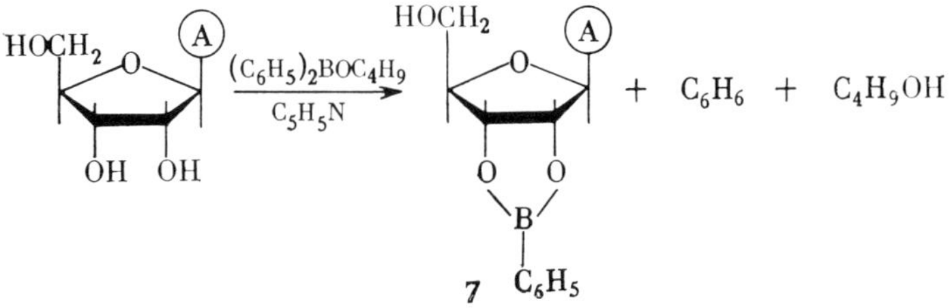

The formation of the cyclic esters *7* by this method takes place without the evolution of water, which is of considerable importance in many cases, for example, in the preparation of 2′,3′-phenylborates of 5′-*O*-tosylinosine, 5′-*O*-tosylguanosine, and 5′-*O*-tosyluridine [95, 96].

1.2.2 Esters of alkyl- and arylboronic acids

a) Preparation

Derivatives of boron with one alkyl or aryl radical (RBX_2) are prepared by the following methods: 1) by the action of organometallic compounds on borates or boron trihalides; 2) by the exchange reactions $R_3B + BX_3$; 3) by the cleavage of two radicals from R_3B or of one radical from R_2BX (Chapter 6); 4) by reactions of ArH with BX_3; 5) by the action of BI_3 on aryl iodides; 6) by the transformation of an organic radical in monoorganoboranes RBX_2; 7) by reactions of trialkyl- or dialkylboron derivatives with some organic and inorganic compounds, e.g., with carbon monoxide or bromine; 8) by the hydroboration of olefins by means of boranes HBX_2 (Chapter 2).

A general method for the synthesis of alkyl- and arylboronic acids or their esters consists of the reaction of the corresponding Grignard reagent and trimethyl borate or tributyl borate at low temperature:

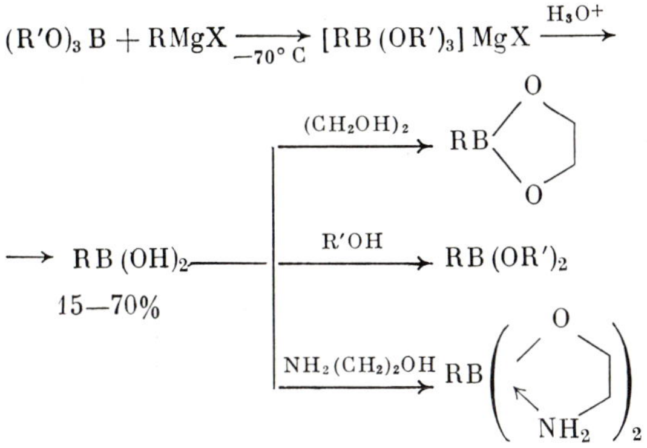

After the addition of the organomagnesium compound to a solution of the trialkyl borate in ether at −70°C, the reaction mass is heated to room temperature and is treated with 10% hydrochloric acid. The organoboron acid is then isolated or is esterified with butanol, ethylene glycol, ethanolamine, diethanolamine, or other higher alcohols [12, 13].

This method has been used for the synthesis of alkyl- (C_1–C_{16}), 3-methoxypropyl-, cyclohexyl-, benzyl-, phenyl-, and 1- and 2-naphthylboronic acids and also of a number of nuclearly substituted phenyl- and naphthylboronic acids and their esters and anhydrides [9, 12, 13]. The diethanolamine esters of 1-phenylethyl- and α-phenylisoprophyl- [97, 98], 1- and 2-naphthylmethylboronic acids [99] have been prepared from the corresponding compounds RMgX, with subsequent treatment with diethanolamine.

Tetramethylene-, decamethylene-, *p*-phenylene-, 4,4′-diphenyl-, and 2,5-thiophenediboronic acids and their 2,2-dimethyltrimethylene esters have been prepared from the dimagnesium derivatives and trimethyl borate [100]. The hydrogenation of 2,5-thiophenediboronic acid over nickel leads to 1,4-tetramethylenediboronic acid.

The dibutyl [101] and diethanolamine [99] esters of butylboronic acid have been synthesized in 60% yields by the reaction of butyllithium with tributoxyborane. In the case of other lithium alkyls, the reaction is complicated by the formation of esters R_2BOR' [101]. At the same time, lithium aryls are widely used for the preparation of arylboronic acids and esters.

$$RLi + B(OR')_3 \xrightarrow{-70°C} [RB(OR')_3]\,Li \xrightarrow{H_3O^+} RB(OH)_2 \xrightarrow{R'OH} RB(OR')_2.$$

It has been established [79, 102] that, in the action of phenyllithium on trialkyl borates, a complex salt $[C_6H_5B(OR')_3]Li$ is initially formed, from which the ester $C_6H_5B(OR')_2$ can be isolated by the action of an excess of borate or of a mineral acid.

Isomeric hydroxyphenyl-, methyl-, and methoxynaphthylboronic acids, 2- and 3-thienyl-, 4-phenoxythienyl-, 2-thianthrene-, and 8-quinolineboronic acids and a number of nuclearly substituted phenylboronic acids have been synthesized using lithium reagents [9, 12, 13]. *o,o'*-Tolanediboronic [103], *o*-(2-α-pyridylethynyl)phenylboronic [104], *p*-phenylenediboronic [105, 106], ferrocenylboronic and 1,1'-ferrocenylenediboronic [107, 108], 2-(*N,N*,-dimethylaminomethyl)ferrocenylboronic [109], 2-(α-pyridyl)-ferrocenylboronic [110, 111], and cyclopentadienylmanganese-tricarbonylboronic [112] acids and a number of other compounds have been prepared by an analogous method [13].

2-(1-Adamantyl)-1,3,2-dioxaborinane has been synthesized by the reaction of 1-bromoadamantane, 2-chloro-1,3,2-dioxaborinane, and sodium [163a].

For the preparation of esters of monosubstituted acids, the reaction of tetraalkyl- [113, 114] and tetraarylstannanes [115, 116] with boron halides can be used:

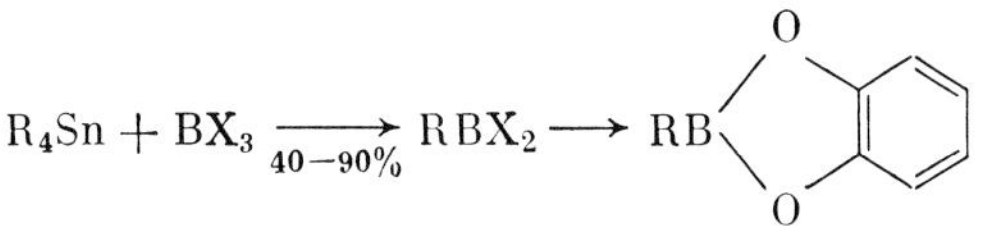

(X = F, Cl, Br).

All four radicals are transferred from the tin to the boron in the case of the aryl compounds, while only two radicals are transferred in the case of tetraalkylstannanes. The initially formed dihalo(organo)-

boranes are converted to the corresponding esters under the action of alcohols and diols.

Many aryl compounds can be conveniently prepared by heating arylmercury chlorides with BCl_3 or BBr_3 in benzene, followed by alcoholysis [117, 118].

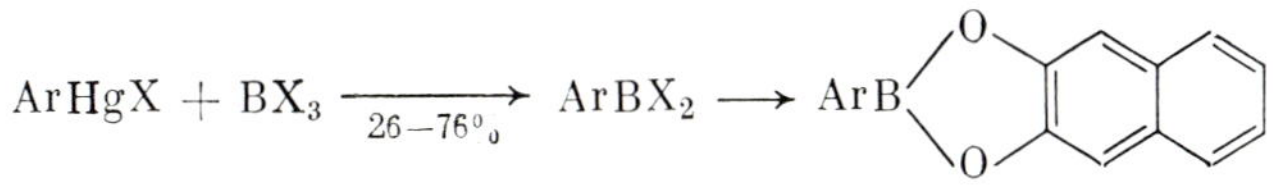

During the course of the reaction of *o*-tolylmercury chloride with BCl_3, a partial rearrangement takes place, as a result of which a mixture of 95% *o*- and 5% of *m*-$CH_3C_6H_4BCl_2$ [119] is obtained. An isomerization of a similar nature is also observed in some reactions of *o*-substituted arylboranes [119].

A number of arylboronic acids, including α-furyl- and α-thienylboronic acid, have been prepared by the reaction of diborane with Ar_2Hg, ArHgX [169], or $ArTl(OCOCF_3)_2$ [170]. The treatment of cyclic esters of α-chloro- [171] or α-bromoalkylboronic acids [172], $R'R''CXB(OR''')_2$, with RLi or RMgX produces the esters $RR'R''CB(OR''')_2$, while $(C_4H_9)_2CHB(OR)_2$ has been synthesized from $Cl_2CHB(OR)_2$ and butyllithium [173].

One of the most convenient and economical methods for the synthesis of esters of alkylboronic acids is the exchange reaction of trialkylboranes with borates [48, 81, 83].

$$R_3B + 2B(OR')_3 \xrightarrow{\rangle B\text{—}H,\ 100^\circ C} \underset{70\text{—}90\%}{3RB(OR')_2}.$$

In the presence of tetraalkyldiboranes and other compounds with B—H bonds, the exchange of radicals proceeds quite rapidly at 100°C [83]. The cyclic ethylene and propylene glycol esters are prepared analogously from the corresponding borates [80, 120]. Secondary trialkylboranes exchange radicals with much greater difficulty than the primary compounds; the redistribution of radicals is accompanied by partial isomerization [121].

The anhydrides $(RBO)_3$ are formed by boiling triphenyl-, tricyclohexyl-, and trialkylboranes with B_2O_3 [9]. It should, however, be kept in mind that alkylboranes isomerize under these conditions (Chapter 4).

Two methods of synthesis of aryl- and ferrocenylboronic acids without the use of organometallic compounds have been developed. The first of these consists of the reaction of aromatic compounds with boron halides BX_3 (X = Cl, Br, I) in a continuous system (platinum black, 500–600°C) [122] or stationary system [10].

$$ArH + BX_3 \longrightarrow \underset{70\%}{ArBX_2} \xrightarrow{H_2O} ArB(OH)_2.$$

In the presence of activated aluminum or $AlCl_3$, reactions of this type proceed at 30–150°C [10, 123, 124]; phenylboronic acid (70%) [123] and ferrocenylboronic acid (37%) [125, 126] are obtained in this way. Toluene and other benzene derivatives form a mixture of isomeric compounds with BCl_3; in the reaction of *p*-xylene, a migration of methyl groups over the benzene ring takes place, as a result of which a mixture of 90% of 3,5- and 10% of 3,4-$(CH_3)_2C_6H_3BCl_2$ is formed [119, 124]. Boron tribromide and boron triiodide are more active in reactions with ArH [10, 127]. Thus, ferrocene reacts with BI_3 in CS_2 at 10°C with the formation of ferrocenylboron diiodide (92%) [128].

The UV-initiated reactions of BBr_3 and BI_3 with aromatic hydrocarbons probably take place by a radical mechanism ($\cdot BX_2$) [129]. Boron trifluoride does not react with benzene even at 250°C [130].

The second method is based on redox reactions of aryl iodides with boron triiodide:

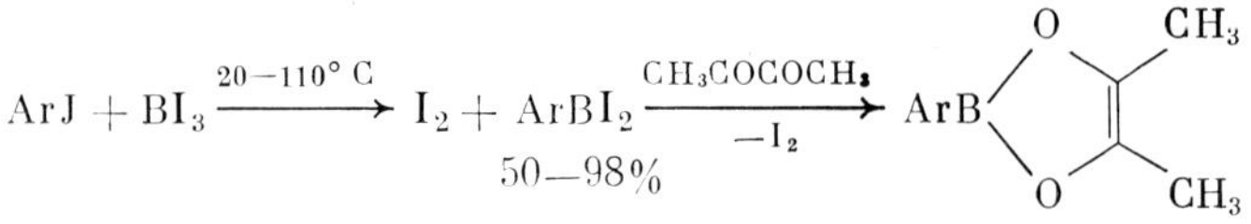

Derivatives of phenyl- [131], *m*- and *p*-tolyl- [132], *m*- and *p*-chlorophenyl-, *p*-fluorophenyl-, 3,5- and 3,6-xylyl-, mesityl-, 2,3,5,6-tetramethylphenyl- [133], and 2-thienylboronic [41] acids have been synthesized by this method, as well as 2,5-bis-(diiodoboryl)thiophene [41]. The reactions of *o*-tolyl iodide and other aryl iodides containing substituents in the ortho position are accompanied by rearrangement and lead to a mixture of the meta- and para-isomers [132, 133].

Functionally substituted esters of alkylboronic acids X—$(CH_2)_nB(OC_4H_9)_2$ ($n = 3$–6; X = Cl, Br, I, OH, CN, COOR, NR_2, $COCH_3$) have been synthesized with the use of the corresponding Grignard reagent, for example, in accordance with the following scheme [134, 135]:

$$Cl(CH_2)_nBr \xrightarrow[2.\ B(OR)_3]{1.\ Mg} Cl(CH_2)_nB(OR)_2 \xrightarrow[130°\ C]{NaCN}$$

$$\longrightarrow NC(CH_2)_nB(OR)_2 \xrightarrow{H_3O^+} HOOC(CH_2)_nB(OR)_2.$$

Numerous functional derivatives of phenylboronic acid have been prepared by bromination, nitration, oxidation, and other conversions of phenyl- and tolylboronic acids [13]. The action of urotropine on the anhydride (p-$BrCH_2C_6H_4BO)_3$ forms p-$NH_2CH_2C_6H_4B(OH)_2$ [136].

b) Some properties. Use for the protection of functional groups

Alkyl- and cycloalkylboronic acids are oxidized in air, while their aryl analogs are generally stable toward atmospheric oxygen. Aryl- and benzylboronic acids are cleaved by heating with water or solutions of acids and alkalis; the acids of the alkyl series are stable under these conditions (Chapter 6). Under the action of dehydrating agents, the acids are converted to the trimeric anhydrides $(RBO)_3$ [13]; with alkalis, they form complex salts with a tetravalent negative boron $[R\bar{B}(OH)_3]M^+$, and they form 1:1 adducts with amines [9]. The acids are esterified by heating with higher alcohols (C_4 and above) with the formation of esters $RB(OR')_2$ [9, 13]; reaction with ethylene glycol or 1,3-propanediols leads to the corresponding cyclic esters [100]; under the action of diethanolamine, even in the presence of water, chelate esters stable in air are obtained [13, 137].

The methyl esters can be obtained from the acids and methanol with the use of a Soxhlet apparatus (with anhydrous copper sulfate) [138].

The alkyl esters $RB(OR')_2$ are hydrolyzed by water and by solutions of acids and alkalis. Their preparative hydrolysis for the purpose of isolating the alcohol is carried out by shaking with alkali solutions, using at least one mole of NaOH or sodium carbonate per mole of borane. The salt $[RB(OH)_3]Na$ obtained is soluble in water [9]. The diethanolamine esters are hydrolyzed only by strong acids (breakdown of the B—N coordination) [137].

Because of their ability to form cyclic esters with 1,2- and 1,3-diols, phenylboronic acid and some of its derivatives are used for the protection of sugars [139–144] (see also Chapter 6). The phenylborate protection is removed by the action of 1,3-propanediol or of a solution of acid (pH 2), sodium carbonate, alkali, pyridine, or diethanolamine.

The methylation of 2,4-phenylborate esters of α- and β-methyl-D-xylopyranosides, with subsequent removal of the protection by 1,3-propanediol, leads to the corresponding methyl ethers [145]. With the use of the phenylborate protective group, 3-*O*-methyl and 2,4-di-*O*-methyl-D-xylose and also the 3-acetates, 3-benzoates, and 3-*N*-phenylcarbamates of partially methylated D-xylopyranose have been obtained [145].

β-Methyl-L-arabinopyranoside and α-methyl-D-lyxopyranoside condense with $C_6H_5B(OH)_2$ to form the 3,4- and 2,3-cyclic esters, respectively; the 2,4-cyclic ester is obtained in 90% yield from β-methyl-D-ribopyranoside [146]. The 3-acyl derivative of D-ribopyranose has been synthesized by acylation of the 2,4-cyclic ester and treatment with 1,3-propanediol; reaction with phenyl isocyanate led to 3-*N*-phenylcarbamate.

3-O-β-D-Glucopyranosyl-D-xylose, 3-O-α-, and β-D-xylopyranosyl-D-xylose have been synthesized by the glycosylation of the 2,4-phenylborate ester of β-benzyl-D-xylopyranoside [147].

By tosylation or tritylation of the air-stable 2′,3′-phenylborate esters of adenosine, uridine, inosine, and cytidine, the 5′-derivatives of the corresponding nucleosides are prepared [148].

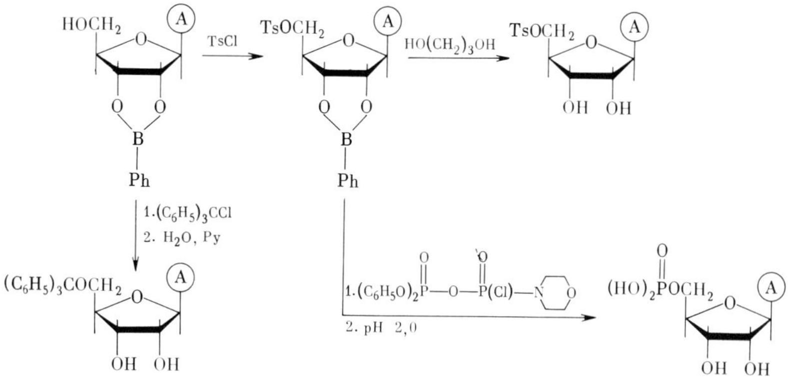

Uridine-5′-phosphate and cytidine-5′-phosphate have been synthesized by the phosphorylation of uridine and cytidine 2′,3′-phenylboronates by P^1-diphenyl-P^2-morpholidopyrophosphochloride [148, 149]. Guanosine-5′-monophosphate was obtained in an analogous manner [93]. The benzoylation of adenosine phenylboronate led to 5′-benzoyladenosine [94]. *p*-Tolyl- and *m*-nitrophenylboronic acids have also been used instead of phenylboronic acid for the protection of diols [94].

When the 2′(3′)-monophosphates of adenosine and cytidine are heated with phenylboronic acid, a specific dephosphorylation takes place, leading to the corresponding 2′,3′-boronate esters; the 5′-monophosphates of the nucleosides have been prepared by phosphorylation of these esters [148, 150].

α-D-Glucofuranose 1:2;3:5-diphenylboronate has been prepared by the treatment of α-D-glucose with phenylboronic anhydride [151]. Subsequent carbodiimide condensation with dimethylglycine hydrochloride leads to 6-*O*-dimethylglcyl-α-D-glucopyranose hydrochloride, which is a derivative of the reduced form of pangamic acid. 6-*O*-Trimethylammoniumacetyl-D-glucopyranose hydrochloride has been synthesized from the same diboronate by successive treatment with chloroacetic anhydride and trimethylamine [151, 152].

The phenylboronate protection of hydroxy groups has been used in the synthesis of analogs of cobamide coenzyme, modified in the corrin ring and in the "lower" axial ligand [95, 153, 154].

Arylboronic acids are used for the chromatographic separation of nucleosides and nucleotides [155–161]. Other methods for the protection of hydroxy groups, and also methods for the separation of *cis*- and *trans*-1,2-diols with the use of organoboron compounds, are considered in Chapter 6.

REFERENCES

1. G. Abeler, H. Bayrhuber and H. Nöth, *Chem. Ber.* **102,** 2249 (1969).
2. G. E. Coates, M. L. H. Green, P. Powell and K. Wade: "Principles of Organometallic Chemistry." London, Methuen, (1968).
3. H. Gilman and L. O. Moore, *J. Am. Chem. Soc.* **80,** 3609 (1958).
4. M. F. Lappert and M. K. Majumdar, *J. Organometal. Chem.* **6,** 316 (1966).
5. H. Nöth and W. Storch, *Synth. Inorg. Metall-Org. Chem.* **1,** 197 (1971).
6. W. Gerrard: "The Chemistry of Organic Boron Compounds" [Russian Edition]. Moscow, Khimiya (1966).

7. P. A. McCusker and L. J. Glunz, *J. Am. Chem. Soc.* **77,** 4253 (1955).
8. M. F. Lappert, *Chem. Rev.* **56,** 959 (1956).
9. B. M. Mikhailov, *Usp. Khim.* **28,** 1450 (1959).
10. M. F. Lappert, In: "The Chemistry of Boron and Its Compounds." Ed. E. L. Muetterties. New York, J. Wiley and Sons (1967), p. 433.
11. K. Smith, *Chem. Soc. Rev.* **3,** 443 (1974).
12. A. N. Nesmeyanov and R. A. Sokolik: "Methods of Elemento-Organic Chemistry. Boron, Aluminum, Gallium, Indium, Thallium." Moscow, Nauka (1964).
13. K. Torssell: In: Progress in Boron Chemistry, vol. 1. Ed. H. Steinberg and A. L. McCloskey. Oxford, Pergamon Press (1964), p. 369.
14. B. M. Mikhailov, Yu. N. Bubnov and V. G. Kiselev, *Zh. Obshch. Khim.* **36,** 62 (1966).
15. G. L. Gal'chenko, N. S. Zaugol'nikova, S. M. Skuratov, L. S. Vasil'ev, Yu. N. Bubnov and B. M. Mikhailov, *Dokl. Akad. Nauk SSSR* **109,** 587 (1966).
16. R. Köster and G. Bruno, in the book: "Organoaluminum Compounds", edited by A. F. Zhigach, Moscow, IL (1962), p. 113.
17. E. Krause and P. Nobbe, *Chem. Ber.* **63,** 934 (1930).
18. B. M. Mikhailov, G. S. Ter-Sarkisyan and N. A. Nikolaeva, *Zh. Obshch. Khim.* **41,** 1721 (1971).
19. R. Ramsey and N. K. Das, *J. Am. Chem. Soc.* **94,** 4227 (1972).
20. Yu. N. Bubnov and B. M. Mikhailov, *Izv. Akad. Nauk SSSR, Ser. Khim.* **1970,** 2156.
21. B. M. Mikhailov, Yu. N. Bubnov, S. A. Korobeinikova and S. I. Frolov, *J. Organometal. Chem.* **27,** 165 (1971).
22. B. M. Mikhailov and V. A Vaver, *Izv. Akad. Nauk SSSR, Otdel. Khim. Nauk* **1960,** 852.
23. B. M. Mikhailov, V. A. Vaver and Yu. N. Bubnov, *Dokl. Akad. Nauk SSSR* **126,** 575 (1952).
24. L. I. Zakharkin and O. Yu. Okhlobystin, *Izv. Akad. Nauk SSSR, Otdel. Khim. Nauk* **1959,** 1135.
25. A. Salvemini, G. Cuneo and F. Smai, *Italian Patent* 758,247 (1967); C. A. **68,** 49728 (1968).
26. R. Köster, P. Binger and W. V. Dahlhoff, *Synth. Inorg. Metal-Org. Chem.* **3,** 359 (1973).
27. H. Lehmkuhl, O. Olbrysch and H. Nehl, *Ann.* **1973,** 708.
28. L. I. Zakharkin and O. Yu. Okhlobystin, *Zh. Obshch. Khim.* **30,** 2134 (1960).
29. R. Köster and G. Benedikt, *Angew. Chem.* **75,** 419 (1963).
30. L. Lardicci, G. P. Giacomelli and L. De Bernardi, *J. Organometal. Chem.* **39,** 245 (1972).
31. P. Binger and R. Köster, *Angew. Chem.* **74,** 652 (1962).
32. D. K. Jenkins and C. Dixon, *Chem. and Ind.* **1966,** 1887.
33. A. Cowley and T. Furtsch, *J. Am. Chem. Soc.* **91,** 39 (1969).
34. B. M. Mikhailov and V. A. Vaver, *Izv. Akad. Nauk SSSR, Otd. Khim. Nauk* **1957,** 812.
35. G. Wittig and G. Herwig, *Chem. Ber.* **88,** 962 (1955).
36. R. Köster, P. Binger and W. Fenzl, *Inorg. Synth.* **15,** 134 (1974).
37. G. Wittig, G. Keicher, A. Rückert and A. Raff, *Ann.* **563,** 110 (1949).
38. A. G. Massay, A. J. Park and F. G. Stone, *J. Organometal. Chem.* **5,** 218 (1966).
39. J. L. W. Pohlmann and F. E. Brinckmann Z. Naturforsch, **20b,** 5 (1965).
40. K. E. Reinhard, R. E. Sherman, H. D. Smith and L. F. Hohnstedt, *Inorg. Synth.* **14,** 52 (1973).

41. W. Siebert, *Chem. Ber.* **103,** 2308 (1970).
42. A. N. Nikitina, V. A. Petukhov, A. F. Galkin, N. S. Fedotov, Yu. N. Bubnov and P. M. Aronovich, *Opt. i Spektroskopiya* **16,** 976 (1964).
43. D. S. Miller and J. E. Leffler, *J. Phys. Chem.* **74,** 2571 (1970).
44. F. Zetter, H. D. Hausen and H. Hess, *J. Organometal. Chem.* **72,** 157 (1974).
45. J. F. Blount, P. Finacchiaro, D. Gust and K. Mislow, *J. Am. Chem. Soc.* **95,** 7019 (1973).
46. J. Hummel, D. Gust and K. Mislow, *J. Am. Chem. Soc.* **96,** 3679 (1974).
47. B. M. Mikhailov and V. A. Dorokhov, *Dokl. Akad. Nauk SSSR* **133,** 119 (1960).
48. B. M. Mikhailov and L. S. Vasil'ev, *Izv. Akad. Nauk SSSR, Otd. Khim. Nauk* **1961,** 531.
49. R. Köster, G. Griasnov, W. Larbig and P. Binger, *Ann.* **672,** 1 (1964).
50. B. M. Mikhailov and T. A. Shchegoleva, *Dokl. Akad. Nauk SSSR* **108,** 481 (1956).
51. G. W. Kramer and H. C. Brown, *J. Organometal. Chem.* **73,** 1 (1974).
52. S. Cabiddu, A. Maccioni and M. Secci, *Gazz. Chim. Ital.* **102,** 555 (1972).
53. J. Soulie and P. Cadiot, *Bull. Soc. Chim. France* **1966,** 1981.
54. B. M. Mikhailov, T. V. Kostroma and N. S. Fedotov, *Izv. Akad. Nauk SSSR, Otd. Khim. Nauk* **1957,** 589.
55. J. C. Doty, B. Bobb, P. J. Grisdale, M. Glogowski and J. L. R. Williams, *J. Organometal. Chem.* **38,** 229 (1972).
56. G. F. Hennion, P. A. McCusker and A. J. Rutkowski, *J. Am. Chem. Soc.* **80,** 617 (1958).
57. G. F. Hennion, P. A. McCusker, E. C. Ashby and A. J. Rutkowski, *J. Am. Chem. Soc.* **79,** 5190 (1957).
58. G. F. Hennion, P. A. McCusker and J. V. Marra, *J. Am. Chem. Soc.* **80,** 3481 (1958).
59. G. Urry, In: "The Chemistry of Boron and Its Compounds." Ed. E. L. Muetterties. New York, J. Wiley and Sons (1967), p. 325.
60. H. C. Brown and M. M. Rogic, *J. Am. Chem. Soc.* **91,** 4304 (1969).
61. A. N. Nesmeyanov, V. A. Sazonova, V. A. Blinova and S. G. D'yachenko, *Dokl. Akad. Nauk SSSR* **200,** 1365 (1971).
62. J. Deberitz, K. Dirscherl and H. Nöth, *Chem. Ber.* **106,** 2783 (1973).
63. H. C. Brown and S. K. Gupta, *J. Organometal. Chem.* **32,** C1 (1971).
64. H. C. Brown, E. Negishi and S. K. Gupta, *J. Am. Chem. Soc.* **92,** 6648 (1970).
65. B. M. Mikhailov and Yu. N. Bubnov, *Zh. Obshch. Khim.* **31,** 160 (1961).
66. B. M. Mikhailov and Yu. N. Bubnov, *Izv. Akad. Nauk SSSR, Ser. Khim.* **1964,** 2248.
67. A. Pelter and D. N. Sharrocks, *Chem. Comm.* **1972,** 566.
68. A. Pelter, K. Rowe and K. Smith, *Chem. Comm.* **1975,** 532.
69. R. L. Letsinger and I. Skoog, *J. Am. Chem. Soc.* **76,** 4174 (1954).
70. B. M. Mikhailov and V. A. Vaver, *Izv. Akad. Nauk SSSR, Otd. Khim. Nauk* **1957**, 989.
71. M. E. Glogowski, P. J. Grisdale, J. L. R. Williams, and T. H. Rodan, *J. Organometal. Chem.* **54,** 51 (1973)
72. V. H. Dodson and V. Fisher, Ohio. *J. Sci.* **58,** 141 (1958).
73. M. Schmidt and E. Gast, Z. Naturforsch. **23b,** 1258 (1968)
74. B. M. Mikhailov and T. A. Shchegoleva, *Izv. Akad. Nauk SSSR, Otd. Khim. Nauk* **1955,** 1124.
75. B. M. Mikhailov and T. A. Shchegoleva, *Zn. Obshch. Khim.* **29,** 3130 (1959).
76. B. M. Mikhailov and P. M. Aronovich, *Izv. Akad. Nauk SSSR, Otd. Khim. Nauk* **1955,** 946.

77. H. A. Staab and B. Meissner, *Ann.* **753,** 80 (1971).
78. I.G.C. Coutts and O. C. Musgrave, *J. Chem. Soc., C* **1970,** 2225.
79. B. M. Mikhailov and P. M. Aronovich, *Izv. Akad. Nauk SSSR, Otd. Khim. Nauk* **1956,** 322.
80. R. Köster, *Angew. Chem.* **71,** 31 (1959).
81. R. Köster, *Angew. Chem.* **73,** 66 (1961).
82. R. Köster and M. A. Grassberger, *Ann.* **719,** 169 (1968).
83. B. M. Mikhailov and L. S. Vasil'ev, *Zh. Obshch. Khim.* **35,** 925 (1965).
84. B. M. Mikhailov and L. S. Vasil'ev, *Izv. Akad. Nauk SSSR, Otd. Khim. Nauk* **1962,** 1756.
85. B. M. Mikhailov and L. S. Vasil'ev, *Dokl. Akad. Nauk SSSR* **139,** 385 (1961).
86. H. C. Brown and S. K. Gupta, *J. Am. Chem. Soc.* **93,** 2802 (1971).
87. R. L. Letsinger and N. Remes, *J. Am. Chem. Soc.* **77,** 2489 (1955).
88. R. Köster, K.-L. Amen and W. V. Dahlhoff, *Ann.* **1975,** 752.
89. A. M. Yurkevich and O. N. Shevtsova, *Zh. Obshch. Khim.* **42,** 1993 (1972).
90. A. M. Yurkevich and G. V. Parkhomenko, *Zh. Obshch. Khim.* **37,** 1977 (1967).
91. A. M. Yurkevich and O. N. Shevtsova, *Zh. Obshch. Khim.* **42,** 1172 (1972).
92. G. Zinner and W. Kliegel, *Chem. Ber.* **99,** 895 (1966).
93. A. M. Yurkevich, L. S. Varshavskaya and I. I. Kolodkina, *Zh. Obshch. Khim.* **38,** 2115 (1968).
94. I. I. Kolodkina, A. S. Guseva, E. A. Ivanova, L. S. Varshavskaya and A. M. Yurkevich, *Zh. Obshch. Khim.* **40,** 2489 (1970).
95. I. P. Rudakova, V. N. Rekunova, L. I. Kuznetsova and A. N. Yurkevich, *Zh. Obshch. Khim.* **43,** 400 (1973).
96. V. N. Rekunova, I. P. Rudakova and A. M. Yurkevich, *Tetrahedron Letters* **1973,** 2811.
97. N. V. Kruglova and R. Kh. Freidlina, *Izv. Akad. Nauk SSSR, Ser. Khim.* **1966,** 1662.
98. N. V. Kruglova and R. Kh. Freidlina, *Izv. Akad. Nauk SSSR, Ser. Khim.* **1965,** 2044.
99. S. O. Lawesson, *Arkiv Kemi* **10,** 171 (1956).
100. I. G. C. Coutts, H. R. Goldschmid and O. C. Musgrave, *J. Chem. Soc. C* **1970,** 488.
101. P. B. Brindley, W. Gerrard and M. F. Lappert, *J. Chem. Soc.* **1955,** 2956.
102. B. M. Mikhailov and P. M. Aronovich, *Dokl. Akad. Nauk SSSR* **98,** 791 (1954).
103. R. Letsinger and J. R. Nazy, *J. Am. Chem. Soc.* **81,** 3013 (1959).
104. R. Letsinger and A. J. Wysocki, *J. Org. Chem.* **28,** 3199 (1963).
105. D. Nielsen and W. McEwen, *J. Am. Chem. Soc.* **79,** 3081 (1957).
106. O. C. Musgrave, *Chem. and Ind.* **1957,** 1152.
107. A. N. Nesmeyanov, V. A. Sazonova and V. N. Drozd, *Dokl. Akad. Nauk SSSR* **126,** 1004 (1959).
108. T. S. Liksandru, *Zh. Obshch. Khim.* **42,** 1991 (1972).
109. G. Marr, R. E. Moore and B. W. Rockett, *J. Chem. Soc., C* **1968,** 24.
110. D. J. Booth, G. Marr, B. W. Rockett and A. Rushworth, *J. Chem. Soc., C* **1969,** 2701.
111. A. N. Nesmeyanov, V. A. Sazonova and V. E. Fedorov, *Izv. Akad. Nauk SSSR, Ser. Khim.* **1970,** 2133.
112. A. N. Nesmeyanov, N. E. Kolobova, Yu. V. Makarov and K. N. Anisimova, *Izv. Akad. Nauk SSSR, Ser. Khim.* **1969,** 1992.
113. W. Gerrard, E. F. Mooney and R. G. Rees, *J. Chem. Soc.* **1964,** 740.
114. K. Niedenzu, J. W. Dawson and P. Fritz, *Inorg. Synth.* **10,** 126 (1967).
115. J. E. Burch, W. Gerrard, M. Hawarth and E. F. Mooney, *J. Chem. Soc.* **1960,** 4916.

116. J. Hooz and J. G. Calzada, *Org. Prep. Proced. Internat.* **4,** 219 (1972).
117. W. Gerrard, M. Hawarth, E. F. Mooney and D. E. Pratt, *J. Chem. Soc.* **1963,** 1582.
118. A. Finch, P. J. Gardner, H. Hill and K. S. Hussian, *J. Chem. Soc., (Dalton)* **1973,** 2543.
119. C. E. Eggers and S. F. A. Kettle, *Inorg. Chem.* **6,** 1975 (1967).
120. H. C. Brown and S. K. Gupta, *J. Am. Chem. Soc.* **92,** 6983 (1970).
121. H. C. Brown and A. B. Levy, *J. Organometal. Chem.* **44,** 233 (1972).
122. E. Pace, *Atti Accad. Nazl. Lincei* **10,** 193 (1929); *C. A.* **24,** 1360 (1930).
123. E. L. Muetterties. *J. Am. Chem. Soc.* **81,** 2597 (1959).
124. E. L. Muetterties and F. N. Tebbe, *Inorg. Chem.* **7,** 2663 (1968).
125. S. McVey, I. G. Morrison and P. E. Pauson, *J. Chem. Soc., C* **1967,** 1847.
126. J. Kotz and E. W. Post, *Inorg. Chem.* **9,** 1661 (1970).
127. B. Asgarauladi, R. Full, K. Schaper and W. Siebert, *Chem. Ber.* **107,** 34 (1974).
128. W. Ruf, M. Fueller and W. Siebert, *J. Organometal. Chem.* **64,** C45 (1974).
129. R. A. Bowie and O. C. Musgrave, *J. Chem. Soc., C* **1970,** 485.
130. E. L. Muetterties, *J. Am. Chem. Soc.* **82,** 4163 (1960).
131. M. Schmidt, W. Siebert and F. R. Rittig, *Chem. Ber.* **101,** 281 (1968).
132. W. Siebert, F. R. Rittig and M. Schmidt, *J. Organometal. Chem.* **25,** 305 (1970).
133. W. Siebert, K. J. Schaper and M. Schmidt, *J. Organometal. Chem.* **25,** 315 (1970).
134. L. S. Vasil'ev, V. P. Dmitrikov, V. S. Bogdanov and B. M. Mikhailov, *Zh. Obshch. Khim.* **42,** 1318 (1972).
135. L. Miginiac and J. Blais, *J. Organometal. Chem.* **29,** 349 (1971).
136. A. M. Pichuzhkina, I. I. Kolodkina and A. M. Yurkevich, *Zh. Obshch. Khim.* **43,** 2275 (1973).
137. H. Weidmann and H. K. Zimmerman, Zimmerman, Jr., *Ann.* **619,** 28 (1958).
138. K. Torssell, *Acta Chem. Scand.* **8,** 1779 (1954).
139. M. L. Wolfrom and J. Solms, *J. Org. Chem.* **21,** 815 (1956).
140. J. M. Sugihara and C. M. Bowman, *J. Am. Chem. Soc.* **80,** 2443 (1958).
141. R. J. Ferrier, D. Prasad and A. Rudowski, *J. Chem. Soc.* **1965,** 858.
142. A. S. Guseva, S. G. Verenikina, S. F. Dymova and A. M. Yurkevich, *Zh. Obshch. Khim.* **44,** 2327 (1974).
143. M. G. Edelev, T. M. Filippova, V. N. Robe, I. K. Shmyrev, A. S. Guseva, S. G. Verenikina and A. M. Yurkevich, *Zh. Obshch. Khim.* **44,** 2321 (1974).
144. J. R. McKinley and H. Weigel, *Chem. Comm.* **1972,** 1051.
145. R. J. Ferrier, D. Prasad, A. Rudowski and I. Sangster, *J. Chem. Soc.* **1964,** 3330.
146. R. J. Ferrier and D. Prasad, *J. Chem. Soc.* **1965,** 7425.
147. R. J. Ferrier and D. Prasad, *J. Chem. Soc.* **1965,** 7429.
148. A. M. Yurkevich, I. I. Kolodkina, L. S. Varshavskaya, V. I. Borodulina-Shvetz, I. P. Rudakova and N. A. Preobrazhenski, *Tetrahedron* **25,** 477 (1969).
149. A. M. Yurkevich, L. S. Varshavskaya, I. I. Kolodkina and N. A. Preobrazhenskii, *Zh. Obshch. Khim.* **37,** 2002 (1967).
150. A. M. Yurkevich, V. I. Borodulina-Shvets, I. I. Kolodkina and N. A. Preobrazhenskii, *Zh. Obshch. Khim.* **37,** 2176 (1967).
151. S. G. Verenikina, A. M. Yurkevich and N. A. Preobrazhenskii, *Zh. Obshch. Khim.* **37,** 2181 (1967).
152. A. M. Yurkevich, S. G. Verenikina, E. G. Chauser and N. A. Preobrazhenskii, *Zh. Obshch. Khim.* **36,** 1746 (1966).
153. E. M. Tkacheva, I. P. Rudakova and A. M. Yurkevich, *Zh. Obshch. Khim.* **43,** 2049 (1973).

154. T. A. Pospelova, I. P. Rudakova and A. M. Yurkevich, *Zh. Obshch. Khim.* **44,** 434 (1974).
155. E. A. Ivanova, I. I. Kolodkina and A. M. Yurkevich, *Zh. Obshch. Khim.* **41,** 455 (1971).
156. I. I. Kolodkina, E. A. Ivanova and A. M. Yurkevich, *Khim. Prir. Soedin.* **6,** 612 (1970).
157. E. A. Ivanova, I. I. Kolodkina and A. M. Yurkevich, *Zh. Obshch. Khim.* **44,** 430 (1974).
158. H. L. Weith, J. L. Weibers and P. T. Gilham, *Biochemistry* **9,** 4396 (1970).
159. H. Schott, E. Rudloff, P. Schmidt, R. Roychoudhury and H. Kössel, *Biochemistry* **12,** 932 (1973).
160. E. A. Ivanova, S. I. Panchenko, I. I. Kolodkina and A. M. Yurkevich, *Zh. Obshch. Khim.* **45,** 208 (1975).
161. S. A. Barker, B. W. Hatt, P. J. Sommers and R. R. Woodbury, *Carbohydr. Res.* **26,** 55 (1973).
162. H. Nöth and T. Taeger, *J. Organometal. Chem.* **142,** 281 (1977).
163. C. G. Whiteley, *Chem. Commun.* **1981,** 5.
163a. T. A. Shchegoleva, E. M. Shashkova and B. M. Mikhailov, *Izv. Akad. Nauk SSSR, Ser. Khim.* **1981,** 1098.
164. P. Jacob, III., *J. Organometal. Chem.* **156,** 101 (1978).
165. P. J. Maddocks, A. Pelter, K. Rowe, K. Smith and C. Subrahmanyam, *J. Chem. Soc., Perkin Trans. I* **1981,** 653.
166. A. Pelter, K. Rowe, D. N. Sharrocks, K. Smith and C. Subrahmanyam, *J. Chem. Soc., Dalton Trans.* **1976,** 2087.
167. E. V. Chistova, V. A. Alyasov, R. F. Galiullina, V. P. Maslennikov, V. A. Dodonov and Yu. A. Aleksandrov, *Zh. Obshch. Khim.* **51,** 1078 (1981).
168. R. F. Galiullina, E. V. Chistova and V. A. Dodonov, *Zh. Obshch. Khim.* **50,** 1657 (1980).
169. S. W. Breuer and F. G. Thorpe, *Tetrahedron Letters* **1974,** 3719.
170. S. W. Breuer, G. M. Pickles, J. C. Podesta and F. G Thorpe, *Chem. Commun.* **1975,** 36.
171. D. S. Matteson and D. Majumdar, *J. Am. Chem. Soc.* **102,** 7588 (1980).
172. H. C. Brown, N. R. De Lue, Y. Yamamoto, K. Maruyama, T. Kasahara, S. Marahashi and A. Sonoda, *J. Org. Chem.* **42,** 4088 (1977).
173. M. Rathke, E. Chao and G. Wu, *J. Organometal. Chem.* **122,** 145 (1976).

CHAPTER 2

The hydroboration of alkenes and cyclenes

A general method for the synthesis of trialkylboranes and tricyclylboranes is the addition of diborane and of other hydroborating agents to alkenes and cyclenes.

Together with the oxidation of organoboranes by an alkaline solution of hydrogen peroxide, which proceeds with retention of the configuration, hydroboration is a convenient preparative method for the hydration of compounds of different types with C=C bonds, including natural compounds.

The ability of diborane to add to the double bond of olefins with the formation of trialkylboranes was discovered by Hurd in 1948 [1]. Later systematic investigations established the general characteristics of the reaction, elucidated the problems of orientation and stereochemistry, and led to the discovery of new methodical routes of hydroboration, which proved to be valuable both in the chemistry of organoboron compounds and in organic chemistry as a whole.

The addition reactions of diborane and its derivatives to multiple carbon–carbon bonds, known under the name of "hydroboration", have been dealt with in monographs [2–5] and in review articles [6, 7].

2.1 PREPARATION OF HYDROBORATING AGENTS

Diborane, tetraalkyl-, dialkyl-, and diaryldiboranes; complex compounds of borane, alkyl-, and arylboranes; alkylmercaptoboranes, and complex compounds of haloboranes are used as hydroborating agents.

2.1.1 Preparation of diborane

Of the different methods for the preparation of diborane [3], the most convenient from the preparative aspect are those based on the use of hydrides and complex hydrides of metals.

The method for the preparation of diborane from lithium hydride and boron trifluoride etherate has been investigated in detail by Elliott and co-workers [8].

The reaction takes place with different stoichiometric ratios, depending on the conditions:

$$6LiH + 8BF_3 \xrightarrow{\text{ether}} B_2H_6 + 6LiBF_4,$$

$$6LiH + 2BF_3 \xrightarrow[\text{ether}]{\text{promoter}} B_2H_6 + 6LiF.$$

With a reagent ratio of 6:8 in diethyl ether, the yield of diborane is 75%. When a smaller amount of boron trifluoride etherate is used (6:2) in the presence of activators [$LiBH_4$ or $B(OCH_3)_3$], the yield of diborane is 85–88%.

Sodium hydride reacts with boron trifluoride etherate analogously to lithium hydride:

$$6NaH + 8BF_3 \longrightarrow B_2H_6 + 6NaBF_4.$$

The yield of B_2H_6 is 40% in diethyl ether, while it reaches a value of 92% in diglyme [9].

Diborane may be obtained from sodium hydride and BCl_3 in the presence of sodium borohydride [10], aluminum chloride [11], a mixture of aluminum chloride and bromide [12], or nitrogenous bases [13].

Esters of metaboric acid may be used instead of boron halides in reactions with alkali metal hydrides [14, 15]:

$$NaH + (ROBO)_3 \longrightarrow B_2H_6 + H_2 + B(OR)_3 + Na_2O.$$

A number of effective methods for the preparation of diborane are based on the use of complex metal hydrides.

In 1947, Finholt, Bond, and Schlesinger [16] found that diborane is obtained in almost quantitative yield by the reaction of lithium aluminum hydride with BCl_3 in an ether medium:

$$3LiAlH_4 + 4BCl_3 \longrightarrow 3LiCl + 3AlCl_3 + 2B_2H_6.$$

Ether is capable of splitting a hydride ion from AlH_4^- by a coordination reaction with AlH_3 [17].

Shapiro, et al. [18] suggested the preparation of diborane from lithium aluminum hydride and boron trifluoride etherate in ether solution:

$$3LiAlH_4 + 4BF_3 \longrightarrow 2B_2H_6 + 3LiF + 3AlF_3.$$

The diborane obtained in this way contains ethyl fluoride as an impurity [19].

Metal borohydrides are successfully used for the preparation of diborane.

Diborane is formed quantitatively from boron trifluoride etherate and lithium borohydride in ether solution [8]:

$$3LiBH_4 + BF_3 \xrightarrow{\text{ether}} 2B_2H_6 + 3LiF.$$

Schlesinger, et al. [20] used sodium borohydride for the preparation of diborane.

$$3NaBH_4 + 4BF_3 \xrightarrow{\text{ether}} 2B_2H_6 + 3NaBF_4.$$

In the above method for the preparation of diborane, diglyme, tetrahydrofuran [6], or ether [21] is used as solvent.

Diborane is prepared by the action of hydrogen chloride [22], sulfuric acid [23], methanesulfonic acid [24], or acetic acid [25, 26] on metal borohydrides.

Diborane is also obtained by the action of methyl iodide, ethyl bromide, or 1,2-dichloroethane on tetraalkylammonium borohydride in methylene chloride [27].

2.1.2 Preparation of tetraalkyl-, dialkyl-, and diaryldiboranes

Preparation from trialkyl- or triarylboranes and diborane. Alkyldiboranes were synthesized for the first time by Schlesinger, et al. [28–30] by the action of diborane on trialkylboranes, using high-vacuum equipment. Symmetrical dimethyldiborane is not obtained directly from trimethylborane and diborane, but is synthesized by the treatment of monomethyldiborane with dimethyl ether [30] or tetrahydrofuran [31] at low temperature; in this process, methyldiborane is converted to borane etherate and methylborane, which then dimerizes:

$$2CH_3B_2H_5 + 2\,(CH_3)_2\,O \longrightarrow 2H_3B{\cdot}O\,(CH_3)_2 + (CH_3BH_2)_2.$$

Ethyldiborane and unsymmetrical diethyldiborane can be obtained by the reaction of triethylborane with diborane (1:5); with excess triethylborane, a mixture of triethyldiborane and tetraethyldiborane is formed [29, 32]. The reaction of diborane with tripropylborane (8:1) leads to propyldiborane and unsymmetrical dipropyldiborane [29].

The alkyldiboranes mentioned above were obtained in high-vaccum equipment in amounts of a few millimoles. Mikhailov, et al. [33] obtained alkyldiboranes in large quantities by the use of the usual methods of organic synthesis. By the introduction of diborane into ether solutions of tripropyl- or tributylborane, they synthesized the thermally relatively stable tetrapropyl- and tetrabutyldiboranes:

$$4BR_3 + B_2H_6 \longrightarrow 3R_4B_2H_2$$

and the unstable unsymmetrical dipropyldiborane.

The infrared spectra of the tetraalkyldiboranes contain absorption bands in the region of 2500 cm^{-1}, indicating their partial disproportionation to trialkylboranes and trialkyldiboranes.

A modification of the above method is the method of preparation of alkyldiboranes from trialkylboranes (trimethyl- and triethylborane), lithium or sodium borohydride, and hydrogen chloride proposed by Long and Wallbridge [34–36].

Wiberg, et al. [37], synthesized 1,2-diphenyldiborane by heating a mixture of triphenylborane and diborane at 40–100°C and 2–4 atmospheres:

$$2B\,(C_6H_5)_3 + 2B_2H_6 \longrightarrow 3\,(C_6H_5BH_2)_2.$$

Mikhailov and Dorokhov [38] found that the reaction between triphenylborane and diborane proceeds smoothly in ether solution, leading to a 74% yield of 1,2-diphenyldiborane.

Preparation from trialkylboranes and hydrogen. Köster, et al. [39, 40] proposed an effective method for the preparation of alkyldiboranes, consisting of heating trialkylboranes with hydrogen to 160–190°C under a pressure of 200–400 atmospheres. The hydrogenolysis of triphenylborane is carried out by heating it with hydrogen to 170–180°C under a pressure of 100 atmospheres, forming different phenyl-boron-hydrogen compounds together with the expected phenylboranes [40].

Preparation from olefins and diborane. A convenient preparative method for the synthesis of tetraalkyldiboranes consists of the action of diborane on olefins in an ether medium, with a reagent ratio of 1:4:

$$4CH_2=CHR + B_2H_6 \longrightarrow (RCH_2CH_2)_4 B_2H_2.$$

Mikhailov, et al. [33] prepared tetrapropyl- and tetrabutyldiboranes for the first time by means of this method, starting from propylene or 1-butene. The reaction is carried out with cooling (−50 to −30°C); after the introduction of diborane is complete, the reaction mass is allowed to heat to room temperature, the solvent is distilled off, and the residue is subjected to fractional distillation. The yield of tetraalkyldiboranes depends on the precision with which the ratio of reagents is maintained. A tetrahydrofuran solution is conveniently used to measure out the diborane.

Brown, et al., in an investigation of the reaction of diborane with α-olefins (1-pentene [41], isobutylene, 2-methyl-1-butene, 2,4,4-trimethyl-1-pentene [42]) in a ratio of 1:4 treated the crude mixture with methanol. On the basis of the yields of the methyl esters of dialkylborinic acids (53–70%), methyl borate (10–13%), and trialkylboranes (20–26%), the authors assumed that the reaction mixture contains R_2BH, BH_3, and BR_3 in the same ratios. However, alcoholysis can not be used as criterion for the disproportionation of tetraalkyldiboranes, since the alkoxy derivatives formed during the reaction, under the action of the compounds with B—H bonds contained in the reaction mixture, enter into exchange reactions [33, 43], the course of which depends on the conditions under which the reaction is carried out and, in particular, on the nature of the hydrolyzing agent. Thus, according to the data of Mikhailov, et al. [33], under the action of alcohols on tetraalkyldiboranes (tetrapropyl- and tetrabutyldiboranes), esters of dialkylborinic acids are obtained in yields of approximately 65%, together with esters of alkylboronic acids and trialkyl borates, whereas, in protolysis by ethylene glycol, the esters of dialkylborinic and alkylboronic acids and trialkyl boranes are obtained in almost identical yields. At the same time, protonolysis with butyl mercaptan leads to butyl dialkylthioborinate in almost quantitative yield (83%).

The hydroboration of 2-methyl-2-butene and 2,3-dimethyl-2-butene produced, respectively, tetra(3-methyl-2-butyl)diborane (called disiamylborane in the monomeric form) and 1,2-bis(2,3-dimethyl-2-butyl)diborane (called thexylborane in the monomeric form) [44]:

$$CH_3\overset{\displaystyle CH_3}{\overset{|}{C}}{=}CH{-}CH_3 \xrightarrow{B_2H_6} (CH_3{-}\overset{\displaystyle CH_3}{\overset{|}{C}H}\overset{\displaystyle CH_3}{\overset{|}{C}H}{-})_4B_2H_2$$

$$CH_3\overset{\displaystyle CH_3}{\overset{|}{C}}{=}\overset{\displaystyle CH_3}{\overset{|}{C}}CH_3 \xrightarrow{B_2H_6} (CH_3\overset{\displaystyle CH_3}{\overset{|}{C}H}{-\!-}\underset{\displaystyle CH_3}{\underset{|}{\overset{\displaystyle CH_3}{\overset{|}{C}}}}{-})_2B_2H_4$$

Tetracyclohexyldiborane and tetra(*trans*-2-methylcyclohexyl)-diborane, respectively, were obtained from diborane and cyclohexene or 2-methylcyclohexene [44]. Bis(2,3-dimethyl-2-butyl)borane does not form a dimer [45].

The hydroboration of α-pinene at 0°C proceeds from the spatially less hindered side and, if excess hydrocarbon is used, leads to symmetrical tetraisopinocampheyldiborane [44, 46, 47]. The product formed from (+)-(1R, 5R)-α-pinene is assigned the absolute configuration (1R, 2S, 3R, 5R) for each isopinocampheyl group [44, 48] and, for simplicity, is called (−)-diisocampheylborane or (−)-di-3-pinanylborane. Tetraisopinocampheyldiborane dissociates to the extent of approximately 5–10% to triisocampheyldiborane and α-pinene, with the degree of dissociation depending on the concentration of the dimer, the amount of excess α-pinene present, and the nature of the solvent.

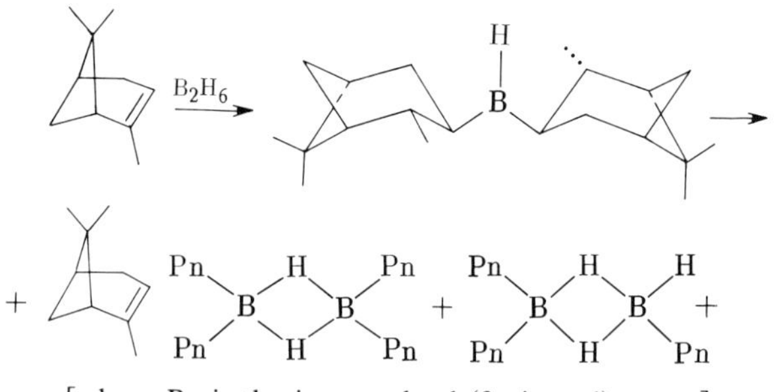

[where Pn is the isocampheyl (3-pinanyl) group]

The reaction of β-pinene with diborane (4:1) in THF, according to the IR spectroscopic data (an intensive absorption band of the bridging bond at 1565 cm^{-1} and a doublet at 2577 and 2518 cm^{-1}), produces mainly tetra(*cis*-myrtanyl)diborane and a small quantity of 1,1-disubstituted diborane [44].

Preparation from esters of alkyl- or arylboronic acids and diborane. Mikhailov and Dorokhov [21, 38, 48, 49] proposed a general preparative method for the synthesis of 1,2-diaryldiboranes, consisiting of the action of diborane on esters of arylboronic or diarylborinic acids in ether solution or in an inert solvent medium. The 1,2-diaryldiboranes are obtained in a yield of 75–80% from esters of diarylborinic acids, according to the following equation:

$$3\,Ar_2BOR + 2\,B_2H_6 \longrightarrow 3\,(ArBH_2)_2 + B(OR)_3$$

$$(Ar = C_6H_5,\ o\text{-}CH_3C_6H_4,\ p\text{-}CH_3C_6H_4,\ p\text{-}ClC_6H_4,\ \alpha\text{-}C_{10}H_7).$$

The action of diborane on esters of arylboronic acids produces 1,2-diaryldiboranes in only 40–50% yields, with 20–25% of the starting ester being recovered [38]. This is explained by the reversible process:

$$6ArB(OR)_2 + 2B_2H_6 \rightleftarrows 3(ArBH_2)_2 + 4B(OR)_3,$$

which has been demonstrated for the example of the reaction of 1,2-diphenyldiborane with *n*-butyl borate. Approximately 40% of diborane was isolated from an ether solution of a mixture of these substances, and dibutyl phenylboronate was obtained in a yield of 44%.

The reaction of diborane with esters of dialkylborinic acids leads to esters of alkylboronic acids and dialkyl- and tetraalkyldiboranes, with the ratio of the latter compounds depending on the ratio of the starting materials and the experimental conditions [43, 50]. Thus, for example, the reaction of hexyl dipropylborinate with diborane (2:1) for 7 days forms dihexylpropylboronate and a mixture of isomeric dipropyldiboranes:

$$2R_2BOR' + B_2H_6 \rightleftarrows RB(OR')_2 + 1{,}5R_2B_2H_4.$$

With an excess of dialkylborinate, the reaction proceeds further, as a result of which esters of alkylboronic acids and tetraalkyldiboranes are obtained:

$$8R_2BOR' + B_2H_6 \rightleftarrows 4RB(OR')_2 + 3(R_2BH)_2.$$

Preparation from alkyl(aryl)chloroboranes and diborane or complex metal hydrides. 1,2-Diaryldiboranes are formed by the reaction of diarylchloroboranes and diborane in an ether medium [38]:

$$Ar_2BCl \xrightarrow{B_2H_6} \underset{60\%}{(ArBH_2)_2}.$$

The reaction of alkyl(aryl)chloroboranes can be carried out with the use of lithium borohydride or lithium aluminum hydride. Thus, 1,2-diphenyldiborane has been obtained by the action of lithium borohydride on phenyldichloroborane [37]:

$$2C_6H_5BCl_2 + 4LiBH_4 \longrightarrow (C_6H_5BH_2)_2 + 4LiCl + 2B_2H_6,$$

and tetraalkyldiboranes have been prepared by the reduction of dialkylchloroboranes with lithium aluminum hydride or sodium borohydride [34, 36], e.g.:

$$4(C_3H_7)_2BCl + LiAlH_4 \longrightarrow 2(C_3H_7)_4B_2H_2 + LiCl + AlCl_3.$$

IR Spectra of alkylboranes. IR spectroscopic data are widely used for the identification of alkylboranes.

The IR spectra of methyl- and ethyldiboranes have been investigated by Lehmann, Wilson, and Shapiro [51–56]. The IR spectra of all alkyldiboranes contain a very intense absorption band in the region of 1580–1610 cm^{-1}, corresponding to antisymmetrical valence vibrations of the BH_2B bridge [57]. An exception is represented by 1,1-dialkyldiboranes, in which this band is displaced in the direction of lower frequencies by 50 cm^{-1}. The intense absorption bands in the region of 2500–2600 cm^{-1} correspond to valence vibrations of terminal B—H bonds. Alkyldiboranes with BH_2 groups (mono- or 1,1-dialkyldiboranes) contain twin or ordinary peaks. Tetraalkyldiboranes naturally do not show absorption bands in this region. Additional characteristics of alkyldiboranes are the absorption bands caused by the bending vibrations of the BH_2 groups (1140–1170 cm^{-1}), the deformational planar vibrations of the B—H groups (1110–1180 cm^{-1}), and the wagging out-of-plane vibrations of the BH_2 groups (930 ± 10 cm^{-1}). Unfortunately, these absorption bands are frequently overlapped by other bands, which limits their analytical value [56]. Trialkyldiboranes frequently show an additional intense peak near the absorption band of the deformational B—H vibration, caused by antisymmetric valence vibrations of the BC_2 groups. The frequencies of the valence terminal B—C bonds are reduced by approximately 10–15 cm^{-1} for each alkyl group, if it is added to another boron atom or to the other end of the molecule. However, when the alkyl group and the hydrogen atom are added to the same boron atom, then the frequency of the B—H bond is reduced by an average of 25 cm^{-1}.

Infrared spectroscopy has been used successfully for studying disproportionation reactions of alkyldiboranes [58].

2.1.3 Preparation of complex compounds of borane and alkyl- and arylboranes

In its reaction with ethers, diborane undergoes dissociation with the formation of coordination compounds. These adducts are very unstable and their existence, as a rule, is established by means of physical methods of investigation.

Schlesinger and Burg isolated neutral complexes of borane [59] and deuteroborane [60] with dimethyl ether in the individual state. These complexes decompose to the starting components within several hours at −78.5°C:

$$B_2H_6 + 2O(CH_3)_2 \rightleftarrows 2H_3B \cdot O(CH_3)_2.$$

As follows from a study of the phase diagrams [61], tetrahydrofuran also forms a complex with borane, which has been isolated in the form of a solid at −78°C, with a saturated vapor pressure of 1.6 mm [62].

A solution of diborane in tetrahydrofuran readily conducts an electric current, from which it can be concluded that the solution, in addition to the neutral complex, also contains an ionic complex of the boronium type [63], similar to the "diammoniate of diborane" [64].

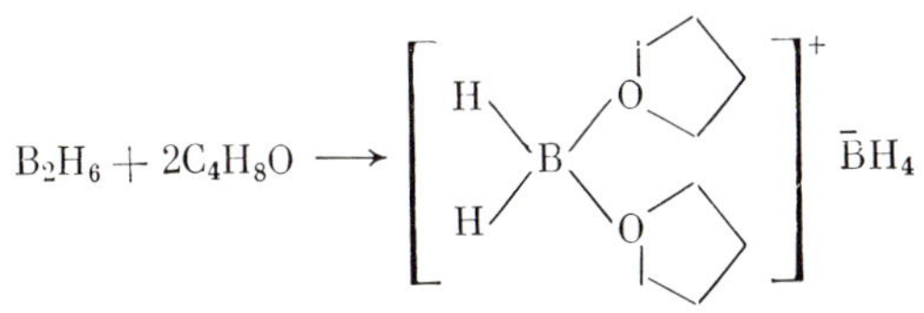

Diborane is significantly less soluble in diethyl ether (1.1 g per 100 g) than in tetrahydrofuran (8 g per 100 g) [65]. Accordingly, the borane etherate is unstable and its existence is demonstrated by a study of the vapor-liquid equilibrium in the ether-diborane system [66], of the phase diagrams [67], and of the Raman spectra [68]. From the data of Rice and Ushida [68], the following order of stability of the ether complexes of borane is obtained:

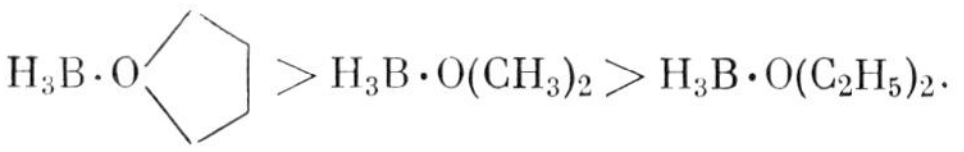

Diborane dissolves relatively well in diethylene glycol dimethyl ether [69].

McAchran and Shore [70], on the basis of a tensimetric study of solutions of diborane and dimethyl sulfoxide in dichloromethane at −78°C, and also of data on the B–11 NMR spectra ($\delta = -38$ ppm and $J_{B-H} = 86$ Hz), suggested the formation of a boronium salt:

$$B_2H_6 + 2(CH_3)_2SO \rightarrow \{H_2B[OS(CH_3)_2]_2\}^+ BH_4^-.$$

Diborane is capable of splitting ether bonds, which is very important, since such compounds as tetrahydrofuran, ethylene glycol dimethyl ether, and polyethylene glycol ethers are used as solvents in hydroboration. According to the data of [71], a solution of diborane in THF (concentration 0.35 M) withstands heating at 50–55°C without marked chemical changes. However, Kollonitsch [72] found that diborane actually reacts with THF upon moderate heating with the formation of tributyl borate, with a yield of 61% of the latter compound, if the solution is maintained at 60°C for 64 hours:

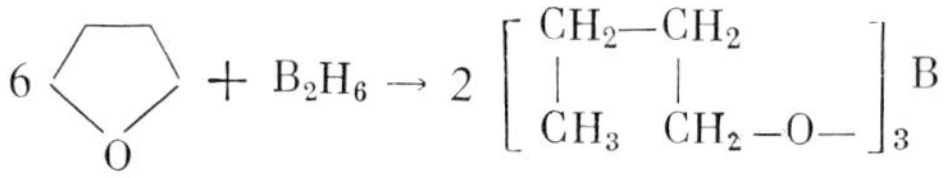

α-Methyltetrahydrofuran is also decomposed by diborane with the formation of 2-pentylborate, the yield of which reaches a value of 30% after 40 hours at 60°C [72].

Diborane is also not inert with respect to diglyme [73]. Thus, a reaction mixture containing $BF_3 \cdot O(C_2H_5)_2$,$NaBH_4$, and an olefin in diglyme contains diethylene glycol monomethyl ether after standing for only 10 minutes at room temperature. The rate of the reaction of diborane with diglyme increases with increasing temperature and pressure [74]. The decomposition of ethers by diborane takes place at room temperature in the presence of iodine [75].

Adducts of borane and alkyl(aryl)boranes with nitrogen and sulfur-containing ligands are used in hydroboration practice.

Alkylamine-boranes are obtained by the direct reaction of diborane with amines in diethyl ether [76, 77] or diglyme [78] in high yields, and also by the action of lithium borohydride on amine hydrochlorides [79–81], for example:

$$[(CH_3)_3NH]Cl + LiBH_4 \longrightarrow H_3B \cdot N(CH_3)_3 + H_2 + LiCl.$$

Amine complexes of alkyldiboranes are prepared by the action of amines on alkyldiboranes [30, 44]. Alkylamine-arylboranes are formed by the reaction of 1,2-diaryldiboranes with primary, secondary, or tertiary amines [50, 76, 82].

Amine complexes of alkylboranes are also obtained by the reduction of anhydrides of alkylboronic acids with lithium aluminum hydride in ether medium in the presence of tertiary amines (40–65%) [83, 84] or of pyridine (11–27%) [85]:

$$(RBO)_3 + 3NR'_3 \xrightarrow[35^\circ\ C]{LiAlH_4} 3RBH_2 \cdot NR'_3.$$

Esters of arylboronic acids or their anhydrides, when reduced with lithium aluminum hydride in ether solution in the presence of pyridine or of tertiary amines, are converted to pyridinates or aminates of arylboranes, respectively [85, 86]:

$$ArB(OR)_2 \xrightarrow[L]{LiAlH_4} ArBH_2 \cdot L \qquad (L = NR_3, \text{ pyridine}).$$

The yields of pyridine-arylboranes range from 17 to 61%, depending on the nature of the aryl group. The yield is 28% for trimethylamine-phenylborane and 51% for pyridine-phenylborane.

The reduction of esters of diarylborinic acids by lithium aluminum hydride in ether solution in the presence of pyridine leads to the formation of pyridine-diarylboranes in yields of 32–73% [85, 87]:

$$Ar_2BOR \xrightarrow[\text{pyridine}]{LiAlH_4} Ar_2BH \cdot NC_5H_5.$$

Pyridinates of dialkylboranes can also be obtained by this method [88].

Trimethylamine-alkylboranes are formed by the reduction of 2-alkyl-1,3,2-benzodioxaboroles with lithium aluminum hydride, followed by treatment of the reaction product with trimethylamine [89].

Triethylamine-alkylborane is obtained by the reaction of (2,3-dimethyl-2-butyl)alkylborane with triethylamine [90]:

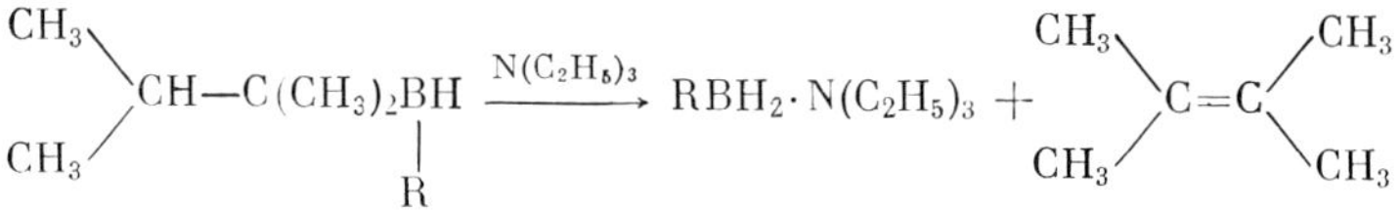

A very convenient hydroborating agent is the complex of borane with dimethyl sulfide, $H_3B \cdot S(CH_3)_2$, which reacts with olefins at room temperature [91–93]. Borane-1,4-oxathiane [327] and the tetrahydrofuran complex of 1-pyrrolylborane [328] have been used as hydroborating agents.

2.1.4 Preparation of alkylmercaptoboranes

Different types of compounds with alkylmercapto groups and hydrogen atoms on the boron are known. With different structural characteristics and properties, these compounds are combined under the general name of "alkylmercaptoboranes".

Monoalkylmercaptoboranes ($RSBH_2$) are unknown in the monomeric form. They exist in the form of polymers, trimers, and complex compounds with secondary or tertiary amines. Of the dialkylmercaptoboranes, only di-*tert*-butylmercaptoborane exists in the monomeric form, while the other representatives of this type of compounds are associated to some degree.

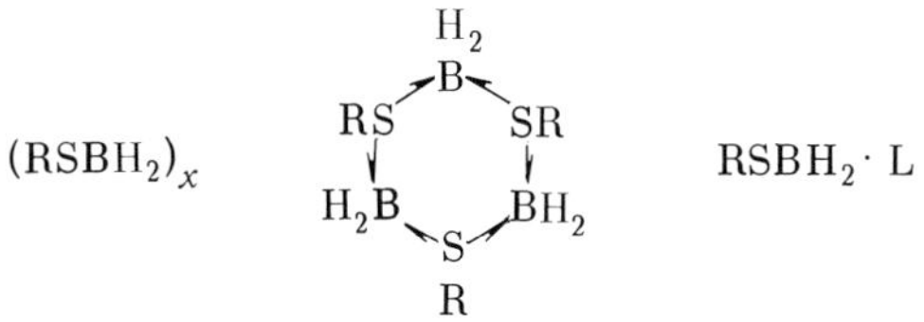

The reaction of diborane with mercaptans was first investigated for the example of methyl mercaptan by Burg and Wagner [94], using high-vacuum equipment. A solid polymer of methylmercaptoborane was obtained.

$$x/2B_2H_6 + x\,CH_3SH \longrightarrow (CH_3SBH_2)_x + xH_2.$$

Treatment of the partially depolymerized reaction product with trimethylamine led to trimethylamine-methylmercaptoborane, $CH_3SBH_2 \cdot N(CH_3)_3$.

The reaction of diborane with the polymer or with trimethylamine-methylmercaptoborane led to the thermally unstable methylmercaptodiborane, which rapidly decomposes at room temperature to diborane and the polymer $(CH_3SBH_2)_x$:

$$(CH_3SBH_2)_x + x/2B_2H_6 \longrightarrow x\,CH_3SB_2H_5,$$

$$CH_3SBH_2 \cdot N(CH_3)_3 + B_2H_6 \longrightarrow CH_3SB_2H_5 + H_3B \cdot N(CH_3)_3.$$

The reaction between diborane and mercaptans was later carried out in ether medium, using conventional laboratory techniques, which made it possible to carry out these reactions on a large scale and to conduct a broad investigation of the properties of the compounds formed.

When one mole of diborane is introduced into an ether solution of two moles of mercaptan, a uniform evolution of hydrogen is observed. At the end of the reaction and removal of the solvent under vacuum, the residue obtained is a polymer in the form of a solid (when using methyl mercaptan or *tert*-butyl mercaptan) or syrupy product (when using ethyl, propyl, or butyl mercaptan) [95, 96]:

$$x\,B_2H_6 + 2x\,RSH \longrightarrow 2\,(RSBH_2)_x + 2x\,H_2 (R = CH_3,\ C_2H_5,\ n\text{-}C_3H_7,\ n\text{-}C_4H_9,\ tert\text{-}C_4H_9).$$

The polymers of ethyl-, propyl-, and butylmercaptoborane are converted upon standing into the trimers of the alkylmercaptoboranes (1,3,5-trialkylcyclotriborothianes); the polymers of methyl- and *tert*-butylmercaptoboranes are converted to the trimers in tetrahydrofuran solution [95, 96]:

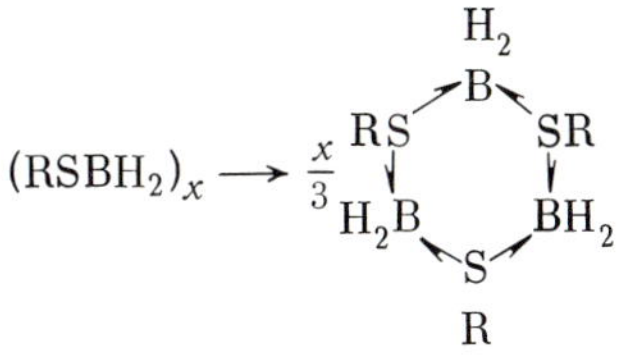

The trimers of alkylmercaptoboranes, with the exception of the trimer of *tert*-butylmercaptoborane (melting point 105–107°C), are liquids, which (except for butylmercaptoborane) distill under vacuum without decomposition.

In the reaction of amyl mercaptan, benzyl mercaptan, and thiophenol with diborane (2:1) in tetrahydrofuran or diglyme, two moles of hydrogen are rapidly evolved. The *RS*-substituted boranes formed in this process have not been isolated from the solutions and have been identified only by their IR and NMR spectra [97, 98].

Polymeric 1,2-bis(borylmercapto)ethane is obtained from diborane and 1,2-dithioethylene glycol [99]:

$$xB_2H_6 + xHSCH_2CH_2SH \longrightarrow (H_2BSCH_2CH_2SBH_2)_x + 2xH_2.$$

Dipropylmercaptoborane and dibutylmercaptoborane have been obtained by the introduction of diborane into an ether solution of the corresponding mercaptan (1:4), followed by distillation of the reaction productions [100, 101]:

$$B_2H_6 + 4RSH \longrightarrow 2\,(RS)_2BH + 4H_2.$$

A general method for the preparation of dialkylmercaptoboranes is the reduction of alkylthioboranes with lithium aluminum hydride in ether medium [102]:

$$4B\,(SR)_3 + LiAlH_4 \longrightarrow 4\,(RS)_2\,BH + LiSR + Al\,(SR)_3.$$

In contrast to the reaction of diborane with alkyl mercaptans, the reaction of diborane with thiophenol in ether solution directly forms the trimer of phenylmercaptoborane, 1,3,5-cyclotriborothiane, which dissociates in tetrahydrofuran with the formation of a complex of the monomeric phenylmercaptoborane [103]:

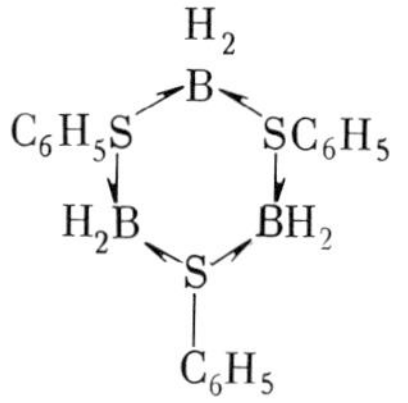

2.1.5 Preparation of complex compounds of haloboranes

The main source for the preparation of amine complexes of haloboranes are amine-boranes, which react with hydrogen halides according to the following equation [30, 104, 105]:

$$H_3B \cdot L + HX \longrightarrow H_2XB \cdot L + H_2 \qquad (L = \text{amine}).$$

Boron halides (BCl_3, BBr_3, but not BF_3) can be used as halogenating agents. Amine complexes of chloroborane [105, 106] and bromoborane [105] have been prepared by this method:

$$3H_3B \cdot N(CH_3)_3 + BX_3 \longrightarrow 3H_2XB \cdot N(CH_3)_3 + 1/2B_2H_6 \qquad (X = Cl, Br).$$

If two moles of BBr_3 are reacted with one mole of the complex, a complex of dibromoborane is formed [104].

Amine-haloboranes are also obtained by the action of halogens on amine-boranes [104]. In benzene solution, chlorine reacts vigorously with trimethylamine-borane, while bromine reacts more moderately and iodine reacts relatively slowly. With a ratio between the complex and iodine of 2:1, the amine-iodoborane is obtained.

$$2H_3B \cdot N(CH_3)_3 + I_2 \longrightarrow 2H_2IB \cdot N(CH_3)_3 + H_2.$$

Complexes of dichloroborane with a dialkylamine have been obtained from chloro(dialkylamino)borane or bis(diethylamino)borane and HCl [107].

$$HBCl(NR_2) + HCl \longrightarrow HCl_2B \cdot NHR_2,$$
$$HB(NR_2)_2 + 3HCl \longrightarrow HCl_2B \cdot NHR_2 + R_2NH_2Cl.$$

Amine-haloboranes can also be prepared from haloborane etherates [105].

Triethylamine-chloroboranes and triethylamine-bromoboranes have been synthesized by the reduction of amine complexes of the corresponding boron trihalides with triethylsilane in the presence of colloidal nickel [108].

Complex compounds of chloroborane with simple and cyclic ethers are formed by the reaction of equimolar quantities of diborane and BCl_3 in a medium of the appropriate ether:

$$B_2H_6 + BCl_3 + 3R_2O \longrightarrow 3H_2BCl \cdot OR_2.$$

To obtain dichloroborane etherates, diborane and BCl_3 should be reacted in a ratio of 1:4:

$$B_2H_6 + 4BCl_3 + 6R_2O \longrightarrow 6BHCl_2 \cdot OR_2.$$

Complexes of dichloroborane [109, 110] and deuterochloroborane [109] with diethyl ether, of dichloroborane with tetrahydropyran [110], and of chloroborane with tetrahydrofuran [111] have been obtained by this method.

Chloroborane etherates are conveniently obtained by the action of HCl on diborane in an ether medium [105]:

$$B_2H_6 + 2HCl + 2R_2O \longrightarrow 2BH_2Cl \cdot OR_2 + 2H_2,$$

and also by the action of BCl_3 on lithium borohydride [104] or sodium borohydride [110] in ether:

$$M[BH_4] + BCl_3 + 2R_2O \longrightarrow 2BH_2Cl \cdot OR_2 + MCl.$$

Dichloroborane etherate is prepared by the addition of an ether solution of $LiBH_4$ to an ether solution of BCl_3 (1:3) [110]:

$$LiBH_4 + 3BCl_3 + 4(C_2H_5)_2O \longrightarrow 4BHCl_2 \cdot O(C_2H_5)_2 + LiCl.$$

The substance is a colorless liquid, unstable at room temperature, but storable for two weeks at 0–5°C.

2.2 REACTIONS OF ALKENES AND CYCLENES WITH DIBORANE AND ITS DERIVATIVES

2.2.1 Methods of carrying out the reaction

Hurd [1] investigated the reaction of diborane with olefins in the gas phase. By heating ethylene with 2% diborane in a sealed tube for four days, he obtained triethylborane:

$$B_2H_6 + 6CH_2=CH_2 \longrightarrow 2B(CH_2CH_3)_3.$$

Isobutylene already reacts slowly with diborane at room temperature and reacts more rapidly at 100°C, forming tri-*tert*-butylborane and triisobutylborane. Bartocha, Graham, and Stone [112] found that a mixture of ethylene and diborane completely reacts over a period of ten days at 60°C to form triethylborane. Stone and Emeleus [113] obtained *tris*(2-phenylethyl)borane by the action of diborane on styrene. Brown and Subba Rao [114] later discovered that, in the presence of ethers, diborane adds smoothly and rapidly to olefins at room temperature.

Diethyl ether, tetrahydrofuran, diglyme, or triglyme are used as solvents. The reaction is carried out by introducing diborane into an ether solution of an unsaturated compound, and the trialkylboranes are then separated by distillation or, as a rule, they are reacted in some way in the crude form. Hydroboration is also carried out by a simplified method, in which diborane is prepared in the same vessel containing the unsaturated compound. The organoboranes obtained in this way are subjected to further chemical conversions.

A third method of hydroboration consists of the use of a solution of diborane in tetrahydrofuran. This method is particularly convenient in those cases where a precise measuring of the reagent is required. It is recommended that a solution of this type be prepared by passing diborane, obtained from sodium borohydride and boron trifluoride etherate in diglyme, into tetrahydrofuran. To collect the etherate entrained by a stream of diborane, the latter is passed through a trap containing a solution of sodium borohydride in diglyme [6].

However, in this method, it is very difficult to remove completely traces of boron trifluoride etherate [115], which causes undesirable side effects in the hydroboration of functional derivatives of olefins. This difficulty can be prevented by preparing diborane from sodium borohydride and iodine in diglyme [116].

In the action of diborane on olefins in an ether medium at 0–25°C, different compounds may be obtained, depending on the ratio of reagents and the nature of the hydrocarbon.

At a specific stoichiometric ratio, when there is one molecule of borane (three B—H bonds) per two molecules of the unsaturated compound, tetraalkyldiboranes are formed as a rule. However, at a hydrocarbon:BH_3 ratio of 3:1, α-olefins, and also such simple disubstituted olefins with a non-terminal double bond (RCH=CHR′) as 2-butene or 2-pentene are rapidly converted to trialkylboranes, which can be judged from the amount of hydrogen evolved in the hydrolysis of the reaction products. However, a disubstituted olefin with a branched chain, *trans*-4,4-dimethyl-2-pentene, reacts rapidly with diborane only to the stage of formation of the tetraalkyldiborane, while the further conversion to the trialkylborane proceeds relatively slowly. *trans*-Di-*tert*-butylethylene, which contains even more highly branched alkyl radicals, undergoes hydroboration only to the dialkyldiborane stage:

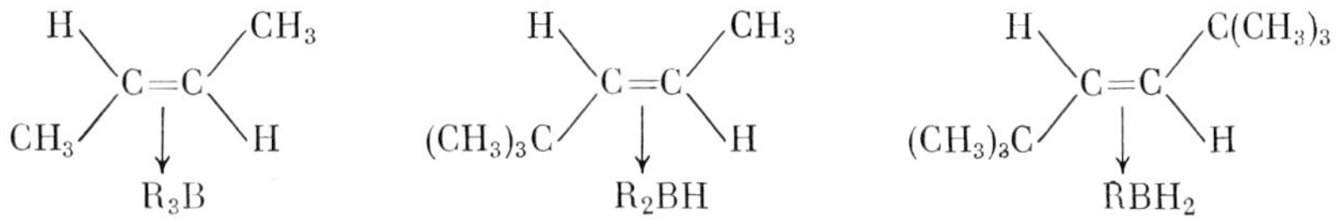

A trisubstituted olefin with a non-terminal double bond, 2-methyl-2-butene, rapidly forms a tetraalkyldiborane; however, the conversion of this to a trialkylborane requires maintaining the reaction mass at 25°C for one day. 2,4,4-Trimethyl-2-pentene and 2,3-dimethyl-2-butene rapidly form only dialkyldiboranes, while the subsequent stage — formation of tetraalkyldiboranes — proceeds slowly [3].

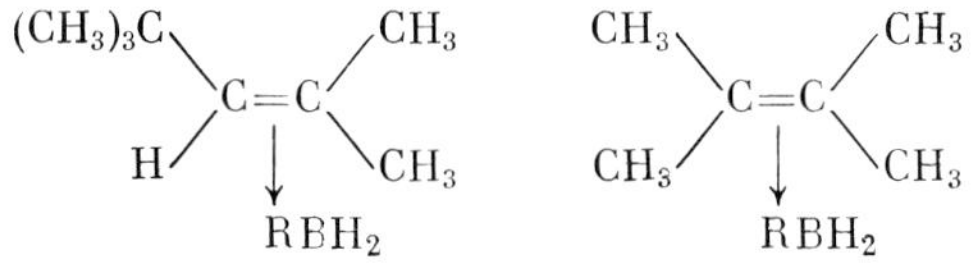

2.2.2 Stereochemistry and mechanism of hydroboration

An investigation of the hydroboration-oxidation of *cis*- and *trans*-2-*p*-anisyl-2-butene led Winstein and co-workers to establish the stereochemistry of the process [117, 118].

The formation of *threo*-3-*p*-anisyl-2-butanol from the *cis* isomer and of *erythro*-3-*p*-anisyl-2-butanol from the *trans* isomer shows that a *cis*-hydration of the double bond takes place:

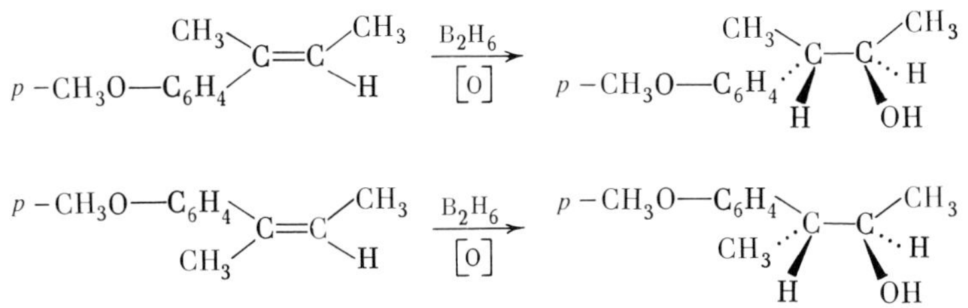

Further proof of the stereospecific *cis*-hydration of simple olefins during oxidative hydroboration has been provided by data on the deuteroboration of *cis*-2-butene and *trans*-2-butene. *erythro*-2-

Butanol-3-*d* was obtained from the former compound and *threo*-2-butanol-3-*d* from the latter [119].

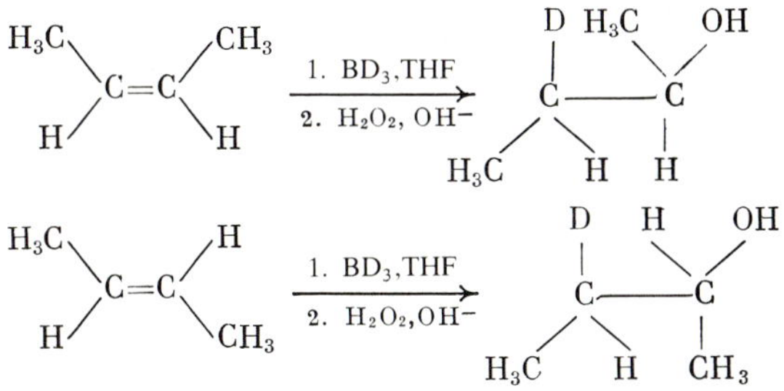

The hydroboration-oxidation of (*Z*)- or (*E*)-3,3-dimethylbutene-*1*-d_1 leads to *erythro*- and *threo*-3,3-dimethylbutan-1-*o*1-*1,2*-d_2 [329].

Brown and Zweifel [120] obtained *trans*-2-methylcyclopentanol and *trans*-2-methylcyclohexanol by the hydroboration-oxidation of 1-methylcyclopentene and 1-methylcyclohexene, respectively, and thereby showed that the sequential use of the indicated operations represents a method for the *cis*-hydration of the double bond of a cyclic system.

Wechter [121] established that the main product of the hydroboration-oxidation of cholesterol is 3β, 6α-cholestanediol, corresponding to *cis*-hydration of the double bond from the α side. A small amount of 3β, 6β-coprostanediol is also formed in this process as a result of *cis*-hydration from the sterically more hindered side.

Wolfe and co-workers [122] showed that *cis*-hydration from the less sterically hindered side takes place in the hydroboration-oxidation of 4-cholestene and 5-cholestene, which are converted to 4α-cholestanol and 6α-cholestanol, respectively.

If we take into consideration the stereochemistry of the addition of a hydroborating agent to an olefin, then the justified assumption arises that the process takes place via a traditional four-center transition state [120]:

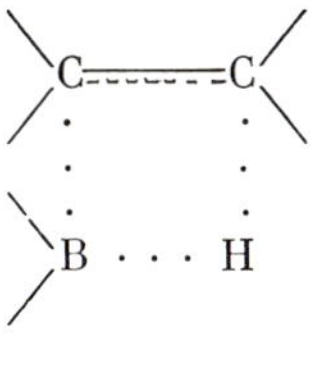

However, this model of the transition state is not in agreement with various experimental data. Thus, it incorrectly predicts the configuration of 1-butanol-1-*d*, which is formed by the asymmetric hydroboration of *cis*-1-butene-*d* by diisopinocampheylborane. In this connection, a mechanism has been proposed according to which the olefin and borane initially form a π-complex, which undergoes a negligible change in its trigonal structure in the transition state [123].

The concept of a four-center transition state is in contradiction with the principle of retention of orbital symmetry in the reaction [124]. A discussion of the orbital symmetry of the molecules participating in the hydroboration reaction shows that the transition state has significant symmetry barriers [125, 126].

On this basis, it has been suggested that the hydroboration reaction be considered as a two-stage process, including the formation of a three-center two-electron complex and a subsequent rate-determining process, with a simultaneous conversion of the complex to the final product [126]:

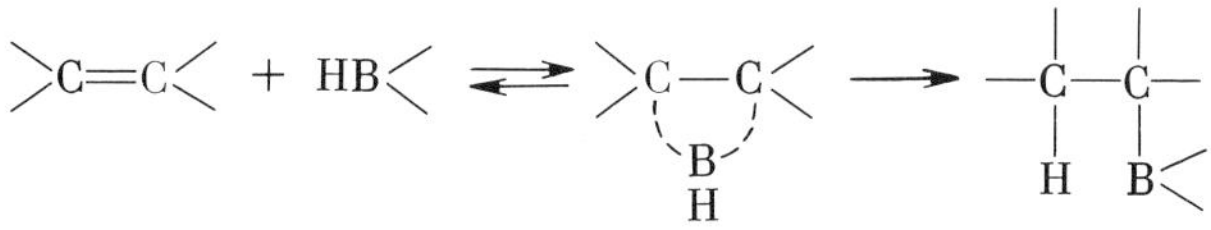

According to [329a], the three-center π-complex formed in the first stage of the process is converted to ethylborane via a four-center transition state.

2.2.3 Kinetics of hydroboration

The kinetics of the reaction of ethylene with diborane over the temperature range 120–175°C has been investigated; the proposed mechanism for the formation of triethylborane includes a dissociation of diborane into two borane molecules in the first stage [127].

The kinetics of the reaction between olefins and aluminum borohydride, which decomposes at 40–140°C with evolution of borane, has been investigated [128].

According to the data of [129], the heat of the reaction of diborane with olefins:

$$B_2H_6 \text{ (gas)} + 6C_nH_{2n} \text{ (liquid)} \rightarrow 2B(C_nH_{2n+1})_3\text{(liquid)}$$

has a value of approximately −136 kcal per mole.

The heat of addition of diborane to 1-hexene, 1-octene, and 2-octene in diglyme has values of −133.1, −132.6, and −118.5 kcal per mole, respectively [130].

The heat of addition of diborane to ethylene has a value of −170.5 kcal per mole [131].

A determination of the relative rate of addition of diborane to olefins has been carried out by the method of competitive reactions in the presence of 1-hexene or cyclopentene [132]. Unbranched α-olefins, such as 1-pentene, 1-hexene, or 1-octene, show an essentially identical reactivity toward diborane. Branching of the alkyl group added at the double bond, such as, for example, in 3-methyl-1-butene or 3,3-dimethyl-1-butene, leads to a negligible decrease in the reaction rate, since the introduction of a methyl group in the 2 position of the α-olefin not only does not decrease the reactivity, but even slightly increases it. A phenyl group at the double bond (styrene) causes a decrease in the reaction rate; olefins with a non-terminal double bond, such as 2-pentene, are also less reactive. On the whole, the reaction rate, at least during the first stages of the hydroboration, shows little change even in the case of significant changes in the structure of the olefin molecule. Nevertheless, for olefins with significant differences in structure, the differences in the rate of reaction with diborane are sufficient for the achievement of selective hydroboration [133]. Thus, for example, in the action of diborane on a mixture of 1-hexene and 2-hexene, the α-olefin is preferentially hydroborated.

Pasto, et al. [134] investigated the relative hydroboration rate of olefins in tetrahydrofuran at different stages of the process. According to the data of [134], dialkylboranes are formed as a result of consecutive addition reactions, and not by a disproportionation of alkylboranes, with the addition rate of borane being less than the addition rate of the alkylborane to the olefin. This conclusion is in contradiction to the data of Fehlner [135] on the gas-phase addition of borane to ethylene, according to which the first stage of the reaction — the addition of borane to the olefin — proceeds 20 times as rapidly as the subsequent stage — the addition of ethylborane, with the formation of diethylborane.

According to the data of Pasto, et al. [134], the rate of addition of the alkylborane to the olefin is greater than the rate of addition of the dialkylborane.

An investigation of the kinetics of hydroboration of tetramethylethylene with a borane-THF complex shows that the rate of formation of 2,3-dimethyl-2-butylborane is first-order with respect to each of the reagents. The activation energy is 9.2 ± 0.4 kcal per mole and the entropy of the reaction -27 ± 1 entropy units [136]. A comparison of the activation enthalpy for the reaction of tetramethylethylene with $H_3B \cdot THF$ and of the gas-phase reaction of borane with ethylene, which has a value of 2 ± 3 kcal per mole [136], indicates that the solvent plays an important role in the hydroboration reaction. On the basis of an analysis and a comparison of the activation parameters of different hydroboration reactions, it can be assumed that they proceed by means of a direct reaction between the borane complex and the olefin. Kinetic data [137] on the reaction of olefins with choroborane in THF, with which it forms a more stable complex than borane, also lead to this conclusion [111].

The rate constants for the addition of chloroborane to olefins ($k \cdot 10^3$, L/mole·minute) at 25°C are shown below [137]:

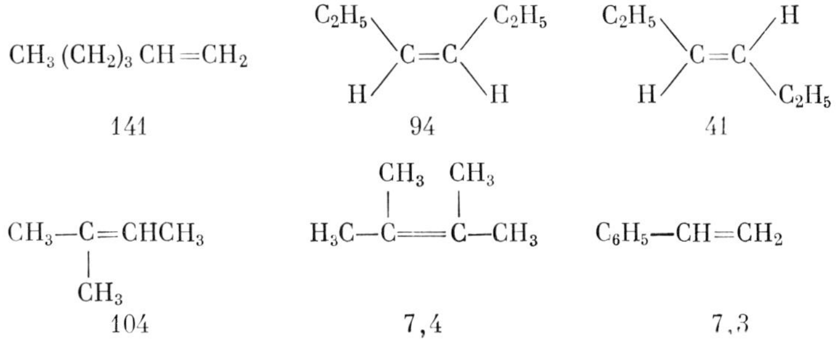

Styrene and its different 4-substituted derivatives show little difference in the rate of addition of chloroborane, and the nature of the substituent in the para position also has little effect on the orientation of the addition (the degree of addition to the terminal hydrocarbon varies from 87% to 93.3%).

There are contradictory data in the literature with respect to the correlation between the effects of orientation and the nature of the substituent in a series of styrene derivatives. Thus, according to the data of [137], the distribution of the boron atoms in the α- and β-positions of styrenes is correlated with the σ scale of the substituents; according to the data of [138], a correlation is observed when the σ^+ scale of substituents is used, whereas Klein and co-workers

[139] found that the hydroboration rate of substituted styrenes is not governed by the Hammett relationship.

The rate of addition of tetraalkyldiboranes depends on the structure of the unsaturated hydrocarbon. Köster, et al. [140] found that tetrapropyldiborane adds most rapidly to linear α-olefins and to cyclopentene, somewhat more slowly to branched α-olefins, considerably more slowly (by a factor of 2.7) to olefins with a non-terminal double bond, and adds to cyclohexene 20 times more slowly than to 1-octene.

The addition to cyclopentene of tetra(3-methyl-2-butyl)diborane, which does not dissociate in tetrahydrofuran, is a first-order reaction with respect to each reagent. This is in agreement with a mechanism according to which the stage determining the reaction rate is the direct interaction of the reagents, leading to the final product and a dialkylborane [141].

With respect to their rate of reaction with tetra(3-methyl-2-butyl)diborane in tetrahydrofuran, the olefins and cyclenes can be arranged in the following order [132, 141] ($k \cdot 10^{-4}$, L/mole · second):

$$\underset{1070}{\text{1-pentene}} > \underset{1020}{\text{1-hexene}} > \underset{580}{\text{3-methyl-1-butene}} > \underset{72}{\text{cyloheptene}}$$

$$> \underset{23}{\textit{cis}\text{-2-butene}} > \underset{21}{\textit{cis}\text{-3-hexene}} > \underset{21}{\textit{cis}\text{-2-pentene}} > \underset{14}{\text{cyclopentene}}$$

$$> \underset{3.9}{\textit{trans}\text{-2-butene}} > \underset{3.0}{\textit{trans}\text{-2-pentene}} > \underset{2.2}{\textit{trans}\text{-3-hexene}}$$

$$> \underset{0.13}{\text{cyclohexene.}}$$

As can be seen, the *cis*-olefins are hydroborated with tetraalkyldiborane considerably more rapidly than the *trans* isomers. The rate of reaction of olefins $RC(CH_3){=}CH_2$ with tetra(3-methyl-2-butyl) diborane decreases with increasing size of the radical R [142, 143]. 1,2-Dialkylethylenes with a *cis* configuration add tetra(3-methyl-2-butyl)diborane more rapidly than the *trans* isomers [144].

The difference in the rate of addition of tetraalkyldiboranes to olefins can be used for the separation of different unsaturated hydrocarbons. Thus, the treatment of an equimolar mixture of 1-pentene and 2-pentene with tetra(3-methyl-2-butyl)diborane leads to the production of 2-pentene with a degree of purity of 99% [145], while the action of tetraethyldiborane on an equimolar mixture of 1-hexene and 2-hexene leads to an almost quantitative yield of the latter hydrocarbon with a degree of purity of 98% [140].

The relative rate of hydroboration of some olefins with 4,4,6-trimethyl-1,3,2-dioxaborinane has been investigated [146], as has the rate of hydroboration of olefins [330] and of styrene derivatives [331] with 9-borabicyclo[3.3.1]nonane.

2.2.4 Scope of the reaction. Synthesis of alcohols and ketones

In this section, we shall consider questions of the orientation of the addition of hydroborating agents to alkenes and cyclenes and the oxidation of the organoboranes formed to alcohols as a method for the hydration of unsaturated compounds based on the hydroboration reaction. Other conversions of organoboranes, obtained both with the use of organometallic compounds and with hydroboration, will be discussed in other sections of this part.

An investigation of the orientation of addition of the boron atoms to an unsaturated system is carried out by oxidation of hydroboration products with hydrogen peroxide in alkaline medium and subsequent analysis of the alcohols formed by gas-liquid chromatography.

a) Alkenes

Monosubstituted α-olefins $RCH{=}CH_2$ with both a linear (from ethylene to 1-dodecene) and branched structure easily undergo hydroboration, with the boron atom adding mainly (93–94%) to the terminal carbon and only to the extent of 6–7% to the second carbon of the double bond [2, 3]:

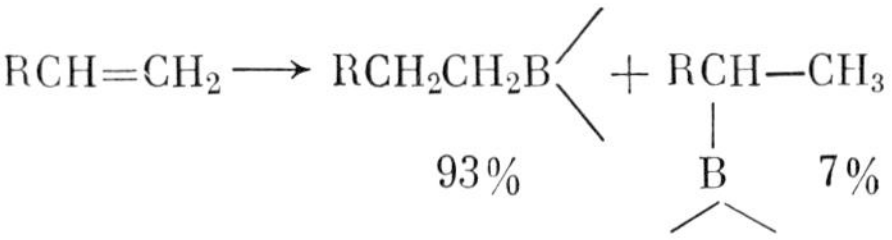

Thus, the hydroboration-oxidation of α-olefins is a method for their anti-Markownikoff hydration although essentially, taking into consideration the polarization of the $\overset{\delta+}{B}-\overset{\delta-}{H}$ bond, the addition of diborane to the olefin proceeds according to the Markownikoff rule.

For disubstituted α-olefins $R_2C{=}CH_2$, the orienting effect of two alkyl substituents shows up very sharply, making it possible to add the boron atom almost exclusively to the terminal carbon. Thus, in the case of 2-methyl-1-butene, the boron atom adds to the extent of 99% to the carbon of the methylene group; analogous results are also obtained in the case of 2,4,4-trimethyl-1-pentene:

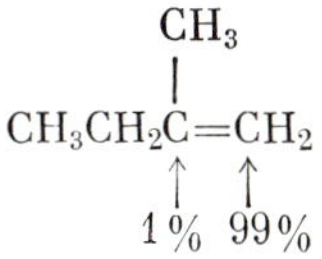

In the hydroboration-oxidation of 2,2-di-*tert*-butylethylene, 2,2-di-*tert*-butylethanol is formed in only 65% yield.

Disubstituted olefins with a non-terminal double bond, $RCH{=}CHR'$, such as 2-pentene, 2-hexene, or 4-methyl-2-pentene, are easily hydroborated to trialkylboranes in such a way that the boron atoms add to approximately identical degrees to both carbon atoms of the double bond. Negligible changes in orientation are observed for the *cis-trans* isomers of hydrocarbons of this type.

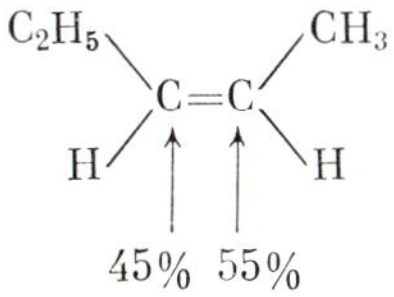

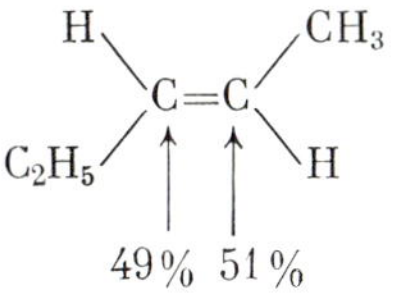

Trisubstituted olefins $R_2C{=}CHR$, such as 2-methyl-2-butene and 2,4,4-trimethyl-2-pentene, add boron to the extent of 98% to the secondary carbon atom and only 2% to the tertiary carbon atom [2, 3]:

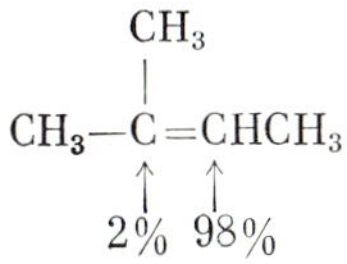

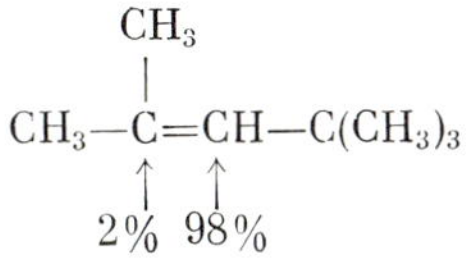

The catalytic hydroboration of ethylene in the gas phase to triethylborane has been investigated [332]. The action of $Me_4B_2H_2$ on C_2H_4 produces Me_2BEt and $MeBEt_2$ [333]. The synthesis of 1-triacontanol, $CH_3(CH_2)_{28}CH_2OH$, from 15-triacontene by hydroboration-isomerization-oxidation has been described [334].

The cyclopropyl radical has a significant effect on the orientation of addition of diborane. Thus, in the action of diborane on cyclopropylethylene, boron adds to the extent of 97% to the terminal carbon [147], whereas it adds to the terminal carbon only to the extent of 93% in the case of unbranched α-olefins:

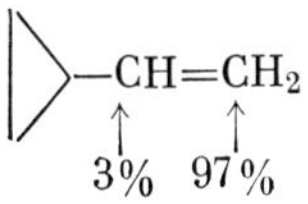

2-Cyclopropylpropylene is converted to 2-cyclopropyl-1-propanol by oxidative hydroboration [148].

p-Allylanisole is converted to 3-(*p*-anisyl)-1-propanol by the above method [149].

Tetraalkyldiboranes add to olefins with the formation of trialkylboranes, with the reaction proceeding more selectively and more slowly than with diborane. Under the action of tetra(3-methyl-2-butyl)diborane, 1-hexene adds a boron atom to the terminal carbon to the extent of 99%. The selectivity of addition of tetra(3-methyl-2-butyl)diborane to olefins with a non-terminal double bond shows up particularly clearly. Thus, in the case of *cis*-2-hexene, 63% of the boron is localized on the C-2 atom instead of 50% under the action of diborane, and branched olefins with a non-terminal double bond, such as *cis*- and *trans*-4-methyl-2-pentene, add boron atoms in the less spatially hindered position to the extent of 85–97%, whereas, with diborane, the addition of boron in the 2 position proceeds to the extent of 57–62% [2, 3].

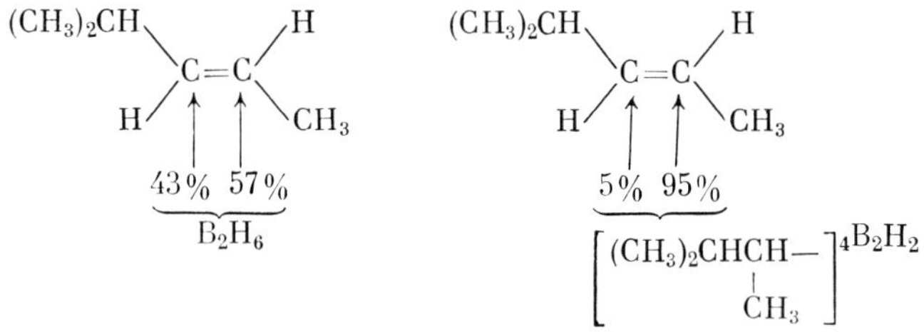

Tetracyclohexyl-, tetra(*trans*-2-methylcyclohexyl)-, and tetraisopinocampheyldiboranes add to 2-hexene and *cis*-4-methylpentene with the same degree of selectivity.

The higher selectivity of addition of the above-mentioned tetrasubstituted diboranes is explained by the spatial effect of the bulky hydrocarbon radicals, but Köster, et al. [140, 150] have shown that branching in the radicals does not play a particularly specific role and that such very simple tetraalkyldiboranes as tetraethyl- and tetrapropyldiborane also add selectively to a multiple carbon–carbon bond.

Tetraalkyldiboranes with primary radicals react with olefins very rapidly at temperatures slightly above −10°C, with the evolution of heat, with an ether medium being optional for this reaction [140]. Tetraalkyldiboranes with secondary radicals react more slowly and also exothermally. In both cases, a mixture of trialkylboranes is obtained, because, under the action of compounds with B—H bonds, an exchange of alkyl radicals takes place, with the establishment of an equilibrium. Thus, in the action of tetra(2-propyl)diborane on 1-butene, the reaction mixture contains 27.4% of triisopropylborane, 54.2% of butyldiisopropylborane, 17.7% of dibutylisopropylborane, and only 0.7% of a trialkylborane containing a 2-butyl radical (1-isopropylbutyl-2-butylborane). Tetraethyldiborane adds to propylene with the formation of the following trialkylboranes, which do not contain isopropyl radicals: 37% R_3B, 41% R_2BR', and 20% RBR_2' (where R = ethyl, R′ = propyl) [140].

A mixture of symmetrical and unsymmetrical trialkylboranes is also formed by the action of tetramethyldiborane on ethylene or propylene [151].

The action of dialkylboranes on (*E*)- and (*Z*)-hexene-*1*,*2*-d_2 produces *threo*- and *erythro*-(1,2-dideuterohexyl)dialkylborane [335].

The hydroboration of α-pinene with tetraethyldiborane at room temperature after disproportionation of a mixed trialkylborane produces pure crystalline tri(*cis*-myrtanyl)borane [140].

The borohydride groups of tetraalkyldiboranes add to α-olefins less selectively in the case where the double bond is under the action of strongly polarizing groups. Thus, styrene, under the action of tetraethyldiborane [140] or tetra(3-methyl-2-butyl)diborane [2, 3], adds boron atoms to the terminal carbon to the extent of 93%.

1,2-Diaryldiboranes react with olefins in ether solution under cooling, with the formation of unstable phenyldialkylboranes which, already at room temperature, undergo relatively rapid symmetrization to triaryl- and trialkylboranes [21]:

$$(ArBH_2)_2 + 4CH_2{=}CHR \longrightarrow 2ArB(CH_2CH_2R)_2,$$
$$3ArB(CH_2CH_2R)_2 \longrightarrow Ar_3B + 2B(CH_2CH_2R)_3$$
$$(Ar = C_6H_5;\ R = CH_3, C_2H_5).$$

Mixed trialkylboranes can be obtained by the reduction of methyl esters of dialkylborinic acids with lithium aluminum hydride in the presence of an olefin. The organoboranes obtained in this way contain impurities of other trialkylboranes. Thus, dibutylpentylborane, containing 13% of tributylborane and 3% of butyldipentylborane, is obtained from methoxydibutylborane and 1-pentene [152]:

$$R_2BOCH_3 + LiAlH_4 \xrightarrow{\text{olefin}} R_2BR'.$$

Alkyldiphenylboranes are obtained in an analogous manner from ethyl diphenylborinate [336].

Unsymmetrical trialkylboranes have been synthesized by the hydroboration of olefins with bis(2,3-dimethyl-2-butyl)diborane [153, 154, 337, 338].

The synthesis of (*Z*)-6-heneicosen-11-one, the sex pheromone of the Douglas fir tussock moth, has been carried out with thexylchloroborane [339].

The data presented above illustrate the important role of steric effects in the orientation of the addition of hydroborating agents to olefins. However, the orientation of addition is affected not only by steric effects, but also by electronic factors: there is a specific correlation between the screening of the carbon atoms of the double bond [155] and the orientation effects.

A phenyl group added directly to a carbon of a double bond (styrene) significantly changes the orientation of addition, increasing the localization of the boron atom on the α-carbon atom to 20%; in the case of *p*-chlorostyrene and *p*-methoxystyrene, the ratio of the α- and β-isomeric alcohols is 27:73 and 7:93, respectively [138, 139, 156].

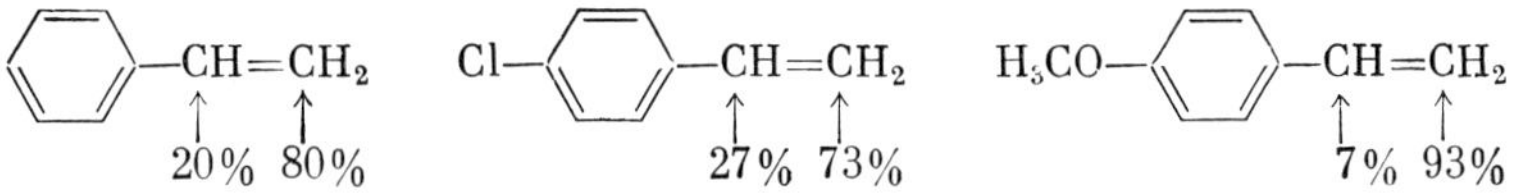

In the hydroboration of styrene, the direction of addition is significantly affected by steric effects in the hydroborating agent. Thus, in the action of tetra(3-methyl-2-butyl)diborane on styrene, the boron atom adds to the terminal carbon to the extent of 98%.

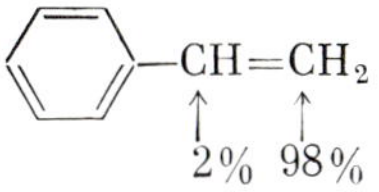

Olefins are hydroborated more rapidly by 9-borabicyclo[3.3.1]-nonane, permitting a selective hydroboration-oxidation of enynes to acetylenic alcohols [340].

The stable complex of thexylborane with *N,N*-diethylaniline hydroborates 1-octene and cyclohexene at 0°C [341].

The hydroboration-oxidation of α-cyclopropyl-β,β-dimethylstyrene (*1*) at room temperature leads to α-cyclopropylbenzyldimethylcarbinol (*6*) (12%) and 5-methyl-4-phenyl-1,3-hexanediol (*5*) (75%). The formation of the diol *5* is explained by cleavage of the three-membered ring in the organoborane *2*, with the formation of an unsaturated organoborane *3*, which is hydroborated to the diboron compound *4* [157]:

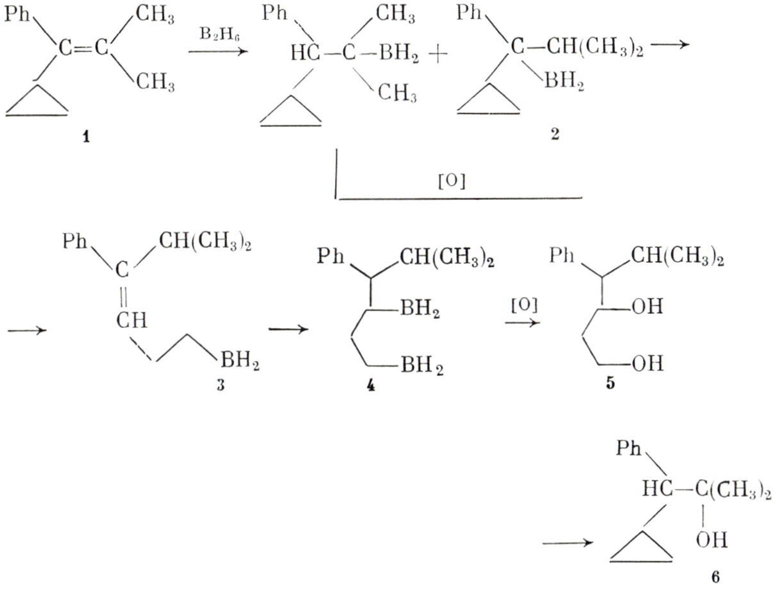

The hydroboration of 4-methyl-3-penten-1-yne and 2,6,6-trimethyl-2-hepten-4-yne with diethylborane leads to alkadienyldiethylboranes [342]. The hydroboration of $FClCHCF_2C{=}CH_2$ has been investigated [343].

The hydroboration of alkenes with chloroboranes, aminoboranes, and alkylthioboranes has been investigated.

The action of chloroborane on olefins in tetrahydrofuran [111, 158, 159, 344], diethyl ether [160, 161], or $H_2BX \cdot S(CH_3)_2$ [345] produces chlorodialkylboranes. The reaction proceeds more slowly than that with diborane, but more selectively. Thus, the hydroboration of 1-hexene leads to 1- and 2-hexanol in a ratio of 96:4 [111]

Olefins react slowly with dichloroborane in tetrahydrofuran [111, 158] or in diethyl ether [162, 346] with the formation of dichloroalkylboranes. It is recommended that the reaction be carried out in the presence of an equimolar quantity of BCl_3 in benzene or pentane [162]:

$$HBCl_2 \cdot O(C_2H_5)_2 + BCl_3 + \text{olefin} \rightarrow RBCl_2 + Cl_3B \cdot O(C_2H_5)_2.$$

The action of $(CH_3)_2S \cdot HBBr_2$ [347], $(CH_3)_2S \cdot HBCl_2$ [348], or $(CH_3)_2S \cdot H_2BHal$ [349, 350] on olefins forms alkyldihaloboranes and dialkylhaloboranes, respectively.

Thexylchloroborane adds to olefins to form thexylalkylchloroboranes [351, 352], which are converted to mixed trialkylboranes by RLi or RMgX [351].

The reaction of dialkylbromoboranes with sodium hydride in the presence of olefins leads to mixed trialkylboranes [353].

The addition of nitrogen-containing borane derivatives has received very little investigation. It is known that tetraethyldiborazane adds to olefins at 120–130°C in the presence of pyridine, with the formation of diethylaminodialkylborane [76]:

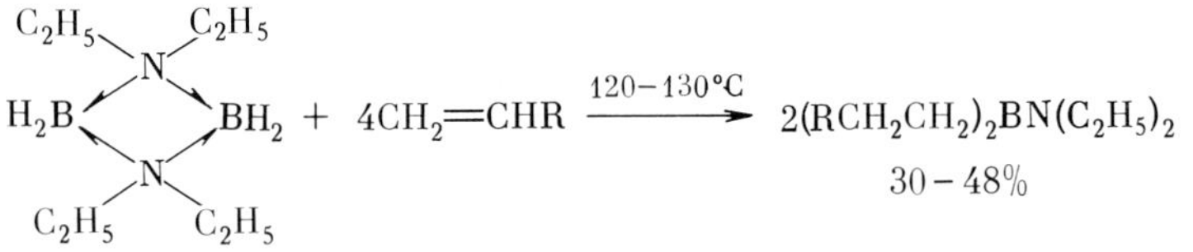

According to patent data [163], dialkylaminoboranes add to dienes at 100–120°C and elevated pressure with the formation of 1-

dialkylaminoborolanes and cyclic diboron compounds:

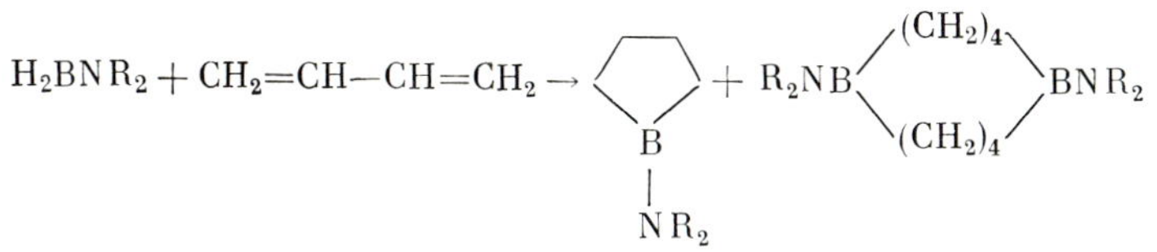

Alkylmercapto(diethylamino)boranes do not add to olefins at 120°C [164].

Polymers of alkylmercaptoboranes react with olefins in an ether medium, with the course of the process depending on the nature of the reagents [165].

In the hydroboration of 1-hexene with phenylmercaptoborane in tetrahydrofuran (25°C, 1–1.5 hours) and subsequent methanolysis of the reaction products, methyl dihexylborinate (45%) and dimethyl hexylboronate (21%) are formed. The reaction of phenylmercaptoborane with styrene proceeds in an analogous manner. Upon oxidation of the hydroboration products, 2- and 1-phenylethanol are obtained in a ratio of 91:9 [98].

Trimers of alkylmercaptoboranes do not react with unsaturated compounds at room temperature. If the reaction is carried out in boiling ether, however, a mixture of substances is obtained. Thus, the reaction of the trimer of ethylmercaptoborane with 1-octene forms triethyl thioborate (10%), diethyl octylthioboronate (20%), and ethyl dioctylthioborinate (15%) [166].

Dialkylmercaptoboranes are less active hydroborating agents than the polymers of alkylmercaptoboranes. The reaction of dibutylmercaptoborane with 1-hexene, 1-octene, or cyclohexene is carried out with heating, forming high yields of the butyl esters of the corresponding alkylthioboronic acids [167]:

$$(n\text{-}C_4H_9S)_2BH + \text{olefin} \longrightarrow RB(SC_4H_9\text{-}n)_2$$

2- and 1-Phenylethanols, in a ratio of 95:5 and a yield of 77%, have been obtained by the reaction of diphenylmercaptoborane with styrene in tetrahydrofuran, followed by oxidation of the reaction products [98].

As a hydroborating agent, one can use 4,4,6-trimethyl-1,3,2-dioxaborinane, which adds to olefins at 100–130°C [146, 168, 169] and also 1,3,2-benzodioxaborole, which reacts with olefins at 100°C [170, 171].

9-Borabicyclo[3.3.1]nonane shows a high regiospecificity in hydroboration reactions of olefins [354].

Olefins and acetylenes are hydroborated with 1,3,2-dithioborolane, forming alkyl- and alkenyl-1,3,2-dithiaborolanes, respectively [355].

b) Cyclic systems with an *exo*-methylene group

Under the action of diborane, methylenecyclopropane is converted to tri(3-butenyl)borane via the intermediate stage of tris(cyclopropylmethyl)borane [172]:

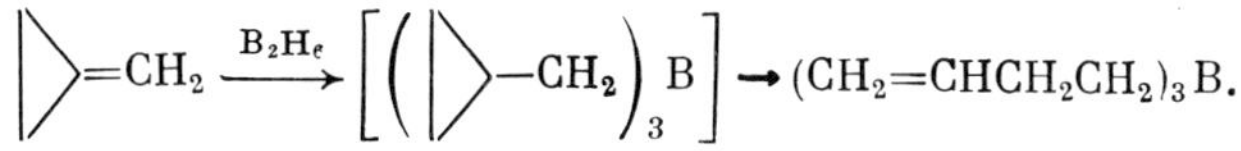

The orientation of the addition of the boron atom in the hydroboration of alkylidenecyclopropanes depends on the number of substituents on the methylene carbon. Diphenylmethylene- (*7*) and α-methylbenzylidenecyclopropanes (*8*) add boron exclusively to the unsaturated carbon of the cyclopropane ring, whereas, in benzylidenecyclopropane (*9*), boron adds to this carbon only to the extent of 24% [173].

The effect of the cyclopropane ring on the orientation of addition of diborane to an olefin system can be seen from a comparison of the results of the hydroboration of compounds *7–9* and compounds *10–11*:

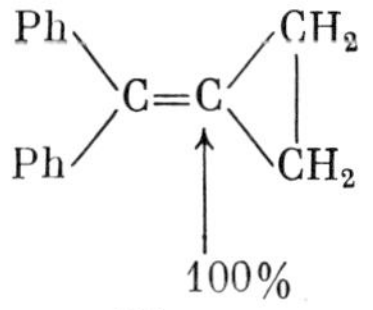

7

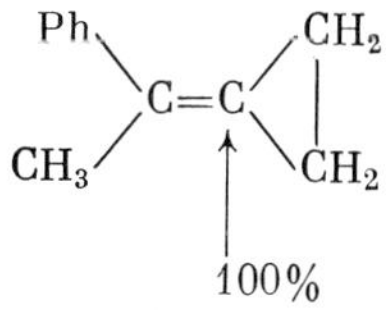

8

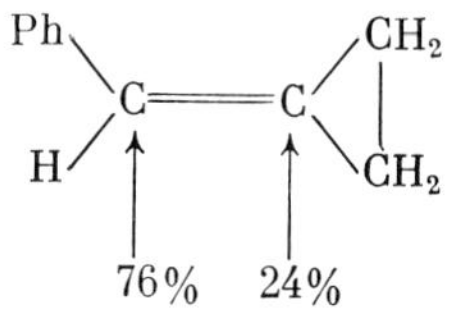

9

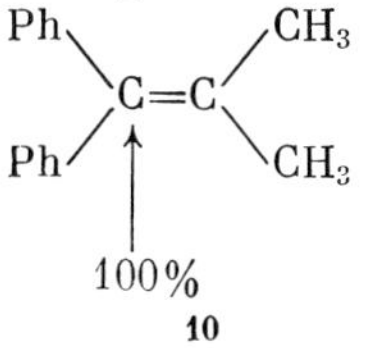

10

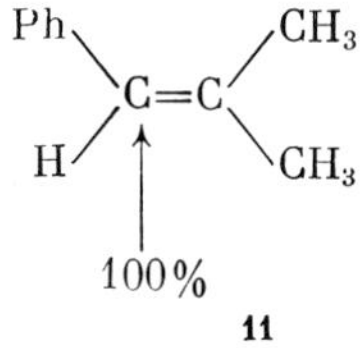

11

The hydroboration-oxidation of 1,1-diphenyl-2-vinylcyclopropane leads to 1,1-diphenyl-2-(2-hydroxyethyl)cyclopropane [356].

Klein and Lichtenberg [174] found that 3-methyl and 4-methylmethylenecyclohexane, upon hydroboration-oxidation, form a mixture of alkyloxymethyl derivatives of cyclohexane, with a predominant content of the less stable axial-equatorial isomers *12* and *12a*. Replacement of a methyl group by a *tert*-butyl group has no significant effect on the course of the reaction.

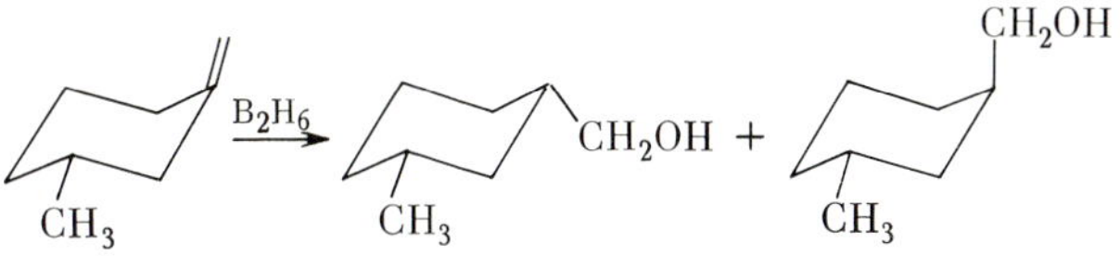

(37%) **12** (63%)

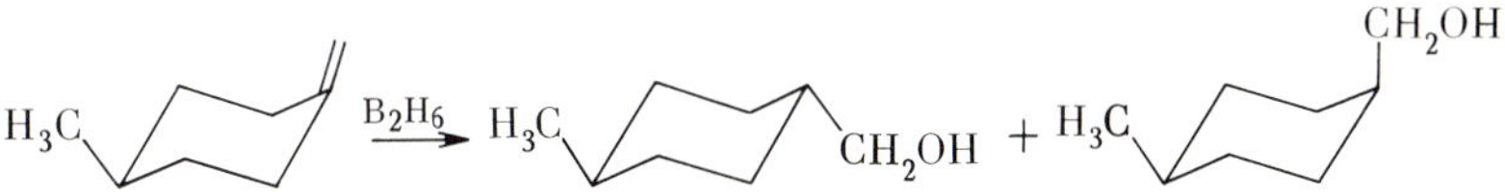

(34%) **12a** (66%)

The preferred attack of the methylene group from the equatorial side is also observed when dichloroborane is used as a hydroborating agent, whereas, with tetracyclohexyldiborane, equal quantities of axial and equatorial products are obtained, which can not be explained, since, with an increase in the volume of the hydroborating agent, an increase in the relative quantities of compounds *12* and *12a* should have been expected.

The hydroboration of 2-substituted derivatives of methylenecyclohexane and methylenecyclopentane proceeds mainly from the side opposite to the substituents on C-2 [357].

The hydroboration of substituted methylenecyclohexanes was studied in [358].

2-Methylenenorbornane is attacked by diborane predominantly from the endo side:

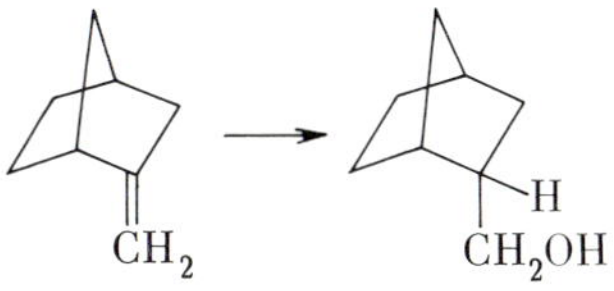

85%

When tetra(3-methyl-2-butyl)diborane is used, the reaction proceeds somewhat more selectively; in this case, 91% of the endo alcohol is obtained. 2-Methylene-7,7-dimethylnorbornane (α-fenchene) is attacked by diborane primarily from the endo side [175].

In the hydroboration-oxidation of camphene, *endo*-camphanol (*13*) is obtained [176, 177] but, if the organoborane is subjected to thermal isomerization before the oxidation, then *exo*-camphanol (*14*) is formed [177]:

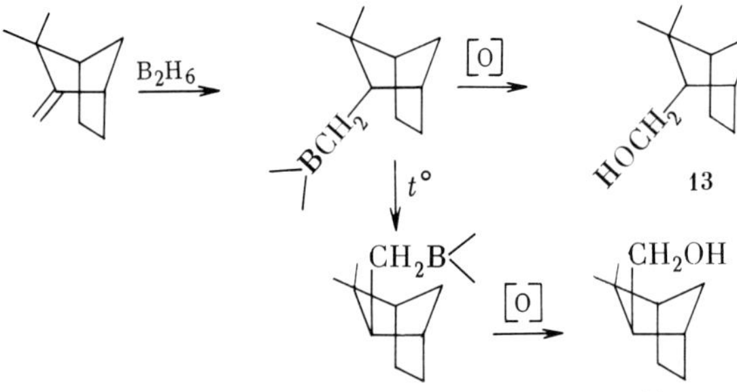

13

14

($-$)-β-Pinene, depending on the hydroboration-oxidation conditions, is converted to ($-$)-*cis*-myrtanol (*17*) or ($-$)-*trans*-myrtanol (*18*) [176, 178, 179].

The compound *18* is obtained by moderate heating of the initially formed organoborane *15*, which has undergone isomerization to the borane *16*:

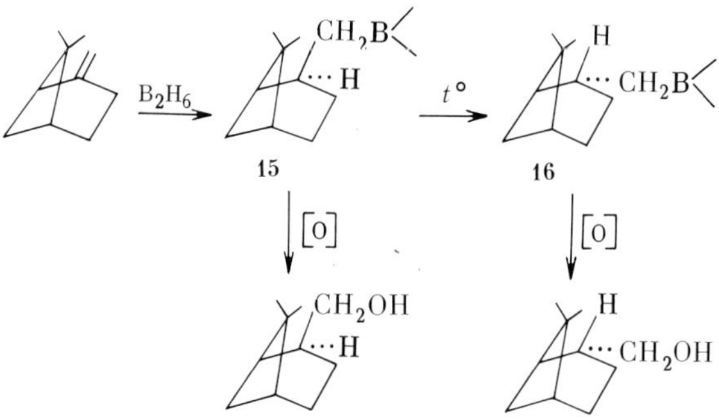

15 16 17 18

The hydroboration-oxidation of 6,6-dimethyl-3-methylenebicyclo[3.1.1]heptane, of its 2-methyl- and 2,4-dimethyl derivatives

[180], and also of various homologs of β-pinene has been investigated [181].

(+)-Sabinene is also hydroborated predominantly from the less hindered side [182]:

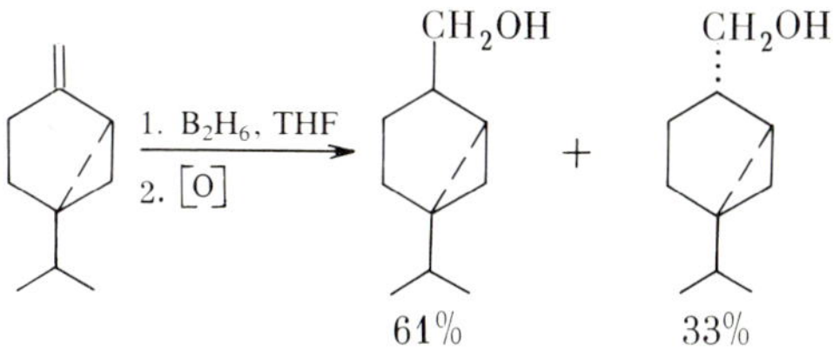

A mixture of olefins (*19*) is converted by hydroboration-oxidation to a mixture of diastereoisomeric alcohols (*20*) [183]:

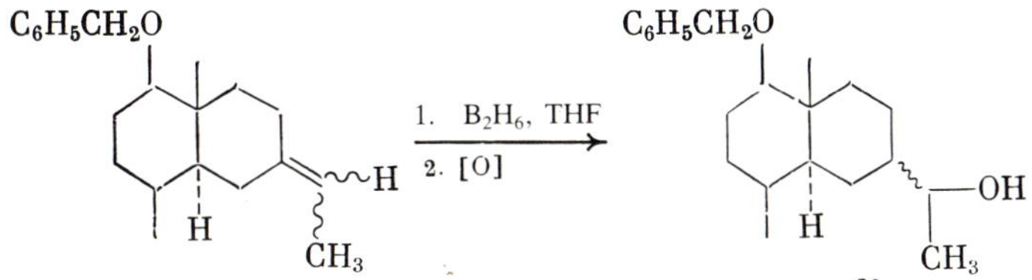

19 **20**

β-Cedrene forms 9-isocedranol [184]:

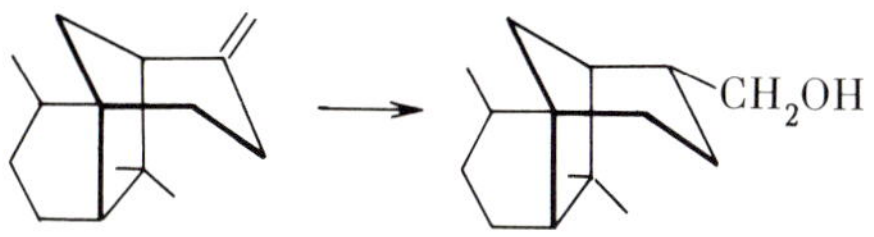

and longifolene forms longifolol [185, 186]:

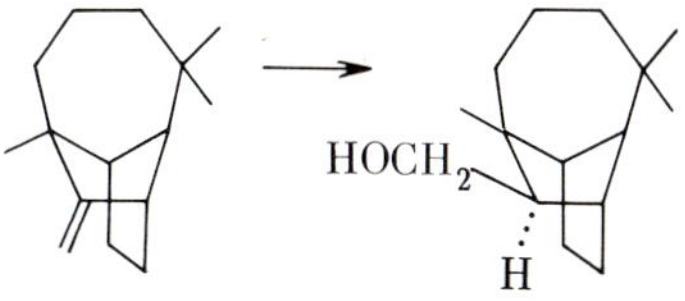

Under the action of diborane, ω-phenyl- or ω,ω-diphenyldibenzofulvene are reduced to 9-benzyl- and 9-benzhydrylfluorene, respectively [187].

c) Monocyclic systems

As found by Köster, et al. [172], in the action of diborane on 1-methylcyclopropene, the boron atom adds mainly in the 2 position:

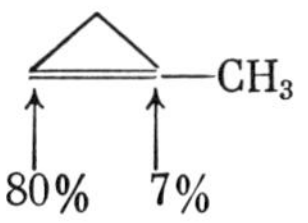

1-Phenylcyclobutene is converted to 2-phenylcyclobutanol by hydroboration-oxidation [188].

The action of diborane on cyclopentene and cyclohexene forms tricyclopentyl- and tricyclohexylborane, respectively, with the final stage of the reaction proceeding slowly in the latter case. The simplest cycloolefins are similar to the unbranched olefins with a non-terminal double bond with respect to their reaction rates with diborane [3].

Substituted cyclenes—1-methylcyclopentene and 1-methylcyclohexene—are rapidly hydroborated only to the tetraalkyldiborane stage, with the boron atoms being localized on the C-2 atom; upon oxidation, *trans*-2-methylcyclopentanol and *trans*-2-methylcyclohexanol, respectively, are obtained. The presence of a methyl group in the 3 position of cyclopentene has a negligible effect on the orientation of addition of the boron atoms [3].

The hydroboration of 1-ethyl- and 1-butylcyclohexenes leads to the corresponding *trans*-2-alkylcyclohexanols [189, 190].

4-Methylcyclohexene forms almost equal quantities (28:27:20:25) of four isomeric alcohols upon oxidative hydroboration [191]:

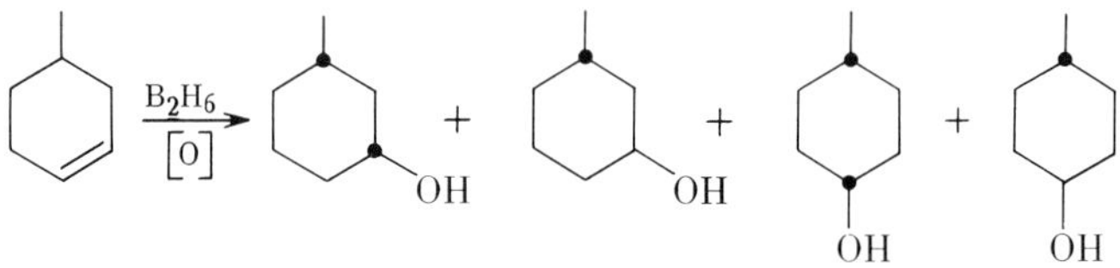

Pasto and Klein [192] investigated the effect of steric factors on the orientation of addition of diborane to different mono- and disubstituted cyclohexenes in tetrahydrofuran with an olefin:BH_3 ratio of 1:2.

In the hydroboration of 4-*tert*-butylcyclohexene, the boron atom adds to the C-1 and C-2 atoms in both the cis and trans positions to the alkyl group, with the cis-addition predominating slightly over the trans-addition (55:45).

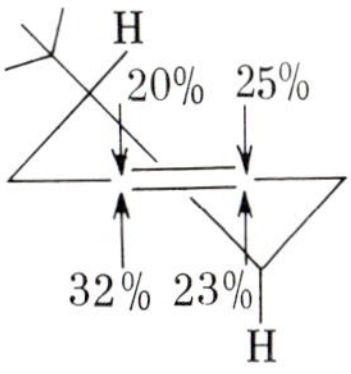

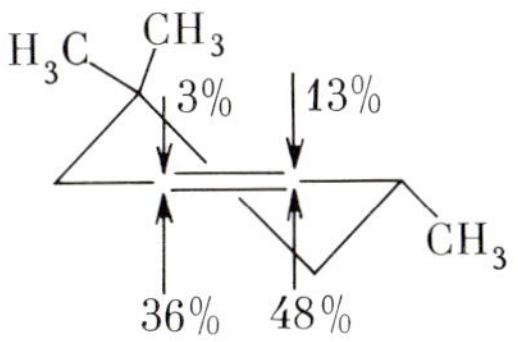

The stereochemical directivity of attack is maintained in the case of 1-methyl- and 1-ethyl-4-*tert*-butylcyclohexene.

In 3,5,5-trimethylcyclohexene, the axial 5-methyl group prevents attack in the cis position to that group, as a result of which the ratio of cis and trans isomeric alcohols is 16:84 [192].

An increase in the volume of the hydroborating agent results in an increase in spatial interactions with the substituents in the axial 4 and 5 positions of cyclohexene. When bis(2,3-dimethyl-2-butyl)diborane and *tert*-butylcyclohexylborane act on 4-*tert*-butylcyclohexene, the following ratios of isomers are obtained [192]:

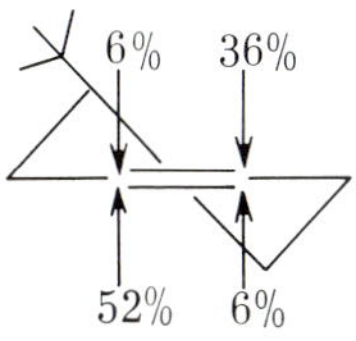

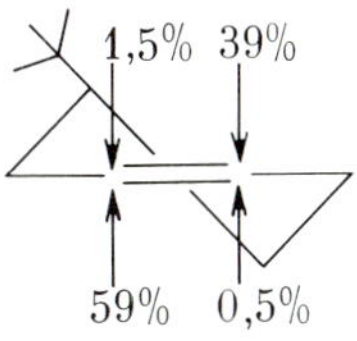

The results of the hydroboration of 3-methylcyclohexene and 3-*tert*-butylcyclohexene, reflecting a direct steric effect of the alkyl group, are as follows:

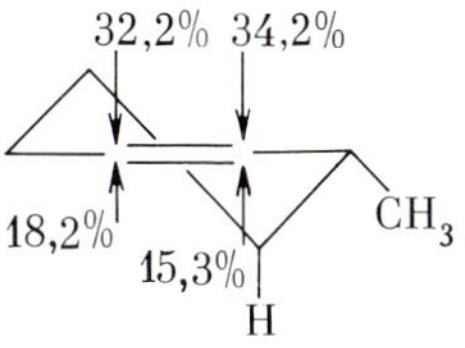

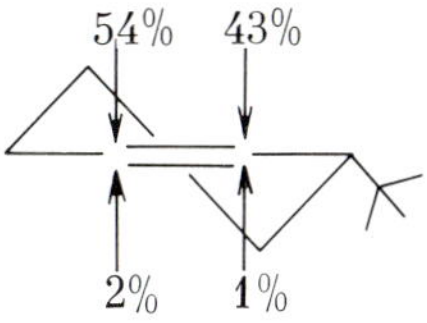

A *tert*-butyl group in the 1 position in 1,4-di-*tert*-butylcyclohexene has a significant effect on the orientation of addition [192, 193]:

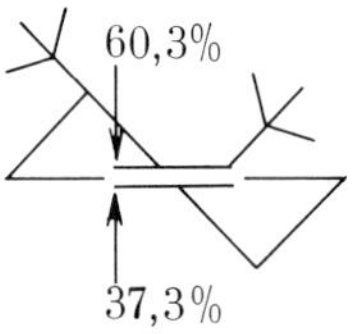

The steric effects caused by a substituent in the 1 position of cyclohexene are also found in the hydroboration-oxidation of 1-*p*-menthene (*21*) [194] and of 3-*p*-menthene (*22*) [195]:

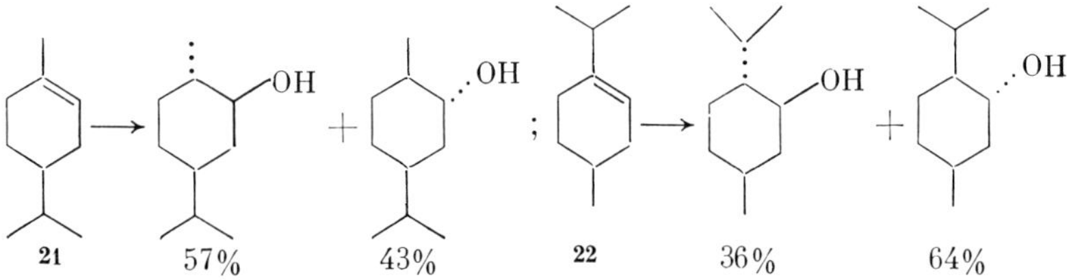

The introduction of a methyl group in the 6 position of 1-methylcyclohexene results in a change in the stereoselectivity of addition of diborane, leading to a 40:60 ratio of *cis* and *trans* isomers [192]:

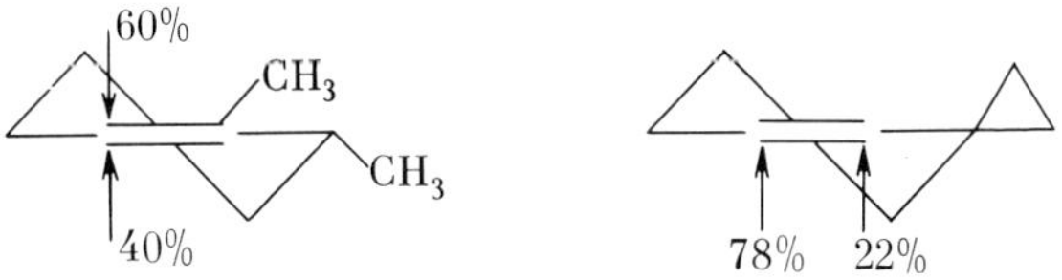

It is interesting that two methyl groups in the 3 position of cyclohexene have no effect on the orientation of addition, whereas the 3,3-spirocyclopropyl group in spiro[2.5]oct-4-ene directs the boron atom predominantly to the 4 position [147].

The stereochemistry of the hydroboration of 1-trimethylsilylcyclohexene (*23*) [196], 1-*tert*-butyl- (*24*), 1-trimethylsilyl- (*25*), and 1-trimethylgermyl-4, 4-dimethylcyclohexenes (*26*) [197] is determined both by the nature of the substituent in the 4 position and by the steric and electronic effects of the $M(CH_3)_3$ groups:

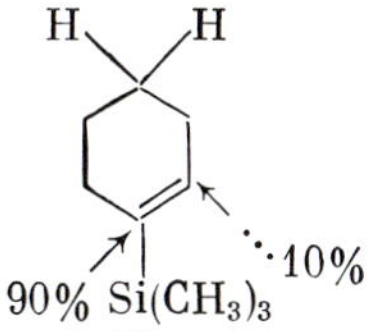

23

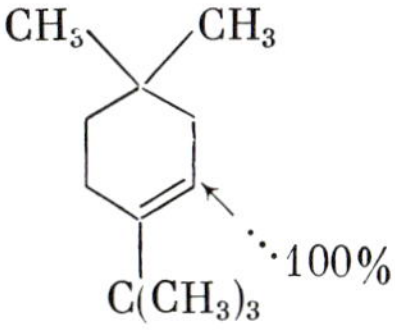

24

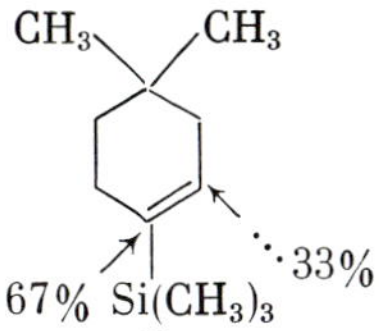

25

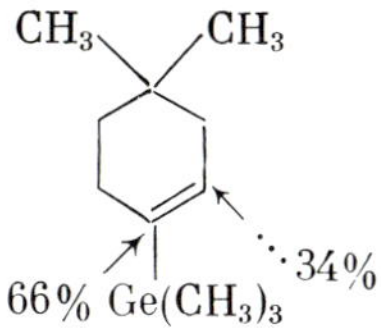

26

In the hydroboration of 6-*tert*-butyl-3,3-dimethylcyclohexene (*27*), only trans alcohols are obtained, whereas 6-trimethylsilyl- (*28*) and 6-trimethylgermyl-3,3-dimethylcyclohexene (*29*) form all four possible isomeric alcohols in different ratios [197]:

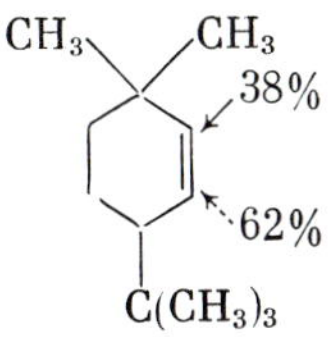

27

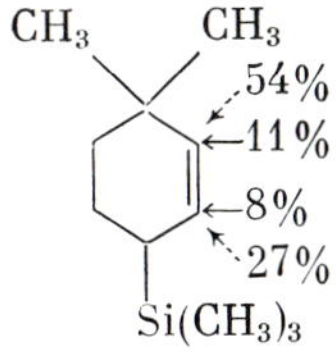

28

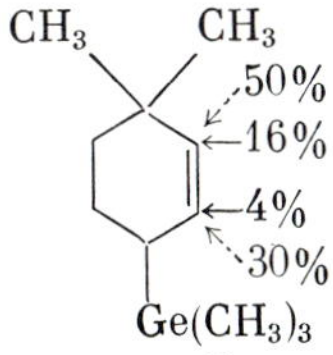

29

The stereoselectivity of the hydroboration of tri- and tetraalkylcyclohexenes [359], the regiochemistry and stereochemistry of the hyhydroboration of cyclic olefin with 9-borabicyclo[3.3.1]nonane [360, 361], and the regioselectivity of the hydroboration of benzocycloalkenes with diborane [362] have been investigated.

The stereochemistry of the hydroboration of bi-1-cyclohexen-1-yl [363, 364] and of bi-1-cyclopenten-1-yl [363] has been studied.

d) Bicyclic systems

Diborane attacks bicyclic hydrocarbons containing cyclohexene or cyclopentene systems connected to a three-membered ring from the α-side. Thus, (+)-carene is converted to (−)-*cis*-caran-*trans*-4-ol

(*31*) by hydroboration-oxidation [198–200]:

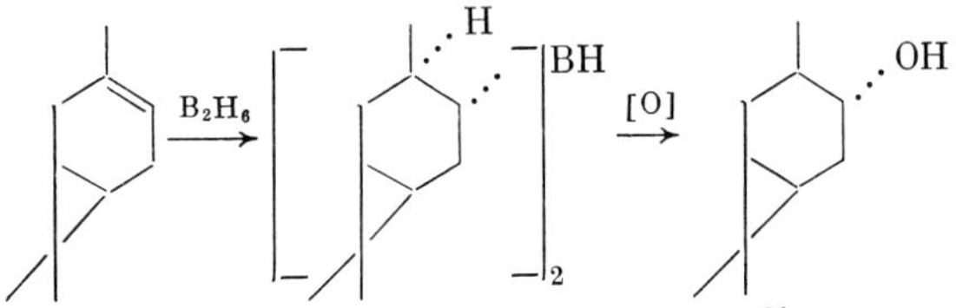

30 31

If bis(*cis*-caran-4-yl)borane (*30*) is heated to 170°C and the isomerized organoboranes are then oxidized, (−)-*cis*-caran-*trans*-5-ol, (−)-*cis*-caran-*trans*-2-ol, (−)-*cis*-*m*-4-menthen-8-ol, and *cis*- and *trans*-caran-10-ol are obtained along with compound *31* [201].

(+)-2-Carene forms (−)-*cis*-caran-*trans*-2-ol upon oxidative hydroboration [198, 199, 202]:

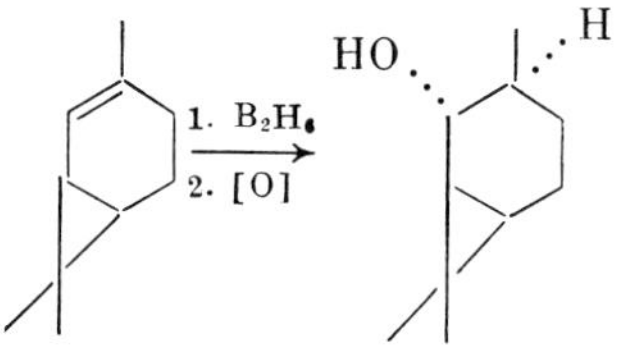

while (−)-*cis*-4-carene is converted to (−)-*cis*-caran-*trans*-4-ol (*31*) and (−)-*cis*-caran-*trans*-5-ol in a ratio of 1:4:

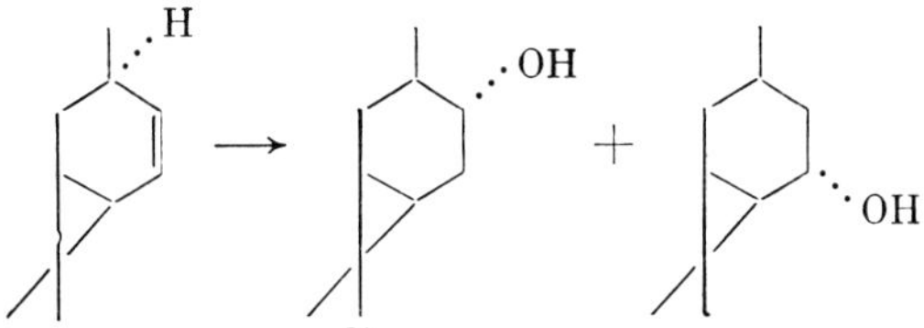

31

(+)-α-Thujene also adds diborane from the side removed from the cyclopropane ring, forming (−)-3-thujyl alcohol [203]:

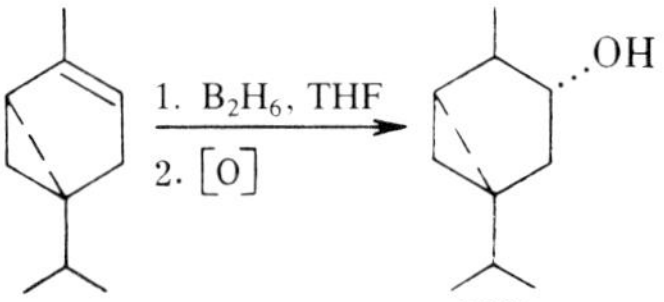

83%

The hydroboration of acyclic terpenes with dichlorodiborane has been investigated [365].

The hydroboration-oxidation of 3-methylbicyclo[4.1.0]hept-2-ene leads to *cis*-3-methylbicyclo[4.1.0]heptan-*trans*-2-ol [366].

The oxidative hydroboration of *cis*-bicyclo[3.3.0]oct-2-ene forms primarily *exo-cis*-bicyclo[3.3.0]octan-2-ol and *exo-cis*-bicyclo-[3.3.0]octan-3-ol, together with traces of the endo-isomeric alcohols [204]:

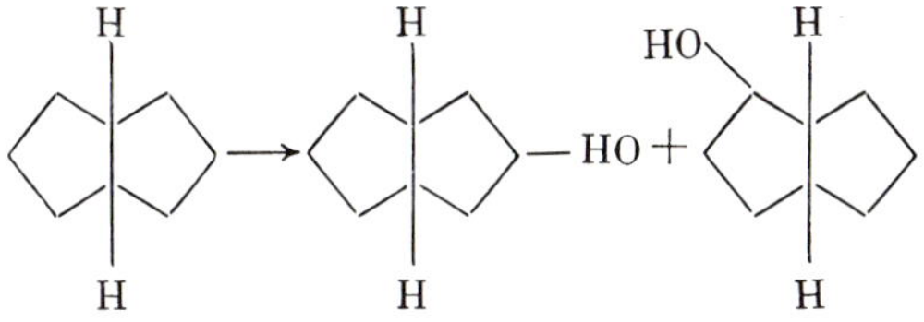

exo-7-Phenylbicyclo[4.2.0]oct-1-ene, containing the endo isomer as an impurity, forms a mixture of isomeric ketones:

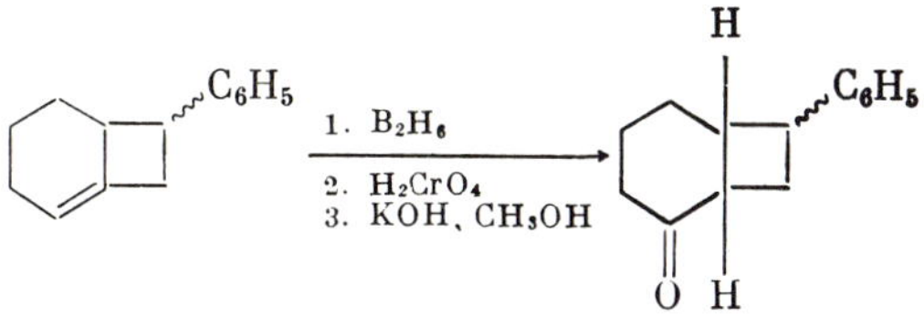

cis-Bicyclo[4.2.0]oct-7-ene forms *cis*-bicyclo[4.2.0]octan-7-one [205]:

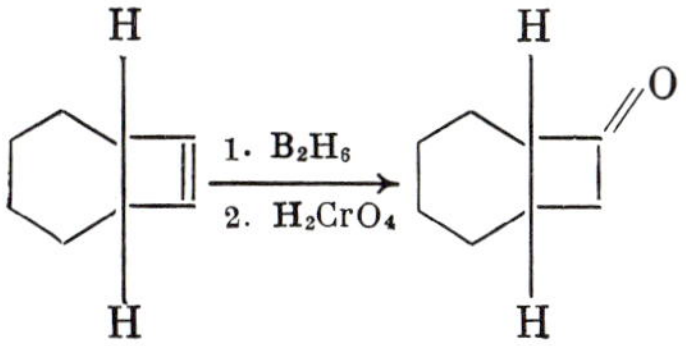

The hydroboration of octalins is interesting from the stereochemical viewpoint.

1(9)-Octalin is considerably superior to 9(10)-octalin in its rate of addition of tetra(3-methyl-2-butyl)diborane. It is therefore possible to carry out a selective hydroboration of an 8:2 mixture of these two hydrocarbons and to obtain pure 9(10)-octalin, free of impurities of the 1(9)-isomer [206]:

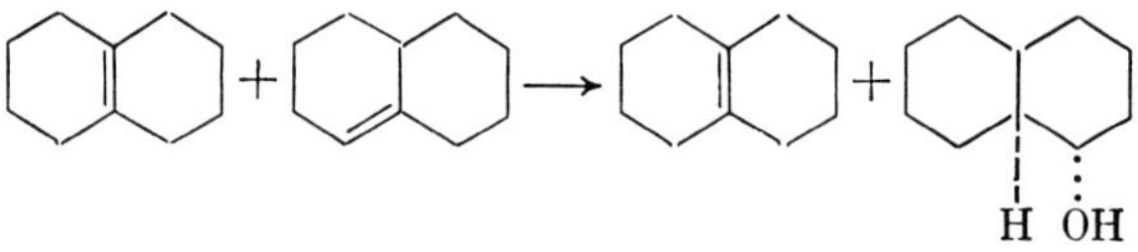

10-Methyl-1(9)-octalin is attacked by diborane from both the β- and α-side (2:1) [207, 208]:

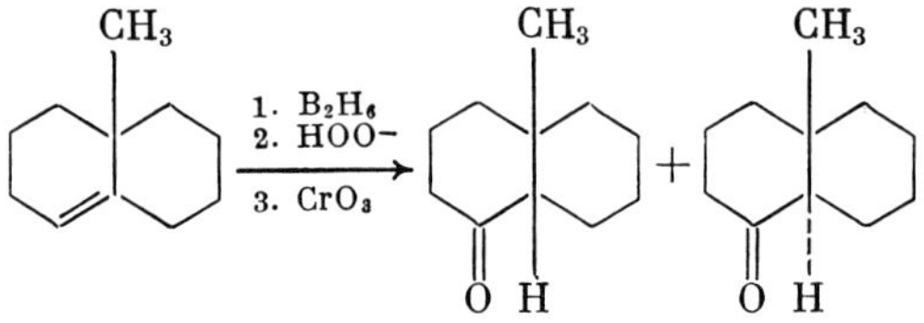

1β- and 1α-alcohols are formed in approximately the same ratio in the hydroboration-oxidation of derivatives of 10-methyl-1(9)-octalin with an equatorial substituent in the 7 position [209–211]:

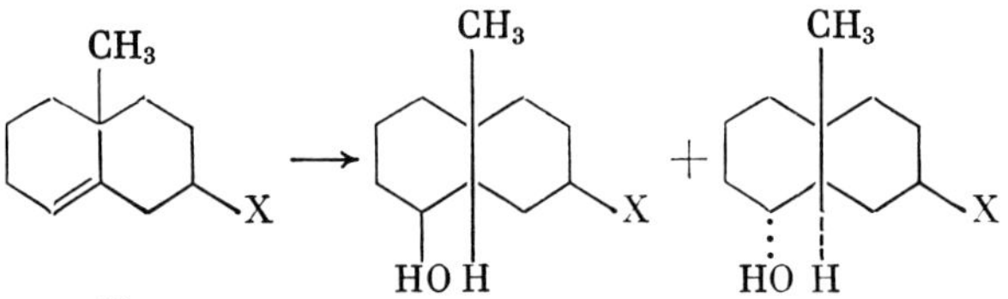

32

$X = \text{iso-}C_3H_7$ 75:25

$X = (CH_3)_2COH$ 67:33

The compound *32* is in the "semichair-chair" conformation (*33*) and the direction of the reaction is determined to a greater degree by the four axial hydrogen atoms than by the angular methyl group and the two vicinal β-axial hydrogen atoms.

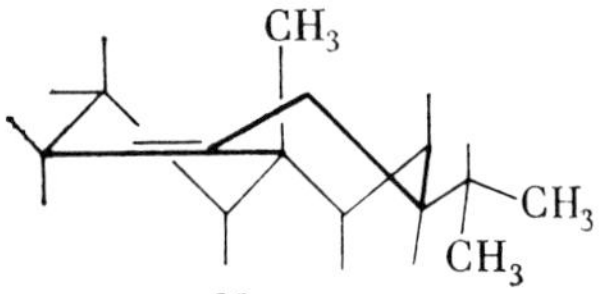

33

With this conformation, steric hindrances from the α-side increase in compound *34* with an α-axial isopropyl group, which should favor an attack from the β-side to an even greater degree. This actually takes place: in the hydroboration-oxidation of compound *34*, only the 1β-alcohol is obtained [212]:

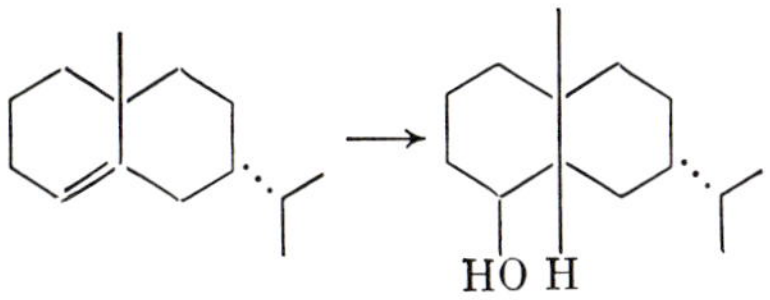

34

The oxidative hydroboration of 2,2-ethylenedioxy-10-methyl-8(9)-octalin, containing an α-axial substituent, also proceeds from the β-side, leading to 2,2-ethylenedioxy-10β-methyl-*cis*-decal-8β-ol [213]:

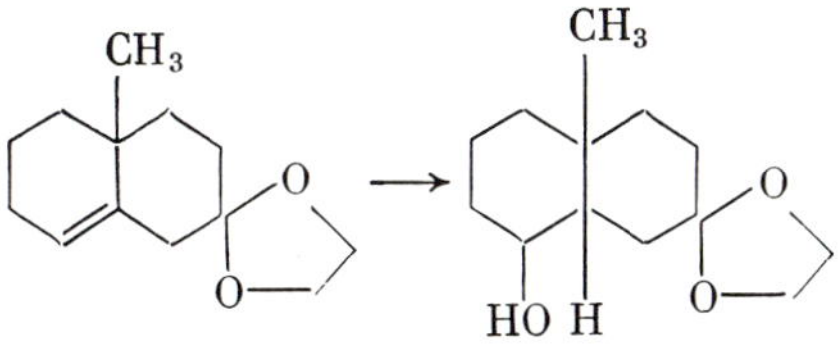

When an equimolar quantity of diborane is used, the cyclic ketal *35* is converted to a mixture of the *cis-trans*-isomeric alcohols *36* and *37*, in a ratio of 2:1 [183]:

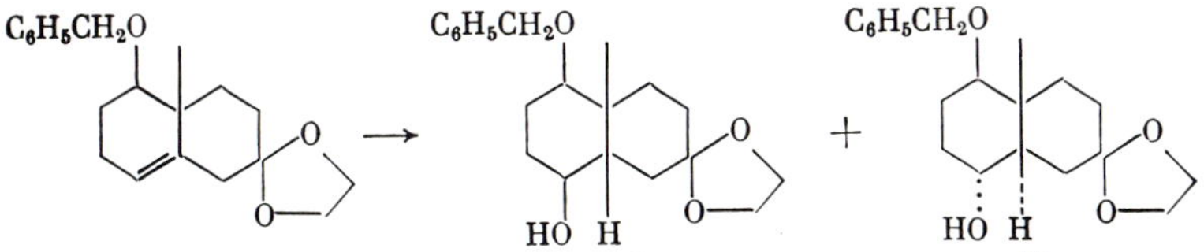

35 **36** **37**

Hydroboration of the hydrocarbon *38* takes place predominantly from the α-side ($\alpha:\beta = 49:1$) [211], which is in agreement with the molecular model *39*, in which the A ring is in a "semiboat" conformation and the B ring, connected to the cyclopropane ring, has an "inverted boat" conformation. In this case, the β-side of the A ring is shielded to a considerable degree.

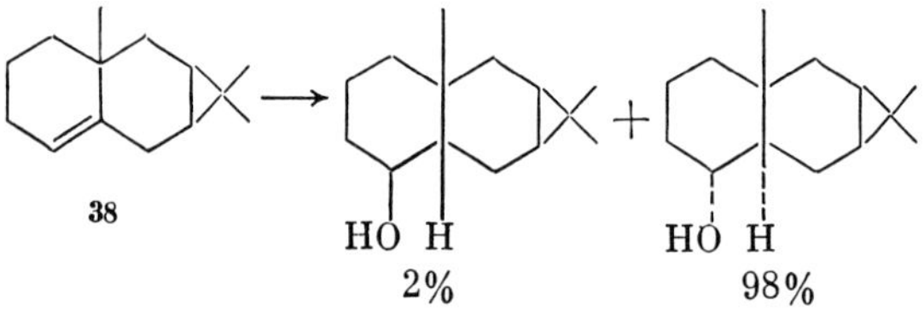

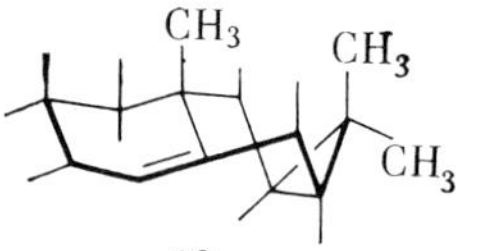

39

8,8,10-Trimethyl-1(9)-octalin is also hydroborated from the α-side, forming the *trans*-connected alcohol [214]:

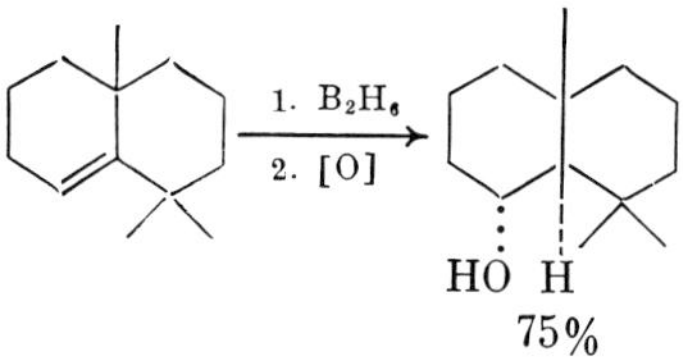

whereas 7-oxo-8,8-dimethyl-10-phenyl-1(9)-octalin does not add diborane, which is related to the steric effect of the phenyl and *gem*-dimethyl groups [215].

7,7,10-Trimethyl-1(9)-octalin is converted to 7,7,10β-trimethyl-*cis*-decal-1β-ol by oxidative hydroboration [207]:

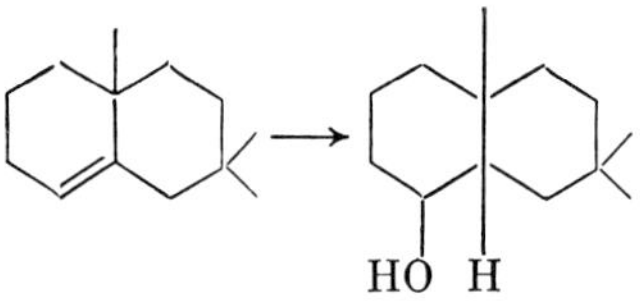

The hydroboration of 2,2,8,8,10-pentamethyl-1(9)-octalin has been investigated [367].

In the hydroboration reaction, 2β-isopropyl-10-methyl-7(8)-octalin is attacked primarily from the α-side ($\alpha:\beta = 9:1$), with the boron atom being localized mainly on the C-7 atom (C-7:C-

8 = 79 : 21) [211]:

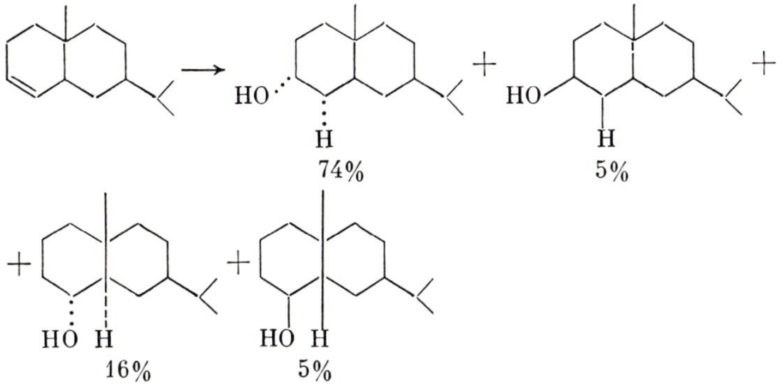

The hydrocarbon *40* also adds diborane primarily from the α-side ($\alpha:\beta = 2.3:1$), with the boron being distributed approximately equally between the C-3 and C-4 atoms (C-3 : C-4 = 49 : 51) [211]:

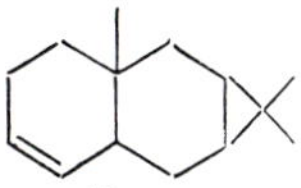

40

The hydroboration of dihydrothujopsene (*41*) proceeds in a characteristic manner [216]. The main reaction product is the tertiary alcohol 2α,8,8,9β,10β-pentamethyl-2β-decalol (*42*), with a small quantity of 8α-hydroxymethyl-2α,8β,9β,10β-tetramethyl-1β-decalol (*43*) also being formed:

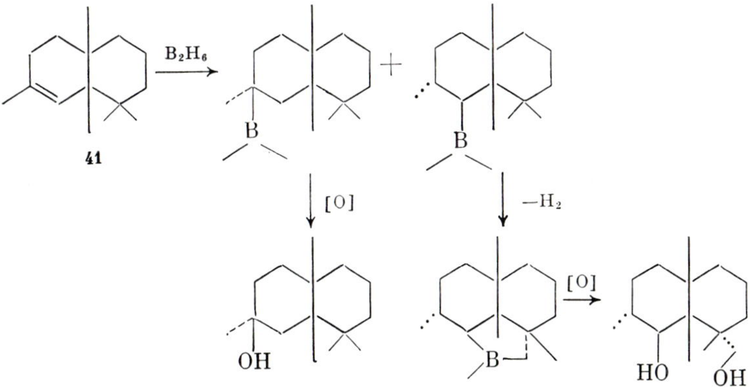

41 **42** **43**

2-Isopropyl-4aβ-methyl-3,4,4a,7,8,8aβ-hexahydronaphthalen-5(6H)-one 2,4-dinitrophenylhydrazone adds diborane from the β-side [217]:

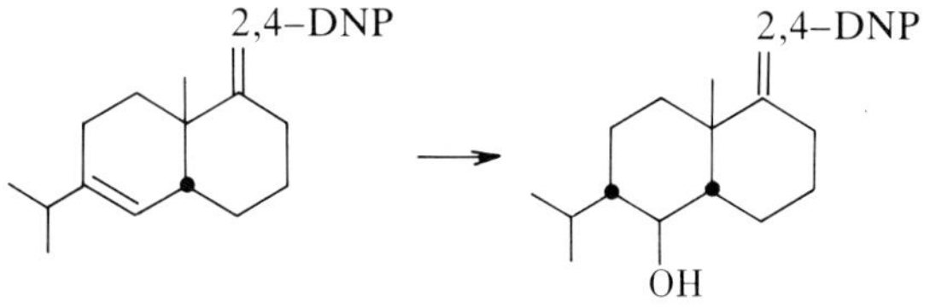

The hydroboration of the norbornene system proceeds with a high stereoselectivity. Norbornene and its 1-methyl and 2-methyl derivatives are attacked by diborane almost completely from the exo side, whereas 7,7-dimethylnorbornene adds diborane mainly from the endo side [175]:

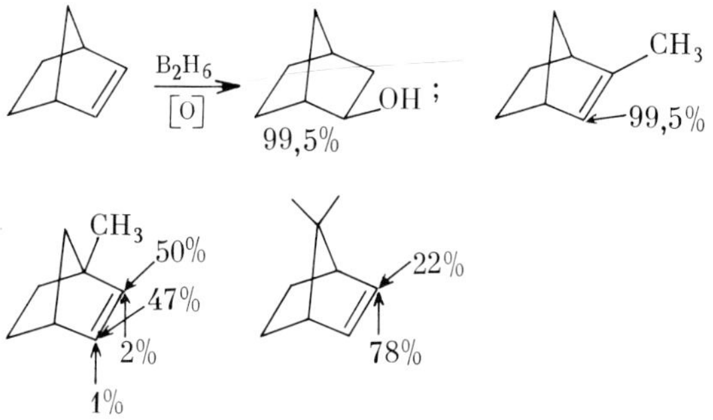

7,7-Dimethylnorbornene is hydroborated more stereoselectively with 9-borabicyclo-[3.3.1]nonane (endo : exo = 97 : 3) [218].

The hydroboration-oxidation of *syn*-7-arylnorbornenes with $BH_3 \cdot$ THF forms a mixture of the exo and endo products [368].

The hydroboration-oxidation of the methyl ester of 2-norbornene-5-*endo*-carboxylic acid leads to a mixture of the esters *44* and *45*:

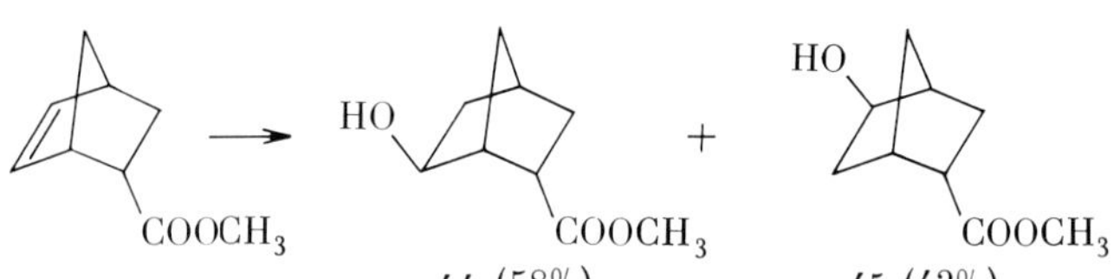

44 (58%) *45* (42%)

while the corresponding alcohols with an exo structure are obtained from the exo isomer [219].

In a hydroboration-oxidation-acetylation, *exo*-isocamphene produces a 59% yield of 6-*exo*-acetoxy-*exo*-isocamphane (*46*) and 5-*exo*-acetoxy-*exo*-isocamphane (*47*) in a ratio of 53:47 [220]:

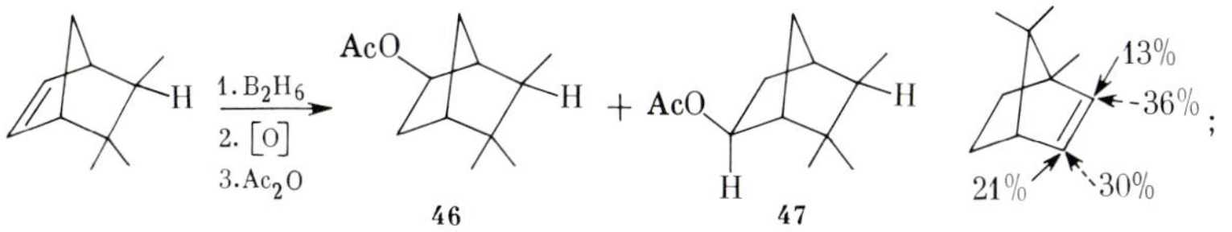

2-Bornene is attacked by diborane primarily from the endo side [221]. When a phenyl group is present in the 2 position, the boron adds mainly to the C-3 atom from both the exo and endo sides [222]:

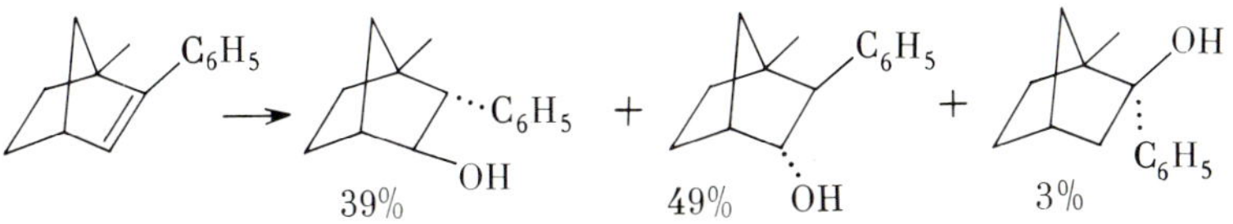

The hydroboration of 1,4,7,7-tetramethylnorbornene produces a mixture of exo and endo alcohols in a ratio of 40:60 [223]:

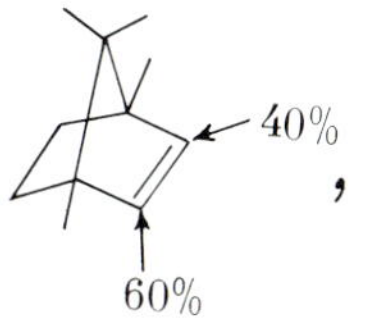

7,7-Dimethoxybicyclo[2.2.1]heptene forms a mixture of exo and endo alcohols in a ratio of 75:20 [224], with a total yield of 74%.

The hydroboration of the anti and syn isomers of 7-trimethyltin-2-norbornene has been described [369].

The effect of a *gem*-dimethyl group on the stereochemistry of addition of diborane is seen in the case of bicyclo[3.1.1]heptene. Thus, α-pinene is converted to isopinocamphanol by hydroboration-oxidation [225] and nopinene forms *trans*-isonopinol [226]:

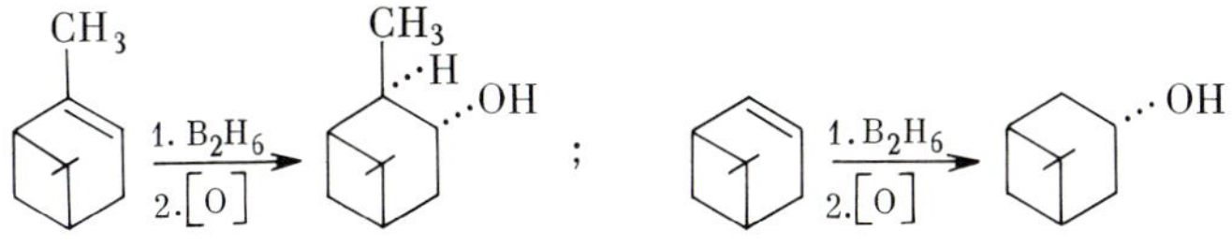

The addition of diborane to 3,7,7-trimethylbicyclo[3.1.1]hept-2-ene also takes place from the side removed from the *gem*-dimethyl group [227]:

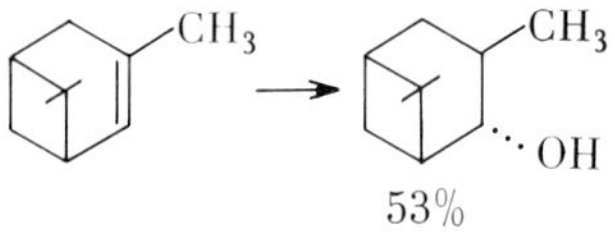

The oxidative hydroboration of bicyclo[3.3.1]non-2-ene forms a 2:3 mixture of *exo*-bicyclo[3.3.1]nonan-2-ol and *exo*-bicyclo[3.3.1] nonan-3-ol, in a yield of 61% [228]:

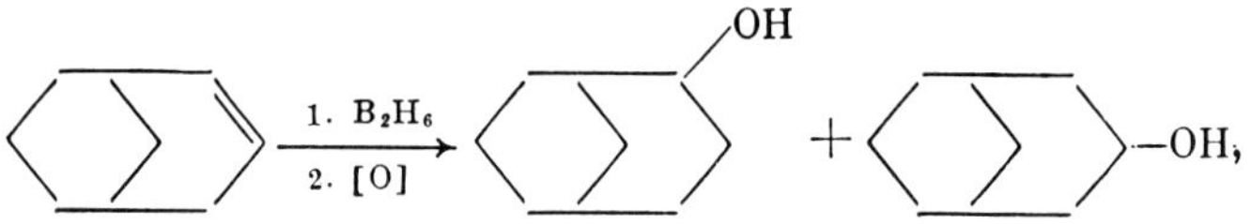

whereas 4β-phenylbicyclo[3.3.1]non-2-ene forms a mixture of the 3-*exo*-alcohol and, probably, the 2-*endo*-alcohol, in a ratio of 3:1 [229].

1,5-Dimethylbicyclo[3.3.1]non-2-ene forms primarily the 3-*exo*-alcohol, together with a small quantity of the 2-*exo*-alcohol [230]:

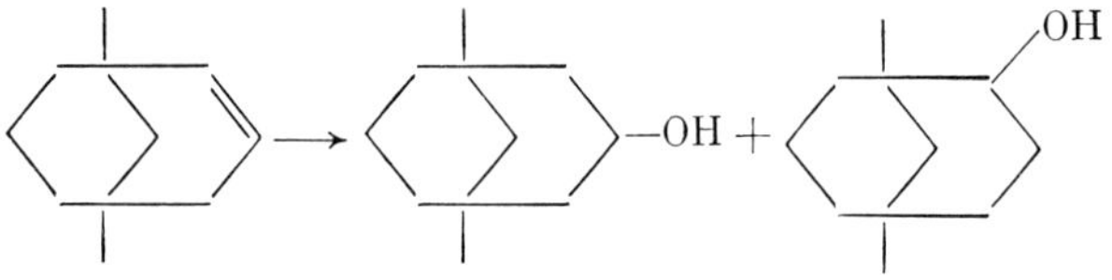

The hydroboration-oxidation of bicyclo[3.3.1]non-1-ene forms 2-*exo*- and 2-*endo*-alcohols and also a compound with a hydroxy group on the tertiary carbon atom [231]:

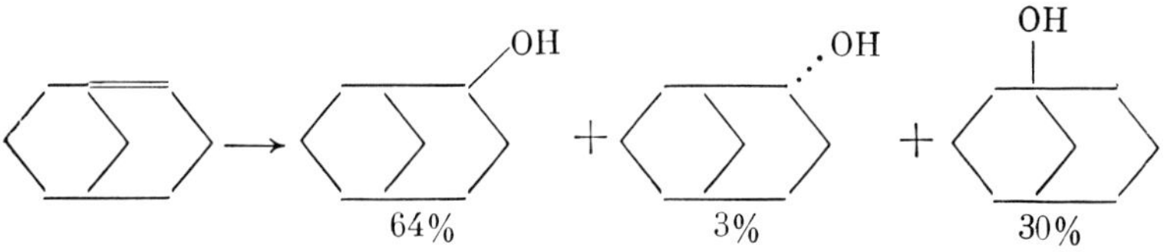

e) Tricyclic systems

The hydroboration of 4,5-*exo*-trimethylene-2-norbornene (*48*) leads to the alcohols *49* and *50*, but, if the reaction is carried out in the presence of BF_3, the alcohol *51* is also obtained [232]:

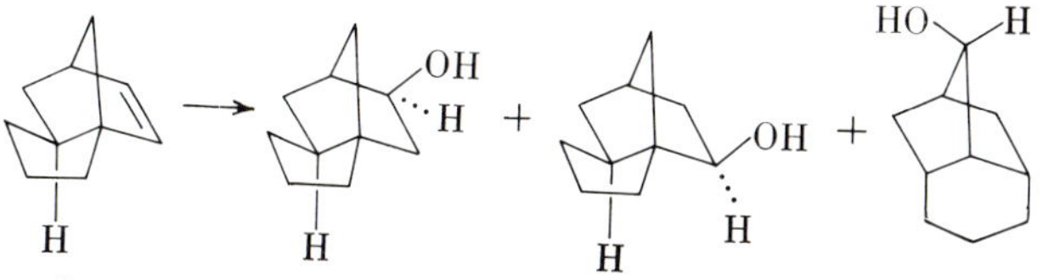

48 49 (42%) 50 (10%) 51 (8%)

exo-Tricyclo[4.2.1.0^{2,5}]non-7-ene (*52*) is converted by oxidative hydroboration to *exo*-7-hydroxy-*exo*-tricyclo[4.2.1.0^{2,5}]nonane (*53*), while the endo isomer (*54*) forms *exo*-7-hydroxy-*endo*-tricyclo [4.2.1.0^{2,5}]nonane (*55*) [233]:

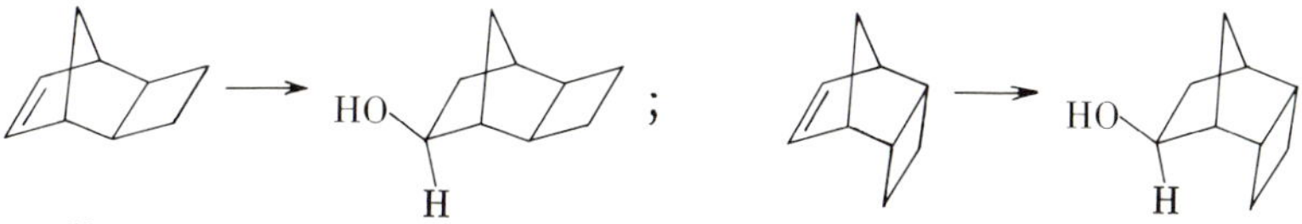

52 53 54 55

The hydration of tricyclo[5.2.1.0^{2,6}]dec-8-ene (*exo*-dihydrodicyclopentadiene) leads to the exo alcohol *56* [234]:

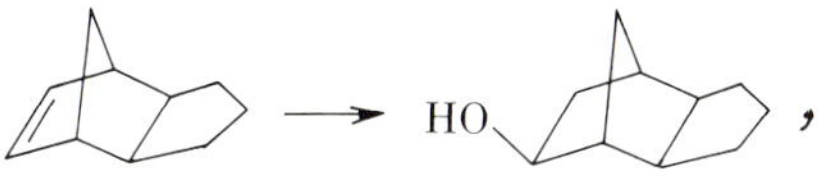

56

and the hydration of isodrin forms the alcohol *57* [235, 236]:

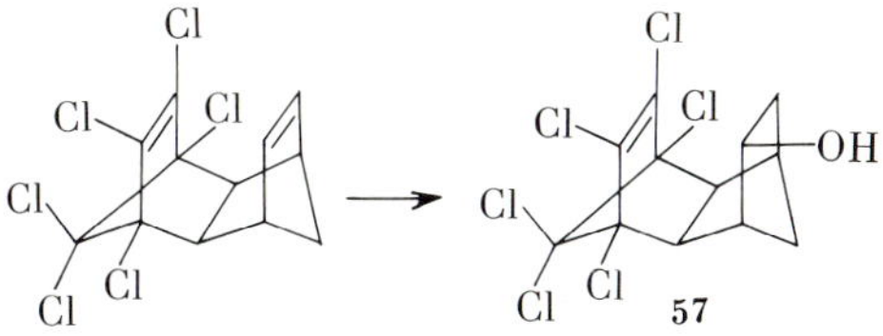

57

endo-Tricyclo[5.2.1.0^{2,6}]dec-3-ene (*endo*-5,6-trimethylene-8-norbornene), upon hydroboration and subsequent oxidation with chromic acid, produces a mixture of the ketones *58* and *59* in a ratio of 2:3 [237]:

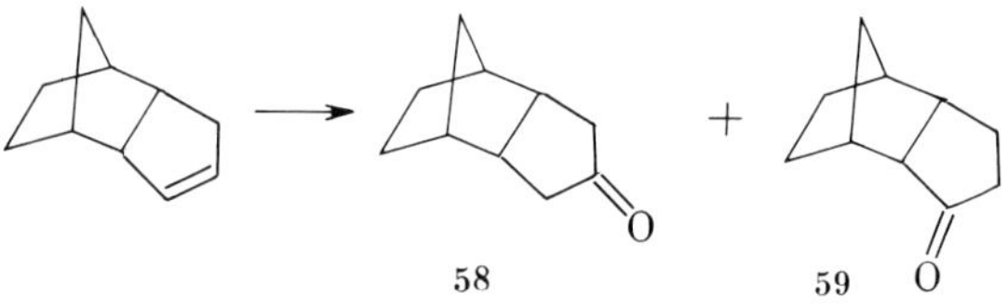

The hydroboration-oxidation of tricyclo[6.2.1.$0^{2,6}$]undec-2(6)-ene forms exclusively *endo*-2-*exo*-6-tricyclo[6.2.1.$0^{2,6}$]undecan-*exo*-3-ol [370].

The tricyclic sesquiterpene α-gurjunene is converted by hydroboration-oxidation to the alcohol *60*, 4α,5α-cyclocoloran-3α-ol (*61*), and the diol *62* [238]:

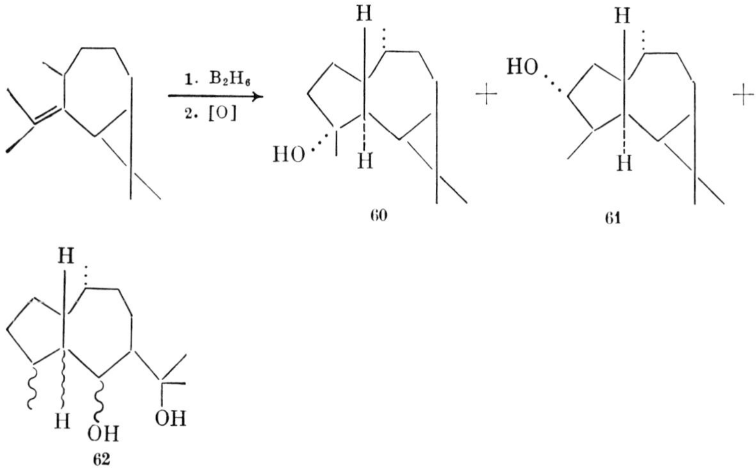

The compound *62* is obtained as a result of cleavage of the three-membered ring, similar to that taking place in norcarane, bicyclo[3.1.0]hexane, and spiro[2.5]octane under the action of diborane at 100°C [239].

It is interesting that the compound *63* adds diborane from the side of the *gem*-dimethyl group, forming the ketone *64* after oxidation with chromic acid [240]:

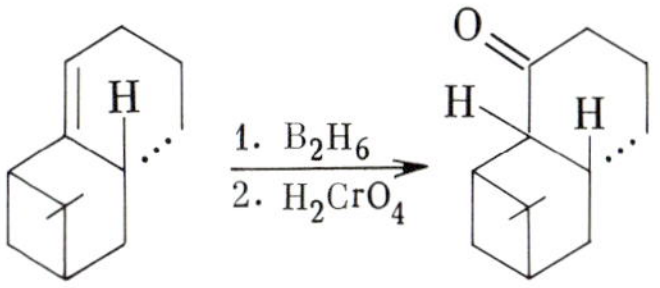

63 64

(−)-α-Cedrene forms (+)-2-isocedranol upon hydration [184, 241]:

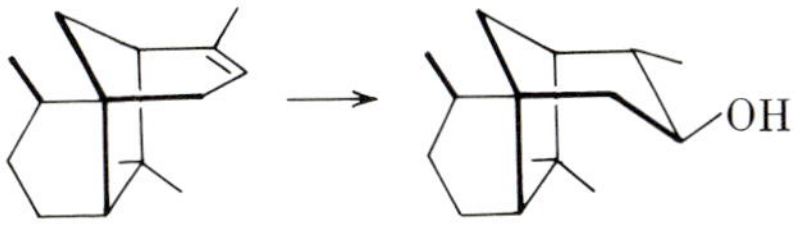

The deuterohydroboration-oxidation of tricyclo[5.3.1.0^{3,8}] undec-2-ene has been investigated [371].

α-Patchoulene is also hydroborated from the side removed from the *gem*-dimethyl group [242]:

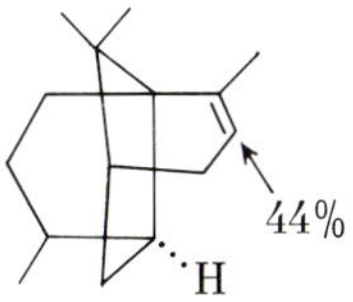

The trimethylsilyl ether of 9-cariolen-1-ol is hydrated to 1,10β-cariolandiol (*65*), the diol *66*, and a mixture of isomeric 1,9-diols (*67*) [243]:

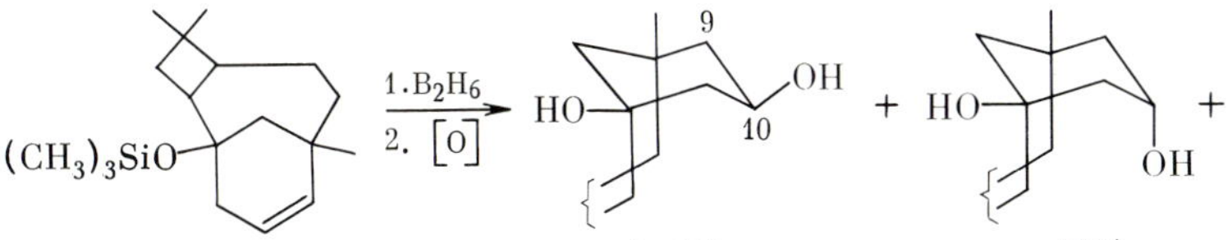

65 (37%) 66 (7%)

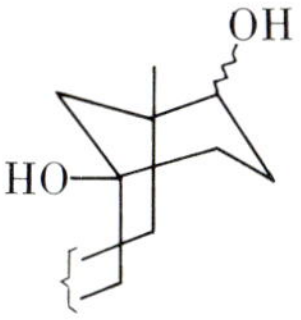

67 (49%)

In contrast to the 2- and 3-carenes, (−)-thujopsene adds diborane exclusively from the β-side with respect to the cyclopropane ring, forming (+)-3-thujopsanol after oxidation [244]:

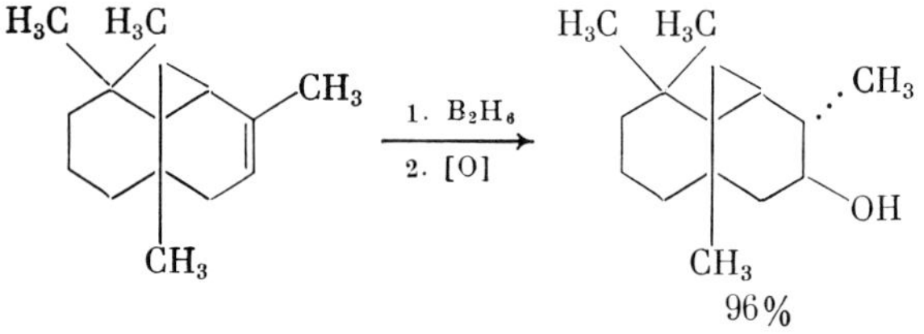

Norrimuene is attacked by diborane from the β-side and is converted to the 9β-hydroxy derivative [245]:

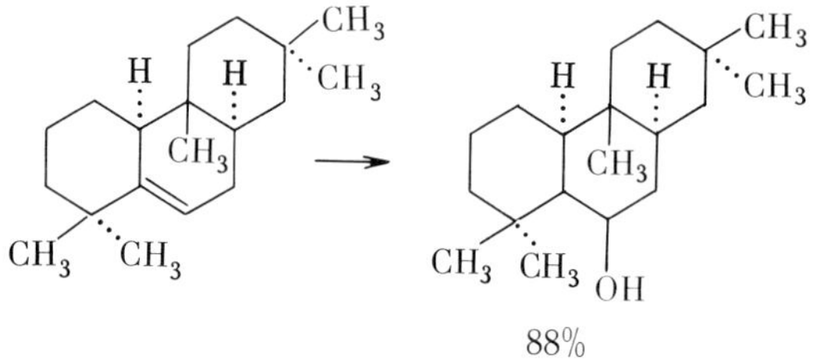

5-Dehydroabietan forms the alcohols *68* and *69* upon oxidative hydroboration [246]:

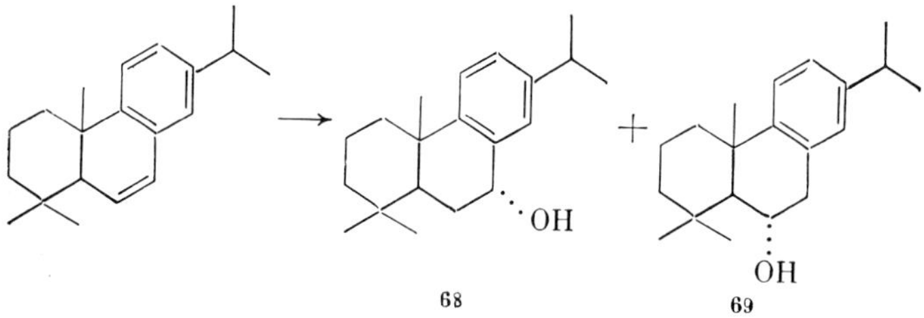

The compound *70*, obtained by the photoaddition of methanol to methyl abietate, forms primarily the 7α-hydroxy compound *71*, with an impurity of the 7β-hydroxy epimer *72*, upon oxidative hydrobora-

In this way, 3,7-dimethyl-1,6-octadiene is converted to tris(3,7-dimethyl-6-octenyl)borane (*84*), the oxidation of which forms citronellol:

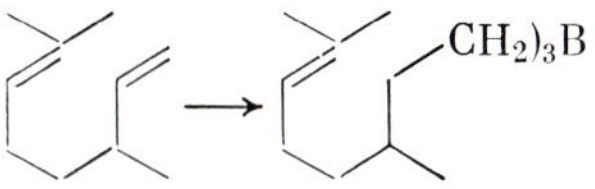

84

The monohydroboration of myrcene with tetra(3-methyl-2-butyl)diborane leads to myrcenol (*85*), while the dihydroboration produces 7-methyl-3-hydroxymethyl-6-octen-1-ol (*86*) [252]:

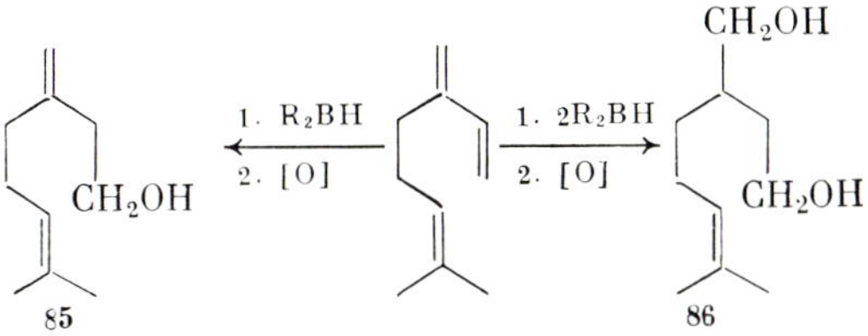

85 86

1-Allyl-3-cyclopentene is converted by partial hydroboration and subsequent oxidation to 3-(3-cyclopentenyl)propanol [253] and α-guaiene is hydrated to the alcohol *87* [254]:

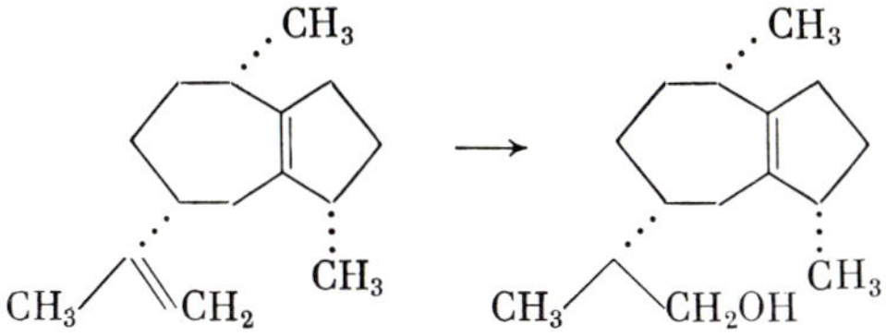

87

Disubstituted *exo*-cyclic diene hydrocarbons add diborane only at the cyclic double bond. Thus, a mixture of terpinolene and isoterpinolene is converted to the alcohol *88* by hydroboration-oxidation [255]:

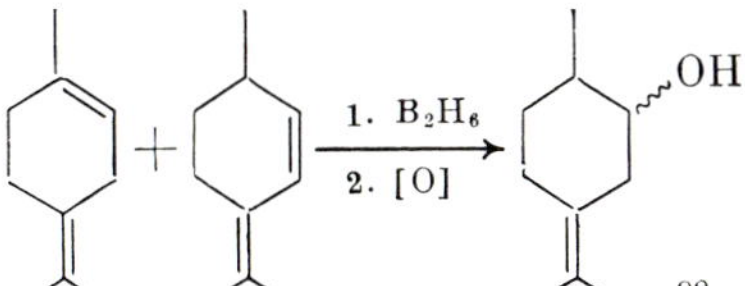

88

The monohydroboration of verbene with tetra(3-methyl-2-butyl)diborane, followed by protolysis, leads to *cis*-δ-pinene [256]:

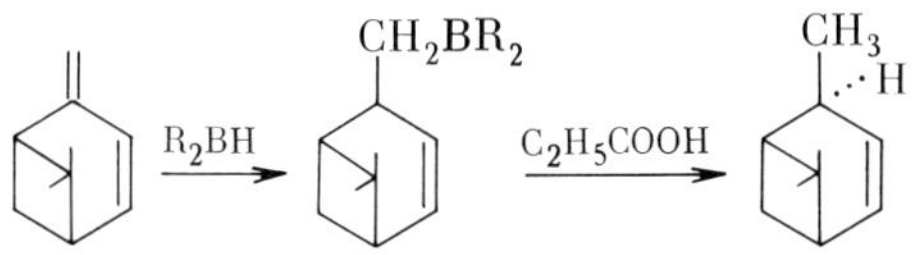

As a result of the reaction of diborane with excess cyclopentadiene and subsequent oxidation, 3-cyclopentenol was isolated in 30% yield [118]:

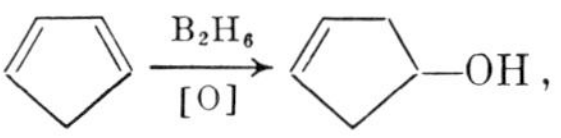

whereas, under dihydroboration conditions, a mixture of *trans*- and *cis*-cyclopentane-1,3-diols (85:15) was obtained in 80% yield [42, 257].

The hydroboration of cyclopentadiene with tetraethyldiborane [140], tetraisopinocampheyldiborane, or tetra(3-methyl-2-butyl) diborane [258] forms predominantly 3-cyclopentenol, together with cyclopentanediols.

The dihydroboration-oxidation of 1,3-cyclohexadiene leads to two glycols of unestablished structure, whereas, when excess hydrocarbon is used (diene:BH_3 = 6:1), the oxidation products are 3-cyclohexen-1-ol and 2-cyclohexen-1-ol (2:3) [259]:

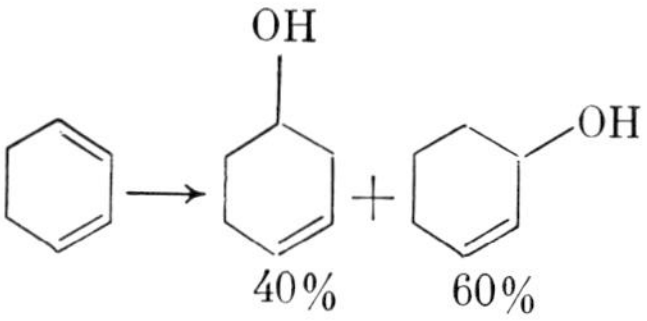

1,4-Cyclohexadiene forms a mixture of *cis*-1,3-, *cis*-1,4-, *trans*-1,3-, and *trans*-1,4-cyclohexanediols in a ratio of 20:28:22:30 as a result of hydroboration-oxidation [260].

Substituted cyclohexadienes are hydroborated at the less substituted double bond, with the formation of unsaturated alcohols.

The main products of the hydroboration-oxidation of 1,5-dimethyl-5-ethyl-1,3-cyclohexadiene are 1,5-dimethyl-5-ethyl-1-cyclohexen-3-

ol and glycols of unestablished structure [261, 262]:

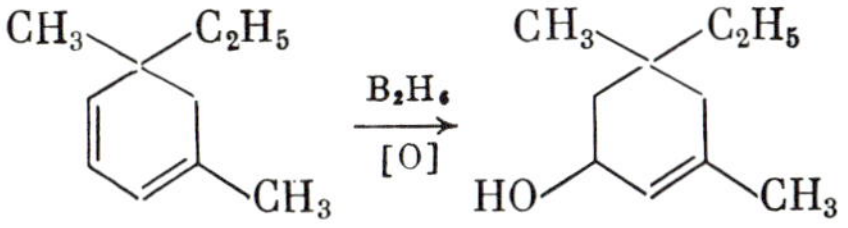

The addition of diborane to the less substituted double bonds, with localization of the boron atom on the carbon most removed from the substituents, is also observed in the case of 1,5,5,6-tetramethyl-1,3-cyclohexadiene [263]:

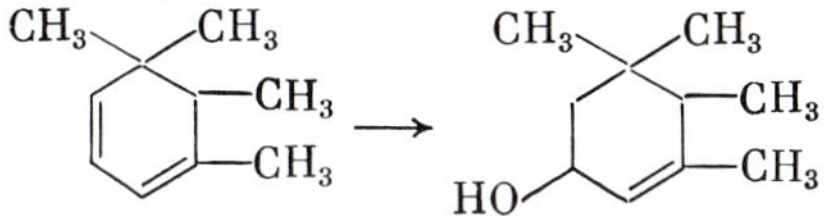

As a result of the hydroboration-oxidation of α-pyronene (1,5,5,6-tetramethyl-1,3-cyclohexadiene) [264], β-pyronene (1,2,6,6-tetramethyl-1,3-cyclohexadiene) [265], or 1,3-dimethyl-1-ethyl-1,3-cyclohexadiene [266], alcohols $C_{10}H_{18}O$ of unestablished structure are formed. The organoboron compounds obtained by the hydroboration of α- and β-pyronenes decompose during vacuum distillation, with elimination of the starting hydrocarbons [267].

The hydration of β-pyronene leads to 2,3,4,4-tetramethyl-2-cyclohexen-1-ol and two glycols [268]:

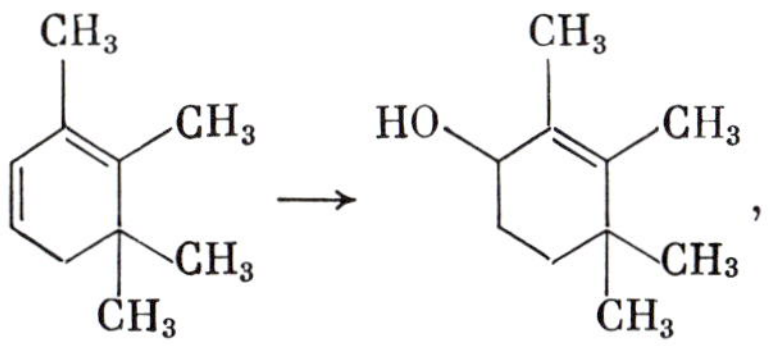

while a mixture of cis and trans alcohols is obtained from α-phellandrene [269]:

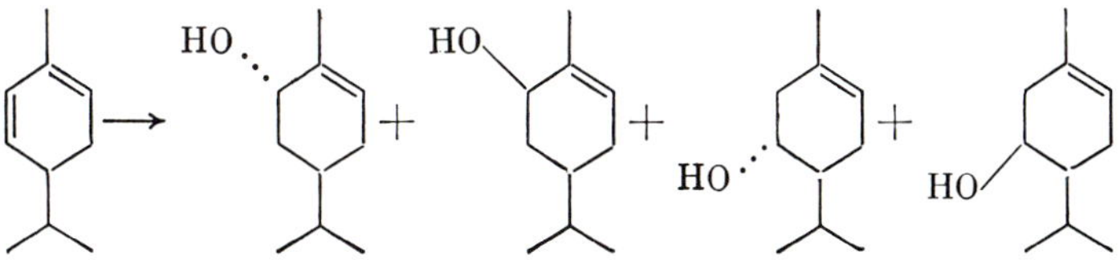

The monohydroboration-protolysis of 1,3-cyclooctadiene forms *cis*-cyclooctene in 35% yield [270].

The oxidative hydroboration of *cis,cis*-1,3-cyclodecadiene leads to *trans,cis*-2-decalol (*89*) (42%), together with a small amount (6%) of cyclodecanone and two isomeric cyclodecanediols (27%) of unestablished structure [271]:

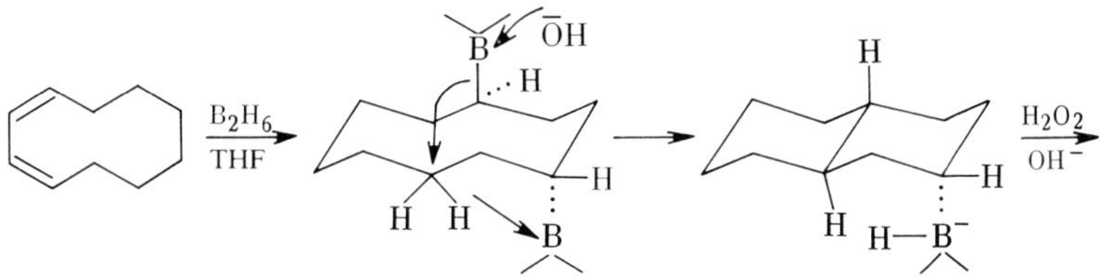

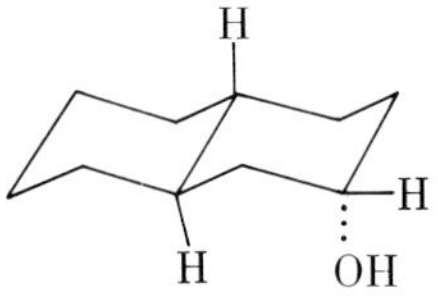

89

The hydroboration-oxidation of *cis,trans*-1,5-cyclodecadiene forms *cis*-1,4-cyclodecanediol (*90*), *cis*-1,5-cyclodecanediol (*91*), cyclodecanone (*92*), and cyclodecanol (*93*), in yields of 61, 10, 4, and 4%, respectively:

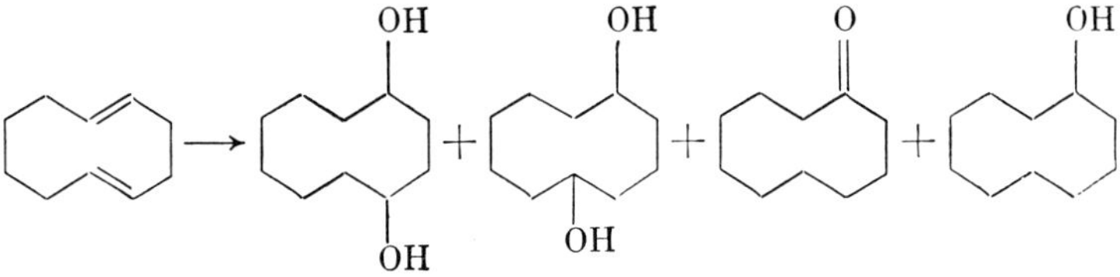

90 91 92 93

However, the thermal isomerization of the organoboranes formed, followed by oxidation, produces a mixture of *92* and *93* in a ratio of 1:1, with a yield of 72%.

The monohydroboration-protonolysis of *cis,trans*-1,5-cyclodecadiene produces *cis*-cyclodecene [272].

1,9-Cyclohexadecadiene, as a result of oxidation with chromic acid of the hydroboration products, is converted to 8-cyclohexadecen-1-one (*94*), cyclohexadecanone (*95*), 1,9-cyclohexadecanedione (*96*), and 1,8-cyclohexadecanedione (*97*) [273]:

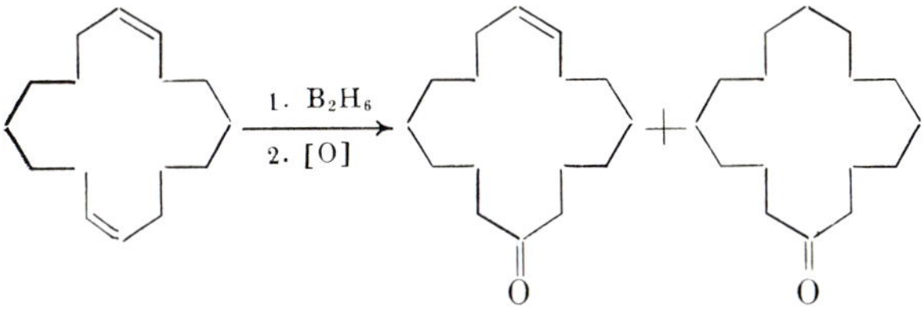

94 (20%) 95 (26%)

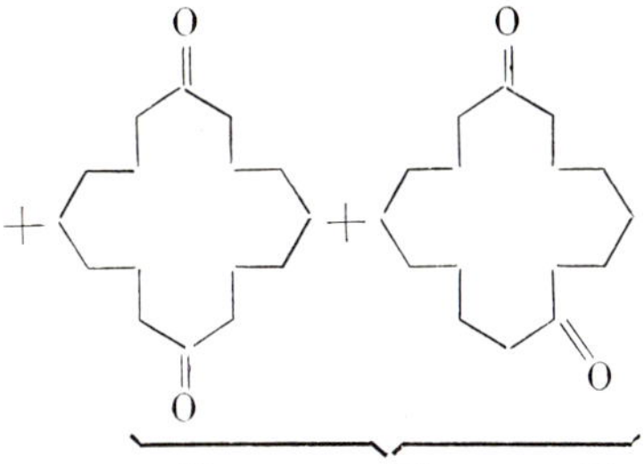

96 (45%) 97

The hydroboration-oxidation of 3,7-diphenylbicyclo[3.3.0]octa-dienes produces derivatives of 3,7-endo-diphenylbicyclo[3.3.0]octane [372].

Bicyclo[2.2.1]hepta-2,5-diene is attacked by diborane predominantly from the exo side. The organoborane *98* obtained is converted by oxidation with hydrogen peroxide in alkaline medium to the alcohol *99* [274, 275]; in the oxidation with chromic acid, the ketone *101* is formed together with the ketone *100* [275]:

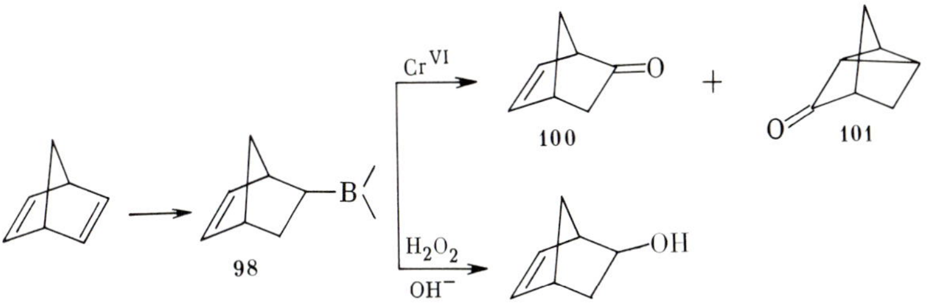

98 100 101 99

The dihydroboration-oxidation of bicyclo[2.2.1]hepta-2,5-diene leads to a mixture of *exo,exo*-bicyclo[2.2.1]heptane-2,5-diol and -2,6-diol in a ratio of 10:1 [373].

The hydrocarbon *102* is hydrated to form a 70% yield of the alcohol *103* [276]:

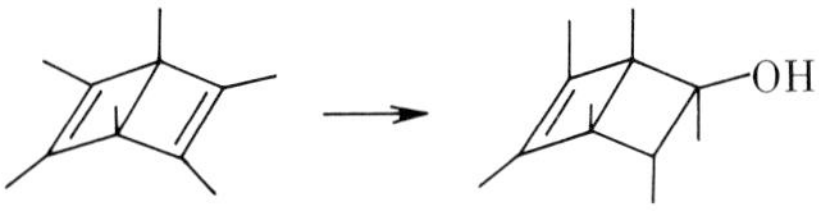

102 103

endo-Tricyclo[5.2.1.2.0^{2,6}]deca-3,8-diene (*endo*-dicyclopentadiene) undergoes hydroboration-oxidation to form four alcohols [277]:

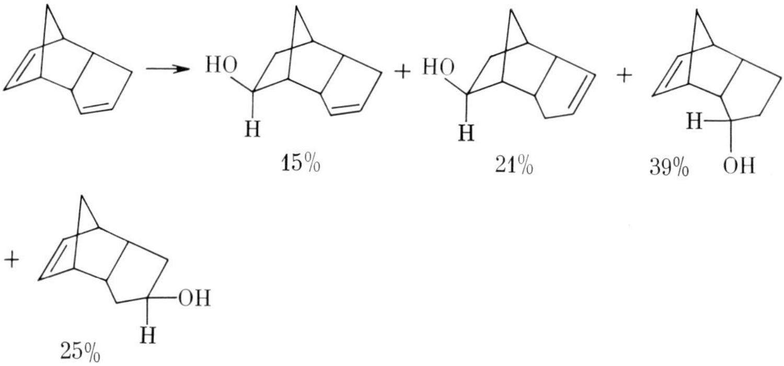

The hydroboration of caryophyllene, isocaryophyllene [374], and *endo*-tricyclo-[5.2.2.0^{2,6}]undeca-3,8-diene has been investigated [375].

Isopimaradiene is hydrated to form the unsaturated alcohol *104* and the diol *105* [278]:

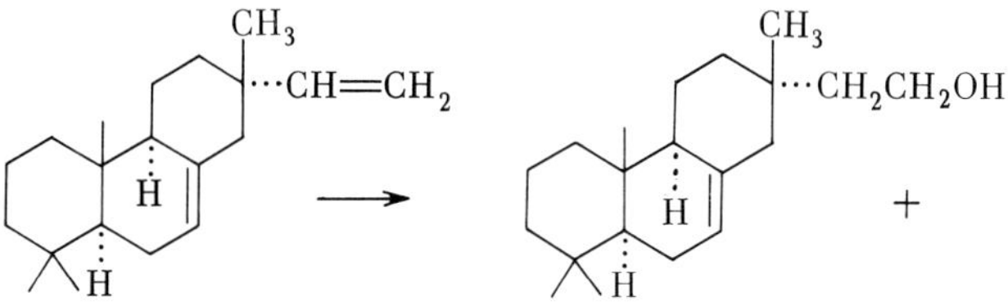

104

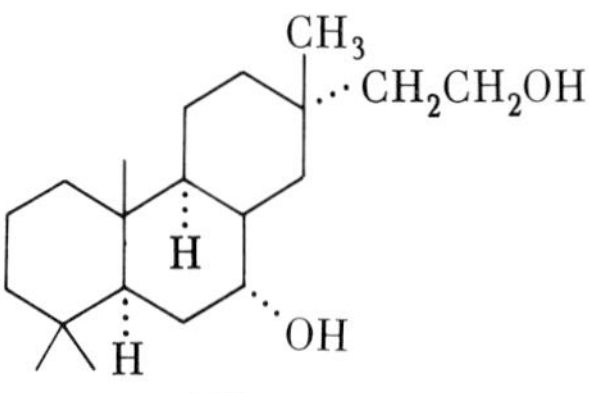

105

Methyl isopimarate (*106*), under the action of tetra(3-methyl-2-butyl)diborane, followed by oxidation, forms the ester of 20-hydroxy-7-dihydroisopimaric acid (*107*):

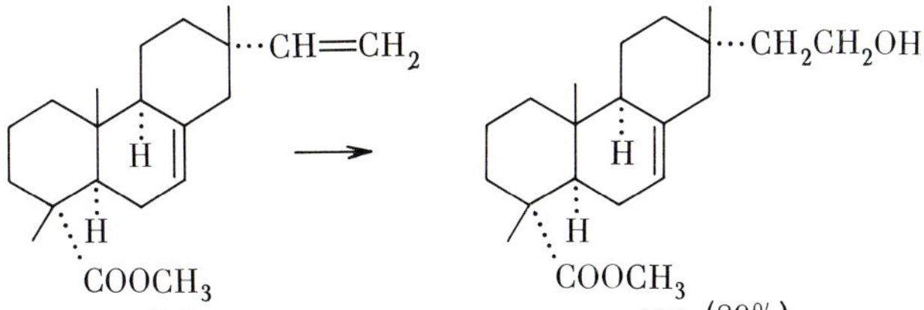

106 107 (89%)

The dihydroboration of compound *106* leads to a mixture of the glycols *108* and *109*, with yields of 17 and 62%, respectively [279]:

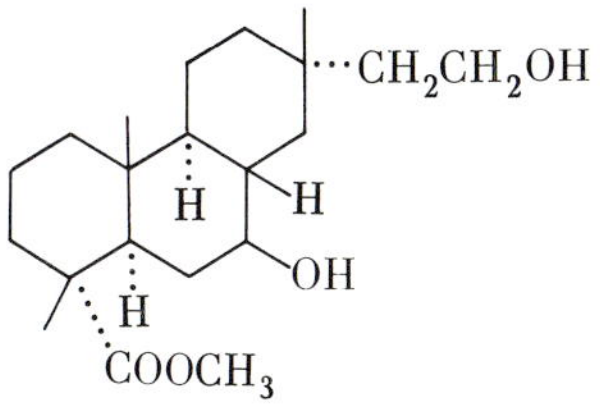

108

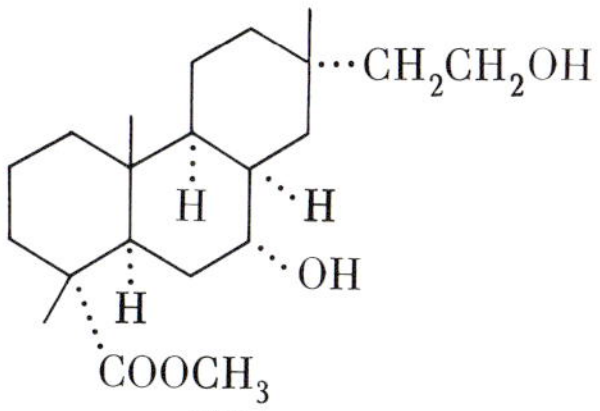

109

Both double bonds in the diene (*110*) are hydroborated from the α-side [250]:

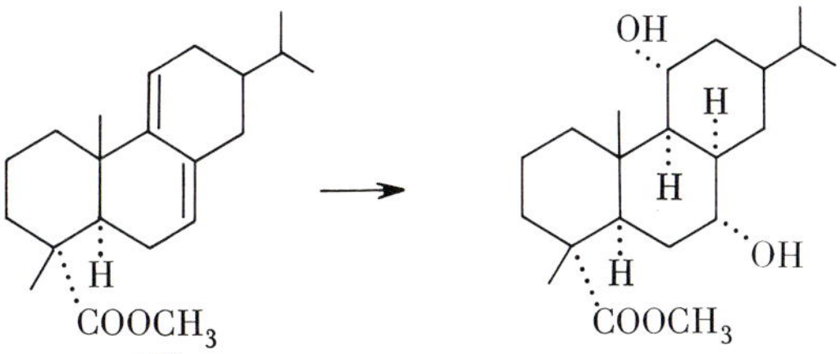

110

Hydroboration has been used in the synthesis of sesquiterpenes [376, 377] and of polyenes related to terpenoids [378].

g) Steroids

The method of hydration with the use of hydroboration and subsequent oxidation has found application in the steroid series.

Sondheimer and Nussim [280, 281] showed that *cis*-hydration from the sterically less hindered side proceeds in the oxidative hydroboration of 5α-cholest-1-ene, which is converted to cholestan-1α-ol (*111*) and cholestan-2α-ol (*112*), with yields of 35 and 40%, respectively:

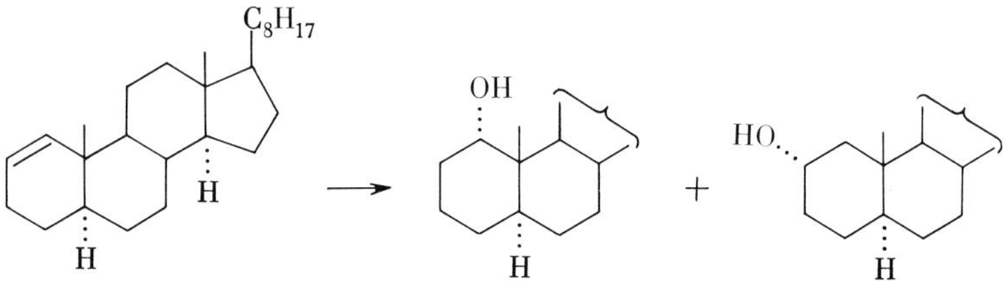

111 112

When tetra(3-methyl-2-butyl)diborane is used as a hydroborating agent, the reaction proceeds selectively and the alcohol *112* is obtained in 75% yield.

From 5α-cholest-2-ene, both alcohols of the α series — 5α-cholestan-3α-ol (*113*) and 5α-cholestan-2α-ol (*114*) [281–283] — and 5α-cholestan-3β-ol (*115*) [282] are obtained:

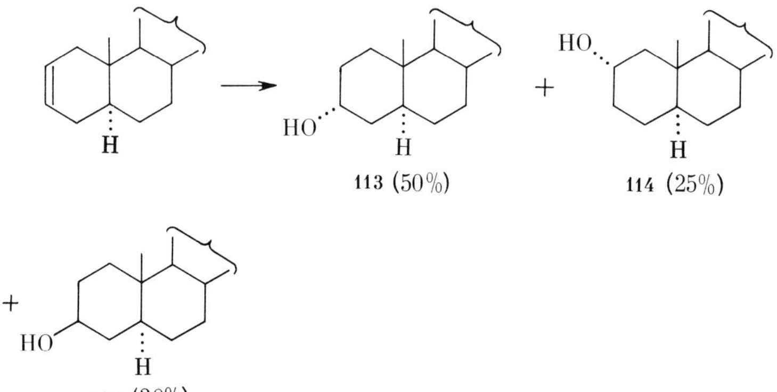

113 (50%) 114 (25%) 115 (20%)

The results of the hydroboration of 5α-cholest-3-ene (*116*) [280–281] and cholest-4-ene (*117*) [122] are shown in the following scheme:

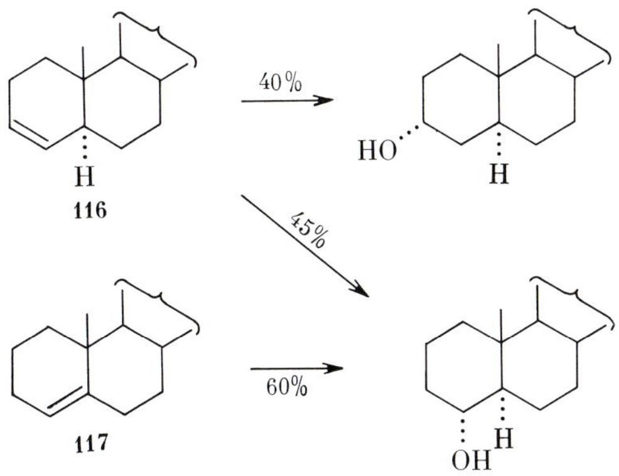

In the hydration of cholest-5-ene, only 5α-cholestan-6α-ol (*118*) is obtained according to some data [122, 281], whereas, according to other data, 5β-cholestan-6β-ol (*119*) is also formed along with *118* [283]:

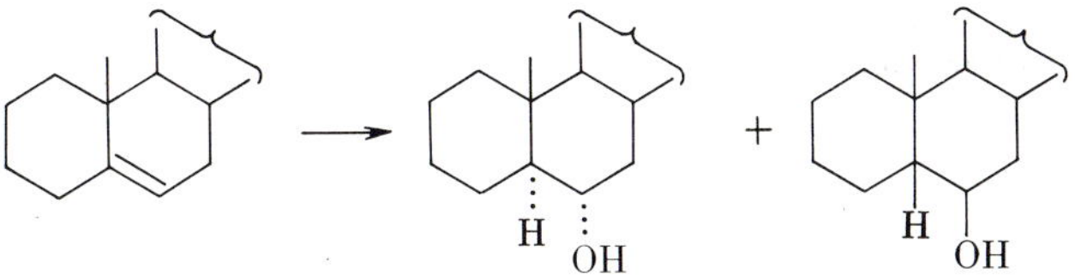

Boranes of the steroid series are capable of undergoing thermal isomerization. Thus, if the products of the hydroboration of 5α-cholest-2-ene are heated for 4 hours at 150°C, a mixture of alcohols containing 5α-cholestan-3β-ol (31%), 5α-cholestan-2α-ol and 5α-cholestan-4α-ol (51%), and also 3α-, 6α-, and 7β-ols (14%) and the 1α-ol is obtained.

An analogous treatment of cholest-5-ene produces 5α-cholestan-3β-ol (18%); 2α- and 4α-ols (30%); 6α-, 3α-, and 7β-ols (40%); 5β-cholestan-6β-ol (9%), and 5α-cholestan-7α-ol (3%) [284].

5β-Cholest-3-ene is converted to 5β-cholestan-3β-ol (75%), with an admixture of an unidentified alcohol, by oxidative hydroboration [285].

In the presence in the 3β position of Δ^5-steroids of such substituents as Cl [286], OAc [286], OCH_3 [287], and $(CH_3)_2N$ [288], the stereochemistry of the addition of diborane remains the same as in the case of cholest-5-ene, whereas, in the case of 3β-OH, a partial β-attack takes place along with the α-attack [282]. If the OAc substituent occupies the 3α position, then the attack proceeds from the β-side to the extent of 92% [289]. 5α-Androst-9(11)-ene is smoothly converted to 5α-androstan-11α-ol by oxidative hydroboration, whereas the compound of the 5β series does not react with diborane [281, 290].

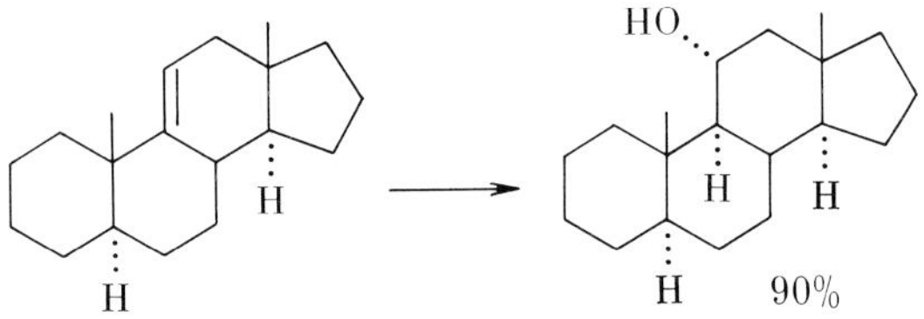

Δ^{14}-Steroids form 15α-alcohols [282] and 5α,6β-dichloro-androst-14-en-3β,17β-diol is converted in low yield to a mixture of 15α- and 15β-alcohols [291].

19-Nor-5α-cholest-1-ene reacts with tetra(3-methyl-2-butyl)-diborane nonselectively, forming, after oxidation, 19-nor-5α-cholestan-2α-ol (*120*), the 2β-ol (*121*), and small quantities of the 1α-ol (*122*) and 1β-ol (*123*) [292]:

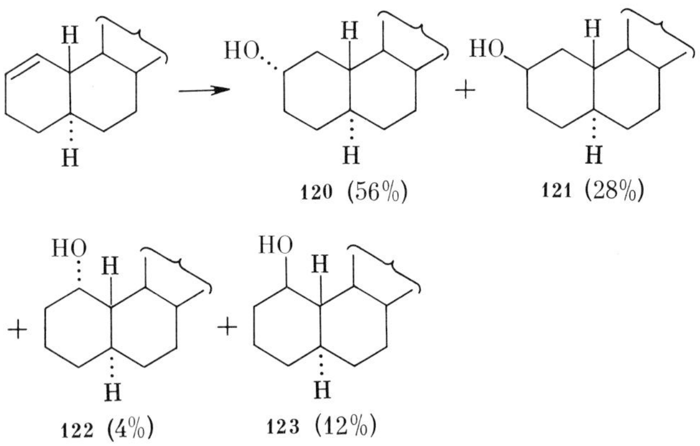

The pentacyclic terpene hydrocarbon serratene is hydroborated stereoselectively, being converted to β-serratan-15-ol [293]:

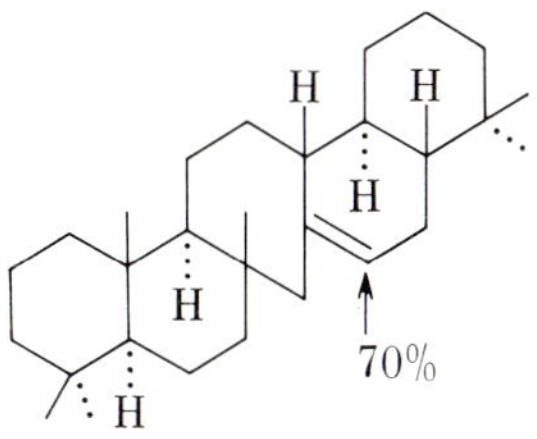

The 17-ethylidene derivative of the methyl ester of estrone (*124a*) is converted to the alcohol *125a* by hydroboration-oxidation [294]. If the hydroboration is carried out at 65°C, then, as a result of isomerization of the initially formed organoborane, the subsequent oxidation leads to the C-21 alcohol *126a* [295]:

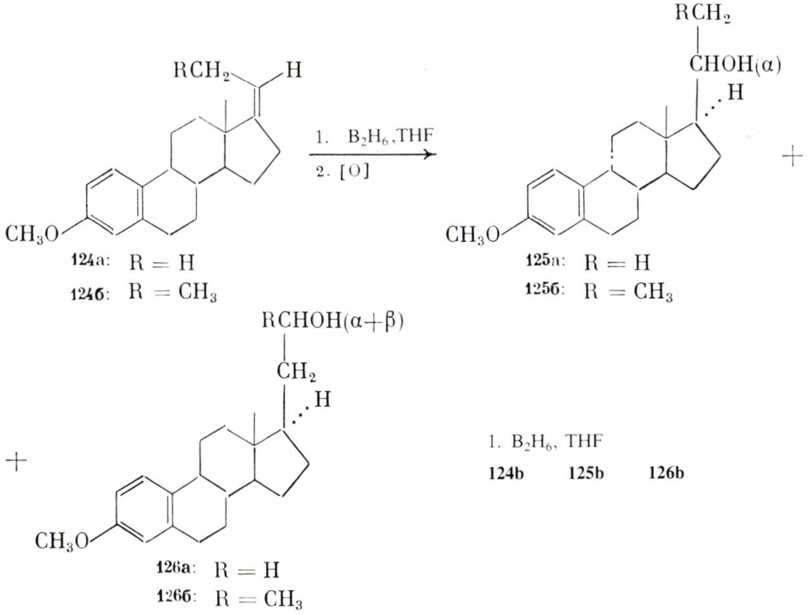

The organoborane formed by the action of diborane on the propylidene derivative *124b* shows an even greater tendency toward isomerization. Thus, the almost pure C-20 alcohol *125b* can be obtained only by carrying out the hydroboration at −5 to 0°C, whereas, at room temperature, a mixture of *125b* and *126b* in a ratio of 3.5:1 is obtained.

The hydroboration-oxidation of cholesta-3,5-diene and cholesta -4,6-diene leads to 5α-cholestan-4α,6α-idol [296]:

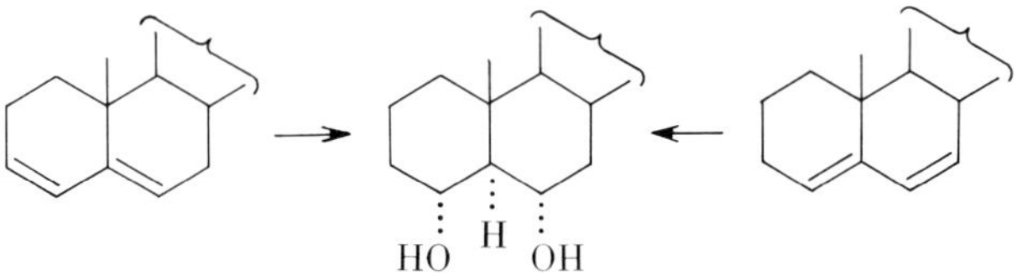

h) Organometallic compounds

Vinylferrocene is hydrated to 2-ferrocenylethanol in a yield of 86%. The regioselectivity of the hydroboration of various alkenylferrocenes has been investigated [379].

Vinylcyclobutadienetricarbonyliron forms 54% of the alcohol *127*; cycloheptatrienetricarbonyliron forms the compound *128* [297]:

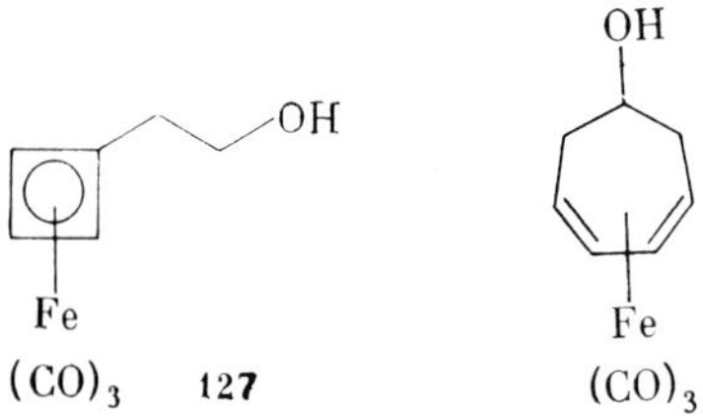

127 **128**

and the triene complex *129* forms the alcohol *130*:

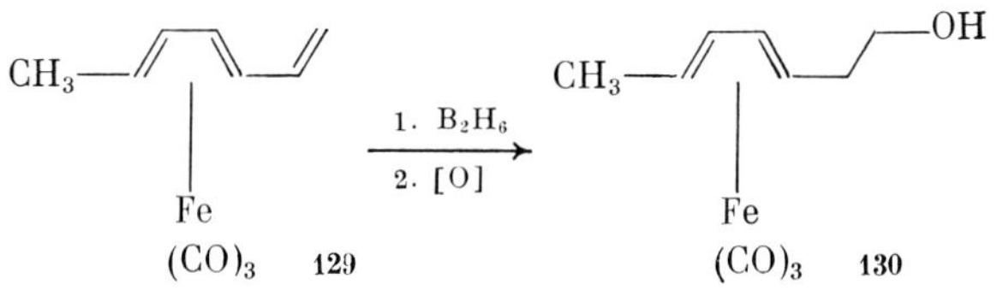

129 **130**

Two hydroxy groups can be introduced into the molecule of a compound with one double bond by the hydroboration-oxidation of allyllithium derivatives. Thus, 1-phenyl-1,2-propanediol is formed from phenylallyllithium, obtained from allylbenzene and butyllithium; the reaction proceeds analogously with *p*-methylallylbenzene [298].

2.3 ASYMMETRIC SYNTHESIS

Hydroboration has been used in the preparative synthesis of optically active compounds.

Brown and Zweifel [299] discovered that tetraisopinocampheyldiborane (referred to in the monomeric form as diisopinocampheylborane or di-3-pinanylborane according to the nomenclature used in *Chemical Abstracts*) adds stereospecifically to olefins. The subsequent oxidation of the organoboron compounds, which proceeds with a retention of the configuration, forms optically active alcohols.

2.3.1 Asymmetric synthesis of alcohols

Sterically unhindered *cis*-olefins react most easily with tetraisopinocampheyldiborane.

The reaction of *cis*-2-butene with (−)-tetraisopinocampheyldiborane, obtained from (+)-α-pinene, followed by oxidation with an alkaline hydrogen peroxide solution, leads to (*R*)-(−)-2-butanol (yield 90%) with a high degree of optical purity (87%). When (+)-tetraisopinocampheyldiborane is used, (*S*)-(−)-2-butanol is formed from *cis*-2-butene [47, 300]:

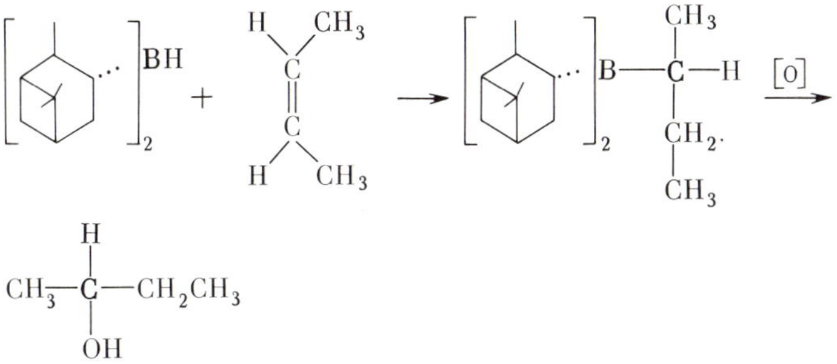

The reactions of (−)- and (+)-tetraisopinocampheyldiboranes with other *cis*-alkenes (norbornene [47, 301, 302], norbornadiene [301], benzonorbornadiene [303, 304], and 5α-cholest-2-ene [305]) proceed analogously; the results of these reactions are shown in Table 1.

It should be noted that the degree of stereoselectivity of the asymmetric synthesis depends to a considerable degree on the opti-

TABLE 1

Asymmetric synthesis of alcohols from *cis*-olefins.

$Pn_4B_2H_2$*	Olefin	Alcohol	Yield, %	Degree of optical purity, %
(−)	*cis*-2-Butene	(*R*)-(−)-2-Butanol	90	87
(+)	*cis*-2-Butene	(*S*)-(+)-2-Butanol	—	86(78% in THF)
(+)	*cis*-2-Pentene	(*S*)-(+)-2-Pentanol	76	82
		3-Pentanol	24	—
(−)	*cis*-3-Hexene	(*R*)-(−)-3-Hexanol	—	91
(+)	*cis*-4-Methyl-2-pentene	(*S*)-(+)-4-Methyl-2-pentanol	92	76
(−)	Norbornene	(1*S*, 2*S*)-(−)-*exo*-Norborneol	—	67–70
(+)	Norbornene	(1*R*, 2*R*)-(+)-*exo*-Norborneol	62	65–68
(+)**	Norbornene	(1*R*, 2*R*)-(+)-*exo*-Norborneol	59	95
(−)***	Benzonorbornadiene	(1*R*, 2*S*)-(+)-*exo*-2-Benzonorborneol	—	46
(−)	5α-Cholest-2-ene	5α-Cholestan-2α-ol	8	—
		5α-Cholestan-3α-ol	88	—
(+)	5α-Cholest-2α-ene	5α-Cholestan-2α-ol	55	—
		5α-Cholestan-3α-ol	26	—
		5α-Cholestan-3β-ol	18	—

Note: Pn = isopinocampheyl (3-pinanyl) group

* $Pn_4B_2H_2$ (tetraisopinocampheyldiborane) was obtained from α-pinene with a 93–95% degree of optical purity.

** From (−)-α-pinene with a 100% degree of optical purity.

*** From (+)-α-pinene with a 75% degree of optical purity.

cal purity of the chiral reagent used. Thus, when (−)- and (+)-tetraisopinocampheyldiborane, obtained from (+)-α- and (−)-α-pinene with a 93–95% degree of optical purity are used, (1*S*, 2*S*)-(−)- and (1*R*, 2*R*)-(+)-*exo*-norborneol, respectively, are formed from norbornene with a 65–70% excess of the enantiomers [47], whereas an increase in the excess of the enantiomer in (+)-tetraisopinocampheyldiborane to 100% leads to (1*R*, 2*R*)-(+)-*exo*-norborneol with a 95% optical purity [302].

The stereoselectivity of the reaction is affected by the nature of the solvent — the degree of asymmetric synthesis is lower in tetrahydrofuran than in diglyme.

trans-Olefins are hydroborated considerably more slowly than *cis*-olefins and form alcohols of low optical purity. Thus, the reaction of tetraisopinocamphenyldiborane with *cis*-2-butene is completed within 2 hours at 0°C, while completion of the reaction with *trans*-2-butene requires 24 hours. The alcohol from *cis*-2-butene contains 75–87% excess of the enantiomer, while the alcohol from *trans*-2-butene contains an excess of only 13% [306]. *Trans*-4-Methyl-2-pentene reacts more slowly than *cis*-4-methyl-2-pentene.

Trisubstituted olefins, such as 2-methyl-2-butene, react particularly slowly with tetraisopinocampheyldiborane, with the alcohol formed after oxidation of the organoborane containing 8–14% excess of the enantiomer [194, 306]. The hydroboration of 1-methylcyclopentene, 1-methylcyclohexene, and bicyclo[2.2.2]oct-2-ene also proceeds slowly; the degree of optical purity of the alcohols obtained is only 14–22%.

The reactions of hindered olefins with tetraisopinocampheyldiborane are accompanied by the formation of α-pinene, with the amount of α-pinene formed depending on the nature of the olefin being hydroborated. Thus, in the case of *trans*-2-butene or cyclohexene, one mole of α-pinene is formed per two moles of reacted olefins, whereas, in the case of 1-methylcyclopentene, one mole of α-pinene is formed for each mole of hydrocarbon reacted.

The characteristics of asymmetric synthesis with the use of hindered olefins are explained by the fact that, in these cases, the hydroborating agent is not tetraisopinocampheyldiborane but triisopinocampheyldiborane (*131*), which is formed from the tetrasubstituted compound by the elimination of α-pinene. In the first stage of the reaction, *131* adds to the olefin with the formation of tris(isopinocampheyl)alkyldiborane (*132*), which then reacts with a second molecule of olefin to form isopinocamphenyldialkylborane (*133*):

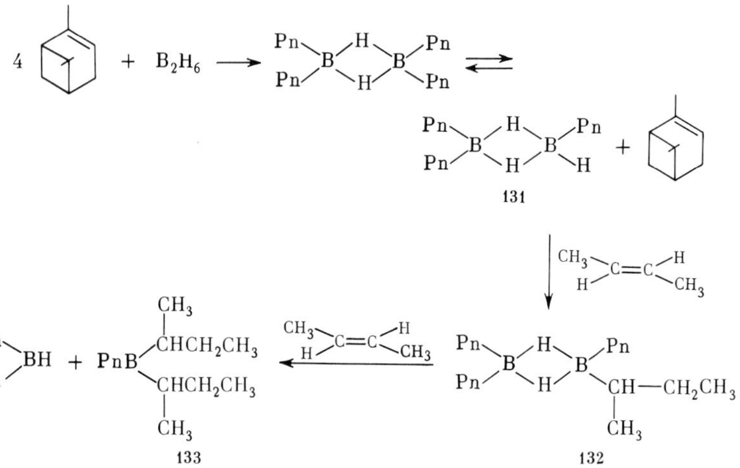

1-Methylcyclopentene is hydroborated in the second stage with bis(isopinocampheyl)alkyldiborane, which shows relatively little steric hindrance and is formed as a result of the displacement of two molecules of α-pinene from tetraisopinocampheylidiborane. In this process, compound *134* is formed.

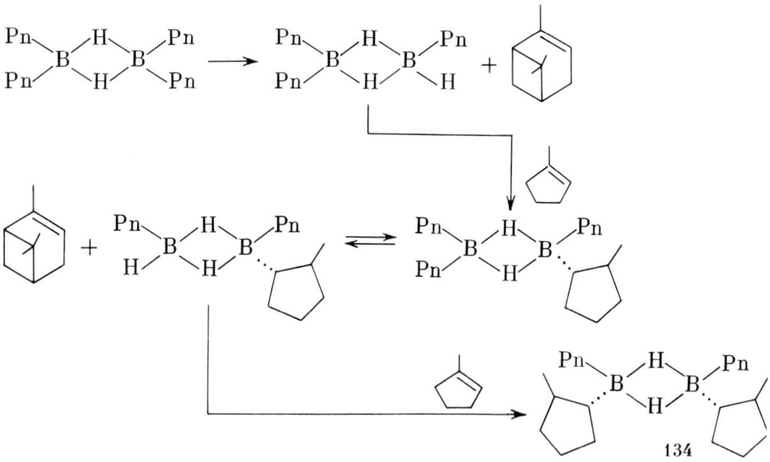

TABLE 2

Asymmetric synthesis of alcohols from trans-olefins and trisubstituted olefins.

Reagent	Olefin	Alcohol	Reaction time, hours	Degree of optical purity, %
$Pn_4B_2H_2$	*trans*-2-Butene	(*S*)-(+)-2-Butanol	24	13
$Pn_4B_2H_2$	2-Methyl-2-butene	(*S*-(+)-3-Methyl-2-butanol	60	14
$Pn_3B_2H_3$	2-Methyl-2-butene	(*S*)-(+)-3-Methyl-2-butanol	3	7
$Pn_4B_2H_2$	1-Methylcyclopentene	(1*S*, 2*S*)-(+)-*trans*-2-Methylcyclopentanol	60	22
$Pn_3B_2H_3$	1-Methylcyclopentene	(1*S*, 2*S*)-(+)-*trans*-2-Methylcyclopentanol	3	17.5
$Pn_3B_2H_3$	1-Methylcyclohexene	(1*S*, 2*S*)-(+)-*trans*-2-Methylcyclohexanol	4	18
$Pn_3B_2H_3$	2-Phenylnorbornene	(2*S*, 3*S*)-(−)-3-*endo*-Phenyl-2-*exco*-borneol	3	26

Note: $Pn_4B_2H_2$ was obtained from (+)-α-pinene and $Pn_3B_2H_2$ from (−)-α-pinene.

The stage determining the rate of the process as a whole is the dissociation of tetraisopinocampheyldiborane to triisopinocampheyldiborane and α-pinene. A confirmation of this scheme is provided by data on the hydroboration of hindered olefins with triisopinocampheyldiborane, which is directly obtained from α-pinene and diborane at a reagent ratio of 3:1. The reaction of 2-methyl-2-butene with this reagent is completed within 3 hours at 0°C, while the completion of the reaction with tetraisopinocampheyldiborane requires 60 hours.

The hydroboration of 3-phenyl-2-norbornene with triisocampheyldiborane, obtained from (−)-α-pinene, leads to (−)-3-*endo*-phenyl-2-*exo*-norborneol, containing a 26% excess of the enantiomer [307].

The results of the asymmetric synthesis of alcohols from hindered olefins are presented in Table 2.

(+)-Diisopinocampheyldiborane has been used as a chiral reagent. Its reaction with racemic 3-*p*-menthene leads to the production of 24% of (−)-menthol and 44% of (−)-isomenthol, with optical purities of 14 and 56%, reespectively [195].

The hydroboration of 2-methyl-1-alkenes with tetraisopinocampheyldiborane leads to optically active 2-substituted 1-alkanols [308]:

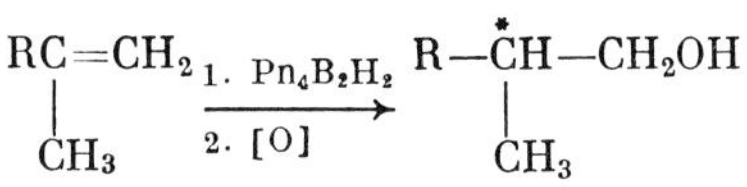

The degree of asymmetric synthesis in this case is low, which is apparently due to the fact that, in the organoborane obtained, the asymmetric carbon atom is not bonded to a boron atom and is therefore further removed from the disymmetric isopinocampheyl group in the transition state.

The results of the hydroboration of 2-substituted alkenes with (−)-tetraisopinocampheyldiborane are shown in Table 3.

Asymmetric hydroboration has been used for the synthesis of deuterated alcohols. Deuterated olefins or $Pn_4B_2D_2$ can be used as the deuterated starting materials [123, 309–311].

The hydroboration-oxidation of an ether of safranol with (+)- and (−)-diisopinocampheylborane leads to (3*R*)-3-hydroxy-β-cyclogeraniol and (3*S*)-3-hydroxy-β-cyclogeraniol, respectively [380].

The reaction of some olefins with a chiral borane can be used for the synthesis of optically active ketones. For this purpose, the chiral triorganoboranes formed by hydroboration are oxidized by chromic

TABLE 3
Asymmetric synthesis of alcohols from 2-methyl-1-alkenes.

Olefin	Alcohol	Degree of optical purity, %
$C_2H_5C(CH_3)=CH_2$	(*R*)-(+)-2-Methyl-1-butanol	21
iso-$C_3H_7C(CH_3)=CH_2$	(*R*)-(+)-2,3-Dimethyl-1-butanol	30
tert-$C_4H_9C(CH_3)=CH_2$	(*R*)-(+)-2,3,3-Trimethyl-1-butanol	—
$C_6H_5C(CH_3)=CH_2$	(*R*)-(+)-2-Phenyl-1-propanol	5

acid. One can also first oxidize the organoboranes to alcohols with hydrogen peroxide and then oxidize the alcohols to ketones with Sarett's reagent. Thus, for example, optically active (1*S*, 4*R*)-(+)-norcamphor with a 21% excess of the enantiomer is obtained from norbornene by hydroboration with tetraisopinocampheyldiborane, followed by oxidation with chromic acid [312].

Optically active hydrocarbons can be obtained by the protolysis of chiral organoboranes obtained by the hydroboration of alkenes with tetraisopinocampheyldiborane. Thus, (S)-(−)-pentane-2*d* has been obtained from 1-pentene-2*d* in this way [311]. Decomposition of the organoboranes from *cis*-2-butene and (−)-tetraisopinocampheyl-diborane with hydroxylamine-*O*-sulfonic acid forms (*R*)-(−)-*sec*-butylamine [300].

The reaction of bis[(*R*)−(+)-2-methylbutyl]borane with 1-hexyne also proceeds stereospecifically; treatment of the intermediately formed vinylborane (Part III) with iodine and alkali leads to (+)−(*R*)-3-methyl-*cis*-5-decene with an optical purity of 99% [313].

The hydroboration of optically active compounds proceeds with a retention of the asymmetric center in the starting substrate.

The monohydroboration of (*R*)-(+)-limonene with tetra(3-methyl-2-butyl)diborane leads to a mixture of (4*R*, 8*R*)-and (4*R*, 8*S*)-*p*-1-menthen-9-ols [314]. D-(−)-(1*R*, 2*R*, 4*R*)-Limonen-2,9-diol and D-(1*R*, 2*R*, 4*R*)-carvomenthol can be obtained by the dihydroboration of D-(+)-limonene with diborane or bis(2,3-dimethyl-2-butyl)-diborane [315].

Monoisopinocampheylborane is a convenient chiral hydroborating agent [381–385]. The preparation of the effective chiral hydroborating agent dilongifolylborane has been described [386].

The hydroboration of *cis*-2-butene and *cis*-3-hexene with diisopinocampheylborane and monoisopinocampheylborane has been investigated [387].

2.3.2 Kinetic separation of racemic alkenes

One of the methods for the preparation of chiral compounds is a method based on kinetically controllable asymmetric reactions. Among such reactions, in which the rates of reaction of the chiral reagent with the enantiomers are different, are hydroboration with tetraisopinocampheyldiborane. The separation of a racemic substrate into the antipodes by hydroboration is not essentially an asymmetric synthesis, since no new asymmetric center is formed in this case, but it will be considered in this section, which deals with the preparative use of chiral borohydrides.

In the hydroboration of a racemic mixture of hydrocarbons with an insufficient quantity of tetraisopinocampheyldiborane, the unconverted part of the hydrocarbon is enriched in the less reactive enantiomer. The reaction of 3-methylcyclopentene with one-half the quantity of chiral reagent (calculated for the monomer) leads to the development of optical activity in the unreacted hydrocarbon (45% of the excess of enantiomer); however, if the amount of reagent is increased to 80% of the theoretical amount, then the stereoselectivity increases considerably (65% of the excess of enantiomer [47]).

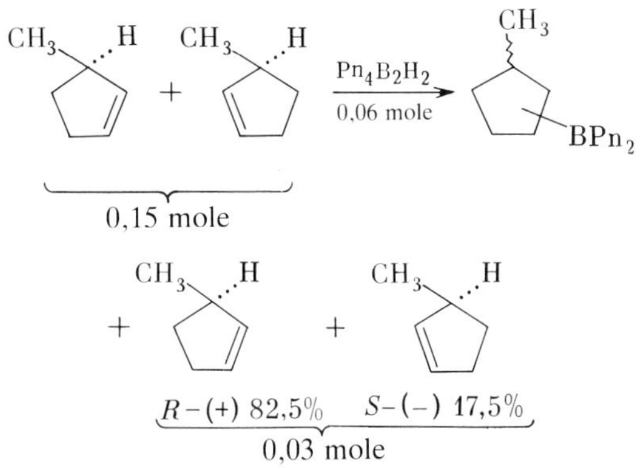

The unreacted hydrocarbon is easily separated from the hydroboration products by distillation.

The above method has been used to carry out the partial kinetic separation into antipodes of 3-ethylcyclopentene, 1-methylnorbornene [47], 4-methylcyclohexene [316], trimethylcyclodecatriene [317], *trans,trans*-2,8-*trans*-bicyclo[8.4.0] tetradecadiene [318], and *trans*-cyclooctene [319].

The method of "kinetic separation" has been used successfully for a series of allene compounds. Racemic 1,3-dimethylallene, taken in twofold excess, reacts rapidly (within five minutes) with (+)-tetraisopinocampheyldiborane. The unreacted hydrocarbon is enriched in (+)-1,3-dimethylallene with the *R* configuration [320–322]. Analogous results are obtained with the use of 1,3-diphenyl- [321], 1,3-diethyl-, 1,3-dipropyl-, and 1,3-di-*tert*-butylallenes [322]. The hydroboration of 1,3-diethyl- and 1,3-dipropylallene proceeds more slowly than the hydroboration of 1,3-dimethylallene, but both hydrocarbons are more reactive than 1,3-di-*tert*-butylallene, the reaction of which is complete only after 24 hours. The use of tri- and diisopinocampheyldiboranes as hydroborating agents leads to a decrease in the degree of optical purity of the unreacted hydrocarbon, but the *R* configuration of the excess enantiomer is retained in this case.

The partial asymmetric hydroboration of 2-phenylisobutenylidenecyclopropane with (+)-tetraisopinocampheyldiborane has been carried out [323]:

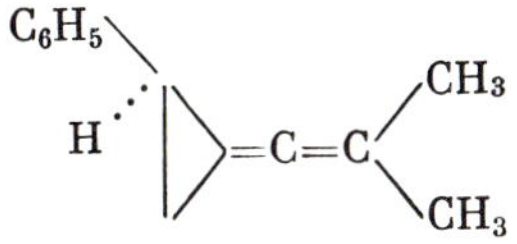

The partial separation of racemic 1,2-cyclononadiene into its antipodes has also been carried out. When (+)-tetraisopinocampheyldiborane is used, the unreacted hydrocarbon is enriched in (+)-1,2-cyclononadiene with the *R* configuration (15% of the excess of enantiomer) [324].

The optical purity of the allene compounds partially separated into the antipodes increases in the following order: 1,3-dimethylallene < 1,2-cyclononadiene < 1,3-di-*tert*-butylallene < 1,3-diethylallene < 1,3-dipropylallene [322].

The orientation of the addition of tetra(3-methyl-2-butyl)diborane and of tetraisopinocampheyldiborane to 1,2-cyclononadiene has been investigated [322]. Oxidation of the hydroboration products leads to cyclononanone, 3-hydroxycyclononene, and cyclononene. From the quantitative ratio of these substances, it follows that the boron atom adds to the central carbon of the allene system to the extent of 72–80%.

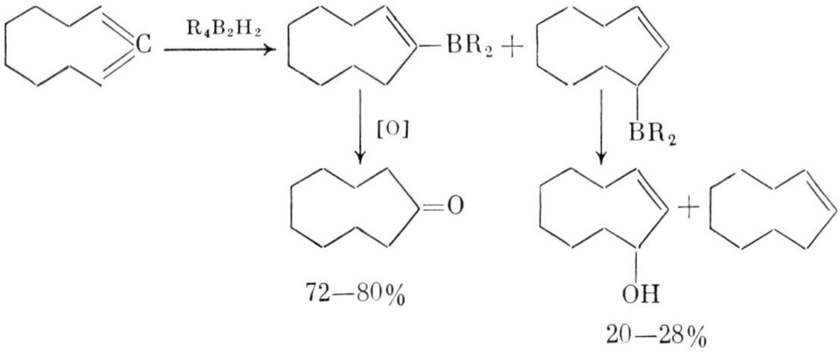

2.3.3 Mechanism of asymmetric hydroboration

To explain the stereochemical results of the hydroboration of alkenes with tetraisopinocampheyldiborane, different models of the transition state have been proposed.

Brown, et al. [47] considered the scheme of asymmetric hydroboration with the participation of the monomeric form of tetraisopinocampheyldiborane, in its most probable conformation *135*, as the hydroborating agent. This conformation has been represented in the form of the model *136*, where S is the smallest group on the asymmetric carbon atom (e.g., H), M is a group of intermediate size (methylene on C-4), and L is a large group (CH_3 on C-2 and C-2′) [325].

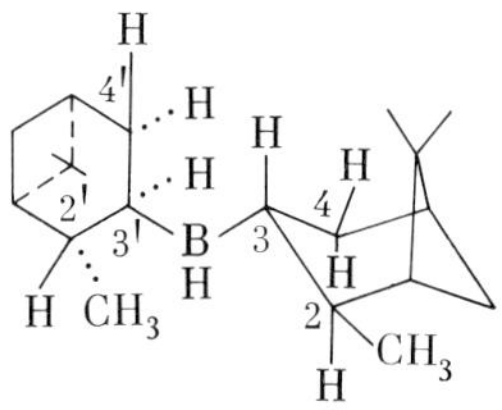

135

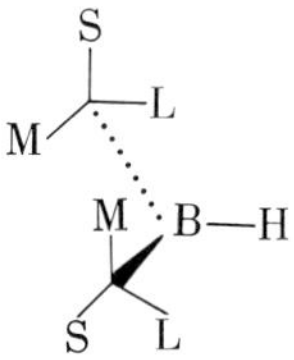

136

The transition state *137* is more preferable than *138*, in which the interaction of the large R and L groups destabilizes the system.

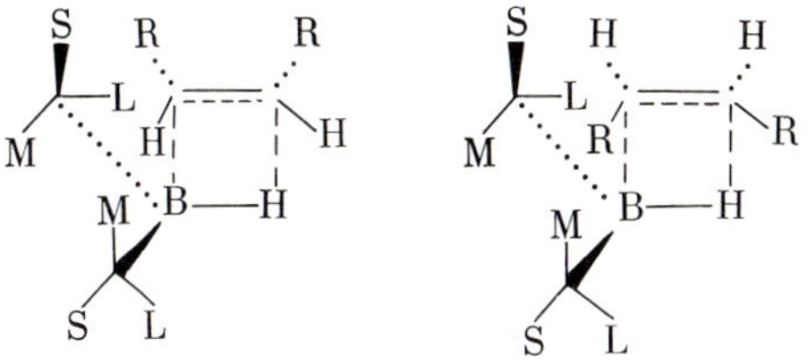

137 138

This model correctly predicts the absolute configuration of the alcohols obtained from acyclic *cis*-olefins and tetraisopinocampheyl-diborane, but its successive application to *trans*-olefins leads to a result that is not in agreement with experiments.

Two variations of the model considered above were later proposed.

To explain the stereochemistry of the hydroboration of *cis*-1-butene-1*d*, Streitwieser, et al. [123] suggested that, in the reaction of an alkene with a chiral reagent, a π-complex is initially formed, and that the subsequent transition state differs to a slight degree from the trigonal structure of this complex. This model satisfactorily describes the results of the hydroboration of other 1-deuteroalkenes, but is not capable of explaining the stereochemistry of the hydroboration of unlabeled 2-methyl-1-alkenes.

In the modification of Varma and Caspi [326], a diisopinocampheylborane conformation *139* is proposed, differing from the conformation of the basic model by the fact that, in it, the "left" isopinocampheyl group is rotated by 90°. In this conformation, the methyl groups on C-2 and C-2′ and, correspondingly, the hydrogen atoms on C-3 and C-3′, are located on different sides of the plane in which the (C-3)-B-(C-3′) and B—H bonds are situated. It is assumed that the environment of the atom which becomes the new asymmetric center has the most significant effect on the course of the reaction.

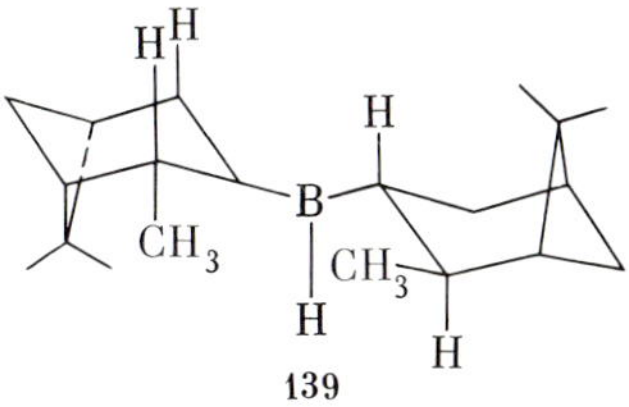

139

Significantly different from the schemes of asymmetric synthesis presented above is the model of McKenna, et al. [325], according to which the hydroboration reaction proceeds by the direct introduction of an alkene into an undissociated tetraisopinocampheyldiborane molecule.

Moore, et al. [322] believed that there is insufficient proof for the McKenna model and proposed a hydroboration scheme with participation of the monomeric form of tetraisopinocampheyldiborane, which is based on the assumption of a non-planar four-center transition state. This model correctly predicts the stereochemistry of the hydroboration of allenes.

REFERENCES

1. D. T. Hurd, *J. Am. Chem. Soc.* **70,** 2053 (1948).
2. H. C. Brown: "Hydroboration", W. A. Benjamin, Inc. (1962).
3. B. M. Mikhailov: "The Chemistry of Borohydrides", Moscow, Nauka (1967).
4. H. C. Brown: "Boranes in Organic Chemistry", New York, Cornell University Press (1972).
5. G. M.L. Cragg: "Organoboranes in Organic Synthesis", New York, M. Dekker (1973).
6. G. Zweifel and H. C. Brown, in: "Organic Reactions", vol. 13, New York, J. Wiley and Sons (1963) p. 1. [Russian Translation: Moscow, Mir (1966), p. 13.]
7. E. Negishi and H. C. Brown, *Synthesis* **1974,** 77.
8. J. R.Elliott, E. M. Boldebuck and G. F. Roedel, *J. Am. Chem. Soc.* **74,** 5047 (1952).
9. H. C. Brown and P. A. Tierney, *J. Am. Chem. Soc.* **80,** 1552 (1958).
10. F. H. May and H. N. Hammar, *U.S. Patent* **3,013,863** (1959); *C. A.* **56,** 8297 (1962).
11. H. D. Batha, C. D. Good and J. P. Faust, *J. Appl. Chem.* **14,** 257 (1964).
12. C. D. Good, *U.S. Patent* **3,019,085** (1962); *C. A.* **56,** 11207 (1962).
13. H. Jenkner, *German Patent* **1,038,019** (1958); *C. A.* **54,** 18908 (1960).
14. G. F. Huff and W. H. Schechter, *U.S. Patent* **2,992,072** (1961); *C. A.* **55,** 23953 (1961).
15. G. N. Tyson, *German Patent* **1,062,687** (1960); *C. A.* **55,** 8788 (1961).
16. A. E. Finholt, A. C. Bond and H. I. Schlesinger, *J. Am. Chem. Soc.* **69,** 1199 (1947).
17. N. L. Paddock, *Nature* **167,** 1070 (1956).
18. I. Shapiro, H. G. Weiss, M. Schmich, S. Skolnik and G. B. L. Smith, *J. Am. Chem. Soc.* **74,** 901 (1952).
19. C. J. Danby, E. Gobbett and J. W. Linnett, *J. Chem. Soc.* **1962,** 2076.
20. H. I. Schlesinger, H. C. Brown, H. R. Hoekstra and L. R. Rapp, *J. Am. Chem. Soc.* **75,** 199 (1953).
21. B. M. Mikhailov and V.A. Dorokhov, *Dokl. Akad. Nauk SSSR* **133,** 119 (1960).
22. H. I. Schlesinger and H. C. Brown, *J. Am. Chem. Soc.* **62,** 3429 (1940).
23. H. G. Weiss and I. Shapiro, *J. Am. Chem. Soc* **81,** 6167 (1959).

24. R. E. Davis and J. A. Gottbrath, *Chem. and Ind.* **1961,** 1911.
25. J. A. Marshall and W. S. Johnson, *J. Org. Chem.* **28,** 595 (1963).
26. V. Hoch, *Synthesis* **1974,** 340.
27. A. Brändstrom, U. Junggren and B. Lamm, *Tetrahedron Letters* **1972,** 3173.
28. H. I. Schlesinger and A. O. Walker, *J. Am. Chem. Soc.* **57,** 621 (1935).
29. H. I. Schlesinger, L. Horvitz and A. B. Burg, *J. Am. Chem. Soc.* **58,** 407 (1936).
30. H. I. Schlesinger, N. W. Flodin and A. B. Burg, *J. Am. Chem. Soc.* **61,** 1078 (1939).
31. A. B. Burg and J. L. Boone, *J. Am. Chem. Soc.* **78,** 1521 (1956).
32. I. J. Solomon, M. J. Klein and K. Hattori, *J. Am. Chem. Soc.* **80,** 4520 (1958).
33. B. M. Mikhailov, A. A. Akhnazaryan and L. S. Vasil'ev, *Dokl. Akad. Nauk SSSR* **136,** 828 (1961).
34. L. H. Long and M. G. H Wallbridge, *Chem. and Ind.* **1959,** 295.
35. L. H. Long, *British Patent* **853, 063** (1960), *C. A.* **55,** 10317 (1961).
36. L. H. Long and M. G. H. Wallbridge, *J. Chem. Soc.* **1963,** 2181.
37. E. Wiberg, J. Evans and H. Nöth, *Z. Naturforsch.* **13b,** 263 (1958).
38. B. M. Mikhailov and V. A. Dorokhov, *Zh. Obshch. Khim.* **31,** 4020 (1961).
39. R. Köster, *Angew. Chem.* **86,** 383 (1956).
40. R. Köster, G. Bruno and P. Binger, *Ann.* **644,** 1 (1961).
41. H. C. Brown, A. Tsukamoto and D. B. Bigley, *J. Am. Chem. Soc.* **82,** 4703 (1960).
42. H. C. Brown and A. W. Moerikofer, *J. Am. Chem. Soc.* **84,** 1478 (1962).
43. B. M. Mikhailov and L. S. Vasil'ev, *Izv. Akad. Nauk SSSR, Otd. Khim. Nauk* **1962,** 628.
44. H. C. Brown and G. J. Klender, *Inorg. Chem.* **1,** 204 (1962).
45. E. Negishi, J.-J. Katz and H. C. Brown, *J. Am. Chem. Soc.* **94,** 4025 (1972).
46. G. Zweifel and H. C. Brown, *J. Am. Chem. Soc.* **86,** 393 (1964).
47. H. C. Brown, N. R. Ayyangar and G. Zweifel, J. Am. Chem. Soc. **86,** 397 (1964).
48. B. M. Mikhailov and V. A. Dorokhov, *Dokl. Akad. Nauk SSSR* **130,** 782 (1960).
49. B. M. Mikhailov and V. A. Dorokhov, *Izv. Akad. Nauk SSSR, Otd. Khim. Nauk* **1962,** 623.
50. B. M. Mikhailov and L. S. Vasil'ev, *Dokl. Akad. Nauk SSSR* **139,** 385 (1961).
51. W. J. Lehmann, C. O. Wilson and I. Shapiro, *J. Chem. Phys.* **32,** 1088 (1960).
52. W. J. Lehmann, C. O. Wilson and I. Shapiro, *J. Chem. Phys.* **32,** 1786 (1960).
53. W. J. Lehmann, C. O. Wilson and I. Shapiro, *J. Chem. Phys.* **33,** 590 (1960).
54. W. J. Lehmann, C. O. Wilson and I. Shapiro, *J. Chem. Phys.* **34,** 476 (1961).
55. W. J. Lehmann, C. O. Wilson and I. Shapiro, *J. Chem. Phys.* **34,** 783 (1961).
56. W. J. Lehmann and I. Shapiro, *Spectrochim. Acta* **17,** 396 (1961).
57. I. Shapiro, C. O. Wilson and W. J. Lehmann *J. Chem. Phys.* **29,** 237 (1958).
58. W. J. Lehmann, C. O. Wilson, J. F. Ditter and I. Shapiro, *Advances in Chemistry, Ser.* **32,** 139 (1961).
59. H. I. Schlesinger and A. B. Burg, *J. Am. Chem. Soc.* **60,** 290 (1938).
60. A. B. Burg, *J. Am. Chem. Soc.* **74,** 1340 (1952).
61. B. Rice, J. A. Livasy and G. W. Schaeffer, *J. Am. Chem. Soc.* **77,** 2750 (1955).
62. T. D. Coyle, H. D. Kaesz and F. G. A. Stone, *J. Am. Chem. Soc.* **81,** 2989 (1959).
63. T. A. Shchegoleva, V. D. Sheludyakov and B. M. Mikhailov, *Dokl. Akad. Nauk SSSR* **152,** 888 (1964).
64. D. R. Schultz and R. W. Parry, *J. Am. Chem. Soc* **80,** 4 (1958).
65. J. R. Elliott, W. L. Roth, G. F. Roedel and E. M. Boldebuck, *J. Am. Chem. Soc.* **74,** 5211 (1952).

66. G. E. McWood and L. J. Paridon, *Nucl. Sci. Abstr.* **11,** 92 (1957).
67. H. E. Wirth, F. E. Massoth and D. X. Gilbert, *J. Phys. Chem.* **62,** 870 (1958).
68. B. Rice and H. S. Ushida, *J. Phys. Chem.* **59,** 650 (1955).
69. S. Yerazunis, J. M. Mullen and B. Steginsky, *J. Chem. Eng. Data* **7,** 337 (1962).
70. G. E. McAchran and S. G. Shore, *Inorg. Chem.* **4,** 125 (1965).
71. H. C. Brown, A. Tsukamoto and D. B. Bigley, *J. Am. Chem. Soc.* **82,** 4703 (1960).
72. J. Kollonitsch, *J. Am. Chem. Soc.* **83,** 1515 (1961).
73. R. E. Lyle and C. K. Spicer, *Chem. and Ind.* **1963,** 739.
74. A. I. Gorbunov, G. S. Solov'eva, I. S. Antonov and M. S. Kharson, Zh. Neorg. Khim. **10,** 1971 (1965).
75. L. H. Long and G. F. Freeguard, *Nature* **207,** 403 (1965).
76. B. M. Mikhailov and V. A. Dorokhov, *Dokl. Akad. Nauk SSSR* **136,** 356 (1961).
77. B. M. Mikhailov and V. A. Dorokhov, *Izv. Akad. Nauk SSSR, Otd. Khim. Nauk* **1961,** 2084.
78. P. A. Chopard and R. F. Hudson, *J. Inorg. Nucl. Chem.* **25,** 801 (1963).
79. G. W. Schaeffer and E. R. Anderson, *J. Am. Chem. Soc.* **71,** 2143 (1949).
80. M. D. Taylor, L. R. Grant and C. A. Sonds, *J. Am. Chem. Soc.* **77,** 1506 (1955).
81. H. Nöth and H. Beyer, *Chem. Ber.* **93,** 928 (1960).
82. B. M. Mikhailov and V. A. Dorokhov, *Izv. Akad. Nauk SSSR, Otd. Khim. Nauk* **1962,** 1213.
83. M. F. Hawthorne, *J. Am. Chem. Soc.* **81,** 5836 (1959).
84. M. F. Hawthorne, *J. Am. Chem. Soc.* **83,** 831 (1961).
85. M. F. Hawthorne, *Chem. and Ind.* **1957,** 1242.
86. M. F. Hawthorne, *J. Am. Chem. Soc.* **80,** 4291 (1958).
87. M. F. Hawthorne, *J. Am. Chem. Soc.* **80,** 4293 (1958).
88. H. C. Brown and S. K. Gupta, *J. Organometal. Chem.* **32,** C1 (1971).
89. H. C. Brown and S. K. Gupta, *J. Am. Chem. Soc.* **93,** 4062 (1971).
90. H. C. Brown, E. Negishi and J.- J. Katz, *J. Am. Chem. Soc.* **94,** 5893 (1972).
91. L. M. Braun, R. A. Braun, H. R. Grissman, M. Opperman and R. M. Adams, *J. Org. Chem.* **36,** 2388 (1971).
92. R. A. Braun, D. C. Brown and R. M. Adams, *J. Am. Chem. Soc.* **93,** 2823 (1971).
93. C. F. Lane, *J. Org. Chem.* **39,** 1437 (1974).
94. A. B. Burg and R. I. Wagner, *J. Am. Chem. Soc.* **76,** 3307 (1954).
95. B. M. Mikhailov, T. A. Shchegoleva, E. M. Shashkova and V. D. Sheludyakov, *Izv. Akad. Nauk SSSR, Otd. Khim. Nauk* **1961,** 1163.
96. B. M. Mikhailov, T. A. Shchegoleva, E. M. Shashkova and V. D. Sheludyakov, *Izv. Akad. Nauk SSSR, Otd. Khim. Nauk* **1962,** 1218.
97. D. J. Pasto, *J. Am. Chem. Soc.* **84,** 3777 (1962).
98. D. J. Pasto, C. C. Cumbo and P. Balasubramaniyan, *J. Am. Chem. Soc.* **88,** 2187 (1966).
99. B. L. Egan, S. G. Shore and J. E. Bonnell, *Inorg. Chem.* **3,** 1024 (1964).
100. B. M. Mikhailov and T. A. Shchegoleva, *Izv. Akad. Nauk SSSR, Otd. Khim. Nauk* **1959,** 1868.
101. B. M. Mikhailov and T. A. Shchegoleva, *Dokl. Akad. Nauk SSSR* **131,** 843 (1960).
102. T. A. Shchegoleva, E. M. Shashkova and B. M. Mikhailov, *Izv. Akad. Nauk SSSR, Otd. Khim. Nauk* **1963,** 494.
103. T. A. Shchegoleva, L. I. Lavrinovich, M. N. Bochkareva and B. M. Mikhailov, *Izv. Akad. Nauk SSSR, Ser. Khim.* **1968,** 1876.
104. H. Nöth and H. Beyer, *Chem. Ber.* **93,** 2251 (1960).

105. K. Borer and J. Dewing, British Patent **881,376** (1961); *C. A.* **57,** 9664 (1962).
106. S. Ratajczak, *Bull. Soc. Chim. France* **1960,** 487.
107. H. Nöth, W. A. Dorochow, P. Fritz and F. Pfab, *Z. Anorg. Chem.* **318,** 293 (1962).
108. L. L. Shchukovskaya, M. G. Voronkov and O. V. Pavlova, *Dokl. Akad. Nauk SSSR* **143,** 887 (1962).
109. T. Onak, H. Landesman and I. Shapiro, *J. Phys. Chem.* **62,** 1605 (1958).
110. H. C. Brown and P. A. Tierney, *J. Inorg. Nucl. Chem.* **9,** 51 (1959).
111. D. J. Pasto and P. Balasubramaniyan, *J. Am. Chem. Soc.* **89,** 295 (1967).
112. B. Barthocha, W. A. G. Graham and F. G. A. Stone, *J. Inorg. Nucl. Chem.* **6,** 119 (1958).
113. F. G. A. Stone and H. J. Emeleus, *J. Chem. Soc.* **1950,** 2755.
114. H. C. Brown and B. C. Subba Rao, *J. Org. Chem.* **22,** 1136 (1957).
115. E. Breuer, *Tetrahedron Letters* **1967,** 1849.
116. K. M. Biswas and A. H. Jackson, *J. Chem. Soc. C* **1970,** 1667.
117. S. Winstein, E. L. Allred and J. Sonnenberg, *J. Am. Chem. Soc.* **81,** 5833 (1959).
118. E. L. Allred, J. Sonnenberg and S. Winstein, *J. Org. Chem.* **25,** 26 (1960).
119. G. W. Kabalka and N. S. Bowman, *J. Org. Chem.* **38,** 1607 (1973).
120. H. C. Brown and G. Zweifel, *J. Am. Chem. Soc.* **81,** 247 (1959).
121. W. J. Wechter, *Chem. and. Ind.* **1959,** 294.
122. S. Wolfe, M. Nussim, Y. Mazur and F. Sondheimer, *J. Org. Chem.* **24,** 1034 (1959).
123. A. Streitwieser, Jr., L. Verbit and R. Bitman, *J. Org. Chem.* **32,** 1530 (1967).
124. R. B. Woodward and R. Hoffman: "The Conservation of Orbital Symmetry", New York, Academic Press (1970) [Russian Translation: Moscow, Mir (1971)].
125. D. J. Pasto, B. Lepeska and T.-C. Cheng, *J. Am. Chem. Soc.* **94,** 6083 (1972).
126. P. R. Jones, *J. Org. Chem.* **37,** 1886 (1972).
127. A. T. Whatley and R. N. Pease, *J. Am. Chem. Soc.* **76,** 835 (1954).
128. R. S. Brokaw and R. N. Pease, *J. Am. Chem. Soc.* **72,** 3237 (1950).
129. J. E. Bennett and H. A. Skinner, *J. Chem. Soc.* **1961,** 2472.
130. H. A. Skinner, J. E. Bennett and J. B. Pedley, *Pure Appl. Chem.* **2,** 17 (1961).
131. E. A. Pope and H. A. Skinner, *J. Chem. Soc* **1963,** 3704.
132. H. C. Brown and A. W. Moerikofer, *J. Am. Chem. Soc.* **85,** 2063 (1963).
133. H. C. Brown and B. C. Subba Rao, *J. Org. Chem.* **22,** 1137 (1957).
134. D. J. Pasto, B. Lepeska and V. Balasubramaniyan, *J. Am. Chem. Soc.* **94,** 6090 (1972).
135. T. P. Fehlner, *J. Am. Chem. Soc.* **93,** 6366 (1971).
136. D. J. Pasto, B. Lepeska and T.-C. Cheng, *J. Am. Chem. Soc.* **94,** 6083 (1972).
137. D. J. Pasto and S.-Z. Kang, *J. Am. Chem. Soc.* **90,** 3797 (1968).
138. H. C. Brown and R. L. Sharp, *J. Am. Chem. Soc.* **88,** 5851 (1966).
139. J. Klein, E. Dunkelblum and M. A. Wolff, *J. Organometal. Chem.* **7,** 377 (1967).
140. R. Köster, G. Griaznow, W. Larbig and P. Binger, *Ann.* **672,** 1 (1964).
141. H. C. Brown and A. W. Moerikofer, *J. Am. Chem. Soc.* **83,** 3417 (1961).
142. R. Fellous and R. Luft, *Tetrahedron Letters* **1970,** 1505.
143. R. Fellous, R. Luft and A. Puiff, *Tetrahedron Letters* **1972,** 248.
144. J. E. Dubois and G. Mauvier, *Tetrahedron Letters* **1965,** 1629.
145. H. C. Brown and G. Zweifel, *J. Am. Chem. Soc.* **82,** 3223 (1960).
146. R. H. Fish, *J. Org. Chem.* **38,** 158 (1973).
147. S. Nishida, I. Moritani, K. Ito and K. Sakai, *J. Org. Chem.* **32,** 939 (1967).
148. M. Hanack and H.-M. Ensslin, *Ann.* **713,** 49 (1968).
149. E. N. Marvell, D. Sturmer and C. Rowell, *Tetraheron* **22,** 861 (1966).
150. R. Köster and G. Griaznov, *Angew. Chem.* **73,** 171 (1961).

151. J. M. Birchall, R. N. Haszeldine and J. F. Marsh, *Chem. and Ind.* **1961,** 1080.
152. H. C. Brown, E. Negishi and S. K. Gupta, *J. Am. Chem. Soc.* **92,** 6648 (1970).
153. H. C. Brown and E. Negishi, *J. Am. Chem. Soc.* **89,** 5285 (1967).
154. C. F. Lane and H. C. Brown, *J. Organometal. Chem.* **34,** C'29 (1972).
155. G. B. Savitsky and K. Namikawa, *J. Phys. Chem.* **68,** 1956 (1964).
156. G. Redeuilh, P. Rumpf and C. Viel, *Bull. Soc. Chim. France* **1973,** 2668.
157. E. Breuer, E. Segall, Y. Stein and S. Sarel, *J. Org. Chem.* **37,** 2242 (1972).
158. G. Zweifel, *J. Organometal. Chem.* **9,** 215 (1967).
159. D. J. Pasto and S.-Z. Kang, *J. Am. Chem. Soc.* **90,** 3797 (1968).
160. H. C. Brown and N. Ravindran, *J. Am. Chem. Soc.* **94,** 2112 (1972).
161. H. C. Brown and N. Ravindran, *J. Org. Chem.* **38,** 182 (1973).
162. H. C. Brown and N. Ravindran, *J. Am. Chem. Soc.* **95,** 2396 (1973).
163. J. Dewing and R. J. Sampson, *German Patent* **1,118,200** (1961); *C. A.* **56,** 14327 (1962).
164. B. M. Mikhailov and V. A. Dorokhov, *Zh. Obshch. Khim.* **31,** 3750 (1961).
165. B. M. Mikhailov, T. A. Shchegoleva, V. D. Sheludyakov and A. N. Blokhina, *Izr. Akad. Nauk SSSR Otd. Khim. Nauk* **1963,** 646.
166. T. A. Shchegoleva and E. M. Belyavskaya, *Dokl. Akad. Nauk SSSR* **136,** 638 (1960).
167. B. M. Mikhailov, T. A. Shchegoleva and A. N. Blokhina, *Izv. Akad. Nauk SSSR, Otd. Khim. Nauk* **1960,** 1307.
168. W. G. Woods and P. L. Strong, *J. Am. Chem. Soc.* **88,** 4667 (1966).
169. R. H. Fish, *J. Am. Chem. Soc.* **90,** 4435 (1968).
170. H. C. Brown and S. K. Gupta, *J. Am. Chem. Soc.* **93,** 1816 (1971).
171. H. C. Brown and S. K. Gupta, *J. Am. Chem. Soc.* **97,** 5249 (1975).
172. R. Köster, S. Arora and P. Binger, *Angew. Chem., Internat. Edn.* **8,** 205 (1969).
173. K. Utimoto, M. Tamura, M. Tanouti and K. Sisido, *Tetrahedron* **28,** 5697 (1972).
174. J. Klein and D. Lichtenberg, *J. Org. Chem.* **35,** 2654 (1970).
175. H. C. Brown and J. H. Kawakami, *J. Am. Chem. Soc.* **92,** 1990 (1970).
176. R. Dulou and Y. Chrétien-Bessière, *Bull. Soc. Chim. France* **1959,** 1362.
177. J. Lhomme and G. Ourisson, *Tetrahedron* **24,** 3201 (1968).
178. R. Dulou and Y. Chrétien-Bessière, *Compt. Rend.* **248,** 416 (1959).
179. J. C. Brown and G. S. Fisher, *Tetrahedron Letters, N* **21,** 9 (1960).
180. Y. Bessière-Chrétien and B. Meklati, *Bull. Soc. Chim. France* **1972,** 2933.
181. M. Barthelemy and Y. Bessière-Chrétien, *Bull. Soc. Chim. France* **1974,** 600.
182. S. P. Acharya, H. C. Brown, A. Suzuki, S. Nozawa and M. Itoh, *J. Org. Chem.* **34,** 3015 (1969).
183. C. H. Heathcock and R. Ratcliffe, *J. Am. Chem. Soc.* **93,** 1746 (1971).
184. S. P. Acharya and H. C. Brown, *J. Org. Chem.* **35,** 196 (1970).
185. J. Lhomme and G. Ourisson, *Tetrahedron* **24,** 3167 (1968).
186. Y. Tanahashi, J. Lhomme and G. Ourisson, *Tetrahedron* **28,** 2655 (1972).
187. M. Rabinovitz, G. Salemnik and E. D. Bergman, *Tetrahedron Letters* **1967,** 3271.
188. C. Beard and A. Burger, *J. Org. Chem.* **26,** 2355 (1961).
189. N. Miyamoto, S. Isiyama, K. Utimoto and H. Nozaki, *Tetrahedron Letters* **1971,** 4597.
190. N. Miyamoto, S. Isiyama, K. Utimoto and H. Nozaki, *Tetrahedron* **29,** 2365 (1973).
191. J. Klein, E. Dunkelblum and D. Avrahami, *J. Org. Chem.* **32,** 935 (1967).
192. D. J. Pasto and F. M. Klein, *J. Org. Chem.* **33,** 1468 (1968).
193. D. J. Pasto and F. M. Klein, *Tetrahedron Letters* **1967,** 963.
194. D. K. Shumway and J. D. Barnhurst, *J. Org. Chem.* **29,** 2320 (1964).

195. J. Katsuhara, H. Watanabe, K. Hashimoto and M. Kobayashi, *Bull. Chem. Soc. Japan* **39,** 617 (1966).
196. W. K. Musker and G. L. Larson, *Tetrahedron Letters* **1968,** 3481.
197. C. Freppel, M.-A. Poirier, J.-C. Richer, Y. Maroni and G. Manuel, *Can. J. Chem.* **52,** 4133 (1974).
198. W. Cocker, P. V. R. Shannon and P. A. Staniland, *Tetrahedron Letters* **1966,** 1409.
199. W. Cocker, P. V. R. Shannon and P. A. Staniland, *J. Chem. Soc., C* **1967,** 485.
200. H. C. Brown and A. Suzuki, *J. Am. Chem. Soc.* **89,** 1933 (1967).
201. W. Cocker, P. V. R. Shannon and P. A. Staniland, *J. Chem. Soc., C* **1967,** 915.
202. S. P. Acharya and H. C. Brown, *J. Am. Chem. Soc.* **89,** 1925 (1967).
203. S. P. Acharya, H. C. Brown, A. Suzuki, S. Nozawa and M. Itoh, *J. Org. Chem.* **34,** 3015 (1969).
204. I. Tabushi, K. Fujita and R. Oda, *J. Org. Chem.* **35,** 2383 (1970).
205. W. R. Moore and W. R. Moser, *J. Org. Chem.* **35,** 908 (1970).
206. R. A. Benkeser and E. M. Kaiser, *J. Org. Chem.* **29,** 955 (1964).
207. F. Sondheimer and S. Wolfe, Can. J. Chem. **37,** 1870 (1959).
208. J. A. Marshall and A. R. Hochstetler, *J. Am. Chem. Soc.* **91,** 648 (1969).
209. A. Taticchi, F. Fringuelli and G. Traverso, *Gazz. Chim. Ital.* **99,** 247 (1967).
210. C. H. Heathcock and T. Ross Kelly, *Tetrahedron* **24,** 1801 (1968).
211. F. Fringuelli and A. Taticchi, *J. Chem. Soc., C* **1971,** 2011.
212. J. A. Marshall, W. I. Fanta and G. L. Bundy, *Tetrahedron Letters* **1965,** 4807.
213. J. Marshall, M. T. Pike and R. D. Carroll, *J. Org. Chem.* **31,** 2933 (1966).
214. G. Defaye, *Compt. Rend.* **260,** 2843 (1965).
215. R. T. Gray and C. Djerassi, *J. Org. Chem.* **35,** 753 (1970).
216. A. R. Hochstetler, *J. Org. Chem.* **37,** 1883 (1972).
217. J. E. McMurry, *J. Am. Chem. Soc.* **90,** 6821 (1968).
218. H. C. Brown, J. H. Kawakami and K.-T. Liu, *J. Am. Chem. Soc.* **95,** 2209 (1973).
219. G. W. Oxer and D. Wege, *Tetrahedron Letters* **1969,** 3513.
220. W. F. Erman, *J. Am. Chem. Soc.* **86,** 2887 (1964).
221. L. Borowiecki and Y. Chrétien-Bessière, *Bull. Soc. Chim. France* **1967,** 2364.
222. J. M. Coxon, M. P. Hartshorn and A. J. Lewis, *Austral. J. Chem.* **24,** 1017 (1971).
223. S. P. Jindal and T. T. Tidwell, *Tetrahedron Letters* **1971,** 783.
224. P. G. Gassman and J. L. Marshall, *J. Am. Chem. Soc.* **88,** 2822 (1966).
225. H. C. Brown and G. Zweifel, *J. Am. Chem. Soc.* **86,** 393 (1964).
226. Y. Chrétien-Bessière, *Compt. Rend.* **255,** 943 (1962).
227. E. Klein and W. Rojahn, *Chem. Ber.* **100,** 1902 (1967).
228. J. P. Schaeffer, J. C. Larc, C. A. Flegal and L. M. Honig, *J. Org. Chem.* **32,** 1372 (1967).
229. R. A. Appleton, K. H. Baggaley, S. C. Egan, J. M. Davies, S. H. Graham and D. O. Lewis, *J. Chem. Soc.* **1968,** 2032.
230. W. D. K. Macrasson, J. Martin and W. Parker, *Tetrahedron Letters* **1965,** 2589.
231. J. A. Marshall and H. Faubl, *J. Am. Chem. Soc.* **92,** 948 (1970).
232. E J. Corey and R. S. Glass, *J. Am. Chem. Soc.* **89,** 2600 (1967).
233. R. R. Sauers, S. B. Schlosberg and P. E. Pfeffer, *J. Org. Chem.* **33,** 2175 (1968).
234. S. J. Cristol, W. K. Seifert and S. B. Soloway, *J. Am. Chem. Soc.* **82,** 2351 (1960).
235. R. C. Cookson and E. Grundwell, *Chem. and Ind.* **1959,** 703.
236. P. Bruck, D. Thompson and S. Winstein, *Chem. and Ind.* **1960,** 405.

237. H. C. Brown, I. Rothberg and D. L. Vander Jagt, *J. Org. Chem.* **37,** 4098 (1972).
238. P. Pesnelle and G. Ourisson, *J. Org. Chem.* **30,** 1744 (1965).
239. B. Bickborn and S. E. Wood, *J. Am. Chem. Soc.* **93,** 3940 (1971).
240. M. Barthelemy and Y. Bessière-Chrétien, *Tetrahedron Letters* **1970,** 4265.
241. P. Teisseire, M. Plattier and W. Wojnarowski, V. G. Ourisson, *Bull. Soc. Chim. France* **1966,** 2749.
242. G. Büchi, W. D. MacLeod, Jr. and J. Padilla O, *J. Am. Chem. Soc.* **86,** 4438 (1964).
243. F. Y. Edamura and A. Nickon, *J. Org. Chem.* **35,** 1509 (1790).
244. S. P. Acharya and H. C. Brown, *J. Org. Chem.* **35,** 3874 (1970).
245. R. E. Ireland and L. N. Mander, *J. Org. Chem.* **34,** 142 (1969).
246. G. Defaye-Duchateau, *Bull. Soc. Chim. France* **1964,** 1469.
247. J. C. Sirgor and G. S. Fischer, *J. Org. Chem.* **34,** 404 (1969).
248. J. W. ApSimon, P. V. Demarco and J. Lemke, *Can. J. Chem.* **43,** 2793 (1965).
249. D. K. Black and G. W. Hedrick, *J. Org. Chem.* **32,** 3758 (1967).
250. W. Herz and J. J. Schmid, *J. Org. Chem.* **34,** 3646 (1969).
251. R. R. Sobti and S. Dev, *Tetrahedron Letters* **1966,** 3939.
252. H. C. Brown, K. P. Singh and B. I. Garner, *J. Organometal Chem.* **1,** 2 (1963).
253. P. D. Bartlett, W. S. Trahanovsky, D. A. Bolon and G. H. Schmid, *J. Am. Chem. Soc.* **87,** 1314 (1965).
254. J. A. Marshall, A. E. Greene and R. Ruden, *Tetrahedron Letters* **1971,** 885.
255. G. Ohloff and G. Uhde, *Helv. Chim. Acta* **48,** 10 (1965).
256. G. Zweifel and C. C. Whitney, *J. Org. Chem.* **31,** 4178 (1966).
257. K. A. Saegebarth, *J. Org. Chem.* **25,** 2212 (1960).
258. H. M. Hess and H. C. Brown, *J. Org. Chem.* **32,** 4138 (1967).
259. G. Zweifel, K. Nagaze and H. C. Brown, *J. Am. Chem. Soc.* **84,** 183 (1962).
260. H. C. Brown, E. Negishi and P. L. Burke, *J. Am. Chem. Soc.* **94,** 3561 (1972).
261. W. Zacharewicz and A. Uzarewicz, *Rocz. Chem.* **36,** 367 (1962).
262. M. Zaidlewicz, A. Uzarewicz and W. Zacharewicz, *Rocz. Chem.* **38,** 591 (1964).
263. A. Uzarewicz, *Rocz. Chem.* **38,** 599 (1964).
264. W. Zacharewicz and A. Uzarewicz, *Rocz. Chem.* **34,** 1521 (1960).
265. A. Uzarewicz, *Rocz. Chem.* **36,** 369 (1962).
266. W. Zacharewicz, A. Uzarewicz and M. Zaidlewicz, *Rocz. Chem.* **36,** 307 (1962).
267. A. Uzarewicz, *Rocz. Chem.* **38,** 1299 (1964).
268. A. Uzarewicz, *Rocz. Chem.* **38,** 385 (1964).
269. I. Uzarewicz, M. Zaidlewicz and A. Uzarewicz, *Rocz. Chem.* **44,** 1403 (1970).
270. D. S. Sethi and D. Devaprabhakara, *Can. J. Chem.* **46,** 1165 (1968).
271. I. Mehrotra and D. Devaprabhakara, *Tetrahedron Letters* **1973,** 1885.
272. J. G. Traynham, G. R. Franzen, G. A. Knesel and D. J. Northington, *J. Org. Chem.* **32,** 3285 (1967).
273. L. G. Wideman, *J. Org. Chem.* **33,** 4541 (1968).
274. G. Zweifel, K. Nagaze and H. C. Brown, *J. Am. Chem. Soc.* **84,** 190 (1962).
275. P. T. Lansbury and E. J. Nienhouse, *Chem. Comm.* **1966,** 273.
276. L. A. Paquetti, S. A. Long, Jr., M. R. Short, B. Parkinson and J. Crardy, *Tetrahedron Letters* **1972,** 3141.
277. P. Wilder, Jr., A. R. Portis, Jr., G. W. Wright and J. M. Shepherd, *J. Org. Chem.* **39,** 1636 (1974).
278. E. Wenkert, P. W. Jeffs and J. R. Mahajan, *J. Am. Chem. Soc.* **86,** 2218 (1964).
279. W. Herz, D. Melchior, R. N. Mirrington and P. J. S. Pauwels, *J. Org. Chem.* **30,** 1873 (1965).
280. F. Sondheimer and M. Nussim, *J. Org. Chem.* **26,** 630 (1961).

281. M. Nussim, Y. Mazur and F. Sondheimer, *J. Org. Chem.* **29,** 1120 (1964).
282. A. Hassner and C. Pillar, *J. Org. Chem.* **27,** 2914 (1962).
283. D. N. Jones and D. E. Kime, *J. Chem. Soc., C* **1966,** 846.
284. J. E. Herz and L. A. Marquez, *J. Chem. Soc., C* **1969,** 2243.
285. G. Cainelli and A. Selva. Gazz, *Chim. Ital.* **97,** 720 (1967).
286. C. W. Shoppe, R. Lack and B. McLean, *J. Chem. Soc.* **1964,** 4996.
287. G. Quinkert, B. Wegemund, Z. Homburg and G. Cimbollek, *Chem. Ber.* **97,** 958 (1964).
288. R. Boutarel, *Bull. Soc. Chim. France* **1964,** 1665.
289. B. Matkovics, Zs. Tegyey and G. Göndos, *Steroids* **5,** 117 (1965).
290. M. Nussim and F. Sondheimer, *Chem. and Ind.* **1960,** 400.
291. R. W. Kelly and P. J. Sykes, *J. Chem. Soc., C* **1968,** 416.
292. C. W. Shoppee, J. C. Coll and R. E. Lack, *J. Chem. Soc., C* **1970,** 1893.
293. Y. Inubushi, T. Tsuda, T. Sano, T. Konita, S. Suzuki, H. Ageta and Y. Otake, *Chem. Pharm. Bull. Tokyo* **15,** 1153 (1967).
294. A. M. Krubiner and E. P. Oliveto, *J. Org. Chem.* **31,** 24 (1966).
295. A. M. Krubiner, N. Gottfried and E. P. Oliveto, *J. Org. Chem.* **33,** 1715 (1968).
296. M. Nussim, Y. Mazur and F. Sondheimer, *J. Org. Chem.* **29,** 1131 (1964).
297. C. H. Mauldin, E. R. Biehl and P. C. Reeves, *Tetrahedron Letters* **1972,** 2955.
298. J. Klein and A. Medlik, *J. Am. Chem. Soc.* **93,** 6313 (1971).
299. H. C. Brown and G. Zweifel, *J. Am. Chem. Soc.* **83,** 486 (1961).
300. L. Verbit and P. J. Heffron, *J. Org. Chem.* **32,** 3199 (1967).
301. K. Mislow and J. G. Berger, *J. Am. Chem. Soc.* **84,** 1956 (1962).
302. R. N. McDonald and R. N. Steppel, *J. Am. Chem. Soc.* **92,** 5664 (1970).
303. D. J. Sandman, K. Mislow, W. P. Giddings, J. Dirlam and G. C. Honson, *J. Am. Chem. Soc.* **90,** 4877 (1968).
304. H. Tanida, H. Ishitobi, T. Irie and T. Tsushima, *J. Am. Chem. Soc.* **91,** 4512 (1969).
305. J. E. Herz and L. A. Marquez, *J. Chem. Soc., C* **1971,** 3504.
306. H. C. Brown, N. R. Ayyangar and G. Zweifel, *J. Chem. Soc.* **86,** 1071 (1964).
307. H. T. Thomas and K. Mislow, *J. Am. Chem. Soc.* **92,** 6292 (1970).
308. G. Zweifel, N. R. Ayyangar, T. Minekata and H. C. Brown, *J. Am. Chem. Soc.* **86,** 1076 (1964).
309. H. Weber, P. Loew and D. Arigoni, *Chimica* **19,** 595 (1965).
310. H. Weber, J. Seibl and D. Arigoni, *Helv. Chim. Acta* **49,** 741 (1966).
311. A. Streitwieser, Jr., I. Schwayer, L. Verbit and H. Rabitz, *J. Org. Chem.* **32,** 1532 (1967).
312. R. K. Hill and A. G. Edwards, *Tetrahedron* **21,** 1501 (1965).
313. G. Giacomelli, A. M. Caporusso and L. Lardicci, *Gazz. Chim. Ital.* **104,** 1311 (1974).
314. B. A. Pawson, H. C. Cheung, S. Gurbaxani and G. Saucy, *Chem. Comm.* **1968,** 1107.
315. H. C. Brown and C. D. Pfaffenberger, *Tetrahedron* **31,** 925 (1975).
316. S. I. Goldberg and F. L. Lam, *J. Org. Chem.* **31,** 290 (1966).
317. J. Furukawa, T. Kakuzen, H. Morikawa, R. Yamamoto and O. Okuno, *Bull. Chem. Soc. Japan* **41,** 155 (1968).
318. P. S. Wharton and R. A. Kretchmer, *J. Org. Chem.* **33,** 4258 (1968).
319. W. L. Waters, *J. Org. Chem.* **36,** 1569 (1971).
320. W. L. Waters and M. C. Caserio, *Tetrahedron Letters* **1968,** 5233.
321. W. L. Waters, W. S. Linn and M. C. Caserio, *J. Am. Chem. Soc.* **90,** 6741 (1968).

322. W. R. Moore, H. W. Anderson and S. D. Clark, *J. Am. Chem. Soc.* **95,** 835 (1973).
323. D. J. Pasto and J. K. Borchardt, *Tetrahedron Letters* **1973,** 2517.
324. L. R. Byrd and M. C. Caserio, *J. Am. Chem. Soc.* **93,** 5758 (1971).
325. D. R. Brown, S. F. A. Kettle, J. McKenna and J. M. McKenna, *Chem. Comm.* **1967,** 667.
326. K. R. Varma and E. Caspi, *Tetrahedron* **24,** 6365 (1968).
327. H. C. Brown and A. K. Mandal, *Synthesis* **1980,** 153.
328. M. Anez, G. Uribe, L. Mendoza and R. Contreras, *Synthesis* **1981,** 214.
329. D. E. Bergbreiter and D. P. Rainville, *J. Org. Chem.* **41,** 3031 (1976).
329a. S. Nagase, N. K. Ray and K. Morokuma, *J. Am. Chem. Soc.* **102,** 5436 (1980).
330. K. K. Wang and H. C. Brown, *J. Org. Chem.* **45,** 5303 (1980).
331. L. C. Vishwakarma and A. Fry, *J. Org. Chem.* **45,** 5306 (1980).
332. T. Matsuda, I. Seto and T. Tanaka, *Bull. Chem. Soc. Japan* **49,** 1713 (1976).
333. E. J. Stampf and J. D. Odom, J. Organomet, *Chem.* **131,** 171 (1977).
334. K. Maruyama, K. Terada and Y. Yamamoto, *J. Org. Chem.* **45,** 737 (1980).
335. G. W. Kabalka, B. J. Newton, Jr. and J. Jacobus, *J. Org. Chem.* **43,** 1567 (1977).
336. P. Jacob, III, *J. Organomet. Chem.* **156,** 101 (1978).
337. H. C. Brown, E. Negishi and J. Katz, *J. Am. Chem. Soc.* **97,** 2791 (1976).
338. H. C. Brown, J. Katz, C. F. Lane and E. Negishi, *J. Am. Chem. Soc.* **97,** 2799 (1975).
339. G. Zweifel and N. R. Pearson, *J. Am. Chem. Soc.* **102,** 5919 (1980).
340. C. A. Brown and R. A. Coleman, *J. Org. Chem.* **44,** 2328 (1979).
341. A. Pelter, D. J. Ryder and J. H. Sheppard, *Tetrahedron Lett.* **1978,** 4715.
342. R. Köster, A. Bussmann and G. Schroth, *Ann. Chem.* **1975,** 2130.
343. V. Dedek, F. Liska, J. Fikar, Z. Chvatal and L. Pohorelsky, *Collect. Czech. Chem. Comm.* **40,** 1008 (1975).
344. H. C. Brown and N. Ravindran, *J. Am. Chem. Soc.* **98** 1785 (1976).
345. H. C. Brown, N. Ravindran and S. U. Kulkarni, *J. Org. Chem.* **44,** 2417 (1979).
346. H. C. Brown and N. Ravindran, *J. Am. Chem. Soc.* **98,** 1798 (1976).
347. H. C. Brown and N. Ravindran, *J. Am. Chem. Soc.* **99,** 7097 (1977).
348. H. C. Brown, N. Ravindran and S. U. Kulkarni, *J. Org. Chem.* **45,** 384 (1980).
349. H. C. Brown, N. Ravindran and S. U. Kulkarni, *J. Org. Chem.* **44,** 2417 (1979).
350. H. C. Brown and N. Ravindran, *Synthesis* **1977,** 695.
351. G. Zweifel and N. R. Pearson, *J. Am. Chem. Soc.* **102,** 5919 (1980).
352. H. C. Brown, J. A. Sikorski, S. U. Kulkarni and H. D. Lee, *J. Org. Chem.* **45,** 4540 (1980).
353. P. J. Maddocks, A. Pelter, K. A. Rowe, K. Smith and C. Subrahmanyam, *J. Chem. Soc., Perkin Trans. 1* **1981,** 653.
354. H. C. Brown, R. Liotta and C. G. Scouten, *J. Am. Chem. Soc.* **98,** 5297 (1976).
355. S. Thaisrivongs and J. D. Wuest, *J. Org. Chem.* **42,** 3243 (1977).
356. N. Shimizu, T. Fujioka, S. Ishizuka, T. Tsuji and S. Nishida, *J. Am. Chem. Soc.* **99,** 5972 (1977).
357. Y. Senda, S. Kamiyama and S. Imaizumi, *Tetrahedron* **33,** 2933 (1977).
358. J. C. Richer and C. Lamarre, *Can. J. Chem.* **53,** 2033 (1975).
359. M. H. Gordon and M. J. T. Robinson, *Tetrahedron Lett.* **1975,** 3867.
360. H. C. Brown, R. Liotta and L. Brener, *J. Am. Chem. Soc.* **99,** 3427 (1977).
361. L. Brener and H. C. Brown, *J. Org. Chem.* **42,** 2702 (1977).
362. P. A. Marshall and R. H. Prager, *Aust. J. Chem.* **32,** 1251 (1979).
363. F. Plénat, F. Pietrasanta, M. R. Darvich and H. Christol, *Bull. Soc. Chim. France* **1976,** 2071.
364. T. W. Bell, *Tetrahedron Lett.* **1980,** 3443.

365. I. Uzarewicz and A. Uzarewicz, *Rocz. Chem.* **50,** 1315 (1976).
366. E. K. Kazakova, L. N. Surkova and Sh. S. Bikeev, *Izv. Akad. Nauk SSSR, Ser. Khim.* **1980,** 875.
367. A. R. Hochstetler, *J. Org. Chem.* **40,** 1536 (1975).
368. J. W. Wilt and V. P. Narutis, *J. Org. Chem.* **44,** 4899 (1979).
369. D. D. Davis and H. T. Johnson, *J. Organomet. Chem.* **86,** 75 (1975).
370. N. Takaishi, Y. Inamoto and K. Aigami, *Chem. Lett.* **1979,** 803.
371. Y. Fujikura, Y. Inamoto, N. Takaishi, H. Ikeda and K. Aigami, *J. Chem. Soc., Perkin Trans, 1* **1976,** 2133.
372. D. Gardette and J. Lhomme, *J. Org. Chem.* **44,** 2315 (1979).
373. R. T. Hawkins, R. S. Hsu and S. G. Wood, *J. Org. Chem.* **43,** 4648 (1978).
374. V. V. R. Rao and D. Devaprabhakara, *Tetrahedron* **34,** 2223 (1978).
375. N. Takaishi, Y. Fujikura and Y. Inamoto, *J. Org. Chem.* **40,** 3767 (1975).
376. C. W. Greengrass, R. Ramage, A. F. Cameron and N. J. Hair, *Tetrahedron* **31,** 679 (1975).
377. C. W. Greengrass and R. Ramage, *Tetrahedron* **31,** 689 (1975).
378. F. Kienzle and R. E. Minder, *Helv. Chim. Acta* **58,** 27 (1975).
379. T. A. Woods, T. E. Boyd, E. R. Biehl and P. C. Reevtes, *J. Org. Chem.* **40,** 2416 (1975).
380. A. Rüttimann and H. Mayer, *Helv. Chim. Acta* **63,** 1456 (1980).
381. H. C. Brown and N. M. Yoon, *J. Am. Chem. Soc.* **99,** 5514 (1977).
382. A. Pelter, D. J. Ryder, J. H. Sheppard, C. Subrahmanyam, H. C. Brown and A. K. Mandal, *Tetrahedron Lett.* **1979,** 4777.
383. H. C. Brown and A. K. Mandal, *Synthesis* **1978,** 146.
384. H. C. Brown, J. R. Schwier and B. Singaram, *J. Org. Chem.* **43,** 4395 (1978).
385. A. K. Mandal, P. K. Jadhov and H. C. Brown, *J. Org. Chem.* **45,** 3543 (1980).
386. P. K. Jadhav and H. C. Brown, *J. Org. Chem.* **46,** 2988 (1981).
387. A. K. Mandal and N. M. Yoon, J. Organomet, *Chem.* **156,** 183 (1978).

CHAPTER 3

Hydroboration of functional alkenes and cyclenes

Organoboron compounds with functional substituents in the hydrocarbon radicals are used in synthetic organic chemistry.

In this chapter, we shall consider the chemical conversions of functionally substituted organoboranes prepared by the hydroboration of acyclic and alicyclic compounds with a carbon-carbon double bond containing different functional groups. On the basis of the hydroboration of unsaturated amines, vinyl ethers, enol acetates, unsaturated alcohols, aldehydes, ketones, halogenated derivatives, and other compounds, methods for the preparation of amino alcohols, alkoxy alcohols, diols, unsaturated alcohols, cyclic 1,5- and 1,6-diene hydrocarbons, and cyclopropane and other compounds have been developed. Hydroboration has found synthetic use in the terpene, steroid, carbohydrate, and heterocyclic series.

In addition to the synthetic aspect, a study of the reactions of functional derivatives of unsaturated systems with diborane and other hydroborating agents is also very important from the theoretical viewpoint, since it throws light on the problem of the effect of a hetero substituent on the orientation and stereochemistry of the addition of borohydrides to unsaturated compounds.

3.1 GENERAL CHARACTERISTICS OF FUNCTIONAL DERIVATIVES OF ALKYLBORANES

The functional groups on the unsaturated carbon atom of an olefin compound have a significant effect on the orientation of the addition of the hydroborating agent. As will be seen from the later presentation, amino and alkoxy groups completely or predominantly promote the addition of a boron atom in a β position, whereas, in the case of chlorinated compounds, it is the α-carbon atom which is attacked primarily and, in certain cases, even exclusively. The ratio between α-substituted (*1*) and β-substituted (*2*) organoboranes is also determined to a considerable degree by the structure of the hydrocarbon chain.

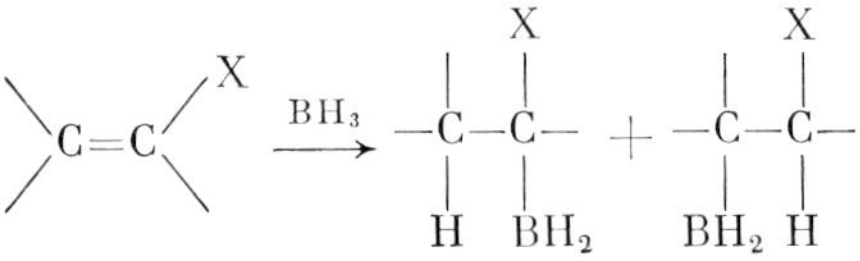

1 2

A characteristic feature of the β-substituted organoboranes *2* is their tendency to undergo decomposition with the elimination of unsaturated hydrocarbons, which are hydroborated to alkylboranes under the conditions of the reaction.

The rate of elimination of β-substituted organoboranes depends on the basicity of the solvent [1, 2]. The process may proceed spontaneously as a *cis*-elimination [3, 4]:

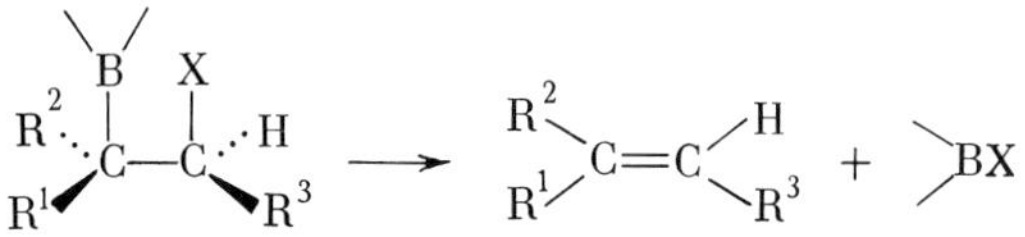

or may take place under the action of acid and basic catalysts as a *trans*-elimination [5]:

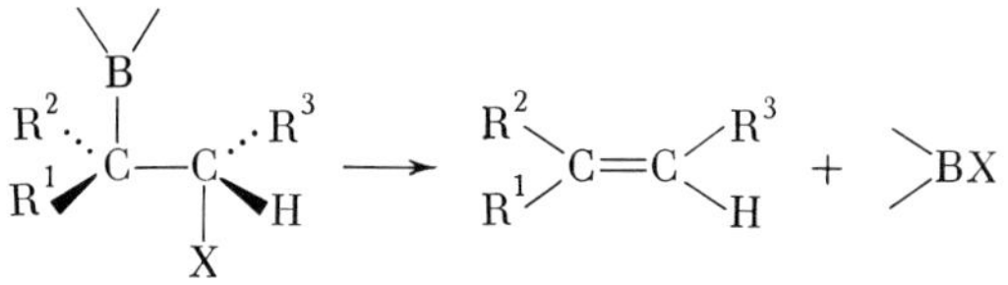

Pasto and co-workers [3, 4, 6, 7] showed that the compounds *1* are capable of undergoing spontaneous rearrangement to the compounds *3*, with a migration of the X atom from the carbon to the boron and of the hydrogen on the boron to the α-carbon:

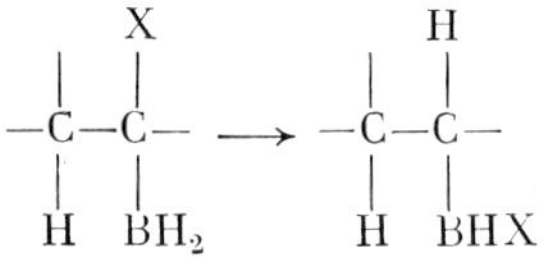

1 3

It is assumed that this rearrangement takes place via a transition state *4* with complete inversion at the carbon atom:

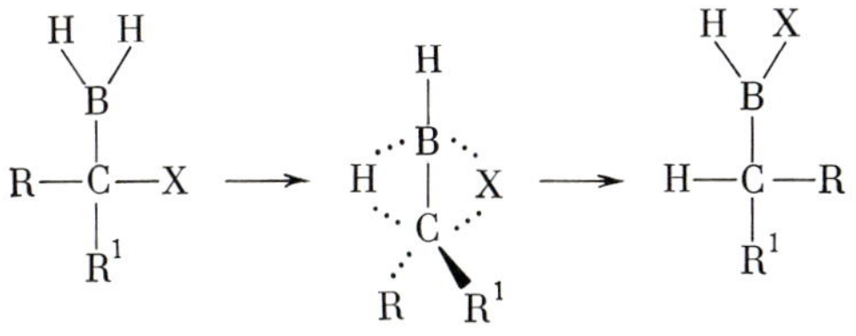

4

This mechanism is similar to the mechanism proposed by Matteson [2, 8] for the analogous rearrangements discovered by him in the series of α-halo-substituted organoboranes, proceeding with the assistance of nucleophiles, which associate with the boron atom to form intermediate tetrahedral borates.

The intramolecular rearrangement mechanism ("α-transfer") has been confirmed by studying the kinetics of the conversion of the α-chloroorganoborane *6* formed from 1-chloro-2-methylpropene (*5*) [9]:

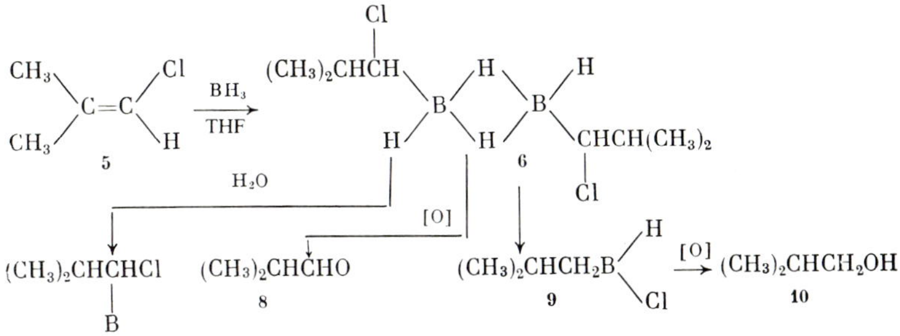

The compound 6, hydrolyzed by water to the acid 7 and oxidized to isobutyraldehyde (*8*), undergoes spontaneous rearrangement within a few hours to the chloroalkylborane *9*, forming isobutanol (*10*) upon oxidation. The process of α-transfer is first-order with respect to the dimer *6*, indicating a dissociation of the dimer and a catalytic action of one monomeric molecule on the rearrangement of the second borane molecule via the transition state *11*:

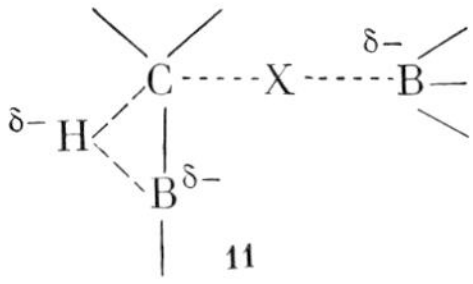

11

The rearrangement of compound *6* to *9* is significantly accelerated by the action of BF_3.

In a study of the hydroboration of functional derivatives of olefins, the reaction products generally undergo direct oxidation with alkaline hydrogen peroxide solution, and the orientation and stereochemistry of the process are judged on the basis of the structure of the alcohols obtained. In a number of studies. the organoboron compounds formed were isolated in the individual state.

3.2 NITROGEN–CONTAINING COMPOUNDS

3.2.1 Hydroboration of enamines. Synthesis of olefins, cyclenes, amino alcohols, and cyclanones

According to the data of Lewis and Pearce [10, 11], the orientation of the addition of diborane to aliphatic enamines depends on their structure. The pyrrolidine enamines of 3-pentanone and 4-heptanone (*12*) add a boron atom in the β-position. Protonolysis of the β-aminoorganoboranes *13* formed by propionic acid in boiling diglyme leads to 2-pentene and 3-heptene, respectively:

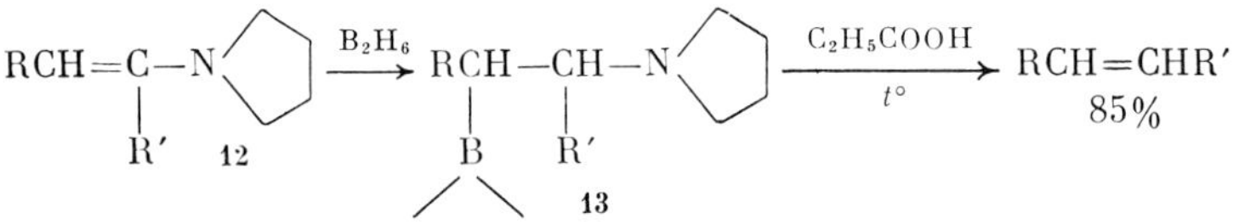

On the other hand, 2-ethyl-1-N-pyrrolidylhexene forms only 42% of 2-ethylhexene together with 52% of 3-methylheptane, indicating the addition of boron atoms in both the β- and α-position:

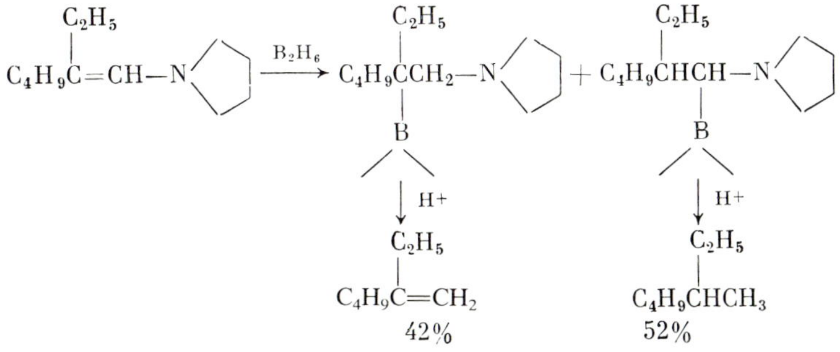

The enamines of cyclic ketones are hydroborated in the β-position in all cases. The protonolysis by propionic acid at 25°C of the β-aminoorganoborane *14* obtained from 1-N-piperidylcyclohexene, followed by hydrolysis, leads to *trans*-2-N-piperidylcyclohexylboronic acid (*15*). The latter is converted to cyclohexene by treatment with propionic acid in boiling diglyme:

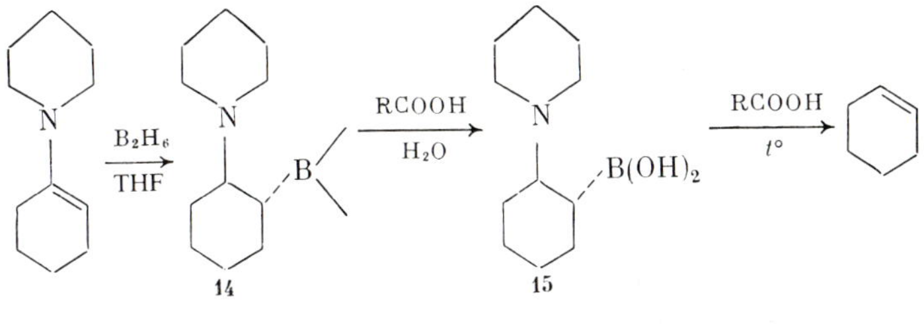

Cyclenes are obtained in yields of 88–98% from enamines of different cyclanones by the above method.

The hydroboration-oxidation of imine derivatives of 2-cyclohexen-1-one leads to *trans*-1-cyclohexylamino-2-hydroxycyclohexane [220]. Acetylated enamines of cyclohexanone are converted to 5-substituted derivatives of 1-acetylcyclohexene by hydroboration-elimination [221].

4-*tert*-Butyl-1-N-pyrrolidylcyclohexene reacts with diborane to form an equimolar mixture of the isomeric β-aminoorganoboranes

16 and *17* [11]. Both isomers are smoothly converted to 4-*tert*-butylcyclohexene by protolysis:

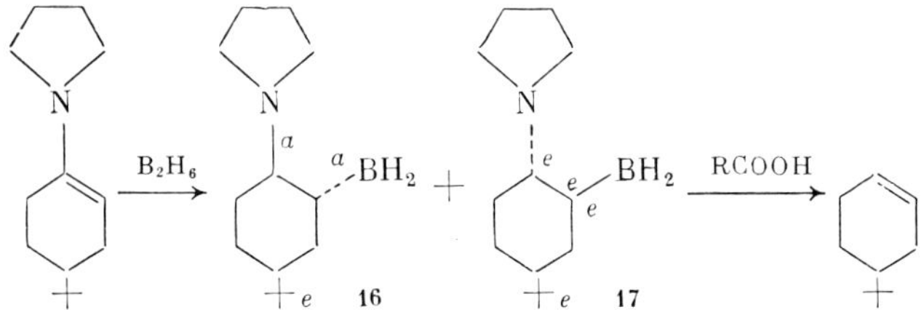

Analogous conversions have been carried out in the steroid series. 3β-N-Pyrrolidyl-2α-cholestanylborane (*19*), obtained by the hydroboration of 3-N-pyrrolidyl-2-cholestene (*18*), is converted to 2-cholestene (*20*) in 40% yield by heating with propionic acid in diglyme:

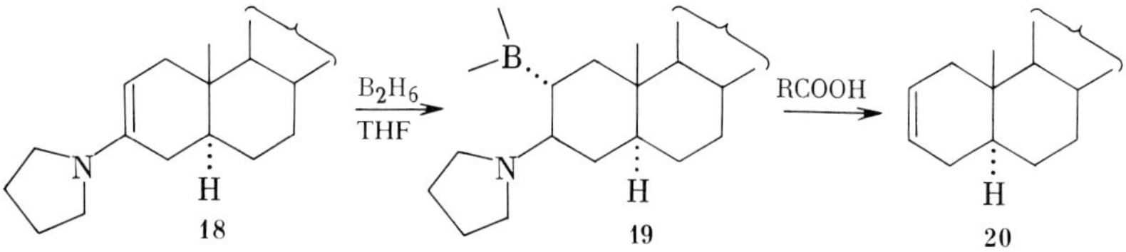

3β-N-Pyrrolidyl-4-cholestene (*21*) forms a β-aminoborane with a *trans*-diequatorial location of the substituents (*22*), which is converted to 3-cholestene (*23*) in moderate yield by protonolysis [11]:

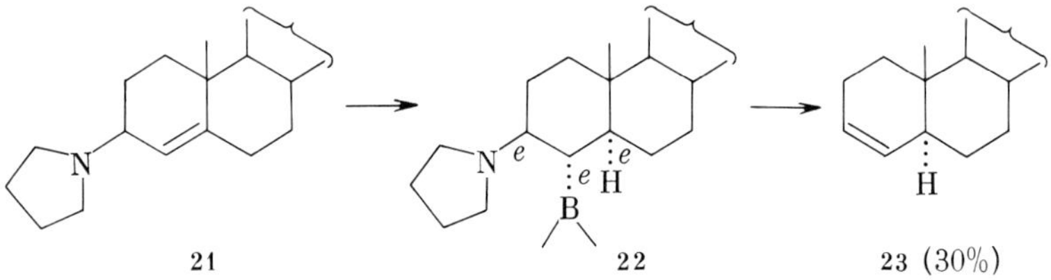

3-N-Pyrrolidyl-3,5-cholestadiene (*24*) is converted to 3β-N-pyrrolidyl-5-cholestene (*25*) by hydroboration-protonolysis [12]:

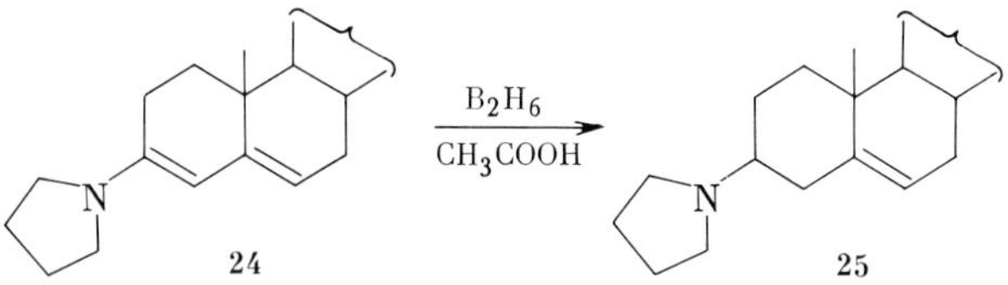

The enamine (*26*), obtained from pyrrolidine and progesterone, forms predominantly 3β-N-pyrrolidyl-4-pregnen-20-one (*27*), with practically no reduction of the carbonyl group being observed during this process [13]:

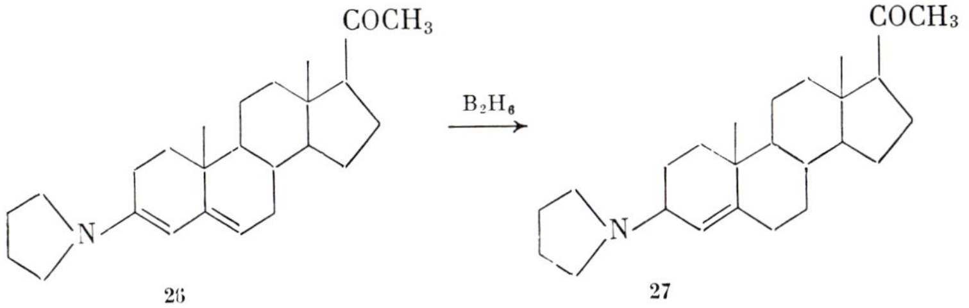

26 27

Trans-Amino alcohols are obtained in high yields by the oxidation of β-aminoorganoboranes with hydrogen peroxide in alkaline media [14]:

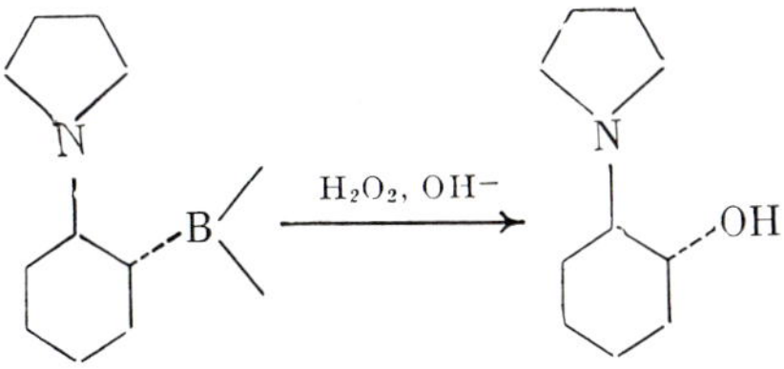

The hydroboration-oxidation of the enamine of α-substituted cyclohexanone forms a mixture of stereoisomeric amino alcohols:

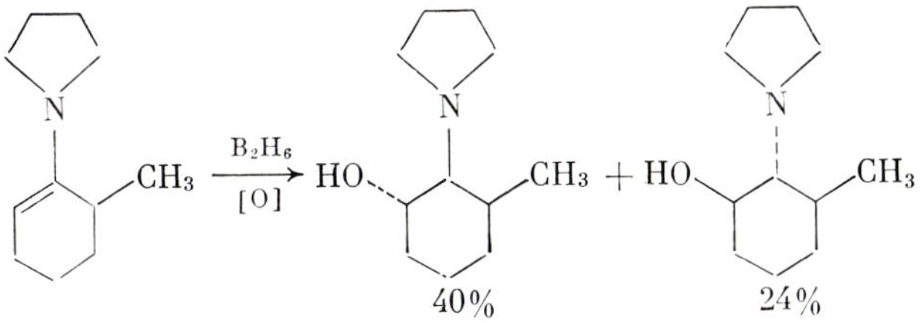

40% 24%

The oxidation of *trans*-amino alcohols of the cyclohexane series with hydrogen peroxide in methanol leads to the corresponding

N-oxides, which decompose at 150°C with the formation of cyclohexen-3-ol or its derivatives [15, 16]:

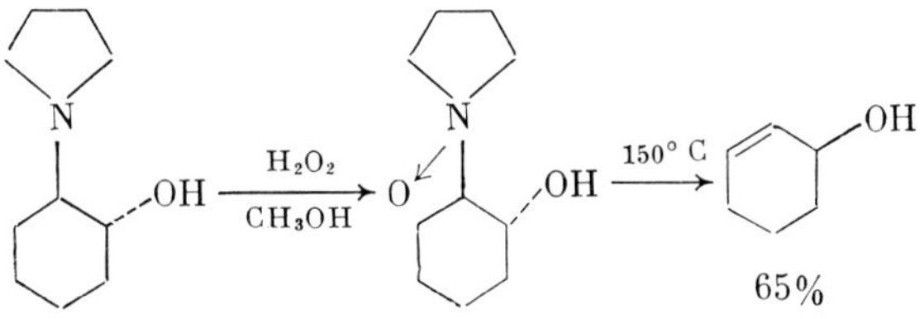

When applied to alkyl-substituted amino alcohols, this method makes it possible to convert symmetrically substituted cyclohexanones (*28*) to the unsymmetrically substituted compounds (*29*):

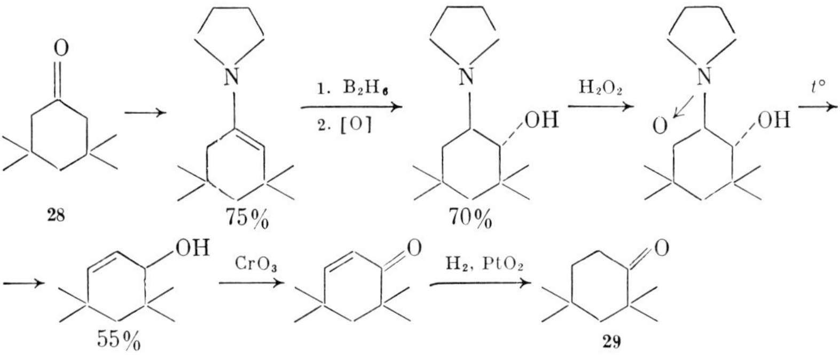

The shift of a keto group into an adjacent position can also be carried out in a series of unsymmetrically substituted cyclic ketones [15, 17]; for example, 2-methylcyclohexanone can be converted to 3-methylcyclohexanone:

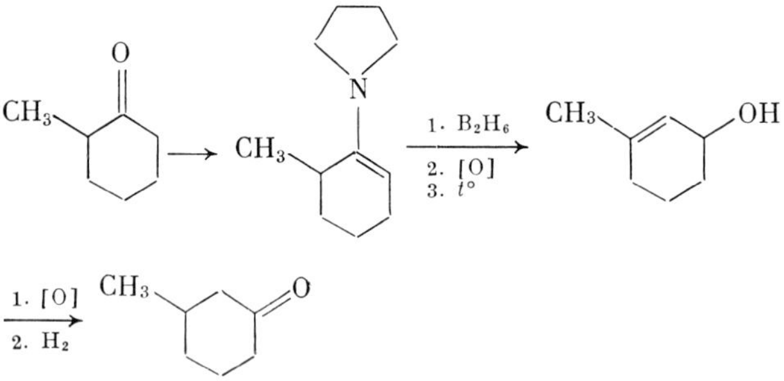

Numerous examples of similar conversions of unsymmetrical derivatives of cyclohexanone, including those of the terpene series, are presented in [17].

It could have been expected that the conversions of enamines indicated above could be carried out in the steroid series. However, it was found that enamines of the steroid series undergo hydroboration-oxidation to form not amino alcohols but the corresponding amino compounds, as the result of a protolytic cleavage of the C—B bond during the oxidation with hydrogen peroxide in alkaline media. Thus, the hydroboration of 3-N-pyrrolidyl-5α-androst-2-en-17β-ol (*30a*), followed by oxidation of the aminoboranes, produces an 80% yield of a mixture of 3α-N-pyrrolidyl-5α-androstan-17β-ol and 3β-N-pyrrolidyl-5α-androstan-17β-ol (*31a*) containing 80% of the 3α epimer [18]:

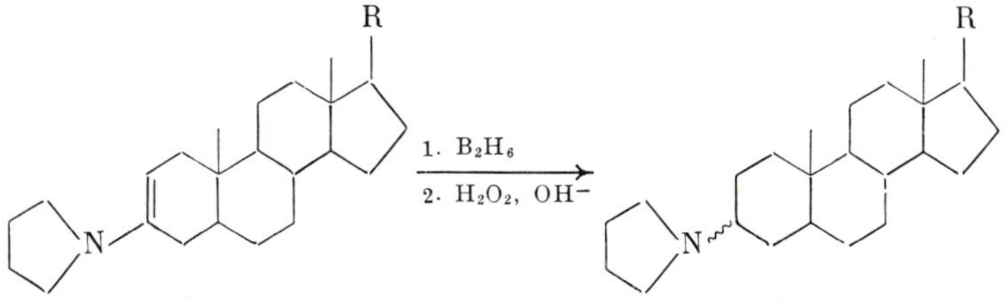

30a: R = OH
30b: R = C_8H_{17}

31a: R = OH
31b: R = C_8H_{17}

The reaction with 3-N-pyrrolidyl-5α-cholest-2-ene (*30b*) proceeds analogously.

The formation of amines in the conversions of enamines of the steroid series considered above is explained by the 1,3-diaxial interaction of the 2β-boryl group with the angular methyl group in the 19 position. This explanation is in agreement with data on the hydroboration-oxidation of enamines of 3-ketosteroids not containing a methyl group in the 19 position or of those compounds in which the amino and BH_2 groups occupy a trans-diequatorial position. Thus, the transition from 3-keto to 2-keto compounds can be carried out in the 19-norsteroid series. The enamine *33* of 5α-estran-17β-ol-3-one benzoate (*32*) is converted in 75% yield to 3α-N-pyrrolidyl-5α-estran-2β,17β-diol (*34*), with an impurity of 13% of 3ξ-N-pyrrolidyl-

5α-estran-17β-ol (*35*) [19]:

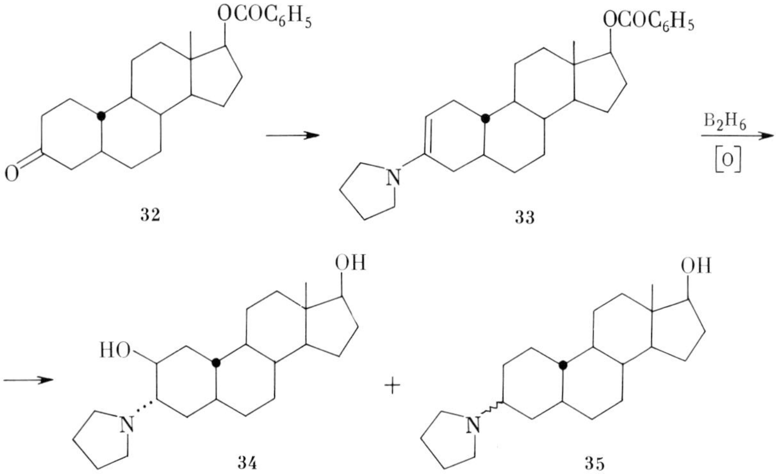

32 33 34 35

The amino alcohol *34* is oxidized with hydrogen peroxide to the N-oxide *36*, which is pyrolyzed to the alcohol *37*, from which 5α-estr-3-en-2,17-dione (*38*) is obtained in 70% yield by oxidation with Jones reagent [20]:

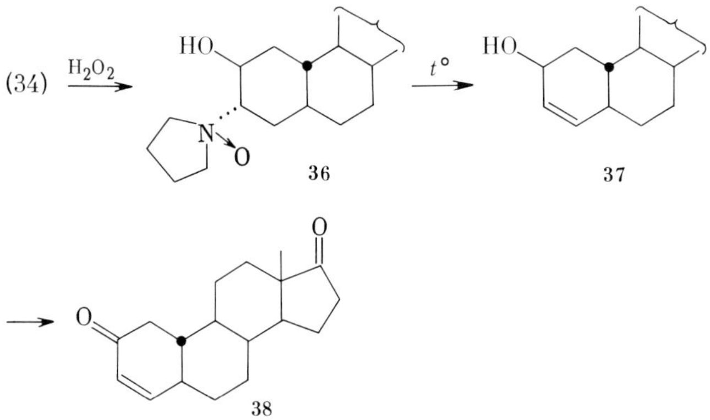

36 37 38

Unsaturated acetamido acids are converted by hydroboration-oxidation to amino alcohols [21]. For example, α-acetamidocinnamic and α-acetamidoacrylic acids produce 2-ethylamino-3-phenyl-1-

propanol and 2-ethylamino-1-propanol, respectively:

$$C_6H_5CH{=}\underset{\displaystyle NHCOCH_3}{\underset{|}{C}}{-}COOH \xrightarrow[\text{[O]}]{B_2H_6} C_6H_5CH_2\underset{\displaystyle NHCH_2CH_3}{\underset{|}{C}}HCH_2OH$$

$$CH_2{=}\underset{\displaystyle NHCOCH_3}{\underset{|}{C}}{-}COOH \xrightarrow[\text{[O]}]{B_2H_6} CH_3\underset{\displaystyle NHCH_2CH_3}{\underset{|}{C}}HCH_2OH$$

The hydroboration of vinyl derivatives of urea, followed by treatment with methanol, leads to dimethyl esters of β-ureido-alkylboronic acids [22]. For example, dimethyl β-ureidoethyl-boronate is obtained from N-vinylurea:

$$NH_2{-}CO{-}NHCH{=}CH_2 \xrightarrow[CH_3OH]{B_2H_6} \underset{80\%}{NH_2{-}CO{-}NHCH_2CH_2B(OCH_3)_2}.$$

N-Propenyl- and N-cyclohexenylurea react analogously with diborane.

In the hydroboration of N-alkenylcarbamates, the boron atom also attacks the β-carbon atom [22]:

$$C_2H_5OCONHCH{=}CH_2 \xrightarrow[CH_3OH]{B_2H_6} C_2H_5OCONHCH_2CH_2B(OCH_3)_2.$$

3.2.2 Hydroboration of ω-aminoalkenes. Synthesis of α,ω-amino alchols

In the hydroboration of allylamines and other ω-aminoalkenes, nitrogen-containing organoboron compounds are formed which, as a rule, have been isolated in the individual state. Their oxidation produces amino alcohols.

Mikhailov and co-workers [23] have found that the course of the reaction between allylamine and tetraalkyldiborane depends on the order of mixing of the reagents. When tetraalkyldiborane is added to an ether solution of allylamine (amine: $R_2BH = 1{:}1$), an adduct *39* is formed which, upon refluxing of the solution, is converted predominantly to allylaminodialkylborane (*40*), with evolution of hydrogen, and partially undergoes intra- or intermolecular hydroboration to a chelated 3-aminopropyldialkylborane, cyclo(N—B)-3-amino-propyldialkylborane (*41*):

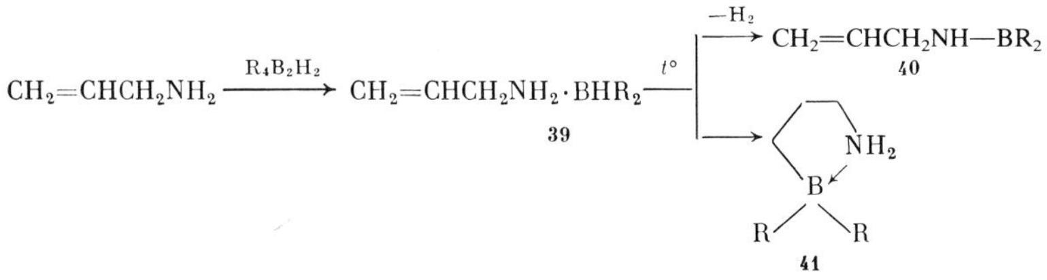

When the reagents are mixed in the reverse order, i.e., when there is an excess of tetraalkyldiborane in the reaction mixture, the adduct *39* adds dialkylborane with the formation of the compound *42*, which splits off the coordinately bonded dialkylborane to form the compound *41*:

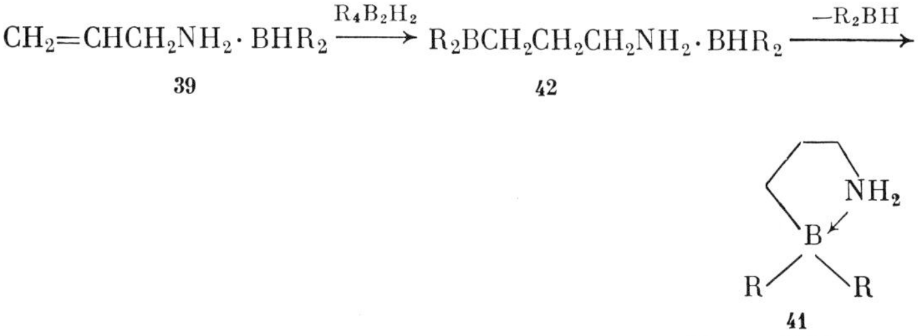

(R = *n*-C_3H_7, *n*-C_4H_9).

The hydroboration of allylamine with a mixture of triethylamine-borane and trialkylborane in a ratio of 3:2:1 at 110–130°C has been investigated [24, 25]. The main reaction products in this case are 2-alkyl-1,2-azaborolidines (*43*) (37–53%), together with cyclo(N—B)-3-aminopropyldialkylboranes (*41*) and allylaminodialkylboranes (*40*):

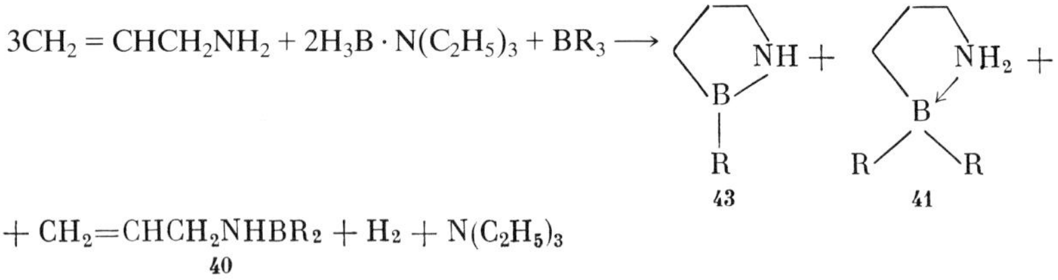

(R = *n*-C_3H_7, *n*-C_4H_9, *iso*-C_5H_{11}, *n*-C_6H_{13}).

At elevated temperature, trialkylamine-borane and trialkylboranes form amine complexes of dialkylboranes and alkylboranes, which also act as hydroborating agents. At the above-indicated ratio of reagents in the reaction mixture, there is a predominance of amine-alkylboranes, the reaction of which with allylamine forms the products *43*, whereas the compounds *41* are obtained as a result of the reaction of allylamine with amine-dialkylboranes.

In the hydroboration of allylamine with a mixture of triethylamine-borane and trialkylborane in a ratio of 1:2 (i.e., when amine-dialkylboranes predominate in the equilibrium mixture), the main reaction products are cyclo(N—B)-3-aminopropyldialkylboranes (*41*) [26].

The hydroboration of allylamine with 1,2-diphenyldiborane leads to 2-phenyl-1,2-azaborolidine [26]:

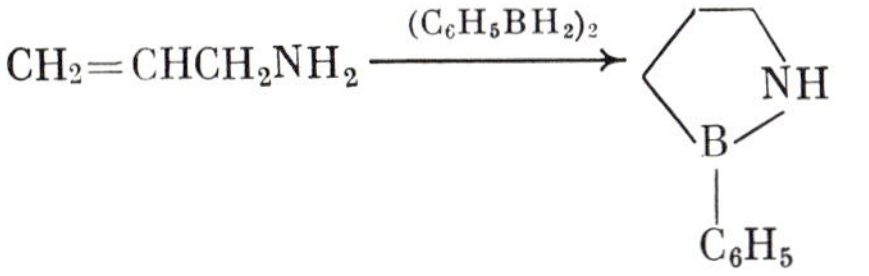

(21%)

When the hydroborating agents used are trialkylboranes, with which allylamine reacts at 180°C, cyclo(N—B)-3-aminopropyl-dialkylboranes (*41*) and 2-alkyl-1,2-azaborolidines (*43*) are formed in almost equal amounts [27].

Secondary allylamines are converted to 1-alkyl-1,2-azaborolidines (*44*) by hydroboration [28]:

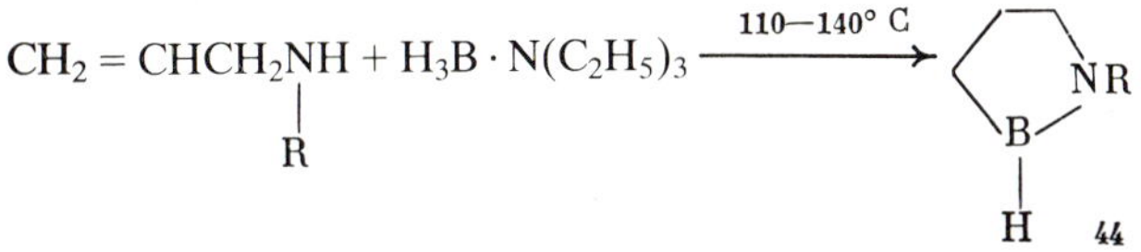

44 (35—40%)

($R = n\text{-}C_3H_7$, $n\text{-}C_4H_9$).

When a mixture of trialkylborane and triethylamine-borane is used as a hydroborating agent, allylalkylamines are converted to 1,2-dialkyl-1,2-azaborolidines (*45*) [26]:

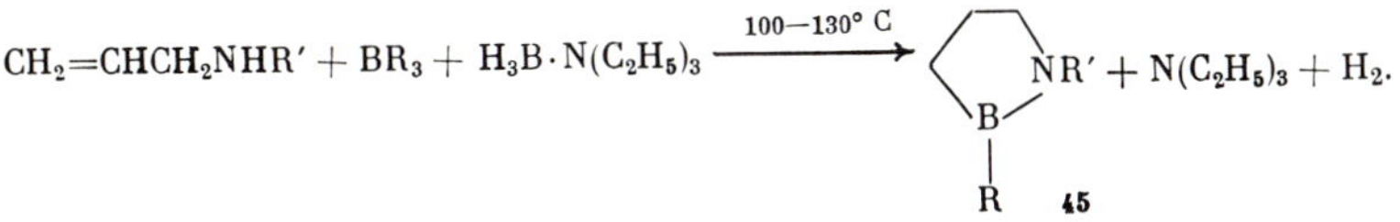

45

The hydroboration of benzalallylamine with triethylamine-borane also leads to 1-substituted 1,2-azaborolidines [28]:

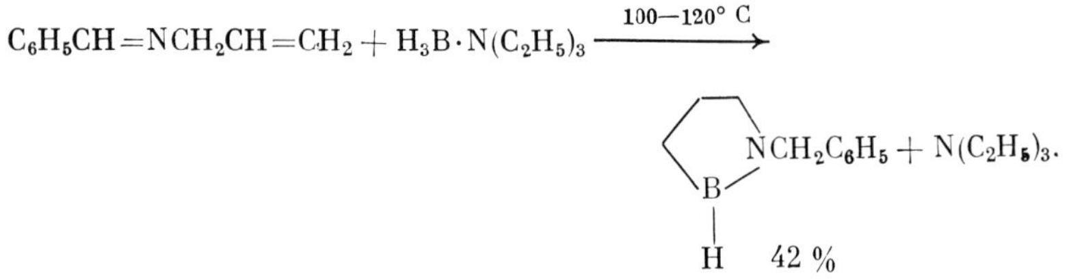

White [*29*] obtained 2-phenyl-1-methyl-1,2-azaborolidine by heating trimethylamine-phenylborane with allylmethylamine in diglyme:

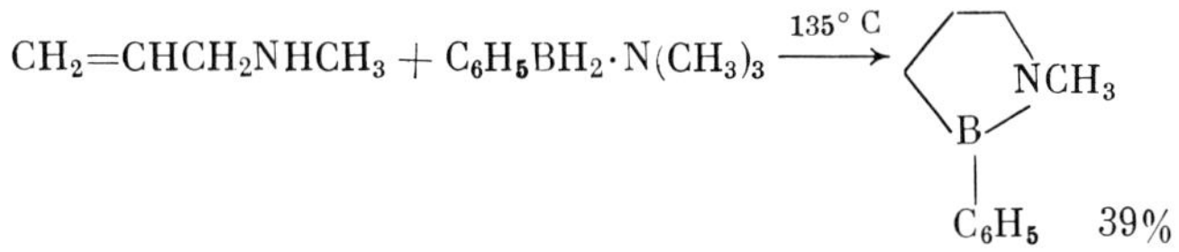

The hydroboration of allyldialkylamines with different reagents has been investigated. Allyldialkylamines react with tetraalkyldiborane to form cyclo(N—B)-3-dialkylaminopropyldialkylboranes (*46*) [30]:

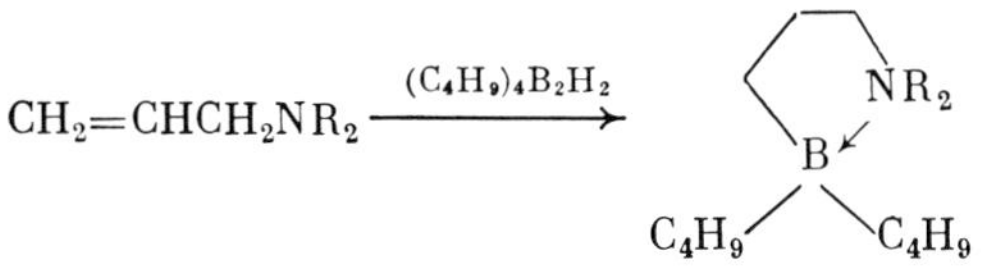

46

The compounds *46* are also obtained by heating allyldialkylamines with trialkylboranes at 180°C [27, 30].

In the reaction of N-allylpiperidine with phenylborane, obtained in the reaction medium from diethyl phenylboronate and lithium aluminum hydride, cyclo(N—B)-3-piperidinopropylphenylborane is formed in small yield [31]:

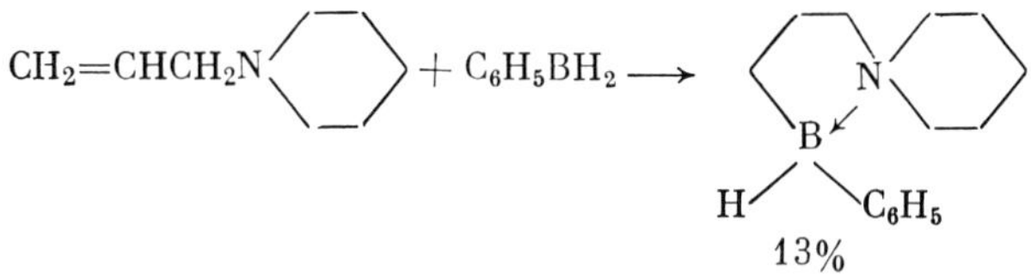

Cyclo(N—B)-3-dimethylaminopropylborane (*47*) has been prepared by heating allyldimethylamine with trimethylamine-borane or triethylamine-borane in toluene [32, 33]:

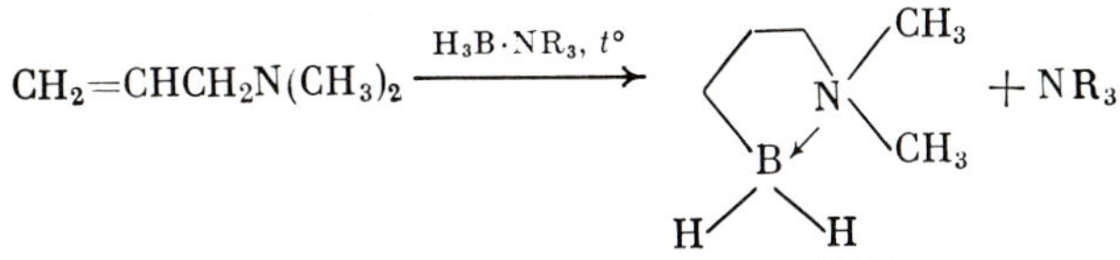

47 (65—86%)

The compound *47* is also obtained by heating allyldimethylamine-borane at 160°C (yield: 61%) [33].

When the enaminoborane *48* is heated with tributylborane, 1-cyclohexenyl-2-butyl-1,2-azaborolidine (*49*) is formed [34]:

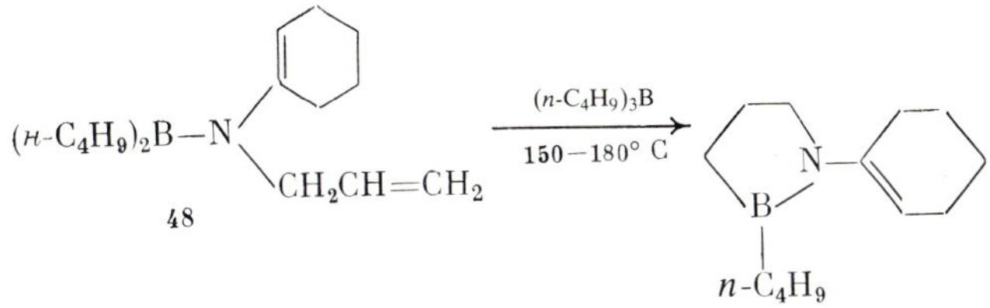

48

49 (75%)

As found by Butler and co-workers [31, 35], diallylalkylamines react with triethylamine-phenylborane to form the cyclic compound *50* and 2-phenyl-1-alkyl-1,2-azaborolidine:

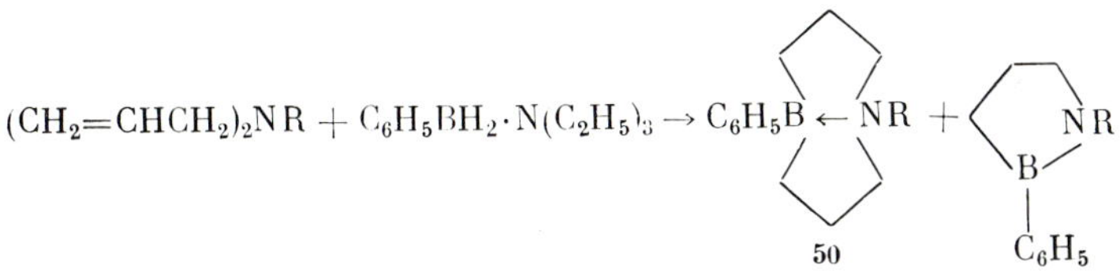

50

Triallylamine hydrochloride is converted to 1-aza-5-bora-tricyclo[3.3.3.0]-undecane by the action of triethylamine-borane [36]:

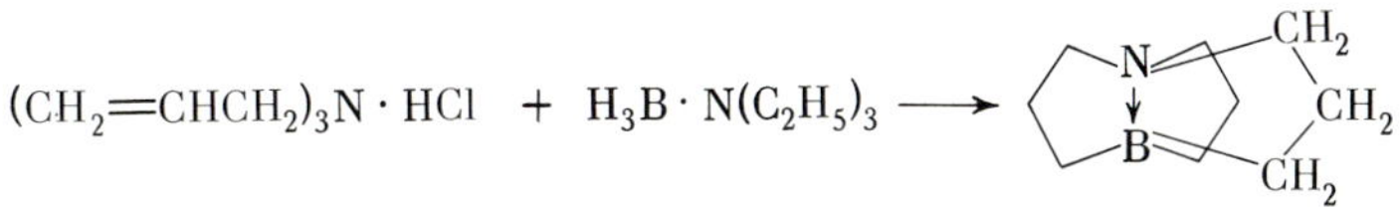

1-Alkyl-1,2-azaborolidines existing in the dimeric form [26, 28, 37] are capable of exchanging hydrogen for alkoxy and alkylmercapto groups [28]. 2-Alkyl-1,2-azaborolidines add organomagnesium and organolithium compounds with the formation of cyclo(N—B)-3-aminopropyldialkylboranes (*41*) [38], are converted to cyclic esters of (3-aminopropyl)alkylboronic acids by the action of alcohols [39], are acylated by acetic anhydride to acetoxy(3-aminopropyl)alkylboranes [40], and form coordination compounds with acetaminodialkylboranes [41].

Cyclo(N—B)-3-dialkylaminopropyldialkylboranes (*46*) react with acetic acid with the formation of cyclo(N—B)-3-dialkylaminopropylacetoxyalkylboranes [42]. Upon treatment with hydrochloric acid, followed by oxidation with an alkaline solution of hydrogen peroxide, cyclo (N—B)-3-dimethylaminopropylboranes are converted smoothly to 3-dimethylaminopropanol [43].

3-Butenylamine is converted to 4-amino-1-butanol (71%) and 4-amino-2-butanol (13%) by the action of diborane and subsequent oxidation of the reaction products; in the hydroboration with tetra(3-methyl-2-butyl)diborane, the yield of primary amino alcohol is 81% (with an impurity of 2% of the isomer) [44]. When heated with trimethylamine-phenylborane, 3-butenylamine forms 2-phenyl-1,2-azaborinane (*51*), which is dehydrogenated over palladium/carbon to 2-phenyl-2,1-borazarene (*52*) [29]:

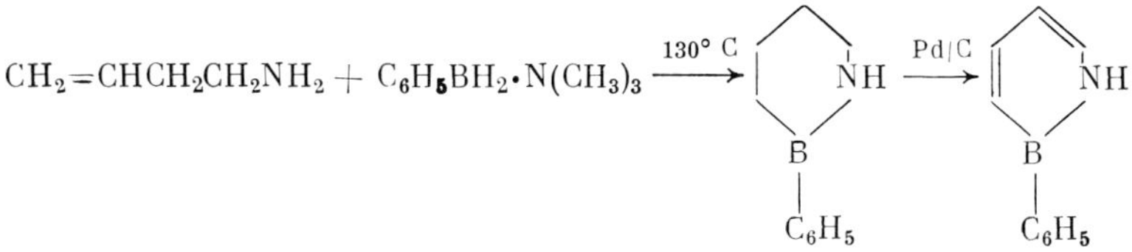

51 (58%) 52

Tritetramethyleneborazole is obtained by heating 3-butenylamine with trimethylamine-borane in diglyme [45]:

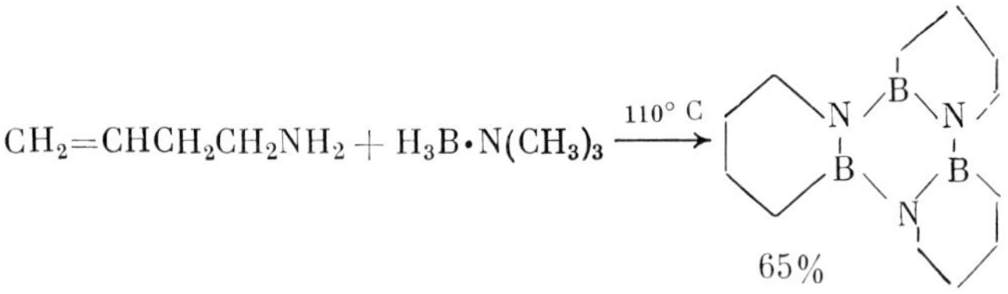

65%

N-Methyl-3-butenylamine reacts with trimethylamine-borane at 150°C with the formation of 1-methyl-1,2-azaborinane [45].

Polívka, Ferles, and co-workers [43, 46–48] carried out a detailed investigation of the hydroboration of ω-dimethylaminoalkenes and of the thermal conversions of the ω-dimethylaminoalkylboranes formed in this process. The orientation of the addition of diborane to 7-dimethylamino-1-heptene is similar to the orientation in the case of α-olefins: the ratio of α- to β-organoboranes is 92:8 [46]. The hydroboration of ω-dimethylaminoalkenes with triethylamine-borane in boiling xylene leads to cyclo(N—B)-ω-dimethylaminoalkylboranes [43, 47, 48]. ω-Dimethylamino-1-butene forms a 70% yield of a mixture of cyclic 4-dimethylamino-1-butylborane (*53*) and 4-dimethylamino-2-butylborane (*54*), in a ratio of 87:13:

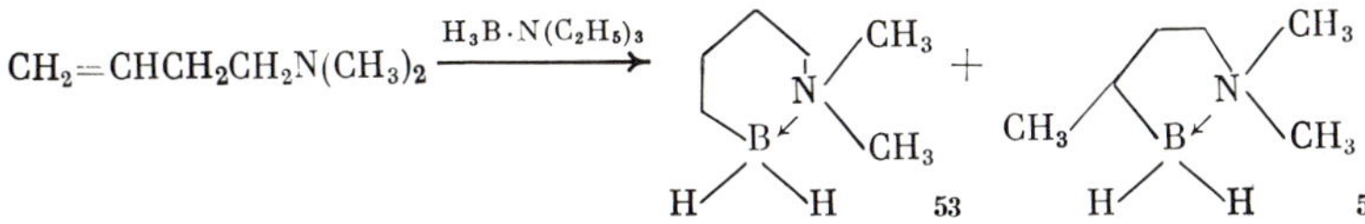

53 54

while 5-dimethylamino-1-pentene forms a mixture of *55* and *56* in a ratio of 96:4 [43]:

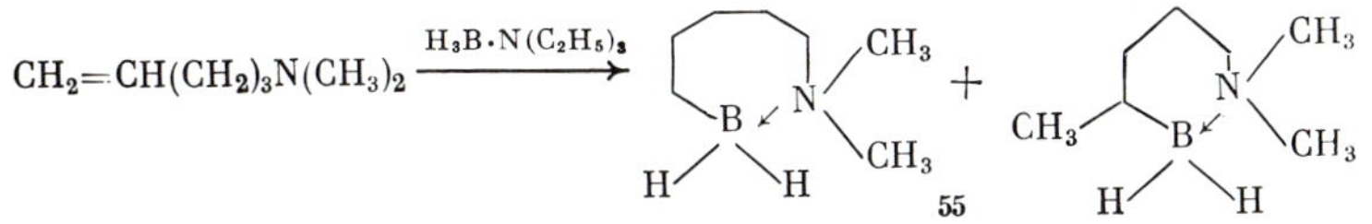

55 56

The higher members of the $(CH_3)_2N(CH_2)_nCH{=}CH_2$ series ($n = 4 - 10$) are converted to a mixture of cyclo(N—B)-ω-dimethylaminoalkylboranes with a predominant content of 6- and 7-membered rings, formed as a result of thermal isomerization of the primary hydroboration products.

The hydroboration-oxidation of dialkylallylamines leads to amino alcohols [222].

The donor-acceptor reaction of δ-dialkylamino- and γ-dialkylaminoalkylboranes with various substituents on the boron atom has been investigated by B-11 and H-1 NMR [49–51]. It has been established that these compounds, both in the individual state and in solutions, consist of an equilibrium mixture of the coordinated (cyclic) form A and the uncoordinated (linear) form B, whose rate of interconversion exceeds 10^3 sec^{-1}:

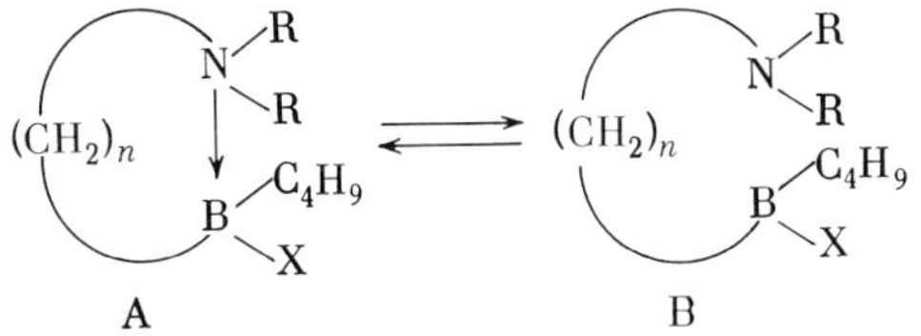

$(X = OR, SR;\ R = CH_3, C_2H_5;\ n = 3,4)$

From measurements of chemical shifts, the equilibrium constants at different temperatures and the ΔH values have been determined.

3.3 OXYGEN-CONTAINING COMPOUNDS

3.3.1 Hydroboration of vinyl and allyl ethers. Synthesis of olefins, cyclenes, and alkoxy alcohols

a) Vinyl ethers

Mikhailov and co-workers [52, 53] have found that the introduction of diborane into an ether solution of vinyl ethyl or vinyl butyl ether (ether: $BH_3 = 3:1$) at −70°C, followed by slow heating of the solution, leads to the formation of thermally unstable tris(2-alkoxyethyl)boranes [57]. Their decomposition starts at a temperature as low as −10°C, with the evolution of ethylene, which accelerates significantly at 0°C and continues for 3–4 hours when room temperature is reached. As a result of the elimination of one mole of ethylene, esters of bis(2-alkoxyethyl)borinic acids (*58*) are formed:

$$3CH_2{=}CHOR + {}^1/_2B_2H_6 \longrightarrow \underset{\mathbf{57}}{B(CH_2CH_2OR)_3} \xrightarrow{t^\circ} CH_2{=}CH_2 + \underset{\mathbf{58}}{ROB(CH_2CH_2OR)_2}$$

$$(R = C_2H_5, n\text{-}C_4H_9).$$

Ethyl bis(2-ethoxyethyl)borinate (*58*, $R = C_2H_5$) is thermally quite stable and can be vacuum-distilled without decomposition (yield: 77%), whereas butyl bis(2-butoxyethyl)borinate (*58*, $R = C_4H_9$) eliminates ethylene upon distillation, being converted to dibutyl 2-butoxyethylboronate (*59*) and partially to tributyl borate:

$$n\text{-}C_4H_9OB(CH_2CH_2OC_4H_9\text{-}n)_2 \longrightarrow C_2H_4 + \underset{\mathbf{59}}{(n\text{-}C_4H_9O)_2BCH_2CH_2OC_4H_9\text{-}n}$$

The hydroboration of vinyl ethyl ether with a 20% excess of diborane (ether: $BH_3 = 2.5:1$), followed by distillation of the reaction products, produces ethyl ethyl(2-ethoxyethyl)borinate (*60*) together with ethyl bis(2-ethoxyethyl)borinate. The formation of the ester *60* may take place as a result of exchange processes between the reaction products and excess of diborane, followed by the addition of $>BH$ to ethylene, for example, in accordance with equations (1) and (2):

$$(C_2H_5OCH_2CH_2)_2BOC_2H_5 \xrightarrow{BH_3} C_2H_5OCH_2CH_2{-}\underset{\displaystyle H}{\underset{|}{B}}OC_2H_5 \qquad (1)$$

$$C_2H_5OCH_2CH_2\underset{\displaystyle H}{\underset{|}{B}}{-}OC_2H_5 + CH_2{=}CH_2 \longrightarrow \underset{60}{C_2H_5OCH_2CH_2\underset{\displaystyle C_2H_5}{\underset{|}{B}}OC_2H_5} \qquad (2)$$

The ester *60*, when heated for 2 hours at 160–170°C, is converted to diethyl ethylboronate with the elimination of ethylene:

$$\underset{60}{C_2H_5OCH_2CH_2\underset{\displaystyle C_2H_5}{\underset{|}{B}}OC_2H_5} \xrightarrow{t^\circ} CH_2{=}CH_2 + C_2H_5B(OC_2H_5)_2.$$

The ester *58* shows a higher thermal stability; it is converted to triethyl borate, with the elimination of ethylene, by heating for 6 hours at 170–180°C.

The heating of dibutyl 2-butoxyethylboronate (*59*) at 150°C for 9 hours leads to tributyl borate.

The tendency of 2-alkoxy-substituted alkylboranes toward β-elimination, like that of other β-substituted compounds, is explained primarily by the inductive effect of the substituent. The physical nature of this effect consists in a rehybridization of the orbitals of the oxygen, carbon, and boron atoms [54–56], as a result of which the polarization of the B—C bond is increased and its heterolytic rupture is facilitated. This is promoted by a coordinative interaction between the vacant boron orbital and the unshared electron pair of the hetero atom. The superposition of the indicated intramolecular effects, acting in the same direction, leads to elimination of the olefin by a consistent mechanism [53, 56].

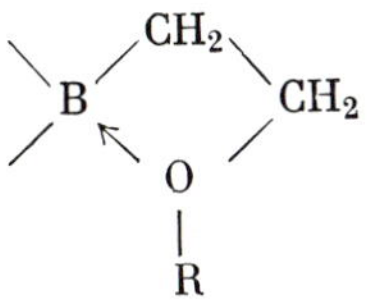

The action of tetrapropyldiborane on butyl vinyl ether in diethyl ether at −20°C produces (2-butoxyethyl) dipropylborane (*61*), which eliminates ethylene at room temperature with the formation of butyl dipropylborinate (*62*) [57]:

$$(n\text{-}C_3H_7)_4B_2H_2 + CH_2{=}CHOC_4H_9\text{-}n \longrightarrow \underset{\mathbf{61}}{(n\text{-}C_3H_7)_2BCH_2CH_2OC_4H_9\text{-}n} \xrightarrow{t^\circ} (n\text{-}C_3H_7)_2BOC_4H_9\text{-}n + CH_2{=}CH_2.$$

Thus, the mono-β-alkoxy-substituted trialkylboranes *61* show a greater tendency toward β-decomposition than $ROCH_2CH_2B(OR)_2$, the esters of β-alkoxyethylboronic acids [53]. This is explained by the fact that, under the action of the less electronegative alkyl radicals (as compared with alkoxy radicals), the *s* character and electronegativity of the boron orbital in the direction of the β-substituted radical decrease, as a result of which the B—C bond in question becomes more highly polarized. Furthermore, the presence of alkyl radicals favors the coordination of the vacant boron orbital with the unshared electron pair of the oxygen atom, which causes an increase in the polarization of both the B—C and the C—O bonds and thereby facilitates their heterolytic rupture.

The hydroboration of β-ethoxystyrene with diborane (1:1) in tetrahydrofuran, followed by oxidation (at 0°C), produces primarily 1-phenyl-2-ethoxyethanol, together with small quantities of 2-phenyl- and 1-phenylethanols [4]. The formation of the latter two alcohols is explained by the hydroboration-oxidation of styrene, which is formed by decomposition of the organoborane *63*:

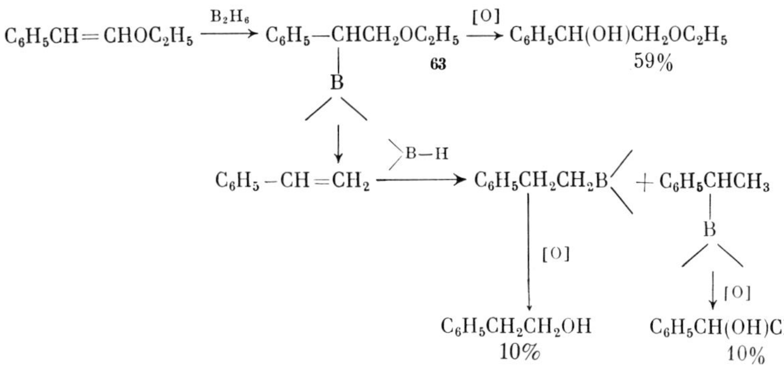

If the hydroboration is carried out at 25°C, then the organoborane *63* almost completely eliminates styrene, as a result of which the main oxidation product is 2-phenylethanol (48%), while only 12% of 2-ethoxy-1-phenylethanol is obtained.

The hydroboration-oxidation of 1-ethoxy-1-butene and 1-ethoxy-2-methylpropylene (1:1) in tetrahydrofuran leads to 1-ethoxy-2-butanol and 1-ethoxy-2-methyl-2-propanol, respectively, indicating localization of the boron in a β-position to the ethoxy group [58].

As indicated above, β-hetero-substituted organoboranes are capable of undergoing a *trans*-elimination catalyzed by acid and basic reagents. Thus, the products of the deuteroboration (*64*) of *cis*-β-ethoxystyrene spontaneously undergo *cis*-elimination with the formation of *trans*-β-deuterostyrene (*65*), whereas, in the presence of a basic (C_4H_9Li) or acid catalyst (BF_3), a *trans*-elimination with the formation of *cis*-β-deuterostyrene (*66*) takes place [5]:

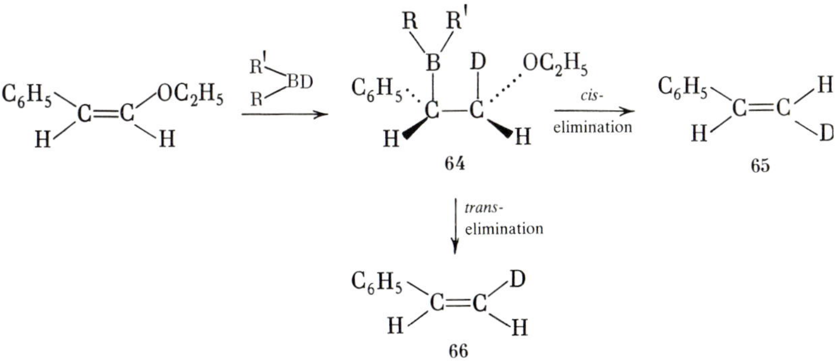

On the other hand, *trans*-β-benzylthiostyrene, as a result of deuteroboration-elimination under the action of C_4H_9Li or BF_3, is converted to *trans*-β-deuterostyrene by *trans*-elimination [5]. In this investigation, conversions of α-substituted organoboranes, obtained together with the β-substituted compounds, were not taken into account.

In connection with the problem of the decomposition of β-hetero-substituted organoboranes, there is considerable interest in a study of the kinetics of decomposition of a number of individual esters of *trans*-2-ethoxycyclohexylboronic acids, *67–69*, and of the ester *70* under the action of different Lewis acids in different solvents [59].

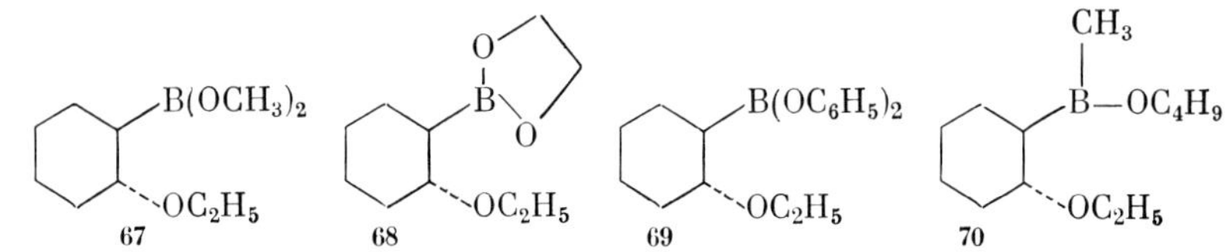

67 68 69 70

The rate constants of the dealkoxyboration of the ester *67*, catalyzed by BF_3, in different ether media are presented below.

Solvent	T, °C	$k_2 \cdot 10^3$, L/mole · sec
Diisopropyl ether	40	2.11
Tetrahydrofuran	42	4.18
Diethyl ether	42	5.72
Tetrahydropyran	37	9.07
Dibutyl ether	40	9.73
Dipropyl ether	40	10.66

The unusually high negative values of ΔS permit the assumption that the effect of the solvent on the elimination rate constant shows up in a solvation stabilization of the transition state of the boronium cation formed, as indicated in the following scheme:

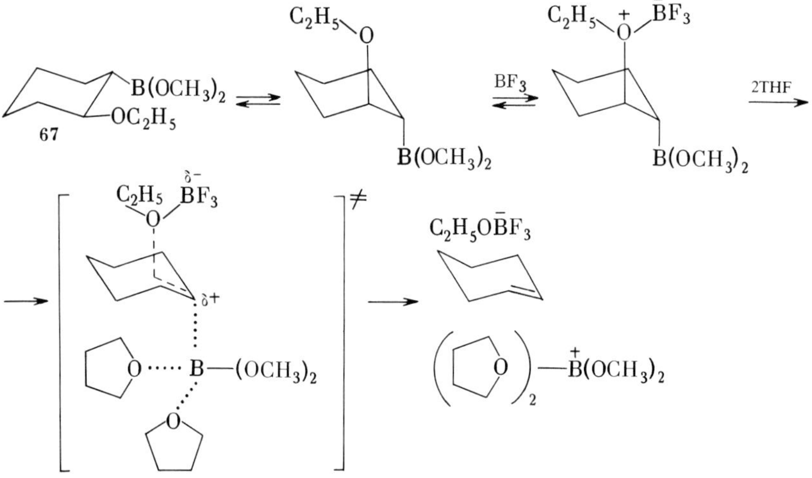

The elimination rate is affected to a considerably greater degree by the strength of the Lewis acid and the nature of the functional group.

As a result of the hydroboration of 1-ethoxycyclohexene in tetrahydrofuran (olefin: $BF_3 = 3:2$), followed by oxidation, *trans*-2-ethoxycyclohexanol (*71*) and traces of cyclohexanol are obtained, indicating the addition of a boron atom in the 2 position and the relative thermal stability of the compound *72* [4]. The addition of boron trifluoride to the hydroboration products causes a decomposition of the compound 72, with the elimination of cyclohexene. It is interesting that, in the hydroboration of 1-ethoxycyclohexene in the presence of an equimolar quantity of BF_3, the boron atom adds to a considerable degree in the 1 position (*73*), as follows from the formation of cyclohexyl ethyl ether (*74*) (21%) as a result of hydrolysis-oxidation of the reaction products:

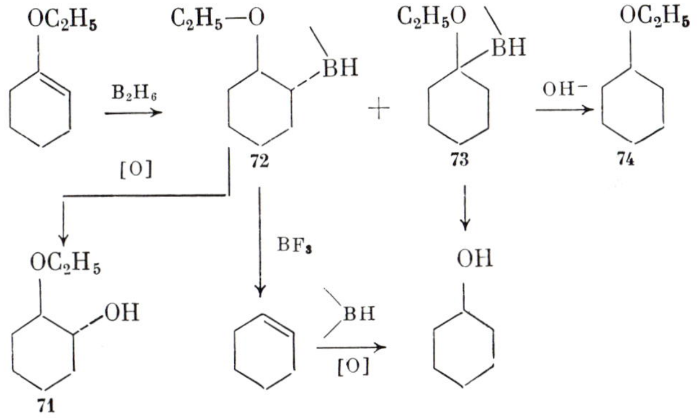

In the hydroboration of 4-*tert*-butyl-1-ethoxycyclohexene, the boron atom adds predominantly to C-2 in the cis (*75*) and trans position (*76*) with respect to the *tert*-butyl group; addition of the boron to C-1 in the cis (*77*) and trans position (*78*) takes place only to a negligible degree [4].

The hydroxy ethers *79* and *80* and *cis*-(*81*) and *trans*-1-ethoxy-4-*tert*-butylcyclohexane (*82*) are obtained by oxidation-hydrolysis.

In the hydroboration of 1-ethoxy-4-*tert*-butylcyclohexene in the presence of BF_3, the diaxial isomer *80* is not obtained, and the yield of the diequatorial isomer *79* is reduced to 21% as a result of β-cleavage of the compounds *75* and *76*. 4-*tert*-Butylcyclohexene (*83*) formed in this process is converted to a mixture of 4- and 3-*tert*-butylcyclohexanol.

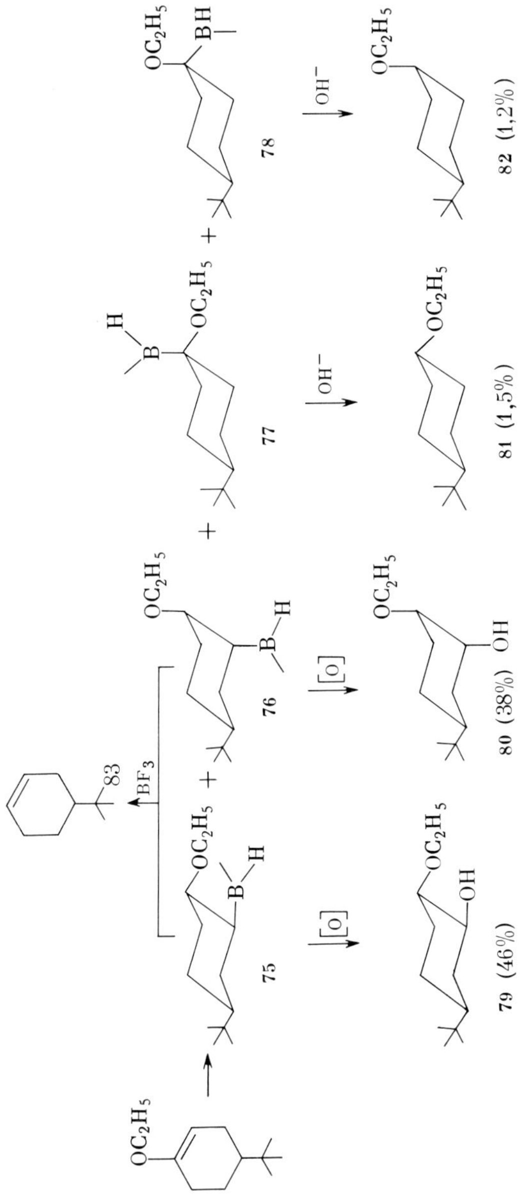
OC2H5
83
BF3
75
76
77
78
[O]
OH−
79 (46%)
80 (38%)
81 (1,5%)
82 (1,2%)

The hydroboration-oxidation of 1-ethoxycyclopentene leads to *trans*-2-ethoxycyclopentanol [58].

3-Methoxy-5α-cholest-2-ene, as a result of hydroboration and subsequent elimination by treatment with NaOH, is converted to 5α-cholest-2-ene [60]:

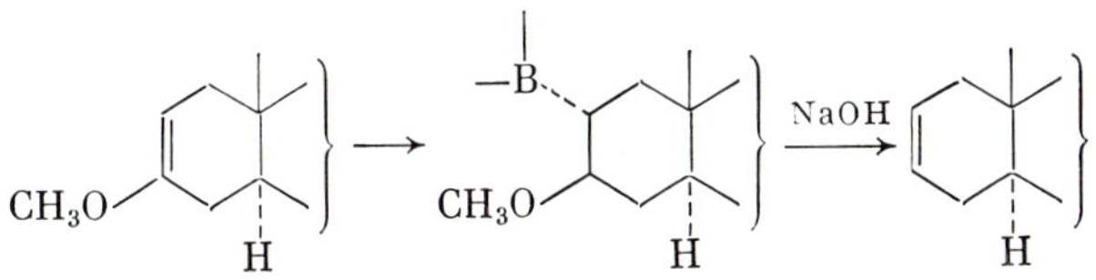

The reaction of trimethylamine-*tert*-butylborane with divinyl ether forms 4-*tert*-butyl-1,4-oxaborinane [61]:

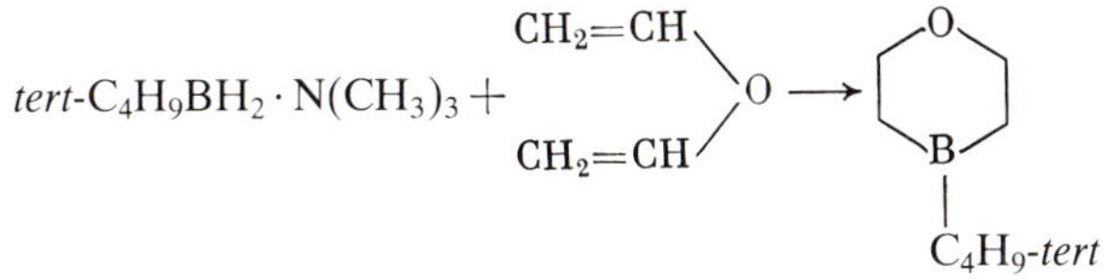

According to patent data [62], 2-vinyloxyethyldimethylborane is obtained from divinyl ether and tetramethyldiborane:

$$^{1}/_{2}(CH_3)_4B_2H_2 + (CH_2{=}CH)_2O \longrightarrow (CH_3)_2BCH_2CH_2OCH{=}CH_2.$$

Pasto and Miesel [3, 63] investigated the hydroboration-oxidation of 2-phenylmercapto-1-phenylethylene, phenyl (1-phenyl-1-propenyl) sulfide, and phenyl 1-propenyl sulfide. The main reaction products are alcohols devoid of a phenylmercapto group and also alkyl phenyl sulfides. Thus, for example, 1-propanol (48%), 2-propanol (4%), methyl phenyl sulfide (31%), and ethyl phenyl sulfide (14%) are formed from phenyl propenyl sulfide. The authors interpret the results obtained with the use of concepts of "α- and β-transfers".

The hydroboration of β-methylmercaptostyrene leads to a mixture of isomeric phenylethanols [64].

The trimethylsilylenol ethers *84–86* prepared from acetophenone, butyraldehyde, and isobutyraldehyde are hydroborated with diborane in tetrahydrofuran or diethyl ether with the primary formation of β-derivatives. These easily eliminate olefinic hydrocarbons which, upon further hydroboration-oxidation, form the corresponding alcohols [65].

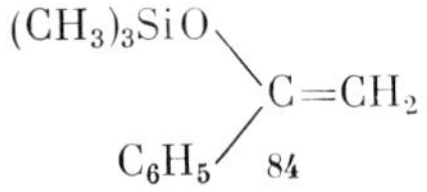

84

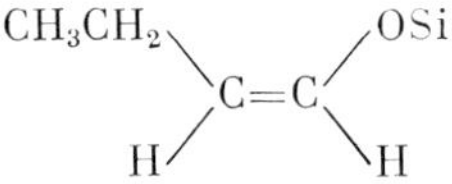

85

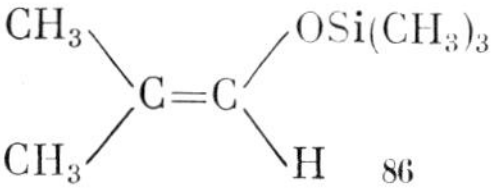

86

Thus, compound *84* in tetrahydrofuran forms 2-phenylethanol (72%), 1-phenylethanol (18%), and small quantities of 2-trimethylsiloxy-2-phenylethane and 1-phenyl-1,2-ethanediol. Alcohols are obtained from aliphatic ketones by the same method [223]. The borane obtained by the hydroboration of 2-trimethylsiloxy-1-phenylpropene undergoes mainly cis-elimination, to form (*Z*)-β-methylstyrene [224].

The trimethylsilyl enol ethers of cyclopentanone, cyclohexanone, and cycloheptanone (*87*) are converted to the corresponding *trans*-2-trimethylsiloxycyclanols (*88*) by hydroboration-oxidation [65]:

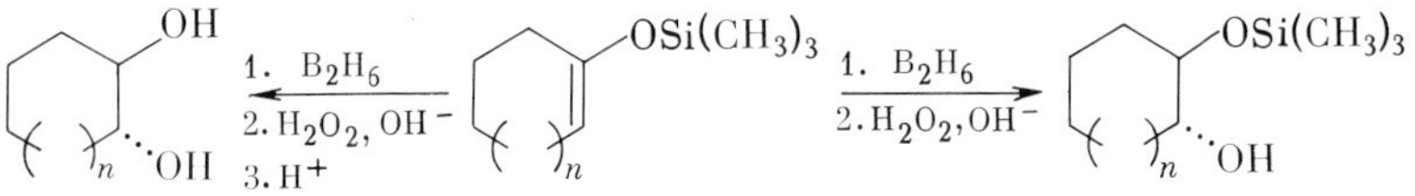

87

88 (30–59%)
(n = 0,1,2)

If ether solutions of the compounds *88* are washed with 10% hydrochloric acid, the corresponding *trans*-glycols are formed in yields of 49–63%.

The hydroboration-oxidation of 1-trimethylsiloxy-4-*tert*-butylcyclohexene leads to a mixture of 4-*tert*-butyl-*trans*-1-*cis*-2- (*89*) and 4-*tert*-butyl-*cis*-1-*trans*-2-cyclohexanediol (*90*) in a ratio of 3:1, with a yield of 73% [65]:

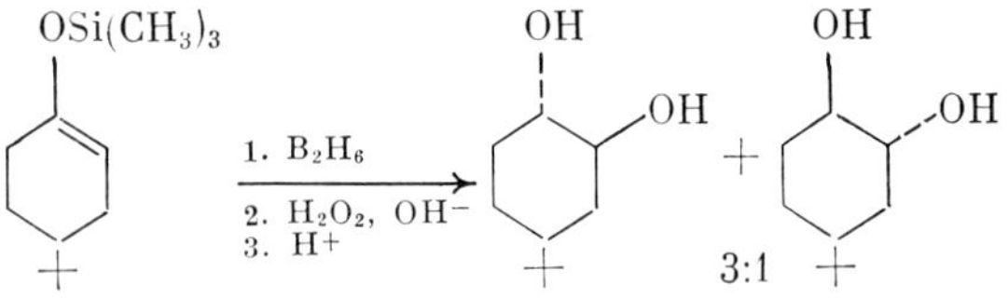

89 90

The products of the hydroboration of cyclic trimethylsilyl enol ethers are converted to cyclenes by heating with 10% HCl [225]. 1-Trimethylsiloxy-4-methylcyclohexene forms the corresponding diols in a ratio of 62:38 [66]. Thus, the trimethylsiloxy group, as compared with the ethoxy group [4], promotes the addition of a boron atom in a cis position with respect to an alkyl group on C-4. On the other hand, a boranate group (*91*) orients the addition of a boron atom primarily in the trans position [66]:

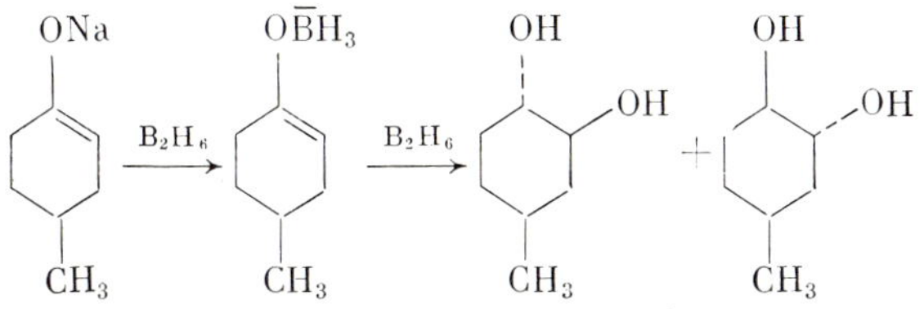

91 2:3

b) Allyl ethers. Synthesis of alkoxy alcohols

Allyl ethers (allyl ethyl and allyl phenyl ethers) [67], and also allyl phenyl sulfide [3] add a boron atom to the extent of 63–78% in the α-position in a hydroboration with diborane, whereas allyl tosylate forms α- and β-substituted organoboranes in almost equal quantities [68]:

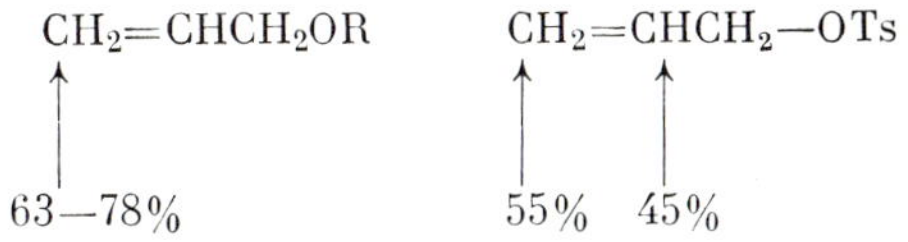

The hydroboration of allyl ethers by tetra(3-methyl-2-butyl)-diborane forms predominantly (86–92%) α-substituted organoboranes [67]. Oxidation of the organoboranes leads to a mixture of the corresponding alkoxy alcohols.

The hydroboration of allyl ethyl ether with a ratio of ether: BH_3 = 3:1 leads to a high yield of tris(3-ethoxypropyl) borane [62].

The introduction of a methyl group into the α-position of an allyl ether sharply changes the orientation of the addition of diborane: crotyl ethyl, crotyl phenyl, and crotyl benzyl ethers add boron predominantly (82–85%) in a β-position to the RO group [69]:

$$CH_3CH{=}CHCH_2OR$$

↑ 82—85%

Alkoxy groups, and also other functional substituents located in a γ-position to an unsaturated carbon atom, have a negligible effect on the orientation of the addition of diborane or tetraalkyldiborane, as compared with unsubstituted olefins. The yields of primary and secondary substituted alcohols formed by the hydroboration-oxidation of different 3-butenyl derivatives with the use of diborane and tetra(3-methyl-2-butyl)diborane are shown below [44].

B_2H_6: $CH_2{=}CHCH_2CH_2X$ — 73—89%; 11—13%

$R_4B_2H_2$: $CH_2{=}CHCH_2CH_2X$ — 81—98%; 1—2%

($X{=}Cl, OCH_3, OC_6H_5, SCH_3, OAc$).

5-Bromo-4-ethoxypentene reacts with tetrabutyldiborane to form (5-bromo-4-ethoxy-1-pentyl) dibutylborane (*92*), on the basis of which a method for the synthesis of 1,2-azaborepanes has been developed [70, 71]:

$$BrCH_2CH(OC_2H_5)CH_2CH{=}CH_2 \xrightarrow{(n\text{-}C_4H_9)_4B_2H_2} BrCH_2CH(OC_2H_5)(CH_2)_3B(C_4H_9)_2 \quad \mathbf{92}$$

Allyl methyl sulfide reacts with dimethyl sulfide-borane with the formation of cyclo(S—B)-3-methylmercaptopropylborane, which hydrolyzes in air [72]:

$$CH_2{=}CHCH_2SCH_3 + H_3B\cdot S(CH_3)_2 \longrightarrow$$

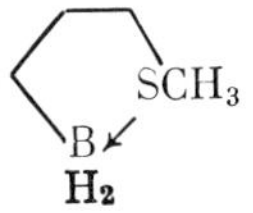

Pasto and Hickman [73] have investigated the hydroboration of 3-methoxycyclohexene. On the basis of the composition of the oxidation products, it was established that a β-addition of borane primarily takes place, with the formation of 81% of the *trans* (*93*) and

10% of the *cis* isomer (*94*), whereas α-addition takes place to the extent of 9%, with the formation of 6% of the *trans* isomer (*95*) and 3% of the *cis* isomer (*96*) (the total yield of the compounds *93–96* is 80–85%):

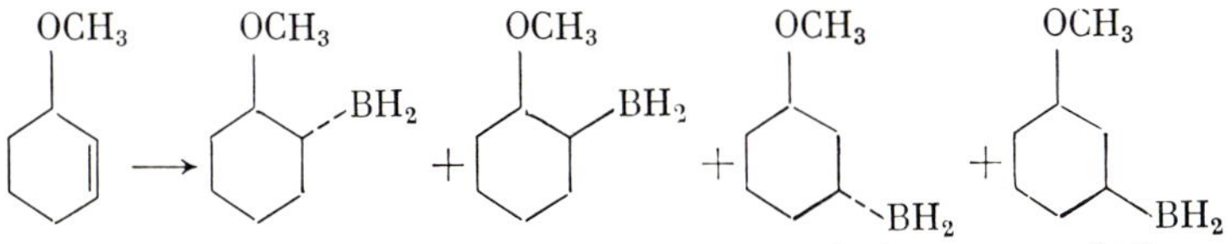

93 (81%) 94 (10%) 95 (6%) 96 (3%)

The methoxy group is similar to a chlorine atom in its orienting effect.

The hydroboration of 4-methoxycyclohexene leads to the following compounds [73]:

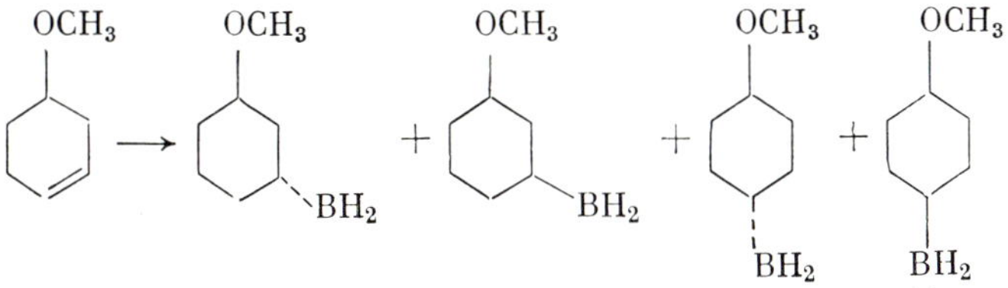

34% 22% 25% 19%

The action of diborane or tetra(3-methyl-2-butyl)diborane on 3-ethoxycyclopentene, followed by oxidation, forms the monoethyl ethers of *trans*-1,2-dihydroxycyclopentane (66%), *cis*-1,3-dihydroxycyclopentane (8%), *trans*-1,3-dihydroxycyclopentane (7–9%), and cyclopentanol (6–18%) [74].

Hydroboration-oxidation as a method for the hydration of unsaturated ethers has been used in carbohydrate chemistry. 5,6-Dideoxy-1,2-O-isopropylidene-α-D-xylohexofuran-5-enose (*97*) has been converted to 5-deoxy-1,2-O-isopropylidene-α-D-xylohexofuranose (*98*) by this method [75]:

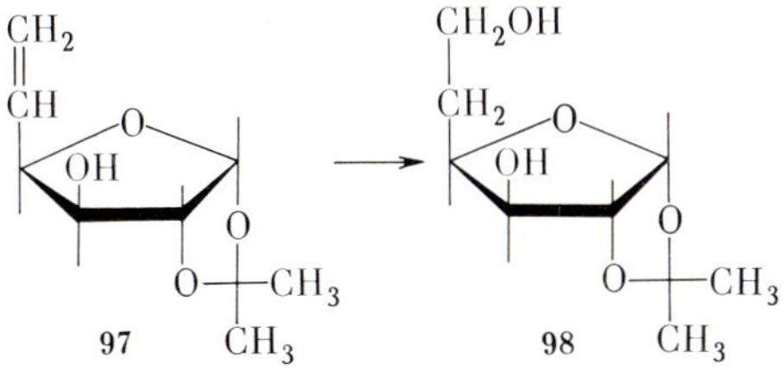

97 98

2,3-O-Isopropylidene-5,6-dideoxy-4,5-dehydro-β-methyl-D-*erythro*-hexofuranoside (*99*) forms 2,3-O-isopropylidene-6-deoxy-β-methyl-D-gulofuranoside (*100*) [76], and 1,2-O-isopropylidene-5,6-dideoxy-3-O-methylsulfonyl-α-D-xylo-5-hexosene (*101*) forms 1,2-O-isopropylidene-3-O-methylsulfonyl-5-deoxy-D-glucofuranose (*102*) [77]:

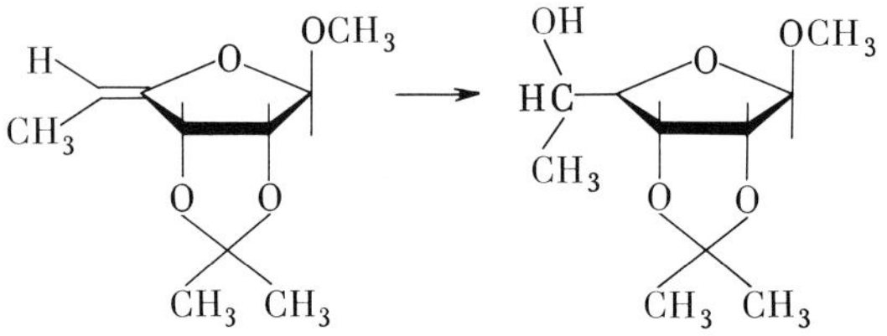

99 100

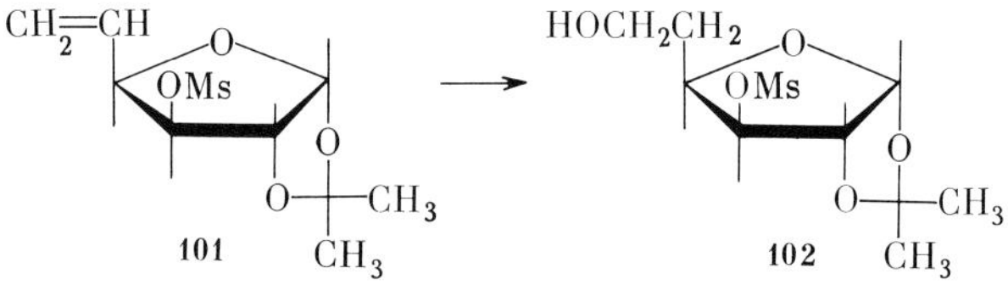

101 102

2,3-O-Isopropylidene-5,6-dideoxy-5,6-dehydro-β-methyl-D-ribo-hexofuranoside (*103*) has been converted to 2,3-O-isopropylidene-5-deoxy-β-methyl-D-ribohexofuranoside (*104*) by hydroboration-oxidation [77]:

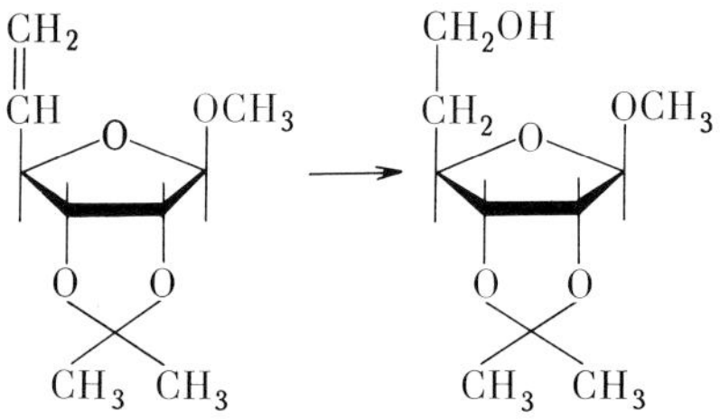

103 104

and 1,2:5,6-di-O-isopropylidene-3-deoxy-3-C-methylene-α-D-ribo-hexofuranose (*105*) has been converted to 1,2:5,6-di-O-isopropylidene-3-deoxy-3-C-hydroxymethyl-α-D-allofuranose (*106*)

[78]:

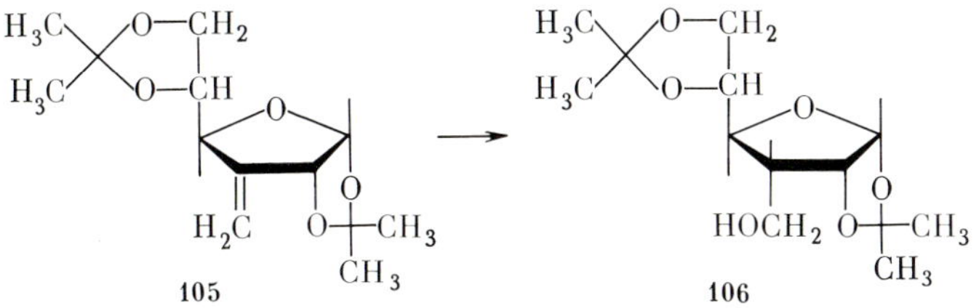

105 106

The conversions of unsaturated thioketals of the steroid series have been investigated [79]. The hydroboration-oxidation of 3-cycloethylenedithio-4-cholestene (*107*) leads to 5β-cholestan-3α-ol (*108*) and 5α-cholestan-4α-ol (*109*):

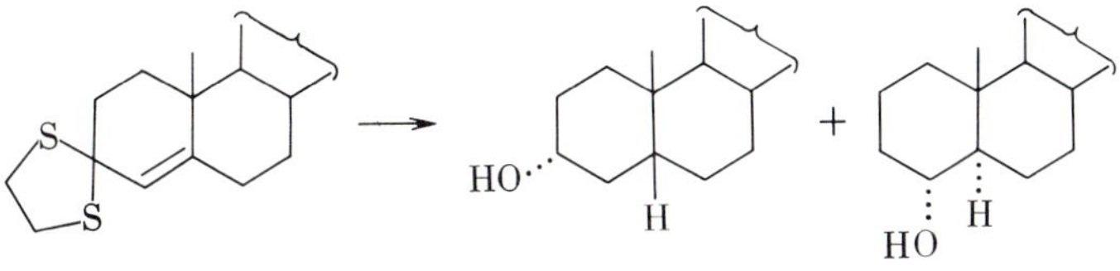

107 108 (32%) 109 (13%)

5α-Cholestan-7β-ol (38%) and 5α-cholestan-6α-ol (16%) are obtained from 7-cycloethylenedithio-5-cholestene.

3.3.2 Hydroboration of alkenyl esters. Synthesis of alcohols, diols, and diene hydrocarbons

The hydroboration-oxidation of unsaturated esters of acids, primarily esters of acetic acid, has been investigated in a number of papers.

The acetoxy group, in its effect on the direction of hydroboration, differs strongly from the alkoxy group, and the orientation of the addition of diborane to enol acetates depends to a considerable degree on the structure of the olefin group.

In the hydroboration of enol acetates (acetate: $BH_3 = 1.67 - 2.7$), 3–4 equivalents of borane are used per mole of the starting substance, depending on the temperature and on the structure of the hydroborating agent.

1-Acetoxybutene adds boron in the α-position to the extent of 30% [58]:

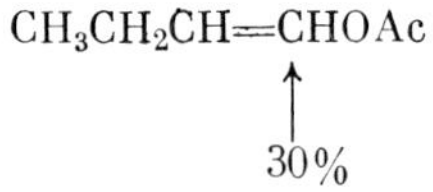

Branching of the chain in a β-position to the acetoxy group leads to a significant change in the orientation of the addition: the boron atom attacks the α-position almost exclusively. Thus, 1-acetoxy-2-methylpropylene (*110*), as a result of hydroboration-oxidation, is converted to isobutyraldehyde in high yield [80]:

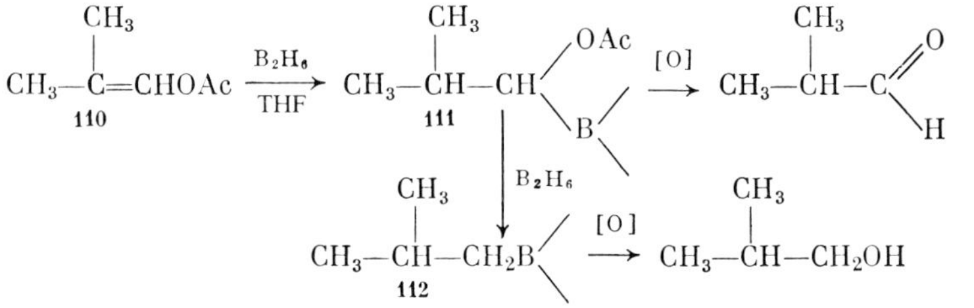

The α-acetoxyorganoborane *111*, at 40°C and in the presence of a large excess of diborane, is converted to the organoborane *112*, which is oxidized to isobutanol in 90% yield. Replacement of the acetoxy group by hydrogen apparently takes place as a result of a nucleophilic reduction of the α-acetoxyorganoborane *111* by diborane.

A number of alkenyl acetates have been subjected to hydroboration at 22°C in tetrahydrofuran with a ratio of acetate:BH_3 = 1:2.7 [81]. As a result of oxidation, isomeric alcohols were obtained in all cases, the ratio of which characterized the distribution of boron between the α- and β-positions:

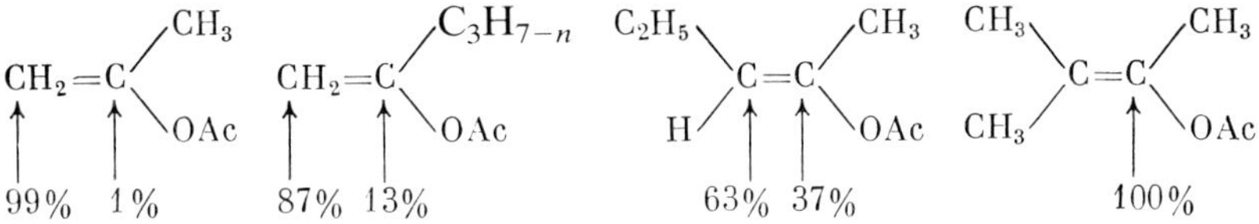

The hydroboration of enol acetates by tetra(3-methyl-2-butyl)-diborane in tetrahydrofuran has been investigated [82]. The enol acetates obtained from normal aldehydes form primary alcohols in a yield of 60–70% as a result of oxidation of the reaction products:

$$RCH{=}CHOAc \xrightarrow{R_2BH} \underset{\displaystyle BR_2}{RCHCH_2OAc} \longrightarrow RCH{=}CH_2 \xrightarrow[{[O]}]{R_2BH} RCH_2CH_2OH.$$

Isopropenyl acetate reacts analogously, whereas the enol acetates obtained from other ketones are inert toward tetraalkyldiborane.

The acetate of (*Z*)-9,11-dodecadien-1-ol (a component of a sex pheromone) has been obtained by the hydroboration-protonolysis of 11-dodecen-9-yn-1-ol acetate [226].

The hydroboration of $ArSO_2CH_2CH{=}CHCH_3$ with diborane, followed by oxidation, leads almost exclusively to $ArSO_2$-$CH_2CHOHCH_2CH_3$, whereas, when 9-borabicyclo-[3.3.1]nonane is used, the main product is $ArSO_2CH_2CH_2CHOHCH_3$ [227].

Cyclohexenyl acetate, under the action of diborane (ether: $BH_3 = 1:2$) in tetrahydrofuran at 0°C, followed by oxidation, is converted to *trans*-cyclohexane-1,2-diol (60%) and cyclohexanol (20%) [83, 84]:

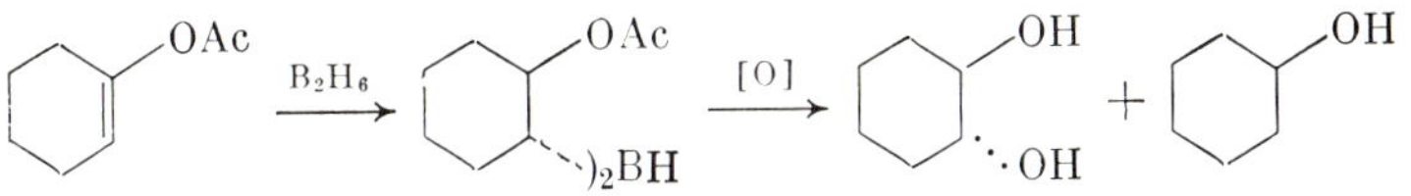

The formation of cyclohexanol is explained by the addition of borane in an α-position and subsequent intra- or intermolecular replacement of the acetoxy group by hydrogen, as has been shown by the production of 1,2-dideuterocyclohexanol (together with cyclohexene) in the deuteroboration of acetoxycyclohexene [85]:

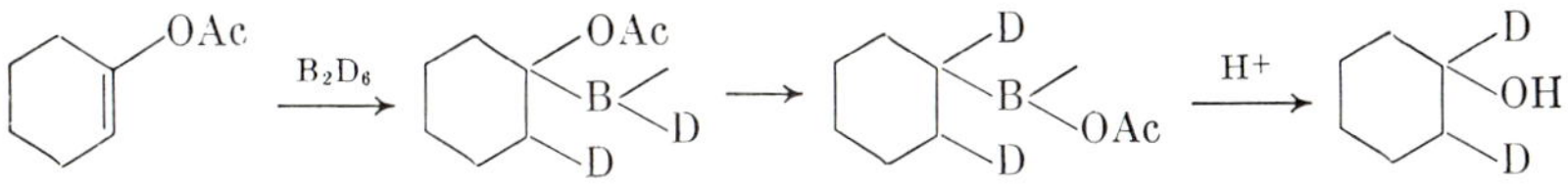

According to some data [84], 1-acetoxycyclopentene forms only cyclopentanol (80%) upon hydroboration-oxidation, whereas, according to other data [58], it forms *trans*-cyclopentanediol (36%), *trans*-2-ethoxycyclopentanol (6%), and cyclopentanol (44%).

The reduction of 16-androsten-3β,17-diol diacetate (*113*) by diborane leads to an 80% yield of androstan-3β,16α,17β-triol (*114*)

[84, 86]:

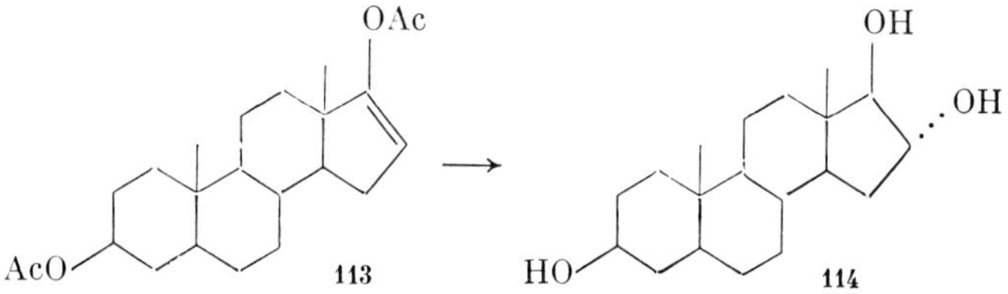

113 → 114

The hydroboration-oxidation of 2-cholesten-3-ol acetate (*115*) forms 3α-cholestanol, 3β-cholestanol, and 2α,3β-cholestanediol [84].

In the protolysis of the hydroboration products of unsaturated esters of the steroid series, elimination of the unsaturated compound takes place. Thus, as a result of the addition of diborane to the compound *115* and subsequent heating of the organoborane with propionic acid, 5α-cholest-2-ene (*116*) is formed [60]:

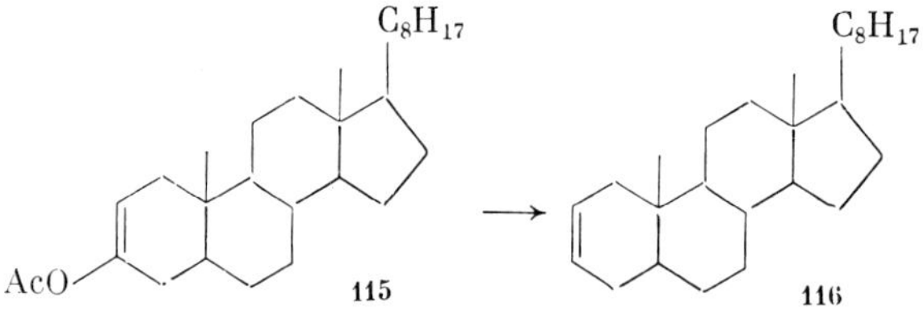

115 → 116

The oxidation of the hydroboration products of compound *115* forms 5α-cholestan-2α,3β-diol [87]. The latter diol is also formed by the hydroboration-oxidation of 3-benzoxy-5α-cholest-2-ene [58].

1,3,5,16-Estratetraene-3,17-diol diacetate (*117*) is converted to estriol (*118*) by hydroboration-oxidation [86]:

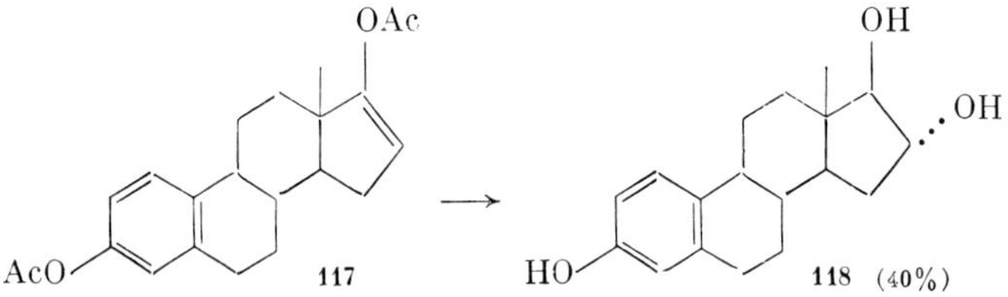

117 → 118 (40%)

The orientation of the addition of diborane to allyl esters of acids depends on the nature of the acid group, as can be seen from the schemes presented [67]:

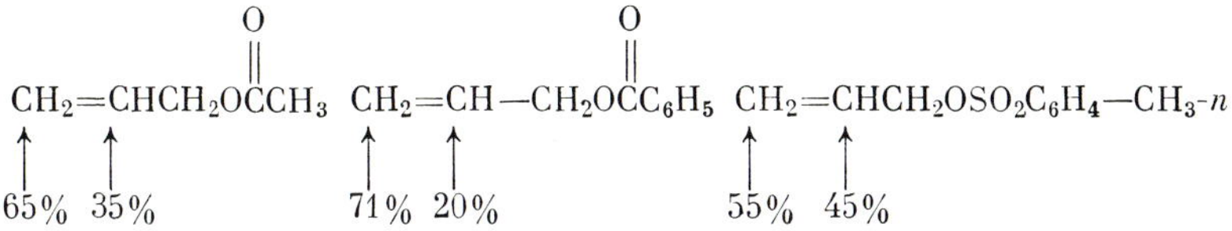

In the hydroboration of allyl acetate by tetra(3-methyl-2-butyl)diborane, the boron atom is almost completely localized on the terminal carbon atom.

Crotyl acetate adds 90% of the boron in the 2 position [69].

The action of diborane or tetra(3-methyl-2-butyl)diborane on phytyl acetate forms 2-phytanol with a small impurity of 1-phytanol (isomer ratio 97:3) [88]:

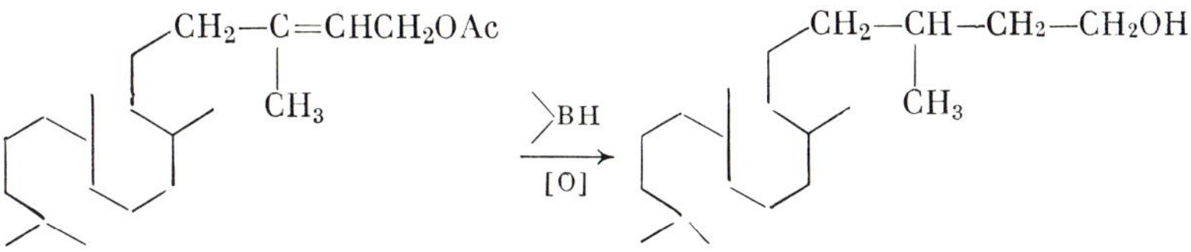

Hydroboration-oxidation has been used in the synthesis of *dl*-α-lipoic acid from 3-acetoxy-1,7-octadiene [228].

The hydroboration of geraniol acetate with diborane leads to 1-acetyl-2-isopropyl-5-methylborepane, which is converted to *trans*-*p*-menthan-4-ol by the action of bromine in the presence of water under exposure to light followed by oxidation [229].

The hydroboration-oxidation of linalyl acetate has been studied [230].

The hydroboration-oxidation of 3-acetoxycyclopentene forms predominantly *trans*-1,2-cyclopentanediol [74].

The hydroboration of *cis*-verbenyl acetate (*119*) proceeds from the side opposite to the *gem*-dimethyl group, with the boron attacking the β-position with respect to the acetoxy group. As a result of oxidation of the reaction products, a mixture of compounds containing $3\alpha,4\beta$-dihydroxy-*cis*-pinane (*120*), 3α-hydroxy-4β-acetoxy-*cis*-pinane (*121*), isopinocampheol (*122*), and *cis*-δ-pinene (*123*) is obtained [89]:

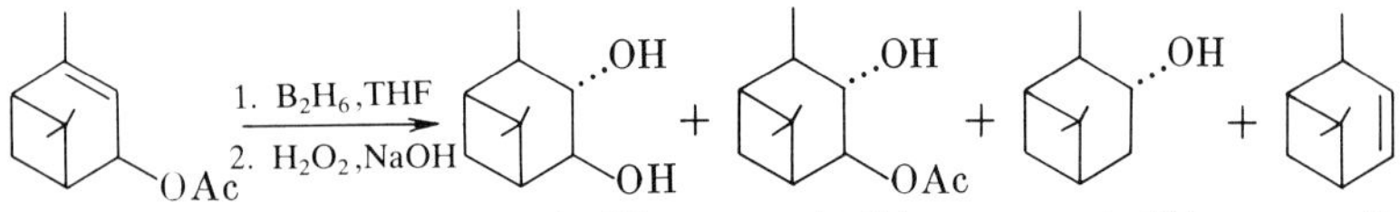

119 120 (33%) 121 (27%) 122 (10%) 123 (12%)

The reaction of *trans*-verbenyl acetate with diborane proceeds very slowly and, after oxidation, leads to low yields of *trans*-verbenol and pinane-3,4-diol monoacetate (6%) [90]. *Cis*-pin-3-en-2-ol acetate is converted by hydroboration-oxidation to *trans*-verbenyl acetate, *trans*-verbenol, and pinane-3,4-diol monoacetate [90]. The hydroboration-oxidation of 3α,5α-etheno-3β-cholestanol acetate produces 3α,5α-ethanocholestan-3β,3aβ-diol (*124*) and 3α,5α-ethanocholestan-3β,5aβ-diol (*125*) [91]:

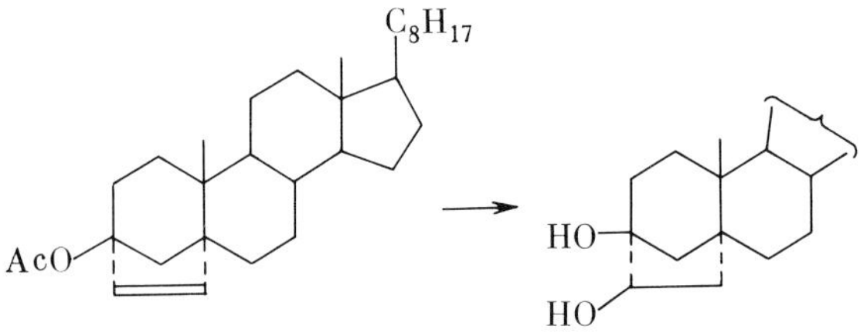

124 (23%)

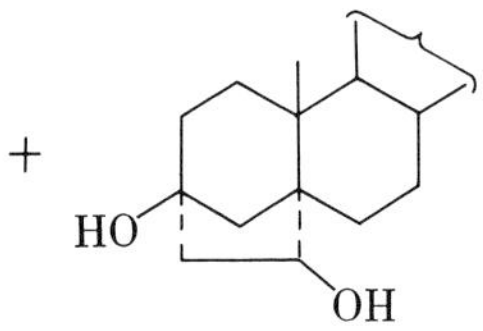

125 (13%)

The characteristic conversions of unsaturated cyclic systems with the use of hydroboration were discovered by Marshall, who summarized the results of his investigations in a review [92]. It was found that the methanesulfonate *126*, as a result of hydroboration, followed by treatment of the hydroboration products with aqueous sodium hydroxide solution, is converted to a mixture (85 : 15) of *trans*,*trans*-1-

methyl-1,6-cyclodecadiene (*127*) and methyltricyclodecane (*128*) in a yield of 90%. These compounds are obtained by the mechanism indicated below, according to which both stereoisomeric secondary borates adopt an antiperiplanar conformation, favoring consistent fragmentation. The tricyclic hydrocarbon *128* is formed as a result of 1,3-elimination. Both reaction centers undergo an inversion of the configuration during the reaction [93, 94]:

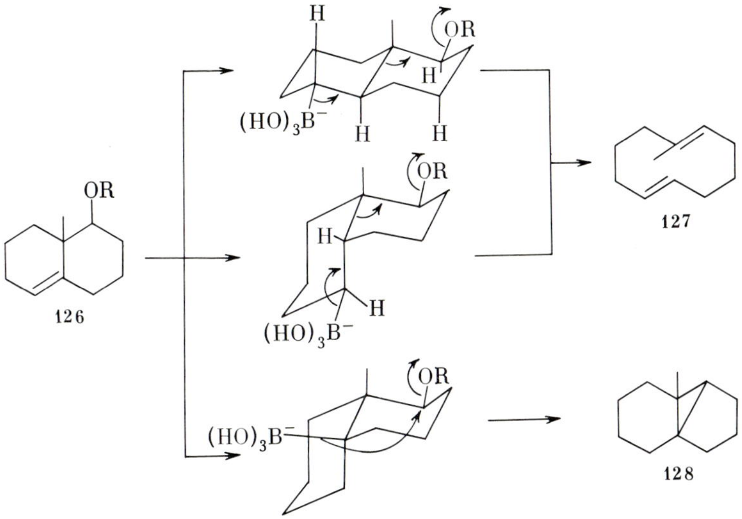

The methanesulfonate *129*, when treated consecutively with diborane and alkali, forms a 93:7 mixture of 1-methyl-*trans*,*trans*-1,5-cyclodecadiene (*130*) and the tricyclic hydrocarbon *128*, with a total yield of 90% [95, 96]:

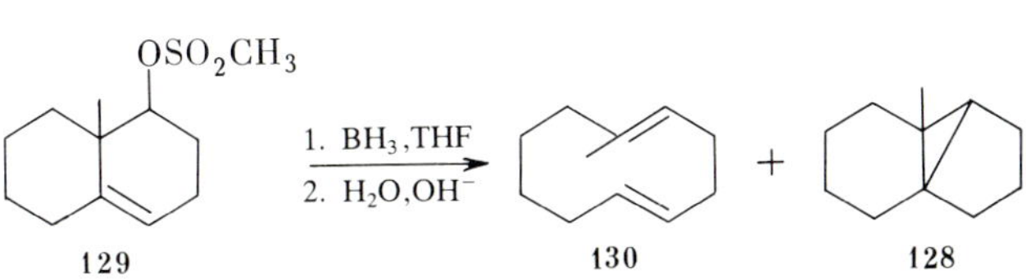

6-Oxo-9β-methyl-5(10)-octalin-1β-ol methanesulfonate is converted by the action of diborane and sodium methoxide to the hydrocarbon *127* in 60% yield [97]:

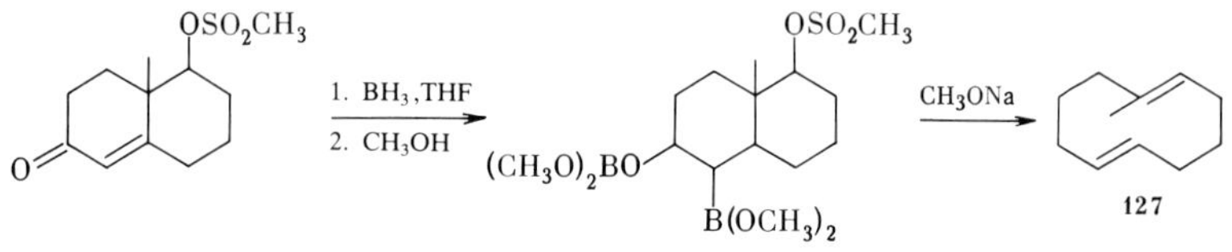

The application of the above method of fragmentation to the methanesulfonates of the cyclohexene series *131–133* leads to the acyclic diene hydrocarbons *134–137* [98]:

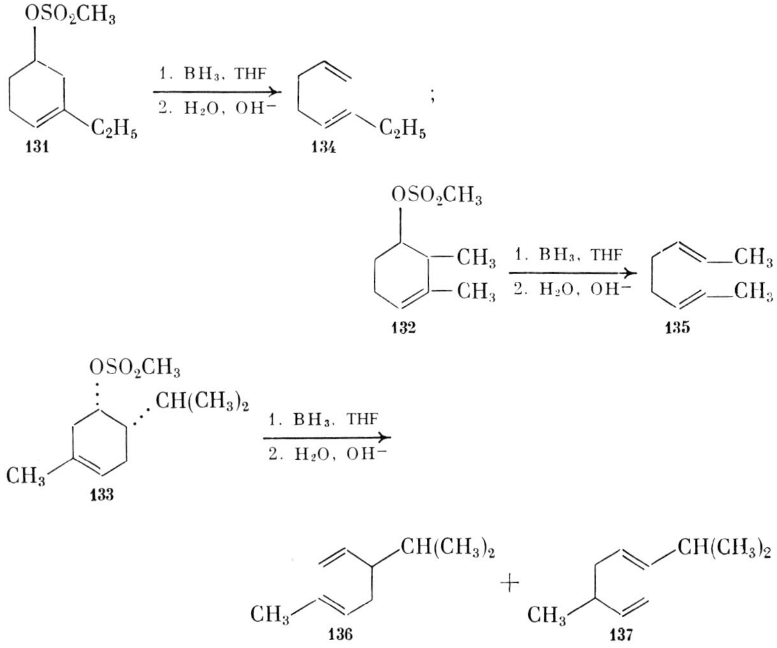

3.3.3 Hydroboration of unsaturated alcohols and mercaptans, aldehydes, ketones, and alkenyl borates. Synthesis of diols and triols

In the action of hydroborating agents on unsaturated alcohols, aldehydes, and ketones, a reaction initially takes place at the functional group, with the formation of the corresponding unsaturated borates, which later add compounds with B—H bonds at the multiple carbon-carbon bond.

In most papers on the hydroboration of unsaturated oxygen-containing compounds, the reaction products undergo direct oxidation and, on the basis of the structure of the alcohols and diols obtained in this process, the orientation and stereochemistry of the addition of the hydroborating agent to the unsaturated system are determined. In some investigations, the organoborane compounds formed were isolated in the individual state.

Hydroboration-oxidation is a valuable preparative method for the hydration of oxygen-containing acyclic and alicyclic olefins, particularly when applied to natural compounds.

a) Compounds of the acyclic and alicyclic series

Köster and Schomburg [99] obtained 2-alkyl-1,2-oxaborolanes (*138*) by the hydroboration of triallyl borate with a mixture of trialkylamine-borane and trialkylborane:

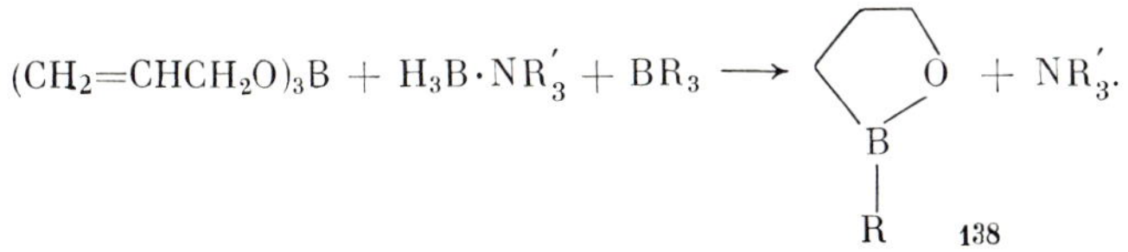

138

Mikhailov and Dorokhov [27, 100] found that the compounds *138* are obtained by heating an allyl dialkylborinate (*139*) with trialkylboranes. A transalkylation of the trialkylborane with the ester *139* (as with the higher olefin) initially takes place, and the borane *140* formed then cyclizes to *138*, with the elimination of BR_3:

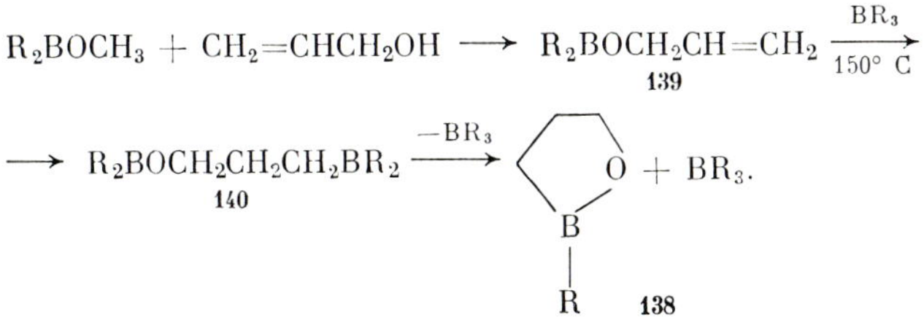

139, 140, 138

2-Alkyl-1,2-oxaborolanes (*138*) are also obtained by heating trialkylboranes with allyl alcohol [27, 100]:

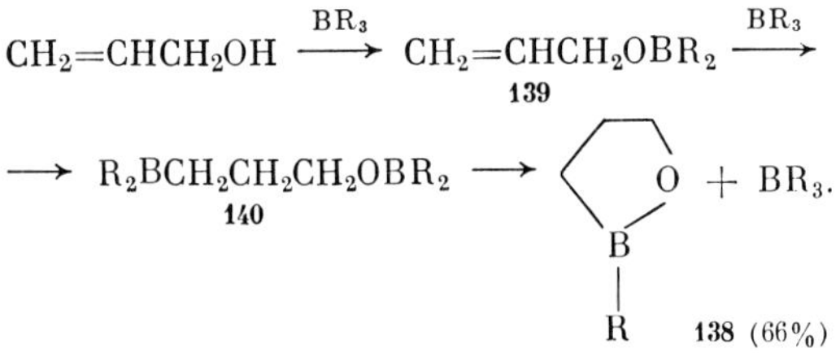

The action of tetrahexyldiborane on 1-hexen-5-ol forms 2-hexyl-7-methyl-1,2-oxaborepane (*141*), with an impurity of 2.5% of 2-hexyl-3,6-dimethyl-1,2-oxaborinane (*142*) [101]:

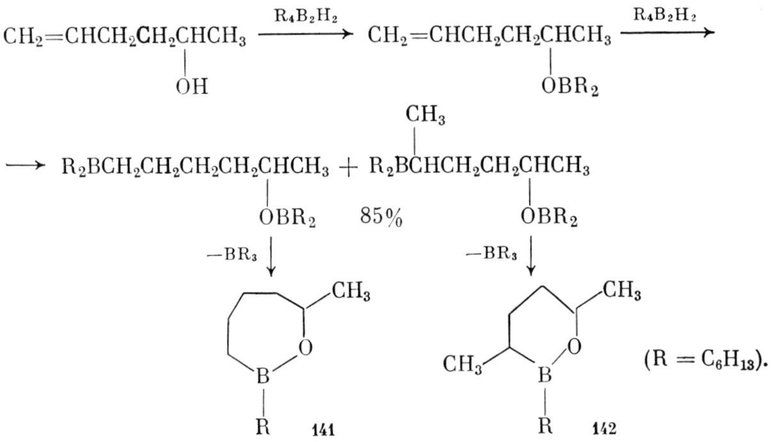

The compounds *141* and *142* are also formed in the hydroboration of allylacetone with tetraalkyldiborane [101]. In this case, the keto group is initially reduced, the unsaturated borate then adds dialkylborane, and the borate-boranes cyclize:

$$CH_2{=}CHCH_2CH_2C({=}O){-}CH_3 \xrightarrow{R_4B_2H_2} \underset{81\%}{(141) + (142)} \qquad (R = C_6H_{13}).$$

Oxidation of a mixture of *141* and *142* forms 1,5-hexanediol with an impurity of 2.5% of 2,5-hexanediol.

2-Alkyl-6,7-benzo-1,2-oxaborepanes (*143*) can be obtained by the action of tetraalkyldiborane on 2-allylphenol [102]:

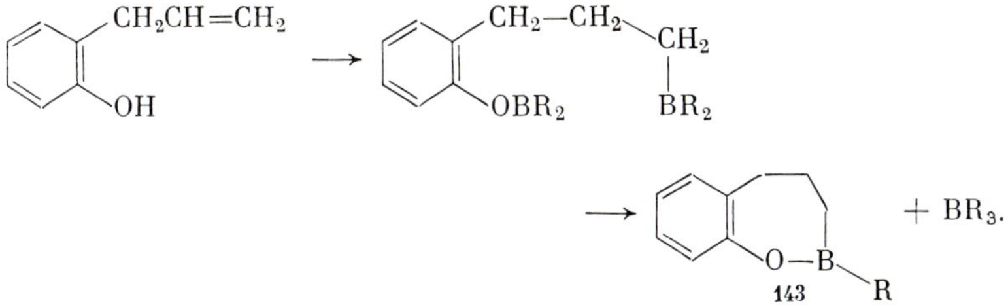

2-Butoxy-1,2-oxaborolane (*144*) has been synthesized by the action of trialkyl borate on the polymeric products formed by the hydroboration of triallyl borate or allyl alcohol [103]:

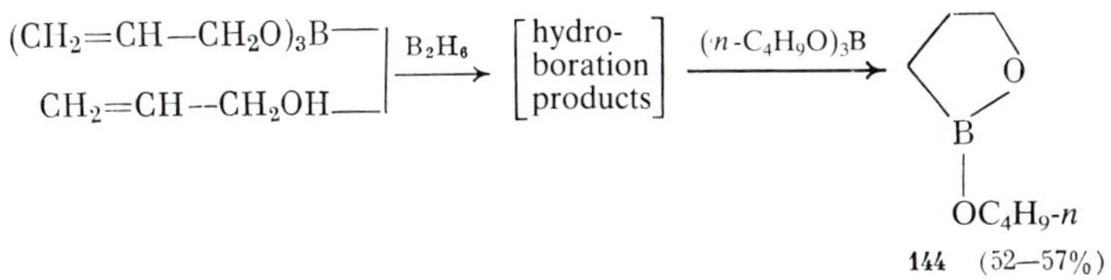

The oxidation of *144* leads to individual 1,3-propanediol.

Hydroboration of 4-penten-1-ol, followed by heating with tributyl borate, leads to a mixture of 2-butyl-1,2-oxaborepane (*145*), 2-butoxy-3-methyl-1,2-oxaborinane (*146*), and 2-butoxy-3-ethyl-1,2-oxaborolane (*147*) [104], in a ratio of 76:23:1 (total yield: 35%):

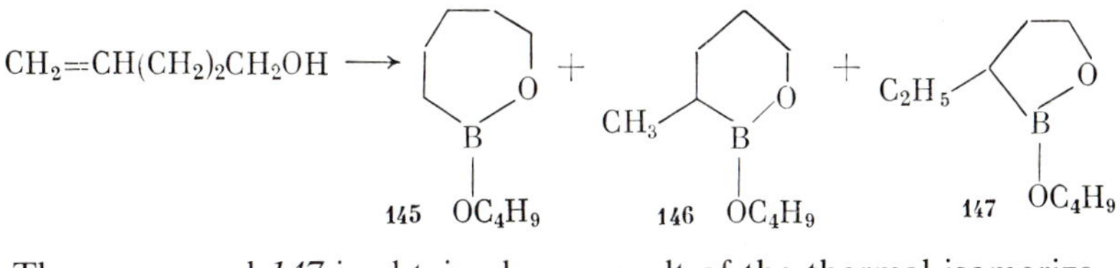

The compound *147* is obtained as a result of the thermal isomerization of *146*.

The hydroboration of 3-, 6-, and 10-undecen-1-ol and the thermal isomerization of the organoboranes obtained in this process have been investigated. The authors assume that the isomerization proceeds with the formation of an intermediate intramolecular π-complex [104].

The action of 1,2-dialkyldiboranes or 1,2-diaryldiboranes on allyl mercaptan forms 2-alkyl- or 2-aryl-1,2-thiaborolanes (*148*) [105, 106]:

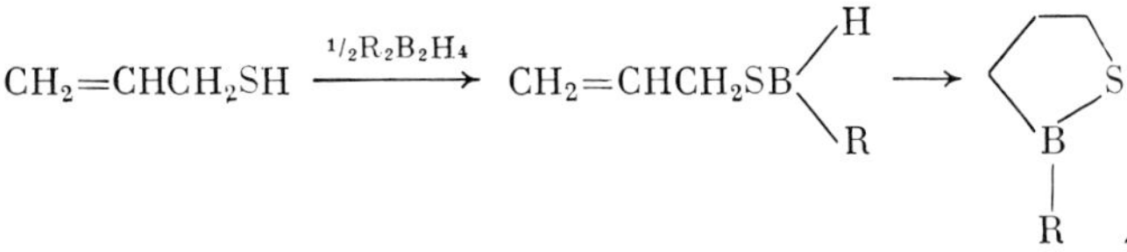

148

($R = n\text{-}C_4H_9$, *iso*-C_5H_{11}, C_6H_5).

The thiaborolanes (*148*) have also been synthesized from allyl mercaptan, triethylamine-borane, and trialkylborane [24]:

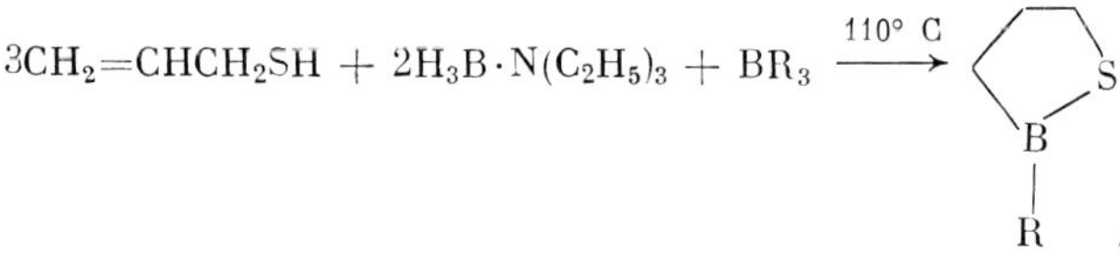

148 (50%)

2-Alkyl-1,2-thiaborolanes react with aldehydes and ketones with the formation of 2,4-disubstituted derivatives of 1,3,4-thiaoxaborepane [107]:

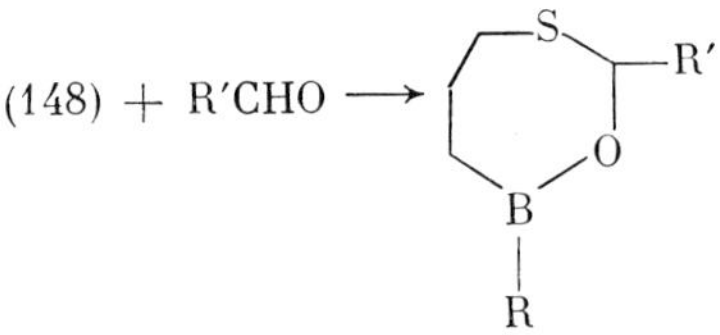

Phytol is converted to the diol *149* and the primary alcohol *150* in an approximate ratio of 3:1 by hydroboration-oxidation [88]:

$$RCH{=}CHCH_2OH \xrightarrow[2.\ [O]]{1.\ B_2H_6,\ THF} \underset{\mathbf{149}\ (75\%)}{RCH_2\overset{OH}{\overset{|}{C}}HCH_2OH} + \underset{\mathbf{150}\ (22\%)}{RCH_2CH_2CH_2OH}$$

($R = C_{17}H_{35}$).

The hydroboration of 4-substituted derivatives of 3-butene ($CH_2{=}CHCH_2CH_2X$, where X = OH, OMe, OAc, Cl, CN) with 9-borabicyclo[3.3.1]nonane has been investigated [231].

Cinnamic alcohol is converted by treatment with diborane and subsequent oxidation to a mixture of 3-phenyl-1,2- and 3-phenyl-1,3-propanediols (2:1) [108]; β-methylcinnamic alcohol forms *threo*-3-

phenyl-2-methyl-1,3-propanediol in 80% yield [109]; crotyl alcohol and crotyl borate form a mixture of 1,2- and 1,3-butanediols and butanol [69]; and crotonaldehyde forms predominantly 1-butanol [110, 111]. Tris(3-cyclopentenyl) borate [69] and 3-cyclopentenyl bis(3-methyl-2-butyl)borinate [74] are converted to cyclopentanediols and cyclopentanol by the above method.

Klein, Dunkelblum, et al. [112–114] carried out a systematic study of the action of different substituents and borate groups on the orientation and stereoselectivity of the hydroboration of substituted 2-cyclohexenones and 2-cyclohexenols. It was found that the borate *151*, obtained by the action of diborane on 2-cyclohexenone, is converted mainly to the *trans*-1,2-diol *152*, together with the *cis*-1,2-diol *153* and the *cis*- and *trans*-1,3-diols *154* and *155* [112].

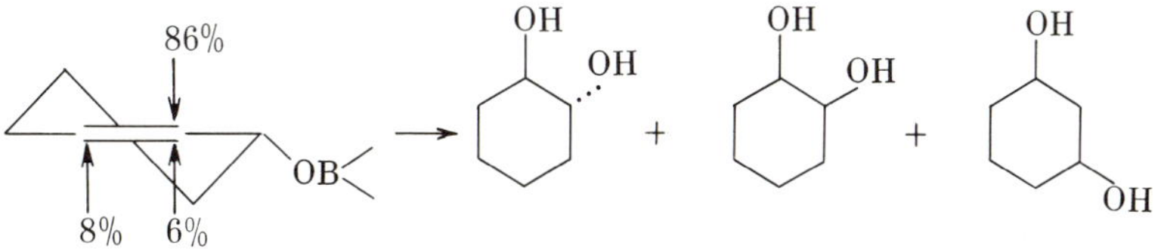

151 152 (86%) 153 (6%) 154 (8%)

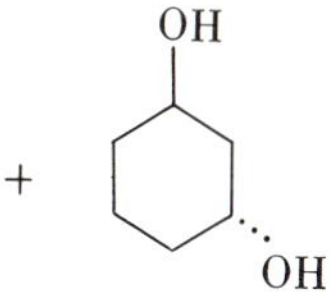

155

The hydroboration of β-substituted 2-cyclohexenones and 2-cyclohexenols proceeds stereospecifically to form *trans*-1,2-diequatorial cyclohexanediols in a yield of 56–70%.

Thus, the hydroboration-oxidation of isophorone (*156*) forms a 65% yield of the *trans*-diequatorial 1,2-diol *157* and a mixture of isomeric monoalcohols (15%) [113–115].

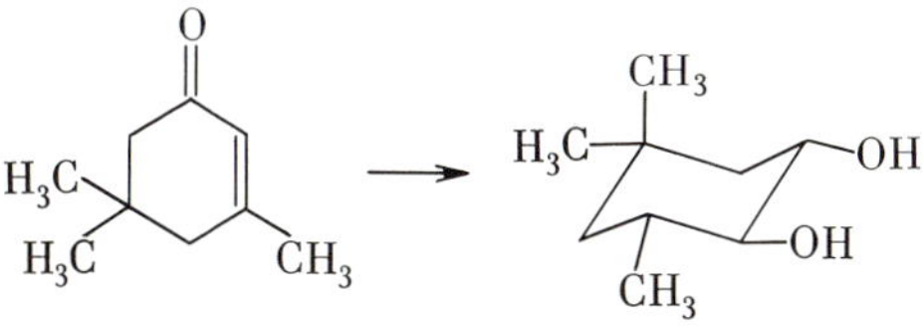

156 157 (65%)

3-Methyl-2-cyclohexenone (*158*) forms 70% of the diol *159* and 17% of monoalcohols [113, 114]:

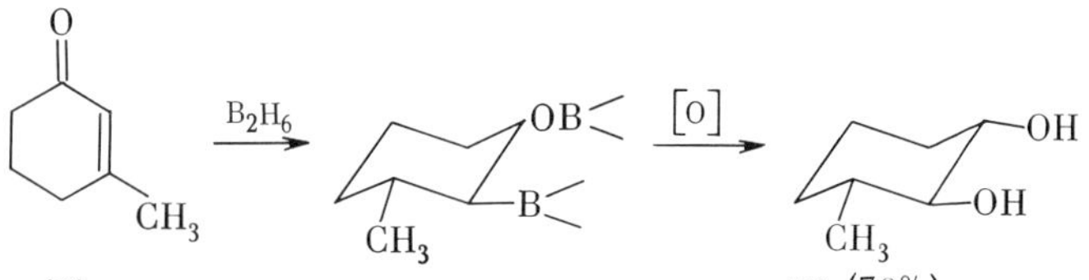

The hydroboration-protonolysis of some alkylcyclohexenones has been investigated [232].

Piperitone (*160*) forms a mixture of the *trans* diequatorial diols *161* and *162* in almost equal quantities [114]:

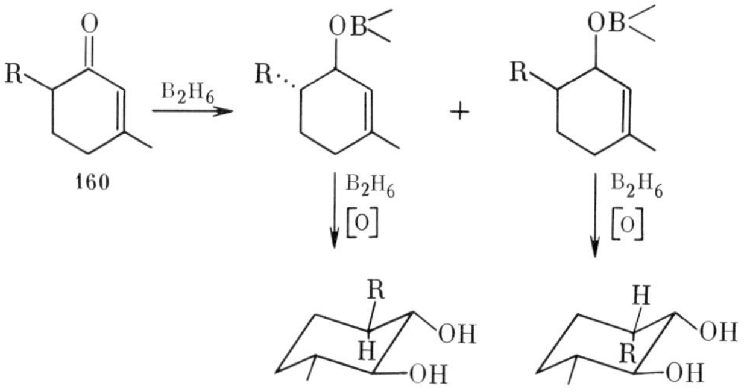

(R = *iso*-C_3H_7).

Ketones and pseudoequatorial alcohols with a substituent in the β-position to a saturated carbon atom form mainly diequatorial *trans*-1,2-diols and a small quantity of 1,3-diols [112].

5-Phenyl-2-cyclohexenone (*163*) and *cis*-5-phenyl-2-cyclohexenol (*164*) form the same pseudoequatorial borate *165* in the first stage of the reaction. The formation of *165* from the ketone *163* is in agreement with the observation that lithium aluminum hydride reduces sterically unhindered cyclohexen-2-ones to the thermodynamically more stable pseudoequatorial alcohols [116]. Further conversions of the borates *165* lead to the *trans*-1,2-diol *166* and the stereoisomeric 1,3-diols *167* and *168*, indicating the high stereospecificity of the attack on the 2 position and the low stereospecificity of that on the 3 position:

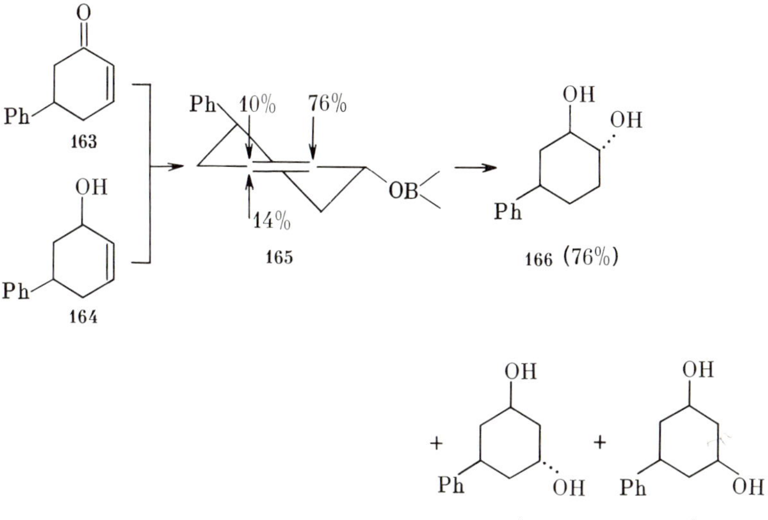

163, 164, 165, 166 (76%), 167 (10%), 168 (14%)

The effect of a pseudoaxially located allylborate group (*170*) on the stereoselectivity of the hydroboration is demonstrated by the example of *trans*-5-phenyl-2-cyclohexenol (*169*), which forms the *trans*-1,2-diol *171* and the 1,3-diols *168* and *172*:

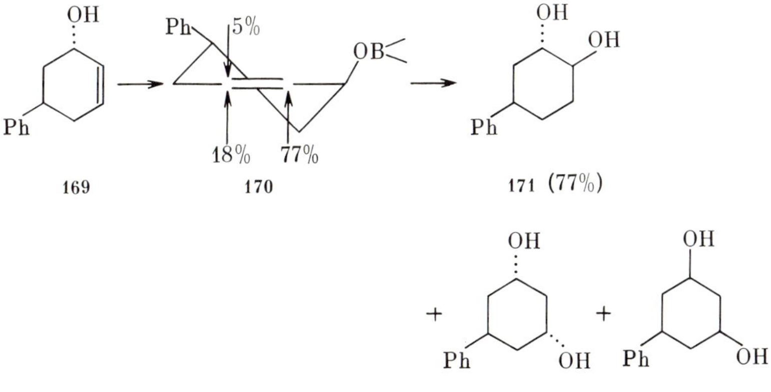

169, 170, 171 (77%), 172 (5%), 168 (18%)

The steric effect of the bulky *tert*-butyl group shows up in the hydroboration of the borate *173* (obtained from 5-*tert*-butyl-2-cyclohexenone or *cis*-5-*tert*-butyl-2-cyclohexenol), which is converted to the *trans*-1,2-diol *174* and only one 1,3-diol (*175*), with a total yield of 56%:

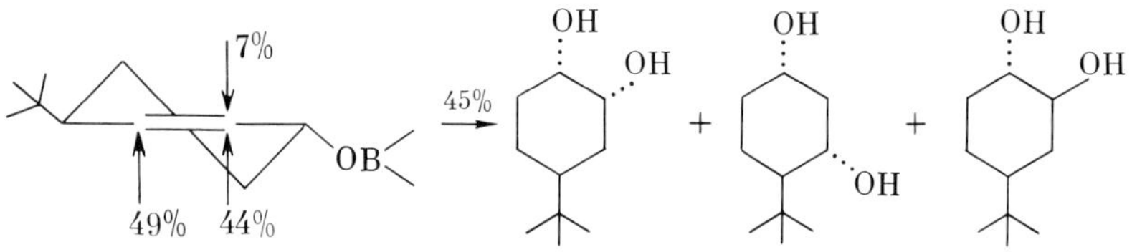

176 177 (44%) 178 (49%) 179 (7%)

The effect of the *tert*-butyl radical shows up particularly strongly in the 4 position. In the case of the borate *176*, obtained from 4-*tert*-butyl-2-cyclohexenone or *trans*-4-*tert*-butyl-2-cyclohexenol, the effect of this radical exceeds the trans-orienting effect of the borate group and, as a result, the main reaction products (total yield: 45%) are *trans*-4-*tert*-butyl-*cis*-1,2-cyclohexanediol (*177*) and *trans*-4-*tert*-butyl-*cis*-1,3-cyclohexanediol (*178*), whereas only 7% of the *trans*-1,2-diol *179* is obtained:

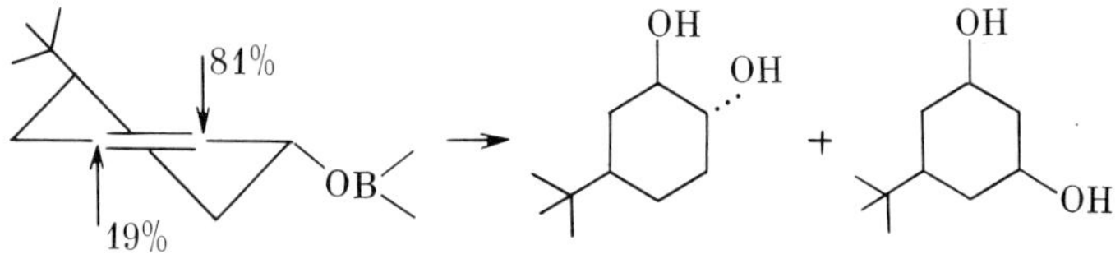

173 174 (81%) 175 (19%)

The hydroboration-oxidation of carvenone forms *trans*-2-hydroxyisomenthol [115]:

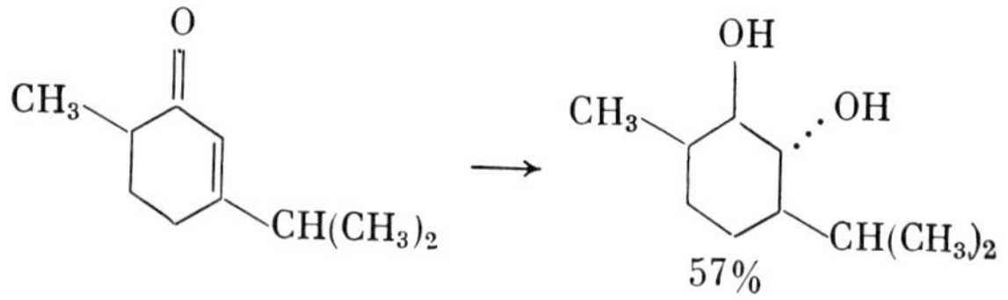

2,3,4,4-Tetramethyl-2-cyclohexenol forms the *trans*-1-2-diol *180* and the *trans*-1,3-diol *181*:

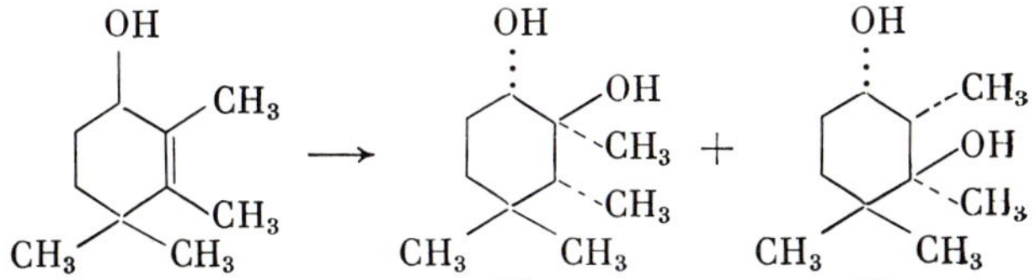

180 181

3,4,5,5-Tetramethyl-2-cyclohexenol (*182*) forms the *trans*-1,2-diols *183* and *184* [117]:

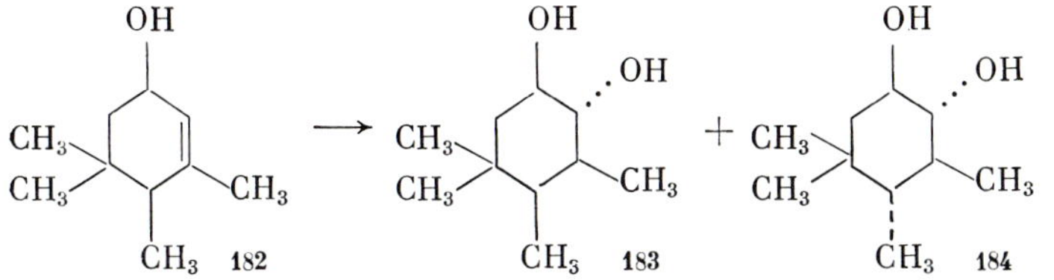

182 183 184

As a result of the conjugate addition of dimethylcopper lithium or methylmagnesium bromide to 2-cyclohexenone, followed by hydroboration-oxidation, the diols *185* and *186* are obtained in a ratio of 87:13, with a yield of 45–55% [118].

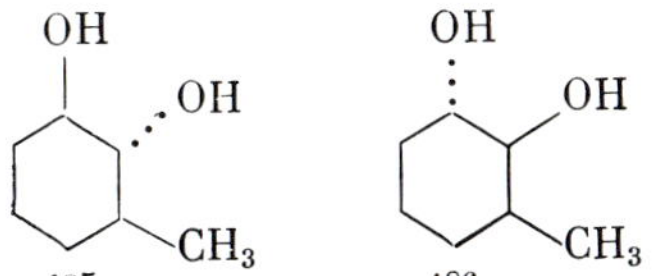

185 186

Analogous conversions have been carried out with isophorone.

A mixture of isomeric 2-α-hydroxybenzylcyclohexanols (7:3) is obtained from 2-benzylidenecyclohexanone by the above method [119]:

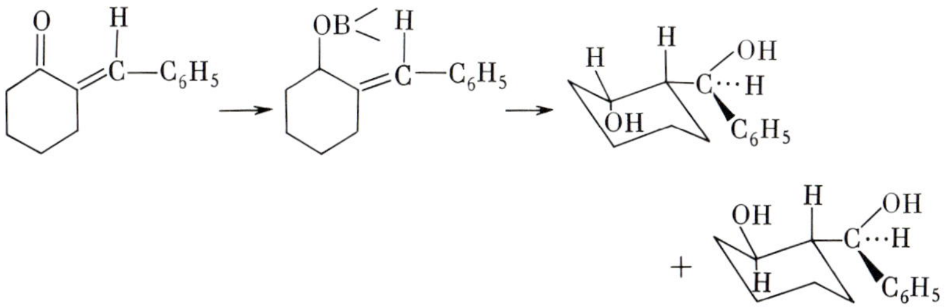

Diphenylcyclopropenone undergoes hydroboration-oxidation to form benzylacetophenone, and protolysis of the hydroboration product with butyric acid produces 1,2-diphenylcyclopropane. A possible reaction mechanism, which is in agreement with all experimental data, is presented in the scheme below [120]:

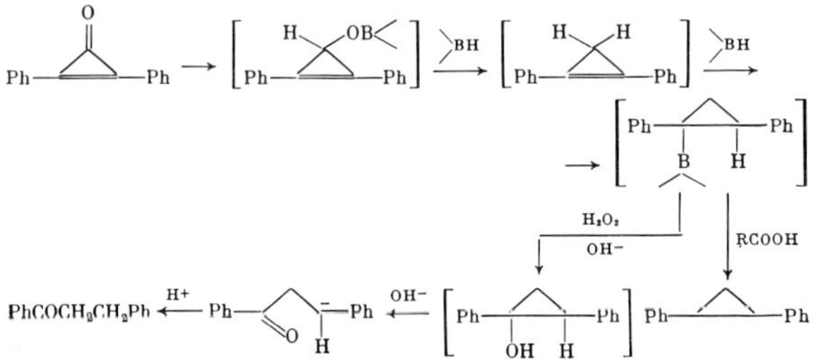

Diphenylketene is converted to diphenylacetaldehyde [121].

10-Methyl-1(9)-octalin-2-one is converted by hydroboration and subsequent treatment with acetic acid and acetic anhydride to the *trans*-diequatorial diacetate *187* [122, 123]:

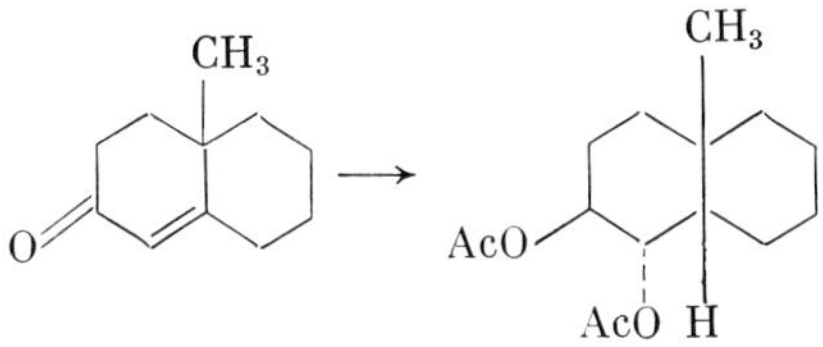

187 (26%)

1β-Acetoxy-9β-methyl-5(10)-octalin-6-one (*188*) is converted by the action of diborane, followed by heating with acetic anhydride, to a 3:1 mixture of *trans*- (*189*) and *cis*-1β-acetoxy-9β-methyl-5-octalins (*190*) [122]:

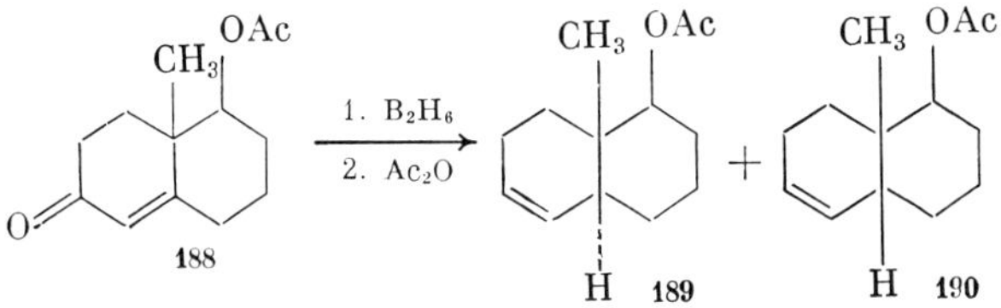

188 189 190

9-Methyl-5(10)-octalin-1,6-dione forms the *cis* isomer *190* under analogous treatment [122], rather than the *trans* isomer *189*, as was assumed in [123].

δ-Cadinol (*191*) is hydrated to cadinan-3,8-diol (*192*) [124]:

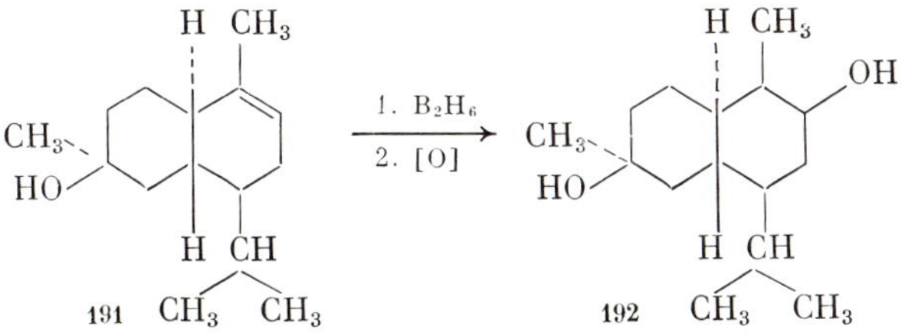

191 192

From the stereochemical viewpoint, the behavior of 2α- (*193*) and 2β-hydroxybicyclo[3.3.1]non-3-ene (*194*) toward diborane is of interest [125]. The equatorial β-isomer *193* is attacked from the β-side only in the 3 position, forming 2α, 3β-dihydroxybicyclo [3.3.1]-nonane (*195*), while the axial β-isomer *194* undergoes attack on the 3 position from the α-side and on the 4 position from the β-side, being converted to 2β,3α- (*196*) and 2β,4β-dihydroxybicyclo[3.3.1]-nonane (*197*), respectively:

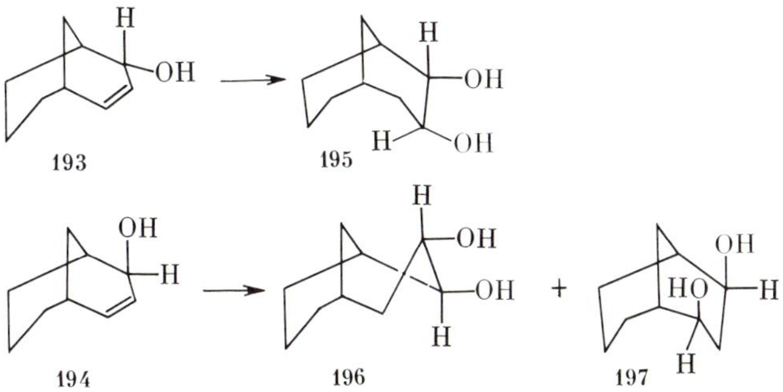

193 195 194 196 197

b) Compounds of the terpene series

A number of investigations of the hydroboration of unsaturated alcohols and ketones of the terpene series have been carried out.

The hydroboration of linalool, geraniol, and citral with diborane and disiamylborane has been investigated [233].

The addition of diborane to myrtenol (*198*) takes place from the side removed from the *gem*-dimethyl group, in a β-position to the substituent, with the formation of the diol *199* [126, 127]:

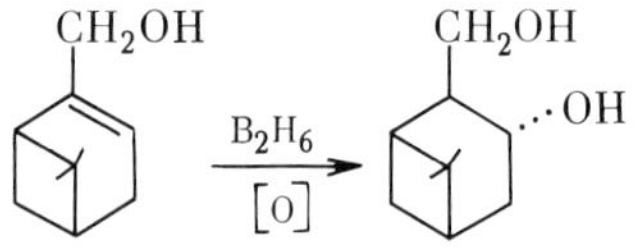

198 199

It is interesting that, in the case of the terpene alcohol *200*, which is isomeric with myrtenol, the boron atom adds to the more substituted carbon of the double bond. The borate-borane *201* obtained undergoes partial elimination, as a result of which the alcohols *203* and *204* are formed along with the diol *202* [128, 129]:

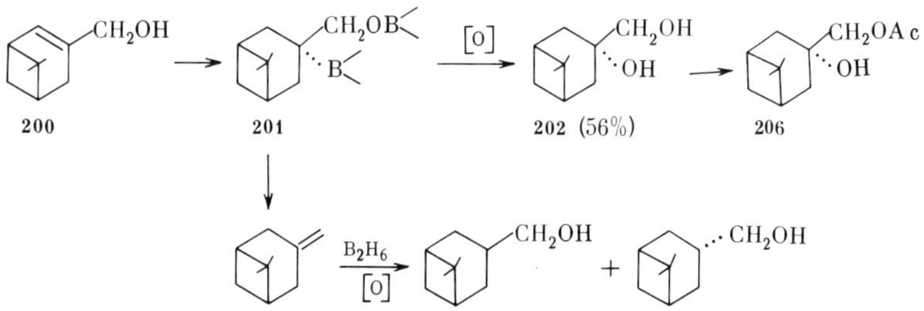

200 201 202 (56%) 206

205 203 (15%) 204

trans-Pinocarveol (*207*) forms two diols, with trans (*208*) and cis (*209*) locations of the substituents, in a ratio of 45:55. From this it follows that the borate group has a significant steric effect, competing with the action of the *gem*-dimethyl group, and the attack of the diborane is no longer stereospecific [127]:

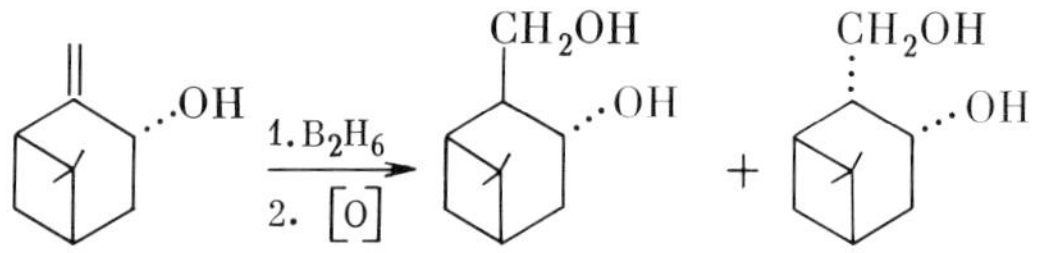

207 208 209

The use of tetra (3-methyl-2-butyl)diborane as a hydroborating agent radically changes the stereochemical course of the reaction: only a diol with a cis location of the substituents (*209*) is obtained.

Stereoselective attack from the side of the *gem*-dimethyl bridge with the use of diborane is observed in the case of the alcohol *210*, which is isomeric with *trans*-pinocarveol [128]:

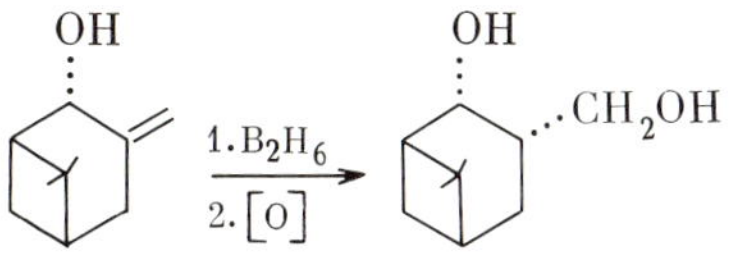

210

The hydration of 3-methylene-*trans*-nopinol (*211*) to the diol *212* also proceeds stereoselectively:

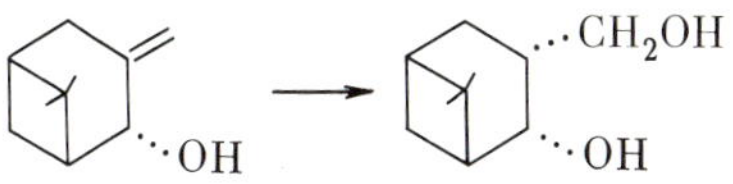

211 212

whereas 3-methylenenopinone (*213*) forms the diols *214* and *215*, since the attack of the carbonyl group proceeds both from the side of the *gem*-dimethyl bridge and from the opposite side [129]:

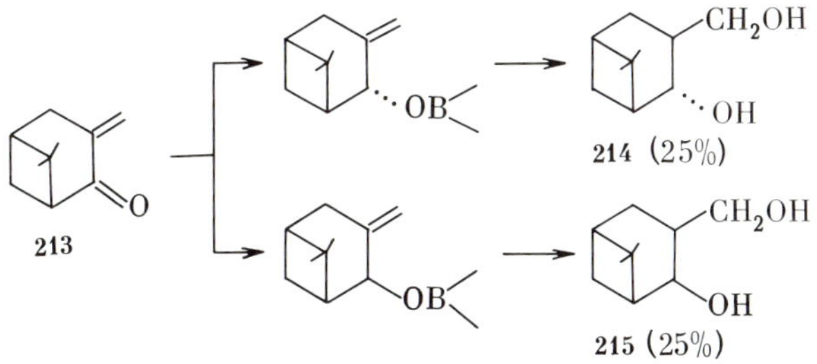

The attack of the carbonyl group of pinocarvone (*216*) is also not stereospecific and leads to the borates *217* and *218* which, as a result of further hydroboration, form the diol *219* and the diols *220* and *221*, respectively [129]:

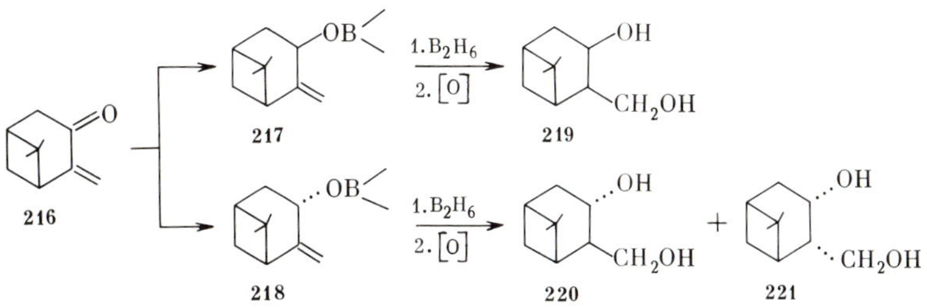

cis-Verbenol (*222*) and verbenone (*223*) are hydrated to *trans*-3,4-pinanediol (*224*) [130]:

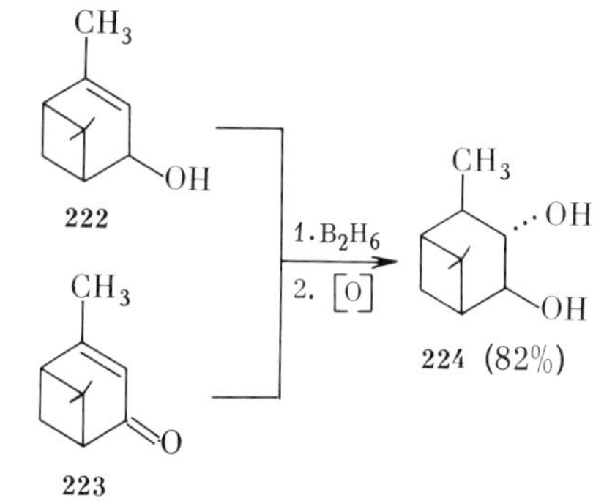

The hydroboration-oxidation of *cis*-2-pinen-3-ol leads to isopinocampheol [131].

3,10-Pinanediol and myrtanol have been obtained from myrtenic acid [234], while α-pinene and isopinocamphanol have been obtained from amines and azides of the pinane series [235].

A detailed investigation has been carried out of the stereochemical course of the hydroboration-oxidation of the diastereoisomeric (1*R*)-isopulegols *225–228*, which are converted to mixtures of 9-hydroxymenthols, with a predominant content of one of the diastereoisomers [132]:

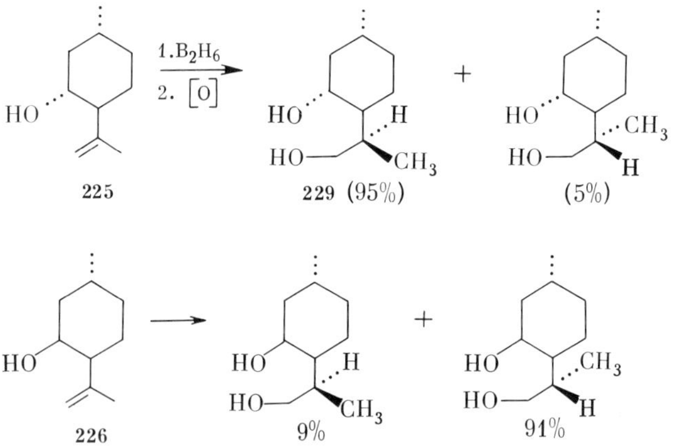

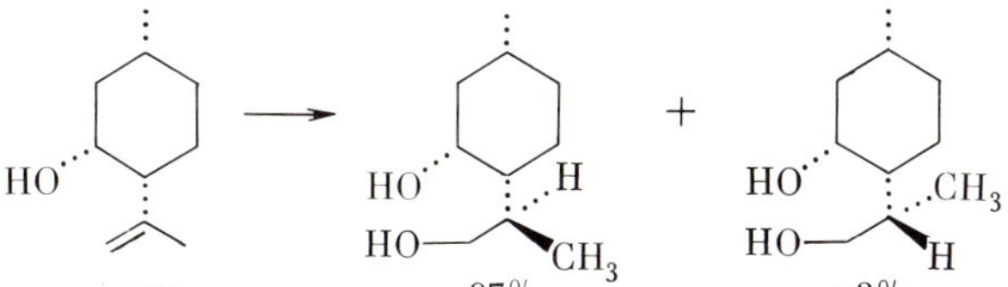

227 97% < 3%

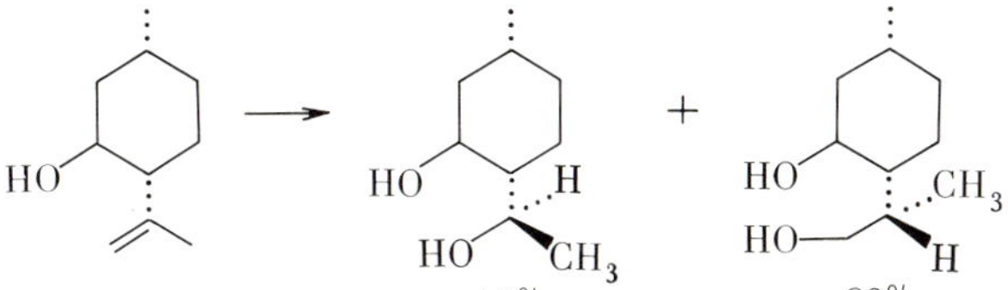

228 18% 82%

The stereochemical ratios observed may be explained by the antiplanar trans-addition of the hydroborating agent to the isopropenyl group of the initially formed borate, for example, *230* (for the case of (−)-isopulegol, *225*):

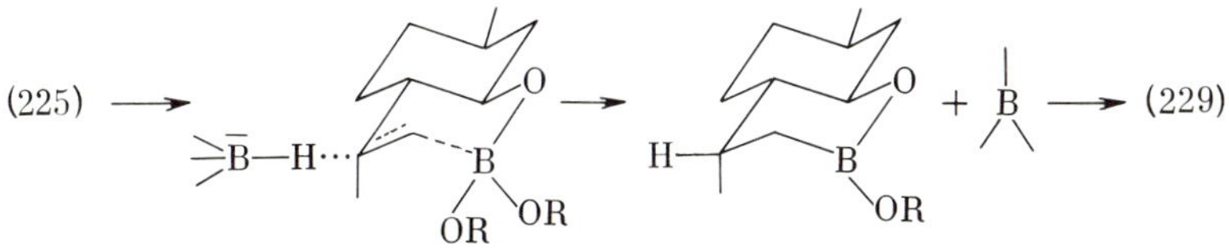

230

On the other hand, the hydroboration of (1*R*, 8*R*)-*p*-3-menthen-9-ol (*231*) probably proceeds by cis-addition [132]:

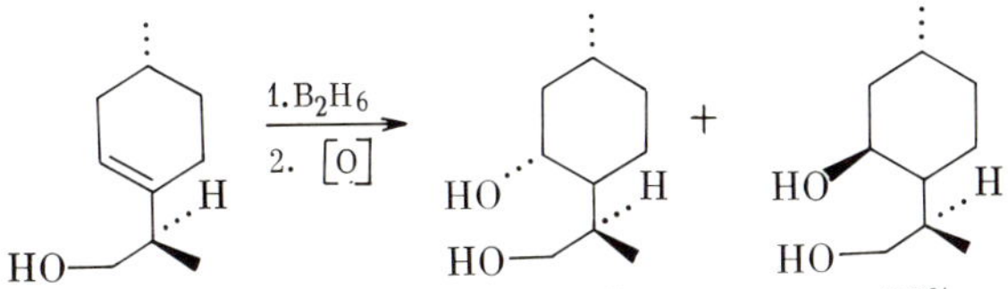

231 67% 33%

The hydroboration-oxidation of *trans*-2-caren-4-ol and *cis*-2-caren-4-ol has been investigated [236].

p-Menthane-9,10-diol is produced by the hydroboration-oxidation of *p*-menth-8-en-10-ol [237].

Hydroboration-oxidation has been used in the synthesis of gymnomitrol [238].

c) Compounds of the steroid series

The hydroboration of alcohols and ketones of the steroid series has been studied.

Wechter [133] established that cholesterol is attacked by diborane primarily from the rear and to a lesser degree from the front. As a result of oxidation of the hydroboration products, 5α-cholestan-3β, 6α-diol (*231a*) is obtained in 45% yield, together with 5β-cholestan-3β, 6β-diol (*232*):

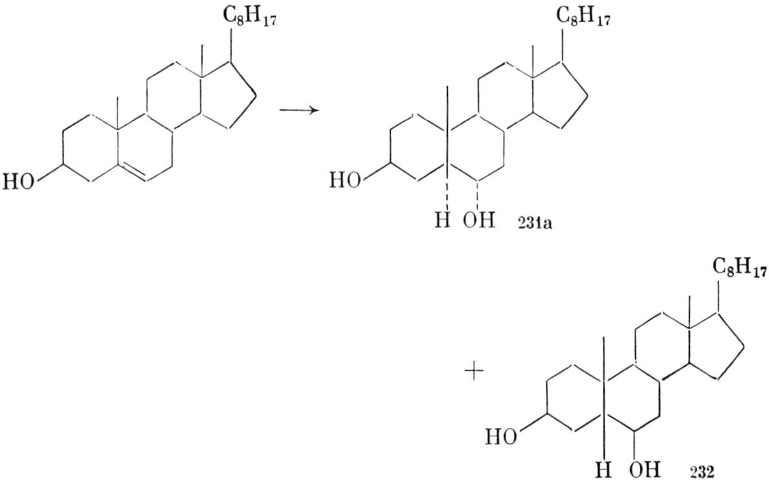

231a

232

Nussim, et al. [134] found that the hydroboration of 4-cholesten-3β-ol (*223*) proceeds stereoselectively, with the formation of 5α-cholestan-3β,4α-diol (*234*):

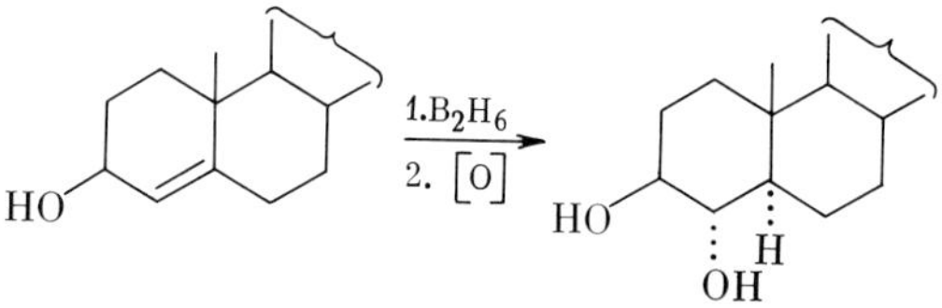

233

234 (55%)

The hydroboration of 4-cholesten-3-one (*235*), followed by oxidation, forms the diol *234* [134, 135] and heating of the hydroboration product with acetic anhydride forms 5α-cholest-3-ene (*236*) [60, 136]:

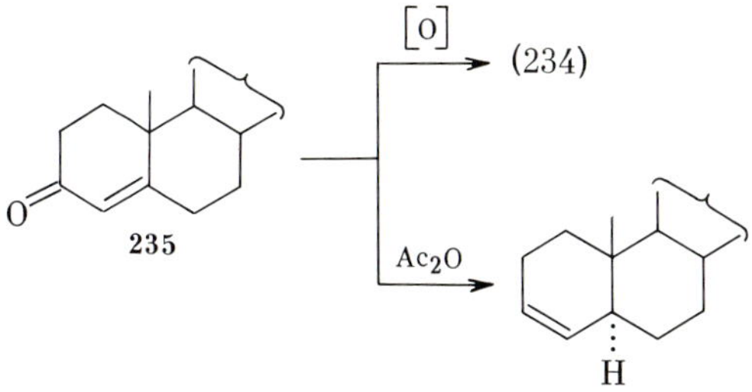

235

236

The treatment of 4-cholesten-3-one (*235*) with excess deuterodiborane and acetic anhydride leads to 3,5-d_2-5α-cholest-3-ene (*237*) whereas 3,5-cholestadiene (*238a*) or 3-*d*-cholesta-3,5-diene (*238b*), respectively, is obtained when an equimolar quantity of diborane or deuterodiborane is used [137]:

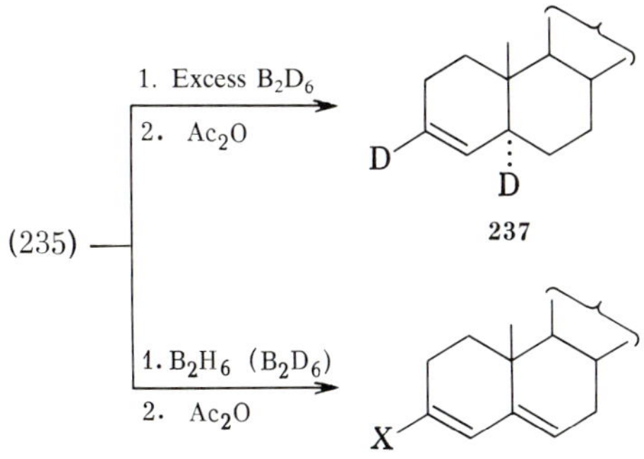

237

238a: X = H
238b: X = D

The compound *239* is converted by excess B_2H_6 or B_2D_6 to a 7,14-diene (*240a*, *240b*) [137]:

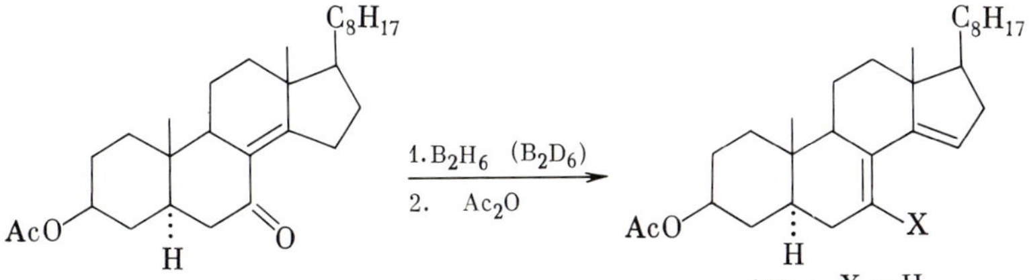

239

240a: X = H
240b: X = D

3β-Acetoxy-7-oxo-5-cholestene (*241*) is converted to 3β-acetoxy-5α-cholest-6-ene (*242*) by the above method:

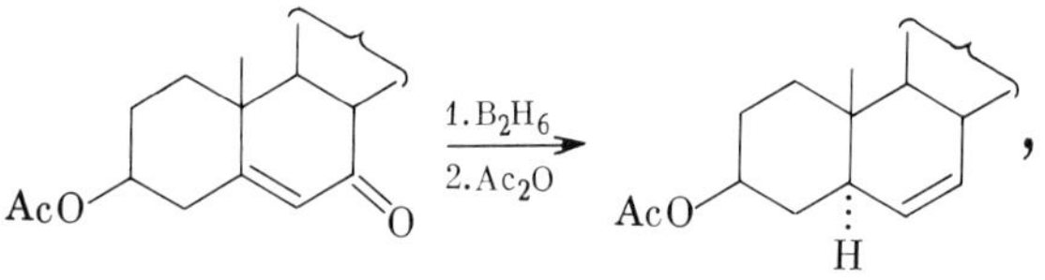

241 242 (80%)

whereas 3-oxo-4-chloro-17β-acetoxy-4-androstene forms 17β-acetoxy-5α-androst-3-ene and 4-chloro-17β-acetoxy-3,5-androstadiene [138].

5α-Cholest-6-en-3β-ol (*243*) adds diborane from the α-side, while boron adds to both C-6 and C-7, as a result of which a 1:1 mixture of 5α-cholestan-3β,6α-diol (*244*) and the 3β,7α-diol *245* is obtained, with a total yield of 57% [134]:

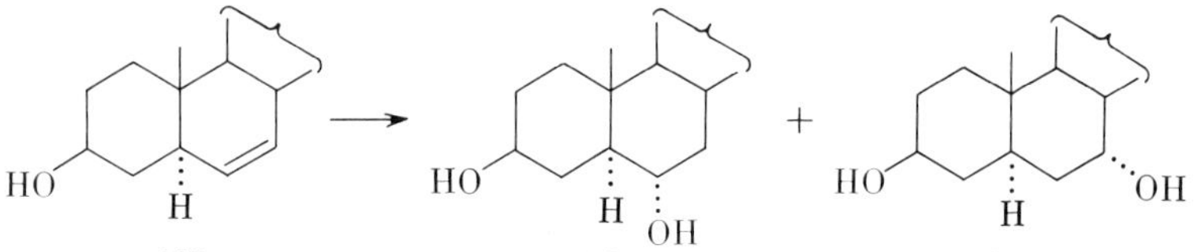

243 244 245

The hydroboration of functionally substituted conjugated dienes of the steroid series has been investigated.

7-Dehydrocholesterol (*246*) reacts with diborane to form the borate-borane *247* which, upon oxidation, forms 5α-cholest-7-en-3β,-6α-diol (*248*) and reacts with acetic anhydride to produce 5α-cholest-6-en-3β-ol (*249*) [139]:

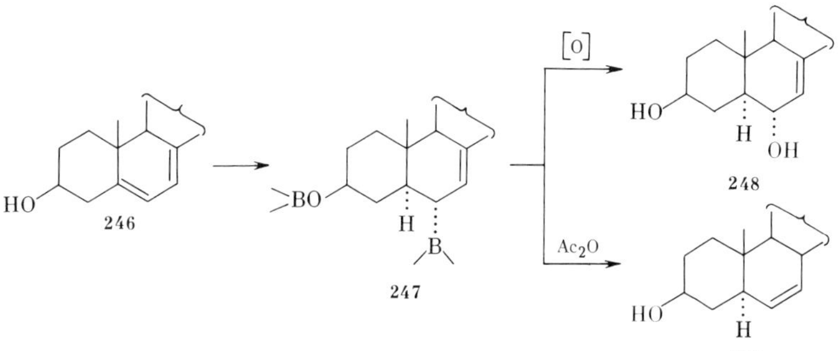

246 247 248 249

5α-Cholesta-7,9(11)-dien-3β-ol (*250*) is hydrated to 5α-cholest-7-en-3β,11α-diol (*251*), which is isolated as the diacetate [140]:

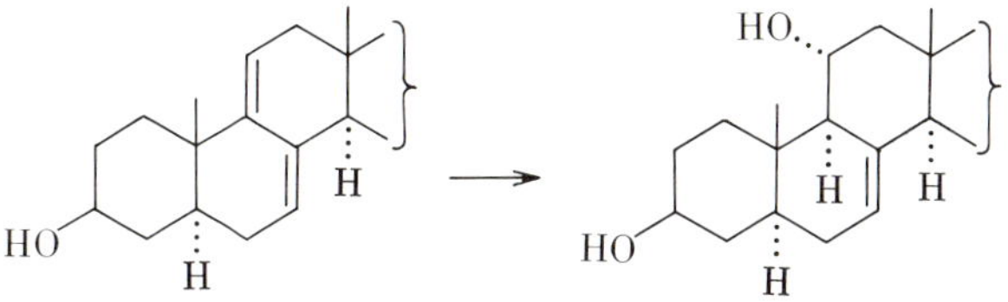

250 251

Cholest-5-ene-3β,26-diol 3-tetrahydropyranyl ether has been obtained by the hydroboration of cholest-5,25-dien-3β-ol tetrahydropyranyl ether with disiamylborane and with (+)- and (−)-diisopinocampheylborane [239].

3β-Propionyloxy-14β-cholest-5-ene has been obtained from 3α,-5α-cyclocholest-8(14)-en-6α-ol by hydroboration [240].

5α-Androstan-3β,4α,17β-triol 17-benzoate (*253*) has been obtained by the hydroboration-oxidation of 4-androsten-3β,17β-diol 17-benzoate (*252*) [141]:

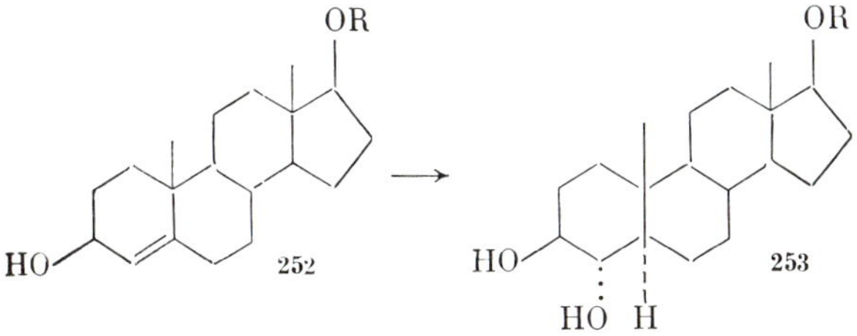

(R = C_6H_5CO).

17,17-Ethylenedioxy-1,5-androstadien-3β-ol (*254*) is hydrated to 17,17-ethylenedioxy-5-androsten-1α,3β-diol (*255*) and 17,17-ethylenedioxy-5-androsten-2α,3β-diol (*256*) [142]:

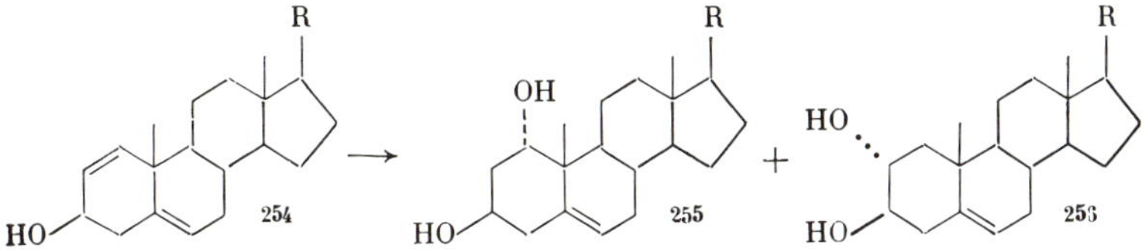

The hydroboration of 20,20-ethylenedioxy-1,5-pregnadien-3β-ol and 3β,20-dihydroxy-1,5-pregnadiene has also been investigated.

Testosterone (*257*) is converted to 17β-acetoxy-5α-androst-3-ene (*258*) by the action of diborane and subsequent heating with acetic anhydride [143]:

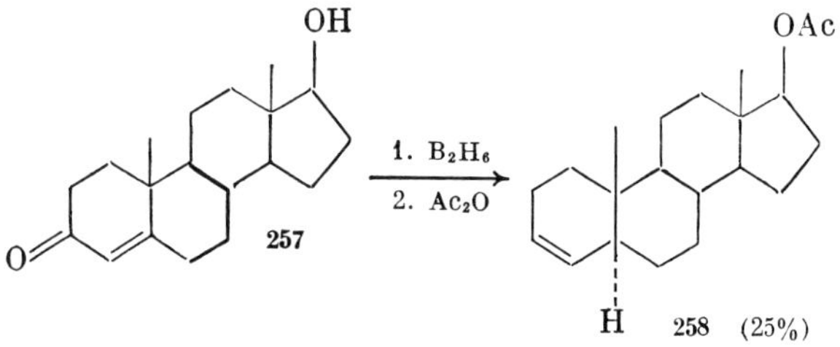

The main product of the hydroboration of progesterone (*259*) is 3β-hydroxy-20-oxo-4-pregnene (*260*); 3β,20β-dihydroxy-4-pregnene (*261*) and 3β,4α,20β-trihydroxy-5α-pregnane (*262*) are also formed [144]:

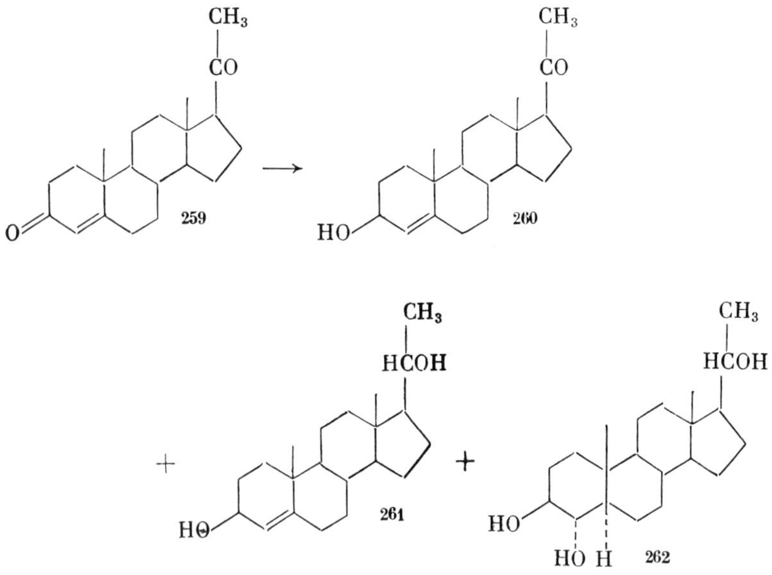

Caglioti, et al. [145] have synthesized androsterone (*264*) from 4-androsten-3,17-dione 17-ethyleneketal (*263*) in accordance with the following scheme:

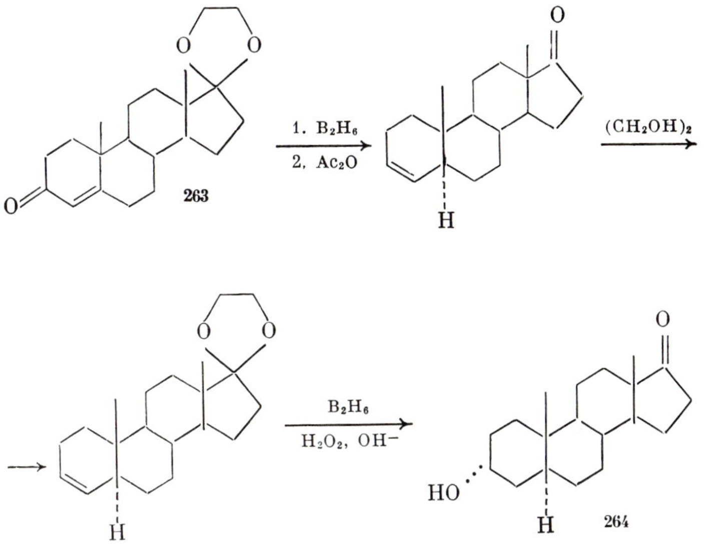

19-Norandrost-4-en-17β-ol-3-one (*265*) has been converted to 19-nor-5α-androst-3-en-17β-ol (*266*):

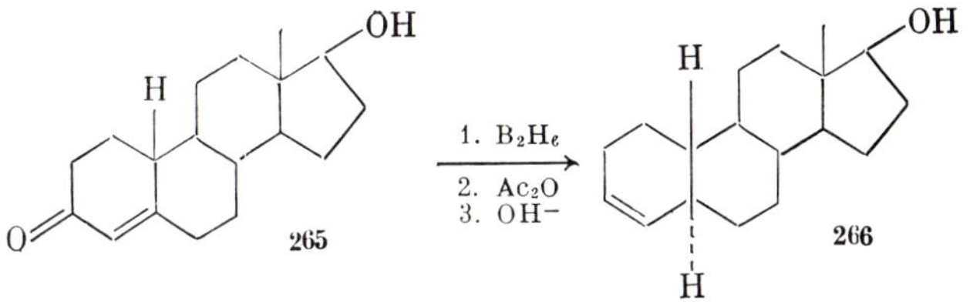

Hydration with the use of diborane has been carried out in the series of functionally substituted 5α-$\Delta^{9(11)}$-steroids.

3,20-Bicycloethylenedioxy-5α-pregn-9(11)ene (*267*) can be converted to 3,20-bicycloethylenedioxy-5α-pregnan-11α-ol (*268*), whereas the 5β compound is inert toward diborane because of steric hindrance caused by cis-fusion of the A and B rings [134]:

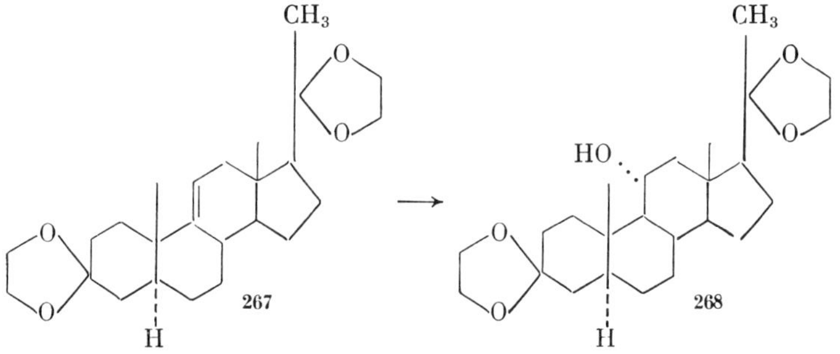

5α,25D-Spirost-9(11)en-3β-ol acetate (*269*) can be converted to 5α,25D-spirostan-3β,11α-diol diacetate (*270*) [134]:

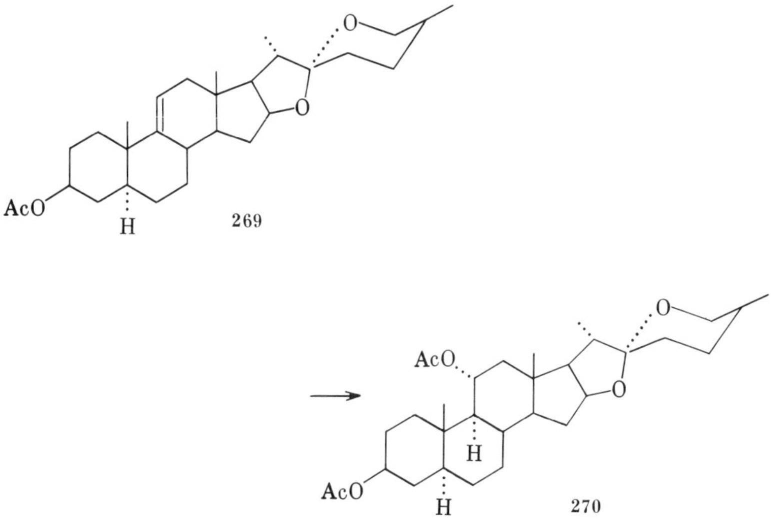

The $\Delta^{13(14)}$-steroid *271* is selectively hydroborated with localization of the boron on the C-13; upon subsequent heating (50–60°C), the boryl group migrates to C-12, with the formation of compounds with *cis*-fusion of the C and D rings (*272*); if the heating is carried out at a higher temperature, then the compound *272* is isomerized in good yield to the 13,14-*trans*-isomer *273* [146]:

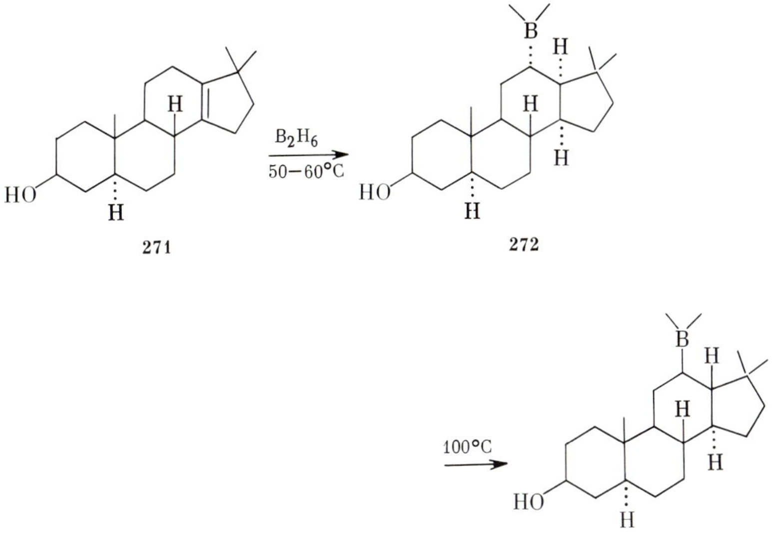

271 272 273 (67%)

Analogous thermal conversions are observed in the series of 8,14- [240a], 5,10- [241], 9,11- [242], and 8,9-unsaturated steroids [243].

The isomerization of a trans-fused borane (*275*) to a cis-fused one (*276*) is observed in the hydroboration of the compound *274* [147]:

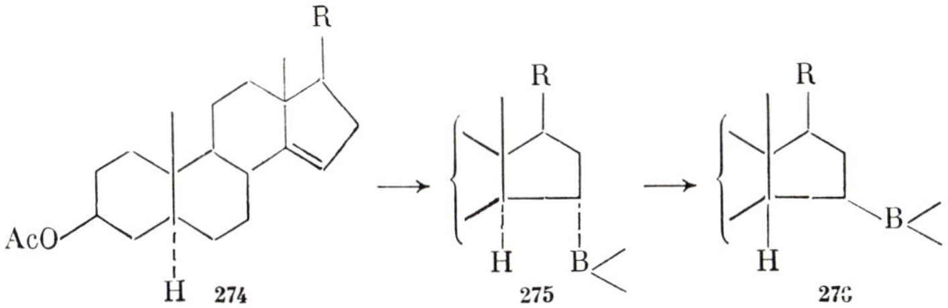

274 275 276

The stereoselective synthesis of 21-hydroxysteroids of the (20*S*) series, including 3*β*,21-dihydroxy-(20*S*)-cholest-4-ene and its conversion to 20-isocholesterol, has been described [148].

3β-Hydroxy-5α-androst-16-ene (*278*) is obtained by the hydroboration of 3β,17-diacetoxy-5α-androst-16-ene (*277*), followed by treatment with acetic anhydride [60]:

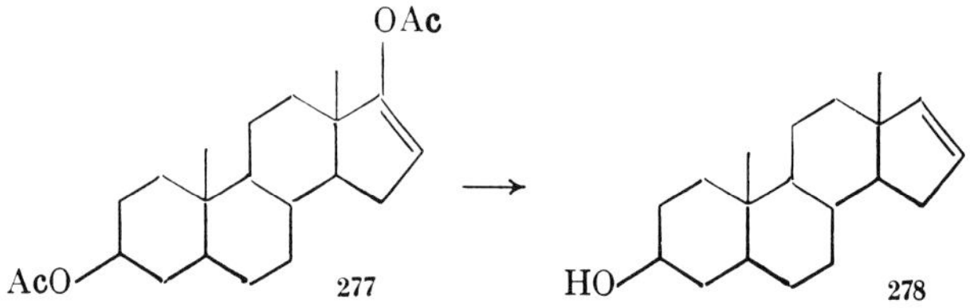

The hydroboration-oxidation of 17-methyl-D-homo-5-androst-17-en-3β-yl acetate has been carried out [244].

5α-Cholest-14-en-3β-ol (*279a*) is hydrated to 5α-cholestan-3β,15α-diol (*280a*), and 5α-androst-14-en-3β,17β-diol (*279b*) to 5α-androstan-3β,15α,17β-triol (*280b*) [134]:

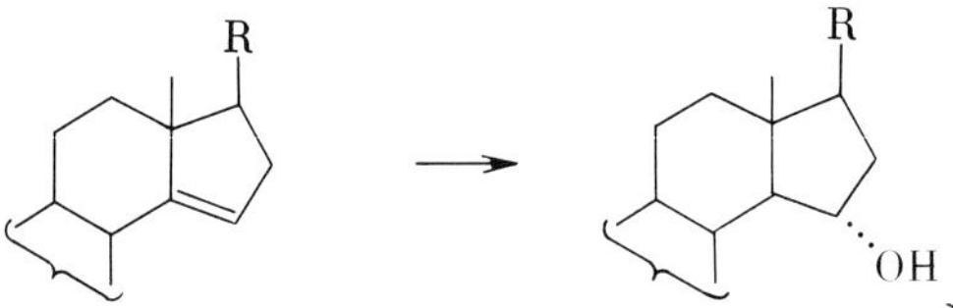

279a: R = C_8H_{17}
279b: R = OH

280a: R = C_8H_{17}
280b: R = OH
} 70–75%

5α-Pregn-16-en-3β,20α-diol (*281*) is smoothly hydrated to 5α-pregnan-3β,16α,20α-triol (*282*) [134]:

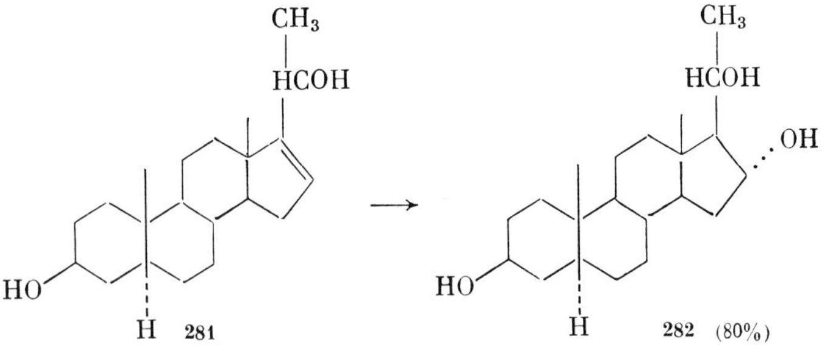

19-Norpregna-1,3,5(10),16-tetraen-3,20α-diol (*283*) is converted to 19-norpregna-1,3,5(10)-trien-3,16α,20α-triol (*284*) by the same method [134]:

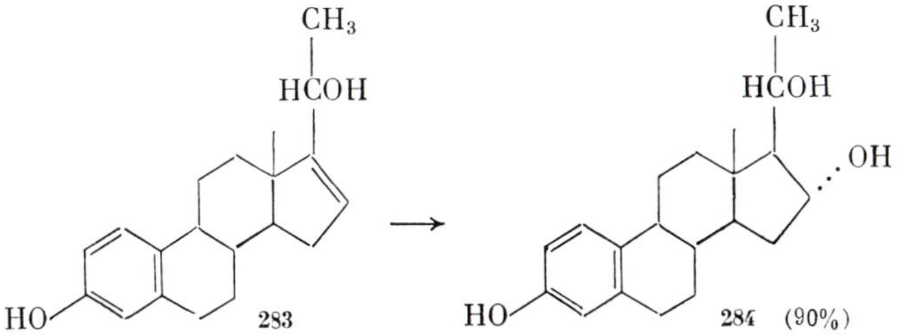

283 → 284 (90%)

The selective hydroboration of $\Delta^{5,20(21)}$-steroids leads to compounds containing a hydroxy group in the side chain [149]. Thus, for example, (20*S*)-3β-tetrahydropyranyloxycholest-5-en-21-ol (*286*) is obtained from 3β-tetrahydropyranyloxycholesta-5,20(21)-diene (*285*):

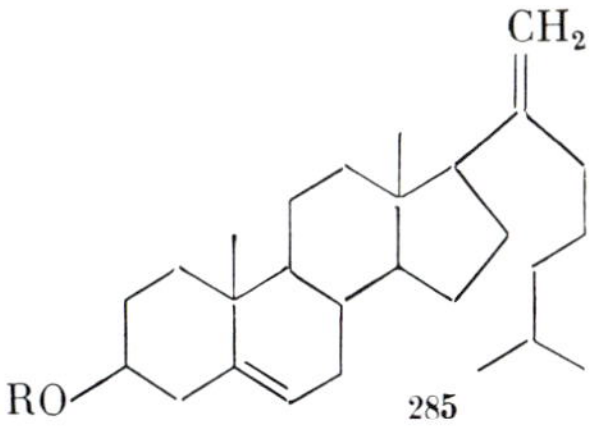

285

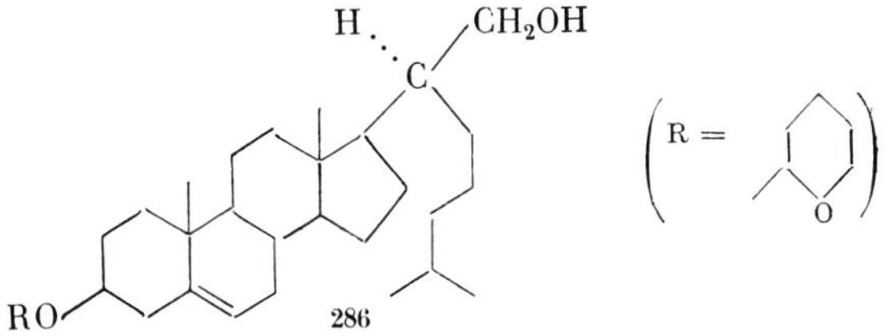

286

3β,16β-Diacetoxy-20-methylene-5α-pregnane (*287*) is converted to 3β,16β-diacetoxy-20-methyl-5α-pregnane (*288*) by hydroboration-protonolysis and 3β,16β-diacetoxy-5α-pregn-20(21)-ene (*289*) is converted to 3β,16β-diacetoxy-5α-pregnane (*290*) [150]:

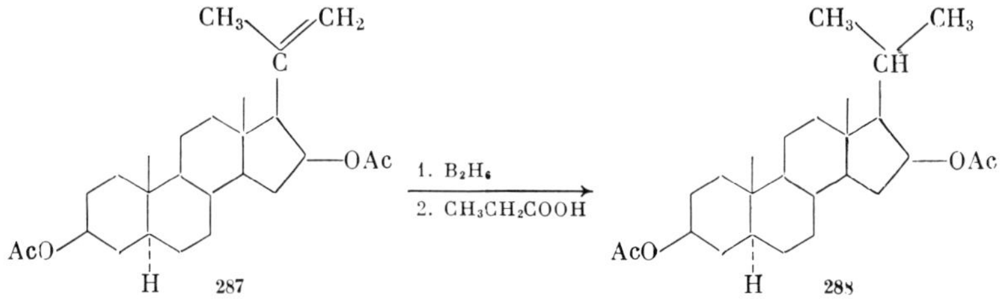

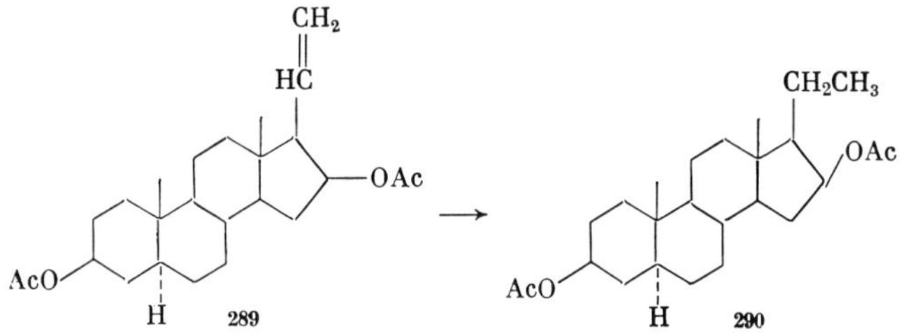

The hydroboration of 3β-acetoxy-22-ergostene has been investigated [151].

Compounds of the lupane type, with a 19αH configuration, have been synthesized from derivatives of 30-nor-19-lupene [245].

Trihydroxylupane has been prepared by the hydroboration-oxidation of betulin [246].

A mixture of 3β-benzoyloxy-22- and -23-hydroxytricarbonylironergosta-5,7-diene has been obtained by the hydroboration of tricarbonylironergosteryl benzoate [247].

($\pm$)-3-Methoxyestra-1,3,5(10)-triene has been synthesized by hydroboration-carbonylation [248].

3.3.4 Hydroboration of esters of unsaturated carboxylic acids. Synthesis of esters of hydroxy acids

As was found by Dulou and Chrétien-Bessière [152] and by Fore and Bickford [153], esters of unsaturated acids with a double bond in the center or at the end of the chain react selectively with diborane, being converted to esters of hydroxycarboxylic acids after oxidation. Thus, methyl undecylenoate forms the ester of 11-hydroxyundecanoic acid (*291*) in 50% yield [152]:

$$CH_2{=}CH(CH_2)_8COOCH_3 \xrightarrow[{[O]}]{B_2H_6} HOCH_2(CH_2)_9COOCH_3,$$

291

while methyl oleate forms an equimolar mixture of esters of 9- and 10-hydroxystearic acids (*292*) and (*293*) in 40% yield [153]:

$$CH_3(CH_2)_7CH{=}CH(CH_2)_7COOCH_3 \xrightarrow[{[O]}]{B_2H_6} CH_3(CH_2)_7CHOHCH_2(CH_2)_7COOCH_3 +$$

292

$$+\ CH_3(CH_2)_7CH_2CHOH(CH_2)_7COOCH_3.$$

293

Esters of the general formula $CH_2{=}CH(CH_2)_nCOOC_2H_5$ ($n = 1$, 2, 8) react analogously to the ester of undecylenic acid, with the boron atom adding predominantly (81–91%) to the terminal carbon [154].

The reaction of diborane with methyl methacrylate in the absence of a solvent proceeds in an involved manner [155], and the action of excess diborane on ethyl acrylate in tetrahydrofuran, followed by oxidation, forms propanediols, propanol, and 1-ethoxy-2-propanol [154].

Methyl phytylate is predominantly converted to 1-phytanol by hydroboration-oxidation [88], while the chlorides of cinnamic and phenylpropionic acids form 1-phenyl-1,2- and 1-phenyl-1,3-propanediol [156].

The hydration of methyl α-cyclogeranate leads to the hydroxy ester *294* [157]:

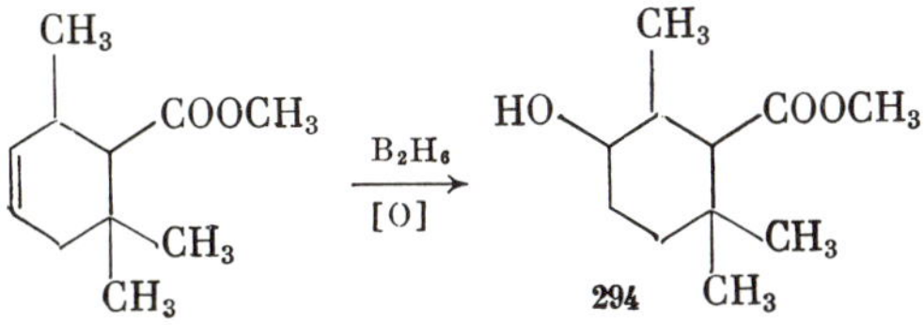

294

Methyl cyclohexene-4-carboxylate is hydrated to a difficult-to-separate mixture of the hydroxy esters *295–298*, in a ratio of 4:2:1:1 [158]:

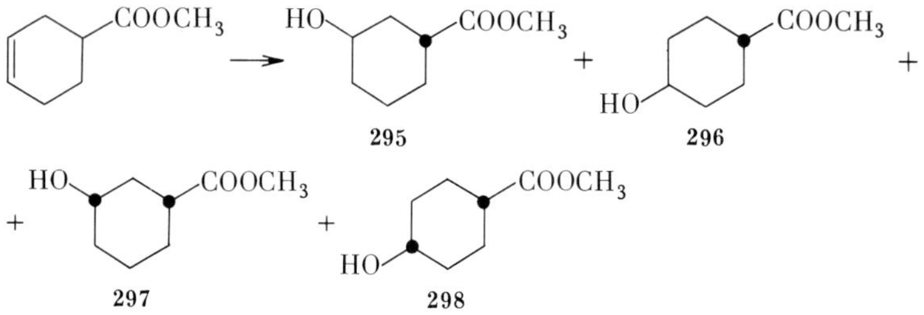

Dimethyl *cis*-tetrahydrophthalate (*299*) forms only the trans alcohol *300*, whereas the trans diester *301* forms a mixture of the two isomeric alcohols *302* and *303* in a ratio of 7:3 [159]:

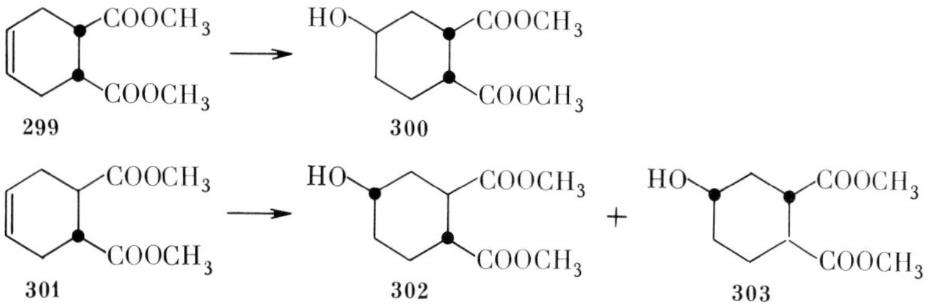

In the hydroboration of the methyl ester of bicyclo [3.1.0]hex-2-en-6-*endo*-carboxylic acid (*304*), followed by oxidation (H_2O_2 in alkaline medium, then CrO_3), the ketones *305* and *306* and methyl 2-cyclopentenylacetate (*307*) are formed [160]:

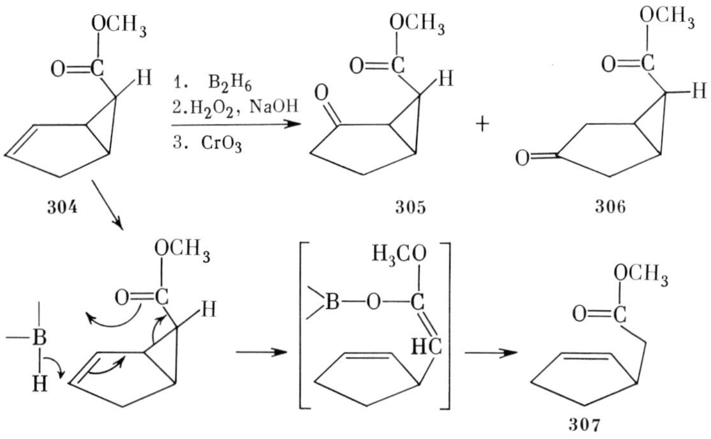

The synthesis of the ethyl ester of cyclopentan-2-octyl-5-oxo-1-heptanoic acid (*308*) and of the ethyl ester of 1-cyclopenten-2-octyl-5-oxo-1-heptanoic acid (*309*), which are related to prostaglandin, has been carried out by the hydroboration of 1-carbethoxy-4-pentene [161].

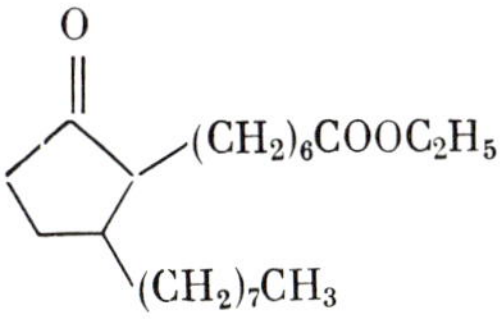

308

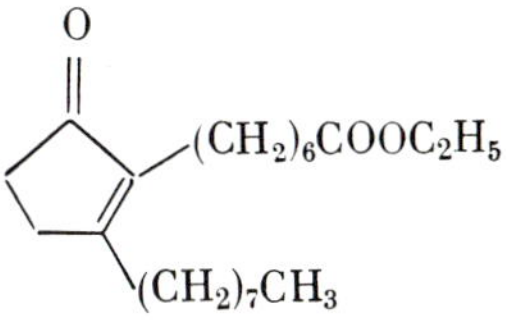

309

The hydroboration-oxidation of trimethylsilyl esters of unsaturated acids proceeds smoothly [249].

3.4 HALOGEN-CONTAINING COMPOUNDS

3.4.1 Hydroboration of vinyl halides. Synthesis of alcohols and halo alcohols

The action of diborane on halogenated derivatives of unsaturated hydrocarbons was investigated for the first time by Stone, et al. [162, 163]. It was established that tetrafluoroethylene reacts with diborane, slowly at room temperature and rapidly at 80°C, with the formation of ethyldifluoroborane, diethylfluoroborane, and a polymeric product. The reactions of diborane with trifluoroethylene, 1,1-difluoroethylene, and fluoroethylene proceed in an equally complex manner [163] and result in products of the partial replacement of fluorine by hydrogen and fluoroethylboranes.

Hawthorne and Dupont [164] obtained a small yield of dichloro-(2-chloroethyl)borane etherate by the reaction of diborane with vinyl chloride in dimethyl ether, indicating a reaction course of the following type:

$$CH_2{=}CHCl \xrightarrow{B_2H_6} B(CH_2CH_2Cl)_3 \xrightarrow{-C_2H_4} ClB(CH_2CH_2Cl)_2 \xrightarrow{-C_2H_4} Cl_2BCH_2CH_2Cl.$$

Pasto and Snyder [6] found that, in the hydroboration of vinyl halides, the boron atom adds completely or predominantly in the

α-position:

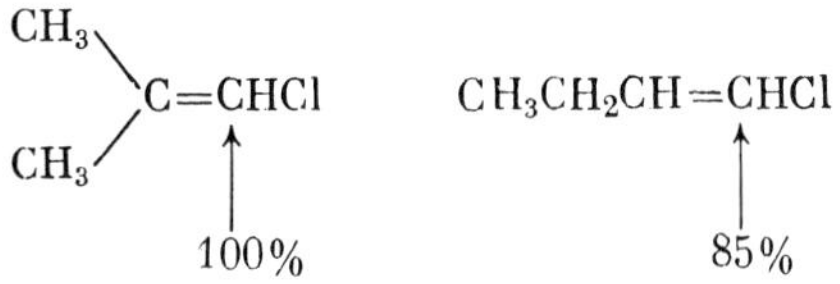

The α-substituted organoboranes formed undergo rearrangement (α-transfer) to alkylhaloboranes. Thus, for example, 2-phenylethanol is obtained from *trans*-β-bromostyrene:

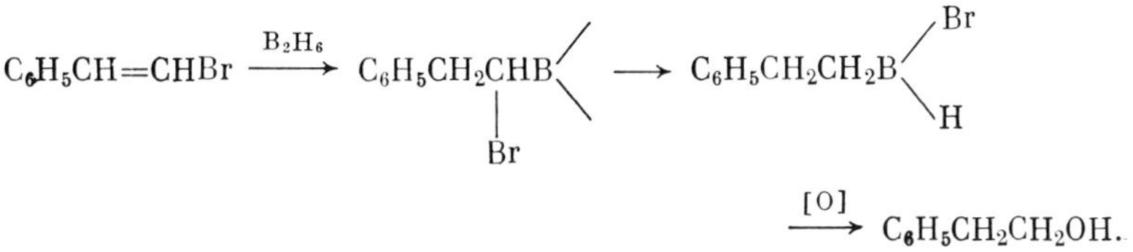

1-Chloro-2-methylpropene and 1-chloro-1-butene are also attacked in the α-position [58]:

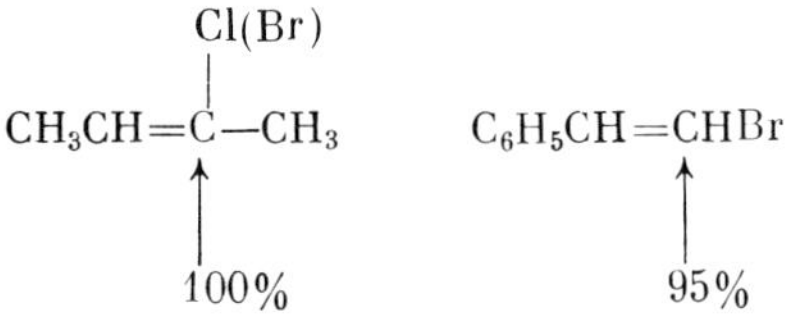

Chloroborane is obtained by the action of BH_3·THF on *cis*-1,2-dichloroethylene [250].

The syntheses of β-phenylethyl alcohol, phenylacetic acid, and their derivatives have been carried out by the hydroboration-oxidation of 1-phenyl-2,2-dihaloethylenes and their nuclearly substituted derivatives [251].

Köster, et al. [165, 166] obtained cyclohexyldialkylborane (*310*) from 1-chlorocyclohexene and tetraalkyldiborane:

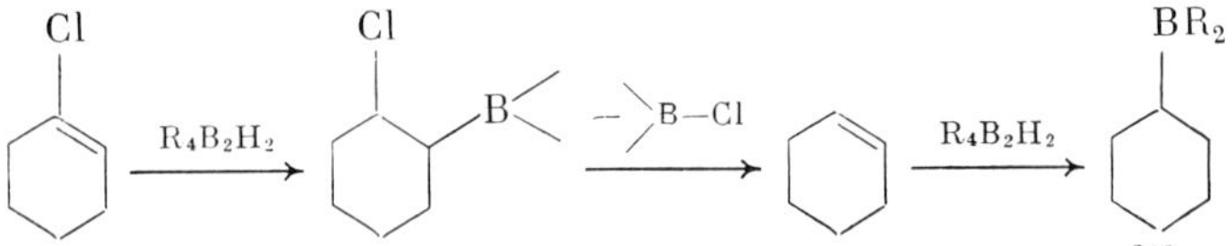

310

Under the action of diborane (1 : 1) in tetrahydrofuran, 1-chloro-4-*tert*-butylcyclohexene forms the organoboranes *311* and *312*:

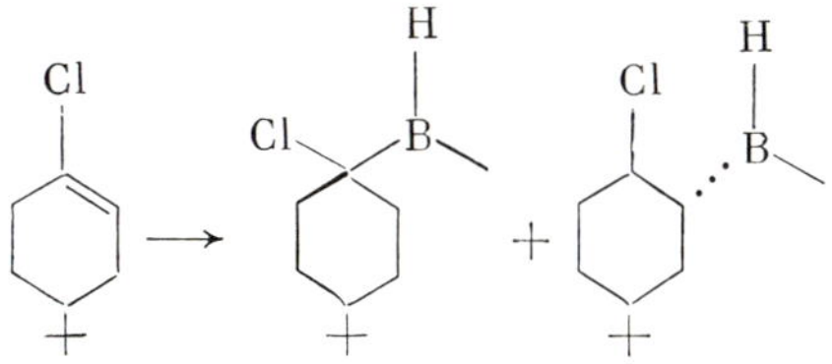

311 312

As a result of their further conversions, including elimination, α-transfer, rehydroboration, and oxidation, 65% of *trans*-4-, 15% of *cis*-4-, 6% of *trans*-3-, and 14% of *cis*-3-*tert*-butylcyclohexanol are obtained [6].

The hydroboration-oxidation of 1-chlorocycloheptene forms cycloheptanol in 57% yield [6].

It is of interest that 7-chlorodibenzobicyclo[2.2.2]octatriene, under the action of diborane in diethyl ether, is converted predominantly to a β-haloborane, which is oxidized by perbenzoic acid to the chlorohydrin *313* [167]:

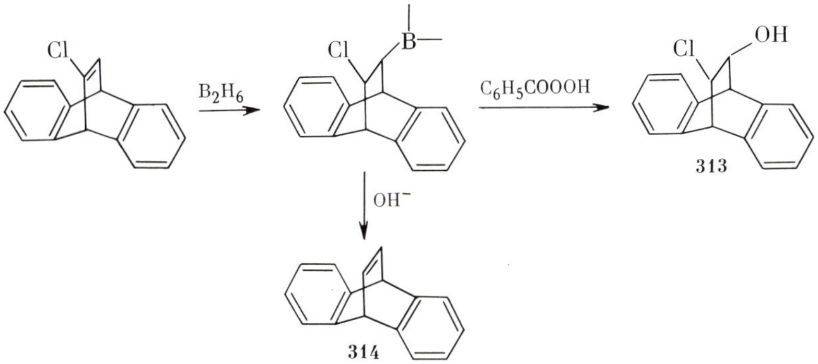

313

314

The oxidation of the β-haloborane with an alkaline solution of hydrogen peroxide forms dibenzobicyclo [2.2.2]octatriene *314*.

3.4.2 Hydroboration of allyl halides. Synthesis of cyclopropane and α, ω-halo alcohols

The orientation of the addition of the boron atom in the hydroboration of allyl halides depends on their structure and on the nature of the hydroborating agent.

Hawthorne and Dupont [164] found that the reaction of diborane with allyl chloride in diethyl ether forms the relatively stable tris(3-chloropropyl)borane (*315*) (36%) and chlorobis(3-chloropropyl)-borane (*316*) (12%):

$$CH_2{=}CHCH_2Cl \xrightarrow{B_2H_6} \underset{\mathbf{315}}{B(CH_2CH_2CH_2Cl)_3} + ClCH_2CH(CH_3)B(CH_2CH_2CH_2Cl)_2 \xrightarrow{-C_3H_6} \underset{\mathbf{316}}{ClB(CH_2CH_2CH_2Cl)_2}$$

The compounds *315* and *316* react with a 10% aqueous sodium hydroxide solution to form cyclopropane in the theoretical yield:

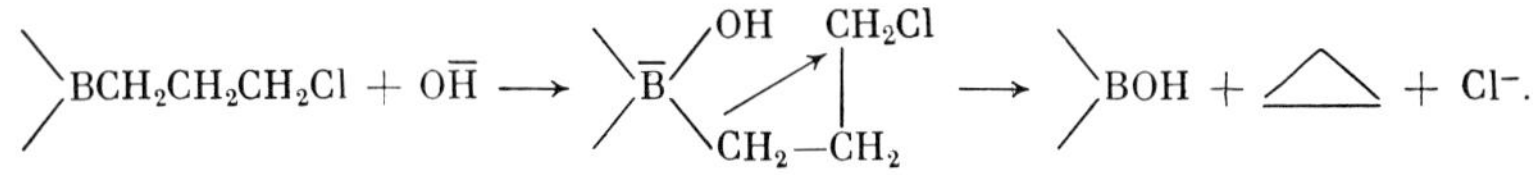

Compound *315* is also smoothly converted to cyclopropane by the action of phenyllithium.

Hawthorne later showed [168] that the capacity for conversion to cyclopropane hydrocarbons is a general property of organoboron compounds with chlorine atoms in the γ position of the alkyl chain. A number of alkyl- and aryl-substituted cyclopropanes have been synthesized in yields of 43–71% by the hydroboration of different β-alkyl- and β-arylallyl chlorides ($RCl:BH_3 = 3:2$), followed by the action of an aqueous solution of NaOH. The moderate yield of cyclopropane from allyl chloride is explained by the fact that the boron atom adds in the β position to the extent of 30%, with the formation of a 2-haloalkylborane:

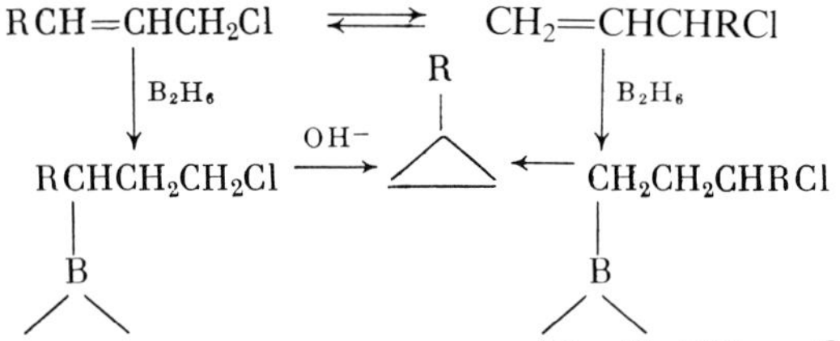

[R=H, CH_3, n-C_3H_7, $C_6H_5CH_2$, C_6H_5]

Tetraalkyldiboranes hydroborate 3-haloolefins more selectively than diborane. Köster, et al. [165, 166] found that, in the reaction of 3-haloolefins with tetraethyl- and tetrapropyldiborane, the boron atom adds predominantly to the terminal carbon atom:

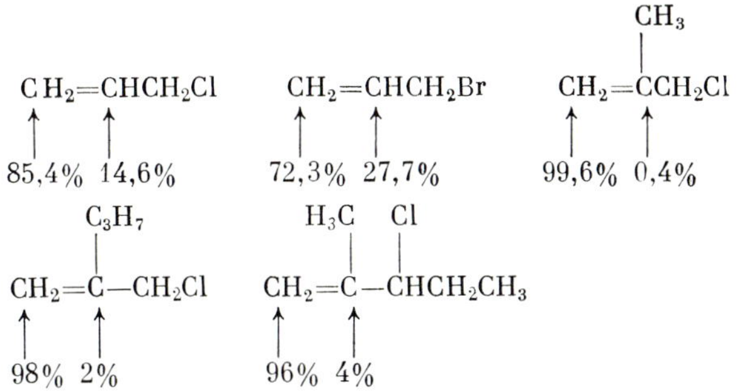

The (3-haloalkyl)dialkylboranes formed by the hydroboration of 3-halo-α-olefins are capable of undergoing symmetrization in the presence of traces of the starting tetraalkyldiborane, and are therefore converted to trialkylborane and tris(3-haloalkyl)borane during distillation. Thus, tris(3-chloro-2-methyl-1-propyl)borane (*317*) and tripropylborane are obtained in the distillation of the products of the reaction of tetrapropyldiborane with 3-chloro-2-methylpropene:

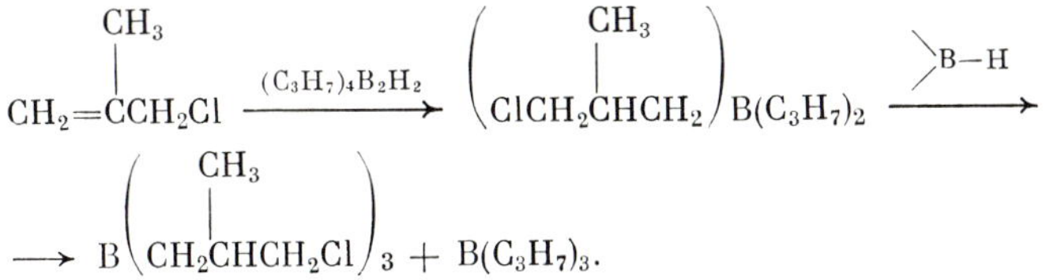

317

The organoboranes obtained from tetraalkyldiboranes and allyl chlorides react with an aqueous solution of NaOH or KOH, alkali metal hydrides, metal alkyls or sodium tetraalkylborates, $NaBR_4$, to form cyclopropane compounds in high yields [165]:

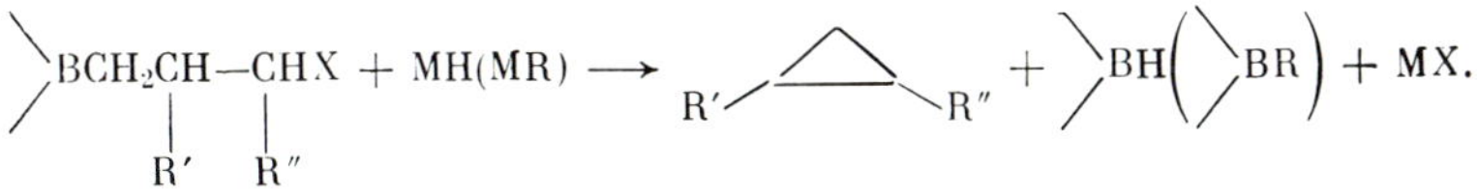

Bis(9-borabicyclo) [3.3.1]nonane can be used in the above method for the synthesis of cyclopropane compounds [169].

The cyclization of organoboranes obtained from allyl chlorides proceeds stereospecifically under the action of bases, with an inversion of the configuration at the carbon atom bonded to the boron [252].

The hydroboration of allyl halides of the type $RCR{=}CHCH_2X$, followed by oxidation of the reaction products, leads to primary alcohols with an impurity of secondary alcohols. Phytyl bromide is converted to 1-phytanol containing approximately 8% of 2-phytanol by this method [88]:

$$\underset{|}{\overset{CH_3}{}}\; RC{=}CHCH_2Br \xrightarrow[\text{[O]}]{B_2H_6,\ THF} \underset{90\%}{R{-}\overset{CH_3}{\overset{|}{C}}HCH_2CH_2OH} + \underset{8\%}{R\overset{CH_3}{\overset{|}{C}}H\underset{\underset{OH}{|}}{C}HCH_3}$$

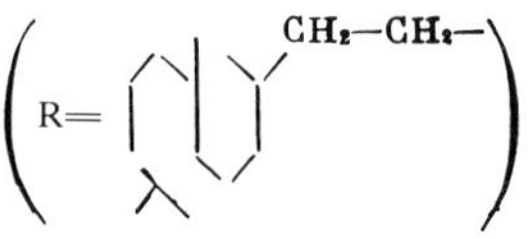

A chlorine atom in a β-position to the double bond of an α-olefin has no effect on the orientation of the addition of tetraalkyldiborane [68, 166]:

$$>BH + CH_2{=}CHCH_2CH_2Cl \longrightarrow \underset{94\%}{>B{-}CH_2CH_2CH_2CH_2Cl} + \underset{6\%}{>B\overset{CH_3}{\overset{|}{C}}HCH_2CH_2Cl}$$

α,ω-Chlorohydrins are obtained by the oxidation of α,ω-chloroalkylboranes.

It should be noted that the degree of branching of the radical in $R_4B_2H_2$ has a negligible effect on the orientation of the addition of tetraalkyldiboranes to 3-choroolefins. Thus, in the hydroboration of allyl chloride by tetrapropyldiborane [166] or tetra(3-methyl-2-butyl)diborane [170], the ratio of γ-chloroorganoborane to β-chloroorganoborane is 86:14 and 94:6, respectively.

In the case of 3-chloro-2-methylpropylene, the degree of branching of the radical in $R_4B_2H_2$ has no effect at all on the direction of the reaction: with the use of both tetraethyldiborane [166] and tetra(3-methyl-2-butyl)diborane [170], the boron atom adds completely in the γ-position. Differences in the orientation of the addition of diborane and tetraalkyldiborane are conditioned by the structure of the 3-chloroolefins. The orientation is markedly different in the case of 3-chloroolefins without a substituent in the 2 position (allyl chloride, 3-chloro-3-methylpropylene), shows little difference in the presence of an alkyl group in the 2 position (2-chloromethylpropylene), and does not differ in the case of crotyl chloride [166, 170]:

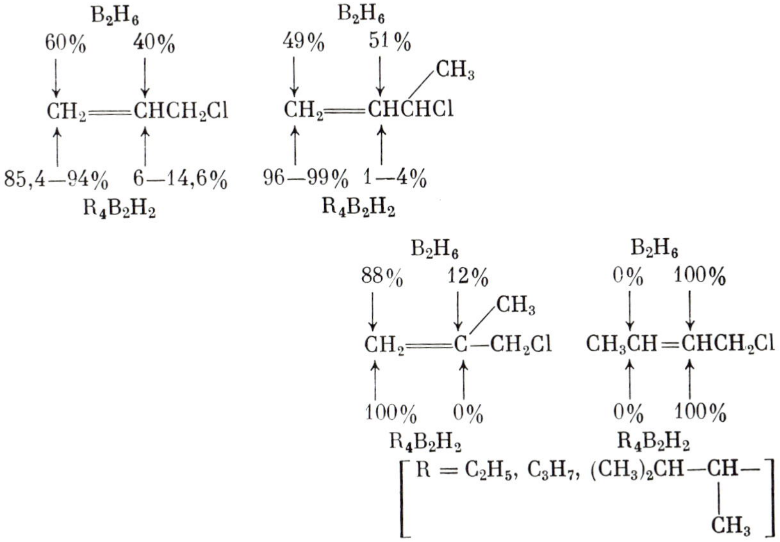

3-Chlorocycloalkenes form primarily 2-chloroorganoboranes under the action of diborane or tetraalkyldiborane.

Köster, et al. [166] found that the hydroboration of 3-chlorocyclopentene with tetraethyldiborane in dibutyl ether at 0–10°C leads to 2- and 3-chlorocyclopentylboranes (*318* and *319*) in a ratio of 90:10:

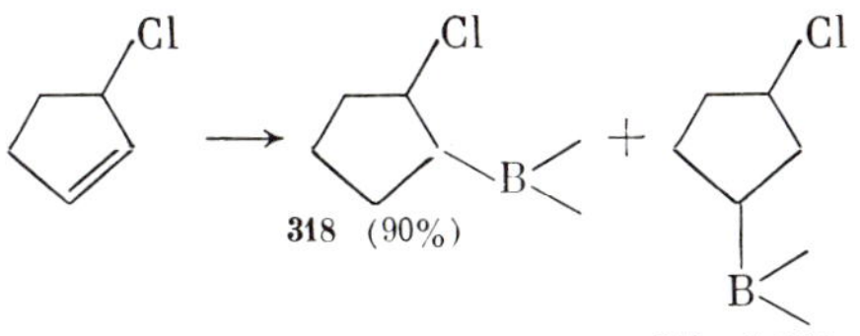

318 (90%) **319** (10%)

This ratio shows little change when diborane or tetra(3-methyl-2-butyl)diborane are used as hydroborating agents [74]. Unfortunately, the stereochemistry of the addition of diborane to 3-chlorocyclopentene has not been investigated.

Pasto and Hickman [73] have carried out a detailed investigation of the hydroboration of 3-chlorocyclohexene with diborane in diethyl ether and tetrahydrofuran. The boron atom adds mainly in the 2 position (95–97%) with the formation of the *trans* isomer (*320*) and *cis* isomer (10%) (*321*) of 2-chlorocyclohexylborane, together with small quantities of the isomeric 3-chlorocyclohexylboranes *322* and *323*. The total yield of the reaction products is 80–85%. Oxidation of the compounds *320–323* leads to the corresponding isomeric chlorocyclohexanols. When the reaction is carried out in tetrahydrofuran, *trans*-2-chlorocyclohexylborane (*320*) undergoes elimination:

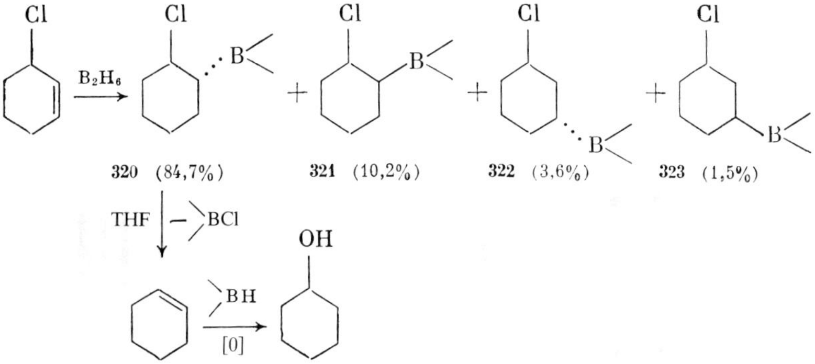

The predominant addition of borane in the trans position is explained by steric hindrances caused by the chlorine atom in the predominantly pseudoaxial conformation of 3-chlorocyclohexene:

According to the data of [165, 166], the hydroboration of 3-chlorocyclohexene with tetrapropyldiborane leads exclusively to cyclohexanol.

The action of diborane on 4-chloronorbornene in tetrahydrofuran forms the hydroxy chlorides *324* and *325* in a ratio of 63:37 [171], and not in the reverse ratio, as was shown earlier [172]:

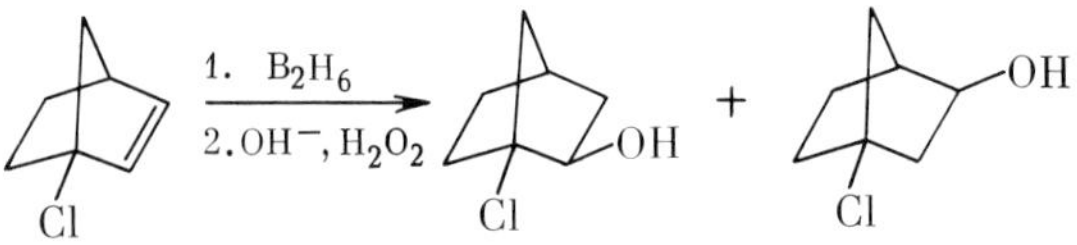

324 (63%) 325 (37%)

Myrtenyl chloride and bromide undergo hydroboration-oxidation to form primarily 2β-halomethyl-3α-hydroxynorpinanes (*326*), along with other products [173]:

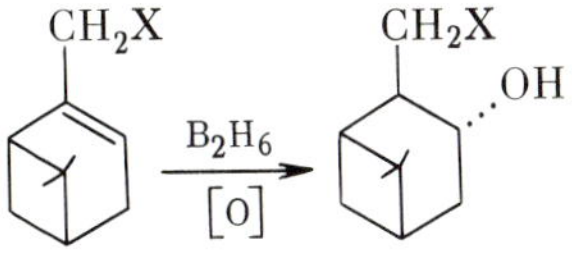

326

3.4.5 Compounds containing silicon, germanium, and phosphorus. Synthesis of silylalkanols

A silicon atom at the carbon-carbon double bond exerts a significant effect on the orientation of the addition of diborane. As shown by Seyferth [174, 175], the hydroboration of trimethylvinylsilane in diglyme leads to a mixture of organoboron compounds, the oxidation of which forms α-trimethylsilylethanol and β-trimethylsilylethanol in a ratio of 1:2:

$$(CH_3)_3SiCH{=}CH_2 \xrightarrow{B_2H_6} [(CH_3)_3Si{-}C_2H_4]_3B \xrightarrow{[O]} \underset{33\%}{(CH_3)_3SiCH(OH)CH_3} + \underset{67\%}{(CH_3)_3SiCH_2CH_2OH}.$$

A mixture of α- and β-alcohols (25:75%) is formed in the hydroboration of $(CH_3)_3Si{-}(CH_2)_2CH{=}CH_2$ [176].

The orientation of the addition of diborane to vinylsilanes changes significantly if a chlorine atom is present on the silicon. In that case, α-isomers are obtained in overwhelming quantities [177, 253]:

$$\underset{\uparrow\;30\%}{CH_2}{=}\underset{\uparrow\;70\%}{CH}Si(CH_3)_2Cl \qquad \underset{\uparrow\;10\%}{CH_2}{=}\underset{\uparrow\;90\%}{CH}Si(CH_3)Cl_2$$

Dialkylboranes (disiamylborane, dicyclohexylborane, 9-borabicyclo[3.3.1]nonane) add to trimethylvinylsilane in a regiospecific manner, forming β-silylethylboranes ($\geqq 95\%$) [254, 255].

1-Trimethylsilycycloalkenes, under the action of $BH_3 \cdot THF$, form mainly 1-boryl-1-trimethylsilylcycloalkanes, while thexylborane produces β-boryl derivatives [256].

Boron-silicon compounds are obtained by the reaction of a trialkylvinylsilane with triisobutylborane [178].

The reaction of diborane with allytrimethylsilane forms tris(3-trimethylsilylpropyl)borane [179], which is also synthesized by the transalkylation of triisobutylborane with allyltrimethylsilane [52]:

$$CH_2{=}CHCH_2Si(CH_3)_3 \xrightarrow{B_2H_6} [(CH_3)_3SiCH_2CH_2CH_2]_3B.$$

Chlorine atoms on the silicon in allylsilanes have a negligible effect on the orientation of the addition of diborane [177].

Dimethyldivinylsilane, when heated with trimethylamine-*tert*-butylborane, forms 4-*tert*-butyl-1,1-dimethyl-1,4-silaborinane [61]:

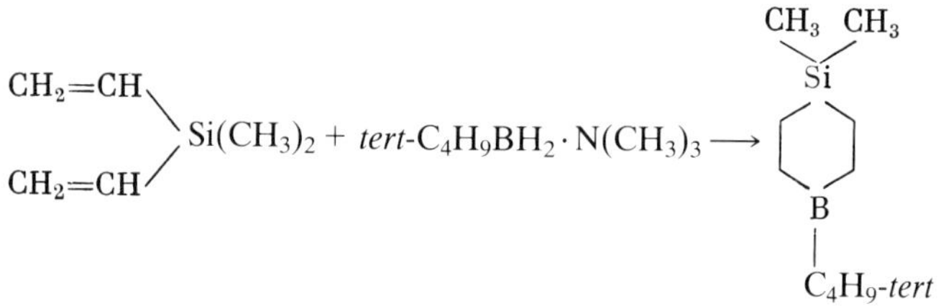

The α,β-unsaturated phosphonate obtained from 5α-cholestan-3-one is converted to the corresponding 3α-hydroxy-3-phosphonate by hydroboration-oxidation [180].

The hydroboration of vinyl phosphonates in oxolane leads to 2-hydroxyethyl phosphonates [257].

3.4.6 Heterocyclic compounds

1. Hydroboration of oxygen–containing heterocyclic compounds

Zaidlewicz and Uzarewicz [181, 182] found that cyclic unsaturated epoxides are converted by the action of diborane and subsequent hydrolysis of the reaction products with an aqueous sodium hydroxide solution to α,β-unsaturated alcohols. The reaction mechanism apparently consists of the addition of diborane to the multiple bond (*327*) and formation of the borate (*328*), which is converted to the final product as a result of elimination with cleavage of the epoxide ring:

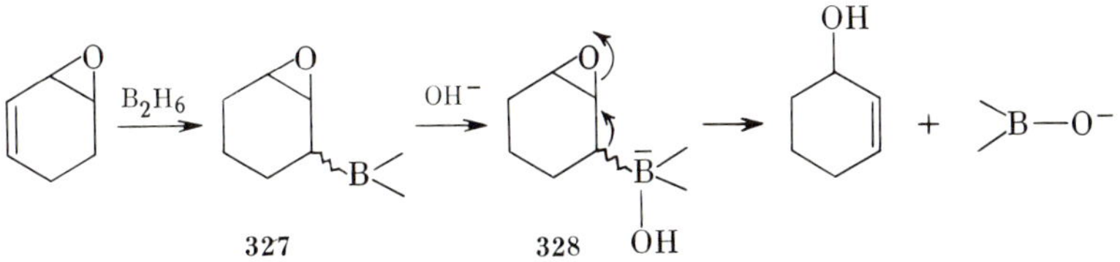

327 328

This mechanism is confirmed by data on the hydroboration of 3,4-epoxy-3,4-dimethylcyclohexene (*329*) and 3,4-epoxy-3,4,5,5-tetramethylcyclohexene (*331*), which are converted to 1,2-dimethyl-2-cyclohexen-1-ol (*330*) and 1,2,6,6-tetramethyl-2-cyclohexen-1-ol (*332*), respectively:

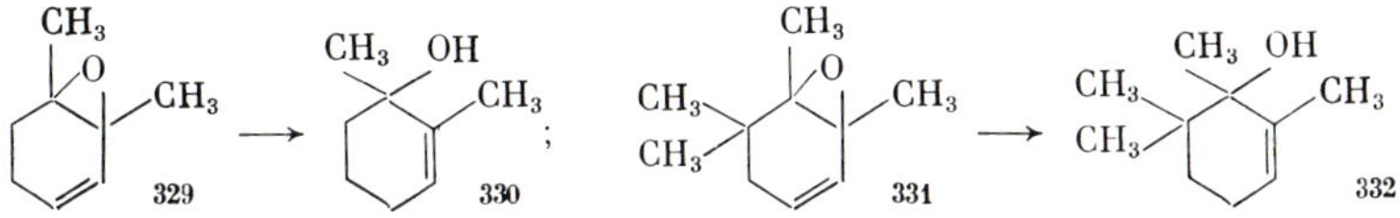

329 330 331 332

The hydroboration of dihydropyrans and dihydrofurans has been investigated.

Δ^2-Dihydropyran is converted to tetrahydro-3-pyranol (*333*) by the action of diborane (olefin: $BH_3 = 3:1$) and subsequent oxidation [183–185], while Δ^3-dihydropyran forms a mixture of tetrahydro-3- (*333*) and tetrahydro-4-pyranol (*334*) [183]:

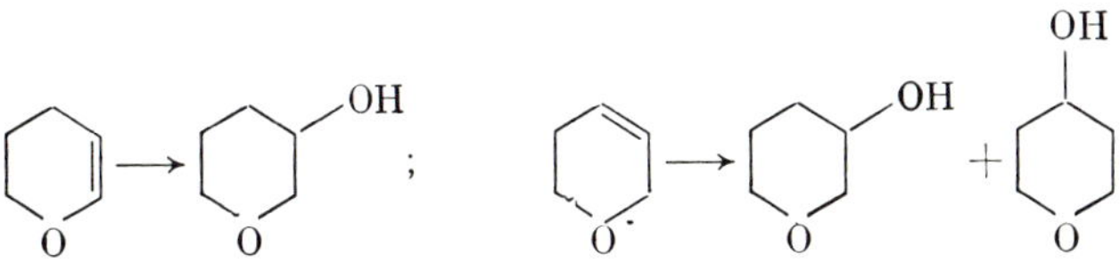

333 (86%) 333 (55%) 334 (30%)

2-Methyl-Δ^2-dihydropyran forms 2-methyltetrahydro-3-pyranol (*335*) [184], while 2-methoxy-3,4-dihydropyran (*336*) is converted to a mixture (1:2) of *cis-* and *trans*-2-methoxy-5-hydroxytetrahydropyrans, *337* and *338* [185]:

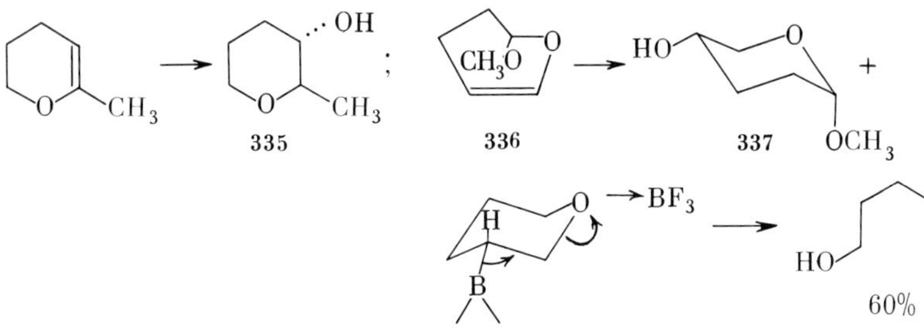

335 336 337 60%

A mixture of *cis*- and *trans*-2-methoxy-5-hydroxy-6-methyltetrahydropyrans, *339* and *340*, is obtained by the hydroboration of 2-methoxy-6-methyl-3,4-dihydropyran [186]:

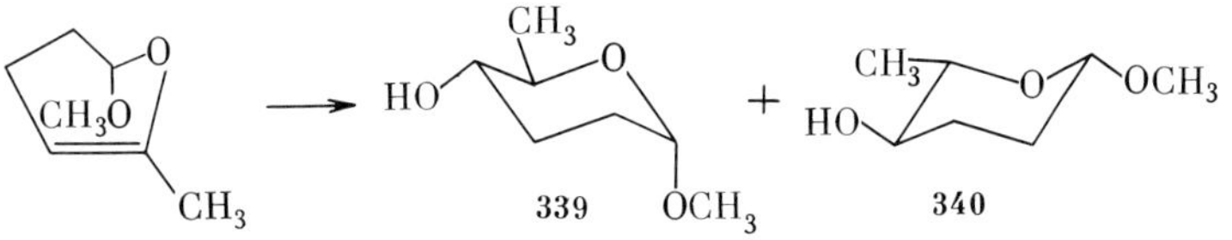

339 340

In the hydroboration of Δ^2-dihydropyran with excess diborane and subsequent oxidation, 1,5- and 1,4-pentanediols are formed along with the compound *333*, as a result of side reactions of elimination and breakdown of the tetrahydropyran ring under the action of diborane [183]:

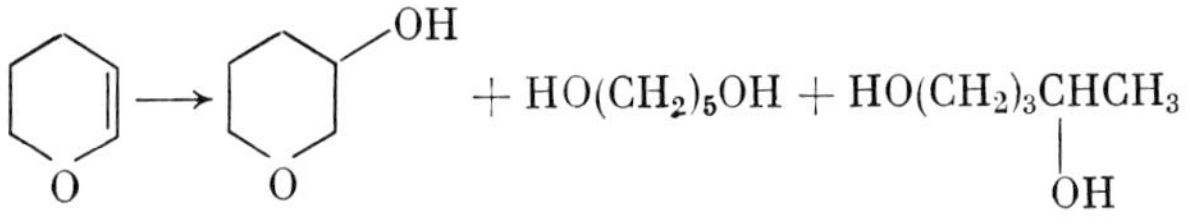

333 (36%) (33%) (22%)

4-Penten-1-ol, a precursor of 1,5-pentanediol, is eliminated in good yield from the product of the hydroboration of Δ^2-dihydropyran under the action of boron trifluoride [183]:

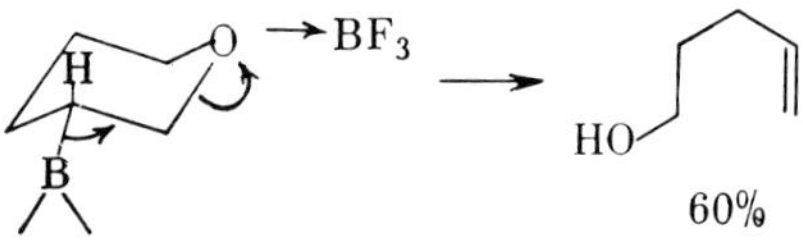

60%

The triorganoboranes *341* and *342*, obtained by the action of diborane on 2,3-dihydrofuran, 2,5-dihydrofuran, and 2-methyl-4,5-dihydrofuran, also undergo elimination under the action of BF_3

[183]:

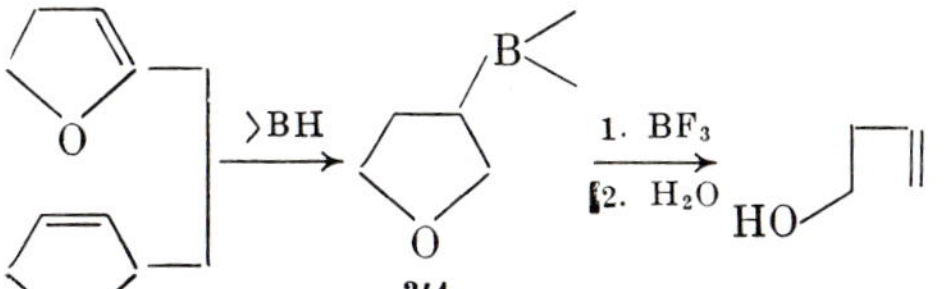

341

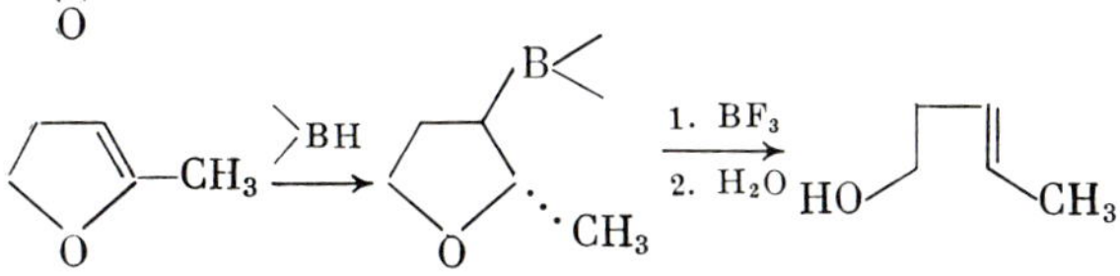

342

3-Chromene undergoes hydroboration-oxidation to form a mixture of 3- and 4-chromanols (*343*, *344*), in a ratio of 85:15 according to some data [187] and 1:8.5 according to other reports [188]:

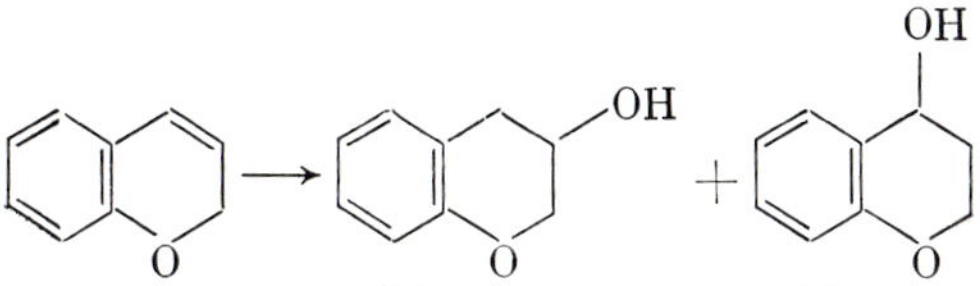

343 344

In the presence of a substituent in the 3 position of 3-chromene, the boron atom attacks only the 4 position. Thus, 3-methyl-3-chromene forms *trans*-3-methyl-4-chromanol (*345*) [189]:

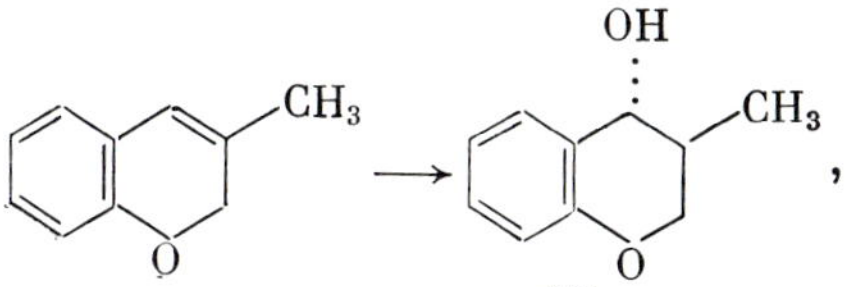

345

while isoflavene forms *trans*-4-isoflavanol (*346*) [190]:

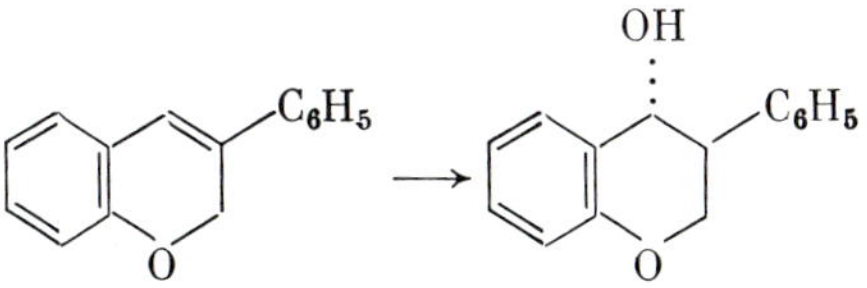

346

4′-Methoxy-3-flavene undergoes hydroboration to form 2,4-*trans*-4′-methoxy-4-flavanol (*347*) (47%) and the products of breakdown of the heterocyclic ring (30%) [191, 192]:

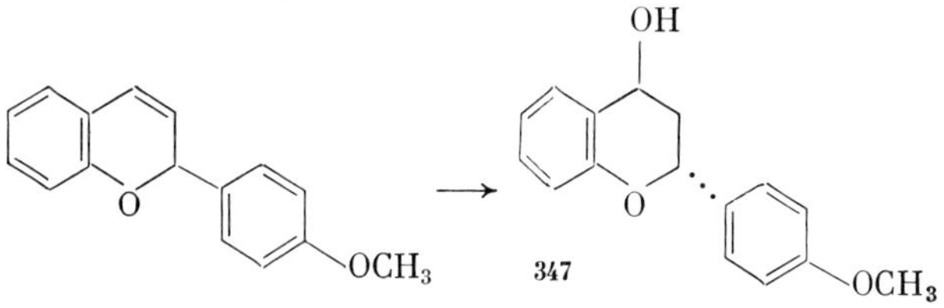

The hydroboration of 7-methoxy-substituted 4′-methoxy-3-flavene proceeds in an analogous manner. The presence of a methoxy group in the A ring promotes boration in the 3 position, and the oxygen-containing ring in the boranes obtained by this process has a tendency to break up, with the formation of phenolic products. Thus, the hydroboration of 4′,7-dimethoxy-3-flavene (*348*) leads to 1,3-diaryl-1-propanol (*349*) and 1,3-diaryl-2-propanol (*350*) [192]:

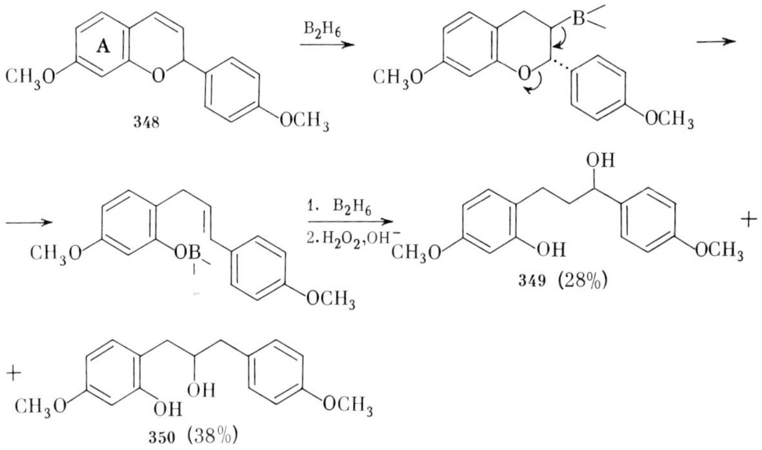

The break-up of a dihydropyran ring also takes place in the borane *351* obtained from 4′-methoxy-2-flavene [188]:

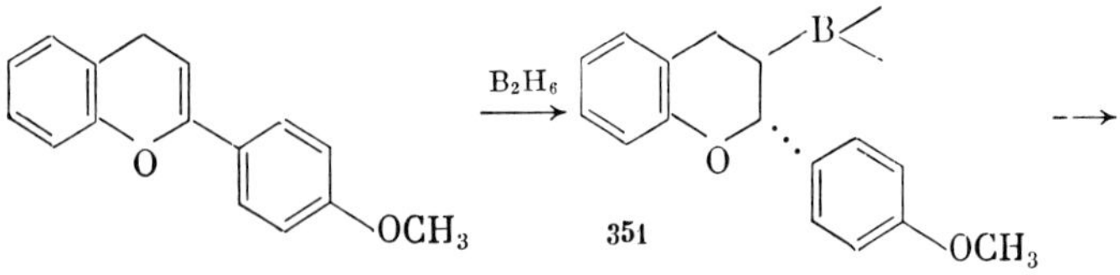

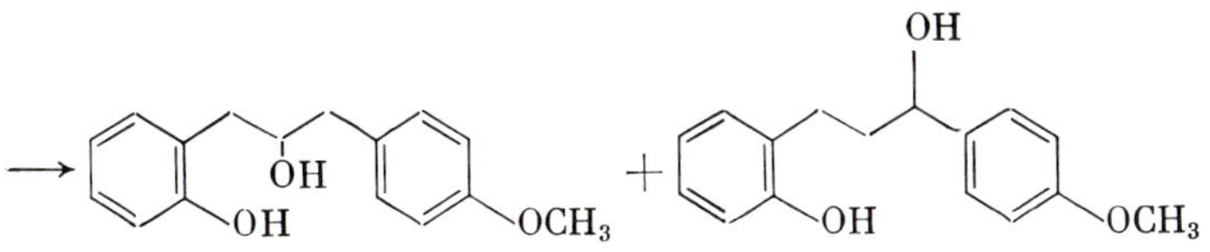

Anhydrotrimethylbraziline is converted to (±)-trimethylbraziline (*352*) by hydroboration-oxidation [193]:

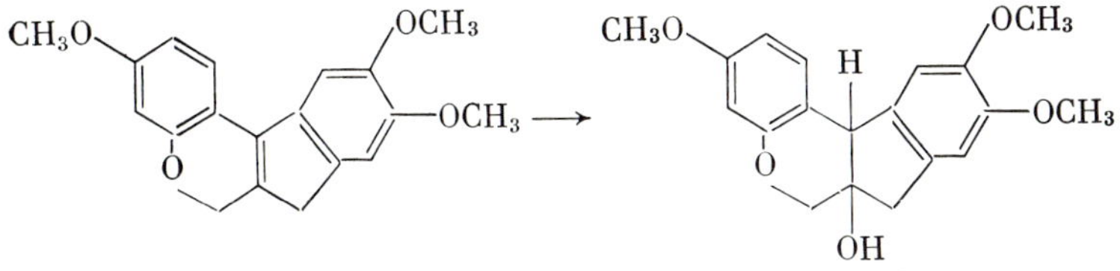

352

The use of hydroboration in the synthesis of isoflavanones is covered in [258, 259].

The hydroboration of coumarin and related compounds has been investigated.

Coumarin (*353*) [185, 194], 3-methylcoumarin (*355*) [185, 194–196], and 4,7-dimethylcoumarin (*357*) [185] form 3-chromanol (*354*), 3-methyl-4-chromanol (*356*), and 4,7-dimethyl-3-chromanol (*358*), respectively, and also compounds with an acyclic side chain:

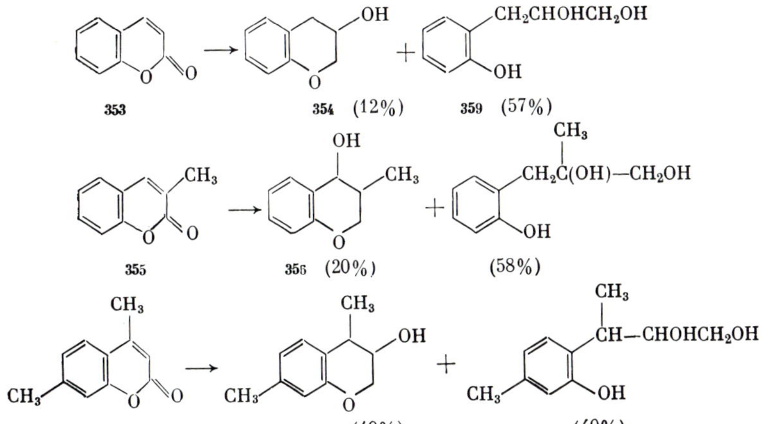

353 354 (12%) 359 (57%)

355 356 (20%) (58%)

357 358 (49%) (40%)

The course of the reactions may be described by a series of consecutive conversions, represented by the following scheme:

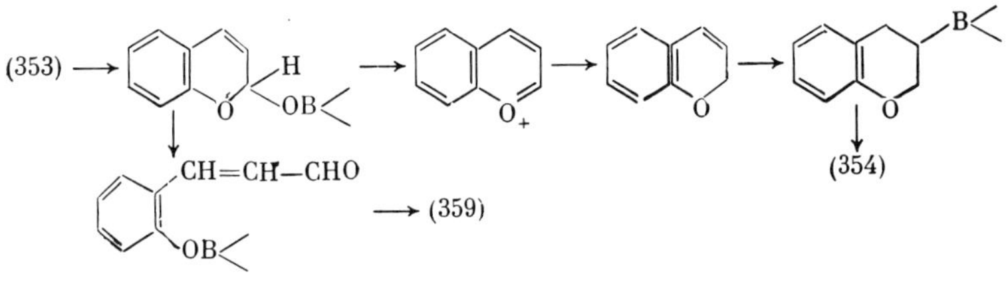

The hydroboration-oxidation of 3-methyl-4-hydroxycoumarin forms 4-hydroxy-3-methylchroman [197], while chromone (*360*) forms 3-chromanol (*361*) [185, 195]:

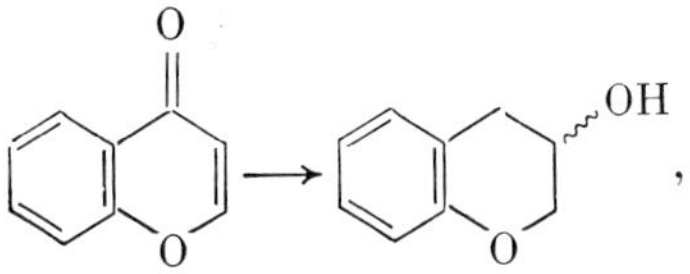

360 361 (79%)

and 3,4-cyclopenteno-4-methoxycoumarin (*362*) forms 3,4-cyclopenteno-3-hydroxy-7-methoxychroman (*363*) and products of the break-up of the heterocyclic ring [198]:

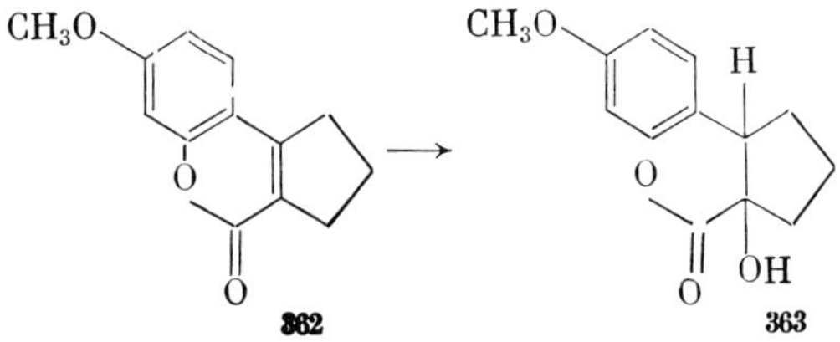

362 363

3-Phenylcoumarin, 4-hydroxy-3-phenylcoumarin [199], and substitution products of the latter compound [200] form the corresponding isoflavanones by hydroboration and subsequent oxidation with chromic acid.

3-Benzyl-4-hydroxycoumarin is converted to homoisoflavanols [260]. The hydroboration-oxidation of angelicin, lomatin, and dihydrooroselone has been investigated [261].

(25*R*)-5α-Spirost-14-ene (*364*) is converted to 14-*d*-(25*R*)-5α,14β-spirostan (*365*) by deuteroboration-protolysis [201]:

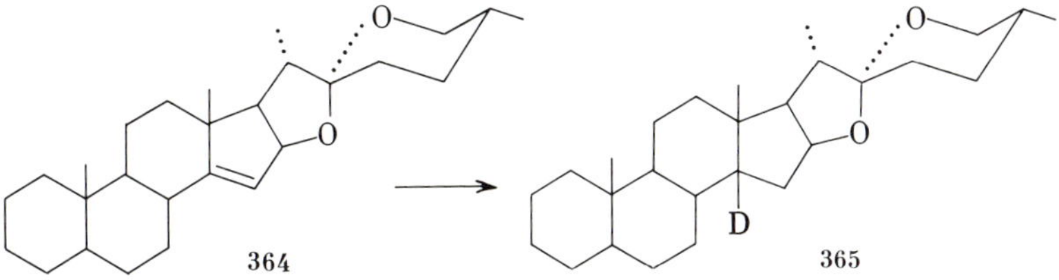

Hydroboration-oxidation has been used in the synthesis of the antibiotic monensin [262].

2. *Hydroboration of nitrogen-containing heterocyclic compounds*

The hydroboration of different substituted tetrahydropyridines has been investigated in a number of papers.

1-Methyl-1,2,3,6-tetrahydropyridine (*366*) is converted to a mixture (71:29) of 1-methyl-3-piperidinol (*367*) and 1-methyl-4-piperidinol (*368*) by the action of diborane and subsequent oxidation [202–204]:

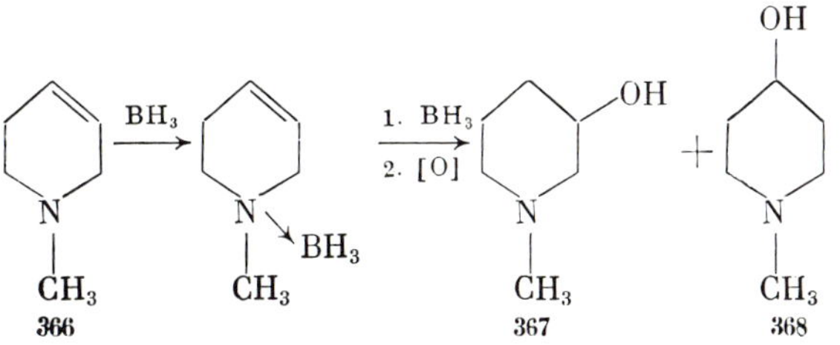

The hydroboration of compound *363* with (−)-diisopinocampheylborane showed that the alcohol *367* has an *R* configuration [203]. Isomeric alcohols of the type *367* and *368* are obtained from 1-propyl-, 1-benzyl-, and 2-phenylethyl-1,2,3,6-tetrahydropyridines in a lower yield (34–65%), but in approximately the same ratios [202].

The hydroboration of *366* with triethylamine-borane in toluene at 130°C leads exclusively to 1-methyl-4-piperidinol (*368*) [204].

trans-1,3,-Dimethyl-4-piperidinol is formed by the action of diborane on 1,3-dimethyl-1,2,3,6-tetrahydropyridine in diglyme at 25°C [204, 205].

1-Methyl-4-phenyl-1,2,3,6-tetrahydropyridine is converted to *trans*-1-methyl-4-phenyl-3-piperidinol (*369*) by hydroboration-oxidation [206]:

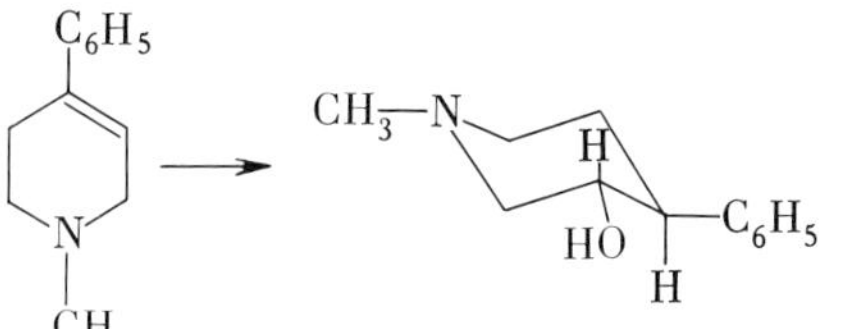

369 (67%)

The reactions of 4-alkyl-substituted 1-methyl-1,2,3,6-tetrahydropyridines with diborane proceed in an analogous manner [207]. 1-Methyl-2-alkyl-1,2,3,6-tetrahydropyridines form approximately 50% of a mixture of isomeric alcohols [208]:

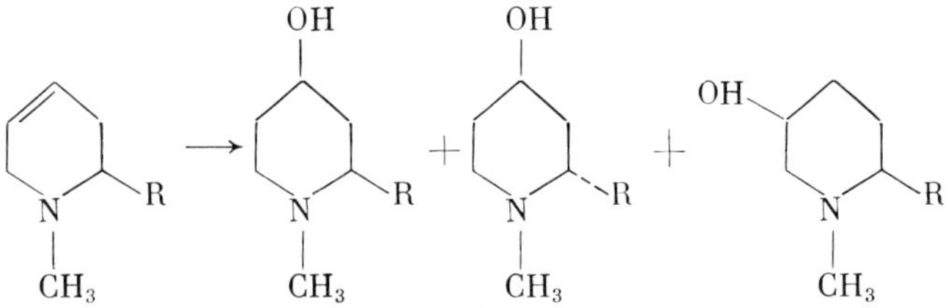

70% 14—28% 4—18%

($R=CH_3$, C_2H_5, *n*-C_3H_7, *iso*-C_3H_7)

The stereochemistry of the hydroboration of 3(or 5)-methyl-4-phenyl-1,2,5,6-tetrahydropyridines has been investigated [263].

The hydroboration of N-carbobenzyloxy-3,4-dehydro-DL-proline methyl ester (*370*), after oxidation and saponification, produces 70% of *trans*-3-hydroxy-DL-proline (*371*) and 10% of *trans*-4-hydroxy-DL-proline (*372*) [209]:

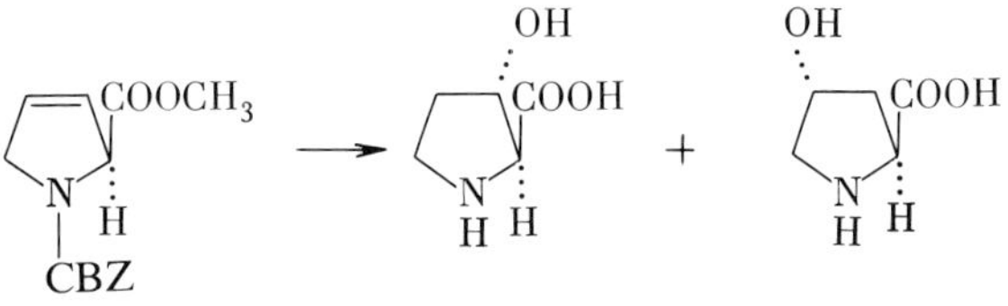

370 **371** (70%) **372** (10%)

DL-Baikiain (1,2,3,6-tetrahydropyridine-2-carboxylic acid) can be converted to *trans*-5-hydroxy- and *trans*-4-hydroxy-DL-pipecolic acids by hydroboration [209].

The hydroboration of derivatives of tropidine (*373*) proceeds primarily from the α side because the β side is shielded by a coordinately added borane molecule [202]. 2-Tropidine (*373*, R = H) forms approximately equal quantities of 2α- (*374*, R = H) and 3α-alcohols (*375*, R = H) and 3% each of the 2β- (*376*, R = H) and 3β-alcohol (*377*, R = H), with a total yield of 68%:

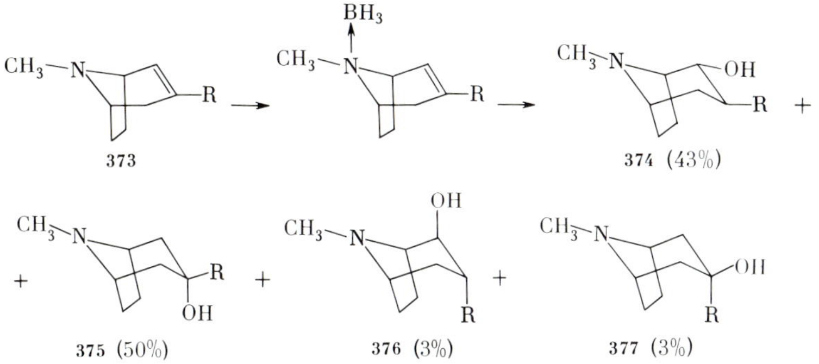

In the case of the aryl derivatives (*373*, R = C_6H_5, *p*-ClC_6H_4, *p*-$CH_3C_6H_4$), 3-aryl-2α-tropanols (*374*, R = Ar) are primarily obtained (70–80%), while the content of 3-aryl-3α-tropanols (*375*, R = Ar) in the mixture is 20–30%. 2,3-Dicarbomethoxy-2,3-diazabicyclo[2.2.1]hept-5-ene (*378*), under the action of diborane, followed by oxidation with hydrogen peroxide in alkaline medium, is converted to *380*. If the borane *379* obtained during the first stage of the reaction is treated with the alkali, β-elimination takes place, with the formation of 4-(N,N′-dicarbomethoxyhydrazino)cyclopentene (*381*) [210–212]:

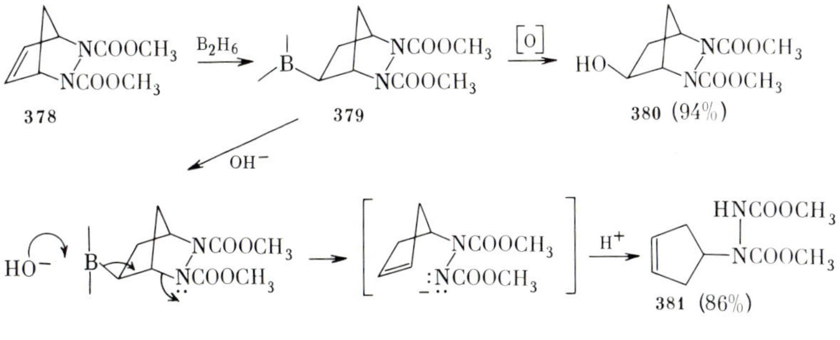

The hydroboration-oxidation of benzocycloazaheptadiene compounds has been investigated [264].

Indole and its homologs, when treated sequentially with diborane and a methanolic solution of sodium methoxide, form the corresponding indolines [213].

Lysergic acid (*382*) is converted to 6-methyl-8β-hydroxymethyl-9α-hydroxy-10α-ergoline (*383*) by hydroboration-oxidation [214]:

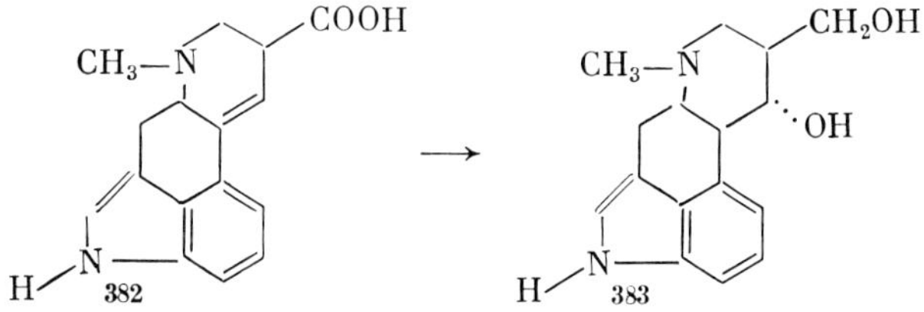

Δ^8-Deoxycodeine (*384*) undergoes hydroboration-oxidation to form *trans*-8α-hydroxydihydrodeoxycodeine (*385*) and small quantities of dihydropseudocodeine and *trans*-dihydrodeoxycodeine [215]:

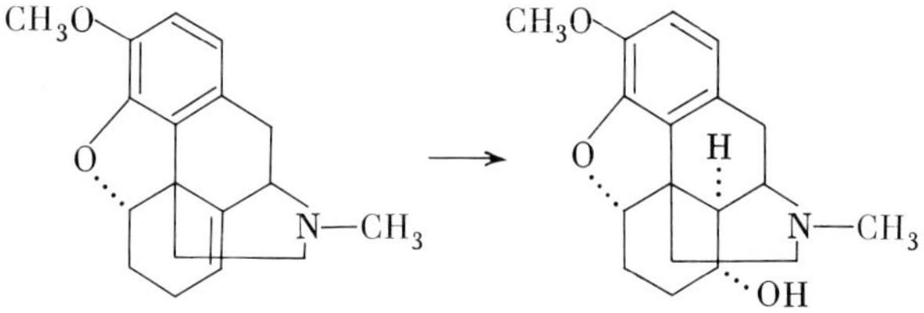

384 385

The hydroboration of isoneopine, neopine, neopinone, and thebaine has been investigated [216].

The hydroboration-oxidation of 9-phenylsulfinyl-9-azabicyclo-[3.3.1]nona-2,6-diene leads to a mixture of the *exo*-2,6- and *exo*-2,7-diols [265].

The hydroboration of vinylaziridines, which proceeds with a high degree of regio- and stereoselectivity, leads to allylamines [266].

1,3,4,7-Tetrahydro-2*H*-1,3-diazepin-2-one is converted to (±)-5-hydroxyperhydro-1,3-diazepin-2-one by hydroboration-oxidation [267].

3. Hydroboration of heterocyclic silicon and germanium compounds

1,1-Diphenylsila-2-cyclohexene is converted to 1,1-diphenylsila-2-cyclohexanol by hydroboration-oxidation [217]:

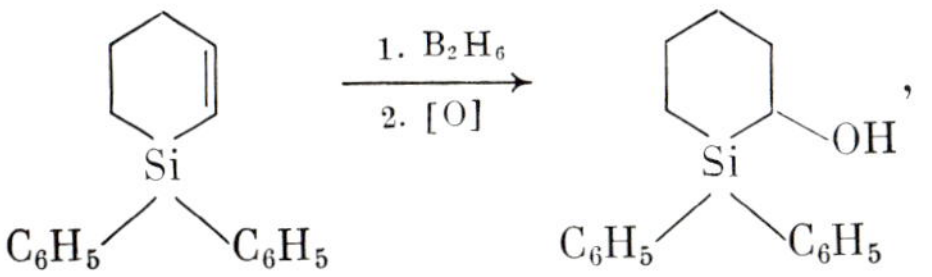

while 1-phenyl-1-methylsila-2-cyclohexene forms a mixture of diastereoisomeric carbinols (*386*) [218]:

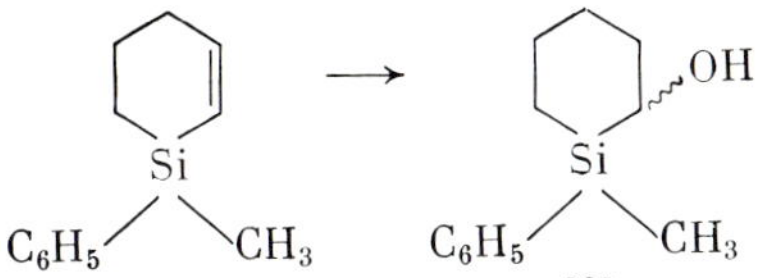

386

1,1-Dimethylsila-3-cyclopentene and 1,1,3-trimethylsila-3-cyclopentene (*387*) have been converted to the corresponding alcohols (*388*) by the above method, while derivatives of germano-3-cyclopentene (*389*) form the alcohols (*390*) [219]:

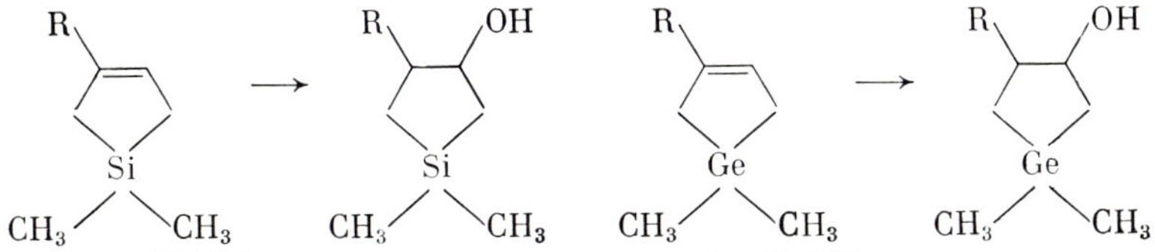

387: R=H, CH_3 388 389: R=H, CH_3 390

REFERENCES

1. D. S. Matteson and J. D. Liedtke, *J. Am. Chem. Soc.* **87,** 1526 (1965).
2. D. S. Matteson, In: "Progress in Boron Chemistry", vol. 3. Ed. R. J. Brotherton and H. Steinberg. Oxford, Pergamon Press, 1970, p. 117.
3. D. J. Pasto and J. L. Miesel, *J. Am. Chem. Soc.* **85,** 2118 (1963).
4. D. J. Pasto and C. C. Cumbo, *J. Am. Chem. Soc.* **86,** 4343 (1964).
5. D. J. Pasto and R. Snyder, *J. Org. Chem.* **31,** 2777 (1966).
6. D. J. Pasto and R. Snyder, *J. Org. Chem.* **31,** 2773 (1966).
7. D. J. Pasto and J. Hickman, *J. Am. Chem. Soc.* **89,** 5608 (1967).
8. D. S. Matteson and R. W. H. Mah, *J. Org. Chem.* **28,** 2171 (1963).
9. D. J. Pasto, J. Hickman and T. Cheng, *J. Am. Chem. Soc.* **90,** 6259 (1968).

10. J. W. Lewis and A. A. Pearce, *Tetrahedron Letters* **1964,** 2039.
11. J. W. Lewis and A. A. Pearce, *J. Chem. Soc. B,* **1969,** 863.
12. J. A. Marshall and W. S. Johnson, *J. Org. Chem.* **28,** 421 (1963).
13. J. Schmitt, J. J. Panouse, A. Hallot, P. J. Cornu, P. Comoy and H. Pluchet. *Bull. Soc. Chim. France* **1963,** 807.
14. I. J. Borowitz and G. J. Williams, *J. Org. Chem.* **32,** 4157 (1967).
15. J. Gore, J. P. Drouet and J. J. Barieux, *Tetrahedron Letters* **1969,** 9.
16. J.-J. Barieux and J. Gore, *Bull. Soc. Chim. France* **1971,** 1649.
17. J.-J. Barieux and J. Gore, *Bull. Soc. Chim. France* **1971,** 3978.
18. J.-J. Barieux and J. Gore, *Tetrahedron* **28,** 1537 (1972).
19. J.-J. Barieux and J. Gore, *Tetrahedron* **28,** 1555 (1972).
20. A. Bowers, T. G. Halsall, E. R. H. Jones and A. J. Lemin, *J. Chem. Soc.* **1953,** 2548.
21. S. Corsano and F. Bombardiere, *Ann. Chim. (Rome)* **54,** 650 (1964).
22. D. N. Butler and A. H. Soloway, *J. Am. Chem. Soc.* **88,** 484 (1966).
23. B. M. Mikhailov, V. A. Dorokhov and N. V. Mostovoi, *Izv. Akad. Nauk SSSR, Ser. Khim.* **1964,** 199.
24. B. M. Mikhailov, V. A. Dorokhov and N. V. Mostovoi, *Izv. Akad. Nauk SSSR, Ser. Khim.* **1964,** 201.
25. B. M. Mikhailov, V. A. Dorokhov, N. V. Mostovoi and I. P. Yakovlev, in the collection: *Zh. Obshch. Khim. Problems of Organic Synthesis,* Leningrad, Nauka (1965), p. 223.
26. B. M. Mikhailov, V. A. Dorokhov, N. V. Mostovoi, O. G. Boldyreva and M. N. Bochkareva, *Zh. Obshch. Khim.* **40,** 1817 (1970).
27. B. M. Mikhailov, V. A. Dorokhov and N. V. Mostovoi, in the collection *Zh. Obshch. Khim. Problems of Organic Synthesis* Leningrad, Nauka (1965), p. 228.
28. V. A. Dorokhov, O. G. Boldyreva and B. M. Mikhailov, *Zh. Obshch. Khim.* **40,** 1528 (1970).
29. D. G. White, *J. Am. Chem. Soc.* **85,** 3634 (1963).
30. N. V. Mostovoi, V. A. Dorokhov and B. M. Mikhailov, *Izv. Akad. Nauk SSSR, Ser. Khim.* **1966,** 90.
31. G. B. Butler, G. L. Statton and V. S. Brey, Jr. *J. Org. Chem.* **30,** 4194 (1965).
32. R. M. Adams and F. D. Poholsky, *Inorg. Chem.* **2,** 640 (1963).
33. Z. Polívka and M. Ferles, *Coll. Czech. Chem. Comm.* **34,** 3009 (1969).
34. V. A. Dorokhov, O. G. Boldyreva, V. S. Bogdanov and B. M. Mikhailov, *Zh. Obshch. Khim.* **42,** 1558 (1972).
35. G. B. Butler and G. L. Statton, *J. Am. Chem. Soc.* **86,** 518 (1964).
36. N. N. Greenwood, J. H. Morris and J. C. Wright, *J. Chem. Soc. (London)* **1964,** 4753.
37. B. M. Mikhailov, M. N. Bochkareva, V. S. Bogdanov, O. G. Boldyreva and V. A. Dorokhov, *Zh. Obshch. Khim.* **41,** 1546 (1971).
38. V. A. Dorokhov, O. G. Boldyreva and B. M. Mikhailov, *Zh. Obshch. Khim.* **43,** 1955 (1973).
39. B. M. Mikhailov, V. A. Dorokhov and N. V. Mostovoi, in the collection: *Zh. Obshch. Khim. Problems of Organic Synthesis*, Leningrad, Nauka (1965), p. 233.
40. V. A. Dorokhov and B. M. Mikhailov, *Izv. Akad. Nauk SSSR, Ser. Khim.* **1970,** 1804.
41. B. M. Mikhailov, V. A. Dorokhov, V. S. Bogdanov, I. P. Yakovlev and A. D. Naumov, *Dokl. Akad. Nauk SSSR* **194,** 595 (1970).
42. M. N. Bochkareva, V. A. Dorokhov, O. G. Boldyreva, V. S. Bogdanov and B. M. Mikhailov, *Zh. Obshch. Khim.* **45,** 780 (1975).

43. Z. Polívka, V. Kubelka, H. Holubová and M. Ferles, *Coll. Czech. Chem. Comm.* **35,** 1131 (1970).
44. H. C. Brown and M. K. Unni, *J. Am. Chem. Soc.* **90,** 2902 (1968).
45. H. Wille and J. Goubeau, *Chem. Ber.* **105,** 2156 (1972).
46. Z. Polivka and M. Ferles, *Coll. Czech. Chem. Comm.* **35,** 1147 (1970).
47. M. Ferles and Z. Polívka, *Coll. Czech. Chem. Comm.* **33,** 2121 (1968).
48. M. Ferles. V. Vošický and *Z. Chem.* **8,** 380 (1968).
49. B. M. Mikhailov, V. S. Bogdanov, L. S. Vasil'ev, V. A. Dorokhov, V. P. Dmitrikov, V. G. Kiselev and A. D. Naumov, *Izv. Akad. Nauk SSSR, Ser. Khim.* **1970,** 1677.
50. L. S. Vasil'ev, V. P. Dmitrikov, V. S. Bogdanov and B. M. Mikhailov, *Zh. Obshch. Khim.* **42,** 1318 (1972).
51. V. S. Bogdanov, V. G. Kiselev, A. D. Naumov, L. S. Vasil'ev, V. P. Dmitrikov, V. A. Dorokhov and B. M. Mikhailov, *Zh. Obshch. Khim.* **42,** 1547 (1972).
52. B. M. Mikhailov and T. A Shchegoleva, *Izv. Akad. Nauk SSSR, Otd. Khim. Nauk* **1959,** 546.
53. B. M. Mikhailov and A. N. Blokhina, *Izv. Akad. Nauk SSSR, Otd. Khim. Nauk* **1962,** 1373.
54. H. A. Bent, *Chem. Rev.* **61,** 275 (1961).
55. B. M. Mikhailov, *Usp. Khim.* **40,** 2121 (1971).
56. B. M. Mikhailov: "Borane Chemistry", Moscow, Nauka (1967).
57. B. M. Mikhailov and E. N. Safonova, *Izv. Nauk SSSR, Ser. Khim.* **1965,** 1487.
58. H. C. Brown and R. L. Sharp, *J. Am. Chem. Soc.* **90,** 2915 (1968).
59. D. J. Pasto and P. E. Timony, *J. Organometal Chem.* **60,** 19 (1973).
60. L. Caglioti, G. Cainelli, G. Maina and A. Selva, *Gazz. Chim. Ital.* **92,** 309 (1962).
61. M. F. Hawthorne, *J. Am. Chem. Soc.* **83,** 2541 (1961).
62. G. F. D'Alelio, *U.S. Patent* 3,115,526 (1963); C. A. **60,** 6866 (1964).
63. D. J. Pasto and J. L. Miesel, *J. Am. Chem. Soc.* **84,** 4991 (1962).
64. G. A. Russel and L. A. Ochrymowycz, *J. Org. Chem.* **35,** 764 (1970).
65. G. L. Larson, D. Hernandez and A. Hernandez, *J. Organometal Chem.* **76,** 9 (1974).
66. J. Klein, R. Levene and E. Dunkelblum, *Tetrahedron Letters* **1972,** 2845.
67. H. C. Brown and O. J. Cope, *J. Am. Chem. Soc.* **86,** 1801 (1964).
68. H. C. Brown and K. A. Keblys, *J. Am. Chem. Soc.* **86,** 1791 (1964).
69. H. C. Brown and B. M. Gallivan, Jr., *J. Am. Chem. Soc.* **90,** 2906 (1968).
70. L. S. Vasil'ev, M. M. Vartanyan and B. M. Mikhailov, *Zh. Obshch. Khim.* **42,** 2675 (1972).
71. L. S. Vasil'ev, M. M. Vartanyan and B. M. Mikhailov, *Izv. Akad. Nauk SSSR, Ser. Khim.* **1972,** 1678.
72. R. A. Braun, D. C. Brown and R. M. Adams, *J. Am. Chem. Soc.* **93,** 2823 (1971).
73. D. J. Pasto and J. Hickman, *J. Am. Chem. Soc.* **90,** 4445 (1968).
74. H. C. Brown, E. F. Knights, *J. Am. Chem. Soc.* **90,** 4439 (1968).
75. M. L. Wolfrom, K. Matsuda, F. Komitsky and T. E. Whiteley, *J. Org. Chem.* **28,** 3551 (1963).
76. H. Arzoumanian, E. M. Aston and L. Goodman, *J. Am. Chem. Soc.* **86,** 74 (1964).
77. K. J. Ryan, H. Arzoumanian, E. M. Aston and L. Goodman, *J. Am. Chem. Soc.* **86,** 2503 (1964).
78. A. Rosenthal and M. Sprinzl, *Can. J. Chem.* **47,** 4477 (1969).
79. D. W. Theobald, *J. Org. Chem.* **30,** 3929 (1965).

80. A. Suzuki, K. Ohmori, H. Takenaka and M. Itoh, *Tetrahedron Letters* **1968,** 4937.
81. A. Suzuki, K. Ohmori and M. Itoh, *Tetrahedron* **25,** 3707 (1969).
82. D. B. Bigley and D. W. Payling, *J. Chem. Soc.* **1965,** 3974.
83. A. Hassner and B. H. Braun, *Univ. Color. Studies, Ser. Chem. Pharm.* **4,** 48 (1962).
84. A. Hassner, R. E. Barnett, P. Catsonlacos and S. H. Wilen, *J. Am. Chem. Soc.* **91,** 2632 (1969).
85. B. C. Uff, M. E. Butcher, J. V. Adams, I. T. Streeting and J. W. Lewis, *J. Chem. Soc. (Perkin I)* **1974,** 566.
86. F. C. Alvarez and M. Arreguin, *Chem. and Ind.* **1960,** 720.
87. L. Caglioti and G. Cainelli, *Atti Accad. Nazl. Lincei, Rend., Classe Sci. Fis., Mat. e Nat.* **30,** 224 (1961); C. A. **56,** 7387 (1962).
88. Y. Bessière-Chrétien and T.-P. Marion, *Bull. Soc. Chim. France* **1970,** 3013.
89. I. Uzarewicz and A. Uzarewicz, *Rocz. Chem.* **44,** 1205 (1970).
90. I. Uzarewicz, A. Uzarewicz, J. Krupowicz and W. Zacharewicz, *Rocz. Chem.* **38,** 1615 (1964).
91. W. Nagata, M. Narisada and T. Wakalayshi,*Chem. Pharm. Bull.* **16,** 875 (1968).
92. J. A. Marshall, *Synthesis* **1971,** 229.
93. J. A. Marshall and G. L. Bundy, *J. Am. Chem. Soc.* **88,** 4291 (1966).
94. J. A. Marshall and J. H. Babler, *Chem. Comm.* **1968,** 993.
95. J. A. Marshall and G. L. Bundy, *Chem Comm.* **1967,** 854.
96. J. A. Marshall and G. L. Bundy, *J. Chem. Soc., D* **1967,** 855.
97. J. A. Marshall and W. F. Huffman, *J. Am. Chem. Soc.* **92,** 6358 (1970).
98. J. A. Marshall and J. H. Babler, *Tetrahedron Letters* **1970,** 3861.
99. R. Köster and G. Schomburg, *Angew. Chem.* **72,** 567 (1960).
100. B. M. Mikhailov and V. A. Dorokhov, *Izv. Akad. Nauk SSSR, Ser. Khim.* **1964,** 946.
101. L. S. Vasil'ev, M. M. Vartanyan, V. S. Bogdanov, V. G. Kiselev and B. M. Mikhailov, *Zh. Obshch. Khim.* **42,** 1540 (1972).
102. L. S. Vasil'ev, M. M. Vartanyan and B. M. Mikhailov, *Izv. Akad. Nauk SSSR, Ser. Khim.* **1976,** 2308.
103. B. M. Mikhailov and V. A. Dorokhov, *Izv. Akad. Nauk SSSR, Ser. Khim.* **1965,** 1661.
104. K. Sisido, M. Naruse, A. Saito and K. Utimoto, *J. Org. Chem.* **37,** 733 (1972).
105. B. M. Mikhailov, N. V. Mostovoi and V. A. Dorokhov, *Izv. Akad. Nauk SSSR, Ser. Khim.* **1964,** 1358.
106. B. M. Mikhailov, V. A. Dorokhov and N. V. Mostovoi, *Dokl. Akad. Nauk SSSR* **166,** 1114 (1966).
107. V. A. Dorokhov, O. G. Boldyreva and B. M. Mikhailov, *Izv. Akad. Nauk SSSR, Ser. Khim.* **1971,** 191.
108. K. Kratzl and P. Claus, *Monatsh.* **94,** 1140 (1963).
109. J. Canceill, J. Gabard and J. Jacques, *Bull. Soc. Chim. France* **1966,** 2653.
110. L. Tai and C. T. Chien, Hua Hsueh Hsueh Pao, **31,** 370 (1965); *C. A.* **64,** 8022 (1966).
111. C. T. Chien, C. Hsieh and L. Tai, Hua Hsueh Hsueh Pao, **31,** 376 (1965); *C. A.* **64,** 8022 (1966).
112. E. Dunkelblum, R. Levene and J. Klein, *Tetrahedron* **28,** 1009 (1972).
113. J. Klein and E. Dunkelblum, *Tetrahedron Letters* **1966,** 6047.
114. J. Klein and E. Dunkelblum, *Tetrahedron* **24,** 5701 (1968).
115. M. Zaidlewicz, I. Uzarewicz and A. Uzarewicz. *Rocz. Chem.* **43,** 937 (1969).
116. E. Toromanoff, *Topics in Stereochemistry* **2,** 157 (1967).

117. A. Uzarewicz, I. Uzarewicz and W. Zacharewicz, *Rocz. Chem.* **39,** 19 (1965).
118. J. Klein, R. Levene and E. Dunkelblum, *Tetrahedron Letters* **1972,** 4031.
119. E. Dunkelblum and D. Hoffer, *J. Chem. Soc., C* **1973,** 1707.
120. E. Dunkelblum, *Tetrahedron* **28,** 3879 (1972).
121. D. S. Sethi, I. Mehrotra and D. Devaprabhakara, *Tetrahedron Letters* **1970,** 2765.
122. J. Schmitz and C. A. Peters, *J. Chem. Soc., C* **1971,** 1905.
123. K. Bailay and T. G. Halsall, *J. Chem. Soc., C* **1968,** 679.
124. W. G. Dauben, B. Weinstein, P. Lim and A. B. Anderson, *Tetrahedron* **15,** 217 (1961).
125. J. M. Davies and S. H. Graham, *J. Chem. Soc., C* **1968,** 2040.
126. Y. Chrétien-Bessière, *Bull. Soc. Chim. France* **1964,** 2182.
127. Y. Chrétien-Bessière and G. Boussac, *Bull. Soc. Chim. France* **1967,** 4728.
128. Y. Chrétien-Bessière and B. Meklati, *Compt. Rend., C* **271,** 318 (1970).
129. Y. Bessière-Chrétien and B. Meklati, *Bull. Soc. Chim. France* **1971,** 2591.
130. I. Uzarewicz, A. Uzarewicz, and W. Zacharewicz, *Rocz. Chem.* **39,** 1051 (1965).
131. I. Uzarewicz, A. Uzarewicz, and W. Zacharewicz, *Rocz. Chem.* **39,** 111 (1965).
132. K. H. Schulte-Elte and G. Ohloff *Helv. Chim. Acta* **50,** 153 (1967).
133. W. J. Wechter, *Chem. and Ind.* **1959,** 294.
134. M. Nussim, J. Mazur, F. Sondheimer, *J. Org. Chem.* **29,** 1120 (1964).
135. L. Caglioti and G. Cainelli, *Atti. Accad. Nazl. Lincei, Rend., Classe Sci. Fis., Mat. e Nat.,* **29,** 255 (1960); *C. A.* **56,** 7386 (1962).
136. L. Caglioti and G. Cainelli, *U.S. Patent* 3,083,199 (1963); *C. A.* **59,** 5238 (1963).
137. I. Midgley and C. Djerassi, *Tetrahedron Letters* **1972,** 4673.
138. L. Caglioti, G. Cainelli and A. Selva, *Chim. Ind.* (*Milan*) **44,** 36 (1962); *C. A.* **60,** 12075 (1964).
139. L. Caglioti, G. Cainelli and G. Maina, *Tetrahedron* **19,** 1057 (1963).
140. M. Nussim, J. Mazur and F. Sondheimer, *J. Org. Chem.* **29,** 1131 (1964).
141. H. Nakata, *Bull. Chem. Soc. Japan* **38,** 378 (1965).
142. C. Kaneko, A. Sigimoto, S. Tamada, M. Ishikawa, S. Sasaki and T. Suda, *Chem. Pharm. Bull.* **22,** 2101 (1974).
143. H. Tada and Y. K. Sawa, *J. Org. Chem.* **33,** 3347 (1968).
144. M. Stefanović and S. Lajšić, *Tetrahedron Letters* **1967,** 1777.
145. L. Caglioti, G. Cainelli, G. Maina and A. Selva, *Tetrahedron* **20,** 957 (1964).
146. E. Mincione and F. Feliziani, *Chem. Comm.* **1973,** 942.
147. E. Mincione and D. Iocco, *Ann. Chim. (Rome)* **62,** 285 (1972); *C. A.* **77,** 114637 (1972).
148. J. Bottin and M. Fetizon, *Chem. Comm.* **1971,** 1087.
149. J. Bottin and M. Fetizon, *Bull. Soc. Chim. France* **1972,** 2344.
150. E. Mincione and C. Iavarone, *Gazz. Chim. Ital.* **101,** 956 (1971).
151. E. Mincione, F. Feliziani and O. Rossi, *Ann. Chim. (Rome)* **63,** 297 (1973); *C. A.* **81,** 37718 (1974).
152. R. Dulou and Y. Chrétien-Bessière, *Bull. Soc. Chim. France* **1959,** 1362.
153. S. Fore and W. Bickford, *J. Org. Chem.* **24,** 920 (1959).
154. H. C. Brown and K. A. Keblys, *J. Am. Chem. Soc.* **86,** 1795 (1964).
155. F. G. A. Stone and H. J. Emeleus, *J. Chem. Soc.* **1950,** 2755.
156. K. Kratzl and P. Claus, *Monatsh.* **94,** 1140 (1963).
157. A. Uzarewicz, *Rocz. Chem.* **38,** 957 (1964).
158. J. Klein, E. Dunkelblum and D. Avrahami, *J. Org. Chem.* **32,** 935 (1967).
159. D. Varech and J. Jacques, *Bull. Soc. Chim. France* **1969,** 3505.
160. D. Döpp, *Chem. Ber.* **102,** 1081 (1969).

161. O. Attanasi, G. Baccolini, L. Caglioti and G. Rosini, *Gazz. Chim. Ital.* **103,** 31 (1973).
162. F. G. A. Stone and W. A. G. Graham, *Chem. and Ind.* **1955,** 1181.
163. B. Bartocha, W. A. G. Graham and F. G. A. Stone, *J. Inorg. Nucl. Chem.* **6,** 119 (1958).
164. M. F. Hawthorne and J. A. Dupont, *J. Am. Chem. Soc.* **80,** 5830 (1958).
165. P. Binger and R. Köster, *Tetrahedron Letters* **1961,** 156.
166. R. Köster, G. Griasnow, W. Larbig and P. Binger, *Ann.* **672,** 1 (1964).
167. S. J. Cristol, F. P. Parungo and D. E. Plorde, *J. Am. Chem. Soc.* **87,** 2870 (1965).
168. M. F. Hawthorne, *J. Am. Chem. Soc.* **82,** 1886 (1960).
169. H. C. Brown and S. P. Rhodes, *J. Am. Chem. Soc.* **91,** 2149 (1969).
170. H. C. Brown and E. F. Knights, *Israel J. Chem.* **6,** 691 (1968).
171. R. J. Müller and B. L. Murr, *J. Org. Chem.* **39,** 2810 (1974).
172. A. J. Fry and W. B. Farnham, *J. Org. Chem.* **34,** 2314 (1969).
173. I. Uzarewicz, E. Zientek and A. Uzarewicz, *Rocz. Chem.* **46,** 1069 (1972).
174. D. Seyferth, *J. Inorg. Nucl. Chem.* **7,** 152 (1958).
175. D. Seyferth, *J. Am. Chem. Soc.* **81,** 1844 (1959).
176. M. Kumado, N. Imake and K. Yamamoto, *J. Organometal. Chem.* **6,** 490 (1966).
177. P. R. Jones and J. K. Myers. *J. Organometal. Chem.* **34,** C9 (1972).
178. B. M. Mikhailov and A. N. Blokhina, *Zh. Obshch Khim.* **30,** 3615 (1960).
179. D. Seyferth, H. Yamazaki and J. Sato, *Inorg. Chem.* **2,** 734 (1963).
180. C. Benezza and G. Ourisson, *Bull. Soc. Chim. France* **1967,** 624.
181. M. Zaidlewicz and A. Uzarewicz, *Rocz. Chem.* **47,** 1433 (1973).
182. M. Zaidlewicz and A. Uzarewicz, *Rocz. Chem.* **48,** 467 (1974).
183. G. Zweifel and J. Plamondon, *J. Org. Chem.* **35,** 898 (1970).
184. J. Core and F. Guigues, *Bull. Soc. Chim. France* **970,** 3521.
185. W. C. Still and D. J. Goldsmith, *J. Org. Chem.* **35,** 2282 (1970).
186. R. M. Srivastava and R. K. Brown, *Can. J. Chem.* **48,** 2334 (1970).
187. B. S. Kirkiacharian and M. Garnier, *Compt. Rend.* **277,** 1037 (1973).
188. J. W. Clark-Lewis and E. J. McGarry, *Austral. J. Chem.* **26,** 819 (1973).
189. K. Hanaya, *Bull. Chem. Soc. Japan* **43,** 442 (1970).
190. S. Yamaguchi, S. Ito, I. Suzuki and N. Inoue, *Bull. Chem. Soc. Japan* **41,** 2073 (1968).
191. J. W. Clark-Lewis and M. I. Baig, *Austral. J. Chem.* **24,** 2581 (1971).
192. J. W. Clark-Lewis and E. J. McGarry, *Austral. J. Chem.* **26,** 809 (1973).
193. B. S. Kirkiacharian, *Compt. Rend. C* **274,** 2096 (1972).
194. B. S. Kirkiacharian and D. Raulais, *Compt. Rend.* **269,** 464 (1969).
195. B. S. Kirkiacharian and D. Raulais, *Bull. Soc. Chim. France* **1970,** 1139.
196. B. S. Kirkiacharian and G. H. Elia and G. Mahuzier, *Compt. Rend. C* **279,** 151 (1974).
197. B. S. Kirkiacharian, *Bull. Soc. Chim. France* **1973,** 999.
198. B. S. Kirkiacharian, D. Billet. *Bull. Soc. Chim. France* **1972,** 3292.
199. B. S. Kirkiacharian, *Chem. Comm.* **1975,** 162.
200. B. S. Kirkiacharian and H. Chidiac, *Compt. Rend.* **280,** 775 (1975).
201. W. H. Faul, A. Failli and C. Djerassi, *J. Org. Chem.* **35,** 2571 (1970).
202. R. E. Lyle, K. R. Carle, C. R. Ellefson and C. K. Spicer, *J. Org. Chem.* **35,** 802 (1970).
203. R. E. Lyle and C. K. Spicer, *Tetrahedron Letters* **1970,** 1133.
204. Z. Polívka and M. Ferles, *Coll. Czech. Chem. Comm.* **35,** 2392 (1970).
205. M. Ferles, J. Hauer and Z. Polívka, *Coll. Czech. Chem. Comm.* **36,** 4099 (1971).

206. R. E. Lyle, D. H. McMahon, W. E. Krueger and C. K. Spicer, *J. Org. Chem.* **31,** 4164 (1966).
207. M. Ferles, J. Hauer, J. Kalář, Z. Polívka and P. Štern, *Coll. Czech. Chem. Comm.* **37,** 2464 (1972).
208. P. Štern. P. Trška and M. Ferles, *Coll. Czech. Chem. Comm.* **39,** 2267 (1974).
209. Y. Fujita, F. Irreverre and B. Witkop, *J. Am. Chem. Soc.* **86,** 1844 (1964).
210. E. L. Allred, C. L. Anderson and R. L. Smith, *Tetrahedron Letters* **1966,** 951.
211. E. L. Allred, C. L. Anderson and R. L. Smith, *J. Org. Chem.* **31,** 3493 (1966).
212. E. L. Allred and R. L. Smith, *J. Org. Chem.* **31,** 3498 (1966).
213. S. A. Monti and R. R. Schmidt, *Tetrahedron* **27,** 3331 (1971).
214. G. Cainelli, L. Caglioti and W. Barbieri, *Farmaco. Ed. Sci.* **22,** 456 (1967); *C. A.* **67,** 117028 (1967).
215. H. Kugita and M. Takeda, *Chem. Pharm. Bull.* **13,** 1422 (1965).
216. M. Takeda, H. Inoue and H. Kugita, *Tetrahedron* **25,** 1839 (1969).
217. A. G. Brouk and J. B. Pierce, *J. Org. Chem.* **30,** 2566 (1965).
218. A. G. Brouk, H. W. Kučera and D. M. MacRae, *Can. J. Chem.* **48,** 818 (1970).
219. G. Manuel, P. Mazerolles and J.-C. Florence, *Compt. Rend., C* **269,** 1553 (1969).
220. M. Montary and J. Gore, *Bull. Soc. Chim. France* **1975,** 2622.
221. M. Montary and J. Gore, *Compt. Rend. C* **282,** 1037 (1976).
222. M. Baboulene, J.-L. Torregrosa, V. Speziole, A. Lattes, *Bull. Soc. Chim. France* **1980,** II-565.
223. G. L. Larson and L. M. Fuentes, *Synth. Comm.* **9,** 841 (1979).
224. G. L. Larson and A. Hernandez, *J. Organomet. Chem.* **102,** 123 (1975).
225. G. L. Larson, E. Hernández, C. Alonso and I. Nieves, *Tetrahedron Lett.* **1975,** 4005.
226. R. Rossi, A. Carpita and M. L. Gaudenri, *Synthesis* **1981,** 359.
227. R. Moreau, Y. Adam and P. Loiseau *Compt. Rend. C* **287,** 287 (1978).
228. J. Tsuji, H. Yasuda and T. Mandai, *J. Org. Chem.* **43,** 3606 (1978).
229. R. Murphy and R. H. Prager, *Aust. J. Chem.* **29,** 617 (1976).
230. R. Murphy and R. H. Prager, *Aust. J. Chem.* **34,** 143 (1981).
231. J. C. Chen, *J. Organomet. Chem.* **156,** 213 (1978).
232. A. Uzarewicz, E. Segiet-Kujawa and I. Uzarewicz, *Rocz. Chem.* **51,** 1537 (1977).
233. J. Walinsky and R. H. Bedoukian, *J. Org. Chem.* **41,** 278 (1976).
234. L. Borowiecki and E. Reca, *Rocz. Chem.* **50,** 1689 (1976).
235. I. Uzarewicz and A. Uzarewicz, *Pol. J. Chem.* **52,** 1907 (1978).
236. I. Uzarewicz, E. Zientek and A. Uzarewicz, *Rocz. Chem.* **50,** 1515 (1976).
237. T. J. Brockson and J. T. B. Ferreira, *Synth. Comm.* **11,** 105 (1981).
238. S. C. Welch and S. Chayabunjonglerd, *J. Am. Chem. Soc.* **101,** 6768 (1979).
239. B. K. Varma, M. Koreeda, B. Yagen, K. Nakanishi and E. Caspi, *J. Org. Chem.* **40,** 3689 (1975).
240. M. Anastasia, A. Scola and G. Galli, *J. Org. Chem.* **41,** 1064 (1976).
240a. E. Mincione and F. Feliziani, *Ann. Chim. (Rome)* **65,** 209 (1975).
241. E. Mincione and A. Stirna, *Ann. Chim. (Rome)* **65,** 421 (1975).
242. E. Mincione and A. Stirna, *Ann. Chim. (Rome)* **65,** 473 (1975).
243. E. Mincione, *Ann. Chim. (Rome)* **67,** 119 (1977).
244. H. Hirschmann, F. B. Hirschmann and Y. Gopichand, *J. Org. Chem.* **44,** 180 (1979).
245. E. Klinotova. S. Besak and A. Vystrcil, Coll. Czech. Chem. Comm. **43,** 2204 (1978).
246. L. G. Matyukhina and I. A. Saltykova, *Zh. Obshch. Khim.* **46,** 2759 (1976).
247. A. A. L. Gunatilaka and A. F. Mateos, *J. Chem. Soc. Perkin Trans. I* **1979,** 935.
248. T. A. Bryson and W. E. Pye, *J. Org. Chem.* **42,** 3214 (1977).

249. G. L. Larson, M. Ortiz and M. R. de Roca, *Synth. Comm.* **11,** 583 (1981).
250. A. Arase, M. Hoshi and Y. Masuda, *Bull. Chem. Soc. Japan* **54,** 299 (1981).
251. S. Takano, K. Ogasawara, I. Nagayama and T. Kutsuma, *Synth. Comm.* **6,** 349 (1976).
252. H. L. Goering and S. L. Trenbeath, *J. Am. Chem. Soc.* **98,** 5016 (1976).
253. T. F. O. Lim, J. K. Myers, G. T. Rogers and P. R. Jones, *J. Organomet. Chem.* **135,** 249 (1977).
254. J. A. Soderquist and A. Hassner, *J. Organomet. Chem.* **156,** C12 (1978).
255. J. A. Soderquist and H. C. Brown, *J. Org. Chem.* **45,** 3571 (1980).
256. M. De Jesus, O. Rosario and G. L. Larson, *J. Organomet. Chem.* **132,** 301 (1977).
257. M. Yamashita, M. Kojima, H. Yoshida, T. Ogata and S. Inokawa, *Bull. Chem. Soc. Japan* **53,** 1625 (1980).
258. B. S. Kirkiacharian and H. Chidiac, *Compt. Rend. C* **280,** 775 (1975).
259. B. S. Kirkiacharian, *J. Chem. Soc., Chem. Comm.* **1975,** 162.
260. B. S. Kirkiacharian, P. Reynaud, and F. W. Wehrli, *Compt. Rend. C* **282,** 907 (1976).
261. K. L. Bhat, V. S. Kamat and G. K. Trivedi, *Ind. J. Chem., Sect. B* **18B,** 11 (1979).
262. G. Schmid, T. Fukuyama, K. Akasaka and Y. Kishi, *J. Am. Chem. Soc.* **101,** 259 (1979).
263. M. A. Iorio, G. N. Barrios, E. Menichini and A. Mazzeo-Farina, *Tetrahedron* **31,** 1959 (1975).
264. P. S. Anderson, *Tetrahedron Lett.* **1976,** 1141.
265. T. Momoze and S. Atarashi, *Heterocycles* **6,** 469 (1976).
266. R. Chaabouni, A. Laurent and B. Marquet, *Tetrahedron Lett.* **1976,** 757.
267. V. E. Marquez, P. S. Liu, J. A. Kelley and J. S. Driscoll, *J. Org. Chem.* **45,** 485 (1980).

CHAPTER 4

Isomerization and transalkylation of alkyl- and cycloalkylboranes

Triorganoboranes show the ability to undergo thermal isomerization, which is of preparative as well as theoretical interest.

A second characteristic of trialkylboranes is their ability to react with olefins upon heating, with replacement of lower by higher radicals. In combination, these two reactions make it possible to carry out counter-thermodynamic isomerization of non-terminal olefins to α-olefins.

4.1 ISOMERIZATION

Hennion, et al. [1, 2] were the first to observe that trialkylboranes with secondary and tertiary hydrocarbon radicals isomerize upon heating to compounds with primary radicals. In an attempt to obtain tri-*tert*-butylborane from *tert*-butylmagnesium bromide and boron trifluoride, organoboron compounds with isobutyl and *tert*-butyl groups were isolated, while the heating of tri-*sec*-butylborane produced tributylborane (82%) and di(*sec*-butyl)butylborane (18%) [2, 3]:

$$(CH_3CH_2\overset{CH_3}{\overset{|}{C}}H)_3B \xrightarrow{120^\circ C} (CH_3CH_2CH_2CH_2)_3B + (CH_3CH_2\overset{CH_3}{\overset{|}{C}}H)_2BCH_2CH_2CH_2CH_3$$

Triisopropylborane [3–5] is converted at 150–160°C to an equilibrium mixture of trialkylboranes containing 83% of tripropylborane,

16% of dipropylisopropylborane, and 1% of propyldiisopropylborane:

$$(iso\text{-}C_3H_7)_3B \xrightarrow{160°C} \underset{83\%}{(n\text{-}C_3H_7)_3B} + \underset{16\%}{(n\text{-}C_3H_7)_2BC_3H_7\text{-}iso} + \underset{1\%}{(iso\text{-}C_3H_7)_2BC_3H_7\text{-}n}.$$

When the temperature is increased, the proportion of boron atoms bonded to secondary carbon atoms in the equilbrium mixture increases slightly. Trialkylboranes with longer alkyl radicals contain predominantly the primary radical and a small quantity of the secondary radical in the equilibrium state.

In the presence of branching in the hydrocarbon radical, the boron atom is primarily bonded to the carbon atom furthest removed from the branching point.

As found by Brown and Zweifel [6], the isomerization of trialkylboranes is accelerated in the presence of alkyldiboranes. Thus, the hydroboration of 2-hexene (with an excess of the starting hydrocarbon) and subsequent heating in diglyme for one hour at 150°C leads, after oxidation, to a mixture of 1-, 2-, and 3-hexanols in a ratio of 13:64:23, whereas, in the presence of compounds with B—H bonds, the above-mentioned alcohols are obtained from 2-hexene in a ratio of 88:7:5 during the same time:

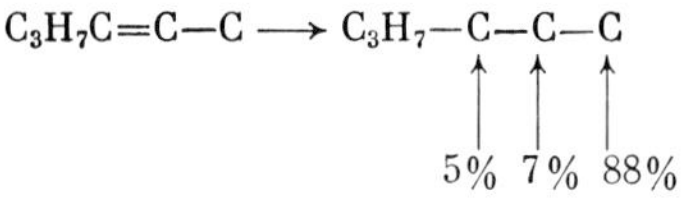

Optically active compounds undergo racemization as a result of thermal isomerization. Thus, tris[(R)-2-methylbutyl]borane is slowly converted to the racemate at 100°C [7]:

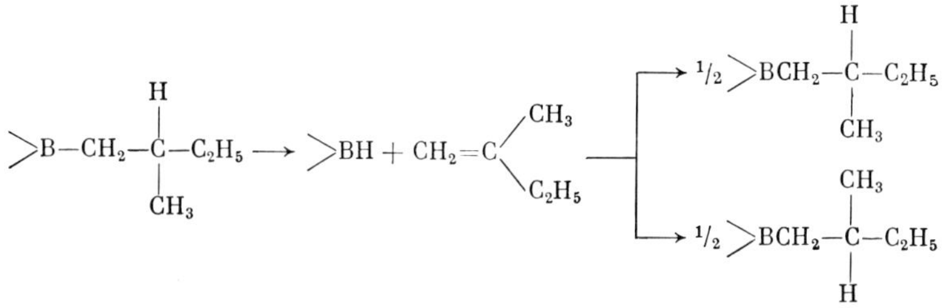

Organoboranes formed from cyclic olefins also undergo thermal isomerization.

Bicyclic organoboranes isomerize particularly easily. The addition of a hydroborating agent to exo-methylenic bicyclic hydrocarbons, which is controlled by a kinetic factor, proceeds from the spatially less hindered side and leads to the less stable axial or exo derivative. Thus, β-pinene forms the borane (*1*), the oxidation of which produces *cis*-myrtanol (*4*). If the organoborane *1* is subjected to brief heating at 125°C, it isomerizes to a thermodynamically more stable compound with an equatorially located substituent (endo derivative) (*2*), which is converted to *trans*-myrtanol (*3*) by oxidation [8]:

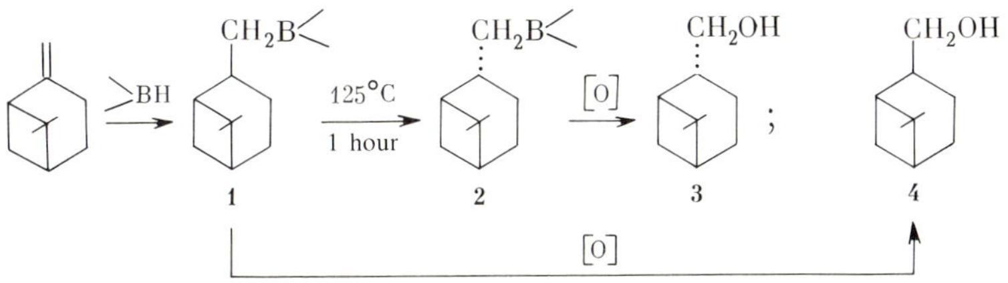

In the thermal isomerization of the organoborane 5, which is obtained from α-pinene, there is not only a transfer of the boron atom into the side chain but also, as in the case of β-pinene, a transition of the substituent from the axial to the equatorial position (*2*), so that the oxidation of the isomerized organoborane forms *trans*-myrtanol (*3*) [9]:

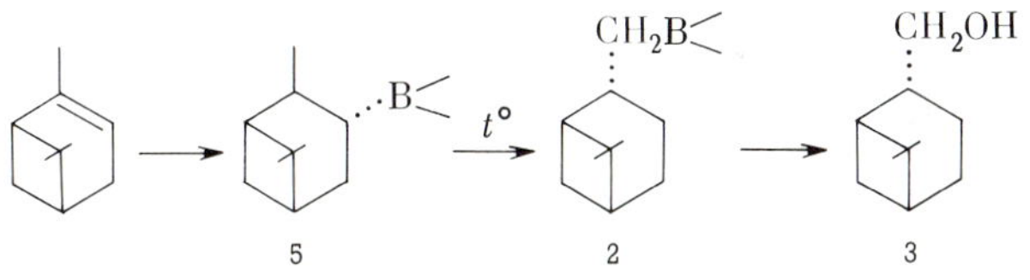

Monocyclic organoboranes isomerize under more severe conditions. The isomerization of 1-alkylcyclohexenes requires heating at 160°C, and, after 4 hours, only a 54–70% transfer of the boron atom to the side chain has taken place. Thus, the organoborane from 1-methylcyclohexene (olefin: BH_3 = 3:1 + 20% excess of hydride), after heating and oxidation, forms a mixture of alcohols containing 54% of cyclohexylcarbinol (*6*) and 46% of secondary and tertiary alcohols:

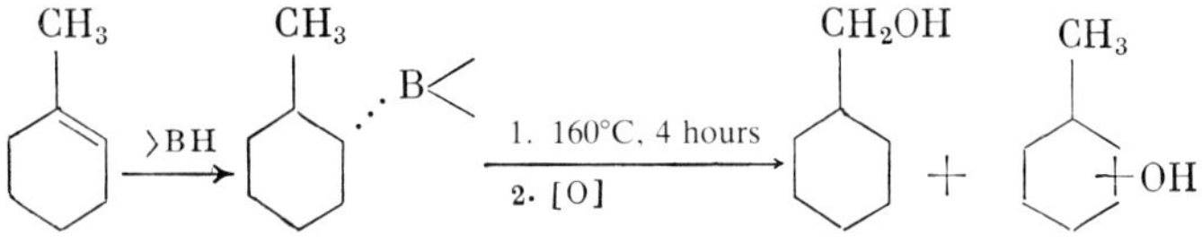

6

The isomerization rate depends on the size of the hydrocarbon ring. Thus, the organoborane from 1-methylcyclopentene isomerizes to an extent of only 30% under the conditions indicated [9].

Boranes of the steroid series, obtained from 5α-cholest-2-ene and 5-cholestene isomerize primarily to the equatorial 3β and 2α derivatives [10].

Other examples of the isomerization of boranes of the steroid series were considered in Chapter 2.

The mechanism of the thermal isomerization of trialkylboranes, according to the concepts of Köster [5] and Brown [11], consists of the elimination of a dialkylborane and its subsequent addition to another atom of the olefin formed.

This mechanism ("elimination-addition") does not explain the role of compounds with B—H bonds. which accelerate the isomerization of trialkylboranes. The catalytic action of compounds with B—H bonds is reflected in the isomerization mechanism proposed by Williams [12]. According to this mechanism, the dialkylborane *7* present in the reaction mixture undergoes isomerization according to the following scheme:

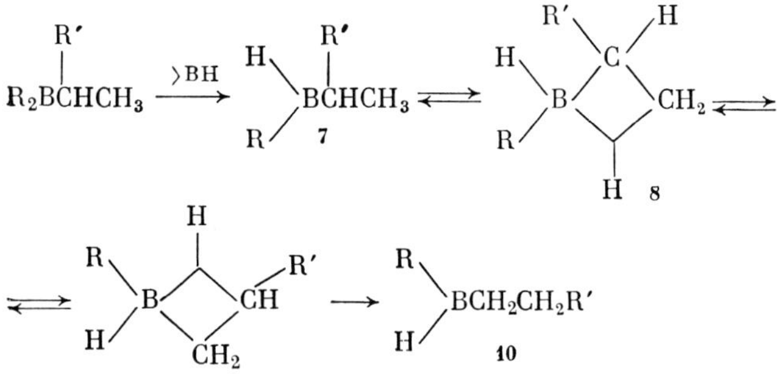

9

The above scheme provides for the formation of a three-center B—H—C bond in the intermediate compound *8*, which rearranges via the thermodynamically more stable intermediate compound *9* to form a dialkylborane with a boron-α-carbon bond (*10*). The later rapid exchange by alkyl groups of the isomerized dialkylborane *10* with the starting trialkylborane leads to a partially isomerized trialkylborane. Repeated cycles of the conversions produce an equilibrium mixture of trialkylboranes with a predominant content of boranes with α-alkyl radicals.

4.2 TRANSALKYLATION

Köster [13] discovered that trialkylboranes, when heated with higher olefins, are capable of splitting off unsaturated hydrocarbons, with the formation of new trialkylboranes containing higher radicals. This transalkylation reaction, which is of interest from both the theoretical and methodical aspect, was later subjected to detailed investigation [14–18]. It proceeds by a consecutive replacement of a lower radical by a higher one and may be represented, in a general form, by the following series of equilibrium reactions:

$$B(C_nH_{2n+1})_3 + C_mH_{2m} \rightleftharpoons B(C_nH_{2n+1})_2(C_mH_{2m+1}) + C_nH_{2n},$$
$$B(C_nH_{2n+1})_2C_mH_{2m+1} + C_mH_{2m} \rightleftharpoons B(C_nH_{2n+1})(C_mH_{2m+1})_2 + C_nH_{2n},$$
$$B(C_nH_{2n+1})(C_mH_{2m+1})_2 + C_mH_{2m+1} \rightleftharpoons B(C_mH_{2m+1})_3 + C_nH_{2n}$$
$$(n \geqslant 3;\ m \geqslant 2).$$

A shift in the equilibrium to the right is achieved by distillation of the low-boiling olefin (C_nH_{2n}).

The formation of symmetrical trialkylboranes also proceeds in the case where the process does not go to completion, since the mixed trialkylboranes initially obtained are unstable at high temperatures [19–21] and rearrange to form symmetrical compounds.

The reaction mechanism consists of consecutive reversible conversion, including decomposition of trialkylboranes to olefins and tetraalkyldiboranes and addition of the latter primarily to olefins, which are present in excess in the reaction mixture [15, 18].

$$2B(C_nH_{2n+1})_3 \rightleftharpoons B_2H_2(C_nH_{2n+1})_4 + 2C_nH_{2n},$$
$$B_2H_2(C_nH_{2n+1})_4 + 2C_mH_{2m} \rightleftharpoons 2B(C_nH_{2n+1})_2(C_mH_{2m+1}).$$

Cleavage of an olefin from the trialkylborane, in most cases, proceeds in the same direction as the addition of the dialkylborane to the olefin. Thus if, for example, 2-methyl-1-butene or 3-ethyl-2-pentene are hydroborated and the reaction products are then treated for 3–6 hours with excess 1-decene at 160°C, the starting hydrocarbon, with an impurity of insignificant quantities of isomeric compounds, distills off in both cases. This indicates that, in the transalkylation process, isomerization takes place to only a negligible degree.

Organoboron compounds from 1-methylcyclenes eliminate an olefin in two directions [22]:

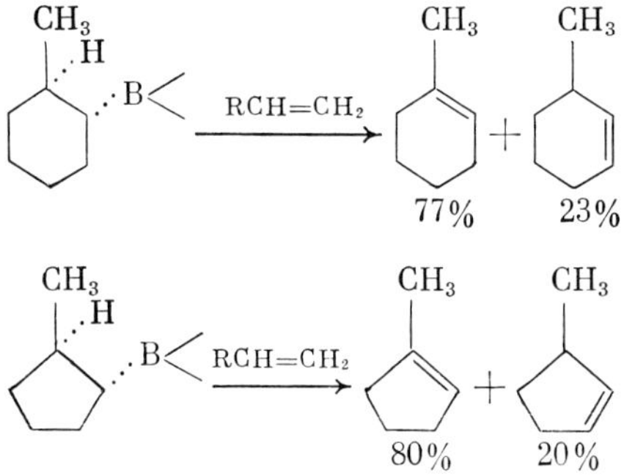

The displacement of a lower olefin by a higher one is used for the synthesis of tritium-labeled 1-hexene [23]. For this purpose, 1-hexene is hydroborated with diborane-H^3 and the labeled trialkylborane is then heated with decene at 160–170°C. The transalkylation of tripropylborane by 1-dodecene has been investigated [24]. To accelerate the transalkylation reaction, the use of catalysts, such as $(C_2H_5)_3PO_4$, $[(CH_3)_2N]_3PO$, $HCON(CH_3)_2$, $(CH_3)_2SO$, $[(CH_3)_2N]_2CO$, and $(C_4H_9)_2S$, is recommended [25]. Higher trialkylboranes with a molecular weight of approximately 300 are obtained from lower ones (tripropylborane) and ethylene in the presence of tripropylaluminum [26].

Trialkylboranes enter into a displacement reaction with paraffin hydrocarbons; for example, trioctylborane produces 1- and 2-octenes when heated with cetane [27].

Functional derivatives of trialkylboranes can be synthesized by transalkylation. Thus tris[3-(2-chloro-1,1,2-trifluoroethoxy)propyl]-borane has been obtained by heating triisobutylborane with 2-

chloro-1,1,2-trifluoroethyl allyl ether:

$$(iso\text{-}C_4H_9)_3B + 3CH_2{=}CHCH_2OCF_2CHFCl \longrightarrow (ClFCHCF_2OCH_2CH_2CH_2)_3B + 3iso\text{-}C_4H_8,$$

and tri(3-trimethylsilylpropyl) borane and tri(3-trichlorosilylpropyl)-borane have been synthesized from allyltrimethylsilane and allyltrichlorosilane, respectively [28]:

$$[(CH_3)_3SiCH_2CH_2CH_2]_3B \quad \text{and} \quad (Cl_3SiCH_2CH_2CH_2)_3B$$

Analogous conversions can be carried out with vinyl derivatives of silicon [29]. 2-Methyl-2-nitrosopropane reacts with trialkylboranes with olefin displacement [30]:

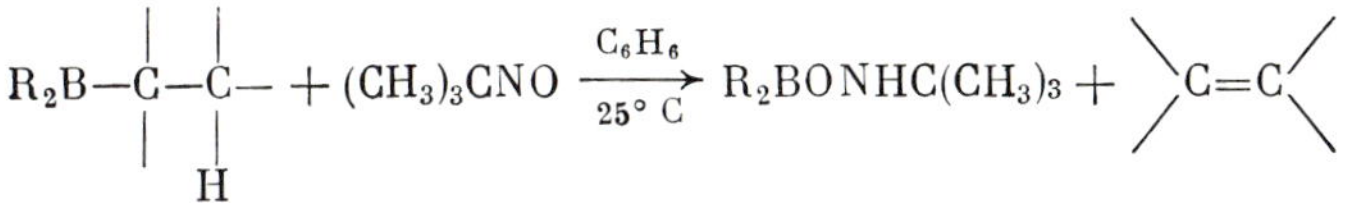

Not only trialkylboranes but also other organoboron compounds containing one or two alkyl radicals and a functional group react with olefins [31, 32]. When chlorodiisobutylborane or bromodiisobutylborane is heated with 1-heptene at 130–150°C, a replacement of both isobutyl radicals take place:

$$(iso\text{-}C_4H_9)_2BX + C_5H_{11}CH{=}CH_2 \longrightarrow (C_5H_{11}CH_2CH_2)_2BX + 2iso\text{-}C_4H_8 \quad (X = Cl, Br)$$

The transalkylation of dibromoisobutylborane by 1-heptene or 1-octene proceeds at 130–140°C at a slightly slower rate:

$$iso\text{-}C_4H_9BBr_2 + RCH{=}CH_2 \longrightarrow RCH_2CH_2BBr_2 + iso\text{-}C_4H_8 \quad (R = C_5H_{11}, C_6H_{13})$$

The transalkylation of ethyl di-*n*-propylthioborinate by 1-heptene proceeds more slowly and under more severe conditions (160–170°C) than that with halides, with only the propyl radical being replaced. The transalkylation rate decreases in the following order:

$$(iso\text{-}C_4H_9)_3B > (iso\text{-}C_4H_9)_2BBr > (iso\text{-}C_4H_9)_2BCl > (C_3H_7)_2BSC_2H_5.$$

The mechanism of the transalkylation reaction, according to Köster [15], consists of consecutive reversible conversions, including decomposition of the trialkylborane to an olefin and a dialkylborane and addition of the latter to the higher olefin.

This mechanism would seem to be confirmed by the data of Rosenblum [33] on the thermal decomposition of tributylborane to butylene and compounds with B—H bonds upon heating under vacuum for a long time (10 days) at 120–130°C, and also by the ability of higher trialkylboranes to split off an olefin during vacuum distillation [15]. However, in these cases, the cleavage of the olefin proceeds either very slowly or under conditions considerably more severe than those of the transalkylation reaction. Thus, McCusker, et al. [3] established that triisobutylborane does not change during heating at atmospheric pressure for 20 hours at 135°C or for 7 hours at 185°C, whereas it undergoes transalkylation at 120–130°C [32, 34].

The mechanism of the displacement of some olefins by other olefins from BR_3 by an elimination-addition mechanism is also not in agreement with other experimental data, specifically with the dependence of the transalkylation rate on the concentration and nature of the higher olefin and also with the fact that the orientation in the transalkylation differs from the orientation of the addition of tetraalkyldiboranes to the same olefins [32, 34].

All of these facts are explained within the scope of the following mechanism [32, 34]. The transalkylation of trialkylboranes starts with the formation of a π-complex involving the vacant orbital of boron and the electron cloud of the carbon-carbon double bond; the further process of replacement of one hydrocarbon radical by another proceeds via a six-center transition state with synchronous rupture and formation of new bonds:

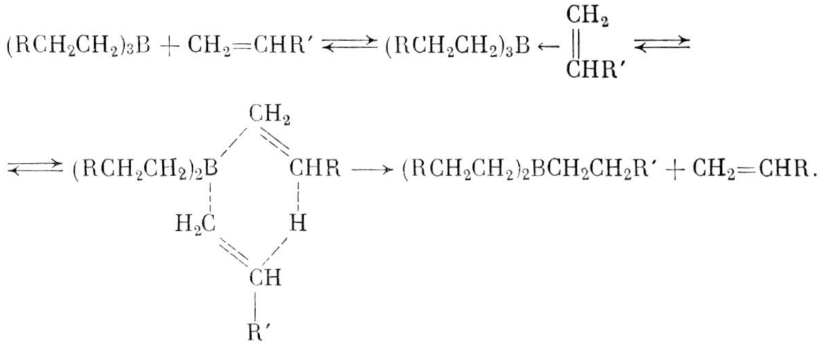

On the basis of data on the kinetics of the thermal decomposition of triisobutylborane in the presence of excess ethylene, Cocks and Egger

[35] concluded that this reaction proceeds via a polar four-center transition state.

The method of counter-thermodynamic isomerization of non-terminal olefins to α-olefins is a combination of hydroboration, isomerization, and transalkylation reactions:

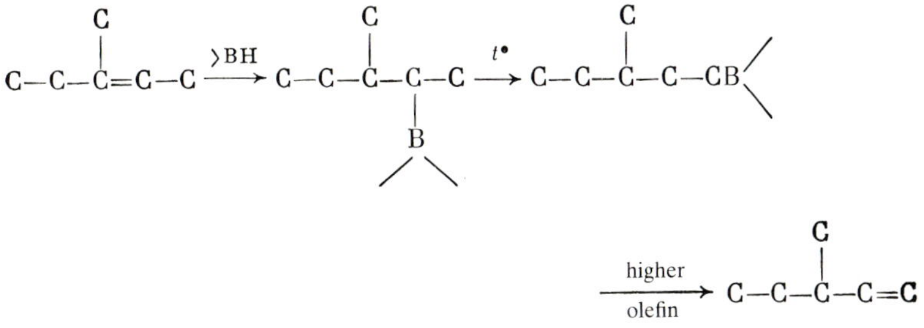

REFERENCES

1. G. F. Hennion, P. A. McCusker, E. C. Ashby and A. J. Rutkowski, Abstracts of Papers, 130th Meeting of the American Chemical Society, September 16–21, 1956.
2. G. F. Hennion, P. A. McCusker, E. C. Ashby and A. J. Rutkowski, *J. Am. Chem. Soc.* **79,** 5190 (1957).
3. P. A. McCusker, F. M. Rossi, J. H. Bright and G. F. Hennion, *J. Org. Chem.* **28,** 2889 (1963).
4. G. Benedikt, R. Köster, W. Larbig, K. Reinert and G. Rotermund, *Angew. Chem.* **75,** 1079 (1963).
5. R. Köster and G. Schomburg, *Angew. Chem.* **72,** 567 (1960).
6. H. C. Brown and G. Zweifel, *J. Am. Chem. Soc.* **88,** 1433 (1966).
7. L. Lardicci, L. Lucarini, P. Palagi and P. Pino, *J. Organometal. Chem.* **4,** 341 (1965).
8. J. C. Braun and G. S. Fisher, *Tetrahedron Letters* **1960,** 9.
9. H. C. Brown and G. Zweifel, *J. Am. Chem. Soc.* **89,** 561 (1967).
10. J. E. Herz and L. A. Marquez, *J. Chem. Soc., C* **1969,** 2243.
11. H. C. Brown: "Hydroboration." New York, W. A. Benjamin Inc. (1962).
12. R. E. Williams, *Inorg. Chem.* **1,** 971 (1962).
13. R. Köster, *Angew. Chem.* **68,** 383 (1956).
14. R Köster. 16e Congrès internationale de chimie pure et appliquée. Paris, 1957, p. 461.
15. R. Köster, *Ann* **618,** 31 (1958).
16. R. Köster, *German Patent* 1,034,176 (1958); *C. A.* **54,** 11995 (1960).
17. H. C. Brown and B. C. Subba Rao, *J. Org. Chem.* **22,** 1137 (1957).
18. H. C. Brown and B. C. Subba Rao, *J. Am. Chem. Soc.* **81,** 6434 (1959).
19. T. D. Parsons and D. M. Ritter, *J. Am. Chem. Soc.* **76,** 1710 (1954).
20. B. M. Mikhailov and T. A. Shchegoleva, *Dokl. Akad. Nauk SSSR* **108,** 481 (1956).

21. G. F. Hennion, P. A. McCusker and A. J. Rutkowski, *J. Am. Chem. Soc.* **80,** 617 (1958).
22. H. C Brown, M. V. Bhalt, T. Munekata and G. Zweifel, *J. Am. Chem. Soc.* **89,** 567 (1967).
23. N. H. Nam, A. T. Russo and R. F. Nystrom, *Chem. and Ind.* **1963,** 1876.
24. T. Ohtaki, K. Fujimori, M. Itoh and A. Suzuki, *J. Chem. Soc. Japan, Ind. Chem. Sect* **74,** 2129 (1971).
25. A. J. Rutkowski, E. J. Inchalik and A. Schriesheim, *U.S. Patent* 3,131,225 (1964); *C. A.* **61,** 4393 (1964).
26. K. Ziegler, R. Köster and W. R. Croll, *U.S. Patent* 2,975,215 (1961); *C. A.* **55,** 25756 (1961).
27. S. J. Brois and A. J. Rutkowski, *U.S. Patent* 3,101,376 (1963); *C. A.* **60,** 548 (1964).
28. B. M. Mikhailov and T. A. Shchegoleva, *Izv. Akad. Nauk SSSR, Otd. Khim. Nauk* **1959,** 546.
29. B. M. Mikhailov and A. N. Blokhina, *Zh. Obshch. Khim.* **30,** 3615 (1960).
30. K. G. Foot and B. P. Roberts, *J. Chem. Soc. A.* **1971,** 3475.
31. B. M. Mikhailov, M. E. Kuimova and E. A. Shagova, *Izv. Akad. Nauk SSSR, Ser. Khim.* **1968,** 548.
32. B. M. Mikhailov, M. E. Kuimova and E. A. Shagova, *Dokl. Akad. Nauk SSSR* **179,** 1344 (1968).
33. L. Rosenblum, *J. Am. Chem. Soc.* **77,** 5016 (1955).
34. B. M. Mikhailov and M. E. Kuimova, *Zh. Obshch. Khim.* **41,** 1714 (1971).
35. A. T. Cocks and K. W. Egger, *J. Chem. Soc., A* **1971,** 3606.

CHAPTER 5

Methods of oxidation of organoboron compounds

In this chapter, different methods of oxidation of organic boron derivatives, used for the transition from organoboranes to several classes of organic compounds, will be discussed.

Many organoboranes are easily oxidized with oxygen, and some of them ignite spontaneously in air. Autoxidation is used for the preparation of alcohols and alkyl hydroperoxides, and also for the initiation of a series of radical processes of addition of organoboranes to unsaturated, aromatic, and carbonyl compounds.

Oxidation with hydrogen peroxide in alkaline or acid medium and with amine oxides leads to alcohols or phenols. Unlike autoxidation, these reactions are strictly stereospecific (retention of the configuration) and, in combination with hydroboration, represent a simple method for the anti-Markownikoff hydration of olefins.

The oxidation of organoboranes with chromic acid and manganese compounds produces ketones.

5.1 AUTOXIDATION

Most alkyl and aryl derivatives of trivalent boron are easily oxidized with atmospheric oxygen, and all operations for their preparation and for the investigation of their properties must therefore be carried out in an inert-gas atmosphere — dry nitrogen, argon, etc.

Lower trialkylboranes, alkyl boracyclanes, and some other compounds ignite spontaneously in air and burn with a green flame. Higher trialkylboranes and triarylboranes do not ignite, but are also easily autoxidized. An exception is presented by trinapthylborane [1, 2] and trimesitylborane [1–4], which are stable toward oxygen owing to steric hindrance.

The autoxidation of boron derivatives has attracted the attention of researchers for a long time. The numerous papers in this area have been generalized to a considerable degree in a number of review articles [5–8] and books [9–12].

The slow introduction of dry air or of one mole of oxygen into a trialkylborane (without a solvent) produces an ester of an alkylboronic acid (C_2—C_6) [13–22]:

$$R_3B + O_2 \longrightarrow RB(OR)_2 \xrightarrow{H_2O} RB(OH)_2 + 2ROH.$$

When a sufficient amount of oxygen is used, trialkylboranes can be oxidized to borates [23, 24].

The first B—C bond in trialkylboranes is oxidized significantly more rapidly than the second, and the second, in turn, is oxidized more rapidly than the third [22–25].

Esters of alkylboronic acids are oxidized very slowly with oxygen [18, 24, 26, 27]. Amines [23, 26–30] and water [23] inhibit the oxidation of organoboranes. Compounds of negative tetravalent boron, as a rule, are not oxidized [31–45].

In 1956, Petry and Verhoek [46] obtained dimethylborylmethyl peroxide by the action of oxygen on trimethylborane in the gas phase:

$$(CH_3)_3B + O_2 \longrightarrow (CH_3)_2BOOCH_3 \xrightarrow{t°C} \underset{90\%}{CH_3B(OCH_3)_2}.$$

The peroxide is a compound that is quite stable at room temperature, with a half-life of 60 days. The same peroxide is formed by the oxidation of trimethylborane in a solution [47].

Peroxide compounds have been isolated in the autoxidation of tributyl- [23, 27, 48, 49], tri-*sec*-butyl- [28], triisobutyl- [30], and tri-2-methylpentylboranes [48], *tert*-butyldisobutylborane [26, 27, 29, 30], dibutylborinic anhydride $[(C_4H_9)_2B]_2O$ [28, 48], alkylboronic anhydrides $(RBO)_3$ [50, 51], $Cl_2BCH_2CH_2BCl_2$ [52], and n-$C_4H_9BX_2$ [53, 54]. It has been shown [23, 24, 27, 28, 30] that, in dilute solutions (in ether or hydrocarbons), trialkylboranes absorb two moles of oxygen, forming diperoxide compounds $RB(OOR)_2$, which can be isolated in the form of viscous liquids by vacuum distillation.

It was initially suggested [12, 27] that the oxidation proceeds by an ionic mechanism with 1,3-migration of the alkyl group from the boron to the oxygen in the complex $R_3B \cdot O_2$ formed [12, 23, 24]:

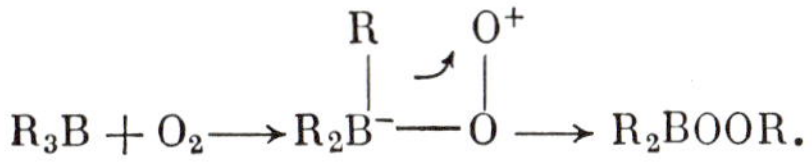

However, up to the present, a quite large amount of data has been obtained indicating the intermediate formation of free alkyl radicals during the autoxidation. This is partially confirmed by the following facts.

From the products of the oxidation of tributylborane, Mirviss [23] isolated approximately 20% of hydrocarbons (butane, butenes, and octane), along with butyraldehyde and a number of other compounds.

At 55°C, triphenylborane absorbs one mole of oxygen, with the formation, after hydrolysis, of 10.1% of biphenyl, 3.5% of terphenyl, and 4.6% of phenol [24]. The oxidation of a mixture of tributylborane and triphenylborane produces butylbenzene along with the above-mentioned compounds.

Kolesnikov [55–57], Furukawa [58–62], Chinese investigators [63, 63a], Hansen [64, 65], Grummit, et al. [51], and other investigators [66–76] have established that trialkylboranes and other organic boron compounds, in the presence of even traces of oxygen, initiate the radical polymerization and copolymerization of methyl methacrylate, acrylonitrile, vinyl acetate, vinyl chloride, and other monomers, while the individual trialkylboranes and their peroxides do not catalyze the polymerization [63, 63a, 66].

The autoxidation of trialkylboranes is completely inhibited by galvinoxyl (*1*) [25, 77–80], which is the most effective free-radical "trap" [81], and is also strongly retarded by iodine [82–84] (see, however, [64]). Less effective inhibitors have practically no effect on the course of the reaction [25].

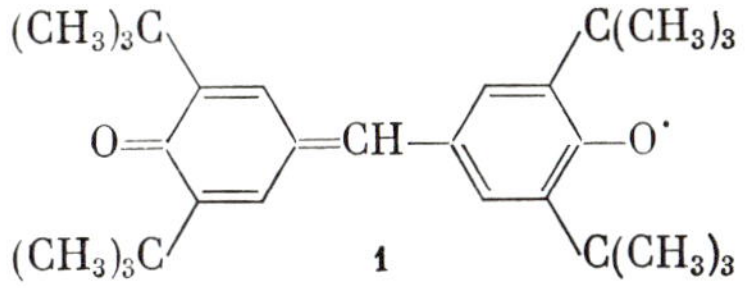

1

It has also been established that the autoxidation of (+)-1-phenylethylboronic acid produces racemic 1-phenylethanol [79]. The reaction of diisopinocampheyl(*sec*-butyl)borane ($[\alpha]_D^{16} + 121°$) with oxygen proceeds primarily with a loss of optical activity [25, 77, 78]. Alcohols of the same composition, 76% *exo*- and 24% *endo*-2-norbornanol, have been obtained from epimeric trinorbornylboranes containing 99.2%, 66%, and 14% of the exo isomer [80].

Finally, in recent years a whole series of preparative radical reactions of triorganoboranes, initiated by oxygen and other peroxide compounds, have been found (Chapters 7–9, 13, 14).

On the basis of all of these data, it has been suggested [25, 77–80, 84, 85] that the autoxidation of organoboranes is a homolytic chain reaction:

$$R_3B + O_2 \longrightarrow R^{\bullet}, \qquad \text{initiation}$$
$$R^{\cdot} + O_2 \longrightarrow R\dot{O}_2, \qquad \text{chain propagation}$$
$$R\dot{O}_2 + R_3B \longrightarrow [RO_2BR_3]^{\cdot} \longrightarrow RO_2BR_2 + R^{\cdot},$$
$$2R\dot{O}_2 (\text{or } R^{\cdot}) \rightarrow \text{stable products (chain termination)}$$

This mechanism shows a considerable similarity to the mechanism of hydrocarbon oxidation [85, 86], but the autoxidation of organoboranes proceeds significantly more easily [25].

As suggested, the initiation stage consists of a slow attack by oxygen, which is a biradical, on the boron atom, as a result of which a free radical $R^{\cdot}$ is formed; the latter then reacts with oxygen to form a peroxy radical. In the third stage, a rapid replacement of the alkyl radical in the trialkylborane by a peroxy group takes place. The kinetic chain is broken by various recombinations of $ROO^{\cdot}$ and $R^{\cdot}$ radicals.

The alkylperoxy(dialkyl)borane formed then either reacts with a second oxygen molecule, being converted to an alkyl(dialkylperoxy)borane $(ROO)_2BR$, or oxidizes a second trialkylborane molecule to form an ester of dialkylborinic acid, R_2BOR [24].

The boron monoperoxide can also be decomposed by oxidation of the second B—C bond, with this oxidation-reduction reaction proceeding intermolecularly [23, 24] or intramolecularly [26, 30]:

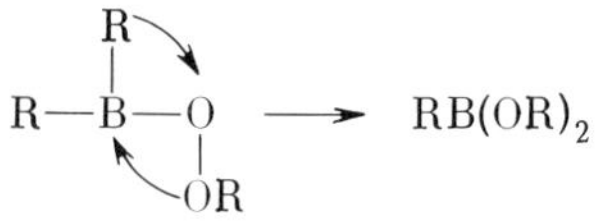

According to kinetic investigations [46, 63, 63a], the conversion of monoperoxides to esters $RB(OR)_2$ is described by a first-order equation, while the decomposition of diperoxides, $RB(OOR)_2$, and of compounds of the type RB(OR)OOR is a second-order reaction [63, 63a, 85].

Although the mechanism for the autoxidation of organoboranes indicated above is generally accepted at this time, it cannot be rigorously verified. In particular, the initiation stage is vague [87]. Furthermore, most of the facts cited in support of a homolytic nature of the initial stage of the process are unconvincing. For example, galvinoxyl (*1*), which inhibits the autoxidation, is a quite complex compound, each molecule of which may add a minimum of two (or three) molecules of organoborane.

The complete or partial absence of stereoselectivity in the autoxidation of optically active organoboranes can also not be used as serious evidence for a radical course of the process, since it is well known [88] that many reactions proceeding by an ionic mechanism (via carbanions) lead to racemates.

Arimoto [70] believes that the alkyl radicals are formed primarily by decomposition of the peroxyboranes $ROOB{<}$ and, to a lesser degree, by the interaction of $R{-}B{<}$ and $ROOB{<}$. At the same time, Maslennikov, et al. [89–97] have established by means of numerous examples that the boron peroxides $(ROO)_3B$, $(RO)_2BOOR'$, and others decompose with the formation of radicals only when heated above 100°C, whereas the reaction of tripropylborane with *tert*-$C_4H_9OOB(OC_4H_9)_2$ [98], which leads to propanol, *tert*-butanol, *tert*-butyl propyl ether, propane, propylene, and other compounds proceeds rapidly at room temperature. According to the data of [244], the peroxide $C_4H_9OOBBu_2$ and its analogs do not decompose by a radical mechanism.

Trialkylboranes and diorganylboranes R_2BX form complexes with peroxide compounds of the type $\geqslant C{-}O{-}O{-}Si{\leqslant}$ [245–247], which decompose to form radicals.

Barton, et al. [99] detected a small quantity of dimethyltrioxadiborolane (*2*) in the products of the gas-phase oxidation of trimethylborane.

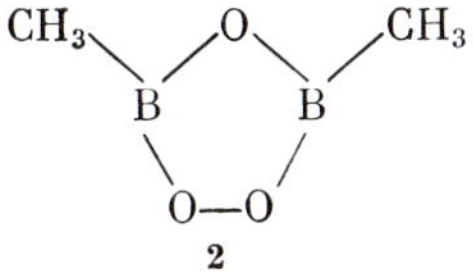

2

Analogous cyclic peroxides are formed as the main product of the oxidation of 1,1-dimethyldiborane [100, 101] and 1-methyldiborane [102] at 125°C.

Lithium and sodium tetrabutylborates are easily oxidized in air with the formation of 48% of butanol, 3% of butylraldehyde, and 15% of butane [103]. When oxygen is passed through an acid solution of $(C_6H_5)_4BNa$, biphenyl is formed [104], whereas, under normal conditions, sodium tetraphenylborate is stable in air.

5.1.1 Synthesis of alcohols

The autoxidation of organoboranes is a convenient method for alcohol synthesis.

Trialkylboranes absorb up to 1.5 mole of oxygen at −20 to +90°C and are converted to the borates $B(OR)_3$ or RB(OR)OOR, the alkaline hydrolysis of which leads to alcohols in a yield of up to 86% [12, 23, 24, 105]:

$$R_3B \xrightarrow[2.\ H_2O,\ OH^-]{1.\ O_2,\ 0°\ C} ROH.$$

The reaction temperature, type of solvent used, and concentration of the solution show little effect on the rate of oxygen absorption; the maximum yield of alcohols is obtained at 0°C [23].

A disadvantage of this method of alcohol synthesis is that the stereochemical configuration of the starting organoborane is not retained, as can be seen from the following two examples [79, 105]:

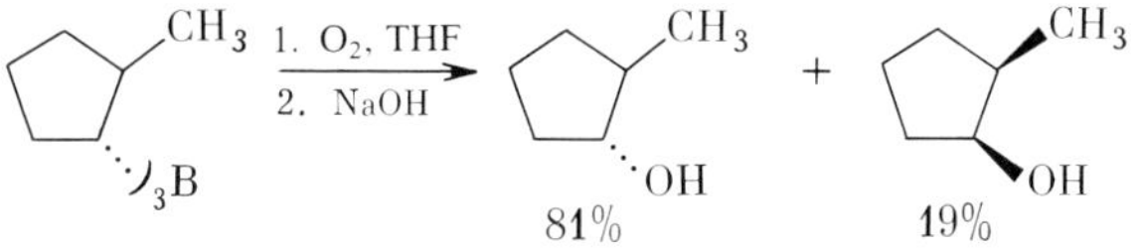

$$(+)C_6H_5\overset{\overset{\displaystyle CH_3}{|}}{C}H\text{—}B(OH)_2 \xrightarrow[2.H_2O]{1.\ O_2} (\pm)C_6H_5\overset{\overset{\displaystyle CH_3}{|}}{C}HOH$$

In some cases, rearrangement is observed during autoxidation. Thus, the oxidation of (7αH)-longifolyl-15-borane (*3*) with air in tetrahydrofuran leads to a mixture of compounds consisting of 33% of (7αH)-longifolol-15 (*6*), 25% of (7αH)-longifolol-3α (*7*), 33% of the 3β-isomer *8*, 3% of the ketone *9*, and also including dimeric compounds [106]:

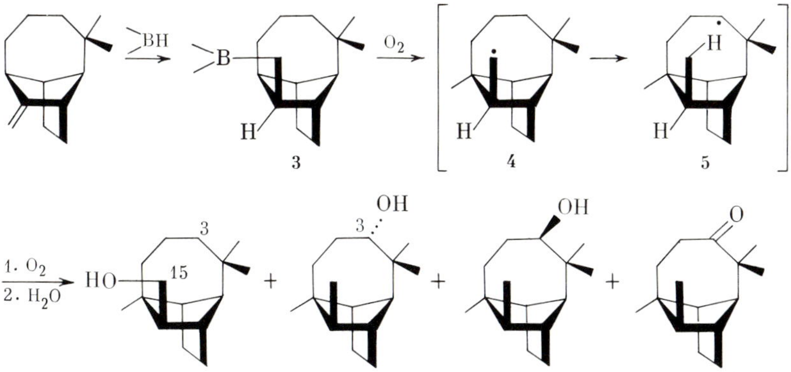

6 (33%) 7 (25%) 8 (33%) 9 (3%)

This course of the reaction is explained by a transannular conversion of the initially formed radical *4* to the radical *5*.

5.1.2 Synthesis of hydroperoxides

Dilute solutions of trialkylboranes (0.01–0.04 M) in paraffin hydrocarbons, benzene, or ethers absorb two moles of oxygen, forming diperoxide compounds, the treatment of which with hydrogen peroxide or peracids leads to hydroperoxides [12, 107]:

$$R_3B \xrightarrow[\text{solvent}]{2O_2} RB(OOR)_2 \xrightarrow{H_2O_2 \text{ or } R'COOOH} ROH + ROOH.$$

The role of the dilution consists of reducing intermolecular oxidation to a minimum.

Wilke and Heimbach [107] have used autoxidation for the synthesis of primary and secondary alkyl hydroperoxides from olefins according to the following scheme:

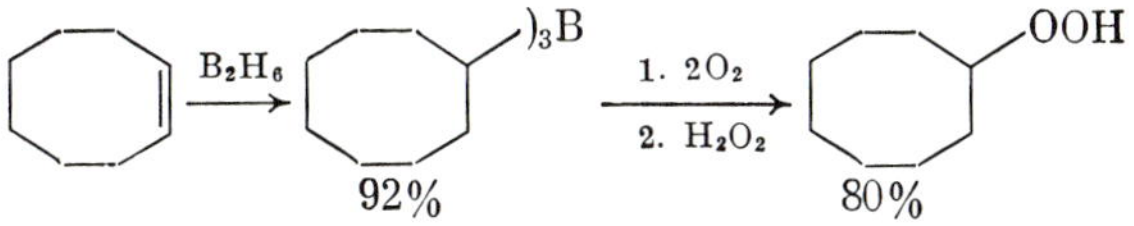

As has been shown by Chinese chemists [63, 63a], triethylborane absorbs one mole of oxygen at −70°C, forming the monoperoxide compound $(C_2H_5)_2BOOC_2H_5$ which, at −20°C, absorbs a second mole of oxygen to form the diperoxide $C_2H_5B(OOC_2H_5)_2$.

This stagewise course of the oxidation is the basis of a convenient method for the conversion of organoboranes to hydroperoxides [108]:

$$R_3B + O_2 \xrightarrow[-78°\,C]{THF} R_2BOOR \xrightarrow[0°\,C]{THF} RB(OOR)_2 \xrightarrow[H_2O]{H_2O_2} \underset{>80\%}{2ROOH} + ROH.$$

Hydroperoxides are also obtained in a yield of approximately 90% in the autoxidation of dichloro(alkyl)boranes $RBCl_2$ in ether [109].

3-Cyclopentenyl hydroperoxide is obtained by the hydroboration-autoxidation of cyclopentadiene [248], and is used further for the synthesis of substituted 2,3-dioxabicyclo[2.2.1]heptanes. The hydrolytic cleavage of boron peroxides can be carried out by the action of peracids [249].

The reaction of trialkylboranes with nitric oxide, which proceeds by a radical mechanism, has been used for the preparation of hydroxylamine derivatives, R_2NOH [110, 111], heterocyclic boron derivatives [112, 113], or alkylamines [114].

5.2 OXIDATION WITH ALKALINE HYDROGEN PEROXIDE. SYNTHESIS OF ALCOHOLS AND PHENOLS

The oxidation of organoboranes by alkaline hydrogen peroxide, which was first carried out by Ainley and Challenger [115], and also by Johnson, et al. [18, 116], is now one of the most important reactions in boron chemistry:

$$R{-}B\langle + H_2O_2 \xrightarrow{OH^-} ROH + \rangle B{-}OH.$$

It is widely used, first, to establish the structures of organoboranes, generally with subsequent chromatographic analysis of the carbinol compounds, and, second, for going from organoboranes to alcohols, polyols, and carbonyl derivatives, for example, as the final stage of the *cis*-hydration of olefins and polyolefins by hydroboration-oxidation. In addition, this reaction represents the basis of the method for determining boron in different derivatives of this element by titration of the boric acid formed with NaOH solution in the presence of mannitol, using phenolphthalein or bromthymol as indicator [117, 118].

The synthetic application of the reaction is considered in almost all sections of this book.

As a result of numerous investigations, it has been shown that the concentration of bases and peroxides, and also the reaction temperature, can be varied within broad limits without a significant change in the alcohol yield [119]. The following method is generally used. An organic boron compound (one mole), in an ether solvent or without a solvent, is mixed with one equivalent of NaOH solution in water (5–20% solution). Homogenization of the mixture is completely unnecessary. After this, 30% hydrogen peroxide is carefully added at 0–35°C (one equivalent plus a 10–15% excess for each B—C bond of the organoborane). The reaction proceeds with vigorous evolution of heat. The mixture is left for 1–2 hours at room temperature or is heated for 10–15 minutes at 50–60°C. The alcohol is extracted with ether. In the final stage of the oxidation, the function of the alkali is to bind the boric acid, as a result of which it remains in the aqueous layer. It should be emphasized that, the milder the oxidation conditions, the higher the alcohol yield.

To isolate the diols formed from 1,2- 1,3- and 1,4-diboron compounds and also from boracyclopentanes, a special method has been developed, consisting of treatment of the oxidation products with mannitol, which "displaces" 1,2- and 1,3-diols from the very stable cyclic boron ethers [120–122].

Alkaline hydrogen peroxide easily oxidizes practically all alkyl-, cycloalkyl-, and arylboranes, independently of the size and structure of the radicals on the boron, including cyclopropylboranes [123–126] (see, however, [127]). The mildness of the conditions makes it possible to use this method for the oxidation of functional organoboranes containing such groups as halogen, RO, CN, a double bond, a carbonyl or carboxy group, or other substituents in the radical, which remain unchanged. Some functional groups are protected by a suitable method. For example, cyclic ketals and acetals are stable in alkaline medium and are not affected by the oxidation [128–131].

In those cases where milder conditions are necessary, the oxidation is carried out in sodium acetate solution [132, 133, 137] or in a phosphate buffer at a pH of approximately 7–8 [134–136]. Thus, α-keto ethers have been obtained in a yield of 61–68% from 3-alkyl-2-boryl acrylates by the action of hydrogen peroxide in 0.5 M NaH_2PO_4 solution, which is adjusted to a pH of approximately 8 by the addition of 6 N NaOH, whereas, in alkaline medium, a protolytic cleavage of the boron-carbon bond takes place [135]:

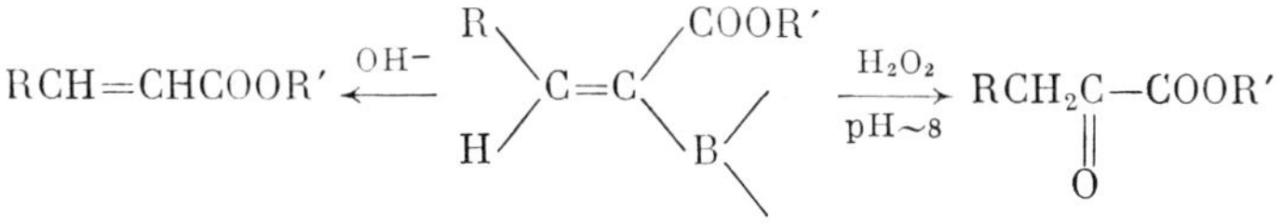

Oxidation in the presence of bases proceeds without a change in the carbon skeleton of the radical, with retention of its configuration. Thus, the oxidation of (+)-tri[(*S*)-2-methylbutyl]borane led to (−)-(*S*)-2-methylbutanol (with an optical purity of 92%), while the (+)-(*S*)-alcohol (4% racemization) was obtained from (+)-tri[(*S*)-3-methylpentyl]borane [138]. The reaction of optically active 2-butylborane with alkaline hydrogen peroxide led to (*R*)-(−)-*sec*-butanol with an optical purity of 76% [139], while the oxidation of (*R*)-(−)-dibutoxy(1-phenylethyl)borane produced (*R*)-(+)-1-phenylethanol [140, 141]. Other examples of retention of the configuration were discussed in Chapters 2 and 3.

On the basis of a study of the kinetics of the oxidation of alkyl- [142] and arylboronic acids [143–147] in the presence of bases, the following reaction mechanism was proposed:

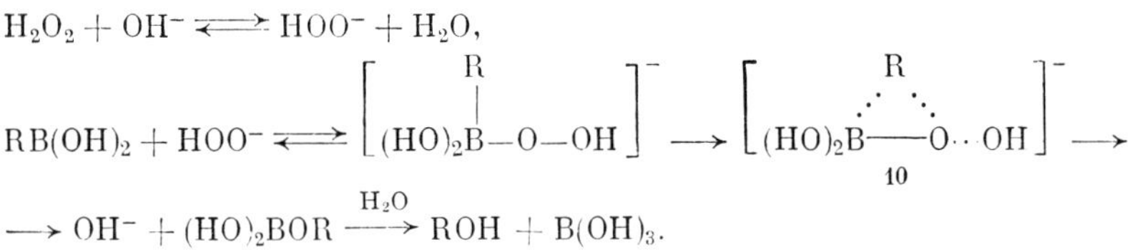

The oxidation proceeds by an S_E2 mechanism via a transition state of the type *10*, with the following order of relative reactivity of the boronic acids $RB(OH)_2$ in alkaline medium being observed [142]: *tert*-C_4H_9(330) > *sec*-C_4H_9(185) > *n*-C_4H_9(52) > $C_6H_5CH_2$(24) > $CH_2{=}CH$ (4.2) > C_2H_5 (2.3) > CH_3 (1).

According to the data of [146], the reaction is second-order with respect to $ArB(OH)_2$ and, consequently, an intermediate complex containing two borane molecules, possibly $(HO)_2Ar\bar{B}OOB(Ar)OH$, is formed in the determining stage. However, this does not significantly change the overall nature of the reaction mechanism. Boric acid catalyzes the oxidation, possibly as a result of the formation of an analogous intermediate product [144–148].

The oxidation of $C_6H_5B(^{18}OH)_2$ in $^{18}H_2O$ medium forms C_6H_5OH containing no ^{18}O [149].

It should be noted that, in the oxidation of trialkylboranes with alkaline hydrogen peroxide, a coupling of the radicals takes place to a slight degree, as a result of which approximately 5% of dimeric hydrocarbons R—R is obtained [150].

Benzylboranes [141, 151, 152], allylboranes (Part IV), and some triarylboranes (see below) are decomposed by alkali solutions at room temperature or on heating, with the formation of hydrocarbons RH. This should be kept in mind in working with compounds of the classes indicated. In the oxidation of 1-phenylethyl-1,2- and 1-phenylethyl-2,2-bisboronates, radical cleavage of a carbon–carbon bond takes place to a considerable extent [153]:

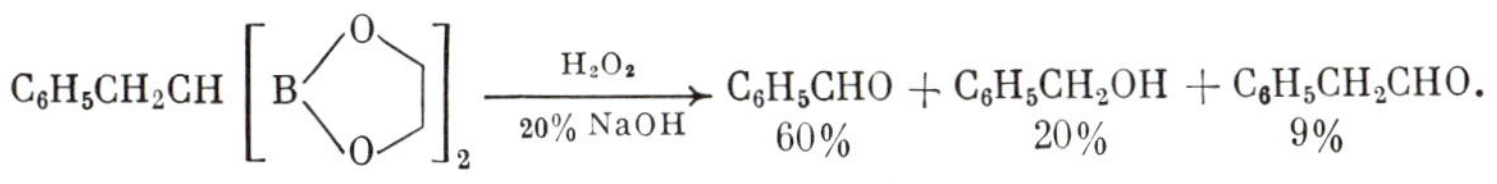

The oxidation of 1,1- and 1,2-bisboronates with hydrogen peroxide in acid medium proceeds normally, without rearrangements or the cleavage of C—C bonds [154, 155].

The oxidation of some l-alkenylboronic acids with hydrogen peroxide is complicated by cleavage of the double bond (approximately 40%), while oxidation with sodium perborate leads smoothly to aldehydes [250, 251].

The oxidation of "ate" complexes of 9-alkyl-9-borabicyclo-[3.3.1]nonane proceeds anomalously to a considerable degree. Thus, the main product obtained from the dibutyl compound is *cis*-bicyclo[3.3.0]octan-1-ol (*11*) [156]:

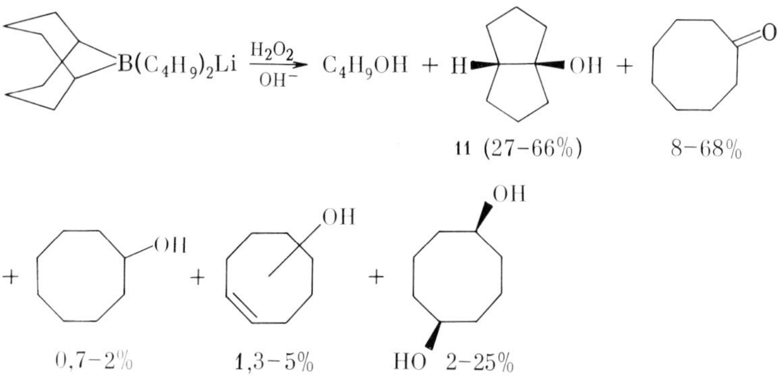

Oxidation in alkaline medium was used by Hawthorne [157] for the conversion of the corresponding aryl halides to phenol, α-naphthol, and *p*-cresol without isolation of the organoboranes:

$$ArX \xrightarrow{Mg} ArMgX \xrightarrow[2.\ H_3O^+]{1.\ B(OCH_3)_3} \left[Ar-B\langle\right] \xrightarrow{H_2O_2,\ OH^-} ArOH.$$

60—80%

Breuer and Broster [158] proposed carrying out the Grignard synthesis in a molar solution of diborane in tetrahydrofuran. The organoborane formed is treated with hydrogen peroxide, without isolation:

$$RBr + Mg + B_2H_6 \xrightarrow{THF} \left[R-B\langle\right] \xrightarrow{H_2O_2,\ OH^-} R-OH.$$

60—95%

In this way, isomeric butanols and octanols, hexadecanol, benzyl alcohol, cyclopentanol and cyclohexanol, phenol, *p*- and *m*-cresols, *p*- and *m*-methoxyphenols, *m*-chlorophenol (from *m*-chloroiodobenzene), and 1-hydroxyadamantane have been obtained from the corresponding bromides.

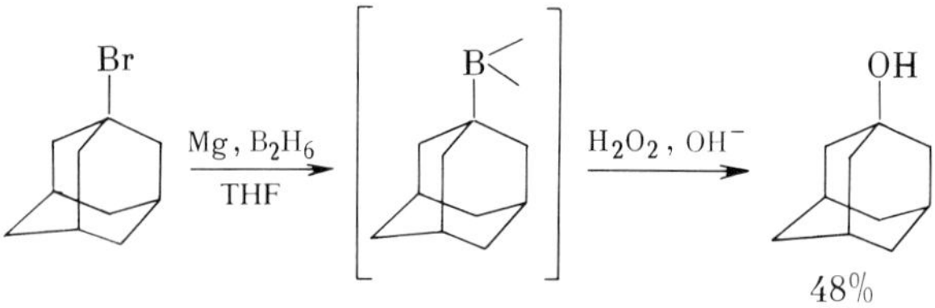

48%

Two of the latter mild methods for the replacement of a halogen by a hydroxy group are of great importance in the aromatic series. For example, 9-hydroxyphenanthrene has been synthesized from 9-bromophenanthrene [158]:

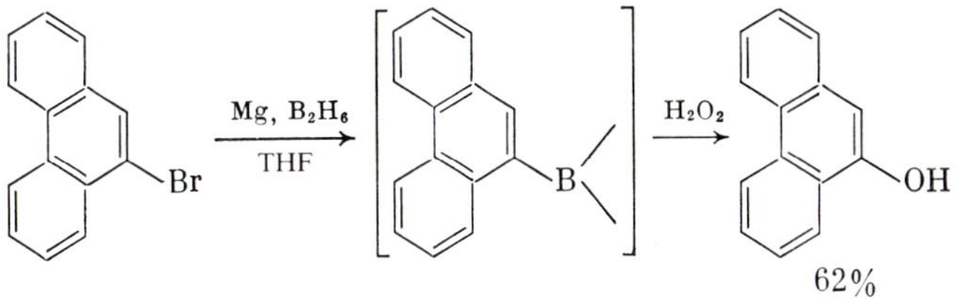

Lithium, potassium, or sodium can be used instead of magnesium for the preparation of phenols from aromatic chlorides, bromides, and iodides in a diborane solution [159]. With calcium, satisfactory yields (28–50%) are observed only when aryl iodides are used.

Hörnfeldt and Gronowitz [160–163] used lithium derivatives of thiophene and butyl borate for the synthesis of 2- and 3-thiolenones:

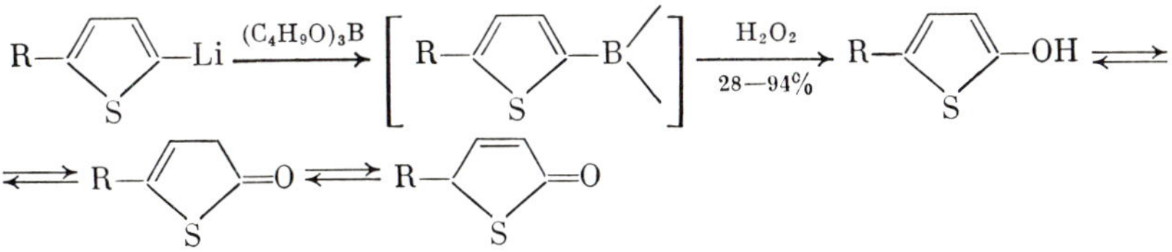

Hawkins [164] used this method for the preparation of thienyl methacrylates in a yield of 63–76% and for the synthesis of polymers based on these.

An analogous method has been used for the mild conversion of alkyl halides to alcohols: the reaction of lithium alkyls with methyl borate has been carried out in pentane at −78°C [165].

Organolithium compounds formed by the action of butyllithium on allylbenzene and *p*-allyltoluene are converted to 1,3-diols by treatment with diborane and subsequent oxidation [166].

Peracids can be used instead of hydrogen peroxide for the preparation of alcohols from organoboranes [18, 167]:

$$R_3B \xrightarrow[CHCl_3]{3C_6H_5COOOH} (RO)_3B \xrightarrow{H_2O} 3ROH.$$

In the reaction of sodium tetraphenylborate with acyl peroxides, heterolytic cleavage of the peroxide bond takes place, producing the phenyl ester and sodium salt of the corresponding acid along with triphenylborane, which undergoes further oxidation [168].

5.3 OXIDATION WITH HYDROGEN PEROXIDE IN ACID MEDIUM. SYNTHESIS OF ALDEHYDES AND DIOLS

As noted above, the alkaline oxidation of 1,1- and 1,2-bisboronates is accompanied by the cleavage of C—C bonds and by other side reactions. At the same time, the reaction proceeds very smoothly in acid medium, leading to aldehydes, ketones, or 1,2-diols [154, 155, 169]:

$$C_2H_5CH[B(OR)_2]_2 \xrightarrow[15-20^\circ C]{H_2O_2,\ H^+} C_2H_5CHO,$$

$$\underset{\displaystyle B(OR)_2}{CH_3\underset{|}{C}H}-CH_2B(OR)_2 \xrightarrow[15-20^\circ C]{H_2O_2,\ H^+} \underset{\displaystyle OH}{CH_3\underset{|}{C}H}-CH_2OH$$

The treatment of vinylboranes with hydrogen peroxide in acid solution leads to aldehydes [170–173], which are usually isolated as the hydrazones.

The reaction in the presence of 2,4-dinitriophenylhydrazine has been used to prove the structures of α-haloalkylboronic acids [174].

Alkyl- and arylboronic acids react with hydrogen peroxide in the presence of acids to form alcohols or phenols [142].

5.4 OXIDATION WITH AQUEOUS HYDROGEN PEROXIDE AND ALKYL HYDROPEROXIDES

In the treatment of trihexylborane with an equimolar quantity of 30% aqueous hydrogen peroxide in tetrahydrofuran, two B—C bonds are cleaved and a mixture of hydrocarbons and n-hexanol is formed [150, 175]:

$$(C_6H_{13})_3B + H_2O_2 \xrightarrow[0^\circ C,\ 24\ \text{hours}]{THF} \underset{6,5\%}{n\text{-}C_6H_{14}} + \underset{28,6\%}{n\text{-}C_{12}H_{26}} + \underset{8\%}{C_4H_9\underset{\underset{\displaystyle CH_3}{|}}{C}HC_6H_{13}} + \underset{5,4\%}{n\text{-}C_6H_{13}OH}$$

The formation of 5-methylundecane is explained by the presence of 2-methylpentyl groups in the starting trihexylborane, which was obtained by the hydroboration of 1-hexene.

Further proof of the free-radical nature of the process was obtained by studying reactions with three epimeric trinorbornylboranes (100% of the exo isomer; a mixture of 65% of the exo and 35% of the endo isomer; a mixture of 24% of the exo and 76% of the endo isomer) in CCl_4. It was found that the ratio of *exo*- and *endo*-norbornyl chlorides formed does not depend on the epimeric composition of the starting organoborane and is equal to 94:6 in all cases [176], just as in other conversions of norbornyl compounds taking place by a radical mechanism [177]. The following reaction mechanism is proposed [150]:

$$R_2BOOH + R_3B \longrightarrow R_2BO^{\bullet} + R^{\bullet} + R_2BOH,$$
$$R_2BO^{\bullet} + H{-}O{-}O{-}H \longrightarrow R_2BOH + HO_2^{\bullet},$$
$$R_2BO^{\bullet} + H{-}O{-}O{-}H \longrightarrow R_2BOOH + HO^{\bullet},$$
$$R_3B + HO_2^{\bullet} \longrightarrow R_2BOOH + R^{\bullet}.$$

Monomeric hydrocarbons are formed by the removal of hydrogen from the solvent molecule by the $R^{\bullet}$ radical. Alcohols may be obtained as a result of the recombination of alkyl and hydroxy (or $HO_2^{\bullet}$) radicals or by the 1,2 migration of the R^- anion from boron to oxygen, which proceeds with a retention of the configuration [176]:

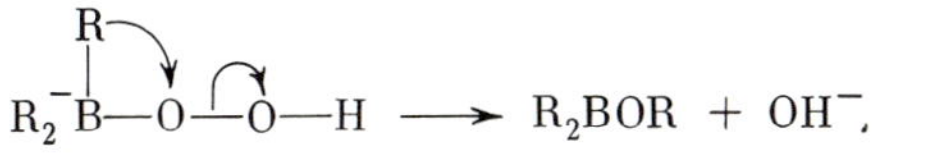

The reaction of tripropylborane with alkyl hydroperoxides produces 40–100% of propane, 95–100% of propanol, 20–30% of hexane, and a number of other products of radical conversions [178]. However, there is practically no protolytic cleavage according to the scheme:

$$R_3B + R'OOH \not\longrightarrow R_2BOOR' + RH$$

because, when R'OOD is used, the total quantity of propane formed contains only 5–6% of C_3H_7D. In the analogous reaction of triphenylborane, a large amount of benzene is obtained, but the content of C_6H_5D does not exceed 6–8% of the total quantity [179]. Consequently, propane and benzene are formed by the abstraction of hydrogen from the solvent molecule by a propyl or phenyl radical.

When the reaction is carried out in CCl_4, C_6H_5Cl and hexachloroethane are obtained, while methylbiphenyl ($CH_3C_6H_4C_6H_5$) and dibenzyl are obtained in toluene. In the action of *tert*-C_4H_9OOH on $(n\text{-}C_5H_{11})_2BOC_4H_9$, oxidation (approximately 30%) is accompanied by protonolysis (approximately 70%) [252]. In the action of hydroperoxides on $(C_6H_5)_2BOSi(C_2H_5)_3$, one boron-phenyl bond is oxidized and $C_6H_5B(OC_6H_5)OSi(C_2H_5)_3$ is obtained [253].

The reactions of organoboranes with Fenton's reagent (H_2O_2—Fe^{+2}) proceed with the intermediate formation of $HOO^{\bullet}$ radicals and lead to a mixture of substances, including products of the cleavage of C—C bonds [153]:

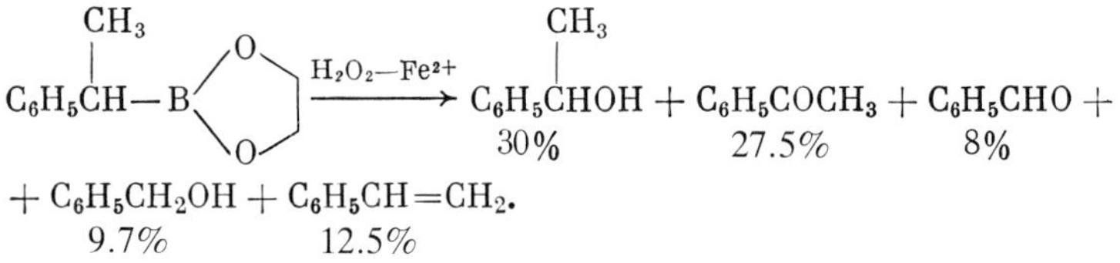

An ester of β-styrylboronic acid, when treated with Fenton's reagent, forms 60% of benzaldehyde and 36% of acetophenone.

Hydrogen peroxide [180] and Fenton's reagent [181] have been used as effective catalysts for radical reactions of trialkylboranes with mercaptans [180] and with α,β-unsaturated ketones [181].

Alkyl- and phenylboronic acids, their anhydrides and esters are oxidized with hydroperoxides to form, after hydrolysis, the corresponding alcohols or phenols [12, 182, 183]:

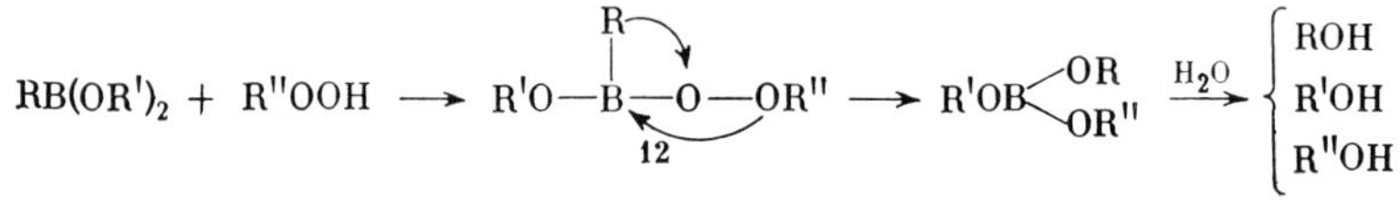

The rearrangement of boron peroxides (*12*) (internal oxidation) proceeds at room temperature or under gentle heating [96, 184, 185]. A kinetic study [182, 183] has shown that the reaction proceeds according to the indicated scheme only in the initial study stages and is later complicated by other radical processes.

1,2,3,4-Tetrahydronaphthyl-1-hydroperoxide is reduced by labeled phenylboronic acid, $C_6H_5B(^{18}OH)_2$, to form a tetralol containing no ^{18}O [149, 186]. Under the same conditions, 1-phenylethyl hydroperoxide forms an alcohol with complete retention of the configuration.

The boron peroxide $(C_6H_5)_2BOOC(CH_3)_2C_6H_5$, obtained from $(C_6H_5)_2BCl$ and the sodium derivative of cumyl hydroperoxide, rearranges at −10°C to form $C_6H_5B(OC_6H_5)OC(CH_3)_2C_6H_5$, phenoxy(cumyloxy)phenylborane [187].

One interesting aspect of the synthetic use of boron compounds should be noted. Oxiranes have been obtained in high yields by heating hydroperoxides with the corresponding olefins in the presence of cyclohexyl metaborate [188]:

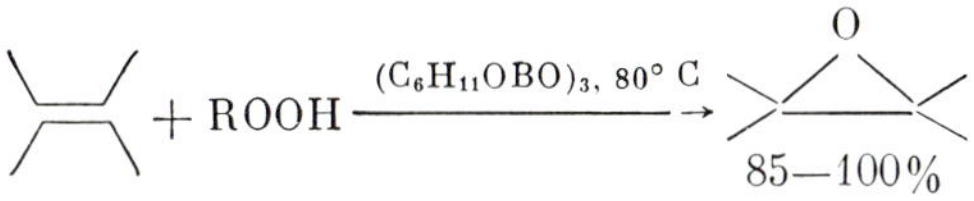

5.5 OXIDATION WITH AMINE OXIDES. SYNTHESIS OF ALCOHOLS

Köster and Morita [189–191] have shown that alkyl, cycloalkyl, cyclic, aryl, alkenyl, and heterocyclic boron derivatives are easily oxidized to the corresponding borates with amine oxides, e.g., the *N*-oxide of trimethylamine or pyridine:

$$R{-}B\Big\langle + R_3NO \xrightarrow[-R_3N]{} ROB\Big\langle \xrightarrow{H_2O} ROH.$$

Alkynylboranes are not oxidized with amine oxides [191]. Complex compounds are slowly oxidized [141, 191].

In triorganoboranes, the first B—C bond is oxidized more rapidly than the second, and the second more rapidly than the third:

$$R_3B + (CH_3)_3NO \longrightarrow R_2BOR + (CH_3)_3N,$$
$$R_2BOR + (CH_3)_3NO \longrightarrow RB(OR)_2 + (CH_3)_3N,$$
$$RB(OR)_2 + (CH_3)_3NO \longrightarrow B(OR)_3 + (CH_3)_3N.$$

The treatment of R_3B with one equivalent of the *N*-oxide forms the pure borinate R_2BOR, while the use of two equivalents leads to $RB(OR)_2$. The oxidation of the third B—C bond generally requires heating.

The selectivity of the reactions is clearly illustrated by the formation of a bis(alkylvinyl)borinate from bis(alkylvinyl)thexylborane

[192, 254]:

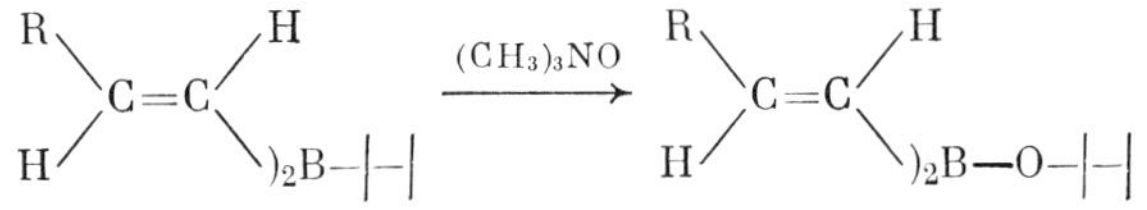

The oxidation with *N*-oxides, is not complicated by side reactions and is apparently a milder method than the reaction with alkaline hydrogen peroxide. The oxidation is usually carried out [190, 191, 193] by adding a solution of the organoborane (1 mole) in dry benzene or toluene to a boiling suspension of trimethylamine oxide (3 equivalents) in the same solvent. The mixture is refluxed for 30 minutes and the $(CH_3)_3N$ evolved is absorbed by sulfuric acid solution. After filtration and removal of the solvent, a pure borate is obtained, from which the alcohol is obtained by the action of methanol, dilute sulfuric acid, or alkali.

This method represents the basis for the quantitative determination of B—C bonds in organoboranes of the type R_3B, R_2BX, RBX_2, and others [191] (titration of the trimethylamine evolved).

The oxidation of β-selenium-containing vinylboranes by means of Me_3NO, followed by hydrogen peroxide, leads to vinyl ketones [255]. Not only anhydrous trimethylamine oxide, but also the dihydrate can be used for the oxidation of alkylboranes [256]. Organoboranes (symmetrical and unsymmetrical) can be arranged in the following order with respect to their ease of oxidation with trimethylamine oxide dihydrate: *tert*-R > *sec*-cyclic > *sec*-acyclic > primary-R > β-branched primary-R > vinyl [257]. In the reaction of R_3B with pyridine, quinoline, isoquinoline, or lepidine *N*-oxides, the reduction is accompanied by alkylation in the aromatic ring [258].

The method of oxidation with amine oxides has been used for the preparation of cyclopropanols from cyclopropylboranes [127]. In the opinion of the authors, alkaline hydrogen peroxide cannot be used in this case, because of decomposition of the cyclopropanols; however, it was later shown [123–126] that this is not true.

The reaction of (+)-dibutoxy-1-phenylethylborane with trimethylamine oxide proceeds to the extent of 98% with retention of the configuration and leads to (−)-1-phenylethanol in 81% yield [141]:

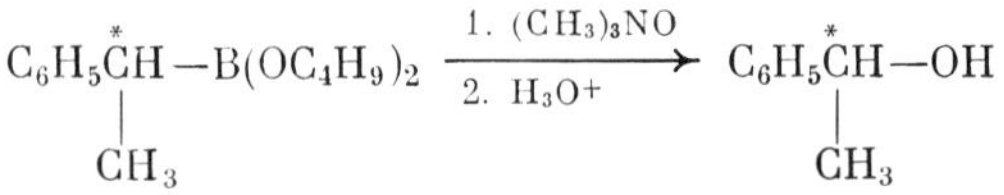

Under analogous conditions, tri-*exo*-norbornylborane forms 99.2% of *exo*-norbornyl alcohol.

The reaction mechanism is apparently similar to that of oxidation with alkaline hydrogen peroxide and includes a nucleophilic 1,2-migration of the $\bar{R}$ anion from the boron to the oxygen:

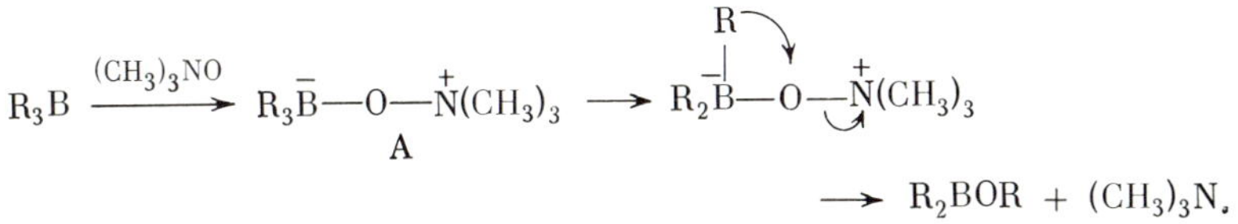

Betaines of the type *A* are quite stable in the case of $(C_6H_5)_3B$ [113], tripropynylborane [194], and, particularly, BF_3 [195–197]. Betaines are intermediate products in a number of conversions of complex systems. For example, BF_3 causes the decomposition of the cubane precursor, 7,8-diazapentacyclo[$4.2.2.0^{2,5}.0^{3,9}.0^{4,10}$]dec-7-ene (diazabasketene) *N*-oxide, leading to benzaldehyde oxime [198]:

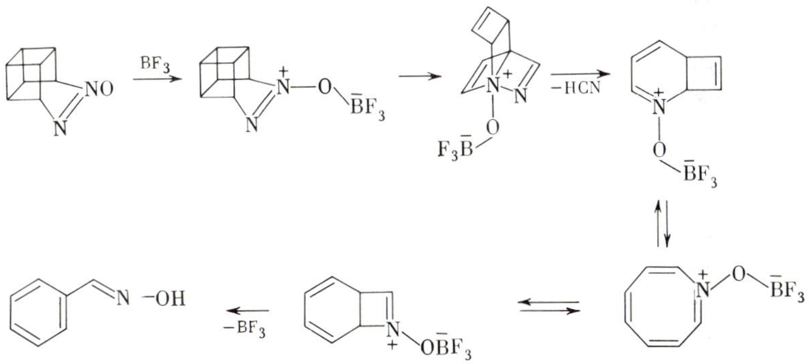

N-Oxides of *N*-substituted diethanolamines form complex compounds of the type *13* with phenylboronic acid [199]:

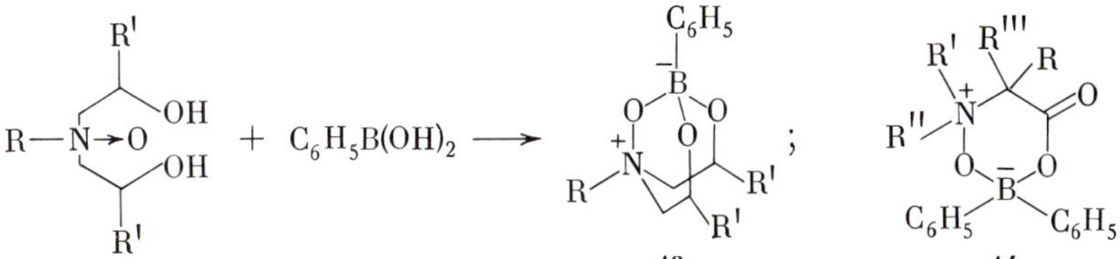

13 14

The action of *N*-oxides of α-dimethylamino acids on phenylboronic acid produces the internal chelates *14* [200].

Triphenylphosphine oxide, when heated with trialkylboranes and other boron derivatives, is converted to triphenylphosphine [201]:

$$(n\text{-}C_3H_7)_3B + (C_6H_5)_3PO \xrightarrow{250°C,\ 5\ \text{hours}} \underset{96\%}{(C_6H_5)_3P}.$$

5.6 OXIDATION WITH SODIUM HYPOCHLORITE

Arylboronic acids are oxidized with sodium hypochlorite to phenols; the action of sodium hypobromite on $C_6H_5B(OH)_2$ produces 2,4,6-tribromophenol, regardless of the ratio of reagents [202].

The treatment of an alkaline solution of organoboranes in diglyme with sodium hypochlorite at 25–30°C leads to the corresponding alcohols in yields of approximately 70% [203].

5.7 OXIDATION WITH CHROMIC ACID AND WITH MANGANESE COMPOUNDS. SYNTHESIS OF KETONES

The oxidation of *sec*-alkylboranes with chromic acid leads to ketones. This reaction, in combination with hydroboration, is widely used for the transition from cyclic olefins to carbonyl compounds, for example:

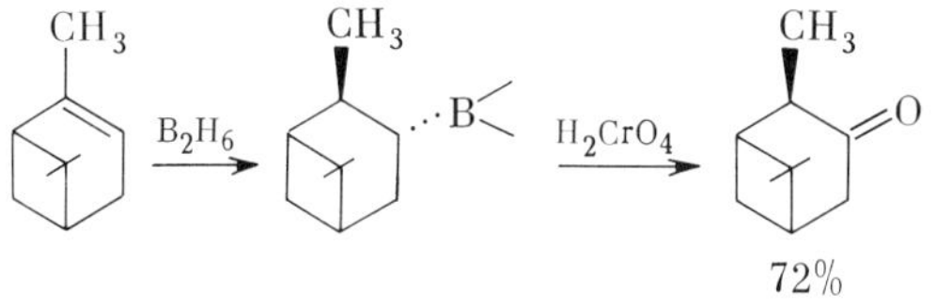

This method was first used by Pappo [204], who converted conessine [3β-dimethylaminocon-5-enine] to 3β-dimethylaminoconanin-6-one.

Ware and Traylor [205] later showed that the direction of the oxidation of alkylboronic acids depends on the acidity of the medium. At low pH values, the reaction leads to ketones, whereas in neutral or weakly acid solutions ($3 < pH < 7$), secondary alcohols are obtained, which are relatively stable toward Cr(VI) within this pH range:

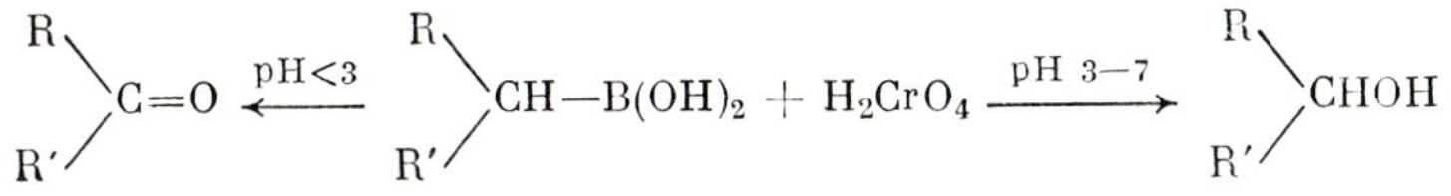

The oxidation of organoboranes with Cr(VI) compounds to produce alcohols has not been used so far in organic chemistry. At the same time, this reaction can be used for the selective oxidation of boron-carbon bonds in mixed organoboranes, since its rate depends on the nature of the R radical in $RB(OH)_2$ to a greater degree than in the case of the oxidation with alkaline hydrogen peroxide [102, 205]. Thus, in 0.114 M perchloric acid at 30°C, the reaction is second-order overall, with the following values of the rate constants: $7.5 \cdot 10^{-2}$($R = tert\text{-}C_4H_9$), $6.6 \cdot 10^{-4}$ ($R = C_2H_5$), and $24 \cdot 10^{-7}$ ($R = CH_3$)$L \cdot mole^{-1} \cdot sec^{-1}$. The following mechanism for formation of the alcohols has been proposed:

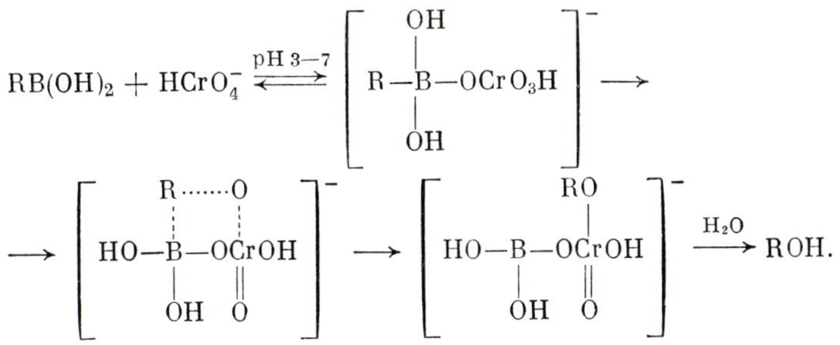

To convert cycloalkylboranes to the corresponding cycloalkanones, 8N H_2CrO_4 is used; the ketone yield is 40–90% [206].

The method has been used for the synthesis of the C_{16} musk compound, 8-cyclohexadecen-1-one, by the monohydroboration-oxidation of 1,9-cyclohexadecadiene [207], and also for the preparation of bicyclic ketones from 5,6-*exo*-dideuteronorbornene [208] and from 4-methyl- [209] and 5-*exo*-6-*endo*-dimethylnorbornenes [210]. The reaction of norbornene with (−)-diisopinocampheylborane, followed by chromic acid oxidation, forms (1*S*, 4*R*)-(+)-norcamphor [211], while *cis*-bicyclo[4.2.0]oct-7-ene, obtained by the photolysis of *cis,cis*-1,3-cyclooctadiene, is converted to *cis*-bicyclo[4.2.0]octan-7-one in 60% yield [212].

In some cases, the oxidation is accompanied by a rearrangement. For example, the product of the monohydroboration of norbornadiene (*15*), when treated with chromic acid, forms a mixture of approximately equal quantities of nortricyclanone (*16*) and dehydronorcamphor (*17*) [213]:

15 16 (51%) 17 (49%)

At the same time, the stepwise oxidation of the organoborane *15* with hydrogen peroxide and chromic acid leads only to dehydronorcamphor. Bicyclo[2.2.1]hepta-2,6-dione is obtained in an analogous manner from the dihydroboration product of norbornadiene [259].

Stepwise oxidation (with H_2O_2 and CrO_3) is used for the preparation of substituted cyclohexanones [214, 215], tricyclic ketones [216], and 6-oxosteroids [217]; *endo*-tricyclo[3.2.2.$0^{2,4}$]nonan-6-one has been synthesized from *endo*-tricyclo[3.2.2.$0^{2,4}$]non-6-ene[218]:

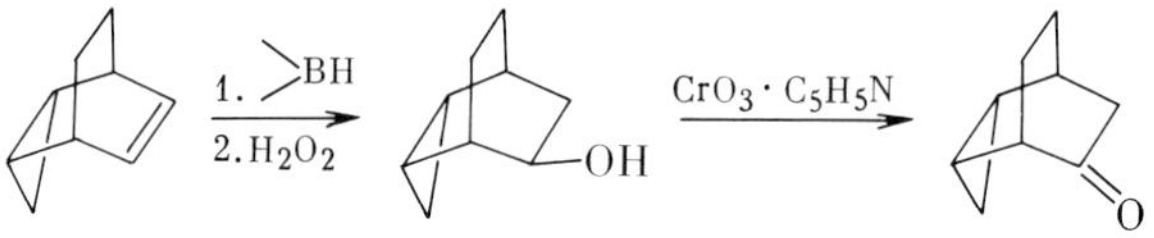

The pyridine complex of chromic anhydride has been used in the second stage of the stepwise oxidation. Pyridinium chlorochromate ($C_5H_5N \cdot HCl \cdot CrO_3$) oxidizes primary alkylboranes to aldehydes [260, 261] and secondary alkylboranes to ketones [260, 262]. The reaction has been used for the conversion of alcohols (RCH_2OH) [263], acids (RCOOH) [264], and terminal olefins, including functional terminal olefins [260, 261] to aldehydes (RCHO) via boron derivatives. The same reagent oxidizes 1,1-diboron compounds (from RC≡CR) to ketones (with an impurity of alcohol), while a mixture of olefins is obtained from 1,2-diboron compounds [265].

Oxidation with chromic anhydride has also been used in the synthesis of *l*-carvone from *l*-α-pinene [219]:

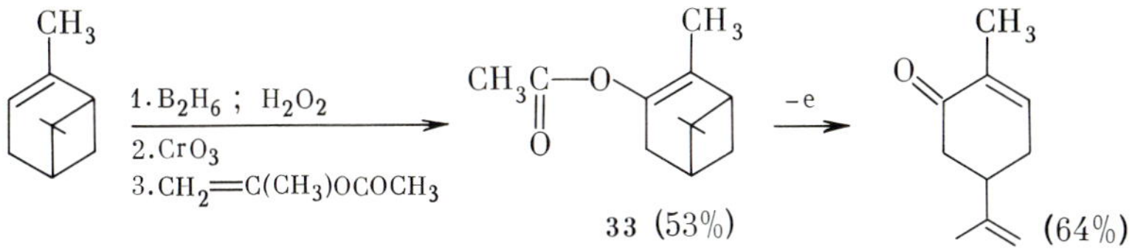

The oxidation of *syn*-7-trimethylstannylnorbornylborane with chromic acid produces *syn*-7-trimethylstannylnorbornan-2-one [266].

Some organic boron derivatives are stable toward the usual oxidants (alkaline hydrogen peroxide, amine oxides, oxygen). For example, the cyclic vinyloxyborane *18* was oxidized only with Jones reagent [220]:

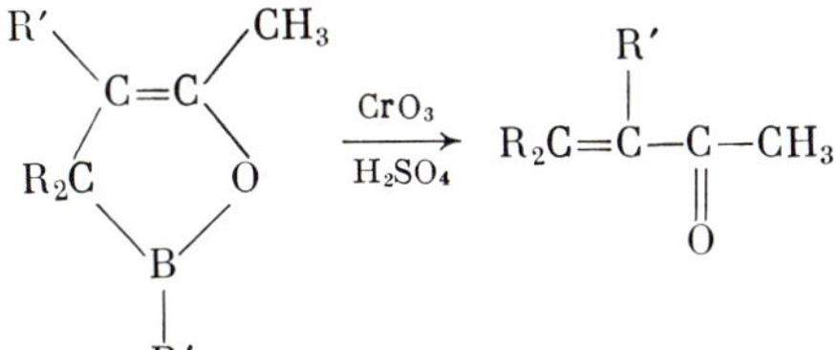

18

1,3-Diaza-2,4-diborolidines and other heterocyclic compounds are oxidized by the action of CrO_3 in acetic acid [221, 222].

In the oxidation of 9-borabicyclo[3.3.1]nonane, 10-borabicyclo[4.3.1]decane, and 11-borabicyclo[5.3.1]undecane with an excess of 8 N H_2CrO_4, monocyclic ketones are obtained instead of the expected dicarbonyl compounds [223]:

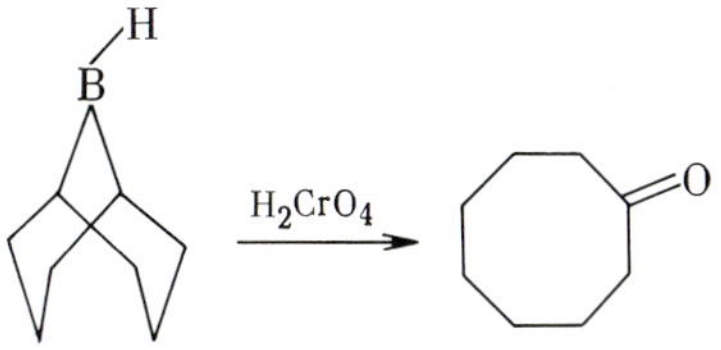

The stagewise oxidation of 9-borabicyclo[3.3.1]nonane produces 5-hydroxycyclooctan-1-one [267], which has been used further in the synthesis of substituted cyclooctyne derivatives. The oxidation of a 1,1-diboron compound, obtained by the hydroboration of diphenylacetylene, with CrO_3 in pyridine led to *trans*-stilbene [268].

Manganese dioxide [217] or sodium periodate in the presence of RuO_4 and sodium acetate in aqueous acetone [269] can be used instead of chromium compounds for the conversion of cycloalkylboranes to ketones:

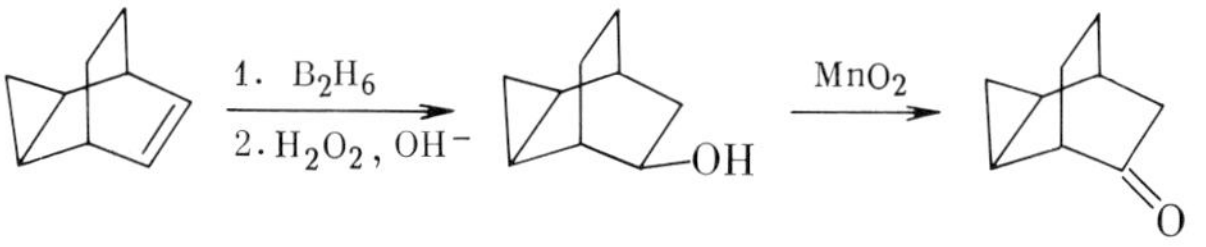

The sequential oxidation of 3-methoxy-3-borabicyclo[3.3.1]nonane with hydrogen peroxide and $KMnO_4$ led to *cis*-1,3-cyclohexanedicarboxylic acid [224].

In alkaline medium, potassium permanganate breaks the aromatic ring in phenylboronic acids; phthalic acid is obtained from α- and β-naphthylboronic acids, while butylboronic acid forms butanol [225]. In *m*- and *p*-tolylboronic acids, the methyl group undergoes oxidation, resulting in the formation of *m*-carboxyphenylboronic acid [227] and *p*-carboxyphenylboronic acid [226, 227, 228], respectively:

$$CH_3C_6H_4B(OH)_2 \xrightarrow[OH^-]{KMnO_4} HOOCC_6H_4B(OH)_2 \xrightarrow[HCl]{CH_3OH} CH_3OCOC_6H_4B(OH)_2.$$

It has not been possible to obtain the ortho isomer by this method [225].

3-Nitro-4-carboxyphenylboronic acid is formed by the oxidation of 3-nitro-4-methylphenylboronic acid [229]; a number of derivatives of this compound (esters, amides, acyl halides, azide, etc.) have been synthesized [230].

Alkyl- and cycloalkylboranes are oxidized to alcohols with $MoO_5 \cdot C_5H_5N \cdot HMPA$ [270–272]. The reaction proceeds with retention of the configuration. The action of lead tetraacetate on trialkylboranes produces alkyl acetates, ROAc [273].

5.8 ANODIC OXIDATION

The electrolysis of trialkylboranes in methanolic KOH solution with the use of platinum electrodes leads to a coupling of the radicals bonded to the boron; e.g., *n*-$C_{16}H_{34}$ is obtained from trioctylborane [231].

In the electrolysis of trialkylboranes in alcoholic $NaClO_4$ and CH_3ONa solution, methyl alkyl ethers are mainly obtained (84–

100%), together with dimerization products R—R [232]:

$$R_3B + 3CH_3O^- \xrightarrow{-3e} 3R^\cdot + B(OCH_3)_3,$$
$$R^\cdot \xrightarrow{-e} R^+ \xrightarrow{CH_3OH} ROCH_3,$$
$$2R^\cdot \longrightarrow R{-}R$$
($R{=}n$-C_5H_{11}, C_6H_{13}, C_8H_{17}, $C_2H_5CH(CH_3)CH_2$, cyclo-C_5H_9, cyclo-C_6H_{11}).

The initially formed carbocation R^+ isomerizes to an appreciable degree. For example, the oxidation of trioctylborane produced a mixture of 1-, 2-, 3-, and 4- methoxyoctanes in a ratio of 8:66:25:1.

The anodic oxidation of trialkylboranes in the presence of CH_3COONa and CH_3COOH leads to alkyl acetates, CH_3COOR.

The electrolysis of trihexylborane in sodium methoxide solution in the presence of 1,3-butadiene leads to a mixture of dodecane (9%), 8-hexadecene (18%), 1-methoxy-2-decene (23%), 8,12-eicosadiene (16%), and 8-vinyl-10-octadecene (13%) [233]. The electrolysis of R_3B in methyl cyanide in the presence of R'_4NX salts leads to higher nitriles, RCH_2CN [274], while nitroalkanes, RCH_2NO_2, are obtained in nitromethane [275].

As shown by Keating and Skell [234], the galvanostatic electrolysis of butylboronic acid in aqueous NaOH (platinum electrodes) produces 1-butene (60%), *cis-* and *trans*-2-butenes (20% each), while sodium propylborate forms propylene (71%), cyclopropane (18%), hexane (2%), ethylene (7%), and propane. Over pyrolytic graphite, the ratio of propane:cyclopropane:ethylene is 39:33:28.

On the other hand, according to the data of Humffray, et al. [235, 236], the electrolysis of $[C_4H_9B(OH)_3]Na$ in NaOH solution, along with a mixture of butenes, forms octane, with the ratio of the products formed depending on the anode potential, the concentrations of the reagents, and the pH of the solution. The yield of octane is 56% at pH 14 and a potential of 2.45 V.

The dimerization of aryl radicals bonded to boron takes place in the electrolysis of arylborates. Thus, for example, biphenyl is obtained as the result of the anodic oxidation of $(C_6H_5)_4BNa$ [237–239] and phenylboronic acid [240]:

$$(C_6H_5)_4B^- \xrightarrow{-2e} (C_6H_5)_2B^+ + (C_6H_5)_2,$$
$$(C_6H_5)_2B^+ + H_2O \longrightarrow (C_6H_5)_2BOH + H^+,$$
$$(C_6H_5)_2BOH + 2H_2O \xrightarrow{-2e} (C_6H_5)_2 + B(OH)_3 + 2H^+$$

The formation of $(C_6H_5)_2BOC_6H_5$ during the electrolysis is explained [238] by the partial autoxidation of triphenylborane [241, 242] (see, however, [24]).

Diaryl compounds are obtained in high yield by the oxidation of alkali metal tetraarylborates with bromine, iodine, *N*-bromosuccinimide, dichlorodicyanobenzoquinone, and Fe(III) and Ce(IV) salts [243].

REFERENCES

1. I. I. Lapkin and G. A. Yuzhakova, *Zh. Obshch. Khim.* **32,** 1967 (1962).
2. H. C. Brown and V. H. Dodson, *J. Am. Chem. Soc.* **79,** 2302 (1957).
3. E. Krause and P. Nobbe, *Ber.* **63,** 934 (1930).
4. H. E. Bent and M. Dorfman, *J. Am. Chem. Soc.* **57** 1259 (1935).
5. M. M. Midland, *Intra-Science Chem. Rep.* **7,** 65 (1973).
6. A. Davies and B. P. Roberts, *Acc. Chem. Res.* **5,** 387 (1972).
7. G. Sosnovsky and J. H. Brown, *Chem. Rev.* **66,** 529 (1966).
8. Yu. A. Alexandrov *J. Organometal. Chem.* **55,** 1 (1973).
9. M. F. Lappert, In: "The Chemistry of Boron and Its Compounds," Ed. E. L. Muetterties. New York, John Wiley and Sons, (1967) p. 443.
10. K. U. Ingold and B. P. Roberts, Free Radical Substitution Reactions. New York, Willey — Interscience, (1971)
11. A. G. Davies. In: "Organic Peroxides," vol. 2. Ed. D. Swern. New York, Wiley — Interscience (1972), Ch. 4.
12. A. G. Davies, In: "Progress in Boron Chemistry," vol. 1. Ed. H. Steinberg and A.L. McCloskey. Oxford, Pergamon Press, (1964), p. 265.
13. R. Köster, *Ann.* **618,** 31 (1958).
14. E. Frankland *Proc. Roy. Soc.* **25,** 165 (1877).
15. E. Frankland *Ann.* **124,** 129 (1862).
16. E. Frankland and B. Duppa, *Proc. Roy. Soc.* **10,** 568 (1859).
17. H. Meerwein and H. Sönke, *J. Prakt. Chem.* **147,** 251 (1936).
18. J. Johnson and M. van Campen, *J. Am. Chem. Soc.* **60,** 121 (1938).
19. E. Krause and R. Nitsche, *Ber.* **54,** 2784 (1921).
20. E. Krause and P. Nobbe, *Ber.* **64,** 2112 (1931).
21. D. Hurd, *J. Am. Chem. Soc.* **70,** 2053 (1948).
22. C. H. Bamford and D. M. Newitt, *J. Chem. Soc.* **1946,** 695.
23. S. B. Mirviss, *J. Am. Chem. Soc.* **83,** 3051 (1961).
24. S. B. Mirviss, *J. Org. Chem.* **32,** 1713 (1967).
25. P. G. Allies and P. B. Brindley, *J. Chem. Soc. B* **1969,** 1126.
26. A. G. Davies, D. G. Hare and R. F. M. White, *J. Chem. Soc.* **1961,** 341.
27. M. H. Abraham and A. G. Davies, *J. Chem. Soc.* **1959,** 429.
28. A. G. Davies and D. G. Hare, *J. Chem. Soc.* **1959,** 438.
29. A. G. Davies, D. G. Hare and R. F. M. White, *Chem. and Ind.* **1959,** 1315.
30. A. G. Davies, D. G. Hare and R. F. M. White, *J. Chem. Soc.* **1960,** 1040.
31. B. M. Mikhailov, V. A. Dorokhov, V. S. Bogdanov, I. P. Yakovlev and A. D. Naumov, *Dokl. Akad. Nauk SSSR* **194**, 595 (1970).

32. V. A. Dorokhov, L. I. Lavrinovich and B. M. Mikhailov, *Dokl. Akad. Nauk SSSR* **195,** 1100 (1970).
33. B. M. Mikhailov and Yu. N. Bubnov, *Izv. Akad. Nauk SSSR, Otd. Khim. Nauk* **1960,** 1883.
34. B. M. Mikhailov, V. A. Dorokhov and V. I. Seredenko, *Dokl. Akad. Nauk SSSR* **199,** 1328 (1971).
35. V. A. Dorokhov and B. M. Mikhailov, *Zh. Obshch. Khim.* **41,** 2702 (1971).
36. V. A. Dorokhov and B. M. Mikhailov, *Izv. Akad. Nauk SSSR, Ser. Khim.* **1972,** 1895.
37. B. M. Mikhailov, V. A. Dorokhov and V.I. Seredenko, *Zh. Obshch. Khim.* **43,** 1102 (1973).
38. V. A. Dorokhov, L. I. Lavrinovich, M. N. Bochkareva, V. S. Bogdanov and B. M. Mikhailov, *Zh. Obshch. Khim.* **43,** 1115 (1973).
39. K. L. Letsinger and I. H. Skoog, *J. Am. Chem. Soc.* **77,** 5176 (1955).
40. V. A. Dorokhov, O. G. Boldyreva and B. M. Mikhailov, *Zh. Obshch. Khim.* **43,** 1955 (1973).
41. B. M. Mikhailov and V. A. Dorokhov, *Izv. Akad. Nauk SSSR, Ser. Khim.* **1973,** 2649.
42. B. M. Mikhailov, V. A. Dorokhov and L. I. Lavrinovich, *Zh. Obshch. Khim.* **44** 1888 (1974).
43. V. A. Dorokhov, V. I. Seredenko and B. M. Mikhailov, *Zh. Obshch. Khim.* **45,** 1772 (1975).
44. W. Gerrard, M. F. Lappert and R. Shafferman, *Chem. and Ind.* **1958,** 722.
45. W. Gerard, M. F. Lappert and R. Shafferman, *J. Chem. Soc.* **1958,** 3648.
46. R. C. Petry and F. H. Verhoek, *J. Am. Chem. Soc.* **78,** 6416 (1956).
47. L. Parts and J. T. Miller, *Inorg. Chem.* **3,** 1483 (1964).
48. A. G. Davies, D. G. Hare and O. R. Khan, *J. Chem. Soc.* **1963,** 1125.
49. N. R. Zutty and F. J. Welch, *J. Org. Chem.* **25,** 861 (1960).
50. E. C. F. Coffee and A. G. Davies, *J. Chem. Soc. C* **1966,** 1493.
51. O. Grummit, *J. Am. Chem. Soc.* **64,** 1811 (1942).
52. A. K. Holliday and A. G. Massay, *J. Chem. Soc.* **1960,** 43.
53. H. Hock and F. Ernst, *Chem. Ber.* **92,** 2716 (1959).
54. H. Hock, A. Kropf and F. Ernst, Angew. Chem **71,** 541 (1959).
55. G. S. Kolesnikov and L. S. Fedorova, *Izv. Akad. Nauk SSSR, Otd. Khim. Nauk* **1957,** 236; **1958,** 906.
56. G. S. Kolesnikov and N. V. Klimentova, *Izv. Akad. Nauk SSSR, Otd. Khim. Nauk* **1957,** 652.
57. G. S. Kolesnikov and L. S. Fedorova, *Vysokomol. Soedin.* **1,** 1266 (1959).
58. J. Furukawa, T. Tsuruta and S. Inoue, *J. Polymer Sci.* **26,** 234 (1957).
59. J. Furukawa and T. Tsuruta, *J. Polymer Sci.* **28,** 227 (1958).
60. J. Furukawa, T. Tsuruta and S. Shiotani, *J. Polymer Sci.* **40,** 234 (1959).
61. J. Furukawa, T. Tsuruta, T. Imada and H. Fukutani, *Makromol. Chem.* **31,** 122 (1959).
62. I. Nakayama, T. Tsuruta and J. Furukawa, *Makromol. Chem.* **40,** 79 (1960).
63. T.-H. Lu, P.-T. Huang and H.-C. Chao, *Hua Hsueh Hsueh Pao* **32,** 34 (1966); *C. A.* **65,** 15185 (1966).
63a. T.-H. Lu, P.-T. Huang and H.-C. Chao, *Hua Hsueh Hsueh Pao* [*Acta Chim. Sinica*] **32,** 34 (1966); *Ref. Zh. Khim.* 4Zh, 351 (1967).
64. R. L. Hansen and R. R. Hamann, *J. Phys. Chem.* **67,** 2868 (1963).
65. R. L. Hansen, *J. Polymer Sci.* **2A,** 4215 (1964).
66. C. H. E. Bawn and D. Margerison and N. M. Richardson, *Proc. Chem. Soc.* **1959,** 397.

67. G. Talamini and G. Vidatto, *Makromol. Chem.* **50,** 129 (1961).
68. F. J. Welch, *J. Polymer Sci.* **61,** 243 (1962).
69. N. Zutty and F. Welch, *J. Polymer Sci.* **43,** 445 (1960).
70. F. S. Arimoto, *J. Polymer Sci.* **4A,** 275 (1966).
71. J. Grotewold and E. A. Lissi, *Chem. Comm.* **1965,** 21.
72. J. Grotewold, E. A. Lissi and E. A. Villa, *J. Polymer Sci.* **6A,** 3157 (1968).
73. E. Arancibia, J. Grotewold, E. A. Lissi and E. A. Villa, *J. Polymer Sci.* **7A,** 3430 (1969).
74. G. Borsini and M. Cipolla, *J. Polymer Sci.* **2B,** 291 (1964).
75. K. Kojima, Y. Iwata, M. Napayama and S. Iwabuchi, *J. Polymer Sci.* **8B,** 541 (1970).
76. K. Utimoto, T. Tanaka, T. Farubayashi and H. Nozaki, *Tetrahedron Letters* **1973,** 787.
77. P. G. Allies and P. B Brindley, *Chem. and Ind.* **1968,** 1439.
78. P. G. Allies and P. B. Brindley, *Chem. and Ind.* **1967,** 319.
79. A. G. Davies and B. P. Roberts, *J. Chem. Soc. B* **1967,** 17; *Chem. Comm.* **1966,** 298.
80. A. G. Davies and B. P. Roberts, *J. Chem. Soc. B* **1969,** 311.
81. P. D. Bartlett and T. Funahashi, *J. Am. Chem. Soc.* **84,** 2596 (1962).
82. H. C. Brown and M. M. Midland, *Chem. Comm.* **1971,** 699.
83. M. M. Midland and H. C. Brown, *J. Am. Chem. Soc.* **93,** 1506 (1971).
84. J. Grotewold, J. Hernandez and E. Lissi, *J. Chem. Soc. B* **1971,** 182.
85. A. G. Davies, K. U. Ingold, B. P. Roberts and R. Tudor, *J. Chem. Soc. B* **1971,** 698.
86. N. M. Emanuel, E. T. Denisov and Z. K. Maizus: "Chain Reactions of Hydrocarbon Oxidation in the Liquid Phase", Moscow, Nauka (1965).
87. P. B. Brindley and J. C. Hodgson, *Chem. Comm.* **1972,** 202.
88. D. J. Cram. "Fundamentals of Carbanion Chemistry," New York, Academic Press (1965),
89. G. P. Gerbert, V. P. Maslennikov and V. A. Shushunov, *Zh. Obshch. Khim.* **40,** 131 (1970).
90. V. P. Maslennikov, G. P. Gerbert and G. F. Khodalev, *Zh. Obshch. Khim.* **40,** 245 (1970).
91. G. P. Gerbert and V. P. Maslennikov, *Zh. Obshch. Khim.* **40,** 1105 (1970).
92. V. P. Maslennikov, G. P. Gerbert and V. A. Shushunov, *Zh. Obshch. Khim.* **40,** 1107 (1970).
93. V. P. Maslennikov, M. B. Nedoshivina and V. A. Shushunov, *Zh. Obshch. Khim.* **41,** 487 (1971).
94. V. P. Maslennikov, G. P. Gerbert, G. F. Khodalev and V. A. Shushunov, *Zh. Obshch. Khim.* **41,** 592 (1971).
95. G. B. Sadikov, V.P. Maslennikov and G. I. Manin, *Zh. Obshch. Khim.* **42,** 1571 (1972).
96. V. A. Shushunov, V. P. Maslennikov and G. P. Gerbert, *Dokl. Akad. Nauk SSSR* **199,** 373 (1971).
97. V. F. Kupriyanov, V. N. Alyasov, V. G. Tsvetkov, M. S. Sheiman, V. I. Buzulukov, V. P. Maslennikov and Yu. A. Aleksandrov, *Tr. Khim. Khim. Tekhnol.* **1974,** No. 1, 80.
98. G. I. Manin, V. P. Maslennikov and Yu. A. Aleksandrov, *Zh. Obshch. Khim.* **44,** 1278 (1974).
99. L. Barton and G. T. Bohn, *Chem. Comm.* **1971,** 77.
100. L. Barton and J. N. Grump, *Inorg. Chem.* **12,** 2252 (1973).
101. L. Barton and J. N. Grump, *Inorg. Chem.* **12,** 2506 (1973).

102. L. Barton, J. N. Grump and J. B. Wheatley *J. Organometal. Chem.* **72,** C1 (1974).
103. R. Damico, *J. Org. Chem.* **29,** 1971 (1964).
104. J. N. Cooper and R. E. Powell, *J. Am. Chem. Soc.* **85,** 1590 (1963).
105. H. C. Brown, M. M. Midland and G. W. Kabalka, *J. Am. Chem. Soc.* **93,** 1024 (1971).
106. Y. Tanahashi, J. Lhomme and G. Ourisson, *Tetrahedron* **28,** 2655 (1972).
107. G. Wilke and P. Heimbach, *Ann.* **652,** 7 (1962).
108. H. C. Brown and M. M. Midland, *J. Am. Chem. Soc.* **93,** 4078 (1971).
109. M. M Midland and H. C. Brown, *J. Am. Chem. Soc.* **95,** 4069 (1973).
110. S. J. Brois, *Tetrahedron Letters* **1964,** 345.
111. M. Inatome and L. Kuhn, *Advances in Chemistry, Ser.* **42,** 183 (1964).
112. M. H. Abraham, J. H. N. Garland, J. A. Hill and L. F. Larkworthy, *Chem. and Ind.* **1962,** 1615.
113. W. Kliegel, *Organometal. Chem. Rev., A* **8,** 153 (1972).
114. S. J. Brois and A. J. Rutkowski, *U.S. Patent* 3,147,310 (1964); *C. A.* **31,** 11891 (1964).
115. A. D. Ainley and F. Challenger, *J. Chem. Soc.* **1930,** 2171.
116. H. R. Snyder, J. A. Kuck and J. R. Johnson, *J. Am. Chem. Soc.* **60, 105** (1938).
117. R. Belcher, D. Gibbons and A. Sykes, *Mikrochim. Acta* **40,** 76 (1952).
118. B. M. Mikhailov and T. A. Shchegoleva, *Dokl. Akad. Nauk SSSR* **108,** 481 (1956).
119. G. Zweifel and H. C. Brown in: "Organic Reactions", vol. 13, New York, J. Wiley & Sons (1963), p. 1.
120. B. M. Mikhailov, A. Ya. Bezmenov, L. S. Vasil'ev and V. G. Kiselev, *Dokl. Akad. Nauk SSSR* **155,** 141 (1964).
121. A. Ya. Bezmenov, L. S. Vasil'ev and B. M. Mikhailov, *Izv. Akad. Nauk SSSR, Ser. Khim.* **1965,** 2111.
122. B. M. Mikhailov, A. Ya. Bezmenov and L. S. Vasil'ev, *Dokl. Akad. Nauk SSSR* **167,** 590 (1966).
123. Yu. N. Bubnov, O. A. Nesmeyanova, T. Yu. Rudashevskaya, B. M. Mikhailov and B. A. Kazansky, *Tetrahedron Letters* **1971,** 2153.
124. Yu. N. Bubnov, O. A. Nesmeyanova, T. Yu. Rudashevskaya, B. M Mikhailov and B. A. Kazanskii, *Zh. Obshch. Khim.* **43,** 127 (1973).
125. Yu. N. Bubnov, O. A. Nesmeyanova, T. Yu. Rudashevskaya, B. M. Mikhailov and B. A. Kazanskii, *Zh. Obshch. Khim.* **43,** 135 (1973).
126. K. Utimoto, M. Tamura, M. Tanouti, and K. Sisido, *Tetrahedron* **28,** 5697 (1972).
127. R. Köster, S. Arora and P. Binger, *Angew. Chem.* **81,** 185 (1969).
128. S. A. Korobeinikova, Yu. N. Bubnov and B. M. Mikhailov, *Izv. Akad. Nauk SSSR, Ser. Khim.* **1969,** 2212.
129. B. M. Mikhailov, Yu. N. Bubnov, S. A. Korobeinikova and V. S. Bogdanov, *Zh. Obshch. Khim.* **40,** 1321 (1970).
130. B. M. Mikhailov, Yu. N. Bubnov and M. Sh. Grigoryan, *Izv. Akad. Nauk SSSR, Ser. Khim.* **1971,** 1842.
131. B. M. Mikhailov, Yu. N. Bubnov, M. Sh. Grigoryan and V. S. Bogdanov, *Zh. Obshch. Khim.* **44,** 2715 (1974).
132. H. C. Brown, G. W. Kabalka and M. W. Rathke, *J. Am. Chem. Soc.* **89,** 4530 (1967).
133. E. Negishi and H. C. Brown, *Synthesis* **1972,** 196.
134. H. C. Brown, R. A. Coleman and M. W. Rathke, *J. Am. Chem. Soc.* **90,** 499 (1968).

135. J. Plamondon, J. T. Snow and G. Zweifel, *Organometal. Chem. Synth.* **1,** 249 (1971).
136. H. C. Brown, E. F. Knights and R. A. Coleman, *J. Am. Chem. Soc.* **91,** 2144 (1969).
137. H. C. Brown and S. P. Rodes, *J. Am. Chem. Soc.* **91,** 4306 (1969).
138. P. Pino, L. Lardicci and A. Stefani, *Ann. Chim. (Rome)* **52,** 456 (1962).
139. L. Verbit and P. J. Heffron, *J. Org. Chem.* **32,** 3199 (1967).
140. D. S. Matteson and R. A. Bowie, *J. Am. Chem. Soc.* **87,** 2587 (1965).
141. A. Davies and B. Roberts, *J. Chem. Soc., C* **1968,** 1474.
142. H. Minato, J. C. Ware and T.G. Traylor, *J. Am. Chem. Soc.* **85,** 3024 (1963).
143. H. G. Kuivila, *J. Am. Chem. Soc.* **76,** 870 (1954).
144. H. G. Kuivila, *J. Am. Chem. Soc.* **77,** 4014 (1955).
145. H. G. Kuivila and R. A. Wiles, *J. Am. Chem. Soc.* **77,** 4830 (1955).
146. H. G. Kuivila and A. G. Armour, *J. Am. Chem. Soc.* **79,** 5659 (1957).
147. W. J. Wechter, *Chem. and Ind.* **1959,** 294.
148. I. R. Wilson, *Austral. J. Chem.* **13,** 583 (1960).
149. A. G. Davies and R. B. Moodie, *J. Chem. Soc.* **1958,** 2372.
150. D. B. Bigley and D. W. Payling, *J. Chem. Soc., B* **1970,** 1811.
151. A. J. Weinheimer and W. E. Marsico, *J. Org. Chem.* **27,** 1926 (1962).
152. N. V. Kruglova and R. Kh. Freidlina, *Izv. Akad. Nauk SSSR, Ser. Khim.* **1966,** 1662.
153. D. J. Pasto, S. K. Arora and J. Chow, *Tetrahedron* **25,** 1571 (1969).
154. B. M. Mikhailov, P. M. Aronovich and V. G. Kiselev. *Izv. Akad. Nauk SSSR, Ser. Khim.* **1968,** 146.
155. P. M. Aronovich, V. S. Bogdanov and B. M. Mikhailov, *Izv. Akad. Nauk SSSR, Ser. Khim.* **1969,** 362.
156. G. W. Kramer and H. C. Brown, *J. Organometal. Chem.* **90,** C1 (1975).
157. M. F. Hawthorne, *J. Org. Chem.* **22,** 1001 (1957).
158. S. W. Breuer and F. A. Broster, *J. Organometal. Chem.* **35,** C5 (1972).
159. G. M. Pickles and F. G. Thorpe, *J. Organometal. Chem.* **76,** C23 (1974).
160. A.-B. Hörnfeldt and S. Gronowitz, *Acta Chem. Scand.* **16,** 789 (1962).
161. A.-B. Hörnfeldt and S. Gronowitz, *Arkiv Kemi* **21,** 239 (1963).
162. A.-B. Hörnfeldt, *Arkiv Kemi* **22,** 211 (1964).
163. A.-B. Hörnfeldt, *Acta Chem. Scand.* **19,** 1249 (1965).
164. R. T. Hawkins, *J. Heterocycl. Chem.* **11,** 291 (1974).
165. G. W. Kramer and H. C. Brown, *J. Organometal. Chem.* **73,** 1 (1974).
166. J. Klein and A. Medlic, *J. Am. Chem. Soc.* **93,** 6313 (1971).
167. G. Zweifel and H. Arzoumanian, *J. Am. Chem. Soc.* **89,** 291 (1967).
168. G. I. Makin, V. P. Maslennikov and V. A. Shushunov, *Zh. Obshch. Khim.* **42,** 135 (1972).
169. D. S. Matteson and J. G. Shdo, *J. Org. Chem.* **29,** 2742 (1964).
170. R. L. Letsinger and I. H. Skoog, *J. Org. Chem.* **18,** 895 (1953).
171. S. I. Frolov, Yu. N. Bubnov and B. M. Mikhailov, *Izv. Akad. Nauk SSSR, Ser. Khim.* **1969,** 1996.
172. Yu. N. Bubnov. S. I. Frolov, V. G. Kiselev, V. S. Bogdanov and B. M. Mikhailov, **40,** 1311 (1970).
173. Yu. N. Bubnov, S. I. Frolov, V. G. Kiselev, V. S. Bogdanov and B. M. Mikhailov, *Organometal. Chem. Synth.* **1,** 37 (1970/1971).
174. D. Matteson, *J. Am. Chem. Soc.* **82,** 4228 (1960).
175. D. B. Bigley and D. W. Payling, *Chem. Comm.* **1968,** 938.
176. A. G. Davies and R. Tudor, *J. Chem. Soc., B* **1970,** 1815.
177. F. R. Jensen, L. H. Gale and J. E. Rodgers, *J. Am. Chem. Soc.* **90,** 5793 (1968).

178. V. P. Maslennikov, G. I. Makin and Yu. A. Aleksandrov, *Zh. Obshch. Khim.* **43,** 2289 (1973).
179. V. A. Shushunov, V. P. Maslennikov, G. I. Makin and A. V. Gorbunov, *Zh. Obshch. Khim.* **42,** 1577 (1972).
180. A. B. Mikhailov and Yu. N. Bubnov. *Izv. Akad. Nauk SSSR, Ser. Khim.* **1964,** 2248.
181. A. Arase, Y. Masuda and A. Suzuki, *Bull. Chem. Soc. Japan* **47,** 2511 (1974).
182. G. I. Makin, V. P. Maslennikov and V. A. Shushunov, *Zh. Obshch. Khim.* **42,** 834 (1972).
183. V. A. Shushunov, V. P. Maslennikov, G. I. Makin and S. I. Kulakov, *Zh. Org. Khim.* **8,** 1409 (1972).
184. V. A. Shushunov, V. P. Maslennikov and G. I. Makin, *Tr. Khim. Khim. Tekhnol.* **1970,** No. 2, 214.
185. G. I. Makin, V. P. Maslennikov and Yu. A. Aleksandrov, *Zh. Obshch. Khim.* **43,** 2270 (1973).
186. A. G. Davies and R. B. Moodie, *Chem. and Ind.* **1957,** 1622.
187. G. P. Gerbert, V. P. Maslennikov, V. A. Shushunov and L. N. Yankitova, *Tr. Khim. Khim. Tekhnol.* **1968,** No. 3, 41.
188. P. F. Wolf and R. K. Barnes, *J. Org. Chem.* **34,** 3441 (1969).
189. R. Köster, *German Patent* 1,294,964 (1969); *C. A.* **71,** 39139 (1969).
190. R. Köster and Y. Morita, *Angew. Chem.* **78,** 589 (1966).
191. R. Köster and Y. Morita, *Ann.* **704,** 70 (1967).
192. G. Zweifel, N. L. Polston and C. C. Whitney, *J. Am. Chem. Soc.* **90,** 6243 (1968).
193. G. W. Kabalka and H. C. Hedgecock, *J. Org. Chem.* **40,** 1776 (1975).
194. R. Köster, H. J. Horstschäfer and P. Binger, *Ann.* **717,** 1 (1968).
195. A. Burg and J. H. Bickerton, *J. Am. Chem. Soc.* **67,** 2261 (1945).
196. U. Wanagat and R. Pfeifenschneider, *Z. Anorg. Chem.* **297,** 151 (1958).
197. I. G. Ryss and S. L. Idel's, *Zh. Neorg. Khim.* **10,** 786 (1965).
198. J. R. Snyder, L. Lee and D. G. Farnum, *J. Am. Chem. Soc.* **93,** 3816 (1971).
199. W. Kliegel, *Z. Chem.* **10,** 437 (1970).
200. W. Kliegel, *Z. Chem.* **9,** 112 (1969).
201. R. Köster and Y. Morita, *Angew. Chem.* **77,** 589 (1965).
202. H. G. Kuivila, L. Benjamin, C. Murphy, A. Price and J. Polevy, *J. Org. Chem.* **27,** 825, (1962).
203. H. C. Brown, *U.S. Patent* 3,439,046 (1969); *C. A.* **71,** 50273 (1969).
204. R. Pappo, *J. Am. Chem. Soc.* **81,** 1010 (1959).
205. J. C. Ware and T. G. Traylor, *J. Am. Chem. Soc.* **85,** 3026 (1963).
206. H. C. Brown and C. P. Garg, *J. Am. Chem. Soc.* **83,** 2951, 2952 (1961).
207. L. G. Wideman, *J. Org. Chem.* **33,** 4541 (1968).
208. K. D. Berlin and R. Ranganathan, *Tetrahedron* **25,** 793 (1969).
209. G. J. Dufresne and M. Blanchard, *Bull. Soc. Chim. France* **1968,** 385.
210. A. F. Thomas and B. Willhalm, *Helv. Chim. Acta* **50,** 826 (1967).
211. R. K. Hill and A. G. Edwards, *Tetrahedron* **21,** 1501 (1965).
212. W. R. Moore and W. R. Moser, *J. Org. Chem.* **35,** 908 (1970).
213. R. T. Lansbury and E. J. Nienhause, *Chem. Comm.* **1966,** 273.
214. N. Miyamoto, S. Isiyama, K. Utimoto and H. Nozaki, *Tetrahedron* **29,** 2365 (1973).
215. D. Pasto and C. C. Cumbo *J. Am. Chem. Soc.* **86,** 4343 (1964).
216. N. Takaishi, Y. Fujikura, Y. Inamoto, H. Ikeda and K. Aigami, *Chem. Comm.* **1975,** 372.
217. J. F. Bagli, P. F. Morand and R. Gaudry, *J. Org. Chem.* **27,** 2938 (1962).
218. P. E. Schueler and Y. E. Rhodes, *J. Org. Chem.* **39,** 2063 (1974).

219. T. Shono, I. Nishiguchi, T. Yokoyama and M. Nitta, *Chem. Letters* **1975,** 433.
220. M. Naruse, T. Tomita, K. Utimoto and H. Nozaki, *Tetrahedron Letters* **1973,** 795.
221. J. Casanova and H. B. Kiefer, *J. Org. Chem.* **34,** 2579 (1969).
222. G. Hesse and H. Witte, *Angew. Chem.* **75,** 791 (1963).
223. M. M. Bhagwat, I. Mehrotra and D. Devaprabhakara, *J. Organometal. Chem.* **82,** C27 (1974).
224. B. M. Mikhailov, Yu. N. Bubnov and S. I. Frolov, *Izv. Akad. Nauk SSSR, Ser. Khim.* **1967,** 2290.
225. B. Bettman and G. E. K. Branch, *J. Am. Chem. Soc.* **56,** 1616 (1934).
226. A. Michaelis, *Ann.* **315,** 33 (1901).
227. B. Bettman, G. E. K. Branch and D. L. Yabroff, *J. Am. Chem. Soc.* **56,** 1865 (1934).
228. K. Torssell, *Arkiv Kemi* **10,** 473 (1957).
229. K. Torssell, H. Meyer and B. Zacharias, *Arkiv Kemi* **10,** 497 (1957).
230. K. Torssell, *Arkiv Kemi* **10,** 513 (1957).
231. T. Taguchi, M. Itoh and A. Suzuki, *Chem. Letters* **1973,** 719.
232. T. Taguchi, Y. Takahashi, M. Itoh and A. Suzuki, *Chem. Letters* **1974,** 1021.
233. H. Schaefer and D. Koch, *Angew. Chem.* **84,** 32 (1972).
234. J. T. Keating and P. S. Skell, *J. Org. Chem.* **34,** 1479 (1969).
235. A. A. Humffray and L. F. G. Williams, *Electrochimica Acta* **17,** 401 (1972).
236. A. A. Humffray and L. F. G. Williams, *Chem. Comm.* **1965,** 616.
237. D. H. Geske, *J. Phys. Chem.* **66,** 1743 (1962).
238. D. H. Geske, *J. Phys. Chem.* **63,** 1062 (1959).
239. W. R. Turner and P. J. Elving, *J. Phys. Chem.* **69,** 1067 (1965).
240. W. R. Turner and P. J. Elving, *Anal. Chem.* **37,** 207 (1965).
241. G. A. Razuvaev and T. G. Brilkina, *Zh. Obshch. Khim.* **24,** 1415 (1954).
242. G. A. Razuvaev and T. G. Brilkina, *Dokl. Akad Nauk SSSR* **85,** 815 (1952).
243. J. J. Fisch and R. J. Wilcsek, *J. Organomtal. Chem.* **71,** C21 (1974).
244. P. B. Brindley, J. C. Hodgson and M. J. Scotton, *J. Chem. Soc., Perkin II* **1979,** 45.
245. G. A. Razuvaev, M. A. Lopatin and V. A. Dodonov, *Zh. Obshch. Khim.* **48,** 2494 (1978).
246. G. A. Razuvaev, V. A. Dodonov, D. F. Grishin and V. K. Cherkasov, *Dokl. Akad. Nauk SSSR* **253,** 113 (1980).
247. V. A. Dodonov, O. S. Morozov, D. F. Grishin, E. G. Lyutin and N. N. Vyshinskii, *Dokl. Akad. Nauk SSSR* **255,** 1123 (1980).
248. A. J. Bloodworth and H. J. Eggelte, *Tetrahedron Letters* **21,** 2001 (1980).
249. V. N. Glushkov, V. N. Alyasov, V. P. Maslennikov, Yu. A. Aleksandrov and G. A. Razuvaev, *Zh. Obshch. Khim.* **50,** 1541 (1980).
250. D. S. Matteson and R. Moody, *J. Org. Chem.* **45,** 1091 (1980).
251. D. S. Matteson, R. Moody and R. K. Jesthi, *J. Am. Chem. Soc.* **97,** 5608 (1975).
252. V. N. Alyasov, V. P. Maslennikov and Yu. A. Aleksandrov, *Zh. Obshch. Khim.* **50,** 362 (1980).
253. E. V. Chistova, V. N. Alyasov, R. F. Galiullina, V. P. Maslennikov, V. A. Dodonov and Yu. A. Aleksandrov, *Zh. Obshch. Khim.* **51,** 1078 (1981).
254. N. Miyaura and A. Suzuki, *Chem. Letters* **1981,** 879.
255. J. Hooz and R. D. Mortimer, *Can. J. Chem.* **56,** 2786 (1978).
256. G. W. Kabalka and H. C. Hedgecock, *J. Org. Chem.* **40,** 1776 (1975).
257. G. W. Kabalka and S. W Slayden, *J. Organometal. Chem.* **125,** 273 (1977).
258. T. Kudo, A. Nose and M. Hamana, *Yakugaku Zasshi* **95,** 521 (1975).
259. R. T. Hawkins, R. S. Hsu and S. G. Wood, *J. Org. Chem. 43,* 4648 (1978).

260. C. G. Rao, S. U. Kulkarni and H. C. Brown, *J. Organometal. Chem.* **172,** C20 (1979).
261. H. C. Brown, S. U. Kulkarni and C. G. Rao, *Synthesis* **1980,** 151.
262. V. V. Ramana Rao, D. Devaprabhakara and S. Chandrasekaran, *J. Organometal. Chem.* **162,** C9 (1978).
263. H. C. Brown, S. U. Kulkarni and C. G. Rao, *Synthesis* **1979,** 702.
264. H. C. Brown, C. G. Rao and S. U. Kulkarni, *Synthesis* **1979,** 704.
265. C. G. Rao, S. K. Agarwal, I. Mehrotra and D. Devaprabhakara, *J. Organometal. Chem.* **166,** 9 (1979).
266. D. D. Davis and H. T. Johnson, *J. Organometal. Chem.* **86,** 75 (1975).
267. H. Meier and H. Peterson, *Synthesis* **1978,** 596.
268. M. M. Bhagwat, I. Mehrotra and D. Devaprabhakara, *Tetrahedron Letters* **1975,** 167.
269. R. H. Mueller and R. M. DiPardo, *Chem. Commun.* **1975,** 565.
270. G. Schmitt and B. Olbertz, *J. Organometal. Chem. 152,* 271 (1978).
271. D. A. Evans, E. Vogel and J. V. Nelson, *J. Am. Chem. Soc.* **101,** 6120 (1979).
272. M. M. Midland and S. B. Preston, *J. Org. Chem.* **45,** 4514 (1980).
273. Y. Masuda and A. Arase, *Bull. Soc. Chem. Japan* **51,** 901 (1978).
274. Y. Takahashi, M. Tokuda, M. Itoh, A. Suzuki, *Chem. Letters* **1975,** 523.
275. Y. Takahashi, M. Tokuda, M. Itoh, A. Suzuki, *Synthesis* **1976,** 616.

CHAPTER 6

Methods of protolysis of organoboron compounds

Organoboron compounds are considerably more stable toward hydrolytic agents than organomagnesium and organoaluminum compounds. In this case, the capacity of the B—C bonds for protolytic cleavage is determined mainly by the hybridization of the carbon atom bonded to the boron. Compounds of the same class are arranged in the following series with respect to their ease of protolysis: $B—C_{sp} > B—C_{sp^2} > B—C_{sp^3}$[1]. Alkylboranes are the most stable, vinyl and aryl derivatives are less stable, and alkynyl derivatives are cleaved most easily. Although the energy of the boron-vinyl bond is 10–15 kcal/mole greater than the energy of the bond between boron and alkyl radicals [2, 3], carbon atoms in the sp and sp^2 state are more electronegative than C_{sp^3}, causing a greater polarity of the $B—C_{sp}$ and $B—C_{sp^2}$ bonds and promoting their heterolytic rupture [1, 4].

Trialkylboranes ($B—C_{sp^3}$) are stable toward water, alcohols, amines [5], and aqueous solutions of acids and alkalies at temperatures of up to 100–120°C; triphenylborane reacts with water at 20°C (more rapidly upon heating) with the rupture of two boron-carbon bonds [6].

Vinyl boron derivatives (Parts III and IV) undergo cleavage considerably more readily than the alkyl compounds [7–11]. Derivatives of di- and monovinylboranes are decomposed by water or alcohols with the formation of an olefin at 100–120°C [12–14], whereas their alkyl analogs are stable under these conditions. The $B—C_{sp}$ bond is ruptured even more easily in bis(dialkylboryl)-acetylenes [15] and ethynylboronic acids (by water or $KHCO_3$ solution) [16, 17].

Allylboranes (Part IV), 1-boracyclopentanes (Part II), benzyl-, and also cyclopropylboranes [18], 1,1-diboron, and some other aliphatic compounds containing "active" boron-carbon bonds show an increased reactivity and decompose under milder conditions than the simple alkylboranes.

The protolytic cleavage of organoboranes is used for the synthesis of boron derivatives of the type R_2BY and RBY_2 from triorganoboranes, for the conversion of organoboron compounds to organic compounds:

$$R{-}B\langle + H{-}Y \longrightarrow RH + \rangle B{-}Y,$$

and also for the protection and quantitative determination of hydroxy groups in alcohols, phenols, and polyols.

The protonolysis of trialkylboranes and of mono- and dialkyl compounds for the synthesis of hydrocarbons RH is usually carried out with higher carboxylic acids (anhydrous alkalies and HF can also be used), while alcohols and alkali solutions are used for the protonolysis of benzylboranes. The most suitable method for the preparative protonolysis of diaryl- and arylboron compounds is their heating with aqueous acetic acid or with 10% alkali solution.

The protolytic cleavage of organoboranes is the final stage in the *cis*-hydrogenation of olefins to paraffin hydrocarbons and of acetylenes to *cis*-olefins, and also in the reduction of halides to hydrocarbons via organoboranes.

The monodealkylation of trialkylboranes by protonic reagents is catalyzed by mercaptans and by pivalic acid and its derivatives.

Questions of the protonolysis of organoboranes are partially discussed in reviews [2, 19] and monographs [20].

6.1 ACTION OF WATER, ALCOHOLS, AND NITROGENOUS COMPOUNDS

At 20–100°C, trialkylboranes are not hydrolyzed with water, acid or alkali solutions and are not cleaved by alcohols, diols, amines, and many other reagents of the type HY. These reactions proceed at a higher temperature, with the cleavage of one B—C bond.

Trimethylborane reacts with water, D_2O, H_2S, ethylene glycol, pinacol, pyrocatechol, *o*-aminophenol, and *o*-phenylenediamine in an autoclave at 180–200°C with the formation of dimethyl derivatives, $(CH_3)_2BY$, and methane [21]. At 300–340°C, monomethyl compounds are obtained. Reactions with ammonia [22, 23], methylamine

[23], aniline [24], and diamines [21, 25] at 200–340°C lead to the corresponding amino(methyl)boranes, aminodimethylboranes, or *B*-methylborazoles. *B*-Triethylborazole was obtained from the complex $Et_3B \cdot NH_3$ at 440–450°C [26].

Higher trialkylboranes (tripropylborane and others) react with water [5], alcohols [5, 27], ammonia [28], amines [5, 29], and hydrazines [30, 31] at 150–180°C, forming dialkylboron compounds containing oxygen or nitrogen:

$$R_3B + HY \longrightarrow R_2B{-}Y + RH + R_{-H} + H_2.$$

The reaction is accompanied by the formation of a saturated hydrocarbon RH, an olefin, and hydrogen, the ratio of which depends on the nature of the nucleophilic agent HY. In this case, the cleavage proceeds along two routes, the first of which (*a*) is a direct cleavage of the boron-carbon bond by a cyclic mechanism or via a transition state with three-center bonds:

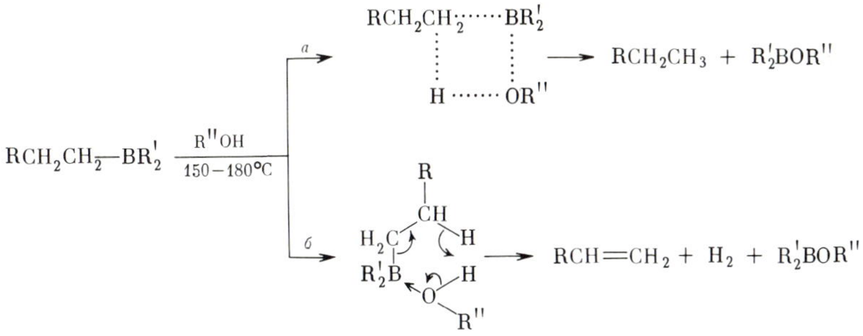

1

The formation of an olefin and hydrogen can be explained by a reaction course in the direction *b*, via a cyclic transition state *1*. A different reaction mechanism, including the formation of boron hydrides R_2BH, is less probable.

A large number of different oxygen, sulfur, and nitrogen derivatives of boron R_2BY have been obtained by the monodealkylation of trialkylboranes (Y = OH, OR, SH, SR, NH_2, NHR, NR_2, $NHNH_2$, NHNHR, NHCOR, OCH=CR_2, —O—SO_2—O—, RSO_2O—, etc.).

Reagents of the type H—Y̤—C=A̤—, as a rule, cleave one B—C bond in trialkylboranes under milder conditions than HY.

2-Aminopyridine [32–34], pyrrole [35–37], pyrazole [38], 2-pyridone, 2-oxopyrrolidine, and 8-hydroxyquinoline [32] react with trialkylboranes at room temperature or on heating, also with the cleavage of one B—C bond and the elimination of only RH. Tributylborane is monodealkylated by β-keto alcohols in tetrahydrofuran at 20°C within two hours. The chelated ethers formed are stereoselectively reduced by $NaBH_4$ to 1,3-diols [161].

Trialkylboranes react with 2-isopropyl-6-methyl-4-pyrimidone to form 4-diorganylboryloxy-2-isopropyl-6-methylpyrimidines [162].

The reactions of trialkylboranes with mercaptans leading to thioesters R_2BSR' are considered in Part VI.

As a result of the monodealkylation of trialkylboranes with acetylacetone [32, 39–43], other β-diketones [41, 43], acetoacetic ester and 4-imino-2-pentanone [41], benzoin [42], diacetamide [44], tetrasubstituted oxamidines [45], and nickel dimethylglyoxime [46, 47], chelated dialkylboron compounds such as the following have been obtained:

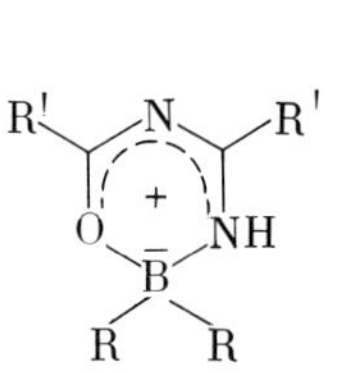

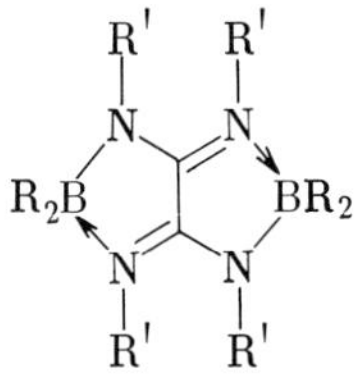

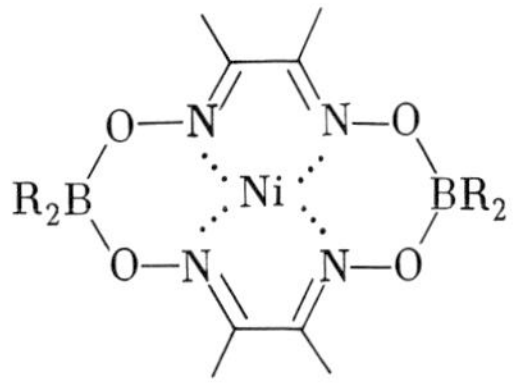

Acetylacetonates have also been synthesized from tribenzyl-triphenyl-, tri-p-tolyl,-, and tri-α-naphthylborane [42].

In contrast to the amines, amides react with trialkylboranes upon heating in tetrahydrofuran to form acylaminodialkylboranes [32, 48, 49]. The latter add easily to nitriles, forming chelated dialkylboryl-acylamidinates (*2*) [49–52]:

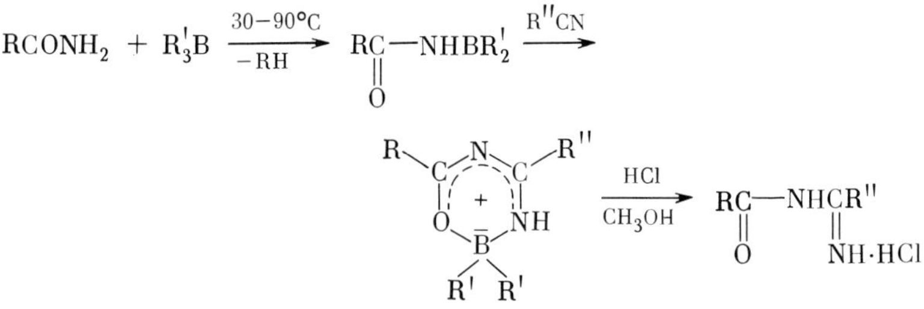

2

Since the chelates *2* are cleaved by the action of a methanolic HCl solution, a combination of these reactions serves as the basis of a method for converting amides and nitriles to acylamidines in a yield of 50–70% [49, 53]:

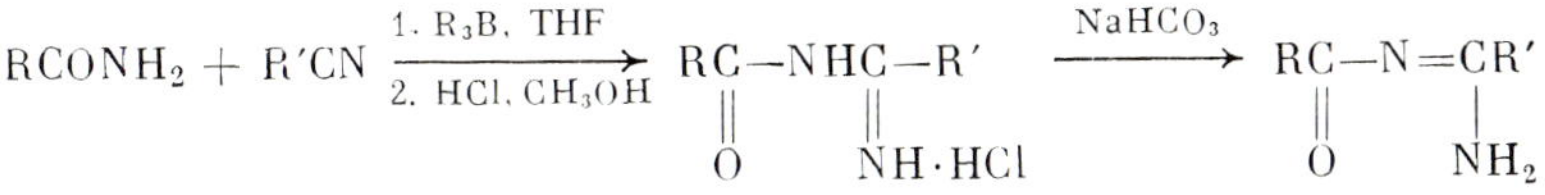

This method is an interesting example of the use of chelation in organic synthesis. The trialkylboranes play an auxiliary role, although they do participate in the process.

Chelated dialkylborylimidoylamidinates are obtained by heating acylamides with trialkylboranes and nitriles [54, 55].

N,N'-Diphenylacetamidine reacts with trialkylboranes with the formation of *N*-dialkylboryl-*N,N'*-diphenylacetamidines which, upon heating, split off RH and are converted to 3-phenyl-4-alkyl-2-methyl-3,4-dihydro-4-boraquinazolines [56]:

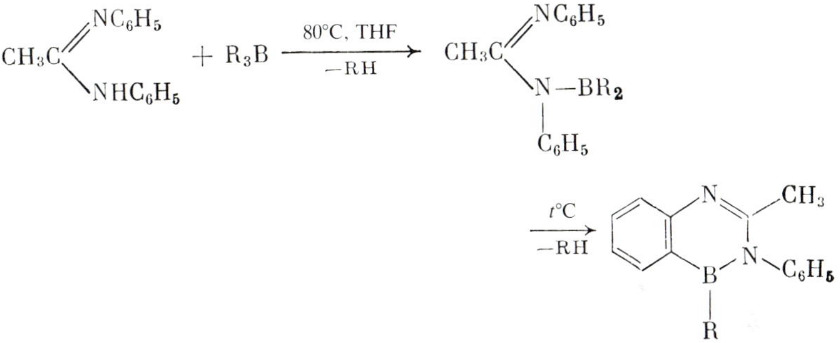

The reaction of diphenylboron chloride with methyl iminotrifluoroacetate forms CF_3CN, C_6H_6, CH_3Cl, and the anhydride $(C_6H_5BO)_3$ [57].

Irradiation of a complex of tribenzylborane with NH_3 in the presence of alcohol leads to the cleavage of three B—C bonds to form toluene [163]. Bibenzyl is formed together with toluene in an aprotic medium.

Tribenzylborane is cleaved by water at room temperature to form dibenzylborinic acid [58–60]. Its reaction with deuteromethanol proceeds in two directions and leads to α-deuterotoluene and o-deuterotoluene in a ratio of 75:25 [61–63]:

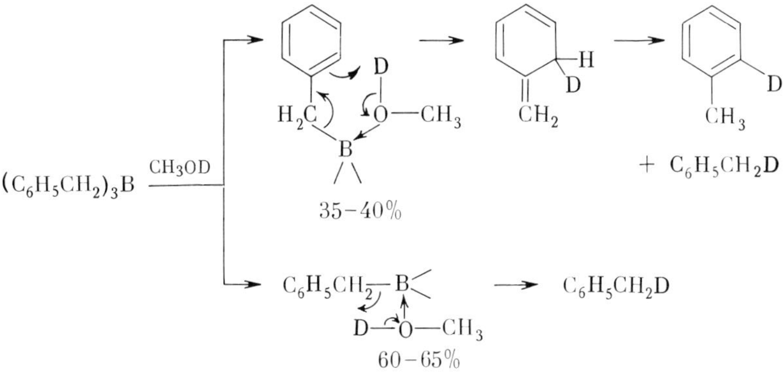

Boranes of the benzyl type are rapidly and quantitatively cleaved under the action of alkalies (Section 4). Substituted triphenylmethyl-(phenyl)borinic acids are hydrolyzed with water with the formation of derivatives of triphenylmethane and phenylboronic acid [64], indicating that the boron-benzyl bond is less stable than the boron-phenyl bond.

Triphenylborane reacts quantitatively with water at 20°C within 8 hours and at 100°C within 20 minutes, being converted to $C_6H_5B(OH)_2$ [6]. Tri-α-naphthylborane is slowly decomposed by hot water, whereas trimesitylborane is stable under these conditions.

The reaction of triphenylborane with ethanolamine led to the chelated β-aminoethyl diphenylborinate, $(C_6H_5)_2BOCH_2CH_2NH_2$ (flavognost) [65]. In amine complexes of triphenylborane, water and alcohols cleave two B—C bonds [66].

Phenylboronic and diphenylborinic acids are completely cleaved by water and alcohols at 150–200°C [67]; the protonolysis is catalyzed by salts of zinc, cadmium, silver, and other metals [68–70] and also by alkalies and acids [68, 69], while a minimal rate is observed at pH 4.85 [71]. At 200°C, phenylboronic acid is cleaved with the formation of benzene and HOBO, while $(C_6H_5BO)_3$ and C_6H_6 are obtained from diphenylborinic acid [67]. The pyrolysis of di-α-naphthylborinic acid at 120–130°C led to naphthalene and $(\alpha\text{-}C_{10}H_7BO)_3$ [72].

The ammonium salts $R_3NH[B(C_6H_5)_4]$ decompose at 200–240°C with the formation of R_3N, benzene, and triphenylborane [73]. The hydrolysis of sodium tetraphenylborate proceeds in stages and leads to phenylboronic acid [74, 75].

6.2 ACTION OF CARBOXYLIC ACIDS. SYNTHESIS OF HYDROCARBONS

Depending on the conditions, organic acids cleave one, two, or three boron-carbon bonds in trialkylboranes:

$$R_3B + R'COOH \xrightarrow{20^\circ\,C} RH + R_2BOCOR',$$

$$R_2B\text{—} + R'COOH \xrightarrow{20\text{—}100^\circ\,C} RH + RB\Big\langle\,,$$

$$RB\Big\langle + R'COOH \xrightarrow{140\text{—}200^\circ\,C} RH + \Big\rangle B\text{—}OCOR'.$$

This method of deboration is widely used for the synthesis of hydrocarbons and functional organic compounds via organoboranes, usually of the aliphatic series.

The first B—C bond in the trialkylboranes is ruptured at 20°C (with the evolution of heat) [76, 77] and the second at 20–100°C [78–81]. The diacyloxy(alkyl)boranes $RB(OCOR')_2$ formed as a result of the elimination of two moles of RH decompose at 20°C or upon distillation, being converted to the acid anhydrides $(R'CO)_2O$ and the alkylboron pyroacetates $[R(R'COO)B]_2O$ [82–84].

Monoalkyl boron derivatives (or the third B—C bond in trialkylboranes) undergo protonolysis upon refluxing for several hours with propionic or butyric acids without a solvent or in diglyme [79, 85]. In most cases, it is better to use higher organic acids (decanoic, stearic, and others), which makes it possible to heat the reaction mass to 160–200°C or higher temperatures [79, 86, 87].

For the complete cleavage of trialkylboranes and of di- and monoalkyl boron derivatives, 2.5 to 4 moles of acid are used per mole of organoborane. The reaction is generally carried out with simultaneous distilling off the hydrocarbon formed at atmospheric pressure or under vacuum. The mixture is heated to 170°C and is then kept for 10 to 30 minutes at 170–200°C [86]. In the case of some sterically hindered organoboranes, more prolonged heating or an increase in the temperature to 250°C is required [88]. The yield of RH is 70–95%. By this method, for example, derivatives of cyclohexane and cyclohexene have been obtained from bicyclic boron derivatives and cyclopropylboranes have been converted to cyclopropane hydrocarbons (Part IV):

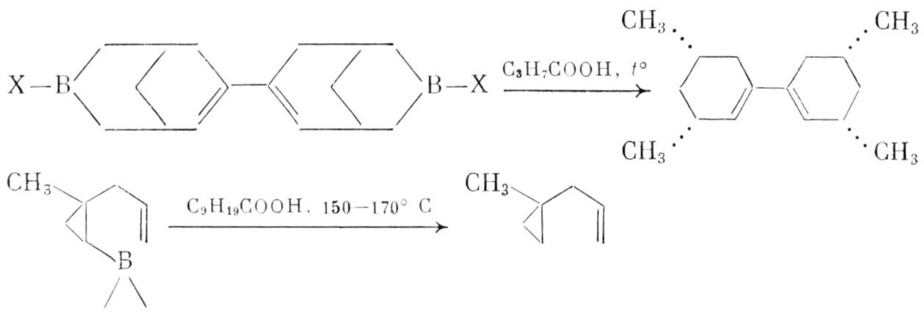

Heating methyl 4-bromobutyl (n-butyl)borinate with benzoic acid at 150–170°C led to butane and butyl bromide [89].

Benzoic acids are suitable for the cleavage of only saturated organoboranes; when ω-unsaturated boron derivatives are subjected to acidolysis by these acids, isomerization of the olefin is observed.

The action of (1*S*)-menthylacetic acid on tributylborane forms (1*S*)-menthoxydibutylborane which, upon irradiation, adds to cyclohexene to form (after oxidation) optically active *cis*-2-butylcyclohexanol [90].

In unsymmetrical trialkylboranes, the $B—C_{prim}$ bond is cleaved slightly more easily than the $B—C_{sec}$ bond, but no significant selectivity is observed [91]. At the same time, *tert*-alkyl groups are split off with significantly greater difficulty than primary and secondary alkyl groups, as a result of steric hindrance.

An investigation of the kinetics of the reaction of triethylborane with different aliphatic and aromatic acids in diglyme has shown that the rate of protonolysis of the first B—C bond decreases with increasing pK_a of the acid. The reaction is first order in each reagent and shows a significant isotope effect ($K_{CH_3COOH}/K_{CH_3COOD} = 3.3$, $\rho^* = -0.94$). It is interesting that trichloroacetic and trifluoroacetic acids react very slowly, while the others show a K_2 value of approximately $5 \cdot 10^{-3}$ $L \cdot mole^{-1} \cdot sec^{-1}$ [32].

The protolytic cleavage of alkylboranes by organic acids proceeds with a retention of the configuration. This is confirmed, for example, by the formation of (−)-1-deuteroethylbenzene by heating (−)-dibutyl 1-phenylethylboronate with deuterocaprylic acid in diglyme [92]. *exo*-Deuteronorborane has been obtained by heating tri-*exo*-norbornylborane with deuteropropionic acid [93]. Retention of the configuration has also been observed in the hydroboration-acetolysis of (*Z*)- and (*E*)-1-phenyl-3,3-dimethyl-1-butene-1,2-d_2 [164]

The last two examples demonstrate the applicability of organoboranes for the synthesis of deuterated compounds. The deuterolysis of tribenzylborane under the action of CH_3COOD led to a high yield of α-deuterotoluene, $C_6H_5CH_2D$ [61–63]. Dideuterated hydrocarbons can be obtained by deuteroboration-deuterolysis [93]:

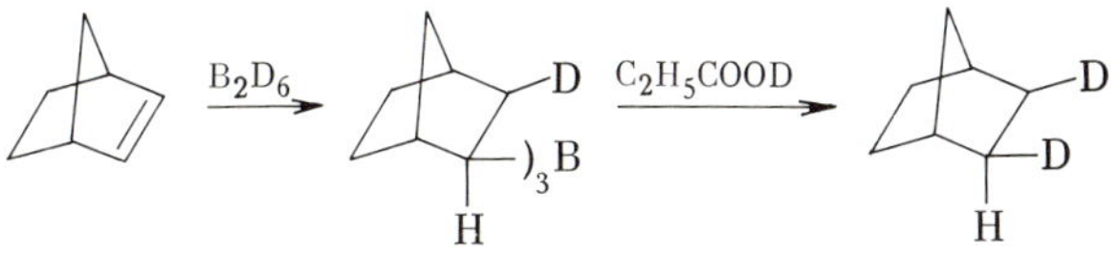

The retention of the radical configuration, and also the kinetic data, indicate the cyclic nature of the acid protonolysis of organoboranes [32, 91, 92]:

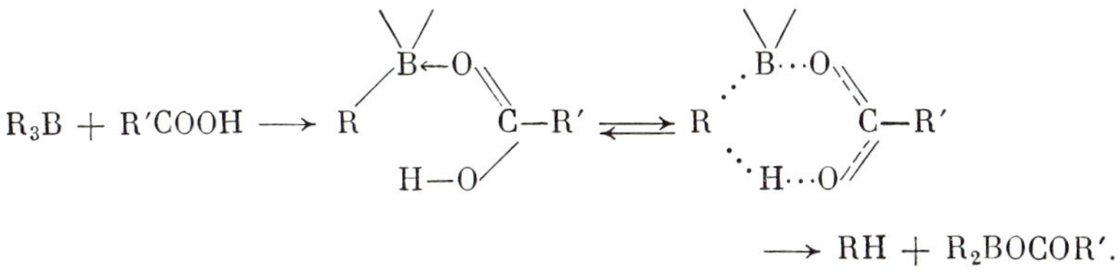

The first stage is coordination of carbonyl oxygen and of boron. This coordination increases the electrophilic nature of the hydrogen of the acid and, on the other hand, promotes a weakening of the B—C bond and the formation of a carbanion [32]. The decrease in the protonolysis rate of the second and third B—C bonds is explained by a decrease in the complexing capacity of di- and monoalkyl acylates as compared with trialkylboranes.

The protonolysis of alkyl- and cycloalkylboranes with carboxylic acids generally proceeds smoothly and stereospecifically. Rearrangements and side processes are observed only in some cases.

Acenaphthenylborane (*3*) forms a mixture of acenaphthene and 3-acylacenaphthene in a ratio of 1:1.5 when heated with acetic and propionic acids [94]:

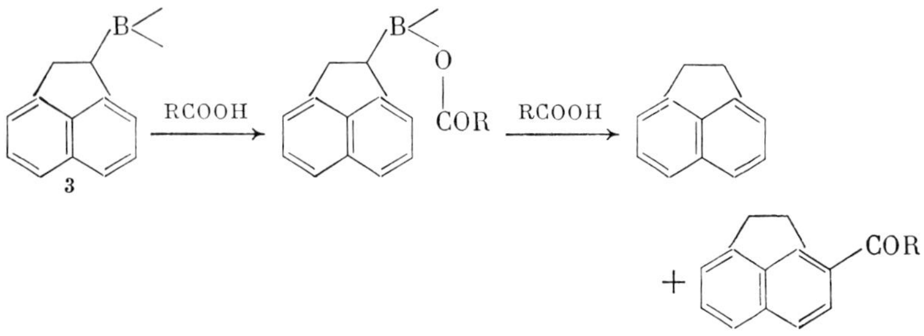

Formolysis of the compound *3* led to 3-formylacenaphthene, the oxidation of which with Ag_2O produced 3-acenaphthenecarboxylic acid. The acylation proceeds intramolecularly, which is confirmed by the data of the deuterolysis and deuteroboration-protonolysis of acenaphthylene.

The hydroboration-protonolysis of 3-phenyl-1,2,4-trimethyl-3-pyrazoline produced a mixture of *cis*- and *trans*-pyrazolidines in a ratio of 3:2 [95]. Acidolysis of the hydroboration product of (+)-2- carene led to a hydrocarbon mixture consisting of 16 compounds [96].

In the acidolysis of *trans*-2-aminocyclopentyl-, -cyclohexyl-, -cycloheptyl-, and -cyclooctylboranes, the main process is β-elimination, as a result of which cycloolefins are obtained in a yield of 85–98% [97]:

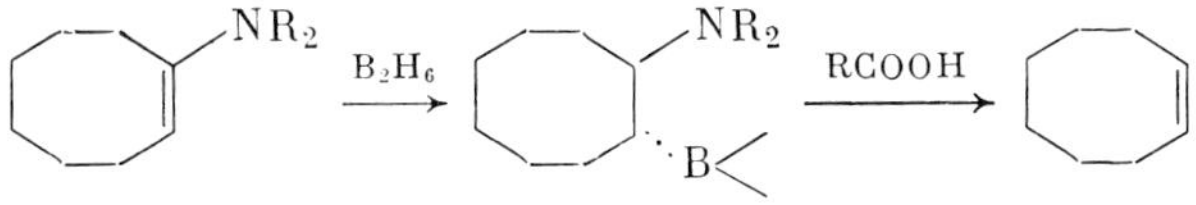

The cleavage of *trans*-2-acyloxycyclylboranes proceeds in an analogous manner.

The protonolysis of boranes of the allyl type proceeds with allyl rearrangement (Part IV).

Heating (100°C) benzylboronic acid with 3 M tartaric acid in D_2O at pH 0.7 led to dideuterotoluene, $C_6H_4DCH_2D$, with a ratio of 1.1 between the ortho and para isomers [98].

In chelated dialkylborylacetylacetonates, -imidoylamidinates [99], and -acylamidinates [51], acetic and trimethylacetic acids cleave one B—C bond at 120–140°C, i.e., at a higher temperature than in the case of the dialkyl compounds of trivalent boron:

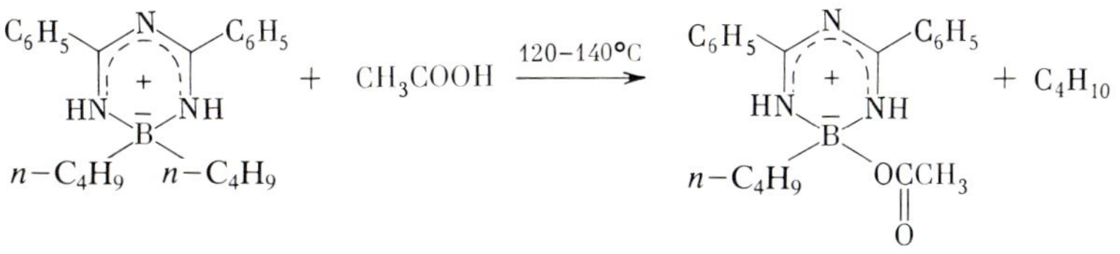

Arylboronic and diarylborinic acids are easily deborated by heating with aqueous acetic acid, alcoholic picric acid, or a 20% sodium bisulfite solution [100]:

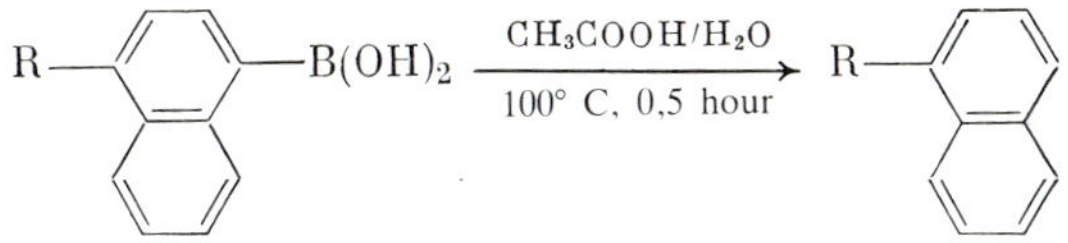

6.3 ACTION OF INORGANIC ACIDS

Hydrogen fluoride quantitatively cleaves trialkylboranes at room temperature in an autoclave, with the formation of boron trifluoride and an alkane [101]:

$$R_3B + 3HF \xrightarrow[10\ \text{minutes}]{20^\circ\ C} 3RH + BF_3.$$

Under the same conditions, the reaction of HF with lithium tetramethylborate led to LiF, BF_3, and four moles of methane.

Other hydrogen halides HX (X=Cl, Br, I) show a lower reactivity toward trialkylboranes. The reaction with HCl proceeds only at 110°C [102, 103], that with HBr at 20°C [104] or 60°C [103, 105], and that with HI at room temperature [103]. In these processes, halodialkylboranes R_2BX and one mole of alkane are obtained. Under more forced conditions (180–210°C), HCl cleaves two B—C bonds; the reaction is catalyzed by $AlCl_3$ [102]. The heats of reaction of hydrogen halides with R_3B have been investigated [103].

Dealkylation may possibly proceed by a cyclic four-center mechanism. Olah, et al. [101] used concepts relating to two-electron three-center bonds to explain the reaction mechanism:

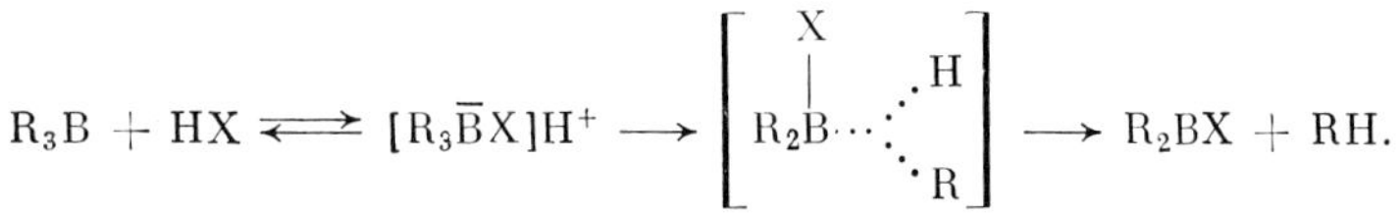

Anhydrous sulfuric acid [106, 107] and alkyl- and arylsulfonic acids [106] react with trialkylboranes upon heating (to about 100°C) with the formation of bis(dialkylboryl) sulfates $(R_2BO)_2SO_2$ or dialkylboryl sulfonates $R_2BOSO_2R_1$, respectively. The latter compounds, which are highly reactive, have been used for the synthesis of *N*-dialkylboryl-*sym*-tetraphenyloxamidine and 4,4-dialkylpyrazabole [106]. Trialkyl- and tricycloalkylboranes are monodealkylated to $R_2BOSO_2CF_3$ by the action of CF_3SO_2OH [165–167].

Aqueous hydrobromic acid, upon boiling, cleaves only one boron-carbon bond in trialkylboranes, producing dialkylborinic acids, which are isolated as the anhydrides in 80% yield [105]. The evolution of one mole of ethane on heating triethylborane with concentrated hydrochloric acid was first observed by Frankland [8].

The acid hydrolysis of benzylboronic acid is accompanied by a rearrangement [98].

Inorganic acids catalyze the protonolysis of arylboronic acids at pH$<$ 4 [68, 69, 71]. At pH 4–5, a non-catalytic reaction takes place at a minimal rate [71]. In a series of papers by Kuivila, et al. [70, 71, 109–112], the results of which were summarized in a monograph [20], the effect of substituents in the phenyl ring, the strength and nature of the acids, solvents, salts, and also of other factors on the kinetics of this reaction of electrophilic replacement of a boryl group by hydrogen (A-S_E2) was investigated. The rate of hydrolysis of thienylboronic acids in the presence of $HClO_4$ is significantly higher than that of phenylboronic acid: (C_6H_5 : 3-thienyl : 2-thienyl $= 1:7.1 \cdot 10^3:8.5 \cdot 10^5$) [113]. 2-Thienylboronic acid is cleaved 450 times more rapidly then *p*-methoxyphenylboronic acid.

6.4 ACTION OF ALKALIES. SYNTHESIS OF HYDROCARBONS

Trialkylboranes containing no "active" B—C bonds are stable toward alkali solutions up to 100°C. Anhydrous NaOH cleaves all three bonds in tributylborane at 140–200°C. The protonolysis with alkali proceeds in a milder manner in ethylene glycol [114]:

$$(C_4H_9)_3B + NaOH \xrightarrow{(CH_2OH)_2,\ 175-180^\circ C} \underset{86\%}{3C_4H_{10}}.$$

In this process, no significant differences in the conditions of cleavage of the boron-carbon bonds are observed. The first bond is cleaved at 135–150°C, the second at 150–170°C, and the third at 170–180°C. Alkylboronic and dialkylborinic acids, when heated with NaOH in ethylene glycol, form the corresponding alkanes in a yield of 85–87%; 2-butyl-1,2-oxaborinane forms 82% of butane and 87% of butanol [114]. The formation of butane by the fusion of butylboronic acid with alkali was noted by Johnson, et al. [115].

Although the alkaline cleavage of trialkylboranes proceeds at high temperatures, this reaction can be used in a number of cases instead of protonolysis with carboxylic acids, particularly when functional groups sensitive to acids are present in the radical.

Boranes of the benzyl and aryl series are less stable toward alkalies.

Triphenylborane is rapidly cleaved, and trinaphthylborane is slowly cleaved in dilute solutions, with the formation of benzene or naphthalene and of the sodium salts of the corresponding arylboronic acids [6]. Trimesitylborane is stable under these conditions.

Benzylboronic acid [98] and its derivatives [116, 117] are hydrolyzed by aqueous alkali at room temperature or with gentle heating (50°C). The reaction of $C_6H_5CH_2B(OH)_2$ with heavy water in alkali led to $C_6H_5CH_2D$ [98].

The cleavage of boron-carbon bonds under the action of alkali solutions (0–5°C) has been observed in 1-alkylboracyclopentanes [118] and 1,1-diboron compounds [119–121].

Phenyl-, *o*-tolyl-, *α*-furyl-, and *α*-thienylboronic acids are rapidly cleaved by hot alkali [68, 69] and by an ammoniacal solution of silver oxide with the formation of ArH [69, 122]. Ferrocenyl- [123, 124], benzyl-, and alkylboronic acids [69] react with Ag_2O in ammonia to form products of coupling of the radicals; *tert*-butylboronic acid forms isobutane and isobutene (see also Chapter 14).

Nitrophenyl- and pyridylboronic acids are cleaved by methanol and by alcoholic alkali under UV irradiation, whereas the photoprotonolysis of ferrocenylboronic acid is accelerated in acid medium [125].

Weinheimer and Marsico [116] have established that the alkaline protonolysis (pH 9–10, 20°C) of boranes of the benzyl type proceeds with retention of the configuration. The product of the hydroboration of *cis*-dimethylstilbene produced a 69% yield of pure *meso*-2,3-diphenylbutane, while the trans isomer formed pure D,L-2,3-diphenylbutane. On the other hand, according to the data of [92], the deuterodeboration of diethanolamine esters of optically active 1-phenylethylboronic acids under the action of NaOD in boiling D_2O takes place with 54% inversion.

Insufficient attention has been devoted to the mechanism of the alkaline cleavage of organoboranes. At the same time, it is well known [126–128] that esters of organoboron acids form the corresponding borates under the action of alkalies. The cleavage of the latter probably proceeds intramolecularly [92, 114, 116]:

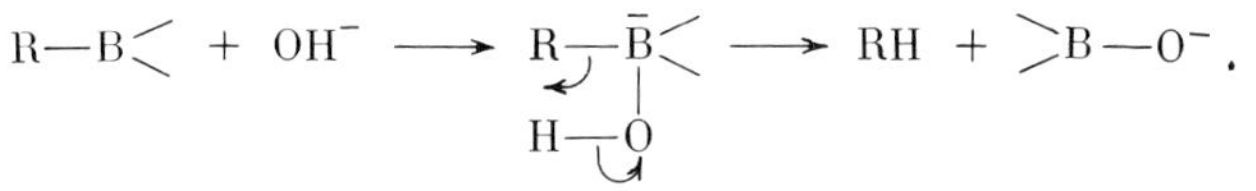

The dissociation of the B—C bond in a complex anion with the formation of an R^- anion does not take place. It is suggested [20, 71] that the transition state in the case of the alkaline deboration of phenylboronic acid has the structure *4*:

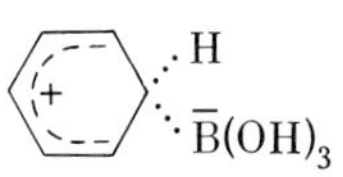

4

Alkylboranes containing electronegative groups, such as a halogen [129–131], in the β position undergo β-elimination with the formation of olefins in alkaline solution; under these conditions, γ-halopropylboranes form cyclopropane [131]. At the same time, the ester $CH_3O(CH_2)_3B(OH_3)_2$ is not cleaved by concentrated alkali even at 100°C but undergoes protonolysis with the formation of methyl propyl ether (yield 75%) upon heating (150–210°C) with solid alkali [114].

The silicon-containing organoborane $[(CH_3)_3SiOSi\,(CH_3)_2CH_2]_3B$ undergoes cleavage under the action of alcoholic alkali to form ethoxytrimethylsilane and hexamethyldisiloxane [132]. It is suggested that the intermediate product is 1,1-dimethyl-1-silaethylene, $(CH_3)_2Si{=}CH_2$.

6.5 CATALYTIC DEALKYATION OF TRIALKYLBORANES

The monodealkylation of trialkylboranes under the action of water, alcohols, amines, and other reagents is catalyzed by mercaptans [133–135] and by trimethylacetic acid and its derivatives [107]:

$$R_3B + H{-}Y \xrightarrow[(CH_3)_3CCOOH]{R'SH(O_2)\text{ or}} R_2BY + RH$$

$$\left(Y = OH,\ RO,\ NH_2,\ NHR,\ NR_2,\ RSO_3,\ R_2C{=}\overset{\cdot}{N}O{-},\ \rangle C{=}\overset{R}{\overset{|}{C}}{-}O,\ \text{etc.}\right).$$

The use of catalysts makes it possible to lower the reaction temperature from 120–180 to 20–80°C and thereby to prevent isomerization of the radicals in the organoborane (Chapter 4).

The reaction of dialkylboranes with mercaptans, which takes place at 20°C (with evolution of heat) [5, 134, 136], is a chain radical process [137, 138] and is catalyzed by oxygen, peroxides and other initiators (Part VI). The thioethers R_2BSR' obtained in this process are highly reactive and exchange the R'S group for OH, OR, NH_2, NR_2, and others with evolution of heat, while the mercaptan liberated reacts again with R_3B, forming R_2BSR', etc., for example [133]:

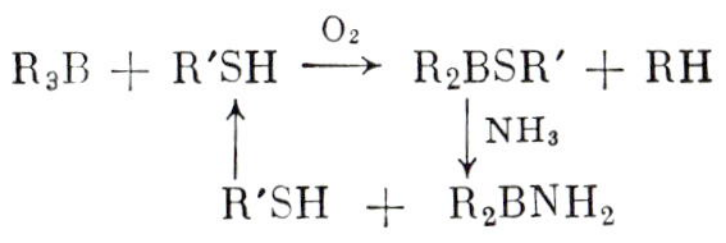

cis-Bicyclo[3.3.0]octane has been obtained by the action of 1-propanethiol and methanol on *cis*-bicyclo[3.3.0]oct-1-yldimethylborane [168].

As was established by Köster, et al. [107], the B—O bond in dialkylboryl pivalates, $R_2BOCOC(CH_3)_3$, unlike the analogous bond in other acyl derivatives of boron, is very reactive, as a result of which the pivalate group is easily replaced under the action of water [104, 107], alcohols and phenols [107, 139], polyols [104, 140], mercaptans, hydrogen halides and amides [104], primary amines [107, 141], sulfonic acids, H_2SO_4 and hydroxylamine derivatives [107], amino acids [112, 142], enols (vinyloxyboranes, $R_2BOCH{=}C\lessdot$, are obtained from aldehydes and ketones) [143], and a number of other nucleophilic reagents HY. In turn, pivalic acid reacts with R_3B at room temperature, splitting off one alkyl group (see below). The combination of these two properties is the basis of a simple method for the dealkylation of trialkylboranes with the reagents indicated above. The reactivity of R_3B decreases in the following series: $C_2H_5 \geqslant n\text{-}C_3H_7 > iso\text{-}C_3H_7 > iso\text{-}C_4H_9$ [107, 139]. C—H Acids do not react with R_3B in the presence of pivalates [104], but triphenylborane reacts with water at 100°C with the formation of boric acid [107].

Triethylborane reacts with *N*-alkylketimines in the presence of pivaloyloxy(diethyl)borane at approximately 100°C to form vinylamino(diethyl)boranes, $(C_2H_5)_2BNRCR'{=}CHR''$ [169].

The protonolysis of trialkylboranes, catalyzed by mercaptans and pivalates, is used for the preparation of organoboranes of the type R_2BY, and also for the protection and quantitative determination of hydroxy groups and for the determination of the water content and dehydration of crystal hydrates of organic and inorganic compounds (see below). Furthermore, the system triethylborane/pivalic acid is used as a dehydrating agent in the condensation of aldehydes and ketones (Chapter 9).

6.5.1 Quantitative determination of hydroxy groups and water

The reaction of triethylborane, activated by a small amount of pivalic acid, is the basis for a rapid gasometric method for the determination of OH groups in alcohols, phenols, diols [139], polyols, sugars, and other compounds [140, 170] and for the determination of the water content in crystal hydrates of inorganic, complex, and organic salts [107, 144]:

$$-\overset{|}{\underset{|}{C}}-OH + (C_2H_5)_3B \xrightarrow{(CH_3)_3CCOOH} (C_2H_5)_2B-O-\overset{|}{\underset{|}{C}}- + C_2H_6,$$

$$MX \cdot n\,H_2O + 2n(C_2H_5)_3B \xrightarrow{(CH_3)_3CCOOH} 2nC_2H_6 + [(C_2H_5)_2B]_2O.$$

The mean error is less than ±2%. Because of the availability of triethylborane and its high accuracy, this method should occupy a valuable place among the methods of analysis according to Chugaev-Zerevitinov [45].

Pure diethylboryl pivalate or a mixture of triethylborane and chlorodipropylborane is used for the determination of hydrogen in acidic, basic, and ammonium salts and also in metal hydroxides [144].

The method has been used for the preparative dehydration of hydrates of salts [144] and sugars [140].

6.5.2 Protection of hydroxy and amino groups

The ability of triethylborane, in the presence of pivalic acid, to react with compounds containing an OH group at 20–80°C is used for the protection of hydroxy groups in alcohols, phenols, diols, and polyols, including their functional derivatives (C=C, C≡C, halogen derivatives, etc.) [139, 171]:

$$R-OH \underset{CH_3OH}{\overset{(C_2H_5)_3B + 1\%\ (CH_3)_3CCOOH}{\rightleftarrows}} R-O-B(C_2H_5)_2 + C_2H_6.$$

The diethylboryl group is easily removed by methanol or acetylacetone (and, probably, by water).

Diethylborylation has been used successfully for the quantitative protection of all OH groups in polyols (glycerol, erythritol, pentaerythritol, ribitol, mannitol, *meso*-inositol), monosaccharides (D-arabinose, D-ribose, D-xylose, D-galactose, D-glucose, D-mannose and D-fructose, L-sorbose and L-rhamnose) and their derivatives (*C*-amino-, *C*-nitro-, *O*-methyl-, *O*-isopropylidene-, *O*-tosyl-, etc.), saccharic acids, their lactones and salts (ascorbic, galactaric, galacturonic, γ-lactone of gluconic acid, etc.), disaccharides (sucrose, maltose, lactose, cellobiose, trehalose, raffinose, cycloamylose), polysaccharides (starch, cellulose, glycogen, inulin, etc.) [140] and other compounds. For example, the diethylborylation of D-glucopyranose led to 1,2,3,4,6-penta-*O*-diethylboryl-D-glucose containing the α and β forms in a ratio of 7:3 [140]:

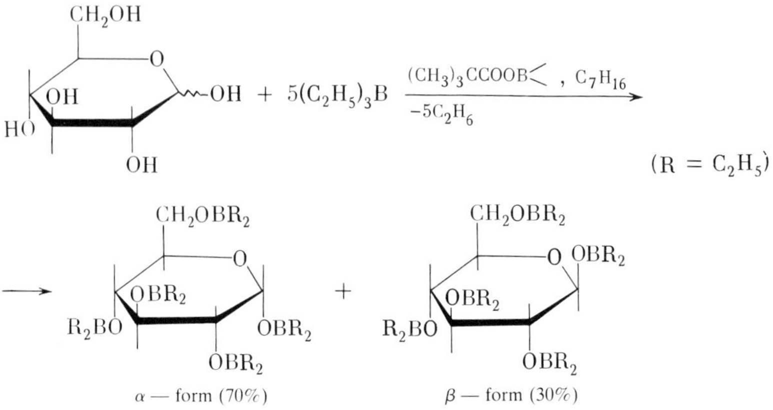

The diethylborylation is carried out with excess triethylborane in heptane or another hydrocarbon (at 20°C or with heating) until the polyol is completely dissolved. The method has been used for the separation of starches of different types into amylose and amylopectin components; for the separation of mono- and disaccharides, since the protected polyols (up to C_6) may be distilled under vacuum; and also for the removal of water of crystallization from polyhydroxy compounds (D-glucose, cycloamylose, etc.).

The ability of diethylboryl derivatives of sugars to dissolve in hydrocarbons is very important for their physicochemical investigation (by mass, IR, and NMR spectroscopy).

Phenylboronic acid [146–149], its anhydride [150], esters of propyl- and butylboronic acids [151–155], and also trialkyl borates [156–158] are used in other methods for the protection and physicochemical analysis of diols, polyols, sugars, and hydroxy acids. Methyl- and butylboronic acids are used for the separation of *cis*- and *trans*-diols [159–160] (see also Chapter 1). The ease of introduction and removal of an ethylboron protecting group has been demonstrated for the examples of xylitol [172, 173], D-mannitol [174], dulcitol [175], a number of methylglycosides [176], L-rhamnose [177], α-D-mannofuranosyl bromide [178], α-D-lyxofuranosyl bromide [179], and 2-deoxy-D-*arabino*-hexose (2-deoxy-D-glucose) [180]. Ethylboron protection has been used for the isolation of pure L-glucose and

D-*gluco*-hexodialdose from D-glucurono-6,3-lactone [181] and D-gulono-1,4-lactone [182].

Amino acids of the type $NH_2(CH_2)_nCOOH$ ($n = 1-3, 5$; glycine, alanine, β-alanine, sarcosine, 4-aminobutyric acid, 6-aminocaproic acid; 2-, 3-, and 4-aminobenzoic acids) react with triethylborane in acetonitrile at 80°C to form *O*-diethylboryl derivatives [142]:

$$R_3B + NH_2(CH_2)_nCOOH \xrightarrow{80°\,C} NH_2(CH_2)_nCOOBR_2 \xrightarrow{R_3B,\ \text{catalyst}}$$
$$\longrightarrow \underset{99\%}{R_2BNH(CH_2)_nCOOBR_2}.$$

The amino group can be protected by heating with triethylborane in the presence of diethylboryl pivalate.

REFERENCES

1. Yu. N. Bubnov, S. A. Korobeinikova, G. V. Isagulyants and B. M. Mikhailov, *Izv. Akad. Nauk SSSR, Otd. Khim. Nauk* **1970,** 2023.
2. M. F. Lappert, in: "The Chemistry of Boron and its Compounds", ed. E. L. Muetterties, New York, J. Wiley and Sons (1967), pp. 500–504.
3. W. C. Steele, L. D. Nickols and F. G. A. Stone, *J. Am. Chem. Soc.* **84,** 1154 (1962).
4. Yu. N. Bubnov, V. S. Bogdanov, I. P. Yakovlev and B. M. Mikhailov, *Zh. Obshch. Khim.* **42,** 1313 (1972).
5. B. M. Mikhailov, V. A. Vaver and Yu. N. Bubnov, *Dokl. Akad. Nauk SSSR* **126,** 575 (1959).
6. H. C. Brown and V. H. Dodson, *J. Am. Chem. Soc.* **79,** 2302 (1957).
7. B. M. Mikhailov and T. K. Baryshnikova, *Zh. Obshch. Khim.* **41,** 1303 (1971).
8. S. I. Frolov, Yu. N. Bubnov and B. M. Mikhailov, *Izv. Akad. Nauk SSSR, Ser. Khim.* **1969,** 1996.
9. B. M. Mikhailov, Yu. N. Bubnov and S. A. Korobeinikova, *Izv. Akad. Nauk SSSR, Ser. Khim.* **1969,** 2465.
10. H. C. Brown and G. Zweifel, *J. Am. Chem. Soc.* **83,** 3834 (1961).
11. Yu. N. Bubnov, S. I. Frolov, V. G. Kiselev, V. S. Bogdanov and B. M. Mikhailov, *Zh. Obshch. Khim.* **40,** 1311 (1970).
12. F. E. Brinckman and F. G. A. Stone, *J. Am. Chem. Soc.* **82,** 6218 (1960).
13. H. D. Kaesz and F. G. A. Stone, Ch. 3 in: "The Chemistry of Organometallic Compounds", ed. H. Zeiss, Reinhold Publishing Corporation, N. Y., Chapman and Hall Ltd., London (1960).
14. B. M. Mikhailov, Yu. N. Bubnov, S. I. Frolov, and S. A. Korobeinikova, *Izv. Akad. Nauk SSSR, Ser. Khim.* **1969,** 1631.
15. H. H. Hartmann and K. Birr, *Z. Anorg. Chem.* **299,** 174 (1959).
16. G. L. Sharanina, V. S. Zavgorodnii and A. A. Petrov, *Zh. Obshch. Khim.* **36,** 1275 (1966).

17. B. S. Matteson and K. Peacock, *J. Am. Chem. Soc.* **82,** 5760 (1960).
18. Yu. N. Bubnov. O. A. Nesmeyanova, T. Yu. Rudashevskaya, B. M. Mikhailov and B. A. Kazanskii, *Zh. Obshch. Khim.* **43,** 135 (1973).
19. B. M. Mikhailov, *Usp. Khim* **28,** 1450 (1959).
20. O. A. Reutov, I. P. Beletskaya and V. I. Sokolov: "The Mechanisms of the Reactions of Organometallic Compounds", Moscow, Khimiya (1972).
21. D. Ulmschneider and J. Goubeau, *Chem. Ber.* **90,** 2733 (1957).
22. E. Wiberg, K. Hertwig and A. Bolz, *Z. Anorg. Chem.* **256,** 177 (1948).
23. E. Wiberg and K. Hertwig, *Z. Anorg. Chem.* **255,** 141 (1947).
24. E. Wiberg and K. Hertwig, *Z. Anorg. Chem.* **257,** 138 (1948).
25. J. Goubeau and A. Zappel, *Z. Anorg. Chem.* **279,** 38 (1955).
26. A. F. Zhigach, E. B. Kazakova and E. S. Krongauz, *Dokl. Akad. Nauk SSSR* **111,** 1029 (1956).
27. B. M. Mikhailov and V. A. Vaver, *Izv. Akad. Nauk SSSR. Otd. Khim. Nauk* **1960,** 852.
28. B. M. Mikhailov and Yu. N. Bubnov, *Zh. Obshch. Khim.* **32,** 1969 (1962).
29. B. M. Mikhailov, V. A. Dorokhov and N. V. Mostovoi, *Izv. Akad. Nauk SSSR, Ser. Khim.* **1964,** 201.
30. B. M. Mikhailov and Yu. N. Bubnov, *Izv. Akad. Nauk SSSR, Otd. Khim. Nauk* **1960,** 370.
31. H. Nöth, *Z. Naturforsch,* **16b,** 147 (1961).
32. L. H. Toporcer, R. E. Dessy and S. I. E. Green, *J. Am. Chem. Soc.* **87,** 1236 (1965).
33. V. A. Dorokhov and B. M. Mikhailov, *Izv. Akad. Nauk SSSR, Ser. Khim.* **1972,** 1895.
34. V. A. Dorokhov and B. M. Mikhailov, *Zh. Obshch. Khim.* **44,** 1281 (1974).
35. H. Becher, *Z. Anorg. Chem.* **289,** 262 (1957).
36. R. Köster, H. Bellut and S. Hattori, *Ann.* **720,** 1 (1969).
37. H. Bellut and R. Köster, *Ann.* **738,** 86 (1970).
38. S. Trofimenko, *J. Am. Chem. Soc.* **89,** 3165 (1967).
39. B. M. Mikhailov and Yu. N. Bubnov. *Izv. Akad. Nauk SSSR, Otd. Khim. Nauk* **1960,** 1883.
40. M. F. Hawthorne and M. Reintjes, *J. Am. Chem. Soc.* **86,** 5016 (1964).
41. M. F. Hawthorne and M. Reintjes, *J. Org. Chem.* **30,** 3851 (1965).
42. R. Köster and G. Rotermund, *Ann.* **689,** 40 (1965).
43. L. H. Toporcer, R. E. Dessy and S. I. E. Green, *Inorg. Chem.* **4,** 1649 (1965).
44. V. A. Dorokhov, L. I. Lavrinovich and B. M. Mikhailov, *Dokl. Akad. Nauk SSSR* **195,** 1100 (1970).
45. S. Trofimenko, *J. Am. Chem. Soc.* **89,** 7014 (1967).
46. G. N. Schrauzer, *Chem. Ber.* **95,** 1438 (1962).
47. F. Umland and D. Theirig, *Angew. Chem.* **75,** 685 (1963).
48. V. A. Dorokhov, L. I. Lavrinovich, I. P. Yakovlev and B. M. Mikhailov, *Zh. Obshch. Khim.* **41,** 2501 (1971).
49. B. M. Mikhailov, V. A. Dorokhov and L. I. Lavrinovich, *Zh. Obshch. Khim.* **42,** 1553 (1972).
50. B. M. Mikhailov and V. A. Dorokhov, *Izv. Akad. Nauk SSSR, Ser. Khim.* **1970,** 1446.
51. V. A. Dorokhov, L. I. Lavrinovich, M. N. Bochkareva, V. S. Bogdanov and B. M. Mikhailov, *Zh. Obshch. Khim.* **43,** 1115 (1973).
52. B. M. Mikhailov, V. A. Dorokhov and L. I. Lavrinovich, *Zh. Obshch. Khim.* **44,** 1888 (1973).

53. B. M. Mikhailov and V. A. Dorokhov, *Izv. Akad. Nauk SSSR. Ser. Khim.* **1971,** 201.
54. B. M. Mikhailov, V. A. Dorokhov and V. I. Seredenko, *Dokl. Akad. Nauk SSSR* **199,** 1328 (1971).
55. V. A. Dorokhov and B. M. Mikhailov, *Zh. Obshch. Khim.* **41,** 2702 (1971).
56. B. M. Mikhailov and V. A. Dorokhov, *Izv. Akad. Nauk. SSSR, Ser. Khim.* **1973,** 2649.
57. E. L. Zaitseva, *Zh. Obshch. Khim.* **36,** 327 (1966).
58. R. Köster and G. Bruno, *Ann.* **629,** 89 (1960).
59. G. Bruno, Dissertation, Techn. Hochschule, Aachen, (1958).
60. B. M. Mikhailov, G. S. Ter-Sarkisyan and N. A. Nikolaeva, **41,** 1721 (1971).
61. Yu. N. Bubnov and B. M. Mikhailov, *Izv. Akad. Nauk SSSR, Ser. Khim.* **1970,** 2156.
62. B. M. Mikhailov, Yu. N. Bubnov and S. A. Korobeinikova, *J. prakt. Chem.* **312,** 998 (1970).
63. B. M. Mikhailov, Yu. N. Bubnov, S. A. Korobeinikova and S. I. Frolov, *J. Organometal. Chem.* **27,** 165 (1971).
64. G. Wittig, L. Gonsior and H. Vogel, *Ann.* **688,** 1 (1965).
65. C. S. Rondestvedt, Jr., R. M. Scribner and C. E. Wulfman, *J. Org. Chem.* **20,** 9 (1955).
66. H. J. Roth and Ch. Schwenke, *Archiv Pharmazie* **298,** 648 (1965).
67. E. W. Abel, W. Gerrard and M. F. Lappert, *J. Chem. Soc.* **1958,** 1451.
68. A. D. Ainley and F. Challenger, *J. Chem. Soc.,* **1930,** 2171.
69. J. R. Johnson, M. G. Van Campen, and O. Grummit, *J. Am. Chem. Soc.* **60,** 111 (1938).
70. H. G. Kuivila, J. F. Reuwer and J. A. Mangravite, *J. Am. Chem. Soc.* **86,** 2666 (1964).
71. H. G. Kuivila, J. F. Reuwer and J. A. Mangravite, *Can. J. Chem.* **41,** 3081 (1963).
72. R. L. Letsinger and N. Remes, *J. Am. Chem. Soc.* **77,** 2489 (1955).
73. G. Wittig, G. Keicher, A. Rückert and P. Raff, *Ann.* **563,** 110 (1949).
74. R. Neu, *Chem. Ber.* **87,** 802 (1954).
75. J. N. Cooper and R. E. Powell, *J. Am. Chem. Soc.* **85,** 1590 (1963).
76. H. Meerwein and H. Sönke, *J. Prakt. Chem.* **147,** 251 (1936).
77. J. Goubeau, R. Epple, D. Ulmschneider and H. Lehmann, *Angew. Chem.* **67,** 710 (1955).
78. B. M. Mikhailov and V. A. Vaver, *Zh. Obshch. Khim.* **31,** 574 (1961).
79. H. C. Brown and K. Murry, *J. Am. Chem. Soc.* **81,** 4108 (1959).
80. G. S. Kolesnikov, L. S. Davydova, M. A. Yampol'skaya and N. V. Klimentova, *Izv. Akad. Nauk SSSR, Otd. Khim. Nauk* **1962,** 841.
81. J. Goubeau and H. Lehmann, *Z. Anorg. Chem.* **322,** 224 (1963).
82. B. M. Mikhailov and N. S. Fedotov, *Izv. Akad. Nauk SSSR, Otd. Khim. Nauk* **1958,** 857.
83. B. M. Mikhailov and T. A. Shchegoleva, *Izv. Akad. Nauk SSSR, Otd. Khim. Nauk* **1958,** 860.
84. B. M. Mikhailov and T. A. Shchegoleva, *Izv. Akad. Nauk SSSR, Otd. Khim. Nauk* **1958.** 1393.
85. J. Crighton, A. K. Holliday, A. G. Massay and N. R. Thompson, *Chem. and Ind.* **1960,** 347.
86. B. M. Mikhailov, Yu. N. Bubnov and S. I. Frolov, *Izv. Akad. Nauk. SSSR, Ser. Khim.* **1967,** 2290.

87. Yu. N. Bubnov, S. I. Frolov, V. G. Kiselev and B. M. Mikhailov, *Zh. Obshch. Khim.* **40,** 1316 (1970).
88. Yu. N. Bubnov, M. Sh. Grigoryan and B. M. Mikhailov, *Zh. Obshch. Khim.* **42,** 1738 (1972).
89. L. S. Vasil'ev, V. P. Dmitrikov and B. M. Mikhailov, *Zh. Obshch. Khim.* **42,** 1015 (1972).
90. N. Miyamoto, S. Isiyama, K. Utimoto and H. Nozaki, *Tetrahedron,* **29,** 2365 (1973).
91. D. B. Bigley and D. W. Payling, *J. Inorg. Nucl. Chem.* **33,** 1157 (1971).
92. A. G. Davies and B. P. Roberts, *J. Chem. Soc., C* **1968,** 1474.
93. H. C. Brown and K. Murry, *J. Org. Chem.* **26,** 631 (1961).
94. H. W. Whitlock, C. Y. Hsu and K. Sundaresan, *Tetrahedron Letters* **1965,** 4821.
95. J. L. Aubagnac, J. Elguero and R. Jacquier, *Bull. Soc. Chim. France* **1969,** 3316.
96. S. P. Acharya and H. C. Brown, *J. Am. Chem. Soc.* **89,** 1925 (1967).
97. J. W. Lewis and A. A. Pearce, *Tetrahedron Letters* **1964,** 2039.
98. W. Hanstein and T. G. Traylor, Tetrahedron Letters, **1967,** 4451.
99. B. M. Mikhailov, V. A. Dorokhov and V. I. Seredenko, *Zh. Obshch. Khim.* **43,** 862 (1973).
100. H. R. Snyder and F. W. Wyman, *J. Am. Chem. Soc.* **70,** 234 (1948).
101. G. Olah, P. Westerman, Y. K. Mo and G. Klopman, *J. Am. Chem. Soc.* **94,** 7859 (1972).
102. R. Booth and C. Kraus, *J. Am. Chem. Soc.* **74,** 1415 (1952).
103. H. Skinner and T. Tees, *J. Chem. Soc.* **1953,** 3378.
104. R. Köster, H. Bellut and W. Fenzl, *Ann.* **1974,** 54.
105. J. Johnson, H. Snyder and M. G. Van Campen, *J. Am. Chem. Soc.* **60,** 115 (1938).
106. S. Trofimenko, *J. Am. Chem. Soc.* **91,** 2139 (1969).
107. R. Köster, K. Amen, H. Bellut and W. Fenzl, *Angew, Chem.* **83,** 805 (1971).
108. E. Frankland, *Ann.* **124,** 129 (1862).
109. H. G. Kuivila and K. V. Nahabedian, *Chem. and Ind.* **1959,** 1120.
110. H. G. Kuivila andd K. V. Nahabedian, *J. Am. Chem. Soc.* **83,** 2159 (1961).
111. H. G. Kuivila and K. V. Nahabedian, *J. Am. Chem. Soc.* **83,** 2164 (1961).
112. K. V. Nahabedian and H. G. Kuivila, *J. Am. Chem. Soc.* **83,** 2167 (1961).
113. R. D. Brown, A. S. Buchanan and A. A. Humffray, *Austral. J. Chem.* **18,** 1521 (1965).
114. L. S. Vasil'ev, V. V. Veselovskii and B. M. Mikhailov, *Izv. Akad. Nauk SSSR, Ser. Khim.* **1977,** 1126.
115. H. R. Snyder, J. A. Kuck and J. R. Johnson, *J. Am. Chem. Soc.* **60,** 105 (1938).
116. A. J. Weinheimer and W. E. Marsico, *J. Org. Chem.* **27,** 1926 (1962).
117. H. C. Brown and R. L. Sharp, *J. Am. Chem. Soc.* **88,** 5851 (1966).
118. B. M. Mikhailov, A. Ya Bezmenov, L. S. Vasil'ev and V. G. Kiselev, *Dokl. Akad. Nauk SSSR,* **155,** 141 (1964).
119. G. Zweifel and H. Arzoumanian, *J. Am. Chem. Soc.* **89,** 291 (1967).
120. D. Pasto, J. Chow and S. Arora, *Tetrahedron Letters* **1967,** 723.
121. B. M. Mikhailov, P.M. Aronovich and V. G. Kiselev, *Izv. Akad. Nauk SSSR, Ser. Khim.* **1968,** 146.
122. A. Michaelis, *Ber.* **27,** 248 (1894).
123. A. N. Nesmeyanov, V. N. Drozd, V. A. Sazonova, V. I. Romanenko, A. K. Prokof'ev and L. A. Nikonova *Izv. Akad. Nauk SSSR, Otd. Khim. Nauk* **1963,** 667.
124. A. N. Nesmeyanov, V. A. Sazonova and V. N. Drozd, *Dokl. Akad. Nauk SSSR* **126,** 1004 (1959).

125. H. C. H. A. Van Riel, F. G. Fischer, J. Lugtenburg and E. Havinga, *Tetrahedron Letters* **1969**, 3085.
126. B. M. Mikhailov and V. A. Vaver, *Dokl. Akad. Nauk SSSR* **102**, 531 (1955).
127. B. M. Mikhailov and V. A. Vaver, *Izv. Akad. Nauk SSSR, Otd. Khim. Nauk* **1956**, 451.
128. B. M. Mikhailov, T. K. Kozminskaya, A. N. Blokhina and T. A. Shchegoleva, *Izv. Akad. Nauk SSSR, Otd. Khim. Nauk* **1956**, 692.
129. B. M. Mikhailov and P. M. Aronovich, *Izv. Akad. Nauk SSSR, Otd. Khim. Nauk* **1961**, 927.
130. D. S. Matteson and J. D. Liedtke, *J. Org. Chem.* **28**, 1924 (1963).
131. M. F. Hawthorne and J. A. Dupont, *J. Am. Chem. Soc.* **80**, 5830 (1958).
132. S. P. Hopper and J. S. Fine, *J. Organometal. Chem.* **80**, C21 (1974).
133. B. M. Mikhailov and Yu. N. Bubnov. *Izv. Akad. Nauk SSSR, Otd. Khim. Nauk* **1960**, 1872.
134. B. M. Mikhailov and Yu. N. Bubnov. *Zh. Obshch. Khim.* **31**, 160 (1961).
135. B. M. Mikhailov and Yu. N. Bubnov, *Zh. Obshch. Khim.* **31**, 577 (1961).
136. B. M. Mikhailov and A. N. Blokhina, *Zh. Obshch. Khim.* **30**, 3615 (1960).
137. B. M. Mikhailov and Yu. N. Bubnov, *Izv. Akad. Nauk SSSR, Ser. Khim.* **1964**, 2248.
138. A. G. Davies and B. P. Roberts, *J. Chem. Soc. B* **1971**, 1830.
139. R. Köster, W. Fenzl and G. Seidel, *Ann,* **1975**, 352.
140. R. Köster, K.-L. Amen and W. Dahlhoff, *Ann.* **1975**, 752.
141. E. Rothgery and R. Köster, *Ann.* **1974**, 101.
142. R. Köster and E. Rothgery, *Ann* **1974**, 112.
143. W. Fenzl and R. Köster, *Angew. Chem.,* **83**, 807 (1971).
144. R Köster and W. Fenzl, *Ann.* **1974**, 69.
145. V. A. Drozdov, A. P. Kreshkov and S. I. Petrov, *Usp. Khim.* **38**, 113 (1969).
146. R. J. Ferrier, D. Prasad, A. Rudowski and I. Sangster, *J. Chem. Soc.* **1964**, 3330.
147. A. J. Gorin and M. Mazurek, *Can. J. Chem.* **51**, 3277 (1973).
148. F. Shafizadeh, G. D. McGinnis and P. S. Chin, *Carbohydr. Res.* **18**, 357 (1971).
149. S. G. Batrakov, A. G. Panosyan, A. N. Ushakov, B. V. Rozynov and L. D. Bergel'son, *Izv. Akad. Nauk SSSR, Ser. Khim.* **1974**, 1739.
150. R. J. Ferrier; in: "Methods in Carbohydrate Chemistry," vol. VI. New York—London, Academic Press, (1972) p. 419.
151. F. Eisenberg, *Carbohydr. Res.* **19**, 135 (1971).
152. P. J. Wood and I. R. Siddiqui, *Carbohydr. Res.* **19**, 283 (1971).
153. C. J. W. Brooks and I. MacLean, *J. Chromatogr. Sci.* **9**, 19 (1971).
154. C. J. W. Brooks and J. Watson, *Chem. Comm.* **1967**, 952.
155. G. M. Anthony, C. J. W. Brooks, J. MacLean and I. Sangster, *J. Chromatogr. Sci.* **7**, 623 (1969).
156. V. V. Gertsev and L. A. Frolova, *Dokl. Akad. Nauk SSSR* **200**, 1107 (1971).
157. V. V. Gertsev, *Dokl. Akad. Nauk SSSR* **208**, 595 (1973).
158. E. M. Shvarts: "Complexes of Boron with Polyoxy Compounds", Riga, Zinatne (1968).
159. H. C. Brown and G. Zweifel, *J. Org. Chem.* **27**, 4708 (1962).
160. M. F. Rathke and H. C. Brown, *J. Am. Chem. Soc.* **88**, 2606 (1966).
161. K. Narasaka and H. C. Pai, *Chem. Letters* **1980**, 1415.
162. W. Fenzl and R. Köster, *Ann.* **1976**, 1370.
163. V. V. Chung, K. Inagaki, M. Tokuda and M. Itoh, *Chem. Letters* **1976**, 209.
164. G. W. Kabalka, R. J. Newton and J. Jacobus, *J. Org. Chem.* **44**, 4185 (1979).
165. D. A. Evans, E. Vogel and J. Nelson, *J. Am. Chem. Soc.* **101**, 6120 (1979).
166. M. Hirama and S. Masamune, *Tetrahedron Letters* **1979**, 2225.

167. T. Inoue and T. Mukaiyama, *Bull. Soc. Chem. Japan* **53,** 174 (1980).
168. G. W. Kramer and H. C. Brown, *J. Org. Chem.* **42,** 2832 (1977).
169. R. Köster, W. Fenzl and F. J. Levelt, *Ann.* **1981,** 734.
170. R. Köster, *Fuel* **57,** 489 (1978).
171. W. V. Dahlhoff and R. Köster, *Ann.* **1975,** 1914.
172. W. V. Dahlhoff and R. Köster, *Ann.* **1975,** 1926.
173. W. Fenzl, W. V. Dahlhoff and R. Köster, *Ann.* **1980,** 1176.
174. W. V. Dahlhoff, W. Schüssler and R. Köster, *Ann.* **1976,** 387.
175. W. V. Dahlhoff and R. Köster, *J. Org. Chem.* **41,** 2316 (1976).
176. R. Köster and W. V. Dahlhoff, *Ann.* **1976,** 1925.
177. W. V. Dahlhoff and R. Köster, *J. Org. Chem.* **42,** 3151 (1977).
178. W. V. Dahlhoff, A. Geisheimer and R. Köster, *Synthesis* **1980,** 935.
179. W. V. Dahlhoff and R. Köster, *Synthesis* **1980,** 936.
180. W. V. Dahlhoff and R. Köster, *Angew. Chem.* **92,** 552 (1980).
181. W. V. Dahlhoff, P. Idelmann and R. Köster, *Angew. Chem.* **92,** 552 (1980).
182. W. V. Dahlhoff, P. Idelmann and R. Köster, *Angew. Chem.* **92,** 553 (1980).

CHAPTER 7

Reactions of alkyl(aryl)boranes with chloramine, hydroxylamine derivatives, and azides

The boron atom in organoboranes of different types can be replaced by an amino group. Metal phthalimides, chloramines, hydroxylamine derivatives, and organic or organoboron azides are used as the source of the nitrogen function.

In this chapter, the reactions of organoboranes leading to primary, secondary, and tertiary amines, aziridines, and amides will be considered.

7.1 SYNTHESIS OF PRIMARY AND TERTIARY AMINES

For the replacement of boron by an amino group in aryl-, styryl-, and ferrocenylboronic acids, Nesmeyanov, Sazonova, et al. [1] have used the reaction with copper(II) phthalimide, leading initially to the corresponding aryl- or styrylphthalimides:

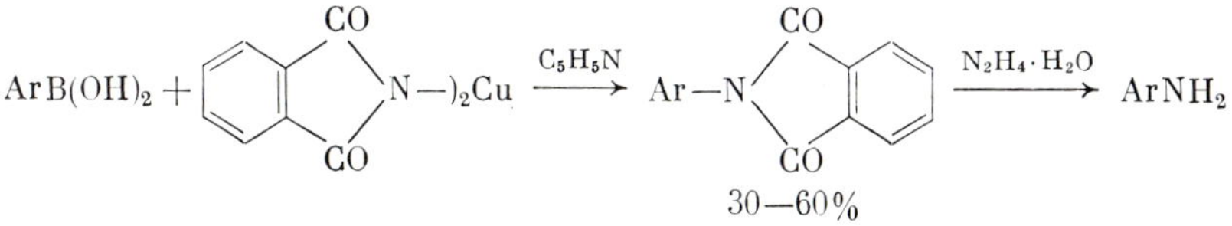

($Ar=C_6H_5$, n-$CH_3OC_6H_4$, $C_6H_5CH=CH$—, $C_5H_5FeC_5H_4$)

The subsequent hydrolysis of *N*-ferrocenylphthalimide obtained by this method produced aminoferrocene [2].

Primary amines are obtained from trialkylboranes by the action of hydroxylamine-*O*-sulfuric acid, chloramine in alkaline medium [3, 4, 41], or *O*-mesitylenesulfonyl hydroxylamine in tetrahydrofuran [5]:

$$R_3B + NH_2{-}X \longrightarrow \underset{50-60\%}{2R{-}NH_2} + RB(OH)_2$$

$$(X = Cl,\ OSO_3H,\ 2{,}4{,}6\text{-}(CH_3)_3C_6H_2SO_3;\ R = \text{alkyl or cycloalkyl})$$

In combination with hydroboration, this reaction represents a method for the anti-Markownikoff amination of olefins. For example, *cis*-myrtanylamine (*1*) has been obtained from β-pinene and α-pinene has been converted to isopinocampheylamine (*2*):

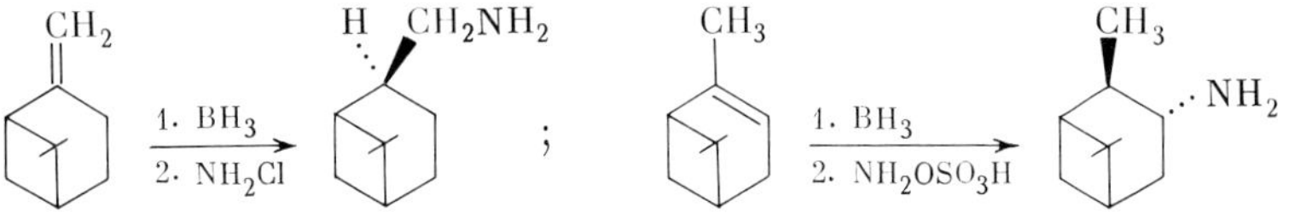

1 (48%) ; 2 (45%)

Cyclopentylamine, cyclohexylamine, and cyclooctylamine have been synthesized in an analogous manner from the corresponding cycloolefins. 11-Aminoundecanoic acid has been obtained in 30% yield from ethyl undecanoate by this method.

The migration of radicals from boron to nitrogen during the amination of organoboranes proceeds with retention of the configuration. This has been demonstrated by a whole series of examples. Thus, *exo*-norbornylamine has been obtained from tris-(*exo*-norbornyl)borane and *trans*-2-methylcyclohexylamine has been synthesized from tris(*trans*-2-methylcyclohexyl)borane.

Retention of the radical configuration is observed not only in cyclic but also in linear structures. *R*-(−)-*sec*-Butylamine has been synthesized by the action of hydroxylamine-*O*-sulfuric acid on the unsymmetrical trialkylborane *3*, which is prepared by the hydroboration of *cis*-2-butene with diisopinocampheylborane [6]:

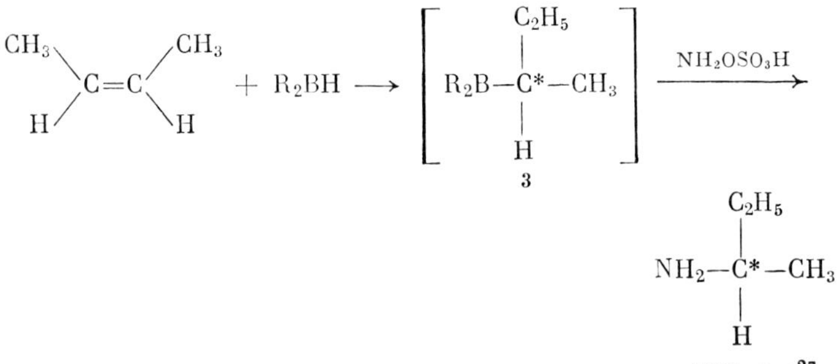

13%, $[\alpha]_D^{27}$ −3,90°

The deuteroboration-amination of *cis*- and *trans*-cyclooctenes has been used for the preparation of the corresponding *cis*- and *trans*-deuterocyclooctylamines (*4*) and (*5*) [7]:

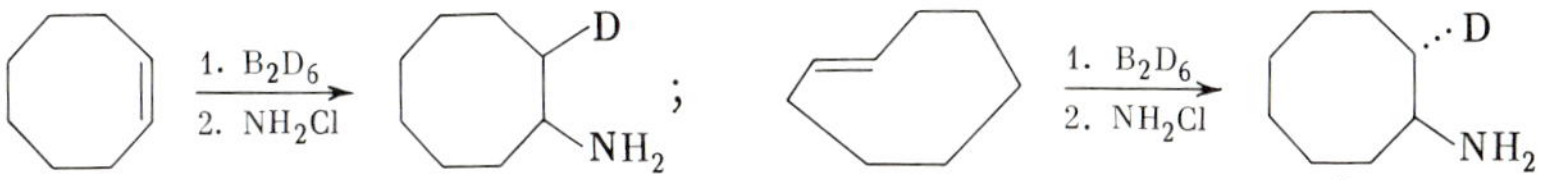

4 5

exo-3-Deutero-2-*exo*-aminonorbornene has been synthesized in an analogous manner from norbornene [8]. Amines deuterated in the 2 position have been used to clarify the mechanism of the Hofmann elimination. The reaction of triphenylborane with NH_2OSO_3H forms aniline and diphenylborinic acid [42]. *p*- and *m*- Tolylamines, and also *p*-ethylaniline, have been obtained by the action of hydroxylamine-*O*-sulfuric acid on a mixture of the corresponding arylboranes, formed by the reaction of Ar_2CuLi with diborane [43].

The mechanism of the amination of organoboranes is apparently analogous to the mechanism of oxidation with hydrogen peroxide in alkaline medium and involves a polar 1,2-migration of an alkyl anion from boron to nitrogen:

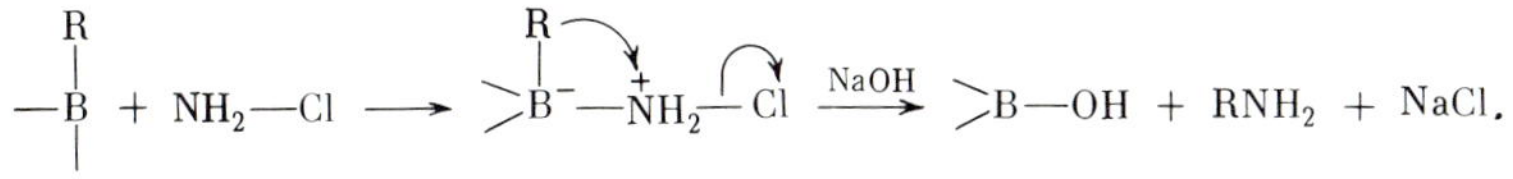

An attempt to expand this reaction to substituted boranes of the benzyl type led to a very interesting result. The treatment of the products of the hydroboration of 2,4,5-trimethoxypropenylbenzene with hydroxylamine-*O*-sulfuric acid produced 2,4,5-trimethoxyaniline [9].

Dimethylbutylamine and dibutylboron chloride have been obtained in quantitative yield (direction *a*) by the action of chlorodimethylamine on tributylborane in the presence of galvinoxyl [10]:

$$(n\text{-}C_4H_9)_3B + Cl{-}N(CH_3)_2 \xrightarrow[100\%]{a} n\text{-}C_4H_9N(CH_3)_2 + (n\text{-}C_4H_9)_2BCl$$

$$(n\text{-}C_4H_9)_3B + Cl{-}N(CH_3)_2 \xrightarrow{L} n\text{-}C_4H_9Cl + (n\text{-}C_4H_9)_2B{-}N(CH_3)_2$$

In the absence of galvinoxyl, which is a powerful inhibitor of radical processes, the reaction proceeds in two directions (*a* and *b*), producing approximately equal quantities of butyl chloride and dimethylbutylamine. Direction *b*, leading to an alkyl halide, is a chain radical process and is initiated by UV light.

In the presence of aqueous alkali, the reaction of trioctyl- or tricyclohexylborane with *N*-chlorodiethylamine or *N*-chloropiperidine proceeds only in direction *b* and leads to alkyl halides (32–53%) and diethylamine or piperidine (54–96%). The reaction of *N*-chlorobutylamine with trialkylboranes is complicated by side processes [11]. The reaction of R_3B with NCl_3 leads to RCl and $>BNCl_2$ [44]. Tosyl derivatives of amines, RNHTs, are obtained by the action of chloramine-T on R_3B [45]. The reaction of R_3B with *p*-$NO_2C_6H_4SO_3NHCO_2C_2H_5$ in the presence of bases and of an interphase catalyst leads to esters of the corresponding *N*-alkylcarbamic acids, $RNHCO_2C_2H_5$ [46].

Another method for the preparation of primary amines consists of the pyrolysis of azidodiorganoboranes, R_2BN_3. In reactions of this type, the migration of a radical from boron to nitrogen proceeds simultaneously with the evolution of N_2, and results in the formation of new B—N and C—N bonds.

When dimesitylborazide (obtained from dimesitylboron fluoride and aluminum azide) is heated in tetrahydrofuran, the cyclic borane *6* is formed, the hydrolysis of which leads to mesidine [12]:

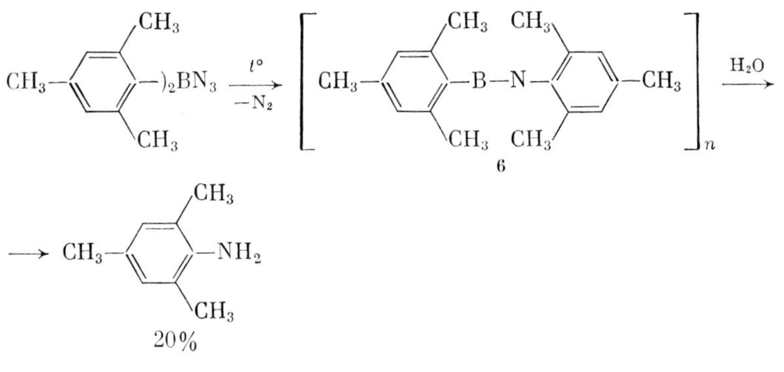

Azidodiphenylborane is converted to tetraphenyl-1,3,2,4-diazadiboretidine (*7*) at elevated temperatures [13–19]:

$$(C_6H_5)_2BCl + LiN_3 \longrightarrow (C_6H_5)_2BN_3 \xrightarrow{t^\circ} \underset{\mathbf{7}}{(C_6H_5BNC_6H_5)_2} \xrightarrow{H_2O} C_6H_5NH_2.$$

Four-membered heterocyclic compounds of the type *7* are easily hydrolyzed to form the corresponding amines [20, 21].

The pyrolysis of the pyridine adduct of azidodiphenylborane leads to hexaphenylborazene in quantitative yield. In the presence of dimethylamine, the main product of the reaction is dimethylhydrazino(diphenyl)borane [15], by the hydrolysis of which *N*,*N*-dimethylhydrazine can be obtained:

$$R_2BN_3 \cdot C_5H_5N \xrightarrow[-N_2]{t^\circ} [R_2BN] \xrightarrow{(CH_3)_2NH} R_2BNHN(CH_3)_2 \xrightarrow{H_2O} (CH_3)_2NNH_2$$

$$(R = C_6H_5).$$

Alkyl radicals are also capable of migrating from boron to nitrogen [14]. For example, by heating azidomethylethylborane, *B*-trimethyl-*N*-triethylborazole is obtained [22, 23], which can easily be converted to ethylamine.

According to the data of [24], the action of NO on tributylborane in heptane at 65°C produces dibutylamine in 35% yield.

7.2 SYNTHESIS OF SECONDARY AMINES AND AZIRIDINES

Secondary amines with identical (R_2NH) or different substituents ($RR'NH$) are obtained by the action of organic azides on triorganoboranes R_3B [25], dialkylboron chlorides [26], and alkyl- and arylboron dichlorides [27], with the reactivity of the organoboranes investigated increasing in the order $R_3B < R_2BCl < RBCl_2$. Some other classes of boron derivatives can apparently also be used for the preparation of amines.

Trialkylboranes and triphenylborane react with alkyl, cycloalkyl, and aryl azides in boiling xylene with the evolution of nitrogen and formation of substituted aminodialkylboranes, which form secondary amines upon hydrolysis or alcoholysis:

$$R_3B + R'N_3 \xrightarrow[-N_2]{\text{xylene, } 135^\circ C} R_2B{-}N\begin{matrix} R \\ R' \end{matrix} \xrightarrow{H_2O,\ OH^-} \underset{70\text{–}80\%}{RR'NH} + R_2BOH.$$

In the presence of bulky radicals in the trialkylborane or azide, the reaction rate decreases considerably and additional heating of the mixture of reagents is required. This difficulty is prevented by the use of dialkylboron chlorides instead of trialkylboranes:

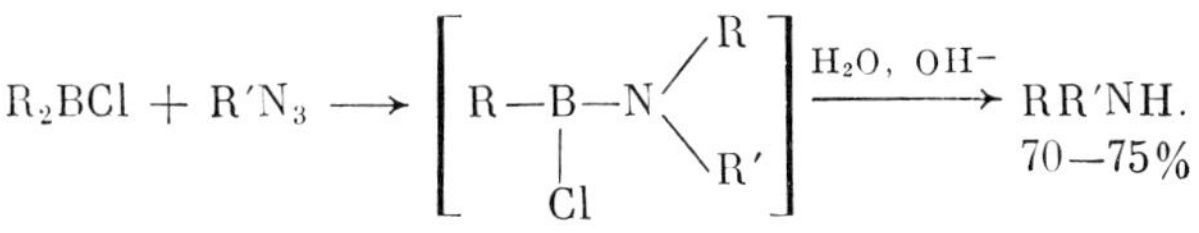

The first stage of the reaction of organic azides with organoboranes apparently consists of the reversible coordination of the organoborane with a molecule of azide, as a result of which a 1:1 complex is formed:

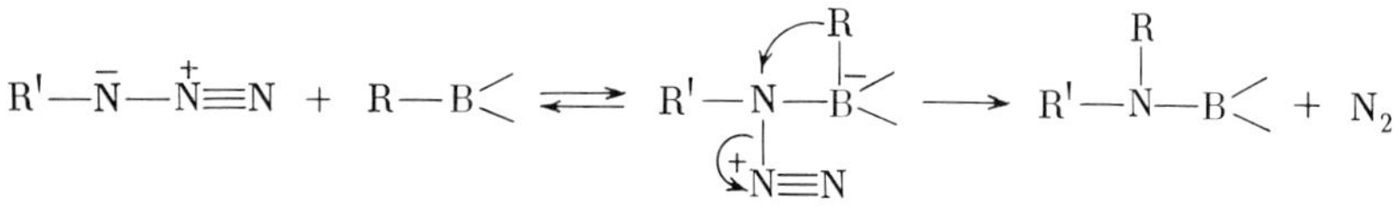

The subsequent migration of the radical from boron to nitrogen proceeds simultaneously with the evolution of a nitrogen molecule (synchronous process). In the transfer of the radical from boron to nitrogen, its configuration is retained. This has been established for the example of the reaction of dichloro-*exo*-norbornylborane with phenyl azide, which led to pure *N*-*exo*-norbornylaniline. *N*-Butyl-*exo*-norbornylamine has been obtained in an analogous manner. *N*-(*trans*-2-Methylcyclopentyl)-cyclohexylamine has been synthesized by the reaction of *trans*-2-methylcyclopentylboron dichloride with cyclohexyl azide [27]:

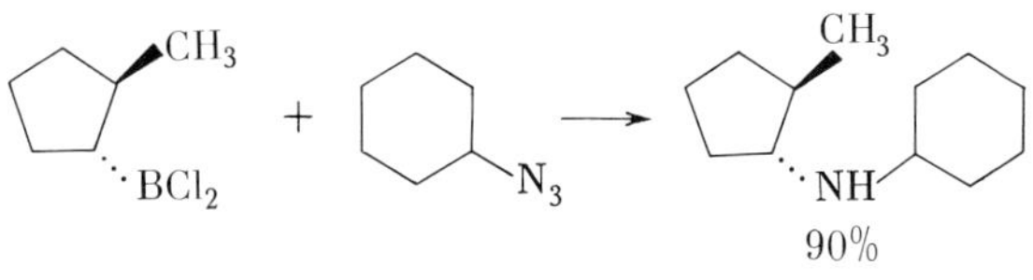

No formation of inversion products was observed in any of the cases. At the same time, the reduction of *N*-cyclohexyl-2-methylcyclopentylimine with lithium aluminum hydride produces a mixture of amines consisting of 78% of the cis isomer and 22% of the trans isomer.

2-Iodoalkyl azides, obtained by the trans-addition of iodine azide to olefins [28], react at room temperature with phenyl- and alkylboron dichlorides with the formation of secondary 2-iodoalkylamines, the treatment of which without isolation with bases (solutions of alkalies, K_2CO_3 or butyllithium) leads to the corresponding *N*-phenyl- or *N*-alkylaziridines [29]:

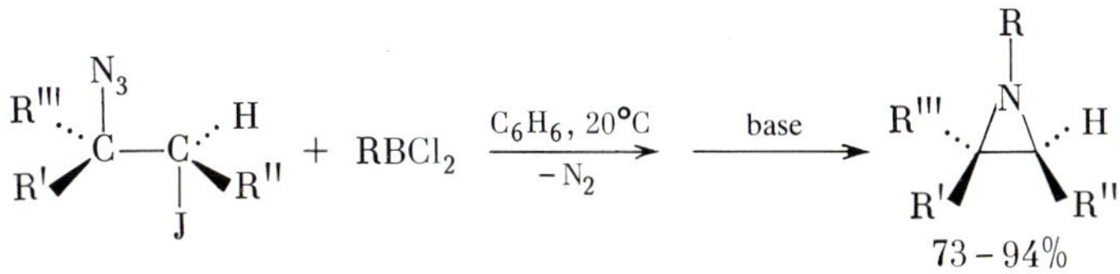

73 – 94%

When tricyclohexylborane is heated with nitrosyl sulfate to 50–80°C, caprolactam and cyclohexylhydroxylamine or dicyclohexylamine are obtained [30]. Dicyclohexylamine is formed in 26% yield by the action of nitrosocyclohexane on tricyclohexylborane [31]. Aromatic nitroso compounds are reduced to the corresponding amines under the same conditions.

A simple method for the synthesis of indoline from β-anilinoethanol has been proposed [32]:

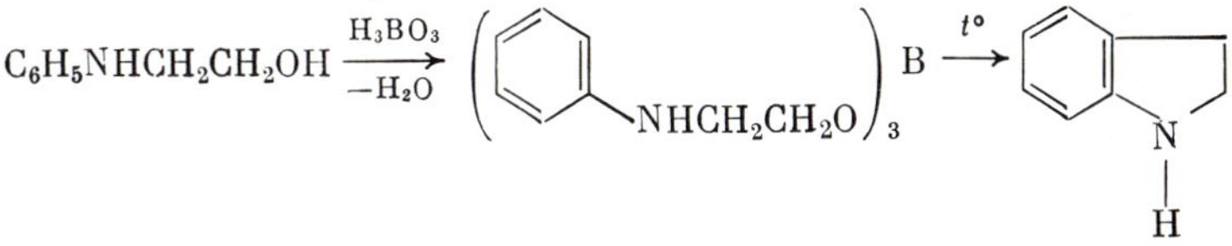

40%

Tris(β-anilinoethyl) borate cyclizes during vacuum distillation.

7.3 SYNTHESIS OF ACID AMIDES AND ENAMINES

Trialkylboranes are used as auxiliary substances in the synthesis of amides from carboxylic acids and amines [33]:

$$RCOOH + R'_3B \xrightarrow[-R'H]{} R'_2BOCOR \xrightarrow{R''NH_2} \underset{90\%}{RCONHR''}.$$

The treatment of an organoborane with one equivalent of acid produces an acyloxydialkylborane which, under the action of primary and secondary amines, is smoothly converted to an amide. The reaction has been used for the synthesis of amides and of macrocyclic (up to 32 atoms) lactams from ω-amino acids [47].

Other methods for the preparation of amides use inorganic boron derivatives. The mixed anhydride of boric and formic acids, formed from BCl_3 and anhydrous HCOOH, produces the corresponding formamides in quantitative yield under the action of primary and secondary amines in benzene at 20°C [34]:

$$HCOOH + BCl_3 \longrightarrow (HCOO)_3B \xrightarrow[-B(OH)_3]{RNH_2} HCONHR.$$

The action of tris(alkylamino)boranes and tris(dialkylamino)boranes on mono- and dicarboxylic acids produces their amides [33, 35, 36]:

$$3RCOOH + B(NR'_2)_3 \longrightarrow [(RCOO)_3B + 3R'_2NH] \longrightarrow \underset{62-87\%}{3RCONR'_2}.$$

The initial reaction product is triacyloxyborane, which, under the action of the amine, forms the amide [37]. Esters of organic acids react slowly with $B(NR_2)_3$; at the same time, the reaction of esters of β-keto acids proceeds rapidly and leads to β-enaminoamides [35].

When triaminoboranes are replaced by their oxygen analogs (trialkyl borates) in the reaction with carboxylic acids, esters are obtained [34, 38, 39]:

$$RCOOH + B(OR')_3 \longrightarrow RCOOR'.$$

This method is convenient for the esterification of polyhydroxy acids, such as *xylo*-trihydroxyglutaric acid [38].

The action of glycine ethyl ester on the borate *8*, prepared from the sodium salt of benzoyl-L-leucine, produced benzoyl-L-leucineglycine ethyl ester with an optical purity of over 90% [33, 40]:

$$C_6H_5CONH\underset{\substack{|\\CH_3}}{C}HCOONa + ClB(OCH_3)_2 \xrightarrow[-NaCl]{C_6H_6,\ 20°C}$$

$$\longrightarrow \underset{8}{C_6H_5CONH\underset{\substack{|\\CH_3}}{C}H-COOB(OCH_3)_2} \xrightarrow{NH_2CH_2COOC_2H_5}$$

$$\underset{31\%}{C_6H_5CONH\underset{\substack{|\\CH_3}}{C}H-CONHCH_2COOC_2H_5}$$

Amides are also obtained by the reaction of acids with amines in the presence of trimethyl borate and toluenesulfonic acid [33].

The reaction of ketones with tris(dialkylamino)boranes is used for the synthesis of enamines [35]:

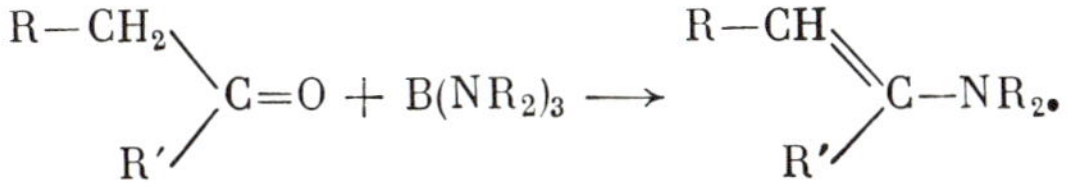

Enamines have been obtained in this way from tripyrrolidylborane and diisobutyl ketone, acetophenone, cyclohexanone, and cholestanone. β-Diketones form enamino ketones under these conditions.

REFERENCES

1. A. N. Nesmeyanov, V. A. Sazonova, A. V. Gerasimenko and V. G. Medvedeva, *Izv. Akad. Nauk SSSR, Otd. Khim. Nauk* **1962,** 2073.
2. A. N. Nesmeyanov, V. A. Sazonova and V. N. Drozd, *Chem. Ber.* **93,** 2717 (1960).
3. H. C. Brown, W. R. Heydkamp, E. Breuer and W. S. Murphy, *J. Am. Chem. Soc.* **86,** 3565 (1964).
4. M. W. Rathke, N. Inoue, K. R. Varma and H. C. Brown, *J. Am. Chem. Soc.* **88,** 2870 (1966).
5. Y. Tamura, J. Minamikawa, S. Fujii and M. Ikeda, *Synthesis* **1974,** 196.
6. L. Verbit and P. J. Heffron, *J. Org. Chem.* **32,** 3199 (1967).
7. J. Coke and M. C. Mourning, *J. Am. Chem. Soc.* **90,** 5561 (1968).
8. J. Coke and M. P. Cooke, *J. Am. Chem. Soc.* **89,** 2779 (1967).
9. L. Levy and L. Fishbein, *Tetrahedron Letters* **1969,** 3773.
10. A. G. Davies, S. C. W. Hook and B. P. Roberts, *J. Organometal. Chem.* **23,** C11 (1970).
11. J. G. Sharefkin and H. D. Banks, *J. Org. Chem.* **30,** 4313 (1965).
12. J. E. Leffler and L. J. Todd, *Chem. and Ind.* **1961,** 512.
13. P. Paetzold. *Z. Anorg. Chem.* **326,** 53 (1963).
14. P. Paetzold *Z. Anorg. Chem.* **326,** 58 (1963).
15. P. Paetzold, *Z. Anorg. Chem.* **326,** 64 (1963).
16. P. Paetzold, P. Habereder and R Müllbauer, *J. Organometal. Chem.* **7,** 45 (1967).
17. P. Paetzold, P. Habereder and R. Müllbauer, *J. Organometal. Chem.* **7,** 51 (1967).
18. P. Paetzold and P. Habereder, *J. Organometal. Chem.* **7,** 61 (1967).
19. P. Paetzold, *Fortschr. Chem. Forsch.* **8,** 437 (1967).
20. Yu. N. Bubnov and B. M. Mikhailov, *Izv. Akad. Nauk SSSR, Ser. Khim.* **1967,** 472.
21. Yu. N. Bubnov, V. S. Bogdanov and B. M. Mikhailov, *Zh. Obshch. Khim.* **38,** 260 (1968).
22. R. Schaeffer and L. J. Todd, *J. Am. Chem. Soc.* **87,** 488 (1965).
23. J. Rathke and R. Schaeffer, *Inorg. Chem.* **11,** 1150 (1972).
24. S. J. Brois and A. J. Rutkowski, U.S. Patent 3,147,310 (1964): *C.A.* **61,** 11891 (1964).

25. A. Suzuki, S. Sono, M. Itoh, H. C. Brown and M. M. Midland, *J. Am. Chem. Soc.* **93,** 4329 (1971).
26. H. C. Brown, M. M. Midland and A. B. Levy, *J. Am. Chem. Soc.* **94,** 2114 (1972).
27. H. C. Brown, M. M. Midland and A. B. Levy, *J. Am. Chem. Soc.* **95,** 2394 (1973).
28. F. W. Fowler, A. Hassner and L. A. Levy, *J. Am. Chem. Soc.* **89,** 2077 (1967).
29. A. Levy and H. C. Brown, *J. Am. Chem. Soc.* **95,** 4067 (1973).
30. T. Okushi, O. Manabe, H. Hiyata and Z. Yoshida, *Kogyo Kagaku Zasshi* **72,** 1660 (1969); *C. A.* **72,** 12802 (1970).
31. T. Okushi, O. Manabe, H. Hiyata and Z. Yoshida, *Kogyo Kagaku Zasshi* **72,** 1665 (1969); *C. A.* **72,** 11797 (1970).
32. V. V. Gertsev and Ya. Ya. Makarov-Zemlyanskii, *Zh. Org. Khim.* **5,** 178 (1969).
33. A. Pelter, T. E. Levitt and P. Nelson, *Tetrahedron* **26,** 1539 (1970).
34. V. V. Gertsev and Ya. Ya. Makarov-Zemlyanskii, *Zh. Prikl. Khim.* **43,** 1633 (1970).
35. P. Nelson and A. Pelter, *J. Chem. Soc.* **1965,** 5142.
36. T. Levitt and A. Pelter, *Nature*, **211,** 299 (1966).
37. A. Pelter and T. E. Levitt, *Tetrahedron* **26,** 1899 (1970).
38. V. V. Gertsev, *Zh. Obshch. Khim.* **37,** 1481 (1967).
39. V. A. Kuskov and V. A. Zhukova, *Izv. Akad. Nauk SSSR, Otd. Khim. Nauk* **1956,** 733.
40. A. Pelter and T. E. Levitt, *Tetrahedron* **26,** 1545 (1970).
41. M. W. Rathke and A. A. Millard, *Org. Synthesis* **58,** 32 (1978).
42. G. W. Kabalka and J. W. Ferrel, *Synth. Commun.* **9,** 443 (1979).
43. N. P. Bullen, K. S. Chiheru and F. G. Thorpe, *J. Organometal. Chem.* **195,** 147 (1980).
44. H. C. Brown and N. R. De Lue, *J. Organometal. Chem.* **135,** C 57 (1977).
45. V. B. Jigaginni, A. Pelter and K. Smith, *Tetrahedron Letters* **1978,** 181.
46. I. Akimoto and A. Suzuki, *Synth. Commun.* **11,** 475 (1981).
47. D. Collum, S.-C. Chen and B. Ganem, *J. Org. Chem.* **43,** 4393 (1978).

CHAPTER 8

Halogenation of alkyl(aryl)boranes

The reactions of organoboranes with halogens may proceed by a radical or electrophilic mechanism (with the participation of Br_2 or Br^+).

In the case of the chlorination and bromination of alkylboranes by a radical mechanism, an α-hydrogen atom is replaced by a halogen:

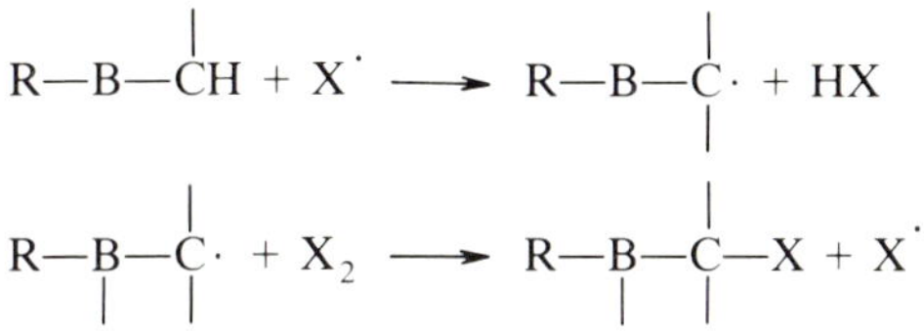

The α-haloalkylborane formed, depending on the conditions, is either cleaved by the hydrogen halide HX liberated or undergoes rearrangement with the transfer of an R group from boron to an α-carbon atom, and of a halogen to the boron:

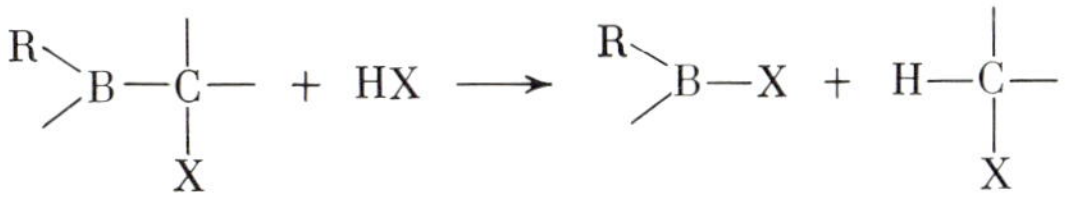

(2)

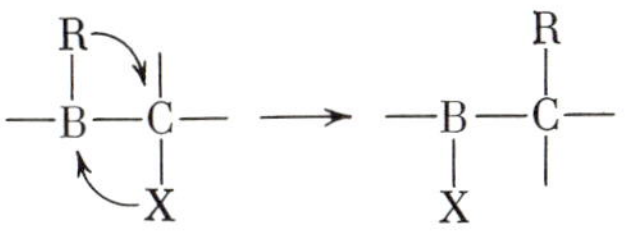

(3)

The direction of reactions (2) or (3) is determined to a considerable degree by the nature of the halogen and of the hydrogen halide liberated. As is known, hydrogen halides (HCl, HBr, and HI) are capable of cleaving one boron-carbon bond in trialkylboranes, with the ease of protolysis depending on the nature of the HX (Chapter 6). Thus, tributylborane reacts with hydrogen chloride at 110°C [1, 2], with hydrogen bromide at 60°C [2, 3], and with hydrogen iodide at room temperature [2]. Although the B—C bond in α-haloalkylboranes is more strongly polarized than that in R_3B, and is therefore cleaved more easily by hydrogen halides, in the chlorination and bromination it is possible to isolate the α-haloalkylboranes or to use them in further conversions.

Rearrangements of type (3), which are accompanied by the formation of new C—C bonds and are typical in boron chemistry, were first observed by Pasto [4–7] and Matteson [8, 9] and their co-workers for the example of α-halo-, α-alkoxy-, and α-alkylthioalkylboranes $R'CHXBR_2$ (R = H, alkyl, aryl).

The halogenation of alkylboranes is used for the synthesis of organic and organometallic compounds of different classes. A combination of reactions (*1*) and (*2*) is used for the preparation of alkyl and cycloalkyl halides and also of ω-substituted alkylboranes and 4-substituted alkanols. A combination of reactions (*1*) and (*3*) is used for the formation of carbon–carbon bonds in acyclic and carbocyclic systems, particularly for the synthesis of higher acyclic and alicyclic alcohols and ketones (Part II).

Depending on their structure and on the conditions, arylboranes are halogenated in the ring or side chain or undergo halodeboronation.

The halogenation of organoboranes is partially discussed in review articles and books [10–19].

8.1 CHLORINATION OF ALKYLBORANES

The chlorination of trimethylborane proceeds easily at temperatures from −95° C to 25° C and leads to chloromethyldimethylborane (*1*) [20–23]:

$$(CH_3)_3B + Cl_2 \xrightarrow{-95°\,C} \underset{\mathbf{1}}{ClCH_2B(CH_3)_2} + HCl.$$

Chloromethyldimethylborane reacts with nucleophilic reagents in two directions. Under the action of mercaptans, dimethylphosphine, and dimethylarsine, chlorine is replaced by a mercapto, phospino, or arsino group, respectively [23]:

$$(CH_3)_2BCH_2P(CH_3)_2 \xleftarrow{(CH_3)_2PH} (CH_3)_2BCH_2Cl \xrightarrow{CH_3SH} (CH_3)_2BCH_2{-}SCH_3.$$

The reaction with KI leads to iodomethyldimethylborane, $(CH_3)_2BCH_2I$.

On the other hand, when compound *1* is treated with such strong nucleophilic agents as water, alcohols, ammonia, amines, hydrazines, and also with lithium azide or antimony trifluoride or trichloride, methylethylborane derivatives of the type *2* or *3* are obtained, i.e., the replacement of chlorine is accompanied by migration of the methyl group from boron to carbon and the formation of a carbon–carbon bond:

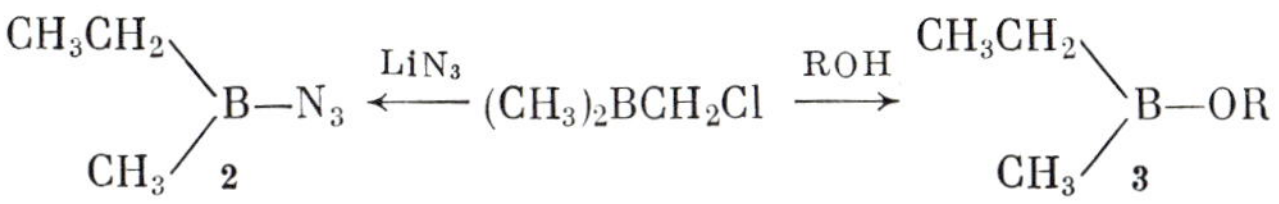

Under the action of trimethylamine, compound *1* undergoes rearrangement with the formation of chloro(ethyl)methylborane. The replacement of chlorine in esters $RCH(Cl)B(OR')_2$ by alkyl, cycloalkyl, phenyl, or the groups C_6H_5S, C_6H_5O, *tert*-$C_4H_9O_2CCH_2$, etc. under the action of the corresponding metal derivatives has been described in [113].

The chlorination of butylboronic anhydride is accompanied by the evolution of hydrogen chloride; the structures of the other products have not been determined [24]. From the complex mixture of products of the reaction of di-*tert*-butyl methylboronate and *tert*-butyl hypochlorite, the ester of chloromethylboronic acid, $ClCH_2B(OR)_2$ was isolated in low yield [25]. The reaction of α-phenylthioalkyl(dialkyl)boranes and of cyclic esters of α-phenylthioalkylboronic acids with *N*-chlorosuccinimide in methanol is a selective method for the synthesis of the monothioacetals $RCH(SC_6H_5)OCH_3$ and of the acetals $RCH(OCH_3)_2$ [114, 115]. The reaction of $C_6H_5SCH_2B(OR)_2$ with thionyl chloride leads to $C_6H_5SCH_2Cl$ [115].

It should be noted that the direct chlorination of alkylboranes has not been used so far for the preparation of alkyl chlorides. At the same time, trialkylboranes can be converted to alkyl chlorides by a radical reaction with chlorodialkylamine [26, 27], $CuCl_2$ [28], $FeCl_3$ or $FeCl_2$ in the presence of hydrogen peroxide [29, 116]. Arylboranes form aryl chlorides under the action of $CuCl_2$ [30].

8.2 BROMINATION OF ALKYLBORANES

The structure of the compounds obtained by the action of bromine on alkylboranes is strongly dependent on the nature of the radicals on the boron and, in particular, on the conditions under which the reaction is carried out. Bromination is used for the preparation of alkyl and cycloalkyl bromides and a combination of this reaction with hydroboration represents a method for the anti-Markownikoff hydrobromination of olefins.

Under specific conditions, a coupling of the radicals takes place in the bromination of di- and trialkyl boron derivatives [reaction (3)]. Processes of this type are used for the synthesis of higher ketones and alcohols and also for the transition from boracyclanes and, consequently, from dienes to carbocyclic compounds (Part II).

8.2.1 General characteristics of the reaction

The bromination of trialkylboranes was carried out for the first time by Johnson, Snyder, and Van Campen [3], who showed that tributylborane reacts slowly with bromine at 20°C to form butyl bromide, butane, and dibutylboron bromide:

$$R_3B + Br_2 \xrightarrow{20^\circ\ C} R_2BBr + RBr + RH + HBr.$$

The main direction of the above reaction consists of the cleavage of a B—C bond by an electrophilic mechanism (S_EC); there is also a partial replacement of hydrogen atoms by bromine to form HBr.

In an investigation of the kinetics of the gas-phase bromination of triethylborane, it was established that this reaction is initiated by light and proceeds by a chain radical mechanism [31]. The main reaction products are hydrogen bromide and α-bromoethyldiethylborane; ethane and ethyl bromide, which are formed in very small quantities, have also been identified. Consequently, the bromination of trialkylboranes proceeds mainly by the replacement of an α-hydrogen atom. It was found that the abstraction of an α-hydrogen from alkylboranes requires considerably less energy than the rupture of the C_{sec}-H Bond in alkanes (the difference is approximately 14.5 kcal/mole). This makes it possible to use a very wide variety of solvents in the bromination, including acyclic and alicyclic hydrocarbons. The bromination rate depends strongly on the nature of the radicals in the organoborane; for example, tri-*sec*-butylborane is brominated much more easily than tributylborane and the latter is brominated more rapidly than triisobutylborane [32]: (sec-$C_4H_9)_3B > (n$-$C_4H_9)_3B > (iso$-$C_4H_9)_3B$. *tert*-Alkylboranes are not cleaved by bromine.

In the bromination of triethylborane under light, with simultaneous rapid removal of the hydrogen bromide formed, α-bromoethyldiethylborane (*4*) can be isolated; this is quite stable and can be stored at 0° C [33]:

$$CH_3CH_2B(C_2H_5)_2 + Br_2 \xrightarrow{h\nu,\ \text{pentane}} \underset{\displaystyle Br}{CH_3CH}-B(C_2H_5)_2 + HBr.$$

4 (70%)

The radical bromination of R_3B is carried out by the action of bromotrichloromethane [34] or 1,2-dibromo-1,1,2,2-tetrachloroethane [35].

α-Bromoalkyldialkylboranes undergo rearrangement upon heating or under the action of nucleophilic reagents [33, 34, 36] and Lewis acids [37] by migration of an alkyl group from boron to carbon. Nucleophilic agents initially coordinate with the boron atom, while electrophilic agents attack the bromine atom:

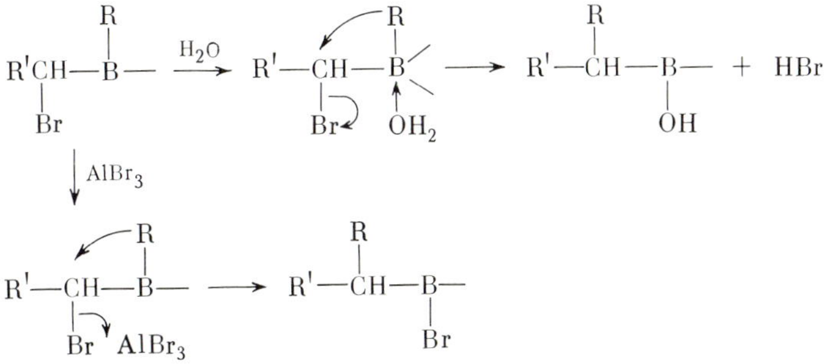

The rearrangement proceeds instantaneously in the presence of water, methanol, pyridine, AlX_3, or $ZnCl_2$. In acetonitrile, tetrahydrofuran, trimethylene oxide, and 2,6-lutidine it proceeds considerably more slowly and is accompanied by cleavage of the ring of cyclic ethers. In diethyl ether and 2,6-di-*tert*-butylpyridine, α-bromoalkylboranes are stable and undergo practically no rearrangement.

N-Bromosuccinimide brominates trialkylboranes under light with the formation of α-bromoalkylboranes which, however, immediately rearrange under the action of the succinimide formed [34].

The bromination of tributylborane pyridinate proceeds by an electrophilic mechanism and leads to bromodibutylborane pyridinate, which, under the action of ethanol, is converted to the ester $(C_4H_9)_2BOC_2H_5$ (81%) [38].

The reactions of esters of dialkylborinic acids with bromine proceed by replacement of an α-hydrogen atom (Section 3).

Esters of *n*-alkylboronic acids do not react with bromine [39] or *N*-bromosuccinimide [25], while their *sec*-alkyl analogs are easily brominated under light in the α-position with the formation of α-bromoalkylboranes in quantitative yield [39, 40]:

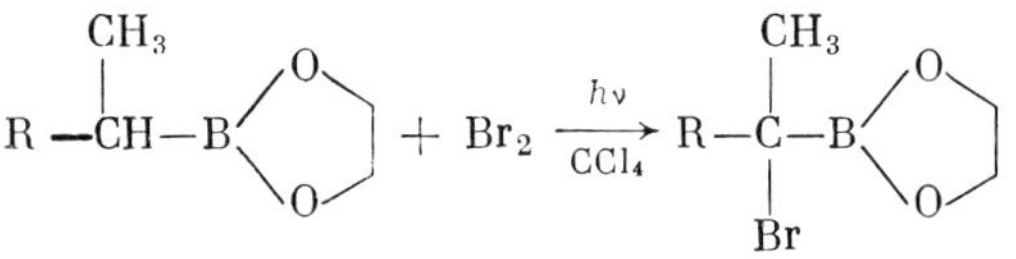

$(R = n\text{-}C_4H_9,\ C_6H_5)$.

By means of these examples, it was established experimentally for the first time that the liquid-phase bromination of organoboranes may proceed by a radical mechanism. The trimethylene esters of benzyl- and *sec*-alkylboronic acids react with bromine in an analogous manner [117].

N-Bromopyridinium perbromide $(C_6H_5N + 2\ Br_2)$ splits a B—C bond in esters of 1-phenylethylboronic acid, as a result of which α-bromoethylbenzene is obtained [40].

The radical bromination of the glycol ester of 2-phenylethylboronic acid, which is initiated by daylight, leads to an ester of 2-bromo-2-phenylethylboronic acid, indicating a stronger activating effect of a phenyl radical on the mobility of an α-hydrogen as compared with a dihydroxyboryl group. Esters of 1,1- and 1,2-(2-phenylethyl)diboronic acid are brominated in the 2 position; these reactions have been used for the synthesis of α- or β-*trans*-bromostyrene.

Matteson, et al. have prepared a number of esters of α-bromoalkylboronic acids by the radical addition of hydrogen halides [41–44], bromohalomethanes [8, 9, 45, 46], bromomalononitrile and butyl bromomalonate [47] to esters of α,β-unsaturated boronic acids and also by the bromination of tetraboronic esters of the type $C[B(OR)_2]_4$ and their derivatives [48–52].

8.2.2 Synthesis of alkyl and cycloalkyl bromides

As previously noted, the reaction of trialkylboranes with bromine under light without a solvent proceeds with the cleavage of one or two boron-carbon bonds and leads to alkyl bromides, but their yield does not exceed 50% [3]. At the same time, bromination in methylene chloride in the dark proceeds smoothly enough. At room temperature, the reaction is completed within 24–48 hours and leads to dialkylboron bromides and organic bromides [32]:

$$R_3B + Br_2 \xrightarrow{CH_2Cl_2,\ 25°\,C} \underset{80-99\%}{RBr} + R_2BBr$$

(R = *sec*-C_4H_9, *iso*-C_4H_9, $(CH_3)_3CCH_2CH(CH_3)CH_2$, cyclo-$C_5H_9$, cyclo-$C_6H_{11}$, *exo*-norbornyl).

As a method for the synthesis of alkyl bromides, the bromination of trialkylboranes in the dark has a very limited application and can be used only for the preparation of *sec*-alkyl or cycloalkyl bromides. Organoboranes prepared by the hydroboration of terminal olefins and containing approximately 7% of secondary alkylboranes, which are brominated much more easily than *n*-alkylboranes, form a mixture of isomeric bromides containing more than 10% of 2-bromoalkanes. For example, the hydroboration-bromination of 1-butene forms 80% of butyl bromide and 11% of 2-bromobutane.

Bromination in the dark proceeds by a radical mechanism [10]. In this case, the chain reaction is probably initiated by traces of oxygen or of organoboron peroxides [53].

In [54], 9-*sec*-alkyl-9-borabicyclo[3.3.1]nonanes are used for the conversion of olefins with a non-terminal double bond to the corresponding *sec*-alkyl halides. Alkyl bromides can be obtained by the action of bromine, BrCl, or sodium bromide and chloramine-T on trialkylboranes and alkyldicycloalkylboranes in aqueous tetrahydrofuran. In this reaction, only one B—C bond is cleaved, with primary organoboranes being brominated more easily than secondary ones [118].

The reaction of tri-*exo*-norbornylborane or 9-*exo*-norbornyl-9-borabicyclo[3.3.1]nonane with bromine leads to *exo*-norbornyl bromide, in which the admixture of the endo isomer does not exceed 1% [32, 54].

In the presence of sodium methoxide, the bromination of trialkylboranes proceeds with the cleavage of all three B—C bonds and alkyl bromides are obtained in high yield [55]:

$$R_3B + 3Br_2 + 4CH_3ONa \xrightarrow{\text{THF, 0°C}} \underset{60\text{—}90\%}{3RBr} + 3NaBr + NaB(OCH_3)_4$$

$R = n\text{-}C_6H_{13},\ n\text{-}C_8H_{17},\ n\text{-}C_3H_7CH(CH_3)CH_2,\ \text{cyclo-}C_5H_9,\ \text{cyclo-}C_6H_{11}$

This reaction is used for the synthesis of acyclic, cyclic, and functional organic bromides:

$$CH_2{=}CH(CH_2)_7CH_2COOCH_3 \xrightarrow[\text{2. } Br_2,\ CH_3ONa]{\text{1. } BH_3} \underset{85\%}{BrCH_2(CH_2)_9COOCH_3}$$

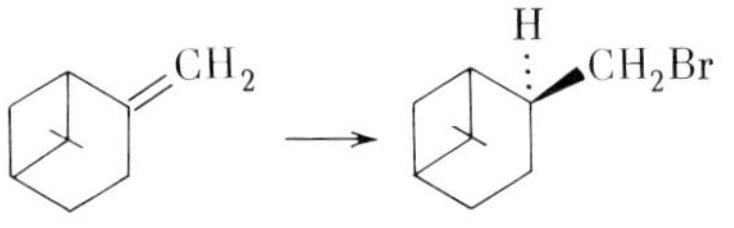

59%

When alkali is used instead of sodium methoxide, the yield of bromides does not exceed 45%, since the hypobromite formed oxidizes the organoboron compound to an alcohol.

The reaction of bromine with alkylboranes in the presence of sodium methoxide proceeds mainly with inversion of the configuration, which is confirmed by the formation of primarily (75%) *endo*-norbornyl bromide (5) from tri-*exo*-norbornylborane and one mole of bromine [56]:

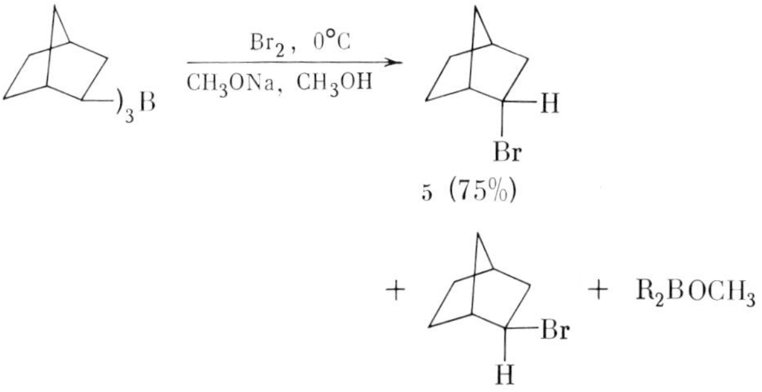

(25%) (R = *exo*-2-norbornyl)

An inversion of the configuration (over 95%) has been observed in the bromination of the *threo* and *erythro* isomers of tris(3,3-dimethyl-1-butyl-1,2-d_2)borane in the presence of sodium methoxide [119]. Three moles of phenyl bromide are obtained from triphenylborane and bromine in the presence of sodium methoxide [120].

Aliphatic bromides can also be prepared by the action of mercuric acetate and bromine [57] or of $CuBr_2$ [28, 29, 116] on trialkylboranes (Chapter 14).

8.2.3 Synthesis of ketones and branched alcohols

Pasto and McReynolds [39] found that an ester of *n*-hexylmethylborinic acid absorbs two moles of bromine under light, with the formation of a boron compound, the oxidation of which forms 2-heptanone:

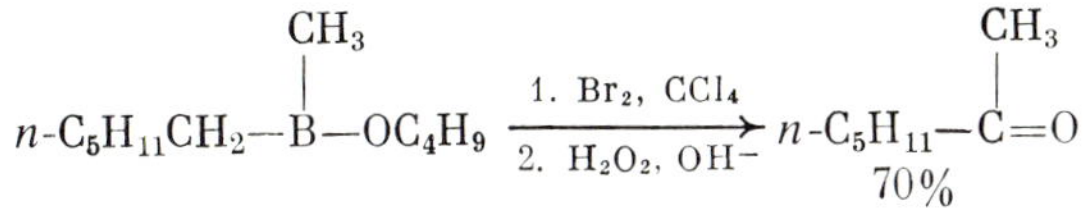

The reaction probably proceeds in the following manner:

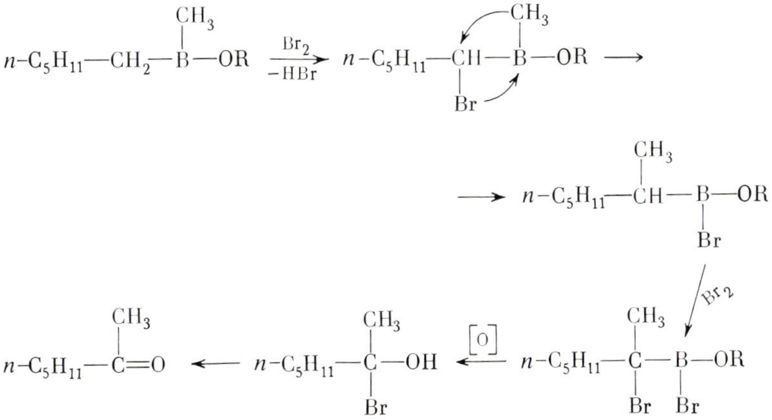

The monobromination of di-*n*-alkylborinic acids in CH_2Cl_2 containing water, followed by oxidation of the rearrangement products, leads to secondary alcohols and ketones [58]:

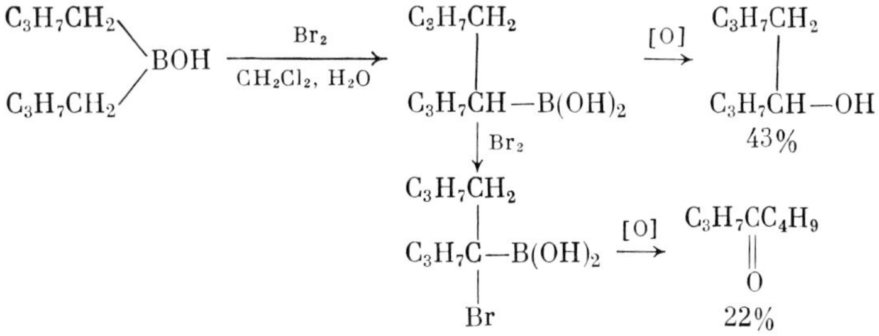

As a result of the bromination-oxidation of di-*sec*-alkyl- and dicycloalkylborinic acids, which react more selectively, branched aliphatic and alicyclic alcohols have been synthesized:

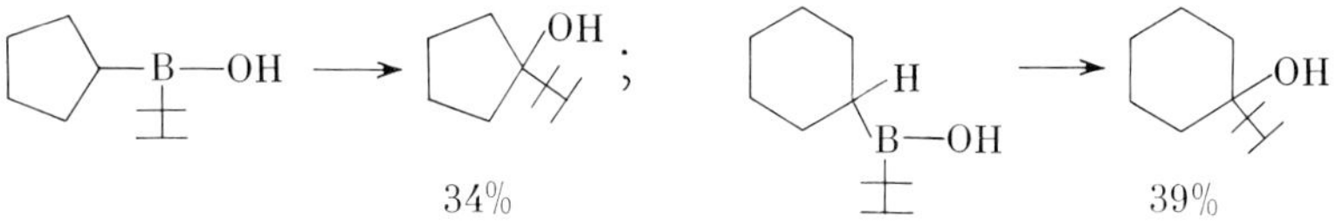

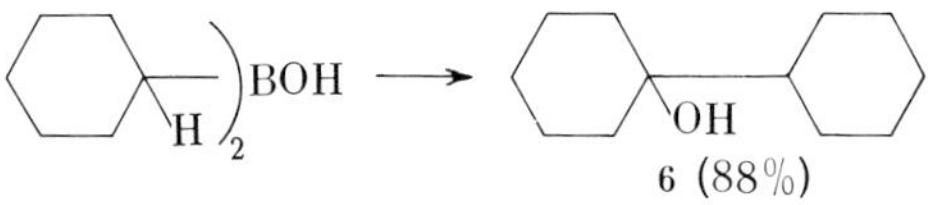

6 (88%)

Alcohols of an analogous type have also been obtained by the monobromination-oxidation of tri-*sec*-alkylboranes [36], e.g., 1-cyclohexylcyclohexanol (*6*) has been synthesized from tricyclohexylborane in 89% yield.

1-Substituted cyclopentanols and cyclohexanols, including alcohols containing functional groups in the hydrocarbon radical, have been obtained from mixed trialkylboranes under light ($CH_2Cl_2 + H_2O$) [59]:

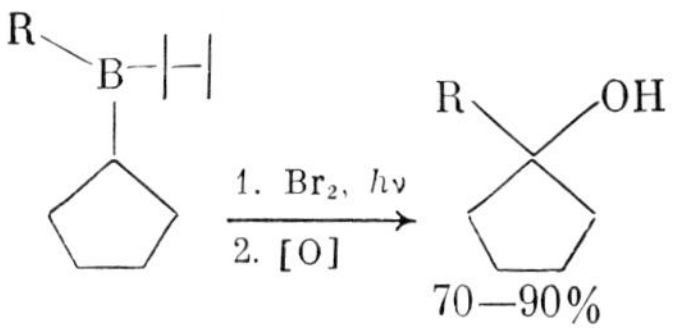

($R = n\text{-}C_4H_9$, *iso*-C_4H_9, $n\text{-}C_6H_{13}$, $ClCH_2(CH_2)_3CH_2$, $C_2H_5OCO(CH_2)_2CH_2$)

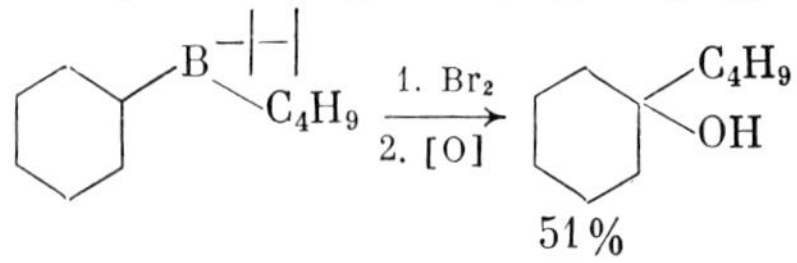

No migration of the bulky 2,3-dimethylbutyl group takes place.

The bromination of tri-*n*-alkylboranes under light in aqueous methylene chloride is accompanied by the migration of two alkyl groups to the α-carbon atom of the third radical. As an example, one can cite the reaction of triethylborane with bromine in a ratio of 1:2, leading to (3-methyl-3-pentyl)boronic acid (*7*), the oxidation of which in aqueous alcoholic alkali produces 3-methyl-3-pentanol (*8*) [35, 36]:

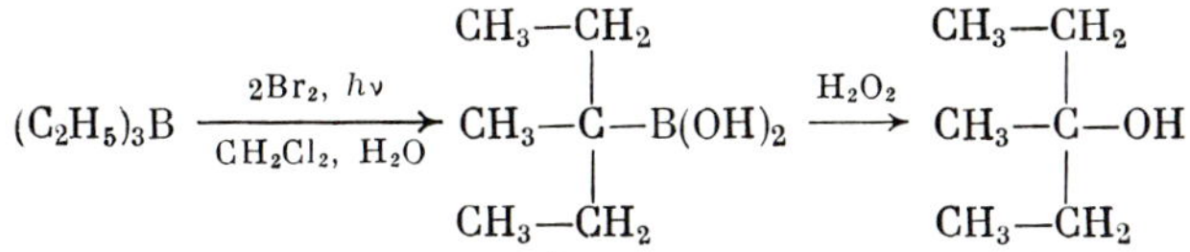

7 **8**

The replacement of an α-halogen by a radical in an α-haloalkyldialkylborane, which takes place as the result of a rearrangement initiated by bases, proceeds with inversion at the α-carbon. α-Haloalkylboranes are readily epimerized under the action of tetrahydrofuran and dimethyl sulfide [121].

8.3 IODINATION OF ALKYLBORANES. SYNTHESIS OF ALKYL IODIDES, OLEFINS, AND 1,5-DIENES

According to [3, 60, 61], trialkylboranes do not react with iodine. However, in 1953 Long and Dollimore [62] showed that the reaction of tripropylborane with iodine proceeds at temperatures above 140°C and leads to propyl iodide and iododipropylborane. The replacement of a second radical by iodine proceeds very slowly [62, 63]:

$$R_3B + I_2 \xrightarrow{155\text{–}165°C} R\text{—}I + R_2BI \xrightarrow{I_2} I\ RBI_2 + RI$$
$$(R{=}n\text{-}C_3H_7,\ iso\text{-}C_3H_7,\ n\text{-}C_4H_9,\ \text{cyclo-}C_6H_{11}).$$

The dialkylboron iodides obtained in this way have been used for the synthesis of acetylenic boron compounds, $R_2BC{\equiv}CBR_2$ [63, 64]. The iodination of $(C_4H_9)_2BOZnC_2H_5$ proceeds at 20°C and leads to a mixture of ethyl and butyl iodides [122].

As has been shown by an investigation of the kinetics of the thermal (100–140°C) [65, 66] or photochemical [65, 67] iodination of triethylborane in cyclohexane, this is a radical process proceeding as three consecutive stages, of which the second—the attack on a trialkylborane molecule by an I• radical—is the determining one:

$$I_2 \rightleftarrows 2I^{\bullet},$$
$$I^{\bullet} + (C_2H_5)_3B \longrightarrow (C_2H_5)_2BI + C_2H_5^{\bullet},$$
$$C_2H_5^{\bullet} + I_2 \longrightarrow C_2H_5I + I^{\bullet}$$

It is possible, however, that the reaction with iodine proceeds analogously to the chlorination and bromination, i.e., by replacement of the α-hydrogen atom. The hydrogen iodide liberated in this process immediately cleaves a boron-carbon bond (as previously noted, HI reacts with trialkylboranes at room temperature [2]).

The ability of iodine to trap radicals is the basis for its use as an oxidation inhibitor for trialkylboranes [68].

In the presence of alkalies, the reaction of trialkylboranes with iodine proceeds very rapidly even at 25°C. In this process, two boron-carbon bonds in the primary trialkylboranes are cleaved and the corresponding alkyl iodides are obtained [69]:

$$R_3B + 2I_2 + 2NaOH \xrightarrow{\text{THF, 25°C}} \underset{60—67\%}{2RI} + 2NaI + RB(OH)_2$$

[R = n-C_4H_9, i-C_4H_9, n-C_6H_{13}, n-$C_{10}H_{21}$, $(CH_3)_3CCH_2CH(CH_3)CH_2$, cyclo-$C_5H_9$ (33%), *sec*-C_4H_9 (30%)].

Secondary trialkylboranes react significantly more slowly than the primary compounds, with the cleavage of only one B—C bond, and this reaction can therefore be used for the conversion of terminal olefins (by hydroboration) to quite pure (approximately 99%) primary alkyl iodides. By this method, 2-phenylpropyl iodide is obtained from α-methylstyrene, β-pinene is converted to the iodide *9*, and vinyl-3-cyclohexene forms 2-cyclohexenylethyl iodide (*10*):

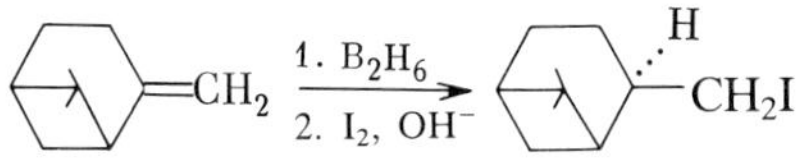

9 (40%)

10 (60%)

For the conversion of some difficulty accessible olefins to primary iodides, they are initially hydroborated with di-*sec*-alkylboranes and the mixed trialkylboranes formed are then iodinated in tetrahydrofuran in the presence of alkali:

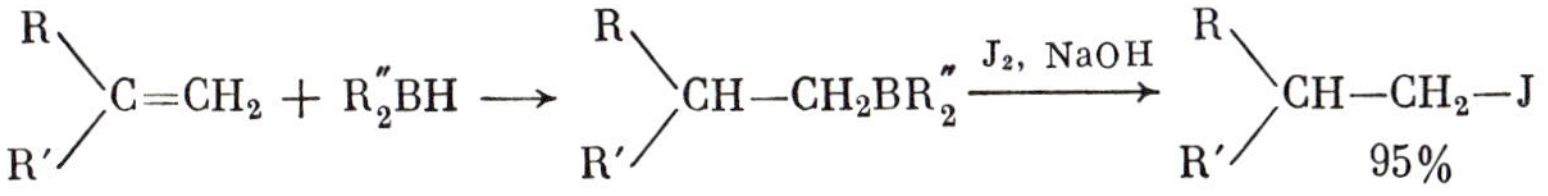

(R″ = *sec*-alkyl or cycloalkyl).

Iodination proceeds more smoothly in the presence of sodium methoxide, with three moles of RI being obtained from primary R_3B [123].

Alkyl iodides are easily obtained by the action of iodine monochloride [124, 125] or sodium iodide and chloramine-T [126] on triorganoboranes in the presence of sodium acetate. Compounds containing radioactive iodine have been obtained by this method [127].

The base-induced replacement of boron by iodine in organoboranes proceeds with inversion of the configuration [123, 124, 128], possibly by an S_E2 mechanism [126].

In triphenylborane under the action of iodine [120], iodine monochloride [129], or sodium iodide and an oxidant [126], one B—C bond is cleaved and iodobenzene is obtained.

Alkyl iodides RI can be obtained by the electrolysis of R_3B in nitromethane in the presence of R'_4NI [130].

According to the data of [131], the reaction of trialkylarylborates with iodine leads quantitatively to an alkyl iodide and a dialkylarylborane. The borates $(RBR'_3)Li$ react with iodine to form RI and BR'_3 [130a].

Alkyl iodides are also obtained as a result of the oxygen-initiated reaction of trialkylboranes with benzyl or allyl iodide [70]:

$$R_3B + CH_2{=}CH{-}CH_2I \xrightarrow[75-100\%]{O_2} RI + (CH_2{=}CHCH_2)_2$$

$(R = C_2H_5,\ n\text{-}C_4H_9,\ n\text{-}C_5H_{11},\ n\text{-}C_6H_{13},\ \text{-cyclo-}C_5H_9)$.

This reaction proceeds by a radical mechanism but is not a chain reaction. The formation of one mole of RI requires the introduction of approximately one mole of oxygen. Under analogous conditions, allyl and benzyl bromides form alkyl bromides with a yield of approximately 20%, while allyl chloride does not react.

Apparently, a more valuable aspect of the reaction is the coupling of the radicals of the iodide used. As a result of the introduction of air into a mixture of triethylborane and 2-methylallyl iodide in a ratio of 2:1, 2,5-dimethyl-1,5-hexadiene is formed. In an analogous manner, 97% of diallyl was obtained from allyl iodide, 87% of dibenzyl from benzyl iodide, and *p*-nitrobenzyl iodide was converted to 4,4′-dinitrodibenzyl:

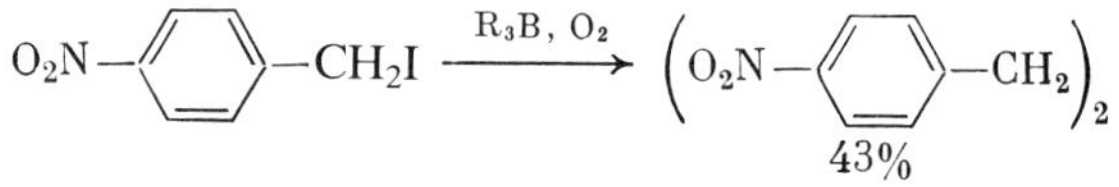

A similar reaction of two different iodides of the allyl or benzyl type leads to mixed products of radical coupling:

$$C_6H_5CH_2I + CH_2{=}CHCH_2I \xrightarrow{(C_2H_5)_3B,\ O_2} \underset{72\%}{C_6H_5CH_2CH_2CH{=}CH_2}.$$

α-Amino derivatives $R_2'NCH_2B(OR)_2$ [132] and the boron analog of hippuric acid, $C_6H_5CONHCH_2B(OH)_2 \cdot 2H_2O$ [133], have been synthesized from the esters $ICH_2B(OR)_2$. 1,2-Diphenylethanol has been obtained by the action of $C_6H_5M_gX$ on an ester of 1-iodo-2-phenylethylboronic acid, followed by oxidation [134].

8.4 HALOGENATION OF ARYLBORANES

Depending on the conditions under which the reactions are carried out and on the reagents used, the halogenation of aromatic boron derivatives proceeds in one of four directions: 1) cleavage of a boron-carbon bond (so-called halodeboration), 2) halogenation in the ring (replacement of hydrogen), 3) halogenation in the side chain, and 4) coupling of the radicals. The B—C bond in arylboranes is not cleaved by halogen radicals.

The reaction of arylboronic and diarylborinic acids with chlorine, bromine, and iodine in aqueous, weakly alkaline, or acid solution, as a rule, proceeds with the cleavage of boron-carbon bonds and leads to aryl halides [71–84]:

$$ArB(OH)_2 \xrightarrow[H_2O]{X_2} ArX \xleftarrow[H_2O]{X_2} Ar_2BOH$$

An investigation of the kinetics of the bromination and iodination of phenylboronic acid in aqueous acetic acid or acetate buffer carried out by Kuivila and co-workers [85–89] showed that these reactions proceed by an electrophilic substitution mechanism and are second-order overall and first-order in each of the components. They are catalyzed by bases, salts, and complexing agents and are inhibited by acids.

It is suggested that the replacement of boron by halogen proceeds analogously to the bromination of aromatic compounds by a dipole-ion interaction between the halogen molecule and the arylborate anion (*11*) via a transitional σ-complex of the type *12*:

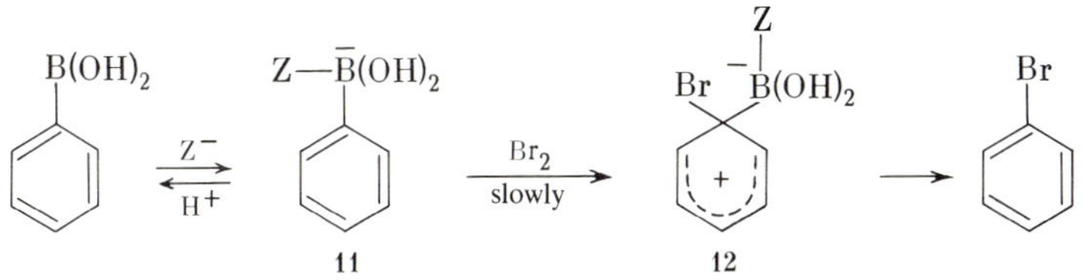

Reagents (bases, complexing agents, salts, etc.) promoting the formation of the borate anion *11* accelerate the halodeboronation process.

Potassium tetraphenylborate reacts with bromine in methanol with the formation of bromobenzene and biphenyl [90] and potassium tetratolylborate forms ditolyl [91], indicating the radical nature of the process.

The action of bromine, bromosuccinimide, and other brominating agents on 2-aminoethyl esters of aryl(2-thienyl)borinic acids produces the corresponding 2-arylthiophenes [135].

In the action of halogens on aryl boron derivatives in organic solvents, a halogenation in the aromatic ring usually takes place [79–81, 92], with the BCl_2 and $B(OH)_2$ groups showing a meta-orienting effect; the values of σ for the meta and para positions of the ring in $C_6H_5B(OH)_2$ have values of -0.01 and $+0.13$, respectively [93]. In the presence of another substituent along with $B(OH)_2$ in the benzene ring, the direction of electrophilic substitution is determined by the nature of the substituent. The reaction of 2- and 3-thienylboronic acids with *N*-bromosuccinimide or bromine in CCl_4 leads to the corresponding 5-bromothiopheneboronic acids [94].

The electrophilic halogenation of heterocyclic boron derivatives of the following types has been described: 3,2-borazaropyridine [95–100], 4-methyl-4,3-borazaroisoquinoline [101, 102], 4,5-borazarothieno(2,3-*c*)pyridine [103], 10,9-borazaro- [104, 105] and 10,9-boroxarophenanthrene [106].

The radical bromination of tolylboronic acids proceeds in the side chain with the replacement of one or two hydrogen atoms by bromine [107–111, 136]. The reaction of *p*-ethylphenylboronic acid with *N*-bromosuccinimide, followed by the elimination of HBr, leads to *p*-vinylphenylboronic acid [112].

REFERENCES

1. R. Booth and C. Kraus, *J. Am. Chem. Soc.* **74,** 1415 (1952).
2. H. Skinner and T. Tees, *J. Chem. Soc.* **1953,** 3378.
3. J. Johnson, H. Snyder and M. van Campen, *J. Am. Chem. Soc.* **60,** 115 (1938).
4. D. Pasto and J. Miesel, *J. Am. Chem. Soc.* **84,** 4991 (1962).
5. D. Pasto and J. Miesel, *J. Am. Chem. Soc.* **85,** 2118 (1963).
6. D. Pasto and J. Hickman, *J. Am. Chem. Soc.* **89,** 5608 (1967).
7. D. Pasto and R. Snyder, *J. Org. Chem.* **31,** 2773 (1966).
8. D. Matteson and R. Mah, *J. Am. Chem. Soc.* **85,** 2599 (1963).
9. D. Matteson, *Organometal. Chem. Rev.* **1,** 1 (1966).
10. C. F. Lane, *Intra-Science Chem. Rep.* **7,** 133 (1973).
11. C. F. Lane, *Aldrichimica Acta,* **6,** 21 (1973).
12. P. I. Paetzold and H. Grundke, *Synthesis* **1973,** 635.
13. E. Negishi and H. C. Brown, *Synthesis* **1974,** 77.
14. H. C. Brown: "Boranes in Organic Chemistry", Cornell University Press, Ithaca, New York (1972).
15. M. F. Lappert, *Chem. Rev.* **56,** 959 (1956).
16. M. F. Lappert. in: "The Chemistry of Boron and Its Compounds", Ed. E. L. Muetterties. New York, J. Wiley and Sons (1967), p. 433.
17. O. A. Reutov, I. P. Beletskaya and V. I. Sokolov: "Mechanisms of Reactions of Organometallic Compounds", Moscow, Khimiya (1972).
18. A. Suzuki, *Kagaku No Ryoiki, Zokan* **89,** 213 (1970); *C. A.* **73,** 14886*n* (1970).
19. K. Torssell, in: "Progress in Boron Chemistry", vol. 1. Ed. H. Steinberg and A. McCloskey. Oxford, Pergamon Press (1964), p. 369.
20. L. Zeldin and P. R. Girardot, Abstracts of the 14th National Meeting of the American Chemical Society. Chicago, Ill. (1961), p. 15N.
21. P. R. Girardot and L. Zeldin. *U.S. Patent* 3,083,230 (1963); *C. A.* **59,** P10116g (1963).
22. R. Schaeffer and L. J. Todd, *J. Am. Chem. Soc.* **87,** 488 (1965).

23. J. Rathke and R. Schaeffer, *Inorg. Chem.* **11,** 1050 (1972).
24. J. C. Perrine and R. N. Keller, *J. Am. Chem. Soc.* **80,** 1823 (1958).
25. D. Matteson, *J. Org. Chem.* **29,** 3399 (1964).
26. J. G. Sharefkin and H. D. Banks, *J. Org. Chem.* **30,** 4313 (1965).
27. A. G. Davies, S. C. W. Hook and B. P. Roberts, *J. Organometal. Chem.* **23,** C11 (1970).
28. C. F. Lane, *J. Organometal. Chem.* **31,** 421 (1971).
29. A. Arase, Y. Masuda and A. Suzuki, *Bull. Chem. Soc. Japan.* **47,** 2511 (1974).
30. R. C. Larock, *Intra-Science Chem. Rep.* **7,** 95 (1973).
31. J. Grotewold, E. A. Lissi and J. C. Scaiano, *J. Organometal. Chem.* **19,** 431 (1969).
32. C. F. Lane and H. C. Brown, *J. Am. Chem. Soc.* **92,** 7212 (1970).
33. H. C. Brown and Y. Yamamoto, *J. Am. Chem. Soc.* **93,** 2796 (1971).
34. H. C. Brown and Y. Yamamoto, *Chem. Comm.* **1971,** 1535.
35. H. C. Brown and Y. Yamamoto, *Synthesis* **1972,** 699.
36. C. F. Lane and H. C. Brown, *J. Am. Chem. Soc.* **93,** 1025 (1971).
37. H. C. Brown and Y. Yamamoto, *Chem. Comm.* **1972,** 71.
38. L. S. Vasil'ev, V.P. Dmitrikov and B. M. Mikhailov, *Zh. Obshch. Khim.* **42,** 1015 (1972).
39. D. J. Pasto and K. McReynolds, *Tetrahedron Letters* **1971,** 801.
40. D. J. Pasto, J. Chow and S. K. Arora, *Tetrahedron* **25,** 1557 (1969).
41. D. Matteson, *Acc. Chem. Res.* **3,** 186 (1970).
42. D. Matteson, R. A. Bowie and G. Srivastava, *J. Organometal. Chem.* **16,** 33 (1969).
43. D. Matteson and J. D. Liedtke, *Chem. and Ind.* **1963,** 1241.
44. D. Matteson and J. D. Liedtke, *J. Org. Chem.* **28,** 1924 (1963).
45. D. Matteson, *J. Am. Chem. Soc.* **82,** 4428 (1960).
46. D. Matteson and R. W. Mah, *J. Org. Chem.* **28,** 2171 (1963).
47. D. Matteson and G. D. Schaumberg, *J. Org. Chem.* **31,** 726 (1966).
48. R. B. Castle and D. Matteson, *J. Organometal. Chem.* **20,** 19 (1969).
49. D. Matteson, R. A. Davis and L. A. Hagelee, *J. Organometal. Chem.* **69,** 45 (1974).
50. D. Matteson and R. J. Wilcsek, *J. Organometal. Chem.* **57,** 231 (1973).
51. D. Matteson, *Synthesis* **1975,** 147.
52. D. Matteson, *Intra-Science Chem. Rep.* **7,** 147 (1973).
53. B. M. Mikhailov and Yu. N. Bubnov, *Izv. Akad. Nauk SSSR, Ser. Khim.* **1964,** 2248.
54. C. F. Lane and H. C. Brown, *J. Organometal. Chem.* **26,** C51 (1971).
55. H. C. Brown and C. F. Lane, *J. Am. Chem. Soc.* **92,** 6660 (1970).
56. H. C. Brown and C. F. Lane, *Chem. Comm.* **1971,** 521.
57. J. J. Tufariello and M. M. Hovey, *J. Am. Chem. Soc.* **92,** 3221 (1970).
58. H. C. Brown and C. F. Lane, *Synthesis* **1972,** 303.
59. H. C. Brown, Y. Yamamoto and C. F. Lane, *Synthesis* **1972,** 304.
60. E. Frankland and B. Duppa, *Proc. Roy. Soc.* **12,** 123 (1862).
61. E. Frankland, *Ann.* **124,** 129 (1862).
62. L. H. Long and D. Dollimore, *J. Chem. Soc.* **1953,** 3902.
63. H. Hartmann and K.-H. Birr, *Z. Anorg. Chem.* **299,** 174 (1959).
64. H. Hartmann and K.-H. Birr, *Angew. Chem.* **68,** 247 (1956).
65. E. Lissi and E. Sanhueza, *J. Organometal. Chem.* **26,** C59 (1971).
66. E. Lissi and E. Sanhueza, *J. Organometal. Chem.* **32,** 285 (1971).
67. M. Abufhele, C. Andersen, E. Lissi and E. Sanhueza, *J. Organometal. Chem.* **42,** 19 (1972).

68. M. M. Midland and H. C. Brown, *J. Am. Chem. Soc.* **93,** 1506 (1971).
69. H. C. Brown, M. W. Rathke and M. Rogič, *J. Am. Chem. Soc.* **90,** 5038 (1968).
70. A. Suzuki, S. Nozawa, M. Harada, M. Itoh, H. C. Brown and M. M. Midland, *J. Am. Chem. Soc.* **93,** 1508 (1971).
71. N. N. Mel'nikov, *Zh. Obshch. Khim.* **6,** 636 (1936).
72. R. Letsinger and N. Remes, *J. Am. Chem. Soc.* **77,** 2489 (1955).
73. R. L. Bruce and A. A. Humffray, *Austral. J. Chem.* **24,** 1085 (1971).
74. E. W. Abel, W. Gerrard and M. Lappert, *J. Chem. Soc.* **1957,** 5051.
75. R. D. Brown, A. S. Buchanan and A. A. Humffray, *Austral. J. Chem.* **18,** 1527 (1965).
76. J. R. Johnson, M. G. van Campen and O. Grummit, *J. Am. Chem. Soc.* **60,** 111 (1938).
77. A. D. Ainley and F. Challenger, *J. Chem. Soc.* **1930,** 2171.
78. N. N. Mel'nikov, *Zh. Obshch. Khim.* **8,** 1766 (1938).
79. B. Serafin and M. Makosza, *Tetrahedron* **19,** 821 (1963).
80. B. Serafin, H. Duda and M. Makosza, *Rocz. Chem.* **37,** 765 (1963).
81. B. Serafin, M. Makosza and I. Szczere, *Rocz. Chem.* **36,** 889 (1962).
82. F. R. Been and J. R. Johnson, *J. Am. Chem. Soc.* **54,** 4415 (1932).
83. D. Nielsen and W. McEwen, *J. Am. Chem. Soc.* **79,** 3081 (1957).
84. O. C. Musgrave, *Chem. and Ind.* **1957,** 1152.
85. H. Kuivila and E. K. Easterbrook, *J. Am. Chem. Soc.* **73,** 4629 (1951).
86. H. Kuivila and A. Hendrickson, *J. Am. Chem. Soc.* **74,** 5068 (1952).
87. H. Kuivila and E. Soboczenski, *J. Am. Chem. Soc.* **76,** 2675 (1954).
88. H. Kuivila and R. W. Williams, *J. Am. Chem. Soc.* **76,** 2679 (1954).
89. H. Kuivilva and L. Benjamin, *J. Am. Chem. Soc.* **77,** 4834 (1955).
90. G. A. Razuvaev and T. G. Brilkina, *Zh. Obshch. Khim.* **24,** 1415 (1954).
91. A. N. Nesmeyanov, V. A. Sazonova, G. S. Liberman and L. I. Emel'yanova, *Izv. Akad. Nauk SSSR, Otd. Khim. Nauk* **1955,** 48.
92. K. Niedenzu and J. W. Dawson, *J. Org. Chem.* **26,** 1671 (1961).
93. K. Torssell, *Arkiv Kemi* **10,** 513 (1957).
94. S. Lawesson, *Arkiv Kemi* **11,** 387 (1957).
95. S. Gronowitz and A. Bugge, *Acta Chem Scand.* **19,** 1271 (1965).
96. S. Gronowitz and J. Namtvedt, *Acta Chem. Scand.* **21,** 2151 (1967).
97. S. Gronowitz, P. Gassne and B. Yom-Tov, *Acta Chem. Scand.* **23,** 2927 (1969).
98. J. Namtvedt and S. Gronowitz, *Acta Chem. Scand.* **22,** 1373 (1968).
99. S. Gronowitz and A. Maltesson, *Chem. Scripta* **2,** 79 (1972).
100. A. Maltesson, Ph. D. Thesis. Lund (1974).
101. J. Namtvedt and S. Gronowitz, *Acta Chem. Scand.* **20,** 1448 (1966).
102. M. J. S. Dewar and J. Rosenberg, *J. Am. Chem. Soc.* **88,** 358 (1966).
103. S. Gronowitz and J. Namtvedt, *Tetrahedron Letters* **1966,** 2967.
104. M. J. S. Dewar and V. P. Kubba, *J. Org. Chem.* **25,** 1722 (1960).
105. M. J. S. Dewar and V. P. Kubba, *Tetrahedron* **7,** 213 (1959).
106. M. J. S. Dewar and R. Dietz, *J. Chem. Soc.* **1960,** 1344.
107. K. Torssell, *Acta Chem. Scand.* **8,** 1779 (1954).
108. H. R. Snyder, A. J. Reedy and W. J. Lennarz, *J. Am. Chem. Soc.* **80,** 835 (1958).
109. K. Torssell, *Arkiv Kemi* **10,** 507 (1957).
110. M. J. S. Dewar and R. C. Dougherty, *J. Am. Chem. Soc.* **86,** 433 (1964).
111. R. T. Hawkins, W. J. Lennarz and H. R. Snyder, *J. Am. Chem. Soc.* **82,** 3053 (1960).
112. K. Hoffmann and W. Thomas, *J. Am. Chem. Soc.* **81,** 580 (1959).
113. D. S. Matteson and D. Majumdar, *J. Am. Chem. Soc.* **102,** 7588 (1980).

114. A. Mendoza and D. S. Matteson, *Chem. Commun.* **1978,** 357.
115. A. Mendoza and D. S. Matteson, *J. Organometal. Chem.* **156,** 149 (1978).
116. Y. Masuda, M. Hoshi and A. Arase, *Bull. Soc. Chem. Japan* **52,** 271 (1979).
117. H. C. Brown, N. R. De Lue, Y. Yamamoto and K. Maruyama, *J. Org. Chem.* **42,** 3252 (1977).
118. G. W. Kabalka, K. A. R. Sastry, H. C. Hsu and M. D. Hylarides, *J. Org. Chem.* **46,** 3113 (1981).
119. D. E. Bergbreiter and D. P. Rainville, *J. Organometal. Chem.* **121,** 19 (1976).
120. G. W. Kabalka and J. W. Ferrel, *Synth. Commun.* **9,** 443 (1979).
121. M. M. Midland, A. R. Zolopa and R. L. Halterman, *J. Am. Chem. Soc.* **101,** 248 (1979).
122. R. F. Galiullina, E. V. Chistova and V. A. Dodonov, *Zh. Obshch. Khim.* **50,** 1657 (1980).
123. N. R. De Lue and H. C. Brown, *Synthesis* **1976,** 114.
124. G. W. Kabalka and E. E. Gooch, *J. Org. Chem.* **45,** 3578 (1980).
125. E. E. Gooch and G. W. Kabalka, *Synth. Commun.* **11,** 521 (1981).
126. G. W. Kabalka and E. E. Gooch, *J. Org. Chem.* **46,** 2582 (1981).
127. G. W. Kabalka and E. E. Gooch, *Chem. Commun.* **1981,** 1011.
128. H. C. Brown, N. R. De Lue, G. W. Kabalka and H. C. Hendecock, *J. Am. Chem. Soc.* **98,** 1290 (1976).
129. G. W. Kabalka and E. E. Gooch, *Synth. Commun.* **11,** 247 (1981).
130. Y. Takahashi, M. Tokuda, M. Itoh and A. Suzuki, *Synthesis* **1976,** 616.
130a. E. Negishi, M. J. Idacavage, K.-W. Chiu, T. Yoshida, A. Abramovitch, M. Goettel, A. Silveira and D. H. Bretheric, *J. Chem. Soc., Perkin II* **1978,** 1225.
131. E. Negishi, *J.Organometal. Chem.* **108,** 281 (1976).
132. D. S. Matteson and D. Majumdar, *J. Organometal. Chem.* **170,** 259 (1979).
133. R. N. Lindquist and A. C. Nguyen, *J. Am. Chem. Soc.* **99,** 6435 (1977).
134. D. S. Matteson and K. Arne, *J. Am. Chem. Soc.* **100,** 1325 (1978).
135. G. M. Davies, P. S. Davies, W. E. Paget and J. M. Wardleworth, *Tetrahedron Letters* **1976,** 795.
136. T. Burgmeister, R. Grobe-Einsler, R. Grotstollen, A. Mannschreck and G. Wulff, *Chem. Ber.* **114,** 3403 (1981).

CHAPTER 9

Reactions of alkyl(aryl)boranes with carbonyl compounds

The direction of the reactions of organic boron derivatives with aldehydes and ketones is determined by the nature of the radicals on the boron and by the structure of the carbonyl compound.

Unlike allylboranes, which react with aldehydes and ketones similarly to a Grignard reagent (Part IV), alkyl- and arylboranes do not usually add to a C=O double bond, which is probably due to the negligible polarity of the boron-carbon bond in these compounds. An exception is represented by the reaction of R_3B with formaldehyde, which, in the presence of oxygen, leads to esters of the type R_2BOCH_2R (lengthening the radical's carbon chain by one carbon atom).

Aromatic aldehydes, when heated above 100°C, are quantitatively reduced with trialkylboranes to the corresponding benzyl alcohols, while aliphatic aldehydes and ketones form mainly enol derivatives — vinyloxyboranes. The reaction of alkylboranes with aromatic aldehydes is used for the synthesis of individual α-olefins and for the isomerization of internal to terminal olefins and of methylcyclenes to methylenecyclanes.

In the reaction of carbonyl compounds with 1,1-diboron compounds, higher olefins are obtained under certain conditions. This method supplements the well-known method for the formation of a double bond via phosphoranes and, in a number of cases, may compete with the Wittig reaction.

Under the action of oxygen or other radical initiators, organoboranes react with linear and cyclic α,β-unsaturated aldehydes and ketones to form the corresponding products of 1,4-addition. These reactions are used for lengthening the carbon chain of the radical on the boron or of the olefin by three or more carbon atoms, with the simultaneous introduction of a functional group. 1,4-Addition reactions, which proceed under mild conditions, are a general method for the alkylation of conjugated carbonyl compounds with organoboranes.

9.1 REACTIONS WITH ALDEHYDES AND KETONES. SYNTHESIS OF ALCOHOLS AND OLEFINS

Meerwein and co-workers [1] found that the reaction of triethylborane with some aldehydes is accompanied by the elimination of ethylene and leads to esters of diethylborinic acid:

$$(C_2H_5)_3B + RCHO \longrightarrow (C_2H_5)_2BOCH_2R + C_2H_4$$

$$(R{=}CCl_3,\ CBr_3,\ p\text{-}ClC_6H_4).$$

Higher trialkylboranes also reduce aromatic aldehydes with the elimination of an olefin; depending on the structure of the radicals in the organoborane, the reaction starts at 80–150°C and proceeds with the evolution of heat [2, 3]. The first B—C bond is cleaved relatively easily, while the second is cleaved only by prolonged heating at 150–200°C, i.e., the esters of di-*n*-alkylborinic acids are weaker reducing agents than the trialkylboranes:

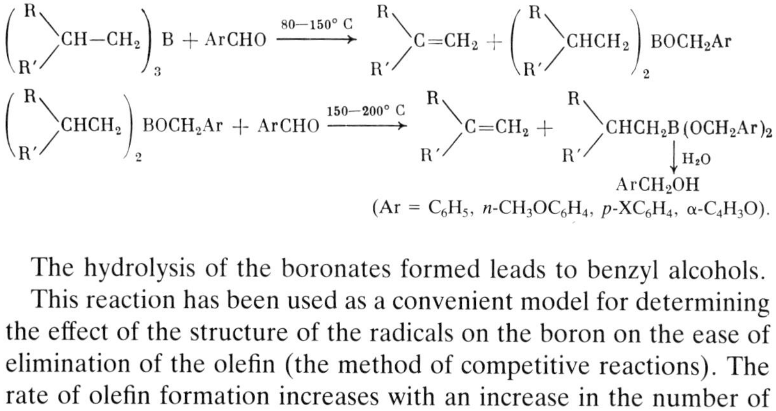

$(Ar = C_6H_5,\ n\text{-}CH_3OC_6H_4,\ p\text{-}XC_6H_4,\ \alpha\text{-}C_4H_3O).$

The hydrolysis of the boronates formed leads to benzyl alcohols.

This reaction has been used as a convenient model for determining the effect of the structure of the radicals on the boron on the ease of elimination of the olefin (the method of competitive reactions). The rate of olefin formation increases with an increase in the number of alkyl groups on the β-carbon atoms, whose inductive effect promotes the abstraction of a hydride ion (H^-):

$$CH_3{-}\underset{}{\overset{CH_3}{\overset{|}{C}H}}{-}CH_2{-} > CH_3CH_2CH_2{-} \geqslant C_2H_5CH_2CH_2{-} > CH_3CH_2{-} > CH_3{-}\overset{CH_3}{\overset{|}{C}H}{-}$$

The reaction with aldehydes probably proceeds by a mechanism of synchronous cyclic electron transfer [3]:

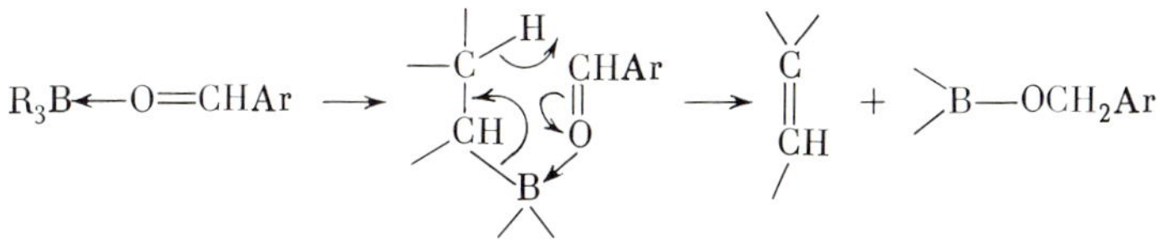

When tricyclohexylborane is heated with benzaldehyde, *p*-chloro-, *p*-methyl-, *p*-isopropyl, and *o*, *p*-dichlorobenzaldehyde, cyclohexene, and the corresponding benzyl alcohols are obtained [4, 5]. According to the data of [6], the prolonged heating of benzaldehyde results in the cleavage of all three boron-carbon bonds in trihexylborane, which is undoubtedly incorrect. In the reaction of trialkylboranes with furfural, two moles of olefin are eliminated at a temperature above 120°C. Acetophenone, dimethyl sulfoxide, and dimethylformamide cleave only one boron-carbon bond [7].

It is known that alkyl boron derivatives in which the boron is bonded to an internal carbon atom of the chain isomerize upon heating to compounds with boron on the terminal carbon (Chapter 4). This property has been used for the contra-thermodynamic isomerization of internal to terminal olefins via hydroboration-isomerization-elimination. In this manner, 1-hexene has been obtained from 2- and 3-hexenes [2]:

$$CH_3CH_2CH{=}CHCH_2CH_3 \xrightarrow{B_2H_6} CH_3CH_2CH_2\underset{\underset{\displaystyle B<}{|}}{C}HCH_2CH_3 \xrightarrow{200^\circ\ C}$$

$$\longrightarrow (CH_3CH_2CH_2CH_2CH_2CH_2{-})_3B \xrightarrow{C_6H_5CHO} CH_3CH_2CH_2CH_2CH{=}CH_2.$$

Methylcyclopentene is converted to methylenecyclopentane by an analogous method:

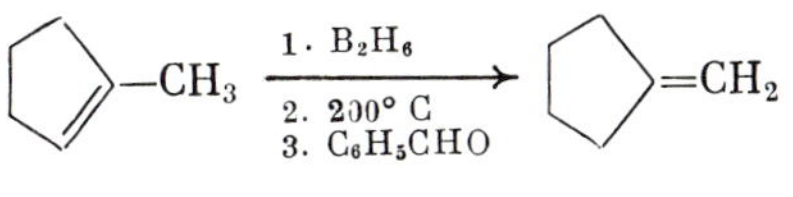

Diborane or triisobutylborane is used for the hydroboration of 1-methylcyclenes:

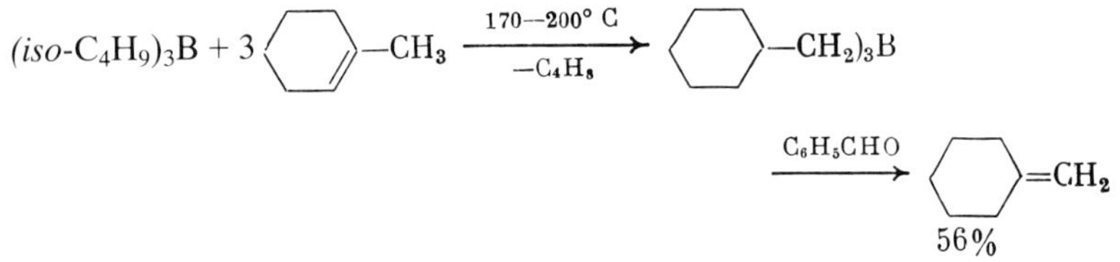

Unlike the aromatic aldehydes, formaldehyde reacts with trialkylboranes at 0°C. In an inert atmosphere, the reaction proceeds by olefin elimination and leads to methyl dialkylborinate, R_2BOCH_3 [8]. In the presence of atmospheric oxygen, however, the addition of a boron-alkyl fragment to the carbonyl bond takes place and a higher ester of dialkylborinic acid is obtained, i.e., the radical is lengthened by one carbon atom:

$$R_3B + CH_2O \xrightarrow{O_2} R_2BOCH_2R \xrightarrow[H_2O_2]{OH^-} 2ROH + RCH_2OH.$$

The reaction proceeds by a radical mechanism, as indicated by the formation of α-tetrahydrofurylcarbinol when tetrahydrofuran is used as a solvent:

$$R_3B \xrightarrow{O_2} R_2BO_2^{\cdot} + R^{\cdot},$$
$$R^{\cdot} + CH_2O \longrightarrow RCH_2O^{\cdot},$$
$$RCH_2O^{\cdot} + R_3B \longrightarrow R_2BOCH_2R + R^{\cdot}.$$

The reaction of triphenylborane with formaldehyde and tertiary amines leads to betaine compounds [9]:

$$(C_6H_5)_3B + CH_2O + R_3N \longrightarrow (C_6H_5)_3\overline{B}{-}OCH_2{-}\overset{+}{N}R_3.$$

Cyclic betaines are obtained by the reaction of formaldehyde with secondary or cyclic amines and halodiphenylboranes [9, 10]:

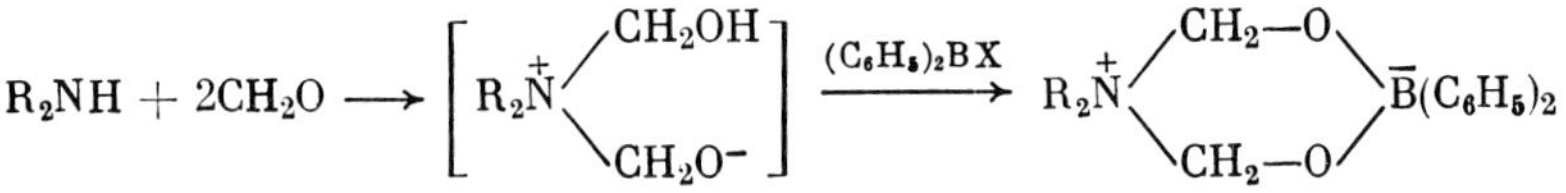

Aliphatic ketones and alkyl aryl ketones react with trialkylboranes at 145–170°C to form a mixture of substances consisting of the product of reduction of the carbonyl group (*1*) and a boron ester of the enol—vinyloxyborane (*2*) [11]:

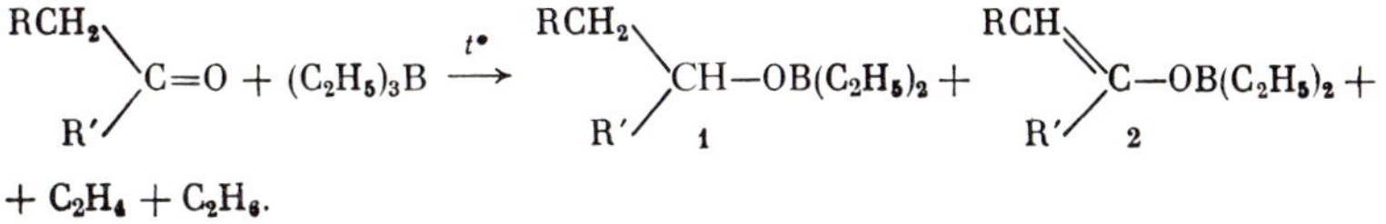

The yield of the esters *2* is 90% from the ketones and 40% from the aldehydes. At the same time, triphenyl- and tribenzylboranes quantitatively form the enol compounds *2* with aldehydes of the type R'_2CHCHO.

In the presence of 1–10% of diethylboryl pivalate, alkenyloxydiethylboranes (*4*) are formed in good yields from triethylborane and acyclic or cyclic ketones (*3*) [12, 12a, 94]:

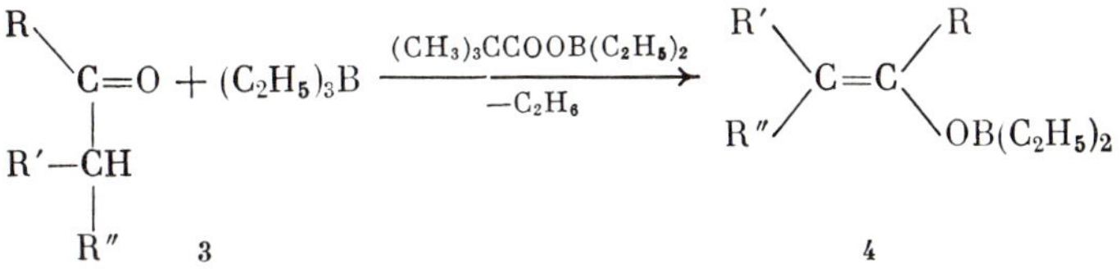

The catalytic action of diethylboryl pivalate is explained by the following two-stage process:

$$(3) + (CH_3)_3CCOOB(C_2H_5)_2 \rightleftharpoons (4) + (CH_3)_3CCOOH,$$
$$(C_2H_5)_3B + (CH_3)_3CCOOH \longrightarrow C_2H_6 + (CH_3)_3CCOOB(C_2H_5)_2.$$

The structural and stereochemical direction of the reaction is determined by the nature of the radicals in the starting ketone. Thus, diethyl ketone and ethyl isopropyl ketone are converted at 80–100°C to a mixture of *Z* and *E* isomers in 75% yield, with a predominant content of the *Z* isomer (90–95%).

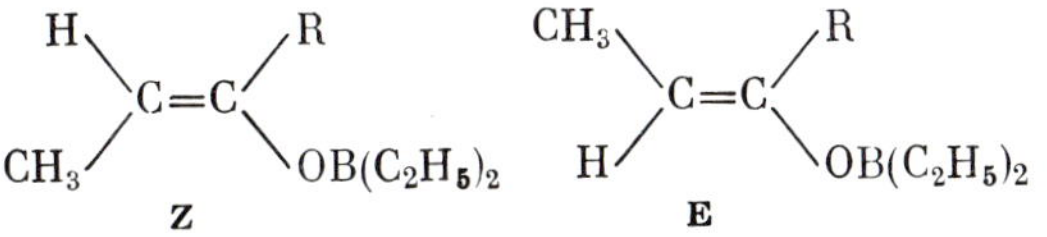

($R = C_2H_5$, *iso*-C_3H_7).

In the case of benzyl ethyl ketone, two structurally isomeric esters are formed, each of which is a mixture of *Z* and *E* isomers.

With the use of diethylboryl pivalate, a number of conjugated ketones has been obtained from alkyl phenyl ketones, including the *E* isomer of dyprone, 1,3-diphenyl-2-buten-1-one [13].

In the excited triplet state ($n \rightarrow \pi^*$), ketones react with R_3B at $-80°C$ to form alkyl and ketyl radicals, which have been recorded by EPR spectroscopy [14–16].

The reaction of acetylacetone or of acetoacetic ester with trialkylboranes at room temperature or on heating is accompanied by elimination of the acyclic hydrocarbon and leads to boron acetylacetonates [17–22]:

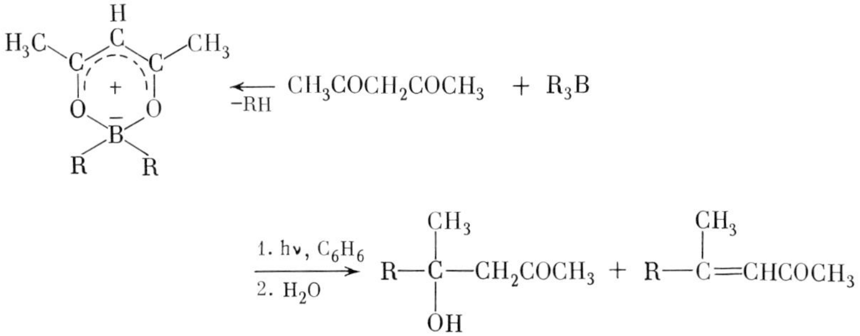

On the other hand, the photoreaction of trialkylboranes with β-dicarbonyl compounds leads, after hydrolysis, to β-keto alcohols and unsaturated ketones [23].

Dialkylboryl diacetamidinates are formed by the action of diacetamide on trialkylboranes in tetrahydrofuran [24]:

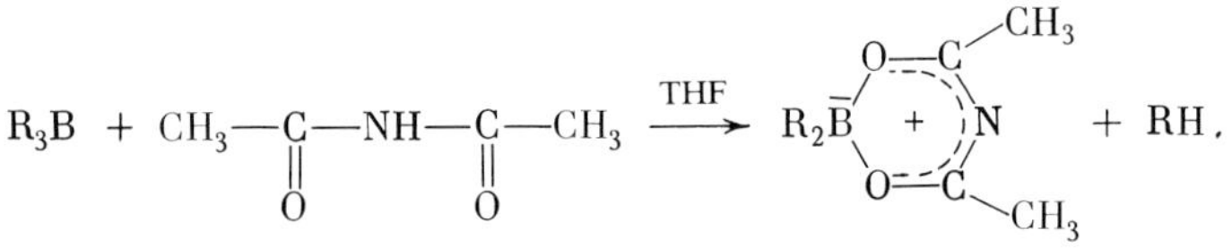

Amides react with R_3B to form acetaminodialkylboranes [25, 26]. Azodibenzoyl reacts with trialkylboranes to form a reduction product and a chelated diboron compound in quantitative yield [27]:

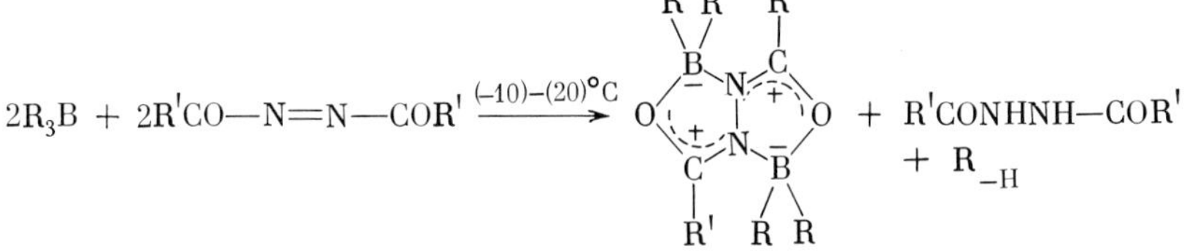

($R = C_2H_5$, n-C_4H_9; $R' = C_6H_5$, C_2H_5O).

In this reaction, it is not the carbonyl group that is reduced but the N=N double bond, just as in the case of the action of R_3B on *cis*-azobenzene [28, 29]. *trans*-Azomethane does not react with triethylborane in the gas phase even at 90°C [30].

In the action of phenacyl bromide on $(C_4H_9)_2BK(Na)$, 15% of acetophenone and 2% of caprophenone are formed (before hydrolysis) [31].

The salts $[(CH_3)_2N{=}C(CH_3)_2][B(C_6H_5)_4]$ are formed by the reaction of dimethylammonium tetraphenylborate with aldehydes or ketones [32, 33].

A simple method for the synthesis of olefins with a terminal and internal double bond has been developed on the basis of the reaction of *gem*-bimetallic compounds with aldehydes and ketones. In the general form, it is described by the following scheme [34–40]:

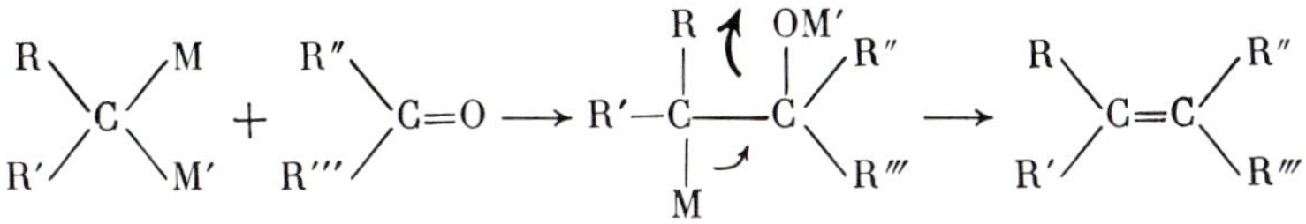

(M and M'=Li[36], Mg[34, 35], B[38, 39], Al[37, 39, 40]; R, R', R'' and R'''=H, alkyl, aryl).

The diboron compounds obtained by the hydroboration of terminal acetylenes form olefins by consecutive treatment with a lithium alkyl and a carbonyl compound:

$$n\text{-}C_4H_9C{\equiv}CH \xrightarrow{R_2BH} n\text{-}C_4H_9CH_2CH(BR_2)_2 \xrightarrow[2.\ (C_6H_5)_2CO]{1.\ C_4H_9Li} n\text{-}C_4H_9CH_2CH{=}C(C_6H_5)_2.$$

By this method, hexylidenecyclohexane has been obtained from 1-hexyne and cyclohexanone, cholestanone has been converted to 3-*n*-hexylidene-5α-cholestane, and tripentylethylene has been synthesized from dipentyl ketone [38, 39].

A method for the homologation of aldehydes and ketones (RR'CO→RR'CHCHO) using lithium bis(ethylenedioxyboryl)-methide has been developed [89, 90]. The reaction of the latter compound with benzoyl chloride and methyl benzoate, followed by hydrolysis, produces acetophenone [89].

The reactions of 1-lithium-1,1-bis(trimethylenedioxyboryl)hexane with RCHO and $RCOOCH_3$, followed by oxidation, produce ketones of the type $RCH_2COC_6H_{13}$ and $RCOC_6H_{13}$, respectively [91].

The α-carbanions generated from cyclic esters of 1-(phenylthio)alkylborinic acids [92] or 1,1-diphenylthiomethylborinic acid [93] react with aldehydes and ketones to form substituted vinyl sulfides [92] or thioacetals of the corresponding ketones [93].

9.2 REACTIONS WITH α, β-UNSATURATED ALDEHYDES AND KETONES

Trialkylboranes react with α,β-unsaturated aldehydes and ketones by 1,4-addition to form alkenyloxydialkylboranes (enol esters of dialkylborinic acids):

$$R_3B + CH_2{=}CHCOR' \longrightarrow RCH_2CH{=}C\begin{matrix} R' \\ OBR_2 \end{matrix}$$

Vinyloxydialkylboranes are stable in an inert atmosphere and can be distilled at reduced pressure [41–44]. In most studies, these compounds were not isolated in the pure state but were immediately subjected to further conversions.

The boron esters of enols, which show a high reactivity, are used for the preparation of aldehydes, ketones, esters of carboxylic acids, 1,3-keto alcohols, Mannich bases, conjugated ketones with several double bonds, and organic compounds of other classes. The reactivity of vinyloxyboranes is apparently higher than that of their silicon analogs, $>C{=}C{-}OSiR_3$, which are widely used in preparative organic chemistry.

9.2.1 Synthesis of aldehydes and ketones

Alkenyloxydialkylboranes are easily hydrolyzed to the corresponding aldehydes and ketones. In this way, the transition from α,β-unsaturated carbonyl compounds to higher saturated carbonyl compounds can be achieved with the use of organoboranes.

The reactions of acrolein [45], 2-methylacrolein [46], 2-bromoacrolein [46], and methyl vinyl ketone [47, 94, 95] with triorganoboranes, followed by hydrolysis of the intermediately formed alkenyloxyboranes, lead to the corresponding aldehydes or ketones:

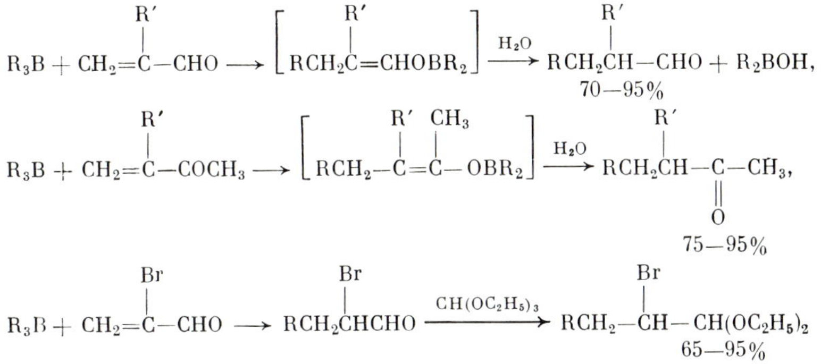

(R = alkyl, cyclo-C_5H_9, cyclo-C_6H_{11}, 2-norbornyl, R′ = H, CH_3).

Most trialkylboranes react with the above-mentioned unsaturated aldehydes and ketones at room temperature or with gentle heating (40°C). An exception is represented by triisobutylborane and its structural analogs, which add slowly to vinyl ketones.

The α-bromoaldehydes obtained from 2-bromoacrolein are generally not isolated, but are immediately converted to diethyl acetals by treatment of the reaction mixture wih orthoformic ester in the presence of methylsulfonic acid as a catalyst.

3-Deutero-2-hexanone has been obtained by the reaction of triethylborane with methyl vinyl ketone in the presence of heavy water [48].

Tri-*sec*-trialkylboranes react with unsaturated aldehydes and ketones significantly more rapidly than the primary compounds. Individual trialkylboranes should therefore be used for the preparation of the individual compounds, and the products of hydroboration of terminal olefins cannot be used. For example, the reaction of acrolein with tributylborane, obtained from 1-butene and diborane and containing 94% of *n*-butyl and 6% of *sec*-butyl groups, leads to a mixture of 85% of heptanal and 15% of 4-methylhexanal, with a total yield of 86–99% [45]:

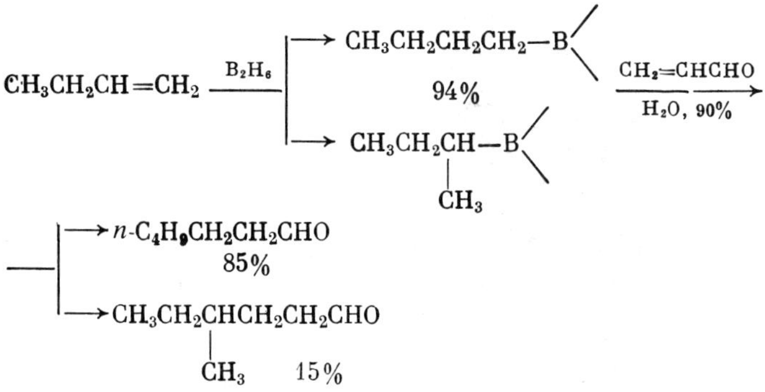

In the action of isopropenyl methyl ketone on tripentylborane, synthesized by the hydroboration of 1-pentene, 73% of 3-methyl-2-nonanone and 27% of 3,5-dimethyl-2-octanone are formed [49]. At the same time, the reaction of methyl vinyl ketone with trihexylborane, obtained by the hydroboration of 1-hexene with borane in tetrahydrofuran and preliminarily treated with a small quantity of anisole and dimethyl sulfoxide (at 170°C for two hours), led to a mixture of 2-decanone and 5-methyl-2-nonanone in a ratio of 98:2 [96].

Alkyl(diphenyl)boranes react with methyl vinyl ketone to form higher 2-alkanones [97]. Unlike trialkylboranes, triphenylborane reacts with 2-methylacrolein by 1,2-addition to the carbonyl group [95].

α, β-Unsaturated aldehydes and ketones containing substituents in the 3 position (crotonaldehyde, 3-penten-2-one, mesityl oxide, 1-acetylcyclohexene, 2-cyclopenten-1-one, 2-cyclohexen-1-one, etc.) add trialkyl- and tricycloalkylboranes only under UV irradiation [49, 50] or in the presence of oxygen, peroxide catalysts of the azobisisobutyronitrile type, diacyl peroxides [50, 51], or Fenton's reagent (H_2O_2 — $FeCl_2$) [52]:

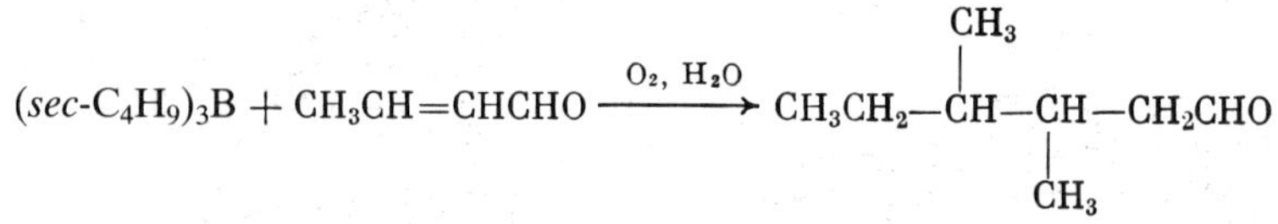

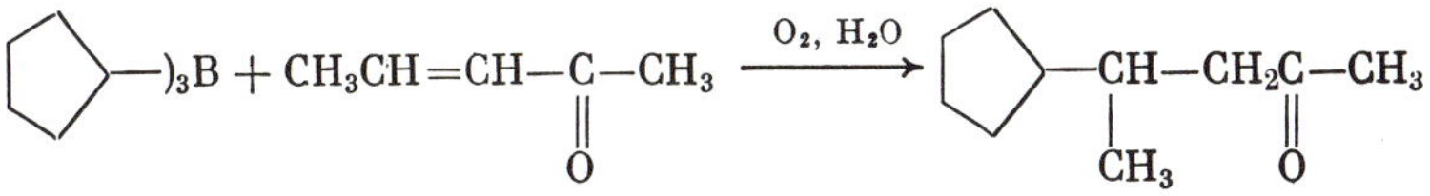

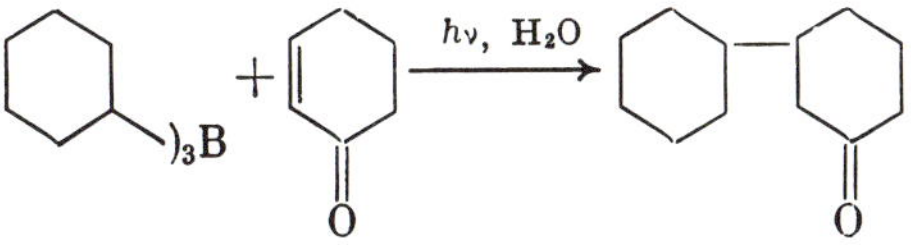

In the oxygen-induced 1,4-addition of R_3B to *d*-carvone and *l*-perillaldehyde, the alkyl group adds stereoselectively in a *trans* position with respect to the isopropenyl group [98].

The available data indicate that the 1,4-addition of organoboranes to α, β-unsaturated aldehydes and ketones proceeds by a chain radical mechanism [50]:

$$R_3B \xrightarrow[\text{or } h\nu]{O_2} R^{\cdot},$$

$$R^{\cdot} + CH_2{=}CH{-}CHO \longrightarrow RCH_2\dot{C}H{-}CHO \longrightarrow RCH_2CH{=}CH{-}O^{\bullet},$$

$$RCH_2CH{=}CH{-}O^{\cdot} + R_3B \longrightarrow R^{\cdot} + RCH_2CH{=}CHOBR_2$$

$$\xrightarrow{H_2O} RCH_2CH_2CHO + R_2BOH.$$

To initiate and complete the reaction of acrolein, methyl vinyl ketone, and other aldehydes and ketones with a terminal double bond, the traces of oxygen present in the starting substances and in the solvent are sufficient. In this case, the length of the kinetic chain is apparently considerably greater than in reactions involving α, β-unsaturated carbonyl compounds substituted in the 3 position. In the presence of galvinoxyl, which is an active inhibitor of radical processes [53, 54], no alkylation of conjugated aldehydes or ketones takes place.

In the transfer of the radical from boron to carbon in the 1,4-addition reactions, its configuration is not retained, which serves as additional evidence of the radical nature of the process. Thus, the reaction of *trans*-2-methylcyclohexylborane with methyl vinyl ketone leads to a mixture of the trans and cis isomers of 4-(2-methylcyclohexyl)butan-2-one in a ratio of 82:18 [55].

Previous data [56] on the retention of the configuration in the course of the 1,4-addition are erroneous.

Triisopropyl- and tributylboranes react at 20°C with diphenylcyclopropenone to form 2-phenyl-3-alkyl-1-indenone [57]:

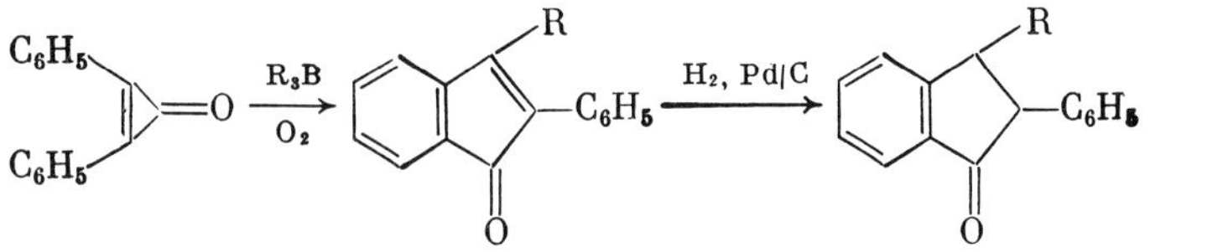

The reaction is catalyzed by oxygen, its radical nature being confirmed by the formation of a significant amount (11%) of 2-phenyl-3-(1-ethoxyethyl)-1-indenone when the process is carried out in the presence of ether.

Irradiation with a mercury lamp at −78°C converts *cis*-cycloocten-2-one to the trans isomer *5* with a strained double bond. The latter reacts readily with triethyl- or triisopropylborane to form, after hydrolysis, 3-ethyl- or 3-isopropylcyclooctanone (*6*) in yields of 70 and 83%, respectively [58, 59]:

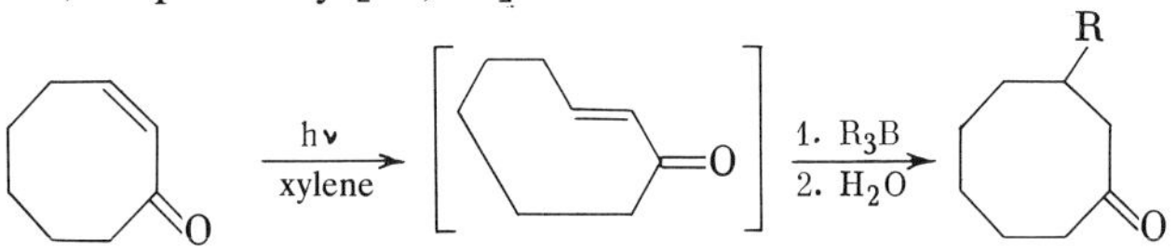

5 6

Alkylated ketones *6* are also obtained by the reaction of *cis*-2-cyclooctenone and trialkylboranes in the presence of oxygen.

The oxygen-catalyzed reaction of cycloocta-2,7-dien-1-one (*7*) with trialkylboranes leads to a mixture of 3-alkylcycloocten-7-one (*8*) and *trans*-3,7-dialkylcyclooctanone (*9*), which are formed as a result of mono- and diaddition:

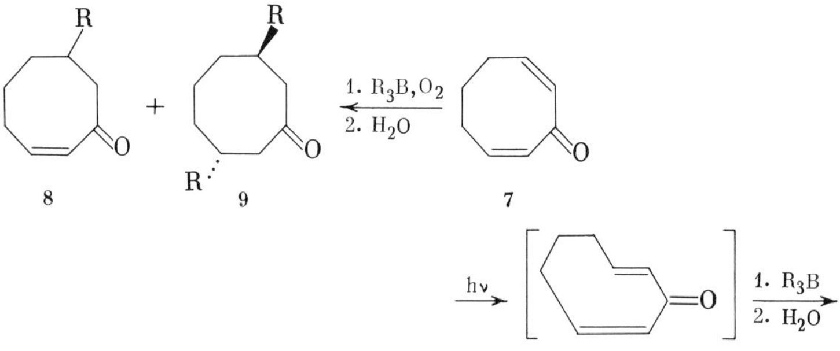

8 9 7

10 (8)

At the same time, the photoreaction of the ketone *7* with trialkylboranes forms only the monoaddition products *8*. The increased reactivity of the previously irradiated ketone *7* is explained by its isomerization to the compound *10* with an active trans double bond.

The action of a three-to-fivefold excess of triethyl-, tributyl-, or triisobutylborane on 16-dehydropregnenolone acetate (*11*) formed (after hydrolysis) the corresponding 16α-alkyl-5-pregnen-3β-ol-20-ones (*12*) in a yield of approximately 40% [60]:

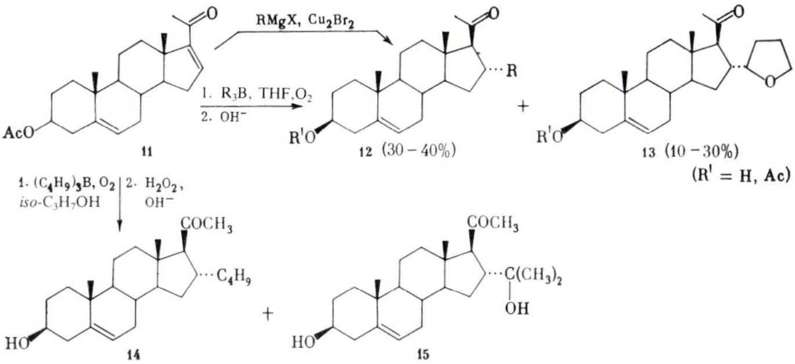

In all cases, along with the products of 1,4-addition of the trialkylboranes (*12*), 10–30% of 16α-tetrahydrofuryl-5-pregnen-3β-ol-20-one (*13*) is obtained, which is formed as a result of the addition of tetrahydrofuran to the starting steroid. Other solvents capable of forming radicals also add to conjugated ketones. For example, the oxygen-initiated reaction of 16-dehydropregnenelone acetate (*11*) with tributylborane in isopropanol leads to 40% of 16α-butyl-5-pregnen-3β-ol-20-one (*14*) and 28% of 16α-(2-hydroxy-2-propyl)-5-pregnen-3β-ol-20-one (*15*).

Tricycloalkylboranes (C_5, C_6, C_8, and C_{12}) and tributylborane alkylate crotonaldehyde in the presence of tertiary amines (triethylamine, pyridine; 2-, 3-, 4-picoline, 2,6-lutidine, and 2,2′-dipyridyl) in toluene at 20°C [61].

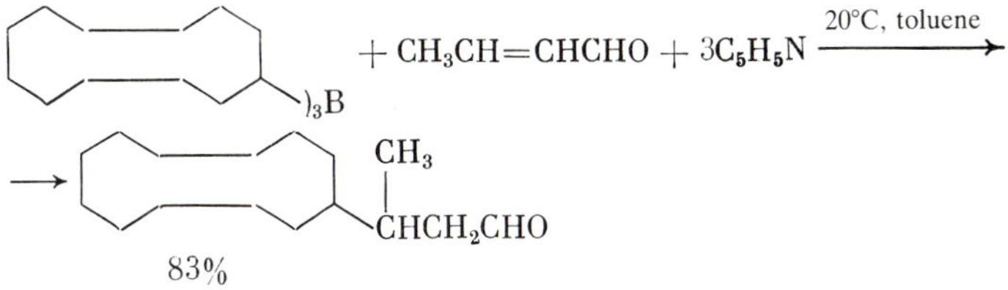

The Mannich base *16*, when treated with methyl iodide and potassium carbonate, generates 2-methylenecyclohexanone (*17*), which forms 2-alkylcyclohexanones with trialkylboranes [62]:

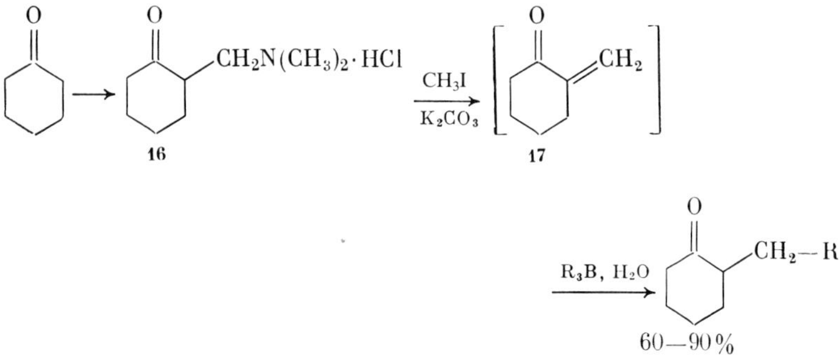

2-Methylenecyclopentanone and 3-methylene-2-norbornanone, obtained from the corresponding ketones via the Mannich base, are alkylated in an analogous manner.

One of the methods for the generation of methylenecyclohexane using organoboranes consists of the cleavage of spiro[10-hydroxy-2-methyldecahydroisoquinoline-4,1′-cyclohexan-2′-one] (a "bis-Mannich base", obtained from cyclohexanone, formaldehyde, and methyl iodide) [63]:

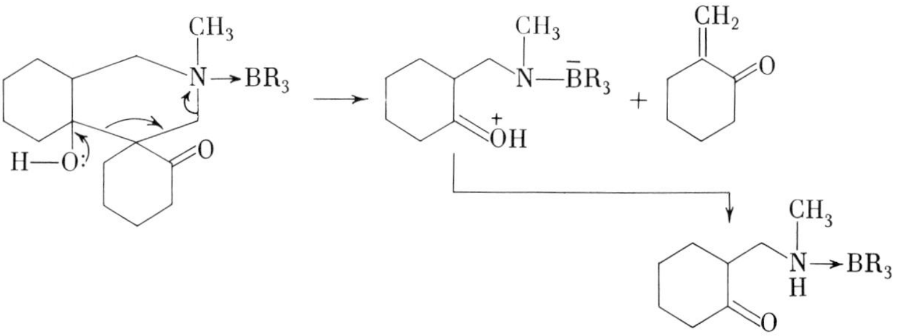

A fragmentation of this type is probably general in nature and can be used for the cleavage of aminopropanols, 4-hydroxytetrahydropyrroles, and 4-hydroxypiperidines. The alkylation of Mannich bases has been used for the preparation of compounds of the prostaglandin series [64]:

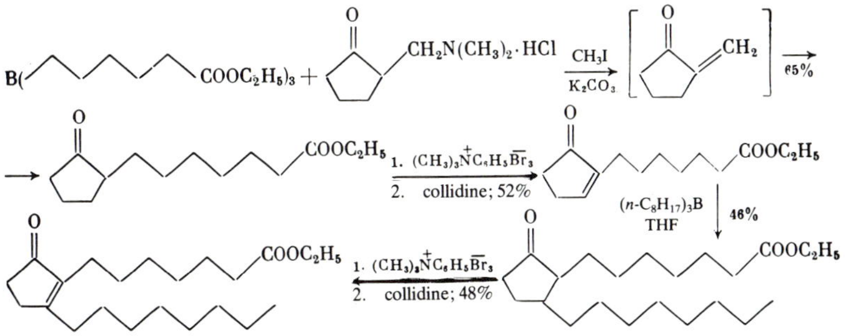

Esters of α,β-unsaturated acids polymerize under the action of trialkylboranes and oxygen [43, 67–70], but are alkylated by trialkylboranes during electrolysis in acetonitrile to form the esters $RCH_2CHR'COOC_2H_5$ [99].

Under the action of oxygen, a 1,6-addition of trialkylboranes to 1-acetyl-2-vinycyclopropane and 1-benzoyl-2-vinylcyclopropane takes place, leading, after hydrolysis to γ,δ-unsaturated ketones [71]:

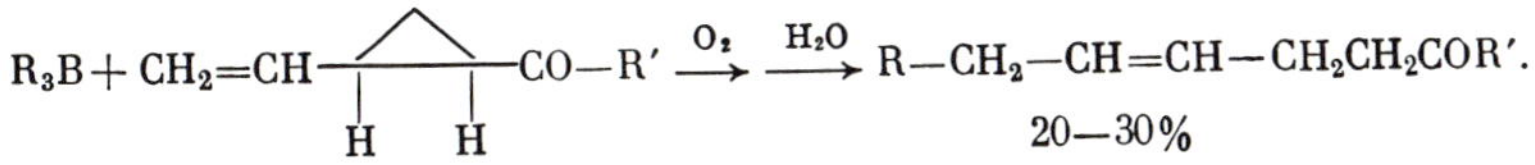

Acetylenic ketones, particularly methyl ethynyl ketone, also add trialkylboranes in the 1,4 position. This reaction proceeds only in the presence of oxygen or of radical initiators of the azobisisobutyronitrile type or under irradiation [72]:

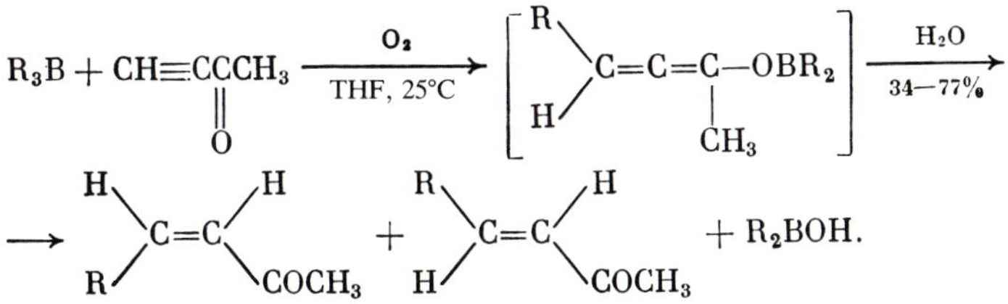

The ratio between cis and trans ketones is 1.1:1.

Acetylenic ketones are capable of adding two molecules of a trialkylborane and can be used for the preparation of branched ketones with identical or different radicals in the 3 position with respect to the carbonyl:

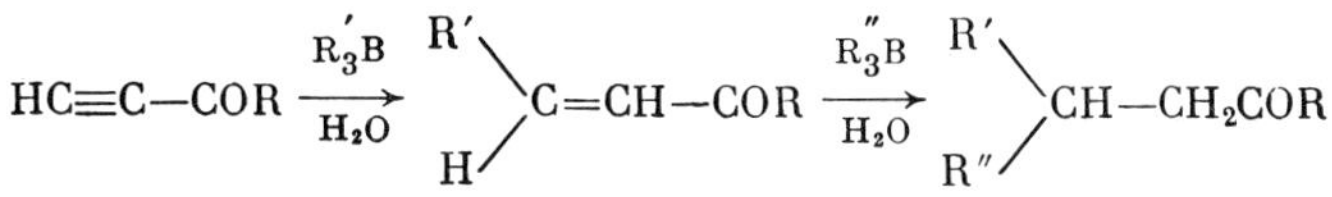

The reactions of unsaturated aldehydes and ketones with organoboranes have been considered in part in review articles [73, 74]; the reactions of boracyclanes with α, β-unsaturated compounds are described in Part II. The addition of organoboranes to unsaturated α-epoxides and other functional olefins is discussed in Chapters 13 and 15.

9.2.2 Synthesis of α-halo ketones, esters of α-halo acids, branched ketones, Mannich bases, β-hydroxy ketones, and esters of β-hydroxy acids

Considered below are some reactions of vinyloxyboranes obtained both by the 1,4-addition of organoboranes to α,β-unsaturated aldehydes and ketones and by other methods, for example, by the reaction of R_3B with ketones or aldehydes (see Section 1 of the present chapter), diazocarbonyl compounds or ylides (Chapter 11), 2-isopropyl-6-methyl-4-pyrimidiloxy(dialkyl)boranes [94, 100, 101]; by the reaction of ketones with dialkylboryltriflates, $R_2BOSO_2CF_3$ [102–105] or BCl_3 [106] in the presence of tertiary amines; by the addition of R_2BBr [107] or thioboranes $\gt$BSR to ketenes or α,β-unsaturated ketones (Part VI); by the reaction of $(R_2N)_2BBr$ with $(CH_3OCOCH_2)_2Hg$ [108], of 1,1-diboron compounds with $C_6H_5COOCH_3$[109], or of triflates and halides R_2BX with silyl esters of enols [110, 111]. Methods for the stereoselective synthesis of boron enolates have been developed. For example, *Z*- or *E*-$CH_3CH{=}CR'OBR_2$ are mainly obtained from ethyl ketones $R'COCH_2CH_3$ and $R_2BOSO_2CF_3$, depending on the type of R and R′, the nature of the base, and the reaction temperature [112]. A *trans*-enolate is formed from *tert*-butyl thiopropionate.

(*E*)-$C_4H_9CH{=}CHROB(C_4H_9)_2$ is converted to the *Z*-isomer by the action of catalytic quantities of lithium phenoxide or *tert*-butoxide, pyridine, and (in some cases) water [113]. The vinyloxyboranes $(CH_3)_2C{=}C(Br)OBR_2$, obtained by the addition of R_2BBr to dimethyl ketene, decompose to the components upon heating [107].

A method for the preparation of enolates $R'CH{=}C(OBR_2)_2$, based on the reaction of two equivalents of $R_2BOSO_2CF_3$ with propionic or phenoxyacetic acid in the presence of diisopropylethylamine, has been developed [112]. Most reactions of boron esters of enols proceed with an allylic rearrangement.

The boron esters of enols are halogenated at 0°C with bromine in the dark with the use of CCl_4 as a solvent [43] or with *N*-chloro- or *N*-bromosuccinimide in tetrahydrofuran [75], as a result of which α-halo ketones (*18*) or esters of α-halo acids (*18*, R′ = OR) are obtained:

$$RCH{=}C(R')(OBR_2) \xrightarrow{X_2} RCH(X){-}COR' \quad \textbf{18}\ (56{-}98\%)$$

$$(R'{=}CH_3, iso\text{-}C_3H_7, C_6H_5, OR;\ X{=}Cl, Br)$$

The bromination of vinyloxyboranes, obtained from aldehydes, followed by methanolysis, leads to α-bromoaldehydes [101].

Bromine initially adds to the double bond to form the corresponding dibromide, the hydrolysis or alcoholysis of which forms an α-bromocarbonyl compound.

This method of halogenation is of particularly great importance in the chemistry of the α-halo ketones, since it can be used to prepare compounds with a fixed position of the halogen (in no case was the formation of isomers observed).

The examples presented below partially demonstrate the scope of application of the reaction:

$$(n\text{-}C_4H_9)_3B + CH_2{=}CH{-}COCH_3 \longrightarrow [n\text{-}C_4H_9CH_2CH{=}C(CH_3)OB(C_4H_9)_2] \xrightarrow{\text{bromination}} n\text{-}C_4H_9CH_2CH(Br){-}COCH_3 \quad 85\%$$

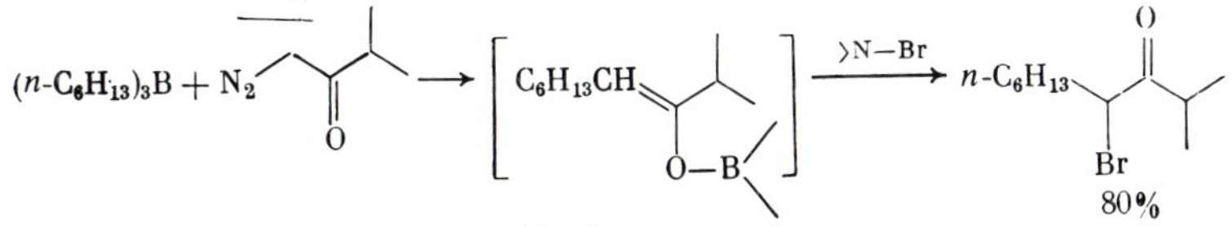

$$(n\text{-}C_4H_9)_3B + N_2CHCOOC_2H_5 \longrightarrow \xrightarrow{\rangle N{-}Br} n\text{-}C_4H_9CH(Br){-}COOC_2H_5 \quad 76\%$$

At the same time, the bromination of ketones always leads to a mixture of isomers with a predominant formation of the ketone with halogen in the most substituted α-position [75–78].

The reaction of alkenyloxydialkylboranes with C_6H_5SeCl leads to α-phenyl selenoketones of the type $C_6H_5CH_2COCH(SeC_6H_5)C_2H_5$ which, upon oxidation (H_2O_2 in pyridine), form substituted α,β-unsaturated acids [114]. A combination of these reactions represents a convenient method for replacement of the β-hydrogen in vinyl ketones by an organic group or a method for the introduction of an alkylidene group into diazo ketones.

Alkenyloxydialkylboranes react with two moles of a lithium alkyl to form lithium enolates (*19*); the latter are easily alkylated in the β position by treatment with methyl iodide or benzyl chloride [44]:

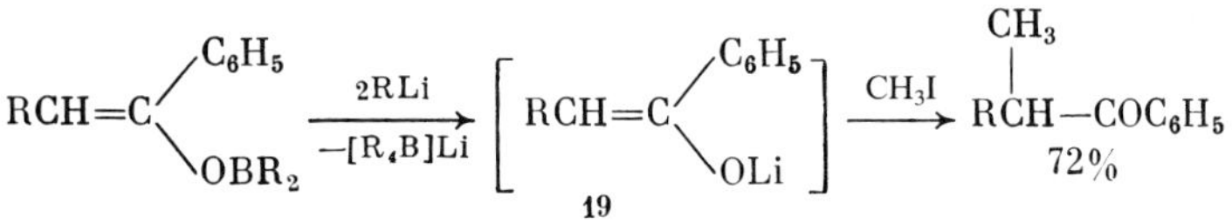

Reactions of this type represent a method for the double α,α-alkylation of diazocarbonyl compounds and the addition of two alkyl groups in the α and β positions of conjugated aldehydes and ketones:

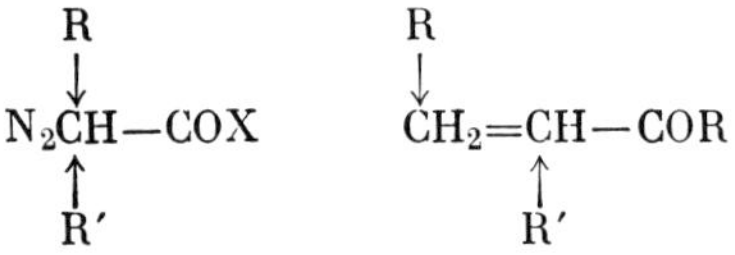

(R from R_3B; R′ from R′I)

As a result of the consecutive treatment of methyl vinyl ketone with tributylborane, two moles of butyllithium, and methyl iodide, 3-methyl-2-octanone [44] with an impurity of 11% of 3-nonanone [79] is formed:

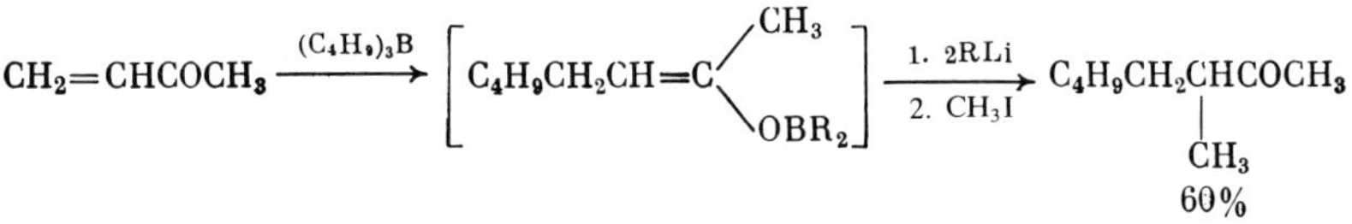

The method can also be used for the synthesis of α,α-dialkylated cyclic ketones. For example, 2-butyl-2-methylcyclohexanone is obtained from diazocyclohexanone [44]:

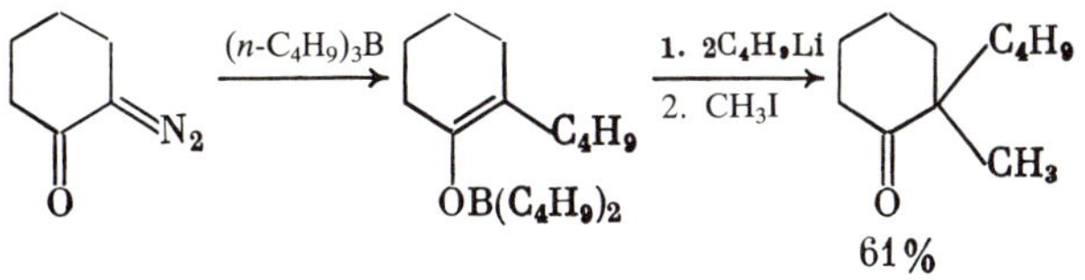

The methods for the preparation of substituted ketones via organoboranes considered above are characterized by a high selectivity and, in a number of cases, show certain advantages over the base-catalyzed alkylation of ketones. As is known [78], ketones with hydrogen atoms in both α positions are alkylated in both of these positions with the formation of a mixture of isomers.

The alkylation, allylation, and benzylation of alkenyloxydialkylboranes proceeds readily and regiospecifically with the use, instead of RLi, of lithium dimethylaminoethoxide, $(CH_3)_2NCH_2CH_2OLi$ [114a].

The alkylation of borates of the potassium enoxytrialkylborate type proceeds more smoothly and more selectively than that of the corresponding potassium enolates [114b]. Thus, the reaction with propargyl bromide leads only to an acetylenic product, without an allenic product as an impurity.

Enol esters react with dimethyl(methylene)ammonium iodide ("Mannich reagent") [80] in dimethyl sulfoxide to form Mannich bases [81]:

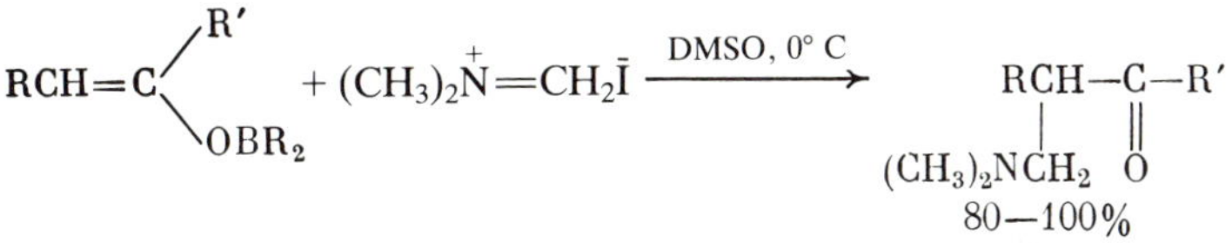

By this method, using diazomethyl isopropyl ketone as a starting material for the generation of the corresponding vinyloxyborane, 4-dimethylaminomethyl-2-methyl-3-hexanone (*20*) has been obtained:

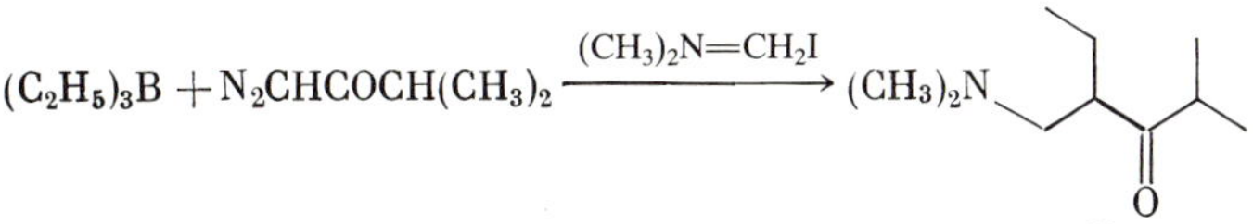

20

An isomer of the compound *20*, the ketone *21*, has been synthesized from the product of radical addition of triethylborane to

3-methyl-3-buten-2-one:

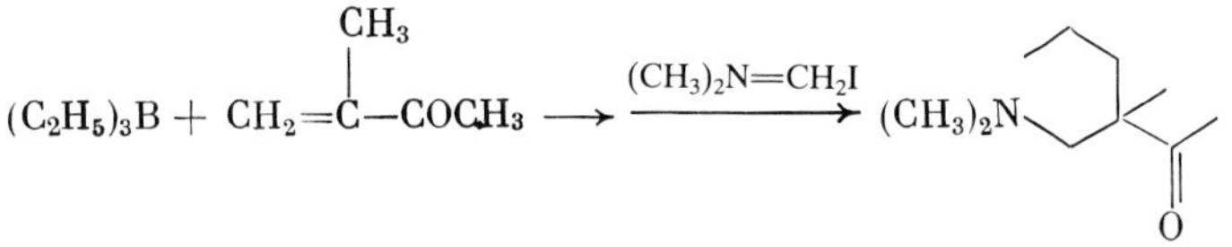

21 (87%)

The reactions of vinyloxyboranes with carbonyl compounds have been broadly investigated. The first reactions of this type were carried out by Köster and Fenzl [11], but the authors initially took the adducts formed to be the dimerization products of vinyl ethers. The reactions of vinyloxyboranes with carbonyl compounds actually proceed in the manner of an aldol condensation and, after hydrolysis, lead to β-keto alcohols [82], for example:

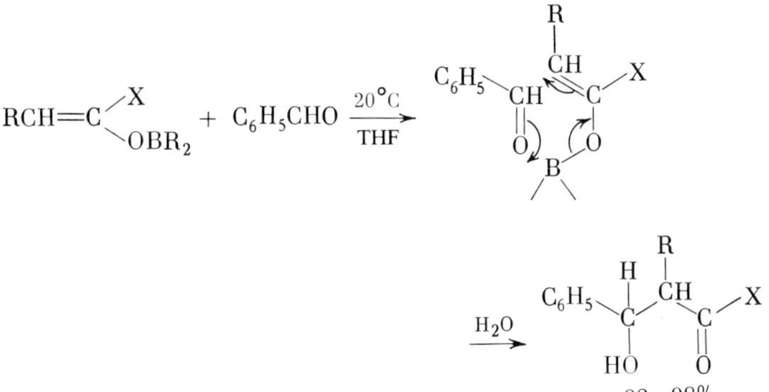

92–98%

($R = H, C_4H_9, C_5H_{11}, C_6H_5SCH_2$; $X = CH_3, C_6H_5, OC_2H_5, SC_6H_5$).

Reactions of this type proceed regio- [102, 103, 111] and stereospecifically [105, 111, 112, 115, 116] and are apparently pericyclic reactions [112]. *erythro*-Ketols are obtained almost exclusively from *Z*-vinyloxyboranes, while *threo*-ketols [104, 112, 113, 117, 118] are mainly obtained from the *E*-isomers. The reactivity of the *E*-isomers is 2-4 times greater than the activity of the corresponding *Z*-isomers [112].

The stereochemistry of the products of the condensation of cyclopentenyl- and cyclohexenyloxyboranes with aldehydes [106, 116] are determined by the nature of the radicals on the boron and of the solvent [105, 110, 112].

A stereoselective synthesis of thioesters of β-hydroxy-α-methylcarboxylic acids from *Z*- and *E*-$CH_3CH{=}C(SC_4H_9\text{-}tert)OBR_2$ has been carried out [117].

The reaction of $CH_3CH{=}C(SC_6H_5)OBR_2$ with OHCCH-$(CH_3)CH_2CH(CH_3)COOCH_3$ is the basis for the elegant synthesis of Prelog-Djerassi lactamic acid [104].

An enantioselective aldol condensation of chiral boron enolates obtained from $R_2BOSO_2CF_3$ and chiral *N*-propionyloxazolidones [115], (*S*)-(−)-1-[1-(4-toluenesulfonyl)-1-azacyclopentan-2-yl]-1-propanone [112, 119], or 1-cyclohexyl-1-(*tert*-butyldimethylsilyloxy)-butan-2-ones [120] has been carried out. The latter enolates have been used in the total synthesis of 6-deoxyerythronolide B [121]. When chiral boron enolates are used, the asymmetric induction is higher than that in the case of the lithium analogs [112]. The mechanism of asymmetric induction has been considered in [112]. A method for the synthesis of optically active β-hydroxy acids based on the reactions of chiral boron enolates with achiral aldehydes has been developed [112, 115, 119].

The dienolate $C_6H_5CH_2OCH{=}C(OBR_2)_2$ condenses with benzaldehyde to form only *threo*-3-hydroxy-3-phenyl-2-phenylmethoxypropionic acid, whose subsequent fragmentation under the action of dimethylformamide dimethylacetal led to *Z*-2-phenylmethoxy-1-phenylethylene [112].

According to the data of [83, 84, 122], the enol ester *22*, formed by the reaction of ketene with dimethylamino(dimethyl)borane, reacts further with ketene to form the chelated compound *23*:

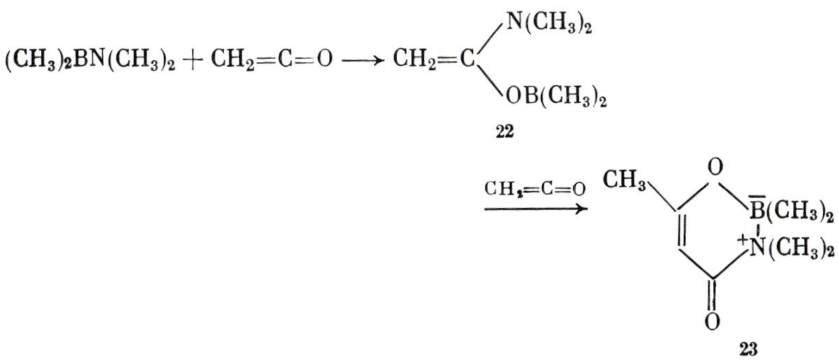

The ester $(CH_3)_2C{=}CBrOB(CH_3)_2$ condenses with aldehydes RCHO to form, after hydrolysis, the β-hydroxy acids $RCH(OH)C(CH_3)_2COOH$ [107].

Köster and Fenzl [11] synthesized boron chelates of the type *24*, which are resistant to hydrolysis, by heating alkenyloxydiethylboranes with nitriles:

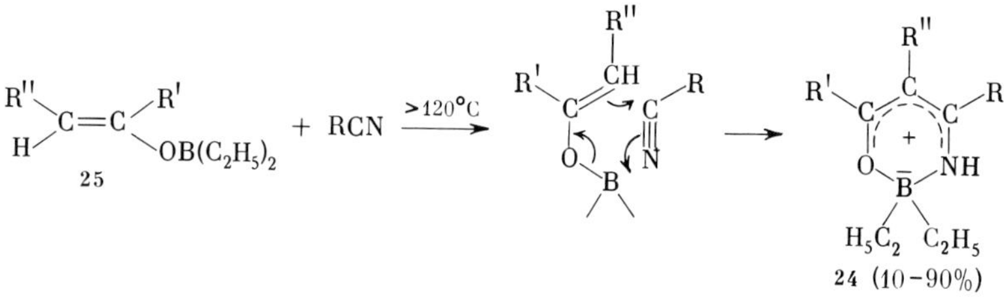

Chelated boron compounds of the type *24*, formed by reaction of the esters *25* (R′ = SC_2H_5) with nitriles, form esters of β-iminothio acids upon hydrolysis (Part VI).

The action of the vinyloxyborane $CH_2{=}C(SC_4H_9\text{-}tert)OB(C_4H_9)_2$ on Schiff bases produces derivatives of β-amino acids [123]. This method has been used in the synthesis of a fragment of the antitumor antibiotic bleomycin.

The *N*-analogs of alkenyloxyboranes — substituted vinylamino(dichloro)- [124] and vinylamino(dialkyl)boranes [125] — react with aldehydes and ketones in the manner of an aldol condensation to form boron chelates [125] or (after hydrolysis) β-keto alcohols [124]. Chiral boron azaenolates (vinylaminoboranes) have been used in an aldol-type enantioselective condensation [126, 127].

The reaction of acetonitrile with benzaldehyde and R_2NBCl_2 in the presence of triethylamine leads mainly to 3-phenyl-3-*sec*-aminopropionitrile, $C_6H_5CH(NR_2)CH_2CN$. It is assumed that dialkylamino(chloro)keteniminoborane, $CH_2{=}C{=}NBClNR_2$, is an intermediate in this reaction [128].

9.3 ALKYLATION OF QUINONES

Hawthorne and Reintjes [85, 86] have shown that trialkylboranes react with *p*-quinone in ether solution with a large evolution of heat, forming, after hydrolysis, 2-alkylhydroquinones in high yield:

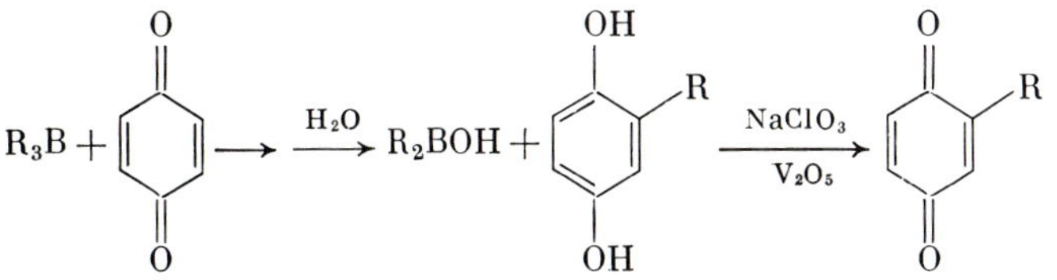

(R = n-C_4H_9, iso-C_4H_9, sec-C_4H_9, n-C_6H_{13}, 3-methyl-2-butyl, cyclohexyl, cycloheptyl, $C_6H_5CH_2$)

Triphenylborane and vinylboranes do not react with benzoquinone.

Since iodine and galvinoxyl inhibit the react of trialkylboranes with benzoquinone, it is suggested [87] that the reaction is catalyzed by oxygen and proceeds by a chain radical mechanism via 1,4-addition:

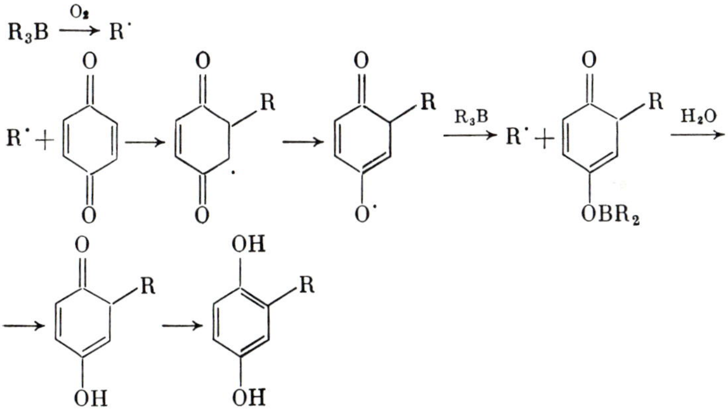

2,5-Dihydroxy-3-n-tridecyl-1,4-benzoquinone (rapanone) has been obtained by the reaction of tri-n-tridecylborane with 2,5-diacetoxy-1,4-benzoquinone [129].

The alkylation of 1,4-naphthoquinone forms 1,4-dihydroxy-2-alkylnaphthalenes (*26*), the oxidation of which leads to 2-alkyl-1,4-naphthoquinones, which show a vitamin K activity [87, 88];

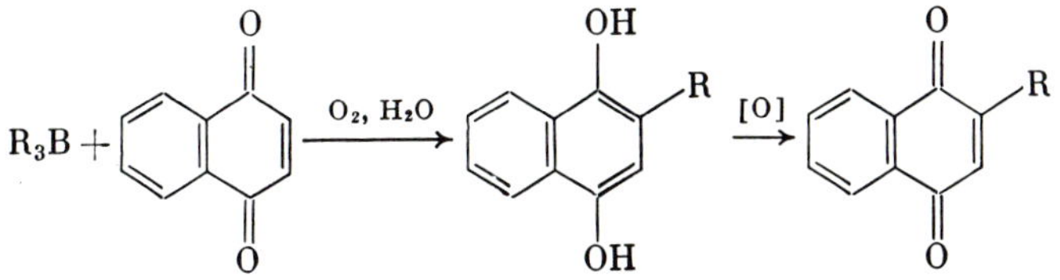

26 (70—95%)

(R = C_2H_5, C_3H_7 C_4H_9, $C_6H_5CH_2$, cyclo-C_5H_9).

According to the data of [87], 1,4-naphthoquinone, recrystallized twice in an inert atmosphere, reacts with organoboranes only when oxygen is slowly passed through the reaction mass.

2-Benzyl compounds of the type *26* are also obtained by the action of dibenzylborinic acid on benzoquinone and 1,4-naphthoquinone [88].

REFERENCES

1. H. Meerwein, G. Hintz, H. Majert and H. Sönke, *J. Prakt Chem.* **147,** 226 (1936).
2. B. M. Mikhailov, V. G. Kiselev and Yu. N. Bubnov, *Izv. Akad. Nauk SSSR, Ser. Khim.* **1965,** 898.
3. B. M. Mikhailov, Yu. N. Bubnov and V. G. Kiselev, *Zh. Obshch. Khim.* **35,** 62 (1966).
4. T. Okushi, O. Manabe, H. Hiyama and Z. Yoshida, *J. Chem. Soc. Japan* **72,** 1702 (1969).
5. T. Okushi, O. Manabe, H. Hiyama and Z. Yoshida, *J. Chem. Soc. Japan* **72,** 1707 (1969).
6. E. A. Timofeeva, G. A. Lutsenko and N. I. Shuikin, *Izv. Akad. Nauk SSSR, Ser. Khim.* **1967,** 1152.
7. A. Arase, Y. Masuda, M. Itoh and A. Suzuki, *Nippon Kagaku Kaishi* **1972,** 395; *C.A.*, **76,** 113285 (1972).
8. N. Miyaura, M. Itoh, A. Suzuki, H. C. Brown, M. Midland and P. Jacob, *J. Am. Chem. Soc.* **94,** 6549 (1972).
9. W. Kliegel, *Organometal. Chem. Rev. A* **8,** 162 (1972).
10. W. Kliegel, *Angew. Chem.* **80,** 614 (1968).
11. R. Köster and W. Fenzl, *Angew. Chem* **80,** 756 (1968).
12. W. Fenzl and R. Köster, *Angew. Chem.* **83,** 807 (1971).

12a. W. Fenzl and R. Köster, *Ann.* **1975,** 1322.

13. R. Köster and A.-A. Pourzal, *Synthesis* **1973,** 674.
14. A. G. Davies, B. P. Roberts and J. C. Scaiano, *Chem. Comm.* **1971,** 196.
15. A. G. Davies, B. P. Roberts and J. C. Scaiano, *J. Chem. Soc. B* **1971,** 2171.
16. A. G. Davies and J. C. Scaiano, *J. Chem. Soc. Perkin II* **1972,** 2234.
17. B. M. Mikhailov and Yu. N. Bubnov, *Izv. Akad. Nauk SSSR, Otd. Khim. Nauk* **1960** 1883.
18. W. Gerrard, M. F. Lappert and R. Shafferman, *Chem. and Ind.* **1958,** 722.
19. M. F. Hawthorne and M. Reintjes, *J. Org. Chem.* **30,** 3851 (1965).
20. V. A. Dorokhov, L. I. Lavrinovich, M. N. Bochkareva, V. S. Bogdanov and B. M. Mikhailov, *Zh. Obshch. Khim* **43,** 1115 (1973).
21. L. H. Toporcer, R. E. Dessy and S. J. Green, *Inorg. Chem.* **4,** 1649 (1965).
22. R. Köster and G. W. Rotermund, *Ann.* **689,** 40 (1965).
23. K. Utimoto, T. Tanaka and H. Nozaki, *Tetrahedron Letters* **1972,** 1167.
24. V. A. Dorokhov, L. I. Lavrinovich and B. M. Mikhailov, *Dokl. Akad. Nauk SSSR* **195,** 1100 (1970).
25. B. M. Mikhailov and V. A. Dorokhov, *Izv. Akad. Nauk SSSR, Ser. Khim.* **1971,** 201.

26. V. A. Dorokhov, L. I. Lavrinovich, I. P. Yakovlev and B. M. Mikhailov, *Zh. Obshch, Khim.* **41,** 2501 (1971).
27. A. Haag and H. Baudisch, *Tetrahedron Letters* **1973,** 401.
28. A. G. Davies, B. P. Roberts and J. C. Scaiano, *J. Chem. Soc. Perkin II* **1972,** 803.
29. A. G. Davies, K. G. Foot, B. P. Roberts and J. C. Scaiano, *J. Organometal. Chem.* **31,** C1 (1971).
30. J. Grotewold, E. A. Lissi and J. C. Scaiano, *J. Chem. Soc. B* **1971,** 1187.
31. D. Pasto and P. W. Wojtkowski, *J. Organometal. Chem.* **34,** 251 (1972).
32. G. Wittig and W. Tochterman, *Chem. Ber.* **94,** 1692 (1961).
33. N. Wiberg and K. H. Schmid, *Angew. Chem.* **76,** 381 (1964).
34. G. Cainelli, F. Bertini, P. Grasselli and G. Zubiani, *Tetrahedron Letters* **1967,** 5153.
35. F. Bertini, P. Grasselli, G. Zubiani and G. Cainelli, *Tetrahedron* **26,** 1281 (1970).
36. G. Cainelli, A. Ronchi, F. Bertini, P. Grasselli and G. Zubiani, *Tetrahedron* **27,** 6109 (1971).
37. G. Cainelli, F. Bertini, P. Grasselli and G. Zubiani *Tetrahedron Letters* **1967,** 1581.
38. G. Cainelli, G. Dal Bello and G. Zubiani, *Tetrahedron Letters* **1966,** 4315.
39. G. Cainelli, *Corsi Semin. Chem.* **1968,** 16; *C. A.* **72,** 31880c (1970).
40. A. Bongini, D. Savoia and A. Umani-Ronchi, *J. Organometal Chem.* **72,** C4 (1974).
41. B. M. Mikhailov, G. S. Ter-Sarkisyan and N. A. Nikolaeva, *Izv. Akad. Nauk SSSR, Ser. Khim.* **1968,** 1655.
42. D. Pasto and P. Wojtkowski, *Tetrahedron Letters* **1970,** 215.
43. N. Miyaura, M. Harada, M. Itoh and A. Suzuki, *Chem. Letters* **1973,** 1145.
44. D. Pasto and P. Wojtkowski, *J. Org.Chem.* **36,** 1790 (1971).
45. H. C. Brown, M. Rogič, M. Rathke and G. Kabalka, *J. Am. Chem. Soc.* **89,** 5709 (1967).
46. H. C. Brown, G. Kabalka, M. Rathke and M. Rogic, *J. Am. Chem. Soc.* **90,** 4165 (1968).
47. A. Suzuki, A. Arase, A. Matsumoto, M. Itoh, H. C. Brown, M. Rogic and M. Rathke, *J. Am. Chem. Soc.* **89,** 5708 (1967).
48. J. Hooz and D. Gunn, *J. Am. Chem. Soc.* **91,** 6195 (1969).
49. G. Kabalka, H. C. Brown, A. Suzuki, S. Honma, A. Arase and M. Itoh, *J. Am. Chem. Soc.* **92,** 710 (1970)
50. H. C. Brown and G. Kabalka, *J. Am. Chem. Soc.* **92,** 712 (1970).
51. H. C. Brown and G. Kabalka, *J. Am. Chem. Soc.* **92,** 714 (1970).
52. A. Arase, Y. Masuda and A. Suzuki, *Bull. Chem. Soc. Japan* **47,** 2511 (1974).
53. P. G. Allies and P. B. Brindley, *Chem. and Ind.* **1968,** 19.
54. P. D. Bartlett and T. Funahashi, *J. Am. Chem. Soc.* **84,** 2596 (1962).
55. E. Negishi and H. C. Brown, *J. Am. Chem. Soc.* **95,** 6757 (1973).
56. H. C. Brown, M. M. Rogič, M. W. Rathke, and G. Kabalka, *J. Am. Chem. Soc.* **91,** 2150 (1969).
57. N. Naruse, K. Utimoto and H. Nozaki, *Tetrahedron Letters* **1972,** 1605.
58. N. Miyamoto, S. Isiyama, K. Utimoto and H. Nozaki, *Tetrahedron* **29,** 2365 (1973).
59. N. Miyamoto, S. Isiyama, K. Utimoto and H. Nozaki, *Tetrahedron Letters* **1971,** 4597.
60. A. A. Akhrem, I. S. Levina, Yu. A. Titov, V. A. Khripach, Yu. N. Bubnov and B. M. Mikhailov, *Zh. Obshch. Khim.* **43,** 2565 (1973).

61. K. Utimoto, T. Tanaka, T. Furubayashi and H. Nozaki, *Tetrahedron Letters* **1973,** 787.
62. H. C. Brown, M. W. Rathke, G. W. Kabalka and M. M. Rogič, *J. Am. Chem. Soc.* **90,** 4166 (1968).
63. H. J. Roth and Ch. Schwenke, *Arch. Pharm.* **298,** 648 (1965).
64. O. Attanasi, G. Baccolini, L. Caglioti and G. Rosini, *Gazz. Chim. Ital.* **103,** 31 (1973).
65. U. S. Johnson, D. J. Boss and K. L. Williamson, *Tetrahedron* **19,** 861 (1963).
66. A. Marquet and J. Jaques, *Bull. Soc. Chim. France* **1962,** 90.
67. G. K. Kolesnikov and L. S. Fedorova, *Vysokomol. Soedin.* **1,** 1266 (1959).
68. G. K. Kolesnikov and N. V. Klimentova, *Izv. Akad. Nauk SSSR, Otd. Khim. Nauk* **1957,** 652.
69. F. Welch, *J. Polymer Sci.* **61,** 243 (1962).
70. G. K. Kolesnikov, A. P. Suprun and T. A. Solvoleva, *Vysokomol. Soedin.* **1,** 627 (1959).
71. N. Miyaura, M. Itoh, N. Sasaki and A. Suzuki, *Synthesis,* **1975,** 317.
72. A. Suzuki, S. Nozawa, M. Itoh, H. C. Brown, G. Kabalka and G. Holland, *J. Am. Chem. Soc.* **92,** 3503 (1970).
73. H. C. Brown and M. Midland, *Angew. Chem.* **84,** 702 (1972).
74. G. Kabalka, *Intra-Science Chem. Rep.* **7,** 57 (1973).
75. J. Hooz and J. N. Bridson, *Can. J. Chem* **50,** 2387 (1972).
76. J. Catch, D. Hey, E. Jones and W. Wilson, *J. Chem. Soc.* **1948,** 276.
77. C. Rappe, L. Knutson, N. Turro and P. Gagosian, *J. Am. Chem. Soc.* **92,** 2032 (1970).
78. H. O. House: "Modern Synthetic Reactions," New York, W. A. Benjamin Inc. (1972).
79. L. E. Hightower, L. R. Glasgow, K. Stone, D. Albertson and H. Smith, *J. Org. Chem.* **35,** 1881 (1970).
80. J. Schreiber, H. Maag, N. Hashimoto and A. Eschenmoser, *Angew. Chem.* **83,** 355 (1971).
81. J. Hooz and J. Bridson, *J. Am. Chem. Soc.* **95,** 602 (1973).
82. T. Mukaiyama, K. Inamata, M. Muraki; *J. Am. Chem. Soc.* **95,** 967 (1973).
83. S. Kosma: Dissertation, Technische Hochschule Aachen (1971); cited in [84], p. 638.
84. P. I. Paetzold and H. Grundke, *Synthesis,* 1973, 635.
85. M. F. Hawthorne and M. Reintjes, *J. Am. Chem. Soc.* **86,** 951 (1964).
86. M. F. Hawthorne and M. Reintjes, *J. Am. Chem. Soc.* **87,** 4585 (1965).
87. G. Kabalka, *J. Organometal. Chem.* **33,** C25 (1971).
88. B. M. Mikhailov, G. S. Ter-Sarkisyan and N. A. Nikolaeva, *Zh. Obshch. Khim.* **41,** 1721 (1971).
89. D. S. Matteson and R. J. Moody, *J. Org. Chem.* **45,** 1091 (1980).
90. D. S. Matteson, *Synthesis* **1975,** 147
91. D. S. Matteson and R. J. Moody, *J. Am. Chem. Soc.* **99,** 3196 (1977).
92. D. S. Matteson and K. Arne, *J. Am. Chem. Soc.* **100,** 1325 (1978).
93. A. Mendoza and D. S. Matteson, *J. Org. Chem.* **44,** 1352 (1979).
94. W. Fenzl, R. Köster and H.-J. Zimmermann, *Ann* **1975,** 2201.
95. R. Köster, H.-J. Zimmermann and W. Fenzl, *Ann.* **1976,** 1116.
96. Y. Masuda, M. Hoshi and A. Arase, *Bull. Chem. Soc. Japan* **52,** 271 (1979).
97. P. Jacob, *J. Organometal. Chem.* **156,** 101 (1978).
98. A. Arase, Y. Masuda and A. Suzuki, *Bull. Chem. Soc. Japan* **49,** 2275 (1976).
99. Y. Takahashi, K. Yuasa, M. Tokuda, M. Itoh and A. Suzuki, *Bull. Chem. Soc. Japan* **51,** 339 (1978).

100. R. Köster, *Pure Appl. Chem.* **49,** 765 (1977).
101. W. Fenzl, H. Kosfeld and R. Köster, *Ann.* **1976,** 1370.
102. T. Mukaiyama and T. Inoue, *Chem. Letters* **1976,** 559.
103. T. Inoue, T. Uchimaru and T. Mukaiyama, *Chem. Letters* **1977,** 153.
104. M. Hirama, D. S. Garvey, L. D.-L. Lu and S. Masamune, *Tetrahedron Letters* **1979,** 3937.
105. D. A. Evans, E. Vogel and J. V. Nelson, *J. Am. Chem. Soc.* **101,** 6120 (1979).
106. T. Sugasawa, T. Toyoda and K. Sasakura, *Synth. Commun.* **9,** 583 (1979).
107. P. Paetzold and M. Lasch, *Chem. Ber.* **112,** 663 (1979).
108. P. Paetzold and H.-P. Biermann, *Chem. Ber.* **110,** 3678 (1977).
109. T. Mukaiyama, M. Murakami, T. Oriyama and M. Yamagushi, *Chem. Letters,* **1981,** 1193.
110. M. Wada. *Chem. Letters* **1981,** 153.
111. I. Kuwayima, M. Kato and A. Mori, *Tetrahedron Letters* **1980** 4291.
112. D. A. Evans, J. V. Nelson, E. Vogel and T. R. Taber, *J. Am. Chem. Soc.* **103,** 3099 (1981).
113. S. Masamune, S. Mori, D. Van Horn and D. W. Brooks, *Tetrahedron Letters* **1979,** 1665.
114. J. Hooz and J. Oudenes, *Synth. Commun.* **10,** 667 (1980).
114a. J. Hooz and J. Oudenes, *Synth. Commun.* **10,** 139 (1980).
114b. E. Negishi, M. J. Idacavage, F. Dipasquale and A. Silveira, jr. *Tetrahedron Letters* **1979,** 845.
115. D. A. Evans, J. Bartroli and T. L. Shih, *J. Am. Ghem. Soc.* **103,** 2127 (1981).
116. T. Inoue and T. Mukaiyama, *Bull. Chem. Soc. Japan* **53,** 174 (1980).
117. M. Hirama and S. Masamune, *Tetrahedron Letters* **1979,** 2225.
118. D. E. Van Horn and S. Masamune, *Tetrahedron Letters* **1979,** 2229.
119. D. A. Evans and T. R. Taber, *Tetrahedron Letters* **1980,** 4675.
120. S. Masamune, W. Choy, F. A. J. Kerdesky and B. Imperiali, *J. Am. Chem. Soc.* **103,**1566 (1981).
121. S. Masamune, M. Hirama, S. Mori, S. A. Ali and D. S. Garvey, *J. Am. Chem. Soc.* **103,** 1568 (1981).
122. P. Paetzold and S. Kosma, *Chem. Ber.* **112,** 654 (1979).
123. M. Otsuka, M. Yoshida, S. Kabayashi, M. Ohno, Y. Umezawa and H. Morishima, *Tetrahedron Letters* **22,** 2109 (1981).
124. T. Sugasawa, T. Toyoda and K. Sasakura, *Synth. Commun.* **9,** 515 (1978).
125. R. Köster, W. Fenzl and F. J. Levelt, *Ann.* **1981,** 734..
126. A. I. Meyers and Y. Yamamoto, *J. Am. Chem. Soc.* **103,** 4278 (1981).
127. T. Sugasawa and T. Toyoda, *Tetrahedron Letters* **1979,** 1423.
128. T. Sugasawa and T. Toyoda, *Synth. Commun.* **9,** 553 (1979).
129. K. Maruyama, K. Saimoto and Y. Yamamoto, *J. Org. Chem.* **43,,** 4895 (1978).

CHAPTER 10

Reactions of alkyl(aryl) boranes with carbon monoxide, isonitriles, and metal cyanides

The common feature of the reactions considered in this chapter is the migration of one, two, or three groups from the boron to the carbon atom of the unsaturated fragment in the initially formed adduct of R_3B with CO, RN≡C, or $\bar{C}$≡N:

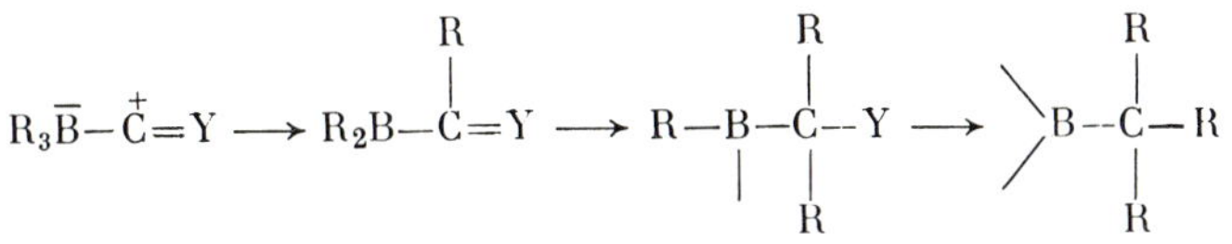

(C=Y = CO, RNC, $\bar{C}$N).

1,2-Anionotropic migrations of this type proceed spontaneously or are caused by nucleophilic or electrophilic reagents and protonic acids.

The fundamental importance of these processes consists of the formation of new C—C bonds and the increase in the length of the carbon chain of the radicals on the boron by one or more carbon atoms. From the synthetic aspect, they can be considered as a method for the replacement of boron by a carbon atom carrying a functional group.

The reactions of organoboranes with CO, isonitriles, and metal cyanides represent the basis of simple methods for the synthesis of higher alcohols, dialkyl- and trialkylcarbinols, ketones, amines, branched olefins, bifunctional and heterocyclic compounds, and organic compounds of a number of other classes.

10.1 REACTION WITH CARBON MONOXIDE

The reaction of organoboron compounds with carbon monoxide is used for the synthesis of heterocyclic derivatives of boron, dialkyl- and trialkylcarbinols, ketones, and also alcohols and aldehydes containing one carbon atom more than the radical of the initial organoborane:

$$R_3B \xrightarrow{CO} \begin{cases} \rightarrow RCH_2OH \text{ or } RCHO \\ \rightarrow R_2CHOH \text{ or } R_2CO \\ \rightarrow R_3COH \end{cases}$$

Carbocyclic compounds are obtained from diene and polyene hydrocarbons and also from higher olefins by hydroboration-carbonylation-oxidation (Parts II and IV).

The conversions of alkylboranes under the action of carbon monoxide have been considered in review articles [1, 2] and, in part, in [3–5].

The carbonylation of trialkylboranes was carried out for the first time by Reppe and Magin [6]. According to their data, the reaction proceeds at 100–200 atmospheres and 10–20°C and leads primarily to two products, $(R_3BCO)_2$ and $(R_3BCO)_3$, the structure of which has not been determined.

Hillman [7–11] studied the carbonylation of trialklboranes over a broad range of temperatures (20–155°C) and pressures (from 1 to 1000 atmospheres). He established the structure of the compounds formed, determined the fundamental characteristics of the process, and showed that the direction of the reaction is almost independent of the pressure and is determined primarily by the temperature and, to a lesser degree, by the nature of the solvent.

The reaction at 25–75°C in the presence of water leads to 2,3,3,5,6,6-hexaalkyl-2,5-dibora-1,4-dioxane (*1*) in almost quantitative yield; at 140–175°C, only a trialkylmethylboronic acid (*2*) is obtained, which is isolated as the trimeric anhydride *3* upon distillation [7]:

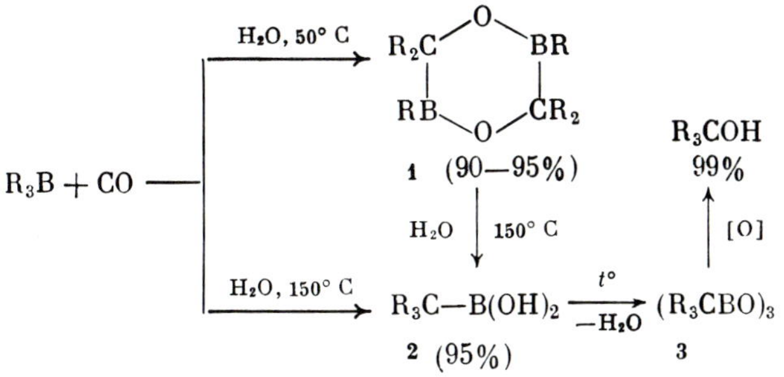

The compounds *1* and *3* are also obtained smoothly (93–95%) with the use as solvents of alcohols, phenols, dioxane, monoglyme, orthoformic ester, or aqueous acetone, whereas their yield is reduced in ether, tetrahydrofuran, or hydrocarbons or in the absence of a solvent. 2,5-Diboradioxanes (*1*), in the presence of water at 150°C, are quantitatively converted to the acids *2*; without water, this reaction does not proceed even at 200°C. In the oxidation of the anhydrides *3* with hydrogen peroxide in alkaline medium, trialkylcarbinols are obtained.

The diboradioxanes *1* are hydrolyzed by heating with solutions of alkalies to dialkylcarbinols and form ketones and alcohols upon oxidation with hydrogen peroxide:

$$R_2CO + ROH \xleftarrow{H_2O_2,\ OH^-} (1) \xrightarrow{OH^-,\ 100°\ C} \underset{75\%}{R_2CHOH}$$

The intracyclic bond B—C is hydrolyzed more readily than the B—C bonds without an oxygen in the α position (Chapter 6).

Carbonylation in a medium of ethylene glycol, 1,3-propanediol; 1,4- 2,3-, or 1,3-butanediol; pinacol, 1,5-pentanediol, pentaerythritol, or polyvinyl alcohol (1 hour at 50°C, followed by 2 hours at 150°C) leads to cyclic trialkylmethylboronate esters of the type *4* and *5* in yields of approximately 95% [8, 10]:

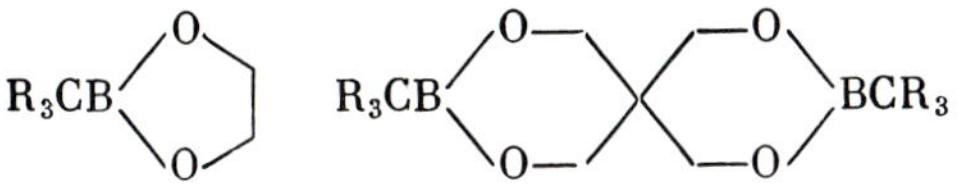

4 5

A large number of different 1,3,2-dioxaborolanes (*4*), including those with functional groups (halogen, RO, RCOO, etc.) in the radicals have been synthesized by this method from the corresponding symmetrical and unsymmetrical trialkylboranes [12–15].

The cyclic esters *4* and *5* are stable toward oxygen and atmospheric moisture and are soluble in organic solvents but insoluble in water. The esters *4* do not hydrolyze even during steam distillation; upon oxidation, they form the corresponding trialkylcarbinols R_3COH.

According to Hillman [7–9], the carbonylation of trialkylboranes proceeds intramolecularly and consists of the consecutive migration of three alkyl anions R^- from the negatively charged boron to the carbon atom carrying a positive charge. The initially formed complex *6* isomerizes to an acyldialkylborane (*7*), which probably dimerizes to form *8*, with the latter rearranging to *1*:

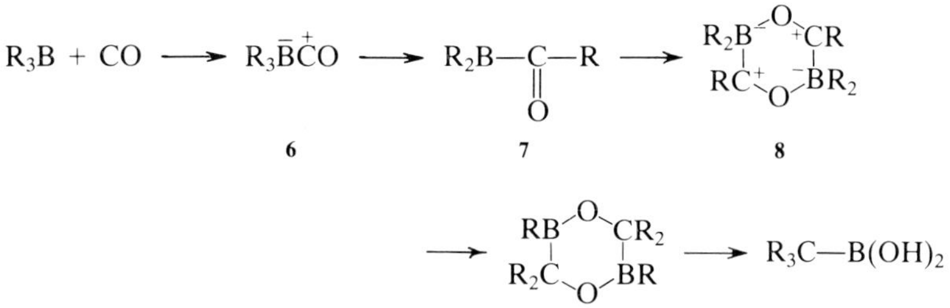

6 7 8 1 2

The hexaalkyldiboradioxane *1* is stable to 50–80°C but completely isomerizes by migration of the third radical from boron to carbon at 140–150°C in the presence of hydroxy-containing reagents. The catalytic action of hydroxy-containing compounds (water, alcohols, glycols) on the carbonylation is explained by their coordination with the boron atom.

The reaction of trialkylboranes with carbon monoxide and aldehydes leads to substituted 4-bora-1,3-dioxolanes (*9*), the alkaline hydrolysis of which produces dialkylcarbinols [9, 11]:

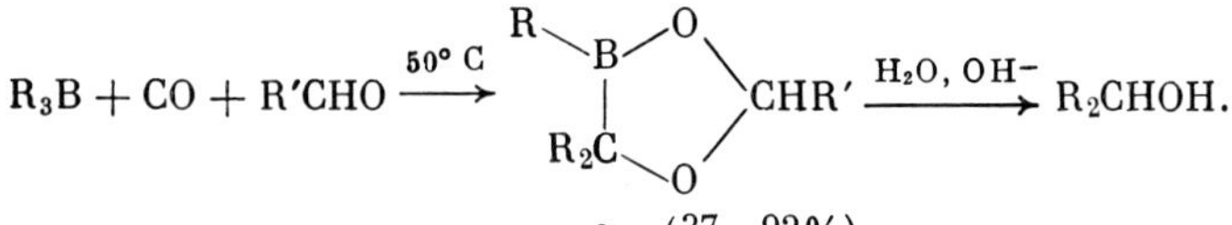

9 (37—92%)

These data confirm the intermediate formation of acyldialkylboranes (*7*), which react with aldehydes by a 1,3-dipolar addition (*10*), with the simultaneous migration of a radical from the boron to the carbon:

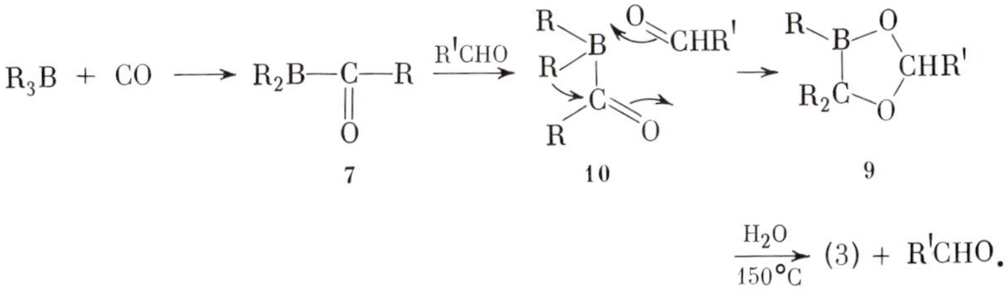

The 4-bora-1,3-dioxolanes *9* are converted by heating with water to the anhydrides of the trialkylmethylboronic acid *3* and the corresponding aldehydes.

The conversions of triarylboranes under the action of CO have received little investigation. The reaction of triphenylborane with CO proceeds only in the presence of $LiAlH(OCH_3)_3$, forming benzaldehyde after oxidation [98]. The carbonylation of $[(C_6H_5)_3BH]Li$ proceeds in the presence of a small quantity of triphenylborane and leads to diphenylmethane after hydrolysis [99]. As a result of the acid hydrolysis of the complex salt formed by the reaction of triphenylborane, carbon monoxide, and triphenylmethylsodium, *p*-benzhydryltriphenylmethane is obtained [16]:

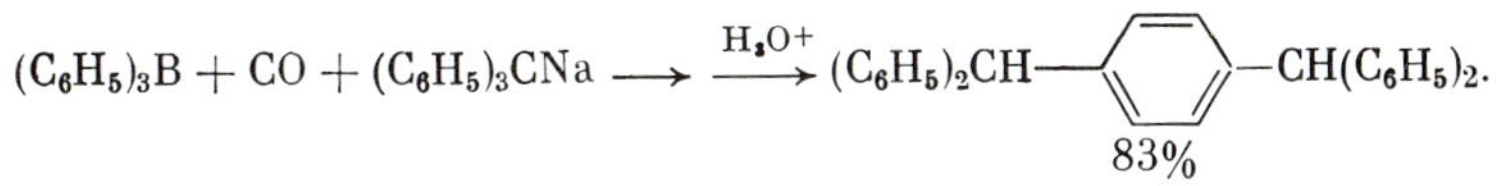

During this reaction, two phenyl groups migrate from the boron to the carbon atom of the CO molecule.

The migration of radicals from boron also takes place in the reaction of triphenylboronsodium with benzoyl chloride; the compound $(C_6H_5)_2B—C(C_6H_5)_2ONa$ formed during this process produces 72% of benzhydrol and 22% of triphenylmethane under the action of aqueous alkali.

The reaction of diborane with carbon monoxide leads to the complex $H_3B \cdot CO$. The synthesis and properties of carbonyl(C—B)borane have been investigated in [17–23] and have been described in a book [24]. Carbonyl-borane is converted under the action of catalytic quantities of $NaBH_4$ in tetrahydrofuran to methylboronic anhydride, $(CH_3BO)_3$ [25]. At −78°C, the reaction of carbonyl-borane with water [26] and alcohols [27] leads to hydroxymethylboronic acid, $HOCH_2B(OH)_2$, or its esters, which exist in the form of the dimeric cyclic esters $(ROB—CH_2O—)_2$. At room temperature, the complex $H_3B \cdot CO$ shows considerable dissociation and the reaction with alcohols therefore produces borate, hydrogen, and carbon monoxide.

10.1.1 Synthesis of trialkylcarbinols

The carbonylation of trialkylboranes at 150°C and atmospheric or elevated pressure is used for the preparation of trialkylcarbinols:

$$R_3B + CO \xrightarrow[150^\circ\,C]{H_2O \text{ or } (CH_2OH)_2} \left[R_3C—B\langle \right] \xrightarrow{H_2O_2,\ OH^-} R_3COH.$$

80—98%

The intermediately formed anhydrides *3* or cyclic boron esters *4* generally are not isolated but are immediately subjected to oxidation with hydrogen peroxide in alkaline medium. In this way, Hillman obtained a 98% yield of tributylcarbinol from tributylborane [7]. Brown and Rathke [28, 29] improved the method of carbonylation and significantly expanded the range of application of this reaction. Triisobutyl-, tri-*sec*-butyl-, and trioctylcarbinols and also tricyclopentyl-, tricyclohexyl-, and tri-2-norbornylcarbinols have been synthesized from the corresponding trialkylboranes:

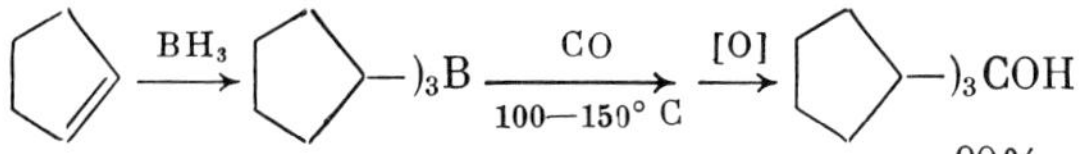

90%

Tri-*exo*-norbornylcarbinol, $R_3{}^{14}COH$, has been synthesized by the reaction of tri-*exo*-norbornylborane with ^{14}CO. Carbon-labeled aldehydes and ketones, $R^{14}CHO$ and $R_2{}^{14}CO$, have been obtained from the corresponding organoboranes and ^{14}CO [100].

The carbonylation of an equimolar mixture of triethyl- and tributylboranes led, after oxidation, to an equimolar mixture of triethyl- and tributylcarbinols, indicating the intramolecular nature of the migration of the radicals.

The carbonylation of organoboranes apparently represents one of the best methods for the synthesis of tertiary alcohols. It should be noted that the reaction of organomagnesium compounds with carbon monoxide proceeds in a complex manner and leads to a mixture of different compounds [30].

The carbonylation-oxidation of unsymmetrical trialkylboranes (R_2BR') produces tertiary alcohols with different radicals on the carbon atom ($R_2R'COH$) [12, 13, 31, 32]. These reactions are generally carried out in an autoclave at 70 atmospheres and a temperature of 150°C.

Unsymmetrical trialkylboranes containing functional groups (halogen, RO, RCOO, CN, and others) in the hydrocarbon chain of the radicals have also been used for the preparation of tertiary alcohols. One of the examples of syntheses of this type is the preparation of the carbinol *11* from 4-penten-1-ol acetate:

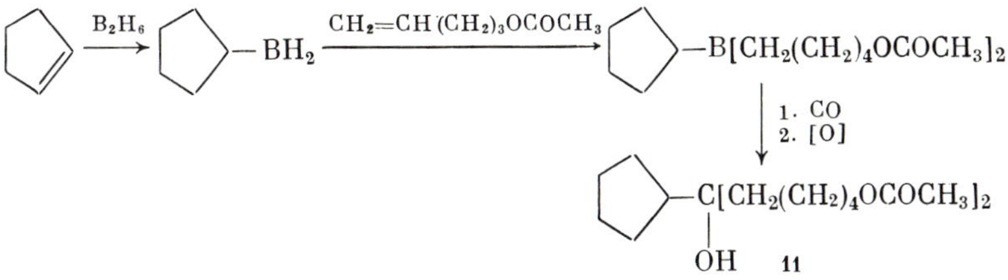

11

ω-Chlorocarbinols of the type $[Cl(CH_2)_n]_2C(OH)R$ have been synthesized in an analogous manner from the corresponding unsaturated ω-chlorides.

(*cis*-Bicyclo[3.3.0]oct-1-yl)dialkylcarbinols have been synthesized by the carbonylation of (*cis*-bicylo[3.3.0]oct-1-yl)dialkylboranes [101].

10.1.2 Synthesis of dialkylcarbinols and ketones

The carbonylation of trialkylboranes in water, dioxane, monoglyme, or orthoformic ester at 40–75°C leads to 2,5-dibora-1,4-dioxanes (*1*) in quantitative yield (migration of two radicals from boron to carbon) [7]. The compounds *1* are cleaved by heating with alkali solutions to form dialkylcarbinols. In the preparation of the latter, the dibora-dioxanes *1* are generally not isolated, but are immediately subjected to alkaline hydrolysis.

The reaction of trialkylboranes with CO in a mixture of diglyme and ethylene glycol with the "addition" of sodium leads to products of the migration of only two alkyl groups—dialkylcarbinols—even at 150°C [14, 15]. Apparently, one equivalent of ethylene glycol sodium alcoholate should be used in the reaction per mole of R_3B.

Hillman [7] obtained dialkyl ketones with an impurity of dialkylcarbinols by the oxidation of the diboradioxanes *1* with hydrogen peroxide in 6 N alkali. Brown and co-workers [33–36] increased the yield of ketones by subjecting the crude products of carbonylation in a solution of sodium acetate or 3 N alkali to oxidation.

The unsymmetrical trialkylboranes form mixed ketones. Thus, dicyclohexyl(alkyl)boranes, formed by the hydroboration of linear and cyclic olefins with dicyclohexylborane, can be easily converted to cyclohexyl alkyl ketones (*12*) with an impurity of dicyclohexyl ketone (*13*) [29, 33]:

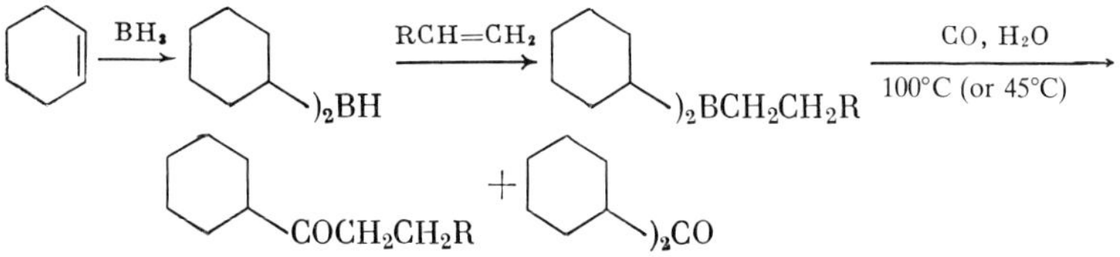

12 (72—92%) **13** (8—20%)

The yield of the compound *13* is reduced when the carbonylation is carried out at 45°C, indicating an increased selectivity of migration of the primary alkyl groups with decreasing temperature.

The ability of alkyl groups to migrate from boron to carbon increases in the following series: tertiary < secondary < primary. The cyclopentyl radical migrates more easily than cyclohexyl and the latter migrates more easily than norbornyl. Tertiary groups migrate only under severe conditions (150°C).

The hydroboration of functionally substituted olefins with diclohexylborane leads to mixed boranes with functional groups in the hydrocarbon chain. Carbonylation-oxidation of the latter produces substituted ketones, such as 4-oxo-4-cyclohexylbutyl cyanide [33]:

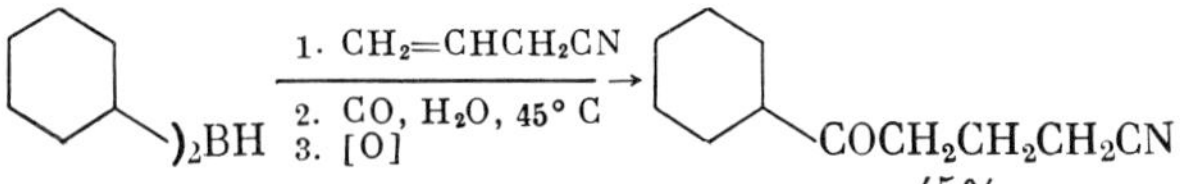

45%

In an analogous manner, methyl 12-oxo-12-cyclohexyldodecylate and 12-oxo-12-cyclohexyldodecyl benzoate have been synthesized from methyl 10-undecenoate and 10-undecenylbenzoate; allyl benzoate is converted to 4-oxo-4-cyclohexylbutyl benzoate. All carbonylation reactions of functional organoboranes are carried out at 45°C in aqueous tetrahydrofuran.

According to the data of Hawthorne, et al. [37, 38], the oxidation of cyclohexyl alkyl ketones with peracids by the Bayer–Williger method proceeds mainly by cleavage of the bond between the carbonyl carbon and the ring. This makes it possible to achieve a transition from cyclohexyl alkyl ketones and, therefore, from the corresponding olefins, to substituted carboxylic acids with an increase by one carbon atom in the chain length. For example, nonanoic acid was obtained by the oxidation of cyclohexyl octyl ketone [33]:

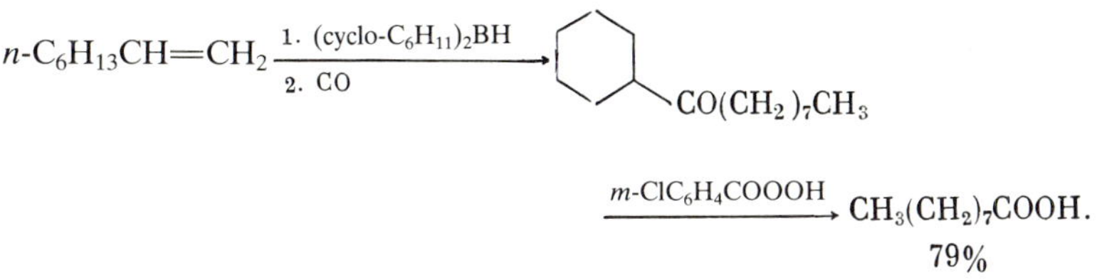

By an analogous method, 1-dodecene was converted to tridecanoic acid, methyl 10-undecenoate to 1,10-decanedicarboxylic acid, and 10-undecenyl alcohol to 12-hydroxydodecanoic acid.

As previously noted, at temperatures below 100°C, *tert*-alkyl groups in mixed alkylboranes undergo practically no migration from boron to carbon. This makes it possible to use trialkylboranes of the type *14*, obtained as a result of consecutive hydroboration of two different olefins with (2,3-dimethyl-2-butyl)borane, for the synthesis of the unsymmetrical ketones *15* [35, 36]:

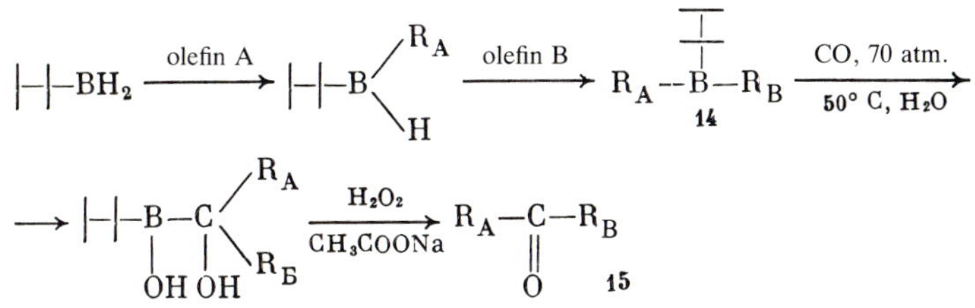

Thus, ethyl 5-oxo-7-methyloctanoate has been synthesized from isobutylene and ethyl vinylacetate:

$$(CH_3)_2C{=}CH_2 + CH_2{=}CHCH_2COOC_2H_5 \longrightarrow (CH_3)_2CHCH_2{-}CO{-}(CH_2)_3COOC_2H_5. \quad 84\%$$

The 12-oxoester *16*, 5-oxonitrile *17*, and 4-oxochloride *18* have been obtained from the corresponding olefins by an analogous combination of reactions:

$$C_6H_5CH{=}CH_2 + CH_2{=}CH{-}(CH_2)_8{-}COOCH_3 \longrightarrow C_6H_5CH_2CH_2{-}CO{-}(CH_2)_{10}COOCH_3$$

16 (73%)

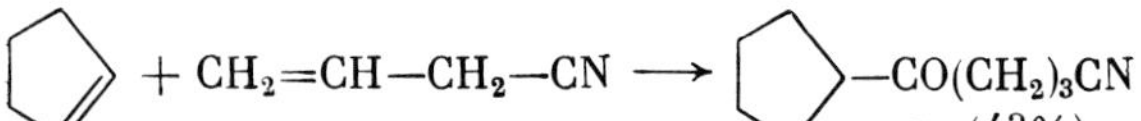

17 (43%)

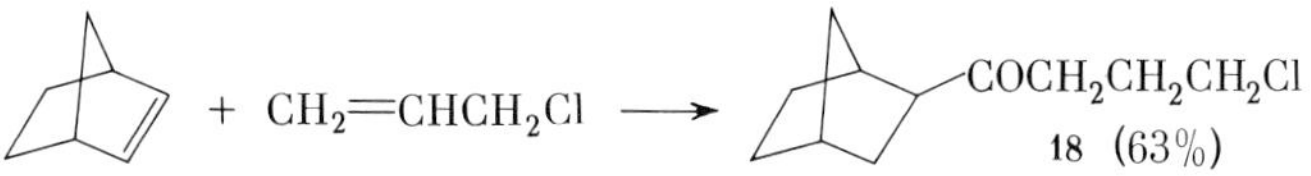

18 (63%)

Juvabione and epijuvabione have been synthesized from methyl perillate and isobutylene by this method [102].

10.1.3 Synthesis of aldehydes and alkylcarbinols

The carbonylation of trialkylboranes in the presence of 0.05–0.2 mole of sodium borohydride in diglyme or of lithium borohydride in tetrahydrofuran, at 45°C and atmospheric pressure, leads, after alkaline hydrolysis, to alcohols containing one more carbon atom than the radical of the starting organoborane [39]:

$$R_3B + CO \xrightarrow[\text{LiBH}_4,\ \text{THF}]{\text{NaBH}_4,\ \text{diglyme}} \xrightarrow{\text{KOH}} RCH_2OH. \quad 70{-}80\%$$

A combination of this reaction with hydroboration represents a method for the conversion of olefins to higher alcohols. Thus, 1-nonanol has been synthesized from 1-octene while cyclopentene, cyclohexene, and norbornene have been converted, respectively, to cyclopentyl-, cyclohexyl-, and 2-*exo*-norbornylcarbinols:

$$n\text{-}C_6H_{13}CH{=}CH_2 \xrightarrow{BH_3} (n\text{-}C_8H_{17})_3B \xrightarrow{\text{CO, LiBH}_4} \xrightarrow{\text{KOH}} n\text{-}C_8H_{17}CH_2OH. \quad 70\%$$

The role of the hydrides in these reactions apparently consists of the reduction of the intermediate acyl(dialkyl)borane *7* to the corresponding α-oxyboryl compound *19*:

$$\underset{\mathbf{6}}{R_3B\cdot CO} \longrightarrow \underset{\mathbf{7}}{R_2B{-}\overset{}{\underset{\underset{O}{\|}}{C}}{-}R} \xrightarrow{MH} \underset{\mathbf{19}}{R_2B{-}\underset{\underset{OM}{|}}{CH}{-}R} \xrightarrow{KOH} RCH_2OH.$$

However, it is not impossible that they reduce the complex *6* and that the process stops after the migration of one group [99]. Compounds of the type *19* may possibly be dimeric "ate" complexes.

Other hydrides used are lithium tri-*tert*-butoxyaluminum hydride, $LiAlH(OC_4H_9\text{-}tert)_3$, or $LiAlH(OCH_3)_3$ (obtained by the treatment of lithium aluminum hydride with three moles of methanol). *trans*-2-Methylcyclopentylmethanol (*20*) has been obtained by the carbonylation of 9-(*trans*-2-methylcyclopentyl)-9-borabicyclo[3.3.1]nonane, followed by alkaline hydrolysis [40, 103]:

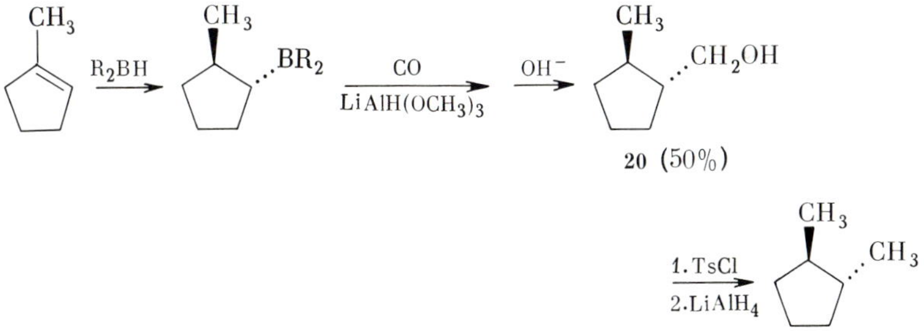

20 (50%)

Tris(*trans*-2-methylcyclopentyl)borane does not react with carbon monoxide at 20–50° C and atmospheric pressure.

The reaction of organoboranes with carbon monoxide at 0–25° C in the presence of $LiAlH(OCH_3)_3$ and subsequent oxidation of the carbonylation product leads to aldehydes [41]:

$$R_3B + CO \xrightarrow{LiAlH(OCH_3)_3} \underset{\mathbf{21}}{\left[R_2B{-}\underset{\underset{OAl(OCH_3)_3Li}{|}}{CHR}\right]} \xrightarrow{H_2O_2,\ pH\ 7} \underset{87\text{—}98\%}{RCHO + 2ROH.}$$

In order to keep the hydrolysis of the intermediate compound *21* and condensation of the aldehyde formed to a minimum, the oxidation is carried out not in alkali but in a phosphate buffer

(NaH_2PO_4-K_2HPO_4) at pH 7. By this method, *n*-heptanal has been synthesized from trihexylborane in 97% yield. Treatment of *21* with acetic or hydrochloric acid causes migration of a second group and, after oxidation, secondary alcohols R_2CHOH are obtained [104].

The reaction with carbon monoxide is one of the stages of the conversion of olefins to aldehydes:

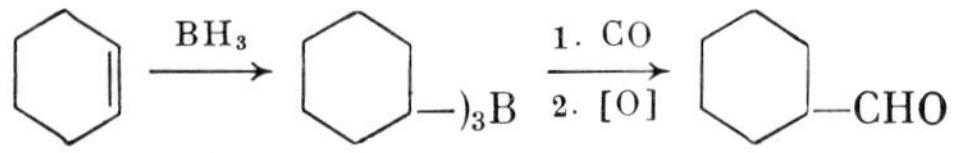

9-Borabicyclo[3.3.1]nonane is used for the synthesis of aldehydes from olefins and cycloolefins [42]. Alcohols or aldehydes can be conveniently obtained by the carbonylation of R_3B or 9-R-9-borabicyclo[3.3.1]nonane in the presence of K[(*iso*-$C_3H_7O)_3BH$] [103].

In the case of functional olefins, $LiAlH(OC_4H_9\text{-}tert)_3$ is used as a reducing agent. This does not react with nitriles or esters [the properties of $LiAlH(OCH_3)_3$ are similar to those of $LiAlH_4$]. The carbonylation is carried out at −25° C and atmospheric pressure and the oxidation at pH 7. By this method, 4-benzoxybutanal has been synthesized from allyl benzoate, acetoxypentanal from 4-butenyl acetate, and acetoxydodecanal from 10-undecenyl acetate. 11-Carbomethoxyundecanal, 10-cyanoundecanal, 4-carbethoxybutanal, and 4-cyanobutanal have also been obtained [43]:

$$CH_2{=}CH(CH_2)_8COOCH_3 \longrightarrow \underset{85\%}{OCH(CH_2)_{10}COOCH_3},$$

$$CH_2{=}CHCH_2CN \longrightarrow \underset{72\%}{OCH(CH_2)_3CN}.$$

Alcohols RCH_2OH and aldehydes RCHO can be obtained by the carbonylation of some compounds of the type [R_3BH]M [99].

10.2 REACTION WITH ISONITRILES

The reaction of boron compounds with isonitriles is of considerable interest in two aspects. First, it is used for the synthesis of heterocyclic boron compounds of several types, including those very stable toward oxygen and atmospheric moisture and even toward acids and alkalies. Secondly, this reaction can be used to prepare ketones, trialkylcarbinols, and a number of organic heterocyclic compounds from organoboranes.

The conversions of boron derivatives under the action of isonitriles have been partially discussed in review articles [5, 44–46].

10.2.1 General characteristics of the reaction

In 1963, Hesse and Witte [47] obtained substituted α-iminoboranes (*23*), which exist in the dimeric form *24*, by the reaction of phenyl isocyanide with triethyl- or tributylborane in ether at room temperature. A further study of the reaction of organoboranes with isonitriles was carried out by several research groups from 1964 to 1970. In establishing the structure of the compounds formed, a whole series of errors was committed, and only as a result of almost seven years of effort has the mechanism of this complex multistage conversion been understood in general outline.

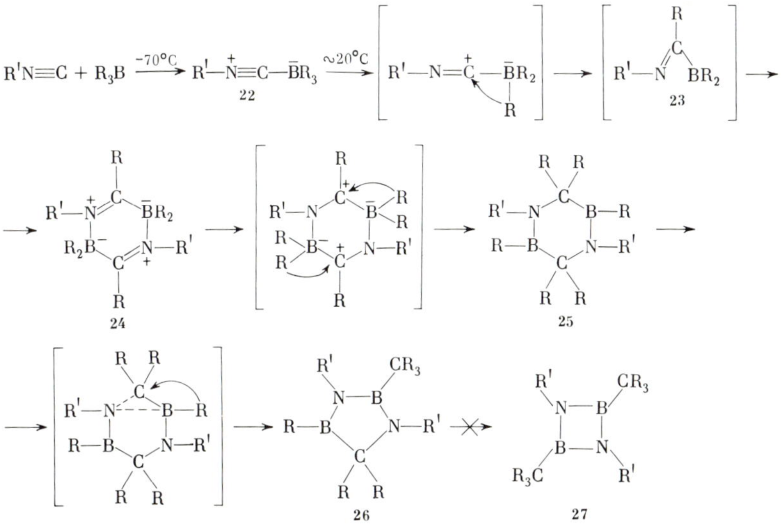

In essence, the main stages of the reaction are analogous to a Wagner–Meerwein rearrangement and consist of a series of consecutive migrations of alkyl or aryl groups from the negatively charged boron to an adjacent carbon atom with an electron deficiency (intramoleular anionotropic rearrangements).

The stability and reactivity of the initially formed 1:1 monomeric adducts *22* depends mainly on the nature of the radical in the isonitrile. At room temperature or by heating to 150°C, they are all converted to cyclic diboron compounds — 2,5-diboradihydropyrazines (*24*) [47–49] — which, in turn, isomerize to 2,5-diborapiperazines (*25*) at 200°C [47, 50] or under irradiation [51]. The latter compounds rearrange to five-membered boron-nitrogen heterocyclic rings, 1,3-diaza-2,4-diborolidines (*26*), under the action of aluminum chloride [52] or as a result of heating for several minutes at 300–310° C [51, 53]. Attempts to use the reaction of organoboranes with isonitriles for the preparation of 1,3-diaza-2,4-diboretidines (*27*) have been unsuccessful. The data of Casanova et al. [54], who reported the synthesis of *27* by this method, were found to be erroneous [44, 46, 51, 53].

The complex of trimethylborane with a *tert*-butyl isonitrile of the type *22*, obtained at −60° C, is gradually converted to the dimer *24* at room temperature; the authors assigned the structure of the iminoborane *23* to this [48]. The reaction of higher trialkylboranes with isonitriles proceeds with the evolution of heat at 20° C and leads to 2,5-diboradihydropyrazines (*24*) [50, 55]. 1:1 Adducts of triphenylborane and aliphatic isonitriles are obtained only in tetrahydrofuran [49, 56], while the reaction does not take place in benzene or ether; they are stable in air but decompose to the starting components upon fusion. The complexes of triphenylborane with aromatic isonitriles are smoothly converted to the dimeric compounds *24* by heating to above 100° C [49]. The esters of dialkylborinic acids, R_2BOR', do not react with phenyl isocyanide in ether or tetrahydrofuran [57]. A simple method for the synthesis of some complex compounds of the type *22* without the use of isonitriles has been described, consisting of the reaction of *tert*-butyl- or triphenylmethyl halides with potassium triphenylcyanoborate [58]:

$$K[(C_6H_5)_3BCN] + R{-}X \xrightarrow[-KX]{} (C_6H_5)_3\overset{-}{B}{\cdot}C{\equiv}\overset{+}{N}{-}R.$$

Monomeric α-imino(alkyl)boranes (*23*) have not been isolated, although their formation is indicated by a whole series of reactions, which will be considered below.

2,5-Diboradihydropyrazines (*24*) are stable toward alcohols, organic acids, and boiling aqueous acids and alkalies and are not oxidized with hydrogen peroxide.

Unlike most aminoboranes $RB(NR_2)_2$, 2,5-diborapiperazines (*25*) do not react with water, alcohols, acetic acid, or acetone; their cyclic structure is not broken down even by the action of strong oxidants. For example, 1,4-diphenyl-2,3,3,5,6,6-hexaethyl-2,5-diborapiperazine is nitrated with concentrated nitric acid with the formation of the corresponding 1,4-dinitrophenyl derivative [51].

According to the data of [51, 52], 1,3-diaza-2,4-diborolidines of the type *26* do not react with HCl and CrO_3 in ether or with bromine in CCl_4 and are not changed by refluxing for 3 hours with a 50% methanolic solution of KOH or with concentrated hydrochloric acid. The *N,N'*-diphenyl compound (*26*, $R = C_2H_5$) is nitrated with fuming nitric acid in CF_3COOH at −5°C to form the *N, N'*-di-*p*-nitrophenyl derivative. The oxidation of diazadiborolidines (*26*) with alkaline hydrogen peroxide leads to ketones, trialkylcarbinols, and aniline. Ketones are also obtained by heating compound *26* with CrO_3 in acetic acid at 90° C.

The unusually high stability and low reactivity of the heterocyclic boron derivatives *24*, *25*, and *26* are apparently due to steric factors making access of most reagents to the boron or nitrogen atom difficult [44], as is the case for the sterically hindered borazoles [59].

The interatomic distances in diazadiborolidine have been determined by X-ray structural analysis [53].

The treatment of organic diisocyanides with trialkylboranes or diborane leads to polymeric compounds containing 2,4-diboradihydropyrazine rings in the unit, and isomerizing upon heating to polymers with 2,5-diborapiperazine rings [60].

As noted above, monomeric α-imino(dialkyl)boranes (*23*) have not been isolated in the pure state. Their formation as intermediate products can be recorded by carrying out the reaction of organoboranes with isonitriles in the presence of compounds with multiple bonds or of organic acids. Thus, as a result of the reaction of *N*-phenylbenzalimine, phenyl isocyanide, and triethylborane or tributylborane, 1,4-diaza-2-boracyclopentanes were obtained (by dipolar 1,3-addition) [61, 62]. The α-imino(dialkyl)boranes *23* also react with phenyl isocyanate and phenyl isothiocyanate by 1,3-dipolar addition, as a result of which 1,4-diphenyl-2,3,3-trialkyl-1,4-

diaza-2-boracyclopentan-5-ones (*28*, X = O) or 1,4-diaza-2-boracyclopentan-5-thiones (*28*, X = S) are obtained.

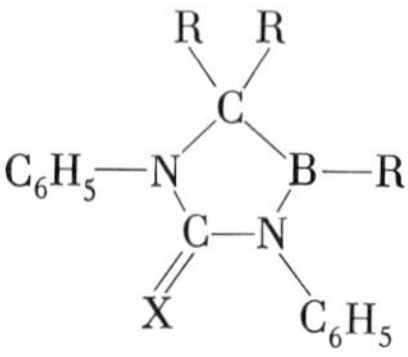

28

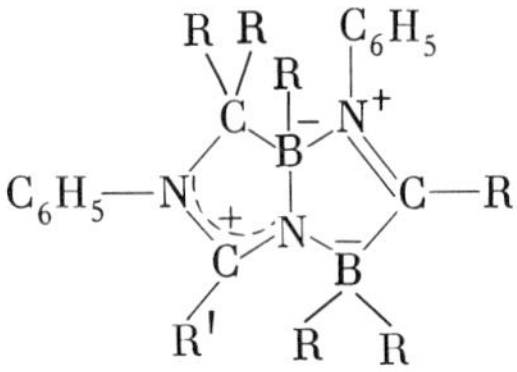

29

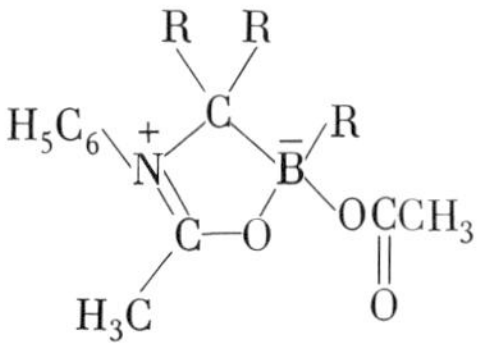

30

N,N'-Dicyclohexylcarbodiimide adds α-iminoboranes at one double bond with the formation of 5-cyclohexylimino-4-phenyl-1-cyclohexyl-2,3,3-trialkyl-1,4-diaza-2-boracyclopentane (*28*, X= cyclo-$C_6H_{11}N$). When nitriles are used as a third component in the reaction of R_3B with phenyl isonitrile, 2,5,7-triaza-1,4-diborabicyclo[3,3,0]octanes (*29*) are obtained, which are formed by the reaction of two molecules of the compound *23* with one molecule of the nitrile. Acetanilide and the cyclic compound *30* have been isolated from the mixture of products of the reaction of acetic acid with $C_6H_5\overset{+}{N}{\equiv}C{-}\overset{-}{B}(C_2H_5)_3$ [63]. All of these compounds form ketones upon oxidation with hydrogen peroxide.

The reaction of trialkylboranes with hydrocyanic acid proceeds, in general, analogously to the reaction with isonitriles, but is greatly complicated by side reactions [64]. Japanese investigators [65] suggested erroneously that the complex acid $H[(C_2H_5)_3BCN]$ is formed by the reaction of HCN with triethylborane.

Thioboranes [66], diborane [67–69], and tetraalkyldiboranes [69] form 2,5-diboradihydropyrazines or 2,5-diborapiperazines by reaction with isonitriles. The reaction of symmetrical and unsymmetrical borazoles containing a hydrogen on the boron atom with isocyanides at 160–170°C leads to *B*-carbiminoborazoles, $(R'NBCH{=}NR)_3$ [69].

According to the data of [70–72], boron halides polymerize isonitriles even at low temperatures. However, Meller and Batka [73] have shown that BCl_3 and BBr_3 form complex compounds with isocyanides at −70° C which, upon heating, are converted to derivatives of 2,5-diboradihydropyrazine (*24*, R = Cl, Br).

The reaction of diborane with hydrocyanic acid in tetrahydrofuran leads to boronium compounds of the type $L{\cdot}\overset{+}{B}H_2C{\equiv}N\overset{-}{B}H_3$, the structure of which has been demonstrated by physicochemical methods and by exchange of the ligand L [74].

10.2.2 Synthesis of ketones, trialkylcarbinols, and heterocyclic compounds

On the basis of the reaction of organoboranes with phenyl isonitrile, Witte, Mischke, and Hesse [63, 75] have developed methods for the synthesis of dialkyl ketones and trialkylcarbinols. As has been shown, in the presence of H-acids, rearrangements of the initially formed adducts $C_6H_5\overset{+}{N}{\equiv}C{-}\overset{-}{B}R_3$ take place under very mild conditions — at room temperature or on cooling. In this process, depending on the nature of the protonic reagent, two or all three alkyl groups migrate from the boron to the carbon of the isonitrile.

In the presence of methanol, thiophenol, diethylamine, and aniline, the main reaction products are (α-phenylaminodialkylmethyl)alkylboranes of the type *31*, containing oxygen, sulfur, or nitrogen, which form ketones upon oxidation:

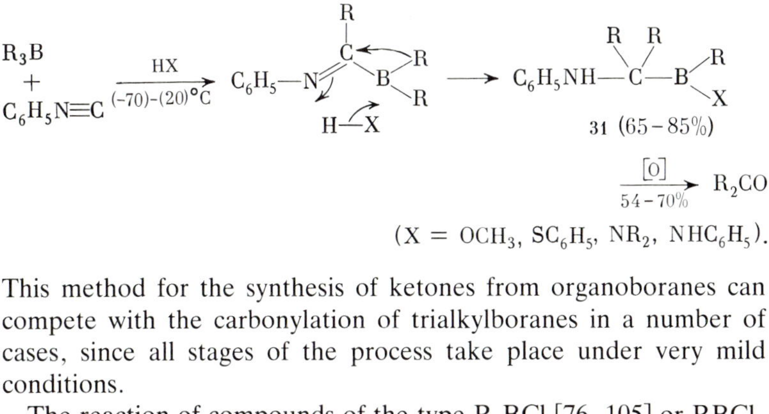

(X = OCH_3, SC_6H_5, NR_2, NHC_6H_5).

This method for the synthesis of ketones from organoboranes can compete with the carbonylation of trialkylboranes in a number of cases, since all stages of the process take place under very mild conditions.

The reaction of compounds of the type R_2BCl [76, 105] or $RBCl_2$ [105] with lithium aldimines is used for the preparation of unsymmetrical ketones:

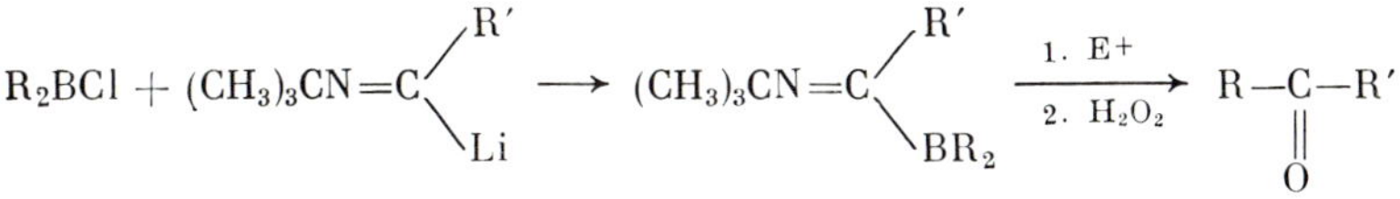

Trifluroacetic acid or its anhydride, dimethyl sulfate, hydrogen sulfide, thiophenol, or thioglycolic acid can be used as the electrophilic agent E. Lithium aldimines are prepared by the addition of RLi to isonitriles [77].

In the presence of water or phenol, the reaction of triethylborane with phenylisonitrile leads (after oxidation) to triethylcarbinol [63]:

$$R_3B + C_6H_5NC \xrightarrow{R'OH,\ 20^\circ C} \underset{88\%}{R_3C{-}B(OR')_2} \xrightarrow{[O]} \underset{>90\%}{R_3COH}.$$

The migration of all three radicals from boron to carbon also takes place when the adduct $R_3B{-}C{\equiv}NC_6H_5$ is treated with dry HCl in ether.

Yoshida, et al. [78] obtained 4.2% of cyclohexanecarboxylic acid *N*-cyclohexylamide by heating a mixture of tricyclohexylborane and cyclohexyl isonitrile, followed by oxidation with air.

An interesting synthetic application of the adducts of triphenylborane with isonitriles has been described by Bittner, Witte, and Hesse [56]. Treatment of the lithium imide *32* with an ether solution of HCl, and then with sulfuric acid at 160° C, led to diphenylmethylenecyclohexane (73%):

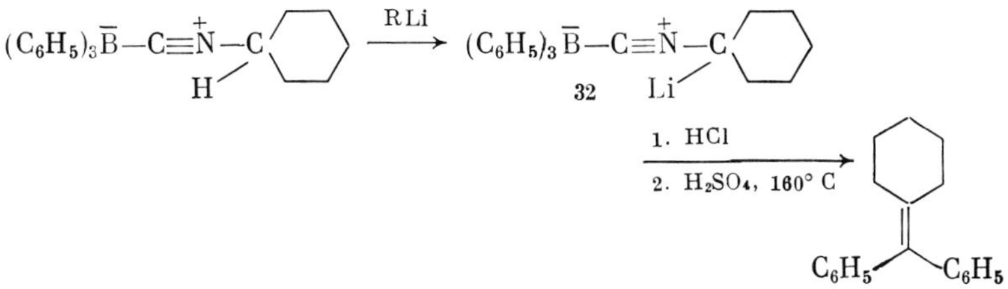

Substituted β-amino alcohols are obtained in a yield of 32–37% by the consecutive treatment of a benzhydrylisocyanide-triphenylborane complex with phenyllithium, a carbonyl compound, and an acid:

$$(C_6H_5)_2CH\overset{+}{N}{\equiv}C{-}\overline{B}(C_6H_5)_3 \xrightarrow[\text{2. RR'CO; 3. } H_2SO_4]{\text{1. } C_6H_5Li} (C_6H_5)_2\underset{NH_2}{C}{-}\underset{OH}{C}RR'$$

Imidazolines have been obtained from the adducts of triphenylborane with isopropyl- or cyclohexylisonitrile:

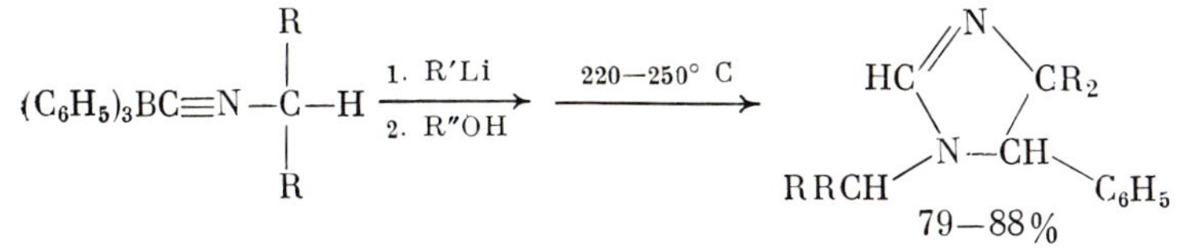

The reaction of trialkylboranes with phenylisocyanide in the presence of aromatic aldehydes leads to substituted oxazolidines [62, 79]:

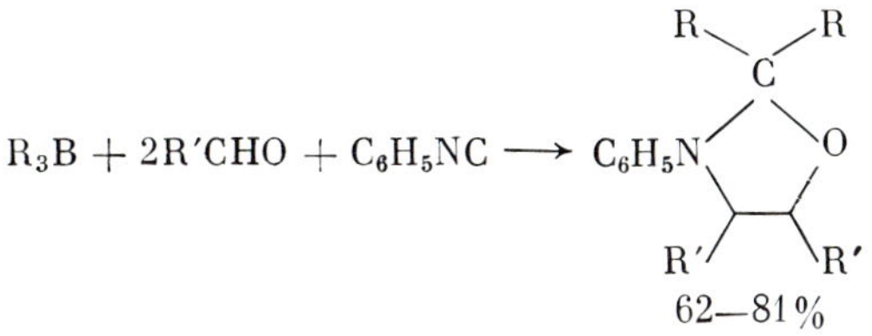

10.3 REACTION WITH METAL CYANIDES

Sodium or potassium trialkyl- and triarylcyanoborates, $M[R_3BCN]$, are used for the synthesis of symmetrical and unsymmetrical ketones, cycloalkanones, amines, trialkylcarbinols, and a number of heterocyclic boron derivatives:

$$M[R_3BCN] \longrightarrow R_2CO, \quad R_2CHNH_2 \text{ or } R_3COH.$$

The synthetic use of cyanoborates has been partially discussed in review articles [80, 81].

Wittig and Raff obtained sodium triphenylcyanoborate, $Na[(C_6H_5)_3BCN]$, by heating triphenylborane with sodium cyanide at 150° C [82, 83]. Sodium triarylcyanoborates have been synthesized by an analogous method from isomeric tritolylboranes [84], tri-*p*-fluoro-, *p*-chloro-, and *p*-bromophenylboranes [85], and also from tri-1-naphthylborane [86].

Sodium trialkylcyanoborates, $Na[R_3BCN]$, were obtained for the first time by Witte, et al. [87, 88] by the reaction of trialkylboranes or their pyridine complexes with NaCN in solvents of the ether type or in water at room temperature. The same authors have synthesized a number of potassium, tetraethylammonium, and phosphonium salts.

Electrophilic reagents cause a rearrangement of the cyanoborate anions, with the migration of organic radicals from the boron to the carbon of the cyano group, involving the migration of two or three radicals, depending on the nature of the electrophilic agent and on the conditions [88, 89]:

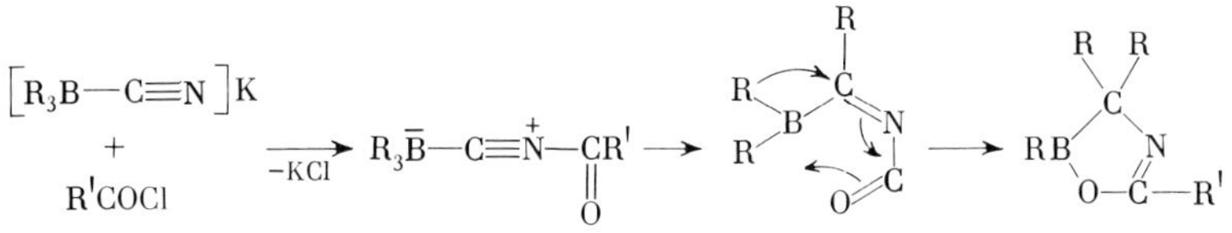

Analogous rearrangements are observed in the acylation of sodium 1-propynyltrialkylborate [90].

Trialkylcyanoborates are used for the synthesis of ketones, amines, and trialkylcarbinols.

10.3.1 Synthesis of ketones and secondary alkylamines

When the reaction of trialkylboranes with cyanides is used for the synthesis of ketones, the salts $M[R_3BCN]$ are generally not isolated but are obtained *in situ* by the addition of trialkylborane to a suspension of NaCN in diglyme. The solution is then treated at −70° C with 1–1.1 moles of an electrophilic reagent (trifluoroacetic anhydride, benzyl chloride, or *N*-phenylbenzimidoyl chloride); the mixture is heated to room temperature or 45° C (in the presence of C_6H_5COCl), after which it is oxidized with hydrogen peroxide in alkaline medium [89, 91, 92]:

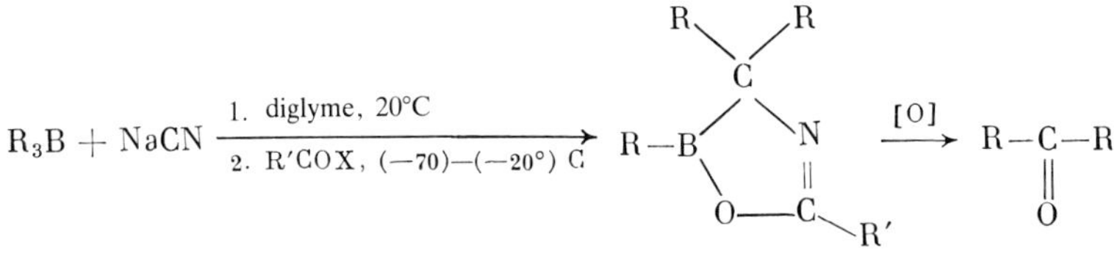

The ketone yields depend on the nature of the electrophilic agent and are almost quantitative when $(CF_3CO)_2O$ is used with diglyme as the solvent (see Table 1). In diethyl ether, methylene chloride, and hydrocarbons, a high yield of ketones is obtained only when a threefold excess of the anhydride is used. In tetrahydrofuran, a mixture of the ketone and trialkylcarbinol is generally obtained from symmetrical trialkylboranes.

TABLE 1
Yield of ketones R_2CO from trialkylboranes and NaCN in diglyme [89, 91] (in %).

R	Electrophilic reagent			
	$H^+ + HCN$	$C_6H_5CCl{=}N{-}C_6H_5$	$(CF_3CO)_2O$	C_6H_5COCl
n-C_4H_9	68	86	94	82
n-C_6H_{13}	*	80	*	*
n-C_8H_{17}	65	86(68)**	95	74
Cyclopentyl	69	88(79)**	84	87
Cyclohexyl	50	80	100	65
exo-Norbornyl	62	*	99	85

*No data **Reaction in tetrahydrofuran

The use of methylsulfonic acid, dimethyl sulfate, triethyloxonium borofluoride, acetic anhydride, or acetyl chloride as an electrophilic reagent does not produce satisfactory results. The yield of ketones in these cases, as a rule, is below 50% [88, 93]. When a cyanoborate is treated with hydrocyanic acid, ketones are obtained in a yield of 50–70% [91].

Unsymmetrical and cyclic ketones, and also ketones whose radicals contain functional groups that are not changed in the hydroboration, can be obtained with the use of cyanoborates [89, 94, 106]. The stepwise hydroboration of two different olefins with bis(2,3-dimethyl-2-butyl)diborane leads to organoboranes with three or two different radicals on the boron, which are easily converted to ketones:

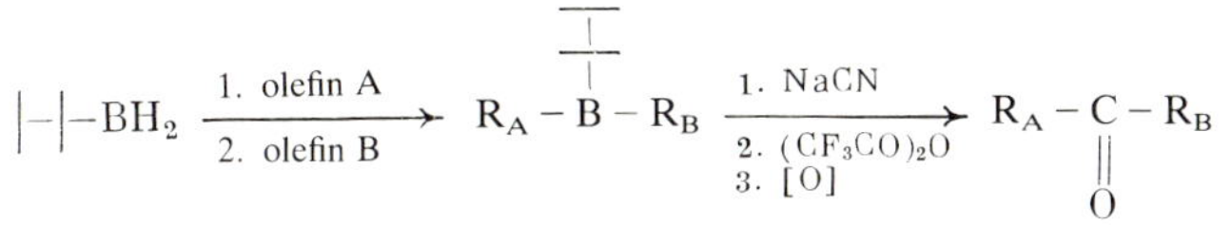

To retain the functional groups, the oxidation is carried out with hydrogen peroxide in $NaHCO_3$ solution at 23° C or with *p*-chloroperbenzoic acid at 0° C. According to the data of [80, 89, 94, 95], the alkyl groups can be arranged in the following order with respect to their ease of migration from boron to carbon: primary > secondary > tertiary. Thus, in the cyanidation of cyclohexyldiethyl- and cyclohexyldibutylboranes, only diethyl or dibutyl ketones are obtained, while dicyclohexylethylborane forms only cyclohexyl ethyl ketone [96]. In no case was the migration of a *tert*-alkyl group observed. On the other hand, the migration capacity of alkenyl groups is higher than that of *n*-alkyl groups [97].

The hydroboration of (+)-limonene (*33*) with bis(2,3-dimethyl-2-butyl)diborane led to 4,8-dimethyl-2-borabicyclo[3.3.1]nonane (*34*), which, as a result of the action of NaCN, $(CF_3CO)_2O$ and oxidation formed 4,8-dimethylbicyclo[3.3.1]-2-nonanone (*35*) in the form of a mixture of the exo and endo isomers in a ratio of 1:1 [89]:

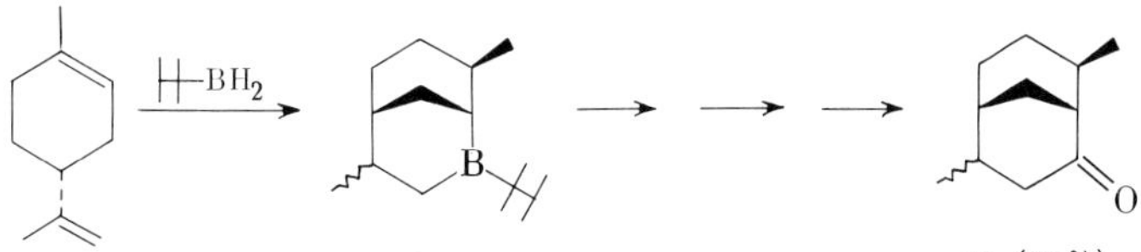

33 34 35 (85%)

1-(2-Methylenecyclohexyl)cyclohexene was converted to *cis,-anti,trans*-perhydrophenanthren-9-one by a similar method:

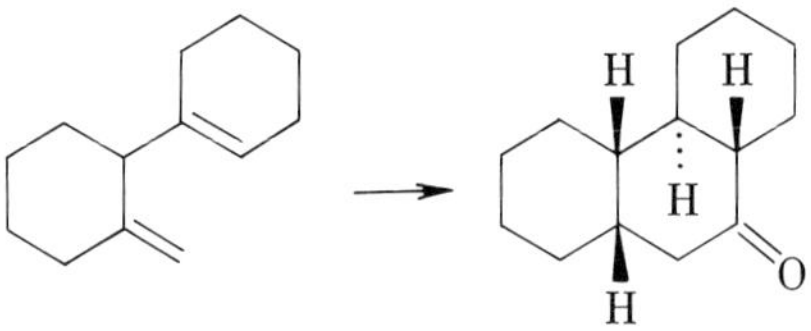

80%

Bicyclo[3.3.1]nonan-2-one and a number of aliphatic ketones have been synthesized from 4-vinyl-1-cyclohexene in an analogous manner [107]. The reaction has been used for the synthesis of *trans*-2-acylcyclopentanols and cyclohexanols from trimethylsilyl esters of the corresponding cycloalkenols [108] and the sex pheromone of the Douglas fir tussock moth, (*Z*)-$C_5H_{11}CH{=}CH(CH_2)_3COC_{10}H_{21}$ [109].

The transfer of radicals from boron to the carbon of the cyanide, initiated by hydrocyanic acid. $(CF_3CO)_2O$, and benzoyl chloride, takes place with a retention of the configuration, as has been demonstrated by several examples with the use of different cyclic systems. Thus, di-*exo*-norbornyl ketone was obtained from tri-*exo*-norbornylborane [91] and the cyanidation of tris(*trans*-2-methyl-cyclopentyl)borane led to bis(*trans*-2-methylcyclopentyl) ketone

[95, 96]:

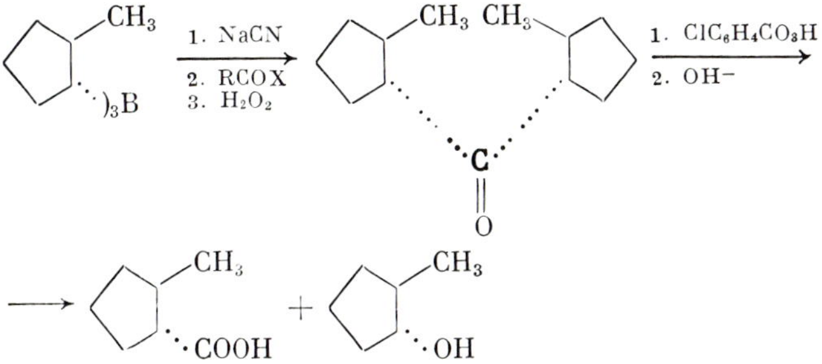

The reaction of trialkylcyanoborates with electrophiles, in which two radicals migrate from the boron to the carbon of the cyanide, is used for the preparation of some amines. According to the data of [89], 5-phenyl-2,3,3-tricyclopentyl-1-oxa-4-aza-2-boracyclopent-4-ene is hydrolyzed by heating with concentrated alkali, being converted to (dicyclopentylmethyl)amine and its *N*-benzoyl derivative:

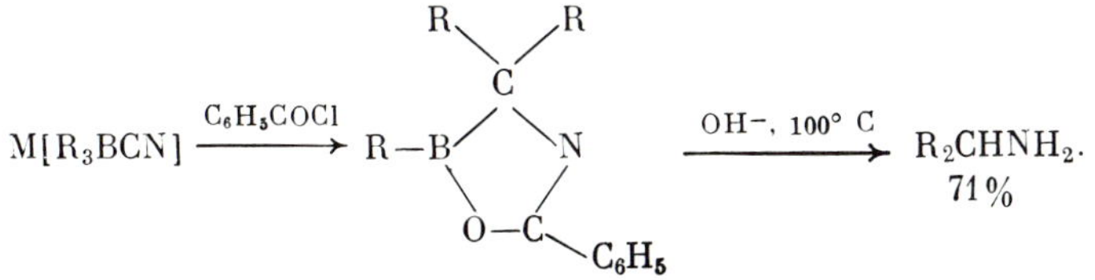

The analogous reaction of vinyldialkylboranes with metal cyanides leads to substituted allylamides [97].

10.3.2 Synthesis of trialkylcarbinols

The reaction of trialkylboranes with potassium or sodium cyanide has been used for the synthesis of trialkylcarbinols [92, 94, 95]:

$$R_3B + MCN \xrightarrow{(CF_3CO)_2O} \xrightarrow{H_2O_2} \underset{73\text{–}80\%}{R_3COH.}$$

As noted above, the migration of two radicals from boron to carbon in a trialkylcyanoborate is easily achieved under the action of trifluoroacetic anhydride at room temperature. The migration of the third radical in the case of *n*-alkylboranes proceeds at 45° C with excess (at least three moles) anhydride; secondary alkyls migrate only in the presence of dimethylformamide or pyridine, while tertiary alkyl radicals are incapable of migration.

Some salts (NaI, NaCN) accelerate the reaction. In the migration of the third group from boron to the cyanide carbon, as in the case of the migration of the first two radicals, the sterochemical configuration is completely retained. This is indicated by the formation of tris-(*trans*-2-methylcyclopentyl)carbinol as the only product from tris-(*trans*-2-methylcyclopentyl)borane [95]:

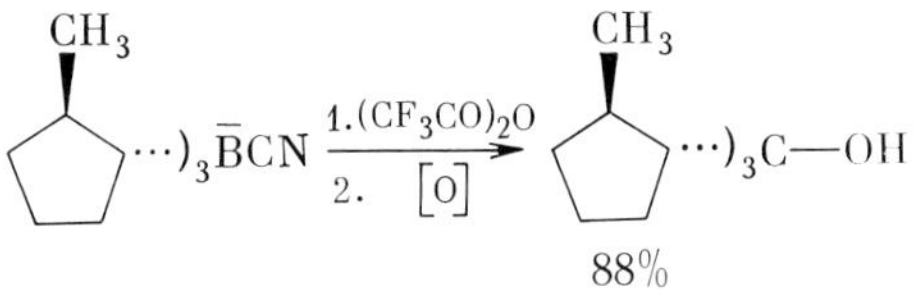

Tri-*exo*-norbornylcyanoborate has been converted to tri-*exo*-norbornylcarbinol in a yield of 79% [92, 95]. *Tris*(4-bromobutyl)-methanol has been obtained from *tris*(4-bromobutyl)borane, KCN, and $(CF_3CO)_2O$ [110] and triphenylmethane from triphenylborane (during oxidation, a protonolysis of the product of migration of three phenyl groups takes place) [98].

The conversion of cycloalkylboranes to tertiary alcohols is accompanied by the formation of methylene compounds, whose yield reaches 20%. For example, the olefin *36* is obtained from dicyclohexyl-*n*-octylcyanoborate and compound *37* from tri-*exo*-norbornylborane:

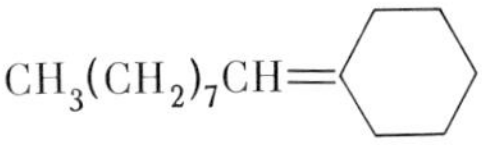

36

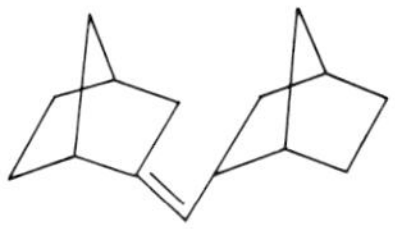

37

REFERENCES

1. E. Negishi, *Intra-Science Chem. Rep.* **7,** 81 (1973).
2. H. C. Brown, *Acc. Chem. Res.* **2,** 65 (1969).
3. H. C. Brown and E. Negishi, *Pure Appl. Chem.* **29,** 527 (1972).
4. E. Negishi and H. C. Brown, *Synthesis* **1974,** 77.

5. P. I. Paetzold and H. Grundke, *Synthesis* **1973,** 635.
6. W. Reppe and A. Magin, *German Patent* 1,076,130 (1960); *C. A.* **55,** P10386i (1961).
7. M. E. D. Hillman, *J. Am. Chem. Soc.* **84,** 4715 (1962).
8. M. E. D. Hillman, *J. Am. Chem. Soc.* **85,** 982 (1963).
9. M. E. D. Hillman, *J. Am. Chem. Soc.* **85,** 1626 (1963).
10. M. E. D. Hillman, *U.S. Patent* 3,206,446 (1965); *C. A.* **63,** 16381e (1965).
11. M. E. D. Hillman, *U.S. Patent* 3,189,627 (1965); *C. A.* **63,** 9984g (1965).
12. H. C. Brown and S. K. Gupta, *J. Am. Chem. Soc.* **93,** 1818 (1971).
13. H. C. Brown and S. K. Gupta, *J. Am. Chem. Soc.* **93,** 4062 (1971).
14. K. V. Puzitskii, S. D. Pirozhkov, K. G. Ryabova, I. V. Pastukhova and Ya. T. Eidus, *Izv. Akad. Nauk SSSR, Ser. Khim.* **1973,** 1817.
15. K. V. Puzitskii, S. D. Pirozhkov, K. G. Ryabova, I. V. Pastukhova and Ya. T. Eidus, *Izv. Akad. Nauk SSSR, Ser. Khim.* **1972,** 1998.
16. G. Wittig, L. Gonsior and H. Vogel, *Ann.* **688,** 1 (1965).
17. A. B. Burg and H. I. Schlesinger, *J. Am. Chem. Soc.* **59,** 780 (1937).
18. J. C. Carter, A. Moe and G. Luther, *J. Am. Chem. Soc.* **96,** 3071 (1973).
19. M. Letmeisl and L. J. Malone, *Inorg. Chem.* **11,** 1245 (1972).
20. A. Kaldor and R. Porter, *J. Am. Chem. Soc.* **93,** 2140 (1971).
21. D. R. Lloyd and N. Linaugh, *Chem. Comm.* **1970,** 1545.
22. D. R. Armstrong and P. G. Perkins, *J. Chem. Soc., A* **1969,** 1044.
23. M. Moireau and A. Veillard, *Theor. Chim. Acta* **11,** 344 (1968).
24. B. M. Mikhailov: "The Chemistry of Bozohydrides", Moscow, Nauka (1967).
25. M. W. Rathke and H. C. Brown, *J. Am. Chem. Soc.* **88,** 2606 (1966).
26. L. J. Malone and M. R. Manley, *Inorg. Chem.* **6,** 2260 (1967).
27. L. J. Malone, *Inorg. Chem.* **7,** 1039 (1968).
28. H. C. Brown and M. W. Rathke, *J. Am Chem. Soc.* **89,** 2737 (1967).
29. H. C. Brown and M. W. Rathke, *J. Am. Chem. Soc.* **89,** 4528 (1967).
30. T. S. Ioffe and A. N. Nesmeyanov: "Methods of Element-Organic Chemistry. Magnesium", Moscow, Publishing House of the Academy of Sciences of the USSR (1963).
31. H. C Brown, E. Negishi and S. K. Gupta, *J. Am. Chem. Soc.* **92,** 6648 (1970).
32. E. Negishi and H. C. Brown, *Synthesis* **1972,** 197.
33. H. C. Brown, G. Kabalka and M. W. Rathke, *J. Am. Chem. Soc.* **89,** 4530 (1967).
34. H. C. Brown and M. W. Rathke, *J. Am. Chem. Soc.* **89,** 2738 (1967).
35. H. C. Brown and E. Negishi, *J. Am. Chem. Soc.* **89,** 5285 (1967).
36. E. Negishi and H. C. Brown, *Synthesis* **1972,** 196.
37. M. F. Hawthorne, W. D. Emmons and K. S. McCallum. *J. Am. Chem. Soc.* **80,** 6393 (1958).
38. M. F. Hawthorne and W. D. Emmons, *J. Am. Chem. Soc.* **80,** 6398 (1958).
39. M. W. Rathke and H. C. Brown, *J. Am. Chem. Soc.* **89,** 2740 (1967).
40. H. C. Brown, M. M. Rogic, M. W. Rathke and G. W. Kabalka, *J. Am. Chem. Soc.* **91,** 2150 (1969).
41. H. C. Brown, R. A. Coleman and M. W. Rathke, *J. Am. Chem. Soc.* **90,** 499 (1968).
42. H. C. Brown, E. F. Knights and R. A. Coleman, *J. Am. Chem. Soc.* **91,** 2144 (1969).
43. H. C. Brown and R. A. Coleman, *J. Am. Chem. Soc.* **91,** 4606 (1969).
44. A. Haag, G. Hesse, *Intra-Science Chem. Rep.* **7,** 105 (1973).
45. W. Kliegel, *Organometal. Chem. Rev., A* **8,** 153 (1972).

46. J. Casanova, Jr. In: Organic Chemistry — Series of Monographs. Isonitrile Chemistry, vol. 20. New York — London, Academic Press (1971), p. 109.
47. G. Hesse and H. Witte, *Angew. Chem.* **75,** 791 (1963).
48. J. Casanova, and R. E. Schuster, *Tetrahedron Letters* **1964,** 405.
49. G. Hesse, H. Witte and G. Bittner, *Ann.* **687,** 9 (1965).
50. G. Hesse and H. Witte, *Ann.* **687,** 1 (1965).
51. J. Casanova and H. R. Kiefer, *J. Org. Chem.* **34,** 2579 (1969).
52. H. Witte, *Tetrahedron Letters* **1965,** 1127
53. C. Tsai and W. Streib, *Tetrahedron Letters* **1968,** 669.
54. J. Casanova, H. R. Kiefer, D. Kuwada and A. H. Boulton, *Tetrahedron Letters* **1965,** 703.
55. S. Bresadola, G. Carraro, C. Pecile and A. Turco, *Tetrahedron Letters* **1964,** 3185.
56. G. Bittner, H. Witte and G. Hesse, *Ann.* **713,** 1 (1968).
57. A. Grote, unpublished data, cited in [44].
58. E. Brehm, A. Haag and G. Hesse, *Ann.* **737,** 80 (1970).
59. K. Nagasawa, *Inorg. Chem.* **5,** 442 (1966).
60. S. Bresadola, F. Rossetto and G. Puosi, *Gazz. Chim. Ital.* **96,** 1397 (1966)
61. G. Hesse, H. Witte and W. Gulden, *Tetrahedron Letters* **1966,** 2707.
62. H. Witte, W. Gulden and G. Hesse, *Ann.* **716,** 1 (1968).
63. H. Witte, P. Mischke and G. Hesse, *Ann.* **722,** 21 (1969).
64. G. Hesse, H. Witte and H. Haussleiter, *Angew. Chem.* **78,** 748 (1966).
65. W. Nagata, M. Yoshioka and S. Hirai, *Tetrahedron Letters* **1962,** 461.
66. A Grote; A. Haag and G. Hesse, *Ann.* **755,** 67 (1972).
67. S. Bresadola, F. Rossetto and G. Puosi, *Tetrahedron Letters* **1965,** 4775.
68. J. Tanaka and J. Carter, *Tetrahedron Letters* **1965,** 329.
69. A. Meller and H. Batka, *Monatsh* **101,** 648 (1970).
70. F. Klages and K. Mönkemeyer, *Chem. Ber.* **85,** 126 (1952).
71. H. J. Kabbe, *Angew. Chem.* **80,** 406 (1968).
72. T. Saguesa, N. Takaishi and Y. Ito. *J. Org. Chem.* **34,** 4040 (1969).
73. A. Meller and H. Batka, *Monaish,* **100,** 1823 (1969).
74. E. A. Lavrent'eva, G. V. Lagodzinskaya, M. L. Khidekel, O. P. Shitov, S. L. Ioffe, V. V. Negrebetskii and V. A. Tartakovskii, *Zh. Obshch. Khim.* **43,** 295 (1973).
75. H. Witte, P. Mischke and G. Hesse, *Z. Naturforsch.* **22b,** 677 (1967).
76. Y. Yamamoto, K. Kondo and I. Moritani, *Tetrahedron Letters* **1974,** 793.
77. H. M. Walborsky and G. E. Niznik, *J. Am. Chem. Soc.* **91,** 7778 (1969).
78. T. Okushi, O. Manabe, H. Hiyama and Z. Yoshida, *J. Chem. Soc. Japan* **72,** 1934 (1969); *C. A.* **72,** 21445a (1970).
79. G. Hesse, H. Witte and W. Gulden, *Angew. Chem.* **77,** 591 (1965).
80. A. Pelter, *Intra-Science Chem. Rep.* **7,** 73 (1973).
81. A. Pelter, *Chem. and Ind.* **1973,** 206.
82. G. Wittig and P. Raff, *Ann.* **573,** 195 (1951).
83. G. Wittig and P. Raff, *Z. Naturforsch.* **6b,** 225 (1951).
84. G. Wittig and W. Herwig, *Chem. Ber.* **88,** 962 (1963).
85. H. Holzapfel, O. Waldner and P. Nenning, *Z. Chem.* **5,** 467 (1965).
86. H. Holzapfel, P. Nenning and P. Döpel, *Z. Chem.* **5,** 315 (1965).
87. H. Witte, E. Brehm and G. Hesse, *Z. Naturforsch.* **22b,** 1083 (1967).
88. E. Brehm A. Haag, G. Hesse and H. Witte, *Ann.* **737,** 70 (1970).
89. A. Pelter, K. Smith, M. Hutchings and K. Rowe, *J. Chem. Soc., Perkin I* **1975,** 129.
90. P. Binger and R. Köster, *Tetrahedron Letters* **1965,** 1901.

91. A. Pelter, M. G. Hutchings and K. Smith, *Chem. Comm.* **1970,** 1529
92. A. Pelter, K. Smith, K. Rowe and M. Hutchings, *J. Chem. Soc., Perkin I* **1975,** 138.
93. A. Pelter, M. Hutchings and K. Smith, *J. Chem. Soc., Perkin I* **1975,** 142.
94. A. Pelter, M. Hutchings and K. Smith, *Chem. Comm.* **1971,** 1048.
95. A. Pelter, M. Hutchings and K. Smith, *Chem. Comm.* **1973,** 186.
96. A. Pelter, M. Hutchings, K. Smith and D. Williams, *J. Chem. Soc., Perkin I* **1975,** 145.
97. A. Pelter, A. Arase and M. Hutchings, *Chem. Comm.* **1974,** 346.
98. G. W. Kabalka J. W. Ferrel, *Synth. Commun.* **9,** 443 (1978).
99. H. C. Brown and J. L. Hubbard, *J. Org. Chem.* **44,** 467 (1979).
100. G. W. Kabalka, E. E. Gooch, C. J. Collins and V. F. Raaen, *Chem. Commun.* **1979,** 607.
101. G. W. Kramer and H. C. Brown *J. Org. Chem.* **42,** 2832 (1977).
102. E. Negishi, M. Sabansky, J. J. Katz and H. C. Brown, *Tetrahedron* **32,** 925 (1976).
103. H. C. Brown, J. L. Hubbard and K. Smith, *Synthesis* **1979,** 701.
104. J. L. Hubbard and H. C. Brown, *Synthesis* **1978,** 676.
105. Y. Yamamoto, K. Kondo and I. Moritani, *Bull. Chem. Soc. Japan* **48,** 3682 (1975).
106. A. Pelter, D. J. Ryden and J. H. Sheppard, *Tetrahedron Letters* **1978,** 4715.
107. S. U. Kulkarni, H. D. Lee and H. C. Brown, *J. Org. Chem.* **45,** 4542, (1980).
108. J. A. Prieto and G. L. Larson, *Synth. Commun.* **10,** 675 (1980).
109. G. Zweifel and N. R. Pearson, *J. Am. Chem. Soc.* **102,** 5919 (1980).
110. J. H. Hallgren and G. M. Lucas, *Tetrahedron Letters* **1980,** 3951.

CHAPTER 11

Reactions of alkyl(aryl)boranes with diazo compounds and ylides

Aliphatic diazo compounds and ylides are sources of carbanions under specific conditions. Their reactions with organoboranes proceed with the elimination of a neutral molecule (N_2, R_3N, R_2S, etc.) and the migration of an organic group from the boron to an α-carbon atom, as a result of which new carbon-carbon and boron-carbon bonds are formed, i.e., the chain length of the radical on the boron atom increases:

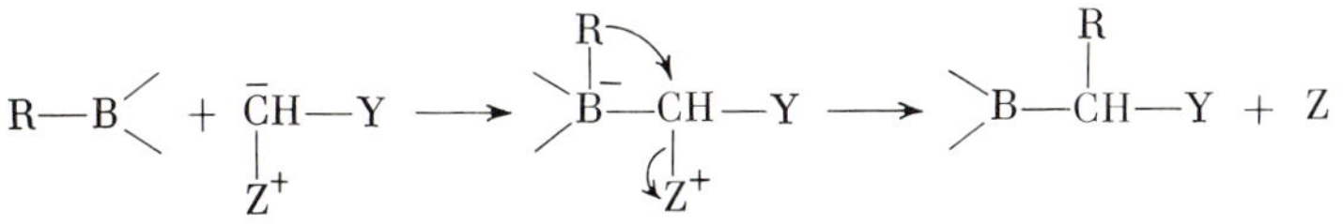

The subsequent deboration of the new organoborane by simple methods (oxidation or protonolysis) leads to various organic compounds.

11.1 REACTIONS WITH DIAZO COMPOUNDS. SYNTHESIS OF ALDEHYDES, KETONES, DIKETONES, NITRILES, AND ESTERS

Diazomethane is converted to polymethylene by the action of trialkylboranes, borates, boron halides [1–8], or thioborates [9]. In some cases, when an excess of a nucleophile (amines or water) is used, higher alcohols are obtained in low yields as a result of the reaction of diazomethane with triorganoboranes, followed by oxidation [5, 6]:

$$(sec\text{-}C_4H_9)_3B \xrightarrow[2.\ [O]]{1.\ CH_2N_2} sec\text{-}C_4H_9CH_2OH,\quad 30\%$$

$$(C_6H_5)_3B \xrightarrow[2.\ [O]]{1.\ CH_2N_2} C_6H_5CH_2OH.\quad 20\%$$

The insertion of a methylene group into a B—Cl bond is observed in the reaction of diazomethane with *B*-trichloroborazole, as a result of which *B*-tris(chloromethyl)borazole is obtained [10], and also in the gas-phase reaction of CH_2N_2 with BF_3, leading to FCH_2BF_2 [11].

A diazoalkane — 2-phenyl-2-methyldiazopropane — is converted to olefinic hydrocarbons by the action of BF_3, BCl_3, or methyl borate [12].

The reactions of boron compounds with functionally substituted diazoalkanes are of preparative significance. These reactions, discovered by Hooz, are used for the synthesis of aldehydes, ketones, diketones, nitriles, esters, their mono- and dideuterated derivatives, and a number of other compounds.

It is assumed that the alkylation of carbonyl diazo compounds proceeds according to the following scheme [13]:

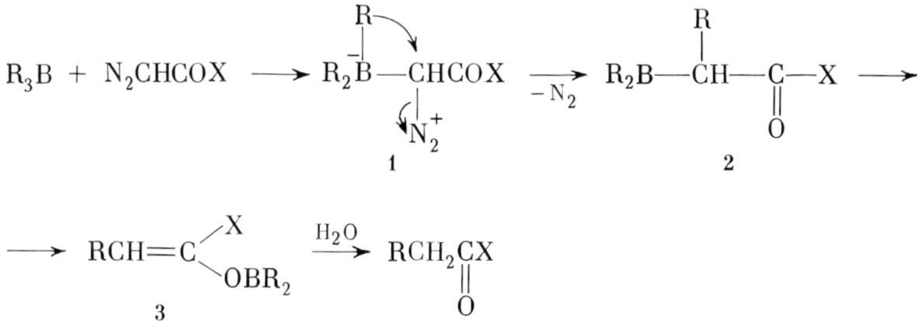

The complex *1* eliminates a nitrogen molecule with simultaneous migration of the radical from the boron to the α-carbon atom, and the β-oxoborane *2* formed then rearranges to the vinyl ester *3*, which is hydrolysed to a carbonyl compound (the properties of vinyloxyboranes (*3*) were considered in Chapter 9).

In aqueous tetrahydrofuran, the reaction of diazoacetaldehyde [14] with trialkyl- or triarylboranes leads to higher aldehydes, the yield of which depends strongly on the steric hindrances in the organoborane and amounts to 33–98% [15]:

$$R_3B + N_2CHCHO \xrightarrow[-N_2]{H_2O,\ THF} \underset{33\text{—}98\%}{RCH_2CHO}$$

$$(R = C_2H_5,\ n\text{-}C_3H_7,\ iso\text{-}C_4H_9,\ sec\text{-}C_4H_9,\ \text{cyclo-}C_5H_9,\ C_6H_5).$$

The reaction of normal trialkylboranes proceeds smoothly at 0–10°C; triisobutylborane reacts at 40°C; and, in the case of secondary trialkylboranes, prolonged heating at 66°C is required. As a result of this difference in the reactivities of primary and secondary organoboranes, trialkylboranes obtained by the hydroboration of terminal olefins can be used in the synthesis of higher aldehydes. For example, tributylborane (containing 7% of secondary butyl groups) prepared from 1-butene reacts with diazoacetaldehyde to form pure *n*-hexanal without an impurity of 3-methylpentanal. Phenylacetaldehyde is obtained in 76% yield from triphenylborane.

The alkylation of diazoketones with trialkylboranes is used for the synthesis of higher ketones [16]:

$$R_3B + N_2CHCOR' \xrightarrow{THF,\ 20°C} RCH{=}C\begin{matrix} R' \\ OBR_2 \end{matrix} \xrightarrow{H_2O} \underset{55\text{—}90\%}{RCH_2COR'}.$$

Reactions of this type proceed readily at room temperature or on heating.

When the reaction of diazoketones with trialkylboranes is carried out in the presence of heavy water, pure α-deuteroketones are obtained [17]:

$$n\text{-}C_6H_{13}CD_2COCH_3 \xleftarrow[D_2O]{N_2CDCOCH_3} (n\text{-}C_6H_{13})_3B \xrightarrow[D_2O]{N_2CHCOR} n\text{-}C_6H_{13}\underset{\substack{|\\D}}{CH}\text{—}COR$$

α,α-Dideuteroketones can be obtained from deuterodiazoacetone by an analogous method.

As a result of the alkylation of cyclic diazoketones with triethylborane, a series of different cyclanones is obtained; several of these are shown in the scheme below [18]:

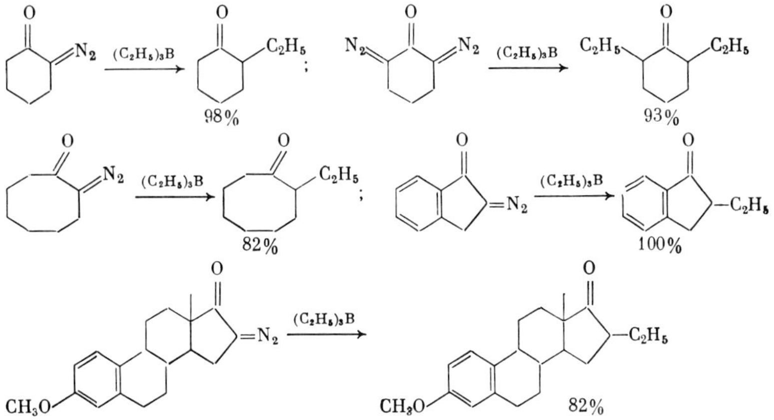

2-Ethyl-5-methylcyclohexanone (93%) and 2-ethylcyclopentanone (92%) have been synthesized by an analogous method.

Hexanol and *cis*-5-hydroxycyclooctylacetone have been obtained by the oxidation of the products of the reaction of 9-hexyl-9-borabicyclo[3.3.1]nonane with diazoacetone [19]:

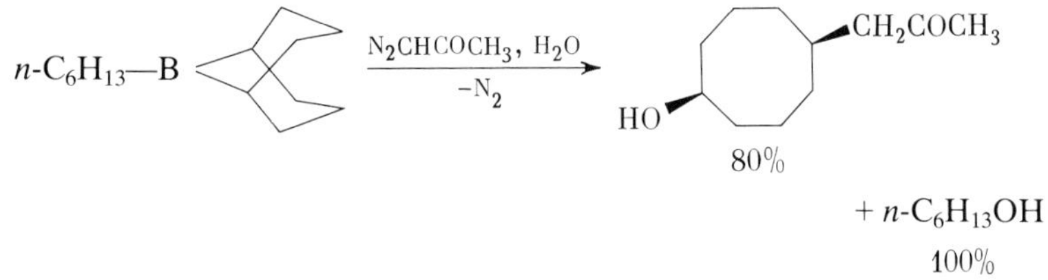

The expected 2-nonanone was not detected, indicating the exclusive participation of the intracyclic B—C bond in the reaction.

Bis-diazoketones, obtained from diacyl chlorides and diazomethane [20], react easily with trialkylboranes at 25°C in tetrahydrofuran to form the corresponding symmetrical 1,4- and 1,5-diketones [21].

Since the latter compounds undergo intramolecular cyclization under the action of alkalies, the reaction has been used for the synthesis of substituted cycloalkenones. Thus, the treatment of the products of the reaction of triethylborane and bis-1,6-diazohexane-2,5-dione with KOH solution leads to 3-propyl-2-ethyl-2-cyclopentenone:

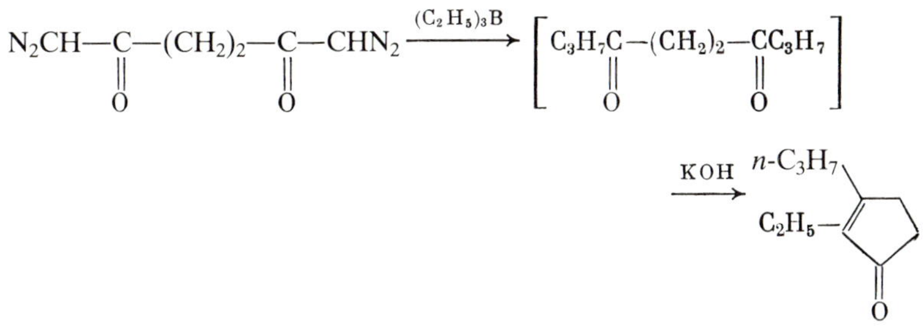

87%

3-Propyl-2-ethyl-2-cyclohexenone has been obtained from bis-1,7-diazoheptane-2,-6-dione by an analogous method.

Most trialkylboranes and triphenylborane react with diazoacetonitrile at 0–20°C, forming, after hydrolysis, nitriles of higher acids in quantitative yield (in the case of triphenylborane, in 52% yield) [22]:

$$R_3B + N_2CHCN \longrightarrow \underset{\sim 100\%}{RCH_2CN}.$$

The structure of the initially formed organoborane has not been established ($RCH{=}C{=}N{-}BR_2$, its dimer, or $R_2BCHRCN$). Pure trialkylboranes, containing no isomeric compounds, should be used in this reaction, since the reactivity of primary and secondary alkylboranes is approximately identical. The sterically hindered alkylboranes with an open chain (tri-*sec*-alkylboranes) react with greater difficulty than normal and cyclic alkylboranes.

Diazoacetic ester is alkylated with trialkylboranes at 0°C or on heating, with the formation (after hydrolysis) of esters of higher acids [22]:

$$R_3B + N_2CHCOOC_2H_5 \xrightarrow[-N_2]{} \underset{4}{RCH{=}C\begin{matrix} OC_2H_5 \\ OBR_2 \end{matrix}} \xrightarrow{H_2O} \underset{40\text{—}85\%}{RCH_2COOC_2H_5}$$

($R = n\text{-}C_6H_{13}$, $n\text{-}C_8H_{17}$, cyclo-C_5H_9, 2-methylpentyl).

Primary alkyl groups migrate from boron to carbon more easily than secondary groups [22, 23]. Thus, only an ester of caprylic acid has been obtained from trihexylborane, prepared from diborane and 1-hexene (7% of *sec*-C_6H_{13}—B<) [22].

Since the initially formed vinyl boron esters *4* are highly active substances, the reaction should be carried out in an aqeuous tetrahydrofuran medium. In this case, the esters *4* undergo hydrolysis immediately after formation. In an anhydrous medium, along with the ester (*4*) ($R = n\text{-}C_3H_7$, 90%), 10% of compound *5* is obtained, the protonolysis of which forms an ester of *n*-propylsuccinic acid (*6*) [16]:

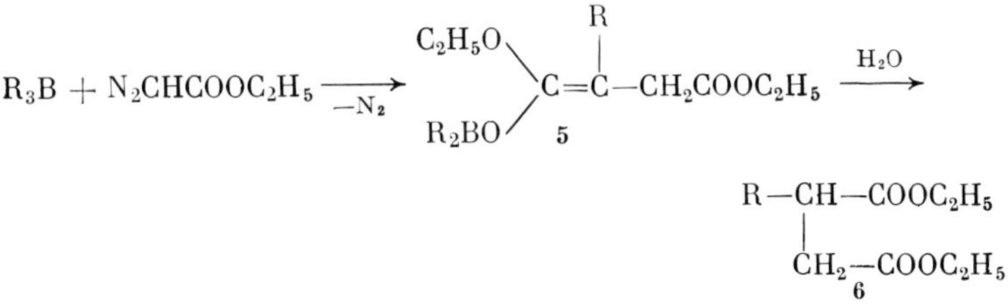

($R = n\text{-}C_3H_7$).

When heavy water is used as a hydrolyzing agent, esters of α-deuterated carboxylic acids or their α,α-dideutero analogs are obtained [17]:

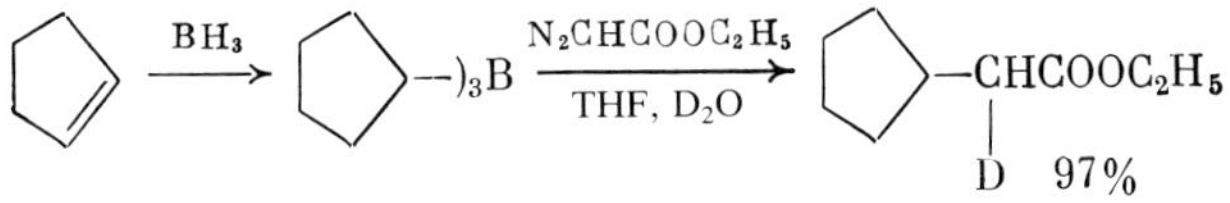

$$(n\text{-}C_6H_{13})_3B + N_2CDCOOC_2H_5 \xrightarrow{D_2O} n\text{-}C_6H_{13}CD_2COOC_2H_5$$

As in other reactions with diazo compounds, only one of the three radicals of the triorganoborane is used in the synthesis of the esters. In the reaction of 9-butyl-[24] and 9-hexyl-9-borabicyclo[3.3.1]nonanes [19] with diazoacetic ester, a quantitative migration of the cyclic bond is observed, as a result of which ethyl *cis*-5-hydroxycyclooctylacetate (*7*) and the corresponding alcohol are obtained after oxidation:

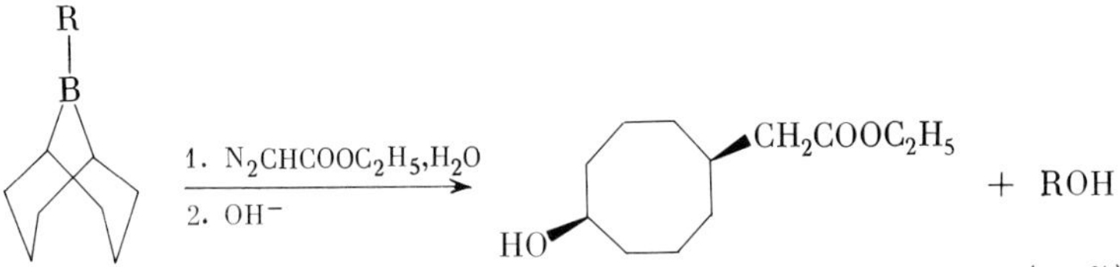

7 (80%) (100%)

This difference in the reactivity of the boron-carbon bonds in acyclic boranes, in which the primary alkyl groups are more active, and in bicyclic compounds is apparently explained by steric factors.

Substituted 9-vinyl-9-borabicyclo[3.3.1]nonanes (*8*), obtained from acetylenes, react with diazoacetic ester to form a mixture of the cyclic ester *7* and a substituted ester of vinylacetic acid (*9*) with a trans location of the substituents with respect to the double bond, indicating a retention of the configuration of the radical during transition from the boron to the carbon [19]:

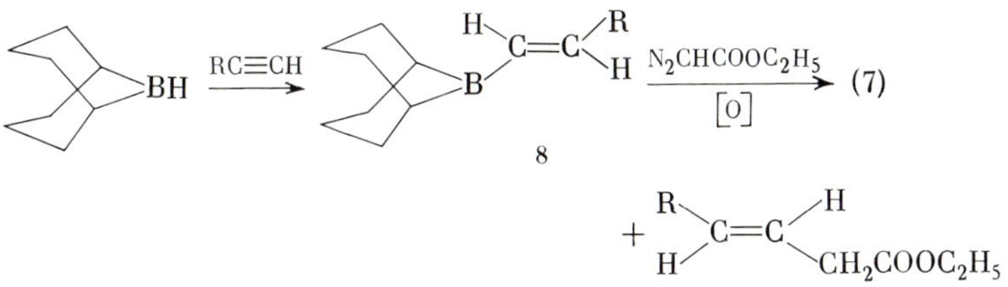

8

9

($R = n\text{-}C_4H_9$, C_6H_5).

Arylboron dichlorides react with diazoacetic ester unambigously, forming the esters of the corresponding arylacetic acids in high yield. At the same time, the reaction of alkylboron dichlorides is complicated by the formation of ethyl chloroacetate (26–32%) [25]:

$$RBCl_2 + N_2CHCOOC_2H_5 \xrightarrow[2.\ H_2O]{1.\ (-60)-(-20)^\circ\ C} \underset{57-98\%}{RCH_2COOC_2H_5} + \underset{0-30\%}{ClCH_2COOC_2H_5}.$$

The three types of radicals investigated can be arranged in the following order with respect to their relative migration capacity: aryl > alkyl > Cl.

Dialkylboron chlorides are active toward diazoacetic ester, reacting at −70°C [26].

11.2 REACTIONS WITH YLIDES. SYNTHESIS OF ALCOHOLS AND ESTERS

The reaction of organoboranes with ylides can be represented by the following scheme:

$$R_3B + \bar{X}-\overset{+}{Y} \rightleftarrows \underset{10}{R_3\bar{B}-X-\overset{+}{Y}} \longrightarrow \underset{11}{R_2\underset{|}{\bar{B}}(R)-X-\overset{+}{Y}} \longrightarrow R_2B-\underset{|}{X}(R) + Y.$$

The initial coordination (*10*) must, to a marked degree, increase the carbanionic nature of the alkyl or aryl groups on the boron and promote reactions in which the R^- anion migrates to the adjacent positively charged atom (*11*).

The reactions of organoboranes with ylides are used for the synthesis of alcohols containing one more carbon atom than the radical of the initial boron compound, and also for the preparation of derivatives of carboxylic acids of the type RCH_2COX (an increase of the chain length by two carbon atoms and introduction of a functional group).

$$R{-}CH_2OH \xleftarrow{(CH_3)_3\overset{+}{N}\overset{-}{C}H_2} R_3B \xrightarrow{(CH_3)_2\overset{+}{S}\overset{-}{C}HCOX} R{-}CH_2COX$$

This method of increasing the length of the carbon chain has two serious disadvantages. First, only one of the three radicals of the organoborane R_3B participates in the reactions, and a second R group is used only by using a large excess of ylide. On the other hand, secondary trialkylboranes are more active with respect to most ylides than the primary compounds. It is therefore necessary to use pure organoboranes, and one should not use trialkylboranes obtained by the hydroboration of terminal olefins (approximately 7% of secondary alkyl groups).

The reactivity of ylides toward organoboranes increases with their basicity. For example, trimethylammonium methylide $(CH_3)_3\overset{+}{N}\overset{-}{C}H_2$ and dimethylsulfonium methylide $(CH_3)_2\overset{+}{S}\overset{-}{C}H_2$ react with trialkylboranes at temperatures below 0°C, whereas triphenylphosphonium methylide, $(C_6H_5)_3PCH_2$, reacts only when heated to above 100°C.

The reaction of trimethylammonium methylide (*12*) with trihexyl- or triphenylborane in a ratio of 1:1 leads (after oxidation) to heptyl or benzyl alcohol [27–29]:

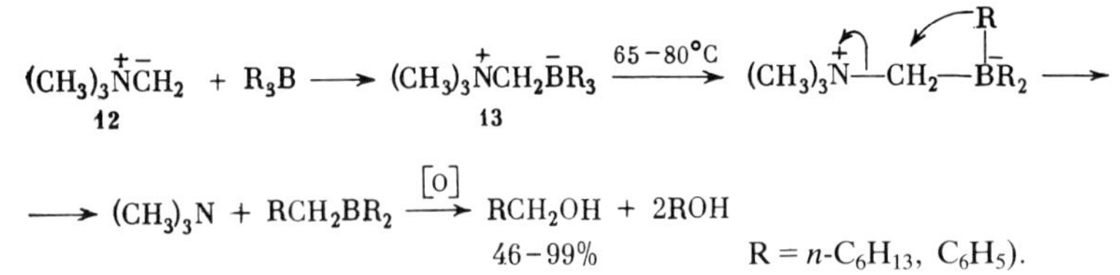

$R = n\text{-}C_6H_{13}, C_6H_5$).

The secondary alkyl groups migrate from the boron to the carbon approximately five times more easily than the primary groups. This is indicated by the formation of 60% of 2,3-dimethylbutanol and only 6% of 1-heptanol by the action of a nitrogenous ylide on hexyldi(1,2-dimethylpropyl)borane.

The hydrazine ylide $(CH_3)_3\overset{+}{N}—\overset{-}{N}H$ forms the betaine $(CH_3)_3\overset{+}{N}—NH\overset{-}{B}(C_6H_5)_3$ with triphenylborane [30].

Dimethylsulfonium methylide reacts with trialkylboranes and triphenylborane in dimethyl sulfoxide under cooling, with the formation of dimethyl sulfide. Subsequent oxidation of the reaction products leads to homologs of alcohols containing one carbon atom more than the radical of the initial organoborane [31]:

$$R_3B + \overset{-}{C}H_2\overset{+}{S}(CH_3)_2 \xrightarrow[-(CH_3)_2S]{\text{DMSO},\ -10°C} \xrightarrow{H_2O_2,\ OH^-} \underset{68-74\%}{ROH} + \underset{26-32\%}{RCH_2OH}.$$

In this way, a mixture of 2,3-dimethyl-1-butanol and octanol in a ratio of 4:1 has been obtained from heptyldi(1,2-dimethylpropyl)borane, i.e., *sec*-alkylboranes are approximately twice as reactive as the primary compounds.

The reaction of R_3B with dimethylsulfoxonium methylide proceeds less smoothly at 0°C: 30% of RCH_2OH and approximately 8% of RCH_2CH_2OH are obtained [32].

The transfer of a radical from the boron to the carbon of the ylide takes place with retention of the configuration. This has been demonstrated for the example of the reaction of tri-*exo*-norbornylborane with dimethylsulfonium [31] and dimethylsulfoxonium [32] methylides — 2-*exo*-hydroxymethylnorbornane is obtained in both cases.

Many known functionally substituted ylides [33] also react with organoboranes. In the reaction of dimethylsulfonium phenacylide with tripropylborane in absolute tetrahydrofuran, *cis*-vinyloxy(dipropyl)borane is obtained almost quantitatively, with a small impurity of the trans isomer [34]:

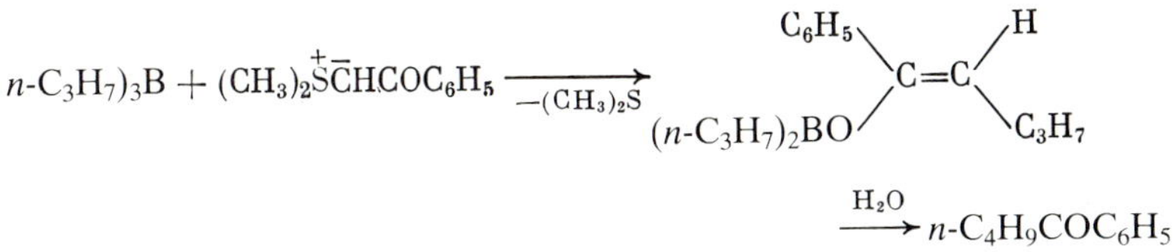

In the reaction of organoboranes with esters and amides of dimethylsulfuranylideneacetic acid, the intermediate products are vinyloxyboranes of the type *14*, which are oxidized in alkaline medium without isolation [35]:

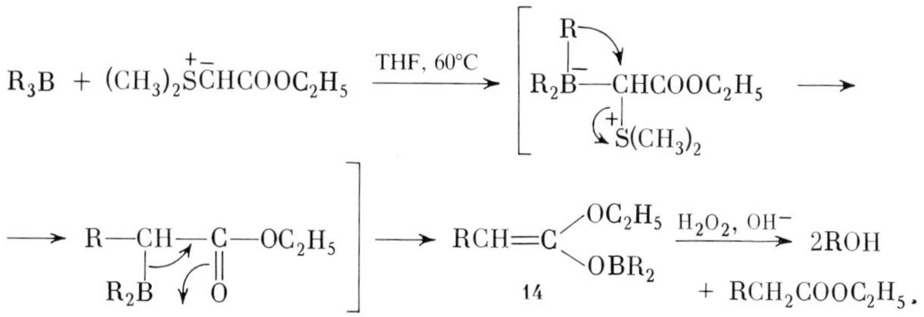

The action of trihexylborane on *N,N*-diethyl(dimethylsulfuranylidene)acetamide produces the corresponding amide *15*:

$$R_3B + (CH_3)_2\overset{+}{S}\overset{-}{C}HCON(C_2H_5)_2 \xrightarrow{THF} \xrightarrow{H_2O_2,\ OH^-} \underset{15\ (100\%)}{RCH_2CON(C_2H_5)_2} + 2ROH.$$

Methylenetriphenylphosphorane, and also ethylidene- and chloromethylenetriphenylphosphoranes form stable complexes of the betaine type, $R_3\overset{+}{P}CH_2\overset{-}{B}X_3$, with triphenylborane, BF_3, BCl_3[36], and B_2H_6 [36, 37]. The adduct $(C_6H_5)_3\overset{+}{P}CH_2\overset{-}{B}(C_6H_5)_3$, obtained from methylenetriphenylphosphorane and triphenylborane or by the action of phenylmagnesium bromide on $(C_6H_5)_3\overset{+}{P}CH_2\overset{-}{B}F_3$ [36], is converted upon heating to triphenylphospine, its complex with $(C_6H_5)_3B$, and tribenzylborane [38].

The reaction of equimolar quantities of triphenylarsonium benzylide (16) with trihexylborane proceeds at room temperature. After hydrolysis, the products of single (*17*) and double (*18*) insertion are obtained in a ratio of 3:1 [35]:

$$\underset{16}{(C_6H_5)_3\overset{+}{A}s\overset{-}{C}HC_6H_5} + (n\text{-}C_6H_{13})_3B \longrightarrow \xrightarrow{H_2O_2,\ OH^-} \underset{17}{C_6H_5CH_2C_6H_{13}} +$$

$$+ \underset{18}{C_6H_5CH_2\underset{\displaystyle C_6H_5}{\underset{|}{C}H}\text{—}C_6H_{13}} + C_6H_{13}OH.$$

The rate of reaction of the initially formed benzyl derivatives of boron, $n\text{-}C_6H_{13}CH(C_6H_5)B\!<$, with the ylide *16* seems to be considerably higher than that of trihexylborane, as a result of which the product of double insertion of the benzylidene fragment is obtained in significant amount. On the other hand, benzylboranes are easily hydrolyzed by alkalies. The oxidation of the reaction mixture in aqueous alkali therefore does not lead to alcohols in this case, but to the hydrocarbons *17* and *18*.

The betaine adduct $(C_6H_5)_3\overset{+}{A}sCH_2\overset{-}{B}F_3$ has been obtained from triphenylarsinomethylene and BF_3 [39].

Two other reactions should also be included in the conversions of organoboranes considered in this section.

The α-lithium derivatives of benzyl sulfides (*19*), which can be easily obtained by the metalation of benzyl phenyl sulfide or benzyl 2-pyridyl sulfide at −70°C [40], react with trialkylboranes at −75°C, forming the alkylbenzenes *20* after alkaline hydrolysis [41]:

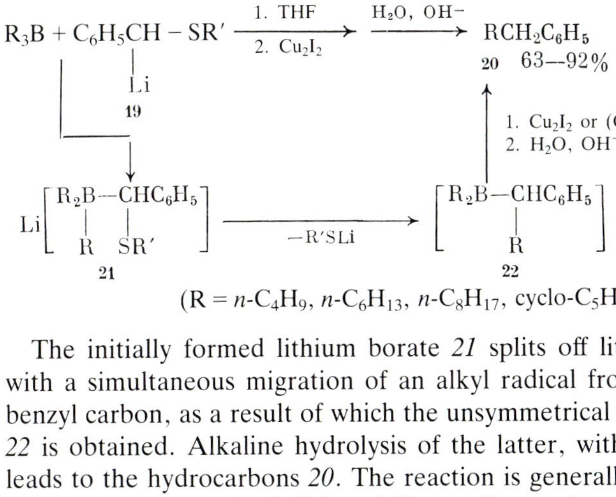

($R = n\text{-}C_4H_9$, $n\text{-}C_6H_{13}$, $n\text{-}C_8H_{17}$, cyclo-C_5H_9, cyclo-C_6H_{11}).

The initially formed lithium borate *21* splits off lithium thiolate with a simultaneous migration of an alkyl radical from boron to a benzyl carbon, as a result of which the unsymmetrical trialkylborane *22* is obtained. Alkaline hydrolysis of the latter, without isolation, leads to the hydrocarbons *20*. The reaction is generally complicated by the formation of 1,2-diphenylalkanes (approximately 25%), the amount of which can be kept to a minimum by treating the reaction mixture with cupric iodide at −75°C or with dimethyl sulfate at room temperature.

Treatment of dibutyl iodomethylboronate with sodium azide in butanol produces formaldehyde and butyl borate [42]:

$$ICH_2B(OC_4H_9)_2 \xrightarrow{NaN_3} [N_3CH_2B(OC_4H_9)_2] \xrightarrow{C_4H_9OH} N{\equiv}\overset{+}{N}NHCH_2\overset{-}{B}(OC_4H_9)_3 \longrightarrow$$
$$\longrightarrow N_2 + B(OC_4H_9)_3 + NH{=}CH_2 \longrightarrow CH_2O.$$

REFERENCES

1. H. Meerwein, *Angew. Chem.* **60,** 78 (1948).
2. C. E. H. Bawn and A. Ledwith: In: *Progress in Boron Chemistry*, vol. 1. Ed. H. Steinberg and A. L. McCloskey. Oxford, Pergamon Press, Ch. 8 (1964).
3. C. E. H. Bawn, A. Ledwith and P. Matties, *J. Polymer Sci.* **34,** 95 (1959).
4. A. Davies, D. G. Hare, O. R. Khan and J. Sikora, *Proc. Chem. Soc.* **1961,** 172.
5. A. Davies, D. G. Hare, O. R. Khan and J. Sikora, *J. Chem. Soc.* **1963,** 4461.
6. J. E. Leffler and B. G. Ramsey, *Proc. Chem. Soc.* **1961,** 117.
7. G. W. Cowell and A. Ledwith, *Quart. Rev.* **24,** 119 (1970).
8. M. G. Krakovyak, E. V. Anufrieva and S. S. Skorokhodov, *Vysokomol. Soedin.* **8,** 1681 (1966).
9. C. D. Gutsche and K. Kinoshita, *J. Org. Chem.* **28,** 1762 (1963).
10. H. S. Turner, *Chem. and Ind.* **1958,** 1405.
11. J. Goubeau and K. H. Rohwedder, *Ann.* **604,** 168 (1957).
12. B. H. Philip, M. C. Lowery and J. Havel, *Tetrahedron Letters* **1967,** 5049.
13. H. O. House: "*Modern Synthetic Reactions*", Second Ed. California. W. A. Benjamin Inc. (1972), p. 138.
14. L. Arnold, *Chem. Comm.* **1967,** 299.
15. J. Hooz and G. Morrison, *Can. J. Chem.* **48,** 868 (1970).
16. J. Hooz and S. Linke, *J. Am. Chem. Soc.* **90,** 5936 (1968).
17. J. Hooz and D. M. Gunn, *J. Am. Chem. Soc.* **91,** 6195 (1969).
18. J. Hooz, D. M. Gunn and H. Kono, *Can. J. Chem.* **49,** 2371 (1971).
19. J. Hooz and D. M. Gunn, *Tetrahedron Letters* **1969,** 3455.
20. E. Fahr, *Ann.* **638,** 1 (1960).
21. J. Hooz and D. Gunn, *Chem. Comm.* **1969,** 139.
22. J. Hooz and S. Linke, *J. Am. Chem. Soc.* **90,** 6891 (1968).
23. J. Hooz and J. N. Bridson, *Can. J. Chem.* **50,** 2387 (1972).
24. H. C. Brown and M. M. Rogič, *J. Am. Chem. Soc.* **91,** 2146 (1969).
25. J. Hooz, J. N. Bridson, J. G. Calzada, H. C. Brown, M. M. Midland and A. B. Levy, *J. Org. Chem.* **38,** 2574 (1973).
26. H. C. Brown, M. M. Midland and A. B. Levy, *J. Am. Chem. Soc.* **94,** 3662 (1972).
27. W. K. Musker and R. R. Stevens, *Tetrahedron Letters* **1967,** 995.
28. W. K. Musker and R. R. Stevens, *Inorg. Chem.* **8,** 255 (1969).
29. F. Bickelhaupt and J. Barnick, *Rec. Trav. Chim.* **87,** 188 (1968).
30. R. Appel, H. Heinen and R. Schölhorn, *Chem. Ber.* **99,** 3118 (1966).
31. J. J. Tufariello, P. Wojtkowski and L. T. C. Lee, *Chem. Comm.* **1967,** 505.
32. J. J. Tufariello and L. T. C. Lee, *J. Am. Chem. Soc.* **88,** 4757 (1966).
33. A. W. Johnson: "Ylide Chemistry", Academic Press, N.Y.-London (1966).
34. D. Pasto and P. Wojtkowski, *Tetrahedron Letters* **1970,** 215.
35. J. J. Tufariello, L. T. C. Lee and P. W. Wojtkowski, *J. Am. Chem. Soc.* **89,** 6804 (1967).
36. D. Seyferth and S. O. Grim, *J. Am. Chem. Soc.* **83,** 1613 (1961).
37. M. F. Hawthorne, *J. Am. Chem. Soc.* **83,** 367 (1961).
38. R. Köster and B. Rickborn, *J Am. Chem. Soc.* **89,** 2782 (1967).
39. D. Seyferth and H. M. Cohen, *J. Inorg. Nucl. Chem.* **20,** 73 (1961).
40. T. Mukaiyama, K. Narasaka, K. Moekawa and M. Furusato, *Bull Chem. Soc. Japan* **44,** 2285 (1971).
41. T. Mukaiyama, S. Yamamoto and M. Shiono, *Bull. Chem. Soc. Japan* **45,** 2244 (1972).
42. D. S. Matteson and T. C. Cheng, *J. Org. Chem.* **33,** 3055 (1968).

CHAPTER 12

Reactions of alkyl(aryl)boranes with halomethanes, thioacetals, α-halocarbonyl compounds, and α-halonitriles

Organoboranes are capable of alkylating some aliphatic halides, thio compounds, and esters. Reactions of alkyl- and arylboranes with haloforms, thioacetals, α-halocarbonyl compounds, and α-halonitriles initiated by bases are considered below. The role of the bases in these processes consists of the generation of carbanions or carbenes, which then react with the organoboranes to form α-thio or α-halo derivatives of boron; the latter, as a rule, are unstable and rearrange rapidly by intramolecular migration of the radical from the boron to the carbon and of the electronegative group to the boron:

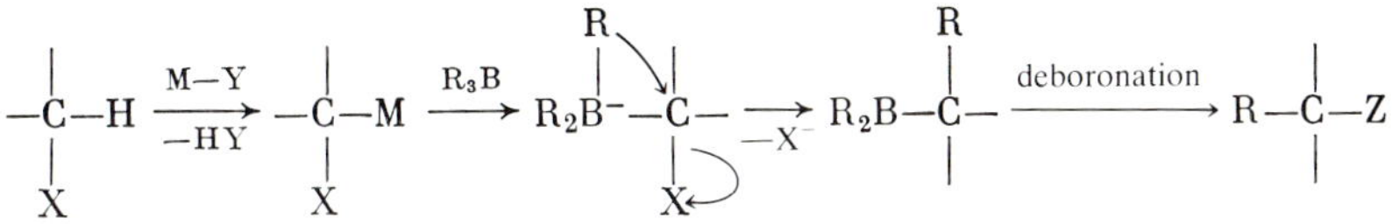

As a result of this rearrangement, a new carbon-carbon bond is formed. The following deboronation of the organoborane obtained by the use of appropriate reagents leads to organic compounds containing an additional one or more extra carbon atoms in the carbon chain, as compared with the radical of the starting organoborane.

12.1 ALKYLATION OF HALOMETHANES AND THIOACETALS

One of the methods of preparation of α-halo-, α-alkylthio-, and α-arylthioalkyltrialkylborates, $-\overset{|}{C}X\text{-}\bar{B}R_3$, which rearrange spontaneously or under the action of various reagents to form a new C—C bond, consists of the reaction of trialkyl- or triarylboranes with derivatives of halomethanes and thioacetals.

It is suggested that carbanions X_3C^- or carbenes $:CX_2$ are formed as intermediate particles. The final result of the conversions is the alkylation of haloforms and thioacetals. These reactions represented the basis of simple methods for the synthesis of secondary and tertiary alcohols, ketones, and olefins:

$$R_3B + HCX_3 \longrightarrow \longrightarrow R_3COH \text{ or } R_2C{=}O,$$

$$R_3B + H_2CX_2 \longrightarrow \longrightarrow R_2CHOH,$$

$$(RCH_2)_3B \xrightarrow{:CX_2} RCH_2CH{=}CHR \qquad (X = \text{Hal}, R'S).$$

The alkylation of haloforms with organoboranes is discussed, in part, in review articles [1, 2].

12.1.1 Synthesis of alcohols and olefins

Köbrich and Merkle [3, 4] found that diphenylcarbinol, with an impurity of benzophenone, is formed by the reaction of triphenylborane with dichloromethyllithium (obtained from C_4H_9Li and CH_2Cl_2) and subsequent oxidation. According to their concepts, the initially formed borate *1* splits off LiCl to form α-chlorobenzylborane(*2*), which rearranges further to the chloroborane *3*, a precursor of diphenylcarbinol:

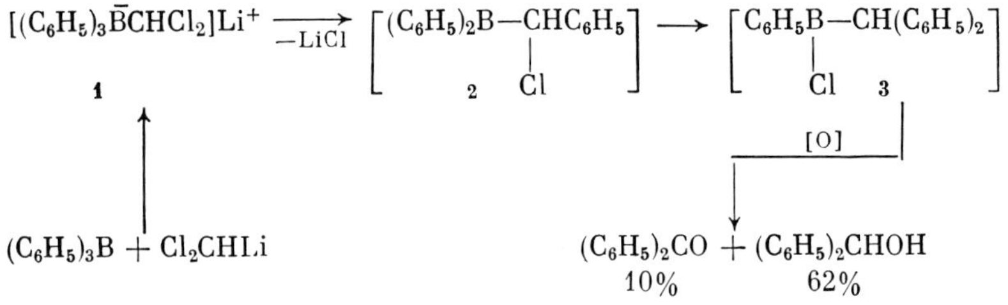

Brown and co-workers [5, 6] developed this method for the synthesis of trialkylcarbinols and dialkyl ketones. Halomethanes and sterically hindered bases were used for the generation of the halomethyl anions. Matteson and Ray [43a], on the basis of the reaction of (+)- or (−)-pinanediol phenylboronates with dichloromethyllithium and methylmagnesium bromide, synthesized (*S*)-1-phenylethanol, (2*S*, 3*S*)-3-phenyl-2-butanol, and (2*R*, 3*S*)-3-phenyl-

2-butanol. A method for the synthesis of aldehydes from RLi (RMgX) and $CHCl_2B(OR')_2$ has been described [43b].

The reaction of trichloromethyllithium [7] with tributylborane led, after oxidation, to a 37% yield of tributylcarbinol [5]. The latter is obtained more smoothly by the reaction of tributylborane with $CHCl_3$ in the presence of two moles of lithium triethylmethoxide:

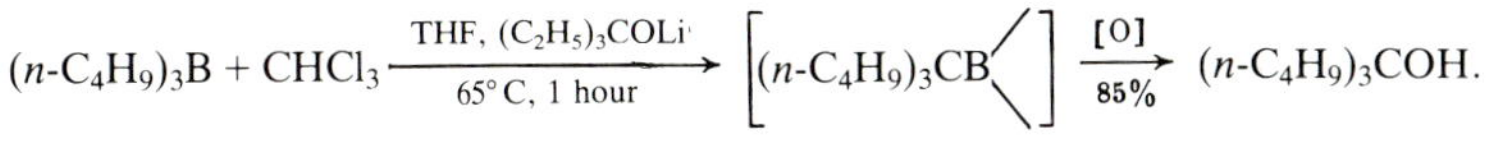

Dichlorofluoromethane, chlorodifluoromethane, or dichloromethyl methyl ether can be used instead of chloroform.

The most convenient method for the synthesis of tertiary alcohols is, apparently, the reaction of dichloromethyl methyl ether with symmetrical [6], mixed [8, 44–46], and cyclic [6, 8] organoboranes, which proceeds at 0–25° C:

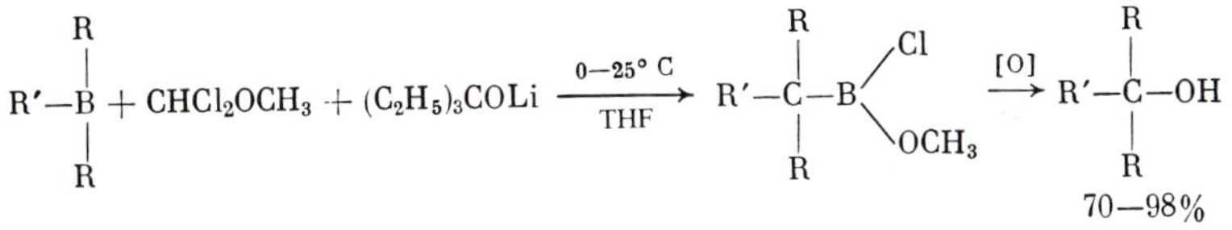

Lithium *tert*-butoxide (with *sec*-R_3B) can be used as a base.

The above method has been used for converting boracyclanes to carbocyclic alcohols. Thus, 1-cyclohexylcyclohexanol has been obtained from 1-cyclohexylborinane and 9-*tert*-butyl-9-borabicyclo[3.3.1]nonane has been converted to 9-*tert*-butylbicyclo-[3.3.1]nonan-9-ol:

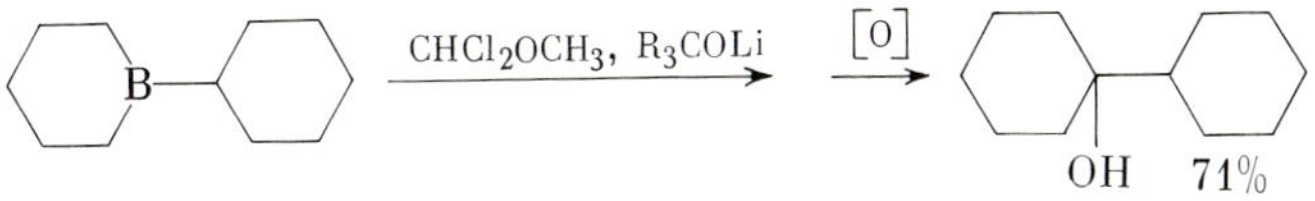

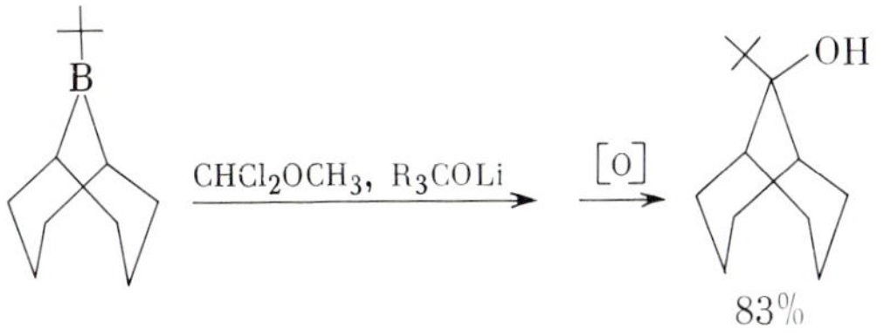

Since the reaction with a dihalo ether proceeds under very mild conditions, the migration of radicals from boron to carbon is not accompanied by isomerization, as is the case in the carbonylation of sterically hindered organoboranes and boracyclanes (at 150°C and 75 atmospheres) [9]. In the synthesis of triarylcarbinols, the oxidation of the initially formed boron derivative $Ar_3CB(Cl)OCH_3$ is carried out in sodium acetate solution.

The reaction of triisobutylborane, and also of tri-*sec*-alkylboranes, with chlorodifluoromethane or dichlorofluoromethane in the presence of lithium triethylmethoxide leads to triethylmethyl esters of fluoro-*tert*-alkylboronic acids (*4*) [10]:

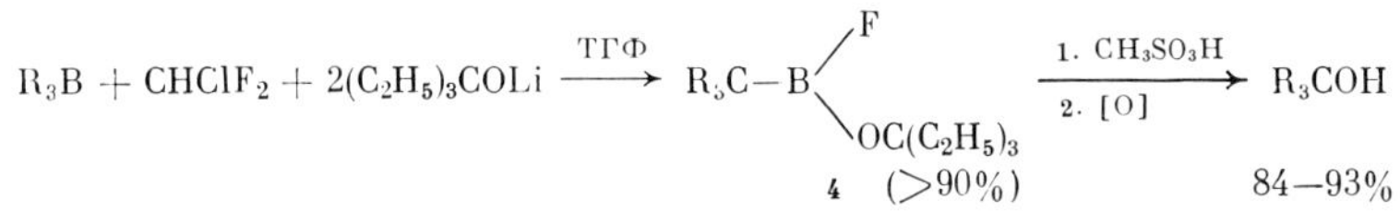

The compounds *4* are resistant to oxidation (O_2, H_2O_2, R_3NO). To convert these esters to trialkylcarbinols, they are initially heated with methylsulfuric acid and are then oxidized with hydrogen peroxide in alkaline medium.

The reactions of R_3B with $(CH_3)_3SiCH(X)Li$ [47, 48] and $(CH_3)_3SiCX_2Li$ [49] have been used for the synthesis of silicon-containing alcohols of the type $(CH_3)_3SiCHROH$ and $(CH_3)_3SiCR_2OH$.

Organoboranes alkylate not only haloforms and halo ethers but also thioketals. The reaction with the lithium derivative of the diphenylthioketal of butyraldehyde, followed by treatment with mercuric chloride [11, 50] or methyl fluorosulfonate [12] and hydrogen peroxide, leads to unsymmetrical alcohols:

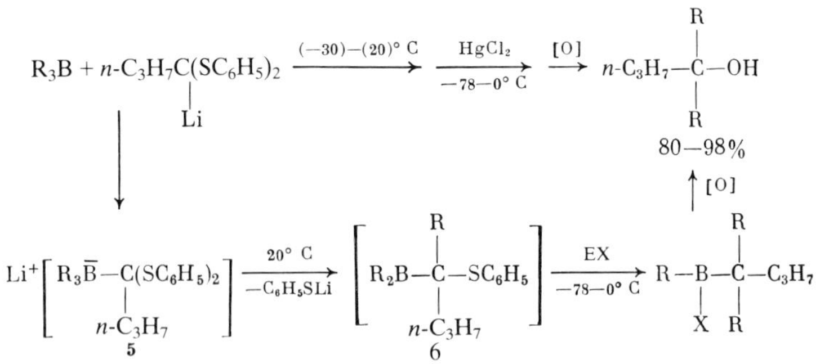

As has been shown by Yamamoto, Shiono, and Mukaiyama [13], the borate *5* formed in the first stage rearranges spontaneously with the formation of α-phenylthioalkylboranes of the type *6*. The migration of a second radical from boron to carbon with the replacement of a mercapto group takes place only under the action of CH_3SO_2F or $HgCl_2$. Higher yields of tertiary alcohols are obtained by the use of 2-lithio-1,3-benzodithioles [51]. The reactions of the α-thioorganoborate complexes obtained from R_3B and $LiCH_2X$ ($X = SCH_3$, $SOCH_3$, $SO_2C_6H_5$) with methyl iodide can be used for the synthesis of primary alcohols, RCH_2OH [52]. Secondary alcohols, RR′CHOH, can be obtained from R_3B and $C_6H_5SO_2CH(M)R'$ (M = Li, MgX) [53].

As a result of the treatment with a halo ether of lithium tetraalkylborate (*7*) and subsequent oxidation, secondary alcohols of the type *9* have been obtained [14]:

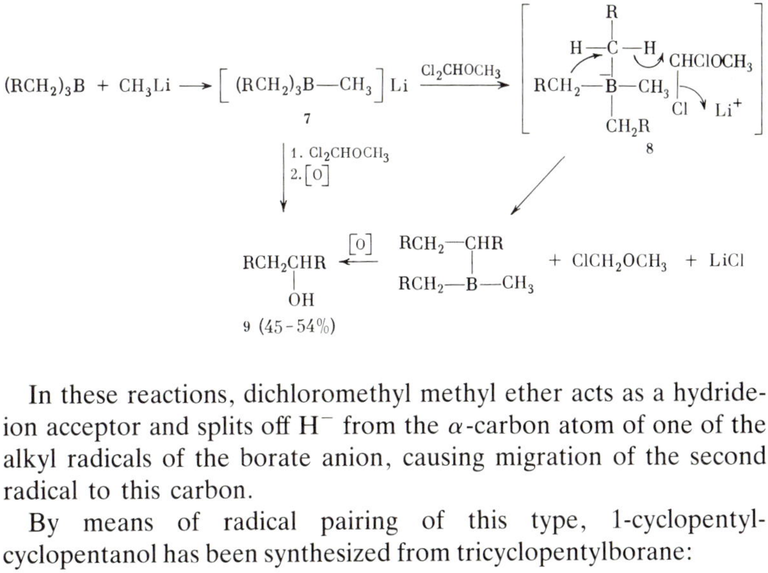

In these reactions, dichloromethyl methyl ether acts as a hydride-ion acceptor and splits off H^- from the α-carbon atom of one of the alkyl radicals of the borate anion, causing migration of the second radical to this carbon.

By means of radical pairing of this type, 1-cyclopentylcyclopentanol has been synthesized from tricyclopentylborane:

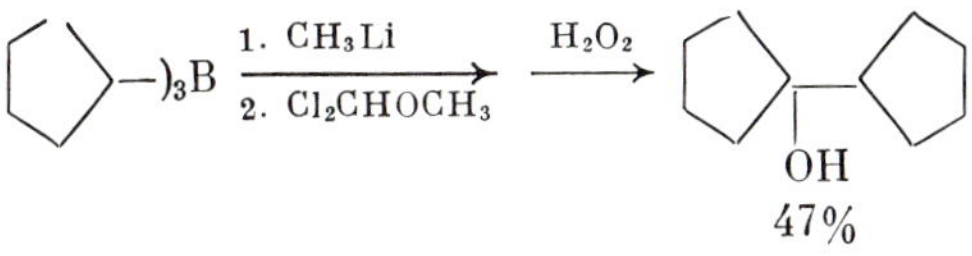

Dichlorocarbene, :CCl_2 [15–17], is an intermediate in the reaction of (bromodichloromethyl)phenylmercury with trialkylboranes, which leads to olefinic hydrocarbons [18]:

$$(RCH_2)_3B + C_6H_5HgCCl_2Br \xrightarrow[60-70°\,C]{C_6H_6} \underset{58-68\%}{RCH_2CH{=}CHR}. \qquad (cis:trans \sim 3:2)$$

In the presence of cyclohexene, the reaction of triphenylborane with (bromodichloromethyl)phenylmercury leads to a mixture of 7-chloro-7-phenylnorcarane and trichlorostyrene, indicating the intermediate formation of phenylchlorocarbene, $C_6H_5\ddot{C}Cl$:

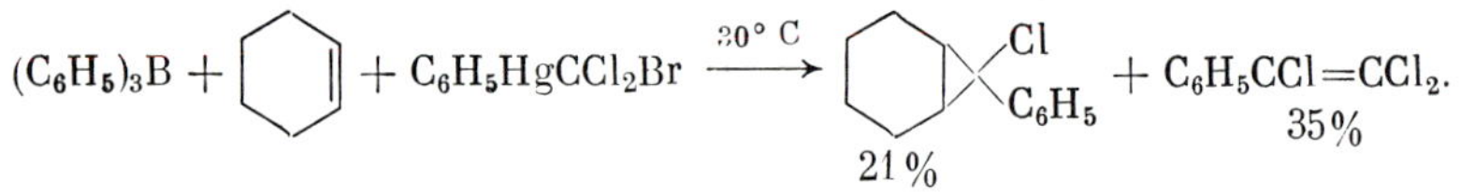

Under analogous conditions, 1,2-dichloro-2-methoxy-1-phenylethylene, $C_6H_5CCl{=}CClOCH_3$, is obtained from dimethyl phenylboronate, together with norcarane (27%) and styrene (23%).

As has been suggested in [19], the electrophilic carbene :$C(NRR')_2$ is an intermediate particle in the reaction of trialkylboranes with bis(1,3-diphenyl-2-imidazolidinylidene), which leads to tertiary alcohols (Chapter 13).

Jäger and Hesse [20] have shown that 4-octene is obtained by heating lithium tetrabutylborate with benzyl chloride in an autoclave:

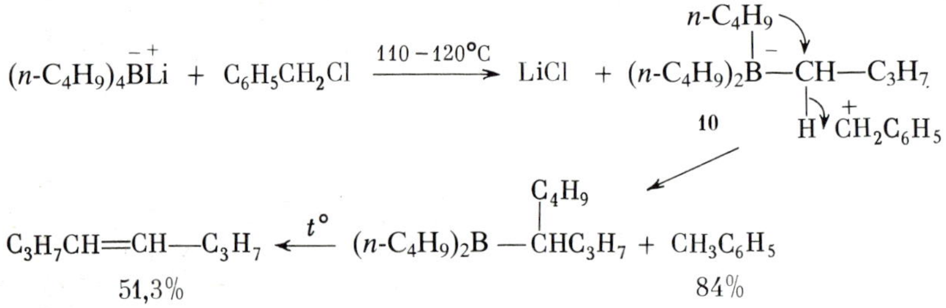

The benzyl cation abstracts a hydride ion from the α-carbon atom of the butyl radical in the borate anion *10*, causing the migration of a second radical from the boron to this carbon.

12.1.2 Synthesis of aldehydes and ketones

The reaction of trialkylboranes or of esters of dialkylborinic acids with α-lithium derivatives of thioacetals and α-halo ethers is used for the synthesis of aldehydes and also of symmetrical and unsymmetrical ketones:

$$R_3B + LiCH(SR')_2 \xrightarrow[75-95\%]{} RCHO.$$

The lithium derivatives of formaldehyde thioacetal (*11*, R′ = H) or of the thioacetals of higher aldehydes (*11*, R′ = Alk) react with trialkylboranes to form borates of the type *12*, which rearrange to form new trialkylboranes *13* by the migration of one alkyl group from boron to carbon, splitting off lithium phenylthiolate [13]:

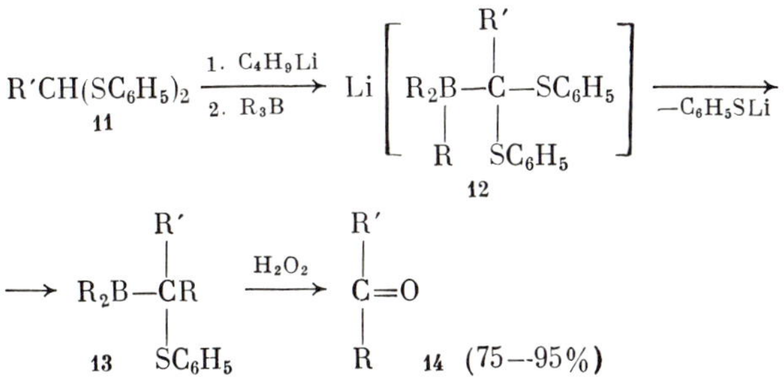

(R′ = H, Alk).

Oxidation of the borane *13* with hydrogen peroxide in sodium acetate or a buffer at pH 7.0 leads to aldehydes (*14*, R′ = H) or ketones (*14*, R′ = Alk). This method may compete with the method for the preparation of aldehydes and ketones from lithium derivatives of thioacetals and alkyl halides [21], since trialkylboranes are easily obtained from olefins.

For the preparation of ketones from R_3B, the compounds *11* can be replaced by 2-lithio-2-alkyl-1,3-benzodithioles [54] or by catechol dichloromethylene ether and methyllithium [55].

Ketones can be obtained from triorganoboranes and $C_6H_5(CH_3)_2SiCCl_2Li$, followed by oxidation with chromic acid [49].

Esters of dialkylborinic acids, $R_2BOR'_2$, are easily converted by the action of dichloromethyl methyl ether and lithium triethylmethylate to esters of chlorodialkylmethylboronic acids, the oxidation of which (without their isolation) leads to the corresponding ketones in high yield [22, 56]:

$$R_2BOR' + Cl_2CHOCH_3 + 2(C_2H_5)_3COLi \xrightarrow[0,5\ \text{hour}]{\text{THF, } 25°\text{ C}} R_2C(Cl){-}B(OR'')_2 \xrightarrow[54-96\%]{H_2O_2,\ OH^-} R_2CO.$$

The transfer of a radical from boron to carbon takes place with retention of the configuration, which is confirmed by the formation of bis(*trans*-2-methylcyclopentyl) ketone from methyl bis(*trans*-2-methylcyclopentyl)borinate.

The conversion of esters of *sec*-dialkylborinic acids to ketones is complicated by the formation of olefins [22]. The reaction may be completely directed toward the formation of a substituted alkene, if the ester $R_2CClB(OR')_2$ formed is treated with a base or with silver nitrate [56].

12.2 ALKYLATION AND ARYLATION OF α-HALOCARBONYL COMPOUNDS AND α-HALONITRILES

Under the action of bases, organoboranes alkylate α-halocarbonyl compounds. This reaction is used for the synthesis of esters of higher organic acids, their nitriles and dinitriles, α-cyano ethers, ketones, and sulfonyl compounds:

$$R_3B + XCH_2COY \xrightarrow{\text{base}} RCH_2COY.$$

In combination with hydroboration, reactions of this type represent a method for increasing the length of the carbon skeleton of olefins by one, two, or more carbon atoms, with the simultaneous introduction of functional groups.

It has been suggested that the alkylation of α-halocarbonyl compounds with organoboranes proceeds in four consecutive stages [2]:

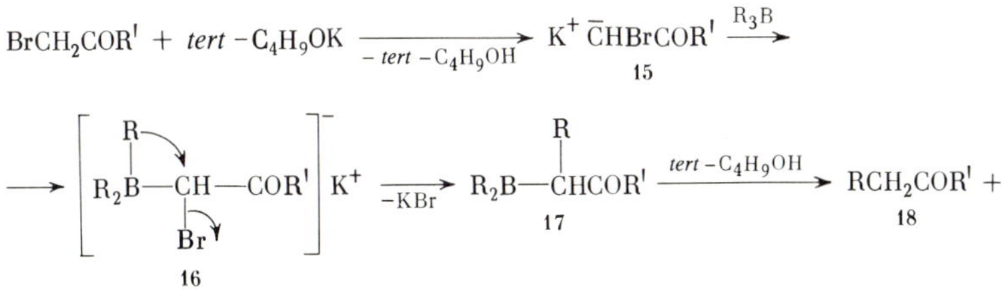

$+ \textit{tert}-C_4H_9OBR_2$ $(R' = \textit{alkyl, aryl,}\ OC_2H_5)$.

In the first stage, as a result of abstraction of an α-proton by a base, an α-halo carbanion *15* is formed, which coordinates with the boron compound, forming the α-halo-substituted borate (*16*). The subsequent rapid rearrangement of the borate, with the migration of one radical from boron to carbon, is accompanied by the elimination of KBr and leads to the compound *17*. In a later stage, protolytic cleavage of the oxoborane *17* takes place, and the alkylation product *18* and a dialkylborinate are formed.

An essential correction, resulting from the work of Pasto and Wojtkowski [23], should be introduced into this scheme. These authors found that β-carbonyl derivatives of boron rearrange rapidly to form substituted vinyloxyboranes (*19*), which are easily converted to carbonyl compounds under the action of water and alcohols:

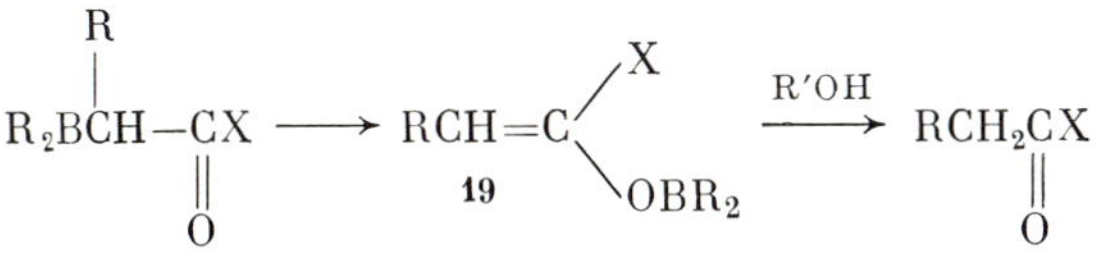

(X = H, SR, OR, Alk, Ar).

As a base in the alkylation, one generally uses potassium *tert*-butoxide or 2,6-di-*tert*-butylphenoxide, obtained from the corresponding phenol by the action of potassium *tert*-butoxide or metallic potassium in tetrahydrofuran. The use of the phenoxide, in a whole series of cases (α-halo ketones, α-halodinitriles) makes it possible to prevent the partial decomposition of the reaction products taking place under the action of potassium *tert*-butoxide.

12.2.1 Synthesis of esters

Esters of higher organic acids are obtained by the treatment of a mixture of a trialkylborane and an ester of chloro- or bromoacetic acid with potassium *tert*-butoxide in *tert*-butanol at O°C [24]:

$$R_3B + BrCH_2COOC_2H_5 + tert\text{-}C_4H_9OK \xrightarrow{0°C,\ THF,\ tert\text{-}C_4H_9OH} RCH_2COOC_2H_5 + tert\text{-}C_4H_9OBR_2.$$
$$80\text{—}99\%$$

Esters of α-halocarboxylic acids have been synthesized from esters of dichloro- or dibromoacetic acid [25]:

$$R_3B + Cl_2CHCOOC_2H_5 + tert\text{-}C_4H_9OK \xrightarrow{0°\ C} \underset{\substack{|\\ Cl}}{R\,CH}COOC_2H_5.$$
$$47\text{—}98\%$$

When two equivalents of trialkylborane and two equivalents of alkoxide are used, a dialkylation of the starting dihalo ester is observed:

$$2(C_2H_5)_3B + Br_2CHCOOC_2H_5 + 2\ tert\text{-}C_4H_9OK \longrightarrow (C_2H_5)_2CHCOOC_2H_5.$$
$$60\%$$

The dialkylation can be carried out in stages, with the use of a different organoborane in each stage. In this way, two different groups can be introduced into the acetic acid molecule.

9-Alkyl-[26–28] or 9-aryl-9-borabicyclo[3.3.1]nonanes [28] can be used as alkylating and arylating agents.

The alkylation of esters of mono- and dihalo acids also proceeds smoothly under the action of potassium 2,6-di-*tert*-butylphenolate in the presence of *tert*-butanol [29].

The transfer of radicals from boron to carbon in these reactions takes place with retention of the configuration. This is confirmed by the formation of only the ester of (*trans*-2-methylcyclopentyl)acetic acid in the alkylation of bromoacetic ester with 9-(*trans*-2-methylcyclopentyl)-9-borabicyclo[3.3.1]nonane [27].

It has been found [30] that organoboranes may be used for the synthesis of β,γ-unsaturated acids. Thus, the alkylation of ethyl 4-bromocrotonoate with trialkyl- and tricycloalkylboranes under the action of potassium 2,6-di-*tert*-butylphenolate leads to esters of substituted *trans*-vinylacetic acids:

$$R_3B + BrCH_2CH{=}CHCOOC_2H_5 \xrightarrow[tert\text{-}C_4H_9OH]{2,6\text{-}(tert\text{-}C_4H_9)_2C_6H_3OK} RCH{=}CH{-}CH_2COOC_2H_5$$

72—89%

($R = C_2H_5$, n-C_4H_9, *iso*-C_4H_9, *sec*-C_4H_9, cyclo-C_5H_9, cyclo-C_6H_{11}).

This reaction represents a method for increasing the length of the radical on the boron atom by four carbon atoms. The authors assume that the shift of the double bond from the 2,3 position to the 3,4 position is due to an allylic rearrangement in the hydrolysis stage (*21*), as takes place in the protolytic cleavage of boranes of the allylic type [31, 32]:

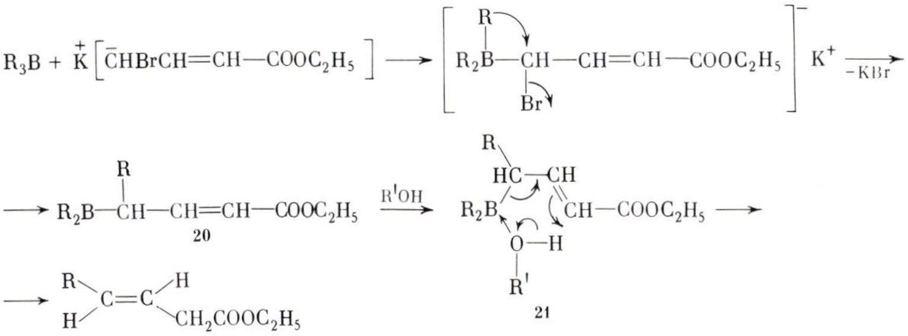

Moreover, it is more probable that the intermediate compound *20,* which is a borane of the allylic type, rearranges by consecutive migration of the boron and a shift of the double bond [31, 33] to the thermodynamically more stable vinyloxyborane *22,* the hydrolysis of which leads to the final unsaturated ester:

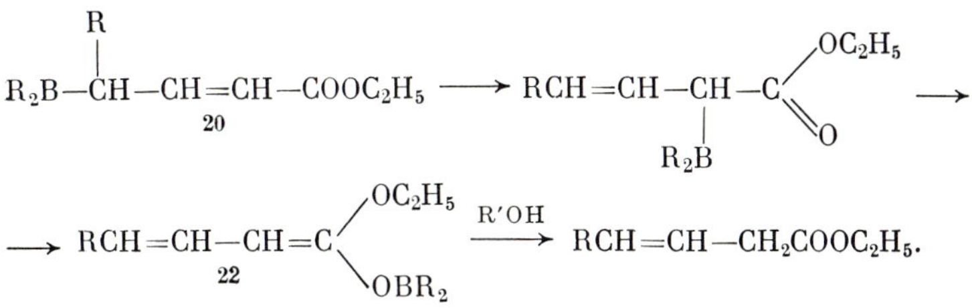

The dianion generated by the action of (*iso*-$C_3H_7)_2NLi$ on phenoxyacetic acid reacts with R_3B to form the acids RCH_2COOH [55].

12.2.2 Synthesis of nitriles and dinitriles

The reaction of α-halonitriles with organoboranes in the presence of potassium *tert*-butoxide does not yield higher nitriles RCH_2CN. At the same time, the alkylation proceeds smoothly under the action of potassium 2,6-di-*tert*-butylphenolate in *tert*-butanol and tetrahydrofuran [34]:

$$RBR_2 + XCH_2CN + 2{,}6\text{-}(tert\text{-}C_4H_9)_2C_6H_3OK \xrightarrow{tert\text{-}C_4H_9OH,\ 20°\,C} \underset{60\text{—}90\%}{RCH_2CN}$$

($R = C_2H_5$, C_4H_9, cyclo-C_5H_9, cyclo-C_6H_{11}, 2-norbornyl, C_6H_5).

Trialkylboranes, 9-alkyl-, 9-cycloalkyl-, or 9-aryl-9-borabicyclo[3.3.1]nonanes are used as alkylating agents. The same organoboranes are used for the alkylation of dichloroacetonitrile. It has been found that, with a ratio of organoborane to Cl_2CHCN of 1:1, α-chloronitriles are obtained which, under the action of a second mole of trialkylborane or 9-alkyl-9-borabicyclo[3.3.1]-nonane, form dialkylation products [35]:

$$RBR'_2 + Cl_2CHCN \xrightarrow{2,6\text{-}(tert\text{-}C_4H_9)_2C_6H_3OK} \underset{70\text{—}90\%}{\underset{|}{RCHCN}\atop Cl} \xrightarrow{R''BR'_2} \underset{|}{R\text{—}CHCN}\atop R''.$$

The alkylation of bromomalodinitrite in the presence of potassium 2,6-di-*tert*-butylphenolate leads to dinitriles of alkylmalonic acids [36].

The reaction with chloromethyl cyanide has been used for the stereo-controlled introduction of a side chain into steroids [57].

12.2.3 Synthesis of ketones and sulfonyl compounds

Primary trialkylboranes and also 9-alkyl-, 9-cycloalkyl-, and 9-aryl-9-borabicyclo[3.3.1] nonanes alkylate or arylate such α-bromo ketones as phenacyl bromide and α-bromopinacolin in the presence of potassium *tert*-butoxide [28, 37, 38] or potassium 2,6-di-*tert*-butylphenoxide [39], forming higher ketones:

$$R_3B + BrCH_2COR' \xrightarrow{R''OK,\ 0°\,C} \underset{60\text{—}100\%}{RCH_2COR'}.$$

An attempt to use this reaction for the synthesis of higher ketones from α,α'-dihalo ketones did not lead to positive results. The alkylation of α,α'-dichloro- and dibromoacetone with triethylborane forms mainly acetone (33–55%) and a small amount (7–23%) of dipropyl ketone [40, 41]. The alkylation of 2-bromocyclopentanone and 2-bromocyclohexanone with trialkylboranes proceeds in the presence of *tert*-C_4H_9OK and $(C_2H_5)_3COK$, but not of di-*tert*-butylphenoxide [58].

The reaction of *cis*-2,6-dibromocyclohexanone, 4-*tert*-butyl- and 3-methyl-*cis*-2,6-dibromocyclohexanones with primary organoboranes in the presence of two equivalents of potassium *tert*-butoxide leads to products of monoalkylation-reduction (*23*), whereas, under the action of one equivalent of this base or of two equivalents of potassium or sodium isopropoxide, complete dehalogenation (reduction) takes place and the corresponding ketone is obtained:

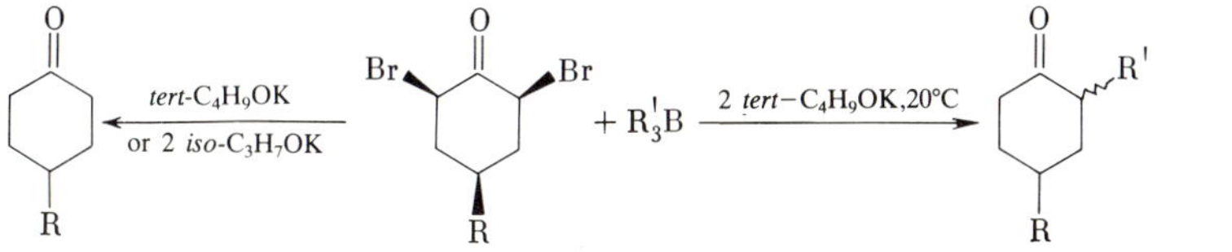

23 (approx. 100 %)

trans-2,6-Dihalocyclohexanones and α,α'-dibromocyclopentanones are not alkylated by the above method.

The mechanism of the reduction-monoalkylation of dihalocyclohexanones with organoboranes has been investigated in [42].

Aliphatic organoboranes are used for the preparation of sulfonyl compounds. The alkylation of α-bromosulfonyl derivatives with trialkylboranes under the action of potassium *tert*-butoxide and *tert*-butanol proceeds under mild conditions and leads to the corresponding sulfones or to esters or amides of sulfonic acids in high yields [43]:

$$R_3B + BrCH_2SO_2Y + tert\ C_4H_9OK \xrightarrow{R'OH,\ -40-0°\ C} \underset{60-99\%}{RCH_2SO_2Y}$$

($R = n\text{-}C_4H_9$, $iso\text{-}C_4H_9$, $n\text{-}C_6H_{13}$,$n\text{-}C_8H_{17}$, cyclo-C_5H_9, cyclo-C_6H_{11}; $Y = C_2H_5$, C_6H_5, $(CH_3)_3CCH_2O$, $(C_2H_5)_2N$).

REFERENCES

1. P. I. Paetzold and H. Grundke, *Synthesis* **1973,** 635.
2. M. M. Rogič, *Intra-Science Chem. Rep.* **7,** 155 (1973).
3. G. Köbrich and H. C. Merkle, *Angew. Chem.* **79,** 50 (1967).
4. G. Köbrich and H. C. Merkle, *Chem. Ber.* **100,** 3371 (1967).
5. H. C. Brown, B. A. Carlson and R. H. Prager, *J. Am. Chem. Soc.* **93,** 2070 (1971).
6. H. C. Brown and B. A. Carlson, *J. Org. Chem.* **38,** 2422 (1973).
7. G. Köbrich, K. Flory and W. Drischel, *Angew. Chem.* **76,** 536 (1964).
8. H. C. Brown, J.-J. Katz and B. A. Carlson, *J. Org. Chem.* **38,** 3968 (1973).
9. E. Negishi, *Intra-Science Chem. Rep.* **7,** 81 (1973).
10. H. C. Brown and B. A. Carlson, *J. Organometal. Chem.* **54,** 61 (1973).
11. R. J. Hughes, A. Pelter and K. Smith, *Chem. Comm.* **1974,** 863.
12. E. Negishi and T. Yoshida: American Chemical Society Meeting, Chicago, August 1973; cited in [11].
13. S. Yamamoto, M. Shiono and T. Mukaiyama, *Chem. Letters* **1973,** 961.
14. A. Suzuki, S. Nozawa, N. Miyaura, M. Itoh and H. C. Brown, *Tetrahedron Letters* **1969,** 2955.
15. W. Kirmse: "*Carbenes*", *N. Y.,* J. Wiley and Sons (1972).
16. W. Kirmse:"Carbene, Carbenoide und Carbenanaloge", Weinheim, Verlag Chemie (1969).
17. D. Seyferth, J. Mui and J. Burlitch, *J. Am. Chem. Soc.* **89,** 4953 (1967).
18. D. Seyferth and B. Prokai, *J. Am. Chem. Soc.* **88,** 1834 (1966).
19. G. Hesse and A. Haag, *Tetrahedron Letters* **1965,** 1123.
20. H. Jäger and G. Hesse, *Chem. Ber.* **95,** 345 (1962).
21. D. Seebach, *Synthesis* **1969,** 17.
22. B. A. Carlson and H. C. Brown, *J. Am. Chem. Soc.* **95,** 6876 (1973).
23. D. Pasto and P. Wojtkowski, *Tetrahedron Letters* **1970,** 215.
24. H. C. Brown, M. M. Rogič, M. W. Rathke and G. W. Kabalka, *J. Am. Chem. Soc.* **90,** 818 (1968).
25. H. C. Brown, M. M. Rogič, M. W. Rathke and G. W. Kabalka, *J. Am. Chem. Soc.* **90,** 1911 (1968).
26. H. C. Brown and M. M. Rogič, *J. Am. Chem. Soc.* **91,** 2146 (1969).
27. H. C. Brown, M. M. Rogič, M. W. Rathke and G. W. Kabalka, *J. Am. Chem. Soc.* **91,** 2150 (1969).
28. H. C. Brown and M. M. Rogič *J. Am. Chem. Soc.* **91,** 4304 (1969).
29. H. C. Brown, H. Nambu and M. M. Rogič, *J. Am. Chem. Soc.* **91,** 6855 (1969).
30. H. C. Brown and H. Nambu, *J. Am. Chem. Soc.* **92,** 1761 (1970).
31. B. M. Mikhailov, *Organometal. Chem. Rev. A* **8,** 1 (1972).
32. B. M. Mikhailov and A. Ya. Bezmenov, *Izv. Akad. Nauk SSSR, Ser. Khim.* **1965,** 931.
33. V. S. Bogdanov, V. F. Pozdnev, Yu. N. Bubnov and B. M. Mikhailov, *Dokl. Akad. Nauk SSSR* **193,** 586 (1970).
34. H. C. Brown. H. Nambu and M. M. Rogič, *J. Am. Chem. Soc.* **91,** 6854 (1969).
35. H. Nambu, H. C. Brown, *J. Am. Chem. Soc.* **92,** 5790 (1970).
36. H. Nambu and H. C. Brown, *Organometal. Chem. Synth.* **1,** 95 (1970/1971).
37. H. C. Brown, M. M. Rogič and G. W. Kabalka, *J. Am. Chem. Soc.* **90,** 6218 (1968).
38. H. C. Brown, M. M. Rogič, H. Nambu and M. W. Rathke, *J. Am. Chem. Soc.* **91,** 2147 (1969).

39. H. C. Brown, H. Nambu and M. M. Rogič, *J. Am. Chem. Soc.* **91,** 6852 (1969).
40. R. H. Prager and J. M. Tippett, *Tetrahedron Letters* **1973,** 5199.
41. R. H. Prager and J. M. Tippett, *Austral. J. Chem.* **27,** 1457 (1974).
42. R. H. Prager and J. M. Tippett, *Austral. J. Chem.* **27,** 1467 (1974).
43. W. E. Truce, L. A. Mura, P. J. Smith and F. Young, *J. Org. Chem.* **39,** 1449 (1974).
43a. D. S. Matteson, and R. Ray, *J. Am. Chem. Soc.* **102,** 7590 (1980).
43b. M. W. Rathke, E. Chao and G. Wu, *J. Organometal. Chem.* **122,** 145 (1976).
44. G. W. Kramer and H. C. Brown, *J. Org. Chem.* **42,** 2832 (1977).
45. H. C. Brown and S. U. Kulkarni, *J. Organometal. Chem.* **218,** 299 (1981).
46. P. Jacob, *J. Organometal. Chem.* **156,** 101 (1978).
47. O. Rosario, A. Olíva and G. L. Larson, *J. Organometal. Chem.* **146,** C8 (1978).
48. G. L. Larson, R. Arguelles, O. Rosario and S. Sandoval, *J. Organometal. Chem.* **198,** 15 (1980).
49. O. Rosario, S. Sandoval and G. L. Larson, *Synth. Commun.* **10,** 813 (1980).
50. R. J. Hughes, S. Ncube, A. Pelter, K. Smith, E. Negishi and T. Yoshida, *J. Chem. Soc., Perkin I* **1977,** 1172.
51. S. Ncube, A. Pelter and K. Smith, *Tetrahedron Letters* **1979,** 1895.
52. E. Negishi, T. Yoshida, A. Silveira, Jr. and B. L. Chiou, *J. Org. Chem.* **40,** 814 (1975).
53. D. Nguen, *Bull. Soc. Chim. France* **1981,** II-99.
54. S. Ncube, A. Pelter and K. Smith, *Tetrahedron Letters* **1979,** 1893.
55. S. Hara, K. Kishimura and A. Suzuki, *Tetrahedron Letters* **1978,** 2891.
56. H. C. Brown, J. J. Katz and B. A. Carlson, *J. Org. Chem.* **40,** 813 (1975).
57. M. M. Midland and Y. C. Kwon, *J. Org. Chem.* **46,** 229 (1981).
58. R. H. Prager and P. A. Reece, *Austral. J. Chem.* **28,** 1775 (1975).

CHAPTER 13

Reactions of alkyl(aryl)boranes with unsaturated, aromatic, and cyclopropane compounds

Addition to multiple carbon-carbon bonds is uncharacteristic for alkyl- and arylboranes. However, there are several exceptions of interest from the practical viewpoint, in which the corresponding reactions proceed extremely smoothly and lead to important classes of organic compounds.

The reactions of primary trialkylboranes with cyclohexene and cycloheptene proceeding under irradiation represent a method for the synthesis of *cis*-2-alkylcyclohexanols and -heptanols and also of the corresponding ketones. Triphenylborane can be used for the preparation of substituted olefinic hydrocarbons and for the hydrogenation, dehydrogenation, and cyclization of some organic compounds.

A number of methods for the synthesis of aromatic ketones and olefins are based on the ability of benzylboranes to add to acetylenes (with benzyl rearrangement).

The reaction of trialkylboranes with 1-methylcyclopropene is the basis for a simple method for increasing the length of the carbon chain of olefins and of alkyl halides by a methallyl group.

Substituted 3-butenylboranes, formed by the action of alkyl- and allylboranes on bicyclo[1.1.0]butane, undergo an allylcarbynyl-cyclopropylcarbynyl rearrangement upon heating. Vinyl- and aryl(triorgano)borates are used for the preparation of different olefins and 1,4-diols and for the alkylation-reduction of aromatic compounds.

Trialkylboranes are used for the alkylation of functional olefins, acetylenes, and aromatic compounds and also for the synthesis of substituted aromatic ketones.

13.1 ALKYLATION AND ARYLATION OF OLEFINS, DIENES, AND ACETYLENES

Trialkylboranes, as a rule, do not add to olefins [1–9], dienes [10–14], or acetylenes [15, 16] but hydroborate these with the formation of new trialkylboranes, boracycloalkanes, or vinylboranes:

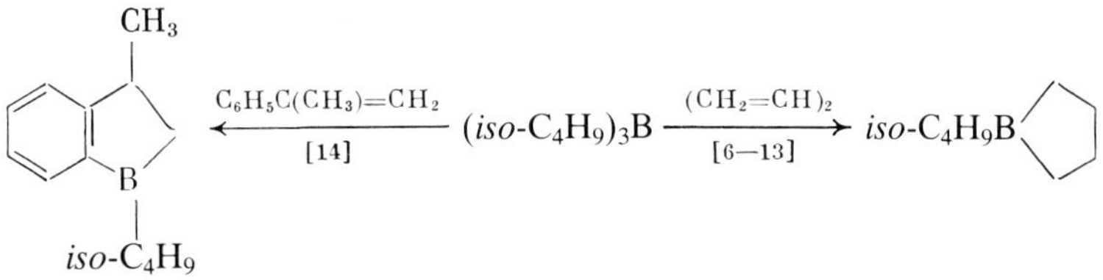

Triethylborane, unlike other trialkylboranes, is able to add partially to decene [17]:

$$B(C_2H_5)_3 + RCH{=}CH_2 \xrightarrow{170°C} (C_2H_5)_2\ B{-}CH_2{-}CHR{-}C_2H_5.$$

When triethylborane and ethylene are heated above 200°C in an autoclave, a mixture of normal butenes is formed [18]:

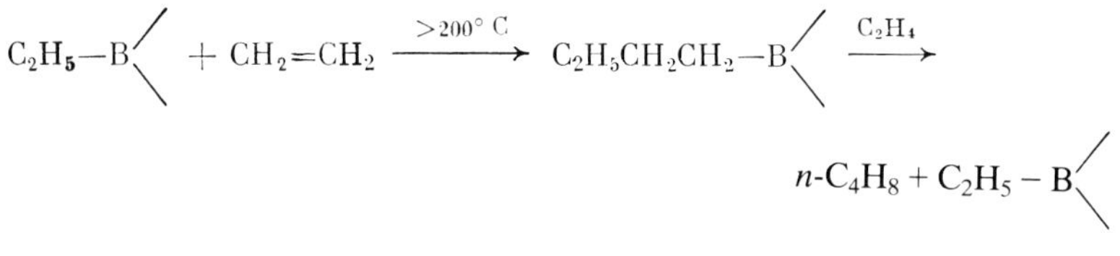

When irradiated with a mercury lamp in the presence of p-xylene as a sensitizer, primary trialkylboranes react with cyclohexene and cycloheptene by *cis*-addition of a boron-alkyl fragment, forming substituted cycloalkylboranes (*3*), the oxidation of which (by H_2O_2,

OH^-) leads to the corresponding *cis*-alcohols (*4*) [19, 20]:

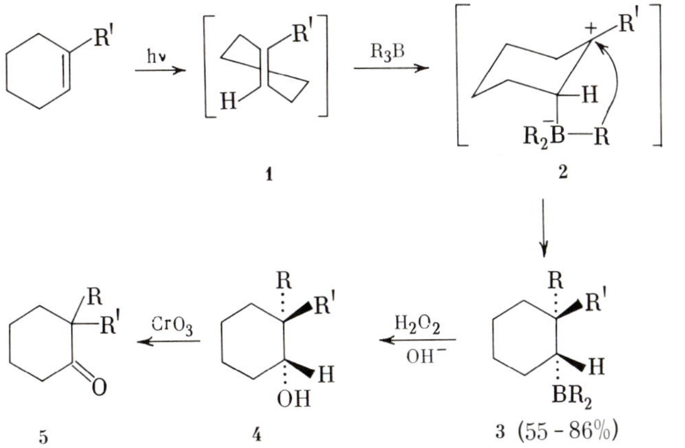

(R = C_2H_5, *n*-C_4H_9, $CH_3OCH_2CH_2CH_2$; R′ = R or H).

By the above method, *cis*-2-ethyl-, *cis*-2-butyl-, and *cis*-2-methoxypropylcyclohexanols (*4*, R′ = H) have been synthesized from cyclohexene in 55–86% yield, while 2,2-diethylcyclohexanol and 2-ethyl-*cis*-2-butylcyclohexanol have been obtained from ethylcyclohexene. The latter two compounds are oxidized with chromic anhydride to the corresponding 2,2-dialkylcyclohexanones (*5*).

The irradiation of mixtures of triethyl- or tributylborane with cycloheptene in a benzene/*p*-xylene medium, followed by oxidation, leads to *cis*-2-ethyl- and *cis*-2-butylcycloheptanols in yields of 10 and 18%, respectively.

Cyclopentene, cyclooctene, and cyclododecene do not react with trialkylboranes even under irradiation. As suggested [19–22], the peculiar behavior of six- and seven-membered cyclenes is due to the fact that, upon irradiation, they form the extremely reactive *trans*-cyclohexene (*1*) and *trans*-cycloheptene, with a strained double bond, which adds R_3B to form an intramolecular ion of the type *2*. The subsequent migration of the alkyl anion from the boron to the positively charged carbon of the ring, according to the authors, proceeds by the shortest route, as a result of which a stereospecific cis-addition is observed. The ionic mechanism of the reaction is confirmed by the formation of optically active *cis*-2-butylcyclohexanol ($[\alpha]_D^{20} = -6.2°$) upon irradiation of a mixture of cyclohexene and dibutylboryl-(*1S*)-menthoxyacetate [20]:

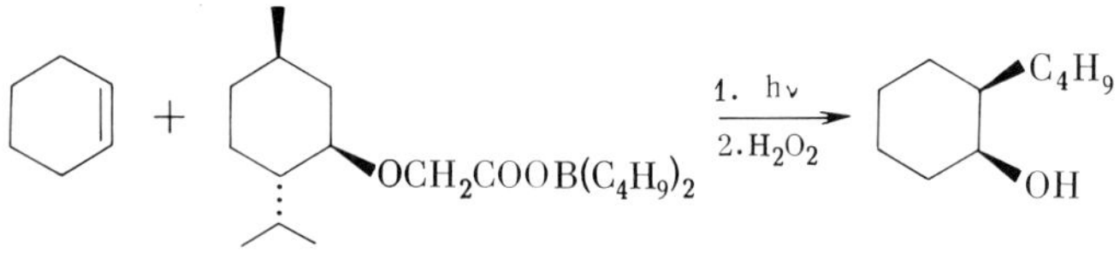

The reaction of trialkylboranes with bis(1,3-diphenyl-2-imidazolidinylidene) proceeds in a distinctive manner. Heating these reagents in a ratio of 1:2 produces a high yield of 1,3-diphenyl-2-boraimidazolidines, the oxidation of which leads to trialkylcarbinols [23]:

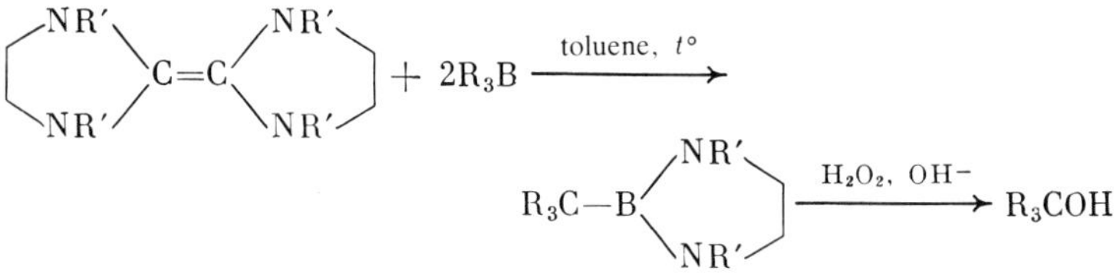

(86—95%) 55—82%

($R = CH_3$, C_4H_9, $C_6H_5CH_2$; $R' = C_6H_5$).

Aryl boron derivatives are capable of adding to some unsaturated compounds. Thus, as a result of the reaction of dichlorophenylborane with norbornadiene, dichloro-(6-phenyl-2-nortricyclyl)borane was obtained in the form of a mixture of the *cis* and *trans* isomers [24, 25]:

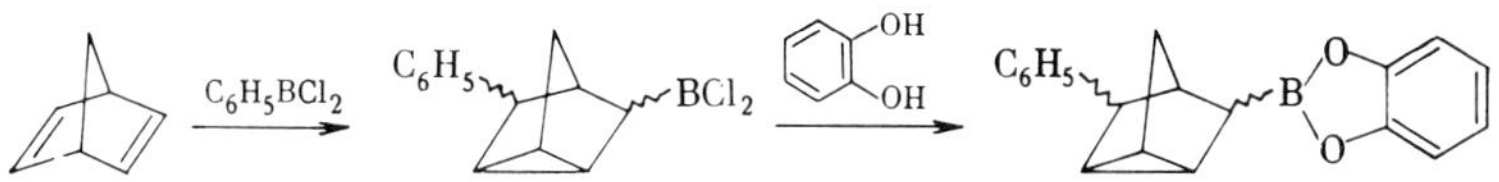

Triphenylborane, diphenylboron chloride, dibutoxy(phenyl)borane, and triamylborane do not react with norbornadiene [24].

The data of Lappert and Prokai [26] on the addition of diphenylboron chloride to ketene to form a boron-containing ketone:

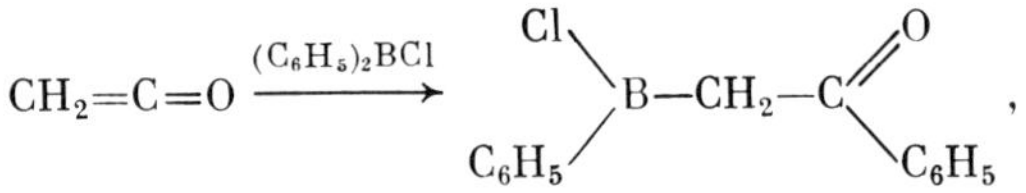

require more precise definition, since β-oxoalkylboranes are unstable and rearrange rapidly to boron esters of enols (Chapter 9).

According to the data of Wittig, et al. [27, 28], the action of triphenylborane and triphenylmethylsodium on styrene forms the borate $[(C_6H_5)_3BCH(C_6H_5)CH_2C(C_6H_5)_3]Na$, the hydrolysis of which forms 1,1,1,3-tetraphenylpropane. Under the same conditions, 1-triphenylmethylacenaphthene has been obtained from acenaphthylene [29] and 1,1,1-triphenyl-4-pentene and its derivatives from 1,3-diene hydrocarbons [27, 28, 30]:

$$(C_6H_5)_3CNa + CH_2{=}CHCH{=}CH_2 + (C_6H_5)_3B \longrightarrow \xrightarrow{H_2O} \underset{73\%}{(C_6H_5)_3CCH_2CH_2CH{=}CH_2}.$$

Triarylboranes are capable of splitting off sodium, potassium, or lithium hydride from some organic compounds of these metals. Thus, 1,1-diphenylethylene (73%) and $[(C_6H_5)_3BH]K$ (83%) have been obtained by the action of triphenylborane on 1,1-diphenylethylpotassium [31] and 1,4-disodium-1,1,4,4-tetraphenylbutane (prepared from sodium and 1,1-diphenylethylene) have been converted to 1,1,4,4-tetraphenylbutadiene [30]. 9,10-Dimethyleneanthracene is formed by treating 9,10-dilithium-9,10-dimethylanthracene with triphenylborane [31].

As a result of the dehydrometalation of the disodium compound *6* by this method, 5,8-diphenyl-1,2,3,4-dibenzocyclooctatetraene (*7*) has been obtained [32]:

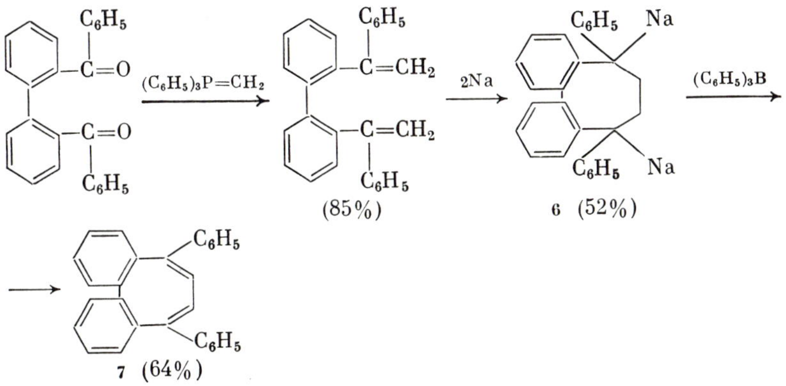

Cyclooctatriene (31%) and hexaphenylethane have been obtained by heating a mixture of cyclooctatetraene, triphenylborane, and triphenylmethylsodium, followed by acid hydrolysis [28].

As noted above, trialkylboranes do not add to acetylenic hydrocarbons. At the same time, some acetylenes containing an activated triple bond are capable of adding organoboranes.

According to [88], triorganoboranes react with bis(trimethylstannyl)acetylene by *cis*-addition of the fragments R and BR_2 to the triple bond. However, it was later found [89] that the compounds obtained have a different structure, $[(CH_3)_3Sn]_2C{=}CR(BR_2)$.

The reaction of tribenzylborane with ethoxyacetylene in a ratio of 1:1 or 1:2 proceeds at 0–5°C with the evolution of heat and, after alcoholysis, leads to 1-ethoxy-1-*o*-tolylethylene, the acid hydrolysis of which produces *o*-methylacetophenone [33–35]:

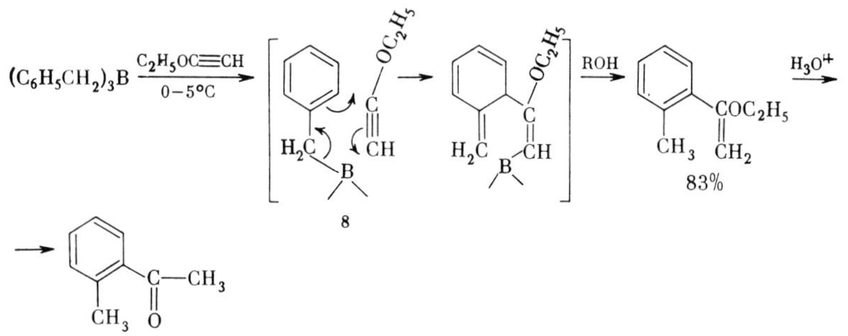

The addition of a boron-benzyl fragment to a triple bond takes place with rearrangement according to a mechanism of cyclic electron transfer (*8*).

When some unsymmetrical boranes of the benzyl type are heated with phenylacetylene in toluene, 2-aryl-4-phenyl-1,2-dihydro-2-boranaphthalenes (*9*) are obtained [36]:

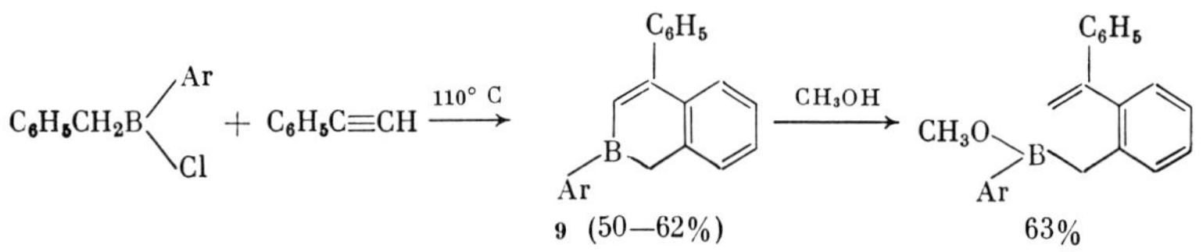

Some arylaminodialkylboranes [37–39] and arylboranes are capable of adding to acetylenic compounds. When butylacetylene is heated to 140°C with phenylboron dichloride or diphenylboron chloride, phenylboration of the triple bond takes place, resulting in the formation of the corresponding 2-phenylhexenylboranes, the acidolysis of which leads to 1-butylstyrene [40]. Diphenylacetylene adds phenylboron dibromide with the formation of dibromo-(triphenylvinyl)borane [41]. At the start of this reaction, the main process is a reversible haloboronation, leading to the *cis*-β-bromovinylborane *10*:

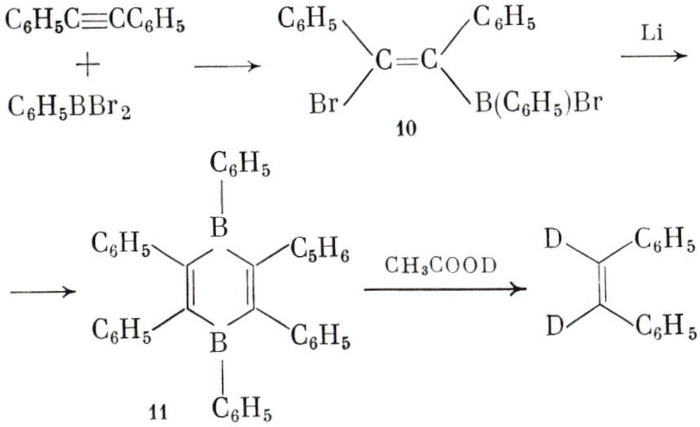

The action of lithium on the compound *10* forms hexaphenyl-1,4-diboracyclohexadiene (*11*), the acidolysis of which by deuteroacetic acid leads to *cis*-dideuterostilbene.

The irradiation of sodium or lithium tetraphenylborate in the presence of diphenylacetylene leads (after acidolysis) mainly to *cis*-stilbene [90].

The reactions of trialkyl- and tricycloalkylboranes R_3B with 1-alkynes $R'C{\equiv}CH$ and lead tetraacetate represent a regio- and stereospecific method for the synthesis of $R'C{\equiv}CR$ and the enol acetates (Z)-$RCH{=}CR'OCOCH_3$ [91].

The electrochemical reaction of R_3B with phenylacetylene leads to phenylalkylacetylenes, $C_6H_5C{\equiv}C{-}R$ [92], while alkylation proceeds with retention of the configuration [93].

13.2 ALKYLATION OF FUNCTIONAL OLEFINS AND ACETYLENES

In this section, the conjugated radical addition of organoboranes to α-oxides of olefins and acetylenes, acrylonitrile, and nitroalkanes and also the cleavage of α,β-unsaturated sulfones and sulfoxides are discussed. These reactions are used for the preparation of higher allylic and allenic alcohols, nitriles, substituted nitroalkanes, and olefins. (The alkylation of α,β-unsaturated carbonyl compounds was described in Chapter 9.)

1,3,-Butadiene monoxide adds organoboranes to form, after hydrolysis, substituted alcohols of the allylic type [42]:

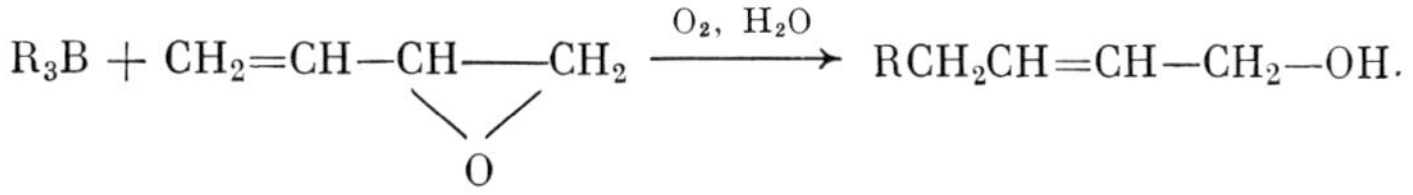

The reaction takes place only in the presence of oxygen or of the other radical initiators. The addition mechanism is represented in the following manner:

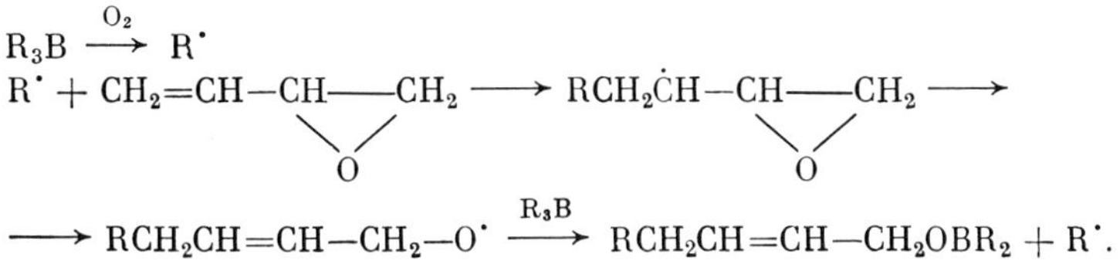

In combination with hydroboration, the reaction represents a method for increasing the carbon skeleton of olefins by four carbon atoms; for example, 4-cyclopentyl-2-buten-1-ol has been obtained from cyclopentene:

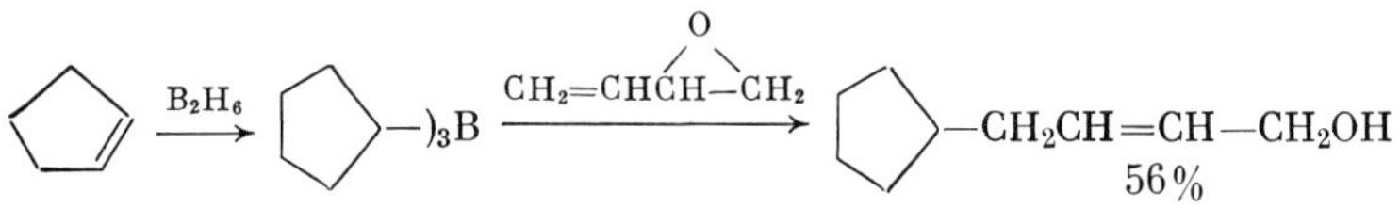

The oxygen-initiated 1,5-addition of trialkylboranes to α-oxides of acetylene derivatives in benzene leads to allenic boron esters, the hydrolysis of which forms allenic alcohols in yields of approximately 60% [43]:

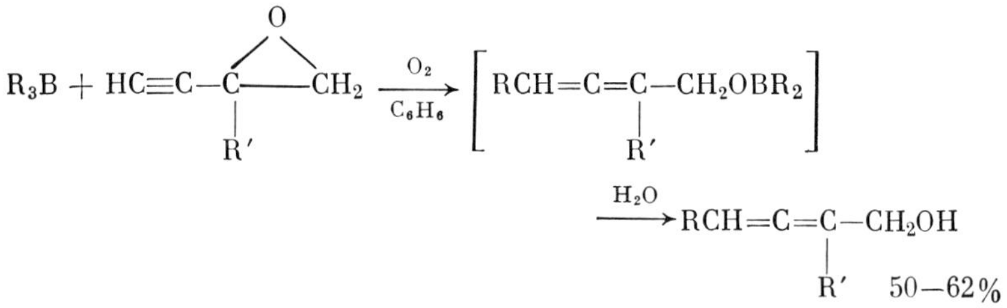

Copper *N*,*N*-diethyldithiocarbamate completely inhibits the reaction, which confirms its chain free-radical nature. The selection of a solvent is very important for processes of this type. For example, the yield of alkylallenic alcohols in tetrahydrofuran is reduced because of the formation of 4-α-tetrahydrofuryl-2,3-butadien-1-ol:

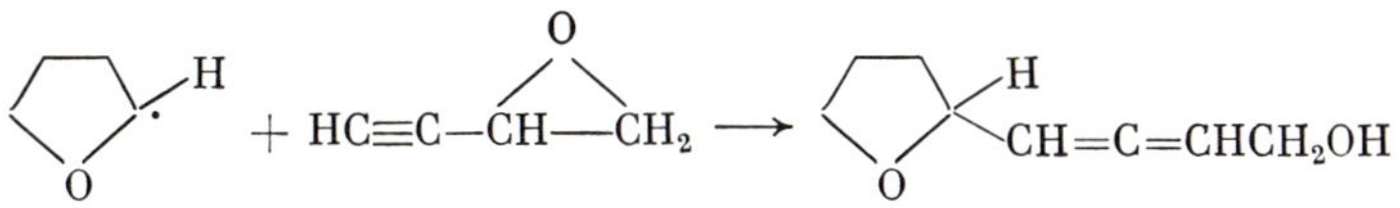

According to the data of [44], in the presence of oxygen, tricyclohexylborane adds to acrylonitrile with the formation, after hydrolysis, of 3-cyclohexylpropionitrile:

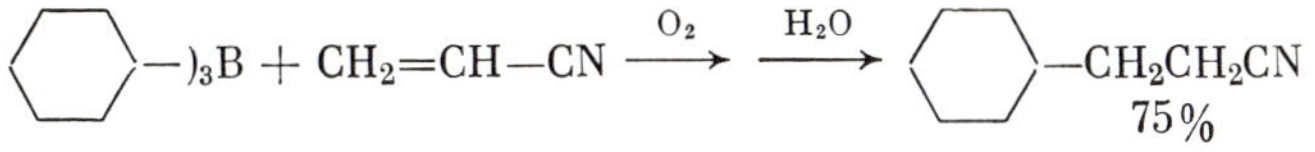

It is noted that the reaction of tri-*n*-alkylboranes is complicated by side reactions, since the intermediately formed organoboranes are extremely active compounds.

The reaction of triethylborane with α-nitroolefins proceeds via a 1,4-addition and leads to dimeric diethyl esters of alkanenitronic acids, the alkaline hydrolysis of which produces branched nitroalkanes [45]:

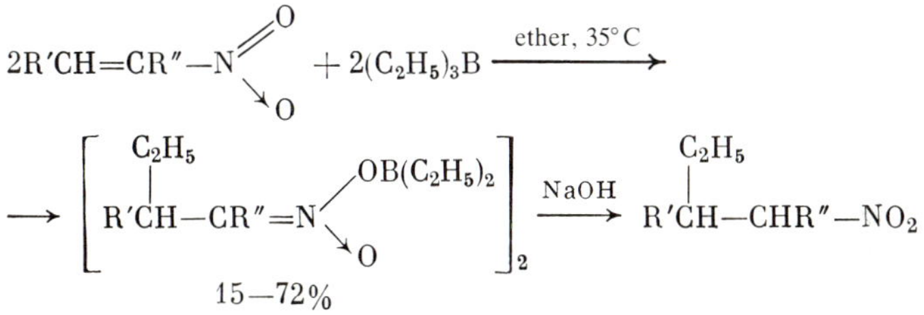

Esters of alkanenitronic acids have also been synthesized by the reaction of diethylboron halides with sodium salts of nitroalkanes, $CH_3CR{=}NOONa$ [46]. The reaction of the silver salt of methylnitramine with diethylboron chloride leads to an equilibrium mixture of *O*- and *N*-diethylboryl derivatives [47].

In α,β-unsaturated sulfones and sulfoxides, organoboranes cleave the C—S bond, resulting in the formation of olefinic hydrocarbons. When heated in tetrahydrofuran or toluene, *trans*-styryl methyl sulfoxide reacts with trialkylboranes to form *trans*-2-alkylstyrene; the reaction of the cis isomer leads to a mixture of cis and trans olefins [48]:

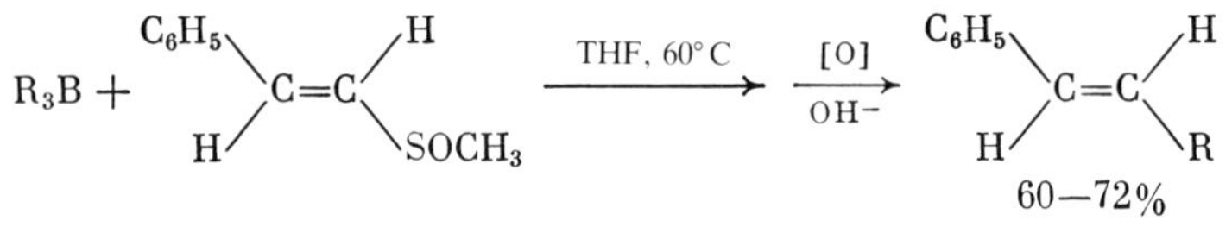

The reaction of R_3B with *cis*- and *trans*-styryl methyl sulfones forms only *trans*-alkylphenylethylenes (70–80%). According to the data of [49, 50], γ,γ-disubstituted allylamines isomerize to the corresponding enamines when heated with tributylborane at 150°C for 30 hours:

$$R_2NCH_2CH{=}C(CH_3)_2 \xrightarrow{(C_4H_9)_3B,\ 150^\circ\ C} \underset{60\%}{R_2N{-}CH{=}CH{-}CH(CH_3)_2}.$$

Tetrachlorodiborane(4), B_2Cl_4, adds to cyclopentene to form *trans*-1,2-bis(dichloroboryl)cyclopentane, the oxidation of which leads to *trans*-1,2-cyclopentanediol [51]. The trans-addition of tetrachlorodiborane(4) to other unsaturated compounds (olefins, dienes, acetylenes, and polynuclear aromatic compounds) is partially discussed in a review [52] and described in numerous papers [53–66].

Triphenyl- and tri-*p*-tolylboranes and phenylboron halides add to aryl isocyanates and aryl isothiocyanates to form diamidoboranes and their thio analogs [67]:

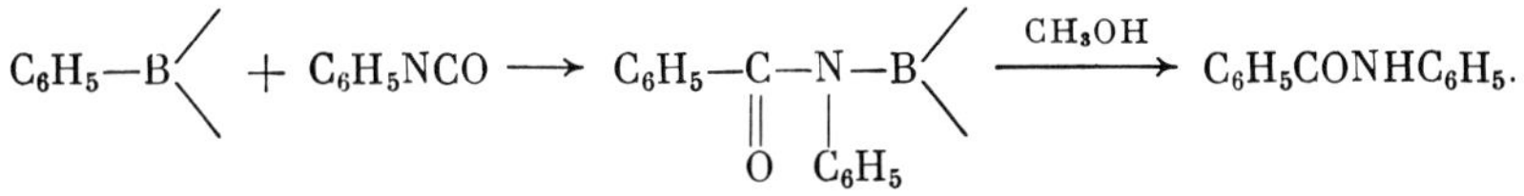

The reaction of triphenylborane and chlorodiphenylborane with di-*p*-tolylcarbodiimide leads to borylenediamidines [68, 69].

13.3 REACTIONS WITH CYCLOPROPANE COMPOUNDS. SYNTHESIS OF OLEFINS

Cyclopropane hydrocarbons are inert toward organoboranes at temperatures up to 100°C [70].

The reaction of primary trialkylboranes with 1-methylcyclopropene proceeds with the evolution of heat and, in the presence of alcohols, leads to 2-methylolefins (an increase in the chain length of the radical on the boron by three carbon atoms). In combination with hydroboration or organometallic synthesis, this reaction represents a method for the introduction of a methallyl fragment into the carbon chain of olefins and alkyl halides [70]:

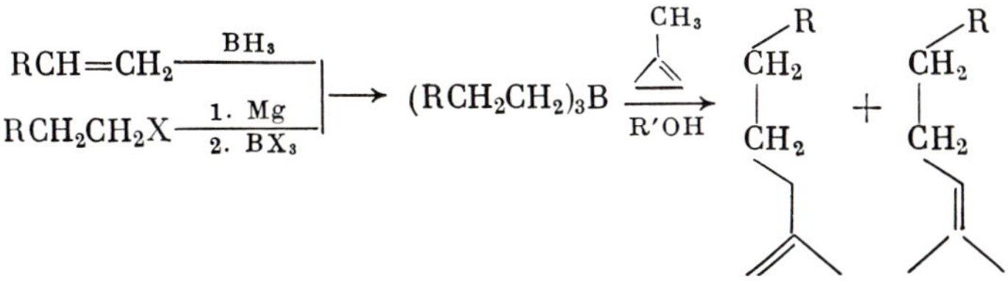

96—98% 2—4%

This method of increasing the chain length has a number of advantages over, for example, the Wurtz reaction, since many important functional groups do not interfere with hydroboration (Hal, OR, SR, NR_2, and others) or react very slowly with diborane (COCl, NO_2, SO_2, etc.); other groups (>C=O) may be protected by an appropriate method. Thus, the hydroboration of allylacetone ketal with diborane led to the trialkylborane *12*, which, under the action of 1-methylcyclopropene in the presence of ethylene glycol, is converted to 2-methyl-1-nonen-8-one ketal (*13*):

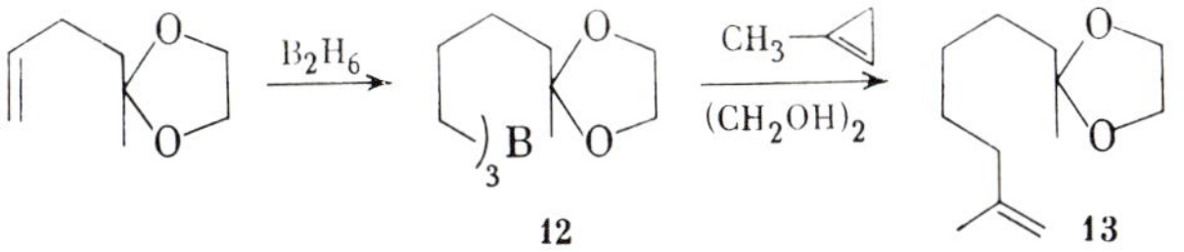

12 13

When the reaction of trialkylboranes with 1-methycyclopropene (1:2) is carried out in the absence of alcohols, substituted cyclopropylboranes of the type *14* are formed, the protolytic cleavage of which by acid leads to the cyclopropane compounds *15* in a yield of over 50%:

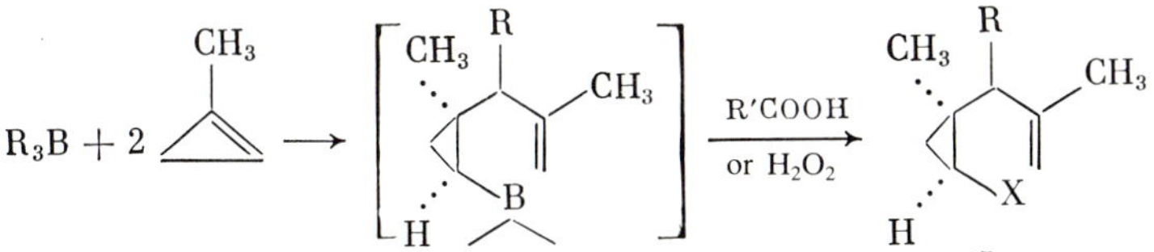

14 15

(X = H, OH).

As suggested, cleavage of the cyclopropane ring proceeds synchronously (*16*). Simultaneously with the rupture of the single C_2—C_3 bond and the formation of a double bond between the C_1 and C_3 atoms, an alkyl radical and a boryl group add to the same C_2 carbon atom. The borane *17* formed immediately rearranges [71] to the thermodynamically more stable allylic compound *18*. The further behavior of the allylborane *18* is determined by the conditions under which the reaction is carried out. In the presence of an alcohol or glycol in the reaction mixture, *18* is immediately cleaved, with protolysis taking place with allylic rearrangement (*19*), resulting in the formation of an olefin with a terminal double bond:

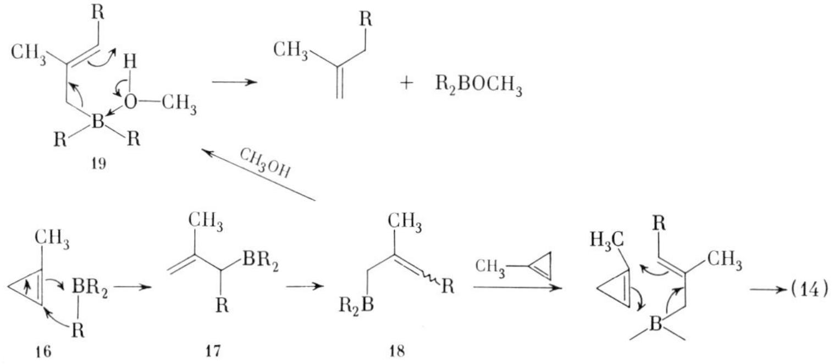

In aprotic solvents, the allylborane *18* formed reacts further with 1-methylcyclopropene by cis-addition to the double bond of the ring [72–74], forming the cyclopropyl compound *14* [70].

Methylenecyclopropane and cyclobutene do not react with triallyl- or trialkylboranes in an inert-gas atmosphere at 20–104°C [72, 73]. At the same time, the reaction of benzylidenecyclopropane with R_3B, which proceeds upon heating in tetrahydrofuran, leads, after treatment with alkali, to a 1-phenyl-4-alkylcyclobutene [75]:

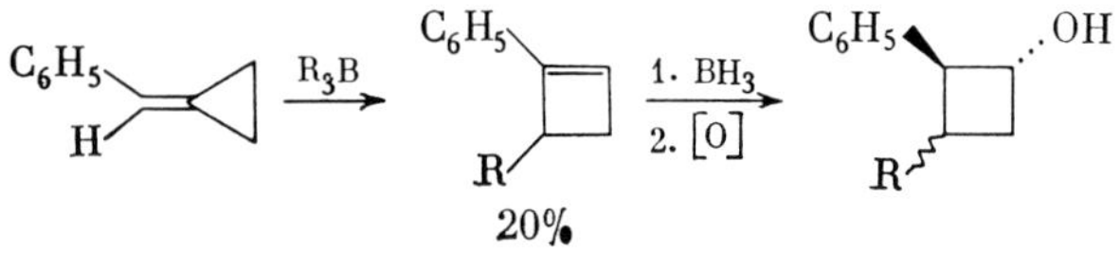

In the reaction of trialkylboranes with bicyclo[1.1.0]butane, a cleavage of the cyclic system takes place and (1-alkyl-3-buten-1-yl)dialkylboranes (*20*) are obtained, the oxidation of which forms 1-alkyl-3-buten-1-ols [76]:

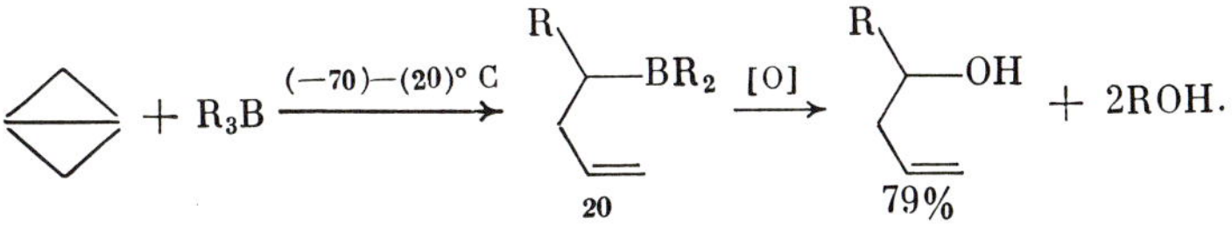

20

In contrast to most other conversions of bicyclobutane, which proceed with cleavage of the central bond, it is a side bond that is broken in the reaction with trialkylboranes.

When butenylboranes of the type *20* are heated to 95°C, they rearrange to form an equilibrium mixture of the compounds *20* and *21* in a ratio of 1:3.

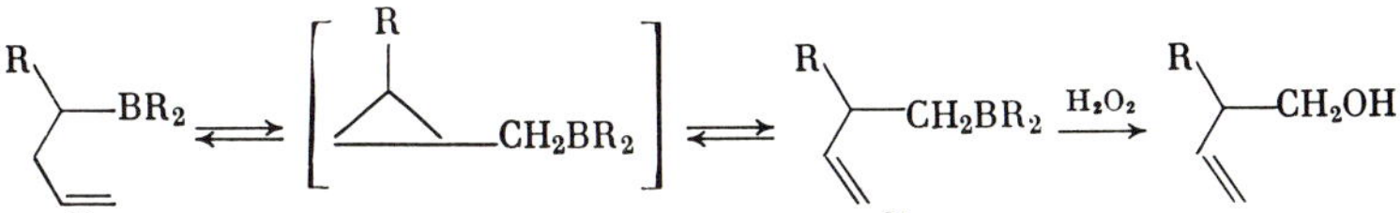

20 21

A similar isomerization is also observed in the case of other cyclopropylmethyl [77] and butenyl [78] derivatives of boron and also of other metals [79].

The reaction of boron tribromide with cyclopropane (1:3) leads to tris(3-bromopropyl)borane, which reacts with acetic acid to form propyl acetate [69]:

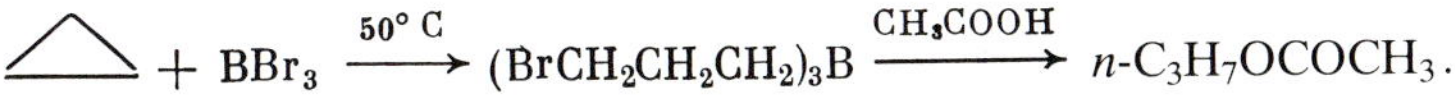

In norcarane, BBr_3 cleaves one of the bonds of the three-membered ring and forms tris(3-bromoalkyl)borane, the structure of which has not been determined.

Cyclopropane, propylcyclopropane, phenylcyclopropane, and norcarane isomerize under the action of phenylboron dibromide and boron chlorides, being converted to the corresponding olefins or their mixtures:

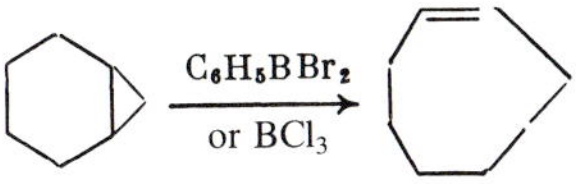

Cyclopropane hydrocarbons are cleaved by tetrachlorodiborane(4) with the formation of 1,3-bis(dichloroboryl)propanes, the oxidation of which leads to 1,3-propanediol or its homologs [57, 80]:

$$\triangle + B_2Cl_4 \longrightarrow Cl_2BCH_2CH_2CH_2BCl_2 \xrightarrow{H_2O_2} HOCH_2CH_2CH_2OH.$$

B_2X_4 (X = F, Cl) adds to the double bond of methylene- and vinylcyclopropane [94].

13.4 ALKYLATION OF AROMATIC COMPOUNDS

Primary and secondary trialkylboranes add to acridine and its derivatives upon heating or irradiation with a mercury lamp in benzene solution. The main reaction products after hydrolysis are 9-alkylacridanes (65–90%) with an impurity of 3–8% of 9-alkylacridines and 3–4% of 9,9′-biacridane [19, 20]:

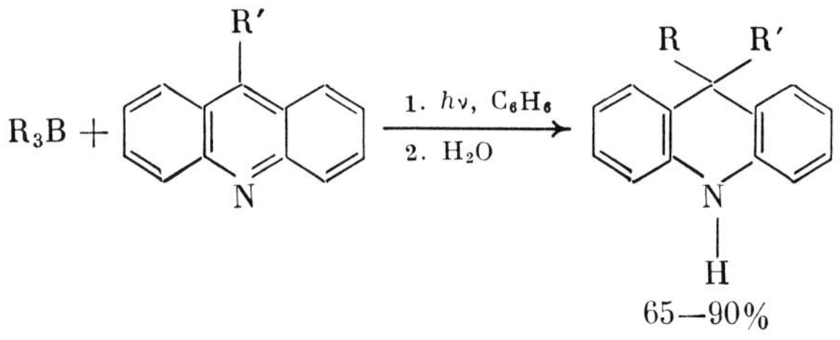

9-Ethyl-, 9-isopropyl-, 9-butyl-, and 9-cyclohexylacridanes have been obtained by this method. 9-Ethyl-9-methylacridane (75%) has been synthesized by the photoreaction of 9-methylacridine with triethylborane and 9,9-diethylacridane (72%) has been obtained from 9-ethylacridine. The reaction proceeds very slowly in the dark. The accelerating action of irradiation is explained by the high basicity of the acridines in the first excited singlet state. Qunioline, isoquinoline, and indole do not react with trialkylboranes upon heating or irradiation.

Lithium 1-naphthyltrialkylborate reacts with methyl fluorosulfonate to form compounds of the type *22*, the hydrolysis of which leads to *trans*-1-alkyl-2-methyl-1,2-dihydronaphthalenes(*23*) [81]:

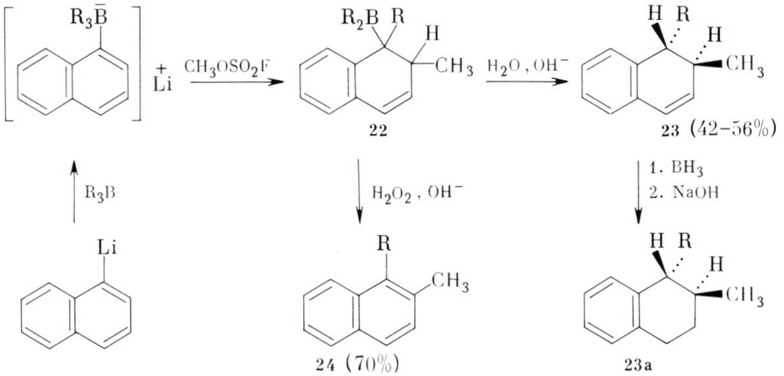

22

23 (42–56%)

24 (70%)

23a

($R = C_2H_5$, n-C_4H_9)

The impurity of cis isomers in hydrocarbons of the type *23*, which cannot be obtained by a Birch reduction, does not exceed 2%. As a result of the hydroboration of the compound *23* ($R = C_4H_9$) and subsequent protonolysis with aqueous alkali, a substituted tetrahydronaphthalene has been synthesized (*23a*, $R = n$-C_4H_9) with a trans location of the methyl and butyl groups.

If the intermediate mixed borane *22* is subjected to oxidation with alkaline hydrogen peroxide or trimethylamine oxide, the main reaction product is 1-alkyl-2-methylnaphthalene (*24*).

1-Butyl-2-methylbenzene (52%) and 9-butyl-10-methylphenanthrene have been synthesized by an analogous series of conversions from phenyllithium or 9-phenanthryllithium and tributylborane:

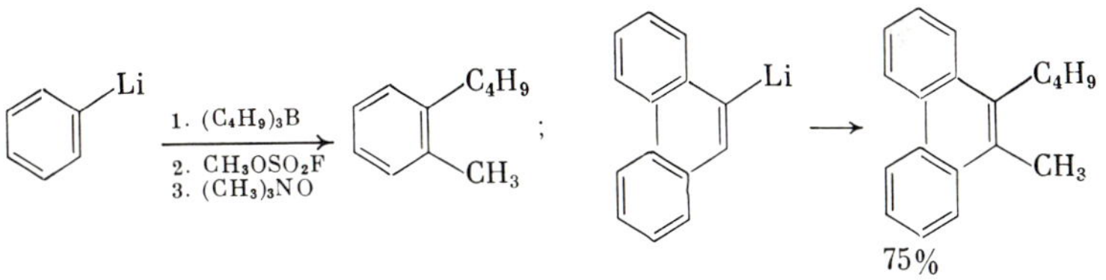

75%

The reactions of aryl(trialkyl)borates with allylic bromides generally proceed in an analogous manner [95].

The sequential treatment of 2-lithiofuran with R_3B ($R =$ alkyl, cycloalkyl), iodine [96, 97], or *N*-chlorosuccinimide (NCS) [97] leads to 2-*R*-furans. 2-Methylfuran [96], thiophene [98], 1-methylpyrrole [97, 98], and *N*-methylindole [99] are alkylated (arylated) in an analogous manner. The action of other electrophiles (methyl iodide, allyl bromide, iodomethyl cyanide, $BrCH(OC_2H_5)_2$, etc.) on lithium 2-indenyl(trialkyl)borate produces 2,3-disubstituted derivatives of *N*-methylindole [100].

3-Alkylfurans and 3-alkylthiophenes are obtained by the treatment of the corresponding "ate" complexes with bromine or iodine [101].

The reaction of halides of aliphatic and aromatic acids with lithium aryltrialkylborates leads to ketones [82]:

$$\mathrm{Ar{-}Li + R_3B \longrightarrow Li[ArBR_3] \xrightarrow[0-25°\,C]{R'COCl} \xrightarrow{H_2O_2,\ CH_3COONa} ArCOR'} \quad 50\text{–}80\%$$

($Ar{=}C_6H_5$, *m*-$CH_3C_6H_4$, *m*-ClC_6H_4, *p*-BrC_6H_4, *m*-$CF_3C_6H_4$, *o*-$CH_3OC_6H_4$, 1-naphthyl; $R'{=}CH_3$, C_6H_5, *o*-$CH_3C_6H_4$, *m*-FC_6H_4, *p*-NCC_6H_4).

The trialkylboranes in this reaction act as a transfer agent for the aromatic radical; high yields of ketones are obtained by the use of tricyclopentyl- or trinorbornylborane.

Aliphatic lithium tetraorganoborate complexes react with R″COX in an analogous manner [102], whereas magnesium bromide trialkylphenylborate complexes, $[C_6H_5BR_3]MgBr$, react with R′COX to form *o*-alkylacylbenzenes, *o*-R—C_6H_4COR′ [103].

The reaction of primary trialkylboranes with α-lithiofuran in a ratio of 1:2 at 20°C leads, after treatment with acetic acid, to 2,3,3-trialkyl-1,2-oxaboracyclohex-4-enes (2,3,3-trialkyl-1,2-oxaborin-4-enes) (*25*). Oxidation of the latter produces 1,1-dialkyl-*cis*-2-butene-1,4-diols (*26*) [83]:

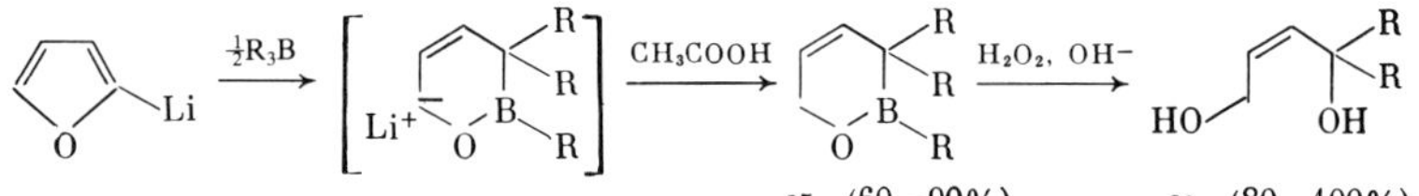

25 (60—90%) **26** (80—100%)

With a ratio of 1:1 between the organoborane and the lithium compound, the yield of cyclic boron derivatives is reduced by a factor of approximately two. Secondary trialkylboranes practically do not react with α-furyllithium in the above manner. Hydroboration-protonolysis of the compound *25* ($R = C_2H_5$) led to a hydrogenated analog (*26*, $R = C_2H_5$).

The reaction of R_3B with 2-bromo-6-lithiopyridine proceeds with ring opening and, depending on the conditions of the later protonolysis, leads selectively to different unsaturated nitriles [104].

13.5 SYNTHESIS OF OLEFINS AND 1,4-DIOLS FROM TRIALKYLBORANES AND VINYL DERIVATIVES OF LITHIUM

Ate complexes of isopropenyllithium with triorganoboranes (*27*), when treated with iodine, followed by hydrolysis-oxidation, are converted to the corresponding 2-methylolefins [84]:

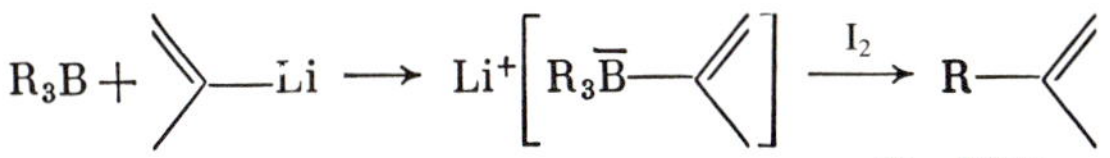

27 75—80%

(R = cyclo-C_5H_9, cyclo-C_6H_{11},C_6H_5).

The reaction represents a method for introducing an isopropenyl group into the molecule of olefins and has been used for the preparation of dihydrolinalool:

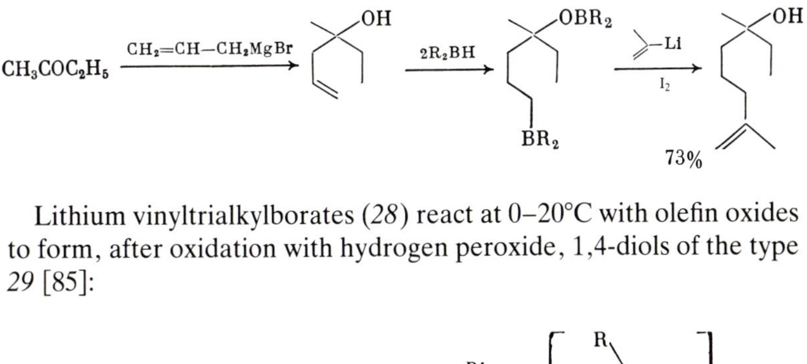

73%

Lithium vinyltrialkylborates (*28*) react at 0–20°C with olefin oxides to form, after oxidation with hydrogen peroxide, 1,4-diols of the type *29* [85]:

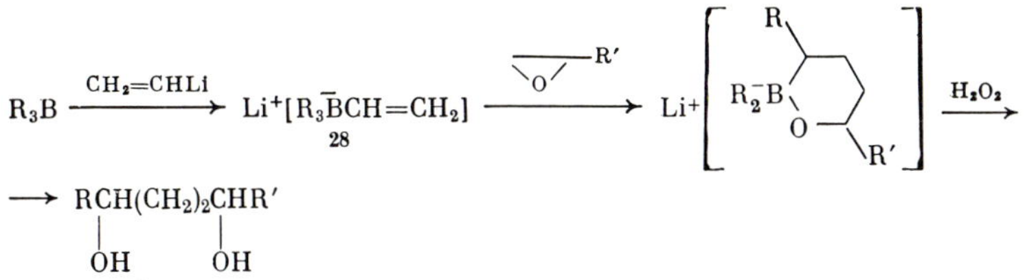

28

29 (85—100%)

This reaction opens up a new route to cyclopentenones. Thus, dihydrojasmone (*30*) has been synthesized by the Jones oxidation of 2,5-undecanediol, which is obtained from trihexylborane and propylene, followed by treatment with alkali:

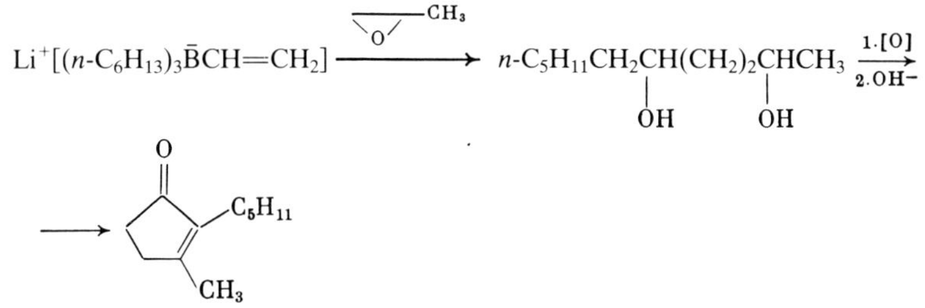

30 (51%)

The reaction of α-halovinyllithium (*31*) with triphenylborane led, after hydrolysis, to α-methyl-*cis*-stilbene (*32*), while α-methyl-*trans*-stilbene (*34*) was synthesized from the cis isomer (*33*) [86, 87]:

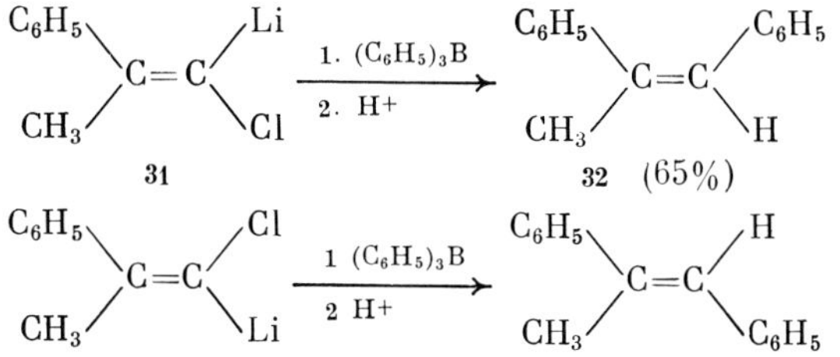

31 32 (65%)

33 34 (72%)

These data indicate that the removal of halogen and migration of a phenyl group occur synchronously and the reaction proceeds stereospecifically:

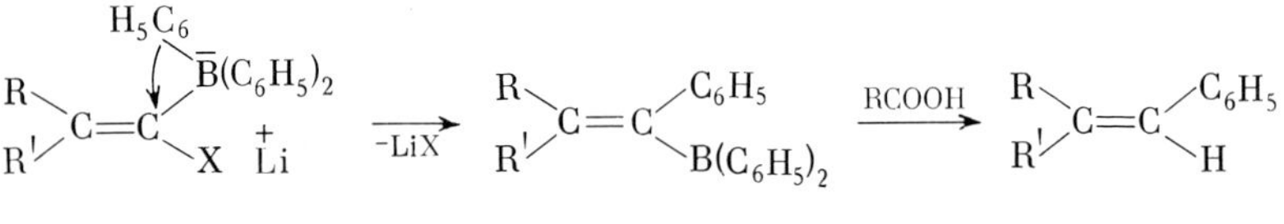

The transfer of radicals from boron to carbon is also observed in the reaction of 1,2-dichlorovinyllithium with triphenylborane.

The borates $[R_3BC(OCH_3){=}CHOCH_3]Li$, when treated with HCl in ether or trichloroacetic acid, $CH_3COONa/(CH_3CO)_2O$, and

$TiCl_4/Ti(OC_3H_7\text{-}iso)_4$ form olefins, $R_2C{=}CH_2$, as a result of the migration of two groups and β-elimination. β-Methoxy alcohols, $CH_3OCH_2CR_2OH$, are obtained from the same borates by sequential treatment with aqueous HCl and hydrogen peroxide [105].

REFERENCES

1. R. Köster, *Angew. Chem.* **68,** 383 (1956).
2. B. M. Mikhailov: "The Chemistry of Borohydrides", Moscow, Nauka (1967), p. 286.
3. M. F. Lappert, in:"The Chemistry of Boron and Its Compounds", ed. E. L. Muetterties, New York, J. Wiley and Sons (1967), p. 443.
4. B. M. Mikhailov and T. A. Shchegoleva, *Izv. Akad. Nauk SSSR, Otd. Khim. Nauk* **1959,** 546.
5. B. M. Mikhailov, M. E. Kuimova and E. A. Shagova, *Dokl. Akad. Nauk SSSR* **179,** 1344 (1968).
6. B. M. Mikhailov and M. E. Kuimova, *Zh. Obshch. Khim.* **41,** 1714 (1971).
7. B. M. Mikhailov, V. G. Kiselev and Yu. N. Bubnov, *Izv. Akad. Nauk SSSR, Ser. Khim.* **1965,** 898.
8. B. M. Mikhailov, Yu. N. Bubnov and V. G. Kiselev, *Zh. Obshch. Khim.* **36,** 62 (1966).
9. B. M. Mikhailov, V. A. Dorokhov and N. V. Mostovoi, in the collection "*Zh. Obshch. Khim.* Problems of Organic Synthesis", Leningrad, Nauka (1965), p. 228.
10. R. Köster, *German Patent* 1,034,176 (1958); *C. A.* **54,** 11995h (1960).
11. R. Köster, *German Patent* 1,109,682 (1960); *C. A.* **56,** 10188i (1962).
12. R. Köster, *U.S. Patent* 2,886,599 (1959); *C. A.* **53,** 19880b (1959).
13. R. Köster, and G. Schomburg, *Angew. Chem* **72,** 567 (1960).
14. R. Köster, *German Patent* 1,089,384 (1960); *C. A.* **56,** 1478c (1962).
15. A. J. Hubert, *J. Chem. Soc.* **1965,** 6669.
16. A. J. Hubert, *J. Chem. Soc.* **1965,** 6679.
17. R. Köster, *Ann.* **618,** 31 (1958).
18. R. Kọster, *Angew. Chem.* **70,** 413 (1958).
19. N. Miyamoto, S. Isiyama, K. Utimoto and H. Nozaki, *Tetrahedron Letters* **1971,** 4597.
20. N. Miyamoto, S. Isiyama, K. Utimoto and H. Nozaki, *Tetrahedron* **29,** 2365 (1973).
21. J. A. Marshall, *Acc. Chem. Res.* **2,** 33 (1969).
22. J. A. Marshall, *Science* **170,** 137 (1970).
23. G. Hesse and A. Haag, *Tetrahedron Letters* **1965,** 1123.
24. F. Joy, M. F. Lappert and B. Prokai, *J. Organometal. Chem.* **5,** 506 (1966).
25. F. Joy and M. F. Lappert, *Proc. Chem. Soc. (London)* **1960,** 353.
26. M. F. Lappert and B. Prokai, Unpublished Data, cited in [3].
27. G. Wittig and H. Schroeder, *Ann.* **592,** 38 (1955).
28. G. Wittig and D. Wittenberg, *Ann.* **606,** 1 (1957).
29. G. Wittig, H. G. Reppe and T. Eicher, *Ann.* **643,** 47 (1961).
30. G. Wittig and A. Rückert, *Ann.* **566,** 101 (1950).
31. G. Wittig and W. Stilz, *Ann.* **598,** 85 (1956).

32. G. Wittig and W. Stilz, *Ann.* **598,** 93 (1956).
33. Yu. N. Bubnov and B. M. Mikhailov, *Izv. Akad. Nauk SSSR, Ser. Khim.* **1970,** 2156.
34. B. M. Mikhailov, Yu. N. Bubnov and S. A. Korobeinikova, *J. Prakt. Chem.* **312,** 998 (1970).
35. B. M. Mikhailov, Yu. N. Bubnov, S. A. Korobeinikova and S. I. Frolov, *J. Organometal. Chem.* **27,** 165 (1971).
36. P. Paetzold and H. G. Smolka, *Chem. Ber.* **103,** 289 (1970).
37. V. A. Dorokhov and B. M. Mikhailov, *Izv. Akad. Nauk SSSR, Ser. Khim.* **1972,** 1895.
38. V. A. Dorokhov and B. M. Mikhailov, *Zh. Obshch. Khim.* **44,** 1281 (1974).
39. P. Paeztold, G. Stohr, H. Maisch and H. Lenz, *Chem. Ber,* **101,** 2881 (1968).
40. M. F. Lappert and B. Prokai, *J. Organometal. Chem.* **1,** 384 (1964).
41. J. J. Eisch and L. J. Gonsior, *J. Organometal. Chem.* **8,** 53 (1967).
42. A. Suzuki, M. Miyaura, M. Itoh, H. C. Brown, G. W. Holland and E. Negishi, *J. Am. Chem. Soc.* **93,** 2702 (1971).
43. A. Suzuki, M. Miyaura, M. Itoh, H. C. Brown and P. Jacob, *Synthesis* **1973,** 305.
44. G. W. Kabalka, *Intra-Science Chem. Rep.* **7,** 57 (1973).
45. O. P. Shitov, S. L. Ioffe, L. M. Leont'eva and V. A. Tartakovskii, *Zh. Obshch. Khim.* **43,** 1127 (1973).
46. O. P. Shitov, S. L. Ioffe, L. M. Leont'eva and V. A. Tartakovskii, *Zh. Obshch. Khim.* **43,** 1266 (1973).
47. S. L. Ioffe, L. M. Leont'eva, A. L. Blyumenfel'd, O. P. Shitov and V. A. Tartakovskii, *Izv. Akad. Nauk SSSR, Ser Khim.* **1974,** 1659.
48. N. Miyamoto, D. Fukuoka, K Utimoto and H. Nozaki, *Bull. Chem. Soc. Japan* **47,** 503 (1974).
49. G. B. Bagdasaryan, K. S. Badalyan, S. M. Markaryan and M. G. Indzhikyan, *Arm. Khim. Zh.* **26,** 929 (1973).
50. G. B. Bagdasaryan and M. G. Indzhikyan, *Arm. Khim. Zh.* **24,** 741 (1971).
51. H. K. Saha, L. J. Glicenstein and G. Urry, *J. Organometal. Chem.* **8,** 37 (1967).
52. G. Urry, in: "The Chemistry of Boron and Its Compounds", Ed. E. L. Muetterties, New York, J. Wiley and Sons (1967) p. 359.
53. J. J. Ritter, T. D. Coyle and J. M. Bellama, *J. Organometal. Chem.* **42,** 25 (1972).
54. G. Urry, J. Kerrigan, T. Parsons and H. Schlesinger, *J. Am. Chem. Soc.* **76,** 5299 (1954).
55. M. Zeldin and T. Wartik, *J. Am. Chem. Soc.* **88,** 1336 (1966).
56. E. Schram and G. Urry, *Inorg. Chem.* **2,** 405 (1963).
57. P. Ceron, A. Finch, J. Frey, J. Kerrigan, T. Parsons, G. Urry and H. Schlesinger, *J. Am. Chem. Soc.* **81,** 6368 (1959).
58. M. Zeldin, T. Gatti and T. Wartik, *J. Am. Chem. Soc.* **89,** 4217 (1967).
59. T. Rudolph, *J. Am. Chem. Soc.* **89,** 4216 (1967).
60. T. D. Coyle and J. J. Ritter, *J. Am. Chem. Soc.* **89,** 5739 (1967).
61. A. Rosen and M. Zeldin, *J. Organometal. Chem.* **31,** 319 (1971).
62. A. Holliday and A. G. Massey, *J. Inorg. Nucl. Chem.* **18,** 108 (1961).
63. J. Feeney, A. Holliday and F. J. Marsden, *J. Chem. Soc.* **1961,** 356.
64. N. B. Fox and T. Wartik, *J. Am. Chem. Soc.* **83,** 498 (1961).
65. C. Chambers, A. K. Holliday and S. M. Walker, *Proc. Chem. Soc. (London)* **1964,** 286.
66. E. B. Moore and W. N. Lipscomb, *Acta Cryst.* **9,** 668 (1956).
67. M. F. Lappert and B. Prokai, *J. Chem. Soc.* **1963,** 4223.
68. R. Jefferson, M. F. Lappert and B. Prokai, *J. Chem. Soc. A,* **1966,** 1584.

69. R. Jefferson and M. F. Lappert, *Intra-Science Chem. Rep.* **7,** 123 (1973).
70. B. M. Mikhailov, Yu. N. Bubnov, O. A. Nesmeyanova, V. G. Kiselev, T. Yu. Rudashevskaya and B. A. Kazansky, *Tetrahedron Letters* **1972,** 4627.
71. V. S. Bogdanov, V. F. Pozdnev, Yu. N. Bubnov and B. M. Mikhailov, *Dokl. Akad. Nauk SSSR* **193,** 586 (1970).
72. Yu. N. Bubnov, O. A. Nesmeyanova, T. Yu. Rudashevskaya, B. M. Mikhailov and B. A. Kazansky, *Tetrahedron Letters* **1971,** 2153.
73. Yu. N. Bubnov, O. A. Nesmeyanova, T. Yu. Rudashevskaya, B. M. Mikhailov and B. A. Kazanskii, *Zh. Obshch. Khim.* **43,** 127 (1973).
74. Yu. N. Bubnov, O. A. Nesmeyanova, T. Yu. Rudashevskaya, B. M. Mikhailov and B. A. Kazanskii, *Zh. Obshch. Khim.* **43,** 135 (1973).
75. M. Naruse, K. Utimoto and H. Nozaki, *Tetrahedron Letters* **1972,** 1605.
76. B. A. Kazansky, Yu. N. Bubnov, S. V. Zotova, N. M. Abramova, V. G. Kiselev and B. M. Mikhailov, *Tetrahedron Letters* **1974,** 567.
77. R. Köster, S. Arora and P. Binger, *Angew. Chem.* **81,** 186 (1969).
78. E. Breuer, E. Segall, Y. Stein and S. Sarel, *J. Org. Chem.* **37,** 2242 (1972).
79. D. J. Cram: "Fundamentals of Carbanion Chemistry" Academic Press, N.Y, London (1965).
80. M. Zeldin and A. Rosen, *J. Organometal. Chem.* **34,** 259 (1972).
81. E. Negishi and R. Merrill, *Chem. Comm.* **1974,** 860.
82. E. Negishi, A. Abramovitch and R. Merrill, *Chem. Comm.* **1975,** 138.
83. A. Suzuki, N. Miyaura and M. Itoh, *Tetrahedron* **27,** 2775 (1971).
84. N. Miyaura, H. Tagami, M. Itoh, A. Suzuki, *Chem. Letters,* **1974,** 1411.
85. K. Utimoto, K. Uchida and H. Nozaki, *Tetrahedron Letters* **1973,** 4527.
86. G. Köbrich and H. Merkle, *Chem. Ber.* **100,** 3371 (1967).
87. G. Köbrich and H. Merkle, *Angew. Chem.* **79,** 50 (1967).
88. B. Wrackmeyer and H. Nöth, *J. Organometal. Chem.* **108,** C21 (1976).
89. G. Menz and B. Wrackmeyer, Z. Naturforsch. **32b,** 1400 (1977).
90. J. J. Eisch, K. Tamao and R. J. Wilcsek, *J. Am. Chem. Soc.* **97,** 895 (1975).
91. Y. Masuda, M. Hoshi and A. Arase, *Chem. Letters* **1980,** 413.
92. Y. Takahashi, M. Tokuda, M. Itoh and A. Suzuki, *Chem. Letters* **1977,** 999.
93. Y. Takahashi, M. Tokuda, M. Itoh and A. Suzuki, *Chem. Letters* **1980,** 461.
94. W. Haubold and K. Stanzl, *Chem. Ber.* **111,** 2108 (1978).
95. E. Negishi, *J. Organometal. Chem.* **108,** 281 (1976).
96. I. Akimoto and A. Suzuki, *Synthesis* **1979,** 146.
97. E. R. Marinelli and A. B. Levy, *Tetrahedron Letters* **1979,** 2313.
98. T. Sotoyama, S. Hara and A. Suzuki, *Bull. Chem. Soc. Japan* **52,** 1865 (1979).
99. A. B. Levy, *J. Org. Chem.* **43,** 4684 (1978).
100. A. B. Levy, *Tetrahedron Letters* **1979,** 4021.
101. I. Akimoto, M. Sano and A. Suzuki, *Bull. Chem. Soc. Japan* **54,** 1587 (1981).
102. E. Negishi, K. W. Chiu and T. Yosida, *J. Org. Chem.* **40,** 1676 (1975).
103. K. Utimoto, K. Okuda and H. Nozaki, *Tetrahedron Letters* **1975,** 4239.
104. K. Utimoto, N. Sakai, M. Obayashi and H. Nozaki, *Tetrahedron* **32,** 769 (1976).
105. T. Yogo and A. Suzuki, *Chem. Letters* **1980,** 591.

CHAPTER 14

Reactions of alkyl(aryl)boranes with metal salts

The direction of the reactions of organic boron derivatives with metal salts is determined by the nature of the two components and the conditions under which the process is carried out (solvent, temperature, etc.). Reactions of this type proceed by a heterolytic or radical mechanism.

The reactions of organoboranes with salts are used for the synthesis of some types of organometallic (Ag, Hg, Cu, Tl, Pb, etc.) and *metallocene compounds* and also for transition from organoboranes to different classes of organic substances (hydrocarbons, halides, esters, etc.).

14.1 REACTIONS WITH SILVER SALTS. SYNTHESIS OF HYDROCARBONS AND ORGANIC SILVER COMPOUNDS

Most organoboranes form products of radical coupling under the action of alkaline silver oxide. Compounds of the type R—R and R—R′ can be obtained from organic halides and olefins via organoboranes by this method.

The reaction is also used for the synthesis of some classes of organic silver derivatives. It can also be used as a method for the generation of alkyl radicals (from R_3B).

Radical coupling and the formation of a silver mirror by the action of an ammoniacal silver oxide solution (Tollens reagent) on alkylboronic acids were observed for the first time by Johnson, et al. [1, 2], who obtained *n*-octane, *n*-dodecane, and dibenzyl:

$$2RB(OH)_2 + 2[Ag(NH_3)_2]^+ + 2H_2O \longrightarrow \underset{70-94\%}{R{-}R} + 2Ag + 2H_3BO_3 + 2NH_4^+ + 2NH_3.$$

Under analogous conditions, *n*-hexane was synthesized from propylboron difluoride [3]. At the same time, the main product of the reaction of *tert*-$C_4H_9B(OH)_2$ is *tert*-butanol [2].

Trialkylboranes react with Tollens reagent very slowly, forming a mixture of R—R and RH [4, 5]. Radical coupling proceeds more smoothly with the use of an alkaline silver nitrate solution [5]:

$$2R_3B + 6AgOH + 2KOH \xrightarrow{\text{THF, 0–10° C}} \underset{35\text{–}80\%}{3R{-}R} + 6Ag + 2KB(OH)_4$$

(R = alkyl, cycloalkyl)

The yield of coupling products is 60–80% from primary trialkylboranes and 35–50% from the secondary compounds.

The above method makes it possible to obtain compounds of the type R—R in high yield directly from organic halides in accordance with the following scheme [6]:

$$R{-}X + Mg + H_3B\cdot THF \xrightarrow{\text{THF}} \left[R{-}B\langle\right] \xrightarrow[\text{2. AgNO}_3]{\text{1. KOH, CH}_3\text{OH}} R{-}R$$

52—90%

(R = alkyl, cycloalkyl, $C_6H_5CH_2$, C_6H_5, *p*-$CH_3C_6H_4$, *p*-ClC_6H_4, *m*-$CH_3OC_6H_4$, 2-$C_6H_5OC_2H_4$; X=Br, I)

The organoborane is prepared from an organic bromide or iodide and magnesium in a solution of BH_3 in tetrahydrofuran and, without isolation, is treated with alcoholic potassium hydroxide solution and $AgNO_3$. In the case of *o*-substituted aryl halides, the product yield is significantly reduced, which is explained by steric hindrance.

Compounds R—R and R—R′ can be obtained from cyclenes and olefinic hydrocarbons by hydroboration, followed by reaction with AgOH [7, 8]:

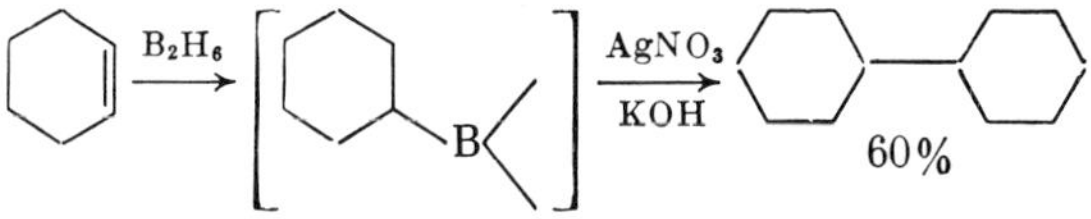

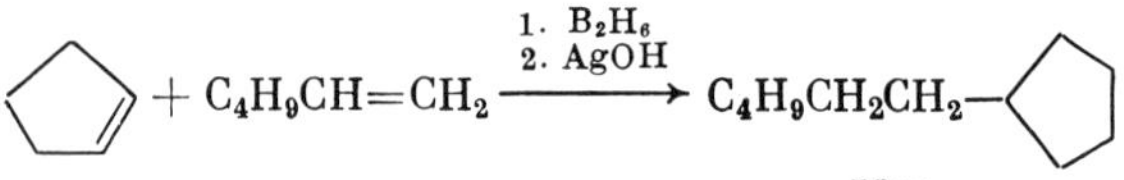

58%

This relatively simple method for the formation of carbon-carbon bonds is very important in organic synthesis, since hydroboration can be used to obtain organoboranes containing some functional groups in the radical, such as halogen, OH, NR_2, COOR, and others.

An interesting example of the application of the "coupling" reaction is the synthesis of 1,10-decanediol from dihydropyran [9]:

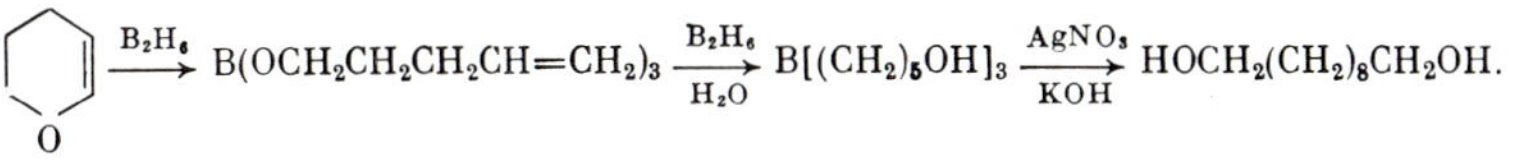

Reactions with silver salts serve as a basis for the synthesis of cyclic hydrocarbons from dienes [115].

Arylboronic acids [10–15], α-furylboronic and α-thienylboronic acids [2], and allylboron difluoride [3] form protonolysis products under the action of an aqueous-ammoniacal solution of silver salts:

$$ArB(OH)_2 + H_2O \xrightarrow{Ag^+} ArH + B(OH)_3.$$

In diarylborinic acids, both B—C bonds are cleaved [16]. The difference between these data and those cited above [6] is apparently due to the different reaction conditions: in [6], a solution of alcoholic alkali in tetrahydrofuran was used.

Vinyl boron derivatives react with Ag_2O in NH_3 solution to form a mixture of the products of protonolysis and radical coupling (1,3-diene compounds) [4, 17–20], with hydrolytic cleavage predominating as a rule. Ferrocenylboronic acid forms 52% of diferrocenyl and 31% of ferrocene under these conditions [21–23].

The mechanism of the reactions of organoboranes with silver salts, which was already proposed by Johnson, et al. [2], involves the formation of organometallic compounds which, as a rule, are unstable [5, 24] and either break down to metallic silver and free radicals or hydrolyze to form hydrocarbons:

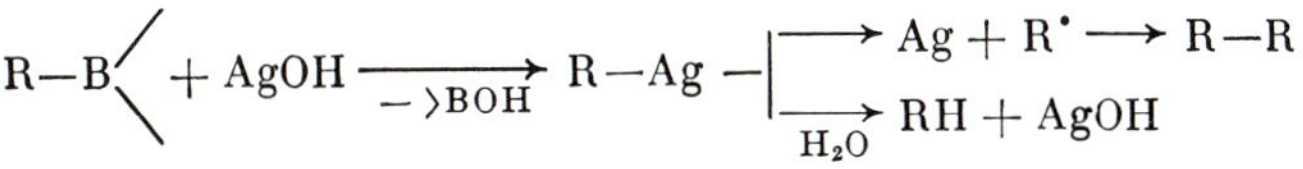

Aryl and vinyl silver derivatives are more stable toward homolytic cleavage than the alkyl and benzyl analogs [24]. At the same time, their hydrolytic cleavage is facilitated, since the Ag—C_{sp^2} bond is more highly polarized than the Ag—C_{sp^3} bond. Similar regularities are observed in the protonolysis of boron derivatives [25–27].

The radical nature of the reactions is confirmed by the following data. Thirty-six per cent of ethyl chloride has been found in the products of the reaction of triethylborane with AgOH in the presence of CCl_4 [5]. The reactions of optically active diethanolamine esters of (+)- and (−)-l-phenylethylboronic acids with $AgNO_3$ in KOH lead to a mixture of *threo*- and *erythro*-2,3-diphenylbutanes in a ratio of 1:1 [28], i.e., they proceed with racemization of the radicals.

Unlike the alkyl, aryl, and vinyl derivatives of silver, some of its cyclopentadiene compounds are quite stable substances. All of these have been synthesized from the corresponding boron derivatives. 1-(2-Chloroferrocenyl)silver (*1*), which is stable at 20°C, has been synthesized in a yield of 37% by the action of Ag_2O on 2-chloroferrocenylboronic acid in aqueous ammonia [29, 30].

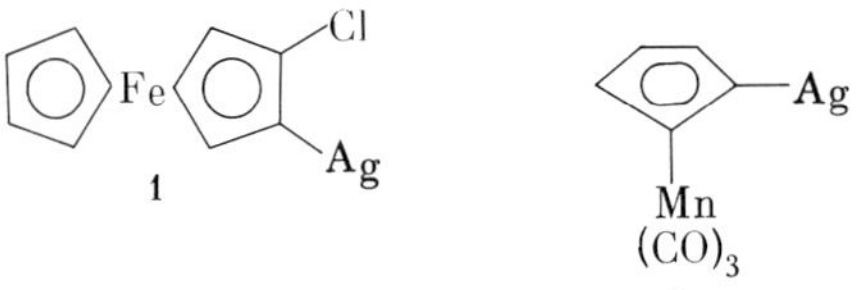

1-(1-Chloroferrocenyl)-[30] and 1-(1-bromoferrocenyl)silver [31] have been obtained in an analogous manner; cyclopentadienylmanganesetricarbonylboronic acid [32] has been converted to manganesetricarbonylcyclopentadienylsilver (*2*) [33]. Silver ferrocenes decompose on heating to form metallic silver and other products.

The reaction of aliphatic boranes with silver oxide is sometimes complicated by the formation of alcohols [28]. (7αH)-Longifolyl-15-borane is oxidized under the action of Ag_2O, forming 25% of rearrangement products along with (7αH)-longifolol-15 [34].

Cyclooctanone (59%) and cyclooctene (10%) are obtained by treating 9-borabicyclo[3.3.1]nonane with one equivalent of $AgNO_3$ in aqueous alkaline tetrahydrofuran [35]:

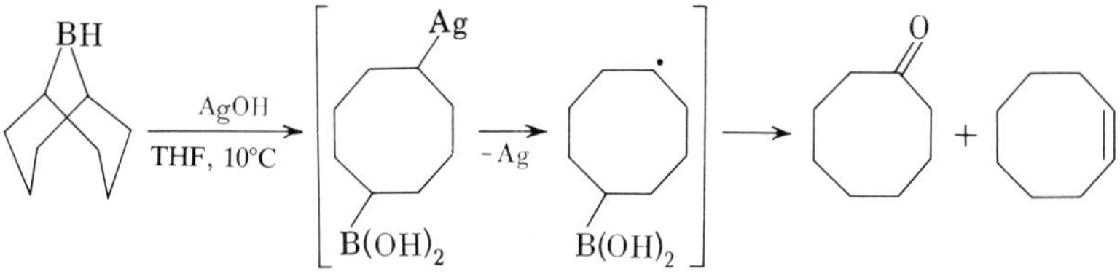

When a threefold excess of $AgNO_3$ is used, the main product (90%) is cyclooctene. The action of $AgNO_3$ in ethanol at 25°C on the esters $RR'CClB(OCH_3)_2$ forms olefins [116]. The 1,2-diboron compounds formed by the hydroboration of internal and cyclic acetylenes with borane in tetrahydrofuran, upon treatment with $AgNO_3$ and a base, also form olefins with a predominance (78–98%) of the *E* isomer [117].

The water-insoluble salt $AgB(C_4H_4N)_4$ is obtained by the action of silver nitrate on potassium or pyridinium tetrapyrrylborate [36].

14.2 REACTIONS WITH COPPER SALTS. SYNTHESIS OF ALKYL AND ARYL HALIDES

Trialkyl- and tricycloalkylboranes react with cupric chloride and bromide upon heating in aqueous tetrahydrofuran to form the corresponding alkyl and cycloalkyl halides [37]:

$$R_3B + 6CuX_2 + 3H_2O \xrightarrow[54-90\%]{THF} 3RX + 3Cu_2X_2 + 3HX + B(OH)_3.$$

Cyclohexyl bromide (87%) and hexyl bromide (3%) have been obtained by the action of two equivalents of $CuBr_2$ on hexyldicyclohexylborane, indicating the higher activity of *sec*-alkylboranes as compared with the primary compounds [38]. The configuration of the radical of the starting organoborane is not retained during the reaction. Thus, a mixture of 2-*exo*- and 2-*endo*-norbornyl bromides in a ratio of 4:1 is formed from tri-*exo*-norbornylborane. It is suggested that the reaction represents a redox process, during which no organocopper compounds are formed:

$$R_3B + CuX_2 + H_2O \longrightarrow R^{\bullet} + R_2BOH + CuX + HX,$$
$$R^{\bullet} + CuX_2 \longrightarrow RX + CuX.$$

Alkylboronic acids do not react with copper halides in aqueous solution even at 100°C [1] while phenyl- [39], α-furyl- and α-thienylboronic acids [2], ferrocenylboronic acid and its derivatives with different substituents in the ring [21, 22, 29, 40], and cyclopentadienylmanganesetricarbonylboronic acid [41] form the corresponding aryl halides in yields of 75–87% under these conditions:

$$ArB(OH)_2 + CuX_2 + H_2O \longrightarrow ArX + CuX + HX + B(OH)_3.$$

One or two boryl groups are replaced in 1,1′-ferrocenylene-diboronic acid, depending on the conditions [21, 42].

The direction of the reaction of arylboronic acids with cupric acetate is determined by the structure of the aryl radical. Phenol is obtained from phenylboronic acid, while *o*- and *m*-nitrophenylboronic acids form biphenyl derivatives [43].

In the case of ferrocenylboronic acid, a mixture of approximately 60% of ferrocenyl acetate and approximately 20% of diferrocenyl is formed [21, 22, 44]. When 1-(2-chloroferrocenyl)boronic acid is treated with cupric acetate, both the boron and the chlorine are replaced by the acyl groups and 1,2-ferrocenylene diacetate is obtained [29]; under the same conditions, 1,1′-ferrocenylene-diboronic acid forms 1,1′-ferrocenylene diacetate, which was later converted to 1,1′-dihydroxyferrocene [45].

Cyclopentadienylmanganesetricarbonylcopper, which is stable under ordinary conditions, has been synthesized by the action of cupric acetate on cyclopentadienylmanganesetricarbonylboronic acid in water at 20°C [46].

The reactions of arylboronic acids with cupric phthalimide, leading to amino derivatives, were considered in Chapter 7. Chlorobenzene has been found in the mixture of products of the thermal reaction (275°C) of the anhydride $(C_6H_5BO)_3$ with anhydrous $CuCl_2$ [47].

Lithium tetraphenylborate reacts with cupric chloride in ether to form cuprous chloride, triphenylborane, and biphenyl [48]. The pyridine complex of copper tetra-*p*-tolylborate decomposes upon acidification to form copper and *p,p*′-ditolyl [49]. An ammoniacal solution of Cu_2Cl_2 in alcohol decomposes ammonium tetraphenylethynylborate with the quantitative formation of copper phenylacetylenide [50].

The reactions of organoboranes in the presence of copper salts are also described in Chapter 15.

14.3 REACTIONS WITH MERCURY SALTS

14.3.1 Synthesis of organomercury compounds

Michaelis and Becker [10] found that phenyl- and *p*-tolylboronic acids are rapidly and quantitatively converted to arylmercury chlorides by the action of mercuric chloride:

$$ArB(OH)_2 + HgCl_2 + H_2O \longrightarrow ArHgCl + B(OH)_3 + HCl.$$

This reaction is widely used in organomercury chemistry. A large number of mercury salts has been obtained by this method [51], including those of the benzene [11, 12, 52–58], naphthalene [55, 56], α-furan [2], α-thiophene [2, 59], ferrocene [21, 22, 42, 60], and cyclopentadienylrheniumtricarbonyl [32] series. Arylboronic acids containing electronegative groups (NO_2, COOR, etc.) in the nucleus do not react with $HgCl_2$ in water, whereas, in glacial acetic acid or aqueous perchloric acid, the reaction proceeds readily [55]. Mercuric bromide [38, 55], and mercuric oxide [63], acetate [56], nitrate and perchlorate [61, 62] can be used in mercuration reactions instead of to mercuric chloride.

Diarylborinic acids exchange both radicals under the action of two equivalents of mercuric acetate or chloride [56, 58, 64].

Boron compounds are used in the synthesis of symmetrical, Ar_2Hg, [63] and unsymmetrical, RHgAr, [65–67] derivatives of mercury:

$$RHgI + ArB(OH)_2 \longrightarrow RHgAr.$$

The latter compounds have been used to refine the Kharasch series.

Vinylboranes show approximately the same reactivity (Part III) as their aryl analogs in mercuration reactions. At the same time, alkylboronic acids are much less active, because of the lower polarity of the boron-carbon bond.

The reaction of alkylboronic acids with mercury salts in water proceeds slowly [1, 63] but, in acetic acid at 100°C, it is completed within a few minutes [55]:

$$n\text{-}C_8H_{17}B(OH)_2 \xrightarrow[CH_3COOH]{(CH_3COO)_2Hg} \xrightarrow{NaCl,\ H_2O} \underset{81\%}{n\text{-}C_8H_{17}HgCl}.$$

Apparently, only primary alkylboronic acids are capable of forming salts RHgCl; only Hg_2Cl_2 was isolated from the products of the reaction of *sec*-$C_4H_9B(OR)_2$ [68].

Benzylboronic acid [54] and its derivatives [68–70], which show a high reactivity, are easily converted to the corresponding benzylmercury chlorides by treatment with mercuric chloride solution. In methane- [71] and ethane-1,1-diboronic acids [72], the two boryl groups are exchanged consecutively, as a result of which salts of the type $ClHgCH(R))B(OH)_2$ or $RCH(HgCl)_2$ are obtained. An ester of 1-phenylethane-1,2-diboronic acid breaks down under these conditions, with the formation of polystyrene [68]. The reaction of $C_6H_5CH_2CH[B(OR)_2]_2$ with $HgCl_2$ leads to the ester $C_6H_5CH_2CH$-$(HgCl)B(OR)_2$ [118].

Butyl esters of α-bromo- and α-butoxybenzylboronic acids react readily with mercuric chloride, but the mercury salts obtained in this process decompose with the formation of $HgCl_2$ and acetophenone [68].

From the ester of methanetetraboronic acid $C[B(OCH_3)_2]_4$ and mercuric acetate, Matteson and co-workers [73] obtained salts of the type $C(HgX)_4$, in which the carbon atom is strongly shielded by mercury atoms, so that any attack on the C—Hg bond becomes difficult.

In contrast to the reactions considered above, in the action of mercuric chloride on 2-*exo*- and 2-*endo*-(5-norbornenyl)boronic acids [61, 74, 75] and on 5-*exo*- and 5-*endo*-(bicyclo[2.2.2]octen-2-yl)boronic acid [76], the initial attack takes place not at the B—C bond but at the C═C bond, with the activity of the endo isomer being 10^2–10^3 times lower than that of the exo isomer:

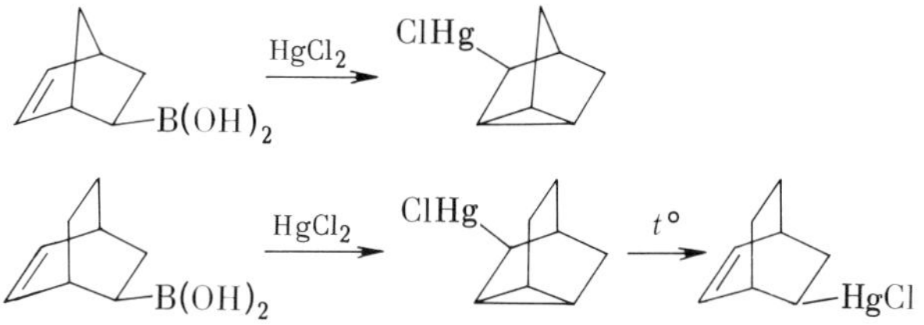

Both compounds R_2Hg and the salts RHgX can be obtained from trialkyl- and triarylboranes.

The reaction of triphenylborane with mercuric chloride was carried out for the first time by Wittig, et al. [77], who obtained phenylmercury chloride.

Honeycutt and Riddle [78–84] developed a method for the synthesis of dialkyl- and diarylmercuric compounds from trialkyl- or triarylboranes:

$$R_3B + HgX_2 \longrightarrow R_2Hg.$$

When mercuric chloride and oxide are used, the reaction is carried out in aqueous alkali [78–84]; with mercuric acetate, it is carried out in tetrahydrofuran or diglyme [78, 79, 84]. In aqueous medium, two of the three radicals of the trialkylborane are transferred to the mercury. This method for the synthesis of dialkylmercury derivatives was developed in [85].

The reaction of primary trialkylboranes with mercuric acetate in a ratio of 1:3 in solvents of the ether type leads to alkylmercuric acetates in almost quantitative yield; the latter are converted to the corresponding halides without isolation [86]:

$$R_3B + 3(CH_3COO)_2Hg \longrightarrow 3RHgOCOCH_3 \xrightarrow[H_2O]{NaX} RHgX.$$

Triisobutylborane and secondary trialkylboranes react much more slowly than the primary compounds [87].

Mercury salts with functional substituents in the alkyl radical can be synthesized with the use of organoboranes. For example, methyl 11-chloromercuriundecanoate has been obtained from methyl 10-undecanoate [86]:

$$CH_3OOC(CH_2)_8CH{=}CH_2 \xrightarrow{B_2H_6} [R_3B] \xrightarrow[2.\ NaCl]{1.\ (CH_3COO)_2Hg} \underset{97\%}{CH_3OOC(CH_2)_{10}HgCl.}$$

In reactions of trialkylboranes with mercury dimethoxide, only one radical migrates from the boron to the mercury [88].

The process of replacement of boron by mercury is a typical reaction of electrophilic substitution and probably proceeds via a transition state with three-center bonds of the type *A* [73, 89, 90] or *B*:

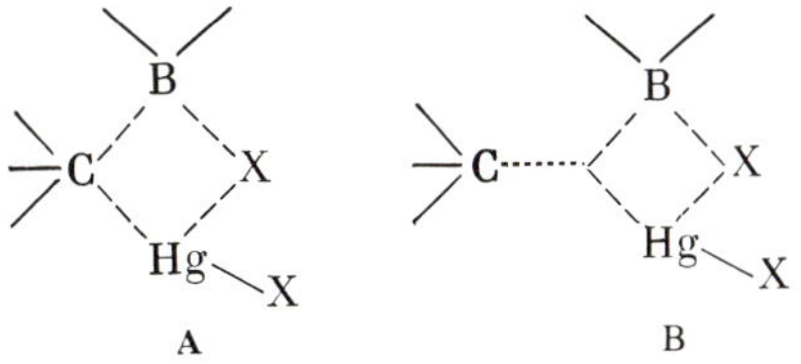

The kinetics of the replacement of boron by mercury [62, 69, 89] have been considered in detail in a book [91].

The stereochemistry of the mercuration of organoboranes has been investigated for the example of the reactions of the ester of (*R*)-(−)-1-phenylethylboronic acid with mercuric chloride [70] and of tri-*exo*-norbornylborane with mercuric acetate [87]. In the former case, the formation of 11–14% of a product with the starting configuration was observed, whereas, in the latter case, after bromination of the mercury salt, 2-norbornyl bromide containing 7–12% of the endo isomer was obtained (it should be noted that, in radical reactions of tri-*exo*-norbornylborane, only approximately 20% of the endo isomer is obtained). On the basis of these data, it has been concluded [70, 87, 92] that the replacement of boron by mercury proceeds with retention of the configuration, while the formation of inversion products is explained by the instability of the mercury salts formed. For the example of the reactions of the *threo* and *erythro* isomers of tris(3,3-dimethyl-1-butyl-1,2-d_2)borane with mercury(II) acetate in tetrahydrofuran, it was established that the mercurodeboroination of primary trialkylboranes proceeds mainly (over 90%) with an inversion of the configuration [119, 120].

Alkoxides of monovalent mercury, $Hg_2(OR)_2$, react with trialkylboranes by a radical mechanism [93]. In this process, two radicals of R_3B are transferred to the mercury, with secondary alkylboranes being approximately seven times more active in this reaction than the primary compounds.

Tetraphenyl- [77], tetratolyl- [94], and tetra-α-furylborates [95] of alkali metals exchange all four radicals in reactions with HgX_2. The borates $M[B(C_2H_5)_3Ar]$ arylate metallic mercury, as a result of which Ar_2Hg [96, 97] and triphenylborane, which does not react with mercury, are obtained.

o,o'-Dichloromercuribiphenyl, which is converted to tetrameric *o,o'*-biphenylenemercury [98] by heating, has been obtained from tetramethylammonium bis-2,2'-biphenyleneborate and mercuric chloride [94].

14.3.2 Synthesis of halides, acetates, and hydrocarbons

Tufariello and Hovey [99, 100] have developed a convenient three-stage method for the anti-Markownikoff hydrobromination of olefins using organoboron and organomercury compounds (hydroboration-mercuration-bromination):

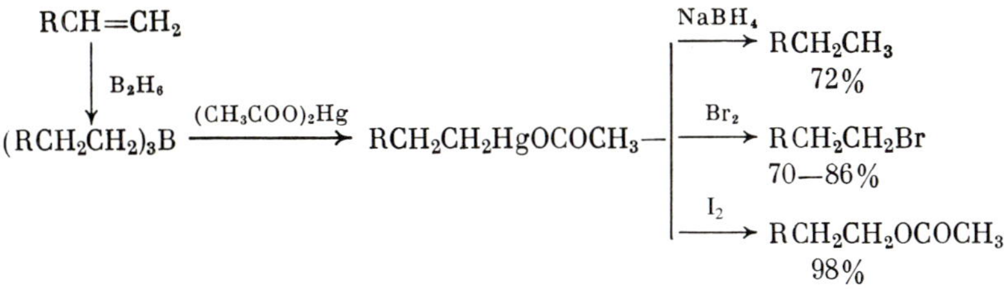

It is very important that primary organoboranes are converted to bromides much more easily than the secondary compounds. The boron and mercury derivatives formed are usually not isolated but are subjected to further treatment.

In contrast to the results of the bromination, alkyl acetates are obtained as a result of hydroboration-mercuration-iodination [101]. The essence of this method consists of an anti-Markownikoff acetylation of olefins. It should be noted that the reaction of carboxylic acids with terminal olefins in the presence of Lewis acids leads to secondary alkyl acetates [102].

When sodium borohydride is used in the final stage instead of halogens, the reaction product is a hydrocarbon [99]. This method of "protonolysis" of alkylboranes is evidently the mildest and may be used instead of acid cleavage (Chapter 6).

In a slight modification of the mercuration-iodination of trialkylboranes, consisting of the addition of lithium iodide to the mixture (before iodination), alkyl iodides are obtained instead of esters, in a yield of approximately 98% [101]:

$$R_3B \xrightarrow[\text{2. LiI; 3. } I_2]{\text{1. } (RCOO)_2Hg} R\text{—}I.$$

Reaction with $HgCl_2$ and subsequent bromination have been used for establishing the orienting effect of the dihydroxyboryl group in the nitration of phenylboronic acid under different conditions [13, 39, 53]:

$$C_6H_5B(OH)_2 \xrightarrow{NO_2^+} O_2NC_6H_4B(OH)_2 \xrightarrow[H_2O]{HgCl_2} O_2NC_6H_4HgCl \xrightarrow{Br_2} O_2NC_6H_4Br.$$

2-Iodo(dimethylaminomethyl)ferrocene has been obtained by the mercuration-iodination of the corresponding ferrocenylboronic acid [60].

14.4 REACTIONS WITH IRON SALTS. SYNTHESIS OF ALKYL HALIDES AND THIOCYANATES

Trialkyl- and tricycloalkylboranes react with ferric chloride in aqueous tetrahydrofuran to form the corresponding alkyl or cycloalkyl halides [37]:

$$R_3B + 6FeCl_3 + 3H_2O \xrightarrow{\text{THF, 55° C}} \underset{50-70\%}{3RCl} + B(OH)_3 + 6FeCl_2 + 3HCl.$$

Both the first and second alkyl groups of the R_3B molecule are split off quite easily, while removal of the third radical proceeds slowly. The reaction represents a redox process and apparently proceeds analogously to the reaction of R_3B with CuX_2. Fenton's reagent ($FeCl_2$-H_2O_2) can be used instead of $FeCl_3$.

The action of ferric thiocyanate on triorganoboranes in aqueous tetrahydrofuran forms the corresponding alkyl thiocyanates with a small impurity of isothiocyanates [37]:

$$(n\text{-}C_6H_{13})_3B + Fe(SCN)_3 \longrightarrow \underset{80\%}{n\text{-}C_6H_{13}SCN}$$

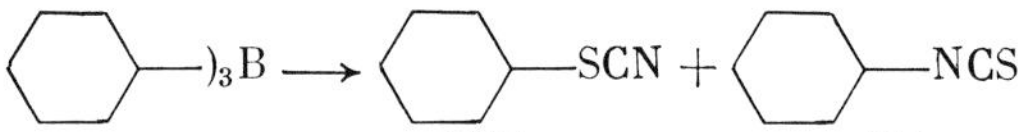

67% 5%

The analogous reaction of R_3B with KSeCN in the presence of ferric salts leads to selenocyanates [121]. *tert*- and *sec*-Alkylboranes react with KSCN and KSeCN more easily than primary alkylboranes [122]. Alkyl and cycloalkyl azides, RN_3, are obtained by the radical reaction of R_3B with sodium azide, ferric sulfate, and hydrogen peroxide [123].

Biphenyl has been isolated from the products of the reaction of lithium tetraphenylborate, $(C_6H_5)_4BLi$, with ferric chloride [103].

Diethylcyclopentadienylborane reacts with ferrous chloride to form ferrocene [104]. The reaction of $(C_6H_5)_4BLi$ with $FeCl_2$ in ether leads to biphenyl, whereas hydrolysis takes place in water [48].

14.5 SYNTHESIS OF ORGANOTHALLIUM COMPOUNDS

Aromatic boronic acids, when heated in aqueous solution with thallium chloride or bromide, form the corresponding organothallium compounds $ArTlX_2$ or Ar_2TlX [63, 105–108]:

$$2ArB(OH)_2 + TlX_3 + 2H_2O \longrightarrow Ar_2TlX + 4HX + 2H_3BO_3,$$
$$ArB(OH)_2 + TlX_3 + H_2O \longrightarrow ArTlX_2 + H_3BO_3 + HX.$$

Vinyl boron derivatives have been used for the synthesis of the corresponding divinylthallium chlorides [50, 109]. An attempt to obtain alkylthallium compounds by this method proved unsuccessful [63].

The reaction of diethanolamine ester of 1-phenylethylboronic acid with $TlCl_3$ in aqueous solution leads to 2,3-diphenylbutane (78%) with an impurity of 1-phenylethanol [28]. The intermediately formed organothallium compound is unstable and breaks down into radicals, which recombine.

$$C_6H_5CH(CH_3)-B(OR)_2 \xrightarrow[OH^-]{TlCl_3} C_6H_5CH(CH_3)-CH(CH_3)C_6H_5$$

Hexyl chloride and hexyl acetate are obtained by treating trihexylborane with thallium triacetate in the presence of lithium chloride [92].

14.6 SYNTHESIS OF ORGANOLEAD COMPOUNDS

In connection with the fact that tetraethyllead is one of the most widely used antiknock compounds, attempts have been made to use ethyl derivatives of boron for the synthesis of this compound.

One of the methods consists of heating metallic lead with sodium tetraethylborate and ethyl chloride in ether or hexane [110]:

$$2(C_2H_5)_4BNa + Pb + 2C_2H_5Cl \xrightarrow{70°\,C} \underset{52-58\%}{(C_2H_5)_4Pb} + 2NaCl + 2(C_2H_5)_3B.$$

The same compound is obtained in approximately 98% yield by heating $NaB(C_2H_5)_4$ with lead(II) oxide, acetate, chloride, or sulfate in a ratio of 2:1 [111].

Triethylborane alkylates PbO and lead(II) naphthenoate in water, aqueous alkali, and ethylene glycol esters with the formation of tetraethyllead, in a yield of over 40% [79]. Trihexylborane reacts analogously.

The treatment of R_3B with lead tetraacetate or (diacetoxyiodo)benzene leads to the alkyl acetates CH_3COOR [124].

Of interest is the bimetallic methane derivative $[(C_6H_5)_3Pb]_2C[B(OCH_3)_2]_2$, obtained by Matteson and Larson [112, 113] as a result of the sequential treatment of the tetraborate ester $C[B(OCH_3)_2]_4$ with a base and chlorotriphenylplumbane.

The action of tetraalkylammonium bromide on lithium tetraalkylborate produces ammonium salts of borates. Compounds of this type, containing one or more higher alkyl radicals, have low melting points (<35°C). Thus, triethyl-*n*-hexylammonium triethyl-*n*-hexylborate, $[(C_2H_5)_3\overset{+}{N}C_6H_{13}][(C_2H_5)_3\overset{-}{B}C_6H_{13}]$, designated as $N_{2226}B_{2226}$, is used as a distinctive aprotic highly polar solvent for some organic reactions of the S_N1 and S_N2 types [114].

REFERENCES

1. H. R. Snyder, J. A. Kuck and J. R. Johnson, *J. Am. Chem. Soc.* **60,** 105 (1938).
2. J. R. Johnson, M. G. Van Campen and O. Grummitt, *J. Am. Chem. Soc.* **60,** 111 (1938).
3. F. E. Brinckman and F. G. A. Stone, *J. Am. Chem. Soc.* **82,** 6218 (1960).
4. T. D. Parsons, M. B. Silverman and D. M. Ritter, *J. Am. Chem. Soc.* **79,** 5091 (1957).
5. H. C. Brown, N. C. Herbert and C. H. Snyder, *J. Am. Chem. Soc.* **83,** 1001 (1961).
6. S. W. Breuer and F. A. Broster, *Tetrahedron Letters* **1972,** 2193.
7. H. C. Brown and C. H. Snyder, *J. Am. Chem. Soc.* **83,** 1002 (1961).
8. H. C. Brown, C. Verbrugge and C. H. Snyder, *J. Am. Chem. Soc.* **83,** 1001 (1961).
9. J. H. Murib and C. A. Bonecutter (to National Distillers and Chemical Corp.), *U.S. Patent* 3,324,146 (1967); *C. A.* **68,** 78115 (1968).
10. A. Michaelis and P. Becker, *Ber.* **15,** 180 (1882).
11. A. Michaelis, *Ber.* **27,** 244 (1894).
12. A. Michaelis and E. Richter, *Lieb. Ann.* **315,** 26 (1901).
13. W. Seaman and J. R. Johnson, *J. Am. Chem. Soc.* **53,** 711 (1931).
14. W. J. Lennarz and H. R. Snyder, *J. Am. Chem. Soc.* **82,** 2172 (1960).
15. A. H. Soloway and P. Szabady, *J. Org. Chem.* **25,** 1683 (1960).
16. E. E. Van Tamelen, G. Brieger and K. G. Untch, *Tetrahedron Letters* **1960,** No. 8, 14.
17. H. D. Kaesz and F. G. A. Stone. Ch. 3 in: "Organometallic Chemistry," ed. H. Zeiss, Reinhold Publishing Corp, N. Y., Chapman and Hall Ltd, London (1960).
18. T. D. Parsons and D. M. Ritter, *J. Am. Chem. Soc.* **76,** 1710 (1954).
19. B. Bartocha, F. E. Brinckman, H. D. Kaesz and F. G. A. Stone, *Proc. Chem. Soc.* **1958,** 116.
20. R. L. Letsinger and I. H. Skoog, *J. Org. Chem.* **18,** 895 (1953).

21. A. N. Nesmeyanov, V. A. Sazonova and V. N. Drozd, *Dokl. Akad. Nauk SSSR* **126,** 1004 (1959).
22. A. N. Nesmeyanov, V. A. Sazonova and V. N. Drozd, *Chem. Ber.* **93,** 2717 (1960).
23. A. N. Nesmeyanov, V. N. Drozd, V. A. Sazonova, V. I. Romanenko, A. K. Prokof'ev and L. A. Nikonova, *Izv. Akad. Nauk SSSR, Otd. Khim.* **1963,** 667.
24. A. N. Nesmeyanov: "Methods of Elemento-Organic Chemistry. The Copper Subgroup", Moscow, Nauka (1974).
25. Yu. N. Bubnov, S. A. Korobeinikova, G. V. Isagulyants and B. M. Mikhailov, *Izv. Akad. Nauk SSSR, Ser. Khim.* **1970,** 2023.
26. Yu. N. Bubnov, V. S. Bogdanov, I.P. Yakovlev and B. M. Mikhailov, *Zh. Obshch. Khim.* **42,** 1313 (1972).
27. B. M. Mikhailov, Yu. N. Bubnov and M. Sh. Grigoryan, *Zh. Obshch. Khim.* **44,** 2466 (1974).
28. A. G. Davis and B. P. Roberts, *J. Chem. Soc. C* **1968,** 1474.
29. A. N. Nesmeyanov, V. A. Sazonova and N. S. Sazonova, *Dokl. Akad. Nauk SSSR* **176,** 598 (1967).
30. A. N. Nesmeyanov, V. A. Sazonova, N. S. Sazonova and V. N. Plyukhina, *Dokl. Akad. Nauk SSSR* **177,** 1352 (1967).
31. A. N. Nesmeyanov, V. A. Sazonova, N. S. Sazonova and L. M. Meskhi, *Izv. Akad. Nauk SSSR, Ser. Khim.* **1969,** 1827.
32. A. N. Nesmeyanov, N. E. Kolobova, Yu. V. Makarov and K. N. Anisimov, *Izv. Akad. Nauk SSSR* **1969,** 1992.
33. A. N. Nesmeyanov, V. A. Sazonova and N. N. Sedova, *Dokl. Akad. Nauk SSSR* **198,** 590 (1971).
34. Y. Tanahashi, J. Lhomme and G. Ourisson, *Tetrahedron* **28,** 2655 (1972).
35. I. Mehrotra and D. Devaprabhakara, *J. Organometal. Chem.* **82,** C1 (1974).
36. V. A. Sazonova and L. P. Sorokina, *Dokl. Akad. Nauk SSSR* **105,** 993 (1955).
37. A. Arase, Y.Masuda and A. Suzuki, *Bull. Chem. Soc. Japan* **47,** 2511 (1974).
38. C. F. Lane, *J. Organometal. Chem.* **31,** 421 (1971).
39. A. D. Ainley and F. Challenger, *J. Chem. Soc.* **1930,** 2171.
40. A. N. Nesmeyanov, V. A. Sazonova and V. E. Fedorov, *Izv. Akad. Nauk SSSR, Ser. Khim.* **1970,** 2133.
41. A. N. Nesmeyanov, V. A. Sazonova and N. N Sedova, *Dokl. Akad. Nauk SSSR* **194,** 825 (1970).
42. A. N. Nesmeyanov, V. A. Sazonova, V. N. Drozd and L. A. Nikonova, *Dokl. Akad. Nauk SSSR* **131,** 1088 (1960).
43. Z. Holzbecher, *Chem. Listy* **46,** 17 (1952).
44. A. N. Nesmeyanov, V. A. Sazonova and V. N. Drozd, *Dokl. Akad. Nauk SSSR* **129,** 1060 (1959).
45. A. N. Nesmeyanov, V. A. Sazonova, V. N. Drozd and L. A. Nikonova, *Dokl. Akad. Nauk SSSR* **133,** 126 (1960).
46. A. N. Nesmeyanov, V. A. Sazonova and N. N. Sedova, *Dokl. Akad. Nauk SSSR* **202,** 362 (1972).
47. R. N. Keller and E. M. V. Wall, *Advances in Chemistry, Ser.* **32,** 221 (1961).
48. G. Wittig and P. Raff, *Ann.* **573,** 195 (1951).
49. A. N. Nesmeyanov, V. A. Sazonova, G. S. Liberman and L. I. Emel'yanova, *Izv. Akad. Nauk SSSR, Otd. Khim. Nauk* **1955,** 48.
50. V. A. Sazonova and N. Ya. Kronrod, *Zh. Org. Khim.* **26,** 1878 (1956).
51. L. G. Makarova and A. N. Nesmeyanov: "Methods of Elemento-Organic Chemistry. Mercury", Moscow, Nauka (1965).

52. K. Torssell, *Arkiv Kemi* **10,** 473 (1957).
53. F. R. Been and J. R. Johnson, *J. Am. Chem. Soc.* **54,** 4415 (1932).
54. E. Khotinsky and M. Melamed, *Ber.* **42,** 3090 (1909).
55. K. Torssell, *Acta Chem. Scand.* **13,** 115 (1959).
56. W. König and W. Scharrnbeck *J. Prakt. Chem.* **128,** 153 (1930).
57. H. R. Snyder and C. Weaver, *J. Am. Chem. Soc.* **70,** 232 (1948).
58. R. Neu, *Chem. Ber.* **88,** 1761 (1955).
59. E. Krause and G. Renwanz *Ber.* **65,** 777 (1932).
60. D. W. Slocum and T R. Engelman, *J. Organometal. Chem.* **24,** 753 (1970).
61. D. S. Matteson, J. O. Waldbillig and S. W. Peterson, *J. Am. Chem. Soc.* **86,** 3781 (1964).
62. H. G. Kuivila and T C. Muller, *J. Am. Chem. Soc.* **84,** 377 (1962).
63. F. Challenger and O. V. Richard, *J. Chem. Soc.* **1934,** 405.
64. K. Torssell, *Acta Chem. Scand.* **9,** 239 (1955).
65. R. Kh. Freidlina, A. N. Nesmeyanov and K. A. Kocheshkov, *Zh. Obshch. Khim* **5,** 1171 (1935).
66. R. Kh. Freidlina, A. N. Nesmeyanov and K. A. Kozeschkov, *Ber* **68,** 565 (1935).
67. H. Gilman and H. L. Yale, *J. Am. Chem. Soc.* **72,** 8 (1950).
68. D. S. Matteson, R. A. Bowie and G. Srivastava, *J. Organometal. Chem.* **16,** 33 (1969).
69. D. S. Matteson and E. Krämer *J. Am. Chem. Soc.* **90,** 7261 (1968).
70. D. S. Matteson and R. A. Bowie, *J. Am. Chem. Soc.* **87,** 2587 (1965).
71. D. S. Matteson and P. Allies, *J. Organometal. Chem.* **54,** 35 (1973).
72. D. S. Matteson and J. G. Shdo *J. Org. Chem.* **29,** 2742 (1964).
73. D. S. Matteson, R. B. Castle and G. L. Larson, *J. Am. Chem. Soc.* **92,** 231 (1970).
74. D. S. Matteson and J. O. Waldbillig, *J. Am. Chem. Soc.* **86,** 3778 (1964).
75. D. S. Matteson and M. L. Talbot, *J. Am. Chem. Soc.* **89,** 1119 (1967).
76. D. S. Matteson and M. L. Talbot, *J. Am. Chem. Soc.* **89,** 1123 (1967).
77. G. Wittig, G. Keicher, A. Rückert and P. Raff, *Ann.* **563,** 110 (1949).
78. J. B. Honeycutt and J. M. Riddle, *J. Am. Chem. Soc.* **81,** 2593 (1959).
79. J. B. Honeycutt and J. M. Riddle, *J. Am. Chem. Soc* **82,** 3051 (1960).
80. A. N. Nesmeyanov and R. A. Sokolik. "Methods of Elemento-Organic Chemistry. Boron, Aluminum, Gallium, Indium, Thallium", Moscow, Nauka (1964).
81. J. M. Riddle, *U.S. Patent* 2,950,305 (1960); *C. A.* **55,** 3434 (1961).
82. J. M. Riddle, *U.S. Patent* 2,950,303 (1969); *C. A.* **55,** 3434 (1961).
83. J. M. Riddle, *U.S. Patent* 2,950,304 (1960); *C. A.* **55,** 3434 (1961).
84. S. Blitzer and T. Pearson, *U.S. Patent* 2,969,381 (1961); *C. A.* **55,** 9282 e(1961).
85. J. D. Buhler and H. C. Brown, *J. Organometal. Chem.* **40,** 265 (1972).
86. R. C. Larock and H. C. Brown, *J. Am. Chem. Soc.* **92,** 2467 (1970).
87. R. C. Larock and H. C. Brown, *J. Organometal. Chem.* **26,** 35 (1971).
88. R. C. Larock, *J. Organometal. Chem.* **67,** 353 (1974).
89. D. S. Matteson and P. Allies, *J. Am. Chem. Soc.* **92,** 1801 (1970).
90. D. S. Matteson, *Organometal. Chem. Rev., A* **4,** 263 (1969).
91. O. A. Reutov, I. P. Beletskaya and V. I. Sokolov: "Mechanisms of the Reactions of Organometallic Compounds", Moscow, Khimiya (1972).
92. R. C. Larock, *Intra-Science Chem. Rep.* **7,** 95 (1973).
93. R. C. Larock, *J. Organometal. Chem.* **72,** 35 (1974).
94. G. Wittig and W. Herwig, *Chem. Ber.* **88,** 962 (1955).

95. A. N. Nesmeyanov, V. A. Sazonova and V. N. Drozd, *Izv. Akad. Nauk SSSR, Ser. Khim.* **1959,** 163.
96. G. A. Razuvaev and T. G. Brilkina, *Dokl. Akad. Nauk SSSR* **91,** 861 (1953).
97. G. A. Razuvaev and T. G. Brilkina, *Dokl. Akad. Nauk SSSR* **85,** 815 (1952).
98. G. Wittig and G. Lehman, *Chem. Ber.* **90,** 875 (1957).
99. J. J. Tufariello and M. Hovey, *J. Am. Chem. Soc.* **92,** 3221 (1970).
100. J. J. Tufariello and M. Hovey, *Chem. Comm.* **1970,** 372.
101. R. C. Larock, *J. Org. Chem.* **39,** 834 (1974).
102. J. Guenzet and M. Camps, *Tetrahedron Letters* **1972,** 2647.
103. G. A. Razuvaev and T. G. Brilkina, *Zh. Obshch. Khim.* **24,** 2415 (1954).
104. H. Grundke and P. I. Paetzold, *Chem. Ber.* **104,** 1136 (1971).
105. F. Challenger and B. Parker, *J. Chem. Soc.* **1931,** 1462.
106. S. S. Nametkin, N. N. Mel'nikov and G. P. Gracheva, *Zh. Obshch. Khim.* **5,** 1455 (1935).
107. N. N. Mel'nikov and M. S. Rokitskaya, *Zh. Obshch. Khim.* **8,** 1768 (1938).
108. N. N. Mel'nikov and M. S. Rokitskaya, *Zh.Obshch. Khim.* **7,** 1472 (1937).
109. A. E. Borisov, *Izv. Akad. Nauk SSSR, Otd. Khim. Nauk* **1951,** 402.
110. F. W. Fray, P. Kobetz, G. C. Robinson and T. O. Sistrunk, *J. Org. Chem.* **26,** 2950 (1961).
111. J. B. Honeycutt and J. M. Riddle, *J. Am. Chem. Soc.* **83,** 369 (1961).
112. D. S. Matteson and G. L. Larson, *J. Am. Chem. Soc.* **91,** 6541 (1969).
113. D. S. Matteson and G. L. Larson, *J. Organometal. Chem.* **57,** 225 (1973).
114. W. T. Ford and R. J. Hauri, *J. Am. Chem. Soc.* **95,** 7381 (1973).
115. R. Murphy and R. H. Prager, *Tetrahedron Letters* **1976,** 463.
116. H. C. Brown, J. J. Katz and B. A. Carlson, *J. Org. Chem.* **40,** 813 (1975)
117. K. Avasthi, S. S. Ghosh and D. Devaprabhakara, *Tetrahedron Letters* **1976,** 4871.
118. D. S. Matteson and P. K. Jesthi, *J. Organometal. Chem.* **114,** 1 (1976).
119. D. E. Bergbreiter and D. P. Rainville, *J. Organometal. Chem.* **121,** 19 (1976).
120. M. Gielen and R. Fosty, *Bull. Soc. Chim. Belg.* **83,** 333 (1974).
121. A. Arase and Y. Masuda, *Chem. Letters* **1976,** 785.
122. A. Arase and Y. Masuda, *Chem. Letters* **1976,** 1115.
123. A. Suzuki, M. Ishidoya and M. Tabata, *Synthesis* **1976,** 687.
124. Y. Masuda and A. Arase, *Bull. Chem. Soc. Japan* **51,** 901 (1978).

CHAPTER 15

Syntheses using copper and palladium compounds

Methods for the preparation of different organic compounds with the use of copper(I) and palladium compounds have been developed.

Copper borates $[R_3BCH_3]Cu$ and complexes $RCu \cdot BF_3$ are used in the syntheses. Both reagents alkylate acrylonitrile and α,β-unsaturated ketones, acids, and esters in the β-position [1, 2] and both react with allyl halides, $R'R''C{=}CHCH_2X$, to form $RR'R''CCH{=}CH_2$ [3, 4]. The latter compounds can be obtained from allyl alcohols and $RCu \cdot BF_3$ [5]. The reagents $[R_3BCH_3]Cu$ alkylate ethyl propiolate [6, 7], α,β-acetylenic aldehydes and ketones, esters of acetylenedicarboxylic acid [7], l-acyl-2-vinylcyclopropane [1], and 1-(1-pyrrolidyl)-6-chloro-1-cyclohexene [8] and replace bromine by R in ethyl β-bromoacrylates [9]; the reaction with propargyl halides leads to substituted allenes [3], that with aroyl chlorides to alkyl aryl ketones [10], and the reaction with benzyl halides to alkylbenzenes and bibenzyls [11]. The reaction of $RCu \cdot BF_3$ with chiral *trans*-8-phenylmenthyl esters of α,β-unsaturated acids proceeds as an enantioselective 1,4-addition [12]. Sorbic acid is alkylated mainly in the 1,6-position while its methyl ester is alkylated in the 1,4-position [2].

The Grignard reagents formed by the action of 1,4-pentamethylenedimagnesium bromide on R_3B in benzene react by coupling of the radicals with allyl bromide, alkenyl iodides, and acyl chlorides in the presence of the catalysts Li_2CuCl_4, $Pd[P(C_6H_5)_3]_4$, and CuI, respectively [13].

Phenylboronic acid reacts with aromatic bromides in the presence of $Pd(PPh_3)_4$ and aqueous sodium carbonate to form biaryls [14].

REFERENCES

1. N. Miyaura, M. Itoh and A. Suzuki, *Tetrahedron Letters* **1976,** 255.
2. Y. Yamamoto and K. Maruyama, *J. Am. Chem. Soc.* **100,** 3240 (1978).
3. N. Miyaura, M. Itoh and A. Suzuki, *Bull. Chem. Soc. Japan* **50,** 2199 (1977).
4. K. Maruyama and Y. Yamamoto, *J. Am. Chem. Soc.* **99,** 8068 (1977).
5. Y. Yamamoto and K. Maruyama, *J. Organometal. Chem.* **156,** C9 (1978).
6. K. Yamada, N. Miyaura, M. Itoh and A. Suzuki, *Bull. Chem. Soc. Japan* **50,** 3431 (1977).
7. Y. Yamamoto, H. Yatagai and K. Maruyama, *J. Org. Chem.* **44,** 1744 (1979).
8. K. Yamada, T. Yano, N. Miyaura and A. Suzuki, *Bull. Chem. Soc. Japan* **52,** 275 (1979).
9. N. Miyaura, N. Sasaki, M. Itoh and A. Suzuki, *Tetrahedron Letters* **1977,** 3369.
10. N. Miyaura, N. Sasaki, M. Itoh and A. Suzuki, *Tetrahedron Letters* **1977,** 173.
11. N. Miyaura, M. Itoh and A. Suzuki, *Synthesis* **1976,** 618.
12. W. Opolzer and H. J. Löher, *Helv. Chim. Acta* **64,** 2808 (1981).
13. K. Kondo and S. Murahashi, *Tetrahedron Letters* **1979,** 1237.
14. N. Miyaura, T. Yanagi and A. Suzuki, *Synth. Commun.* **11,** 573 (1981).

PART II

BORACYCLANES

Heterocyclic compounds with boron and carbon atoms in the ring are of interest from both the theoretical and practical viewpoint. The first representatives of this class of organoboranes were obtained with the use of dilithium and dimagnesium derivatives of saturated hydrocarbons. The method using organometallic compounds was later extended to the synthesis of boracyclanes coupled to aromatic rings, in which the boron atom, in principle, may be included in the overall aromatic system at the expense of its vacant orbital.

A second method for the synthesis of boracyclanes is based on thermal conversions of alkylboranes and alkylarylboranes. This method, which closely links the chemistry of acyclic and aromatic boron compounds with the chemistry of cyclic boron compounds, makes it possible to obtain systems with one, two, and three rings in the molecule, including mixed compounds also containing aromatic nuclei. Boracyclanes coupled to a carborane system can also be obtained by the thermal decomposition of organoboran compounds of the *o*-carborane series.

The method of synthesis of boracyclanes from diene and polyene acyclic and alicyclic hydrocarbons by their hydroboration with diborane and its derivatives proved to be equally fruitful. Five-, six- and seven-membered boracyclanes, bis(boracycloalkyl)alkanes, perhydro-9*b*-boraphenalenes, and compounds of the 9-borabicyclo[3.3.1]nonane, 2-borabicyclo[3.3.1]nonane, and 2-boraadamantane series are obtained by this method.

By means of simple methods, boracyclanes are converted to different organic compounds, including those which are difficult to obtain by the usual methods of organic synthesis. Among the compounds obtained from boracyclanes are α, ω-bifunctional derivatives of aliphatic hydrocarbons; monocyclic, bicyclic, and tricyclic alcohols; monocyclic and bicyclic ketones, and a number of other substances.

The chemistry of the boracyclanes has been covered in review articles [1–4].

PART II

CHAPTER 1

Methods for the preparation of boracyclanes

1.1 PREPARATION OF BORACYCLANES WITH THE USE OF ORGANOMETALLIC COMPOUNDS

Boracyclanes are obtained by the reaction of haloorganoboranes with dimetallic alkane derivatives.

Torssell [5] was the first to obtain 1-arylborolane and 1-arylborinane by the action of difluorophenylborane on α,ω-dilithioalkanes:

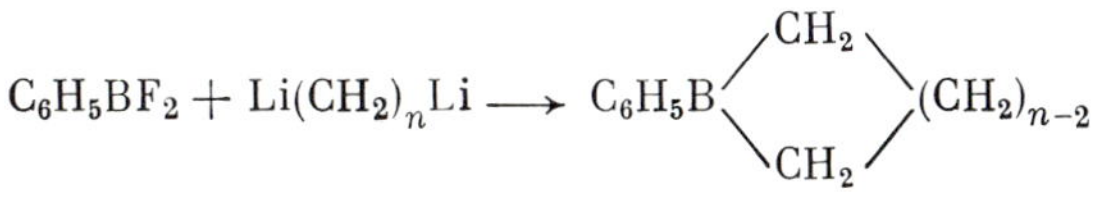

$(n = 4, 5)$

1-Propenylborolane and 1-propenylborinane have been synthesized from dichloropropenylborane with the use of a Grignard reagent [6]:

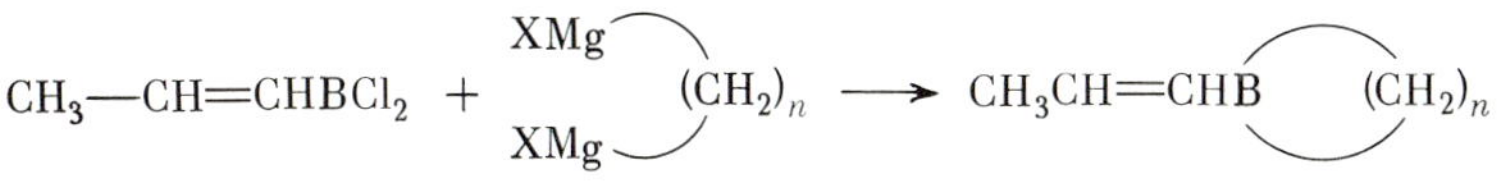

$(n = 4, 5)$.

Halodialkylboranes can be used as a starting material for the preparation of boracyclanes. In this case, the initially formed α,ω-bis(dialkylboryl)alkane undergoes cyclization with the cleavage of a trialkylborane [7, 8]:

$$BrMg(CH_2)_nMgBr + (n - C_4H_9)_2BCl \longrightarrow (n - C_4H_9)_2B(CH_2)_nB(C_4H_9 - n)_2 \longrightarrow$$

$$\longrightarrow 3n\text{-}C_4H_9B(CH_2)_n) + n\text{-}C_4H_9)_3B \qquad (n = 4,5).$$

A boracyclane coupled with aromatic rings — 1-butoxy-2,3;6,7-dibenzodihydroborepin (*1*)—has been obtained from *o,o'*-dilithiodibenzyl and tributyl borate [9]:

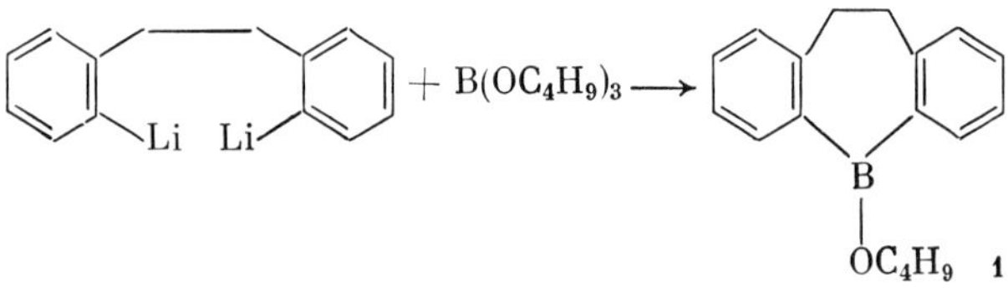

1

The anhydride of the acid obtained from the ester *1* is converted to 1-hydroxy-2,3;6,7-dibenzoborepin by the action of bromosuccinimide, followed by treatment with sodium methoxide [10].

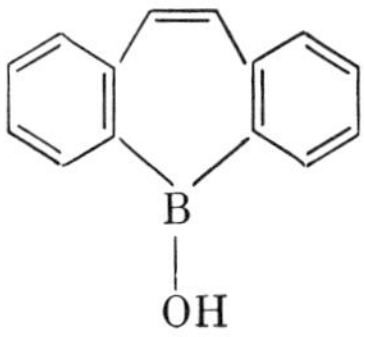

The original lithium borate *2* has been synthesized by the action of boron trifluoride etherate on *o,o'*-dilithiodiphenyl [11]:

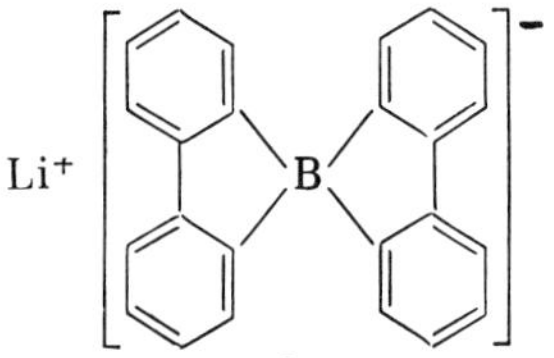

2

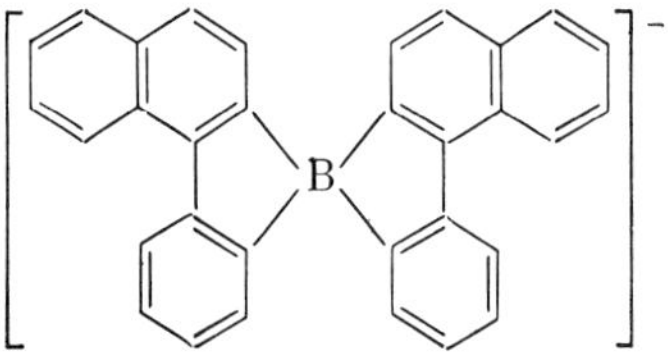

3

The asymmetric spiro anion *3* has been separated into its optical antipodes with the use of the optically active methylpropylphenylbenzylammonium salt [12].

Compounds of the 2,3;5,6-dibenzoborine series have been obtained by the action of diethylamino- or diphenylaminodichloroborane on the Grignard reagent from *o,o'*-bis(bromophenyl)-methane [13]. 2,3;7,8-Dimethylenedioxy-5-phenyldibenzoborole has been obtained with the use of the dilithium reagent [14].

1-Substituted 2,3;5,6-dibenzoborines have been synthesized from 9,9-dimethyl-9,10-dihydro-9-stannaanthracene and haloboranes [15]:

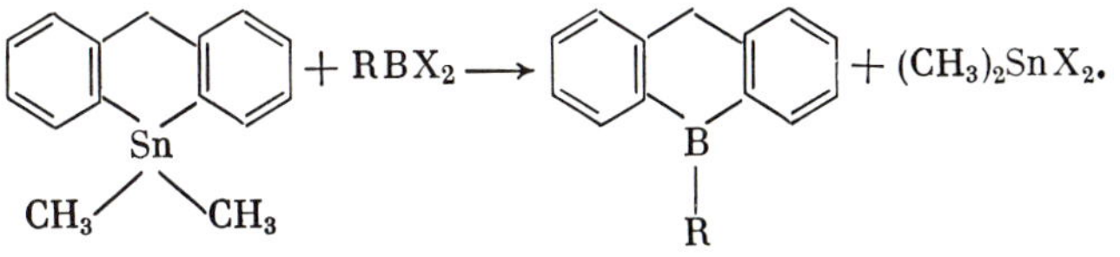

($R = C_6H_5, CH_3, Cl$; $X = Br, Cl$)

1.2 PYROLYSIS OF ALKYL(ARYL)BORANES

Different types of boracyclanes are obtained by the pyrolytic decomposition of alkyl- and alkylarylboranes. Review articles [1–3] have been devoted to a consideration of this method for the synthesis of boracyclanes, developed by Köster.

Trialkylboranes containing four or more carbon atoms in the chain are converted to boracyclanes in high yield at 200–300°C. Dialkylboranes, which are capable of cyclizing to boracyclanes with the elimination of H_2, are formed as intermediate compounds in this process [16–18]:

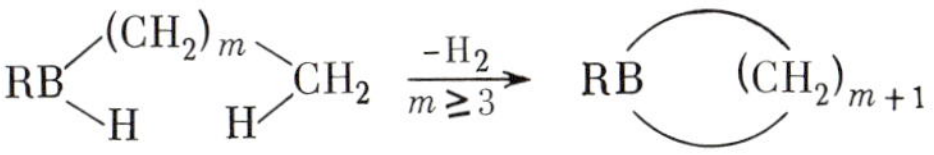

The pyrolysis of tributylborane at 300°C results in the evolution of hydrogen and butenes, primarily 2-butene, with the formation of a mixture of 1-(1-butyl)borolane and 1-(2-butyl)borolane (85:15) in a yield of 85% [16, 19]:

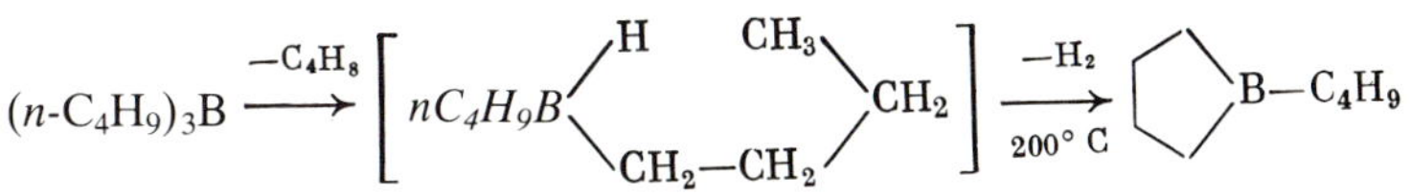

With increasing chain length, the thermal decomposition of trialkylboranes proceeds in a more complex manner, leading to isomeric compounds of the borolane and borinane series [16, 20].

Higher trialkylboranes, containing alkyl radicals with eight or more carbon atoms, are converted to bicyclic compounds by pyrolysis.

Thus, trioctylborane forms 8-borahydrindan [20, 21]:

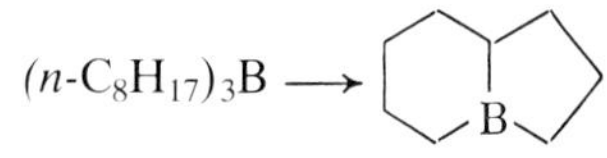

Trinonylborane is converted at 250–300°C to 9-boradecalin in 80% yield [20, 21]:

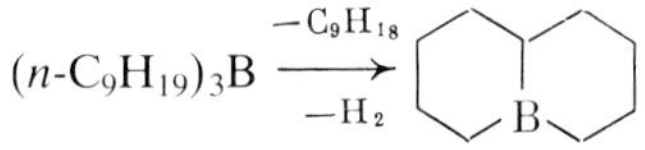

The pyrolysis of tricyclooctylborane leads to the elimination of *cis*-cyclooctene and hydrogen, forming an almost quantitative yield of 9-cyclooctyl-9-borabicyclo[3.3.1]nonane (*4*) [20, 22]:

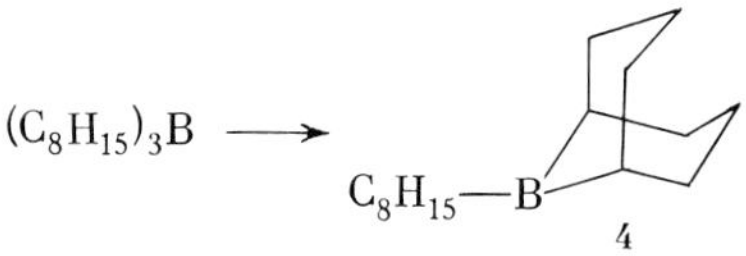

4

In compound *4*, the cyclooctyl radical on the boron atom is replaced by a propyl radical by a reaction with tripropylborane in the presence of compounds with B—H bonds:

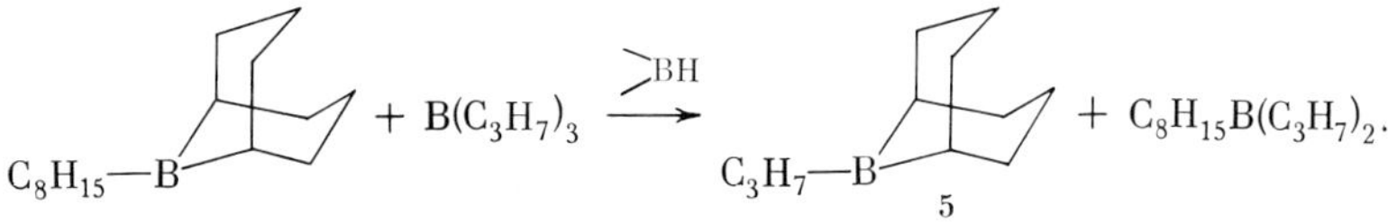

5

The action of tetrapropyldiborane on 9-propyl-9-borabicyclo[3.3.1]-nonane (*5*) forms 9-borabicyclo[3.3.1]nonane, which exists in the form of the dimer *6* [23]:

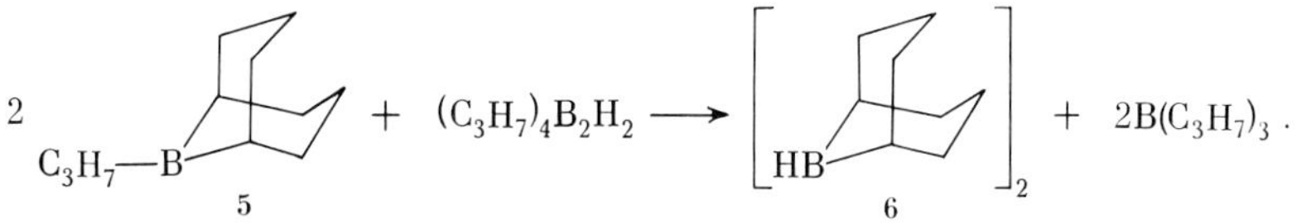

5 6

When tricyclododecylborane is heated, *cis*- and *trans*-cyclododecenes and hydrogen are initially split off. The compound 13-cyclododecyl-13-borabicyclo[7.3.1]-tridecane (*7*) obtained is converted by further heating to *cis*, *cis*, *cis*-perhydro-9b-boraphenalene

(*8*) [20,22]:

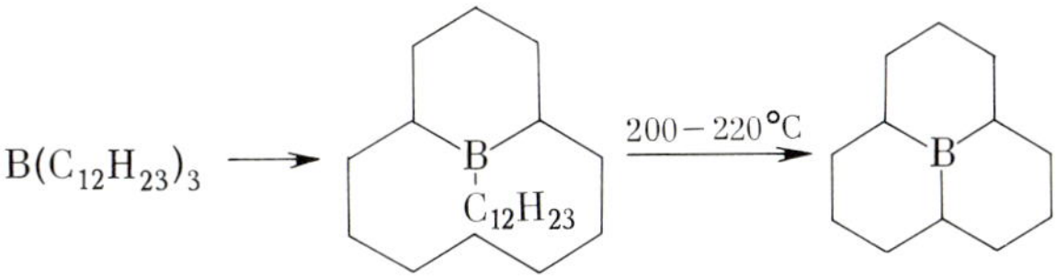

7 8

At 180–200°C, tris(2-phenylpropyl)borane splits off α-methylstyrene and hydrogen and is converted to 1-(2-phenylpropyl)-3-methyl-l-boraindan [24, 25]:

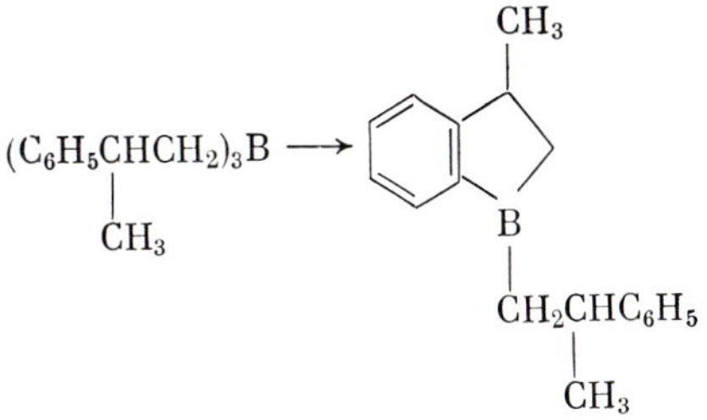

Boratetralines can be prepared by the same method. Thus, tris (3-phenylpropyl)borane forms 1-(3-phenylpropyl)-1-boratetralin at 200°C [25]:

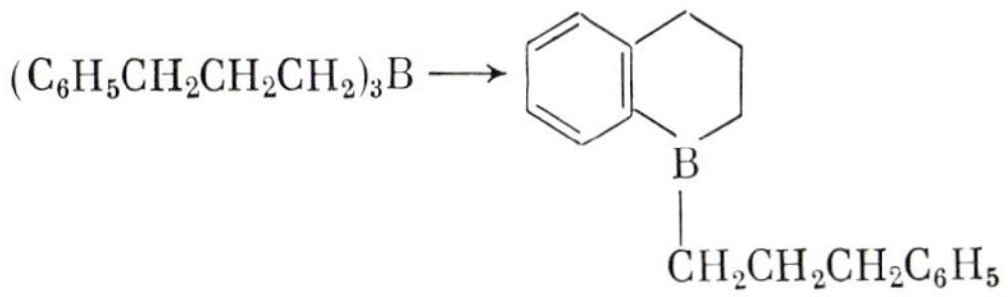

1-Alkylboraindans and 1-alkyboratetralins can be converted by the action of hydrogen under pressure at 160°C to boraindan (*9*) and borateralin (*10*), which exist in the form of dimers [24].

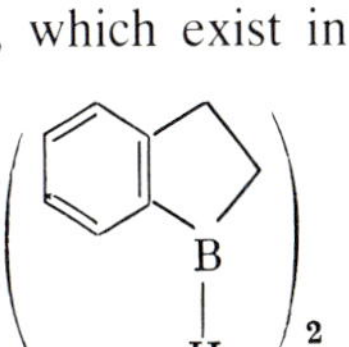

9

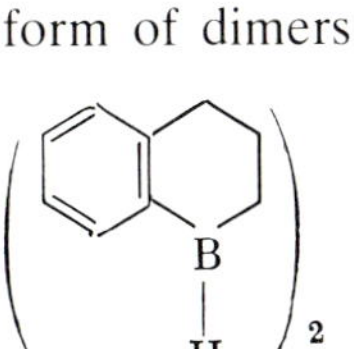

10

2-Biphenylyldialkylboranes prepared from 2-lithiobiphenyl and R_2BCl are converted to 1-alkydibenzoboroles (9-alkylborafluorenes) at 180–200°C [25, 26]:

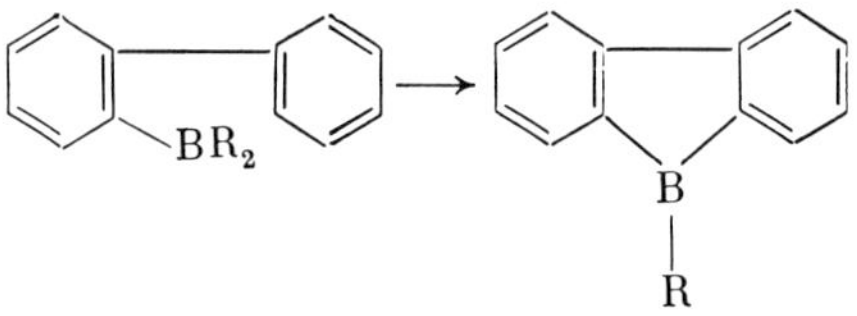

40–65%

In the pyrolysis of a mixture of 2-lithiobiphenyl and dialkylchloroboranes or diarylchloroboranes in a ratio of 2:1, lithium bis(2,2′-biphenylyl)spiroborate is formed [11]:

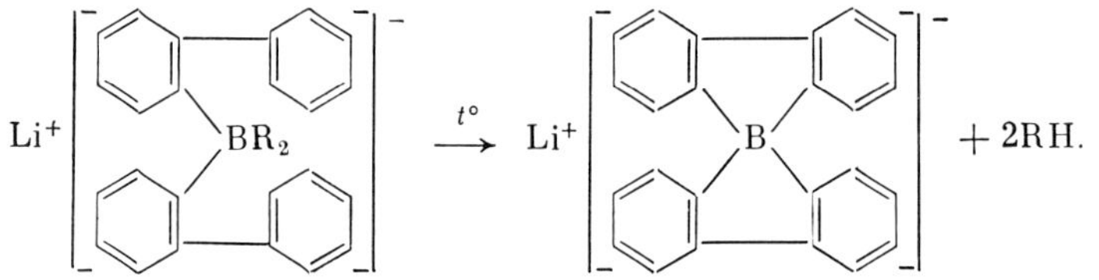

Tricyclic diboron compounds with two borolane rings coupled to a benzene nucleus can be obtained by a pyrolytic method [25]:

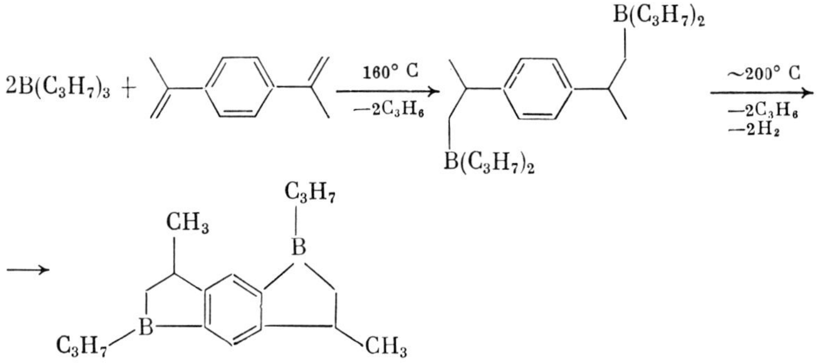

The compound *11* is smoothly formed from 3,3-dimethyl-1-indanylethylborane, which is easily obtained by the hydroboration of 1-vinyl-3,3-dimethylindan:

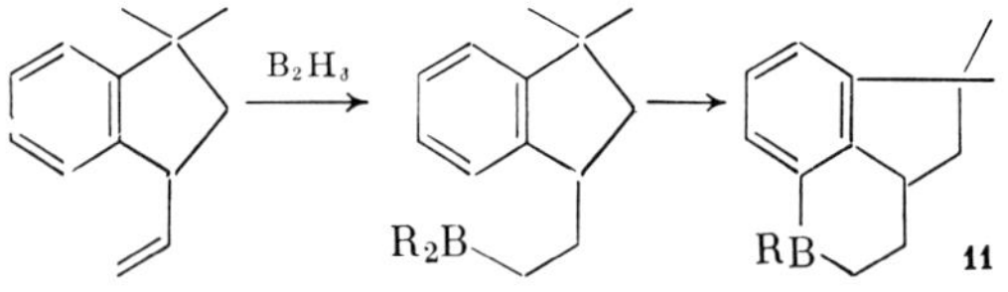

11

The tricyclic boron compounds *12–15* have been synthesized by an analogous method from the corresponding substituted cyclopentenes and cyclohexenes [25]:

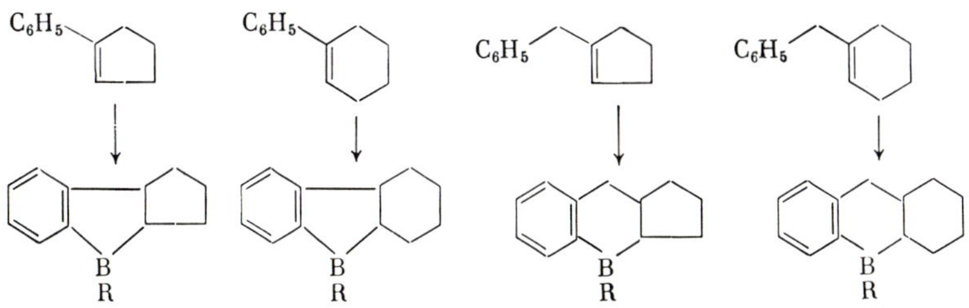

12 13 14 15

1,2-(2,2′-Biphenylylene)-1,2-diethyldiborane (*16*) is formed by the action of tetraethyldiborane on (2-biphenylyl)diethylborane [27]:

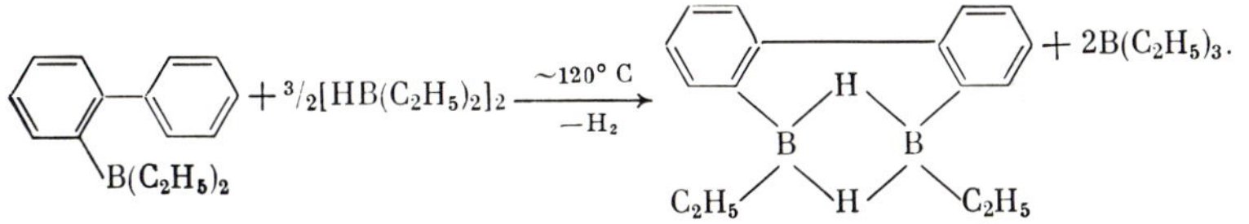

16

1,2-(2,2′-Biphenylylene)diborane (*17*) and 1,2;1,2-bis(2,2′-biphenylylene)diborane (*18*) are obtained by the action of diborane and boron trifluoride etherate, respectively, on the compound *16*.

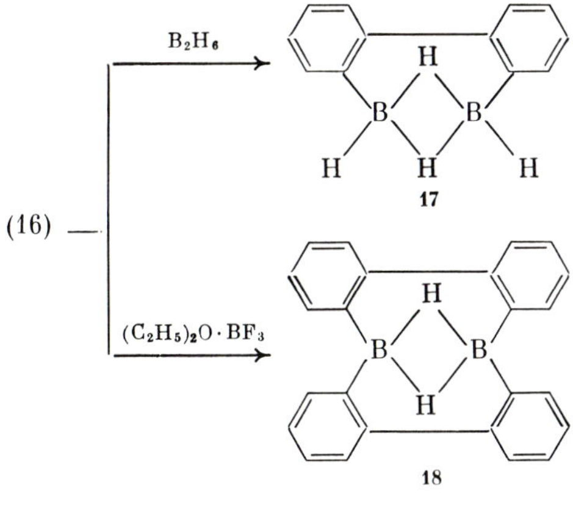

17

18

The structure of the compound *18* has been confirmed by X-ray analysis.

Boracyclanes coupled to an *o*-carborane nucleus have been synthesized from organoboron compounds obtained by the action of haloorganoboranes on lithium compounds of alkyl- and alkenyl-*o*-carborane [28, 29].

(1-Isopropyl-2-*o*-carboranyl)dibutylborane (*19*) cyclizes to 1,2-(3-butyl-5-methyl-3-boracyclopentano)-*o*-carborane (*20*) at 200–230°C [29, 30]:

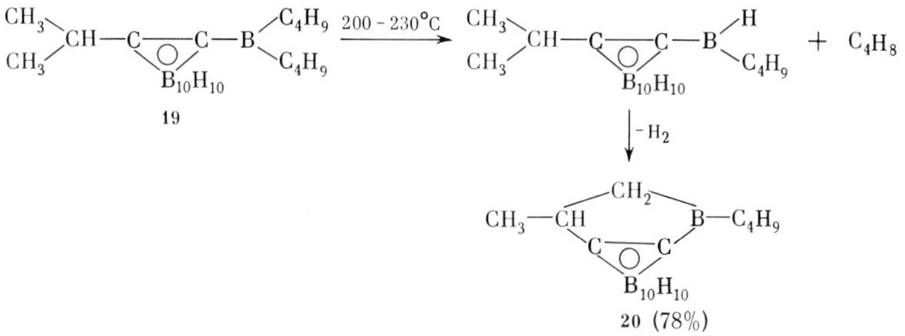

An analogous compound—1,2-(3-butyl-3-boracyclopentano)-*o*-carborane (*23*) — is obtained by the pyrolysis of (1-vinyl-2-*o*-carboranyl)dibutylborane (*21*) [29]. In this case, the intermediately formed borane *22* is converted to the final product by intramolecular hydroboration or transalkylation:

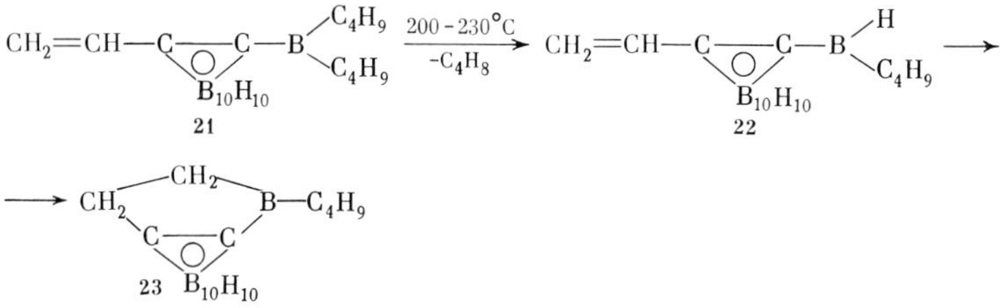

It is interesting that (1-isopropenyl-2-*o*-carboranyl)dibutylborane (*24*) is converted by heating to 1,2-(3-butyl-5-methyl-3-bora-4-cyclopenteno)-*o*-carborane (*25*). Thus, in the presence of a CH_3 substituent in the α-position of the vinyl group, instead of cyclization by intramolecular hydroboration, ring closure takes place by the removal of hydrogen atoms from the boron and of a methylene group from the isopropylene radical:

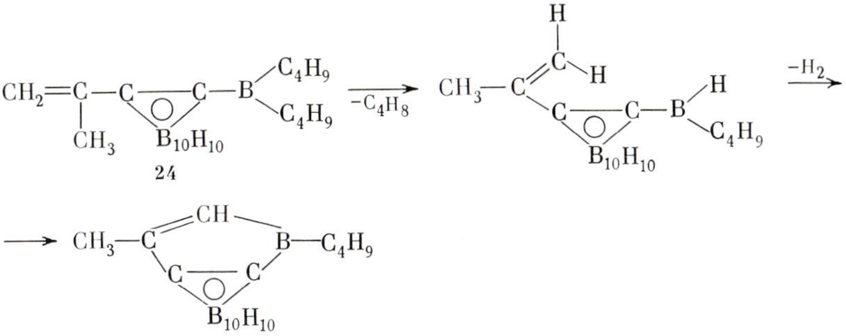

24

25 (60%)

The cyclization of (1-isopropenyl-2-*o*-carboranyl)diisobutylborane proceeds in an analogous manner.

2,4,6,8,9,10-Hexamethyl-2,4,6,8,9,10-hexaboraadamantane is formed by the pyrolysis of trimethylborane [122].

1.3 HYDROBORATION OF DIENE AND POLYENE HYDROCARBONS

Cyclic boron compounds are obtained by the hydroboration of diene hydrocarbons.

A study of the reactions of diborane with diene hydrocarbons was a difficult task, requiring numerous investigations over a period of many years. The reactions of conjugated diene hydrocarbons with diborane are particularly complex. The presence of two active centers in the molecule causes a reaction course in different directions in the first and subsequent stages of the conversions and thereby leads to the formation of different organoboron compounds with a tendency toward association, polymerization, and isomerization.

1.3.1 Hydroboration of acyclic diene hydrocarbons with diborane

Hydroboration of 1,3-butadiene Particular attention has been devoted to the reaction of 1,3-butadiene with diborane, the course of which depends on the ratio of the starting reagents and on the nature of the solvent.

The hydroboration of 1,3-butadiene was first investigated by Köster. According to his brief communications [31, 32], which do not contain any experimental data, 1,4-bis(1-borolyl)butane (*26*) is obtained by the action of triethylamine-borane or diborane on 1,3-butadiene:

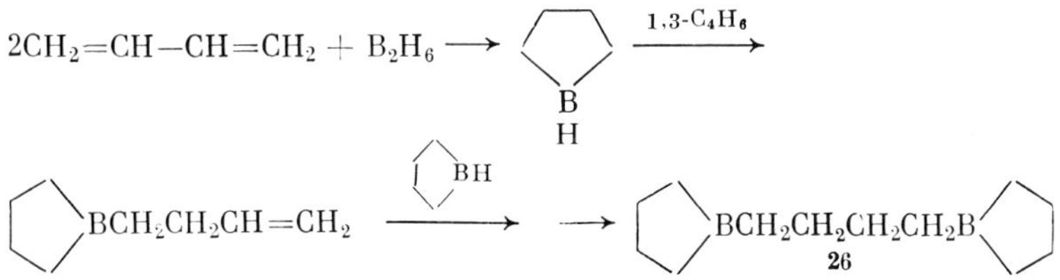

The reaction between 1,3-butadiene and diborane was later subjected to detailed investigation, during which its characteristics were interpreted more fully.

Mikhailov, et al. [33, 34] established the fact that the hydroboration of 1,3-butadiene in ether (diene : $BH_3 = 5:2$) proceeds in a complex manner. Fractional distillation of the reaction products produces two fractions, of which the first (yield 11%, calculated for boron, boiling point 61.5–66.5°C/40 mm Hg) is a mixture of isomeric 1-butenylborolanes and the second (the main fraction; yield 70%, boiling point 75.5–81°C/2 mm Hg), which corresponds in its physical properties to the product previously assumed to be the pure compound *26*, actually contains 55% of *26*, 39% of 1,3-bis(1-borolyl)-butane (*27*), and 6% of 1,2-bis(1-borolyl)butane (*28*).

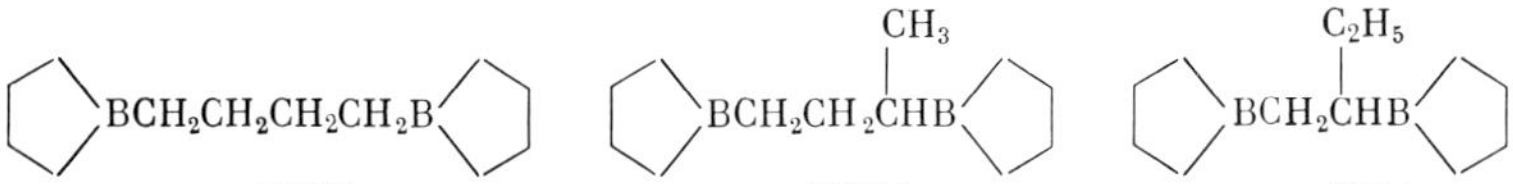

26 (55%) 27 (39%) 28 (6%)

An investigation of the IR spectra and of the products of hydrolysis and oxidation of the first fraction showed that it is almost pure 1-crotylborolane (*30*), with an impurity of approximately 2% of 1-(3-buten-2-yl)borolane (*29*). An analysis of the whole set of experimental data permitted the conclusion that the hydroboration of 1,3-butadiene forms the 1-butenylborolanes *29* and *31*, with *29* being converted to *30* by allylic rearrangement. The compound *30* reacts slowly with diborane and therefore accumulates in the reaction mass, whereas *31*, which is an α-olefin, reacts rapidly with diborane to form the bis(1-borolyl)butanes *26* and *27*.

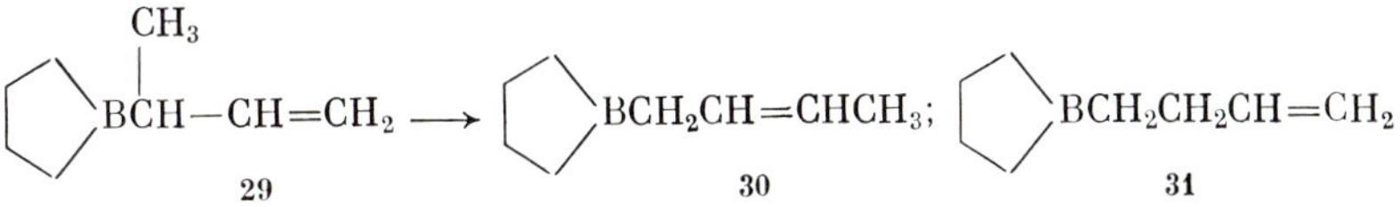

The mechanism of the hydroboration of 1,3-butadiene involves the following sequential and parallel conversions:

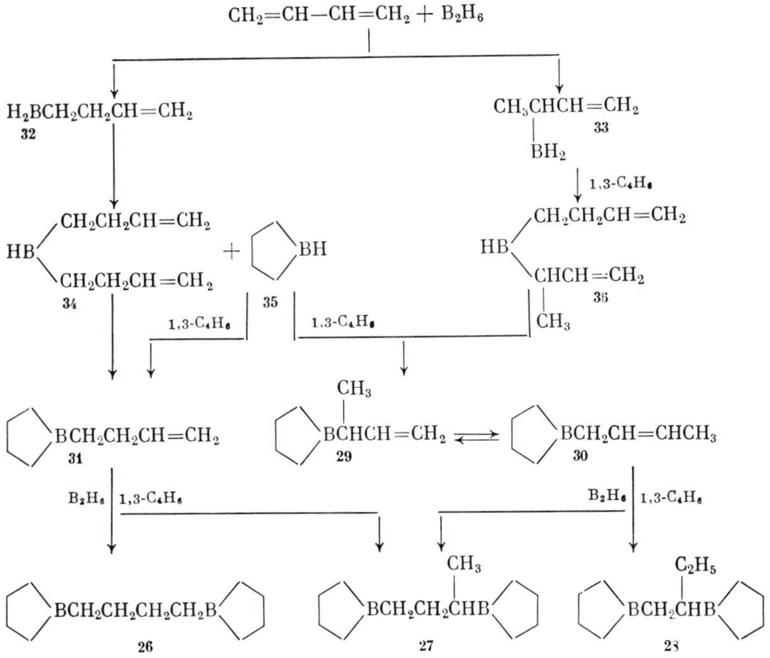

In the first stage of the process, 3-buten-1-ylborane (*32*) and a smaller amount of 3-buten-2-ylborane (*33*) are formed.

The compound *32*, by addition to the starting diene and cyclization, forms the dialkenylborane *34* and borolane *35*, respectively, which are converted to 1-(3-buten-1-yl)borolane (*31*). The compound *33*, going through the dialkenylborane *36*, forms 1-(3-buten-2-yl)borolane (*29*), which undergoes allylic rearrangement to 1-crotylborolane (*30*).

The borane *27* is also formed by addition of *35* to the starting diene. The subsequent reactions of *30* and *31* with diborane and the starting diene lead to *26*, *27*, and *28*.

The addition of diborane to 1,3-butadiene in the 1,4-position, as assumed by Zweifel, et al. [35], actually does not take place.

When they carried out the reaction between diborane and 1,3-butadiene in tetrahydrofuran solution with a diene : BH_3 ratio of 3 : 2, Zakharkin and Kovredov [36] obtained a mixture of compounds *26* and *27* in an approximate ratio of 40 : 60, while Brown, et al. [37] obtained them in a ratio of 30 : 70. The two papers mentioned above contain contradictory data on the behavior of a mixture of compounds *26* and *27* upon heating. In [36], it is indicated that, at 160°C, compound *27* isomerizes to *26* and, after five hours, the mixture contains 76% of compound *26* and 24% of compound *27*, whereas, according to [37], compound *27* isomerizes at 140–170°C to 1,1-bis(1-borolyl)butane (*37*), with the content of *27* in the mixture decreasing to 40%.

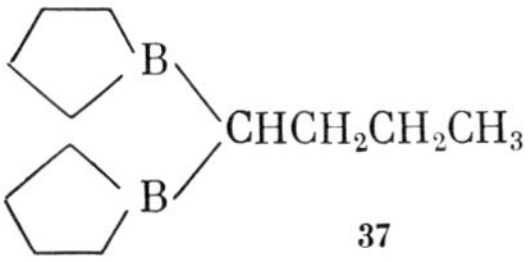

37

By reacting butadiene with diborane in a ratio of 2 : 1, Köster [23] obtained a compound to which he assigned the structure of a dimer of borolane (*38*):

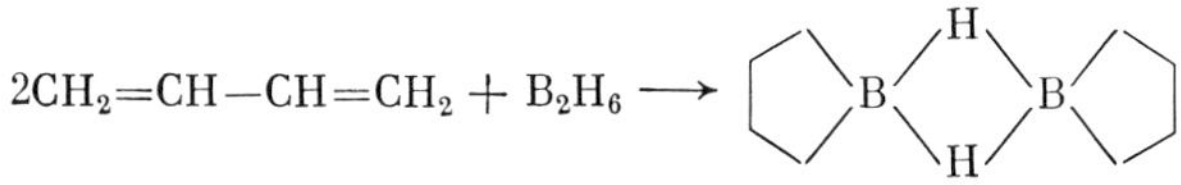

38

According to the data of [2], compound *38* reacts with diborane to form the monocyclic diborane derivative *39*:

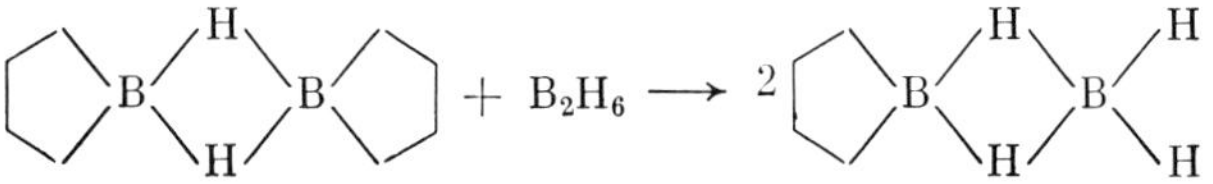

38 **39**

However, even before [2], Weiss and co-workers [38] showed that the reaction of 1,3-butadiene with diborane in the gas phase at a diene : BH_3 ratio of 1 : 2 forms 1,2-tetramethylenediborane (*40*) and 1,2-(1-methyltrimethylene)diborane (*41*).

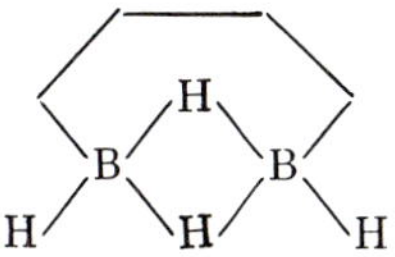

40

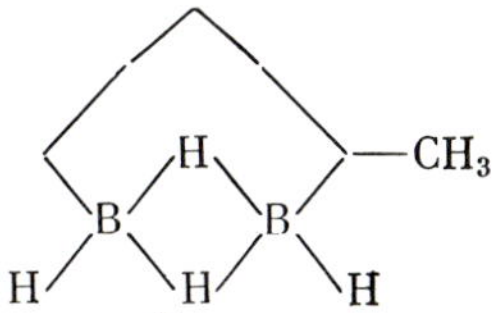

41

Young and Shore [39] later showed that the reaction of 1,3-butadiene with diborane in ether solutions proceeds similarly to the reaction in the gas phase and leads mainly to the compound *40* together with *41*:

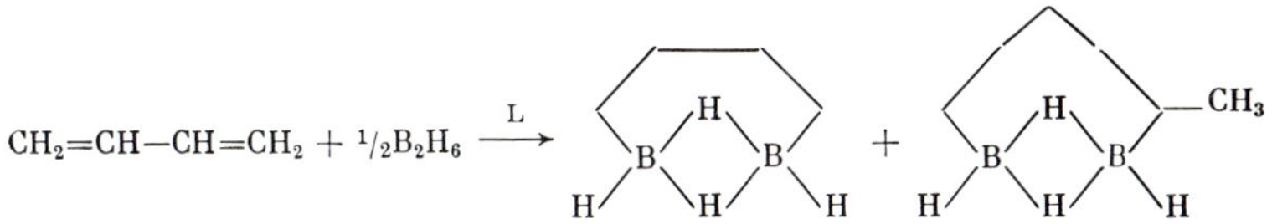

40; 60% **41**

(L = $(C_2H_5)_2O$, $(iso\text{-}C_3H_7)_2O$, $(n\text{-}C_4H_9)_2O$)

Young and Shore [39] also cleared up the question of the structure of the substance formed by the reaction of 1,3-butadiene and diborane at a diene:BH_3 ratio of 1:1, which was found not to have the structure of the borolane dimer *38* [2] but turned out to be 1,2-bis(tetramethylene)diborane (*42*). This compound is obtained by the action of 1,3-butadiene on 1,2-tetramethylenediborane (*40*) in tetrahydrofuran or by the direct reaction of 1,3-butadiene and diborane (2:1) in ether solution. When the reaction is carried out in tetrahydrofuran, a polymer is formed, which is converted to the compound *42* by heating:

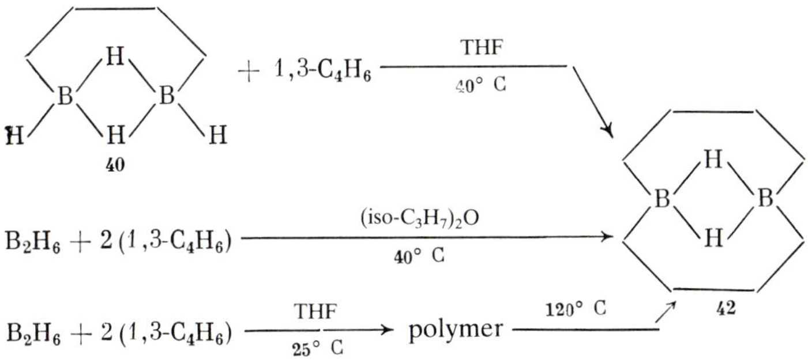

The structure of compounds *40–42* has been demonstrated by studying their conversions under the action of amines.

1-Alkylborolanes can be obtained in yields of 50–60% from the polymer formed in tetrahydrofuran at a diene:BH_3 ratio of 1:1 by heating with olefins [40].

It should be noted that the simplest cyclic diboron compound is 1,2-trimethylenediborane (*43*), which is formed in the gas phase from allene and diborane [41]:

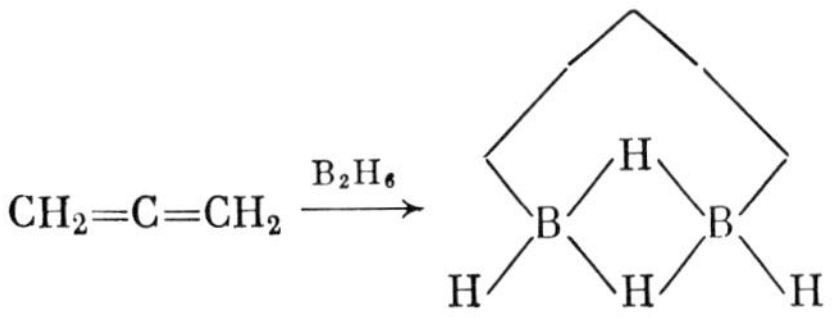

43

In the liquid phase, compound *43* is converted to a polymer, which quantitatively depolymerizes to the starting compound under vacuum at 60°C.

The possible structures of the polymer are shown in the following scheme:

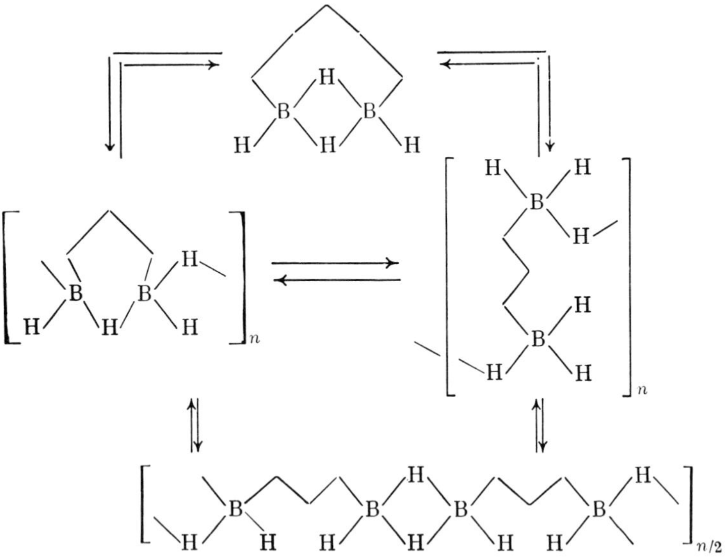

Hydroboration of isoprene According to reports by Saegebarth [42, 43] and by Zakharkin and Kovredov [44], the hydroboration of isoprene (diene:BH_3 = 3:2) forms 1,4-bis(3-methyl-1-borolyl)-2-methylbutane (*44*). However, as was found by Mikhailov, et al. [45], the hydroboration of isoprene leads not only to the compound *44* but also to its isomer, 1,3-bis(3-methyl-1-borolyl)-2-methylbutane (*45*):

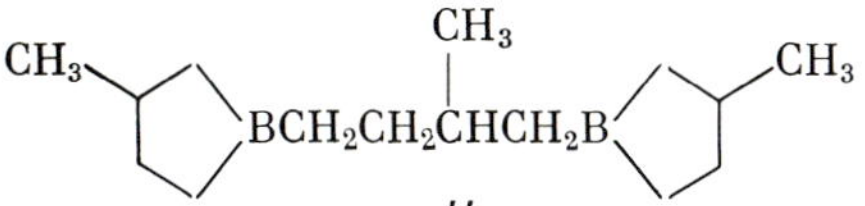

44

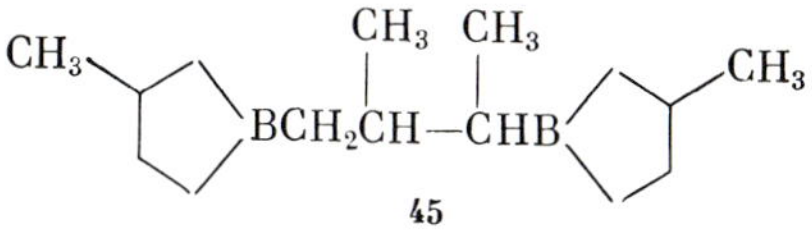

45

and, when an excess of isoprene is used, 1-pentenyl-3-methylborolanes are obtained in addition to compounds *44* and *45*. In the distillation of the reaction products (diene:BH_3 = 3:1), two fractions are isolated, the first of which (yield 22%, calculated on boron, boiling point 61–68°C/9 mm Hg) contains 60% of 1-(2-methyl-2-buten-1-yl)-3-methylborolane (*46*), 35% of 1-(3-methyl-3-buten-1-yl)-3-methylborolane (*47*), and a small amount of 1-(2-methyl-3-buten-1-yl)-3-methylborolane (*48*).

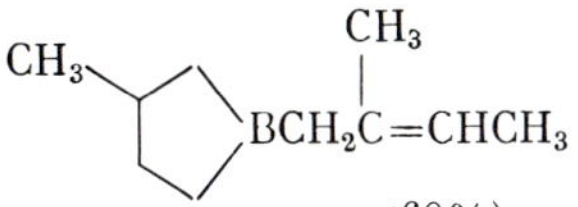

46 (60%)

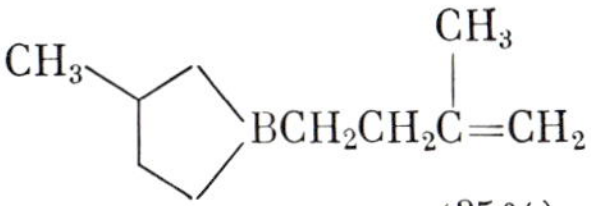

47 (35%)

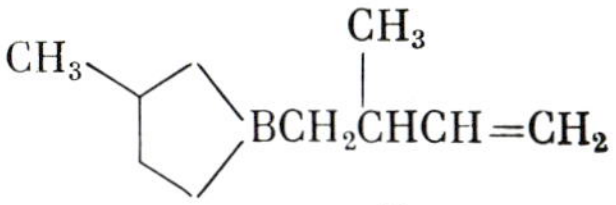

48

The second fraction (yield 68%, boiling point 101–104°C/1.5 mm Hg) contains 90% of compound *44* and 10% of compound *45*.

The hydroboration of isoprene with a mole ratio diene:BH_3 = 1:1 leads to a substance corresponding in composition to the dimer of 3-methylborolane [22] which, by analogy with the product of the hydroboration of 1,3-butadiene [39], has the structure of 1,2-bis(2-methyltetramethylene)diborane (*49*).

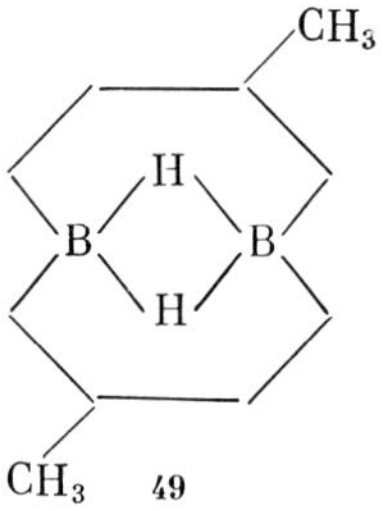

49

Hydroboration of 1,3-pentadiene Saegebarth [46] carried out the hydroboration of 1,3-pentadiene in diglyme (diene: $BH_3 = 3:2$). He did not isolate the hydroboration products but subjected them to oxidation and, on the basis of the ratio between the 1,3- and 1,4-pentanediols obtained, concluded that 1,4-bis(2-methyl-1-borolyl)pentane (*50*) 1,3-bis(2-methyl-1-borolyl)pentane (*51*) are formed in the reaction.

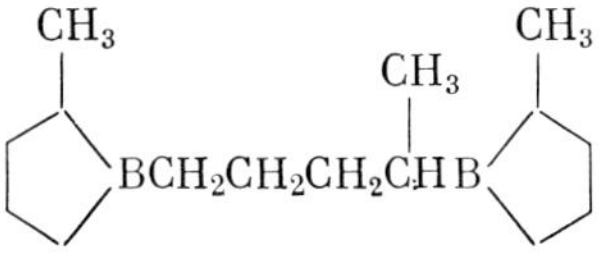

50

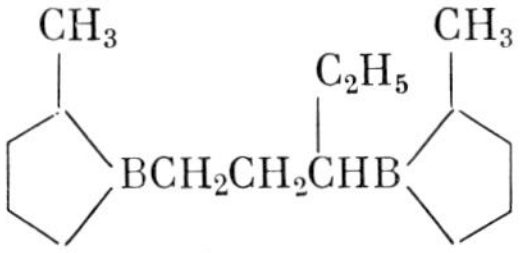

51

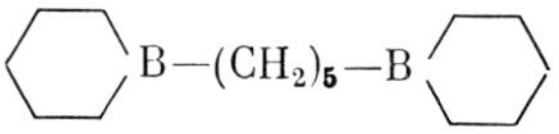

52

According to the data of [46], a mixture of compounds *50* and *51*, when heated to 160–175°C for six hours, undergoes isomerization to a compound with six-membered rings and a normal chain (*52*), so that almost pure 1,5-pentanediol is obtained in the subsequent oxidation. According to Zakharkin and Kovredov [47], the yield of the isomerized compound, 1,5-bis(1-borinyl)pentane (*52*), is 62%.

Mikhailov and co-workers [45] investigated the hydroboration of *cis*- and *trans*-1,3-pentadiene in ether solution at a diene: BH_3 ratio of 3:2. It was found that the structural characteristics of the diene have no significant effect on the course of the process. In the fractional distillation of the hydroboration products, a fraction containing 1-alkenyl-2-methylborolanes (43%, calculated on the boron) and a viscous polymer (33–44%) are obtained. The mixture of 1-alkenyl-2-methylborolanes contains the alkenylboranes *53*, *54*, and *55* in a ratio of 50:35:15.

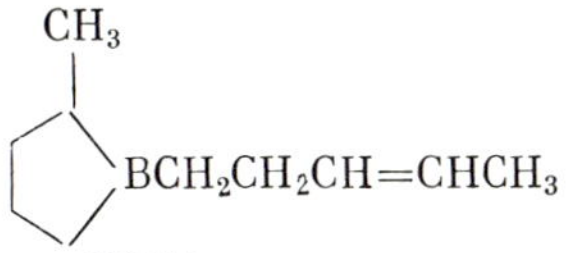

53 (50%)

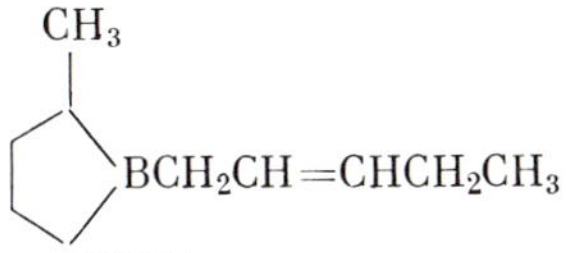

54 (35%)

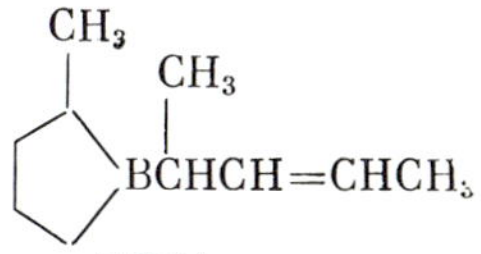

55 (15%)

By thermal degradation of the polymeric residue (160–175°C, six hours) and subsequent distillation, a 60% yield of bicyclic compounds is obtained, the oxidation of which forms 77% of 1,5-pentanediol, 18% of 1,3-pentanediol, and 5% of other diols, indicating incomplete isomerization of the bicyclic compounds to the compound *52*. The formation of significant quantities of polymer in the hydroboration of piperylenes is explained by the fact that, in the initial stages of the process, a large quantity of alkenylboranes with non-terminal double bonds is formed, for which an intramolecular hydroboration is difficult, resulting in their being cross-linked to polymeric chains by diborane.

Hydroboration of 1,4-pentadiene According to the data of [46], the hydroboration of 1,4-pentadiene (*56*) (diene:BH_3 = 3:2) and subsequent isomerization of the hydroboration products at 160–175°C for six hours forms 1,5-bis(1-borinyl)pentane (52):

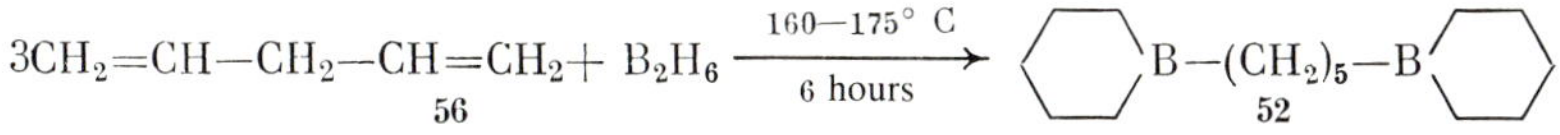

The hydroboration of 1,4-pentadiene in tetrahydrofuran (diene:BH_3 = 3:2) and subsequent distillation of the reaction products produce a mixture of isomeric 1-pentenyl-2-methylborolanes (*57*) and 1-pentenylborinanes (*58*) in a ratio of 66:34 and also a mixture of isomeric bis(boracycloalkyl)pentanes (*59*), containing five- and six-membered rings connected by normal and branched C_5H_{10} hydrocarbon chains [48].

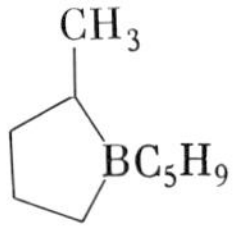

57

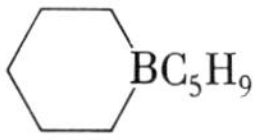

58

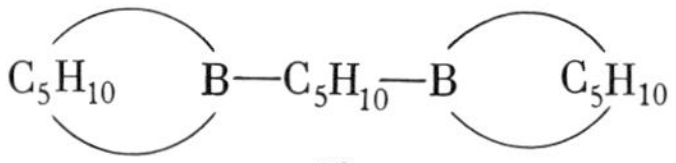

59

1,5-Bis(1-borinyl)pentane (*52*), which is formed by the thermal isomerization of the mixture of compounds *59* [2, 23, 46], is converted by treatment with diborane or trialkylamine-borane to diborinane (*60*) [2, 23], which smoothly reacts with olefins to form 1-alkylborinanes (*61*) [49, 50] and is converted by the action of methanol to 1-methoxyborinane (*62*) [50]:

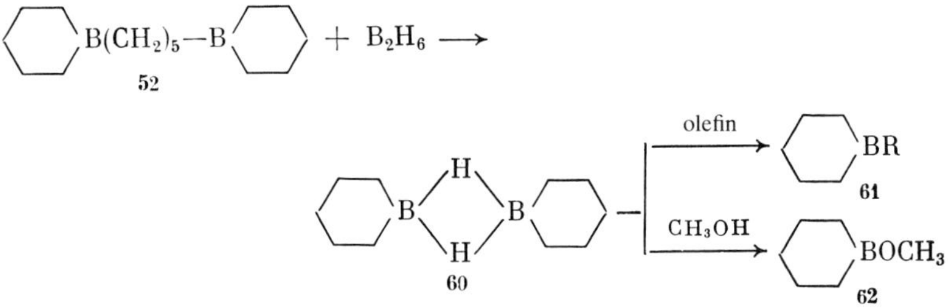

Diborinane (*60*) is also formed by the reaction of 1,4-pentadiene with diborane in a ratio of 1:1 [23].

When the polymeric products of the hydroboration of 1,4-pentadiene (diene: BH_3 = 1:1) are heated to 70°C in tetrahydrofuran for six hours and are then subjected to methoxylation, 1-methoxyborinane (*62*) and 1,2-bis(1-methyltetramethylene)diborane (*49a*) are obtained [50]:

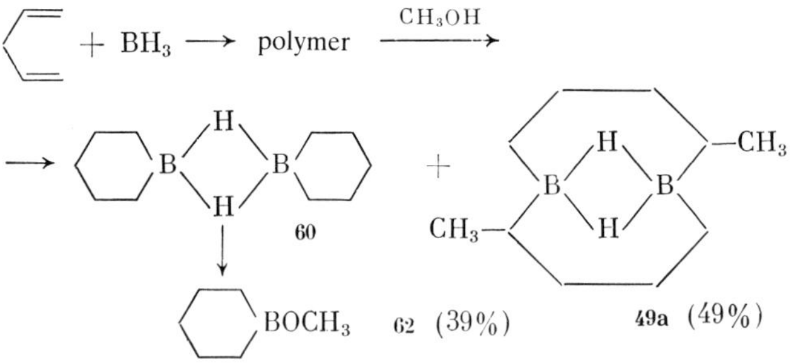

Hydroboration of 1,5-hexadiene According to Saegebarth [46], the hydroboration of 1,5-hexadiene leads to an 82% yield of 1,6-bis(1-borepyl)hexane (*63*):

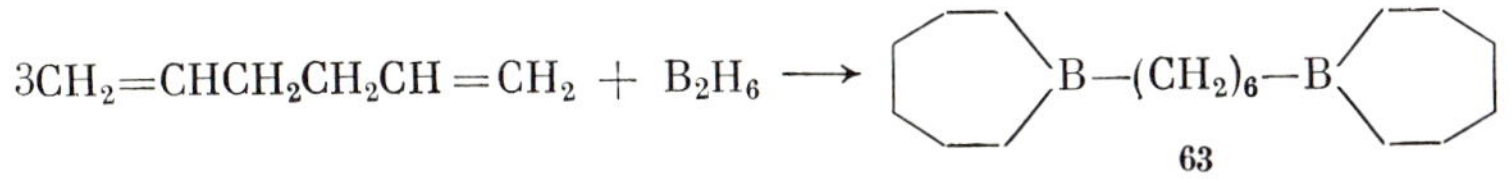

63

which isomerizes at 160–175°C to 1,6-bis(2-methyl-1-borinyl)hexane.

In [47, 51], the product of the hydroboration of 1,5-hexadiene was also taken to be the pure compound *63*. However, oxidation of the products of the hydroboration of 1,5-hexadiene (diene : $BH_3 = 3:1$) indicated a more complex orientation of the addition reaction. Under "monohydroboration" conditions, i.e., with a diene : BH_3 ratio of 3 : 1, 35% of the diene entering into the reaction adds diborane at only one double bond, and it was assumed that the final reaction product is a polymer [35].

The question of the nature of the organoboron compounds obtained by the reaction of 1,5-hexadiene with diborane was later cleared up by Mikhailov and co-workers [48], who investigated the process in ether and tetrahydrofuran with different ratios of the reagents. With a ratio of 1,5-hexadiene : BH_3 of over 1.5, three groups of products are formed in different proportions: a) a mixture of monoboracyclic compounds with hexenyl radicals on the boron atom (*64*); b) a mixture of bis(boracycloalkyl)hexanes (*65*), and c) a polymer.

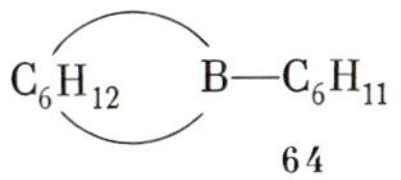

64

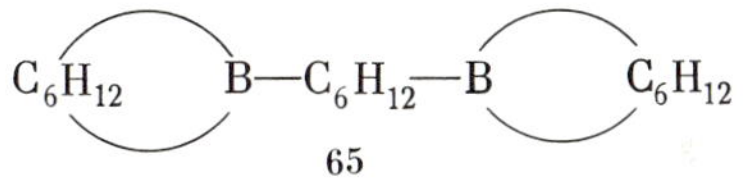

65

As the diene : BH_3 ratio is increased, the content of the compounds *64* in the reaction products increases, and that of the compounds *65* decreases, whereas, with a diene : BH_3 ratio of 1.5 or lower, a polymer is formed.

With a diene : BH_3 ratio of 2 : 1, 29% of compounds *64* and 64% of compounds *65* is obtained.

From data on the oxidation, it follows that the mixture *64* contains seven-membered (*66*), six-membered (*67*) and five-membered (*68*) boracyclanes with a normal hexenyl chain. The mixture *64* contains only 7% of compounds with a branched chain.

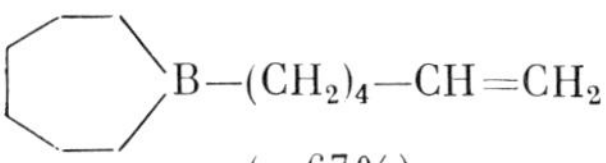

66 (~67%)

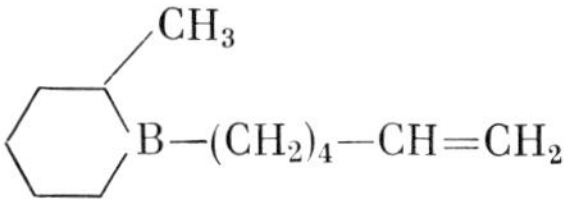

67 (~22%)

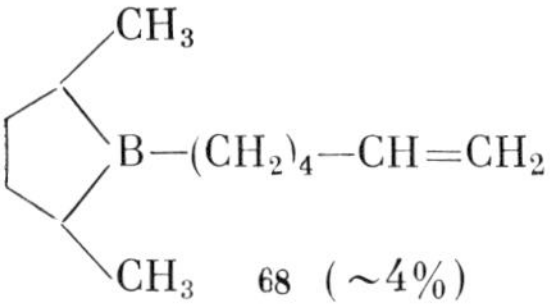

68 (~4%)

Heating for six hours at 160–190°C leads to isomerization of the seven- and five-membered rings *66* and *68* to the six-membered ring *67*, as a result of which the mixture *64* is converted practically completely to the compound *67*.

The mixture of compounds *65* contains isomeric bicyclic compounds with seven-, six-, and five-membered rings, connected by carbon chains with both a normal and an iso structure.

The thermal isomerization of the compounds *65* is accompanied by a contraction of the seven-membered rings, an expansion of the five-membered rings, and a "straightening" of the bridge. The almost complete isomerization of the mixture of compounds *65* to 1,6-bis(2-methyl-1-borinyl)hexane (*69*) is achieved by fourfold vacuum distillation or by heating at 160–190°C for six hours.

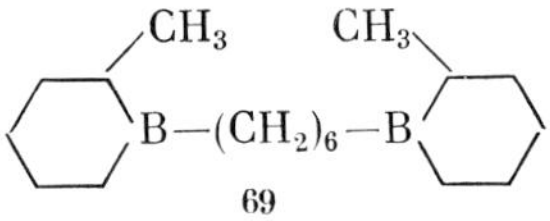

69

The polymer obtained with a diene : BH_3 ratio of 3 : 2 undergoes degradation upon heating to convert to the mixture of compounds *65*.

The small content of the polymer in the reaction mixture at a diene : BH_3 ratio of 2 : 1, and also its formation in the treatment of the mixture of compounds *64* and *65* with diborane, permits the conclusion that the polymer is not a primary reaction product, but is formed as a result of subsequent conversions of bis(boracycloalkyl)hexanes under the action of diborane.

The formation of the bis(boracycloalkyl)hexanes *65* in the case of a diene : BH_3 ratio of 2 : 1 indicates the higher rate of hydroboration of the hexenylboracyclanes *64* as compared with the starting diene.

The most probable mechanism of the hydroboration of 1,5-hexadiene may be represented by the following scheme:

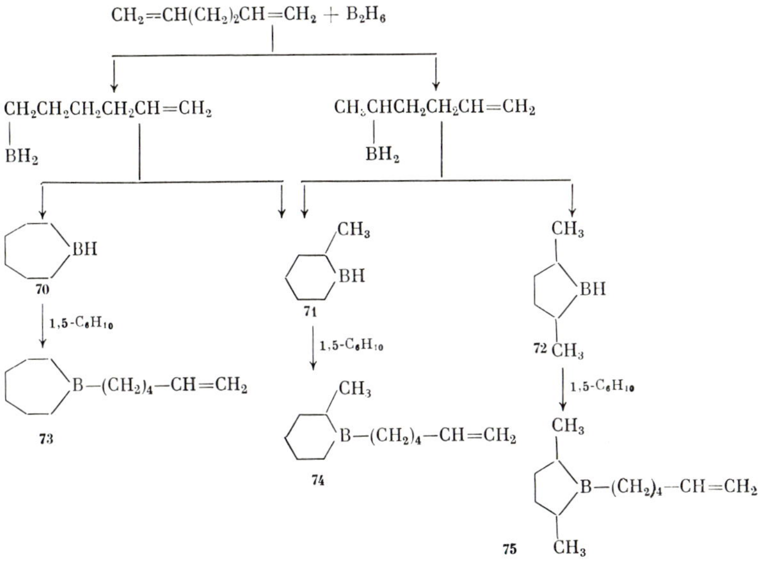

Further sequential reactions of the compounds *73–75* with diborane and the diene or with the compounds *70–72* lead to the formation of different isomeric bis(boracycloalkyl)hexanes (*65*).

The experimental data of [33, 34, 45, 48] indicate that both polymeric and low-molecular-weight mono- and bicyclic compounds may be obtained by the hydroboration of diene hydrocarbons, with their ratio depending both on the nature of the diene and on the conditions under which the reaction is carried out.

The cyclic compounds present in the reaction mixture distill relatively easily even when polymer products are present. However, when cyclic compounds are formed as a result of thermal degradation of the polymer, the reaction mass must be heated for a long time at a temperature considerably higher than the boiling point of the cyclic compounds. The distillation of the latter proceeds slowly in this case and is accompanied by the transfer to the distillate of part of the polymeric compounds, which can be removed only by further fractionation.

The composition of the isomeric bis(boracycloalkyl)hexanes *65* formed by hydroboration (diene:$BH_3 = 3:2$) may vary within wide limits, depending on the length of the thermal treatment of the reaction products. Thus, in repeating the work of [48], Zakharkin and Kovredov [52] obtained 67% of 1,6-hexanediol, 28% of 1,5-hexanediol, and 2% of 2,5-hexanediol in the oxidation of the compounds *65*, whereas Brown and co-workers [53] obtained a ratio of the above-mentioned diols of 49:50:1, indicating a greater extent of thermal isomerization of the rings.

The hydroboration of 1,5-hexadiene in tetrahydrofuran with a diene:BH_3 ratio of 1:1, followed by heating of the mixture at 70°C and methanolysis of the reaction products, leads to a difficult-to-separate mixture of 1-methoxyborepane, 1-methoxy-2-methylborinane, and 1-methoxy-2,5-dimethylborolane in a ratio of 73:20:7 [54].

Hydroboration of 2,4-dimethyl-1,4-pentadiene and 2,5-dimethyl-1,5-hexadiene 2,4-Dimethyl-1,4-pentadiene and diborane (diene: $BH_3 = 1:1$), when refluxed for one hour in tetrahydrofuran, form bis(3,5-dimethyl)borinane, which is converted by methanolysis to 1-methoxy-3,5-dimethylborinane:

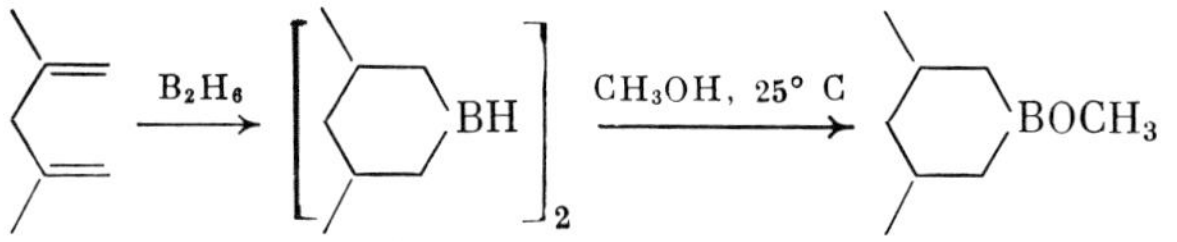

87% 98%

The hydroboration of 2,5-dimethyl-1,5-hexadiene under the same conditions produces bis(3,6-dimethylborepane), which is identified in the form of 1-methoxy-3,6-dimethylborepane [55]:

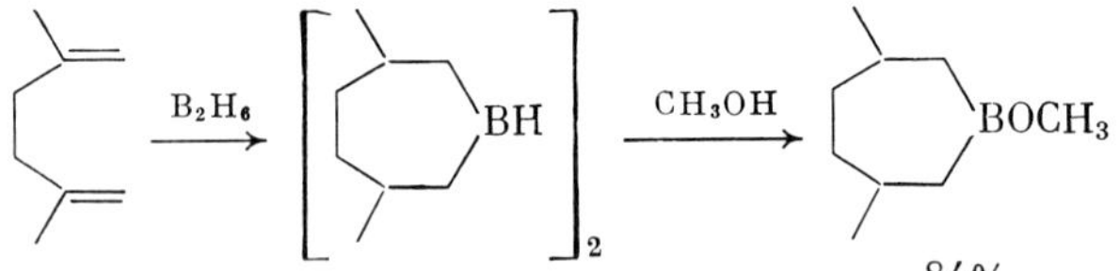

84%

1.3.2 Hydroboration of acyclic diene hydrocarbons by diborane derivatives

The action of amine complexes of alkylboranes; dialkyl-, diaryl-, and tetralkyldiboranes, chloroborane, and alkylmercaptoboranes on diene hydrocarbons has been studied.

Hawthorne [56, 57] showed that the hydroboration of 1,3-diene hydrocarbons can be carried out with alkylboranes in the form of their adducts with amines. Thus, when triethylamine-*tert*-butylborane was used, 1-*tert*-butylborolane was obtained from 1,3-butadiene:

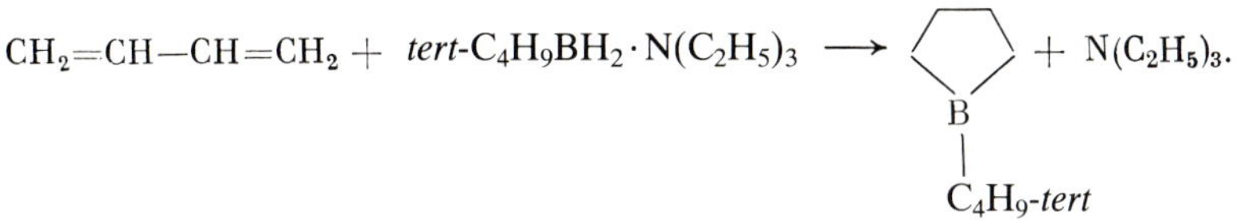

1-Phenylborolane was synthesized from triethylamine-phenylborane and 1,3-butadiene in an analogous manner [58]:

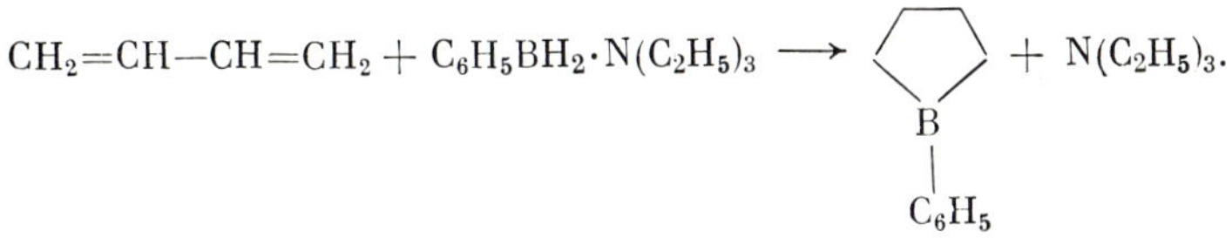

1-Alkylborolanes can be easily obtained by the reaction of isoprene with trialkylamine-borane in the presence of trialkylboranes [59]:

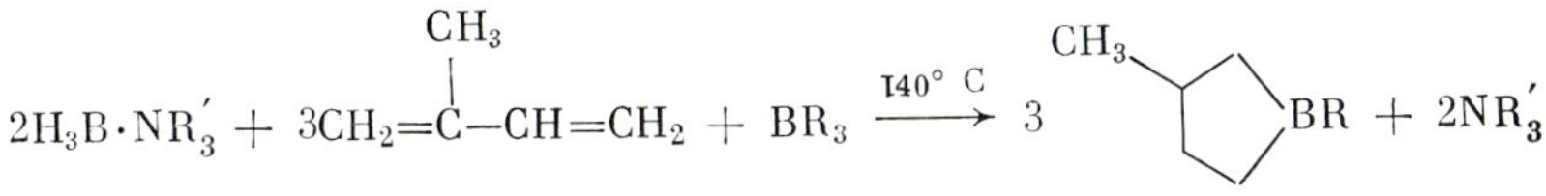

Mikhailov and Dorokhov [60] found that substituted diboranes are convenient hydroborating agents. Thus, 1,2-diaryldiboranes react smoothly with 1,3-butadiene and isoprene in ether to form 1-arylborolanes:

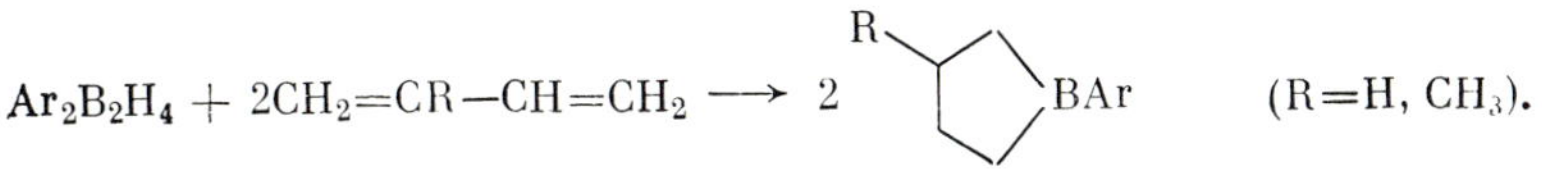

The use of tetralkyldiboranes as hydroborating agents also makes it possible to obtain 1-alkylborolanes from dienes [61], since the intermediate α,ω-bis(dialkylboryl)alkanes formed tend to cyclize with the elimination of trialkylborane:

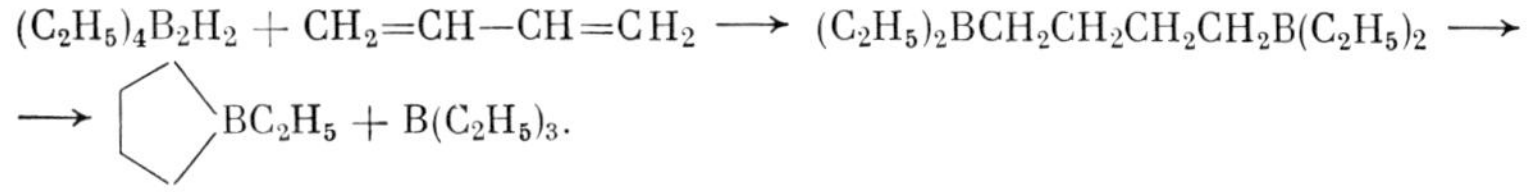

The reaction of 1,4-pentadiene with tetraethyldiborane leads predominantly to 1-ethylborinane, along with 1-ethyl-2-methylborolane (85:15) [61].

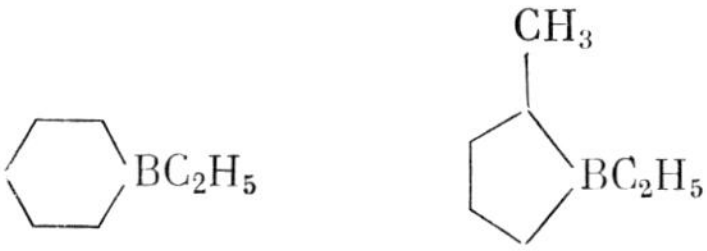

In the hydroboration of 1,5-hexadiene, the main product (97%) is 1-ethylborepane, with a small content of 1-ethyl-2-methylborinane and 1-ethyl-2,5-dimethylborolane:

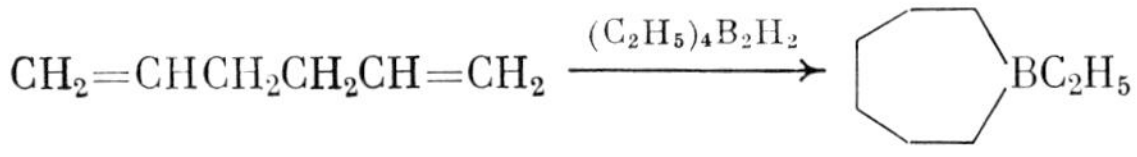

In the hydroboration of 2,5-dimethyl-1,5-hexadiene, the formation of a five- or six-membered ring is made difficult by the presence of methyl groups in the 2 and 5 positions, and a mixture (1:1) of *cis*- and *trans*-1-ethyl-3,5-dimethylborepanes is therefore formed:

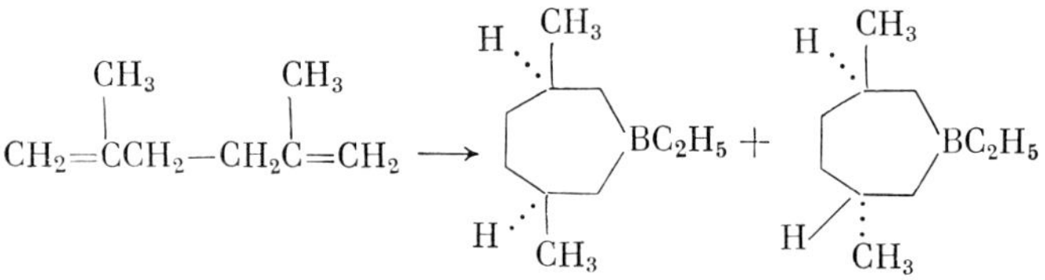

From the data of a chromatographic analysis of the crude hydroboration products, it follows that boracyclanes are already formed at room and even lower temperatures (−10 to 10°C). The yield of boracyclanes can be increased by subsequent thermal isomerization to thermodynamically more stable six-membered rings. Thus, when the isomeric mixture obtained by the hydroboration of 1,5-hexadiene with tetraethyldiborane is heated to 160°C, isomerization of the seven-membered to a six-membered ring takes place and 1-ethyl-2-methylborinane becomes the main component (89%) of the equilibrium mixture [61].

The hydroboration of 2,4-dimethyl-1,3-pentadiene with tetraalkyldiborane, followed by thermal isomerization, produces 1-alkyl-3,5-dimethylborinane (*76*), which is converted under the action of diborane to bis(3,5-dimethylborinane) (*77*) [2]:

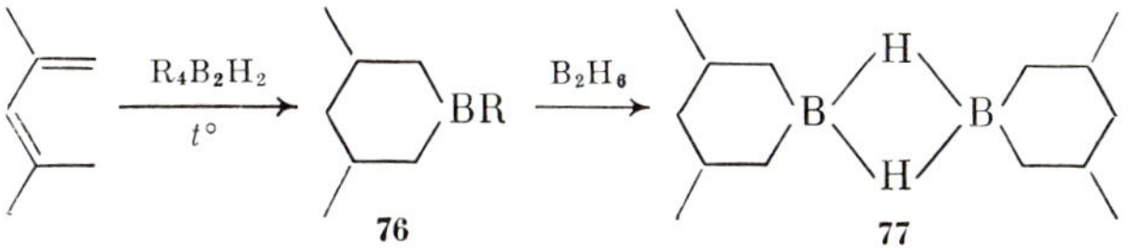

76 77

The diboracyclane (*77*) can be obtained from 2,4-dimethyl-1,3-pentadiene with the use of diborane [62].

When bis(2,3-dimethyl-2-butyl)diborane is used as a hydroborating agent, different diene hydrocarbons are converted to the corresponding boracyclanes, containing 2,3-dimethyl-2-butyl groups on the boron atom. Borolanes are obtained in yields of 54–89% from 1,3-butadiene, isoprene, and 2,3-dimethyl-1,3-butadiene, while a mixture of borolane and borinane is formed from 1,4-pentadiene; the hydroboration of 1,5-hexadiene leads to a mixture of five-, six-, and seven-membered heterocyclic compounds (8:2:90) [63].

1,5-Hexadiene is hydroborated with bis(9-borabicyclo[3.3.1]-nonane) to form 1,6-bis(9-borabicyclo[3.3.1]nonyl)hexane [53, 123].

The hydroboration of 1,3-butadiene, 1,4-pentadiene, and 1,5-hexadiene with chloroborane has been investigated [64, 124]. 1,3-Butadiene reacts with chloroborane to form a mixture of compounds, from which 1-butoxyborolane can be isolated in 38% yield by treatment with butanol. 1,4-Pentadiene adds chloroborane to form 1-chloroborinane and 1-chloro-2-methylborolane (70:30):

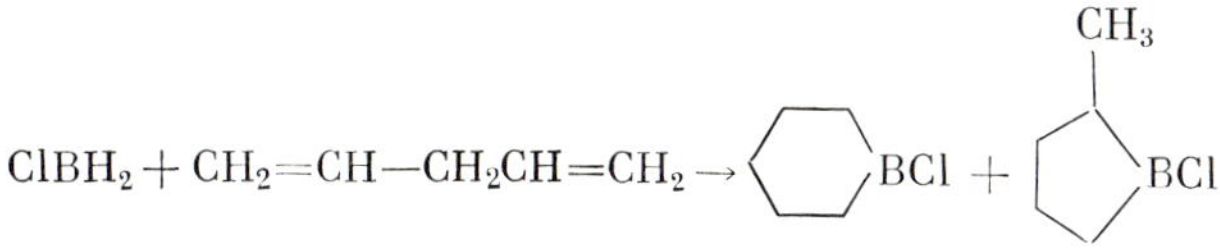

The hydroboration of 1,5-hexadiene with chloroborane leads to a mixture of compounds, from which 1-chloroborepane with a degree of purity of 94% is isolated by distillation:

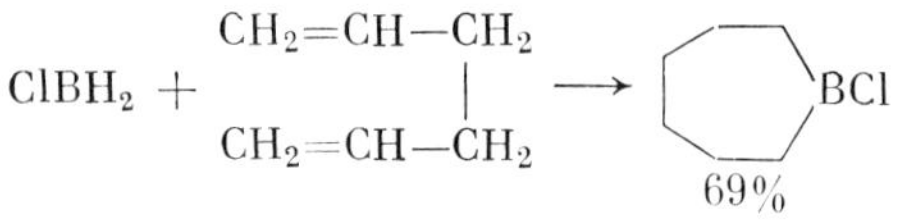

Polymers of alkylmercaptoboranes react with diene hydrocarbons to form 1-alkylmercaptoboracyclanes. From the products of the reaction of polymers of methylmercaptoborane [65, 66] or butylmercaptoborane [67, 68] with 1,3-butadiene or isoprene, 1-alkylmercapto-substituted borolane and 3-methylborolane can be isolated in the pure state:

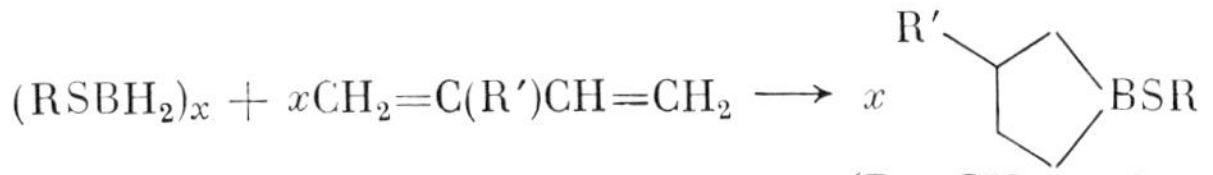

($R=CH_3$, n-C_4H_9; $R'=H$, CH_3)-

The reaction of the polymer of butylmercaptoborane with 1,4-pentadiene produces a mixture of 1-butylmercaptoborinane and 1-butylmercapto-2-methylborolane in a ratio of 55:45 (yield 63%):

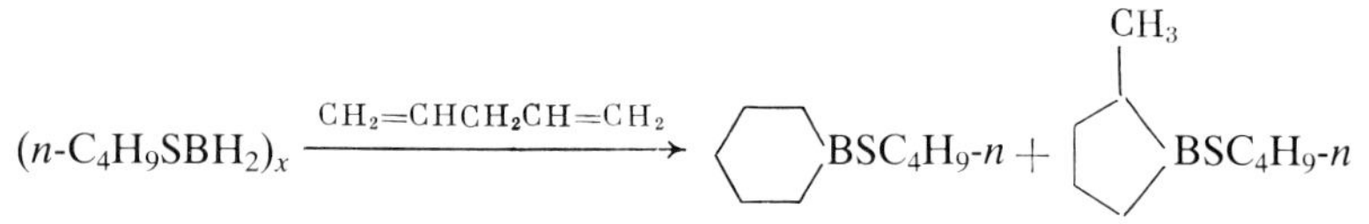

while 1,5-hexadiene predominantly forms 1-butylmercaptoborepane together with 1-butylmercapto-2-methylborinane and, apparently, 1-butylmercapto-2,5-dimethylborolane (80:17:3) [68], with a total yield of 76%:

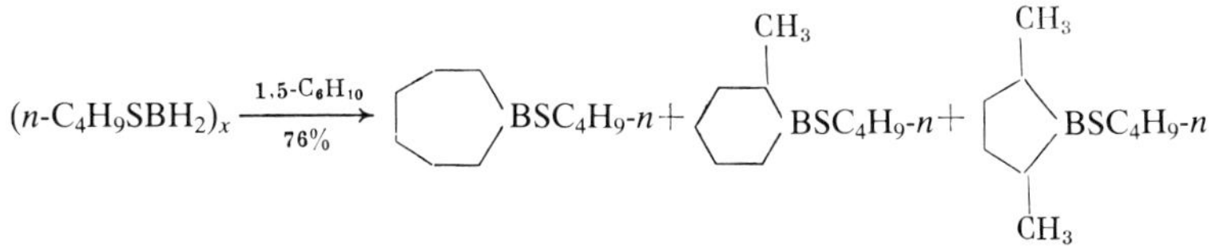

5-(E)-2,6,10-Trimethyl-1,5,9-undecatriene has been converted to 6-cadinol by hydroboration-carbonylation-oxidation [125].

1.3.3 Hydroboration of cyclic diene and polyene hydrocarbons

In this section, methods for the synthesis of boracyclanes based on the hydroboration of cyclic diene and polyene hydrocarbons will be considered.

The dihydroboration of 4-vinylcyclohexene with diborane forms bicyclic boranes in low yield. On the basis of the composition of the diols obtained by the oxidation of the methanolysis products, it can be concluded that the main component of the mixture of *B*-methoxy derivatives is the bicyclic compound *78*. Distillation of the methoxy compounds leads to isomerization, as a result of which 2-methoxy-2-borabibicyclo[3.3.1]nonane (*79*) is found in predominant quantity in the reaction mixture [54]:

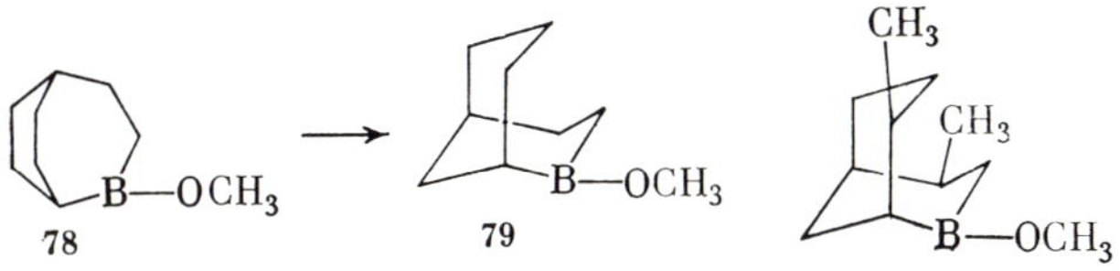

78 79 80

Bicyclic boranes are also obtained in low yield by the hydroboration of D-(+)-limonene. In the distillation of the methanolysis product, the main compound obtained is 2-methoxy-4,8-dimethyl-2-borabicyclo[3.3.1]nonane (*80*), apparently in the form of a mixture of the cis and trans isomers [54].

The hydroboration of D-(+)-limonene with bis(2,3-dimethyl-2-butyl)diborane, followed by distillation of the reaction products, forms 2-(2,3-dimethyl-2-butyl)-4,8-dimethyl-2-borabicyclo[3.3.1]-nonane (*80a*) [63]:

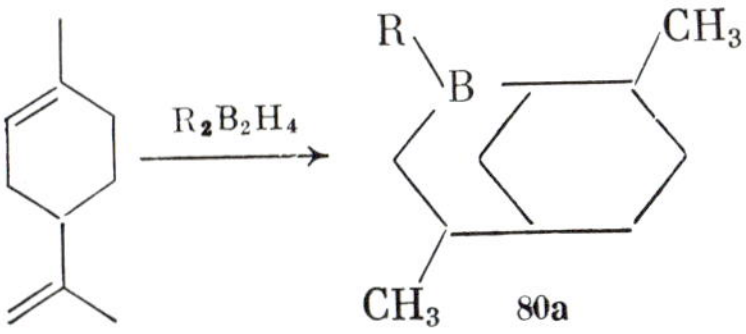

80a

[R = $(CH_3)_2CHC(CH_3)_2$].

Oxidation of the compound *80a* leads to D-(−)-(1*S*, 2*R*, 4*R*)-limonene-2,9-diol (*81*) and protonolysis of the primary B—C bond leads to the borane *82*, the oxidation of which produces (−)-carvomenthol (*83*) [63, 69, 126]:

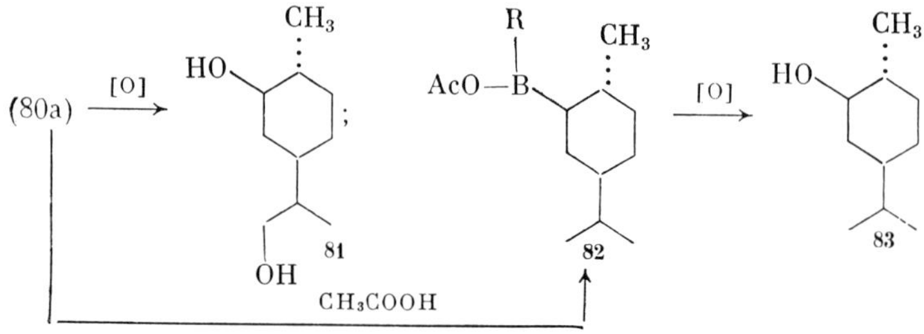

Köster and co-workers [61, 70] obtained 9-ethyl-9-borabicyclo-[3.3.1]nonane by the hydroboration of *cis,cis*-1,5-cyclooctadiene with tetraethyldiborane:

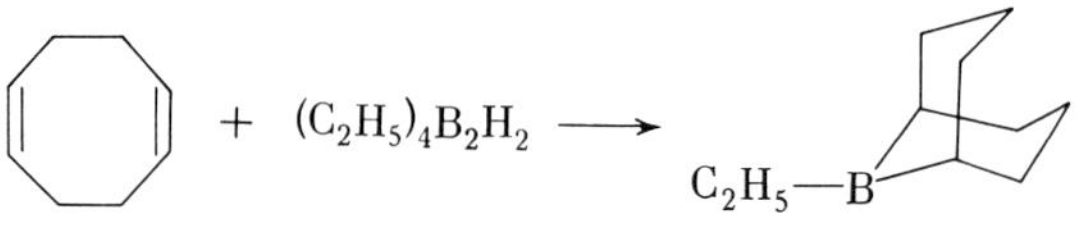

As a result of the hydroboration of 1,5-cyclooctadiene with diborane in tetrahydrofuran, followed by distillation of the reaction products, bis(9-borabicyclo[3.3.1]nonane) is formed [54, 71, 127], which had been previously synthesized by the action of tetraethyldiborane on 9-propyl-9-borabicyclo[3.3.1]nonane [23]:

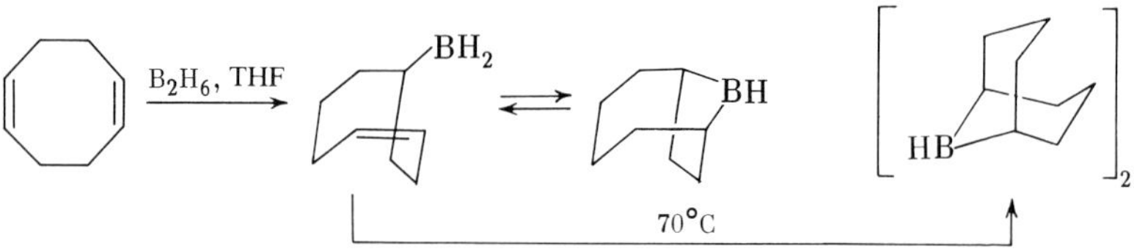

For the methods of preparation and properties of 9-borabicyclo[3.3.1]nonane, see also [128–130]. *B*-Halo-9-borabicyclo-[3.3.1]nonane can be obtained by the action of $H_2BX \cdot SMe_2$ on 1,5-cyclooctadiene [131]. The synthesis of 9-borabicyclo-[3.3.1]nonane-9*d* has been described [132].

1,5-Cyclooctadiene is converted by the action of bis(2,3-dimethyl-2-butyl)diborane to a mixture of *B*-alkyl-9-borabicyclo[4.2.1]nonane

(*84*) and *B*-alkyl-9-borabicyclo[3.3.1]nonane (*85*) in a ratio of 80:20 [63]:

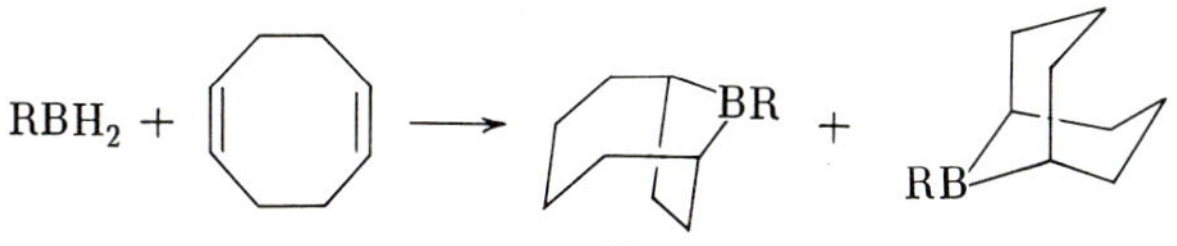

84 85

[R = $(CH_3)_2CH\overset{|}{C}(CH_3)_2$].

The distillation of this mixture is accompanied by partial isomerization of the cyclic system *84*, as a result of which the ratio between the isomers *84* and *85* changes to the reverse (20:80) (yield 91%). *B*-alkyl-9-borabicyclo[3.3.1]nonanes are obtained by the action of lithium dialkylcuprates (R_2CuLi) on 9-borabicyclo[3.3.1]nonane [133].

The hydroboration of bicyclo[3.3.1]nona-2,6-diene (*86*) with tetraethyldiborane leads to the formation of the polymer *87* which, when heated under vacuum (200–250°C), symmetrizes to triethylborane, 2-ethyl-2-boraadamantane (*88*) (27%) and bis(2-boraadamant-2-yl)bicyclo[3.3.1]nonane (*89*). When the compound *89* is refluxed with methyl borate in the presence of catalytic quantities of $(C_2H_5)_4B_2H_2$, 2-methoxy-2-boraadamantane (*90*) is formed [72]:

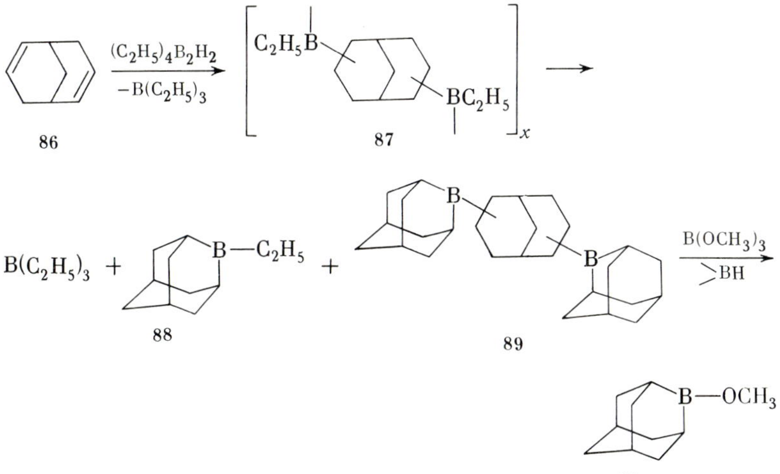

86 87 88 89 90

The hydroboration of bicyclo[3.3.1]nona-2,6-diene (*86*) with diborane in the presence of methyl borate leads to the formation of polymeric products, the pyrolysis of which forms 2-methoxy-2-boraadamantane (*90*) (46%) and bis(2-boraadamant-2-yl)-bicyclo[3.3.1]nonane (*89*) (27%):

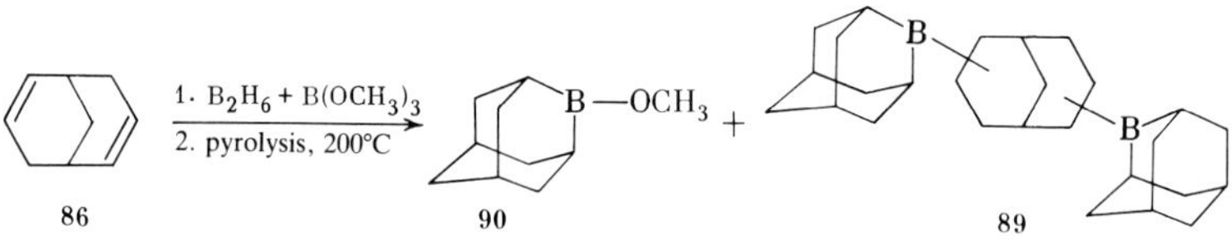

86 90 89

The reduction of compound *90* by diborane forms the dimer of 2-boraadamantane (*91*) in a yield of approximately 80% [73]:

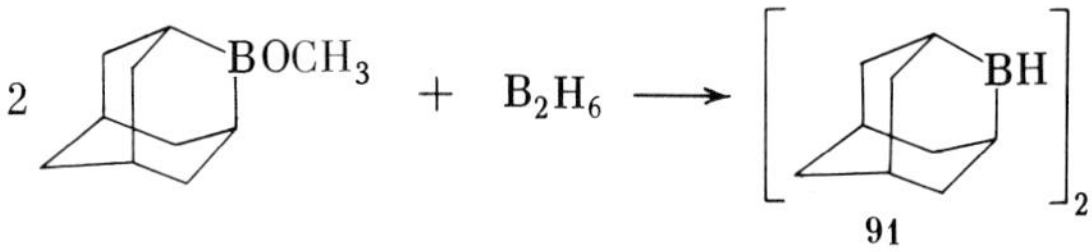

91

The hydroboration of *trans,trans,trans*-1,5,9-cyclododecatriene (*92*) with triethylamine-borane [74, 75] or tetraethyldiborane leads to *cis,cis,cis*-perhydro-9b-boraphenalene (*8*):

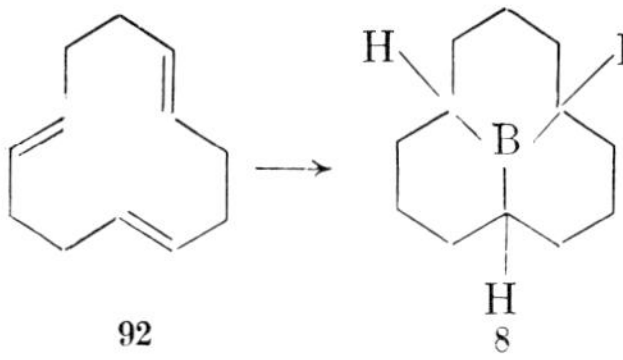

92 8

cis,trans,trans-1,5,9-Cyclododecatriene (*93*) is converted by hydroboration with triethylamine-borane [75] or tetraethyldiborane [70] to *cis,cis,trans*-perhydro-9b-boraphenalene (*8a*):

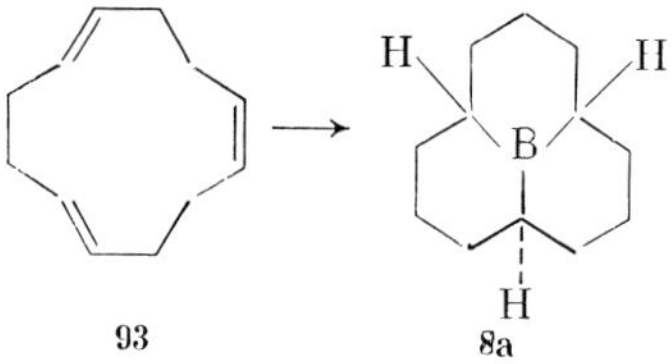

93 8a

In [76], the structure of the *cis,cis,trans* isomer *8a* is assigned to the compound *8* and the structure of the *cis,cis,cis* isomer *8* to the compound *8a*.

2,6-Dichloro-2,6-diboraadamantane has been obtained in the form of an adduct with Me_2S by the hydroboration of 1,3,5,7-cyclooctatetraene with $BH_2Cl \cdot Me_2S$ [134].

CHAPTER 2

Chemical conversions of boracyclanes

2.1 THERMAL ISOMERIZATION

One of the most important characteristics of boracyclanes is their ability to undergo isomerization, with contraction or expansion of the ring, at elevated temperature.

Köster and co-workers [59, 75] found that 1-ethylborepane (*94*), which is obtained (with an impurity of 3% of 1-ethyl-2-methylborinane (*96*)) by the action of tetraethyldiborane on 1,5-hexadiene, isomerizes at 160°C to a mixture of 1-ethylboracyclanes containing 80% of 1-ethyl-2-methylborinane (*95*), 12.5% of 1-ethylborepane (*94*), and 7.5% of 1,2-diethylborolane (*96*).

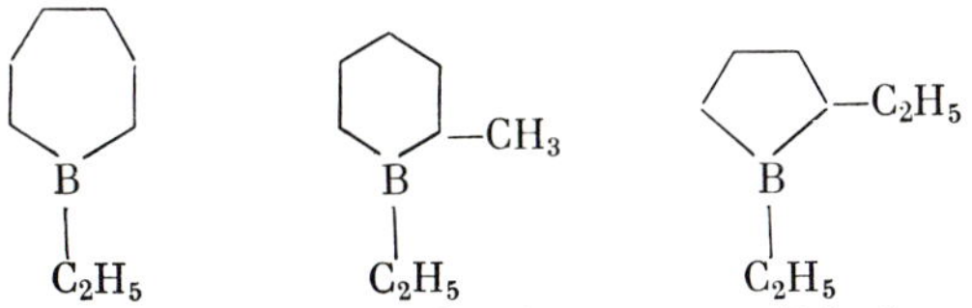

94 (12,5%)   95 (80%)   96 (7,5%)

9-Boradecalin forms the following equilibrium mixture at 160°C [20]:

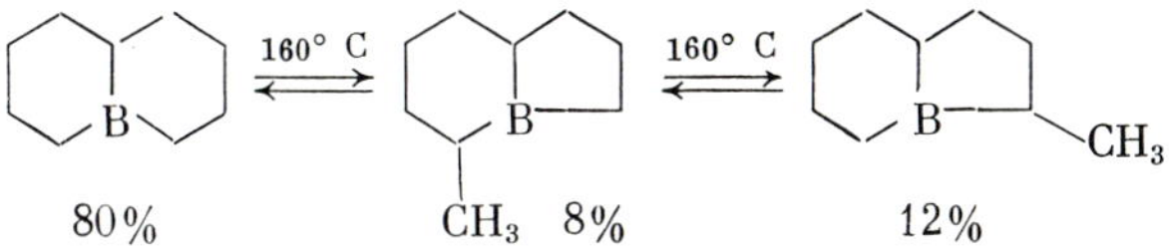

80%   8%   12%

Perhydro-9b-boraphenalene isomerizes at 200°C to the compound *97* [77]:

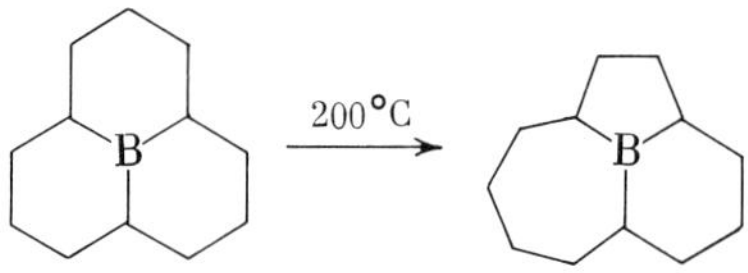

97

B-Alkyl-9-borabicyclo[3.3.1]nonanes show a higher thermal stability than trialkylboranes [135] and, at the same time, are easily transalkylated with olefins upon heating in tetrahydrofuran [136].

2.2 OXIDATION. SYNTHESIS OF DIOLS

The hydroboration of diene hydrocarbons, followed by oxidation with hydrogen peroxide in alkaline medium of the crude reaction products or the products isolated in the form of alkenylboranes or bis(boracycloalkyl)alkanes, may be used as a method for the preparation of alkanediols. However, it is rarely possible to obtain diols in the pure form by this method, such as, for example, 2,3-dimethyl-1,4-butanediol from 2,3-dimethyl-1,3-butadiene; in most cases, a mixture of diols with a predominant content of one particular isomer is obtained. Thus, a mixture of isomeric diols – 1,4-butanediol and 1,3-butanediol in a ratio of 2:1 – is obtained from 1,3-butadiene and the diols from isoprene contain 87% of 2-methyl-1,4-butanediol.

A number of patents have been devoted to the preparation of diols by the hydroboration-oxidation of diene hydrocarbons [42, 43, 78, 79].

In those cases where the isomeric bis(boracycloalkyl)alkanes differ only in the structure of the aliphatic chain, as in the case of the hydroboration of 1,3-butadiene, isoprene, or 1,3-pentadiene, the synthesis of pure diols can be carried out by the conversion of bis(boracycloalkyl)alkanes to 1-methoxyboracyclanes, followed by oxidation of the latter compounds. Thus, for example, 1-methoxyborolane, which forms pure 1,4-butanediol by oxidation, is obtained by treatment with methyl borate of the crude products of the hydroboration of 1,3-butadiene or of an isomerized mixture of 1,4-, 1,3-, and 1,2-di(1-borolyl)butane in the presence of catalytic quantities of tetraalkyldiborane [80].

The hydroboration of α,ω-dienes with thexylborane, followed by oxidation of the boracyclanes formed, leads to a mixture of diastereomeric acyclic diols [137].

Cyclooctadienes and cyclic diene hydrocarbons containing a large number of carbon atoms are hydroborated with the formation of bridged bicyclic compounds, which isomerize upon heating in tetrahydrofuran to the more stable borabicyclic systems with six-membered rings. Oxidation of the isomerized compounds produces cis-diols in high yields. *cis,cis*-1,3-Cyclooctadiene is converted to *cis*-1,5-cyclooctanediol by hydroboration-isomerization-oxidation [81].

In the case of the nine-membered cyclic dienes, the formation of *cis*-1,5-cyclononanediol is observed from both *cis,cis*-1,3-cyclononadiene and *cis,cis*-1,4-cyclononadiene [81].

The hydroboration-oxidation of *cis,cis*-1,3-cyclooctadiene forms *cis*-1,3-cyclooctanediol (23%) and *cis*-1,4-cyclooctanediol (77%), while *cis,cis*-1,5-cyclononadiene forms a 1:4 mixture of *cis*-1,4-cyclononanediol and *cis*-1,5-cyclononanediol. Only *cis*-1,5-cyclodecanediol is obtained from *cis,cis*-1,6-cyclodecadiene, regardless of whether the hydroboration products undergo isomerization or not [81].

cis,cis,cis-Perhydro-9b-boraphenalene is smoothly converted by oxidation with alkaline hydrogen peroxide solution to *cis,cis,cis*-1,5,9-cyclodecanetriol, while the oxidation of the *cis,cis,trans* isomer leads to *cis,cis,trans*-1,5,9-cyclodecanetriol [75].

Bis(9-borabicyclo[3.3.1]nonane) forms *cis*-1,5-cyclooctanediol when oxidized by alkaline hydrogen peroxide solution [71].

9-Borabicyclo[3.3.1]nonane is converted by oxidation with chromic acid to cyclooctanone, which is explained by protonolysis of one of the B—C bonds during the oxidation. Cyclic ketones are also obtained by the chromic acid oxidation of 10-borabicyclo[4.2.1]nonane and 11-borabicyclo[5.3.1]undecane, which are prepared by the hydroboration of *cis,cis*-1,5-cyclooctadiene and *cis,cis*-1,6-cyclodecadiene, respectively [82]. 9-Borabicyclo[3.3.1]-nonane is converted to 1,5-cyclooctanedione by oxidation with Jones reagent [83].

"Ate" complexes of *B*-alkyl-9-borabicyclo[3.3.1]nonane form *cis*-1,5-cyclooctanediol together with *cis*-bicyclo[3.3.0]octan-1-ol upon oxidation with alkaline hydrogen peroxide [138].

2.3 EXCHANGE REACTIONS WITH BORATES, THIOBORATES, BORON HALIDES, AND DIBORANE

Boracyclanes and bis(boracycloalkyl)alkanes, like trialkylboranes [84–86], enter into exchange reactions with alkyl borates in the presence of compounds with B—H bonds, forming 1-alkoxyboracyclanes.

Mikhailov and co-workers [33, 51] found that 1-methoxyborolane is obtained in 80% yield by the reaction of methyl borate with a mixture of di(1-borolyl)butanes in the presence of tetraalkyldiborane:

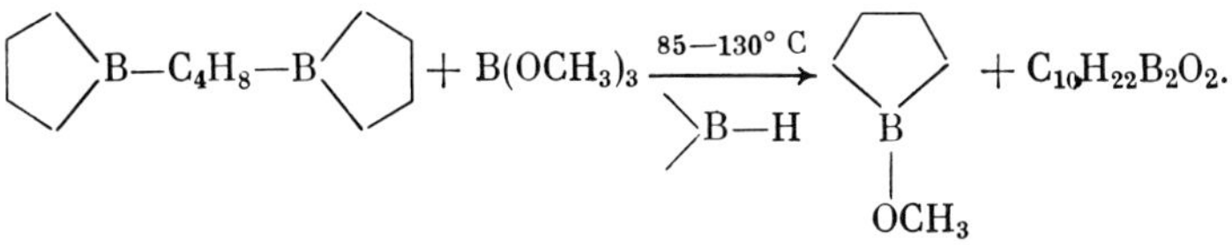

The higher fraction, with the composition $C_{10}H_{22}B_2O_2$, is a mixture of dimethoxy derivatives of diboracyclic compounds (16%), containing the following hydrocarbon chains:

$$-CH_2CH_2CH_2CH_2- \qquad -\underset{|}{\overset{CH_3}{}}\!CHCH_2CH_2- \qquad -\overset{C_2H_5}{\underset{}{CH}}CH_2-$$

1-Methoxyborolane reacts with methyl boráte to form tetramethyl 1,4-butanediboronate (*98*) [51]:

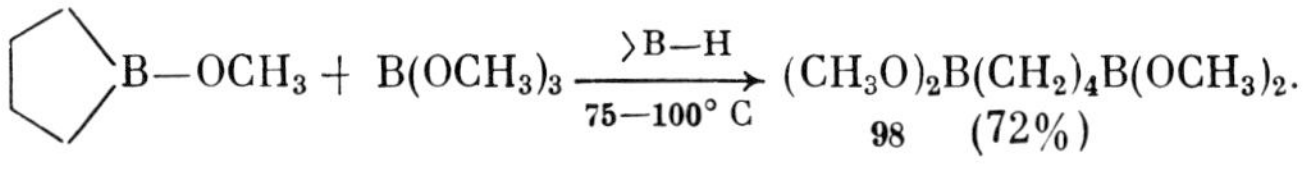

Transesterification of the ester *98* with hexanol forms tetrahexyl 1,4-butanediboronate which, at 300°C, disproportionates to 1-hexyloxyborolane and trihexyl borate:

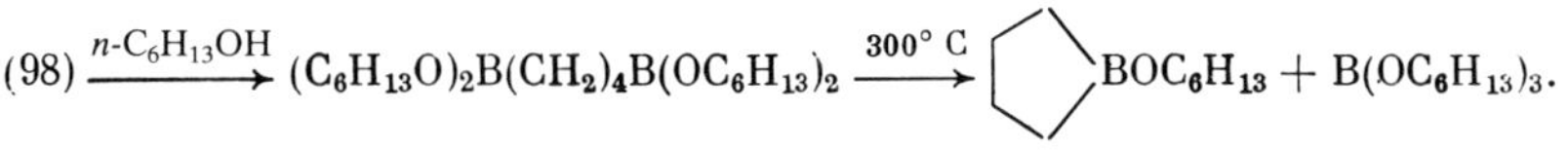

The action of an excess of methyl borate on a mixture of di(1-borolyl)butanes forms a mixture of tetramethyl esters of isomeric butanediboronic acids [33].

A mixture of isomeric di(boracyclyl)hexanes enters into an exchange reaction with methyl borate in the presence of compounds with B—H bonds and is converted to a mixture of 1-methoxyboracyclanes [48], containing 71% of 1-methoxyborepane and 28% of 1-methoxy-2-methylborinane [53]. 1,6-Bis(9-borabicyclo[3.3.1]non-9-yl)hexane (*99*) is converted by treatment with trimethyl borate in the presence of compounds with B—H bonds to 1-methoxyborepane [53]:

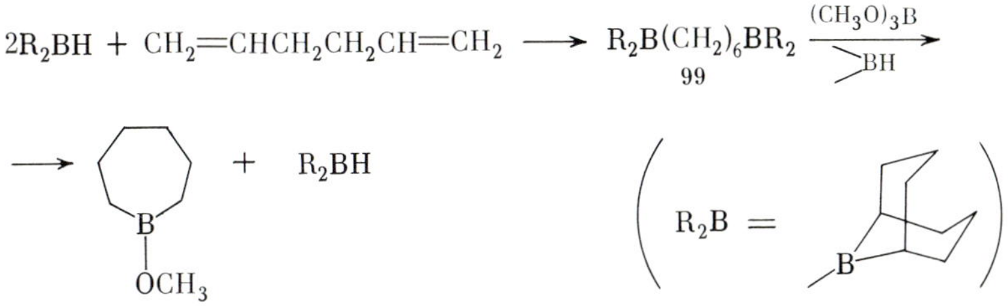

In the presence of compounds with B—H bonds, thioborates also readily enter into exchange reactions with di(boracycloalkyl)alkanes and boracyclanes. Thus, a 1:1 mixture of di(1-borolyl)butanes and ethyl thioborate forms a high yield of 1-ethylmercaptoborolane, which is converted by the action of thioborate to tetraethyl 1,4-butanedithioboronate (*100*) [87]:

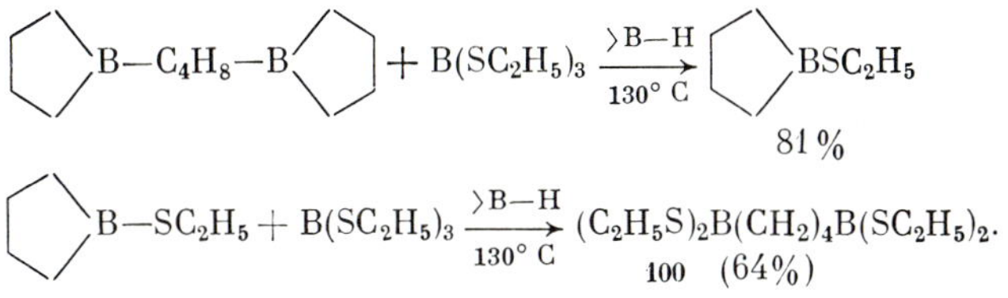

The ester *100*, which can also be obtained by the exchange reaction of tetramethyl 1,4-butanediboronate with ethyl thioborate, disproportionates at 200°C to 1-ethylmercaptoborolane and ethyl thioborate [88].

The mixture of di(1-borolyl)butanes, which was assumed to be pure 1,4-di(1-borolyl)butane by Köster [89], is converted with BCl_3 at 130°C, in the presence of compounds with B—H bonds, to

1-chloroborolane:

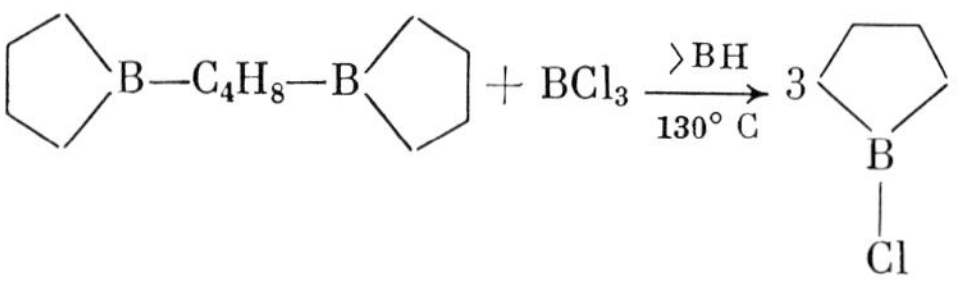

It was found that, in the presence of sodium borohydride, which is a source of compounds with B—H bonds, this reaction already proceeds smoothly at room temperature. In addition to 1-chloroborolane, other cyclic products of the composition $C_8H_{16}B_2Cl_2$, constituting a higher fraction, are formed in this process [87].

1-Chloroborolane reacts with excess BCl_3 at room temperature to form 1,4-bis(dichloroboryl)butane [87]:

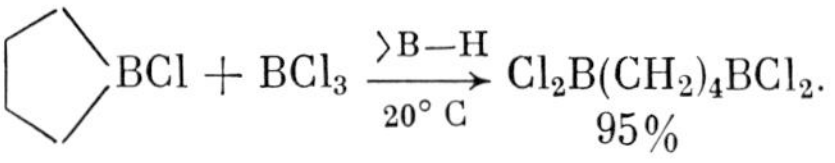

The reaction of a mixture of isomeric di(1-borolyl)butanes with excess BCl_3 forms a mixture of isomeric bis(dichloroboryl)butanes, *101–103* [87], rather than pure 1,4-bis(dichloroboryl)butane (*101*), as was assumed by Zakharkin and Kovredov [90, 91].

$Cl_2B(CH_2)_4BCl_2$ (**101**); $Cl_2BCH(CH_3)CH_2CH_2BCl_2$ (**102**); $Cl_2BCH(C_2H_5)CH_2BCl_2$ (**103**)

The investigation of the reaction of a mixture of isomeric di(1-borolyl)butanes and 1-chloroborolane with BCl_3 [87] was repeated in [36].

The reaction of equimolar quantities of the hydroboration products of isoprene (diene : $BH_3 = 3:2$) with BCl_3 smoothly forms 1-chloro-3-methylborolane [52, 89], which reacts with excess BCl_3 in the presence of tetrapropyldiborane to form 1,4-bis(dichloroboryl)-2-methylbutane [52].

The hydroboration products of isoprene react with excess BCl_3 to form a mixture of bis(dichloroboryl)alkanes, rather than pure 1,4-bis(dichloroboryl)-2-methylbutane, as was erroneously assumed in [44].

Also erroneous are the data on the preparation of 1-chloroborepane and 1,6-bis(dichloroboryl)hexane in the pure state by the reaction of the hydroboration products of 1,5-hexadiene with BCl_3 in a ratio diene:BCl_3 = 1:1 or 1:5 [47, 91]. A mixture of isomeric 1-chloroboracyclanes or bis(dichloroboryl)alkanes is actually formed in this process [52].

The following chlorides of boracyclanes have been synthesized by the replacement of alkyl groups by chlorine with the use of BCl_3 [92]:

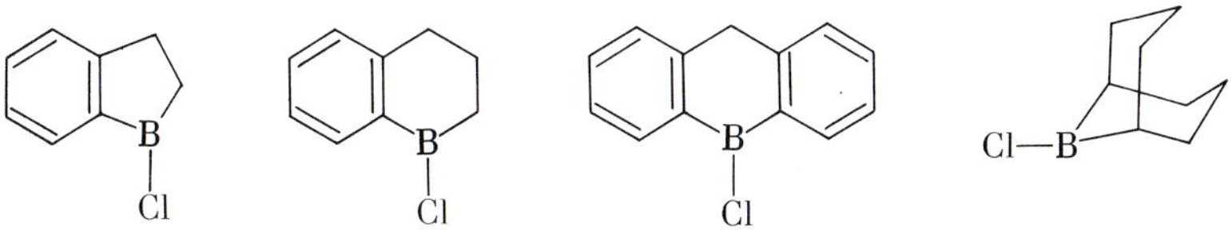

The chlorine atoms in *B*-chloroborolanes are replaced by hydrogen by the action of $LiAlH_4$. By this method, 1-chloro-3-methylborolane [89] and 1-chloroborolane [36] have been converted, respectively, to dimers of 3-methylborolane and borolane, to which, in the light of the data of [39], there should be assigned the structures of 1,2-bis(2-methyltetramethylene)diborane and 1,2-bis(tetramethylene)diborane, respectively.

The deutero analog of 1,2-bis(2-methyltetramethylene)diborane has been obtained from 1-chloro-3-methylborolane and $LiAlD_4$ [61].

The alkyl groups of 1-alkylborinanes are replaced by hydrogen under the action of diborane [2, 93]:

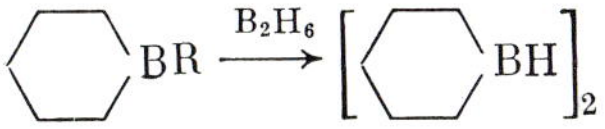

The reactions of α,ω-(diboracycloalkyl)alkanes with diborane proceed in an analogous manner. For example, 1,6-bis(9-borabicyclo[3.3.1]non-g-ylhexane is converted to di(borepane) [53].

1-Butylborolane reacts with diborane with ring opening, being converted to 1-butyltetramethylenediborane [93]:

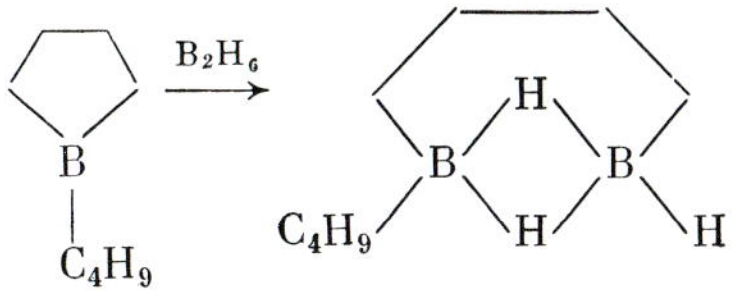

2.4 REACTIONS WITH METAL HYDRIDES AND ORGANOMETALLIC COMPOUNDS

1,2-Bis(tetramethylene)diborane, which reacts with amines with the rupture of both three-center B—H—B bonds [39], reacts with potassium hydride in tetrahydrofuran to form an anion with a transannular B—H—B bridge bond (*104*) [94]:

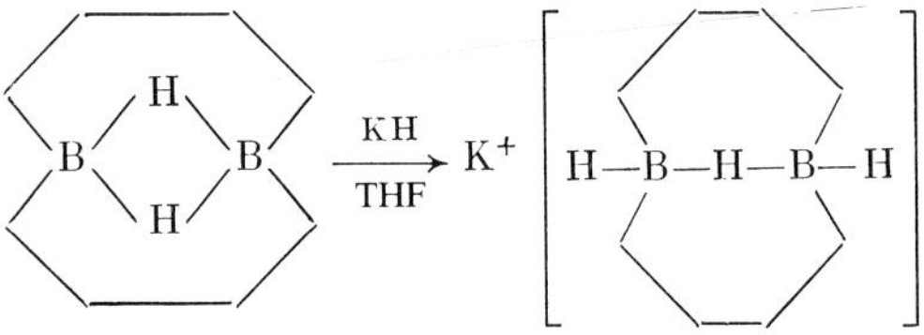

104

The compound *104* enters into an exchange reaction with tetrabutylammonium iodide to form the ammonium salt *105*, which is stable at room temperature in the absence of moisture:

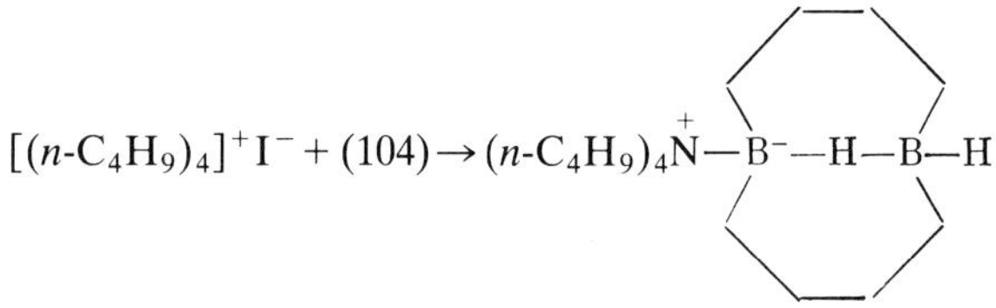

105

X-ray analysis has shown that, in the compound *105*, the length of the C—C bonds varies from 1.507 to 1.527 Å; the anion is in a "chair-chair" conformation, with an angle of approximately 120° between the fused planes of the rings.

According to the data of [95], bis(3-methylborolane) adds sodium hydride in heptane at 100°C to form the complex compound *106*:

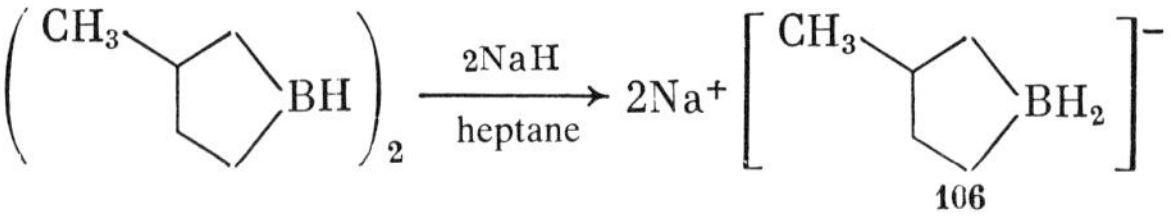

106

The action of lithium hydride on the complex of *B*-chloroborolane with tetrahydrofuran forms lithium tetramethyleneborohydride (*107*) [96]:

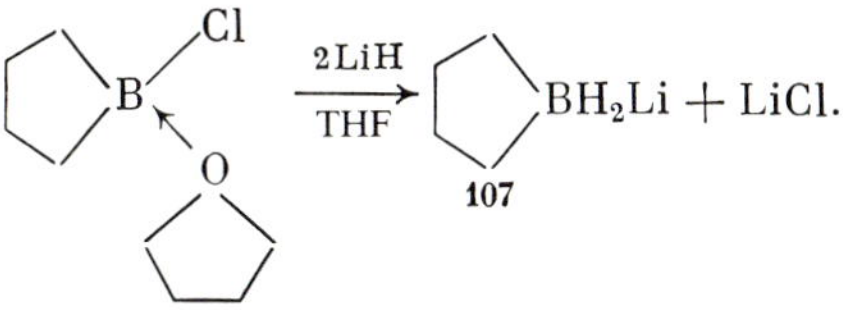

Treatment of the compound *107* with methanesulfonic acid in the presence of 1-octene leads to *B*-(*n*-octyl)borolane. It is assumed that the initial product is the unstable diborolane *108*, which is capable of rapidly adding to the olefin:

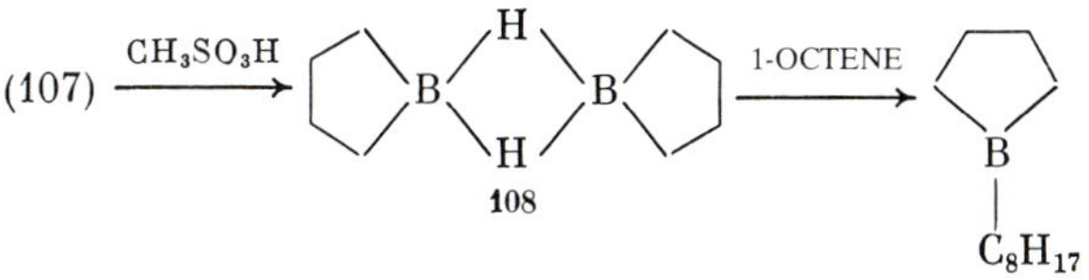

Perhydro-9b-boraphenalene reacts with lithium hydride in boiling tetrahydrofuran to form lithium perhydro-9b-boraphenalylhydride (*109*) [97]:

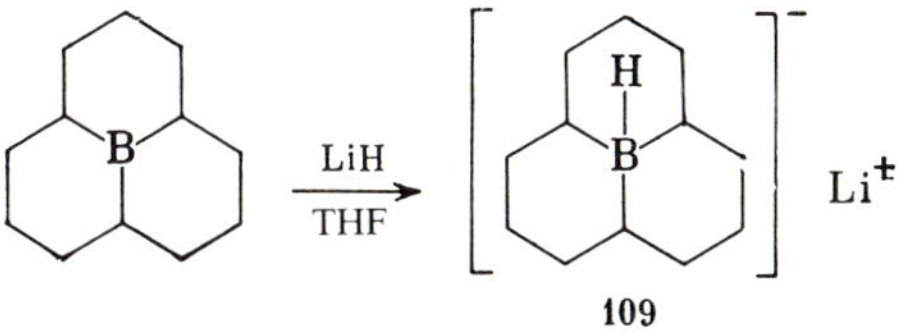

The compound *109* is an active reducing agent, which is stereoselective with respect to cyclic ketones.

Both isomers of perhydro-9b-boraphenalene add acetylene compounds of alkali metals to form complex salts (*110*) [95]:

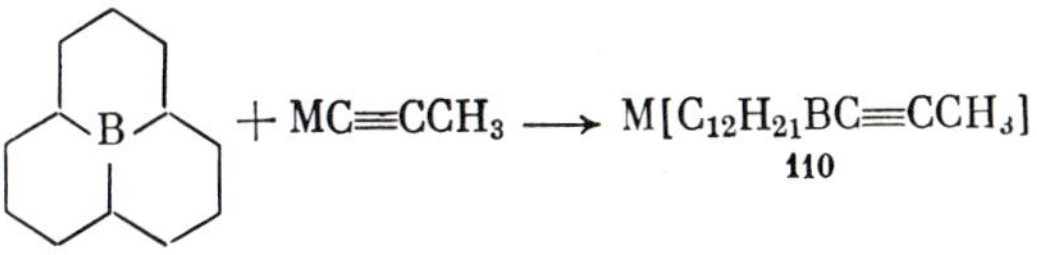

Bis(boracycloalkyl)alkanes enter into exchange reactions with aluminium trialkyls [32, 98]. With a reagent ratio of 1:1, 1-alkylboracyclanes are formed, but the use of excess AlR_3 leads to the formation of 1-alkylalumocyclanes:

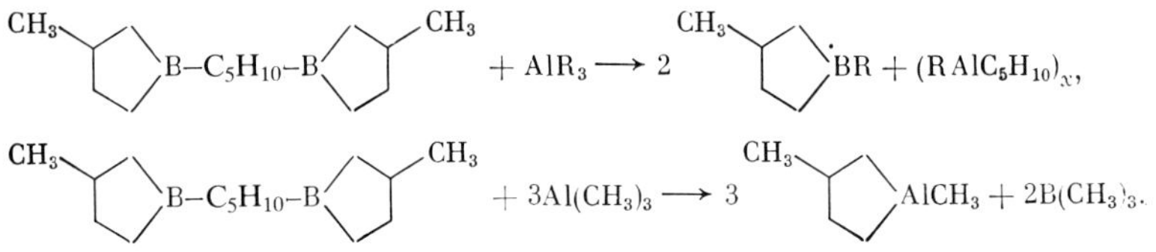

1-Alkylboraindans and 1-alkylboratetralins are converted by the action of aluminum trialkyls to 1-alkylalumoindans and 1-alkyl-alumotetralins, respectively [98]:

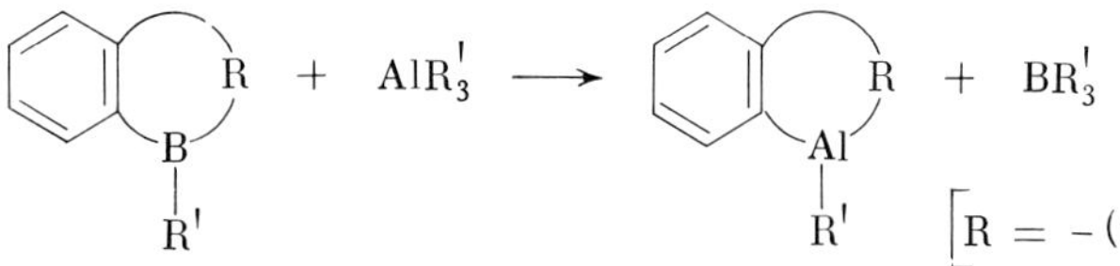

$[R = -(CH_2)_2-, -(CH_2)_3-]$

1-Phenyl-2,3;5,6-dibenzoborine is metalated in the 4 position by the action of *tert*-butyllithium in tetrahydrofuran [99, 100].

The metalation of 1-mesityl-2,3;5,6-dibenzoborine by *tert*-butyllithium, followed by deuterolysis with the use of D_2O or by treatment with CO_2, leads to the compounds (*111*) and (*112*), respectively [101]:

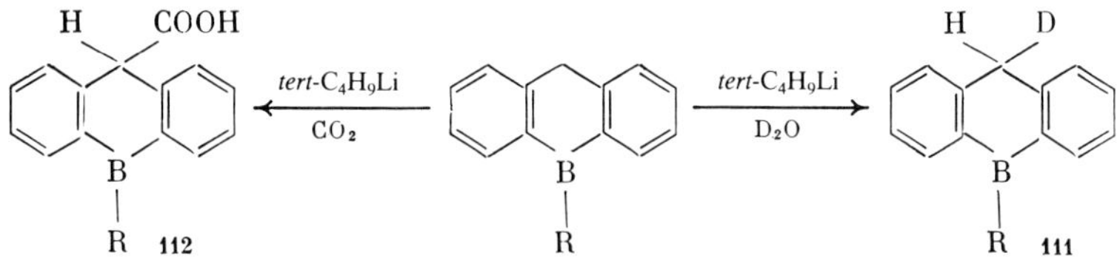

Deuterium has been introduced into 1-mesityl-4-phenyl-2,3;6,7-dibenzoborine by this method [102].

9-Methyl-9-borabicyclo[3.3.1]nonane is metalated by 2,2,6,6,-tetramethylpiperidyllithium in benzene [103].

2.5 CARBONYLATION. SYNTHESIS OF ALICYCLIC ALCOHOLS AND KETONES

Like the trialkylboranes [104], boracyclanes react with carbon monoxide to form, depending on the conditions, tertiary cyclic alcohols or ketones.

The carbonylation-oxidation of 1-alkylborolanes and 1-alkylborinanes in an autoclave (70 atmospheres) at 150°C in the presence of ethylene glycol leads to cyclic tertiary alcohols in 90% yields [105]:

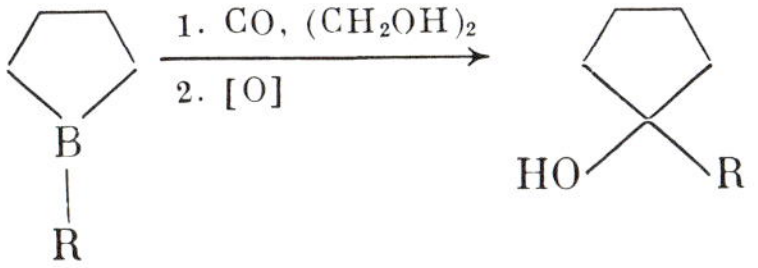

8-Borahydrindan, obtained by the hydroboration-isomerization of 1-octene [20, 21] or by the hydroboration of 1,3,7-octatriene, is converted to *cis*-8-hydrindanol (*113*) (33%) by carbonylation-oxidation. The structure of the latter compound has been confirmed by reverse synthesis from 8(9)-hexahydroindene [106]:

$$n\text{-}C_6H_{13}CH{=}CH_2 \xrightarrow{B_2H_6} (n\text{-}C_8H_{17})_3B$$

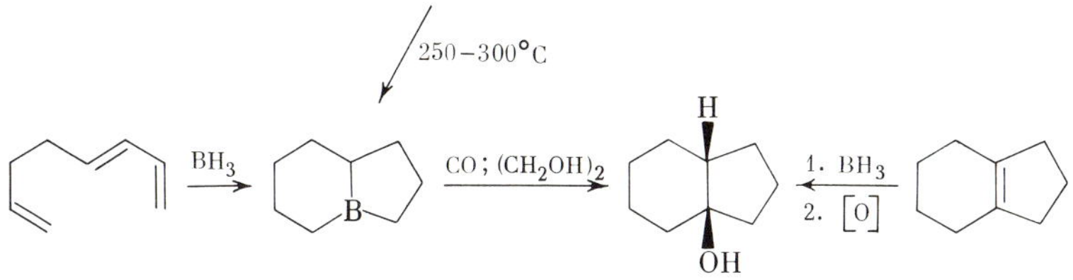

113

The pyrolysis of trinonylborane [20, 21] or hydroboration of 1,4,8-nonatriene leads to 9-boradecalin, from which 9-hydroxydecalin (*114*) is obtained in the form of a mixture of the cis and trans isomers in a ratio of 4:1. *cis*-9-Hydroxy-10-methyldecalin (*115*) has been synthesized by an analogous method [106]:

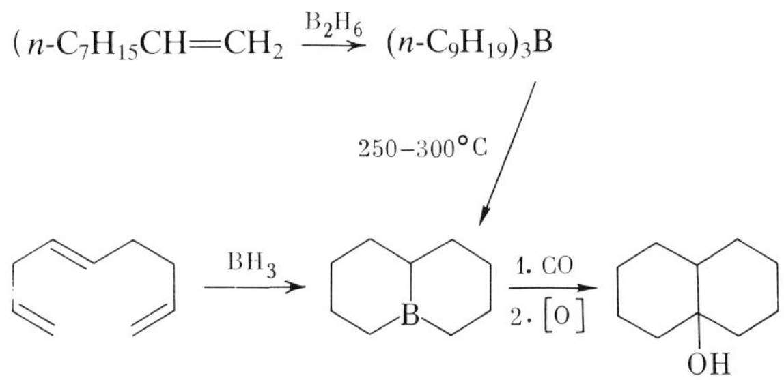

52% 114 (89%)

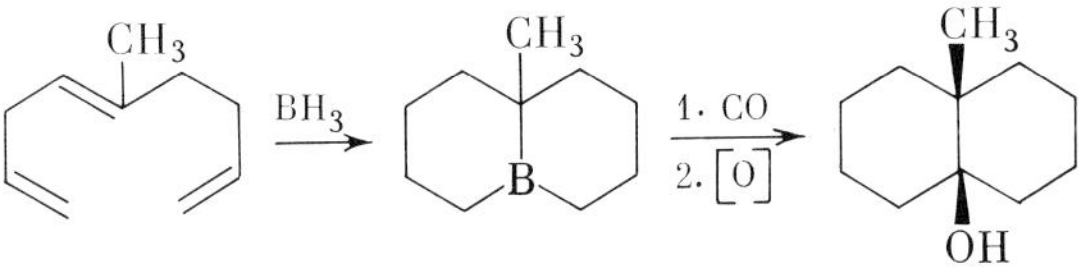

115

A number of 9-hydroxy-9-alkylbicyclo[3.3.1]nonanes (*116*), containing 5% of impurities, have been obtained by the carbonylation of 9-alkyl-9-borabicyclo[3.3.1]nonanes in an autoclave, followed by oxidation of the intermediate organoboranes [107, 108]:

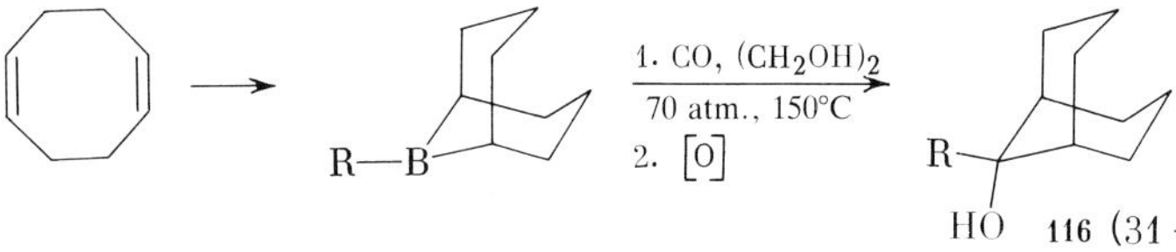

116 (31 - 97%)

Perhydro-9b-boraphenalene is carbonylated to perhydro-9b-phenalenol [76, 109]:

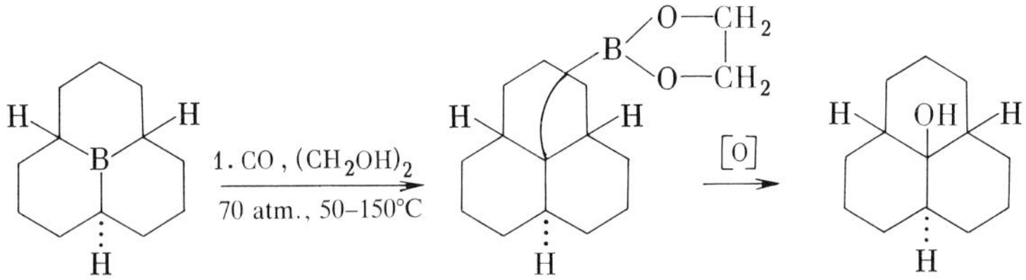

The carbonylation of 1-alkylboracyclanes in the presence of water forms cyclic ketones. Satisfactory yields of cyclanones are obtained only from 1-(2,3-dimethyl-2-butyl)boracyclanes, which are formed by the hydroboration of unsaturated hydrocarbons with the use of bis(2,3-dimethyl-2-butyl)diborane. The reaction is carried out with heating and at elevated pressure (usually at 75 atmospheres of CO and 50°C).

1-(2,3-Dimethyl-2-butyl)borolane is converted by carbonylation-oxidation to 3-methylcyclopentanone in a yield of 86% [110]:

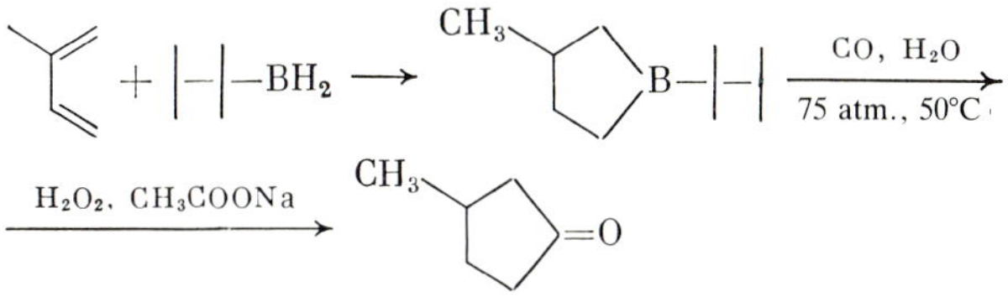

At the same time, as a result of carbonylation-oxidation of the mixture of hydroboration products of isoprene, the ketone can be isolated only in a yield of 46%.

The following conversions have been carried out by the same method:

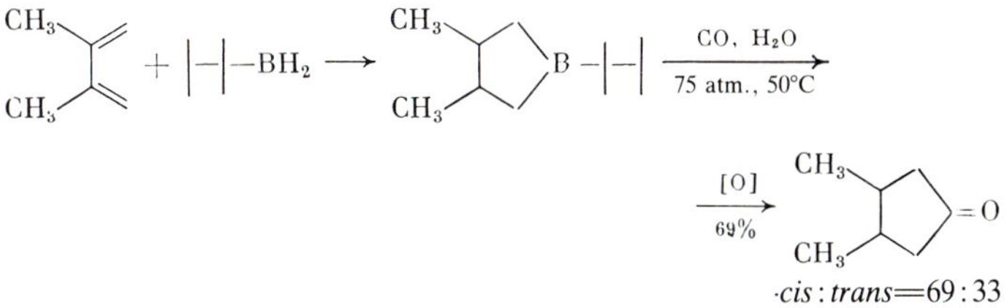

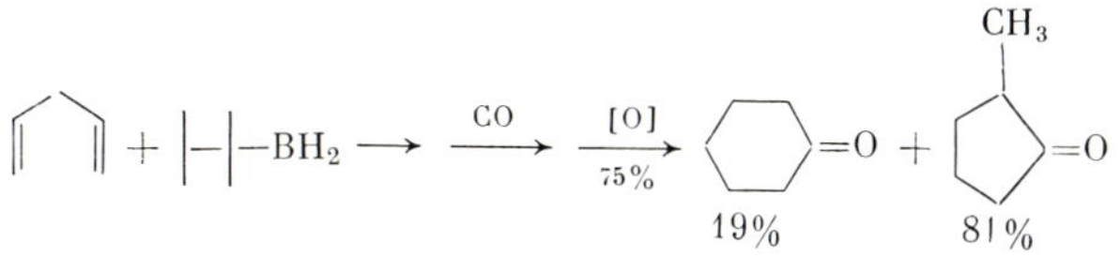

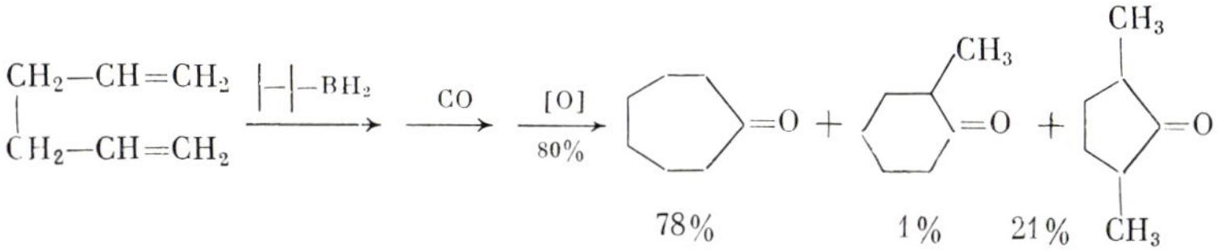

trans-Perhydro-1-indanone (*117*) [108], which is the thermodynamically least stable isomer [111], has been synthesized by the sequential hydroboration-carbonylation-oxidation of 1-vinylcyclohexene:

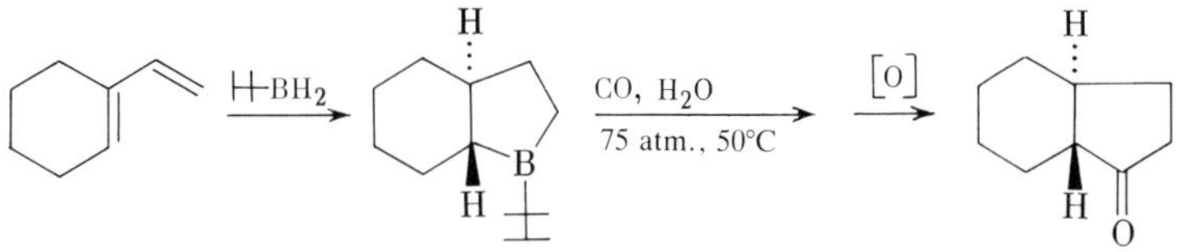

117 (60%)

trans-Perhydro-1-azulenone (*118*), *trans*-perhydro-4-indanone (*119*), *trans*-1-decalone (*120*), and *trans*-bicyclo[5.4.0]undecan-6-one (*121*) have been synthesized by an analogous method from the corresponding vinyl- or allylcyclenes [112].

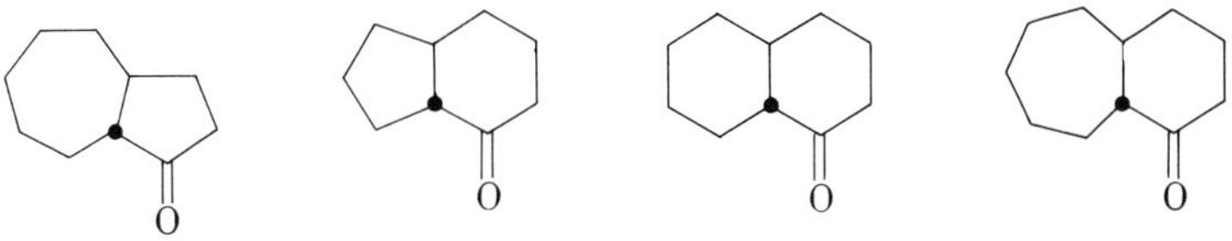

118 119 120 121

The carbonylation of *B*-alkyl-9-borabicyclo[3.3.1]nonanes in the presence of LiAl $(OCH_3)_3H$, followed by reduction of the intermediates formed with $LiAlH_4$, leads to a lengthening of the side chain by one carbon atom [139].

The bicyclic ketones formed by the above method do not contain any cis isomers. This stereospecificity is achieved in the case where the oxidation of the carbonylation products is carried out in sodium acetate solution. In the case of oxidation in alkaline medium, epimerization of the ketones takes place.

The synthesis of *trans*-1-hydrindane and of *cis,anti,trans*-9-ketoperhydrophenanthrene, labeled at the C-1 and C-9, respectively, has been carried out by the reaction of the corresponding boracyclanes with $K^{13}CN$ or $K^{14}CN$ [140].

2.6 BROMINATION. SYNTHESIS OF ALICYCLIC DIBROMOALKANES, BROMOALKANOLS, ALCOHOLS, GLYCOLS, AND KETONES

The reaction of 1-substituted boracyclanes with bromine, which proceeds by a radical mechanism (Chapter 8), has been used for the synthesis of secondary alkyl halides; substituted butanols containing halogen, an alkoxy or amino group in the 4 position; and also for the preparation of some carbocylic compounds.

According to the data of [18], the bromination of 1-hexyl-2-methylborinane in acetic anhydride proceeds with the cleavage of all boron-carbon bonds and leads to hexyl bromide, 1,5-dibromohexane, and boron acetate:

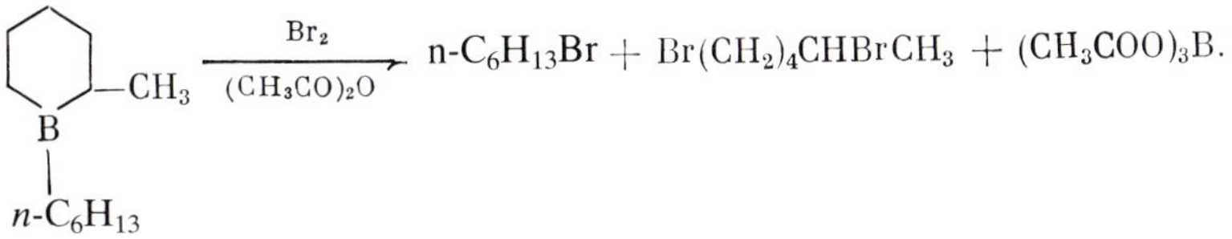

1-Butylborolane reacts with bromine in benzene solution at 5°C and, after methanolysis, is converted to methyl butyl-δ-bromobutylborinate (*122*) in 60% yield:

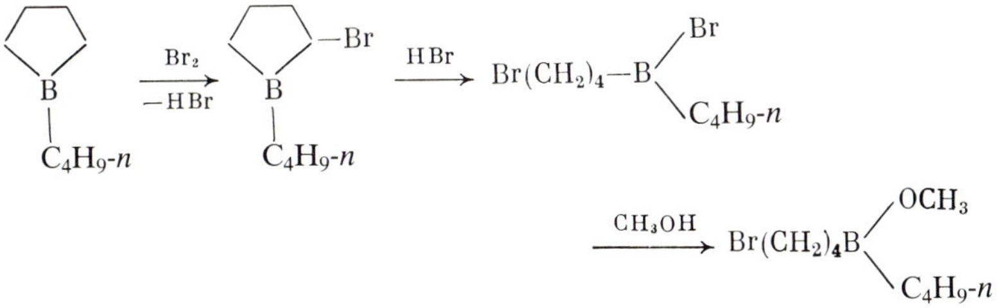

122

The reaction proceeds more smoothly in the case of the bromination of 1-butylborolane pyridinate or in the action of pyridine dibromide on 1-butylborolane. In these cases, a 93% yield of the ester *122* is obtained [113, 114]. Methyl hexyl-δ-bromobutylborinate is obtained from 1-hexylborolane in 81% yield by this method.

The esters of alkyl-δ-bromobutylborinic acids are used in the synthesis of 1,2-azaborinanes [115, 116] and 1,2-oxaborinanes [117, 118, 141].

δ-Bromo-, δ-amino-, and δ-alkoxy derivatives of 1-butanol can be obtained in good yields from esters of the type of *122*.

The bromination of 1-substituted borinanes has been used for the synthesis of five-membered carbocyclic systems. Thus, as a result of the treatment of 1-methoxyborinane with two moles of bromine in the presence of water under ordinary laboratory illumination and subsequent oxidation, cyclopentanone has been obtained in a yield of 48% [119]:

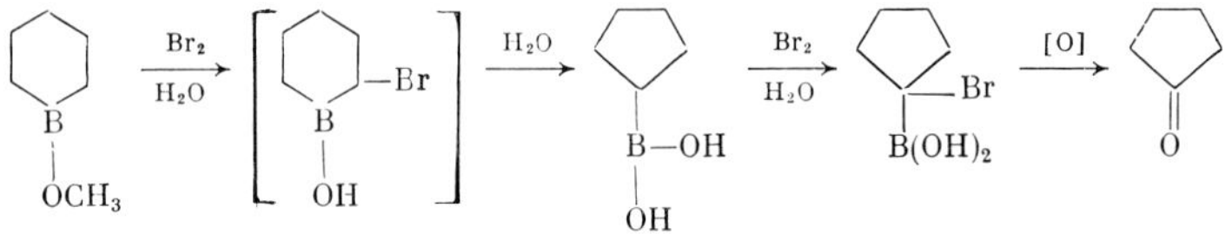

At the same time, it was not found possible to obtain cyclobutane compounds from boracyclopentanes, which is probably a result of the high strain of the cyclobutane compounds.

The bromination-oxidation of 1-(2,3-dimethyl-2-butyl)-2-methylborolane led to 2,3,3,4-tetramethyl-4,7-heptanediol:

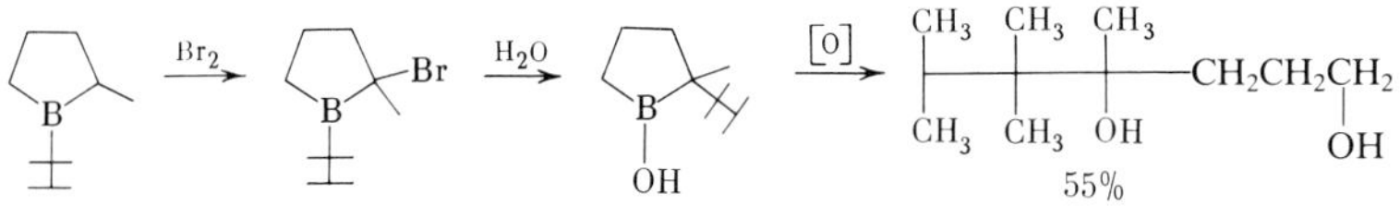

The bromination-oxidation of the bicyclic borane *123* forms *cis*-8-hydrindanol (*124*) in 60% yield [119]:

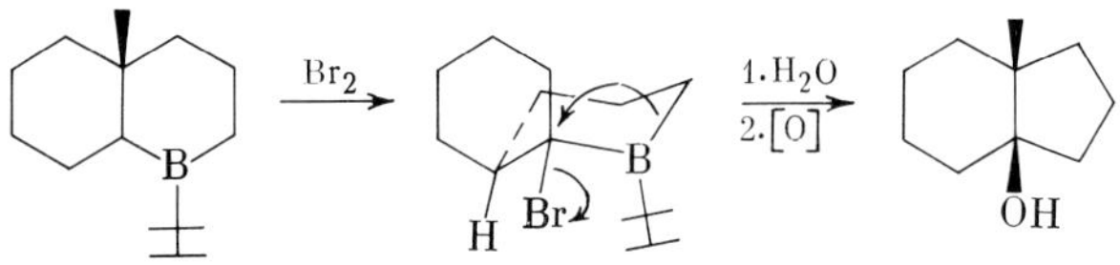

123 124

The treatment of 9-boradecalin with bromine under light in the presence of water, followed by oxidation, leads to 1-(4-hydroxybutyl)cyclopentanol [120]:

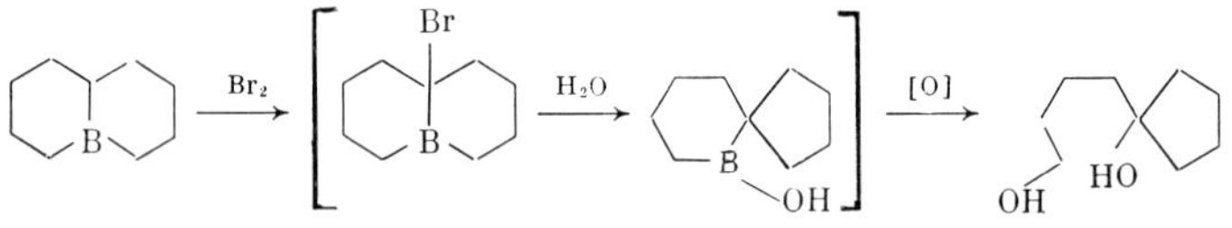

Bicyclo[7.3.0]dodecane-1,5-diol has been synthesized in 70% yield by the α-bromination-oxidation of *cis,cis,trans*-perhydro-9b-boraphenalene:

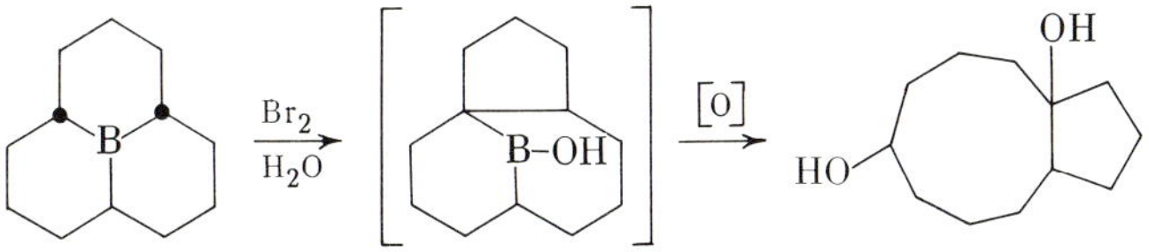

9-Borabicyclo[3.3.1]nonane is converted to *B*-bromo-9-borabicyclo[3.3.1]nonane by the action of bromine, HBr, or BBr_3 [142].

cis-Bicyclo[3.3.0]octan-1-ol (*126*, 55%) and 9-oxabicyclo-[3.3.1]nonane (*127*, 22%) have been obtained by the bromination-oxidation of 9-methoxy-9-borabicyclo[3.3.1]nonane (*125*) [120]:

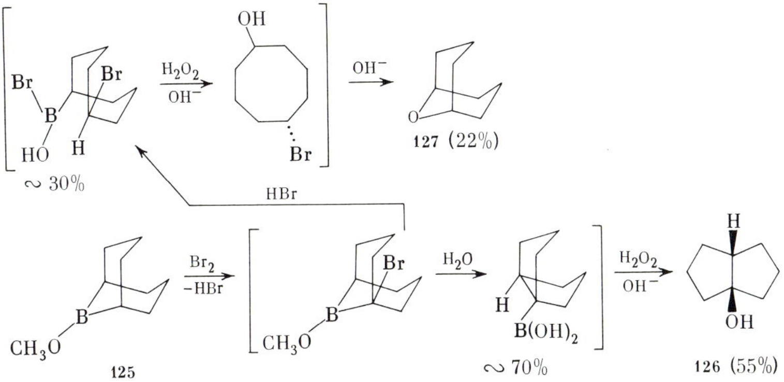

The monobromination of 9-isopropyl-9-borabicyclo[3.3.1]nonane, with simultaneous removal of HBr formed under vacuum, leads to α-bromoisopropylborane (*128*), which, under the action of water, forms 10-hydroxy-9,9-dimethyl-10-borabicyclo[3.3.2]-decane (*129*); the latter compound is oxidized, without isolation, to *cis*-5-(2-hydroxy-2-propyl)cyclooctanol (*130*) [121]:

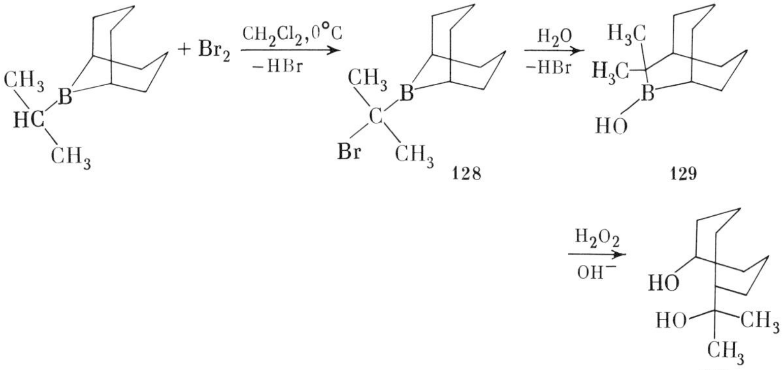

128 129 130

According to [143], the compound *129* is formed by the action of bromine on *B*-isopropyl-9-borabicyclo[3.3.1]nonane in the presence of water under light, while hydroxy(bicyclo[3.3.0]oct-1-yl)isopropylborane is formed in the dark.

The bromination of 2-isopropyl-2-boraadamantane in CH_2Cl_2 proceeds simultaneously by a radical and electrophilic mechanism, leading to 2-(2-bromo-2-propyl)-2-boraadamantane (*131*) and 2-bromo-2-boraadamantane (*132*), the ratio of which depends on the degree of illumination [144]. 2-Isopropyl-2-boraadamantane in CH_2Cl_2 in the presence of water is also brominated by an ionic mechanism, with the participation of solvated Br^+, forming hydroxy(3-noradamantyl)isopropylborane (*133*) [144, 144a]:

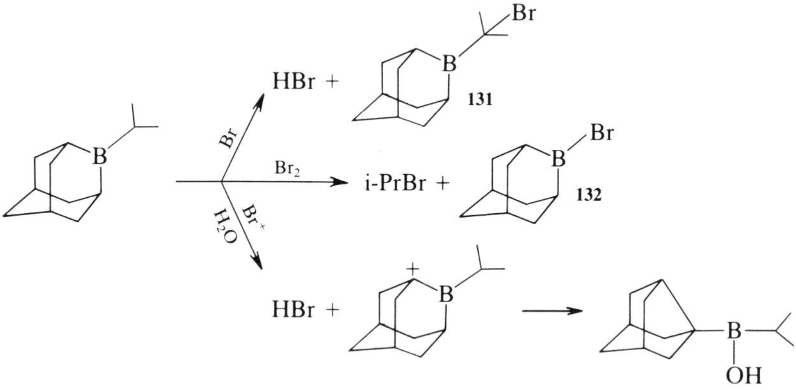

131 132 133

In arylboranes, under the action of bromine in aqueous solution, a rupture of boron-carbon bonds takes place and aryl bromides are obtained (Chapter 8). For example, 2,2-dibromodibenzyl has been obtained by the bromination of 1-(2-aminoethoxy)-2,3;6,7-dibenzodihydroborepin [9]:

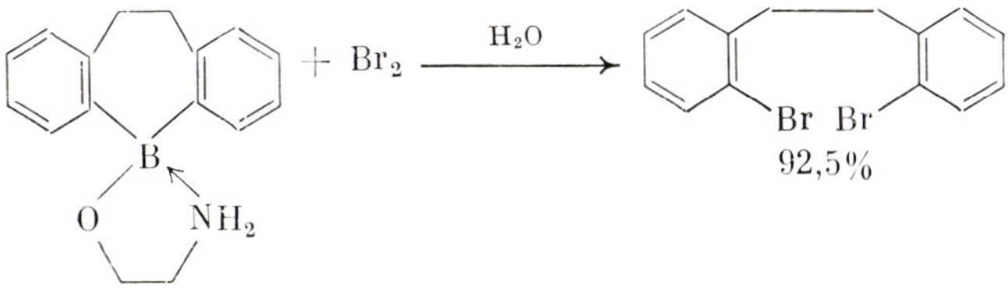

2.7 REACTIONS WITH CARBONYL COMPOUNDS. SYNTHESIS OF KETONES AND ALCOHOLS

1-Alkylborinanes and 1-alkyl-3,5-dimethylborinanes react with methyl vinyl ketone to form 4-substituted derivatives of 2-butanone:

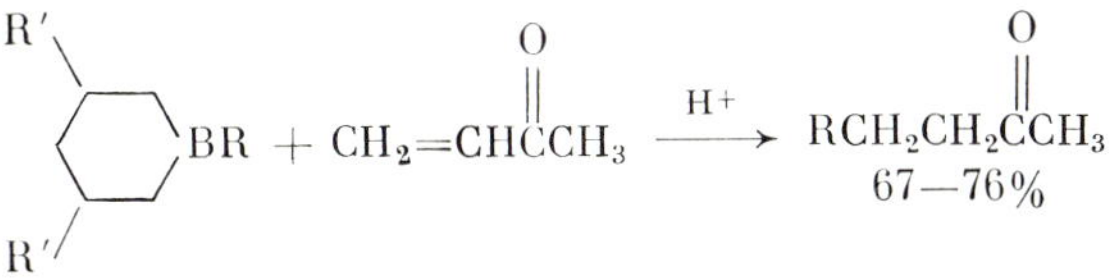

(R = *sec*-C_4H_9, *tert*-C_4H_9, n-C_6H_{13}; R′ = H, CH_3).

l-Alkyl-3,6-dimethylborepanes react analogously with methyl vinyl ketone [55].

B-Alkyl-9-borabicyclo[3.3.1]nonanes react with benzaldehyde under mild conditions to form olefins and benzyl alcohol [145–147]. When *B*-(cycloalkylmethyl)-9-borabicyclo[3.3.1] nonanes are used in this reaction, methylenecycloalkanes are formed [148, 149].

9-Borabicyclo[3.3.1]nonane and its derivatives reduce aldehydes and ketones to alcohols[150–154].

2.8 MISCELLANEOUS REACTIONS

B-Bromo-9-borabicyclo[3.3.1]nonane is capable of cleaving ethers [155]. The adduct of 9-borabicyclo[3.3.1]nonane with NH_3 is converted in boiling benzene to the dimer of *B*-amino-9-borabicyclo[3.3.1]nonane [156].

9-Borabicyclo[3.3.1]nonane and its *B*-substituted derivatives form complex compounds with amines [157–159].

The alkylation of *α*-bromo ketones with *B*-alkyl-9-borabicyclo[3.3.1]nonane has been described [160].

The "ate" complex of *B*-butyl-9-borabicyclo[3.3.1]nonane with butyllithium (*131*) is an active reducing agent for *tert*-alkyl, benzyl, and allyl halides and also for ketones [161–163]. The reaction is accompanied by rearrangement of the bicyclo[3.3.1]nonane system to a bicyclo[3.3.0]octane system (*132*):

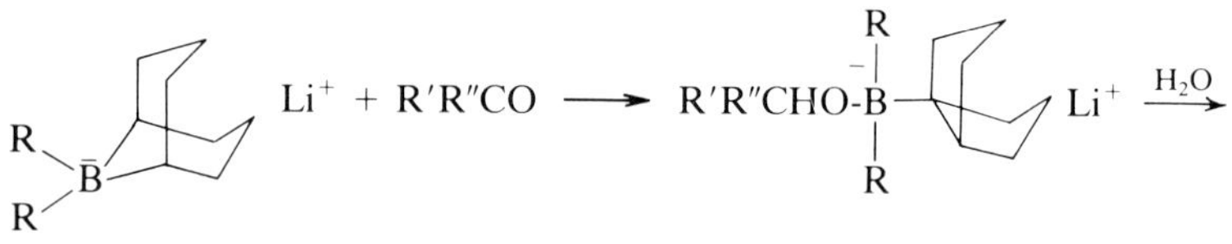

131

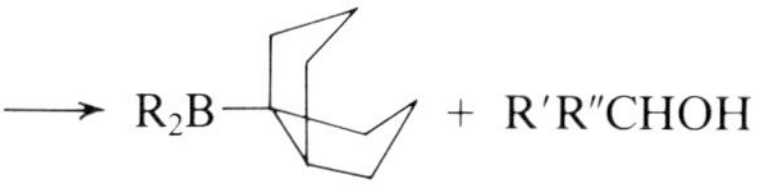

132

The "ate" complexes *131* form bicyclo[3.3.0]oct-l-yl(dialkyl)-boranes under the action of acetyl chloride [138, 164, 165].

Lithium 2,2-dimethyl-2-boraadamantanate (*133*) undergoes intramolecular rearrangement under the action of acetyl chloride to form 3-noradamantyl(dimethyl)borane (*134*), the oxidation of which forms 3-noradamantanol (*135*) [166]:

133 **134** **135**

Perhydro-98-Boraphenalene has been converted to 13-azabicyclo-[7.3.1]tridecan-5-ol and perhydroazaphenalene under the action of *N*-chloro-*O*-2,4-dinitrophenylhydroxylamine [167].

Some dialkylboranes (dicyclohexylborane, disiamylborane, 9-borabicyclo[3.3.1]nonane) react with R′Li to form a mixture of $R_2BR'_2Li$ and R_2BH_2Li [168].

B-Methyl-9-borabicyclo[3.3.1]nonane reacts with *t*-butyllithium to form lithium *B*-methyl-9-borabicyclo[3.3.1]nonylborate and isobutylene, while methyllithium reacts with *B*-*t*-butyl-9-borabicyclo[3.3.1]nonane to form the corresponding "ate" complex [169].

REFERENCES

1. G. Benedikt, R. Köster, W. Larbig, K. Reinert and G. Rotermund, *Angew. Chem.* **75**, 1079 (1963).
2. R. Köster, *Advances Organometal. Chem.* **2**, 257 (1964).
3. R. Köster, In: Progress in Boron Chemistry, vol. 1. Ed. H. Steinberg and A. L. McCloskey. Oxford, Pergamon Press, 1964, p. 289.
4. H. C. Brown and E. Negishi, *Pure and Appl. Chem.* **29**, 257 (1972).
5. K. Torssell, *Acta Chem. Scand.* **8**, 1779 (1954).
6. J. Braun, *Compt. Rend.* **256**, 2422 (1963).
7. S. L. Clark and J. B. Jones, *U.S. Patent* 2,858,339 (1958); *C. A.* **53**, 7198 (1959).
8. S. L. Clark and J. B. Jones, *U.S. Patent* 2,858,340 (1958); *C. A.* **53**, 7206 (1959).
9. R. L. Letsinger and I. H. Skoog, *J. Am. Chem. Soc.* **77**, 5176 (1955).
10. E. E. Van Tamelen, G. Brieger and K. G. Untch, *Tetrahedron Letters* **1960**, No. 8, 14.
11. G. Wittig and W. Herwig, *Chem. Ber.* **88**, 962 (1955).
12. K. Torssell, *Acta Chem. Scand.* **16**, 87 (1962).
13. R. Van Veen and F. Bickelhaupt, *J. Organometal. Chem.* **43**, 241 (1972).
14. F. Dallacker and G. Adolphen, *Ann.* **694**, 110 (1966).
15. P. Jutzi, *J. Organometal. Chem.* **19**, P1 (1969).
16. R. Köster and G. Rotermund, *Angew. Chem.* **72**, 138 (1960).
17. T. J. Logan and T. J. Flautt, *J. Am. Chem. Soc.* **82**, 3446 (1960).
18. P. F. Winternitz and A. A. Carotti, *J. Am. Chem. Soc.* **82**, 2430 (1960).
19. G. Schomburg, R. Köster and D. Hennenberg, *Z. Anal. Chem.* **170**, 283 (1959).
20. R. Köster, W. Larbig and G. W. Rotermund, *Ann.* **682**, 21 (1963).
21. R. Köster and G. Rotermund, *Angew. Chem.* **72**, 563 (1960).
22. R. Köster and G. Rotermund, *Angew. Chem.* **74**, 252 (1962).
23. R. Köster, *Angew. Chem.* **72**, 626 (1960).
24. R. Köster and K. Reinert, *Angew. Chem.* **71**, 521 (1959).
25. R. Köster, G. Benedikt, W. Fenzl and K. Reinert, *Ann.* **702**, 197 (1967).
26. R. Köster and G. Benedikt, *Angew. Chem.* **75**, 419 (1963).
27. R. Köster and H.-G. Willemsen, *Ann.* **1974**, 1843.
28. B. M. Mikhailov and E. A. Shagova, *Izv. Akad. Nauk SSSR, Ser. Khim.* **1972**, 1222.
29. B. M. Mikhailov and E. A. Shagova, *Zh. Obshch. Khim.* **45**, 1052 (1975).
30. B. M. Mikhailov and E. A. Shagova, *Izv. Akad. Nauk SSSR, Ser. Khim.* **1972**, 1450.
31. R. Köster, *Angew. Chem.* **70**, 413 (1958).
32. R. Köster, *Angew. Chem.* **71**, 520 (1959).
33. B. M. Mikhailov, A. Ya. Bezmenov, L. S. Vasil'ev and V. G. Kiselev, *Dokl. Akad. Nauk SSSR* **155**, 141 (1964).

34. B. M. Mikhailov, A. Ya. Bezmenov and L. S. Vasil'ev, *Dokl. Akad. Nauk SSSR* **167,** 590 (1966).
35. G. Zweifel, K. Nagaze and H. C. Brown, *J. Am. Chem. Soc.* **84,** 183 (1962).
36. L. I. Zakharkin and A. I. Kovredov, *Zh. Obshch. Khim.* **26,** 2153 (1966).
37. H. C. Brown, E. Negishi and S. K. Gupta, *J. Am. Chem. Soc.* **92,** 2460 (1970).
38. H. G. Weiss, W. J. Lehman and I. Shapiro, *J. Am. Chem. Soc.* **84,** 3840 (1962).
39. D. E. Young and S. G. Shore, *J. Am. Chem. Soc.* **91,** 3497 (1969).
40. H. C. Brown, E. Negishi and P. L. Burke, *J. Am. Chem. Soc.* **93,** 3400 (1971).
41. H. H. Linder and T. Onak, *J. Am. Chem. Soc.* **88,** 1886 (1966).
42. K. A. Saegebarth, *U.S. Patent* 3,008,997 (1961); *C. A.* **57,** 12534 (1962).
43. K. A. Saegebarth, *U.S. Patent* 3,060,244 (1962); *C. A.* **58,** 5573 (1963).
44. L. I. Zakharkin and A. I. Kovredov, *Izv. Akad. Nauk SSSR, Otd. Khim. Nauk* **1962,** 362.
45. A. Ya. Bezmenov, L. S. Vasil'ev and B. M. Mikahilov, *Izv. Akad. Nauk SSSR, Ser. Khim.* **1965,** 2111.
46. K. A. Saegebarth, *J. Am. Chem. Soc.* **82,** 2081 (1960).
47. L. I. Zakharkin and A. I. Kovredov, *Izv. Akad. Nauk SSSR, Otd. Khim. Nauk* **1962,** 357.
48. B. M. Mikhailov, E. N. Safonova and V. G. Kiselev, in the collection: *"Zh. Obshch. Khim. Problems of Organic Synthesis", Leningrad, Nauka* (1965), p. 213
49. H. C. Brown and E. Negishi, *J. Organometal. Chem.* **26,** C67 (1971).
50. E. Negishi, P. L. Burke and H. C. Brown, *J. Am. Chem. Soc.* **94,** 7431 (1972).
51. B. M. Mikhailov, L. S. Vasil'ev and E. N. Safonova, *Dokl. Akad. Nauk SSSR* **147,** 630 (1962).
52. L. I. Zakharkin and A. I. Kovredov, *Izv. Akad. Nauk SSSR, Ser. Khim.* **1969,** 106.
53. P. L. Burke, E. Negishi and H. C. Brown, *J. Am. Chem. Soc.* **95,** 3654 (1973).
54. H. C. Brown, E. Negishi and P. L. Burke, *J. Am. Chem. Soc.* **94,** 3561 (1972).
55. E. Negishi and H. C. Brown, *J. Am. Chem. Soc.* **95,** 6757 (1973).
56. M. F. Hawthorne, *J. Am. Chem. Soc.* **82,** 748 (1960).
57. M. F. Hawthorne, *J. Am. Chem. Soc.* **83,** 2541 (1961).
58. W. N. Greenwood and J. C. Wright, *J. Chem. Soc.* **1965,** 448.
59. R. Köster and G. Schomburg, *Angew Chem.* **72,** 567 (1960).
60. B. M. Mikhailov and V. A. Dorokhov, *Dokl. Akad. Nauk SSSR* **133,** 119 (1960).
61. R. Köster, G. Griaznov, W. Larbig and P. Binger, *Ann.* **672,** 1 (1964).
62. H. C. Brown and E. Negishi, *J. Organometal. Chem.* **28,** C1 (1971).
63. H. C. Brown and E. Negishi, *J. Am. Chem. Soc.* **94,** 3567 (1972).
64. T. A. Shchegoleva, E. M. Shashkova, V. G. Kiselev and B. M. Mikhailov, *Izv. Akad. Nauk SSSR, Ser. Khim.* **1964,** 365.
65. B. M. Mikhailov, T. A. Shchegoleva, E. M. Shashkova and V. D. Sheludyakov, *Izv. Akad. Nauk SSSR, Otd. Khim. Nauk* **1961,** 1163.
66. B. M. Mikhailov, T. A. Shchegoleva, V. D. Shceludyakov and A. N. Blokhina, *Izv. Akad. Nauk SSSR, Otd. Khim. Nauk* **1963,** 646.
67. B. M. Mikhailov and T. A. Shchegoleva, *Izv. Akad. Nauk SSSR, Otd. Khim. Nauk* **1961,** 1142.
68. T. A. Shchegoleva, E. M. Shashkova, V. G. Kiselev and B. M. Mikhailov, *Zh. Obshch. Khim.* **35,** 1078 (1965).
69. H. C. Brown and C. D. Pfaffenberger, *J. Am. Chem. Soc.* **89,** 5475 (1967).
70. R. Köster and G. Griaznov, *Angew. Chem. Soc.* **90,** 5280 (1968).
71. E. F. Knightsand H. C. Brown, *J. Am. Chem. Soc.* **90**, 5280 (1968).

72. B. M. Mikhailov, T. A. Shchegoleva and E. M. Shashkova, *Izv. Akad. Nauk SSSR, Ser. Khim.* **1975,** 980.
73. B. M. Mikhailov, T. A. Shchegoleva and E.M. Shashkova, *Izv. Akad. Nauk SSSR, Ser. Khim.* **1977,** 894.
74. R. Köster, *Angew. Chem.* **69,** 684 (1957).
75. G. W. Rotermund and R. Köster, *Ann.* **686,** 153 (1965).
76. H. C. Brown and W. C. Dickason, *J. Am. Chem. Soc.* **91,** 1226 (1969).
77. G. Rotermund and R. Köster, *Angew. Chem.* **74,** 319 (1962).
78. K. A. Saegebarth, *U.S. Patent* 3,014,075 (1961); *C. A.* **56,** 13095 (1962).
79. *British Patent* 948,662 **(1964);** *C. A.* **60,** 13280 (1964).
80. B. M. Mikhailov, L. S. Vasil'ev and A. Ya. Bezmenov, *U.S.S.R. Author's Certificate* 172,295 (1964); *Byull. Izobret.* **1965,** No. 13, 14.
81. I. Mehrotza and D. Devaprabhakara, *Tetrahedron Letters*, **1970,** 4493.
82. M. M. Bhagwat, I. Mehrotra and D. Devaprabhakara, *J. Organometal. Chem.* **82,** C27 (1974).
83. R. Bishop, *J. Chem. Soc. (Perkin I)* **1974,** 2364.
84. B. M. Mikhailov and L. S. Vasil'ev, *Izv. Akad. Nauk SSSR, Otd. Khim. Nauk* **1961,** 531.
85. B. M. Mikhailov and L. S. Vasil'ev, *Dokl. Akad. Nauk SSSR* **139,** 385 (1961).
86. R. Köster, *Angew. Chem.* **73,** 66 (1961).
87. B. M. Mikhailov, T. K. Kozminskaya and A. Ya. Bezmenov, *Izv. Akad. Nauk SSSR, Ser. Khim.* **1965,** 355.
88. B. M. Mikhailov, L. S. Vasil'ev and A. Ya. Bezmenov, *Izv. Akad. Nauk SSSR, Ser. Khim.* **1965,** 712.
89. R. Köster and G. Benedikt, *Angew. Chem.* **75,** 346 (1963).
90. L. I. Zakharkin and A. I. Kovredov, *Zh. Obshch. Khim.* **32,** 1421 (1962).
91. A. I. Kovredov and L. I. Zakharkin, *Izv. Akad. Nauk SSSR, Ser. Khim.* **1964,** 50.
92. R. Köster and M. A. Grassberger. *Ann* **719,** 169 (1968).
93. H. C. Brown, E. Negishi and P. L. Burke, *J. Am. Chem. Soc.* **92,** 6649 (1970).
94. W. R. Clagton, J. Laturnino, P. W. R. Corfield and S. G. Shore, *Chem. Comm.* **1973,** 377.
95. P. Binger, G. Benedikt, G. W. Rotermund and R. Köster, *Ann.* **717,** 21 (1968).
96. H. C. Brown and E. Negishi, *J. Am. Chem. Soc.* **93,** 6682 (1971).
97. H. C. Brown and W. C. Dickason, *J. Am. Chem. Soc.* **92,** 709 (1970).
98. R. Köster and G. Benedikt, *Angew. Chem.* **74,** 589 (1962).
99. P. Jutzi, *Angew. Chem.* **83,** 912 (1971).
100. P. Jutzi, *Angew. Chem* **84,** 28 (1972).
101. R. Van Veen and F. Bickelhaupt, *J. Organometal. Chem.* **30,** C51 (1971).
102. R. Van Veen and F. Bickelhaupt, *J. Organometal. Chem.* **43,** 241 (1972).
103. M. W. Rathke and R. Kow, *J. Am. Chem. Soc.* **94,** 6854 (1972).
104. M. E Hillman, *J. Am. Chem. Soc.* **84,** 4715 (1962).
105. H. C. Brown, E. Negishi and S. K. Gupta, *J. Am. Chem. Soc.* **92,** 6648 (1970).
106. H. C. Brown and E. Negishi, *J. Am. Chem. Soc.* **91,** 1224 (1969).
107. E. F. Knights and H. C. Brown, *J. Am. Chem. Soc.* **90,** 5283 (1968).
108. H. C. Brown and E. Negishi, *Synthesis* **1972,** 197.
109. H. C. Brown and E. Negishi, *J. Am. Chem. Soc.* **89,** 5478 (1967).
110. H. C. Brown and E. Negishi, *J. Am. Chem. Soc.* **89,** 5477 (1967).
111. H. O. House and G. H. Rasmusson, *J. Org. Chem.* **28,** 31 (1963).
112. H. C. Brown and E. Negishi, *Chem. Comm.* **1968,** 594.
113. B. M. Mikhailov and L. S. Vasil'ev, *Izv. Akad. Nauk SSSR, Otd. Khim. Nauk* **1967,** 2586.

114. L. S. Vasil'ev, V. P. Dmitrikov and B. M. Mikhailov, *Zh. Obshch. Khim.* **42,** 1015 (1972).
115. B. M. Mikhailov, L. S. Vasil'ev and V. P. Dmitrikov, *Izv. Akad. Nauk SSSR, Ser. Khim.* **1968,** 2164.
116. L. S. Vasil'ev, V. P. Dmitrikov, V. S. Bogdanov and B. M. Mikhailov, *Zh. Obshch. Khim.* **42,** 1318 (1972).
117. B. M. Mikhailov, L. S. Vasil'ev and V. P. Dmitrikov, *Izv. Akad. Nauk SSSR, Ser. Khim.* **1970,** 198.
118. L. S. Vasil'ev, M. M. Vartanyan and B. M. Mikhailov, *Izv. Akad. Nauk SSSR, Ser. Khim.* **1976,** 2308
119. Y. Yamamoto and H. C. Brown, *Chem. Comm.* **1973,** 801.
120. Y. Yamamoto and H. C. Brown, *J. Org. Chem.* **39,** 861 (1974).
121. H. C. Brown and N. R. DeLue, *J. Am. Chem. Soc.* **96,** 311 (1974).
122. M. P. Brown, A. K. Holliday and G. M. Way, *J. Chem. Soc., Dalton Trans.* **1975,** 148.
123. R. Liotta and H. C. Brown, *J. Org. Chem.* **42,** 2836 (1977).
124. H. C. Brown and M. Zaidlewicz, *J. Am. Chem. Soc.* **98,** 4917 (1976).
125. C. F. Reichert, W. E. Pye and T. A. Bryson *Tetrahedron* **37,** 2441 (1981).
126. H. C. Brown and C. D. Pfaffenberger, *Tetrahedron* **31,** 925 (1975).
127. H. C. Brown, E. F. Knights and C. G. Scouten, *J. Am. Chem. Soc.* **96,** 7765 (1974).
128. J. R. Köster and P. Binger, *Inorganic Synthesis, XV,* 141 (1974).
129. H. C. Brown and C. Lane, *Heterocycles* **7,** 454 (1977).
130. H. C. Brown, A. K. Mandal and S. U. Kulkarni, *J. Org. Chem.* **42,** 1392 (1977).
131. H.C. Brown and S. U. Kulkarni, *J. Org. Chem.* **44,** 2422 (1979).
132. M. M. Midland and S. Greer, *Synthesis* **1978,** 845.
133. C. G. Whiteley, *J. Chem. Soc., Comm.* **1981,** 5.
134. S. U. Kulkarni and H. C. Brown, *J. Org. Chem.* **44,** 1747 (1979).
135. H. C. Brown, U. S. Racherla and H. Taniguchi, *J. Org. Chem.* **46,** 4313 (1981).
136. M. M. Midland, J. E. Petre and S. A. Zderic, *J. Organomet. Chem.* **182,** C53 (1979).
137. W. S. Still and K. P. Darst, *J. Am. Chem. Soc.* **102,** 7385 (1980).
138. G. W. Kramer and H. C. Brown, *J. Organomet. Chem.* **90,** C1 (1975).
139. H. C. Brown, T. M. Ford and J. L. Hubbard, *J. Org. Chem.* **45,** 4067 (1980).
140. G. W. Kabalka, *Chem. Comm.* **10,** 93 (1980).
141. L. S. Vasil'ev, M. M. Vartanyan, V. S. Bogdanov and B. M. Mikhailov, *Izv. Akad. Nauk SSSR, Ser. Khim.* **1976,** 2204.
142. H. C. Brown and S. U. Kulkarni, *J. Organomet. Chem.* **168,** 281 (1979).
143. H. C. Brown and N. R. De Lue, *Tetrahedron Lett.* **1977,** 3007.
144. B. M. Mikhailov, T. A. Shchegoleva, E. M. Shashkova and V. G. Kiselev (in press).
144a. T. A. Shchegoleva, E. M. Shashkova and B. M. Mikhailov, *Izv. Akad. Nauk SSSR, Ser. Khim.* **1980,** 2436.
145. M. M. Midland, A. Tramontano and S. A. Zderic. *J. Organomet. Chem.* **134,** C17 (1977).
146. M. M. Midland and A. Tramontano, *J. Org. Chem.* **43,** 1470 (1978).
147. M. M. Midland, A. Tramontano and S. A. Zderic, *J. Am. Chem. Soc.* **99,** 5211 (1977).
148. M. M. Midland, A. Tramontano and S. A. Zderic, *J. Organomet. Chem.* **156,** 203 (1978).
149. H. C. Brown and T. M. Ford, *J. Org. Chem.* **46,** 647 (1981).
150. S. Krishnamurthy and H. C. Brown, *J. Org. Chem.* **40,** 1864 (1975).

151. H. C. Brown, S. Krishnamurthy and N. M. Yoon, *J. Org. Chem.* **41,** 1778 (1976).
152. S. Krishnamurthy and H. C. Brown, *J. Org. Chem.* **42,** 1197 (1977).
153. S. Krishnamurthy, F. Vogel and H. C. Brown, *J. Org. Chem.* **42,** 2534 (1977).
154. H. C. Brown and S. U. Kulkarni, *J. Org. Chem.* **42,** 4169 (1977).
155. M. V. Bhatt, *J. Organomet. Chem.* **156,** 221 (1978).
156. R. Köster and G. Seidel, *Lieb. Ann. Chem.* **1977,** 1837.
157. H. C. Brown and S. U. Kulkarni, *Inorganic Chem.* **16,** 3090 (1977).
158. H. C. Brown and J. A. Soderquist, *J. Org. Chem.* **45,** 846 (1980).
159. H. C. Brown and K. K. Wang, *J. Org. Chem.* **45,** 1748 (1980).
160. J.-J. Katz, J.-E. Dubois and C. Lion, *Bull. Soc. Chim. Fr.* **1977,** 683.
161. Y. Yamamoto, H. Toi, S. Murahashi, and I. Moritani, *J. Am. Chem. Soc.* **97,** 2558 (1975).
162. Y. Yamamoto, H. Toi, A. Sonoda and S. Murahashi, *J. Am. Chem. Soc.* **98,** 1965 (1976).
163. H. Toi, Y. Yamamoto, A. Sonoda and S. Murahashi, *Tetrahedron* **37,** 2261 (1981).
164. G. W. Kramer and H. C. Brown, *J. Am. Chem. Soc.* **98,** 1964 (1976).
165. G. W. Kramer and H. C. Brown, *J. Org. Chem.* **42,** 2832 (1977).
166. T. A. Shchegoleva, E. M. Shashkova and B. M. Mikhailov, *Izv. Akad. Nauk SSSR, Ser. Khim.* **1980,** 2169.
167. R. H. Mueller, *Tetrahedron Lett.* **1976,** 2925
168. J. L. Hubbard and G. W. Kramer, *J. Organomet. Chem.* **156,** 81 (1978).
169. H. C. Brown, G. W. Kramer, J. L. Hubbard and S. Krishnamurthy, *J. Organomet. Chem.* **188,** 1 (1980).

PART III

ALKENYLBORANES

Organoboranes with radicals of the vinyl type are readily available compounds. They show a high reactivity and find various uses in organic synthesis.

The first group of methods for the synthesis of alkenylboranes is based on exchange reactions between organometallic compounds and halo or alkoxy derivatives of boron. 1-Alkenylboron halides and esters of 1-alkenylboronic acids are generally obtained by these methods.

A second group of methods consists of the addition of different types of boron compounds to acetylenes. The hydroboration of acetylenic compounds, leading to 1-alkenylalkylboranes, is of very great preparative importance. Functional derivatives of 1-alkenylboranes are obtained by the addition of alkylmercaptoboranes and haloboranes to acetylenes.

A third group of methods for the synthesis of 1-alkenylboranes is based on conversions of alkali metal 1-alkynyltrialkylborates.

The following types of compounds are obtained from alkenylboranes by various chemical conversions: cis and trans olefins, haloolefins with cis and trans configurations, saturated and unsaturated alcohols, aldehydes, α,β-unsaturated ketones, acetylenic ketones, oxo alcohols, α,β-unsaturated acids, α-oxo acids, β-hydroxy acids, alkylmalonic and alkylglutaric acids; allene, 1,3-diene, 1,2,3-butatriene, and enyne hydrocarbons; cyclopropane and cyclobutane compounds, and others.

PART III

CHAPTER 1

Methods for the preparation of alkenylboranes

In this chapter, methods for the preparation of alkenylboranes using exchange reactions between organometallic compounds and boron compounds and addition reactions of diborane, its derivatives, and alkylmercapto- and halogen-containing boron compounds to acetylenes will be considered.

The method of synthesis based on metal 1-alkynyltrialkylborates, because of its specific characteristics, and the chemical conversions of the alkenylboranes obtained by this method are presented in Chapter 3.

The preparation of 1,4-pentadienyl compounds of boron from allylboranes and acetylenes is considered in Part IV.

1.1 PREPARATION WITH THE USE OF ALKENYL COMPOUNDS OF METALS

Several carefully performed investigations have been devoted to methods of preparation and investigation of the properties of the simplest alkenylalkylboranes, the methylvinylboranes.

Parsons, et al. [1, 2] obtained dimethylvinylborane by the action of bromodimethylborane on vinylsodium:

$$(CH_3)_2BBr + CH_2{=}CHNa \longrightarrow (CH_3)_2BCH{=}CH_2.$$

This compound is unstable and, at room temperature, disproportionates slowly to trimethylborane, methyldivinylborane, and trivinylborane. Thus, after 360 hours, a reaction mixture of 0.78 mmole of dimethylvinylborane contains 0.55 mmole of the starting compound, 0.10 mmole of $B(CH_3)_3$, 0.6 mmole of $CH_3B(CH{=}CH_2)_2$, and 0.002 mmole of $B(CH{=}CH_2)_3$ [3]. Methyldivinylborane symmetrizes to the extent of 10% after 90 hours at 25°C to $(CH_3)_2$-$BCH{=}CH_2$ and $B(CH{=}CH_2)_3$ [3].

By an analogous method, starting from propenyllithium and bromodimethylborane, dimethylpropenylborane has been obtained, which shows a higher thermal stability but also has a tendency to symmetrize to methyldipropenylborane and trimethylborane [1, 2]. Methylvinylboranes and trivinylborane have been obtained from dimethylzinc and chlorovinylboranes [3]. Thus, the reaction of dimethylzinc with chlorodivinylborane leads to a mixture of $(CH_3)_2BCH{=}CH_2$, $CH_3B(CH{=}CH_2)_2$, $B(CH_3)_3$, and $B(CH{=}CH_2)_3$. $B(C_2H_5)_3$ and $B(CH{=}CH_2)_3$ have been obtained in an attempt to synthesize ethylvinylboranes from diethylzinc and chlorodivinylborane.

In each of the cases described, the reaction products have been separated with the use of high-vacuum equipment.

According to [4], trivinylborane is obtained in 70% yield by the reaction of vinylmagnesium bromide and BF_3 in ether.

1-Propenylborolane and 1-propenylborinane have been synthesized from dichloropropenylborane and α,ω-dimagnesium derivatives of alkanes [5]:

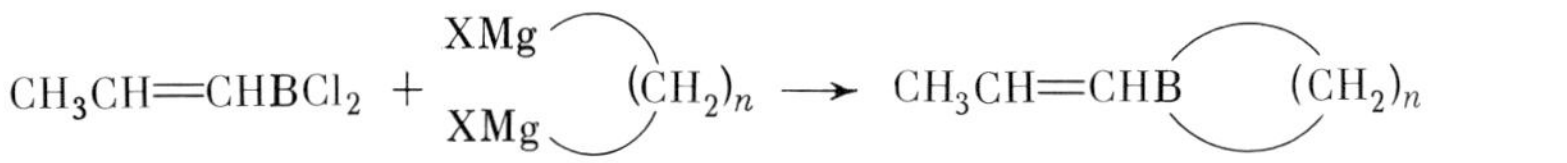

38% ($n = 4, 5$).

Brinckman and Stone [6] have synthesized difluorovinylborane in 30% yield by the action of BF_3 on divinylmercury:

$$BF_3 + Hg(CH{=}CH_2)_2 \longrightarrow CH_2{=}CHBF_2 + HgF_2.$$

The above compound is formed in slightly lower yield by the reaction of BF_3 and divinylzinc, while the use of tetravinyltin makes it possible to increase the yield to 52% [6].

Tetravinyltin and BCl_3, taken in a ratio of 1:2, form dichlorovinylborane in good yield [6, 7]. With an equimolar ratio of the reagents at 60°C, a mixture of substances consisting of 92% of chlorodivinylborane is obtained (yield 35%). Dichlorovinylborane is obtained by the reaction of BCl_3 and divinylzinc, while dibromovinylborane is formed from divinylmercury and BBr_3 [8].

Fluorodivinylborane has been obtained in 41% yield by an exchange reaction between antimony trifluoride and chlorodivinylborane [6].

Boron halides with halovinyl radicals are also obtained by exchange reactions between the corresponding organometallic compounds and haloboranes.

Dichloro(*trans*-2-chlorovinyl)borane has been obtained by the action of boron trichloride on *trans*-2-chlorovinylmercury chloride in an inert solvent:

$$ClCH{=}CHHgCl + BCl_3 \longrightarrow ClCH{=}CHBCl_2 + HgCl_2,$$

and chlorobis(*trans*-2-chlorovinyl)borane has been synthesized in 50% yield from BCl_3 and *trans*,*trans*-bis(2-chlorovinyl)mercury [9].

Dichloroperfluorovinylborane [10] and tris(perfluorovinyl)borane [11] have been synthesized by the action of BCl_3 on bis(perfluorovinyl)mercury.

Stafford and Stone [12] have synthesized difluoro(perfluorovinyl)borane in 59% yield by the action of boron trifluoride on bis(perfluorovinyl)dimethyltin:

$$BF_3 + (CH_3)_2Sn(CF{=}CF_2)_2 \longrightarrow F_2BCF{=}CF_2.$$

Dichloro(perfluorovinyl)borane and chlorobis(perfluorovinyl)-borane have been obtained by the same method, using boron trichloride [12, 13]. Tris(perfluorovinyl)borane is formed by the reaction of chlorobis(perfluorovinyl)borane and bis(perfluorovinyl)dimethyltin [12].

Tris(2-chlorovinyl)borane is obtained as a result of the disproportionation of chlorobis(2-chlorovinyl)borane at 140–190°C [14].

Esters of 1-alkenylboronic acids are synthesized from 1-alkenylmagnesium bromides (or chlorides) and alkyl borates. The corresponding acids are obtained by saponification of the esters. This method was used for the first time by Mel'nikov and Rokitskaya [15] for the synthesis of styrylboronic acid. Dibutyl vinylboronate is obtained in good yields by the action of vinylmagnesium bromide on methyl borate in tetrahydrofuran [16–19]:

$$CH_2{=}CHMgBr + B(OCH_3)_3 \xrightarrow[n\text{-}C_4H_9OH]{H^+} CH_2{=}CHB(OC_4H_9\text{-}n)_2.$$

Esters of propenylboronic [17, 19], isopropenylboronic [17, 19], isobutenylboronic [17], α-styrylboronic [20], and β-styrylboronic [21] acids have been synthesized by this method, starting from the corresponding 1-alkenylmagnesium bromides. Esters of perfluorovinylboronic acid are obtained in yields of 22–30% [11]. The use of isobutenyllithium led to the formation of isobutenylboronic acid in a yield of only 7% [22].

From the different 1-alkenylboronic acids, the corresponding esters, chlorides, esters of 1-alkenylthioboronic acids, alkylamino-(alkenyl)boranes, and other types of compounds have been synthesized [5, 23].

Di-*tert*-butyl vinylboronate is obtained by esterification of the pyridine complex of vinylboronic anhydride [24]. The action of arylmagnesium bromides on dibutyl vinylboronate produces butyl vinylarylboronates [25, 26].

A peculiar method for the preparation of tetramethyl 2-methyl-1,1-propenediboronate consists of the sequential action of methyllithium, acetone, and chlorodimethoxyborane on octamethyl methanetetraboronate [27]:

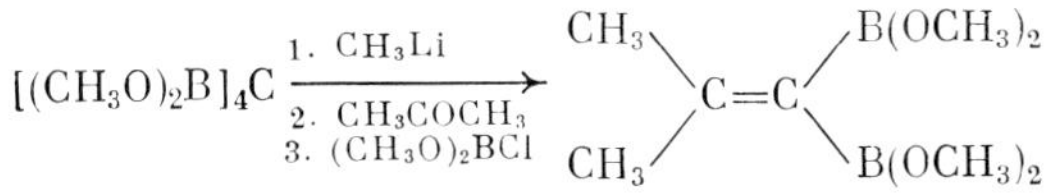

Tetramethyl 1,1-propenediboronate has been obtained by the same method, using acetaldehyde instead of acetone.

Matteson, et al. [167–170] have proposed a method for the synthesis of esters of 1-alkenylboronic acids based on the condensation of tetrakis(trimethyleneboryl)methane or of bis(ethylenedioxyboryl)methane with aldehydes and ketones.

1.2 PREPARATION BY HYDROBORATION OF ACETYLENIC COMPOUNDS

A widely used method for the synthesis of alkenyl boron compounds is based on the hydroboration of acetylenic compounds with diborane, alkyldiboranes, adducts of alkylboranes with nitrogen-containing bases, or with functional derivatives of borane.

A prototype of this method for the synthesis of alkenylboranes was the addition of aluminum hydrides to alkynes, investigated by Wilke and Müller during the period of 1956–1958 [28, 29]. As was found, diisobutylalane adds to 3-hexyne to form predominantly *cis*-3-alkenyldiisobutylalane which, under the action of methanol, undergoes protonolysis to form *cis*-3-hexene, containing only 3% of the *trans* isomer [28]:

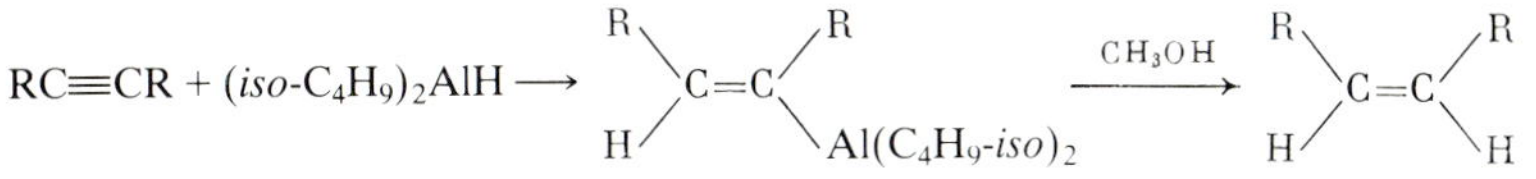

$(R = C_2H_5)$.

Tolane is converted to isostilbene by this method [28].

In the reaction of 1-hexyne with diethylalane in a ratio of 1:1, the dialkylaluminum group adds selectively to the terminal carbon atom. The 1-hexenyldiethylalane formed during this process can be used for the synthesis of 1-deutero-1-hexene by the action of heavy water [29]:

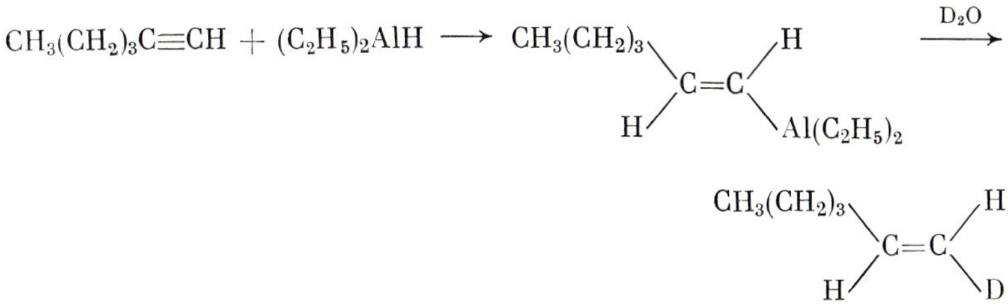

By using $(C_2H_5)_2AlD$ and H_2O as reagents, different deutero-substituted 1-hexene derivatives can be smoothly obtained:

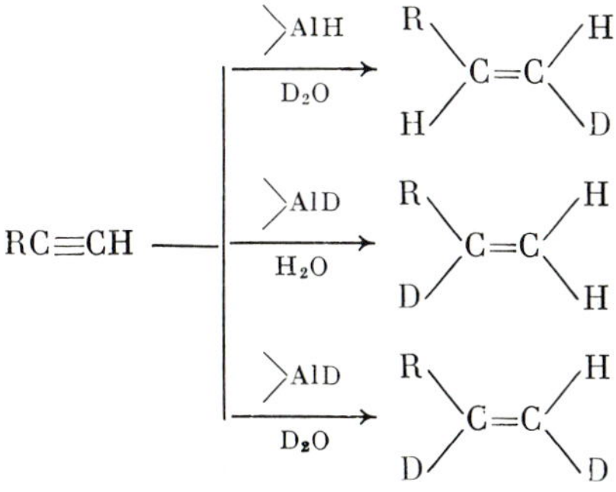

Mono- and dideutero-substituted olefins with a cis configuration are obtained from symmetrical acetylenic hydrocarbons by the use of deuterium-labeled reagents [29]:

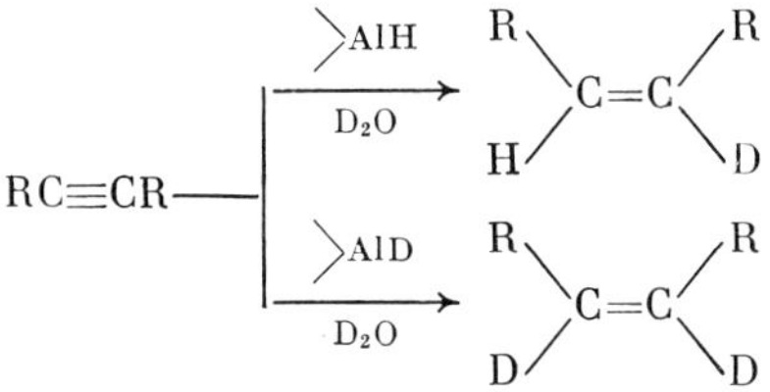

The action of two equivalents of diethylalane on 1-hexyne, followed by protonolysis with heavy water, produces 1,1-dideuterohexane, indicating that the second hydride molecule also adds to the terminal carbon of the olefin system [29]:

$$C_4H_9C{\equiv}CH + (C_2H_5)_2AlH \longrightarrow C_4H_9CH{=}CHAl(C_2H_5)_2 \xrightarrow{(C_2H_5)_2AlH}$$

$$\longrightarrow C_4H_9CH_2{-}CH{\left\langle\begin{matrix}Al(C_2H_5)_2\\ Al(C_2H_5)_2\end{matrix}\right.} \xrightarrow{D_2O} C_4H_9CH_2CHD_2.$$

1.2.1 Hydroboration of acetylenic hydrocarbons

Acetylenic hydrocarbons are capable of adding diborane and its derivatives. In most cases, the process can be stopped at the first stage and, depending on the nature of the acetylenic compound and of the hydroborating agent, 1-alkenyldialkylboranes, di-1-alkenylalkylboranes, or tri-1-alkenylboranes are obtained. As a rule, the organoboron compounds obtained by this method have not been isolated in the pure state, but have been used as reagents for the preparation of different organic compounds.

From alkenylboron compounds of this type, the only products that have been isolated in the pure state and characterized are di-1-alkenylalkylboranes, obtained by Hawthorne [30] by the action of trimethylamine-*tert*-butylborane in tetrahydrofuran at 60°C on acetylenes with a terminal triple bond:

$$tert\text{-}C_4H_9BH_2{\cdot}N(CH_3)_3 + 2RC{\equiv}CH \longrightarrow \underset{41\text{–}70\%}{tert\text{-}\,C_4H_9B(CH{=}CHR)_2} + N(CH_3)_3.$$

By this method, starting from 1-butyne, 1-pentyne, and 1-hexyne, di(1-butenyl)di(1-pentenyl)-, and di(1-hexenyl)-*tert*-butylboranes, which are liquids not undergoing disproportionation during vaccum distillation, have been obtained.

Some alkenylboranes have been obtained in the pure state by the addition of functional borane derivatives to acetylenic hydrocarbons.

The regioselectivity of the hydroboration of 5-methyl-2-hexyne with different R_2BH has been investigated [171].

Woods and Strong [31] obtained esters of alkenylboronic acids by the hydroboration of 1-alkynes with 4,4,6-trimethyl-1,3,2-dioxaborinane in ether at 110–120°C:

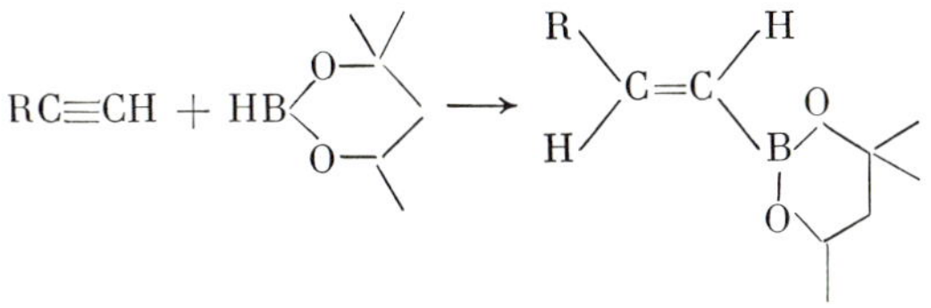

The synthesis of esters of alkenylboronic acids from acetylenic hydrocarbons is also carried out with the use of 1,3,2-benzodioxaborole as a hydroborating agent [32, 172]:

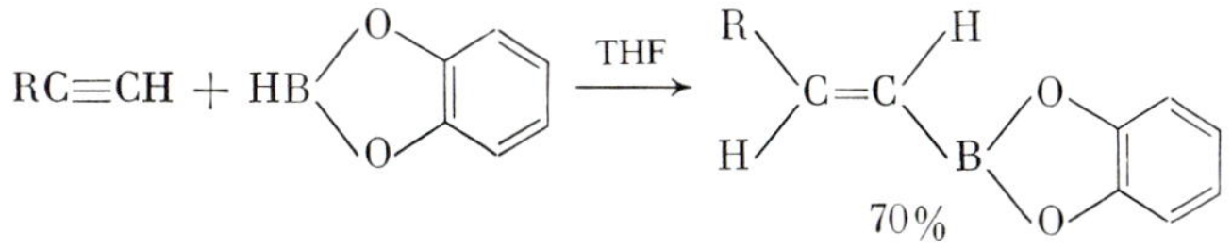

Unsymmetrical disubstituted acetylenes form a mixture of isomeric compounds as a result of the addition of a boron atom to different carbon atoms of the triple bond. Thus, the ratio of isomers is 92:8 in the case of 1-cyclohexenylpropyne and 73:27 for 1-phenylpropyne.

Alkenylboron dichlorides can be obtained by the hydroboration of acetylenic hydrocarbons by dichloroborane etherate in the presence of BCl_3 [33, 173]. By this method, (*trans*-hexenyl) dichloroborane is formed in 69% yield from 1-hexyne; the presence of an isomeric compound as an impurity in the product has not been established:

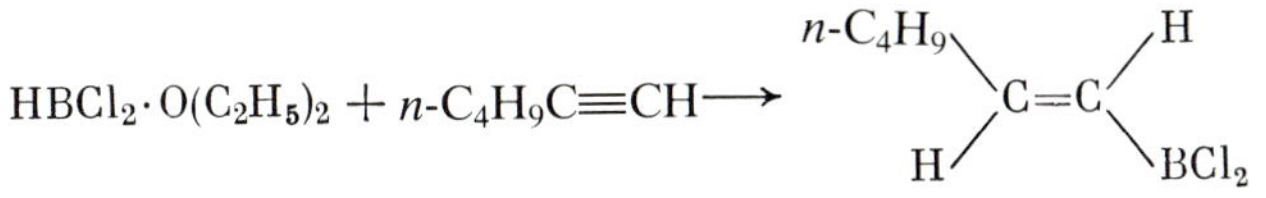

(*cis*-3-Hexenyl)dichloroborane is obtained in an analogous manner from 3-hexyne. Acetylenic hydrocarbons form dialkenylchloroboranes under the action of chloroborane [34, 174]:

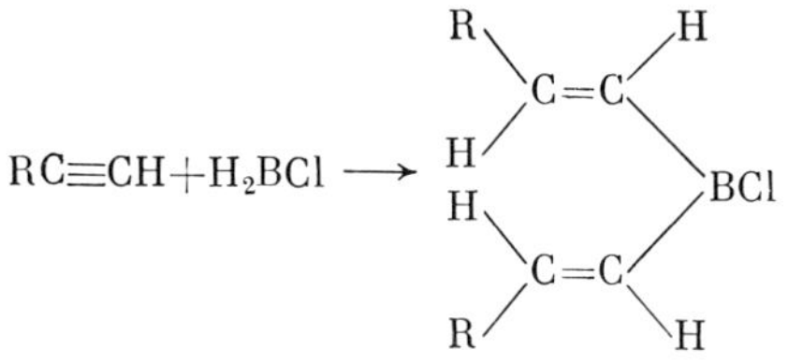

The reactions are complicated by the formation of isomeric alkenyl compounds and by dihydroboration processes.

Acetylenes react with $Br_2BH{\cdot}S(CH_3)_2$ to form 1-alkenyldibromoboranes [175].

The use of diborane itself as a hydroborating agent has not proved acceptable for 1-alkynes from the preparative aspect, because, even with the necessary ratio of reagents (6:1), one-half of the acetylenic hydrocarbon remains unconsumed while the other half undergoes dihydroboration with the predominant formation of the 1,1-diboron compound [35, 36]. A more selective reaction is that of diborane with symmetrically substituted acetylenes, leading to trialkenylboranes, but in this case a considerable quantity of dihydroboration products is formed [36, 37].

Tetraalkyldiboranes, which react in the monomeric form with acetylenic hydrocarbons similarly to the dialkylalanes, have been used successfully for the synthesis of alkenyldialkylboranes. The structure of the hydroboration products has been determined by protonolysis and oxidation; syntheses of alkenyldialkylboranes are considered in the next section.

1.2.2 Hydroboration of functional derivatives of acetylenic hydrocarbons

Bubnov and co-workers [38, 39] have found that ethoxyacetylene reacts with diborane in tetrahydrofuran with the formation of tris(*trans*-2-alkoxyvinyl)borane. Protonolysis of the latter compound with deuteromethanol leads to methoxybis(*trans*-2-ethoxyvinyl)borane and *trans*-1-deutero-2-ethoxyethylene:

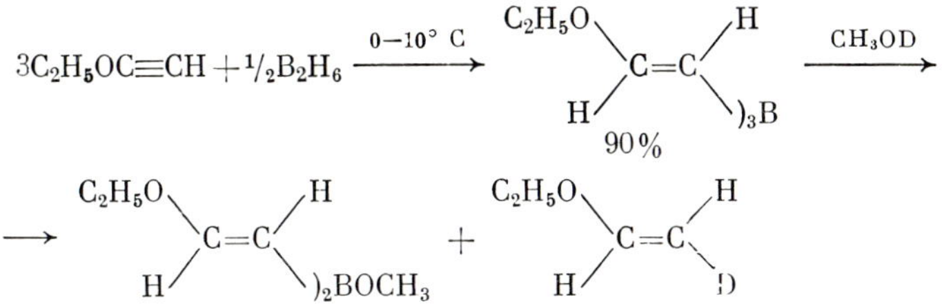

The hydroboration of ethoxyacetylene with tetraalkyldiboranes forms *trans*-2-ethoxyvinyldialkylboranes [38, 39]:

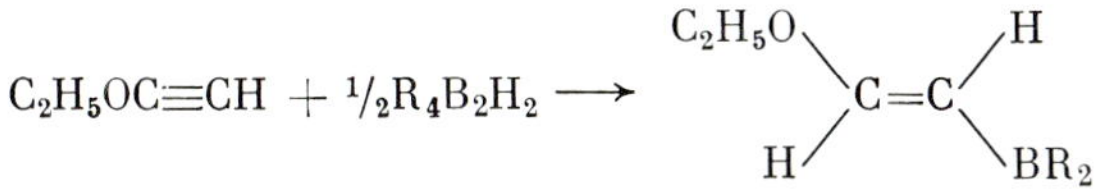

(R = n-C_3H_7, n-C_4H_9).

Tris(*trans*-2-ethoxyvinyl)borane can be converted to methyl esters of bis(*trans*-2-ethoxyvinyl)borinic and *trans*-2-ethoxyvinylboronic acids by an exchange reaction with methyl borate [38, 39]:

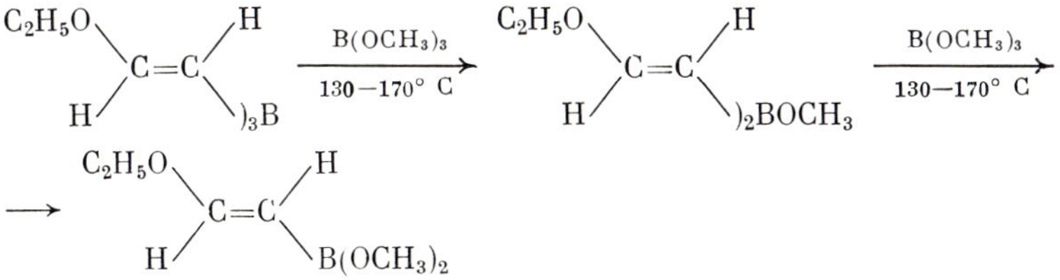

Exchange reactions between alkoxyvinylboranes and an alkyl borate, in which the stereochemical configuration of the olefinic radical is retained, apparently proceed via the formation of intermediate diboron compounds with three-center B—C_{sp^2}—B bonds and semicoordinate B—O bonds [39]:

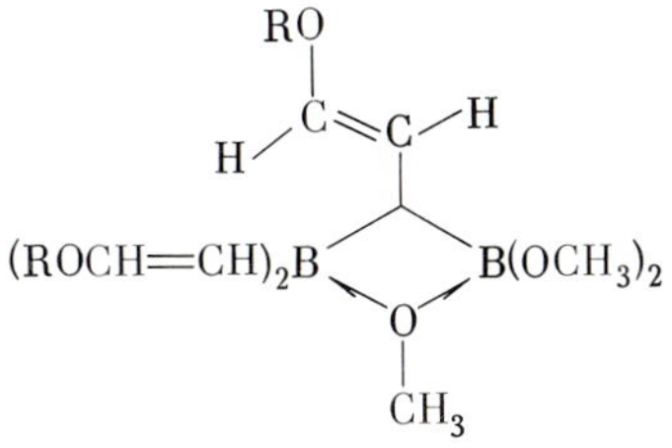

The action of tetracyclohexyldiborane on 1-bromo- and 1-iodo-1-hexyne forms *trans*-1-halo-1-hexenyldicyclohexylboranes [40]:

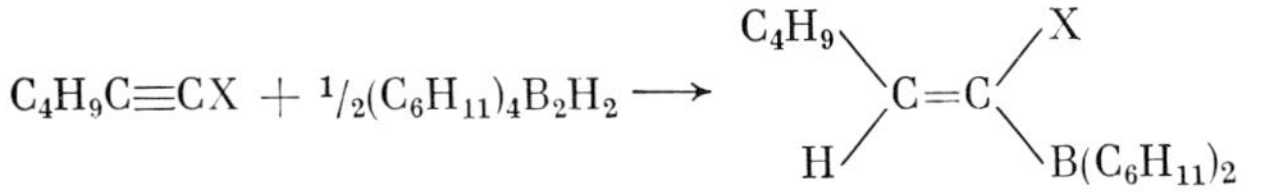

(X=Br, I).

1-Haloacetylenes react analogously with bis(2,3-dimethyl-2-butyl)-diborane [41].

1-Chloro-1-heptyne adds a dialkylboryl group to the terminal carbon atom of the triple bond under the action of tetra(3-methyl-2-butyl)diborane [42].

α-Halo-1-alkenyldialkylboranes, obtained from 1-halo-1-alkynes and R_2BH, are converted to *cis*-alkenylboranes with high stereoselectivity by the action of Et_3BHLi [176] or *tert*-butyllithium [177].

Propargyl chloride reacts with tetraethyldiborane to form diethyl(3-chloro-1-propenyl)borane:

$$HC{\equiv}CCH_2Cl \xrightarrow{(C_2H_5)_4B_2H_2} (C_2H_5)_2BCH{=}CHCH_2Cl,$$

whereas, in the case of alkyl-substituted propargyl chloride, the orientation of addition of the dialkylborane is different [43]:

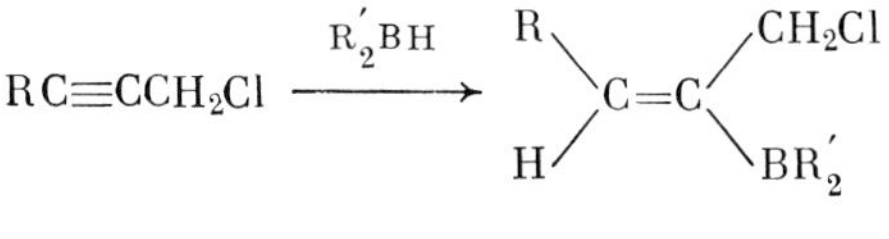

The hydroboration of a number of isomeric methoxy-2-hexynes has been investigated [178].

1.3 THIOBORATION OF ACETYLENIC COMPOUNDS

Mikhailov and co-workers [44, 45] found that alkyl thioborates are capable of adding to ethoxyacetylene under mild conditions (5–25°C). With an equimolar ratio of reagents, esters of (2-alkylthio-2-ethoxyvinyl)thioboronic acids are formed, consisting of a 1:1

mixture of *cis* and *trans* isomers:

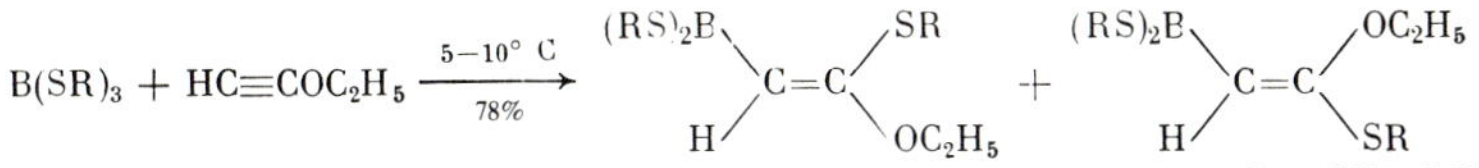

$(R = CH_3, C_2H_5)$,

but, at a ratio of 1:3, tris(2-alkylthio-2-ethoxyvinyl)boranes are obtained, containing a predominant quantity of vinyl groups with a *trans* location of the boron and sulfur atoms (90% in the mixture) [44, 46]:

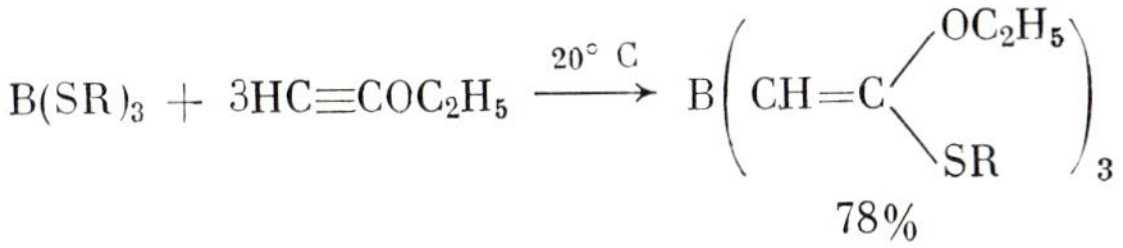

$(R = CH_3, C_2H_5)$.

Thus, the initially formed monovinylboranes add further to ethoxyacetylene in a *trans* position.

Esters of dialkylthioborinic acids add to ethoxyacetylene with the formation of (2-alkylthio-2-ethoxyvinyl)dialkylboranes, consisting of a mixture of *cis* and *trans* isomers in ratios depending on the nature of the thioester [46]:

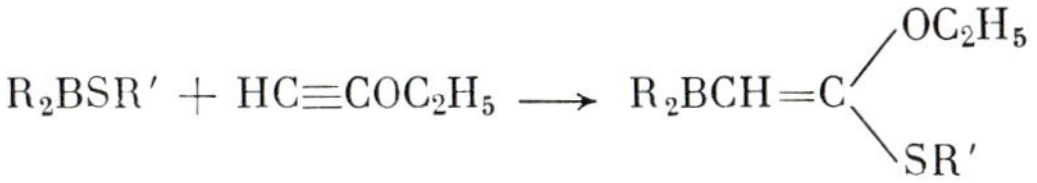

$(R = C_2H_5, n\text{-}C_3H_7;\ R' = C_2H_5,\ tert\text{-}C_4H_9)$.

The addition of 3-methyl-1-ethylthioboracyclopentane to ethoxyacetylene proceeds stereospecifically, with the formation of only a *cis*-addition product [46]:

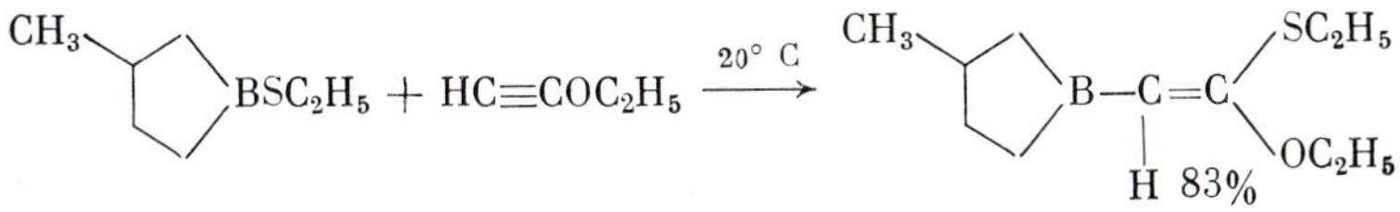

The treatment of dimethyl (2-methylthio-2-ethoxyvinyl)thioboronate with methanol leads to the corresponding boronate with a predominant content (80–90%) of the *trans* isomer [45]:

$$(CH_3S)_2BCH{=}C\begin{matrix}\diagup OC_2H_5\\ \diagdown SCH_3\end{matrix} + 2CH_3OH \longrightarrow (CH_3O)_2BCH{=}C\begin{matrix}\diagup OC_2H_5\\ \diagdown SCH_3\end{matrix} + 2CH_3SH.$$

The conversion of the *cis* to the *trans* form proceeds under the action of the mercaptan evolved during the reaction. It should be noted that *cis-trans* isomerization under the action of a mercaptan is observed in the radical or nucleophilic [47] addition of mercaptans to acetylenic compounds.

The *trans* isomer of the above-mentioned boronate can be isolated in the pure state by crystallization of a mixture of the stereoisomers depleted in the *cis* isomer. The *trans* isomer is partially converted to the *cis* isomer by irradiation with UV light or by heating.

1.4 HALOBORATION OF ACETYLENIC HYDROCARBONS

Alkenylhaloboranes are obtained by the addition of haloboranes to acetylene.

Arnold [48] obtained dichloro(2-chlorovinyl)borane in low yield by the gas-phase reaction of boron trichloride with acetylene at 150–220°C in the presence of mercuric chloride on activated carbon:

$$BCl_3 + HC{\equiv}CH \longrightarrow ClHC{=}CHBCl_2.$$

Gipstein and co-workers [49] carried out the vinylation of boron trichloride over activated carbon and increased the yield of dichloro(2-chlorovinyl)borane to 90%. Chlorodi(2-chlorovinyl)borane (8%) and tri(2-chlorovinyl)borane were obtained as byproducts. The vinylation of dichloro(2-chlorovinyl)borane by passing it with excess acetylene at 100°C over mercuric chloride on activated carbon leads to chlorodi(2-chlorovinyl)borane and tri(2-chlorovinyl)-borane in low yields [49].

Dichloro(2-chlorovinyl)borane, obtained by the addition of BCl_3 to acetylene, forms esters of 2-chlorovinylboronic acid by treatment with alcohols [50].

Dichloroethylborane reacts with acetylene over activated carbon to form chloro(2-chlorovinyl)ethylborane (90%):

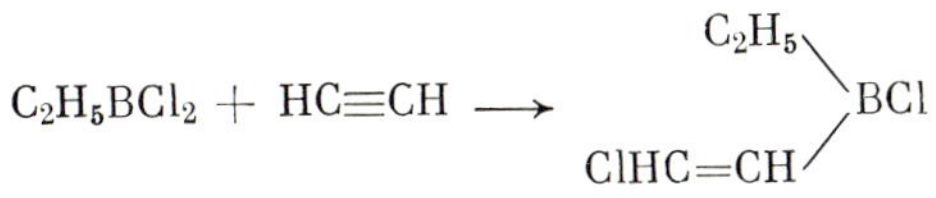

The latter compound undergoes thermal disproportionation and is converted to di(2-chlorovinyl)ethylborane in 90% yield [49].

Chlorodiethylborane does not react with acetylene over activated carbon at 100°C [51].

Different chemical conversions of tri(2-chlorovinyl)borane have been investigated [52].

Lappert and Prokai [53] investigated the reactions of different haloboranes with acetylene, 1-hexyne, and phenyl- and diphenylacetylenes. Boron tribromide reacts with acetylene at 25°C to form a *trans*-addition product. Boron trichloride adds to phenylacetylene in the following manner:

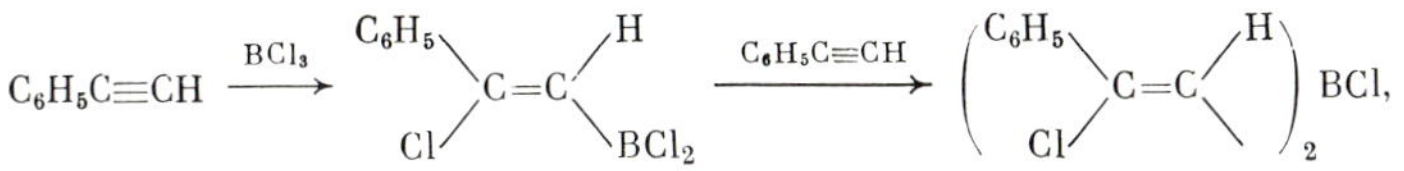

i.e., with the formation of *cis*-addition products.

The reaction of phenylacetylene with phenyldichloroborane proceeds in an analogous manner.

It is interesting that, in the reaction of phenyldichloroborane or diphenylchloroborane with 1-hexyne, arylboration, rather than haloboration, takes place:

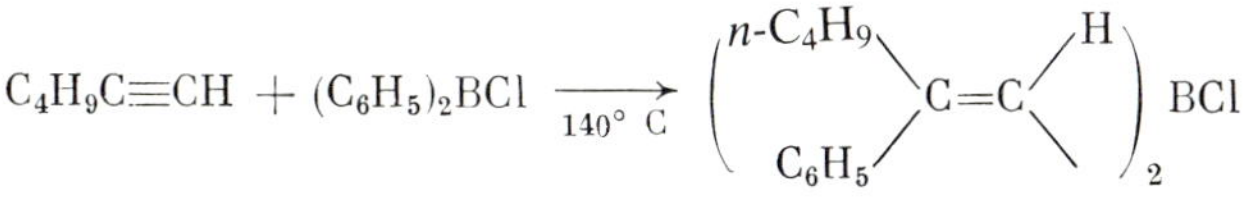

The reactions of phenylacetylene with trihaloboranes (BCl_3, BBr_3, BI_3) have been investigated in detail in [54]. It was found that the action of BCl_3 on phenylacetylene (1:1) forms products of both *cis*- and *trans*-addition and, furthermore, approximately 20% of the products of abnormal addition are obtained, in which the chlorine atom is located on the terminal carbon atom:

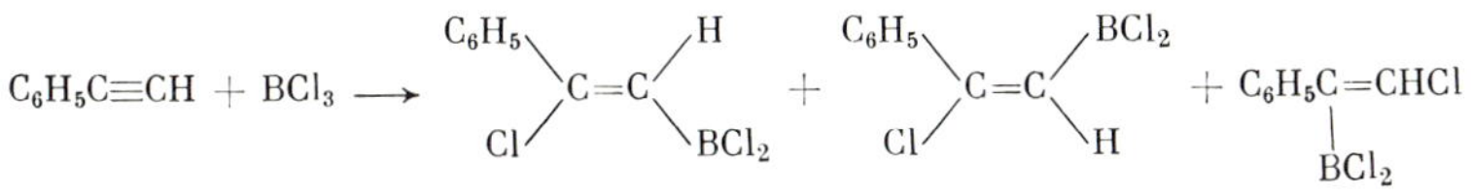

The reactions of BBr_3 and BI_3 with phenylacetylene form approximately equal quantities of the *cis* and *trans* isomers, with the almost complete absence of products of abnormal addition. In the reaction of $BHal_3$ with phenylacetylene in a ratio of 1:3, substituted trivinylboranes are formed, in which 66–92% of the vinyl groups have a *trans* structure. The formation of products of *cis*- and *trans*-addition in an approximate ratio of 1:1 in reactions of $BHal_3$ with phenylacetylene (1:1) indicates that the reaction proceeds via an ionic transition state.

3-Hexyne is capable of adding triiodoborane to form C_2H_5-$(I_2B)C{=}CI(C_2H_5)$ [179].

Tetrachlorodiborane adds to acetylene to form 1,2-bis(dichloroboryl)ethylene [55]:

$$B_2Cl_4 + HC{\equiv}CH \longrightarrow Cl_2BCH{=}CHBCl_2.$$

CHAPTER 2

Chemical properties of alkenylboranes

2.1 SPECTRA

In the IR spectra of alkenylboranes, the absorption bands of the C=C bond show a significant shift into the region of lower frequencies, as compared with the alkenes.

The vibrational frequency of C=C in trivinylborane in the gaseous state is 1602 [56]—1605 cm^{-1} [57].

Halovinylboranes [CH_2=$CHBF_2$, CH_2=$CHBCl_2$, (CH_2=$CH)_2BF$, (CH_2=$CH)_2BCl$] show absorption bands at 2975 and 3100 cm^{-1}, corresponding to vibrations of C—H bonds [6]. The intensity of these bands is relatively higher in halodivinylboranes than in dihalovinylboranes. The absorption bands characterizing the vibrations of the C=C bonds are located in the region 1610–1640 cm^{-1}.

The IR spectra of β-alkoxyvinylboranes contain two very intense absorption bands of the C=C bond (Table 1) [39]:

TABLE 1
IR spectroscopic data of β-alkoxyvinylboranes

Compound	$\nu_{C=C}$, cm^{-1}	Compound	$\nu_{C=C}$, cm^{-1}
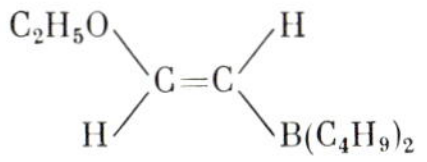	1610, 1585	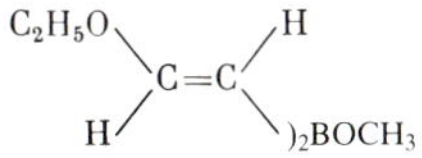	1624, 1601
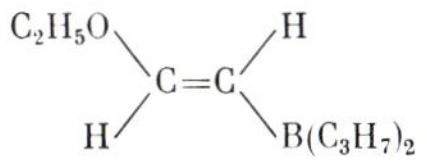	1610, 1587	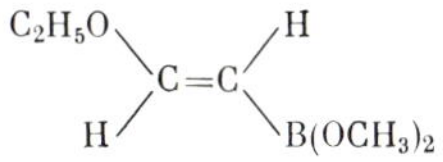	1633, 1610
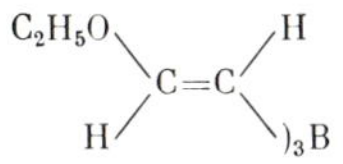	1612, 1590		

The C=C absorption band in dimethoxy(2-methyl-1,4-pentadienyl)borane is located at 1628 cm^{-1} and that in 1,5-diallyl-3-methyl-1-bora-2-cyclohexene at 1600 cm^{-1} [58]. The presence of

Table 2. Data of the ¹H and ¹¹B NMR spectra of some alkenyl boron compounds.

Compound	Chemical shift, ppm				Coupling constant								
	$\delta^{11}B$	$^{\delta}CH_3$	$^{\delta}H_a$	$^{\delta}H_b$	^2J H—H	$^3J_{cis\text{-}trans}$ H—H	$^3J_{vic}$H—H	$^4J_{cis}$ H—H	$^4J_{trans}$ H—H	$^2J_{B—H}$	$^3J_{cis}$B—H	$^3J_{trans}$ B—H	$^3J_{cis}$ B—H
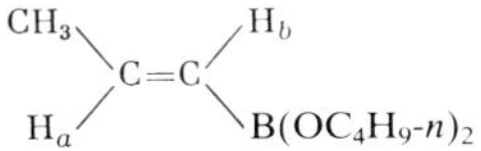	—27.2	1.76	6.60	5.49	—	17.80	6.45	1.60	—	3.5	4.3	—	—
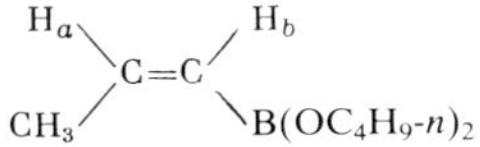	—27.2	1.87	6.34	5.37	—	14.20	6.75	—	1.60	3.5	—	11.5	—
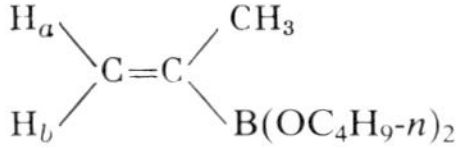	—27.1	1.80	5.46	5.26	3.40	—	—	1.80	1.34	—	6.0	10.7	3.8
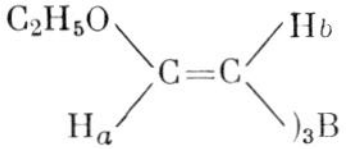	—54.5	—	7.12	5.35	—	14.00	—	—	—	0.0	—	—	—
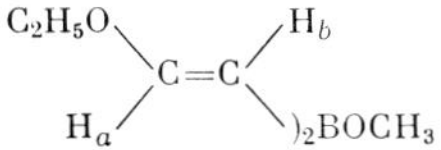	—42.2	3.62	6.95	4.75	—	14.00	—	—	—	0.0	2.9	—	2.9
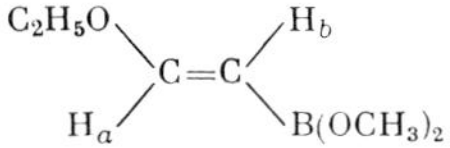	—30.1	3.49	6.95	4.39	—	14.00	—	—	—	0.0	2.9	—	2.4

Note. The chemical shifts of the protons are given on the δ scale; the chemical shifts of ^{11}B were measured relative to $BF_3.O(C_2H_5)_2$.

alkoxy or alkylthio groups in the β-position of trivinylborane significantly reduces the vibrational frequency of the C=C bond. Thus, *tris*(β-ethylthio-β-ethoxyvinyl)borane absorbs at 1532 cm^{-1} [46].

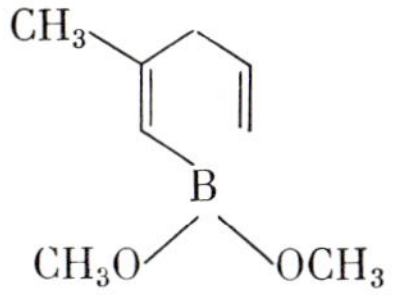

1628 cm^{-1}

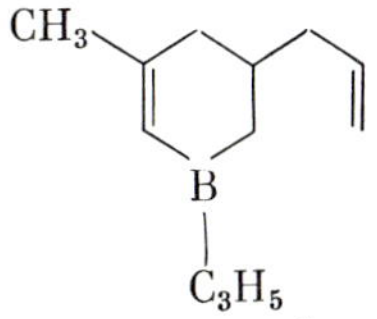

1600 cm^{-1}

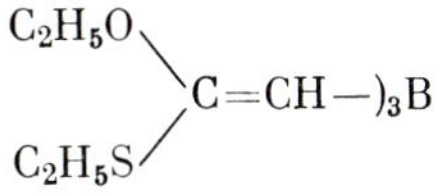

1532cm $^{-1}$

The ^1H and ^{11}B NMR spectra of different types of alkenyl boron compounds have been investigated.

The chemical shift of ^{11}B in trivinylborane has a value of 55.2 ppm relative to boron trifluoride etherate [3].

Table 2 presents data on the chemical shifts of ^1H and ^{11}B and on the coupling constants $J_{H—H}$ and $J_{B—H}$ for esters of *trans-* and *cis*-propenylboronic acids and some β-ethoxyvinylboranes. The values of geminal and vicinal $J_{B—H}$ were obtained by an indirect method from the line broadening in the ^1H and ^{11}B NMR spectra by quadrupole relaxation of ^{11}B [59].

Data on the ^1H NMR spectra of propenyl and isoproprenyl compounds with different groups on the boron atom are presented in [60]. For the ^{13}C NMR spectra of trivinylborane and halovinylboranes, see [180].

2.2 DISPROPORTIONATION

As noted above, lower alkenylalkylboranes undergo disproportionation at room temperature [2], while higher members of this series of compounds are quite stable and withstand vacuum distillation [31].

Halovinylboranes differ considerably in thermal stability. Difluorovinylborane is relatively stable in the gas phase (it remains unchanged during storage for one month). Dichlorovinylborane is not decomposed by brief heating to 80–100°C, but, in the liquid phase, undergoes partial change during storage for several months, possibly as a result of polymerization. Fluorodivinylborane undergoes a relatively fast change in the gas phase at room temperature, with the formation of difluorovinylborane and trivinylborane. The thermal stability increases in the following order:

$$(CH_2{=}CH)_2BF < (CH_2{=}CH)_2BCl < CH_2{=}CHBCl_2 < CH_2{=}CHBF_2 \text{ [6].}$$

Tris(2-chlorovinyl)borane starts to decompose, with the evolution of HCl and acetylene, only at 225°C [49]. Bis(2-chlorovinyl)-ethylborane is partially converted at 195°C over a period of five hours to chlorobis(2-chlorovinyl)borane and *tris*(2-chlorovinyl)-borane [51].

Dichloro(2-chlorovinyl) borane does not change when boiled for 48 hours but, in the presence of activated carbon at 100°C, is converted to BCl_3 and chlorobis(2-chlorovinyl)borane. The latter compound shows a very high thermal stability [49].

Dibutyl vinylboronate undergoes polymerization upon heating or in the presence of oxygen [61].

2.3 PROTONOLYSIS. SYNTHESIS OF OLEFINS, HALOOLEFINS, α, β-UNSATURATED ACIDS, CYCLENES, DIENES, DIENE ALCOHOLS, AND ENYNES

Alkenyl boron compounds are converted to olefins by the action of water, alcohols, or carboxylic acids. The ease with which the reaction proceeds is determined by the nature of the alkenylborane and of the hydrolyzing agent.

Tris(perfluorovinyl)borane is decomposed by water at 120°C with the formation of trifluoroethylene [12]. The relative stability of vinylboranes towards rupture of the B—C bond under the action of water increases in the following order [7]:

$$C_2H_5BF_2 < (CH_2{=}CH)_2BCl < CH_2{=}CHBCl_2 < CH_2{=}CHBF_2 < CF_2{=}CFBF_2.$$

The protonolysis of (1-alkenyl)alkylboranes is generally carried out with acetic acid at room temperature.

Deuterolysis with deuteroacetic acid of the products of the monohydroboration of 1-octyne with tetracyclohexyldiborane or bis(2,3-dimethyl-2-butyl)diborane leads to *trans*-1-deutero-1-octene in a yield of 85–94% [64]. From these data, it follows that tetraalkyldiboranes, like alkylalanes [29], add predominantly to the terminal carbon atom of the triple bond:

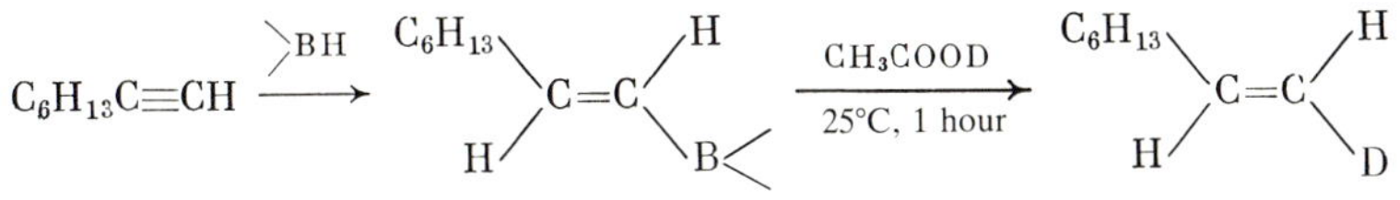

Phenylacetylene is converted to *trans*-β-deuterostyrene in a yield of 87–93% by an analogous method [62].

The hydroboration of 1-hexyne with tetra(3-methyl-2-butyl)-diborane followed by deuterolysis, leads to *trans*-1-deutero-1-hexene (yield 48%); *cis*-1-deutero- or 1,1-dideutero-1-hexene can be obtained from 1-deutero-1-hexyne [63].

1-Hexene, labeled with tritium predominantly in the 1 position, has been synthesized by the hydroboration of 1-hexyne with tetra(3-methyl-2-butyl)diborane, followed by acidolysis with CH_3COOT [64].

3-Hexyne is converted to *cis*-3-hexene (yield 83%) by hydroboration with tetra(3-methyl-2-butyl)diborane, followed by protonolysis with acetic acid [36]:

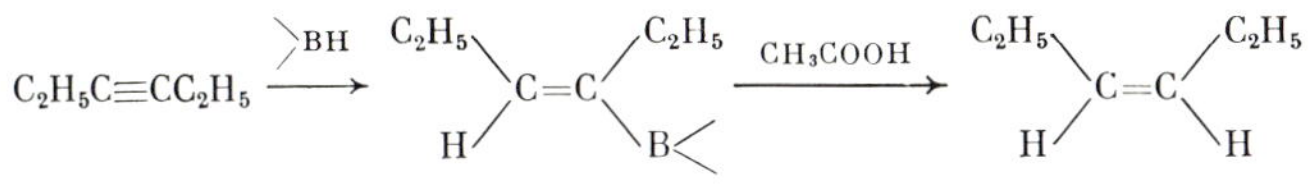

cis-Olefins are formed by the acidolysis of trialkenylboranes, obtained by the transalkylation of triisobutylborane with 5-decyne [65]:

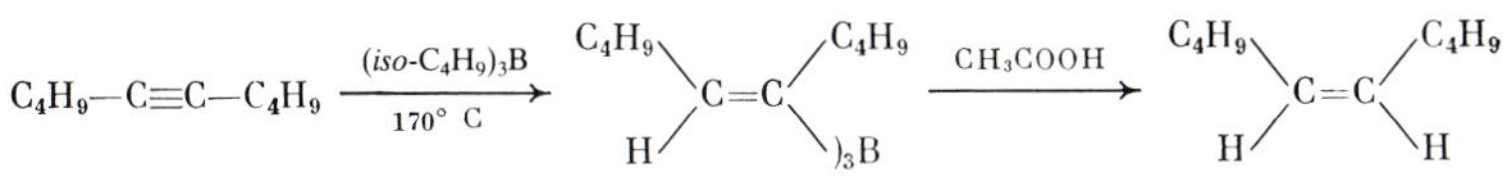

Optically active olefins are obtained from alkenylboranes with optically active substituents by protonolysis with acetic acid [181].

1-Alkenyldialkylboranes undergo protonolysis by the action of catalytic quantities of palladium(II) acetate in acetone and deuteroacetone with the formation of *Z*-olefins [182].

Alkenyldialkylboranes with a terminal alkenyl group (—CH=CHR) undergo ordinary protonolysis upon heating with hydrochloric acid, whereas non-terminal alkenyl compounds

(—CR=CHR) undergo rearrangement under these conditions, with migration of an alkyl group from the boron to the α-carbon [183].

Cyclodecyne is converted to *cis*-1,2-dideuterocyclodecene by hydroboration with hexadeuterodiborane, followed by acidolysis with deuteroacetic acid [66]:

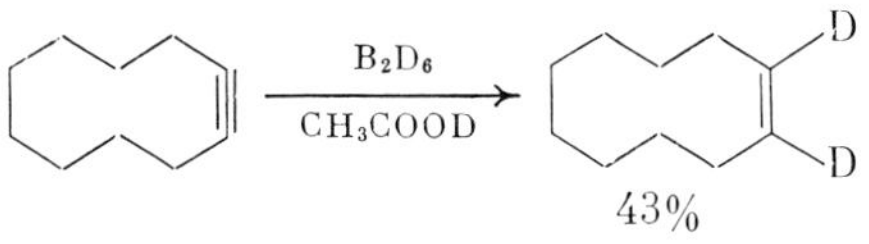

Other cycloalkynes have also been converted to cycloalkenes in the same manner [67–69].

α-Halohexenyl-1-dicyclohexylboranes are converted to *cis*-1-halo-1-hexenes by protonolysis with acetic acid [40]:

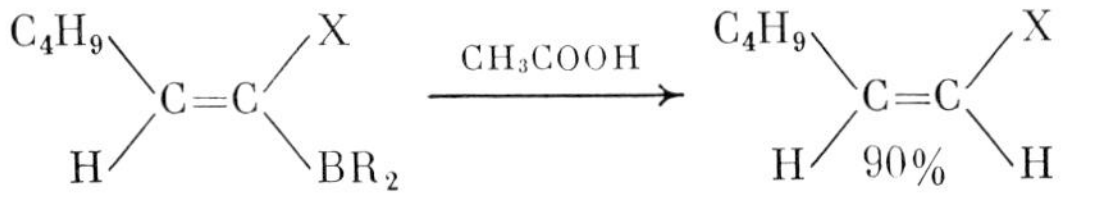

(X = Br, I).

Boiling a methanollic solution of dimethyl (2-alkylthio-2-ethoxyvinyl)boronate leads to 1-alkylthio-1-ethoxyethylene [45]. The action of methanol on tris(2-ethylthio-2-ethoxyvinyl)borane leads to the rupture of all three B—C bonds, with the first two being ruptured very easily (within one hour at room temperature), while the protonolysis of the third bond requires a longer action of the alcohol (approximately 10 hours) or moderate heating [46]:

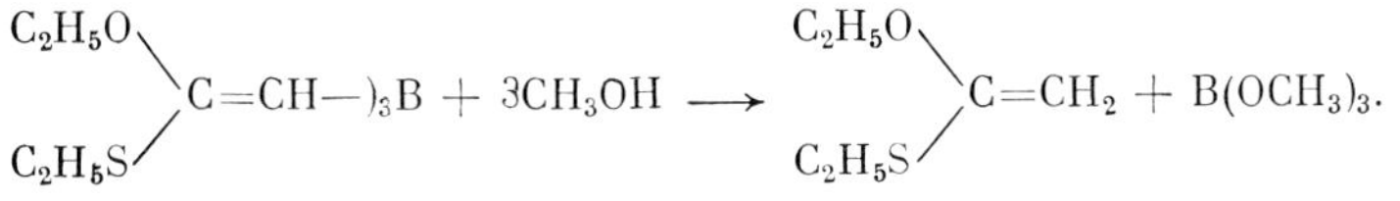

Vinyl boron derivatives are generally characterized by a greater tendency toward protonolysis of the B—C bond than trialkylboranes [70]. A qualitative investigation of the rate of protonolysis of different types of vinylboranes with alcohol indicates that this increases in the following order [45, 46, 70, 71]:

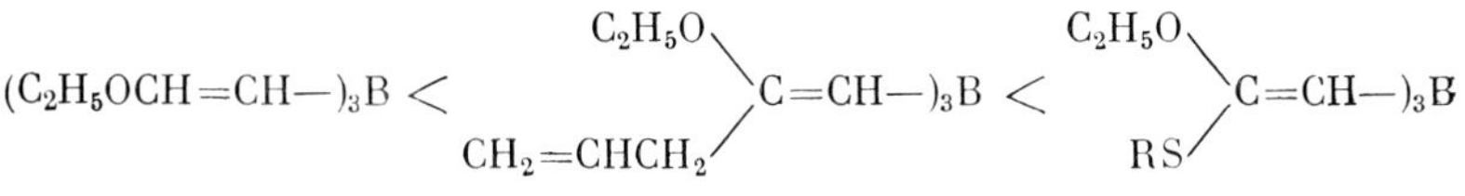

Esters of acetylenic carboxylic acids react with tetra(3-methyl-2-butyl)diborane to form predominantly α-substituted organoboranes ($\alpha:\beta = 90:10$), which hydrolyze upon heating (65°C) with acetic acid to esters of *cis*-α,β-unsaturated carboxylic acids in 80% yield [72]:

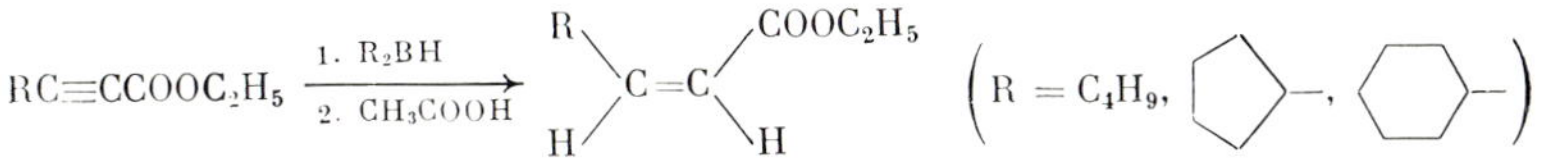

When R is *tert*-C_4H_9, a mixture of the *cis* and *trans* isomeric esters in a ratio of 85:15 is obtained.

In enyne hydrocarbons, the triple carbon-carbon bond is initially attacked in the hydroboration. Petrov, et al. [73] found that the action of tetra(3-methyl-2-butyl)diborane on isopropenylacetylene, followed by acidolysis, leads to a 63% yield of isoprene:

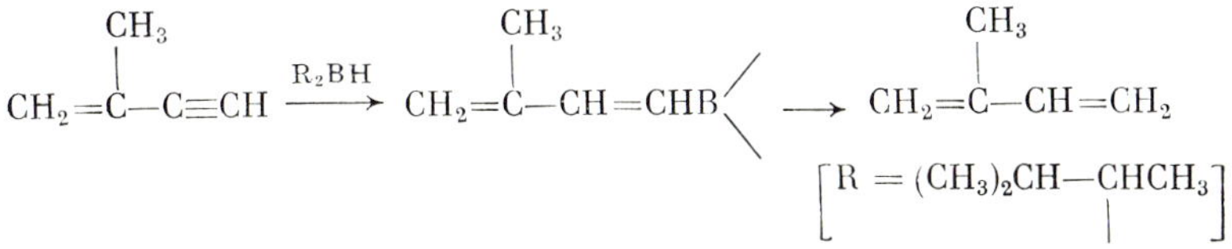

The monohydroboration of isopropenylacetylene with tetracyclohexyldiborane or bis(2,3-dimethyl-2-butyl)diborane, followed by deuterolysis, leads to 1-deutero-3-methyl-1,3-butadiene in high yield [62].

1-Ethynylcyclohexene is converted to *trans*-1-cyclohexenyl-2-deuteroethylene by hydroboration-deuterolysis [62]:

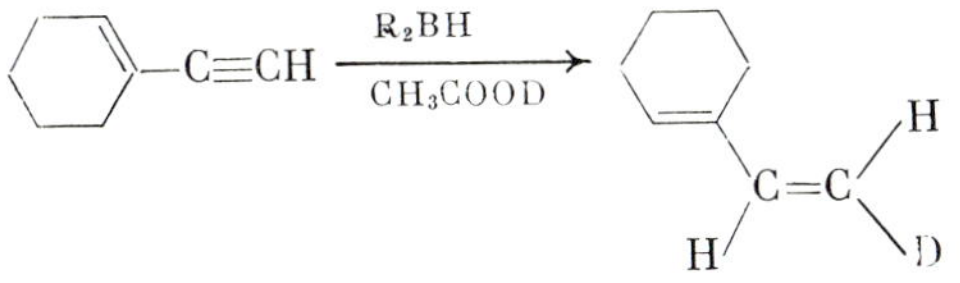

2,2,7,7-Tetramethyl-*cis*-3-octen-5-yne is converted to 2,2,7,7-tetramethyl-*cis*,*cis*-3,5-octadiene in 75% yield by the above method [74]:

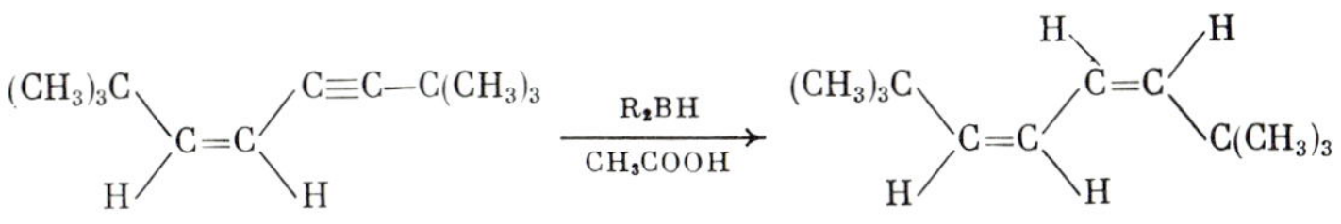

trans-1-Acetoxy-4-methyl-4-hexen-2-yne is converted to the diene alcohol *1* by hydroboration with tetra(3-methyl-2-butyl)diborane, followed by protonolysis [75]:

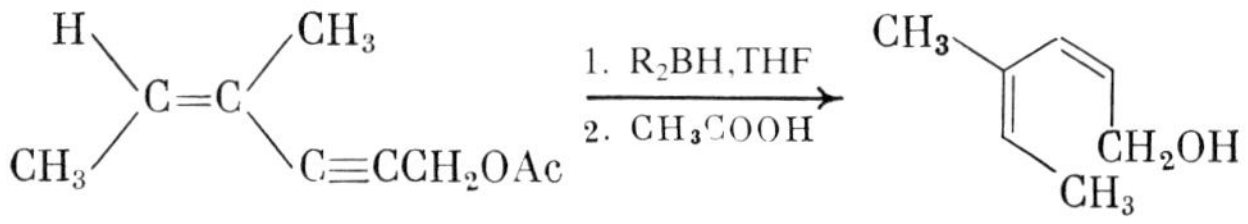

1

Dicyclohexyl-3-methyl-*trans*-1,3-butadienylborane is capable of undergoing photocyclization to 1,2-dicyclohexyl-4-methylboracyclopent-3-ene [76]:

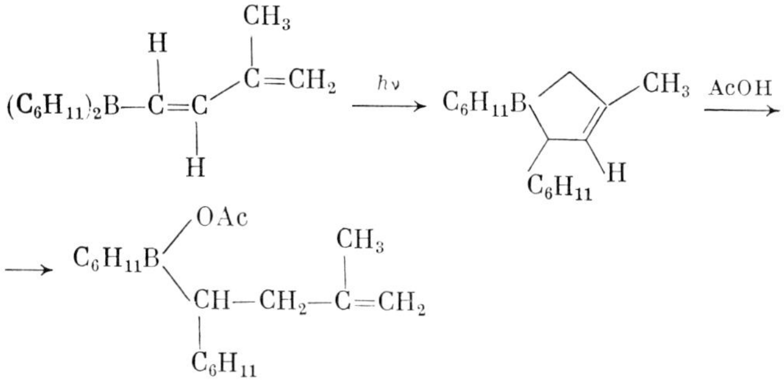

Zweifel and Polston [74] found that, in the hydroboration of conjugated diacetylenic hydrocarbons—5,7-dodecadiyne, 2,7-dimethyl-3,5-octadiyne, and 2,2,7,7-tetramethyl-3,5-octadiyne—with tetra(3-methyl-2-butyl)diborane, the boron atom primarily attacks the internal carbon atom of the diyne system. Protonolysis of the intermediate unsaturated boranes formed leads to the formation of 75% of the corresponding *cis*-enynes, containing diene hydrocarbons as impurities (1–7%):

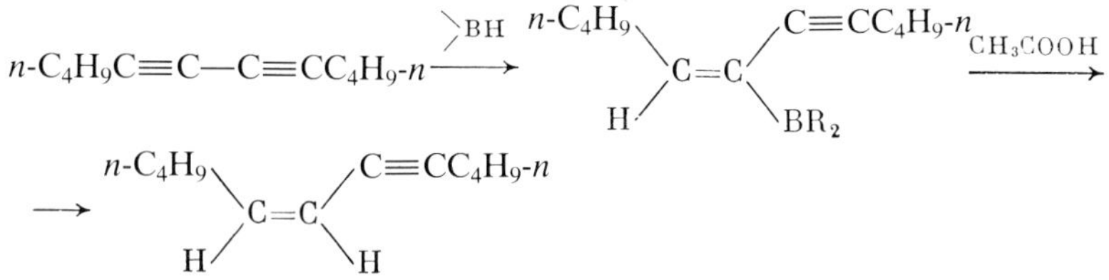

If different radicals are located at the ends of the conjugated carbon chain, such as, for example, in 2,2-dimethyl-3,5-decadiyne, a mixture of the two possible *cis*-enynes is formed:

$$(CH_3)_3CC{\equiv}C{-}C{\equiv}CC_4H_9\text{-}n \xrightarrow[2.\ CH_3COOH]{1.\ R_2BH} (CH_3)_3CCH{=}CHC{\equiv}CC_4H_9\text{-}n +$$
$$+ (CH_3)_3CC{\equiv}C{-}CH{=}CHC_4H_9\text{-}n$$

A small yield of *cis*-5,7-dodecadiene is formed in the hydroboration of 5,7-dodecadiyne with tetra(3-methyl-2-butyl)diborane, followed by protonolysis [65]. The best dihydroborating agent in this case is tetracyclohexyldiborane, the use of which produces a diene yield of 79% [74]:

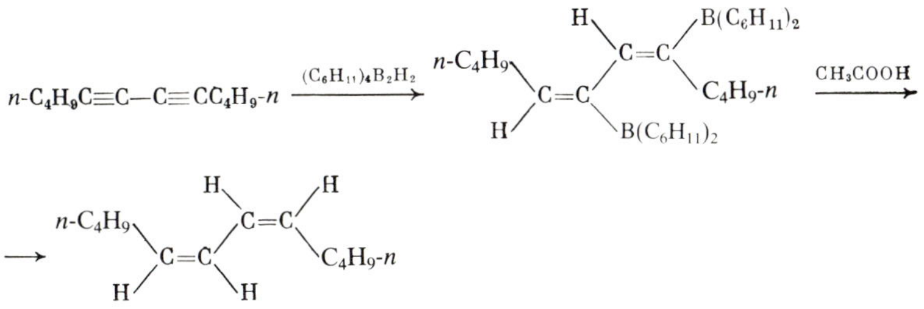

2.4 OXIDATION. SYNTHESIS OF ALDEHYDES, ACETYLENIC KETONES, α-OXO ACIDS, AND SATURATED AND UNSATURATED ACIDS.

Alkenyldialkylboranes, formed by the hydroboration of acetylenes with a terminal C≡C bond with bis(3-methyl-2-butyl)borane, are oxidized with hydrogen peroxide in alkaline medium to the corresponding aldehydes in yields of 60–88% [36, 77]:

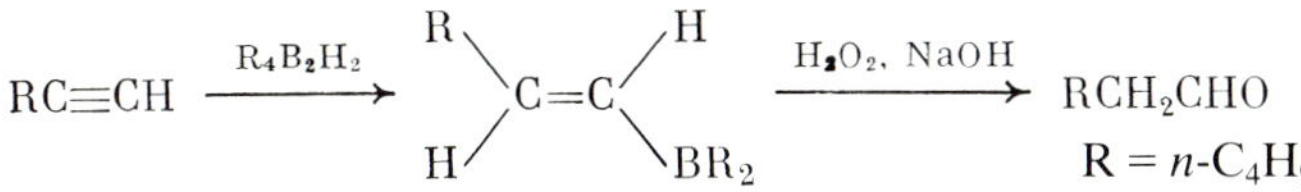

$R = n\text{-}C_4H_9,\ n\text{-}C_6H_{13}$).

1-Hexyne is converted to caproaldehyde in a yield of 82% by the action of bis(2,3-dimethyl-2-butyl)diborane (4:1), followed by oxidation [78].

The monohydroboration-oxidation of terminal acetylenes with the use of 9-borabicyclo[3.3.1]nonane forms aldehydes, while protonolysis leads to *cis*-olefins [184].

The oxidation of alkenyldialkylboranes, obtained by the hydroboration of 3-hexyne [36], or of trialkenylboranes, obtained by the transalkylation of triisobutylborane [65], leads to the corresponding ketones in yields of 50–76%.

Unsymmetrical dialkylacetylenes are converted by hydroboration-oxidation to a mixture of two ketones, the ratio of which is determined by the nature of the starting hydrocarbons and of the hydroborating agents [62].

The esters of acetylenic carboxylic acids are converted to esters of α-oxocarboxylic acids by hydroboration with bis(3-methyl-2-butyl)borane, followed by oxidation with hydrogen peroxide in a buffer solution [72]:

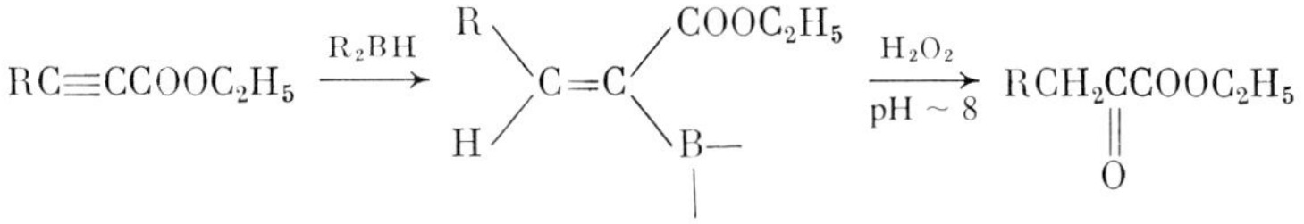

The cyclic vinylboranes obtained in predominant quantity by the action of diborane or of bis(3-methyl-2-butyl)borane on cyclic allenic hydrocarbons are converted to cyclanones by oxidation with alkaline hydrogen peroxide [79, 80].

The acetylenic hydrocarbons *2* and *4* are converted by hydroboration with bis(3-methyl-2-butyl)borane and oxidation to the aldehydes *3* [81] and *5* [82], respectively:

2 3 4 5

Oxidation of the products of monohydroboration of symmetrical disubstituted 1,3-diynes forms α,β-acetylenic ketones in yields of 74–80% (according to GLC data) [74]:

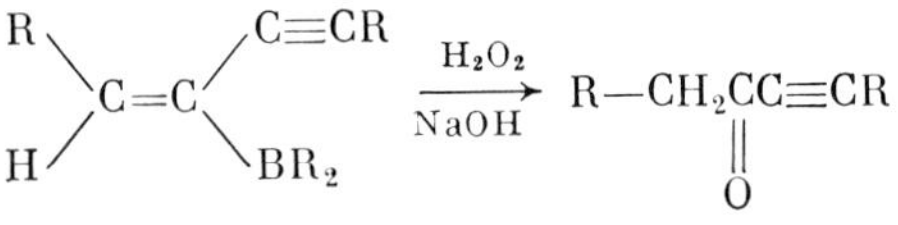

Macrocyclic alkadiynes are converted to cyclic acetylenic ketones by heating with triisobutylborane and oxidation [83]:

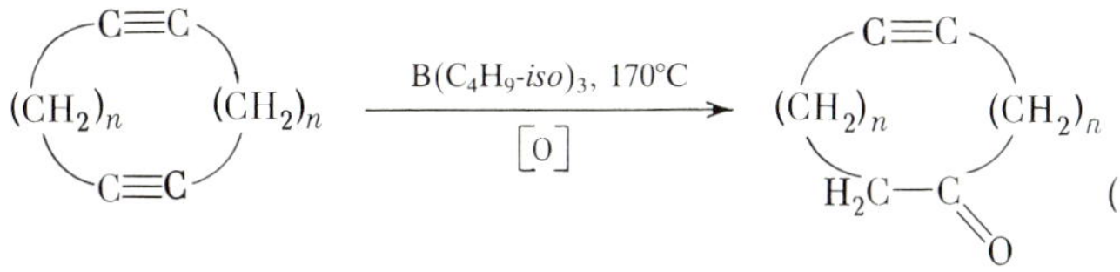

$(n = 5, 6, 7)$.

The hydroboration-oxidation of alkynylsilanes has been used as a basis for the development of methods for the synthesis of saturated and unsaturated carboxylic acids [185] and acylsilanes [186–187].

2.5 REACTIONS WITH HALOGENS

2.5.1 Reactions with chlorine and bromine. Synthesis of bromoolefins and α, β-unsaturated acids

Mikhailov and Aronovich showed that vinyl boron compounds smoothly add chlorine [84] and bromine [85].

The action of chlorine on a hexane solution of dibutyl vinylboronate at −60 to +50°C forms dibutyl 1,2-dichloroethylboronate:

$$CH_2{=}CHB(OC_4H_9\text{-}n)_2 + Cl_2 \longrightarrow ClCH_2CHClB(OC_4H_9\text{-}n)_2.$$

With bromine, the vinylboronate forms a 93% yield of dibutyl 1,2-dibromoethylboronate:

$$CH_2{=}CHB(OC_4H_9\text{-}n)_2 + Br_2 \longrightarrow BrCH_2CHBrB(OC_4H_9\text{-}n)_2.$$

Cadiot and co-workers [86] investigated the bromination in carbon tetrachloride of glycol esters of vinyl-, propenyl-1- and isopropenylboronic acids, propenylboron dichloride, 1-propenylborepane, and some nitrogen-containing vinylboranes; the yields of the corresponding dibromides were 60–90% in most cases. From a mixture of the *cis* and *trans* isomers of glycol propenylboronate (30:70), a mixture of dibromides with threo and erythro configurations is formed, while the dibromide with the erythro configuration is obtained from the *trans* isomer.

Under the action of alkoxides, dilute acids, and aqueous alkali, dibutyl 1,2-dibromoethylboronate eliminates bromoethylene [85]. 2-(1,2-Dibromoethyl)-4,4,6-trimethyl-1,3,2-dioxaborinane reacts analogously with alkali [87].

To investigate the stereochemistry of the elimination process, Matteson and Liedtke [88] prepared the erytho (*6*) and threo (*7*) forms of dibutyl 2,3-dibromo-2-butylboronate by the addition of bromide to dibutyl *trans*- and *cis*-2-butenylboronate, respectively.

Under the action of water, aqueous sodium hydroxide, or pyridine, the esters *6* and *7* undergo β-elimination with *cis*-2-bromo-2-butene being obtained from the erythro form *6* and *trans*-2-bromo-2-butene from the threo form *7*, indicating the stereospecifity of the *trans*-elimination:

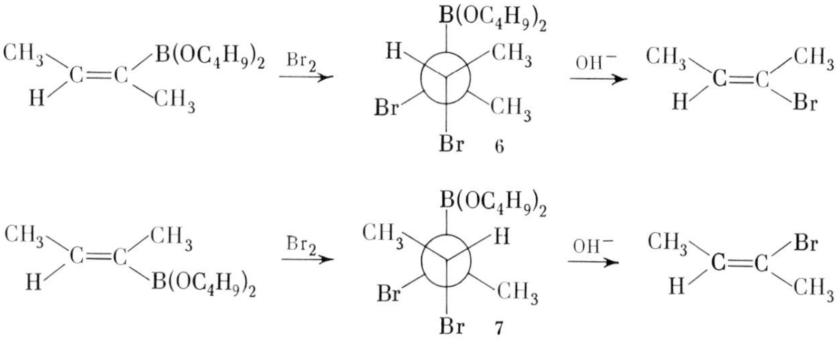

A number of bromoalkenes has been obtained by the above method with the use of pyrocatechol esters of alkenylboronic acids [89].

For the synthesis of bromoalkenes, 1-alkenyldialkylboranes, formed by the action of alkyldiboranes on acetylenic hydrocarbons, can be used as starting materials. 1-Alkynes are hydroborated with bis(3-methyl-2-butyl)borane and the *trans*-alkenyldi(3-methyl-2-butyl)boranes formed, without isolation in the pure state, are then treated sequentially with bromine and aqueous sodium hydroxide solution. The *cis*-bromoalkenes obtained by this method (Table 3) contain 2–5% of the trans isomer; the yields are 65–76%, according to GLC data [90]:

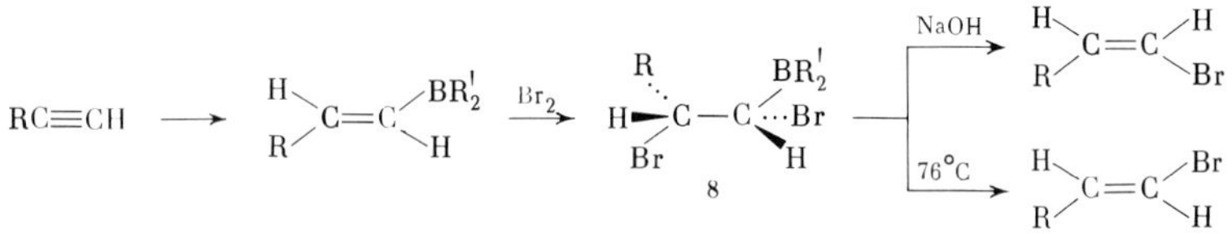

TABLE 3
Synthesis of bromoalkenes from acetylenes.*

Acetylenic hydrocarbon	Bromoalkene	Solvolysis Yield	Solvolysis cis : trans	Thermal decomposition Yield	Thermal decomposition cis : trans
1-Pentyne	1-Bromo-1-pentene	65	97:3	40	0:100
1-Hexyne	1-Bromo-1-hexene	67	95:5	75	12:88
1-Heptyne	1-Bromo-1-heptene	76	97:3	60	12:88
1-Octyne	1-Bromo-1-octene	68	98:2	50	6:94
Phenylacetylene	β-Bromostyrene	88	4:96	78	82:18

* Values in %

If the dibromide *8* is heated in carbon tetrachloride solution, thermal elimination of the *trans*-bromoalkene predominantly takes place (Table 3). It is interesting that the dibromide obtained from phenylacetylene by sequential hydroboration and bromination forms *trans*-β-bromostyrene (96% pure) when treated with alkali but is converted mainly to *cis*-β-bromostyrene by thermal decomposition.

Negishi and co-workers [91] discovered a stereospecific method for the synthesis of esters of (*E*)- and (*Z*)-α,β-unsaturated acids from alkyl(2,3-dimethyl-2-butyl)(β-carbethoxyvinyl)boranes (*9*), obtained by the addition of alkyl(2,3-dimethyl-2-butyl)borane to ethyl propiolate.

Treatment of the borane *9* with bromine, followed by heating, leads to ethyl esters of (*E*)-3-alkylacrylic acids (*11*), whereas, under the action of sodium ethoxide, the dibromides *10* are converted to the *Z* isomers *12*:

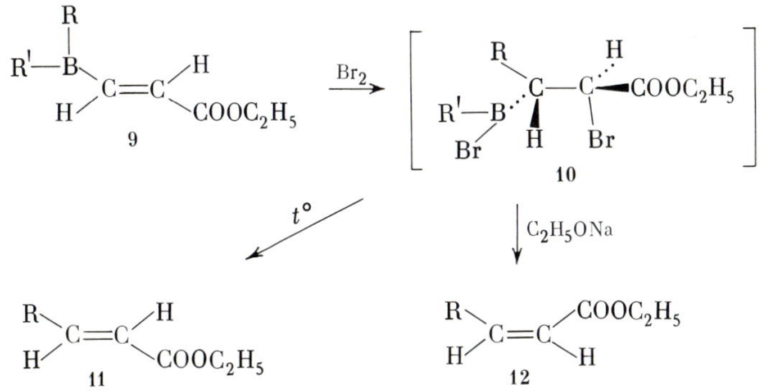

[R′ = $(CH_3)_2CH—C(CH_3)_2—$]

In the above case, the presence of the carboethoxy group promotes the α-transfer of an alkyl group from boron to the α-carbon atom and of a bromine atom from carbon to boron, as a result of which the dibromide *10* is formed instead of the normal dibromide in the bromination of *9*.

The reaction of alkenylboronic acids with bromine in the presence of sodium methoxide and methanol leads to α-bromoacetals [188].

2.5.2 Reactions with iodine. Synthesis of olefins, iodoolefins, and dienes

Zweifel and co-workers [92] found that the reaction of iodine with alkenyldialkylboranes leads not to iodoalkenes but to olefins, formed as a result of the migration of an alkyl group from the boron atom to an adjacent carbon. Thus, in the action of iodine on vinylborane, obtained by hydroboration of 1-hexyne with dicyclohexylborane, *cis*-1-cyclohexyl-1-hexene, containing 8% of the *trans* isomer, is formed in 81% yield. The reaction proceeds in a more stereospecific manner of iodine is reacted with a solution of the starting vinylborane in tetrahydrofuran to which a 6 N aqueous sodium hydroxide solution has been added. Under these conditions, the olefin yield is slightly lower (75%), but it is more than 99% pure:

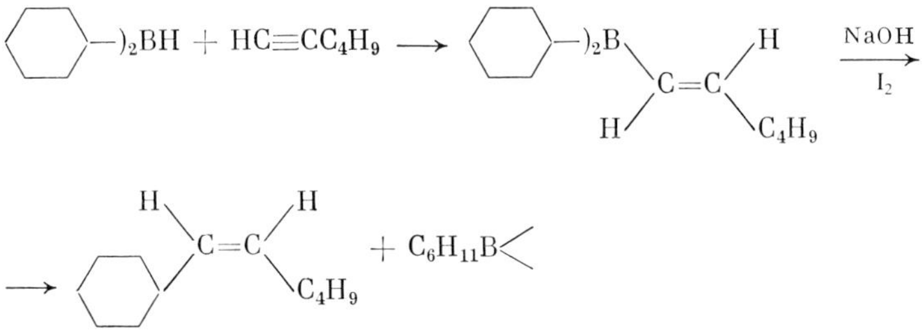

The following reaction mechanism for the iodination of vinylboranes is suggested, including the addition of iodine at the double bond with the formation of a bridging iodonium ion, migration of a hydrocarbon radical from the boron to an adjacent carbon atom with inversion at the migration center, and subsequent *trans*-elimination of the boron-containing group and iodine [92, 93]:

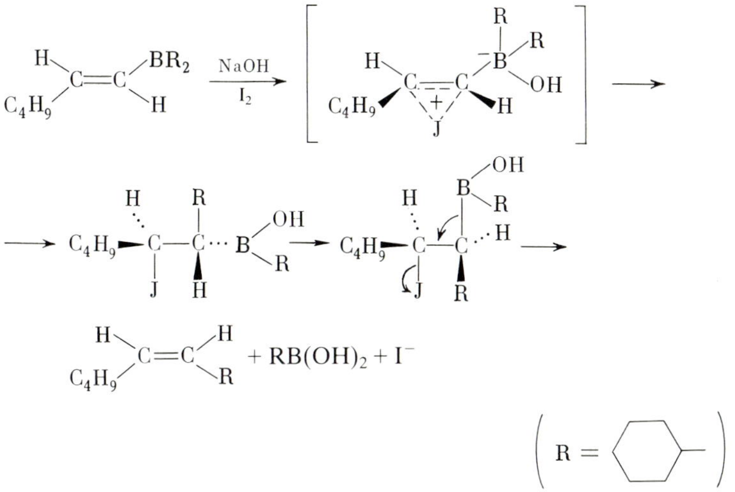

The nature of the solvent has a very significant effect on the course of the reaction of alkenyldialkylboranes with iodine in the presence of NaOH. Thus, if the reaction is carried out not in tetrahydrofuran but in methylene chloride, the stereospecificity of the elimination process increases considerably, with the main product (77%) being a *trans*-olefin rather than a *cis*-olefin [94].

An investigation of the iodination of the vinylborane obtained by the monohydroboration of 1-hexyne with bis(*trans*-2-methylcyclohexyl)borane has been carried out in order to clear up the question of retention or inversion of the configuration of the group migrating from the boron atom to the carbon [95]. It was found that, when this vinylborane is treated sequentially with sodium hydroxide and iodine solution in tetrahydrofuran, *trans*-1-methyl-2-(*cis*-1-hexenyl)cyclohexane is formed in 85% yield, indicating a retention of the configuration of the migrating 2-methylcyclohexyl radical:

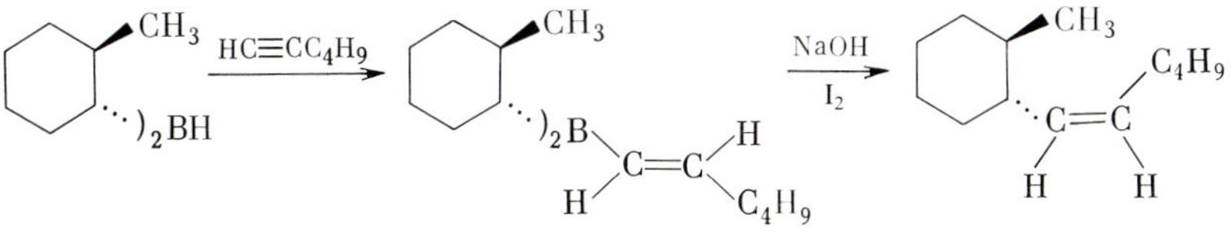

Similar results have also been obtained in other reactions investigated.

The reactions of vinylboranes with iodine in the presence of alkali make it possible to obtain diene hydrocarbons with a *cis,trans* configuration. Di- and trivinylboranes, obtained by the monohydroboration of the corresponding acetylenic hydrocarbons with diborane or dialkyldiborane, are used for this purpose. Thus, tris(*cis*-3-hexen-3-yl)borane, formed by the action of diborane on 3-hexyne, is converted to *cis,trans*-4,5-diethyl-3,5-octadiene in 68% yield by the action of iodine and sodium hydroxide in tetrahydrofuran [96]:

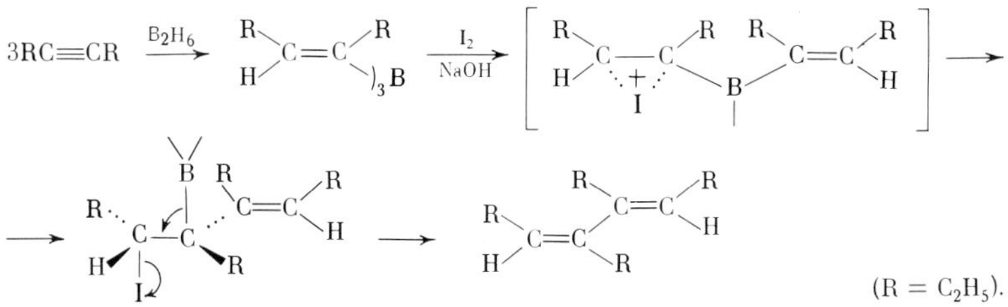

(R = C_2H_5).

In this method for the synthesis of diene hydrocarbons, bis(2,3-dimethyl-2-butyl)diborane is widely used as a hydroborating agent, since acetylenes with a terminal triple bond can be involved in the reaction with its use [96]. 1-Hexyne reacts with this dialkyldiborane to form the corresponding divinylalkylborane which, via the stage of oxidation to a divinylborinate by means of trimethylamine oxide [97], is converted by treatment with iodine and sodium hydroxide to *cis,trans*-5,7-dodecadiene in a yield of 65% (degree of purity 98%):

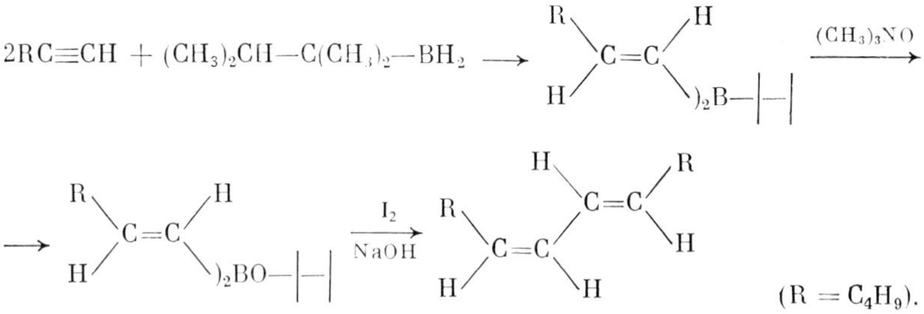

(R = C_4H_9).

In this way, *cis,trans*-1,4-dicyclohexyl-1,3-butadiene has been obtained from cyclohexylacetylene in 64% yield. Bis(2,3-dimethyl-2-butyl)borane has been used with equal success for the conversion of non-terminal acetylenes to dienes. Thus, *cis,trans*-3,4-dimethyl-2,4-hexadiene is formed from 2-butyne in 49% yield and the yield of *cis,trans*-4,5-diethyl-3,5-octadiene from 2-hexyne is 69%.

The alkenyldialkylboranes formed by the protonolysis of alkynyltrialkylborates react with iodine in alkaline medium to form trialkylethylenes [189].

Lithium tricycloisopropenylborates react with iodine to form isopropenylcyclanes. Limonene, diprene, and an isomeric dihydrolinalool have been synthesized by this method [98].

The action of iodine on lithium alkenyltrialkylborates produces trisubstituted olefins [190, 191] and 1,1-dialkylethylenes [192]. Lithium α-methoxyvinyltrialkylborates (*12a*), obtained from α-methoxyvinyllithium and R_3B at −78°C, rearrange under the action of iodine to the enol ethers *12b*, which form methyl ketones by protonolysis. The borates *12a* rearrange at 25°C to the alkenyldialkylmethoxyborates *12c*. Tertiary alcohols are obtained by protonolysis and oxidation of the compounds *12c*, while iodination leads to 1,1-dialkylethylenes [193, 194]:

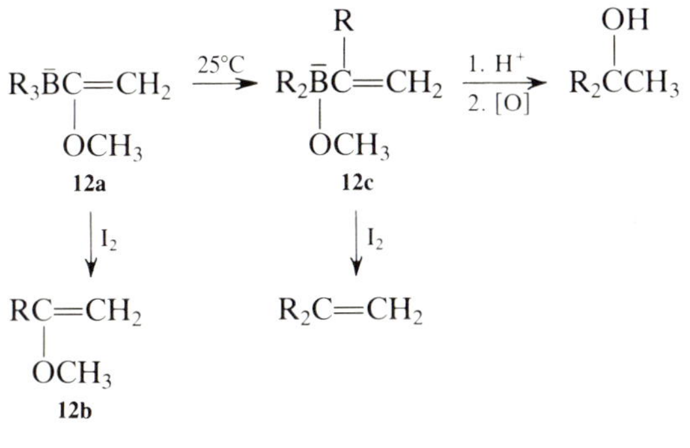

Evans, et al. [195, 196] have used the Zweifel reaction in the synthesis of prostanoids. In the best variation, the starting compound is *trans*-2-ethylcyclopentyldimethoxyborane, which is converted to an *ate* complex with the corresponding vinyllithium reagents. Subsequent rearrangement of the *ate* complex under the action of iodine and alkali leads to a prostanoid:

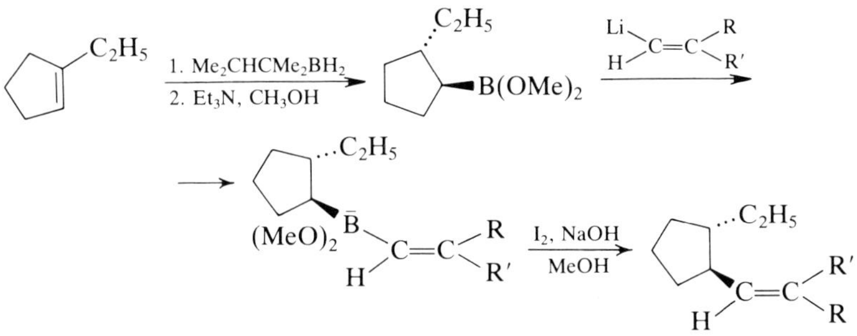

$R = —CH(OSiMe_3\text{-}t\text{-}Bu)—C_5H_{11}$, R′ = H or vice versa

The iodination of alkenylboranes has been used in the synthesis of propylure [197].

trans-1-Iodoalkenes are formed by the action of iodine on alkaline solutions of 1-alkenylboronic acids [198]:

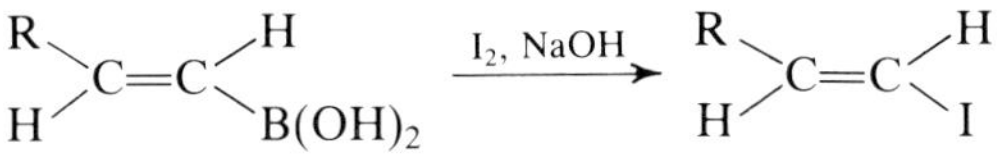

$R{=}C_4H_9$, C_6H_{13}, 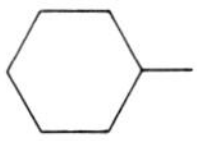, C_6H_5

Iodine monochloride can be used instead of iodine in the synthesis of alkenyl iodides from alkenylboronic acids [199].

The action of ICl in the presence of sodium acetate on the product of the hydroboration of methyl 10-undecynoate with catecholborane forms (*E*)-11-iodo-10-undecenoic acid [200].

2.6 REACTIONS WITH HALOCYANIDES. SYNTHESIS OF OLEFINS

Dicyclohexyl(*trans*-1-hexenyl)borane, when reacted with an equimolar quantity of bromocyanide in tetrahydrofuran or methylene chloride at 0–5°C, is converted to *trans*-cyclohexyl-1-hexene, containing 4–6% of the *cis* isomer, in 74–80% yield [94].

It is assumed that, in this case, as in the reaction with iodine, an electrophilic addition of bromine takes place, with the formation of a bromonium ion and the subsequent migration of an alkyl group from boron to carbon. In the organoboron *13* obtained, a donor-acceptor interaction takes place between the bromine atoms and the boron connected to the strongly electron-drawing nitrile group, resulting in the *cis*-elimination of bromine and the boron-containing group from the eclipsed conformation of the organoborane *14*:

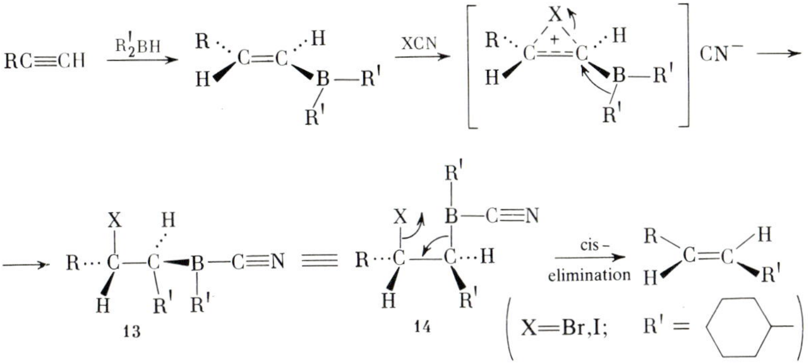

When iodocyanide is used, the highly pure (94%) *trans* isomer is obtained only when the reaction is carried out in methylene chloride; in tetrahydrofuran, a mixture of the *cis* and *trans* isomers of the olefins in a ratio of 46:54 is formed. These data are in good agreement with the scheme presented above, because, in the molecule with the bulkier iodine atom, the eclipsed conformation is energetically less favorable, and in this case a significant effect is exerted by tetrahydrofuran, which competes with the iodine atom in the coordinative interaction with boron. Experiments with the use of bis(*trans*-2-methylcyclohexyl)borane indicate that the migration of the hydrocarbon radical from boron to carbon proceeds with retention of the configuration.

2.7 REACTIONS WITH HYDROGEN HALIDES, MERCAPTANS, AND HALOMETHANES

Matteson and co-workers [99, 100] investigated the heterolytic addition of hydrogen bromide to dibutyl alkenylboronates. A vinylboronate does not react with liquid HBr at atmospheric pressure, but forms a mixture of isomers at elevated pressure [100]:

$$CH_2{=}CHB(OR)_2 + HBr \longrightarrow \underset{60\%}{CH_2BrCH_2B(OR)_2} + \underset{40\%}{CH_3CHBrB(OR)_2}.$$

A propenylboronate adds HBr to form a 2-bromopropylboronate [101]:

$$CH_3CH{=}CH{-}B(OR)_2 + HBr \longrightarrow CH_3CHBrCH_2B(OR)_2.$$

The orienting effect of the dialkoxyboryl group in the electrophilic addition of HBr is completely suppressed by the effect of a methyl group in the α-position. Thus, isopropenylboronates and *trans*-2-butenyl-2-boronates add a hydrogen atom in the β-position:

$$CH_2{=}\overset{\overset{CH_3}{|}}{C}{-}B(OR)_2 + HBr \longrightarrow CH_3\overset{\overset{CH_3}{|}}{C}BrB(OR)_2$$

$$CH_3CH{=}\overset{\overset{CH_3}{|}}{C}{-}B(OR)_2 + HBr \longrightarrow CH_3CH_2\overset{\overset{CH_3}{|}}{C}BrB(OR)_2$$

As found by Cadiot and co-workers [86], the nature of the substituents on the boron atom has an effect on the orientation of the addition of HBr to vinylboranes. Thus, chloro(diisopropylamino)-vinylborane adds boron exclusively in the α-position.

The electrophilic hydrobromination of glycol β-styrylboronate at -70°C in the dark leads to a 2-bromo-2-phenylethylboronate [21]:

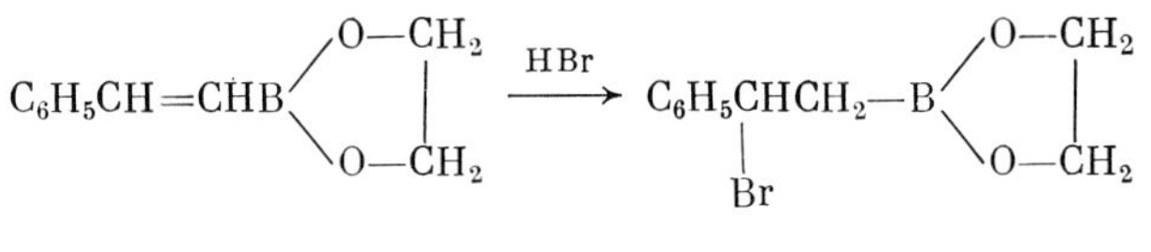

Hydrogen iodide adds to dibutyl vinylboronate to form 1- and 2-iodoethylboronates (60:40) [100].

Dibutyl vinylboronate adds hydrogen bromide [102], bromotrichloromethane [19, 103], bromoform and carbon tetrabromide [101, 103], bromomalononitrile [100], mercaptans and thio acids [19, 103, 104] by a radical mechanism.

Reactions of the indicated reagents with dibutyl propenyl- and isopropenylboronates have also been investigated [19]:

$$RCH{=}C(R')B(OC_4H_9)_2 \xrightarrow{XY} RCH(X){-}CR'(Y)B(OC_4H_9)_2$$

a) $R = R' = H$; b) $R = H$, $R' = CH_3$; c) $R = CH_3$, $R' = H$;

a) $R = R' = H$, $X = CCl_3$, $Y = Cl$;
b) $R = R' = H$, $X = CCl_3$, $Y = Br$;
c) $R = R' = Y = H$, $X = n\text{-}C_6H_{13}S$;
d) $R = H$, $R' = CH_3$, $X = CCl_3$, $Y = Cl$;
e) $R = H$, $R' = CH_3$, $X = CCl_3$, $Y = Br$;
f) $R = CH_3$, $R' = H$, $X = CCl_3$, $Y = Br$.

Review articles by Matteson [101, 105] have been devoted to a consideration of these reactions, which lead to esters of functionally substituted alkylboronic acids.

Butyl divinylboronate adds hydrogen sulfide in the presence of azobisisobutyronitrile to form 4-butoxy-1,4-thiaborinane (4-butoxy-1-thia-4-boracyclohexane) in 12% yield [106]:

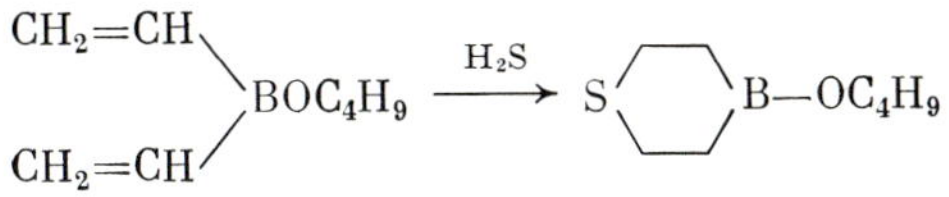

The addition of HBr, $BrCCl_3$, thiophenol [107], and triphenylstannane [108] to *B*-trivinyl-*N*-triphenylborazene has been investigated.

2.8 REACTIONS OF ALKENYLBORATES WITH VARIOUS REAGENTS. SYNTHESIS OF OLEFINS, UNSATURATED ALCOHOLS AND KETONES, β-HYDROXY ACIDS, DIENES, ALLENES, AND 1,2,3-BUTATRIENES

α-Halovinylboranes undergo anionotropic rearrangement during conversion to the corresponding borates. Thus, *trans*-olefins are formed by the action of sodium methoxide on α-halovinylboranes, followed by treatment of the reaction mixture with acetic acid [40, 95]:

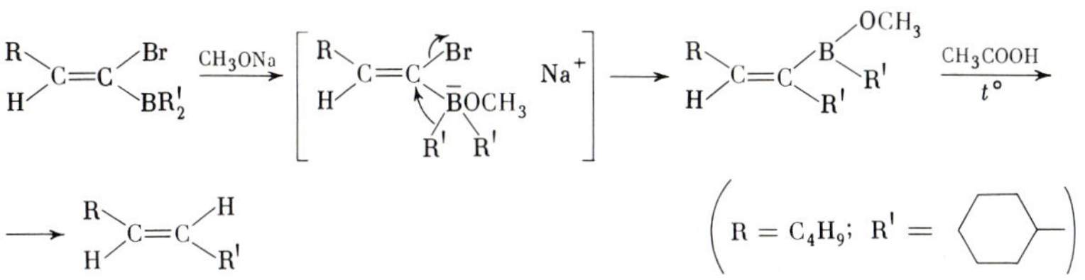

The hydroboration of 1-bromo-1-hexyne with bis(*trans*-2-methylcyclohexyl)borane, followed by treatment of the α-halovinylborane *15* obtained with sodium methoxide, and then with acetic acid, leads to *trans*-1-methyl-2-(*trans*-1-hexenyl)cyclohexane (*16*). The configuration of the migrating group is retained in this process [95]:

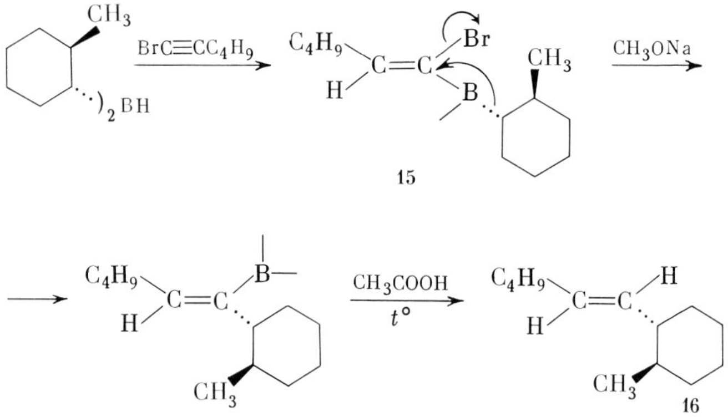

15

16

The above method for the synthesis of *trans*-olefins has been used for the preparation of *trans*-1-cyclohexyl-1-octen-3-ol (*17*) [109].

trans-Olefins have been synthesized from 1-bromoalkynes with the use of (2,3-dimethyl-2-butyl)alkylboranes as hydroborating agents [110]:

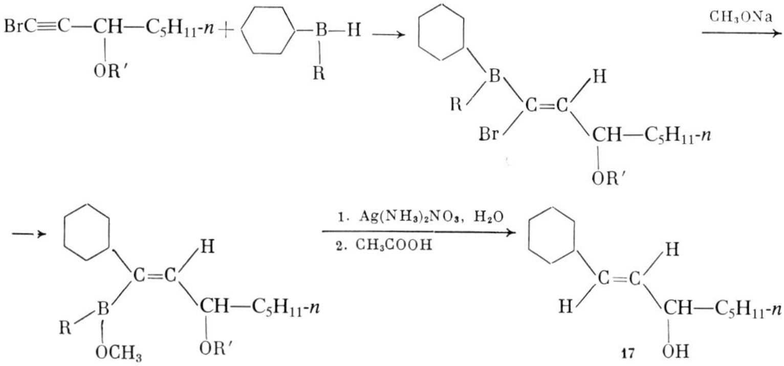

17

The rupture of the boron-carbon bond was carried out under mild conditions with an aqueous solution of $Ag(NH_3)_2NO_3$ at 75–80°C.

This reagent was previously used for the conversion of trivinylborane to ethylene [2, 6]. *trans*-1-(2-Methyl-5-hydroxycyclopentyl)-1-octen-3-ol has been synthesized by an analogous method.

The treatment of α-halovinylboranes with sodium hydroxide also leads to rearranged 1,2-disubstituted vinylboranes which are oxidized with hydrogen peroxide to the corresponding ketones [40]:

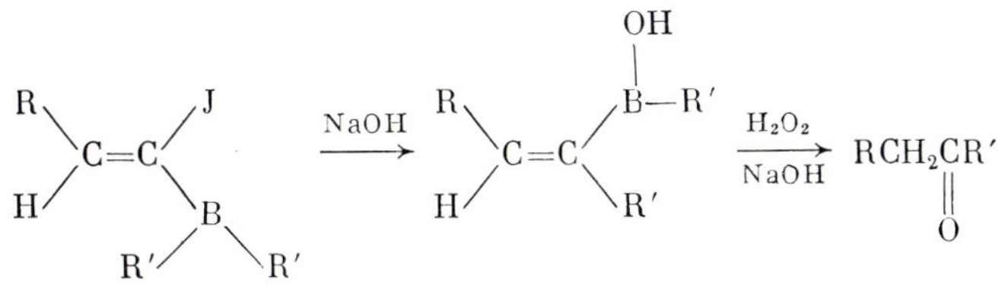

Negishi and Yoshida [41] proposed a stereoselective general method for the synthesis of conjugated *trans,trans*-diene hydrocarbons, based on the hydroboration of haloacetylenes with bis(2,3-dimethyl-2-butyl)diborane. The haloalkenylalkylborane *18* formed during the first stage of the synthesis is then reacted with a 1-alkyne, resulting in the formation of the borane *19*. The latter is converted by the action of sodium methoxide to the compound *20*, the protonolysis of which leads to a conjugated diene with a *trans,trans* configuration (*21*):

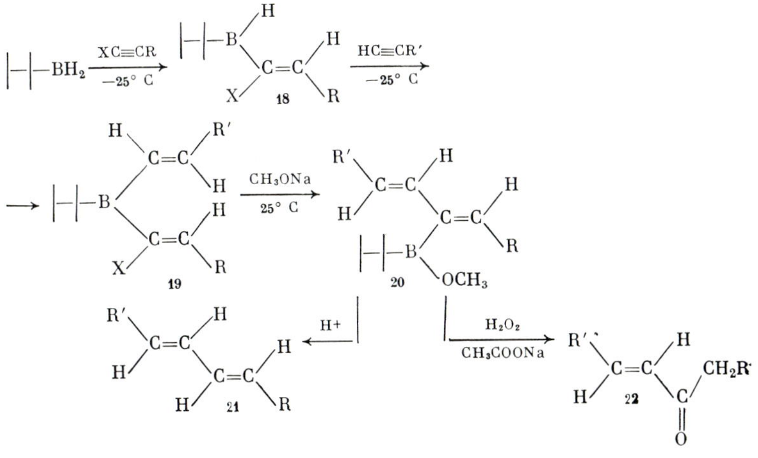

Oxidation of the product of the rearrangement of *20* with hydrogen peroxide forms *trans*-alkenylketones (*22*).

The reaction of bis(2,3-dimethyl-2-butyl)diborane with four equivalents of a 1-iodo-1-alkyne, followed by treatment of the intermediate organoborane *23* with sodium alkoxide, leads to *trans*-1,4-dialkyl-1,2,3-butatriene (*24*) in yields of 29–47% [111]:

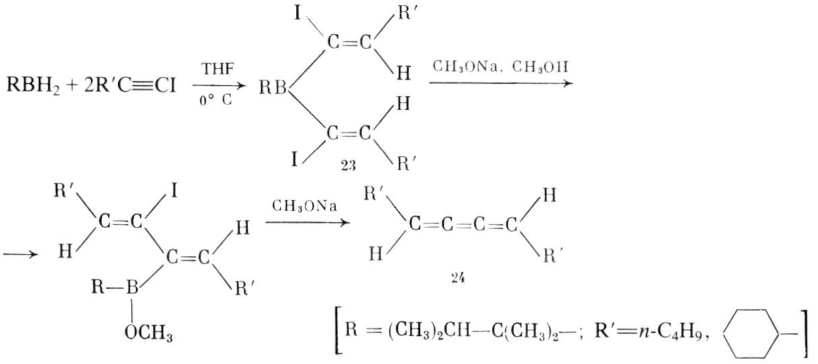

A method for the preparation of β-hydroxycarboxylic acids based on carbethoxy-substituted vinyldialkylboranes has been developed [112]. Ethyl propiolate adds alkyl(2,3-dimethyl-2-butyl)boranes to form predominantly alkyl(2,3-dimethyl-2-butyl)(2-carbethoxy-vinyl)boranes which, when treated with alkali or sodium ethoxide, followed by oxidation with hydrogen peroxide and hydrolysis, are converted to β-hydroxycarboxylic acids (*25*) in yields of 51–69%:

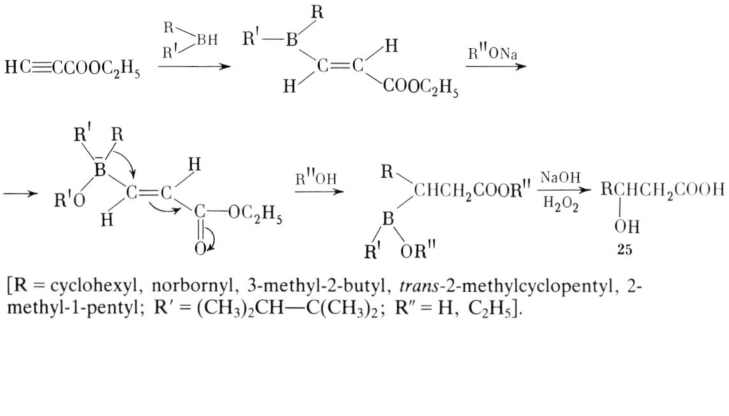

[R = cyclohexyl, norbornyl, 3-methyl-2-butyl, *trans*-2-methylcyclopentyl, 2-methyl-1-pentyl; R' = $(CH_3)_2CH—C(CH_3)_2$; R'' = H, C_2H_5].

In investigating the reaction of propargyl chloride with tetraethyldiborane, Binger and Köster [43] established that, at the monohydroboration stage, dialkyl(3-chloro-1-propenyl)borane is formed which, under the action of alkaline reagents (aqueous sodium hydroxide or $NaBR_4$), is converted to allene with a small content (5%) of methylacetylene:

$$CH{\equiv}CCH_2Cl + {}^1/_2(C_2H_5)_4B_2H_2 \longrightarrow 2(C_2H_5)_2BCH{=}CHCH_2Cl \xrightarrow{OH^-} CH_2{=}C{=}CH_2.$$

The orientation of the addition of a dialkylborane to substituted propargyl chlorides is different, with the boron atom adding to the carbon carrying the chloromethyl group [42]. Treatment of the hydroboration product with aqueous sodium hydroxide causes its decomposition, with the formation of a substituted allene in yields of 64–73%:

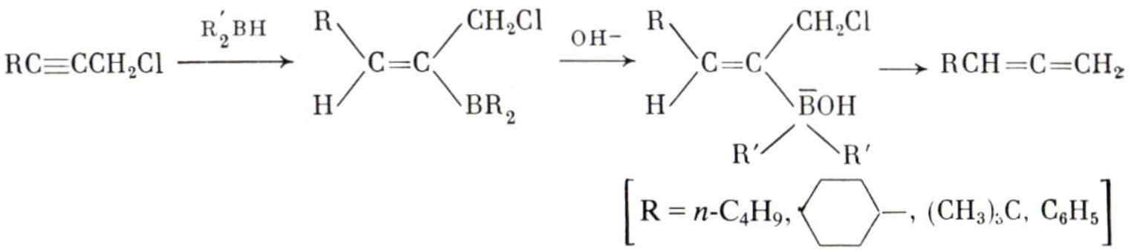

The isolation of substituted allenes from the reaction mass in the pure state involves specific technical problems.

The conversion of 3-chloro-1-heptyne to an allene compound with the use of di(3-methyl-2-butyl)borane can not be carried out, since the product of monohydroboration, under the action of alkali, undergoes intramolecular rearrangement with the splitting of chlorine and, as a result of subsequent oxidation with hydrogen peroxide, 2,3-dimethyl-*trans*-5-decen-4-ol is obtained:

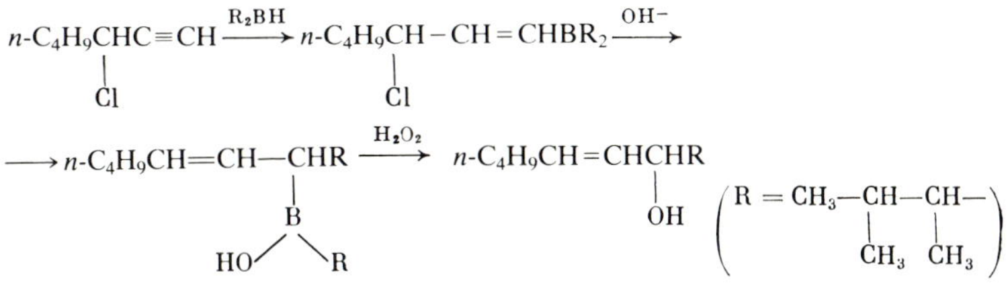

Acidolysis of the product of hydroboration of 1-chloro-2-heptyne leads to pure *cis*-1-chloro-2-heptene [42].

It has been established that the base-induced migration of an alkyl group from boron to carbon in γ-acetylvinyldialkylborane, followed by 1,3-elimination, proceeds predominantly from an anti position with respect to the leaving group, to form *trans*-allyl alcohol after oxidation [201].

trans-1-Hexenyldicyclohexylborane (*26*) reacts with methyllithium to form the borate *27*. When this is treated with an ether solution of hydrogen chloride, lithium chloride splits off, a proton adds to the β-carbon atom of the double bond, and the cyclohexyl radical migrates from the boron to the α-carbon, forming cyclohexyl-α-cyclohexylhexylmethylborane (*28*), the oxidation of which leads to 1-cyclohexyl-1-hexanol (*29*) (70%).

In the reaction of the borate with HCl, there is a partial cleavage of the boron-vinyl bond, with the formation of 1-hexene (10%) [113]:

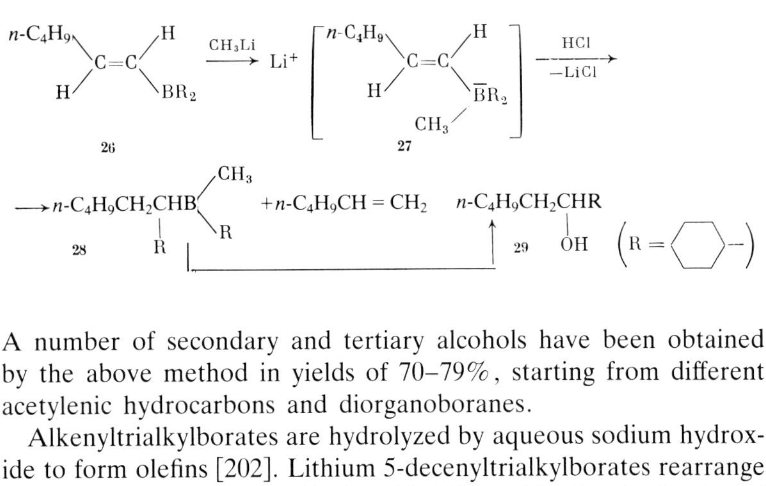

26 27 28 29

A number of secondary and tertiary alcohols have been obtained by the above method in yields of 70–79%, starting from different acetylenic hydrocarbons and diorganoboranes.

Alkenyltrialkylborates are hydrolyzed by aqueous sodium hydroxide to form olefins [202]. Lithium 5-decenyltrialkylborates rearrange under the action of hydrochloric acid, followed by the elimination of a dialkylborane, as a result of which olefins are obtained [203].

In the action of methyllithium on the alkenyl compounds *30*, obtained by the hydroboration of propargyl chloride with tetraalkyldiboranes, the borates *31* are formed as intermediate compounds, which are converted spontaneously to allylboranes (*32*) by elimination of LiCl and migration of an alkyl radical from the boron to the

α-carbon. The compounds *32* rearrange to the allylboranes *33*, which undergo acidolysis with allylic rearrangement to form the olefins *34* [114]:

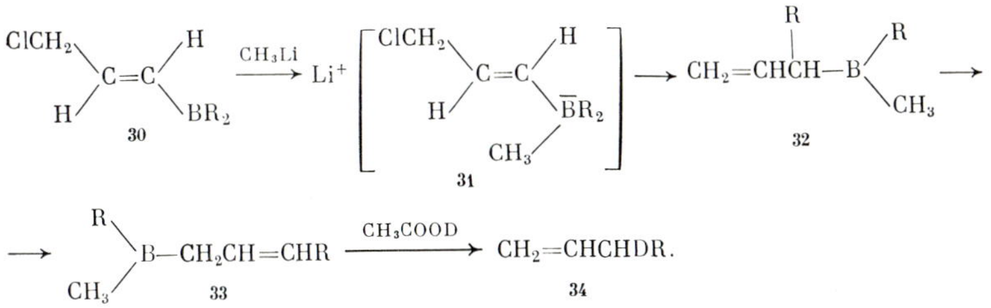

The action of 2,2,6,6-tetramethylpiperidyllithium on 1-alkenylbis(3-methyl-2-butyl)boranes, followed by the addition of trimethylchlorosilane and oxidation of the reaction products with hydrogen peroxide, forms 3-trimethylsilyl-substituted aldehydes in yields of 66–90% [115]. The reaction sequence in this process is expressed by the following scheme: :

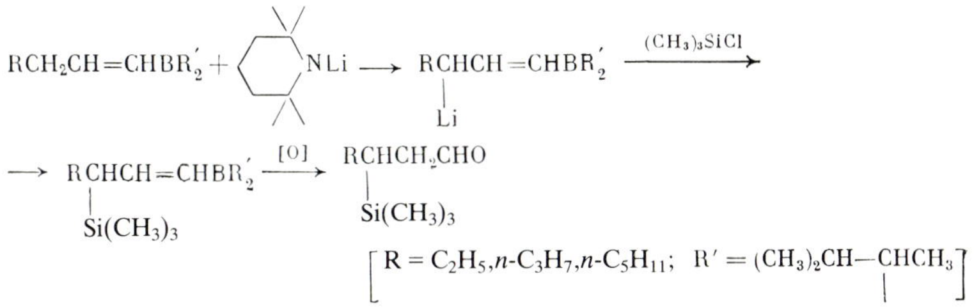

The action of acetone on the product of metalation of octenyl-1-bis(3-methyl-2-butyl)borane, followed by protonolysis, produces an unsaturated alcohol:

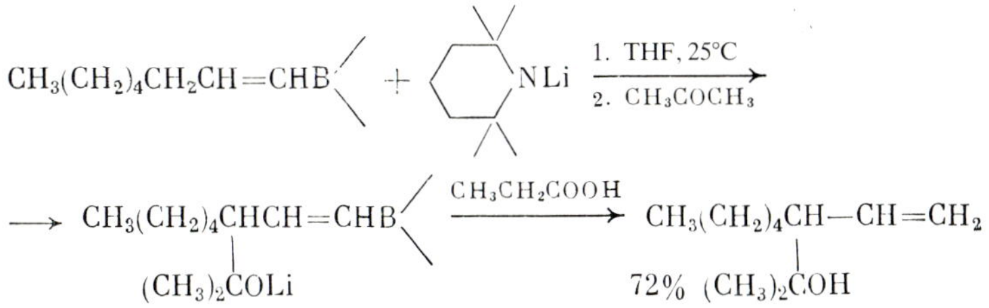

It is interesting that the hydrolysis with water of the product of metalation of 1-octenylbis(3-methyl-2-butyl)borane, followed by oxidation, leads to 2-octen-1-ol (yield:72%), containing 5% of octene and 10% of octanal:

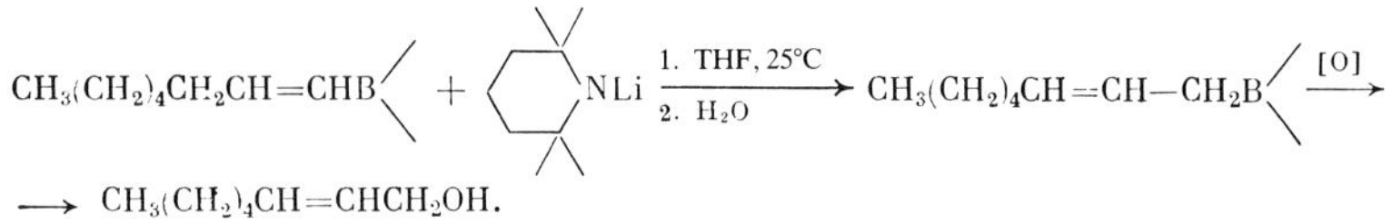

The action of methyl iodide on the metalation product produces 3-nonen-2-ol.

Dialkylalkenylboranes react with potassium cyanide to form the borates *35* which, under the action of trifluoroacetic anhydride, followed by alkaline hydrolysis of the intermediate product *36*, are converted to the allylamides *37* in yields of 67–90% [116]:

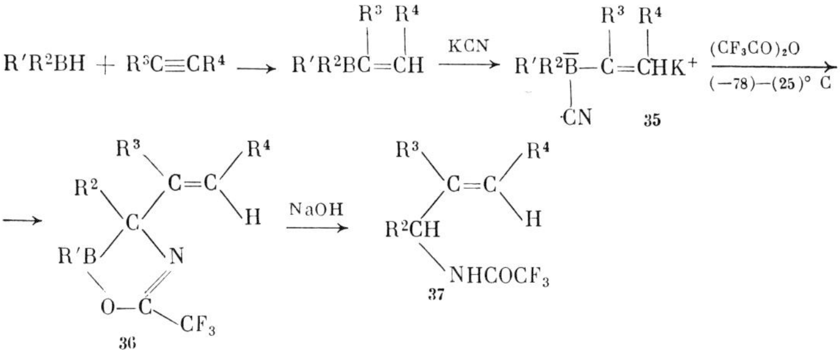

2.9 HYDROBORATION. SYNTHESIS OF ALCOHOLS, OLEFINS, MALONIC ACIDS, CYCLOPROPANE, AND CYCLOBUTANE COMPOUNDS

The action of diborane on acetylenic hydrocarbons at a reagent ratio of 3:2, which is necessary to carry out the dihydroboration, forms polymeric products, the oxidation of which with hydrogen peroxide in the presence of NaOH leads to a complex mixture of organic compounds. Thus, as a result of the dihydroboration-oxidation of 1-hexyne, 1- and 2-hexanols, 1,2-hexanediol, and caproic aldehyde are formed [36, 117, 118]. On the basis of the quantitative ratio of these compounds, it can be concluded that the reaction between 1-hexyne and diborane forms mainly 1,1-diboron compounds and approximately 20–30% of 1,2-diboron compounds. The formation of a large quantity of 1-hexanol (60–80%) indicates that protonolysis of both the 1,1- and the 1,2-diboron compounds takes place during the oxidation [118].

The dihydroboration of phenylacetylene, followed by oxidation, produces a mixture of various compounds (1- and 2-phenylethanols, phenylethanediol, styrene, and others), the relative quantities of which depend both on the ratio of the starting reagents and on the nature of the solvent used [117]. The dihydroboration of diphenylacetylene, followed by oxidation-hydrolysis, forms 1,2-diphenylethanol, benzyl phenyl ketone, and *dl*-hydrobenzoin as the main compounds [117, 119]. As shown by experiments on the deuterolysis of the hydroboration products, a cleavage of the B—C bonds in both the 1,1- and 1,2-diboron compounds takes place in this case.

The action of diborane on 3-hexyne forms almost equal quantities of 1,1- and 1,2-diboron compounds, as was found from the ratio of the organic compounds obtained by oxidation of the reaction mass [117].

The hydroboration of the following individual alkenyl boron compounds with diborane in tetrahydrofuran has been investigated: dibutyl vinyl-, propenyl-, and isopropenylboronates. The action of diborane on the vinylboronate (BH_3 : ester = 1.5), followed by treatment of the reaction products with butanol, forms a mixture of tetrabutyl 1,1- and 1,2-ethanediboronates (yield 60%) in a ratio of approximately 3:[120] (see also [121]):

$$CH_2=CHB(OC_4H_9\text{-}n)_2 \xrightarrow[n\text{-}C_4H_9OH]{B_2H_6} (C_4H_9O)_2B\overset{\displaystyle CH_3}{\overset{|}{C}}HB(OC_4H_9)_2 + (C_4H_9O)_2BCH_2CH_2B(OC_4H_9)_2$$

The introduction of a CH_3 group into the β-position of the vinylboronate exerts a strong effect on the orientation of the addition of diborane to the double bond. The hydroboration of dibutyl propenylboronate leads, after subsequent treatment with butanol, almost exclusively to tetrabutyl 1,1-propanediboronate [120, 122].

When a CH_3 group is present in the α-position of the vinylboronate, the orientation of the addition of diborane is reversed. Dibutyl isopropenylboronate forms a mixture of 1,1- and 1,2-diboron compounds in a ratio of 1:4 [120].

Intramolecular coordination has a significant effect on the orientation of the addition of diborane to alkenylboronates. Thus, 2,2'-iminodiethyl isopropenylboronate forms only a vicinal derivative with diborane, which forms tetrabutyl 1,2-propanediboronate by hydrolysis with hydrochloric acid, followed by esterification with butanol [122]:

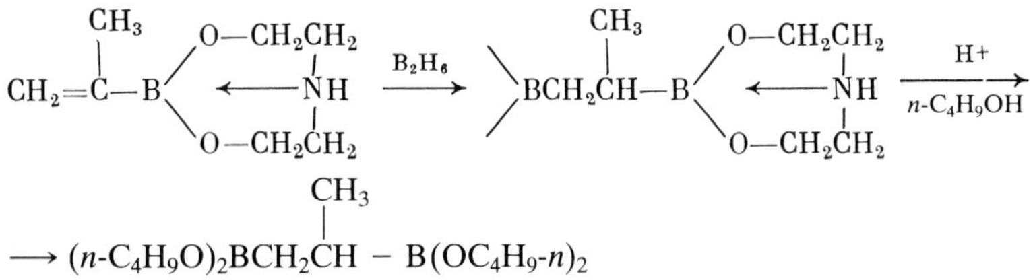

Dibutyl β-styrylboronate reacts with diborane to form a 1,1-diboron compound, while the α-styrylboronate forms a 1,2-diboron compound [21, 123].

Tetracyclohexyldiborane adds to 1-hexenyldicyclohexylborane mainly with the formation of a 1,1-diboron compound, which is converted by oxidation with hydrogen peroxide in alkaline medium to 1-hexanol (89%), with impurities of 1,2-hexanediol (2%) and caproic aldehyde (1%) [118]:

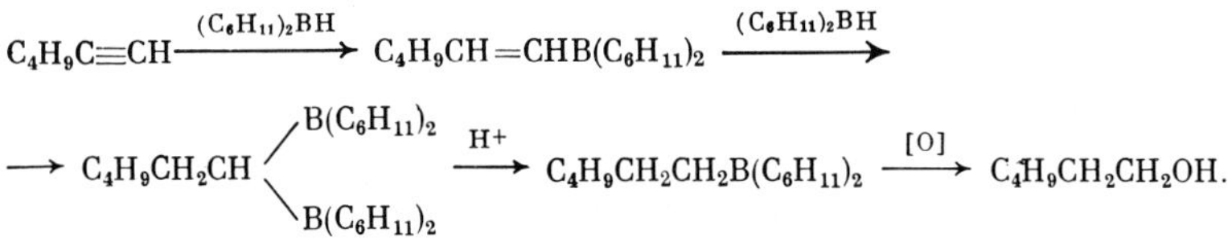

The dihydroboration of 1,2-cyclotridecadiene, followed by oxidation, leads to a mixture of different oxygen derivatives, while hydrolysis of the hydroboration products with aqueous NaOH solution at 80°C, followed by oxidation, forms primarily cyclotridecanol [124].

Cainelli, et al. [125] found that the 1,1-diboron compounds formed by the dihydroboration of 1-hexyne, 1-decyne, or phenylacetylene with tetracyclohexyldiborane undergo transmetalation by the action of two equivalents of butyllithium in heptane at $-70°C$ to form mixed 1-bora-1-lithium organic compounds. The latter are converted by the action of CO_2 to derivatives of malonic acid in yields of 65–70%:

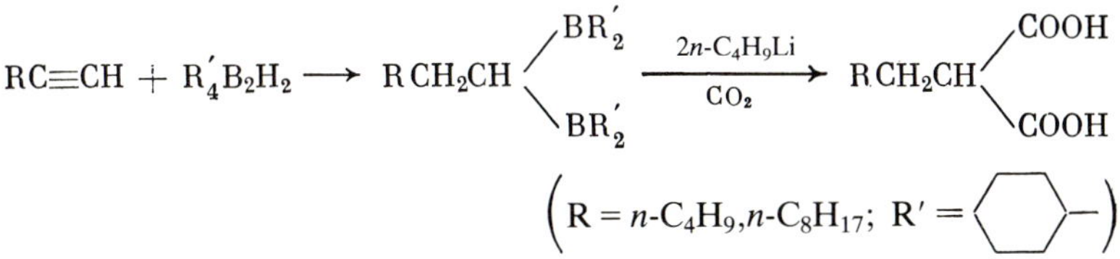

gem-Boron-lithium compounds are derivatives of trivalent boron (*38*) or borate complexes (*39*):

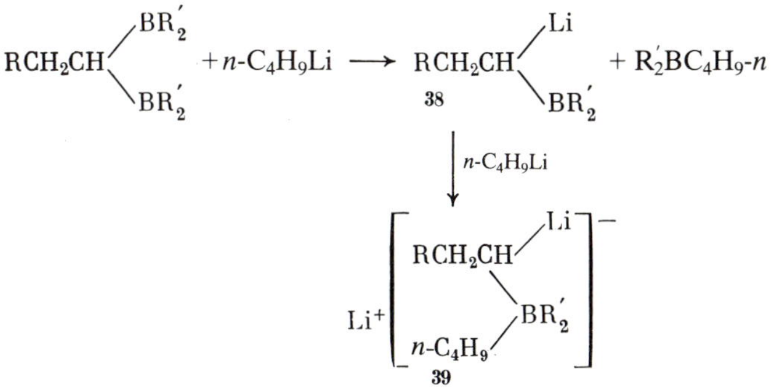

Compounds of the type *38* react with aldehydes and ketones to form olefinic hydrocarbons [126]:

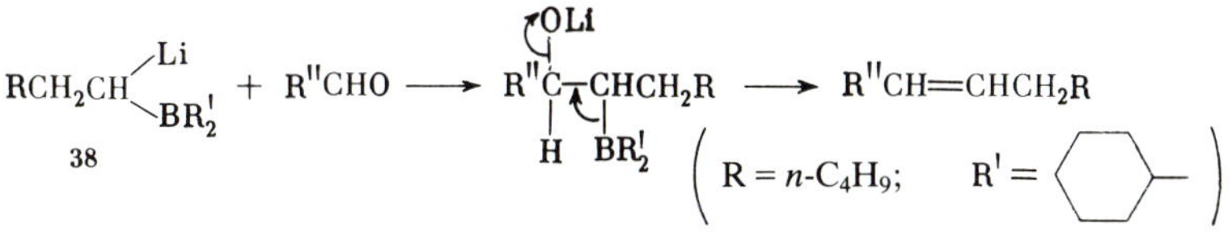

In the reaction with benzaldehyde, the ratio of *cis* to *trans* isomers depends on the temperature conditions: at 40°C, the *trans* isomer is obtained with a 96% degree of purity, while at $-78°C$ a mixture of isomers in the ratio of 7:3 is formed.

The above method may be used for the synthesis of alcohols with increased chain length. The *gem*-boron-lithium compound *40*, formed by the hydroboration of 1-pentyne with tetracyclohexyldiborane and subsequent treatment with butyl- or methyllithium, reacts with ethyl bromide to form the borane *41*, which is converted by oxidation to 3-heptanol (*42*) [127]:

$$C_3H_7C{\equiv}CH \xrightarrow[2.\ R'Li]{1.\ R_4B_2H_2} \underset{40}{C_3H_7CH_2{-}CH\begin{matrix}\diagup Li \\ \diagdown BR_2\end{matrix}} \xrightarrow{C_2H_5Br}$$

$$\longrightarrow \underset{41}{C_3H_7CH_2\underset{\displaystyle |\atop BR_2}{C}HC_2H_5} \xrightarrow{[O]} \underset{42}{C_3H_7{-}CH_2\underset{\displaystyle |\atop OH}{C}H{-}C_2H_5}$$

With the use of 3,6-dimethylborepane (obtained from 2,5-dimethyl-1,5-hexadiene and diborane) as a hydroborating agent and methyllithium, different secondary alcohols have been synthesized by the above method [128].

Binger and Köster [43] have found that (3-chloro-1-propenyl)-dialkylborane (*43*), obtained by the action of tetraethyldiborane on propargyl chloride, is hydroborated to 1,1-bis(dialkylboryl)-3-chloropropane (*44*). When heated with sodium tetraethyl borate, the compound *44* cyclizes to cyclopropyldialkylborane (*45*) in 80% yield:

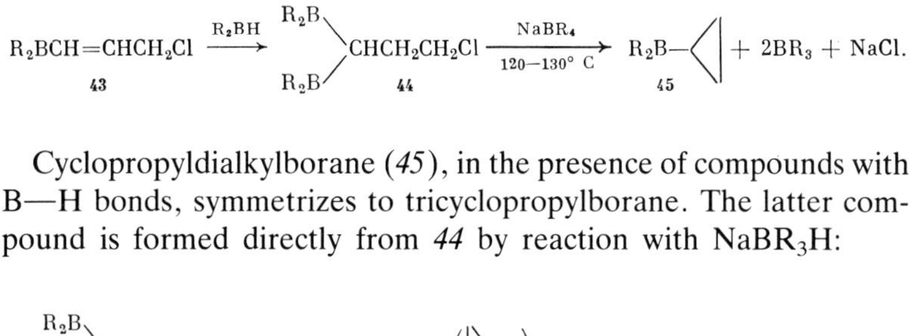

Cyclopropyldialkylborane (*45*), in the presence of compounds with B—H bonds, symmetrizes to tricyclopropylborane. The latter compound is formed directly from *44* by reaction with $NaBR_3H$:

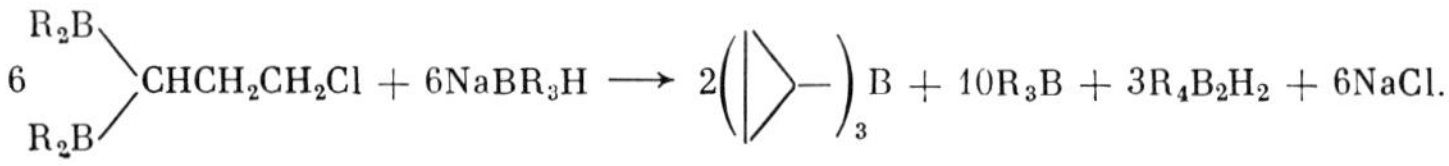

Tricyclopropylborane enters into an exchange reaction with triethylaluminum to form cyclopropylaluminum compounds, which are hydrolyzed with D_2O to deuterocyclopropane.

Cyclopropanol is obtained by the oxidation of cyclopropylboranes.

Bis(9-borabibicylo[3.3.1]nonane) can be used in the synthesis of cyclopropylboranes from propargyl chloride [129]. In this case, the 1,1-diboron compound cyclizes under the action of NaOH or CH_3Li.

3-Butyn-1-ol tosylate is converted by dihydroboration with bis-(9-borabicyclo-[3.3.1]nonane) to 9-cyclobutyl-9-borabicyclo[3.3.1]-nonane (*46*), which is oxidized to cyclobutanol [129]:

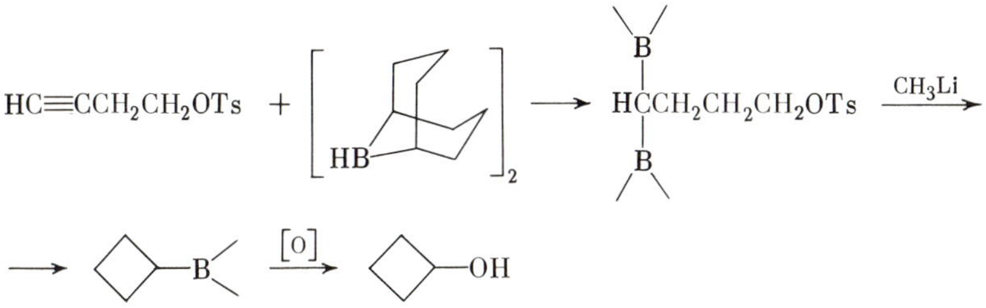

46

1,1-Bis(dimethoxyboryl)alkanes, obtained by the hydroboration of 1-alkynes, react with mercuric chloride to form 1,1-bis(chloromercuri)alkanes [130].

The reaction of $(C_2H_5)_3BHLi$ with α-halovinylboranes leads to *cis*-alkenylboranes [204].

2.10 DIENE CONDENSATION. SYNTHESIS OF SUBSTITUTED NORBORNENES

Dibutyl vinylboronate undergoes a diene condensation with cyclopentadiene at 100°C to form a mixture of 60% of dibutyl *exo*-norborn-2-en-5-ylboronate and 40% of the *endo* isomer [131]:

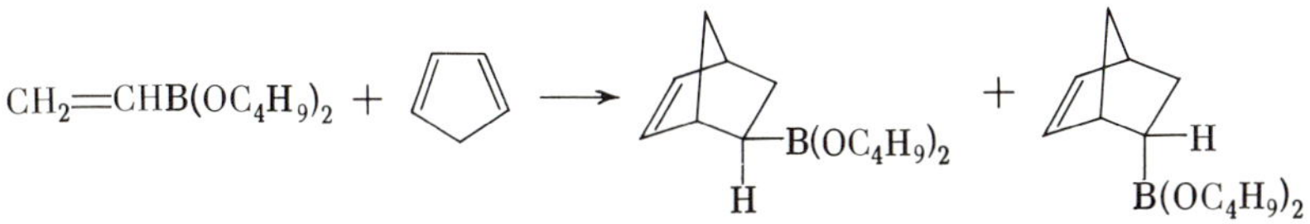

A mixture of the *cis* and *trans* isomers (30:70) of the glycol ester of propenylboronic acid reacts with cyclopentadiene to form a mixture of adducts with a predominant content of *endo* isomers, $(a+b)$:

$(c+d) = 53:47$, whereas 37% $(a+b)$ and 63% $(c+d)$ is obtained from the *trans* esters:

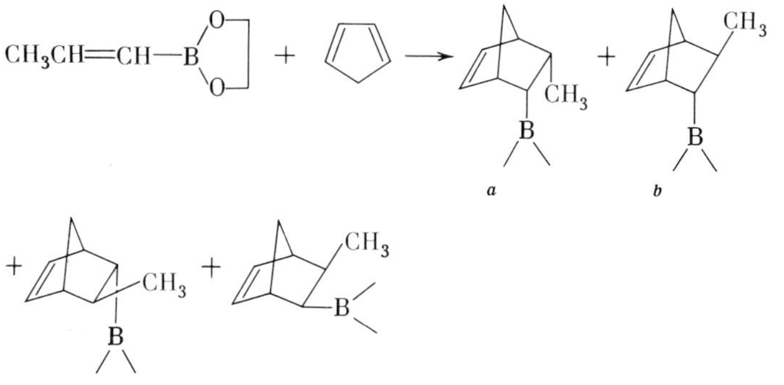

a *b* *c* *d*

Glycol ester of isopropenylboronic acid forms the adducts *e* and *f* in a ratio of 37:63 [132]:

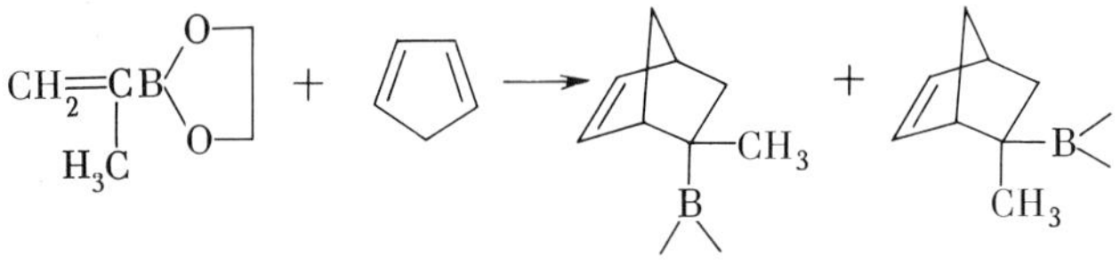

e *f*

The glycol ester of vinylboronic acid reacts with hexachlorocyclopentadiene to form an *exo* adduct, while only an adduct of type *a* is obtained with the ester of propenylboronic acid.

The condensation of dibutyl vinylboronate with 1,3-cyclohexadiene forms a mixture of the *exo* and *endo* isomers (20:80) of dibutyl bicyclo[2.2.2]oct-2-en-5-ylboronate [132a].

2.11 REACTIONS WITH DIAZO ESTERS, ALDEHYDES, ETHYLENE OXIDE, AND ORGANOMETALLIC COMPOUNDS. SYNTHESIS OF PYRAZOLINES, ARYLGLUTARIC ACIDS, ALCOHOLS, 1,3-DIOLS, KETONES, AND COMPOUNDS OF MERCURY, BISMUTH, SILICON, AND TIN.

A vinylboronate reacts with a diazoacetate to form an unstable boron-substituted pyrazoline, which is converted by ethanolysis to

5-carbethoxy-2-pyrazoline [133]:

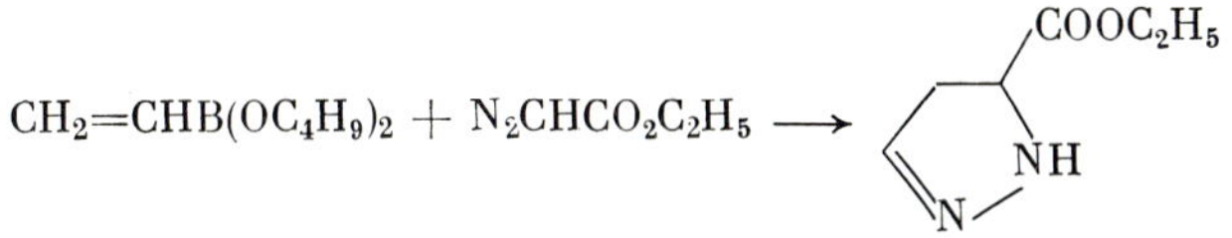

Tris(2-alkylthio-2-ethoxyvinyl)boranes react under mild conditions with aromatic aldehydes to form B_2O_3 and 1,5-dialkylthio-1, 5-diethoxy-3-aryl-1,4-pentadienes, which are converted by hydrolysis to 3-arylglutaric acids [134]:

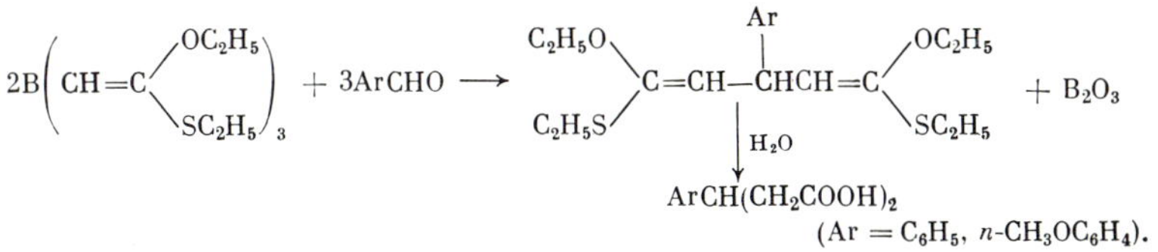

Vinyltrialkylborates react with aldehydes with rearrangement, forming the cyclic borates *47*, which are converted to 1,3-diols by oxidation [205, 206]:

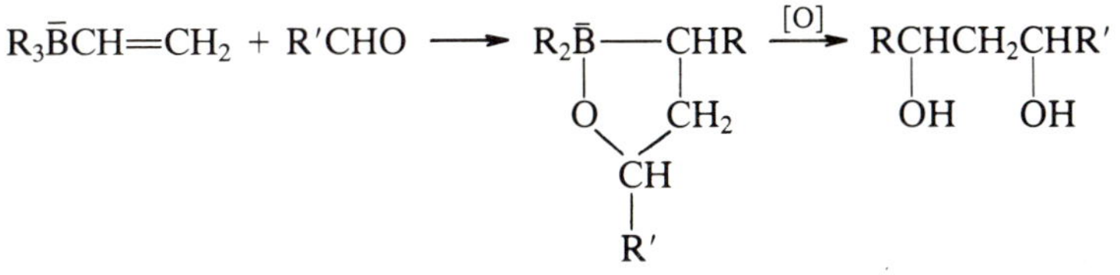

47

Treatment of the borate *47* with phosphorus pentachloride leads to cyclopropane derivatives [206, 207].

Allyl alcohols are obtained from 9-alkenyl-9-borabicyclo-[3.3.1]nonanes and aldehydes [208], while reaction with methyl vinyl ketone forms γ,δ-unsaturated ketones [209].

Dibutyl vinylboronate adds methylchlorosilane and trichlorosilane in the presence of chloroplatinic acid to form borosilico-organic compounds [135].

2-Vinyl-4,4,6-trimethyl-1,3,2-dioxaborinane adds trialkylstannanes at 110°C in the β-position to form borostanno-organic compounds [136]:

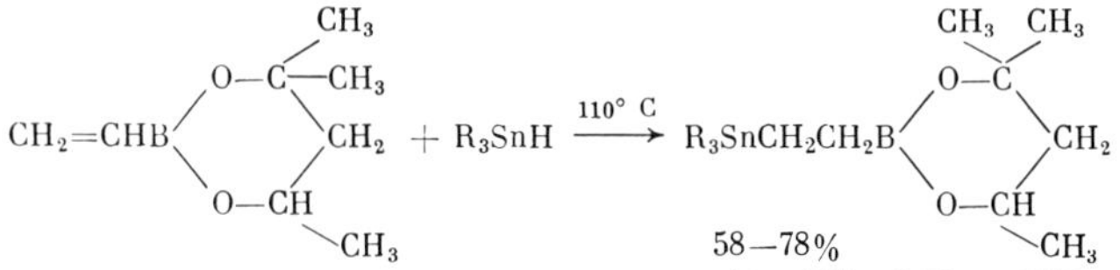

($R=CH_3$, C_2H_5, n-C_3H_7, n-C_4H_9, C_6H_5).

A method for the stereoselective synthesis of (*Z*)-2-alkenyl compounds of silicon and tin using 9-alkenyl-9-borabicyclo[3.3.1]nonanes has been developed [210].

Alkenylboranes undergo an exchange reaction with mercury salts to form alkenyl mercury compounds. The reactions proceed smoothly with alkenylboronic acids [9, 137, 138]:

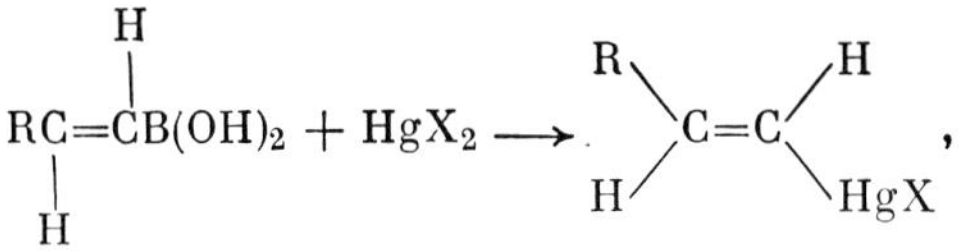

their esters [139], 1-alkenyldialkylboranes [140], and trialkenylboranes [141], with retention of the configuration of the vinyl group.

A rapid electrophilic substitution of boron-ester groups takes place in the action of mercuric chloride and sodium acetate in methanol on 2-methyl-1,1-propenediboronic acid glycol ester, with the formation of a 1,1-dimercury alkenyl compound [27]:

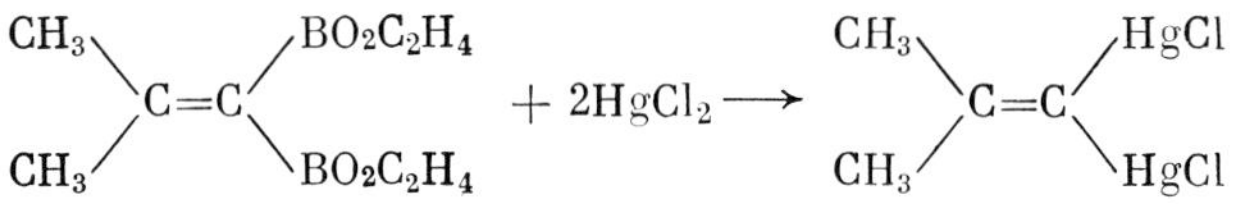

Chlorobis(*trans*-2-chlorovinyl)borane reacts with thallium trichloride to form bis(*trans*-2-chlorovinyl)thallium chloride [9]:

The cleavage of halovinylboranes and trivinylborane under the action of an ammoniacal silver oxide solution has been investigated [6]. Together with ethylene, a negligible quantity of 1,3-butadiene is formed in this reaction.

2.12 REACTIONS INVOLVING COPPER AND PALLADIUM COMPOUNDS. SYNTHESIS OF OLEFINS, DIENES, AND ENYNES.

Different organic compounds can be synthesized from 1-alkenylboranes with the use of copper and palladium compounds.

Yamamoto, et al. [211, 212] have synthesized (*E*,*E*)-1,3-dienes by treating dialkenylchloroboranes with methylcopper at 0°C:

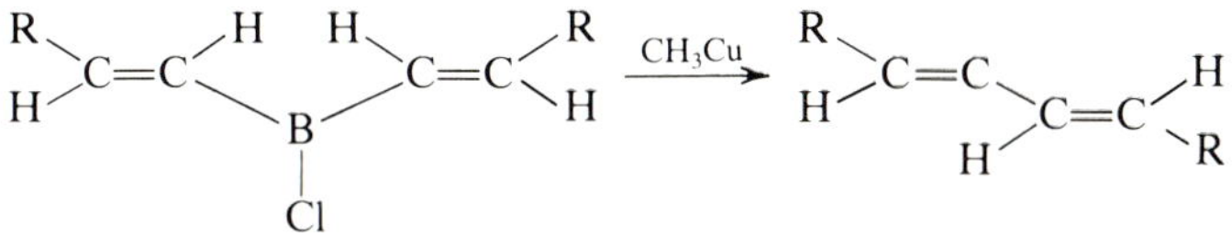

Dialkenylmethylboranes react analogously [213]. The reaction of dialkenylchloroboranes with allyl halides in the presence of methylcopper leads to isomerically pure (4*E*)-1,4-dienes [214–216].

The hydroboration of 1-trimethylsilyl-1-alkynes with dicyclohexylborane proceeds regio- and stereospecifically with the formation of 1-trimethylsilanovinylboranes, which are converted to (*Z*)-1,2-dialkylvinylsilanes by treatment with methyllithium, cuprous iodide, and alkyl halides [217, 218].

1,4-Alkadienes or 1,2,7-alkatrien-5-ynes are formed by the reaction of copper(I) 1-alkenyltrimethylborates with allylic bromides or 2-propynyl bromide [219].

Symmetrical conjugated dienes are formed by the action of cuprous bromide-dimethyl sulfide on borates obtained from alkenyldialkyl-boranes and sodium methoxide [220].

The synthesis of 1,4-dienes [221] and conjugated enynes [222] has been carried out on the basis of the alkenylcopper intermediates obtained from 9-alkenyl-9-borabicyclo[3.3.1]nonanes.

trans-Alkenylboranes are converted to trans-olefins by the action of palladium (II) acetate and triethylamine in tetrahydrofuran [223–224].

(*E*)-1-Alkenyldisiamylboranes and (*E*)-1-alkenyl-1,3,2-benzodioxaboroles react with 1-alkenyl or 1-alkynyl halides in the presence of catalytic quantities of $Pd(PPh_3)_4$ to form the corresponding (*E*)-dienes or (*E*)-enynes with high regio- and stereospecificity [225]. The reactions of esters of 1-alkenylboronic acids with aryl bromides and iodides proceed in an analogous manner [226].

1-Alkenyldisiamylboranes react with allyl and benzyl bromide in the presence of sodium hydroxide and a catalytic quantity of $Pd(PPh_3)_4$ to form 1,4-alkadienes or allylbenzenes [227].

(*Z*)-1-Alkenyldisiamylboranes react with (*Z*)- or (*E*)-1-alkenyl bromides in the presence of a catalytic quantity of $Pd(PPh_3)_4$ and sodium ethoxide to form the corresponding conjugated (*Z*,*Z*)- or (*Z*,*E*)-alkadienes. (*E*)-1-alkenyl-1,3,2-benzodioxaboroles react analogously [228].

1-Alkenyl-1,3,2-benzodioxaboroles enter into reactions, catalyzed by palladium compounds, with phenyl- or 1-alkenyl iodides to form 2-phenyl-1-alkenes or 2-alkyl-1,3-alkadienes [229].

β-Arylvinylboronic acids are converted to 1,4-diaryl-1,3-butadienes in good yield under the action of palladium(II) chloride [230].

A stereospecific synthesis of esters of α,β-unsaturated carboxylic acids, based on the carbonylation of 1-alkenylboranes with carbon monoxide in the presence of palladium(II) chloride and sodium acetate, has been developed [231].

The stereospecific synthesis of 1-halo-1,2-dialkylethylenes is based on reactions of (1-halo-1-alkenyl)dialkylboranes with lead (IV) acetate or diacetoxyiodobenzene [232, 233].

CHAPTER 3

Preparation of alkenylboranes from 1-alkynyltrialkylborates and their properties Synthesis of olefins, unsaturated alcohols, amines, ketones, acids, and allenes

Binger, Köster, and co-workers [142] found that 1-alkynyltrialkylborates *47*, obtained from trialkylboranes and alkali metal acetylenides [143, 144], undergo an anionotropic rearrangement under the action of electrophilic reagents, with a migration of the alkyl group from the boron atom to the α-carbon, resulting in the formation of alkenylboranes. Hydrogen chloride in ether medium reacts with borates to form a mixture of *cis* and *trans* isomers of alkenyldialkylboranes (*48*), which hydrolyze to *cis* and *trans* olefins:

$$\underset{\mathbf{47}}{Na[R_3BC{\equiv}CR']} \xrightarrow[-NaCl]{HCl} \underset{\mathbf{48}}{R_2BC(R){=}CHR'} \xrightarrow{H^+} RCH{=}CHR'.$$

Under the action of water, a rupture of the B—C_{sp} bond takes place in the borates *47*, with the formation of an alkyne and BR_3 [142].

The borates *47* react with chlorodialkylboranes in a sterically selective manner, forming *cis*-1,2-bis(dialkylboryl)alkenes (*49*), the protonolysis of which with water or alcohols produces *cis* olefins [143]:

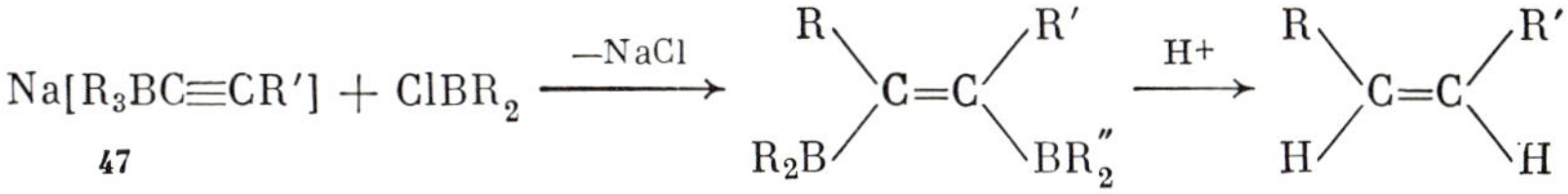

49 (65—80%)

The work of Köster was the beginning of numerous investigations of the chemistry of 1-alkynyltrialkylborates, which led to the development of new methods for the synthesis of different classes of organic compounds.

It was found that, if one starts from a sterically hindered (2,3-dimethyl-2-butyl)dialkylborane, then the acidolysis with propionic acid of the corresponding borate of the type *47* forms yields of 68–82% of a mixture of olefins with a predominant content of the *cis* isomer (92–98%) [145].

This method can also be applied to the synthesis of terminal olefins, using a complex of lithium aceylenide with ethylenediamine (*50*) as the acetylene component [145]:

$$\underset{\mathbf{50}}{HC{\equiv}CLi \cdot EDA} + BR_3 \longrightarrow [R_3BC{\equiv}CH]Li \cdot EDA \xrightarrow{CH_3CH_2COOH} \underset{56-64\%}{RCH{=}CH_2}$$

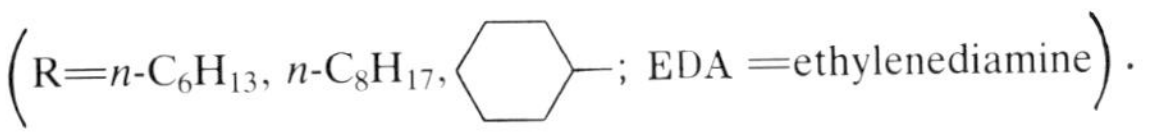

1,2-Bis(dialkylboryl)alkenes (*49*) are converted by pyrolysis (150–180°C) to alkyl-substituted 1,3-diboracyclopent-4-enes and 2,3,4,5-tetracarbahexaboranes(6), the ratio of which depends on the nature of the alkyl groups in the starting compounds [146]. The hydroboration of compounds *49* with tetraethyldiborane and the thermal conversions of the tris(dialkylboryl)alkanes obtained thereby have been investigated [147].

As was found by Binger [148], 1-alkynyltrialkylborates react with chlorides to form 1,2-oxaboracyclopent-4-enes (1,2-oxaborol-4-enes) (*51*):

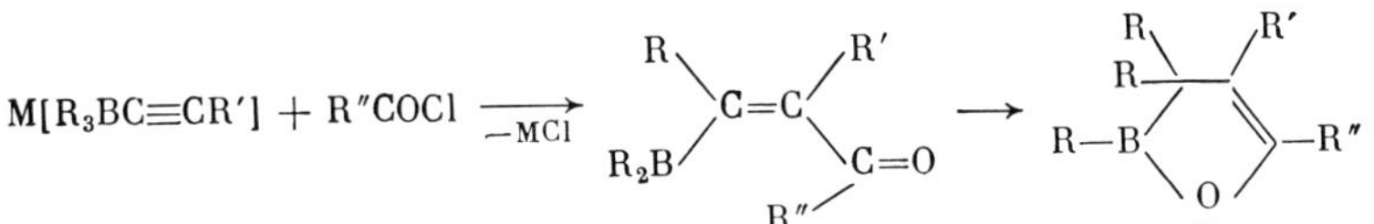

51

(R″ = CH_3, C_6H_5).

Oxidation of the compounds *51* with Jones reagent forms the α,β-unsaturated ketones *52* in yields of 28–42% [149, 150]:

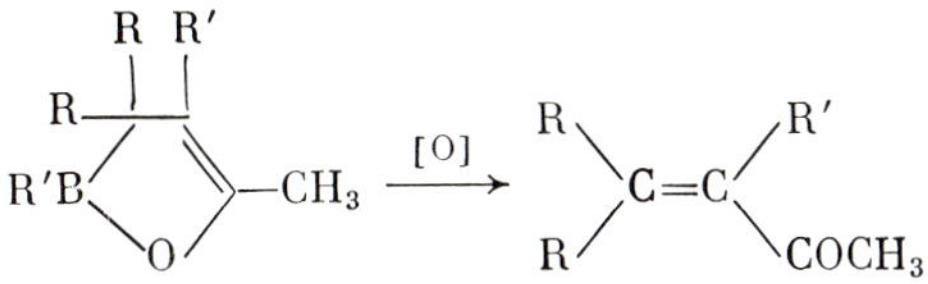

51 52

($R = iso$-C_3H_7, n-C_4H_9, sec-C_4H_9; $R' = C_5H_{11}$, C_6H_5).

The ketones *55* are obtained by the action of pyridine and acetyl chloride on alkynyltrialkylborates at −78°C, followed by oxidation of the intermediate compounds *53* and *54* with hydrogen peroxide. The hydrolysis of the 4-substituted dihydropyridines *53* and *54* forms the compounds *56* and *57*, which are converted by oxidation to the olefins *58* and *59* (65:35) [151]:

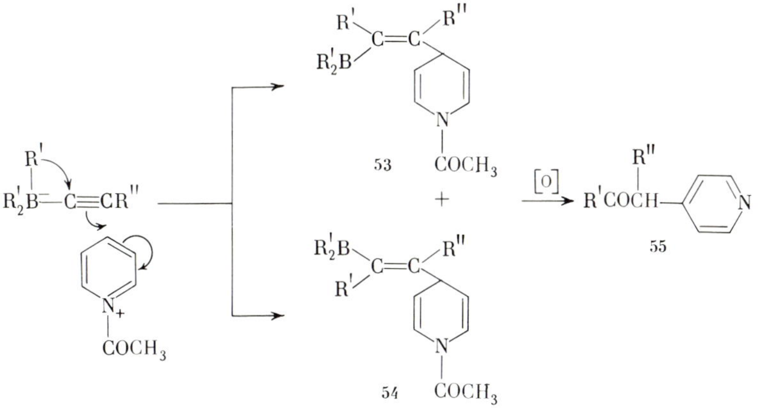

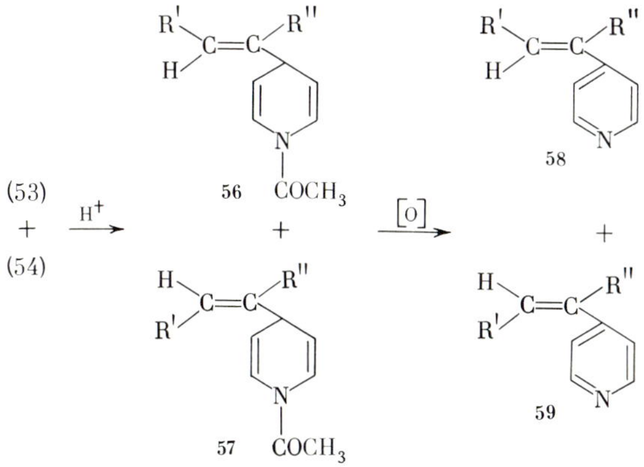

Sodium 1-alkynyltrialkylborates (*47*) react with chloromethyl methyl ether to form the dialkylboryl derivatives of allyl methyl ether *Z-60* and *E-60* (total yield: 45–75%, [152, 234]:

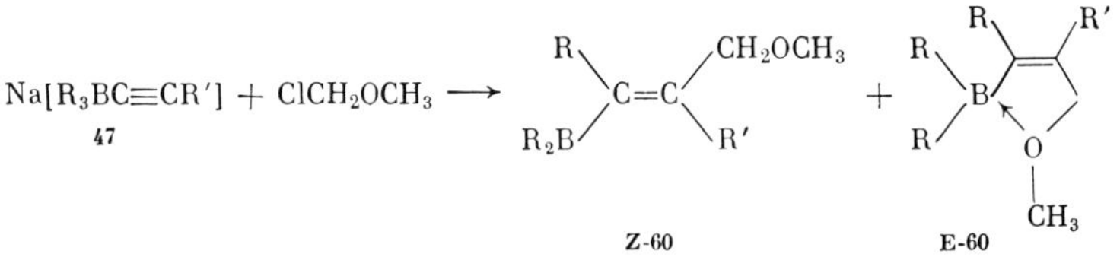

47a: R = R′ = CH_3; *47b*: R = CH_3, R′ = C_2H_5; *47c*: R = R′ = C_2H_5

The ratio of the *Z* and *E* isomers is determined by the structure of the alkyl radicals and by the nature of the solvent. Thus, in diethyl ether, an almost equimolar mixture of *Z* and *E* isomers is formed from the 1-alkynyltrialkylborates (*47b*) and (*47c*), the *E* isomer is predominantly obtained in benzene, while 1-propynyltrimethylborate (*47*a) in ether forms the almost pure isomer *E*-60a. When a mixture of *Z-60* and *E-60* is treated with water, the allyl methyl ether *Z-61* is formed from the isomer *Z-60*, while the isomer *E-60* undergoes protonolysis only under the action of acetic acid, being converted to the isomer *E-61*:

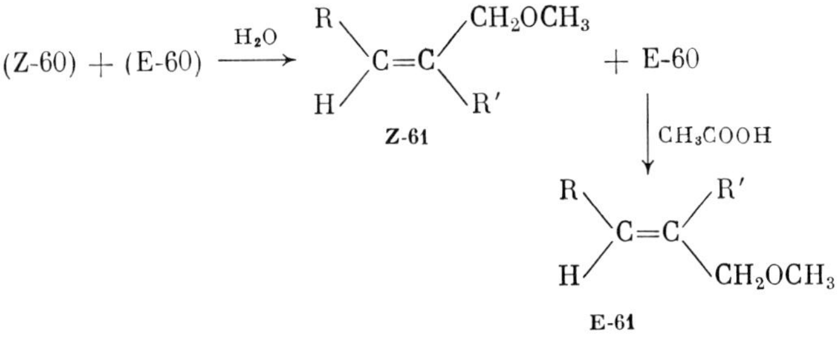

The reaction of sodium propynyltrialkylborates with chlorotrimethylsilane proceeds stereospecifically, with the formation of *E*-3-diethylboryl-2-trimethylsilyl-2-alkene, while in this case trialkylborane and methyltrimethylsilylacetylene are obtained as byproducts [153, 234]:

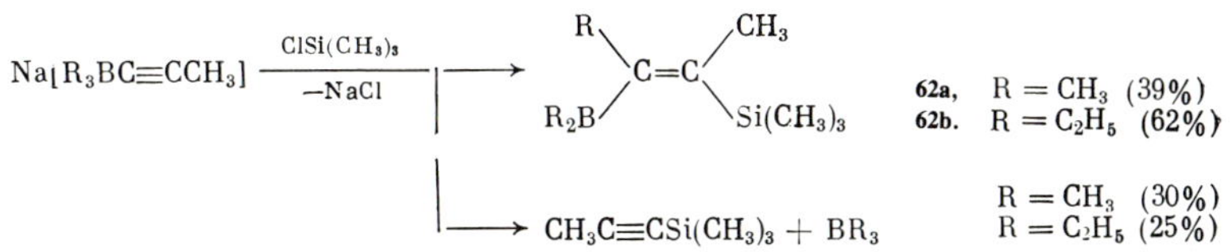

Acetic acid or acetylactone only partially cleaves the compound *62b*, forming *Z*-2-trimethylsilyl-2-alkene (*E-63*); however, these reagents (and also methanol) add mainly to the B and β—C atoms, after which the addition products undergo rearrangement to the compounds *64*:

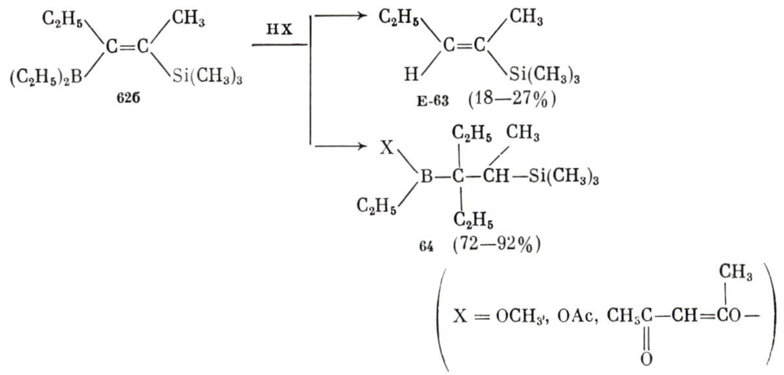

The compounds *64* undergo transmetalation on reaction with triethylalane; the organoaluminum compounds formed in this reaction hydrolyze to form a mixture of alkenylsilanes and alkylsilane:

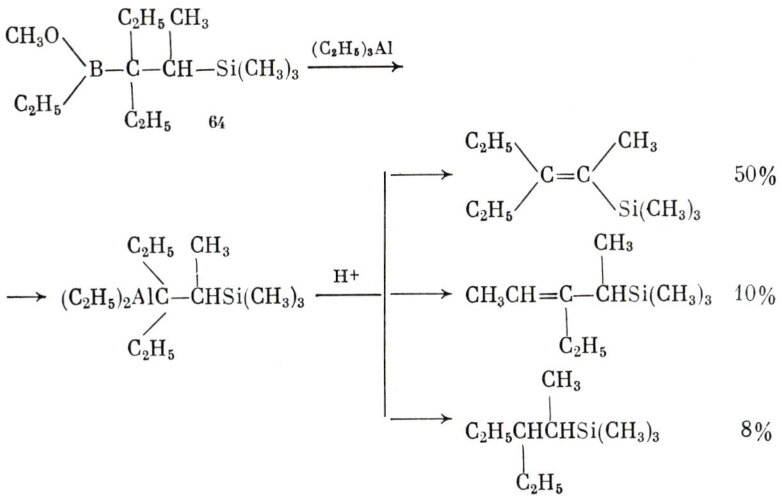

Lithium or sodium 1-alkynyltrialkylborates also react with triethyloxonium borofluoride [142, 154, 234–236], allyl bromide [142, 154, 235, 236], dimethyl sulfate, and other alkylating agents [154, 235–238]. The products of addition-migration thereby formed (*65*) are converted by oxidation with hydrogen peroxide in alkaline solution to the corresponding ketones *66* in high yields [154, 235–239]:

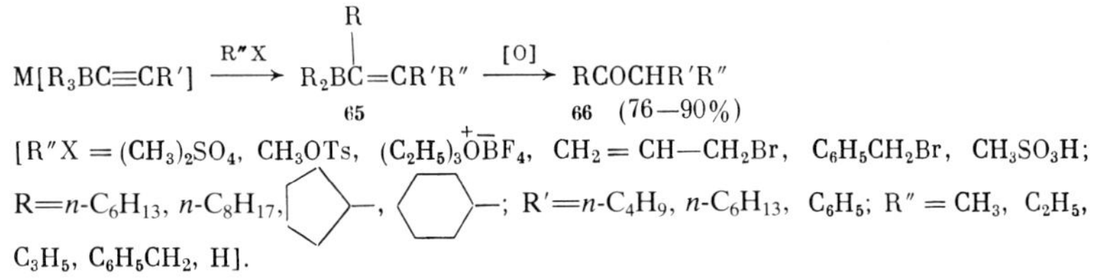

The action of electrophilic halides on $R_3\bar{B}C{\equiv}CSi(CH_3)_3$ forms alkenyldialkylboranes of the type $R_2BCR{=}CR'Si(CH_3)_3$ ($R'{=}CH_3OCH_2$, $(CH_3)_3Si$, CH_3CO, Ph_2P) [240].

Unlike the 1-alkenyltrialkylborates, lithium ethoxyethynyltrialkylborates react with alkylating agents without rearrangement, forming methyl- and allylethoxyacetylene, respectively [241, 242]:

$$R_3\bar{B}C{\equiv}COEt\ Li^+ \xrightarrow{RHal} R^1C{\equiv}COEt + R_3B + LiHal$$

$$R{=}CH_3,\quad CH_2{=}CHCH_2-$$

The action of a bromomethyl ketone or ethyl bromoacetate on metal 1-alkynyltrialkylborates forms alkenyldialkylboranes (*67*), which are oxidized to γ-diketones or esters of γ-oxo acids and are hydrolyzed to substituted allyl ketones or esters of β,γ-unsaturated acids in high yields [155]:

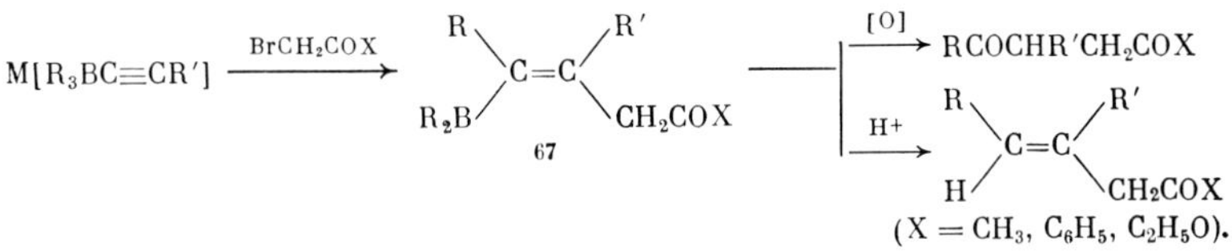

1-Alkynyltrialkylborates react with methanesulfenyl chloride via the intermediate compounds *68* to form disubstituted acetylenes [156, 157]:

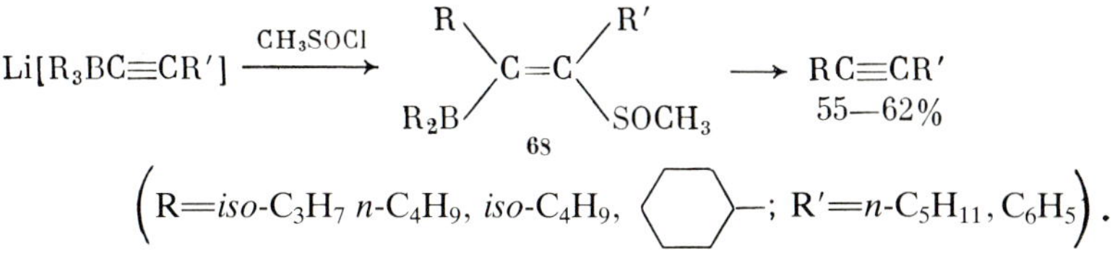

The reaction of CH_3SOCl with the 1-alkynylborate formed from 9-*n*-hexyl-9-borabicyclo[3.3.1]nonane and heptyllithium proceeds in a peculiar manner. As a result of oxidation of the products of this reaction, 6-tridecyne (2%) and 1-(5-hydroxy-1-cyclooctyl)heptyne (*69*) (23%) are obtained [156, 157]:

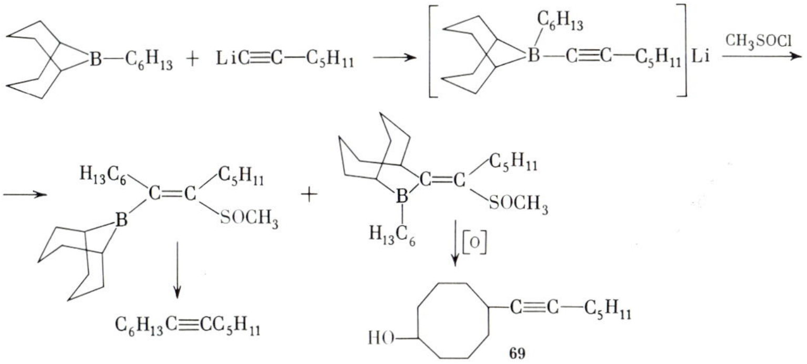

The reaction of alkylnyltrialkylborates with PhSeCl, followed by oxidation, leads to α,β-unsaturated ketones [243].

Oxiranes react with 1-alkynyltrialkylborates with opening of the oxygen ring and the formation of intermediate cyclic borates (*70*). The borates *70* split off lithium alkyls at 150–200°C, being converted to oxaborinenes (*71*). Upon oxidation with hydrogen peroxide, they form γ-hydroxy ketones (*72*) (60–95%); they are hydrolyzed with boiling acetic acid to trisubstituted unsaturated alcohols (*73*) (75–98%); and, when treated with iodine in alkaline medium, form tetrasubstituted unsaturated alcohols (*74*) [158, 159]:

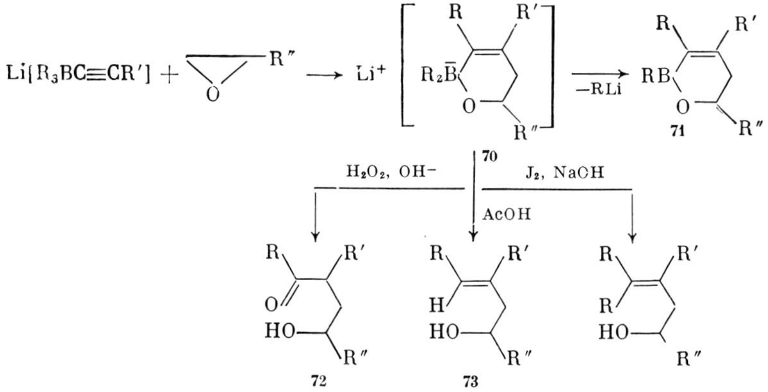

($R = iso\text{-}C_3H_7$, $n\text{-}C_4H_9$; $R' = C_5H_{11}$; $R'' = H, CH_3, C_2H_5, C_6H_5$).

The stereoselective synthesis of (*Z*)- and (*E*)-3-methyl-3-octen-1-ol has been carried out by the reaction of lithium 1-propenyltrialkylborate with ethylene oxide [244].

Sodium 1-propinyltriethylborate and dimethylmethyleneammonium bromide in ether forms a 96% yield of an approximately equimolar mixture of the *Z* and *E* isomers of *N*-(3-diethylboryl-2-methyl-2-pentenyl)dimethylamines, *E-75* and *Z-75* [160]:

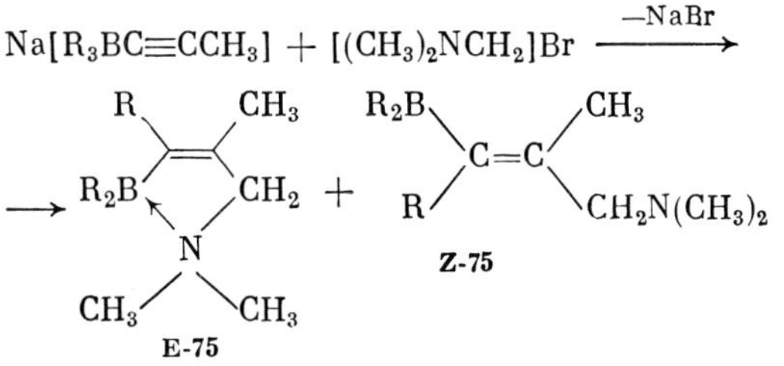

($R = C_2H_5$).

The compounds *E-75* and *Z-75* behave differently with respect to nucleophilic and electrophilic reagents. With water at room temperature, the vinylborane *Z-75* forms *Z-N*-(2-methyl-2-pentenyl)-dimethylamine (*Z-76*), while the B—C bond in the compound

E-75 is cleaved by water only upon heating:

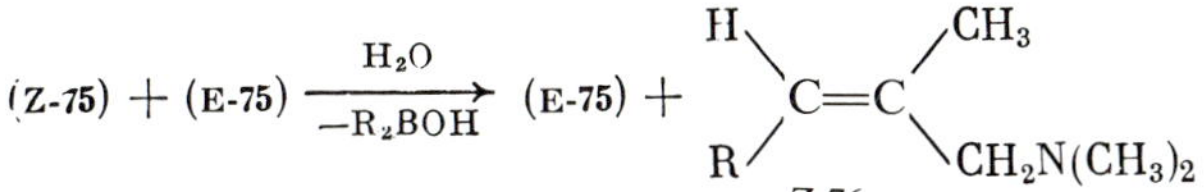

Z-76

Pure *E-N*-(2-methyl-2-pentenyl)dimethylamine (*78*) can be obtained from the compound *E-75* by transmetalation under the action of AlR_3, followed by hydrolysis of the aluminum compound *77*:

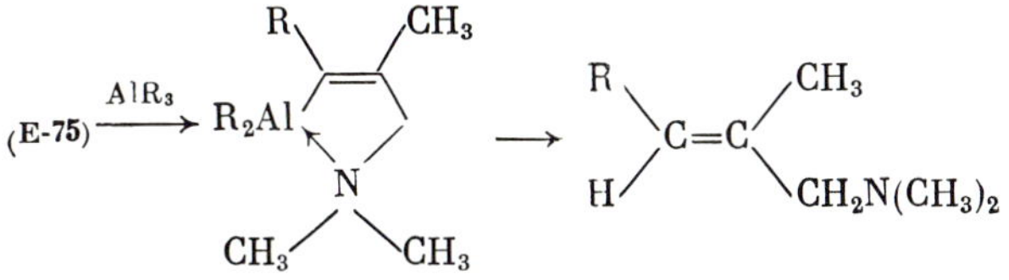

77 78

Under the action of an ether solution of HCl, the compound *Z-75* is converted to the salt *79*, while *E-75* forms the compound *80*:

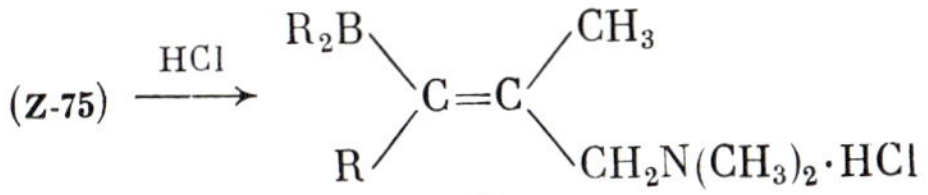

79

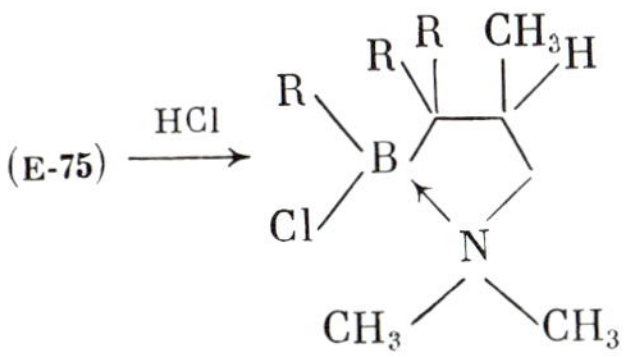

80

Analogs of *E-75* and *Z-76* have been obtained from sodium 1-propynyltrimethylborate.

Trimethylsilylacetylenes are obtained from trialkylboranes and trimethylsilylpropargyl phenyl ether [245].

The 1-alkynyltrialkylborates *81*, obtained with the use of propargyl chloride, spontaneously undergo rearrangement to the allenic boron compounds *82*, which are converted by acidolysis to the allenic compounds *83* [161]:

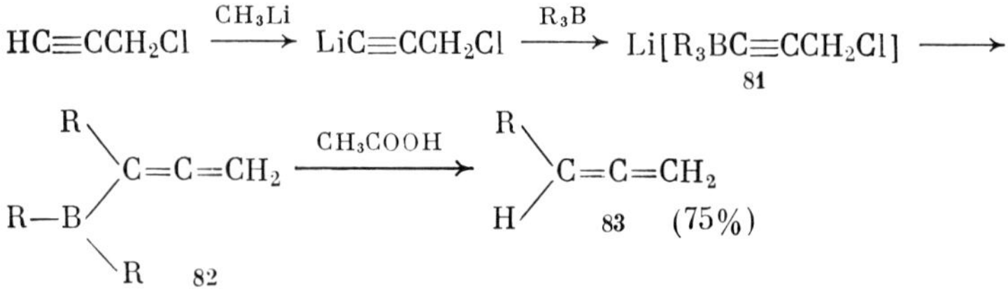

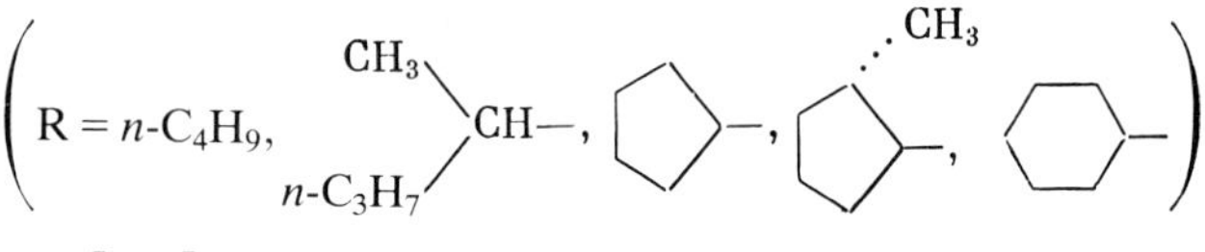

Zweifel, et al. [246] have found that the allenylboranes *82*, formed from *81* at −90°C, isomerize at 25°C to the propargylboranes *83a*. The reactions of *82* and *83a* with aldehydes produce, after oxidation, homopropargyl and α-allenyl alcohol, respectively:

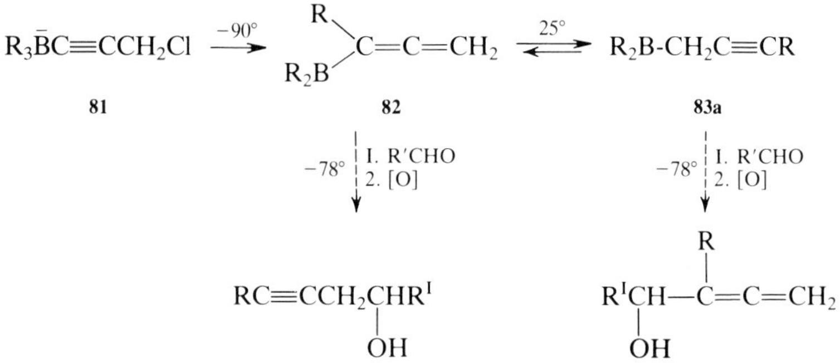

Lithium derivatives of propargyl acetates react with trialkylboranes to form allenylboranes, which are converted to allenes by the action of acetic acid and form acetylenes with water [247].

1-Alkynyltrialkylborates react with chlorodiorganophosphines to form organo-1,2-phosphaboret-3-enes (*84*), which are cleaved selectively in boiling acetic acid to diorganoalkenylphosphines (*85*) [162]:

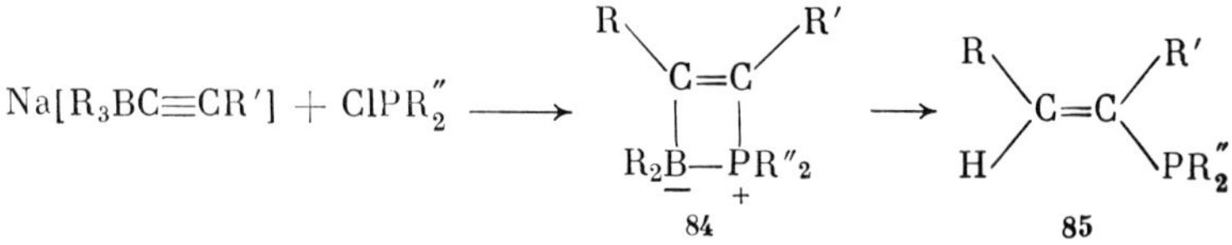

The Zweifel reaction — the rearrangement of alkenyl compounds of tetracoordinated boron under the action of iodine (Part III, Chapter 2.5.2) — has found application among alkynyl boron derivatives. The action of iodine on 1-alkynyltrialkylborates or 1-alkynyltriphenylborate forms acetylenic hydrocarbons [163, 248]:

$$Li[R_3BC{\equiv}CR'] \xrightarrow{I_2,\ -78°C} RC{\equiv}CR'.$$

By this method, 5-decyne, 4-methyl-4-nonyne, 3-methyl-4-nonyne, 1-cyclopentyl-1-hexyne, 1-cyclohexenyl-1-hexyne, 1-phenyl-1-hexyne, and other acetylenic hydrocarbons have been obtained in high yields.

The action of iodine on ethylenediamine complexes of ethynyltrialkylborates forms acetylenic hydrocarbons with a terminal C≡C bond in good yields [164].

The action of iodine on chloroethynyltrialkylborates produces symmetrical alkynes and alkenes [249].

Lithium ethoxyethynyltrialkylborate reacts with iodine in the presence of quinoline to form 1-ethoxyalkylacetylenes [241, 242]. Borates of the type $R'_2\bar{B}(C{\equiv}CR)_2$ are converted to RC≡C—C≡CR by the action of iodine [250, 251].

The reaction of BrCN with dilithium ethynyldi(trialkylborate)leads to di-, tri-, and tetrasubstituted ethylene derivatives [252].

The relative migratory capacity of alkyl groups in the iodination of lithium ethynyltrialkylborates has been investigated [253].

Pelter and Harrison [165] found that 1-alkynyltrialkylborates are capable of reacting with dihalomethanes, with the migration of two alkyl groups from the boron to the α-carbon atom, forming α-olefins after hydrolysis. In the primary reaction product *86*, migration of a second alkyl radical takes place, apparently as the result of attack on the boron by a halogen atom. The borane *87* obtained rearranges to the borane *88*, which has a greater thermodynamic stability, and the latter is converted by protonolysis with allylic rearrangement to the olefin *89*:

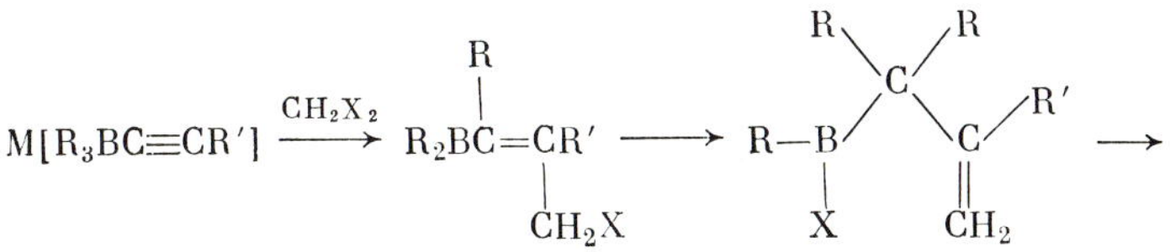

86 87

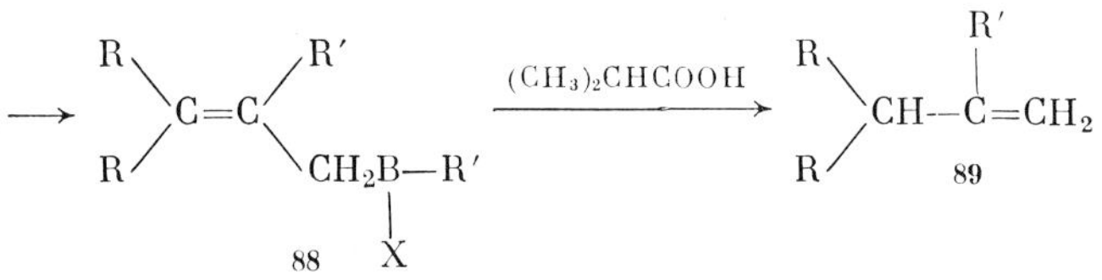

The olefins *89* contain an impurity of the hydrocarbons $R_2C{=}CHR'$ which are formed as a result of side reactions.

The conversion of metal 1-alkynyltrialkylborates under the action of the cationic complexes $[(C_6H_5OCH_3)Fe(CO)_3]^+BF_4^-$ has been investigated [166].

Lithium trialkylalkynylborates react with α,β-unsaturated nitro compounds to form alkenylboranes, which produce γ-nitroalkyl alkyl ketones upon oxidation and nitroolefins by protonolysis [254].

Lithium alkynyltrialkylborates react with alkylidene malonates and alkyldiene acetoacetates to form the corresponding vinylboranes, the oxidation of which leads to 1,5-keto esters, while hydrolysis produces a mixture of stereoisomeric γ, δ-unsaturated esters [255].

Lithium trialkylalkynylborates react with methyl vinyl ketone in the presence of titanium tetrachloride to form δ-diketones [256].

The lithium trialkylalkynylborate obtained from the lithium derivative of *R*-(+)-1-octyn-3-ol acetate and tributylborane forms predominantly the *S*-(+)-allene (5,6-dodecadiene) by protodeboronation, from which it follows that the migration of the butyl group from the boron to the carbon proceeds predominantly form an *anti* position with respect to the leaving acetate group [257].

The synthesis of 1,4-disubstituted (*E*,Z)-1,3-dienes from lithium dicyclohexyl(*trans*-l-alkenyl)(1-alkynyl)borates has been carried out [258].

The reaction of tributyltin chloride with lithium trialkylalkynylborate proceeds stereospecifically — in the vinylborane formed, the migrating group and the entering electrophilic group (R_3Sn) are in a *trans* position. Protonolysis with formic acid produces *cis*-olefins [259].

The consecutive treatment of chloropropargyllithium with thexylalkenylchloroboranes and aldehydes leads to 1,3-enynols or 1,2,4-trienols, depending on the reaction conditions [260].

The alkylation of acetylenic carbonyl compounds has been carried out with the use of *ate* complexes [261].

Trialkylalkynylborates react with benzo-1,3-dithiolium tetrafluoroborates to form vinylboranes, which form 3-oxoaldehydes upon oxidation and α,β-unsaturated aldehydes upon hydrolysis [262].

Trialkylboranes react with $LiC{\equiv}C(CH_2)_2OTs$ to form alkynylboron intermediates which, upon oxidation, furnish cyclopropyl alkyl ketones or acetylenic alcohols [263].

A chemo-, regio-, and stereoselective synthesis of (7*E*, 9*Z*)-dodecadien-1-yl acetate (a sex pheromone) has been carried out with the use of a functionalized organoborate [264].

Lithium trialkylethoxyethynylborates are capable of adding trialkylboranes to form the compounds, which split off R_2BOEt to form the lithium trialkylalkenylborates. The latter forms alkylacetylenes with water and dialkylacetylenes with iodine [242, 265].

REFERENCES

1. T. D. Parsons and D. M. Ritter, *J. Am. Chem. Soc.* **76,** 1710 (1954).
2. T. D. Parsons, M. B. Silverman and D. M. Ritter, *J. Am. Chem. Soc.* **79,** 5091 (1957).
3. C. D. Good and D. M. Ritter, *J. Am. Chem. Soc.* **84,** 1162 (1962).
4. A. V. Topchiev, Ya. M. Paushkin and A. A. Prokhorova, *Dokl. Akad. Nauk SSSR* **129,** 598 (1959).
5. J. Braun, *Compt. Rend.* **256,** 2422 (1963).
6. F. E. Brinckman and F. G. A. Stone, *J. Am. Chem. Soc.* **82,** 6218 (1960).
7. G. C. Brown, B. E. Deuters, W. Gerrard and D. B. Green, *Chem. and Ind.* **1965,** 1634.
8. B. Bartocha, C. M. Duglas and M. Y. Grag, *Z. Naturforsch. B* **1,** 809 (1959).
9. A. E. Borisov, *Izv. Akad. Nauk SSSR, Otd. Khim. Nauk* **1951,** 402.
10. R. N. Sterlin, W. K. Li and I. L. Knunyants, *Zh. Vses. Khim. O-va.* **6,** No. 1, 108 (1961).
11. R. N. Sterlin, V. L. Isaev, G. M. Zakharov, B. Martin and I. L. Knunyants, *Zh. Vses. Khim. O-va..* **12,** 475 (1967).
12. S. L. Stafford and F. G. A. Stone, *J. Am. Chem. Soc.* **82,** 6238 (1960).
13. H. D. Kaesz, S. L. Stafford and F. G. A. Stone, *J. Am. Chem. Soc* **81,** 6336 (1959).
14. B. F. Clark, E. Gipstein, M. A. Higgins and P. R. Kippur, U.S. Patent 2,918,498 (1959); *C. A.* **54,** 7560 (1960).
15. N. N. Mel'nikov and M. S. Rokitskaya, *Zh. Obshch. Khim.* **8,** 1768 (1938).
16. B. M. Mikhailov and F. B. Tutorskaya, *Izv. Akad. Nauk SSSR, Otd. Khim. Nauk* **1959,** 1865.
17. H. Normant and J. Braun, *Compt. Rend.* **248,** 826 (1959).
18. H. Normant and J. Braun, *Izv. Akad. Nauk SSSR, Otd. Khim. Nauk* **1959,** 1397.
19. D. S. Matteson, *J. Am. Chem. Soc* **82,** 4228 (1960).
20. D. S. Matteson and R. A. Bowie, *J. Am. Chem. Soc.* **87,** 2587 (1965)..
21. D. J. Pasto, J. Chow and S. K. Arora, *Tetrahedron* **25,** 1557 (1969).

22. R. L. Letsinger and I. H. Skoog, *J. Org. Chem.* **18,** 895 (1953).
23. J. Braun and H. Normant, *Bull. Soc. Chim. France* **1966,** 2557.
24. D. S. Matteson, *J. Org. Chem.* **27,** 3712 (1962).
25. D. S. Matteson and R. W. H. Mah, *J. Org. Chem.* **28,** 2171 (1963).
26. D. S. Matteson, R. A. Bowie and G. Srivastava, *J. Organometal. Chem.* **16,** 33 (1969).
27. D. S. Matteson and P. B. Tripathy, *J. Organometal. Chem.* **21,** P6 (1970).
28. G. Wilke and H. Müller, *Chem. Ber.* **89,** 444 (1956).
29. G. Wilke and H. Müller, *Ann.* **618,** 267 (1958).
30. M. F. Hawthorne, *J. Am. Chem. Soc.* **83,** 2541 (1961).
31. W. G. Woods and P. L. Strong, *J. Am. Chem. Soc.* **88,** 4667 (1966).
32. H. C. Brown and S. K. Gupta, *J. Am. Chem. Soc.* **94,** 4370 (1972).
33. H. C. Brown and N. Ravindran, *J. Organometal. Chem.* **61,** C5 (1973).
34. H. C. Brown and N. Ravindran, *J. Org. Chem.* **33,** 1617 (1972).
35. R. Dulou and Y. Chrétien-Bessière, *Bull. Soc. Chim. France* **1959,** 1362.
36. H. C. Brown and G. Zweifel, *J. Am. Chem. Soc.* **83,** 3834 (1961).
37. T. J. Logan and T. J. Flaut, *J. Am. Chem. Soc,* **82,** 3446 (1960).
38. Yu. N. Bubnov, V. S. Bogdanov and B. M. Mikhailov, *Izv. Akad. Nauk SSSR, Ser. Khim.* **1970,** 2416.
39. Yu. N. Bubnov, V. S. Bogdanov, I. P. Yakovlev and B. M. Mikhailov, *Zh. Obshch. Khim.* **42,** 1313 (1972).
40. G. Zweifel and H. Arzoumanian, *J. Am. Chem. Soc.* **89,** 5086 (1967).
41. E. Negishi and T. Yoshida, *Chem. Comm.* **1973,** 606.
42. G. Zweifel, A. Horng and J. T. Snow, *J. Am. Chem. Soc.* **92,** 1427 (1970).
43. P. Binger and R. Köster, *Angew. Chem.* **74,** 652 (1962).
44. B. M. Mikhailov and T. A. Shchegoleva, *Izv. Akad. Nauk SSSR, Ser. Khim.* **1971,** 661.
45. B. M. Mikhailov, T. A. Shchegoleva, E. M. Shashkova, L. I. Lavrinovich and V. S. Bogdanov, *Zh. Obshch. Khim.* **44,** 2185 (1974).
46. B. M. Mikhailov, T. A. Shchegoleva and E. M. Shashkova, *Zh. Obshch. Khim.* **44,** 2193 (1974).
47. F. Montanari, *Tetrahedron Letters* **1960,** No. 4, 18.
48. H. R. Arnold, U. S. Patent 2,402,589 (1946); *C. A.* **40,** 5769 (1946).
49. E. Gipstein, P. R. Kippur, M. A. Higgins and B. F. Clark, *J. Org. Chem.* **26,** 943 (1961).
50. W. Lazier and P. Zalzburg, U. S. Patent 2,402,591 (1946); *C. A.* **40,** 5769 (1946).
51. E. Gipstein, P. R. Kippur, M. A. Higgins and B. F. Clark, *J. Org. Chem.* **26,** 2947 (1961).
52. E. Gipstein, P. R. Kippur, J. F. Haffer and B. F. Clark, *J. Org. Chem.* **31,** 1087 (1966).
53. M. E. Lappert and B. Prokai, *J. Organometal Chem,* **1,** 384 (1964).
54. I. B. Blackborow, *J. Chem. Soc. (Perkin II)* **1973,** 1989.
55. C. Chambers and A. K. Holliday, *J. Chem. Soc.* **1965,** 3459.
56. J. D. Odom, L. W. Hall, S. Riethmitler and J. R. During, *Inorg. Chem.* **13,** 170 (1974).
57. A. K. Holliday, W. Reade, K. R. Seddon and I. A. Steer, *J. Organometal. Chem.* **67,** 1 (1974).
58. S. I. Frolov, Yu. N. Bubnov and B. M. Mikhailov, *Izv. Akad. Nauk SSSR, Ser. Khim.* **1969,** 1996.
59. B. M. Mikhailov and V. N. Smirnov, *Izv. Akad. Nauk SSSR, Ser. Khim.* **1974,** 1137.
60. M. P. Simmonin and J. Braun, *Bull. Soc. Chim. France* **1968,** 4918.

61. B. M. Mikhailov and P. M. Aronovich, *Vysokomol. Soedin.* **3,** 861 (1961).
62. G. Zweifel, G. M. Clark and N. L. Polston *J. Am. Chem. Soc.* **93,** 3395 (1971).
63. R. W. Murray and G. J. Williams, *J. Org. Chem.* **34,** 1896 (1969).
64. N. H. Nam, A. T. Russo and R. F. Nystrom, *Chem. and Ind.* **1963,** 1876.
65. A. J. Hubert, *J. Chem. Soc.* **1965,** 6669
66. A. C. Cope, G. A. Berchtold, P. E. Peterson and S. H. Sharman, *J. Am. Chem. Soc.* **82,** 6370 (1960).
67. J. Sicher, M. Svoboda, J. Zavada, R. B. Turner and P. Goebel, *Tetrahedron* **22,** 659 (1966).
68. D. S. Sethi, S. Vaidyanathaswamy and D. Devaprabhakara, *Org. Prep. Proced. Internat.* **2,** 171 (1970).
69. J. Sicher, M. Svoboda, B. J. Mallon and R. B. Turner, *J. Chem. Soc., B* **1968,** 441.
70. B. M. Mikhailov, Yu. N. Bubnov and M. Sh. Grigoryan, *Zh. Obshch. Khim.* **44,** 2466 (1974).
71. Yu. N. Bubnov, S. A. Korobeinikova, G. V. Isagulyants and B. M. Mikhailov, *Ivz. Akad. Nauk SSSR, Ser. Khim.* **1970,** 2023.
72. J. Plamondon, J. T. Snow and G. Zweifel, *Organometal. Chem. Synth.* **1,** 249 (1971).
73. V. V. Markova, V. A. Kormer and A. A. Petrov, *Zh. Obshch. Khim.* **35,** 1669 (1965).
74. G. Zweifel and N. L. Polston, *J. Am. Chem. Soc.* **92,** 4068 (1970).
75. E. J. Corey and D. K. Herron, *Tetrahedron Letters* **1971,** 1641.
76. G. M. Glark, K. G. Hancock and G. Zweifel, *J. Am. Chem. Soc.* **93,** 1308 (1971).
77. B. M. Mikhailov: *The Chemistry of Borohydrides*, Moscow, Nauka (1967).
78. G. Zweifel and H. C. Brown, *J. Am. Chem. Soc.* **85,** 2066 (1963).
79. D. Devaprabhakara and P. D. Gardner, *J. Am. Chem. Soc.* **85,** 1458 (1963).
80. D. S. Sethi, G. C. Joshi and D. Devaprabhakara, *Can. J. Chem.* **46,** 2632 (1968).
81. H. Kretschmar and W. F. Erman *Tetrahedron Letters* **1970,** 41.
82. R. G. Lewis, D. H. Gustafson and W. F. Erman, *Tetrahedron Letters* **1967,** 401.
83. A. J. Hubert, *J. Chem. Soc.* **1965,** 6679.
84. B. M. Mikhailov and P. M. Aronovich, *Izv. Akad. Nauk SSSR, Ser. Khim.* **1963,** 1233.
85. B. M. Mikhailov and P. M. Aronovich, *Izv. Akad. Nauk SSSR, Otd. Khim. Nauk* **1961,** 927.
86. G. Goindard, J. Braun and P. Cadiot, *Bull. Soc. Chim. France* **1972,** 811
87. W. G. Woods and I. S. Bengelsdorf, *J. Org. Chem.* **31,** 2769 (1966).
88. D. S. Matteson and J. D. Liedtke, *J. Am. Chem. Soc.* **87,** 1526 (1965).
89. H. C. Brown, T. Hamaoka and N. Ravindran, *J. Am. Chem. Soc.* **95,** 6456 (1973).
90. H. C. Brown, D. H. Bowman, S. Misumi and M. K. Unni, *J. Am. Chem. Soc.* **89,** 4531 (1967).
91. E. Negishi, G. Lew and T. Yoshida, *J. Org. Chem.* **39,** 2321 (1974).
92. G. Zweifel, H. Arzoumanian and C. C. Whitney, *J. Am. Chem. Soc.* **89,** 3652 (1967).
93. G. Zweifel, *Intra-Science Chem. Rep.* **7,** 181 (1973).
94. G. Zweifel, R. P. Fisher, J. T. Snow and C. C. Whitney, *J. Am. Chem. Soc.* **94,** 6560 (1972).
95. G. Zweifel, R. P. Fisher, J. T. Snow and C. C. Whitney, *J. Am. Chem. Soc.* **93,** 6309 (1971).

96. G. Zweifel, N. L. Polston and C. C. Whitney, *J. Am. Chem. Soc.* **90,** 6243 (1968).
97. R. Köster and Y. Morita, *Angew. Chem.* **78,** 589 (1967).
98. N. Miyaura, H. Tagami, M. Itoh and A. Suzuki, *Chem. Lett.* **1974,** 1411.
99. D. S. Matteson and J. D. Liedtke, *Chem. and Ind.* **1963,** 1241.
100. D. S. Matteson and G. D. Schaumburg, *J. Org. Chem,* **31,** 726 (1966).
101. D. S. Matteson, *Organometal. Chem. Rev.* **1,** 1 (1966).
102. D. S. Matteson and J. D. Liedtke, *J. Org. Chem.* **28,** 1924 (1963).
103. D. S. Matteson, *J. Am. Chem. Soc.* **81,** 5004 (1959).
104. D. S. Matteson, A. H. Soloway, D. W. Tomlinson, J. D. Campbell and G. H. Nixon, *J. Med. Chem.* **7,** 640 (1964).
105. D. S. Matteson, In: Progress in Boron Chemistry, vol. 3. Ed. R. J. Brotherton and H. Steinberg. Oxford, Pergamon Press, 1970, p. 117.
106. D. S. Matteson, *J. Org. Chem.* **27,** 275 (1962).
107. D. Seyferth and M. Takamizawa, *J. Org. Chem.* **28,** 1142 (1963).
108. D. Seyferth and M. Takamizawa, *Inorg. Chem.* **2,** 731 (1963).
109. E. J. Corey and T. Ravindranathan, *J. Am. Chem. Soc.* **94,** 4013 (1972).
110. E. Negishi, J.-J. Katz and H. C. Brown, *Synthesis* **1972,** 555.
111. T. Yoshida, R. M. Williams and E. Negishi, *J. Am. Chem. Soc.* **96,** 3688 (1974).
112. E. Negishi and T. Yoshida, *J. Am. Chem. Soc.* **95,** 6837 (1973).
113. G. Zweifel and R. P. Fisher, *Synthesis* **1974,** 339.
114. G. Zweifel and A. Horng, *Synthesis* **1973,** 672.
115. R. Kow and M. W. Rathke, *J. Am. Chem. Soc.* **95,** 2715 (1973).
116. A. Pelter, A. Arase and M. G. Hutchings, *Chem. Comm.* **1974,** 346.
117. D. J. Pasto, *J. Am. Chem. Soc.* **86,** 3039 (1964).
118. G. Zweifel and H. Arzoumanian, *J. Am. Chem. Soc.* **89,** 291 (1967).
119. A. Hassner and B. H. Braun, J. Org. Chem. **28,** 261 (1963).
120. P. M. Aronovich and B. M. Mikhailov, *Izv. Akad. Nauk SSSR, Ser. Khim.* **1968,** 2745.
121. D. S. Matteson and J. G. Shdo, *J. Am. Chem. Soc.* **85,** 2684 (1963).
122. P. M. Aronovich, V. S. Bogdanov and B. M. Mikhailov, *Izv. Akad. Nauk SSSR, Ser. Khim.* **1969,** 362.
123. D. J. Pasto, J. Chow and S. K. Arora, *Tetrahedron Letters* **1967,** 723.
124. I. Mehrotra and D. Devaprabhakara, *Tetrahedron Letters* **1972,** 4871.
125. G. Cainelli, G. Dal Bello and G. Zubiani, *Tetrahedron Letters* **1965,** 3429.
126. G. Cainelli, G. Dal Bello and G. Zubiani, *Tetrahedron Letters* **1966,** 4315.
127. G. Zweifel and H. Arzoumanian, *Tetrahedron Letters* **1966,** 2535.
128. G. Zweifel, R. P. Fisher and A. Horng, *Synthesis* **1973,** 37.
129. H. C. Brown and S. P. Rhodes, *J. Am. Chem. Soc.* **91,** 4306 (1969).
130. R. C. Larock, *J. Organometal. Chem.* **61,** 27 (1973).
131. D. S. Matteson and J. O. Waldbillig, *J. Org. Chem.* **28,** 366 (1963).
132. G. Goindard and J. Braun, *Bull. Soc. Chim. France* **1972,** 817.
132a. D. S. Matteson and M. L. Talbot, *J. Am. Chem. Soc.* **89,** 1123 (1967).
133. D. S. Matteson, *J. Org. Chem.* **27,** 4293 (1962).
134. B. M. Mikhailov and T. A. Shchegoleva, *Izv. Akad. Nauk SSSR, Ser. Khim* **1971,** 661
135. B. M. Mikhailov, P. M. Aronovich and L. V. Tarasova, *Zh. Obshch. Khim.* **30,** 3624 (1960).
136. R. H. Fish, *J. Organometal. Chem.* **42,** 345 (1972).
137. V. A. Sazonova and N. Ya. Kronrod, *Zh. Obshch. Khim.* **26,** 1876 (1956).
138. A. N. Nesmeyanov, A. E. Borisov and M. A. Osipova, *Dokl. Akad. Nauk SSSR* **169,** 602 (1966).

139. R. C. Larock, S. K. Gupta and H. C. Brown, *J. Am. Chem. Soc.* **94,** 4371 (1972).
140. R. C. Larock and H. C. Brown, *J. Organometal. Chem.* **36,** 1 (1972).
141. Y. M. Riddle, U. S. Patents 2,950,303 and 2,950,305 (1960); *C. A.* **55,** 3434 (1961).
142. P. Binger, G. Benedikt, G. W. Rotermund and R. Köster, *Ann.* **717,** 21 (1968).
143. P. Binger and R. Köster, *Tetrahedron Letters* **1965,** 1901.
144. R. Köster, H. -J. Horstschäfer and P. Binger, *Ann.* **717,** 1 (1968).
145. N. Miyaura, T. Yoshinari, M. Itoh and A. Suzuki, *Tetrahedron Letters* **1974,** 2961.
146. P. Binger, *Agnew. Chem.* **80,** 288 (1968).
147. P. Binger, *Tetrahedron Letters* **1966,** 2675.
148. P. Binger, *Angew. Chem.* **79,** 57 (1967).
149. M. Naruse, T. Tomita, K. Uchimoto and H. Nozaki, *Tetrahedron Letters* **1973,** 795.
150. M. Naruse, T. Tomita, K. Uchimoto and H. Nozaki, *Tetrahedron* **30,** 835 (1974).
151. A. Pelter and K. J. Gould, *Chem. Comm.* **1974,** 347.
152. P. Binger and R. Köster, *Synthesis* **1974,** 350.
153. P. Binger and R. Köster, *Synthesis* **1973,** 309.
154. A. Pelter, C. R. Harrison and D. Kirkpatrick, *Chem. Comm.* **1973,** 544.
155. A. Pelter, C. R. Harrison and D. Kirkpatrick, *Tetrahedron Letters* **1973,** 4491.
156. M. Naruse, K. Uchimoto and H. Nozaki, *Tetrahedron Letters* **1973,** 1847.
157. M. Naruse, K. Uchimoto and H. Nozaki, *Tetrahedron* **30,** 2159 (1974).
158. M. Naruse, K. Uchimoto and H. Nozaki, *Tetrahedron Letters* **1973,** 2741.
159. M. Naruse, K. Uchimoto and H. Nozaki, *Tetrahedron* **30,** 3037 (1974).
160. P. Binger and R. Köster, *Chem. Ber.* **108,** 395 (1975).
161. T. Leung and G. Zweifel, *J. Am. Chem. Soc.* **96,** 5620 (1974).
162. P. Binger and R. Köster, *J. Organometal. Chem.* **73,** 205 (1974).
163. A. Suzuki, N. Miyaura, S. Abiko, M. Itoh, H. C. Brown, J. Sinclair and M. M. Midland, *J. Am. Chem. Soc.* **95,** 3080 (1973).
164. M. M. Midland, J. A. Sinclair and H. C. Brown, *J. Org. Chem.* **39,** 731 (1974).
165. A. Pelter and C. R. Harrison, *Chem. Comm.* **1974,** 828.
166. A. Pelter, K. J. Gould and L. A. P. Kane-Maguire, *Chem. Comm.* **1974,** 1029.
167. D. S. Matteson, *Synthesis* **1975,** 147.
168. D. S. Matteson and P. K. Jesthi, *J. Organometal Chem.* **110,** 25 (1976).
169. D. S. Matteson, M. S. Biernbaum, R. A. Bechtold, J. D. Campbell and R. J. Wilcsek, *J. Org. Chem.* **43,** 950 (1978).
170. D. S. Matteson and R. J. Moody, *J. Am. Chem. Soc.* **99,** 3196 (1977).
171. S. Slayden, *J. Org. Chem.* **45,** 2908 (1980).
172. H. C. Brown and S. K. Gupta, *J. Am. Chem. Soc.* **97,** 5279 (1975).
173. H. C. Brown and N. Ravindran, *J. Am. Chem. Soc.* **98,** 1798 (1976).
174. H. C. Brown and N. Ravindran, *J. Am. Chem. Soc.* **98,** 1785 (1976).
175. H. C. Brown and J. B. Campbell, Jr., *J. Org. Chem.* **45,** 389 (1980).
176. E. Negishi, R. M. Williams, G. Lew and T. Yoshida, *J. Organometal Chem.* **92,** C4 (1975).
177. J. B. Campbell, Jr. and G. A. Molander, *J. Organometal Chem.* **156,** 71 (1978).
178. G. W. Kabalka and S. Slayden, *J. Organometal Chem.* **93,** 33 (1975).
179. W. Siebert, R. Full, J. Edwin and K. Kingberger, *Chem. Ber.* **111,** 823 (1978).
180. L. W. Hall, J. D. Odom and P. D. Ellis, *J. Am. Chem. Soc.* **97,** 4527 (1975).
181. A. M. Caporusso, G. Giacomelli and L. Lardicci, *Gazz. Chim. Ital.* **106,** 879 (1976).
182. H. Yatagai, Y. Yamamoto and K. Maruyama, *Chem. Comm.* **1978,** 702.

183. M. M. Midland and H. C. Brown, *J. Org. Chem.* **40,** 2845 (1975).
184. H. C. Brown, C. G. Scouten and R. Liotta, *J. Am. Chem. Soc.* **101,** 96 (1979).
185. G. Zweifel and S. J. Backlund, *J. Am. Chem. Soc.* **99,** 3184 (1977).
186. A. Hassner and J. A. Soderquist, *J. Organometal Chem.* **131,** C1 (1977).
187. J. A. Miller and G. Zweifel, *Synthesis* **1981,** 288.
188. T. Hamaoka and H. C. Brown, *J. Org. Chem.* **40,** 1189 (1975).
189. G. Zweifel and R. P. Fisher, *Synthesis* **1975,** 376.
190. N. J. La Lima, Jr. and A. B. Levy, *J. Org. Chem.* **43,** 1279 (1978).
191. A. B. Levy, B. Angelastro and E. R. Marinelli, *Synthesis* **1980,** 945.
192. K. Avasthi, T. Baba and A. Suzuki, *Tetrahedron Lett.* **1980,** 9.
193. A. B. Levy and S. J. Schwartz, *Tetrahedron Lett.* **1976,** 2201.
194. A. B. Levy, S. L. Schwartz, N. Wilson and B. Christie, *J. Organometal Chem.* **156,** 123 (1978).
195. D. A. Evans, B. C. Thomas and J. A. Walker, *Tetrahedron Lett.* **1976,** 1427.
196. D. A. Evans, T. C. Crawford, R. C. Thomas and J. A. Walker, *J. Org. Chem.* **41,** 3947 (1976).
197. K. Uchimoto, M. Kitai, M. Naruse and H. Nozaki, *Tetrahedron Lett.* **1975,** 4233.
198. H. C. Brown, T. Hamaoka and N. Ravindran, *J. Am. Chem. Soc.* **95,** 5786 (1973).
199. G. W. Kabalka, E. E. Gooch and H. C. Hsu, *Synthetic Comm.* **11,** 247 (1981).
200. G. W. Kabalka and E. E. Gooch, *J. Org. Chem.* **46,** 2582 (1981).
201. M. M. Midland and S. B. Preston, *J. Org. Chem.* **45,** 747 (1980).
202. E. Negishi and K. W. Chiu, *J. Org. Chem.* **41,** 3484 (1976).
203. K. W. Chiu, E. Negishi, M. S. Plante and A. Silveira, *Jr. J. Organometal Chem.* **112,** C3 (1976).
204. E. Negishi, R. M. Williams, G. Lew and T. Yoshida, *J. Organometal Chem.* **92,** C4 (1975).
205. K. Uchimoto, K. Uchida and H. Nozaki, *Chem. Lett.* **1974,** 1493.
206. K. Uchimoto, K. Uchida and H. Nozaki, *Tetrahedron* **33,** 1949 (1977).
207. K. Uchimoto, K. Uchida and H. Nozaki, *Tetrahedron* **33,** 1945 (1977).
208. P. Jacob, III and H. C. Brown, *J. Org. Chem.* **42,** 519 (1977).
209. P. Jacob, III and H. C. Brown, *J. Am. Chem. Soc.* **98,** 7832 (1976).
210. H. Yatagai, Y. Yamamoto and K. Maruyama, *J. Am. Chem. Soc.* **102,** 4548 (1980).
211. Y. Yamamoto, H. Yatagai and T. Moritani, *J. Am. Chem. Soc.* **97,** 5606 (1975).
212. Y. Yamamoto, H. Yatagai, K. Maruyama, A. Sonoda and S. Murahashi, *J. Am. Chem. Soc.* **99,** 5652 (1972).
213. Y. Yamamoto, H. Yatagai, K. Maruyama, A. Sonoda and S. Murahashi, *Bull. Chem. Soc. Japan* **50,** 3427 (1977).
214. Y. Yamamoto, H. Yatagai, A. Sonoda and S. Murahashi, *J. Chem. Soc. Chem. Comm.* **1976,** 452.
215. H. Yatagai, *J. Org. Chem.* **45,** 1640 (1980).
216. R. Köster and L. A. Hagelee, *Synthesis* **1976,** 118.
217. K. Uchida, K. Uchimoto and H. Nozaki, *J. Org. Chem.* **41,** 2941 (1976).
218. K. Uchida, K. Uchimoto and H. Nozaki, *Tetrahedron* **33,** 2987 (1977).
219. N. Miyaura, T. Yano and A. Suzuki, *Bull. Chem. Soc. Japan* **53,** 1471 (1980).
220. J. B. Campbell, Jr. and H. C. Brown, *J. Org. Chem.* **45,** 549 (1980).
221. H. C. Brown and J. B. Campbell, Jr., *J. Org. Chem.* **45,** 550 (1980).
222. H. C. Brown and G. A. Molander, *J. Org. Chem.* **46,** 645 (1981).
223. H. Yatagai, Y. Yamamoto, K. Maruyama, A. Sonoda and S. Murahashi, *J. Chem. Soc., Chem. Comm.* **1977,** 852.
224. H. Yatagai, *Bull. Chem. Soc. Japan* **53,** 1670 (1980).

225. N. Miyaura, K. Yamada and A. Suzuki, *Tetrahedron Lett.* **1979,** 3437.
226. N. Miyaura and A. Suzuki, *J. Chem. Soc., Chem. Comm.* **1979,** 866.
227. N. Miyaura, T. Yano and A. Suzuki, *Tetrahedron Lett.* **1980,** 2865.
228. N. Miyaura, H. Suginome and A. Suzuki, *Tetrahedron Lett.* **2,** 127 (1981).
229. N. Miyaura and A. Suzuki, *J. Organometal Chem.* **213,** C53 (1981).
230. V. V. R. Rao, C. V. Kumar and D. Devaprabhakara, *J. Organometal Chem.* **179,** C7 (1974).
231. N. Miyaura and A. Suzuki, *Chem. Lett.* **1981,** 879.
232. Y. Masuda, A. Arase and A. Suzuki, *Chem. Lett.* **1978,** 665.
233. Y. Masuda, A. Arase and A. Suzuki, *Bull. Chem. Soc. Japan* **53,** 1652 (1980).
234. R. Köster and L. A. Hagelee, *Synthesis* **1976,** 118.
235. A. Pelter, C. Subrahmanyam, R. J. Laub, K. J. Gould and C. R. Harrison, *Tetrahedron Lett.* **1975,** 1633.
236. A. Pelter, T. W. Bentley, C. R. Harrison, C. Subrahmanyam and R. J. Laub, *J. Chem. Soc., Perkin I* **1976,** 2419.
237. A. Pelter, K. J. Gould and C. R. Harrison, *J. Chem. Soc., Perkin I* **1976,** 2428.
238. A. Pelter, C. R. Harrison, C. Subrahmanyam and D. Kirkpatrick, *J. Chem. Soc., Perkin I* **1976,** 2435.
239. A. Pelter, K. J. Gould and C. R. Harrison, *Tetrahedron Lett.* **1975,** 3327.
240. L. A. Hagelee and R. Köster, *Synth. React, Inorg. Met.—Org. Chem.* **7,** 53 (1977).
241. B. M. Mikhailov, M. E. Gurskii and M. G. Gverdtsiteli, *Izv. Akad. Nauk SSSR, Ser. Khim.* **1977,** 1456.
242. B. M. Mikhailov, M. E. Gurskii, M. G. Gverdtsiteli and V. G. Kiselev, *Izv. Akad. Nauk SSSR, Ser. Khim.* **1979,** 855.
243. J. Hooz and R. D. Mortimer, *Can. J. Chem.* **56,** 2786 (1978).
244. K. Uchimoto, T. Furubayashi and H. Nozaki, *Chem. Lett.* **1975,** 397.
245. T. Yogo, J.Koshino and A. Suzuki, *Synthetic Comm.* **9,** 809 (1979).
246. G. Zweifel, S. J. Backlund and T. Leung, *J. Am. Chem. Soc.* **100,** 5561 (1978).
247. M. M. Midland, *J. Org. Chem.* **42,** 2650 (1977).
248. K. Uchimoto, Y. Yabuki, K. Okada and H. Nozaki, *Tetrahedron Lett.* **1976,** 3969.
249. K. Yamada, N. Miyaura, M. Itoh and A. Suzuki, *Tetrahedron Lett.* **1975,** 1961.
250. A. Pelter, K. Smith and M. Tabata, *J. Chem. Soc., Chem. Comm.* **1975,** 8570.
251. J. A. Sinclair and H. C. Brown, *J. Org. Chem.* **42,** 1078 (1976).
252. N. Miyaura, S. Abiko, M. Itoh and A. Suzuki, *Synthesis* **1975,** 669.
253. S. W. Slayden, *J. Org. Chem.* **46,** 2311 (1981).
254. A. Pelter and L. Hughes, *J. Chem. Soc., Chem. Comm.* **1977,** 913.
255. A. Pelter and J. M. Rao, *Tetrahedron Lett.* **1981,** 797.
256. S. Hara, K. Kishimura and A. Suzuki, *Chem. Lett.* **1980,** 221.
257. M. M. Midland and D. C. McDowell, *J. Organometal Chem.* **156,** C5 (1978).
258. G. Zweifel and S. J. Backlund, *J. Organometal Chem.* **156,** 159 (1978).
259. J. Hooz and B. Mortimer, *Tetrahedron Lett.* **1976,** 805.
260. G. Zweifel and N. B. Pearson, *J. Org. Chem.* **46,** 829 (1981).
261. K. Yamada, N. Miyaura, M. Itoh and A. Suzuki, *Tetrahedron Lett.* **1977,** 679.
262. A. Pelter, P. Rupani and P. Stewart, *J. Chem. Soc., Chem. Comm.* **1981,** 164.
263. R. E. Merril, J. L. Allen, A.Abramovitch and E. Negishi, *Tetrahedron Lett.* **1977,** 1019.
264. E. Negishi and A. Abramovitch, *Tetrahedron Lett.* **1977,** 411.
265. B. M. Mikhailov, M. E. Gurskii, V. G. Kiselev and M. G. Gverdtsiteli, *Izv. Akad. Nauk SSSR, Ser. Khim.* **1977,** 2633.

PART IV

ALLYLBORANES

Allylic boron compounds, showing a high reactivity and a specific behavior with respect to different classes of organic compounds, are valuable and promising reagents in organic chemistry.

A characteristic feature of triallylborane and some of its homologs is the ability to be in a state of *permanent allylic rearrangement*, proceeding at different rates, depending on the nature of the compound and the temperature.

The high reactivity of allylboranes shows up in reactions with water, alcohols, and amines, with which they react even at room temperature, distinguishing them sharply from trialkylboranes, which react with these substances only at elevated temperature.

Allylboranes react with aldehydes, ketones, quinones, and nitriles as organometallic compounds, and their use for synthetic purposes, in many cases, has significant advantages over the use of compounds of magnesium and other metals.

Allylboranes are capable of adding to the double bonds of different unsaturated compounds. The reaction of allylboranes with vinyl ethers is a convenient preparative method for the synthesis of 1,4-pentadiene hydrocarbons and their functional derivatives.

Allylboranes have found use in the chemistry of small rings. With their use, it is possible to convert cyclopropene hydrocarbons to allylcyclopropanes and their functional derivatives and also to hexadiene and heptadiene hydrocarbons, including deutero-substituted compounds.

The allylboration of acetylenic and allenic compounds has proved fruitful from the synthetic viewpoint.

On the basis of the allylboron-acetylene condensation, which takes place via consecutive stages with the formation of acyclic, monocyclic, and bicyclic allylic boron compounds, methods have been developed for the synthesis of diene and triene hydrocarbons and their functional derivatives, unsaturated esters and ketones, unsaturated δ-oxo alcohols and δ-oxo carboxylic acids, and aromatic compounds of the biphenyl series, and a stereospecific method for the synthesis of compounds of the cyclohexene and cyclohexane series has been worked out.

Derivatives of 3-boracyclo[3.3.1]nonane and 3-borabicyclo[3.3.1]-non-6-ene have been used for the synthesis of unique 1-boraadamantane compounds, which have been used as the basis of an original method for constructing cage compounds of the adamantane series. The chemistry of allylic boron compounds has been covered in reviews [1–4, 184, 185].

PART IV

CHAPTER 1

Triallylboranes and alkylallylboranes

1.1 METHODS FOR THE PREPARATION OF TRIALLYLBORANES AND ALKYLALLYLBORANES

Triallylborane was first obtained in 33% yield by the reaction of triallylborate and allylmagnesium bromide [5]. Triallylborane is also formed from allylmagnesium bromide and butyl metaborate [6]. The reaction between allylmagnesium bromide and boron trifluoride etherate [6] is used as a preparative method for the synthesis of triallylborane:

$$CH_2{=}CH{-}CH_2{-}MgBr + BF_3 \longrightarrow \underset{68\%}{(CH_2{=}CH{-}CH_2)_3B} + 3MgBrF.$$

The reaction may be carried out in a single stage, by adding a mixture of allyl bromide and boron trifluoride etherate to magnesium in an ether medium [7, 8].

When allylmagnesium chloride and boron trifluoride etherate are used, the yield of triallylborane is 10–47% [7, 8]. To prepare triallylborane, allylmagnesium chloride and methyl borate can be used [9]. Triallylborane is obtained most conveniently from sesquiallylaluminum bromide and butyl borate or boron trifluoride etherate [10]:

$$(C_3H_5)_3Al_2Br_3 + BF_3 \longrightarrow \underset{78\%}{(C_3H_5)_3B} + AlF_3 + AlBr_3.$$

Tricrotylborane, in 75% yield [11], and tris(2-methylallyl)borane [12] have been synthesized by the above method.

A general method for the synthesis of mixed alkylallylboranes consists of the reaction of esters of organoboron acids with sesquiallylaluminum bromide (from allyl bromide and aluminum). In this way, both dialkylallylboranes and alkyldiallylboranes can be obtained in high yield [13, 14, 186]:

$$R_2BOR' + CH_2{=}CHCH_2Br + Al \longrightarrow \underset{90\%}{R_2BCH_2CH{=}CH_2},$$

$$RB(OR')_2 + 2CH_2{=}CHCH_2Br + Al \longrightarrow \underset{90\%}{RB(CH_2CH{=}CH_2)_2}$$

$$(R = C_2H_5, n\text{-}C_3H_7, n\text{-}C_4H_9, n\text{-}C_6H_{13}).$$

Dialkylallyboranes are also formed by the action of allylmagnesium halides on esters of dialkylborinic acids, but in this case the reaction is complicated by the formation of trialkylboranes, the relative quantity of which depends on the nature of the R_2BOR', the allyl halide, and the method by which the reaction is carried out.

A general method for the synthesis of dialkylallylboranes involves an exchange reaction between esters of dialkylthioborinic acids and triallylborane [15, 186, 187]:

$$R_2BSC_6H_{13} + (C_3H_5)_3B \longrightarrow \underset{95\%}{R_2BCH_2CH{=}CH_2}.$$

7-Alkyl derivatives of 3-allyl-3-borabicyclo[3.3.1]non-6-ene can also be used as a source of allyl groups [15, 186]:

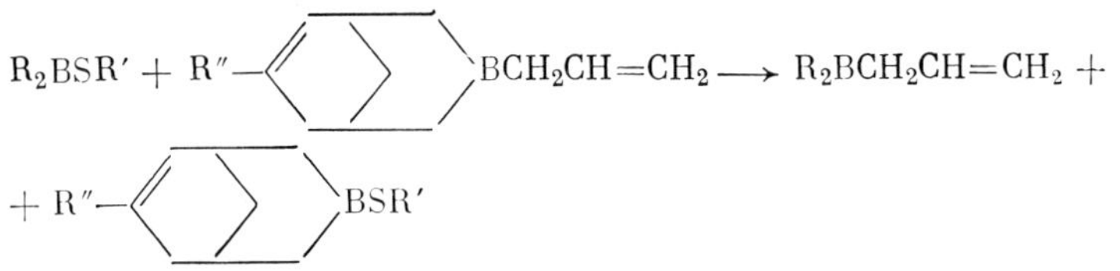

Alkylallylboranes undergo practically no symmetrization when heated to 115°C.

The data of [16, 17] on the preparation of dialkylallylboranes from trialkylboranes and allylmagnesium halides are erroneous – mixed alkylallylboranes are not formed by this method [14].

Allyldiarylboranes are obtained by the action of allylmagnesium bromide on butyl diarylborinates [18]. Allyldimesityl borane has been obtained from allylmagnesium bromide and fluorodimesitylborane [188].

Mixed dialkylallylboron compounds can be obtained by the addition of tetraalkyldiboranes to allenic hydrocarbons. Dialkyl(3-methyl-2-buten-1-yl)boranes with an impurity of approximately 10% of isomeric compounds have been synthesized by this method [19]:

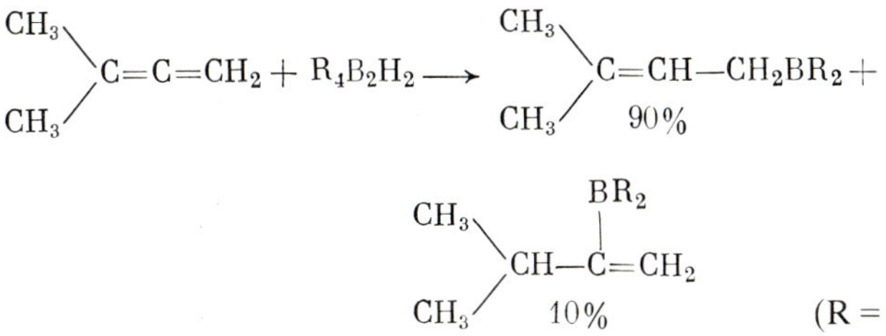

($R = C_2H_5, n\text{-}C_4H_9$).

A number of acyclic [20] and cyclic [21–23] allenic hydrocarbons has been subjected to hydroboration with tetra(3-methyl-2-butyl)diborane, and the reaction products, containing a mixture of the corresponding substituted dialkylallylboranes and dialkylvinylboranes, have been oxidized with hydrogen peroxide.

The hydroboration of allenes with 9-borabicyclo[3.3.1]nonane leads to products of dihydroboration and monohydroboration or merely to 9-allyl derivatives of 9-borabicyclo[3.3.1]nonane, depending on the structure of the hydrocarbons [189–191]. The 9-allyl derivatives can also be obtained from 9-methoxy-9-BBN [190].

Cyclic allenic hydrocarbons add a boron atom to the central carbon of the allenic system under the action of catecholborane [192], whereas, in the case of 9-borabicyclo[3.3.1]nonane, the boron adds mainly to the terminal carbon. The dihydroboration of cyclic allenes with diborane has been investigated in detail [193].

Different 9-allyl derivatives of 9-borabicyclo[3.3.1]nonane are obtained by the monohydroboration of conjugated dienes with the use of 9-BBN [194].

The hydroboration-mercuration of allene leads to $ClHgCH_2CH_2CH_2HgCl$, from which the corresponding Grignard reagent can be obtained [195].

Alkyldiallylboranes may be obtained by the symmetrization of dialkylallylboranes. Thus, (3-methyl-2-buten-1-yl)diethylborane, when heated above 100°C, is converted to a mixture of triethylborane, bis(3-methyl-2-buten-1-yl)ethylborane, and tris(3-methyl-2-buten-1-yl)borane, which are separated by fractional distillation [19].

An interesting type of allylic boron compounds is represented by 2-substituted (1,4-pentadien-1-yl)diallylboranes, formed by the addition of allylboranes to acetylenes. The preparation and properties of these compounds are considered in Chapter 4, which is devoted to the allylboration of acetylenic and allenic compounds.

1-Allylborolane (1-allylboracyclopentane) has been synthesized by the action of allylmagnesium bromide or sesquiallylaluminum bromide on 1-methoxy- or 1-butoxyborolane [24] which, in turn, may be prepared from 1,3-butadiene and diborane [25]:

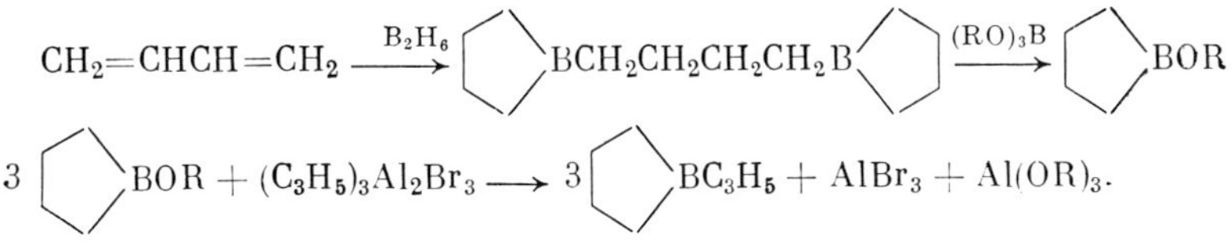

1-Crotylborolane has been obtained from 1-butoxyborolane and sesquicrotylaluminum bromide [26, 27].

The reaction of 1,3-butadiene with diborane, taken in a ratio of 5:1, in an ether medium produces a small yield of a mixture of unsaturated compounds with the composition $C_8H_{15}B$ [together with di(1-borolyl)butanes], which consists almost completely of 1-crotylborolane [28].

The hydroboration of isoprene (hydrocarbon:$BH_3 = 12:1$) leads to a mixture of 3-methyl-1-alkenylborolanes [together with bis(3-methyl-1-borolyl)pentanes], which contains approximately 60% of 3-methyl-1-(2-methyl-2-buten-1-yl)borolane [29].

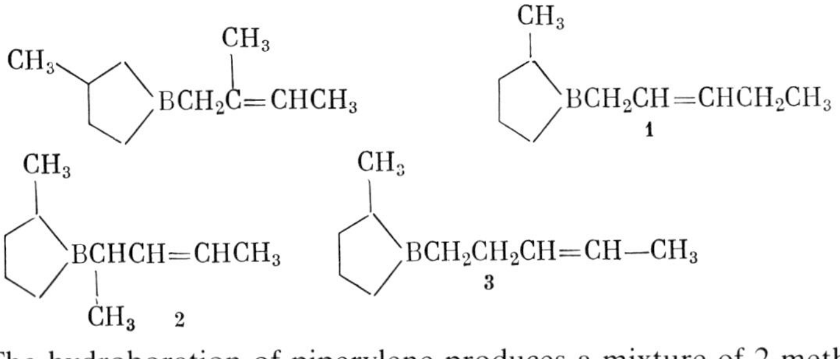

The hydroboration of piperylene produces a mixture of 2-methyl-1-alkenylboracyclopentanes, containing approximately 35% of 2-

methyl-1-(2-penten-1-yl)boracyclopentane (*1*), 15% of 2-methyl-1-(3-penten-2-yl)boracyclopentane (*2*), and 50% of 2-methyl-1-(3-penten-1-yl)boracyclopentane (*3*) [29].

Mixed dialkylaminoallylboranes are obtained from the corresponding chloroorganoboranes and Grignard reagents. The reaction of chloro(dimethylamino)ethylborane with crotylmagnesium bromide produces a mixture of (3-buten-2-yl)(dimethylamino)ethylborane (*4*) and (2-buten-1-yl)(dimethylamino)ethylborane (*5*) in a ratio of 65:35 [30]:

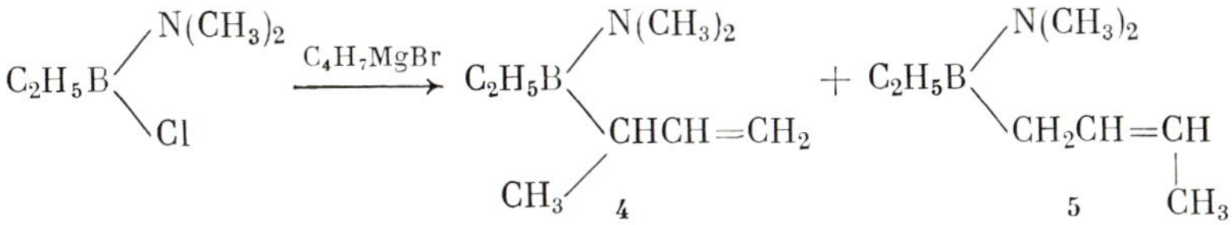

The compound *4* is relatively stable at room temperature, but isomerizes to *5* at 150°C.

The action of crotylmagnesium bromide on chloro(dimethylamino)phenylborane forms a mixture of dimethylamino(phenyl)-1-methylallylborane and dimethylamino(phenyl)crotylborane in a ratio of 8:15 [31]. Bis(dimethylamino)chloroborane reacts with crotylmagnesium bromide to form a mixture of bis(dimethylamino)-(1-buten-3-yl)borane and bis(dimethylamino)(2-buten-1-yl)borane in a ratio of 84:16 [32].

For the IR and Raman spectra of triallylboranes and their pyridine complexes, see [196]. The ^{13}C NMR spectra of allylic compounds of tri- and tetracoordinated boron are presented in [197].

1.2 CHEMICAL PROPERTIES OF TRIALLYLBORANES

A specific feature of triallylborane and of some other allylic boron compounds is their ability to be in a state of permanent allylic rearrangement, which proceeds at different rates, depending on the nature of the compound and the temperature.

Triallylborane and its analogs show a high reactivity, exceeding that of trialkylboranes. They react with nucleophilic reagents even at room temperature, which sharply distinguishes them from trialkylboranes, since the latter react with water, alcohols, and amines only upon heating [33].

The increased reactivity of allylboranes, expressed in the easy rupture of the boron-carbon bond under the action of different reagents, is in accordance with the high chemical activity of other classes of allylic compounds, e.g., allyl halides. The decrease in the energy of the carbon-element bond in the system $CH_2{=}CHCH_2X$, as compared with the saturated system RCH_2X, is explained by an increase in the overlap between the orbitals of the carbon in the $—CH_2X$ group and the carbon in the sp^2 state of the vinyl system as compared with the overlap between the orbitals of two carbon atoms in the sp^3 state, implying a decrease in the s character of the carbon orbitals in the direction of X and thereby a decrease in the overlap between the orbitals of the carbon and the X atom.

1.2.1 Permanent allylic rearrangement

It has been determined by the ^1H NMR method that a permanent allylic rearrangement takes place in triallylborane [34–36].

The ^1H NMR spectrum of triallylborane at low temperatures (-65 to -40°C) is similar to the spectra of allylic organic compounds (allyl alcohol and others). It contains a doublet at 2.05 ppm due to the CH_2 group bonded to the boron atom, signals from terminal protons in the region of 4.70 ppm, and a multiplet from a methine proton in the region of 5.65 ppm. With increasing temperature, the doublet and the signals from the terminal protons broaden, the fine structure disappears, and at room temperature they are converted to two broad signals, which later merge into one. At 135°C, the spectrum of triallylborane is a spectrum of the AX_4 type, containing a normal quintet centered at 5.00 ppm with an intensity ratio of 1:4:6:4:1 and a doublet from four protons of CH_2 groups centered at 3.2 ppm, with $J_{AX} = 10.3 \pm 0.2$ Hz. Consequently, the terminal protons and the protons of the CH_2 groups bonded to the boron atom become equivalent.

The spectral data on the dependence of the exchange rate on the temperature have made it possible to evaluate the activation energy of the process, which was found to have a value of 10 ± 1 kcal/mole. As shown by an analysis of the spectra of triallylborane dissolved in

CCl_4 or cyclohexane, the exchange rate does not depend on the concentration of triallylborane, indicating an intramolecular mechanism of allylic rearrangement.

Allylic rearrangement is also observed in pyridine-triallylborane, but a spectrum of the AX_4 type appears only at a temperature of approximately 160°C, with the exchange rate being approximately one-hundredth of that in triallylborane [35, 36]. An investigation of solutions of pyridine-triallylborane in pyridine and *n*-decane by the ^1H NMR method has shown that allylic rearrangement takes place in the free triallylborane formed as a result of dissociation of the complex.

An equilibrium exchange of CH_2 groups in allylic boron compounds takes place via a transition state *6* with hydrogen atoms in the CH_2 groups symmetrically arranged with respect to the boron atom and with equivalent B—C and B—C bonds.

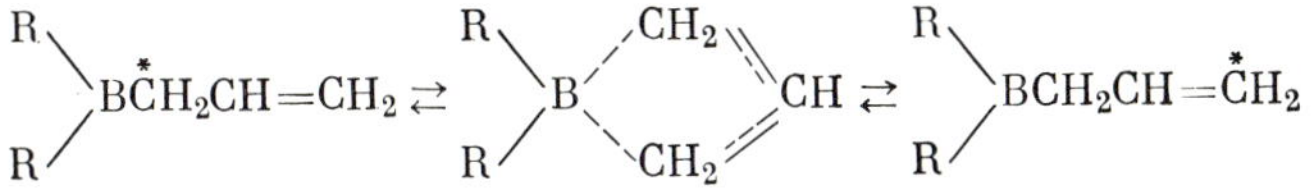

6

An investigation of the ^1H NMR spectrum of tricrotylborane, using the method of double proton-proton resonance at 60 and 100 MHz, has shown that this compound exists in the form of the *cis* (*7*) and *trans* (*8*) isomers of the 2-buten-1-yl moiety [37].

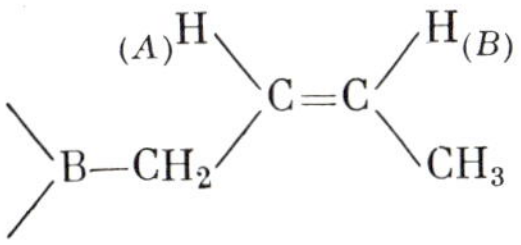

7

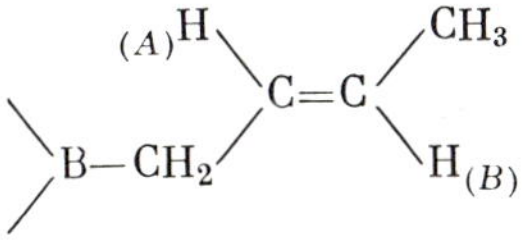

8

In tricrotylborane, unlike triallylborane, permanent allylic rearrangement does not take place at room temperature. From the spectra of the protons of the system—$CH_{(A)}$=$CH_{(B)}$—, obtained by the method of triple proton-proton resonance, it follows that the mixture contains predominantly *trans*-tricrotylborane ($J_{AB} = 15.0$ Hz), the content of which at room temperature is 70%.

The form of the ^1H NMR spectrum of tricrotylborane at temperatures above 150°C indicates a reversible thermal interconversion of the *cis* and *trans* isomers at a rate such that an averaging of the chemical shifts and of the coupling constants of the $H_{(A)}$ and $H_{(B)}$ protons ($J = 14$ Hz) takes place. The observed conversion apparently proceeds via a stage of isomerization of the 2-buten-1-yl fragment of tricrotylborane (*8*) to the 3-buten-2-yl form (*9*), in which a free rotation about the $H_{(A)}C$—$CH_{(B)}$ bond takes place, ensuring a transition to *cis*-tricrotylborane (*7*):

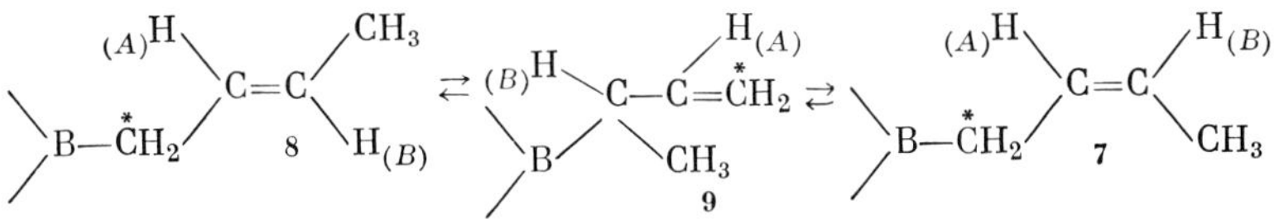

The interconversions *8*⇄*9* and *9*⇄*7* proceed via a transition state of the type *6*.

The study of the ^1H NMR spectra of tri(2-methylallyl)borane [12] has shown that, in this compound also, a permanent allylic rearrangement takes place, which proceeds more readily than in triallylborane. Thus, at 100°C, the rate of rearrangement is 5709 sec^{-1}, whereas the rate in triallylborane at this temperature is 2421 sec^{-1}. The activation energy of the rearrangement has a value of 9.8 ± 0.5 kcal/mole.

The permanent allylic rearrangement has been detected by the ^1H NMR method in tri(3-methyl-2-buten-1-yl)borane, with a free activation energy of 22.8 kcal/mole [38, 39]. The free activation energy of the allylic rearrangement ($\Delta G^{\ddagger}$) increases in the following order (in kcal/mole):

$(CH_2{=}CHCH_2)_3B$	$(CH_2{=}C(CH_3)CH_2)_3B$	$(CH_3CH{=}CH{-}CH_2)_3B$	$[(CH_3)_2C{=}CHCH_2]_3B$
14.6	14.6	19.0–19.7	22.8

The activation energy of the allylic rearrangement in 1-allylborolane has a value of 15 ± kcal/mole [35, 36].

The nature of the disturbance of the proton chemical shifts, the ^{13}C chemical shifts, and the coupling constants J_{HH}, J_{BH}, and J_{CH} in triallylborane, its complexes with trimethylamine, pyridine, and 3-methylpyridine, and in tri(2-methylallyl)borane, determined by ^1H

and ^{11}B NMR spectroscopy and by double heteronuclear magnetic resonance H-{^{11}B} and H-{^{13}C} over the temperature range of −60 to + 200°C, indicates that an allyl-boron σ bond is kept under conditions of rapid permanent allylic rearrangement [39].

An intermolecular permanent rearrangement in the exchange reactions of triallyl-, tricrotyl-, and tris(2-methylallyl)boranes with borates [40] and thioborates [41] has been detected by ^{1}H and ^{11}B NMR methods.

1.2.2 Complexing capacity

Triallylborane forms complexes with pyridine [42, 43], 2-methylpyridine, 3-methylpyridine [43], and diethylamine [44]; an adduct of tricrotylborane with pyridine is known [11].

An investigation of the complexing capacity of triallylborane has been carried out by ^{1}H, ^{11}B NMR, and double boron-proton resonance [43]. The activation energy values characterizing the relative stability of the complexes of triallylborane with trimethylamine, pyridine, 2-methylpyridine, and 3-methylpyridine are 20.6, 25.1, 19.1, and 26.2 kcal/mole, respectively.

It has been found by ^{1}H NMR that the value of ΔH of the dissociation of the complex of triallylborane with 2-methylpyridine is 18.8 kcal/mole. This is in good agreement with the ΔH value obtained by an investigation of the permanent allylic rearrangement. It should be noted that tripropylborane forms a less stable complex with 2-methylpyridine ($\Delta H = 13.1$ kcal/mole).

Substitution of some ligands by others in adducts of triallylborane have been investigated by ^{1}H NMR spectroscopy [43]. It was found that, in accordance with the relative strength of the coordinate bond in pyridine-triallylborane and 3-methylpyridine-triallylborane, 3-methylpyridine is capable of replacing pyridine in its complex with triallylborane, but the displacement of pyridine is not complete. On the other hand, pyridine, which forms a more stable adduct than trimethylamine, completely displaces the latter from its complex with triallylborane.

It has been shown by NMR spectroscopy that, in amine solutions of triallylborane complexes, an exchange between free and complexed amine takes place [43].

Tris(2-methylallyl)borane shows a weaker complexing capacity than triallylborane [12].

1.2.3 Reactions with water, alcohols, and amines

Allylic boron compounds react readily with reagents containing active hydrogen, undergoing cleavage of the boron-carbon bond. The reactions of triallyl- and tricrotylboranes with water and alcohols, leading to allyl- and crotylboronic acids and their esters, will be considered in Chapter 2.

1-Crotylborolane, under the action of water at 20°C, undergoes hydrolysis at the B—C bond, forming 1-hydroxyborolane and 1-butene with a 98% degree of purity, indicating an allylic rearrangement during the hydrolysis [26, 27]. The reaction of triallylborane with mercaptans will be considered in Chapter 2.

According to [42, 45], triallylborane reacts vigorously with acetic or butyric acid to form the corresponding acyloxydiallylboranes.

According to [45], the action of benzoic acid on triallylborane produces benzoyloxydiallylborane and bis(benzoyloxy)allylborane.

In the first stage of the reaction of triallylborane with ammonia and with primary and secondary amines, which proceeds exothermally, complex compounds are formed which, upon heating to 100–130°C, are converted to *N*-substituted aminodiallylboranes [6, 44] in high yields (77–92%).

The adduct obtained by the reaction of triallylborane with diethylamine has been isolated in the pure state; when this is heated, propylene is evolved and diethylaminodiallylborane is obtained [44]:

$$(C_3H_5)_3B + NH(C_2H_5)_2 \longrightarrow (C_3H_5)_3B\cdot NH(C_2H_5)_2 \xrightarrow{100-130^\circ C} \underset{92\%}{(C_3H_5)_2BN(C_2H_5)_2} + C_3H_6.$$

The reaction of triallylborane with butylamine [5] and ethylamine [44] proceeds analogously. Triallylborane reacts with aniline with the evolution of propylene and formation of phenylaminodiallylborane even at room temperature [44]. The reaction of triallylborane with ammonia initially produces an adduct, which is converted to aminodiallylborane by heating [44].

The synthesis of aminodiallylborane and its *N*-alkyl derivatives by the above method was repeated in [46].

1.2.4 Reactions with borates, trialkylboranes, and diborane

Allyl boron compounds are capable of exchange reactions with borates. Triallylborane reacts with methyl borate to form a mixture of methyl esters of allylboronic and diallylborinic acids, with the ratio of the reaction products depending on the ratio of the starting components. Dimethyl allylboronate can be obtained in good yield by heating a mixture of triallylborane with methyl borate in a ratio of 1:2 [47]:

$$(C_3H_5)_3B + 2B(OCH_3)_3 \longrightarrow \underset{78\%}{3C_3H_5B(OCH_3)_2}.$$

It has been shown by ^{1}H NMR that an equilibrium mixture of allylboronates and diallylborinates (4:1) is obtained from an equimolar mixture of triallylborane and methyl borate, whereas, with a 2:1 ratio of the starting reagents, the ratio of these esters is 1:2.3 [48]. Dibutyl allylboronate is obtained by heating triallylborane with tributyl borate [10].

Triallylborane reacts with triisobutylborane at 130–140°C to form butylene and a polymer which may possibly have the following structure [49]:

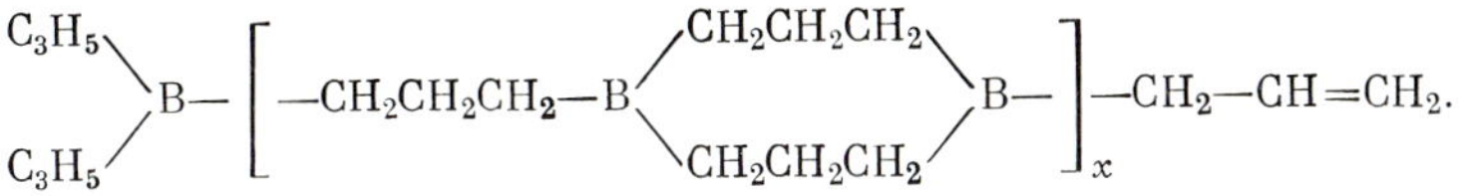

When heated with ethanol or *n*-butanol, rupture of the boron-carbon bonds in the side chain takes place and the polymer is converted to cyclic compounds— 1,5-dialkoxy-1, 5-diborocanes (1,5-dialkoxy-1,5-diboracyclooctanes) (*10*).

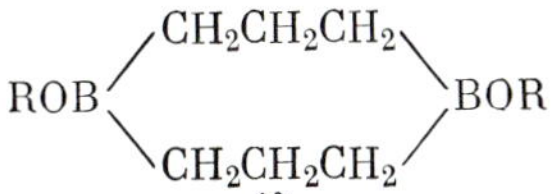

10

$$(n\text{-}C_4H_9O)_2B(CH_2)_3B(OC_4H_9\text{-}n)_2$$

11

The action of diborane on triallylborane (1:2) in ether solution produces a solid polymer [50]. When the polymer is heated with *n*-butanol, hydrogen, propylene, and liquid products consisting mainly of 1,5-dibutoxy-1,5-diboracyclooctane (*10*) and tetrabutyl 1,3-propanediboronate (*11*) are formed.

When heated with methyl borate, the solid polymer enters into exchange reactions, as a result of which tetramethyl 1,3-propanediboronate and methyl bis(3-dimethyoxyborylpropyl)boronate (*12*) are obtained [51–53].

$$(CH_3O)_2B(CH_2)_3\underset{\displaystyle OCH_3}{\underset{|}{B}}(CH_2)_3B(OCH_3)_2$$

12

Heating of the polymer with boron trichloride leads to 1,3-bis(dichloroboryl)propane [54].

The reaction of triallylborane with triethylamine-borane forms a polymer, the pyrolysis of which leads to a complex mixture of products, with 1,5-diborabicylo[3.3.3]undecane as its main component [55].

1.2.5 Reactions with carbonyl compounds. Synthesis of unsaturated alcohols

According to [42, 45], triallylborane reacts with aldehydes to form alkyl diallylborinates and cyclohexadiene or dialkyl allylboronates, cyclohexadiene, and propylene. However, these data have proved erroneous. The reaction of triallylborane with aldehydes actually proceeds as an organometallic synthesis, i.e., by addition of the allyl group to the carbon atom and of the BR_2 group to the oxygen atom of the carbonyl [56]. In this process, depending on the ratio of the starting reagents, diallylborinates (*13*), allylboronates, or borates are obtained:

$$(C_3H_5)_3B \xrightarrow{RCHO} \underset{\mathbf{13}}{(C_3H_5)_2B{-}O\overset{\displaystyle C_3H_5}{\overset{|}{C}}HR} \xrightarrow{RCHO} (C_3H_5)B(O\overset{\displaystyle C_3H_5}{\overset{|}{C}}HR)_2 \xrightarrow{RCHO} B(O\overset{\displaystyle C_3H_5}{\overset{|}{C}}HR)_3$$

$$(R = C_2H_5,\ C_6H_5).$$

The reaction of triallylborane with ketones, even with heating (80°C), stops at the stage of the formation of an allylboronate [56]:

$$(C_3H_5)_3B + 2(CH_3)_2CO \longrightarrow C_3H_5B(O\overset{\overset{CH_3}{|}}{\underset{\underset{CH_3}{|}}{C}}-CH_2CH{=}CH_2)_2$$

Since allylboronates and borates are easily hydrolyzed and transesterified with amino alcohols, the reaction of triallylborane with aldehydes and ketones may be used for the synthesis of unsaturated alcohols.

Tricrotylborane reacts with formaldehyde with allylic rearrangement [11]. Triallylborane adds smoothly to α,β-unsaturated aldehydes in the 1,2 position [57, 58].

In those cases where the reaction was carried out with an equimolar ratio of aldehyde to triallylborane, the intermediate organoboron compounds *13* were not isolated, but were directly hydrolyzed to the corresponding alcohols *14*, the yields of which in some cases exceed the yields of the alcohols obtained with the use of allylmagnesium bromide:

$$B(C_3H_5)_3 + RCHO \longrightarrow \underset{\mathbf{13}}{(C_3H_5)_2BO\overset{\overset{C_3H_5}{|}}{C}HR} \xrightarrow{H_2O} \underset{\mathbf{11}}{R\underset{\underset{OH}{|}}{C}H-C_3H_5}$$

(R = $CH_2{=}CH$, $CH_3CH{=}CH$, $C_6H_5CH{=}CH$,

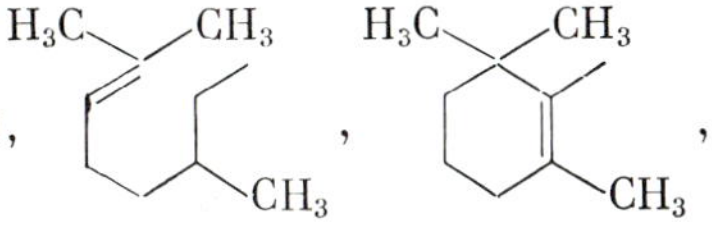

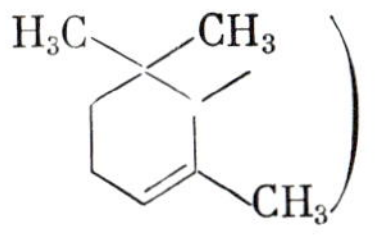
)

The reaction of acrolein with triallylborane (2:1) forms bis(1,5-hexadien-3-yl) allylboronate (*15*) [57, 58]:

$$(C_3H_5)_3B + 2CH_2{=}CHCHO \longrightarrow (CH_2{=}\overset{\overset{H}{|}}{\underset{\underset{C_3H_5}{|}}{C}}CH-O)_2BC_3H_5 \quad \mathbf{15}$$

The reaction of triallylborane with α-cyclocitral or β-cyclocitral (1:3) leads to the corresponding borates *16* and *17*:

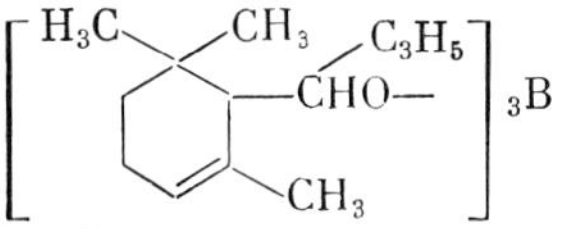

16

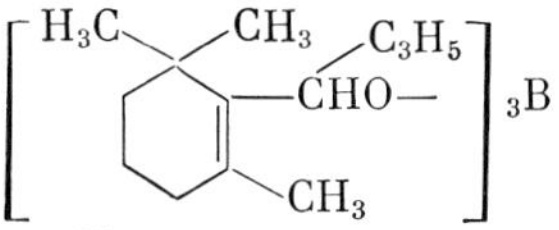

17

The reactions of triallylborane with α,β-unsaturated ketones have also been investigated [57, 59]. Unlike tributylborane, which adds to methyl vinyl ketone in the 1,4 position to form 2-octenyl dialkylborinate (*18*) [57]:

$$(n\text{-}C_4H_9)_3B + CH_2{=}CH\overset{\overset{\large O}{\|}}{C}CH_3 \longrightarrow \underset{\mathbf{18}}{C_4H_9{-}CH_2CH{=}\overset{\overset{\large OB(C_4H_9)_2}{|}}{C}CH_3},$$

triallylborane adds to methyl vinyl ketone in the 1,2 position to form 3-methyl-1,5-hexadien-3-yl diallylborinate (*19*), which forms 3-methyl-1,5-hexadien-3-ol upon hydrolysis:

$$(C_3H_5)_3B + CH_2{=}CH\overset{\overset{\large O}{\|}}{C}CH_3 \longrightarrow CH_2{=}CH{-}\underset{\underset{\large CH_3}{|}}{\overset{\overset{\large OB(C_3H_5)_2}{|}}{C}}CH_2CH{=}CH_2 \quad \mathbf{19}$$

Triallylborane and methyl β-chlorovinyl ketone (1:1) form 1-chloro-3-methyl-1,5-hexadien-3-yl diallylborinate, the hydrolysis of which leads to 1-chloro-3-methyl-1,5-hexadien-3-ol [57].

Triallylborane reacts smoothly with benzalacetone, benzalacetophenone, cinnamylideneacetone, mesityl oxide, and α- and β-ionone to form, after hydrolysis of the organoboron esters, the corresponding unsaturated tertiary alcohols. The yields of the latter reach values of 77–83% which, in some cases, is higher than the yields achieved with the use of allylmagnesium bromide [59].

The use of dialkylallylboranes in organic synthesis is promising. When these react with carbonyl compounds, butenyl dialkylborinates are obtained, which form unsaturated alcohols upon hydrolysis. Thus, the reaction of (3-methyl-2-buten-1-yl)dialkylborane with mesityl oxide forms the ester *20*, the hydrolysis of which leads to 3,3,4,6-tetramethyl-1,5-heptadien-4-ol (*21*) [14, 19]:

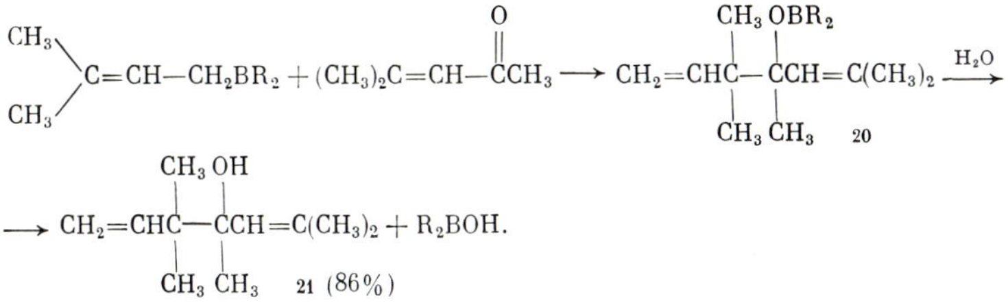

20, 21 (86%)

The reactions of (3-methyl-2-buten-1-yl)dialkylborane with acetone, benzaldehyde, and methyl vinyl ketone proceed analogously. In all cases, the addition takes place with allylic rearrangement, apparently via a six-center transition state. (3-Phenyl-2-butenyl)di(3-methyl-2-butyl)borane [60], dialkylallylboranes [15, 198, 199], alkyldiallylboranes [199], 3-allyl-7-alkyl-3-borabicyclo[3.3.1]non-6-enes [199], and 9-allyl-9-borabicyclo[3.3.1]nonanes [200] react with carbonyl compounds in a similar manner.

The dialkylallylboranes obtained from lithiated 2-butenyloxybenzimidazole and R_3B react regio- and stereospecifically with aldehydes and ketones to form (E)-erythro-γ,δ-unsaturated alcohols [201].

The reaction of α-silyl- or α-stannyl-substituted 9-crotyl-9-borabicyclo[3.3.1]-nonane with aldehydes in the presence of pyridine leads to *threo,cis*-alkenylsilanes or -stannanes [202].

The reaction of lithium trialkylcrotylborate with aldehydes proceeds threoselectively, leading to unsaturated alcohols [203]. In the presence of trialkylborane, the threoselectivity of the reaction of 2-butenyllithium with aldehydes increases [204].

Tri(2-methylallyl)borane adds to the carbonyl group of 16-dehydropregnenolone acetate (*22*) to form 20-(2-methylallyl)-5,16-pregnadien-3β,20ξ-diol 3-acetate (*23*) [61]:

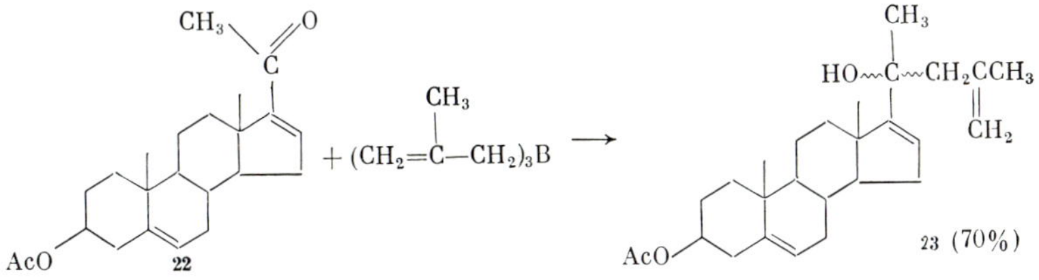

22, 23 (70%)

Dialkylallylboranes and alkyldiallylboranes react with esters and with organic acids to form alkyl(diallyl)carbinols [199].

1.2.6 Reactions with quinones. Synthesis of allylphenols

The reactions of triallylborane with quinones have been investigated [62–64].

The action of triallylborane on 1,4-benzoquinone (1:1), followed by hydrolysis of the organoboron compounds formed, produces high yields of the *cis* and *trans* isomers of 1,4-diallyl-1,4-dihydroxy-2,5-cyclohexadiene (*24*), which are converted by steam distillation in the presence of boric acid to 2,4-diallylphenol (*25*):

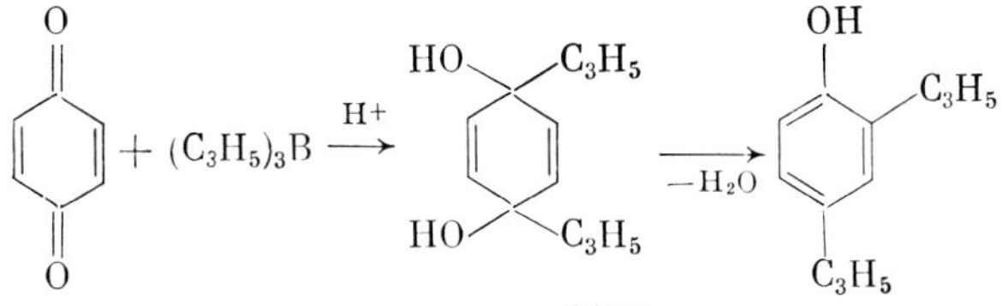

24 (84%) 25

With a reagent ratio of 2:1 and subsequent decomposition of the reaction mixture with water and steam distillation, the main reaction product (61%) is 2-allyl-1,4-hydroquinone (*26*), accompanied by 26% of 2,4-diallylphenol (*25*) [64]:

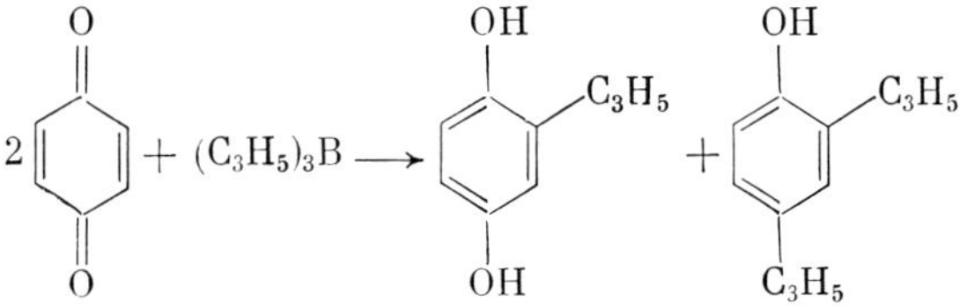

26 (61%) 25 (26%)

Allylboronates and diallylborinates are very active reagents with respect to quinones [63]. Dibutyl allylboronate reacts with 1,4-benzoquinone to form 2-allyl-1,4-hydroquinone in good yield. The mechanism of formation of the latter compound was established by studying the reaction between dimethyl allylboronate and 1,4-naphthoquinone. In this case (when the reaction is carried out under mild conditions), it is possible to isolate an intermediate compound, 1-allyl-1-hydroxy-4-oxo-1,4-dihydronaphthalene (*27*), which, when heated with boron trifluoride etherate, is converted to 2-allyl-1,4-dihydroxynaphthalene (*28*), similarly to the rearrangement of *p*-tolylquinol to 2-methylhydroquinone [65]:

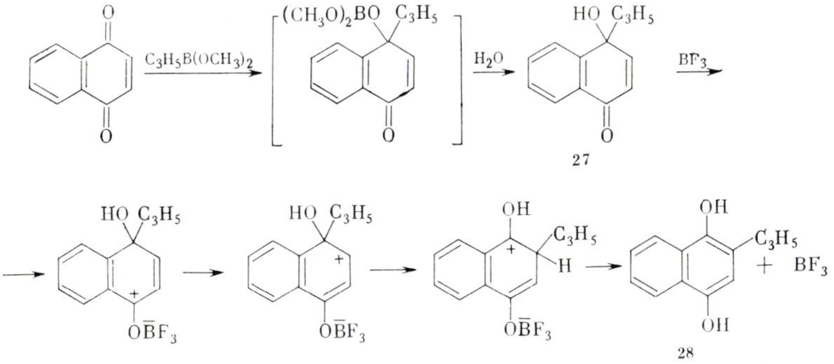

Trialkylboranes [66] react with 1,4-naphthoquinone to form 2-alkyl-1,4-dihydronaphthalene (see Part I, Ch.9.3), while triallylborane adds readily to the two carbonyl groups of 1,4-naphthoquinone in the 1,2 position to form, after hydrolysis, 1,4-diallyl-1,4-dihydroxy-1,4-dihydronaphthalene [63]:

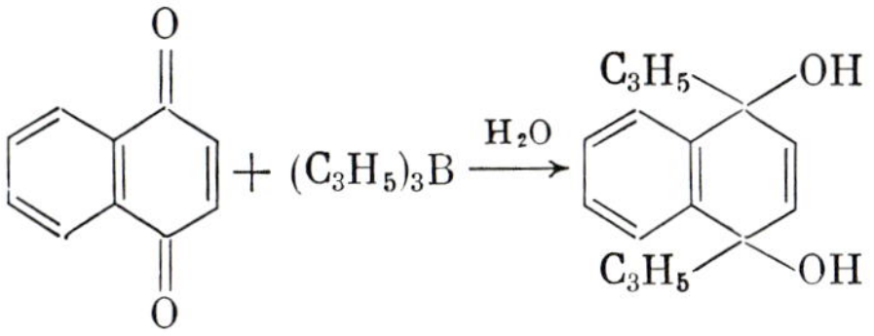

73%

The reactions of triallylborane with substituted 1,4-benzoquinones have been investigated [63]. The reaction of equimolar quantities of 2-allyl-1,4-benzoquinone and triallylborane forms 2,4,6-triallylphenol:

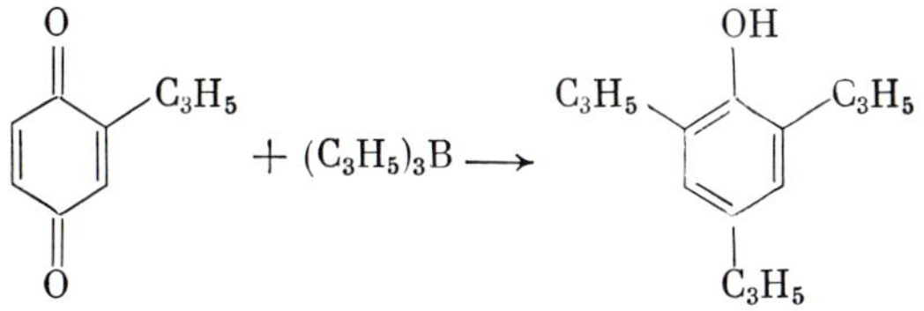

74%

whereas, with a reagent ratio of 2:1, only a small amount of 2,4,6-triallylphenol is obtained, while the main reaction products are 2,5- (*29*) and 2,6-diallylhydroquinones (*30*).

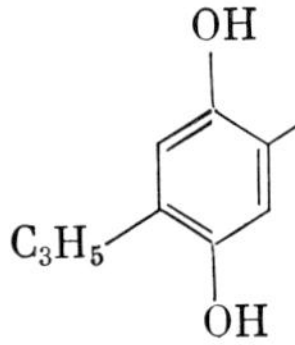

29

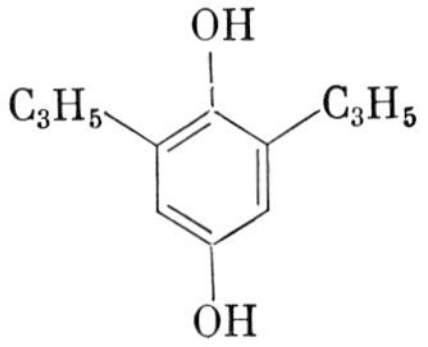

30

2-Butyl-1,4-benzoquinone reacts with triallylborane (1:1) to form 4,6-diallyl-2-butylphenol [64].

The reaction between 9,10-anthraquinone and triallylborane with an equimolar reagent ratio leads to 9-allyl-9-hydroxyanthrone (*31*); when two equivalents of triallylborane are used, *cis*-9,10-diallyl-9, 10-dihydroxy-9,10-dihydroanthracene (*32*) is mainly formed, with an impurity of the *trans* isomer [64]:

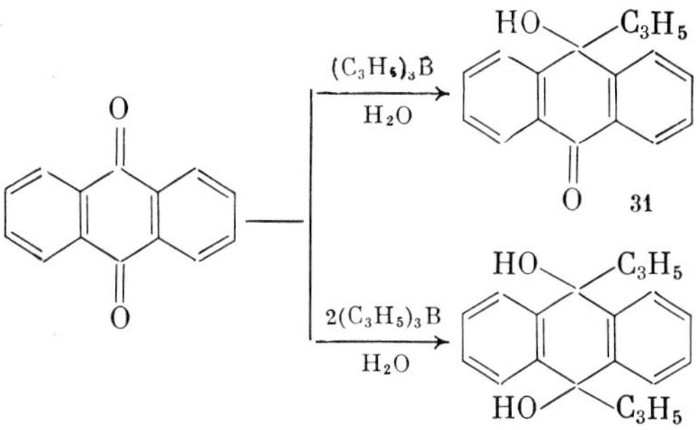

31

32

The reactions of triallylborane with *o*-quinones have been investigated [64]. The reaction of equimolar quantities of 1,2-naphthoquinone and triallylborane, followed by hydrolysis, forms *trans*-1,2-diallyl-1,2-dihydroxy-1,2-dihydronaphthalene (*33*); the same diol is also obtained by the action of dibutyl allylboronate on 1,2-naphthoquinone:

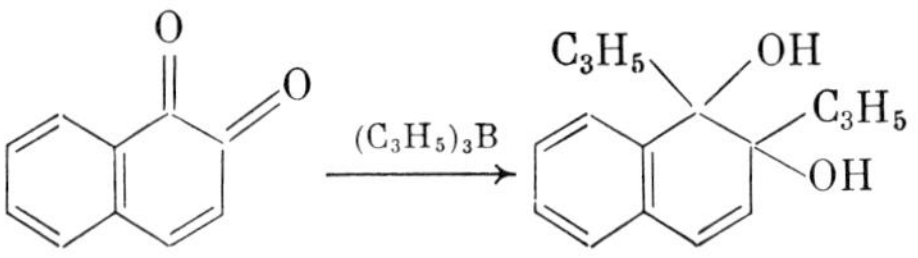

33

9,10-Phenanthrenequinone and acenaphthenequinone react analogously with triallylborane, forming, respectively, (after hydrolysis) *trans*-9,10-diallyl-9,10-dihydroxy-9,10-dihydrophenanthrene and *trans*-1,2-diallyl-1,2-dihydroxyacenaphthene.

1.2.7 Reactions with nitriles. Synthesis of aminodienes

In the first stage of the reaction of triallylborane with nitriles, carried out in isopentane at −70 to −50°C, the complexes *34* are formed which, upon gradual heating of the reaction mass to room temperature, are completely converted to the substituted (3-butenylideneamino)diallylboranes *35*, which dimerize to the substituted 1,3-bis(3-butenylidene)-2,2,4,4-tetraallylcyclodiborazanes *36* [67, 68]:

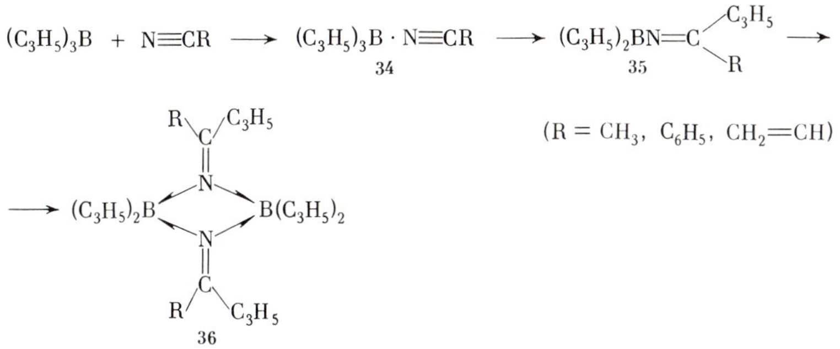

(R = CH_3, C_6H_5, $CH_2{=}CH$)

The dipole moment of cyclodiborazane (*36*, R = C_6H_5) has a value of zero, indicating a planar structure of the system:

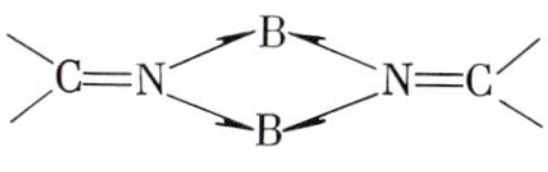

The cyclodiborazanes *36* are converted by heating (100°C) to the substituted 1,3-bis(1,6-heptadien-4-yl)-2,4-diallyl-1,3,2,4-diazadiboretidines *37*, with the formation of covalent B—N bonds instead of the semicoordinate (B←N) bonds between the boron and nitrogen atoms and with the conversion of the C=N bonds to single

N—C bonds:

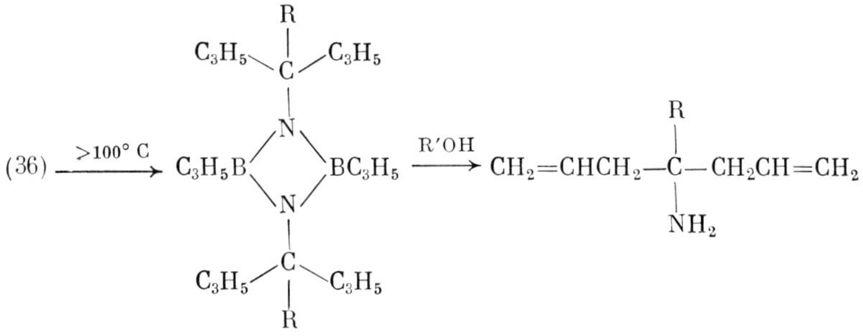

4-Substituted 4-amino-1,6-heptadienes are obtained by alcoholysis of the diazadiboretidines *37*.

Methacrylonitrile reacts with triallylborane to form (2-methyl-1,5-hexadiene-3-imino)diallylborane [69]:

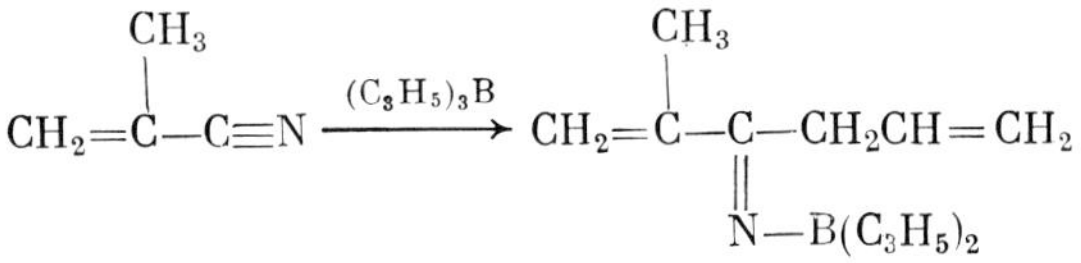

The iminoboranes obtained by the addition of dialkyl(allyl)-boranes to nitriles isomerize to 1-substituted (1,3-butadien-1-yl)aminodialkylboranes under the action of nucleophilic reagents (traces of moisture, ROH, amines) [205].

1.2.8 Reactions with vinyl ethers. Synthesis of 1,4-dienes

Allylboranes are capable of reacting with compounds containing an activated double bond [70, 71].

When triallylborane is heated with butyl vinyl ether (110–140°C), 1,4-pentadiene is formed in a yield of over 80% (calculated for the two boron-allyl bonds of the starting borane). The first stage of the reaction is apparently an addition of the allyl and allylboron fragments of the triallylborane molecule to the double bond of the vinyl ether to form (2-alkoxy-4-penten-1-yl)diallylborane (*38*), which is then converted by β-elimination to the diene *39* and butyl diallylborinate (*40*):

$$(C_3H_5)_3B + CH_2{=}CHOC_4H_9\text{-}n \longrightarrow \underset{38}{(C_3H_5)_2BCH_2\overset{\overset{\displaystyle OC_4H_9\text{-}n}{|}}{C}HCH_2CH{=}CH_2} \longrightarrow$$

$$\longrightarrow \underset{39}{CH_2{=}CHCH_2CH{=}CH_2} + \underset{40}{(C_3H_5)_2BOC_4H_9\text{-}n}$$

$$(C_3H_5)_2BOC_4H_9\text{-}n + CH_2{=}CHOC_4H_9\text{-}n \longrightarrow \begin{matrix} C_3H_5\diagdown \\ \\ n\text{-}C_4H_9O\diagup \end{matrix}\!\!BCH_2\overset{\overset{\displaystyle OC_4H_9\text{-}n}{|}}{C}HCH_2CH{=}CH_2 \longrightarrow$$

$$\longrightarrow CH_2{=}CHCH_2CH{=}CH_2 + C_3H_5B(OC_4H_9\text{-}n)_2.$$

A tendency of β-alkoxyalkylboranes toward olefin elimination has also been observed in an investigation of the hydroboration of alkyl vinyl ethers [72]. The action of tricrotylborane on butyl vinyl ether forms 3-methyl-1,4-pentadiene (*41*):

$$(CH_3CH{=}CHCH_2)_3B + CH_2{=}CHOC_4H_9\text{-}n \xrightarrow{140^\circ\ C} \underset{41\ (50\%)}{CH_2{=}CH\overset{\overset{\displaystyle CH_3}{|}}{C}HCH{=}CH_2}$$

The formation of the diene *41* indicates that the addition of tricrotylborane to the double bond proceeds with allylic rearrangement via a cyclic transition state:

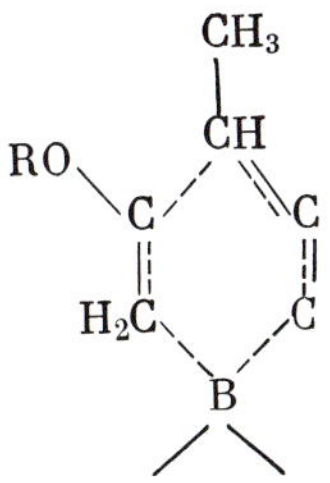

2-Methyl-1,4-pentadiene (*42*) is obtained by the reaction of butyl vinyl ether and tris(2-methylallyl)borane or of ethyl isopropenyl ether and triallylborane:

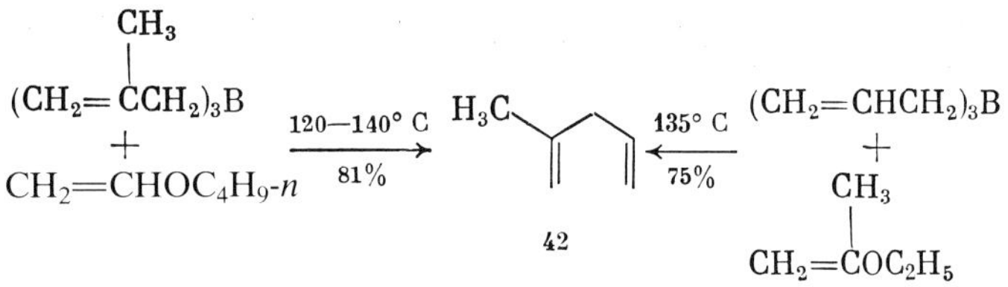

The reaction of triallylborane with 2-methyl-4,5-dihydrofuran proceeds slowly at room temperature and rapidly (within a few minutes) at 80–100°C. Alkaline hydrolysis of the adduct leads to pure 4-methyl-1,4-heptadien-7-ol (*43*):

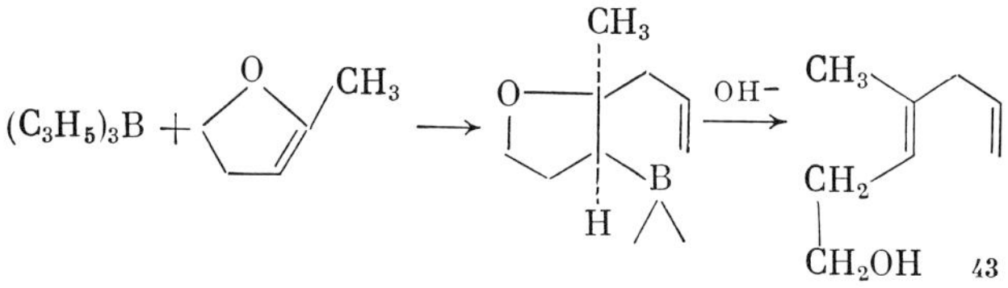

Thus, the reaction of allylboranes with acyclic and cyclic vinyl ethers is a convenient preparative method for the synthesis of difficultly accessible 1,4-diene hydrocarbons and their functional derivatives.

1.2.9 Reactions with cyclopropene hydrocarbons. Synthesis of cyclopropane derivatives and dienes.

Triallylborane reacts under cooling (−70 to 0°C) with 1-methylcyclopropene in two directions (see the reaction scheme on the next page) [73–75]. *Cis*-addition of the organoborane to the double bond of the hydrocarbon takes place predominantly, with the formation of (*cis*-1-allyl-1-methyl-2-cyclopropyl)diallylborane (*44*), but the reaction also proceeds partially with rupture of the cyclopropene ring at the C_2—C_3 bond, forming (2-methyl-2,5-hexadien-1-yl)diallylborane (*45*). Methanolysis of the compound *44* forms dimethoxy(*cis*-1-allyl-1-methyl-2-cyclopropyl)borane (*46*), which is converted to 2-methyl-2-allylcyclopropanol (*47*) by oxidation and to 1-methyl-1-allylcyclopropane (*48*) by acidolysis. The compound *45* was not isolated in the pure state, but its formation was evidenced by the isolation of 2-methyl-1,5-hexadiene (*49*) from the products of methanolysis of the reaction mass.

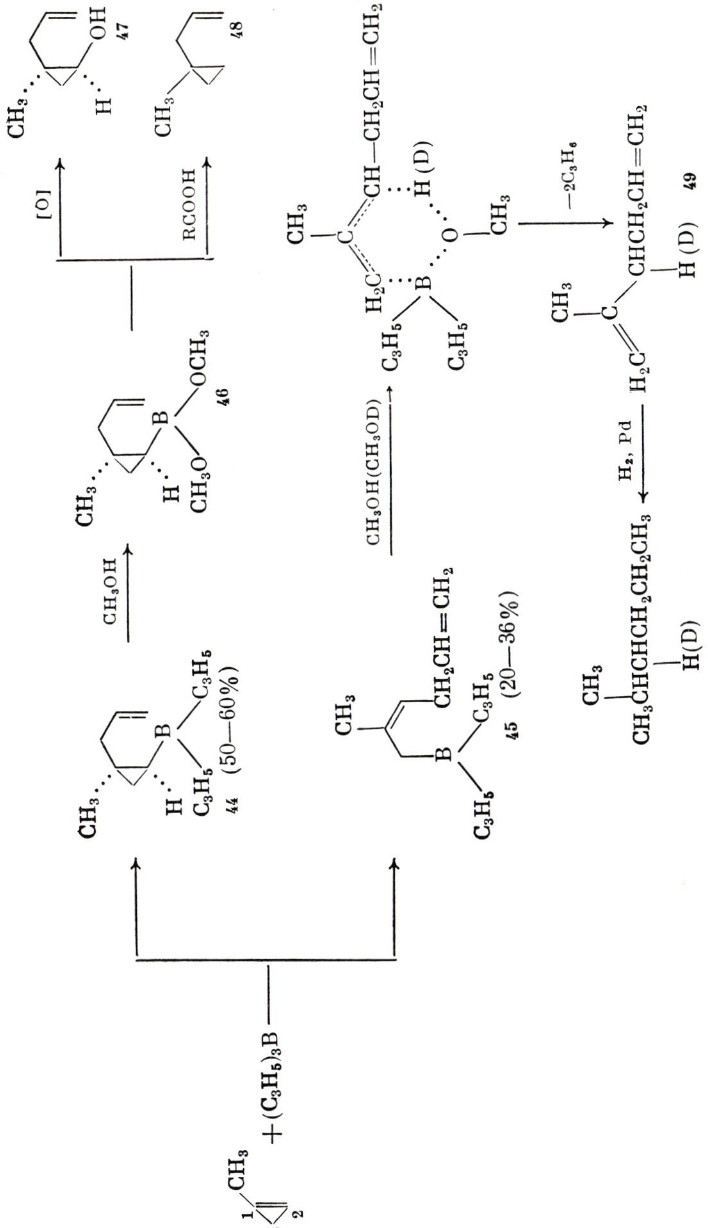

+(C3H5)3B
(50—60%)
44
(20—36%)
45
CH3OH
46
[O]
47
RCOOH
48
CH3OH(CH3OD)
−2C3H6
49
H2, Pd

Protonolysis of the compound *45* with deuteromethanol leads to 2-methyl-3-deutero-1,5-hexadiene (*49*).

The action of triallylborane on 1-methyl-2-deuterocyclopropene, followed by methanolysis of the reaction products, produces dimethoxy(*cis*-1-allyl-1-methyl-2-deutero-2-cyclopropyl)borane (*50*) (50–60%) and 2-methyl-3-deutero-1,5-hexadiene (*51*) (30–35%) [74, 75]:

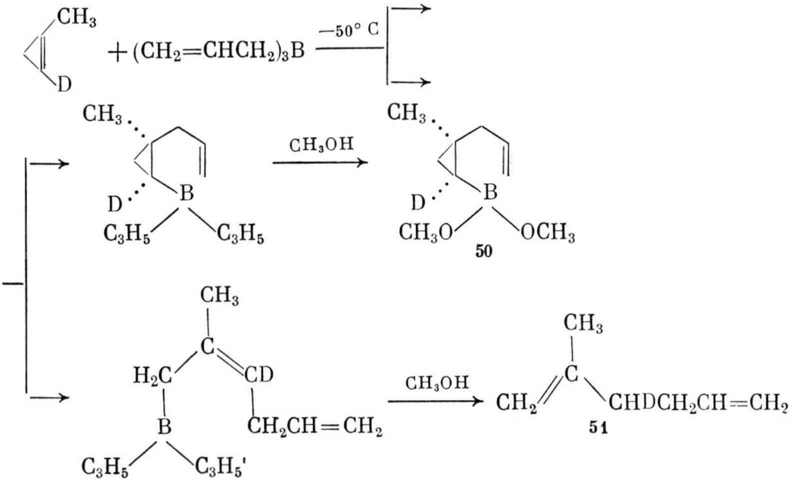

The stereochemical direction of the addition of triallylborane to 1-methylcyclopropene was determined by the conversion of dimethoxy(*cis*-1-allyl-1-methyl-2-cyclopropyl)borane (*46*) on hydroboration to 5-methoxy-1-methyl-5-borabicyclo[4.1.0]heptane (*52*) [75]:

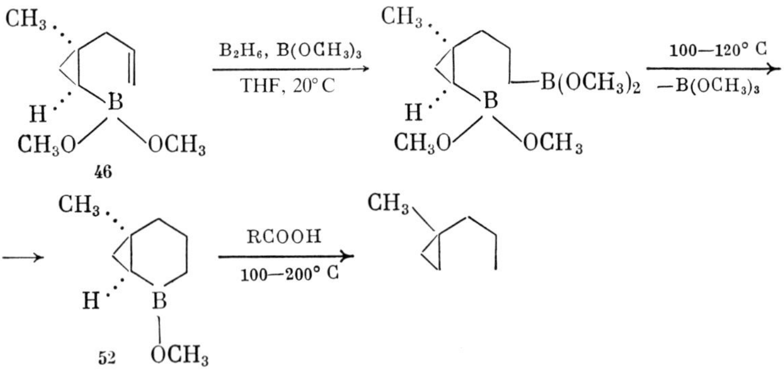

The reaction of 1-methylcyclopropene with tricrotylborane produces predominantly (85–90%) 1-(*cis*-3-buten-2-yl)-1-methyl-2-cyclopropylbis(2-butenyl)borane (*53*), together with 10–15% of (2-methyl-2,5-heptadien-1-yl)bis(2-butenyl)borane (*57*), indicating that the addition of allylborane to the double bond of the three-membered ring proceeds with allylic rearrangement, while the cleavage of a single C—C bond takes place without rearrangement [74, 76].

The methanolysis of compound *53* leads to dimethoxy[1-*cis*-(3-buten-2-yl)-1-methyl-2-cyclopropyl]borane (*54*), which forms 1-*cis*-(3-buten-2-yl)-1-methyl-2-cyclopropanol (*55*) by oxidation and 1-(3-buten-2-yl)-1-methylcyclopropane (*56*) by acidolysis:

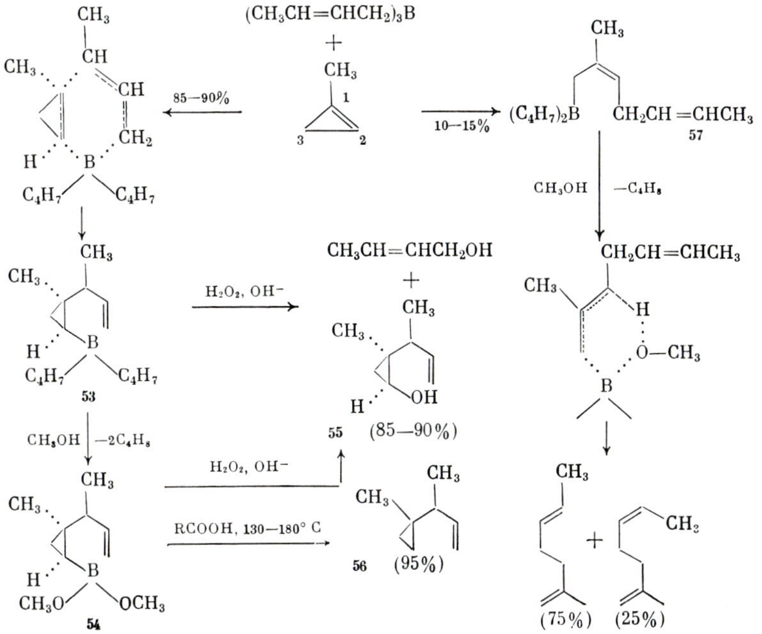

The most probable mechanism for the introduction of allylboranes at the C_2—C_3 bond of 1-methylcyclopropene involves the addition of allyl and diallylboryl groups to the C_2 carbon, with a shift of the double bond from the C_1—C_2 to the C_1—C_3 position (*58*), followed by an allylic rearrangement to the compound *57*:

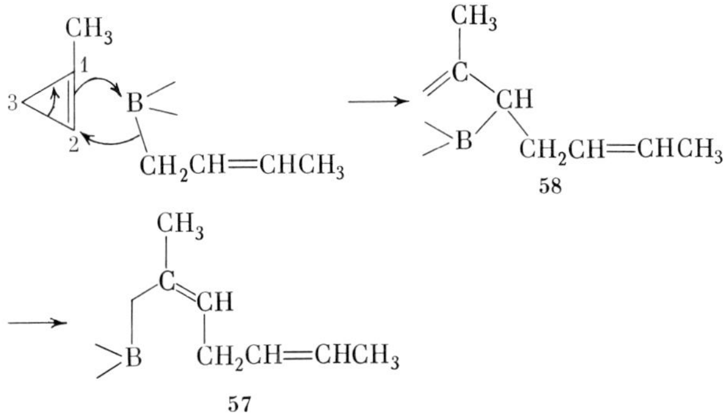

Triallylborane also reacts with 1,3,3-trimethylcyclopropene in two directions, with the formation of the addition product *59* and the product of introduction at the C_2—C_3 bond *60* which, unlike the intermediate compound *58*, does not isomerize to a compound of the type *57* and is converted to the hydrocarbon *61* by protonolysis with allylic rearrangement [77]:

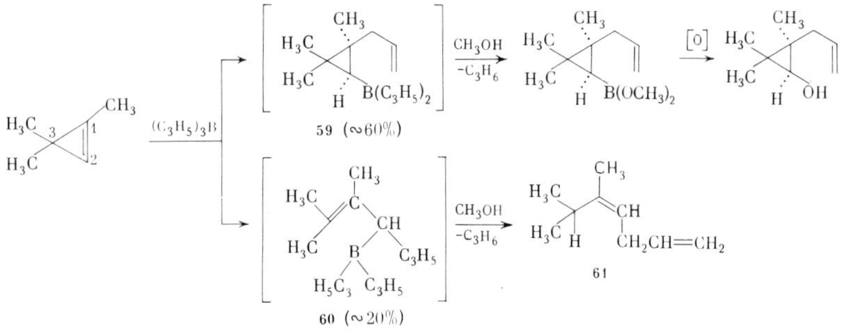

With the use of ^{1}H NMR, ^{11}B NMR, and multi-frequency resonances ^{1}H-{^{13}C}, ^{1}H-{^{1}H}-{^{2}H} and ^{1}H-{^{1}H}-{^{2}H}-{^{11}B}, the coupling constants $^2J_{^{11}BH}$, $^3J_{^{11}BH}$, and $^1J_{^{11}B^{13}C}$ were measured in dimethoxy(1-*cis*-allyl-1,3,3-trimethyl-2-cyclopropyl)borane, dimethoxy-[1-*cis*-(3-buten-2-yl)-1-methyl-2-cyclopropyl]borane, dimethoxy(1-*cis*-allyl-1-methyl-2-deutero-2-cyclopropyl)borane, and 5-methoxy-1-methyl-5-borabicyclo[4.1.0]heptane (*52*) by an indirect method [78].

Triallyl- and tricrotylborane react at −70°C with the saturated but highly strained compound, bicyclo[1.1.0]butane, to form the unsaturated compounds *62* [79]:

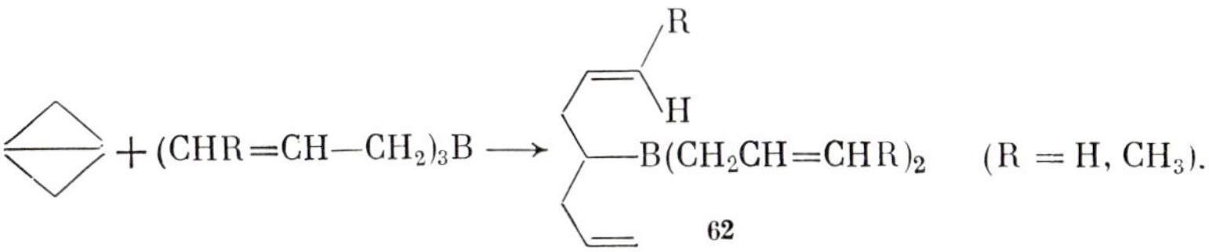

62

1.2.10 Thermal and miscellaneous reactions. Synthesis of higher olefins and unsaturated acids

Different hydrocarbons and boron-containing products are formed by the thermal decomposition of triallylborane [80, 81].

According to some data [42], the action of bromine on triallylborane at room temperature forms tris(2,3-dibromopropyl)borane, while, according to other data [10], triallylborane reacts with bromine at −35°C to form 1,2,3-tribromopropane as the main product.

Triallylborane is converted to tripropylborane by hydrogenation over a skeletal catalyst at room temperature [10].

An investigation of the oxidative polymerization of triallylborane in the presence of atmospheric oxygen at room temperature and also in isopropylbenzene or *tert*-butylbenzene solution at 130°C under the action of an oxygen-nitrogen mixture has been carried out. The polymerization products contained 25–30% of oxygen [82]. It was found that triallylborane is an active catalyst for the polymerization of methyl methacrylate [82].

Allyllithium adds to triallylborane to form lithium tetraallylborate with a ^{11}B chemical shift of +16.8 ppm [83].

Triallyl-, tris(2-methylallyl)-, and tricrotylboranes enter into exchange reactions with dimethylzinc [84–86] and dimethylcadmium [86, 87] to form trimethylborane and diallylzinc or diallylcadmium, respectively. Vanadium tetrachloride reacts with triallylborane to form boron trichloride and allylvanadium trichloride [88].

The oxidation of a tricrotylborane-triethylamine complex in excess amine with hydrogen peroxide leads to crotyl alcohol with an impurity of only 2% of 1-methylallyl alcohol, i.e., no allylic rearrangement takes place during the oxidative cleavage of the B—C bond [11].

Triallyl-, tricrotyl-, and tris(2-methylallyl)boranes react with ethyl diazoacetate to form esters of γ,δ-unsaturated acids [89]:

$$(RCH{=}CR'CH_2)_3B + N_2CHCOOC_2H_5 \xrightarrow[H_2O]{-N_2} RCH{=}CR'CH_2CH_2COOC_2H_5$$

$$(R = R' = H \text{ or } Ch_3).$$

Tricrotylborane enters into an exchange reaction with triethylaluminum; the crotylaluminum compound formed is capable of addition to olefins with a terminal double bond. As a result of hydrolysis of the reaction products, higher olefins are obtained [90]:

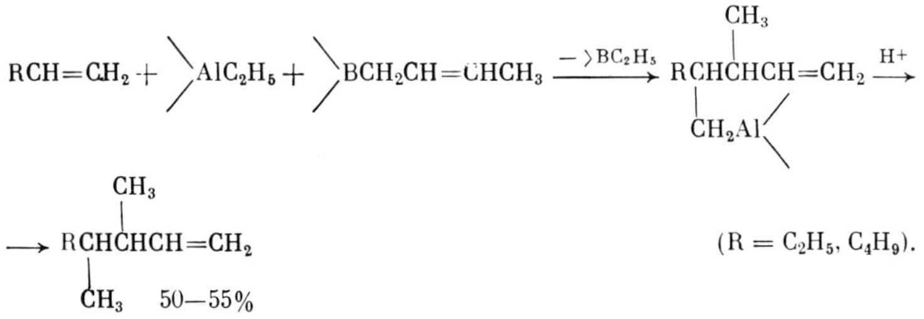

Lithium trialkylallylborates react regioselectively with allyl halides to form 1,5-dienes [206–208]. The reaction of allyl halides with alkylthioallyltrialkylborates proceeds analogously [209].

CHAPTER 2

Esters of allylboronic and allylthioboronic acids

Esters of organoboron acids of the allylic type and their thio analogs are readily available substances. They show a high reactivity and are used successfully both for the production of different classes of organoboron compounds and in organic synthesis.

2.1 METHODS FOR THE PREPARATION OF ESTERS OF ALLYLBORONIC AND ALLYLTHIOBORONIC ACIDS

Triallylborane is converted by the action of water at room temperature to allylboronic acid, which is isolated as the anhydride by distillation [6]:

$$(C_3H_5)_3B + 2H_2O \xrightarrow{-C_3H_6} C_3H_5B(OH)_2 \longrightarrow \tfrac{1}{3}(C_3H_5BO)_3.$$

The action of water on tricrotylborane produces crotylboronic anhydride, with the rupture of the boron-carbon bonds being accompanied by allylic rearrangement [11].

A general method for the synthesis of allyl- and crotylboronates is the reaction of triallylborane or tricrotylborane with alcohols.

Triallylborane reacts with alcohols at room temperature [6]. The action of one equivalent of butanol produces butyl diallylborinate, with the evolution of propylene:

$$(C_3H_5)_3B + n\text{-}C_4H_9OH \longrightarrow (C_3H_5)_2BOC_4H_9\text{-}n + C_3H_6.$$

Butyl diallylborinate, in turn, is capable of reacting with the alcohol, and the action of two equivalents of alcohol on triallylborane therefore forms dibutyl allylboronate:

$$(C_3H_5)_3B + 2n\text{-}C_4H_9OH \longrightarrow C_3H_5B(OC_4H_9\text{-}n)_2 + 2C_3H_6.$$

The action of methanol on triallylborane at a reagent ratio of 1:1 forms dimethyl allylboronate rather than methyl diallylborinate, and part of the triallylborane is recovered unchanged. This is explained by the fact that the initially formed $(C_3H_5)_2BOCH_3$ undergoes rapid symmetrization to $C_3H_5B(OCH_3)_2$ and $(C_3H_5)_3B$ [6].

Diethyl [42], dipropyl [45, 46]; diamyl, diallyl, dibenzyl, and diphenyl [45] allylboronates have been synthesized by the action of the corresponding alcohols on triallylborane.

Tricrotylborane reacts vigorously with methanol at room temperature with an equimolar reagent ratio, forming methyl dicrotylborinate and 1-butene (98%) with a small impurity of 2-butene (2%) [11]. Thus, the cleavage by alcohols of the B—C bond in allylic boron compounds proceeds with allylic rearrangement:

$$(CH_3CH{=}CH{-}CH_2)_3B + CH_3OH \longrightarrow (CH_3CH{=}CH{-}CH_2)_2BOCH_3 + \\ + \underset{98\%}{CH_3CH_2CH{=}CH_2} + \underset{2\%}{CH_3CH{=}CH{-}CH_3}$$

The conversion of tricrotylborane to dimethyl crotylboronate requires heating with methanol [11].

A convenient method for the preparation of allylboronates is based on an exchange reaction between triallylborane and alkyl borates (Chapter 1).

Allylboronates are obtained by the action of allylmagnesium bromide on methyl borate or butyl metaborate, followed by hydrolysis and esterification of the organoboron acid with alcohols [6]. According to the data of [91], as a result of the reaction of crotylmagnesium bromide with methyl borate, followed by hydrolysis and transesterification with *n*-hexanol, a mixture of dihexyl 1-methylallyl- and crotylboronates in a ratio of 77:23, with a total yield of 44%, is obtained. When crotyllithium is used, only the crotylboronate (36%) is obtained. The 1-methylallylboronate is converted slowly upon standing (over a period of several weeks), and more rapidly by heating in the presence of magnesium bromide (18 hours at 55°C), to the crotylboronate. The reaction of 2-hexen-1-ylmagnesium bromide with methyl borate, followed by transesterification with butanol, leads to a mixture of dibutyl 1-hexen-3-yl- and 2-hexen-1-ylboronates in a ratio of 86:14.

The preparation of dimethyl allylboronate by the action of allylzinc or allylaluminum on trimethyl borate has been described [92]. According to patent data [93], diamyl allylboronate is formed by the reaction of triallyl borate, allyl bromide, and sodium, followed by esterification with amyl alcohol.

Ethylene glycol esters of allylboronic acids are obtained by the reaction of trialkylallyltin with 2-chloro-1,3,2-dioxaborolane [210].

A mixture of cyclic esters of allylboronic (*63*) and isopropenylboronic (*64*) acids in a ratio of 86.5:13.5, with a total yield of 25%, is obtained by the prolonged heating (130°C) of allene with 4,4,6-trimethyl-1,3,2-dioxaborinane [94]:

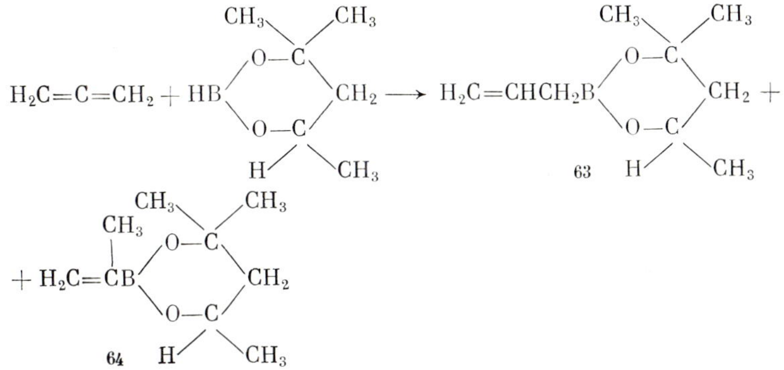

63

64

The addition of 4,4,6-trimethyl-1,3,2-dioxaborinane to 3-methyl-1,2-butadiene proceeds more smoothly, producing a mixture of cyclic esters of 3-methyl-2-buten-1-ylboronic and 3-methyl-1-buten-2-ylboronic acids in a ratio of 78:22, in a yield of 71%.

Triallylborane reacts with ethyl mercaptan at a temperature of −10 to −15°C, forming ethyl diallylthioborinate and propylene when equimolar quantities of the reagents are used [95, 96]:

$$(C_3H_5)_3B + C_2H_5SH \longrightarrow (C_3H_5)_2BSC_2H_5 + C_3H_6.$$

When two equivalents of ethyl mercaptan are used, however, the reaction proceeds in a more complex manner. The main reaction product is diethyl allylthioboronate, along with ethyl diallylthioborinate and ethyl 3-ethylmercaptopropylthioboronate (*65*). The latter ester is obtained as a result of the addition of ethyl mercaptan to the allylthioboronate:

$$(C_3H_5)_3B + C_2H_5SH \xrightarrow[-C_3H_6]{} (C_3H_5)_2BSC_2H_5 \xrightarrow[-C_3H_6]{C_2H_5SH} C_3H_5B(SC_2H_5)_2 \xrightarrow{C_2H_5SH}$$

$$\longrightarrow \underset{65}{C_2H_5SCH_2CH_2CH_2B(SC_2H_5)_2}.$$

The preparation of the ethyl esters of diallylthioborinic and allylthioboronic acids by the above method was repeated in [46].

Allyldifluoroborane is formed by heating boron trifluoride with tetraallyltin [97].

2.2 CHEMICAL PROPERTIES OF ESTERS OF ALLYLBORONIC AND ALLYLTHIOBORONIC ACIDS. SYNTHESIS OF ALCOHOLS, DIOLS, AND SILICON AND GERMANIUM COMPOUNDS

Allylboronate esters are easily saponified with water to allylboronic acid, the distillation of which produces the anhydride, $(C_3H_5BO)_3$ [6].

Diallylborinates are very unstable toward water and are converted to allylboronic acid even in the cold. It is indicated in [98] that the action of allylmagnesium bromide on boron trifluoride, followed by decomposition of the reaction products with water, produces diallylborinic anhydride. However, these data are erroneous, since diallylborinic acid does not withstand contact with water. Also erroneous are the data of [98] on the preparation of methyl diallylborinate from allylmagnesium bromide and methyl borate, followed by treatment of the reaction products with water. It is impossible to produce a diallylborinate by this method [6].

Diallylborinates and dicrotylborinates are converted to allylboronates and crotylboronates by the action of alcohols. As indicated in the preceding section of this chapter, allylthioboronates are capable of adding mercaptans at the carbon–carbon double bond.

Like dialkylborinates and alkylboronates [99–101], allylthioboronates enter readily into different exchange reactions. The action of ammonia on diethyl allylthioboronate forms *B*-triallylborazene [95]:

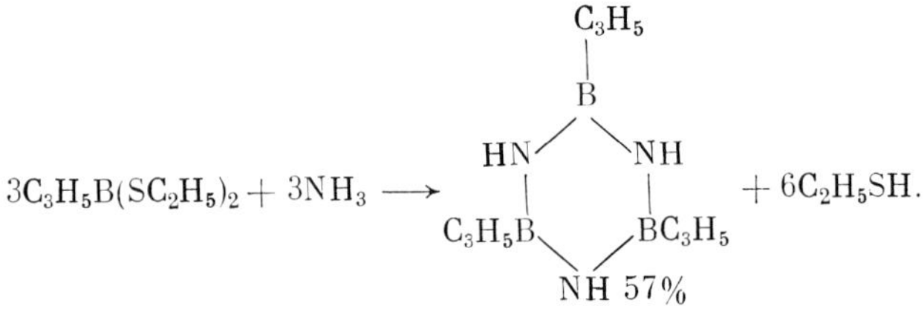

With butylamine, the ester forms allyldibutylaminoborane.

Diethyl allylthioboronate reacts with diethylamine to form approximately equal quantities of ethylthio(diethylamino)allylborane (*66*) and ethylthio(diethylamino)-3-ethylthiopropylborane (*67*):

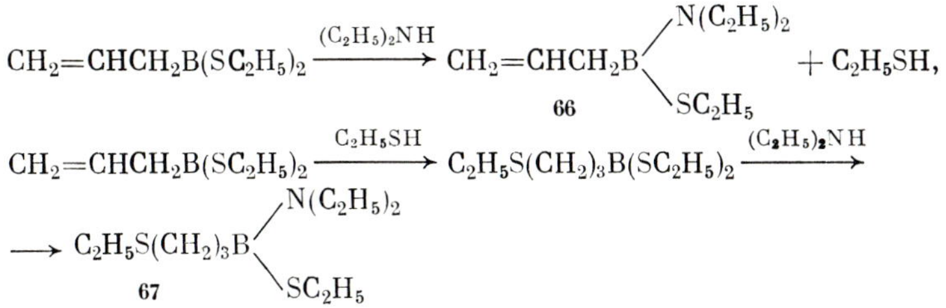

The hydroboration of dimethyl allylboronate, followed by oxidation of the reaction products, leads to a mixture of diols containing 97% of 1,3-propanediol and 3% of 1,2-propanediol [47]. The main hydroboration product is a tetraboron compound — tris(3-methoxyborylpropyl)borane (*68*):

$$3CH_2{=}CHCH_2B(OCH_3)_2 + {}^1/_2B_2H_6 \longrightarrow \underset{68}{B[(CH_2)_3B(OCH_3)_2]_3},$$

which, under the action of alcohol in the presence of a catalytic quantity of mercaptan, is converted to methyl bis(3-dimethoxyborylpropyl)borinate (*69*) and dimethyl propylboronate:

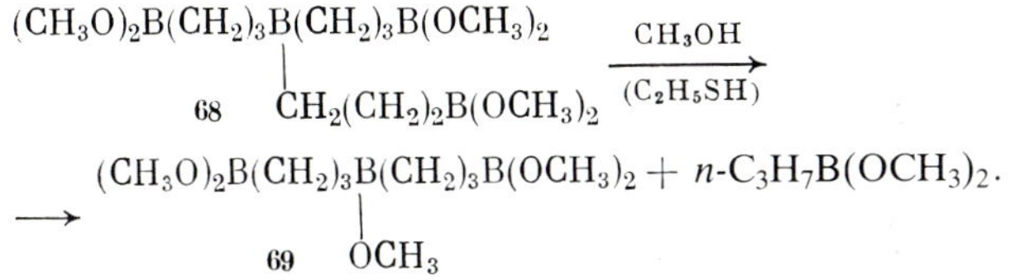

In the reaction of diborane with butyl diallylborinate, boron adds to the extent of 97% to the terminal carbon of the allyl radical [102]. Hydroboration of the ester with a reagent ratio of 1:2 leads to a complex mixture of products, the treatment of which with butanol produces hydrogen, and, as a result of distillation of the alcoholysis products, 1,5-dibutoxy-1,5-diborocane (*70*) and 1,3-bis(dibutoxyboryl)propane (*71*) are obtained:

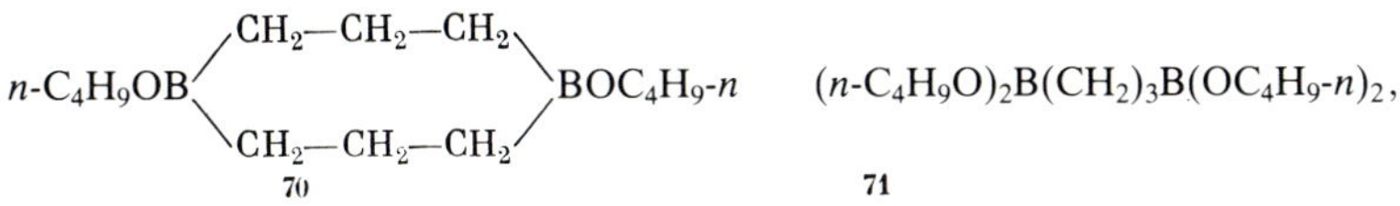

along with butyl borate with an impurity of the butyl esters of propylboronic and dipropylborinic acids.

The reaction of tetrapropyldiborane with dimethyl allylboronate forms the unstable dimethyl 3-dipropylborylpropylboronate (*72*) [103], which decomposes upon heating to form methyl dipropylborinate and the polymer *73*:

$$C_3H_5B(OCH_3)_2 + \tfrac{1}{2}(n\text{-}C_3H_7)_4B_2H \longrightarrow \underset{\mathbf{72}}{(n\text{-}C_3H_7)_2B(CH_2)_3B(OCH_3)_2} \longrightarrow$$

$$\longrightarrow (n\text{-}C_3H_7)_2BOCH_3 + \underset{\mathbf{73}}{1/n[CH_2CH_2CH_2BOR]_n}.$$

When heated with methyl borate in the presence of a small quantity of tetrapropyldiborane, which acts as a catalyst for the reaction [104], the polymer is converted to 1,3-bis(dimethoxyboryl)propane.

Dimethyl allylthioboronate behaves analogously to the allylboronate with respect to tetraalkyldiboranes [103].

Dibutyl allylboronate adds methyl dichlorosilane in the presence of chloroplatinic acid to form dibutyl 3-methyldichlorosilylpropylboronate (*74*) [105]:

$$C_3H_5B(OC_4H_9\text{-}n)_2 + CH_3SiHCl_2 \longrightarrow \underset{\mathbf{74}\quad(64\%)}{Cl_2(CH_3)Si(CH_2)_3B(OC_4H_9\text{-}n)_2}.$$

Triethylsilane adds to an allylboronate in the presence of chloroplatinic acid with considerably greater difficulty.

An investigation of the addition of germanium hydrides — triethylgermane and triphenylgermane — to dimethyl allylboronate [106] showed that these reagents add with greater ease to the double bond of the allyl radical joined to the $B(OR)_2$ group than to the double bond of olefinic hydrocarbons and their γ-functional derivatives. The latter compounds add triethylgermane only in the presence of H_2PtCl_4, while the addition of triethylgermane to dimethyl allylboronate proceeds without a catalyst at 110–190°C and leads to the formation of dimethyl 3-triethylgermylpropylboronate (*75*) in high yield:

$$C_3H_5B(OCH_3)_2 + (C_2H_5)_3GeH \longrightarrow \underset{\mathbf{75}\quad(78\%)}{(C_2H_5)_3Ge(CH_2)_3B(OCH_3)_2}.$$

Triphenylgermane adds to dimethyl allylboronate with even greater ease, forming dimethyl 3-triphenylgermylpropylboronate (*76*):

$$C_3H_5B(OCH_3)_2 + (C_6H_5)_3GeH \longrightarrow \underset{76\quad(87\%)}{(C_6H_5)_3GeCH_2CH_2CH_2B(OCH_3)_2}.$$

Dibutyl allylboronate adds bromine to form dibutyl 2,3-dibromopropylboronate, which partially decomposes during distillation [107].

Allylboronates undergo a diene condensation with cyclopentadiene and its chlorinated derivatives [108]. Dimethyl allylboronate condenses with cyclopentadiene only at high temperatures, forming the endo form of dimethyl (bicyclo[2.2.1]hept-5-en-2-yl)methylboronate (*77*):

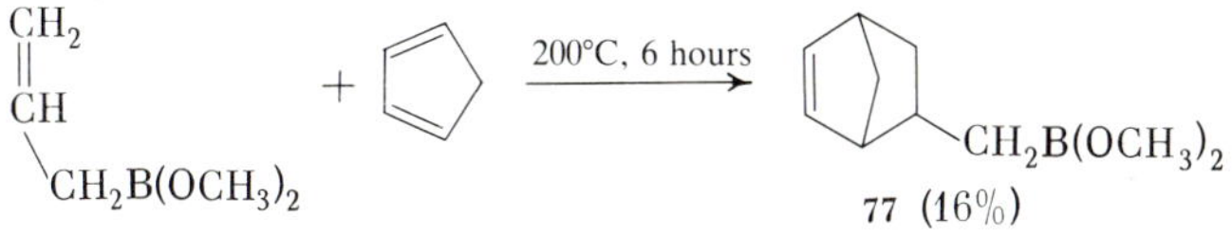

77 (16%)

Dimethyl allylboronate condenses more easily with hexachlorocyclopentadiene to form dimethyl (1,4,5,6,7,7-hexachlorobicyclo-[2.2.1]hept-5-en-2yl)methylboronate (*78*):

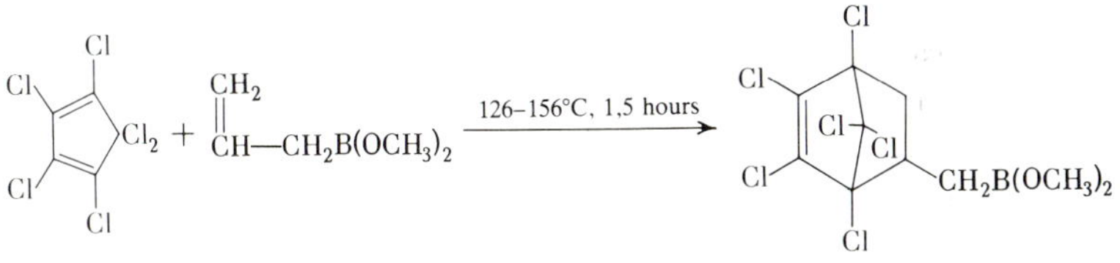

78 (68%)

The allylboronate reacts with 1,2,3,4-tetrachlorocyclopentadiene to form a (1,4,5,6-tetrachlorobicyclo[2.1.1]hept-5-en-2-yl)methylboronate.

Diallylborinates and allylboronates react as organometallic reagents with aldehydes and ketones [57–60, 109].

According to the data of [91, 110], a mixture of dihexyl crotyl- and 1-methylallyboronates (31:69) reacts with butyraldehyde or cyclohexanone to form a mixture of alcholos with the reverse ratio of isomers:

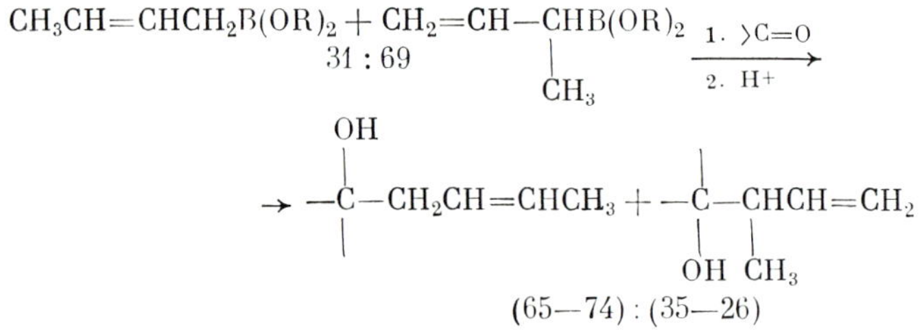

while dihexyl crotylboronate reacts with butyraldehyde or cyclohexanone to form the corresponding 2-methylcarbinols.

The enantioselective synthesis of homoallylic alcohols based on reactions of chiral esters of allylboronic acid with aldehydes has been described [211–213]. Crotylboronate esters react stereoselectively with aldehydes to form homoallylic alcohols [214].

The addition of esters of (*Z*)-2-buten-1-ylboronic [215] and 1-buten-3-ylboronic acids [216] to aldehydes has been investigated.

The pinacol ester of *E*-1-trimethylsilyl-1-propene-3-boronic acid reacts with aldehydes to form (+)-(R^*,S^*)-3-trimethylsilyl-4-hydroxy-alkenes which, upon dehydroxysilylation under different conditions, are converted to substituted *Z*- or *E*-1,3-dienes. This method has been used, in particular, for the synthesis of *Z*- and *E*-$AcO(CH_2)_8CH{=}CH{-}CH{=}CH_2$ [217], a 1:4 mixture of which, according to the data of [218], is the sex pheromone of the red bollworm moth (*Diparopsis castanea*).

The synthesis of stegobinone has been achieved with the use of chiral crotyldi(organyloxy)borane [219].

Diallylborinates and allylboronates also react with quinones (Chapter 1).

CHAPTER 3

Allyl esters of boronic and alkylboronic acids

Borates with radicals of the allylic type are obtained by the action of the corresponding alcohols on boron trichloride:

$$BCl_3 + 3ROH \longrightarrow B(OR)_3 + 3HCl.$$

Triallyl borate [111–113] and tris(2-methylallyl) borate [113] have been synthesized by this method.

Allyl borates are conveniently obtained by the esterification of B_2O_3 or of boric acid by the corresponding alcohols:

$$B(OH)_3 + 3ROH \longrightarrow B(OR)_3 + 3H_2O.$$

Triallyl borate [112, 114, 115], tri(2-methylallyl) borate [116, 117], tri(3-methylallyl) borate [118], and tri(1-methylallyl) borate [118] have been obtained by the above method.

Borates with complex radicals of the allylic type can be obtained by the transesterification of trialkyl borates [119].

Triallyl borate is formed by the cleavage of alkyl allyl ethers by boron trichloride [120]. Allyl dibutylborinate has been obtained by the esterification of dibutylborinic anhydride [121] and by the transesterification of methyl dibutylborinate with allyl alcohol [122].

The ^{11}B chemical shift of triallyl borate is −17.5 ppm (relative to boron trifluoride etherate) [123, 124].

Triallyl borate forms complex compounds with hydrazine, piperidine, and ethylenediamine [125].

The action of octanol on tri(3-methylallyl) borate forms a mixture of 3-methylallyl (91.3%) and 1-methylallyl (8.7%) alcohols [118].

Triallyl borate reacts with hydrogen chloride at 120°C to form allyl chloride, 1,2-dichloropropane, and boric acid. Hydrogen bromide reacts with triallyl borate at 20°C to form allyl bromide and tri(3-bromopropyl) borate [113].

Triallyl borate in CCl_4 adds bromine to form tri(2,3-dibromopropyl) borate [111, 112, 126]. It reacts with BCl_3 to form unstable allyloxydichloroborane and diallyloxychloroborane, depending on the ratio of components [113]:

$$B(OR)_3 \xrightarrow{BCl_3} (RO)_2BCl \xrightarrow{BCl_3} ROBCl_2.$$

Allyloxyboron chlorides react with pyridine to form complexes with a 1:1 and 1:2 composition. 1-Methyl- and 3-methylallyloxyboron dichlorides are unstable and cannot be stabilized even in the form of adducts with pyridine [118]. The thermal decomposition of both 1- and 3-methylallyloxyboron dichloride forms a mixture of 1-chloro-2-butene and 3-chloro-1-butene [118].

Chlorine-substituted 3-methylallyloxyboron dichloride is formed by the reaction of crotonaldehyde and BCl_3 [127].

Triallyl borate is used for the allylation of aromatic hydrocarbons. In the presence of ferric chloride, allylbenzene can be obtained from benzene in a yield of 35% by this method, and *p*-allylcumene can be obtained in 40% yield from cumene [128–130]. When aluminum chloride is used as a catalyst, 1,2-diphenylpropane [131] and isopropylbenzene [131, 132] are obtained from triallyl borate and benzene and 1,2-ditolylpropane and 1,2-dixylylethane are formed from toluene [132].

Triallyl borate can be used for the reduction of cyclohexanone to cyclohexanol [133].

Triallyl borate forms a polymeric product when heated in a stream of air [134] and forms copolymers with vinylidene chloride [135] and methyl methacrylate [136].

The hydroboration of allyl borates has been covered in Part I, Chapter 3.

CHAPTER 4

The allylboration of acetylenic and allenic compounds

Mikhailov and Bubnov [137] found that triallylborane reacts with acetylene at 130°C to form 3-allyl-3-borabicyclo[3.3.1]non-6-ene (*79*):

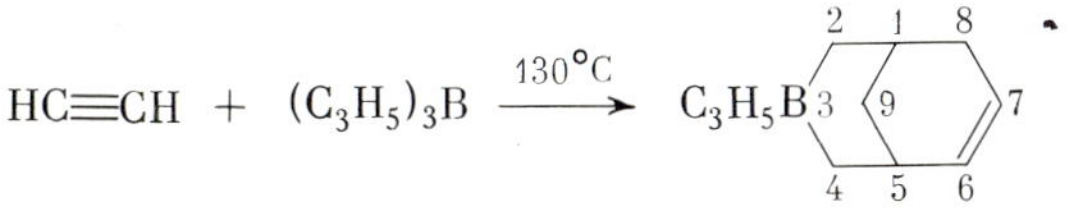

79 (75%)

A detailed investigation of the reaction between triallylborane and acetylenic compounds has shown that it proceeds via three consecutive stages:

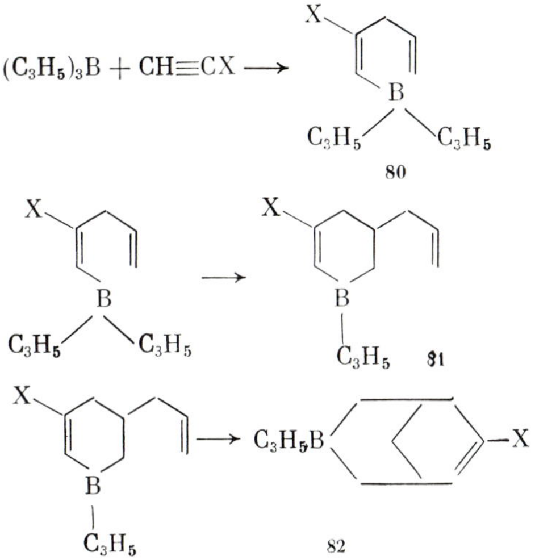

80

81

82

The first stage of allylboron-acetylene condensation consists of a *cis*-addition of allyl and diallylboryl fragments to the C≡C bond to form 2-substituted (1,4-pentadien-1-yl)diallylboranes (*80*). Depending on the nature of the acetylenic compound, the reaction proceeds at room or reduced temperature.

In the second stage, which proceeds at room temperature or with moderate heating, the compounds *80* cyclize to 3-substituted 1,5-diallyl-1-boracyclohex-2-ene compounds (1,5-diallyl-1-borin-2-ene) (*81*).

At elevated temperature (approximately 130°C), the third stage of the allylboron-acetylene condensation takes place: cyclization of the monocyclic compounds *81* to 7-substituted 3-borabicyclo[3.3.1]non-6-enes (*82*). This process involves the rupture of the B—C_2 bond and addition of the C_2 atom to the C_8 atom and of the boron atom to the C_9 atom of the $C_8{=}C_9$ double bond of the allyl radical, which occupies an axial position:

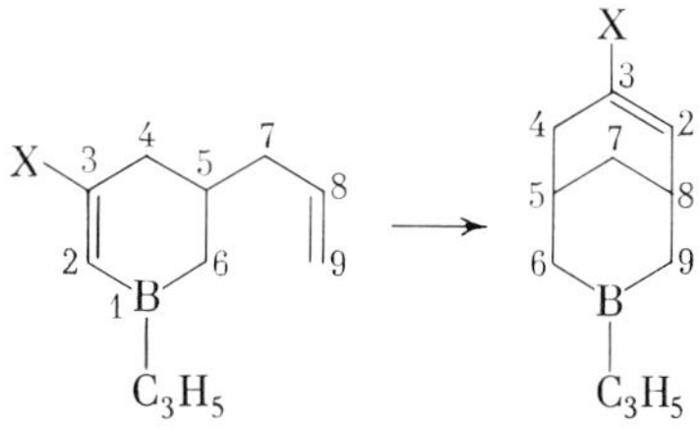

4.1 (1,4-PENTADIEN-1-YL)ALLYLBORANES

4.1.1 Methods for the preparation of (1,4-pentadien-1-yl)allylboranes

Alkylacetylenes (methylacetylene, butylacetylene) react with triallylborane at room temperature to form (2-alkyl-1,4-pentadien-1-yl)-diallylboranes (*83*) [138–141]:

$$(C_3H_5)_3B + RC{\equiv}CH \longrightarrow \underset{\mathbf{83}}{(C_3H_5)_2BCH{=}\overset{\overset{\displaystyle R}{|}}{C}CH_2CH{=}CH_2},$$

which cannot be isolated in the pure state, since they are unstable and cyclize to 3-alkyl-substituted 1,5-diallyl-1-boracyclohex-2-enes, slowly at room temperature and rapidly during vacuum distillation [141]. The formation of the compound *83* is proved by the fact that oxidation of the products of the reaction of alkylacetylenes with triallylborane produce 2-alkyl-4-penten-1-als (*84*):

$$\underset{83}{(C_3H_5)_2BCH{=}\overset{\displaystyle R}{\overset{|}{C}}CH_2CH{=}CH_2} \xrightarrow{H_2O_2} \underset{84}{CH_2{=}CHCH_2\overset{\displaystyle R}{\overset{|}{C}H}\overset{\displaystyle H}{\overset{|}{C}}{=}O}$$

2-Substituted (1,4-pentadien-1-yl)diallylboranes (*80*), which are formed in the first stage of the reaction of triallylborane with propargyloxytrimethylsilane [142], phenylacetylene [143], cyclohexylacetylene [144], or propargyl chloride [145, 146], are quite unstable. However, alkoxyacetylenes form quite stable addition products with triallylborane, which can be isolated in the pure state. The exothermal reaction of triallylborane with alkoxyacetylenes proceeds at −70 to −50°C and leads to (2-alkoxy-1,4-pentadien-1-yl) diallylboranes (*85*), bis(2-alkoxy-1,4-pentadien-1-yl)allylboranes (*86*), or tris(2-alkoxy-1,4-pentadien-1-yl)boranes (*87*), depending on the ratio of reagents [138, 147]:

$$(C_3H_5)_3B + ROC{\equiv}CH \longrightarrow \underset{85}{(C_3H_5)_2BCH{=}\overset{\displaystyle OR}{\overset{|}{C}}CH_2CH{=}CH_2}$$

$$(C_3H_5)_3B + 2ROC{\equiv}CH \longrightarrow \underset{86}{C_3H_5B[CH{=}\overset{\displaystyle OR}{\overset{|}{C}}CH_2CH{=}CH_2]_2}$$

$$(C_3H_5)_3B + 3ROC{\equiv}CH \longrightarrow \underset{87}{B[CH{=}\overset{\displaystyle OR}{\overset{|}{C}}CH_2CH{=}CH_2]_3}$$

The compounds *85* are formed in a yield of approximately 85% with a ratio of triallylborane to alkoxyacetylene of 1:1; with a reagent ratio of 1:2, compounds *86* are obtained in a yield of approximately 80% and, with a ratio of 1:3, compounds *87* are obtained (yield approximately 60%).

(2-Trimethylsilyl-1,4-pentadien-1-yl)diallylborane (*88*) has been obtained from triallylborane and trimethylsilylacetylene [138]:

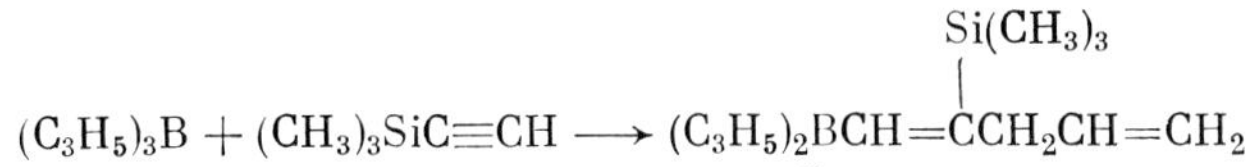

88

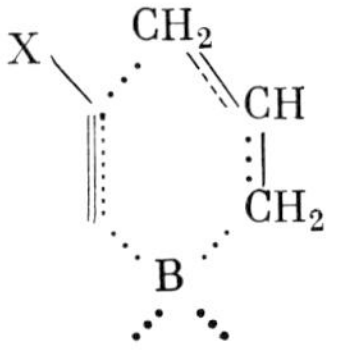

89

The allylboration of acetylenes probably proceeds via a six-center transition state *89* and is accompanied by allylic rearrangement, as has been shown in an investigation of the reaction of 1-crotylboracyclopentane with ethoxyacetylene, as a result of which 1-(2-ethoxy-3-methyl-1,4-pentadien-1-yl)boracyclopentane is formed [148, 149]:

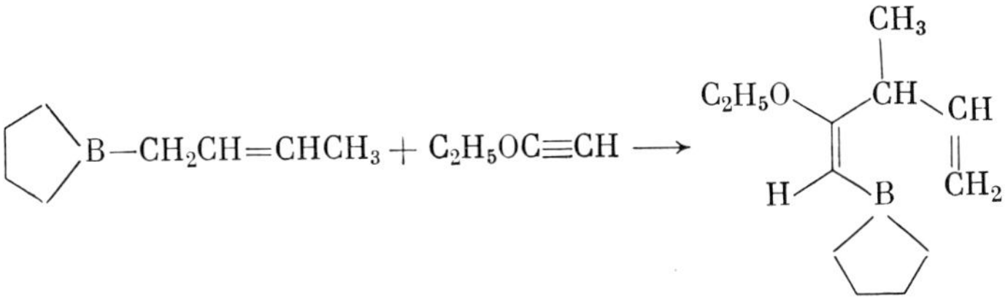

Triallylborane adds to 1-diethylaminopropyne to form 1-diethylamino-1-diallylboryl-2-methyl-1,4-pentadiene [69]:

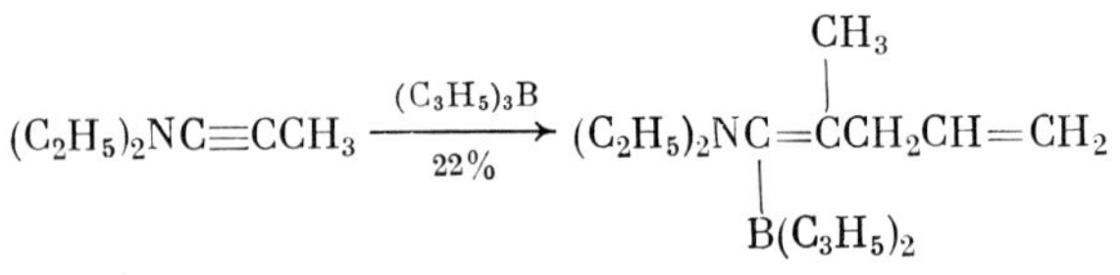

4.1.2 Chemical properties of (1,4-pentadien-1-yl)allylboranes. Synthesis of unsaturated aldehydes and ketones and of alkoxydienes

(2-Alkyl-1,4-pentadien-1-yl)diallylboranes (*83*), under the action of methanol, readily exchange allyl for methoxy groups [138–140, 147]. The dimethyl 2-alkyl-1,4-pentadien-1-ylboronates (*90*) obtained by this process:

$$\underset{\mathbf{83}}{(C_3H_5)_2BCH{=}\overset{\overset{R}{|}}{C}CH_2{-}CH{=}CH_2} + 2CH_3OH \longrightarrow \underset{\mathbf{90}}{CH_2{=}CHCH_2\overset{\overset{R}{|}}{C}{=}CHB(OCH_3)_2} + 2C_3H_6$$

are oxidized with hydrogen peroxide to 2-alkyl-4-pentenals (*91*) and are converted with a mixture of methanol and acetic acid to 2-alkyl-1,4-pentadienes (*92*) [139, 140, 150]:

$$CH_2{=}CHCH_2\overset{\overset{R}{|}}{C}{=}CHB(OCH_3)_2 \xrightarrow{[O]} \underset{\mathbf{91}}{CH_2{=}CHCH_2\overset{\overset{R}{|}}{C}HCHO}$$

$$CH_2{=}CHCH_2\overset{\overset{R}{|}}{C}{=}CHB(OCH_3)_2 \xrightarrow[CH_3COOH]{CH_3OH} \underset{\mathbf{92}}{CH_2{=}CHCH_2\overset{\overset{R}{|}}{C}{=}CH_2}$$

$(R = CH_3,\ n\text{-}C_4H_9)$.

(2-Trimethylsilyl-1,4-pentadien-1-yl)diallylborane reacts with methanol to form dimethyl 2-trimethyl silyl-1,4-pentadien-1-yl boronate [138].

$$CH_2{=}CHCH_2\overset{\overset{Si(CH_3)_3}{|}}{C}{=}CHB(OCH_3)_2$$

In (2-alkoxy-1,4-pentadien-1-yl)diallylboranes (*85*) and bis(2-alkoxy-1,4-pentadien-1-yl)allylboranes (*86*), the action of alcohols leads to the rupture of not only boron-allyl but also boron-vinyl bonds, to form, depending on the ratio of reagents and the duration of heating, 2-alkoxy-1,4-pentadienes (*93*) or ketals of allyl methyl ketone (*94*) [147, 150]. An analogous behavior toward alcohols is shown by tris(2-alkoxy-1,4-pentadien-1-yl)boranes. Acid hydrolysis of the compounds *93* forms allyl methyl ketone (*95*):

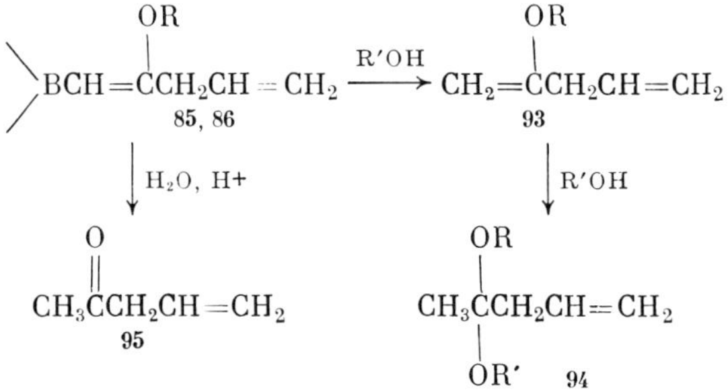

Proceeding from the structure of the compounds 85–87 which are, on the one hand, vinyl boron compounds and, on the other hand, β-boron-substituted vinyl ethers, two schemes for their alcoholysis may be suggested. According to the first of these, the process takes place via the stage of formation of the complex *96*, followed by rupture of the boron-carbon bond:

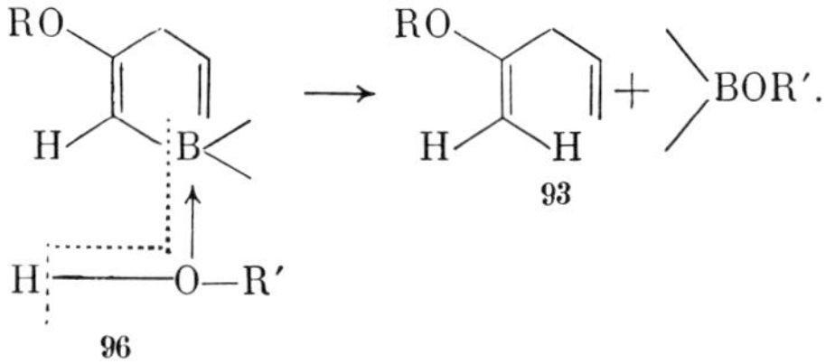

In this case, the alkoxy group of the starting organoboron compound is retained in the vinyl ether *93* formed, regardless of which alcohol is used for the alcoholysis. According to the second possible mechanism, the alcohol adds to the C=C bond of the vinyl ether group, and the organoboron ketal *97* formed then undergoes β-elimination with formation of the ethers *93* and *98*:

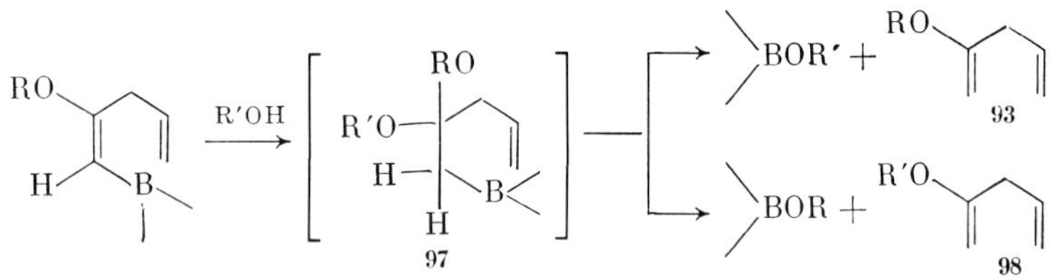

Experiments on the alcoholysis of compounds *85*–87 with different alcohols have shown the correctness of the first mechanism—direct cleavage of the B—C bond. This mechanism was confirmed by an investigation of the reaction of *tris*(2-methoxy-1,4-pentadien-1-yl)-borane (*99*) with methanol containing labeled carbon:

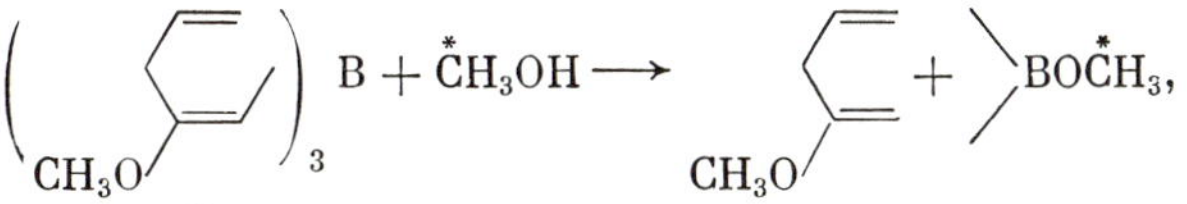

99

and by the reaction of *tris*(2-ethoxy-1,4-pentadien-1-yl)borane with radioactive ethanol [151].

1-(2-Ethoxy-3-methyl-1,4-pentadien-1-yl)boracyclopentane is converted by alcoholysis to 2-ethoxy-3-methyl-1,4-pentadiene (*100*), the hydrolysis of which leads to 3-methyl-4-penten-2-one (*101*) [148, 149]:

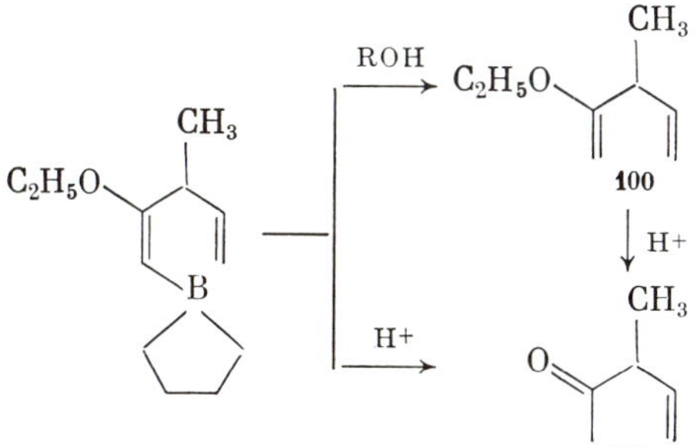

101

The alcoholysis of *tris*(2-alkoxy-4-methyl-1,4-pentadien-1-yl) boranes leads to 2-alkoxy-4-methyl-1,4-pentadienes [152].

In (2-alkoxy-1,4-pentadien-1-yl)diallylboranes (*85*), the terminal double bonds are hydrogenated over technical platinum black, while the vinyl bond on the boron atom is hardly affected [147]. The substituted dialkylvinylboranes *102* obtained by this process do not symmetrize during distillation:

$$\underset{85}{(C_3H_5)_2BCH{=}\overset{\overset{OR}{|}}{C}CH_2CH{=}CH_2} \xrightarrow{H_2/Pt} \underset{102}{(C_3H_7)_2BCH{=}\overset{\overset{OR}{|}}{C}CH_2CH_2CH_3}$$

(2-Ethoxy-1,4-pentadien-1-yl)diallylborane (*103*) reacts with acetone to form the compound *104* [147]:

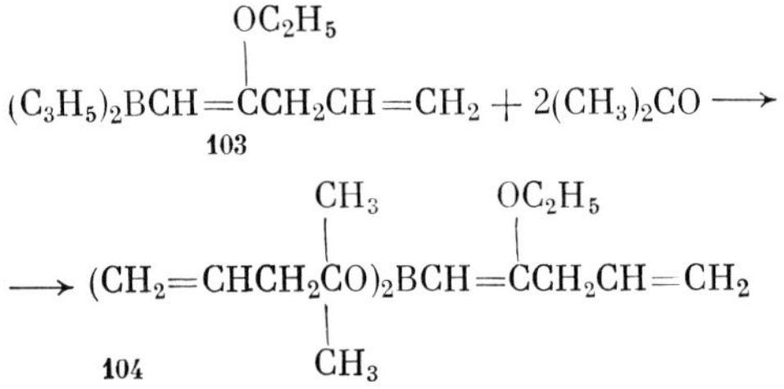

The reactions of alkoxyacetylenes with allylboranes have been used as a basis for developing a method for the synthesis of 2-alkoxy-1,4-pentadienes [220]:

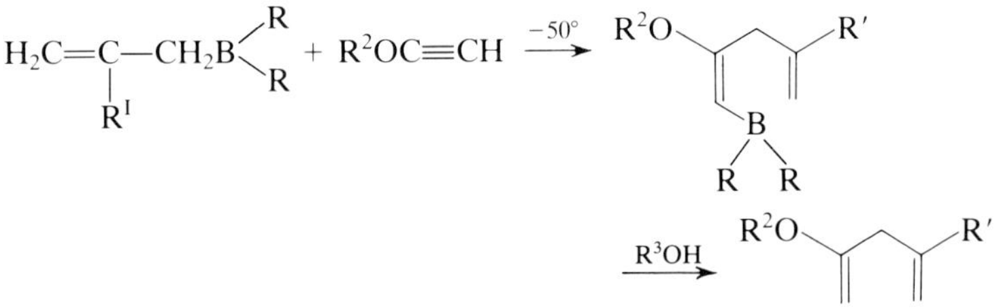

and of 1,4-pentenynes [221]:

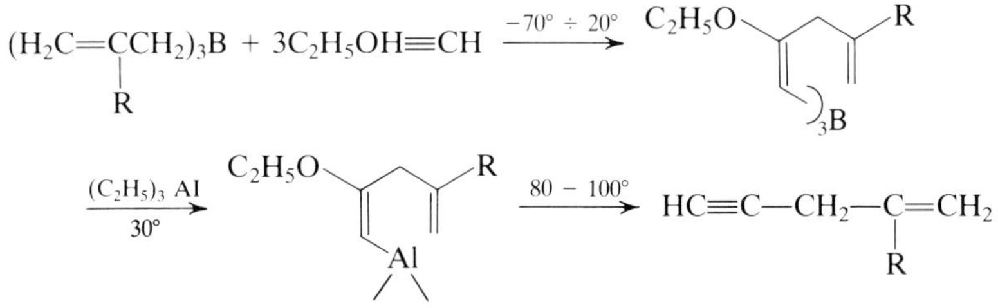

Allylboranes add to 1-ethoxy-2-trimethylsilylacetylene to form 2-ethoxy(1-trimethysilyl-1,4-pentadien-1-yl)boranes, which are converted to the corresponding 1-trimethylsilyl-4-penten-1-ynes under mild conditions, with the elimination of $R'_2BOC_2H_5$ [222].

4.2 ALLYLIC COMPOUNDS OF THE 1-BORACYCLOHEX-2-ENE SERIES

4.2.1 Methods for the preparation of allylic compounds of the 1-boracyclohex-2-ene series

(2-Alkyl-1,4-pentadien-1-yl)diallylboranes (*83*) cyclize to 3-alkyl-substituted 1,5-diallyl-1-boracyclohex-2-ene derivatives (*105*), slowly at room temperature and rapidly under moderate heating (40–100°C) [139–141, 153]:

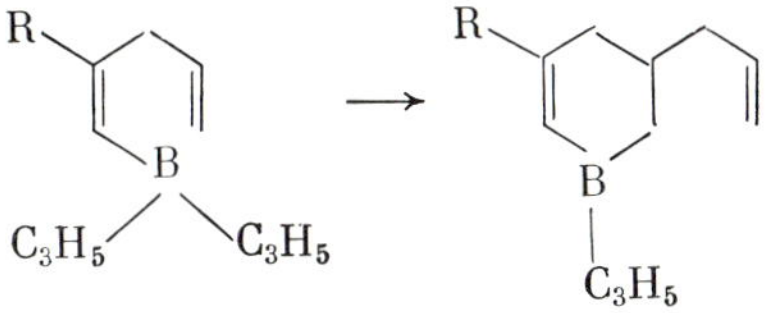

83 105

($R = CH_3, n\text{-}C_4H_9$).

Proparygyloxytrimethylsilane [142], phenylacetylene [143], and cyclohexenylacetylene [144] react with triallylborane similarly to acetylene homologs to form, respectively, the thermally unstable 3-trimethylsiloxymethyl-1,5-diallyl- (*106*), 3-phenyl-1,5-diallyl- (*107*), and 3-cyclohexenyl-1,5-diallyl-1-boracyclohex-2-enes (*108*):

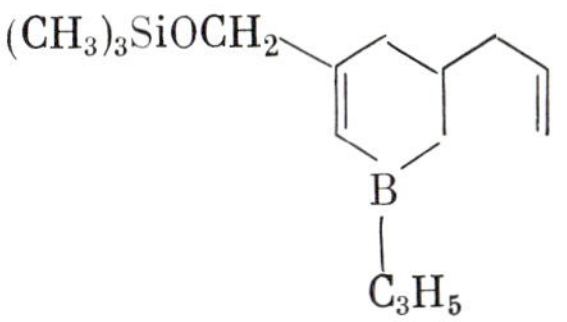

106

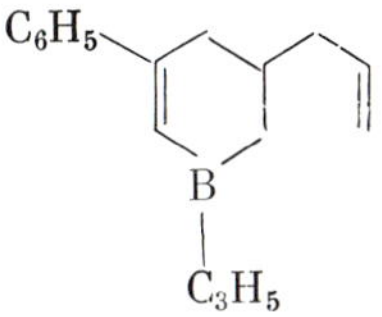

107

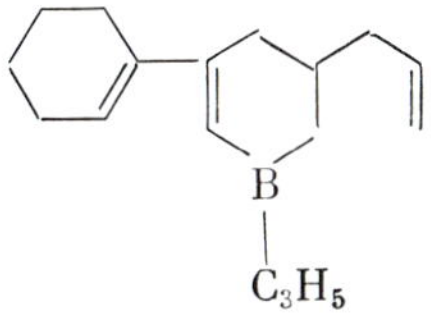

108

(2-Alkoxy-1,4-pentadien-1-yl)diallylboranes (*85*) are converted to 3-alkoxy-1,5-diallyl-1-boracyclohex-2-enes (*109*) by heating at 120–130°C [153, 154]:

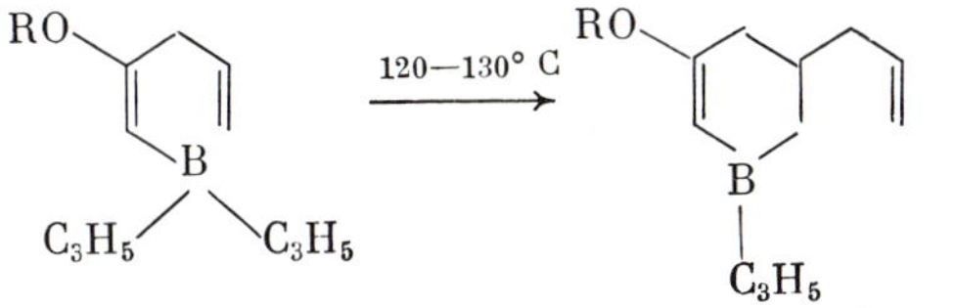

85 109 (80%)

($R = CH_3, C_2H_5$).

Propargyl chloride reacts slowly with triallylborane at room temperature to form 3-chloromethyl-1,5-diallyl-1-boracyclohex-2-ene (*110*), which gradually isomerizes to 2-chloro-3-methylene-1,5-diallyl-1-boracyclohexane (*111*). After twenty days, the reaction mixture contains approximately 80% of compound *110* and 20% of compound *111* [145, 146]:

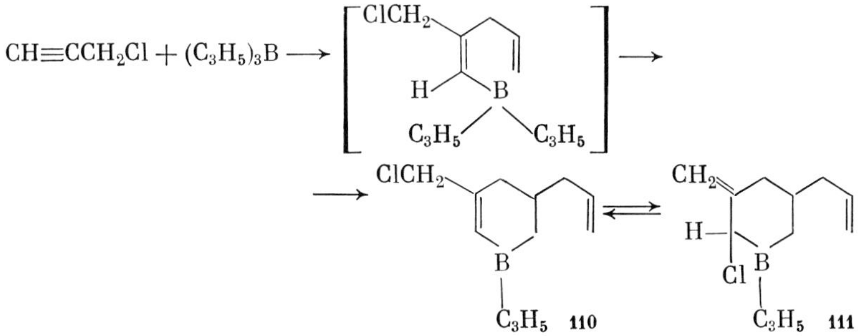

110 **111**

Triallylborane reacts at 20°C with methyl propargyl ether [223, 224], propargyl tetrahydropyranyl ether [225], and (propargyloxy)-dipropylborane [226] to form, respectively, 3-methoxymethyl-1,5-diallyl-1-boracyclohex-2-ene (*105*, R = CH_3OCH_2), 3-tetrahydro-pyranyloxymethyl-1,5-diallyl-1-boracyclohex-2-ene (*105*, R =

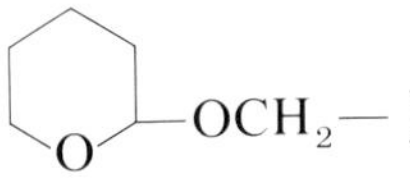

), and 3-dipropylboryloxy-1,5-diallyl-1-boracyclo-hex-2-ene (*105*, R = $CH_2OB(C_3H_7)_2$). The reaction of triallylborane with 1-alkoxypropynes at 20°C leads to 3-alkoxy-1,5-diallyl-2-methyl-1-boracyclohex-2-enes (*111a*) [227].

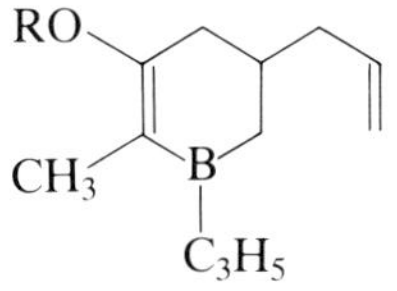

IIIa

4.2.2 Chemical properties of allylic compounds of the 1-boracyclohex-2-ene series. Synthesis of dienes, alkoxydienes, and unsaturated oxo alcohols.

The action of methanol (20–30°C) on 1,5-diallyl-3-alkyl-1-boracyclohex-2-enes (*105*) leads to the formation of propylene and 1-

methoxy-5-allyl-3-alkyl-1-boracyclohex-2-enes (*112*) [139, 140, 153]. Prolonged boiling with excess methanol results in a protolytic cleavage of the boron-vinyl bond in the compounds *112*, forming dimethoxy(2-allyl-4-alkyl-4-penten-1-yl)boranes (*113*):

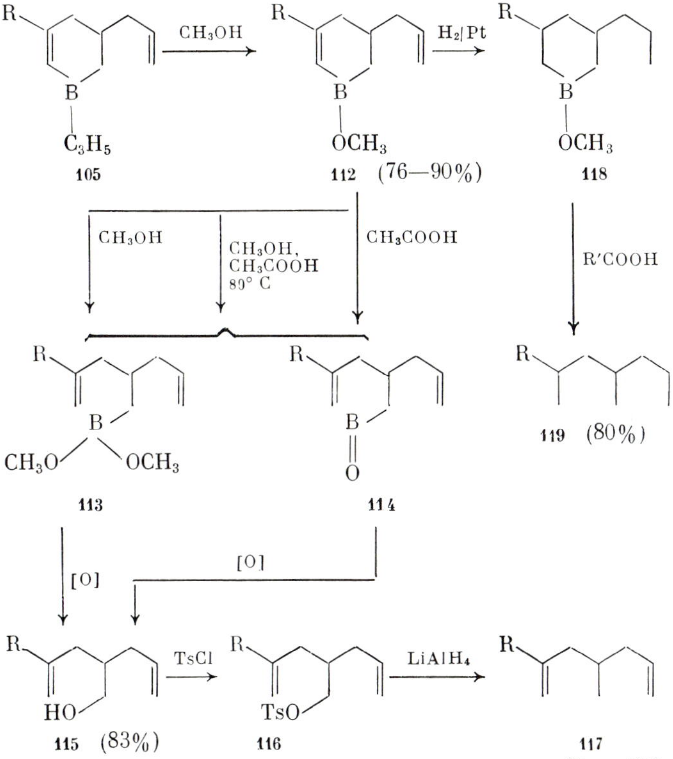

(R = CH_3, *n*-C_4H_9).

Ring opening in the compounds *112* proceeds more rapidly by heating them to 80°C with a mixture of methanol and acetic acid. Under these conditions, the esters *113* and trimers of anhydrides *114* are obtained, or only the latter if the reaction mixture is kept overnight at room temperature.

Oxidation of the esters *113* or anhydrides *114* with alkaline hydrogen peroxide produces 2-allyl-4-alkyl-4-penten-1-ols (*115*) in 80–90% yield. 2,4-Dimethyl-1,6-heptadiene (*117*, R = CH_3) or 2-butyl-4-methyl-1,6-heptadiene (*117*, R = *n*-C_4H_9) can be obtained by reduction of the tosylates *116* with lithium aluminum hydride.

Hydrogenation of the methoxy derivatives *112* over platinum black leads to 1-methoxy-3-alkyl-5-propylborinanes (*118*). 1-Methoxy-3-methyl-5-propylborinane can be converted to 2,4-dimethylheptane (*119*, R = CH_3) by acidolysis.

1-Methoxy-3-phenyl-5-allyl-1-boracyclohex-2-ene (*120*), obtained by the methanolysis of 3-phenyl-1,5-diallyl-1-boracyclohex-2-ene (*107*), is hydrogenated stepwise to 1-methoxy-3-phenyl-5-propyl-1-boracyclohex-2-ene (*121*) (over platinum black) and 1-methoxy-3-phenyl-5-propyl-1-borinane (*122*) (over Pd/$SrCO_3$) [143]:

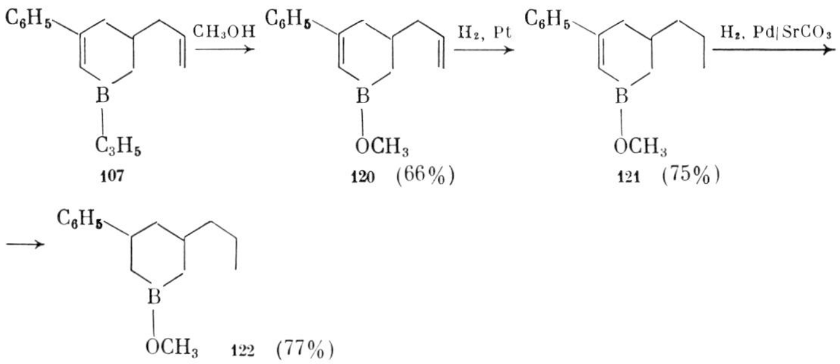

3-Cyclohexenyl-1,5-diallyl-1-boracyclohex-2-ene (*108*) is converted by the action of methanol to 1-methoxy-3-cyclohexenyl-5-allyl-1-boracyclohex-2-ene (*123*), which is converted to 1-methoxy-3-cyclohexenyl-5-propyl-1-boracyclohex-2-ene (*124*) by hydrogenation over platinum black and forms dimethoxy(4-cyclohexenyl-2-allyl-4-penten-1-yl)borane (*125*) when treated with a mixture of methanol and acetic acid [144]:

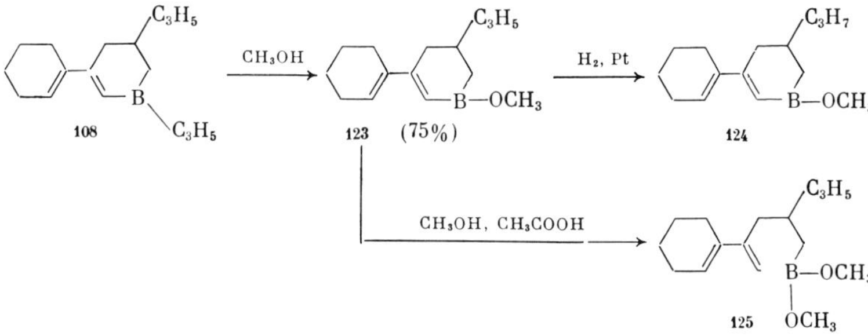

The compound *124* enters into a diene condensation with maleic anhydride, forming 1-methoxy-3-allyl-1-bora-Δ^{4a}-dodecahydrophenanthrenedicarboxylic anhydride (*126*):

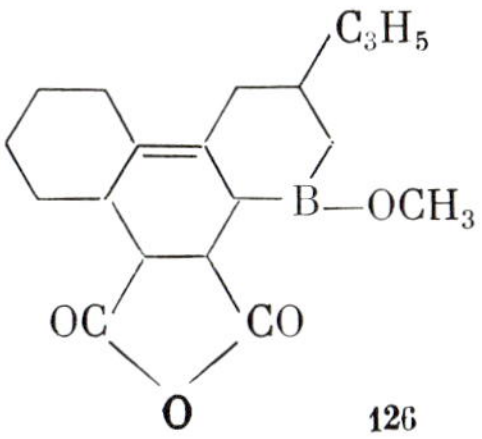

126

1,5-Diallyl-4-trimethylsiloxymethyl-1-boracyclohex-2-ene (*106*) is hydrogenated over Adams catalyst to form 1,5-dipropyl-3-trimethylsiloxymethyl-1-boracyclohex-2-ene (*127*), which is smoothly converted by brief heating with methanol, with cleavage of the Si—O and B—C bonds, to the compound *128*, the subsequent intramolecular condensation of which produces 6-methylene-2,4-dipropyl-1,2-oxaborepane (*129*) [142]:

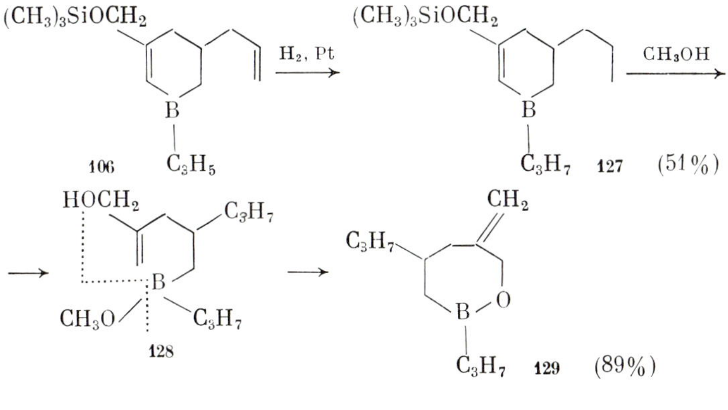

106 127 (51%) 128 129 (89%)

2,4-Dimethylheptane can be obtained from the compound *129* by a series of consecutive conversions.

It is interesting that, in the compound *130*, obtained by methanolysis of the compound *106*, no cleavage of the cyclic boron-carbon bond takes place upon heating with methanol, but only a cleavage of the Si—O bond. The compound *131* obtained in this process undergoes intermolecular condensation to form the compound *132*:

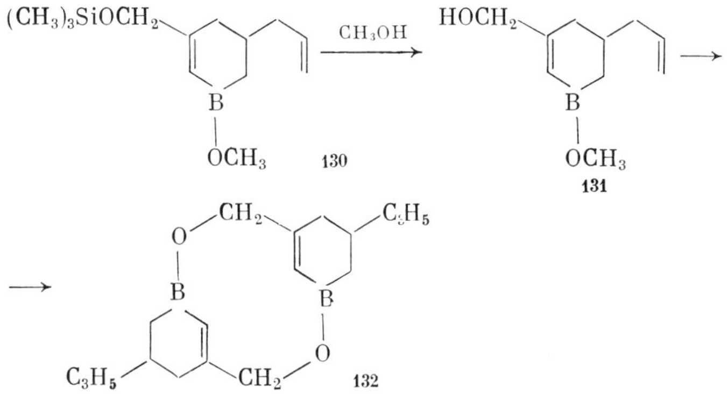

130 131 132

In 3-alkoxy-1,5-diallyl-1-boracyclohex-2-enes (*109*), the cyclic boron-vinyl bond is initially ruptured under the action of alcohols, as has been detected by IR spectroscopy, and the boron-allyl bond in the acyclic organoboron compound formed then undergoes alcoholysis [154]. The final product of the reaction of *109* with methanol is dimethoxy(4-alkoxy-2-allyl-4-penten-1-yl)borane (*133*), from which 4-hydroxymethyl-1-hepten-6-one (*134*), 4-hydroxymethyl-6-heptanone (*135*), and 2-propyllevulinic acid (*136*) can be obtained. When the compound *109* is heated with ethylene glycol, 2-(4-alkoxy-2-allyl-4-penten-1-yl)-1,3,2-dioxaborolane (*137*) is formed, which, upon further heating (100–180°C) with ethylene glycol, forms 2-(4,4-ethylenedioxy-2-allyl-1-pentyl)-1,3,2-dioxaborolane (*138*). The latter compound is converted by catalytic hydrogenation and oxidation with hydrogen peroxide to 2-(4,4-ethylenedioxy-2-propyl-1-pentyl)-1,3,2-dioxaborolane (*139*), 4,4-ethylenedioxy-2-allyl-1-pentanol (*140*), and 4,4-ethylenedioxy-2-propyl-1-pentanol (*141*), from which can be obtained the compounds *134–136*. (See the reaction scheme on the next page.)

A mixture of 1,5-diallyl-3-chloromethyl-1-boracyclohex-2-ene (*110*) and 2-chloro-3-methylene-1,5-diallyl-1-borinane (*111*), in a ratio of 4:1, undergoes methanolysis to form a mixture of 1-methoxy-5-allyl-3-chloromethyl-1-boracyclohex-2-ene (*142*) and 2-chloro-1-methoxy-3-methylene-5-allyl-1-borinane (*143*), which are converted by acidolysis to 2-chloromethyl-4-methyl-1,6-heptadiene (*144*). The latter compound is hydrogenated over Pd/$SrCO_3$ to 2,4-

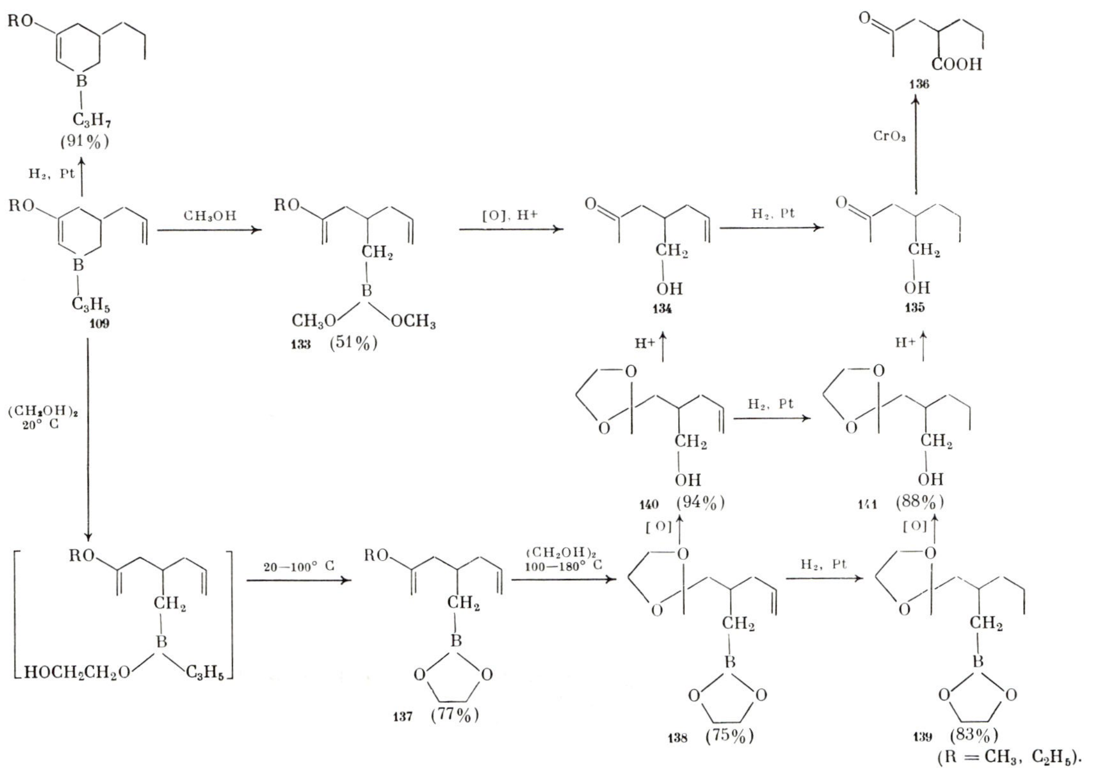

(R = CH3, C2H5).

dimethylheptane (*145*), [145, 146]:

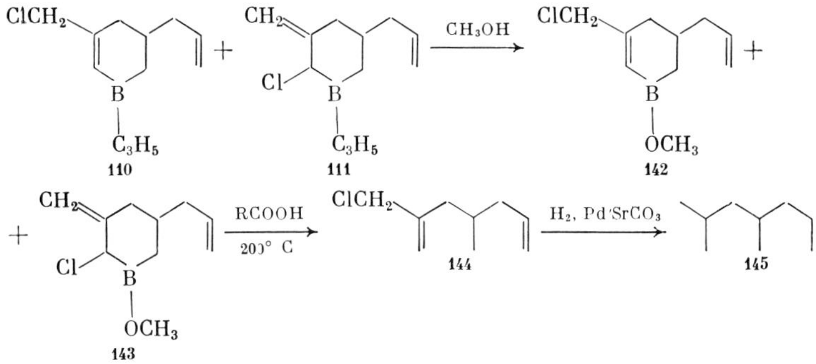

Hydrogenation over platinum of a mixture of compounds *110* and *111* forms 1,5-dipropyl-3-chloromethyl-1-boracyclohex-2-ene (*146*) and 2-chloro-1,5-dipropyl-3-methylborinane (*147*). Compounds *146* and 147 are converted by further hydrogenation over $Pd/SrCO_3$ to chloro(4-methyl-2-propyl-1-pentyl)propylborane (*148*), which reacts with methanol to form methyl (4-methyl-2-propyl-1-pentyl)-propylborinate (*149*). Acidolysis of the chloride *148* forms 2,4-dimethylheptane.

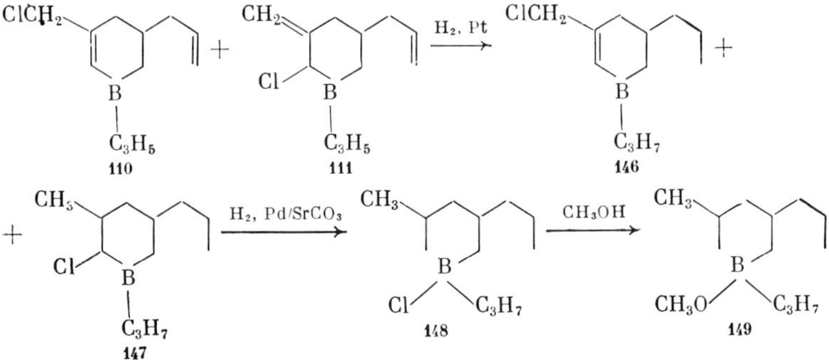

The action of methanol on *105* (R = CH_3OCH_2), *105* (R =

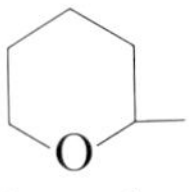

$-OCH_2-$), and *111a* produces, respectively, 1-methoxy-3-methoxymethyl-5-allyl-1-boracyclohex-2-ene [224], 1-methoxy-3-tetrahydropyranyloxymethyl-5-allyl-1-boracyclohex-2-ene [225], and 3-alkoxy-1-methoxy-5-allyl-2-methyl-1-boracyclohex-2-enes [227].

The protonolysis of 1-methoxy-3-methoxymethyl-5-allyl-1-boracyclohex-2-ene with methanolic HCl solution leads to dimethoxy)-4-methoxymethyl-2-allyl-4-penten-1-yl)borane, the oxidation of which smoothly forms 4-methoxymethyl-2-allyl-4-penten-1-ol [224].

The action of one mole of methanol on 3-methoxy-1,5-diallyl-2-methyl-1-boracyclohex-2-ene (*149a*) initially leads to protonolysis at the boron-vinyl bond, with the formation of methoxy(4-methoxy-2-allylhex-4-en-1-yl)borane (*149b*). The action of a second mole of methanol causes cleavages of the B-allyl bond, with the formation of dimethoxy(4-methoxy-2-allylhex-4-en-1-yl)borane (*149c*) [227].

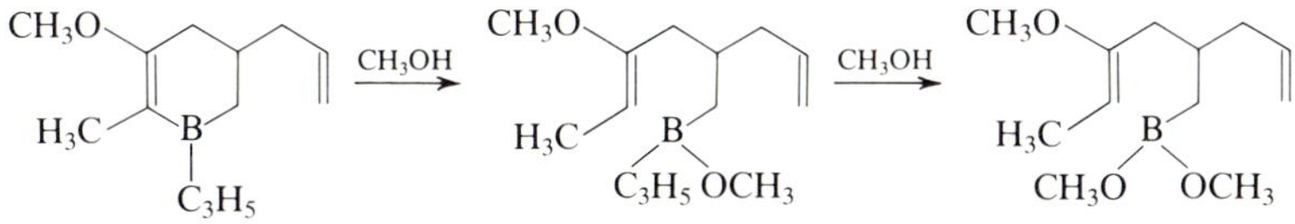

149a 149b 149c

6-Methylene-2-methoxy-4-allyl-1,2-oxaborepane is obtained by the methanolysis of *105* ($R = (C_3H_7)_2BOCH_2$) [226].

4.3 ALLYLIC COMPOUNDS OF THE 3-BORABICYCLO[3.3.1]NONANE SERIES

4.3.1 Methods for the preparation of allylic compounds of the 3-borabicyclo[3.3.1]nonane series

Triallylborane reacts at 100–130°C with acetylene to form 3-allyl-3-borabicyclo[3.3.1]non-6-ene (*79*) [137, 155]:

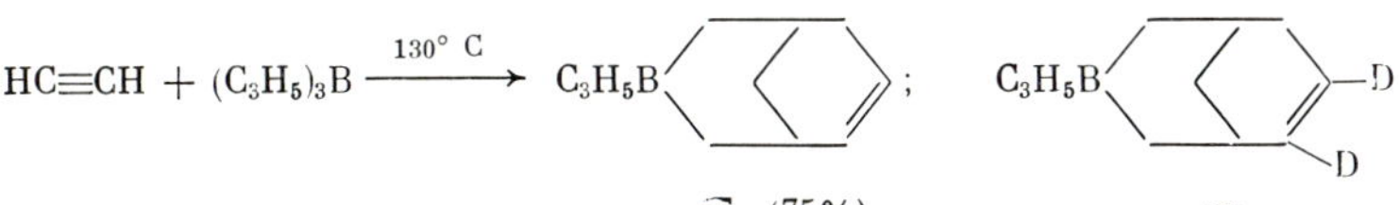

79 (75%) 150

Unlike the case of the reaction of triallylborane with acetylene derivatives, it is not possible in the present case to isolate the products of the first and second stages of the process — (1,4-pentadien-1-yl)diallylborane and 1,5-diallyl-1-boracyclohex-2-ene — since they are converted to the final product *79* at the temperature required for the initial stage of the reaction. The reaction of triallylborane with deuteroacetylene proceeds 1.5 to 2 times more slowly than that with acetylene and leads to 3-allyl-6,7-dideutero-3-borabicyclo[3.3.1]non-6-ene (*150*) [155].

Homologs of acetylene and some of its other derivatives react with triallylborane, as a rule at room temperature, and it is therefore possible in these cases, as was already indicated previously, to isolate 2-substituted (1,4-pentadiene-1-yl)diallylboranes and 3-substituted 1,5-diallyl-1-boracyclohex-2-enes. For the preparation of 7-substituted 3-allyl-3-borabicyclo[3.3.1]non-6-enes, the reaction can initially be carried out by moderate heating, and the 3-substituted 1,5-diallyl-1-boracyclohex-2-enes formed can then be converted to bicyclic compounds, or the reaction can be carried out in a single stage at the temperature necessary for the thermal conversion of the monocyclic compound to a bicyclic one.

To realize the third stage of the allylboron-acetylene condensation in the case of acetylene homologs, i.e., the preparation of 3-allyl-7-alkyl-3-borabicyclo[3.3.1]non-6-enes (*151*), a temperature of 130–140°C is required [155, 156]:

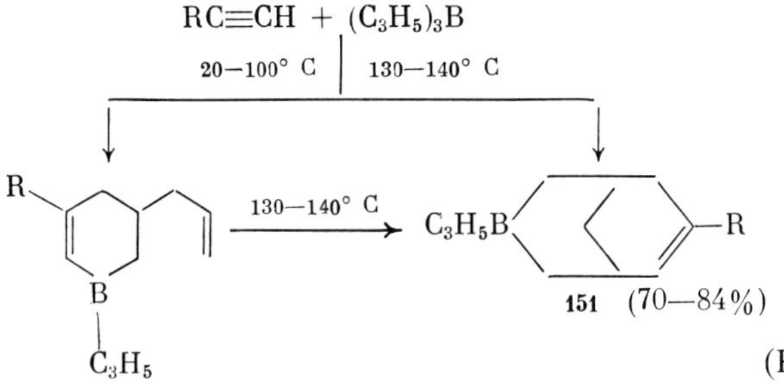

151 (70—84%)

($R = CH_3$, n-C_4H_9).

For the preparation of 7-alkoxy-3-allyl-3-borabicyclo[3.3.1]non-6-enes (*152*), the reaction must be carried out at 170–180°C [157]:

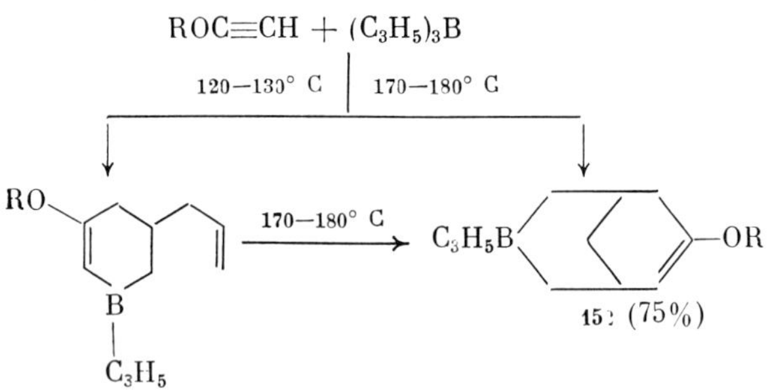

152 (75%)

($R = CH_3$, C_2H_5)

3-Allyl-7-trimethylsiloxymethyl-3-borabicyclo[3.3.1]non-6-ene (*153*) is formed in 80% yield by heating triallylborane with propargyloxytrimethylsilane at 140°C [142]; for the preparation of 7-phenyl-3-allyl-3-borabicyclo[3.3.1]non-6-ene (*154*) (yield: 96%), a mixture of triallylborane and phenylacetylene is kept for one hour at 80°C [143]; 7-cyclohexenyl-3-allyl-3-borabicyclo[3.3.1]non-6-ene (*155*) is formed by the vacuum distillation of 3-cyclohexenyl-1,5-diallyl-1-boracyclohex-2-ene (*108*) [158].

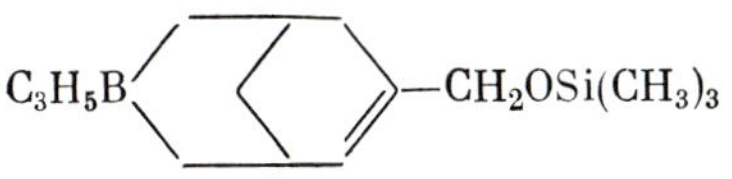

153

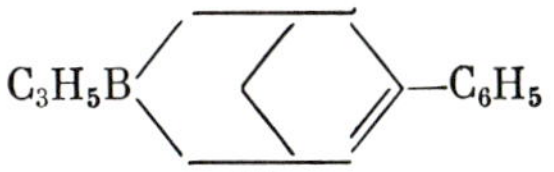

154

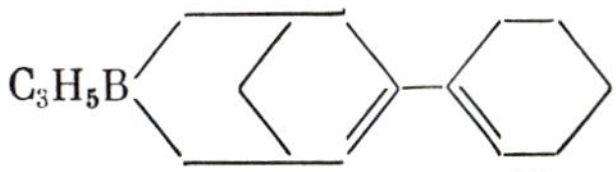

155

When a mixture of 3-chloromethyl-(*110*) and 2-chloro-3-methylene-1,5-diallylborinane (*111*) is kept at room temperature or moderately heated, the former compound cyclizes to 3-allyl-7-chloromethyl-3-borabicyclo[3.3.1]non-6-ene (*156*), whereas the chloride *111* does not cyclize under these conditions.

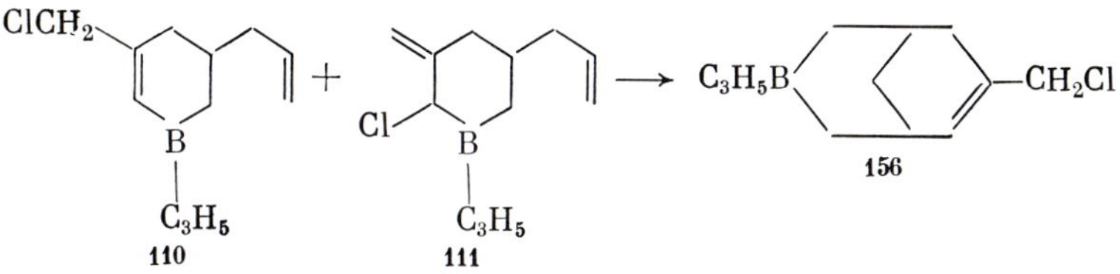

If a mixture of compounds *110* and *111* is heated for several hours at 80–130°C, it is converted to the bicyclic compound *156*. In this case, the equilibrium *110*⇄*111* is displaced to the left, and the compound *111* is converted to *156*, via the stage of isomerization to *110*.

However, if the reaction between triallylborane and propargyl chloride is carried out by addition of the chloride to the borane, heated to 130°C, a mixture of compound *156* and 6-chloro-7-methylene-3-allyl-3-borabicyclo[3.3.1]nonane (*157*) in a ratio of 4:1

is obtained [145, 159, 160]:

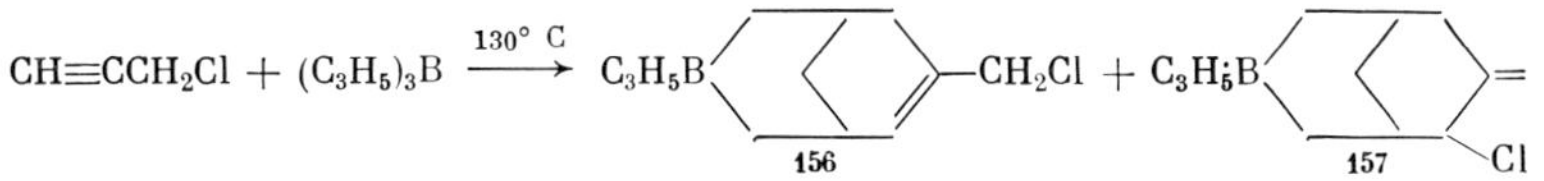

156 157

Triallylborane and propargyl bromide, when heated, form a mixture of 3-allyl-7-bromomethyl-3-borabicyclo[3.3.1]non-6-ene and 6-bromo-7-methylene-3-allyl-3-borabicyclo[3.3.1]nonane [161].

It should be noted that the ability of monocyclic compounds to cyclize depends on the nature of the substituent on the boron atom. Thus, 3-substituted 1-ethoxy-3-allyl-1-boracyclohex-2-enes are not converted to derivatives of 3-borabicyclo[3.3.1]non-6-ene by heating, whereas replacement of the allyl group on the boron atom by an alkyl group does not prevent cyclization. For example, 5-allyl-3-butyl-1-methyl-1-boracyclohex-2-ene has been converted to 3-methyl-7-butyl-3-borabicyclo[3.3.1]non-6-ene by heating at 140–150°C [156].

Tri(2-methylallyl)borane also reacts with acetylene and its derivatives. Heating tri(2-methylallyl)borane with acetylene in an autoclave at 140–150°C leads to a 67% yield of 3-(2-methylallyl)-1,5-dimethyl-3-borabicyclo[3.3.1]non-6-ene (*158*), which is capable of adding to acetylene to form the compound *159* [162, 163]:

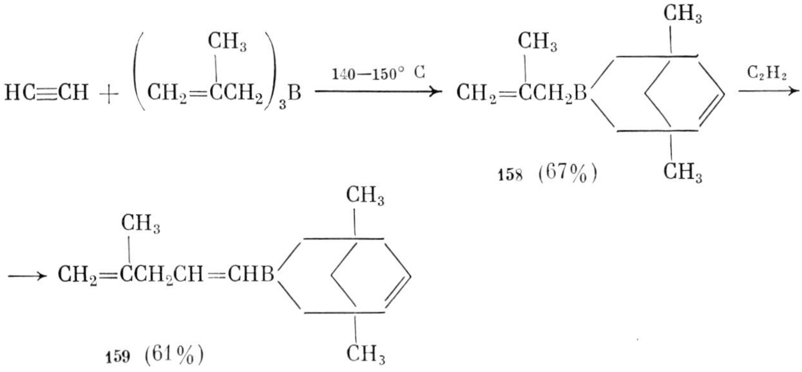

158 (67%)

159 (61%)

Tri(2-methylallyl)borane reacts with butylacetylene to form 3-(2-methylallyl)-7-butyl-1,5-dimethyl-3-borabicyclo[3.3.1]non-6-ene

(*160*) [163]:

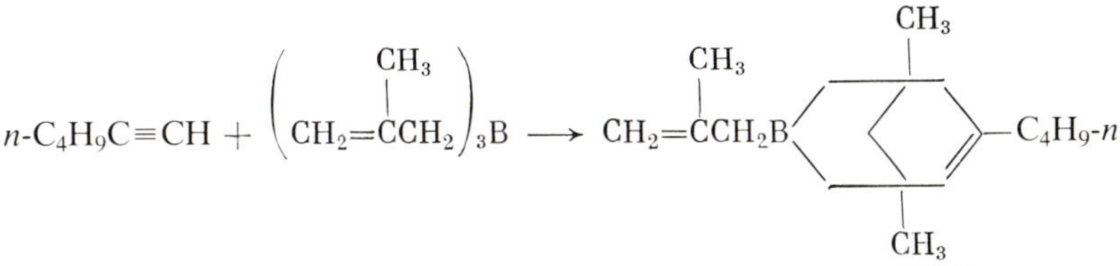

160 (77%)

7-Alkoxy-3-(2-methylallyl)-1,5-dimethyl-3-borabicyclo[3.3.1]non-6-enes (*161*) have been obtained by the condensation of tri(2-methylallyl)borane with alkoxyacetylenes [162, 164]:

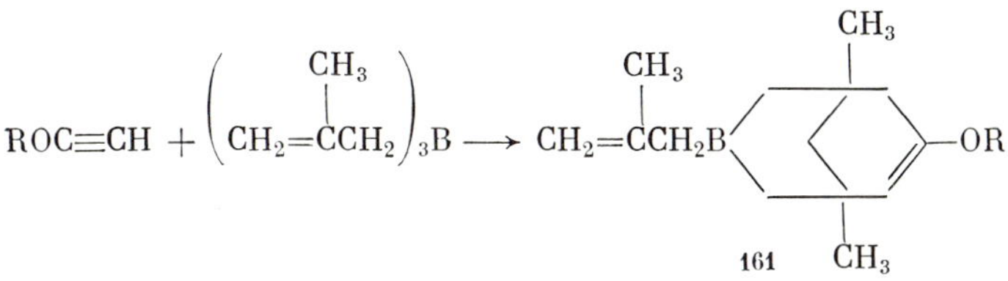

161

The reaction of triallylborane with vinylacetylene or isopropenylacetylene forms, respectively, 3-vinyl-1,5-diallyl-1-boracyclohex-2-ene (*162*, R = H) and 3-isopropenyl-1,5-diallyl-1-boracyclohex-2-ene (*162*, R = CH_3), which are converted by heating to 7-vinyl-3-allyl-3-borabicyclo[3.3.1]non-6-ene (*163*, R = H) and 7-isopropenyl-3-allyl-3-borabicyclo[3.3.1]non-6-ene (*163*, R = CH_3), respectively [165–168]:

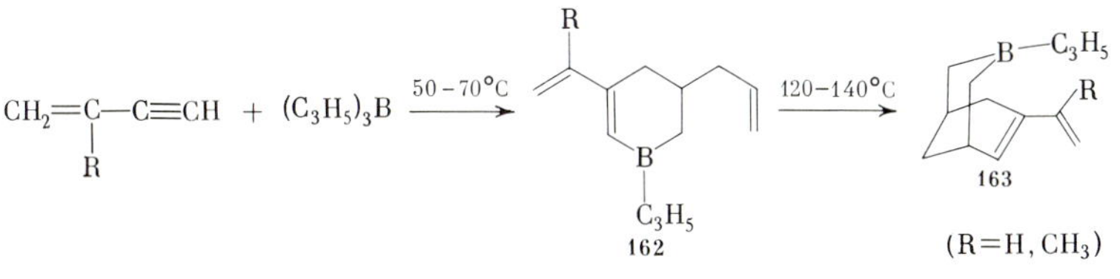

162 163 (R=H, CH_3)

The reaction of diacetylene with triallylborane or tri(2-methylallyl)borane at 120–150°C leads to diboron compounds with a system of conjugated double bonds (*164*) [169]:

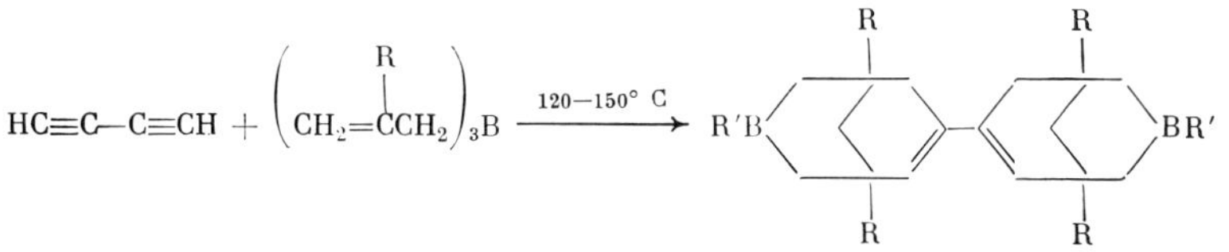

164 (90%)

($R = H$; $R' = CH_2{=}CH{-}CH_2$)
($R = CH_3$; $R' = CH_2{=}C(CH_3){-}CH_2{-}$)

Triallylborane reacts with methyl propargyl ether at 130–140°C to form 7-methoxymethyl-3-allyl-3-borabicyclo[3.3.1]non-6-ene (*164a*) [223, 224] and with propargyl tetrahydropyranyl ether to form 7-tetrahydropyranyloxymethyl-3-allyl-3-borabicyclo[3.3.1]non-6-ene (*164b*) [225].

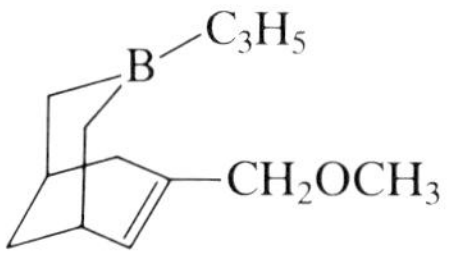

164a

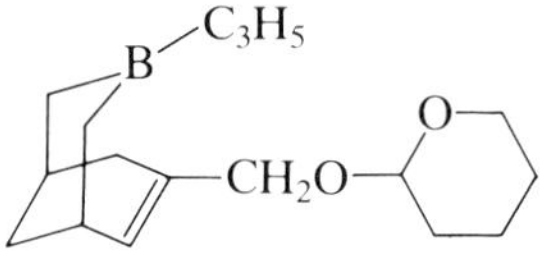

164b

Heating of methyl propargyl ether with tri(2-methylallyl)borane at 140–150°C leads to 7-methoxymethyl-1,5-dimethyl-3-(2-methylallyl)-3-borabicyclo[3.3.1]non-6-ene (*164c*) [228]. The reaction of methyl propargyl ether with tricrotylborane at 120–130°C proceeds stereospecifically and leads to 7-methoxymethyl-3-crotyl-8,9-dimethyl-3-borabicyclo[3.3.1]non-6-ene (*164d*) [229].

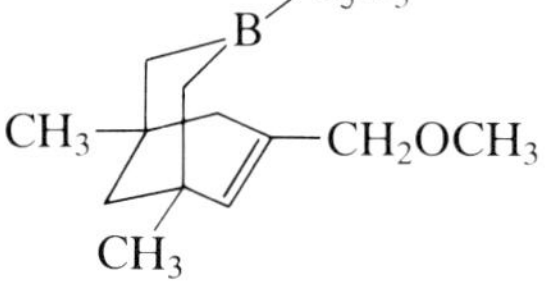

164c

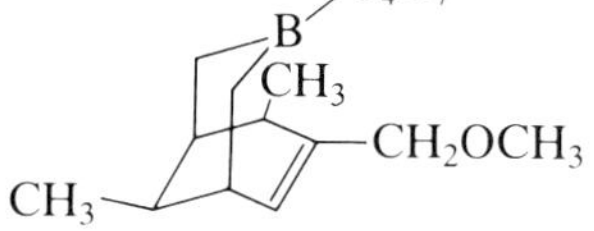

164d

For the example of this reaction, it has been shown that the second stage of an allylboron-acetylene condensation proceeds with allylic rearrangement.

When triallylborane is heated with 1-methoxypropyne to 135–145°C, 7-methoxy-3-allyl-6-methyl-3-borabicyclo[3.3.1]non-6-ene is formed [227].

Mikhailov and Smirnov [170, 171] found that allenic compounds, like acetylenic ones, are capable of reacting with allylboranes to form cyclic compounds, with the rate and direction of the reaction depending on the nature of the allene component.

The reactions of allylboranes with 1-chloro-3-methyl-1,2-butadiene, allene, and 3-methyl-1,2-butadiene have been investigated. It was found that 1-chloro-3-methyl-1,2-butadiene reacts slowly with triallylborane even at room temperature, forming 5-chloromethylene-2-(3-buten-1-yl)-1-allyl-4,4-dimethylborolane (*166*) [170, 171]. By analogy with the allylboron-acetylene condensation, the first stage of the reaction evidently consists of the addition of triallylborane to the allene system, with the diallylboryl group adding to the central carbon atom and the allyl group to the terminal carbon atom, bonded to two methyl groups. (1-Chloromethylene-2,2-dimethyl-4-penten-1-yl)diallylborane (*165*) thus obtained undergoes intramolecular cyclization at room temperature, but the direction of the cyclization in this case is different from that in the case of pentadienyl boron derivatives of the type *80*, with the allyl adding to the terminal carbon atom of the pentene system and the boron atom adding to a non-terminal carbon. The orientation of the cyclization is apparently affected by a steric factor favoring an approach of the boron atom to a closer distance to the C_4 atom than to the C_5 atom in the transition state:

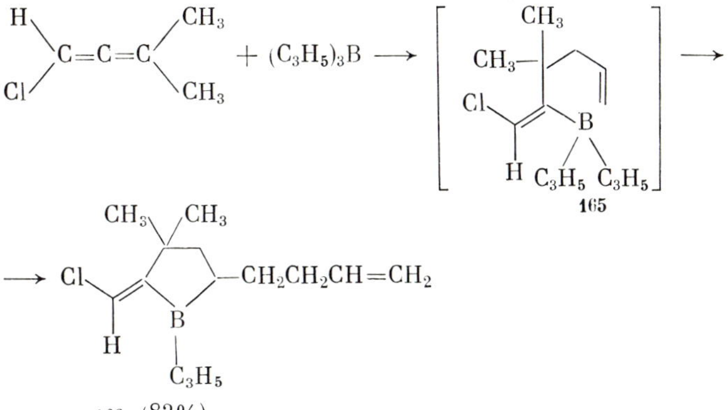

Allene, as a compound with less polarized bonds, is more inert toward allylboranes than 1-chloro-3-methyl-1,2-butadiene. It reacts slowly with triallylborane when heated to 150°C in an autoclave, forming a 60% yield of 7-methylene-3-allyl-3-borabicyclo-[3.3.1]nonane (*167*), the structure of which has been determined by IR and NMR spectroscopy and also by conversion to known compounds [171–173]:

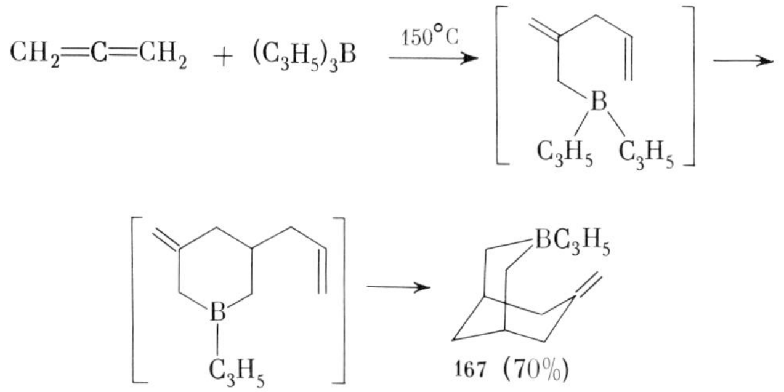

167 (70%)

The reaction of allene with tri(2-methylallyl)borane proceeds analogously, but under more forced conditions [171–173]. When an equimolar mixture of the reagents is heated in an autoclave at 160–170°C, 7-methylene-3-(2-methyl-2-propen-l-yl)-1,5-dimethyl-3-borabicyclo[3.3.1]nonane (*168*) is obtained:

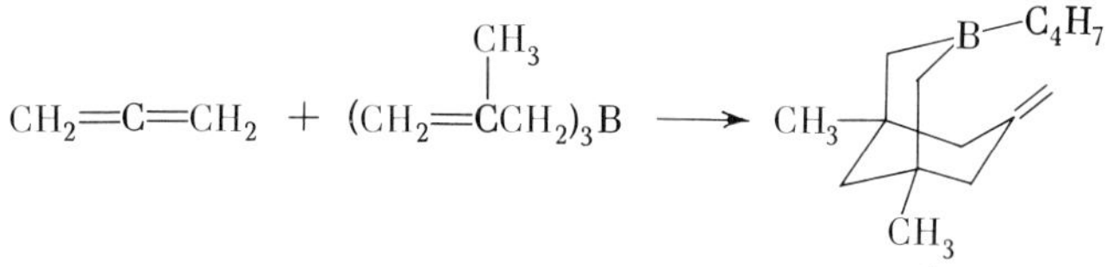

168 (51%)

The action of triallylborane on 3-methyl-1, 2-butadiene produced a bicyclic compound, to which the structure of 3-allyl-6-methylene-7,7-dimethyl-3-borabicyclo[3.3.1]nonane was erroneously assigned [171, 172]. The compound obtained in this reaction is actually 3-allyl-7-methylene-6,6-dimethyl-3-borabicyclo[3.3.1]nonane [230].

4.3.2 Chemical properties of allylic compounds of the 3-borabicyclo[3.3.1]nonane series. Synthesis of derivatives of cyclohexene, cyclohexane, and trienes

Permanent allylic rearrangement has been investigated by ^{1}H NMR spectroscopy in 3-allyl-3-borabicyclo[3.3.1]non-6-ene (*79*), 3-allyl-6,7-dideutero-3-borabicyclo[3.3.1] non-6-ene (*150*), 7-methyl-3-allyl-3-borabicyclo[3.3.1]non-6-ene (*151a*, R = CH_3), and 7-butyl-3-allyl-3-borabicyclo[3.3.1]non-6-ene (*151b*, R = *n*-C_4H_9) [174].

The activation energy of the intramolecular rearrangement was found to have values of 8.4, 5.8, 4.7, and 5.8 ± 1 kcal/mole, respectively. The values obtained are lower, on average, than those of the activation energy for 1-allylborolane (15 kcal/mole) and triallylborane (10 kcal/mole). Thus, from the value of the activation energy of the permanent rearrangement, the compounds under consideration can be arranged in the following order:

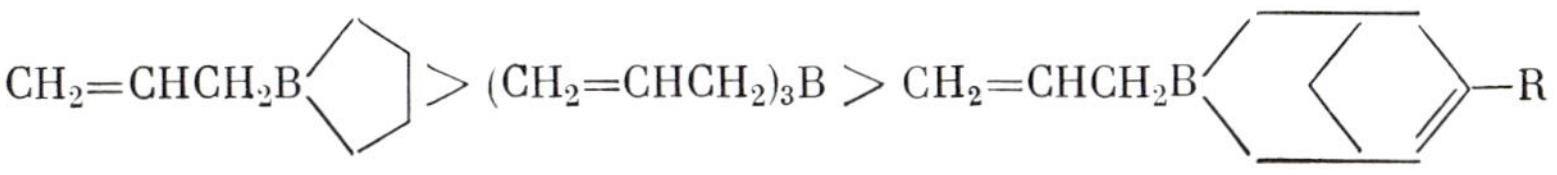

3-Allyl-3-borabicyclo[3.3.1]non-6-ene (*79*) has been subjected to various chemical conversions [137, 155]. Its reaction with methanol leads to 3-methoxy-3-borabicyclo[3.3.1]non-6-ene (*169*), which, when heated with caproic acid, is converted to *cis*-3,5-dimethylcyclohexene (*170*), as a result of the rupture of B—C bonds. When the compound *170* is heated with palladium black in tetrahydrofuran, a mixture of *m*-xylene and *cis*- and *trans*-1,3-dimethylcyclohexanes is formed. The double bonds in the bicyclic compounds *79* and *169* are easily hydrogenated over a palladium catalyst at 20°C, producing, respectively, 3-propyl-3-borabicyclo-[3.3.1]nonane (*171*) and 3-methoxy-3-borabicyclo[3.3.1]nonane (*172*), which, similarly to bicyclo[3.3.1]nonane, have a "chair-chair" conformation. Acidolysis of the compounds *171* and *172* leads to *cis*-1,3-dimethylcyclohexane (*173*), containing only approximately 1% of unidentified impurities. The oxidation of compounds *171* and *172* with alkaline hydrogen peroxide forms *cis*-1,3-bis(hydroxymethyl)cyclohexane (*174*), which is converted to *cis*-1,3-cyclohexanedicarboxylic acid (*175*) by the action of $KMnO_4$:

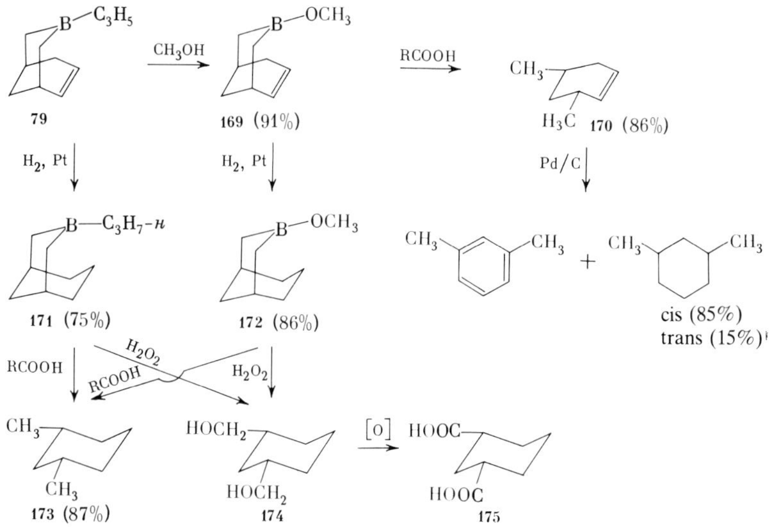

3-Allyl-3-borabicyclo[3.3.1]non-6-ene (*79*) is converted by heating with butylamine to 3-butylamino-3-borabicyclo[3.3.1]non-6-ene (*176*), which can also be obtained by the transamination of 3-amino-3-borabicyclo[3.3.1]non-6-ene (*177*). The latter compound is obtained by the action of ammonia on 3-butylmercapto-3-borabicyclo[3.3.1]non-6-ene (*178*), which is formed by the reaction of tributyl thioborate with the compound *169* [175]:

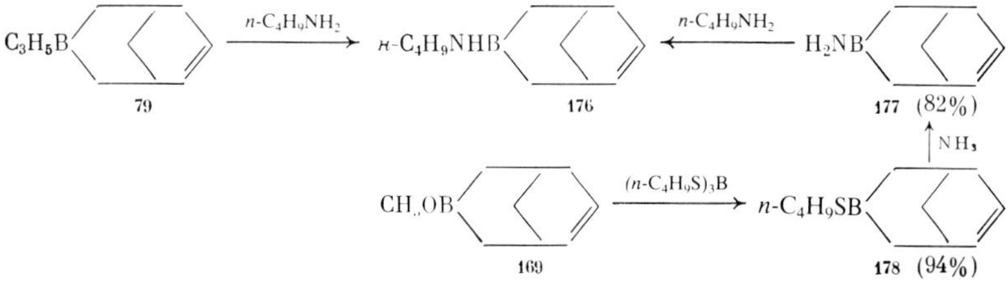

The synthesis of 3-alkylmercapto- and 3-amino-substituted 3-borabicyclo[3.3.1]nonanes has been carried out by an analogous method [175].

3-Allyl-7-methyl-3-borabicyclo[3.3.1]non-6-ene (*151a*) has been subjected to the conversions indicated by the following scheme [155]:

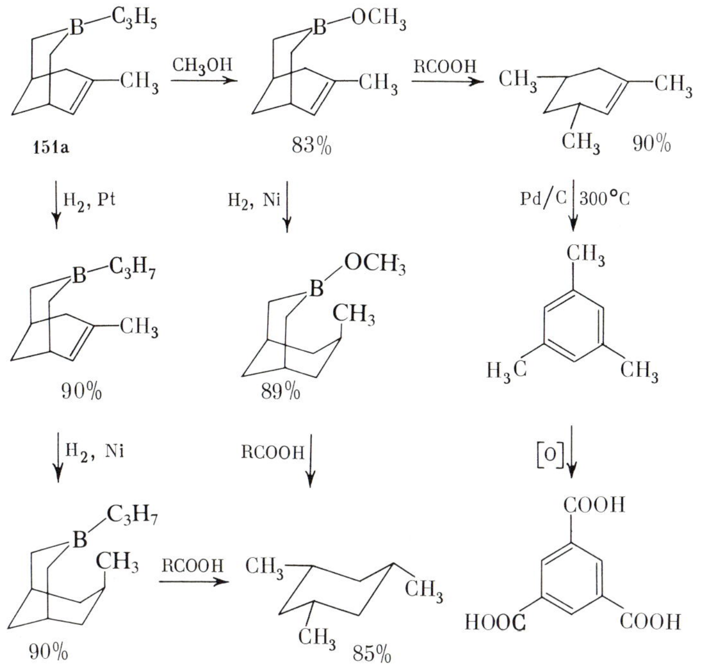

Analogous reactions have been carried out with 3-allyl-7-butyl-3-borabicyclo[3.3.1]non-6-ene (*151b*) [156]:

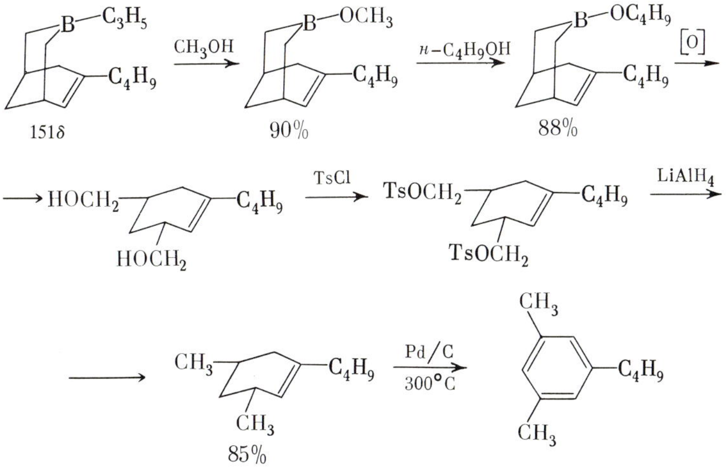

The structure of 7-phenyl-3-allyl-3-borabicyclo[3.3.1]non-6-ene (*154*) has been demonstrated by its conversion to 3,5-dimethylbiphenyl (*180*) by the following method [143]:

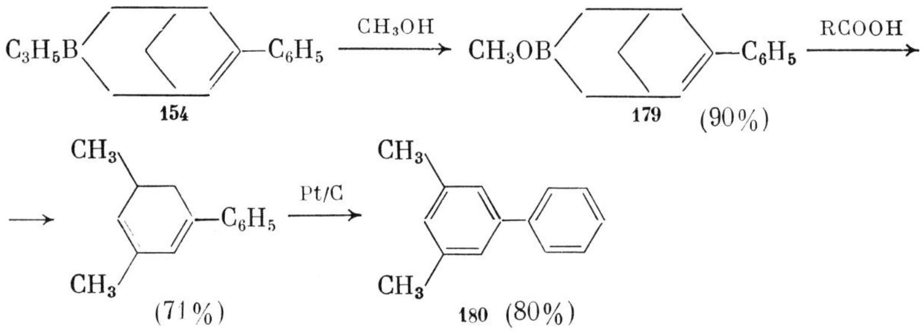

Unlike 3-methoxy-7-methyl-3-borabicyclo[3.3.1]non-6-ene, the compound *179* is not hydrogenated in an autoclave over Raney nickel, but can be hydrogenated to 3-methoxy-7-phenyl-3-borabicyclo[3.3.1]nonane (*181*) in the presence of $Pd/SrCO_3$ at room temperature [143]:

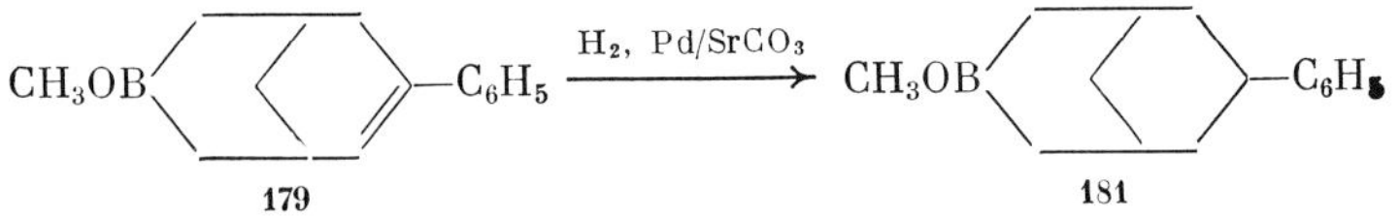

Prolonged heating of the compound *181* with methyl borate in the presence of tetrapropyldiborane leads to cleavage of the boron-containing ring, with the formation of 5-phenyl-1,3-bis-(dimethoxyborylmethyl)cyclohexane (*182*):

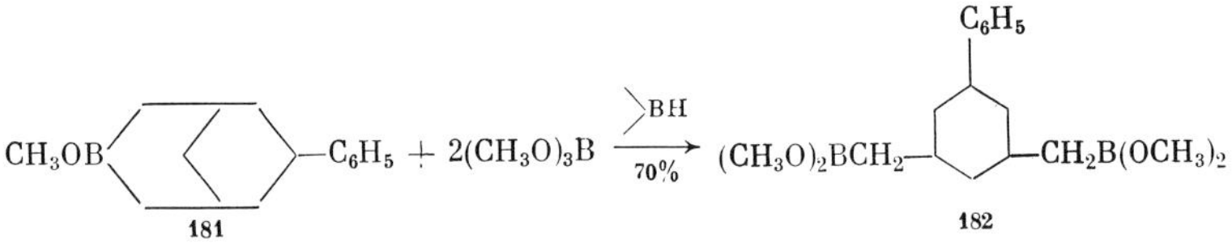

Oxidation of the compound *181* with hydrogen peroxide forms 5-phenyl-1,3-bis(hydroxymethyl)cyclohexane (*183*), while 5-phenyl-*cis*-1,3-dimethylcyclohexane (*184*) can be obtained from *181* by heating it with stearic acid:

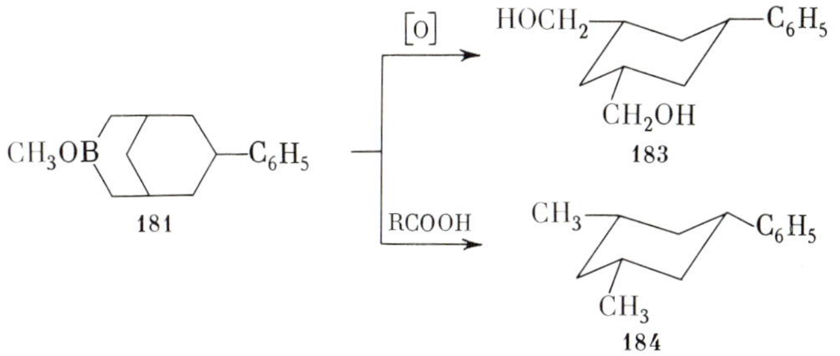

The hydrogenation of 7-alkoxy-3-allyl-3-borabicyclo[3.3.1]non-6-enes (*152*) over a Pd/$SrCO_3$ catalyst proceeds with the absorption of two moles of hydrogen and leads to the compounds *185* [157] (see the scheme on the next page). The action of alcohols on the compounds *152* produces propylene and the dialkoxy derivatives *186*. In the presence of acids, the diesters *186*, containing a vinyl ester group, readily add an alcohol to form the ketals *187*, which, however, are unstable and split off the alcohols to form the starting compounds *186* during vacuum distillation at 100°C. When the bicyclic compounds *152* or *186* are heated with ethylene glycol, the diboron cyclic ketal *188* is obtained. By the alkaline hydrolysis of the ethylene ketal *188*, the bicyclic acid *189* can be obtained, which is converted by esterification with butanol to 3-butoxy-7,7-ethylenedioxy-3-borabicyclo[3.3.1]nonane (*190*). Oxidation of the cyclic ketal *190* or diketal *188* with alkaline hydrogen peroxide leads to 1,1-ethylenedioxy-*cis*-3,5-bis(hydroxymethyl)cyclohexane (*191*). Reduction of the bis-tosylate of *191* with lithium aluminum hydride produces 1,1-ethylenedioxy-*cis*-3,5-dimethylcyclohexane (*192*), which has been converted by the action of dilute acid to *cis*-3,5-dimethylcyclohexanone (*193*) [157].

3-Methoxy-7-cyclohexenyl-3-borabicyclo[3.3.1]non-6-ene enters into a diene condensation with maleic anhydride to form the adduct *194* [158].

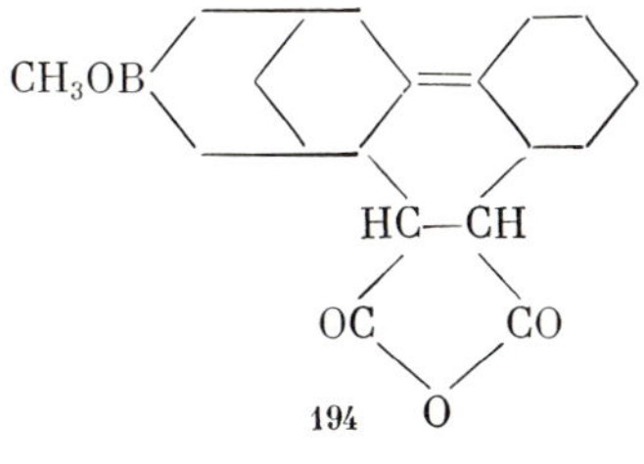

194

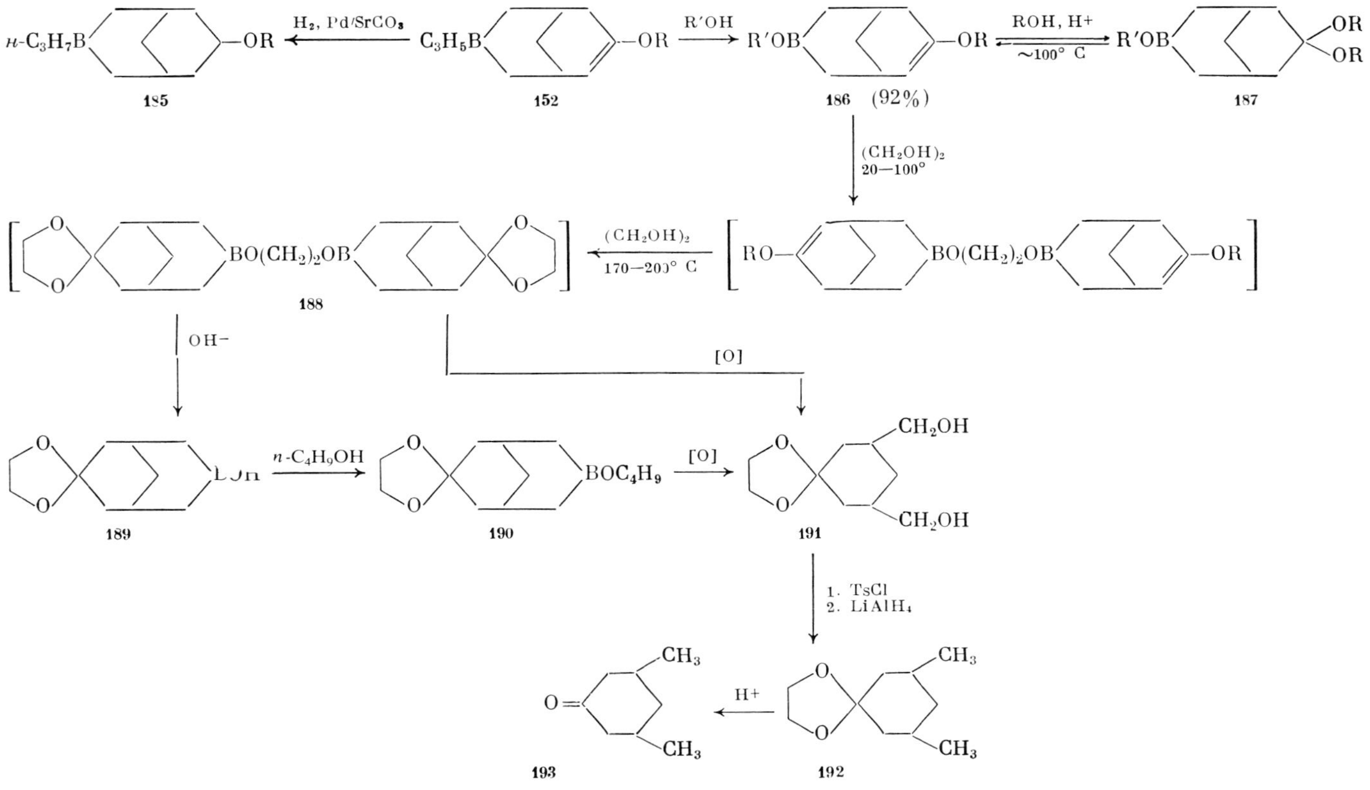

n-C3H7B
—OR
H2, Pd/SrCO3
C3H5B
—OR
R'OH
R'OB
—OR
ROH, H+
~100° C
R'OB
OR
OR
185
152
186 (92%)
187
(CH2OH)2
20—100°
O
O
BO(CH2)2OB
O
O
(CH2OH)2
170—200° C
RO—
BO(CH2)2OB
—OR
188
OH−
[O]
O
O
BOH
n-C4H9OH
O
O
BOC4H9
[O]
O
O
CH2OH
CH2OH
189
190
191
1. TsCl
2. LiAlH4
CH3
O=
CH3
H+
O
O
CH3
CH3
193
192

The action of alcohols on a mixture of 3-allyl-7-chloromethyl-3-borabicyclo[3.3.1]non-6-ene (*156*) and 6-chloro-7-methylene-3-allyl-3-borabicyclo[3.3.1]nonane (*157*) forms a mixture of the esters *195* and *196*, which are converted by hydrogenation over platinum-group catalysts to 3-alkoxy-7-methyl-3-borabicyclo[3.3.1]nonanes (*197*); acidolysis of the latter compounds produces *cis*-1,3,5-trimethylcyclohexane (*198*) with a 99% degree of purity [145, 159]:

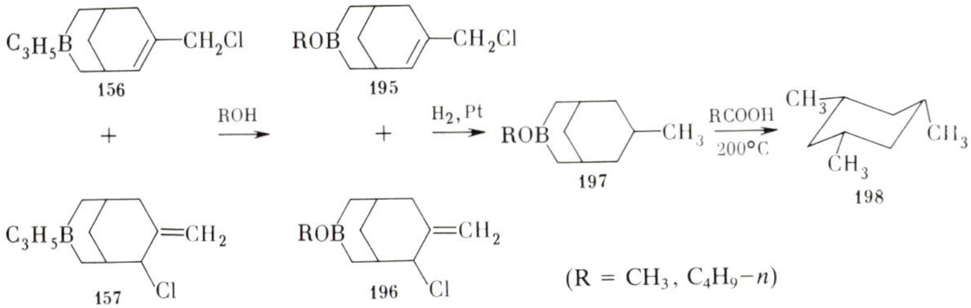

(R = CH_3, C_4H_9-n)

The methanolysis of 3-(2-methylallyl)-1,5-dimethyl-3-borabicyclo[3.3.1] non-6-ene (*158*) leads to 3-methoxy-1,5-dimethyl-3-borabicyclo[3.3.1]non-6-ene (*199*), the oxidation of which with hydrogen peroxide forms *cis*-3,5-bis(hydroxymethyl)-*cis*-3,5-dimethylcyclohexene (*200*), while the acidolysis of *199* produces 3,3,5,5-tetramethylcyclohexene (*201*) [162]:

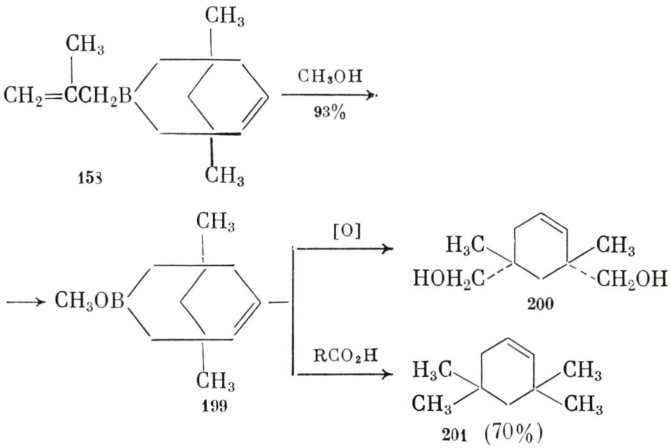

7-Alkoxy-3-(2-methylallyl)-1,5-dimethyl-3-borabicyclo[3.3.1]non-6-ene (*161*) can be converted by the action of ethylene glycol, followed by oxidation with alkaline hydrogen peroxide, to 1,1-ethylenedioxy-*cis*-3,5-bis(hydroxymethyl)-*cis*-3,5-dimethylcyclohexane (*202*), the tosylate of which is reduced with lithium aluminum hydride to 1,1-ethylenedioxy-3,3,5,5-tetramethylcyclohexane (*203*). Hydrolysis of the latter compound forms 3,3,5,5-tetramethylcyclohexanone (*204*) [162, 164]:

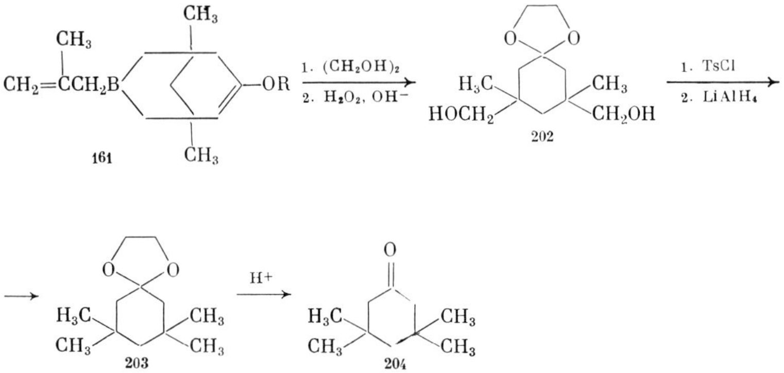

The compound *162* is converted by acidolysis or acidolysis-oxidation to the triene hydrocarbons *205* or triene alcohols *206*.

The compounds *163* are oxidized to 1-vinyl-*cis*-3,5-bis-(hydroxymethyl)cyclohexenes (*207*) [165–168]:

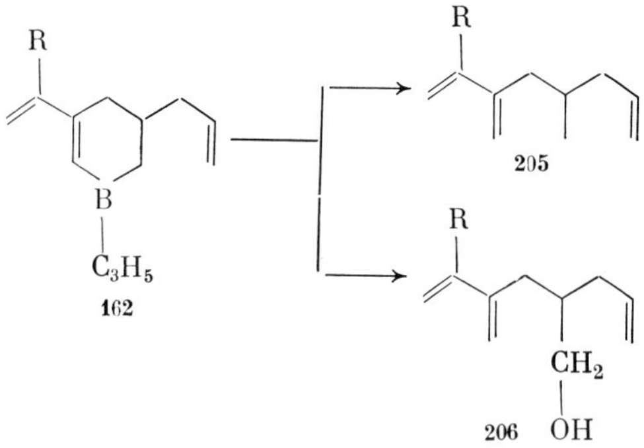

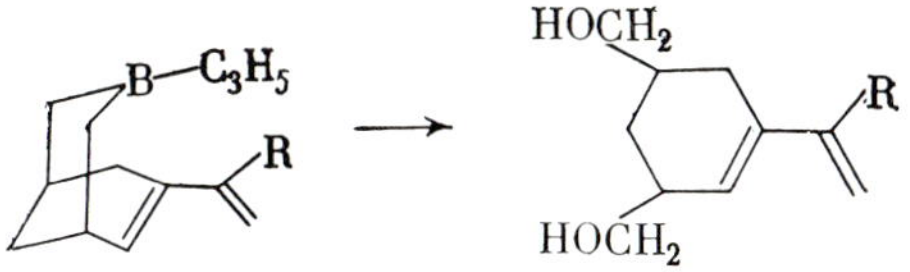

163 207 (R = H, CH$_3$).

3,3′,5,5′-Tetramethylbicyclohexenyl (*208*) and 3,3′,5,5′-tetramethylbiphenyl (*209*) can be obtained by consecutive chemical conversions from the compound *164* (R = H) [169]:

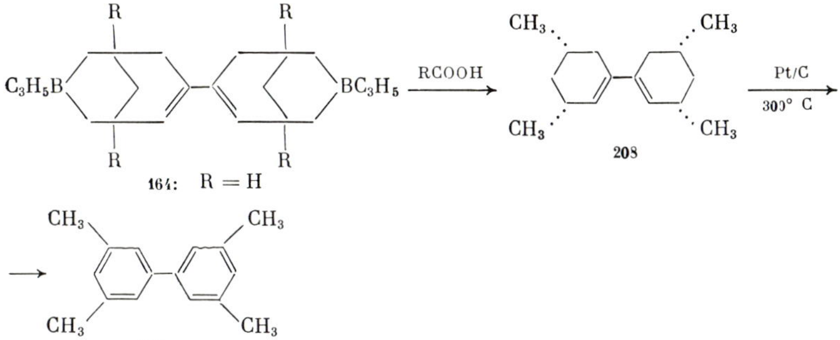

7-Tetrahydropyranyloxymethyl-3-methoxy[3.3.1]nonane is hydrolyzed to the dimer of 4-hydroxy-3-borahomoadamantane (*209a*) [225]:

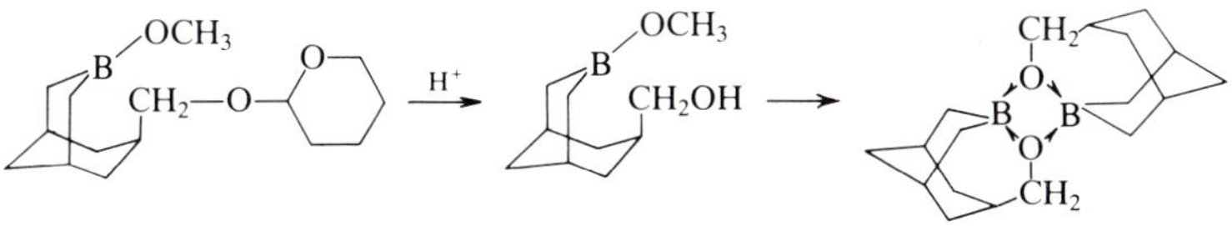

209a

In the dimer *209a*, the oxygen atom, according to X-ray structural data [231], is not in a tetrahedral valence state, as is characteristic for compounds of tricoordinated oxygen, but in a trigonal state. 7-Tetrahydropyranyloxymethyl-3-propyl-3-borabicyclo [3.3.1] non-6-ene reacts with carbon monoxide at 150°C and 50–70 atmospheres to form 3-hydroxy-7-tetrahydropyranyloxymethyl-3-propylbicyclo[3.3.1] non-6-ene in the form of a mixture of stereoisomers [225].

7-Methylene-3-allyl-3-borabicyclo[3.3.1]nonane is smoothly converted by the action of methanol to 3-methoxy-7-methylene-3-borabicyclo[3.3.1]nonane, which forms 3-methoxy-7-methyl-3-borabicyclo[3.3.1]nonane upon catalytic hydrogenation [171–173].

The hydroboration of 3,7-dimethoxy-6-methyl-3-borabicyclo[3.3.1]non-6-ene with the use of $H_3B \cdot THF$ produces 3-methoxy-7β-dimethoxyboryl-6α-methyl-3-borabicyclo[3.3.1]nonane, the protonolysis of which with stearic acid at 260°C leads to 1-*cis*-2-*cis*-4-trimethylcyclohexane, while oxidation forms 3,5-dihydroxymethyl-2-methylcyclohexane [*227*].

The synthesis of different derivatives of 3-borabicyclo[3.3.1]-nonane [232] and of chelated boron compounds on the basis of derivatives of 7-methoxymethyl-3-borabicyclo[3.3.1]non-6-ene has been carried out [233].

Conformational analysis with the use of ^1H and ^{13}C NMR spectroscopy has shown that derivatives of 3-borabicyclo[3.3.1]nonane with

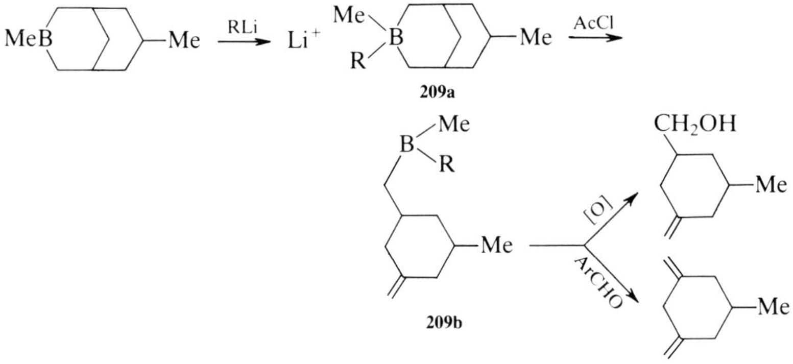

209a

209b

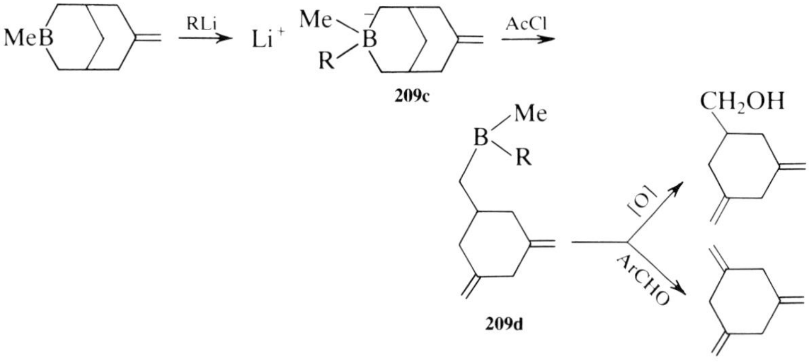

209c

209d

a tricoordinated boron atom are in a flattened double-chair conformation. 3-Borabicyclo[3.3.1]nonane derivatives with a tetracoordinated boron atom and a substituent in the 7α position show a predominant chair-boat conformation, the boat form being characteristic of the cyclohexane ring [234, 235].

The *ate* complexes *209a* and *209c*, obtained by the action of methyllithium or butyllithium on 3,7-dimethyl- and 7-methylene-3-methyl-3-borabicyclo[3.3.1]nonane react by intramolecular β-hydride transfer with acetyl chloride, being converted, respectively, to methylene-(*209b*) or dimethylene-substituted (*209d*) 1-cyclohexylmethyl(methyl)(alkyl)borane derivatives. The oxidation of *209b* and *209d* leads to 1-methylene-3-hydroxymethyl-5-methyl cyclohexane and 3,5-dimethylene-5-hydroxymethylcyclohexane, respectively, while the reactions of *209b* and *209d* with veratraldehyde form 1,3-dimethylene-5-methylcyclohexane and 1,3,5-trimethylenecyclohexane, respectively [236].

Thus, on the basis of the reactions of acetylenic and allenic compounds with allylboranes, the following have been developed: a stereospecific method for the synthesis of cyclohexene and cyclohexane derivatives and methods for the synthesis of unsaturated aldehydes and ketones, dienes, alkoxydienes, trienes, unsaturated oxo alcohols, biphenyl derivatives, and other organic compounds.

CHAPTER 5

1-Boraadamantane and its analogs. The synthesis of adamantanes, homoadamantanes, and 3-azabicyclo[3.3.1]nonanes

In this chapter, methods for the preparation of 1-boraadamantane and its derivatives with the use of 3-borabicyclo[3.3.1]nonanes and 3-borabicyclo[3.3.1]non-6-enes and their chemical behavior under the action of different reagents will be considered.

5.1 METHODS FOR THE PREPARATION OF 1-BORAADAMANTANE AND ITS DERIVATIVES

Derivatives of 3-borabicyclo[3.3.1]nonane with a methylene group in the 7 position, obtained by the condensation of allylboranes with allenes, have been used for the synthesis of 1-boraadamantane compounds. The basis for different modifications of this synthesis consists of the hydroboration of the 7-methylene group, followed by intramolecular cyclization of the diboron compounds formed by this process, which had been observed previously in a number of acyclic diboron [25, 176] and triboron [53] systems.

In the first modification, 7-methylene-3-allyl-3-borabicyclo-[3.3.1]nonane (*167*) was partially hydrogenated by hydrogen over platinum black to 7-methylene-3-propyl-3-borabicyclo[3.3.1]-nonane (*210*), which was then hydroborated with tetrapropyl-diborane to 3-propyl-7-dipropylborylmethyl-3-borabicyclo[3.3.1]-nonane. The latter compound cyclizes with elimination of tripropylborane to 1-boratricyclo[$3.3.1^{3,7}$]decane (1-boraadamantane) (*211*), which forms a stable adduct with pyridine (*212*) [171, 173, 177, 178, 237]:

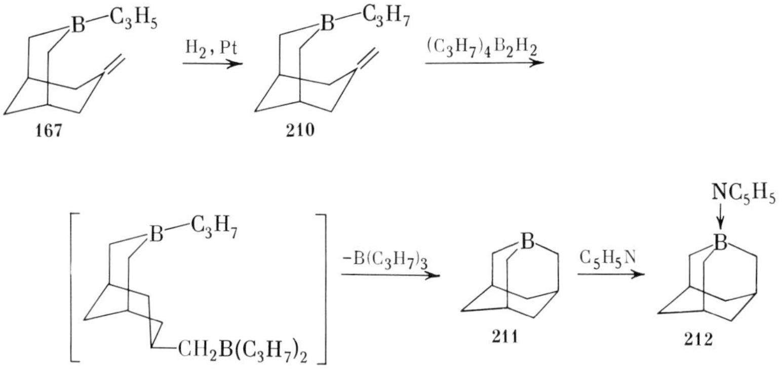

1-Boraadamantane is obtained more conveniently by the hydroboration of 3-methoxy-7-methylene-3-borabicyclo[3.3.1]nonane (*213*), since the latter compound is easier to obtain than *210*. Diborane in ether medium or tetraalkyldiborane can be used as a hydroborating agent. The action of tetraethyldiborane on *213* forms 3-methoxy-7-diethylborylmethyl-3-borabicyclo[3.3.1]nonane (*214*), which is converted, with the elimination of methyl diethylborinate, to 1-boraadamantane. The latter can be isolated by sublimation or in the form of the etherate (*215*) or pyridinate (*212*). In the hydroboration of the compound *213* with diborane in ether medium, the etherate *215* formed is isolated in the form of a colorless viscous liquid after vacuum distilling off the solvent and of the methyl borate formed. If the hydroboration of *213* is carried out with diborane in tetrahydrofuran, a complex of 1-boraadamantane with tetrahydrofuran is obtained in 72% yield.

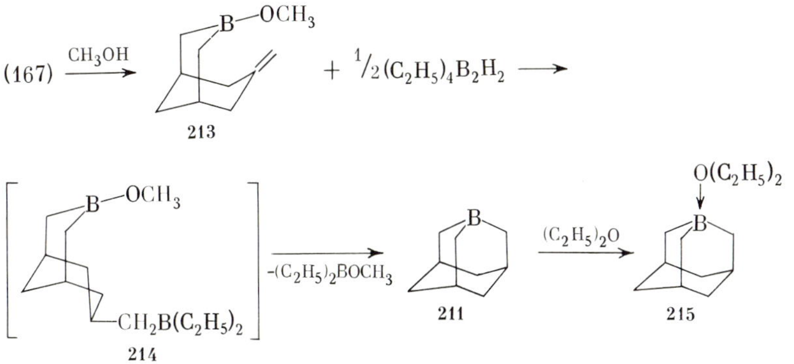

The ether complex of 1-boraadamantane is relatively stable, but is partially dissociated into its components even at room temperature. The etherate *215* has been used for the preparation of complexes with other ligands with a greater donor capacity, simply by mixing the reagents in an inert solvent [171, 178]. A thermally stable complex of 1-boraadamantane with tetrahydrofuran and very stable complexes with pyridine and triethylamine, which do not dissociate even at 200°C, have been obtained in this way. A complex of 1-boraadamantane with ethyl acetate has been synthesized by the direct reaction of these compounds.

3-Methoxy-7-methylene-1,5-dimethyl-3-borabicyclo[3.3.1]nonane (*216*) is smoothly converted to 3,5-dimethyl-1-boraadamantane etherate (*217*) by the action of tetraethyldiborane in an ether medium. The etherate *217* reacts with pyridine to form 3,5-dimethyl-1-boraadamantane pyridinate (*218*) [171, 173]:

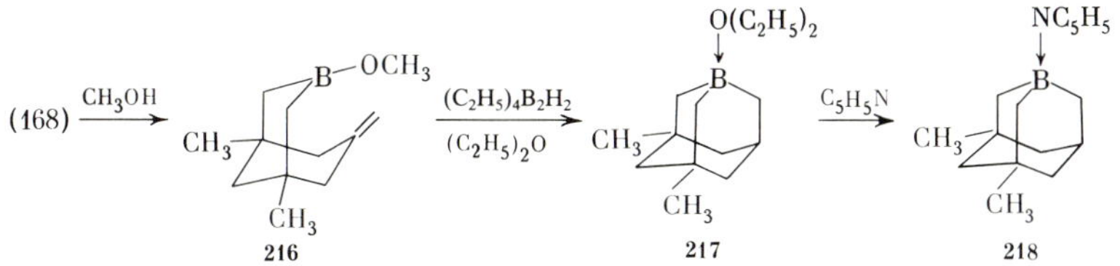

4,4-Dimethyl-1-boraadamantane has been obtained by the hydroboration of 3-methoxy-7-methylene-6,6-dimethyl-3-borabicyclo[3.3.1]nonane with diborane or tetraethyldiborane [230]. This compound had previously been erroneously considered as 4,4-dimethyl-1-boraprotoadamantane [171, 172].

1-Boraadamantane compounds have also been synthesized from compounds formed in the allylboron-acetylene condensation.

3-Methoxy-7-chloromethyl-3-borabicyclo[3.3.1]non-6-ene (*219*), obtained from triallylborane and propargyl chloride [145, 159], reacts with diborane in tetrahydrofuran to add 27% of the boron in the β-position to the CH_2Cl group to form the borane *220*, which eliminates 3-methoxy-7-methylene-3-borabicyclo[3.3.1]nonane (*213*). The latter compound is converted by hydroboration and cyclization to 1-boraadamantane, which crystalizes from the reaction mixture in the form of a complex with tetrahydrofuran (*222*) [160]:

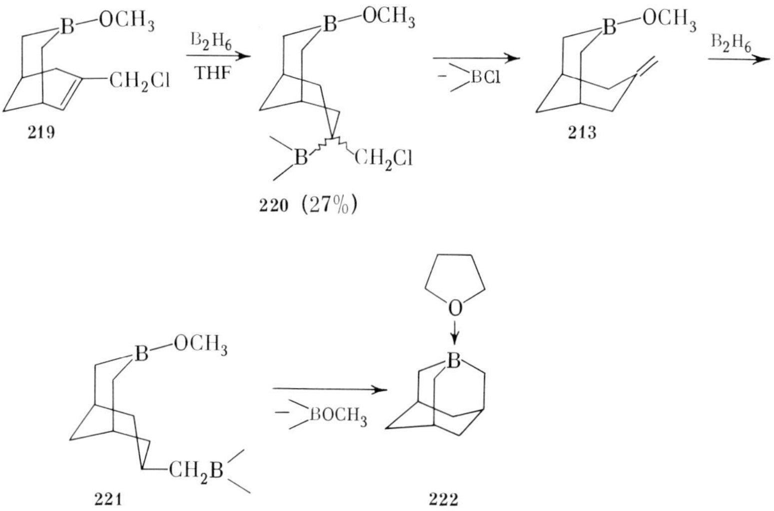

Hydroboration of a mixture of the compounds *219* and *223* (4:1) forms a mixture of the boranes *224–226*. The borane *224* undergoes cyclization to 4-chloro-1-boraadamantane, which is isolated from the reaction mixture in the form of a complex with tetrahydrofuran (*227*) [179]. The borane *225*, through a number of stages indicated in the preceding scheme, is converted to the complex *222*, which can be isolated from the reaction mixture by vacuum distillation [160]:

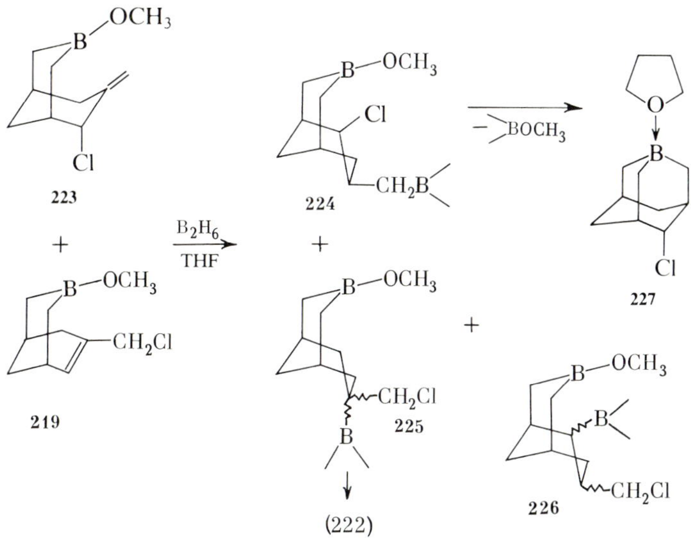

When heated to 120°C, the compound *219* isomerizes to *223*. A 4:1 mixture of *219* and *223* can be considerably enriched in compound *223* (to a ratio of 2:3) by heating. Hydroboration of such a mixture produces the tetrahydrofuran complex of 4-chloro-1-boraadamantane in high yield [238].

A convenient preparative method for the synthesis of 1-boraadamantane is based on the hydroboration of 3-methoxy-7-methoxymethyl-3-borabicyclo[3.3.1]non-6-ene (*228*) [223, 239], which is obtained by the condensation of triallylborane with methyl propargyl ether [223, 224]. The hydroboration is carried out with the use of B_2H_6, $H_3B \cdot THF$, or $H_3B \cdot N(C_2H_5)_3$.

In the hydroboration of *228*, the boron atom adds in both the 6 and 7 positions, with the formation of compounds *229* and *230*. The borane *230* is converted by β-elimination to *213*, which is converted to 1-boraadamantane by further hydroboration. The borane *229* readily isomerizes to *230* and the conversion cycle of the latter is then repeated. It should be emphasized that the isomerization of *229* to *230* proceeds rapidly even at 0°C, particularly with an excess of $H_3B \cdot THF$. The isomerization of organoboron compounds generally proceeds at 100–160°C [240] and takes place under milder conditions (40–60°C) only for some boranes of the steroid type [241].

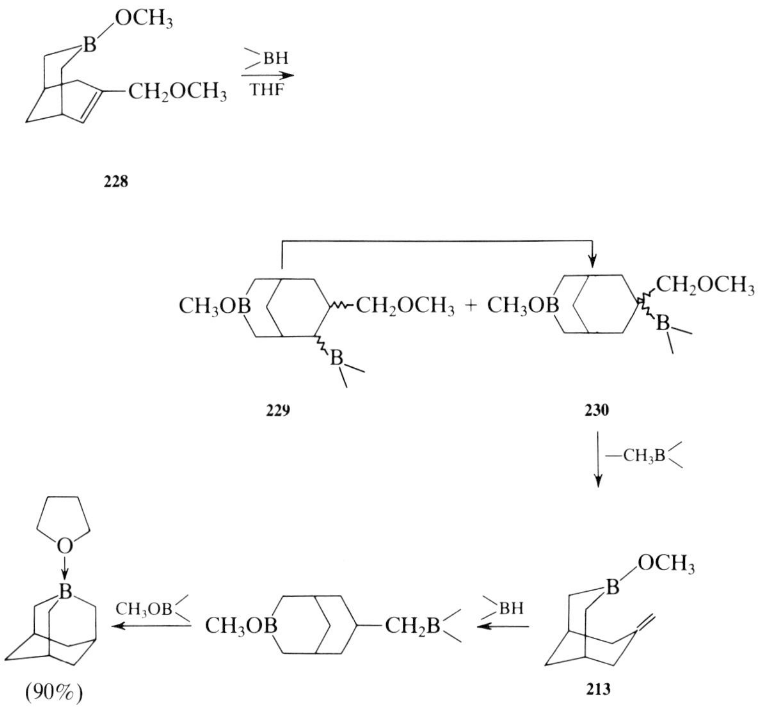

228

229

230

213

(90%)

The observed catalytic effect of compounds with B—H bonds on the rate of isomerization of *229* to *230* is in agreement with a mechanism of "bridge hydrogen tautomerism" [242] and is not explained by an "elimination-addition" mechanism [243]. Free 1-boraadamantane is obtained by heating the etherate *215* under vacuum. In this process, the complex dissociates and 1-bora-adamantane sublimes in crystal form.

1-Boraadamantane is also formed by the hydroboration of 3-methoxy-7-trimethylsilyloxymethyl-3-borabicyclo[3.3.1]non-6-ene and of the cyclic compound obtained by heating 7-dipropyl-boryloxy-3-allyl-3-borabicyclo[3.3.1]non-6-ene [226].

The above method for the synthesis of 1-boraadamantane has been expanded to the preparation of its derivatives.

3-Methoxy-7-methoxymethyl-1,5-dimethyl-3-borabicyclo[3.3.1]-non-6-ene [231], obtained from tri(2-methylallyl)borane and methyl propargyl ether, forms the THF complex of 3,5-dimethyl-1-boraadamantane (*232*) by hydroboration with $H_3B \cdot THF$ [228]:

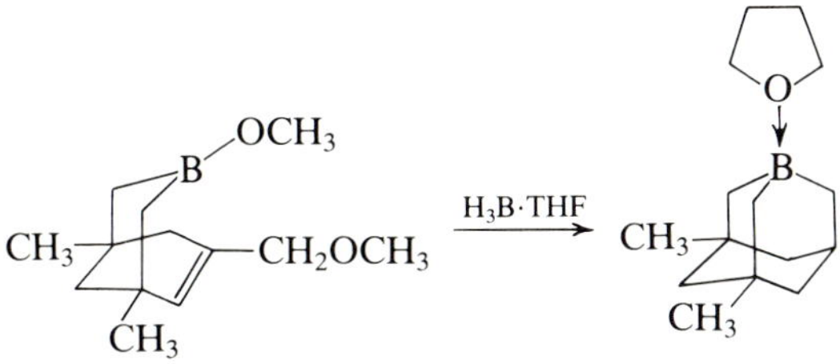

231 **232**

3-Methoxy-7-methoxymethyl-8,9-dimethyl-3-borabicyclo[3.3.1]-non-6-ene (*233*), obtained from tricrotylborane and methyl propargyl ether, reacts with $H_3B \cdot THF$ to form the THF complex of 4,6-dimethyl-1-boraadamantane (*234*) [229].

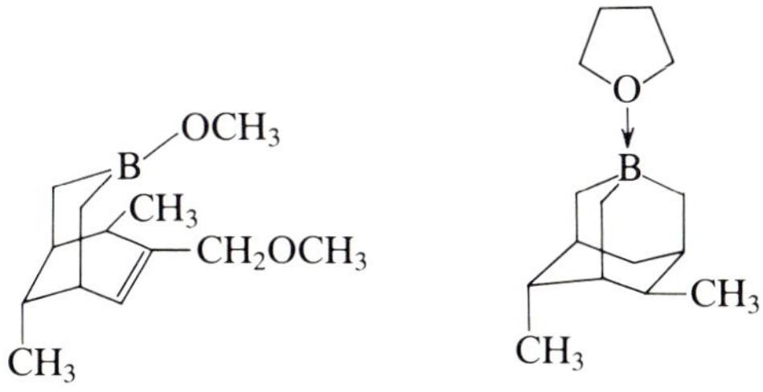

233 **234**

The THF complex of 1-boraadamantane is also obtained by the hydroboration of 3-methoxy-7-tetrahydropyranyloxymethyl-3-bora-bicyclo[3.3.1]non-6-ene with $H_3B \cdot THF$ [225].

5.2 PROPERTIES OF 1-BORAADAMANTANE AND ITS DERIVATIVES

1-Boraadamantane sublimes in the form of well-formed prisms without a sharp melting point [177].

According to NMR spectroscopic data, the HC_2H angles in 1-boraadamantane have a value of 109°54′, while the $C_\alpha BC_\alpha$ angles are close to tetrahedral [185].

5.2.1 Complexing ability

1-Boraadamantane is a unique compound, in which the boron atom is not in a trigonal valence state, as in all of the compounds of trivalent boron known so far, but in a tetrahedral state. This characteristic of 1-boraadamantane is reflected in its chemical properties, particularly in an increased tendency toward complex formation.

The 1-boraadamantane molecule, in which ∠CBC approaches a tetrahedral angle, is ready for complex formation, since its vacant orbital is no longer a pure *p* orbital, as in the trialkylboranes, but an sp^3 orbital. In other words, the energy losses related to rehybridization have already taken place during the synthesis of 1-boraadamantane, and its potential energy is higher than the energy of a trialkylborane by the magnitude of the reorganization energy of the latter. The enthalpy of the reaction of 1-boraadamantane with any ligand is a measure of the actual energy of the coordinate bond, since, in this case, no energy is expended for the compensation of energy losses related to a change in the hybridization of the boron atom, as is the case in the formation of a complex from a normal compound of trivalent boron.

1-Boraadamantane was found to be a unique type of structural model of trivalent boron, permitting an experimental evaluation of the reorganization energy of a trialkylborane during its transition from a planar to a pyramidal configuration.

An investigation of the donor-acceptor interaction in complexes of 1-boraadamantane with diethyl ether and ethyl acetate by means of ^{11}B NMR has made it possible to obtain a value of 6–10 kcal/mole for the reorganization energy [171].

Complexes of 1-boraadamantane with diethyl ether [173], tetrahydrofuran [173, 223], and triethylamine [223] are obtained directly during the synthesis. These complexes can be easily converted to others by exchange reactions with ligands showing a higher donor capacity. In this way, adducts of 1-boraadamantane with different aliphatic amines and heterocyclic bases [178, 223, 239] and pyridinates of 3,5-dimethyl- [228], 4,6-dimethyl- [229], and 4,4-dimethyl-1-boraadamantane [230] have been obtained.

The geometric parameters of complexes of 1-boraadamantane with pyridine and quinoline have been determined by X-ray analysis [244].

3-Butoxy-8-bromomethyl-4,4-dimethyl-3-borabicyclo[4.3.1]decane (*251*, R = C_4H_9) reacts with lithium in ether to form 4,4-dimethyl-3-borahomoadamantane (*255*), the reaction of which with carbon monoxide produces 4,4-dimethylhomoadamantan-3-ol (*356*) [*246*]:

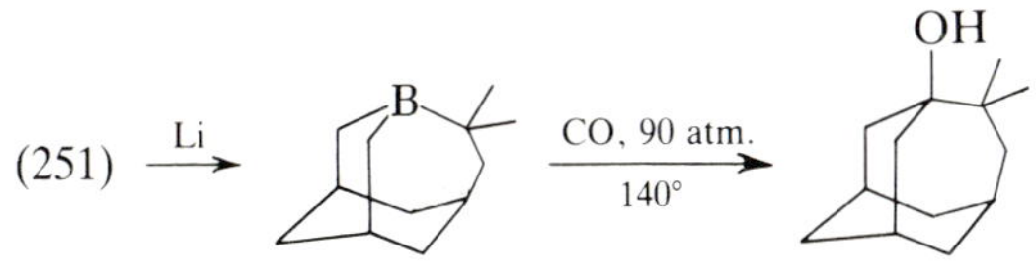

255 **256**

5.2.5 Miscellaneous Reactions

The high complexing capacity of 1-boraadamantane is displayed in its formation of the adducts *257* with carbonyl compounds [248]:

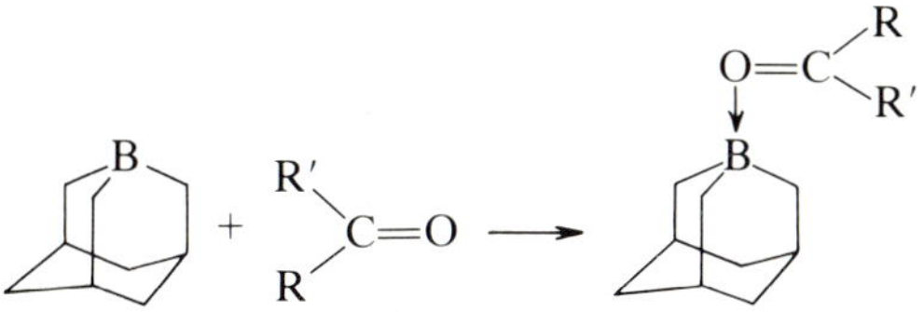

257

Complexes with benzaldehyde, 4-methoxybenzaldehyde, 3,4-dimethoxybenzaldehyde, isobutyraldehyde, and acetone have been described. Upon heating, these adducts rearrange to 5-substituted 4-oxa-3-bora-1,1-bihomoadamantane derivatives, which exist in the form of the dimers *258*:

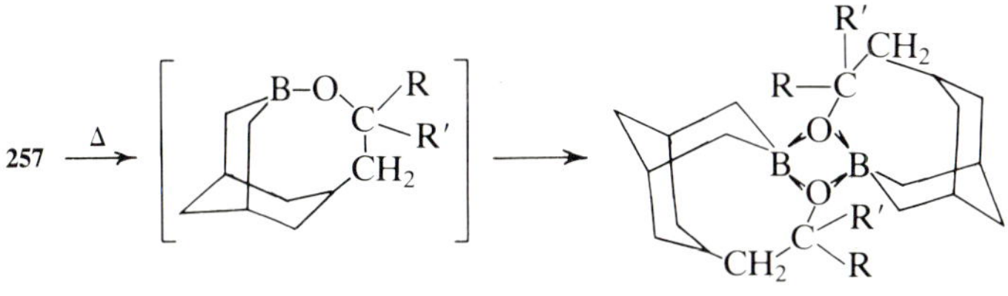

258

The dimers *258*, obtained from 4-methoxybenzaldehyde and 3,4-dimethoxybenzaldehyde, are converted by heating with methanol to

3-methoxy-7α-(2′-methoxy-2′-arylethyl)-3-borabicyclo[3.3.1]-nonanes (*259*), which split off methanol to form 3-methoxy-7α-(2′-arylvinyl)-3-borabicyclo[3.3.1]nonanes (*260*):

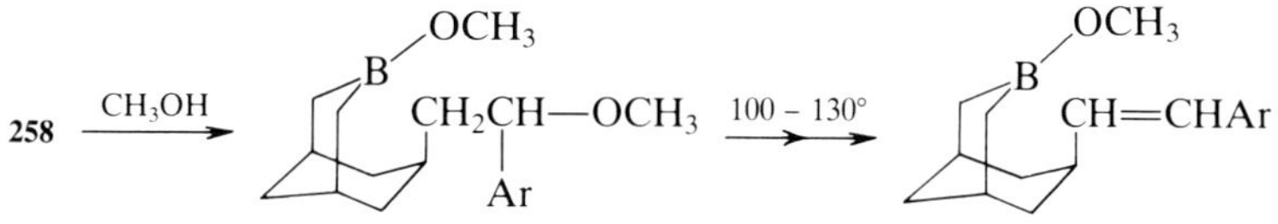

259 **260**

1-Boraadamantane readily reacts with hydroxylamine derivatives and chloramines [249]. The action of hydroxylamine-*O*-sulfonic acid or of *O*-2,4-dinitrophenylhydroxylamine leads to the chelate 3-methoxy-7-aminomethyl-3-borabicyclo[3.3.1]nonane (*261*), the oxidation of which produces *cis*-1,3-bis(hydroxymethyl)-5-aminomethylcyclohexane. With chlorodiethylamine, the chelate 3-chloro-7-diethylaminomethyl-3-borabicyclo[3.3.1]nonane (*262*) is formed, which is converted by treatment with sodium methoxide to 3-methoxy-7-diethylaminomethyl-3-borabicyclo[3.3.1]nonane (*263*):

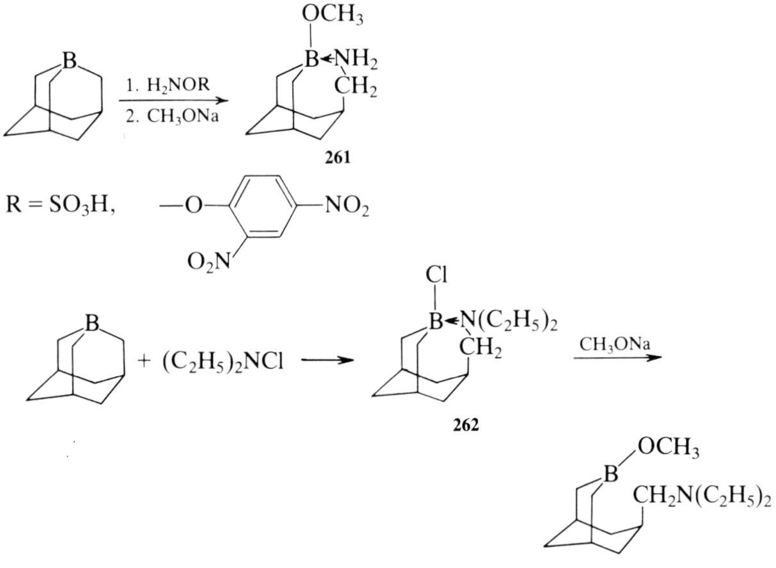

263

The action of alkyldichloramines on 1-boraadamantane forms 3-alkyl-7-dichloroboryl-3-azabicyclo[3.3.1]nonanes (*264*), which furnish 3-alkyl-7-hydroxymethyl-3-azabicyclo[3.3.1]nonanes (*265*) upon oxidation with H_2O_2-NaOH:

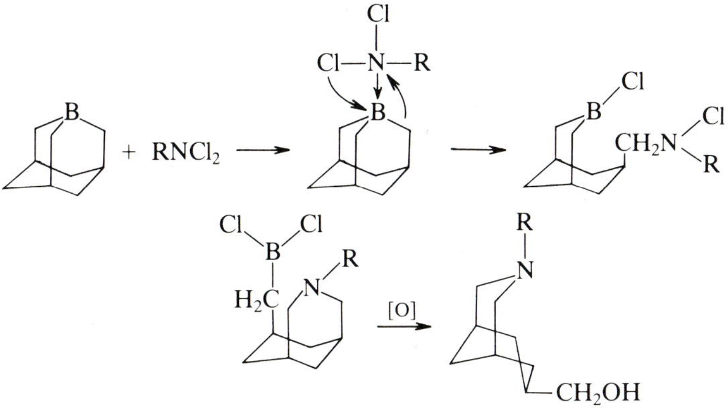

264 **265**

$R = CH_3, n\text{-}C_3H_7,\ n\text{-}C_4H_9$

The reaction of trimethylammonium methylide with the tetrahydrofuran complex of 1-boraadamantane initially forms the betaine *266*, which is converted by heating in dimethoxyethane to a trimethylamine complex of 3-borahomoadamantane (*267*) [250]. The treatment of *267* with boron trifluoride etherate in tetrahydrofuran produces the tetrahydrofuran complex of 3-borahomoadamantane:

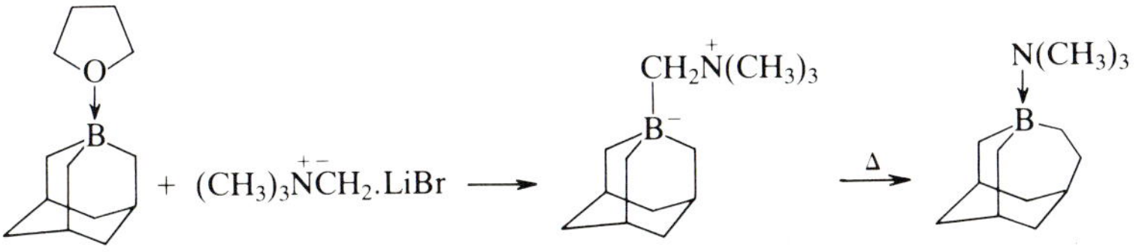

266 **267**

Carbonylation of the latter with carbon monoxide at 120°C and 80 atmospheres forms 3-homoadamantanol.

The *ate* complexes obtained from RLi and 1-boraadamantane or 3,5-dimethyl-1-boraadamantane are converted by the action of acetyl chloride to 3-alkyl-7-methylene-3-borabicyclo[3.3.1]nonane and 1-alkyl-3,5-dimethyl-7-methylene-3-borabicyclo[3.3.1]nonane, respectively. Abstraction of a hydride ion from the 3 position of the 1-boraadamantane system during this reaction has been demonstrated by the formation of 1,3,5-trimethyl-7-methylene-3-borabicyclo[3.3.1]nonane (*269*) from lithium [7-D]-3-5-dimethyl-1-boraadamantane (*268*) [251]:

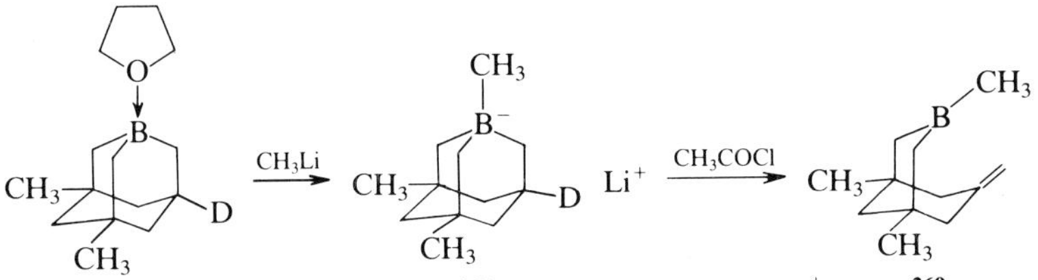

268 **269**

5.2.6 Biological Activity

Polyhedral boron compounds are of interest from not only the chemical but also the biological viewpoint. It has been found [252, 253] that many 1-boraadamantane complexes show a effective therapeutic and prophylactic action against influenza. Complexes with 1-aminoadamantane, remantadine, phenylhydrazine, 4-aminopyridine, ethylenimine, benzimidazole, and other nitrogenous bases show a high antiviral activity in chick embryos and chicks. Some of these complexes protect groups of chicks infected with influenza against death in 75–93% of the cases.

The chemistry of 1-boraadamantane has been discussed in reviews [184, 185].

REFERENCES

1. B. M. Mikhailov, *Organometal. Chem. Rev., A* **8,** 1 (1972).
2. B. M. Mikhailov, *Intra-Science Chem. Rep.* **7,** 191 (1973).
3. B. M. Mikhailov, *Pure Appl. Chem.* **39,** 505 (1974).
4. B. M. Mikhailov, *Usp. Khim.* **45,** 1102 (1976).
5. K. Masao and I. Kunihiko, Japan. Patent 7019 (1954); *C. A.* **50,** 4196 (1956).
6. B. M. Mikhailov and F. B. Tutorskaya, *Dokl. Akad. Nauk SSSR* **123,** 479 (1958).
7. A. V. Topchiev, A. A. Prokhorova, Ya. M. Paushkin and M. V. Kurashev, *Izv. Akad. Nauk SSSR, Otd. Khim. Nauk* **1958,** 370.
8. A. V. Topchiev, Ya. M. Paushkin and A. A. Prokhorova, *Dokl. Akad. Nauk SSSR* **129,** 598 (1959).
9. H. E. Ramsden, Brit. Patent 823,958 (1959); *C. A.* **54,** 17239 (1960).
10. L. I Zakharkin and V. I. Stanko, *Izv. Akad. Nauk SSSR, Otd. Khim. Nauk* **1960,** 1896.

11. B. M. Mikhailov and V. F. Pozdnev, *Izv. Akad. Nauk SSSR, Ser. Khim.* **1964,** 1477.
12. V. S. Bogdanov, Yu. N. Bubnov, M. N. Bochkareva and B. M. Mikhailov, *Dokl. Akad. Nauk SSSR* **201,** 605 (1971).
13. B. M. Mikhailov, Yu. N. Bubnov and A. V. Tsyban, *Izv. Akad. Nauk SSSR, Ser. Khim.* **1975,** 979.
14. B. M. Mikhailov, Yu. N. Bubnov, A. V. Tsyban and K. Base, *Izv. Akad. Nauk SSSR, Ser. Khim.* **1978,** 1586.
15. B. M. Mikhailov, Yu. N. Bubnov and A. V. Tsyban, *Izv. Akad. Nauk SSSR, Ser. Khim.* **1975,** 483.
16. G. E. Kacheishvili, N. I. Pirtskhalava and B. D. Dzhioshvili, *Zh. Obshch. Khim.* **43,** 2402 (1973).
17. G. E. Kacheishvili, N. I. Pirtskhalava and B. D. Dzhioshvili, *Soobshch. Akad. Nauk Gruz. SSR* **73,** 69 (1974).
18. A. T. Po, C.-H. Pan and C.-H Kao, *Hua Hsueh Hsueh Pao* **30,** 79 (1964).
19. B. M. Mikhailov, V. N. Smirnov and O. D. Ryazanova, *Dokl. Akad. Nauk SSSR* **204,** 612 (1972).
20. D. S. Sethi, G. C. Joshi and D. Devaprabhakara, *Can. J. Chem.* **47,** 1083 (1969).
21. D. S. Sethi, G. C. Joshi and D. Devaprabhakara, *Can. J. Chem.* **46,** 2632 (1968).
22. D. S. Sethi, G. C. Joshi and D. Devaprabhakara, *Indian J. Chem.* **6,** 402 (1068).
23. D. Devaprabhakara and P. D. Gardner, *J. Am. Chem. Soc.* **85,** 1458 (1963).
24. B. M. Mikhailov and K. L. Cherkasova, *Zh. Obshch. Khim.* **41,** 2578 (1971).
25. B. M. Mikhailov, L. S. Vasil'ev and E. N. Safonova, *Dokl. Akad. Nauk SSSR* **630** (1962).
26. B. M. Mikhailov and A. Ya. Bezmenov, *Izv. Akad. Nauk SSSR, Ser. Khim.* **1965,** 931.
27. B. M. Mikhailov, A. Ya. Bezmenov and L. S. Vasil'ev, *Dokl. Akad. Nauk SSSR* **167,** 590 (1966).
28. B. M. Mikhailov, A. Ya. Bezmenov, L. S. Vasil'ev and V. G. Kiselev, *Dokl. Akad. Nauk SSSR* **155,** 141 (1964).
29. A. Ya. Bezmenov, L. S. Vasil'ev and B. M. Mikhailov, *Izv. Akad. Nauk SSSR, Ser. Khim.* **1965,** 2111.
30. K. G. Hancock and J. D. Kramer, *J. Am. Chem. Soc.* **95,** 6463 (1973).
31. K. G. Hancock and J. D. Kramer, *J. Am. Chem. Soc.* **95,** 3426 (1973).
32. K. G. Hancock and J. D. Kramer, *J. Organometal. Chem.* **64,** C29 (1974).
33. B. M. Mikhailov, Yu. N. Bubnov and V. A. Vaver, *Dokl. Akad. Nauk SSSR* **126,** 575 (1959).
34. B. M. Mikhailov, V. S. Bogdanov, G. V. Lagodzinskaya and V. F. Pozdnev, *Izv. Akad. Nauk SSSR, Ser. Khim.* **1966,** 386.
35. V. S. Bogdanov, G. V. Lagodzinskaya, V. F. Posdnev and B. M. Mikhailov, *Izv. Akad. Nauk SSSR, Ser. Khim.* **1966,** 944.
36. V. S. Bogdanov, V. F. Pozdnev, G. V. Lagodzinskaya and B. M. Mikhailov, *Teor. Eksp. Khim.* **3,** 488 (1967).
37. V. S. Bogdanov, V. F. Posdnev, Yu. N. Bubnov and B. M. Mikhailov, *Dokl. Akad. Nauk SSSR* **193,** 586 (1970).
38. B. M. Mikhailov, V. N. Smirnov, O. D. Smirnova and V. S. Bogdanov, *Zh. Obshch. Khim.* **45,** 51 (1975).
39. B. M. Mikhailov, V. V. Negrebetskii, V. S. Bogdanov, A. V. Kessenikh, Yu. N. Bubnov, T. K. Baryshnikova and V. N. Smirnov, *Zh. Obshch. Khim.* **44,** 1878 (1974).
40. B. M. Mikhailov, Yu. N. Bubnov and V. S. Bogdanov, *Zh. Obshch. Khim.* **45,** 333 (1975).

41. B. M. Mikhailov, Yu. N. Bubnov and V. S. Bogdanov, *Zh. Obshch. Khim.* **45,** 324 (1975).
42. A. V. Topchiev, Ya. M. Paushkin, A. A. Prokhorova and M. V. Kurashev, *Dokl. Akad. Nauk SSSR* **128,** 110 (1959).
43. V. S. Bogdanov, T. K. Baryshnikova, V. G. Kiselev and B. M. Mikhailov, *Zh. Obshch. Khim.* **41,** 1533 (1971).
44. B. M. Mikhailov and F. B. Tutorskaya, *Izv. Akad. Nauk SSSR, Otd. Khim. Nauk* **1961,** 1158.
45. A. V. Topchiev, Ya. M. Paushkin, A. A. Prokhorova and M. V. Kurashev, *Dokl. Akad. Nauk SSSR* **134,** 364 (1960).
46. J. P. Laurent and R. Haran, *Bull. Soc. Chim. France* **1964,** 2448.
47. B. M. Mikhailov, V. F. Pozdnev and V. G. Kiselev, *Dokl. Akad. Nauk SSSR* **151,** 577 (1963).
48. B. M. Mikhailov, G. Yu. Pek, V. S. Bogdanov and V. F. Pozdnev, *Izv. Akad. Nauk SSSR, Ser. Khim.* **1966,** 1117.
49. B. M. Mikhailov and F. B. Tutorskaya, *Izv. Akad. Nauk SSSR, Otd. Khim. Nauk* **1959,** 1128.
50. B. M. Mikhailov and V. F. Pozdnev, *Izv. Akad. Nauk SSSR, Otd. Khim. Nauk* **1962** 1475.
51. B. M. Mikhailov and V. F. Pozdnev, *Izv. Akad. Nauk SSSR, Otd. Khim. Nauk* **1962,** 1698.
52. B. M. Mikhailov and V. F. Pozdnev, *Izv. Akad. Nauk SSSR, Otd. Khim. Nauk* **1962,** 1861.
53. B. M. Mikhailov and V. F. Pozdnev, in the collection: Journal of Organic Chemistry. *Problems of Organic Synthesis*, Leningrad, Nauka (1965), p. 220.
54. L. I. Zakharkin and A. I. Kovredov, *Izv. Akad. Nauk SSSR, Otd. Khim. Nauk* **1962,** 1564.
55. N. N. Greenwood, J. H. Morris and J. C. Wright, *J. Chem. Soc.* **1964,** 4753.
56. B. M. Mikhailov and Yu. N. Bubnov, *Izv. Akad. Nauk SSSR, Ser. Khim.* **1964,** 1874.
57. B. M. Mikhailov, G. S. Ter-Sarkisyan and N. A. Nikolaeva, *Izv. Akad. Nauk SSSR, Ser. Khim.* **1968,** 1655.
58. G. S. Ter-Sarkisyan, N. A. Nikolaeva and B. M. Mikhailov, *Izv. Akad. Nauk SSSR, Ser. Khim.* **1970,** 876.
59. G. S. Ter-Sarkisyan, N. A. Nikolaeva and B. M Mikhailov, *Izv. Akad. Nauk SSSR, Ser. Khim.* **1968,** 2516.
60. I. Mehrotra and D. Devaprabhakara, *J. Organometal. Chem.* **51,** 93 (1973).
61. A. A. Akhrem, I. S. Levina, Yu. A. Titov, Yu. N. Bubnov and B. M. Mikhailov, *Izv. Akad. Nauk SSSR, Ser. Khim.* **1972,** 1639.
62. B. M. Mikhailov and G. S. Ter-Sarkisyan, *Izv. Akad. Nauk SSSR, Ser. Khim.* **1966,** 380.
63. B. M. Mikhailov, G. S. Ter- Sarkisyan and N. A. Nikolaeva, *Izv. Akad. Nauk SSSR, Ser. Khim.* **1968,** 341.
64. G. S. Ter- Sarkisyan, N. A. Nikolaeva, V. G. Kiselev and B. M. Mikhailov, *Zh. Obshch. Khim.* **41,** 152 (1971).
65. S. Goodwin and B. Witkop, *J. Am. Chem. Soc.* **79,** 179 (1957).
66. G. S. Ter-Sarkisyan, N. A. Nikolaeva and B. M. Mikhailov, *Zh. Obshch. Khim.* **41,** 1721 (1971).
67. Yu. N. Bubnov and B. M. Mikhailov, *Izv. Akad. Nauk SSSR, Ser. Khim.* **1967,** 472.
68. Yu. N. Bubnov, V. S. Bogdanov and B. M. Mikhailov, *Zh. Obshch. Khim.* **38,** 260 (1968).

69. A. Meller and W. Gerger, *Monatsh.* **105,** 684 (1974).
70. B. M. Mikhailov and Yu. N. Bubnov, *Tetrahedron Letters* **1971,** 2127.
71. B. M. Mikhailov and Yu. N. Bubnov, *Zh. Obshch. Khim* **41,** 2039 (1971).
72. B. M. Mikhailov and A. N. Blokhina, *Izv. Akad. Nauk SSSR, Otd. Khim. Nauk* **1962,** 1973.
73. B. M. Mikhailov, Yu. N. Bubnov, O. A. Nesmeyanova, V. G. Kiselev, T. Yu. Rudashevskaya and B. A. Kazansky, *Tetrahedron Letters* **1972,** 4627.
74. Yu. N. Bubnov, O. A. Nesmeyanova, T. Yu. Rudashevskaya, B. M. Mikhailov and B. A. Kazansky, *Tetrahderon Letters* **1971,** 2153.
75. Yu. N. Bubnov, O. A. Nesmeyanova, T. Yu. Rudashevskaya, B. M. Mikhailov and B. A. Kazansky, *Zh. Obshch. Khim.* **43,** 127 (1973).
76. Yu. N. Bubnov, O. A. Nesmeyanova, T. Yu. Rudashevskaya, B. M. Mikhailov and B. A. Kazanskii, *Zh. Obshch. Khim.* **43,** 135 (1973).
77. Yu. N. Bubnov, B. A. Kazansky, O. A. Nesmeyanova, T. Yu. Rudashevskaya and B. M. Mikhailov, *Izv. Akad. Nauk SSSR, Ser. Khim.* **1977,** 2545.
78. V. V. Negrebetskii, V. S. Bogdanov, A. V. Kessenikh, P. V. Petrovskii, Yu. N. Bubnov and B. M. Mikhailov, *Zh. Obshch. Khim.* **44,** 1882 (1974).
79. B. A. Kazansky, Yu. N. Bubnov, S. V. Zotova, N. M. Abramova. V. G. Kiselev and B. M. Mikhailov, *Tetrahedron Letters* **1974,** 567.
80. P. F. Winternitz and A. A. Carotti, *J. Am. Chem. Soc.* **82,** 2430 (1960).
81. P. F. Winternitz and A. A. Carotti, U. S. Patent 3,002,350 (1962); *C. A.* **57,** 2488 (1962).
82. A. V. Topchiev, A. A. Prokhorova, Ya. M. Paushkin and M. V. Kurashev, *Dokl. Akad. Nauk SSSR* **131,** 105 (1960).
83. R. J. Thompson and J. C. Davis, Jr., *Inorg. Chem.* **4,** 1464 (1965).
84. K.-H. Thiele and P. Zdunneck, *J. Organometal. Chem.* **4,** 10 (1965).
85. K.-H. Thiele, G. Ingelhardt, J. Köhler and M. Arnstedt, *J. Organometal. Chem.* **9,** 385 (1967).
86. D. Abenhaim, E. Henry-Basch. and P. Freon, *Bull. Soc. Chim. France* **1969,** 4038.
87. K.-H. Thiele and J. Köhler, *J. Organometal. Chem.* **7,** 365 (1967).
88. K.-H. Thiele and S. Wagner, *J. Organometal. Chem.* **20,** P25 (1969).
89. B. M. Mikhailov and M. E. Gurskii, *Izv. Akad. Nauk SSSR, Ser. Khim.* **1973,** 2644.
90. A. Stefani, *Helv. Chim. Acta* **56,** 1192 (1973).
91. J. Blais, J. Soulié and P. Cadiot, *Compt. Rend. C* **271,** 589 (1970).
92. E. Favre and M. Gaudemar, *Compt. Rend. C* **262,** 1332 (1966).
93. S. J. Groszos, U. S. Patent 2,884,441 (1959); *C. A.* **54,** 298 (1960).
94. R. H. Fish, *J. Am. Chem. Soc.* **90,** 4435 (1968).
95. B. M. Mikhailov and F. B. Tutorskaya, *Zh. Obshch. Khim.* **32,** 833 (1962).
96. B. M. Mikhailov and F. B. Tutorskaya, *Izv. Akad. Nauk SSSR, Otd. Khim. Nauk* **1961,** 2068.
97. F. E. Brinckman and F. G. A. Stone, *J. Am. Chem. Soc.* **82,** 6218 (1960).
98. E. Rothstein and R. Soville, *J. Chem. Soc.* **1952,** 2987.
99. B. M. Mikhailov and Yu. N. Bubnov, *Zh. Obshch. Khim.* **29,** 1648 (1959).
100. B. M. Mikhailov, T. K. Kozminskaya, N. S. Fedotov and V. A. Dorokhov, *Dokl. Akad. Nauk SSSR* **127,** 1029 (1959).
101. B. M. Mikhailov, *Usp. Khim.* **37,** 2121 (1968).
102. B. M. Mikhailov, A. N. Blokhina and V. F. Pozdnev, *Izv. Akad. Nauk SSSR, Ser. Khim.* **1965,** 197.
103. B. M. Mikhailov and E. N. Safonova, *Izv. Akad. Nauk SSSR, Ser. Khim.* **1965,** 1487.

104. B. M. Mikhailov and L. S. Vasil'ev, *Izv. Akad. Nauk SSSR, Otd. Khim. Nauk* **1961,** 531.
105. B. M. Mikhailov, P. M. Aronovich and L. V. Tarasova, *Zh. Obshch. Khim.* **30,** 3624 (1960).
106. B. M. Mikhailov, Yu.N. Bubnov and V. G. Kiselev, *Izv. Akad. Nauk SSSR, Ser. Khim.* **1965,** 68.
107. B. M. Mikhailov and P. M. Aronovich, *Izv. Akad. Nauk SSSR, Otd. Khim. Nauk* **1961,** 927.
108. B. M. Mikhailov and Yu. N. Bubnov, *Izv. Akad. Nauk SSSR, Ser. Khim.* **1964,** 2170.
109. E. Favre and M. Gaudemar, *Compt. Rend., C,* **263,** 1543 (1966).
110. J. Blais, A. L'Honore, J. Soulié and P. Cadiot, *J. Organometal. Chem.* **78,** 323 (1974).
111. C. Counkler, *Ber.* **9,** 485 (1876).
112. C. Counkler, *J. Prakt. Chem.* **18,** 371 (1878).
113. W. Gerrard, M. F. Lappert, and H. B. Silver, *J. Chem. Soc.* **1956,** 3285.
114. S. Z. Haider, M. H. Khundkar and M. Siddiquallah, *J. Appl. Chem.* **4,** 93 (1954).
115. E. P. Irany, U. S. Patent 2,523,433 (1950); *C. A.* **45,** 6654 (1951).
116. L. H. Thomas, *J. Chem. Soc.* **1946,** 820.
117. A. Scattergood, W. H. Miller and J. Gammon, *J. Am. Chem. Soc.* **67,** 2150 (1945).
118. W. Gerrard, M. F. Lappert and H. B. Silver, *J. Chem. Soc.* **1957,** 1647.
119. J. E. Dubois, R. Luft and F. Weck, *Ann. Univ. Saraviensis* **1,** 157 (1952); *C. A.* **47,** 9920 (1953).
120. W. Gerrard, M. F. Lappert and H. B. Silver, *J. Chem. Soc.* **1956,** 4987.
121. W. Gerrard, M. F. Lappert and R. Schafferman, *J. Chem. Soc.* **1957,** 3828.
122. B. M. Mikhailov, V. A. Dorokhov and N. V. Mostovoi, in the collection: Journal of General Chemistry. *Problems of Organic Synthesis,* Leningrad, Nauka (1965), p. 228.
123. T. P. Onak, H. Landesman, R. E. Williams and I. Shapiro, *J. Phys. Chem.* **63,** 1533 (1959).
124. C. D. Good and D. M. Ritter, *J. Am. Chem. Soc.* **84,** 1162 (1962).
125. H. Landesman and R. E. Williams, *J. Am. Chem. Soc.* **83,** 2663 (1961).
126. C. Counkler, *Chem. Ber.* **10,** 1655 (1877).
127. M. F. Frazer, W. Gerrard and M. F. Lappert, *J. Chem. Soc.* **1957,** 739.
128. I. Scriabine and A. Peyrolade, West German Patent 1,014,984 (1957); *C. A.* **54,** 2251 (1960).
129. British Patent 787,615 (1957); *C. A.* **52,** 10171 (1958).
130. French Patent 1,124,561 (1956); *C. A.* **53,** 15006 (1959).
131. V. K. Kuskov and B. M. Sheiman, *Dokl. Akad. Nauk SSSR* **106,** 479 (1956).
132. V. K. Kuskov, B. M. Sheiman and Z. I. Maksimova, *Zh. Obshch. Khim.* **27,** 1454 (1956).
133. H. G. Kuivila, S. C. Slack and P. K. Siliteri, *J. Am. Chem. Soc.* **73,** 123 (1951).
134. S. A. Ballard, U. S. Patent 2,431,224 (1947); *C. A.* **42,** 1960 (1948).
135. E. C. Britton, C. W. Davis and F. L. Taylor, U. S. Patent 2,160,942 (1939): *C. A.* **33,** 7441 (1939).
136. H. S. Rothrock, U. S. Patent 2,276,094 (1942); *C. A.* **36,** 4630 (1940).
137. B. M. Mikhailov and Yu. N. Bubnov, *Izv. Akad. Nauk SSSR, Ser. Khim.* **1965,** 1310.
138. B. M. Mikhailov, Yu. N. Bubnov, S. A. Korobeinikova and S. I. Frolov, *Izv. Akad. Nauk SSSR, Ser. Khim.* **1968,** 1923.

139. S. I. Frolov, Yu. N. Bubnov and B. M. Mikhailov, *Izv. Akad. Nauk SSSR, Ser. Khim.* **1969**, 1996.
140. Yu. N. Bubnov, S. I. Frolov, V. G. Kiselev, V. S. Bogdanov and B. M. Mikhailov, *Zh. Obshch. Khim.* **40,** 1311 (1970).
141. Yu. N. Bubnov, S. I. Frolov, V. G. Kiselev, V. S. Bogdanov and B. M. Mikhailov, *Organometal. Chem. Synth.* **1,** 37 (1970).
142. B. M. Mikhailov and T. K. Baryshnikova, *Zh. Obshch. Khim.* **41,** 1303 (1971).
143. B. M. Mikhailov and K. L. Cherkasova, *Izv. Akad. Nauk SSSR, Ser. Khim.* **1971,** 1244.
144. B. M. Mikhailov and K. L. Cherkasova, *Zh. Obshch. Khim.* **42,** 138 (1972).
145. B. M. Mikhailov, B. I. Bryantsev and T. K. Kozminskaya, *Zh. Obshch. Khim.* **203,** 837 (1972).
146. B. M. Mikhailov and B. I. Bryantsev, *Zh. Obshch. Khim.* **43,** 1102 (1973).
147. B. M. Mikhailov, Yu. N. Bubnov and S. A. Korobeinikova, *Izv. Akad. Nauk. SSSR, Ser. Khim.* **1969,** 2465.
148. B. M. Mikhailov, Yu. N Bubnov and S. A. Korobeinikova, *Izv. Akad. Nauk SSSR, Ser. Khim.* **1970,** 2631.
149. B. M. Mikhailov, Yu. N. Bubnov and S. A. Korobeinikova, *J. Prakt. Chem.* **312,** 998 (1970).
150. B. M. Mikhailov, Yu. N. Bubnov, S. A. Korobeinikova and S. I. Frolov, *J. Organometal. Chem.* **27,** 165 (1971).
151. Yu. N. Bubnov, S. A. Korobeinikova, G. V. Isagulyants and B. M. Mikhailov, *Izv. Akad. Nauk SSSR, Ser. Khim.* **1970,** 2023.
152. B. M. Mikhailov, Yu. N. Bubnov and M. Sh. Grigoryan, *Zh. Obshch. Khim.* **44,** 2466 (1974).
153. B. M. Mikhailov, Yu. N. Bubnov, S. A. Korobeinikova and S. I. Frolov, *Izv. Akad. Nauk SSSR, Ser. Khim.* **1969,** 1631.
154. S. A. Korobeinikova, Yu. N. Bubnov and B. M. Mikhailov, *Izv. Akad. Nauk SSSR, Ser. Khim.* **1969,** 2212.
155. B. M. Mikhailov, Yu. N. Bubnov and S. I. Frolov, *Izv. Akad. Nauk SSSR, Ser. Khim.* **1967,** 2290.
156. Yu. N Bubnov, S. I. Frolov, V. G. Kiselev and B. M. Mikhailov, *Zh. Obshch. Khim.* **40,** 1316 (1970).
157. B. M. Mikhailov, Yu. N. Bubnov, S. A. Korobeinikova and V. S. Bogdanov, *Zh. Obshch. Khim.* **40,** 1321 (1970).
158. B. M. Mikhailov and K. L. Cherkasova, *Zh. Obshch. Khim.* **42,** 1744 (1972).
159. B. M. Mikhailov, B. I. Bryantsev and T. K. Kozminskaya, *Zh. Obshch. Khim.* **43,** 1108 (1973).
160. B. M. Mikhailov and K. L. Cherkasova, *Izv. Akad. Nauk SSSR, Ser. Khim.* **1976,** 2056.
161. B. M. Mikhailov, T. K. Kozminskaya and B. I. Bryantsev, *Zh. Obshch. Khim.* **46,** 87 (1976).
162. B. M. Mikhailov, Yu. N. Bubnov and M. Sh. Grigoryan, *Izv. Akad. Nauk SSSR, Ser. Khim.* **1971,** 1842.
163. B. M. Mikhailov, Yu. N. Bubnov and M. Sh. Grigoryan, *Zh. Obshch. Khim.* **42,** 1738 (1972).
164. B. M. Mikhailov, Yu. N. Bubnov, M. Sh. Grigoryan and V. S. Bogdanov, *Zh. Obshch. Khim.* **44,** 2715 (1974).
165. B. M. Mikhailov, Yu. N. Bubnov and S. I. Frolov, USSR Author's Certificate 431,152 (1972); *Byull. Izobret.* **1974,** No. 21, 81.
166. B. M. Mikhailov, Yu. N. Bubnov and S. . Frolov, USSR Author's Certificate 469,679 (1972); *Byull. Lzobret.* **1975,** No. 17, 62.

167. B. M. Mikhailov, Yu. N. Bubnov and S. I. Frolov, USSR Author's Certificate 466,204 (1972); *Byull. Izobret.* **1975,** No. 13, 50.
168. B. M. Mikhailov, Yu. N. Bubnov and S. I. Frolov, *Izv. Akad. Nauk SSSR, Ser. Khim.* **1981,** 2114.
169. B. M. Mikhailov, Yu. N. Bubnov and M. Sh. Grigoryan, *Izv. Akad. Nauk SSSR, Ser. Khim.* **1973,** 2399.
170. B. M. Mikhailov and V. N. Smirnov, *Dokl. Akad. Nauk SSSR* **193,** 1311 (1970).
171. B. M. Mikhailov and V. N. Smirnov, *Izv. Akad. Nauk SSSR, Ser. Khim.* **1974,** 1137.
172. B. M. Mikhailov, V. N. Smirnov and E. P. Prokof'ev, *Dokl. Akad. Nauk SSSR* **206,** 125 (1972).
173. B. M. Mikhailov, V. N. Smirnov and V. A. Kasparov, *Izv. Akad. Nauk SSSR, Ser. Khim.* **1976,** 2302.
174. V. S. Bogdanov, Yu. N. Bubnov, S. I. Frolov and B. M. Mikhailov, *Izv. Akad. Nauk SSSR, Ser. Khim.* **1968,** 307.
175. Yu. N. Bubnov, S. I. Frolov and B. M. Mikhailov, *Izv. Akad. Nauk SSSR, Ser. Khim.* **1968,** 824.
176. B. M. Mikhailov, L. S. Vasil'ev and A. Ya Bezmenov, *Izv. Akad. Nauk SSSR, Ser. Khim.* **1965,** 712.
177. B. M. Mikhailov and V. N. Smirnov, *Izv. Akad. Nauk SSSR, Ser. Khim.* **1973,** 2165.
178. B. M. Mikhailov and V. N. Smirnov, USSR Author's Certificate 466,237 (1973); *Byull. Izobret.* **1975,** No. 13 (1975).
179. B. M. Mikhailov and K. L. Cherkasova, Abstracts of Papers of the Eleventh Mendeleev Congress on General and Applied Chemistry (Alma-Ata), Moscow, Nauka (1975), p. 100.
180. M. E. Hillman, *J. Am. Chem. Soc.* **84,** 4715 (1962).
181. M. E. Hillman, *J. Am. Chem. Soc.* **85,** 982 (1963).
182. M. E. Hillman, *J. Am. Chem. Soc.* **85,** 1626 (1963).
183. A. Pelter, M. G. Hutchings and K. Smith, *Chem. Comm.* **1971,** 1048.
184. B. M. Mikhailov, *Pure Appl. Chem.* **52,** 691 (1980).
185. B. M. Mikhailov, *Soviet Scientific Reviews, Sec. B. Chem. Rev., Vol. 2,* **(1980),** p. 283.
186. B. M. Mikhailov, Yu. N. Bubnov and A. V. Tsyban, *J. Organomet. Chem.* **154,** 113 (1978).
187. B. M. Mikhailov, Yu. N. Bubnov and A. V. Tsyban, *Izv. Akad. Nauk SSSR, Ser. Khim.* **1978,** 1594.
188. N. M. D. Brown, F. Davidson and J. W. Wilson, *J. Organomet. Chem.* **209,** 1 (1981).
189. L. Chevolot, J. Soulié and P. Cadiot, *Tetrahedron Lett.* **1974,** 3435.
190. G. W. Kramer and H. C. Brown, *J. Organomet. Chem.* **132,** 9 (1977).
191. H. C. Brown, R. Liotta and G. W. Kramer, *J. Am. Chem. Soc.* **101,** 2966 (1979).
192. V. V. R. Rao, S. K. Agarwal, D. Devaprabhakara and S. Shandrasekaran, *Synth. Comm.* **9,** 437 (1979).
193. I. Menrotra, M. M. Bhagwat and D. Devaprabhakara, *Proc. Indian Acad. Sci., Sec. A* **86A,** 15 (1977).
194. H. C. Brown, R. Liotta and G. W. Kramer, *J. Org. Chem.* **43,** 1058 (1978).
195. L. C. Costa and G. M. Whitesides, *J. Am. Chem. Soc.* **99,** 2390 (1977).
196. W. Brüser, S. Schroder and H. Wittke, *Z. Anorg. Allg. Chem.* **421,** 89 (1976).
197. Y. Yamamoto, H. Yatagai, Y. Naruto, K. Maruyama and T. Okamoto, *Tetrahedron Lett.* **21,** 3599 (1980).

198. B. M. Mikhailov, Yu. N. Bubnov and A. V. Tsyban, *Izv. Akad. Nauk. SSSR, Ser. Khim.* **1978,** 1892.
199. B. M. Mikhailov, Yu. N. Bubnov, A. V. Tsyban and M. Sh. Grigoryan, *J. Organomet. Chem.* **154,** 131 (1978).
200. G. W. Kramer and H. C. Brown, *J. Org. Chem.* **42,** 2292 (1977).
201. M. Yamagushi and T. Mukaiyama, *Chem. Lett.* **1980,** 993.
202. Y. Yamamoto, H. Yatagai and K. Maruyama, *J. Am. Chem. Soc.* **103,** 3229 (1981).
203. Y. Yamamoto and K. Maruyama, *J. Am. Chem. Soc.* **103,** 1969 (1981).
204. Y. Yamamoto, H. Yatagai and K. Maruyama, *J. Chem. Soc., Chem. Comm.* **1980,** 1072.
205. Yu. N. Bubnov, A. V. Tsyban and B. M. Mikhailov, *Izv. Akad. Nauk SSSR, Ser. Khim.* **1976,** 2842.
206. Y. Yamamoto and K. Maruyama, *J. Am. Chem. Soc.* **100,** 6282 (1978).
207. Y. Yamamoto, H. Yatagai and K. Maruyama, *Chem. Lett.* **1979,** 385.
208. Y. Yamamoto and K. Maruyama, *J. Am. Chem. Soc.* **103,** 1969 (1981).
209. Y. Yamamoto, H. Yatagai and K. Maruyama, *J. Chem. Soc., Chem. Comm.* **1979,** 157.
210. R. W. Hoffmann, G. Feussner, H.-J. Zeiss and S. Schulz, *J. Organomet. Chem.* **187,** 321 (1980).
211. T. Herold and R. W. Hoffmann, *Angew. Chem.* **90,** 823 (1978).
212. T. Herold, U. Schrott and R. W. Hoffmann, *Chem. Ber.* **114,** 359 (1981).
213. R. W. Hoffmann and T. Herold, *Chem. Ber.* **114,** 375 (1981).
214. R. W. Hoffmann and H.-J. Zeiss, *J. Org. Chem.* **46,** 1309 (1981).
215. R. W. Hoffmann and H.-J. Zeiss, *Angew. Chem.* **91,** 329 (1979).
216. R. W. Hoffmann and U. Weidmann, *J. Organomet. Chem.* **195,** 137 (1980).
217. D. J. S. Tsai and D. S. Matteson, *Tetrahedron Lett.* **22,** 2751 (1981).
218. B. F. Nesbitt, P. S. Beever, R. A. Cole, B. Lester and R. G. Poppi, *Tetrahedron Lett.* **1973,** 4669.
219. R. W. Hoffmann and W. Ladner, *Tetrahedron Lett.* **1979,** 4653.
220. Yu. N. Bubnov, M. Sh. Grigoryan, A. V. Tsyban and B. M. Mikhailov, *Synthesis* **1980,** 902.
221. Yu. N. Bubnov, A. V. Tsyban and B. M. Mikhailov, *Synthesis* **1980,** 904.
222. S. V. Ponomarev, Yu. N. Bubnov and B. M. Mikhailov, *Izv. Akad. Nauk SSSR, Ser. Khim.* **1981,** 1859.
223. B. M. Mikhailov and T. K. Baryshnikova, *Dokl. Akad. Nauk SSSR* **243,** 929 (1978).
224. B. M. Mikhailov and T. K. Baryshnikova, *Izv. Akad. Nauk SSSR, Ser. Khim.* **1979,** 2541.
225. B. M. Mikhailov and K. L. Cherkasova, *Izv. Akad. Nauk SSSR, Ser. Khim.* **1979,** 2729.
226. B. M. Mikhailov and T. K. Baryshnikova, *J. Organomet. Chem.* **219,** 295 (1981).
227. B. M. Mikhailov and M. E. Gurskii, *Izv. Akad. Nauk SSSR, Ser. Khim.* **1979,** 2335.
228. B. M. Mikhailov, T. V. Potapova and A. S. Shashkov, *Izv. Akad. Nauk SSSR, Ser. Khim.* **1979,** 2734.
229. B. M. Mikhailov, M. E. Gurskii and A. S. Shashkov, *Izv. Akad. Nauk SSSR, Ser. Khim.* **1979,** 2551.
230. B. M. Mikhailov, V. N. Smirnov, O. D. Smirnova, E. P. Prokof'ev and A. S. Shashkov, *Izv. Akad. Nauk SSSR, Ser. Khim.* **1979,** 2340.
231. L. G. Vorontsova, O. S. Chizhov, L. S. Vasil'ev, V. V. Veselovskii and B. M. Mikhailov, *Izv. Akad. Nauk SSSR, Ser. Khim.* **1981,** 353.

232. M. E. Gurskii and B. M. Mikhailov, *Izv. Akad. Nauk SSSR, Ser. Khim.* **1981,** 394.
233. B. M. Mikhailov and M. E. Kuimova, *Izv. Akad. Nauk SSSR, Ser. Khim.* **1980,** 1881.
234. M. E. Gurskii, A. S. Shashkov and B. M. Mikhailov, *J. Organomet. Chem.* **199,** 171 (1980).
235. M. E. Gurskii, A. S. Shashkov and B. M. Mikhailov, *Izv. Akad. Nauk SSSR, Ser. Khim.* **1981,** 341.
236. M. E. Gurskii, S. V. Baranin and B. M. Mikhailov, *Izv. Akad. Nauk SSSR, Ser. Khim.* **1980,** 2188.
237. B. M. Mikhailov and V. N. Smirnov, *Izv. Akad. Nauk SSSR, Ser. Khim.* **1972,** 1672.
238. B. M. Mikhailov and K. L. Cherkasova (in press).
239. B. M. Mikhailov, T. K. Baryshnikova, V. G. Kiselev and A. S. Shashkov, *Izv. Akad. Nauk SSSR, Ser. Khim.* **1979,** 2544.
240. B. M. Mikhailov: "The Chemistry of Borohydrides", Nauka, Moscow (1967).
241. E. Mincione and F. Felisiani, *J. Chem. Soc. Chem. Comm.* **1973,** 942.
242. B. E. Williams, *Inorg. Chem.* **1,** 971 (1962).
243. R. Köster and G. Schomburg, *Angew. Chem.* **72,** 567 (1960).
244. L. G. Vorontsova, O. S. Chizhov, V. N. Smirnov and B. M. Mikhailov, *Izv. Akad. Nauk SSSR, Ser. Khim.* **1981,** 595.
245. B. M. Mikhailov, L. S. Vasil'ev and V. V. Veselovskii, *Izv. Akad. Nauk SSSR, Ser. Khim.* **1980,** 1106.
246. L. S. Vasil'ev, V. V. Veselovskii, M. I. Struchkova and B. M. Mikhailov, *J. Organomet. Chem.* **226,** 115 (1982).
247. L. S. Vasil'ev, V. V. Veselovskii and B. M. Mikhailov, *Izv. Akad. Nauk SSSR, Ser. Khim.* **1980,** 2839.
248. B. M. Mikhailov, T. K. Baryshnikova and A. S. Shashkov, *J. Organomet. Chem.* **219,** 301 (1981).
249. B. M. Mikhailov, E. A. Shagova and M. Yu. Etinger, *J. Organomet. Chem.* **220,** 1 (1981).
250. B. M. Mikhailov, N. N. Govorov, Ya. A. Angelyuk, V. G. Kiselev and M. I. Struchkova, *Izv. Akad. Nauk SSSR, Ser. Khim.* **1980,** 1621.
251. B. M. Mikhailov, M. E. Gurskii, T. V. Potapova and A. S. Shashkov, *J. Organomet. Chem.* **201,** 81 (1980).
252. B. M. Mikhailov, V. N. Smirnov, O. D. Smirnova, V. A. Kasparov, N. A. Lagutkin, N. I. Mitin and M. M. Zubairov, *Khim.-Farm. Zh.* **13,** 35 (1979).
253. N. A. Lagutkin, N. I. Mitin, M. M. Zubairov, T. I. Arkhipova, T. K. Petracheva, B. M. Mikhailov, V. N. Smirnov, T. K. Baryshnikova and N. N. Govorov (in press).

PART V

ALKYNYL- AND ALLENYLBORANES

Different types of boron compounds of the acetylene series are known. Compounds with a triple carbon-carbon bond directly adjacent to the boron atom include tris(1-alkynyl)boranes, mixed 1-alkynylalkylboranes and 1-alkynylarylboranes, and also metal 1-alkynylborates, containing from one to four 1-alkynyl groups.

Another type of acetylenic boron compounds comprises esters of propargylboronic acids.

Esters of alkynylboronic acids genetically related to propargylboron compounds will also be considered in this section.

Acetylenic and allenic boron compounds, which show a high reactivity, find use in the synthesis of different classes of unsaturated organic compounds.

Esters of propargylboronic and allenylboronic acids are used successfully for the synthesis of acetylenic and allenic alcohols. In some cases, the methods of synthesis based on boron compounds of these types successfully supplement methods using Grignard reagents.

PART V

CHAPTER 1

Tri(1-alkynyl)boranes and 1-alkynylalkyl(aryl)boranes

Tri(1-alkynyl)boranes are unstable, easily polymerizable compounds and can therefore be obtained only in the form of complexes with amines.

Ashby and Foster [1] investigated a method for the preparation of amine adducts of tris(1-alkylnyl)boranes, based on the reaction of sodium derivatives of acetylenic hydrocarbons with BF_3, BCl_3, and $B(OCH_3)_3$ in different media. Of the modifications of this method tested, the best results were obtained by carrying out the reaction of 1-alkynylsodium with an amine complex of boron trifluoride in an amine medium at room temperature:

$$3RC{\equiv}CNa + F_3B{\cdot}L \rightarrow (RC{\equiv}C)_3B{\cdot}L \qquad (L = \text{amine})$$

$(HC{\equiv}C)_3B \cdot NC_5H_5$, $(HC{\equiv}C)_3B \cdot NH(C_2H_5)_2$, $(HC{\equiv}C)_3B \cdot N(CH_3)_3$, $(C_6H_5C{\equiv}C)_3B \cdot NC_5H_5$, and $(C_4H_9C{\equiv}C)_3B \cdot NC_5H_5$ have been obtained by this method.

The pyridine complex of tri(phenylethynyl)borane is stable to 200°C, while the complex of tri(1-hexynyl)borane dissociates even at room temperature.

Hooz and Layton [2] obtained tri(1-alkynyl)boranes from lithium compounds of acetylenic hydrocarbons and boron trifluoride etherate in tetrahydrofuran at −20°C and, without isolating the products, reacted them with ethyl diazoacetate. In this way, they obtained esters of propargylcarboxylic acids in high yields:

$$3RC{\equiv}CLi + F_3B \cdot O(C_2H_5)_2 \xrightarrow{\text{THF, } -20^\circ C} (RC{\equiv}C)_3B,$$

$$(RC{\equiv}C)_3B + N_2CHCOOC_2H_5 \xrightarrow[-N_2]{\text{THF, } -20^\circ C} \underset{63\text{—}92\%}{RC{\equiv}CCH_2COOC_2H_5}$$

$[R = C_4H_9, C_5H_{11}, C_6H_5, (CH_3)_3C]$.

Mixed 1-alkynylboranes were synthesized for the first time by Soulié and Willemart [3] by the action of 1-alkynylmagnesium bromides on β-aminoethyl esters of diphenylborinic acid in the presence of pyridine in tetrahydrofuran; the pyridinates of different diphenyl-1-alkynylboranes were isolated in yields of 41–77%:

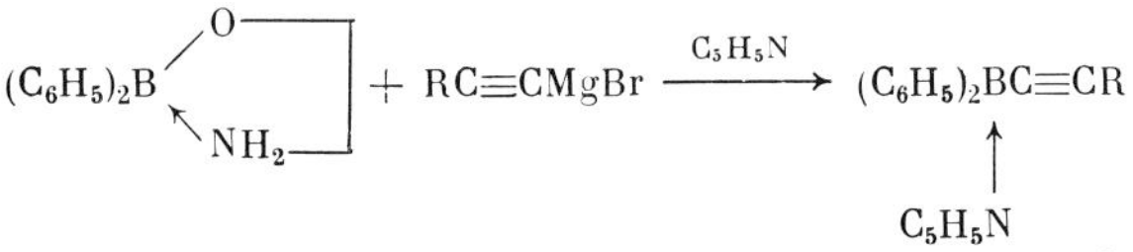

$[R = H, CH_3, C_6H_5, p\text{-}BrC_6H_4, (CH_3)_3Si, (CH_3)_3SiCH_2, C_6H_5S,$

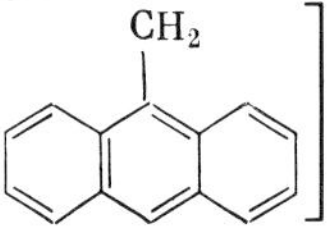

$]$.

Cadiot and co-workers [4, 5] synthesized unsymmetrical 1-alkynyldiarylboranes, using the β-aminoethyl esters of unsymmetrical diarylborinic acids as the starting materials:

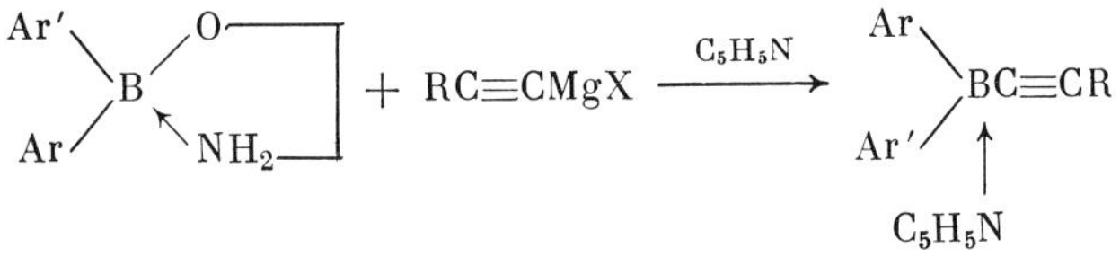

$[Ar = C_6H_5, p\text{-}CH_3C_6H_4; Ar' = o\text{-}CH_3C_6H_4, m\text{-}CH_3C_6H_4, p\text{-}CH_3C_6H_4, p\text{-}CH_3OC_6H_4;$
$R = H, CH_3, C_6H_5, CH_2{=}CH, CH_2{=}C(CH_3), CH_3C{\equiv}C$, etc.].

Numerous adducts of di(1-alkynyl)arylboranes have been synthesized from the 1-alkynylmagnesium reagents and aryldichloroboranes [4, 5]:

$$ArBCl_2 + 2RC{\equiv}MgX \xrightarrow{C_5H_5N} ArB(C{\equiv}CR)_2 \cdot NC_5H_5$$

$[Ar = C_6H_5, p\text{-}CH_3C_6H_4, p\text{-}CH_3OC_6H_4, o\text{-}CH_3C_6H_4;$
$R = CH_3, C_6H_5CH_2{=}C(CH_3)]$.

Dibutylpropynyl- and dibutylphenylethynylboranes have been obtained from the corresponding sodium acetylenides and dibutylchloroborane [6].

Diphenyl(phenylethynyl)- and phenyl(diphenylethynyl)boranes have been synthesized by the action of chlorodiphenylborane and dichlorophenylborane, respectively, on phenylethynyllithium [7]:

$$\underset{74\%}{(C_6H_5)_2BC{\equiv}CC_6H_5} \xleftarrow{(C_6H_5)_2BCl} C_6H_5C{\equiv}CLi \xrightarrow{C_6H_5BCl_2} \underset{78\%}{C_6H_5B(C{\equiv}CC_6H_5)_2}$$

Dimethyl(phenylethynyl)borane has been obtained by the reaction of $(CH_3)_2BI$ with $IC{\equiv}CC_6H_5$ [47].

Ethynyldimesitylborane and allenyldimesitylborane have been obtained from fluorodimesitylborane and ethynyllithium or "propargyl magnesium bromide", respectively [48].

Köster, et al. [8] carried out a systematic investigation of the synthesis of mixed 1-alkynylalkylboranes. Complex compounds of 1-alkynyldialkylboranes with amines are obtained in high yields by the reaction of lithium or sodium compounds of acetylenic hydrocarbons and adducts of dialkylhaloboranes with amines:

$$R_2BX\cdot L + MC{\equiv}CR' \xrightarrow{0-10^\circ} R_2BC{\equiv}CR'\cdot L + MX$$

[$X = F$, Cl; $L = N(CH_3)_3$, pyridine; $R = C_2H_5$, C_3H_7, C_4H_9, $R' = CH{\equiv}C$, $CH_3C{\equiv}C$, $C_2H_5C{\equiv}C$].

To obtain the free 1-alkynyldialkylboranes, their complexes with amines are treated with BF_3 etherate in an ether medium:

$$R_2BC{\equiv}CR'\cdot L + F_3B\cdot O(C_2H_5)_2 \longrightarrow R_2BC{\equiv}CR' + F_3B\cdot L + O(C_2H_5)_2.$$

Adducts of bis(1-alkynyl)alkylboranes with amines are obtained by the same method:

$$RBX_2\cdot L + 2MC{\equiv}CR' \rightarrow RB(C{\equiv}CR')_2\cdot L + 2MX$$

$$[X = F,\ Cl;\ L = (CH_3)_3N,\ \text{pyridine}].$$

The isolation of the free di(1-alkynyl)alkylboranes from their adducts with amines by means of BF_3 etherate does not proceed smoothly. Only isobutyldipropynylborane and di(1-butynyl)isobutylborane have been isolated, in yields of 30–36%.

Adducts of tri(1-alkynyl)boranes are obtained by the action of trimethylamine oxide on the adducts of 1-alkynyldiethylboranes:

$$3(C_2H_5)_2BC{\equiv}CR \cdot N(CH_3)_3 + 6(CH_3)_3NO \longrightarrow 2B(OC_2H_5)_3 + (RC{\equiv}C)_3B \cdot N(CH_3)_3 + \\ + 8N(CH_3)_3.$$

The reaction of isopropenylethynylmagnesium bromide with dibutylchloroborane forms unstable isopropenylethynyl(dibutyl)borane [9].

Dialkylalkynylboranes can be obtained by the action of alkynyllithium on methyl dialkylborinates, followed by treatment of the *ate* complexes formed with boron trifluoride etherate [49].

Acetylenic diboron compounds — bis(dialkylboryl)acetylenes — have been obtained by the action of dimagnesium derivatives of acetylene on dialkyliodoboranes [10]:

$$2R_2BI + BrMgC{\equiv}CMgBr \longrightarrow R_2BC{\equiv}CBR_2$$

$$(R = C_3H_7,\ iso\text{-}C_3H_7,\ C_4H_9,\ iso\text{-}C_4H_9).$$

These substances are easily hydrolyzed with water and form adducts with amines, which are more resistant to hydrolysis.

Bis(diphenylboryl)acetylenes have been synthesized by an analogous method [3].

Soulié and Cadiot obtained a number of diamino-1-alkynylboranes, aminodi(1-alkynyl)boranes [6], and amino(aryl)-1-alkynylboranes [11, 12] by the reaction of sodium acetylenides with aminochloroboranes:

$$RC{\equiv}CNa + ClB(NR'_2)_2 \longrightarrow RC{\equiv}CB(NR'_2)_2,$$

$$2RC{\equiv}CNa + Cl_2BNR'_2 \longrightarrow (RC{\equiv}C)_2BNR'_2$$

Cyclic diaminoboranes of the propargyl and allene series have also been synthesized [13].

The rate constants of internal rotation in dimethylaminophenyl(propynyl)borane and dimethylaminophenyl(ethynyl)borane have been investigated [14].

Diethylaminodipropynylborane reacts with dimethylstannane to form 1-diethylamino-3,4,4,5-tetramethyl-4-stanna-1-bora-2,5-cyclohexadiene [15]:

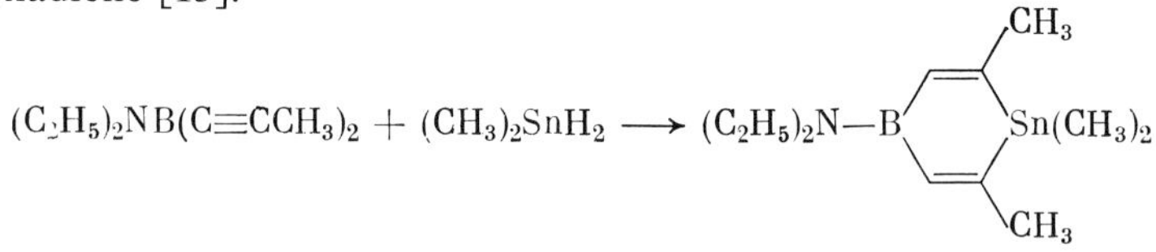

1-Alkynyldialkylboranes form 1,1-bis(dialkylboryl)-1-alkenes under the action of tetraalkyldiboranes [50].

9-(1-Alkynyl)-9-borabicyclo[3.3.1]nonanes are capable of addition to α,β-unsaturated ketones [51], while conjugated enynones are formed with 4-methoxy-3-buten-2-one and related compounds [52].

N,*N'*-Disubstituted *N*-(2-alkynoyl)ureas are formed by the reaction of 9-(1-alkynyl)-9-borabicyclo[3.3.1]nonanes with isocyanates, followed by protonolysis [53].

1-(9-Borabicyclo[3.3.1]nonan-9-yl)oct-4-yne isomerizes to 1-(9-borabicyclo[3.3.1] nonan-9-yl)oct-7-yne in the presence of potassium 3-aminopropylamide [54].

CHAPTER 2

Metal 1-alkynylborates

A compound of tetra-coordinated negative boron with an alkynyl radical was obtained for the first time by Wittig and Raff [16] by the addition of phenylethynyllithium to triphenylborane:

$$(C_6H_5)_3B + C_6H_5C{\equiv}CLi \longrightarrow Li[(C_6H_5)_3BC{\equiv}CC_6H_5].$$

Binger, et al. [17] carried out a detailed investigation of the methods of preparation and properties of metal 1-alkynyltrialkylborates. A general method for their synthesis is based on the reaction of alkali metal acetylenides with trialkylboranes:

$$BR_3 + MC{\equiv}CR' \longrightarrow M[R_3BC{\equiv}CR'] \qquad (M = Li, Na, K).$$

The ease with which the reaction takes place depends on the nature of the metal. Thus, the reaction of triethylborane with propynylsodium takes place at room temperature, while the reaction with propynyllithium proceeds in boiling toluene over a period of two hours. Diethyldipropynyl- or ethyltripropynylborates can be obtained from metallic derivatives of propyne and adducts of diethylpropynyl- or ethyldipropynylborane:

$$(C_2H_5)_2BC{\equiv}CCH_3 \cdot N(CH_3)_3 + MC{\equiv}CCH_3 \longrightarrow M[(C_2H_5)_2B(C{\equiv}CCH_3)_2] + N(CH_3)_3,$$
$$C_2H_5B(C{\equiv}CCH_3)_2 \cdot N(CH_3)_3 + MC{\equiv}CCH_3 \longrightarrow M[C_2H_5B(C{\equiv}CCH_3)_3] + N(CH_3)_3.$$

Borates have been obtained from 1-propyl-3-methyl-1-boraindan [18] and perhydro-9b-boraphenalene [19] by the above method.

Sodium 1-alkynyltrialkylborates are also obtained by another method. Binger and Köster [8, 20] found that alkali metal trialkylhydroborates react with acetylenic hydrocarbons to form 1-alkynyltrialkylborates:

$$Na[R_3BH] + HC{\equiv}CR' \longrightarrow Na[R_3BC{\equiv}CR'] + H_2.$$

With acetylene, the trialkylhydroborates form the disodium salts *1*:

$$2Na[(C_2H_5)_3BH] + HC\equiv CH \longrightarrow \underset{1}{Na_2[(C_2H_5)_3BC\equiv CB(C_2H_5)_3]} + 2H_2.$$

The action of tetraethyldiborane on sodium triethylpropynylborate leads to replacement of one ethyl group by hydrogen:

$$Na[(C_2H_5)_3BC\equiv CCH_3] + (C_2H_5)_2BH \xrightarrow{20^\circ C} \underset{2}{Na[(C_2H_5)_2\underset{|}{B}(H)C\equiv CCH_3]} + B(C_2H_5)_3.$$

In the borate *2* obtained, the B-H group reacts with propyne upon heating to above 100°C.

Potassium tetrafluoroborate and phenylethynylmagnesium bromide form potassium tetraphenylethynylborate which, when treated with ammonium chloride, forms the ammonium salt [21]:

$$KBF_4 + 4C_6H_5C\equiv CMgBr \longrightarrow \underset{27\%}{K[(C_6H_5C\equiv C)_4B]}.$$

Magnesium tetraphenylethynylborate is obtained in higher yield (60%) by the action of boron trifluoride etherate on phenylethynylmagnesium bromide in benzene at room temperature. The salt decomposes at 40°C, while the ammonium salt is more stable [22]. When a solution of this salt in tetrahydrofuran is treated with an ethereal HCl solution, an adduct of triphenylethynylborane with tetrahydrofuran is formed, which is hydrolyzed slowly with cold water but rapidly at 50°C. The adduct with tetrahydrofuran reacts with pyridine or dimethylamine to form the corresponding stable amine complexes of triphenylethynylborane [23].

1-Alkynyltrialkylborates are readily hydrolyzed with water, with the elimination of an acetylenic hydrocarbon, for example:

$$Na[(C_2H_5)_3BC\equiv CCH_3] \xrightarrow{H_2O} Na[(C_2H_5)_3BOH] + CH_3C\equiv CH.$$

Dialkynyl- and trialkynylborates split off only one acetylenic group under the action of water [17]:

$$Na[C_2H_5B(C\equiv CCH_3)_3] \xrightarrow{H_2O} Na\left[C_2H_5\overset{OH}{\overset{|}{B}}(C\equiv CCH_3)_2\right] + CH_3C\equiv CH.$$

Sodium triethylborate reacts with alkenynes to form alkenynyetriethylborates [55].

The lithiation of ethynyltrialkylborates, followed by treatment with butyl bromide, leads to 1-hexynyltrialkylborates [56].

The use of 1-alkynyltrialkylborates in organic synthesis was described in Part III.

CHAPTER 3

Esters of 1-alkynylboronic acids

Matteson and Peacock [24, 25] obtained dibutyl ethynylboronate by the action of ethynylmagnesium bromide on methyl borate:

$$CH{\equiv}CMgBr + B(OCH_3)_3 \xrightarrow[2.\ C_4H_9OH]{1.\ H^+} \underset{63\%}{HC{\equiv}CB(OC_4H_9)_2}.$$

The dibutyl esters of propynyl- and vinylethynylboronic acids have been synthesized by an analogous method [26, 27].

Ethynylboronates have also been obtained with the use of ethynylsodium [28]. Dipropyl ethoxyethynylboronate is obtained in 44% yield by the reaction of ethoxyethynylmagnesium bromide and dipropyl 2-bromoethylboronate [29]:

$$C_2H_5OC{\equiv}CMgBr + BrCH_2CH_2B(OC_3H_7)_2 \longrightarrow C_2H_5OC{\equiv}CB(OC_3H_7)_2 + C_2H_4 + MgBr_2.$$

The boron-carbon bond in ethynylboronic acid is not ruptured by the action of neutral or acid hydrolyzing agents, but is rapidly broken under the action of such weak bases as sodium bicarbonate [25].

Esters of 1-alkynylboronic acids react with aliphatic organometallic compounds to form esters of alkylboronic acids, according to the following scheme:

$$RC{\equiv}CB(OR)_2 + R'MgBr \longrightarrow R'B(OR)_2 + RC{\equiv}CMgBr.$$

Thus, Zavgorodnii and Petrov [26] found that dibutyl vinylethynylboronate is converted to dibutyl ethylboronate under the action of ethyllithium.

The reaction of dibutyl ethynylboronate with ethylmagnesium bromide produces dibutyl ethylboronate in moderate yield. The reaction of dibutyl ethynylboronate with vinylmagnesium bromide results in the formation of dibutyl vinylboronate (47%) and butyl borate [30].

The reaction of ethynylboronates with aromatic Grignard reagents proceeds in a different manner and leads to aryl(ethynyl)borinates

[30]:

$$HC{\equiv}CB(OC_4H_9)_2 + ArMgBr \xrightarrow[H^+]{-70°\ C} \underset{79\%}{HC{\equiv}C{-}B}\begin{matrix} OC_4H_9 \\ Ar \end{matrix}$$

The difference between the behavior of ethynylboronates toward alkyl- and arylmagnesium reagents is explained by the relative electronegativity of the radicals, determining their tendency to split off from the intermediately formed borate anion during hydrolysis. The effect of the nature of the hydrocarbon radical and of the hydrolysis conditions on the direction of the decomposition of borates was investigated in [31–34].

Alkynylboronates are capable of adding halomethanes and mercaptans.

Dibutyl ethynylboronate reacts with bromotrichloromethane to form dibutyl 1-bromo-3,3,3-trichloropropenylboronate [24, 25] and also adds thiophenol and hexyl mercaptan:

$$HC{\equiv}CB(OC_4H_9)_2 \xrightarrow{CCl_3Br} Cl_3CCH{=}C(Br)B(OC_4H_9)_2 \quad 91\%$$

$$HC{\equiv}CB(OC_4H_9)_2 \xrightarrow{RSH} RSCH{=}CHB(OC_4H_9)_2$$

Butyl ethynyl-2,5-dimethylphenylborinate adds bromotrichloromethane to form butyl (2,5-dimethylphenyl)-1-bromo-3,3,3-trichloro-1-propenylborinate [29]:

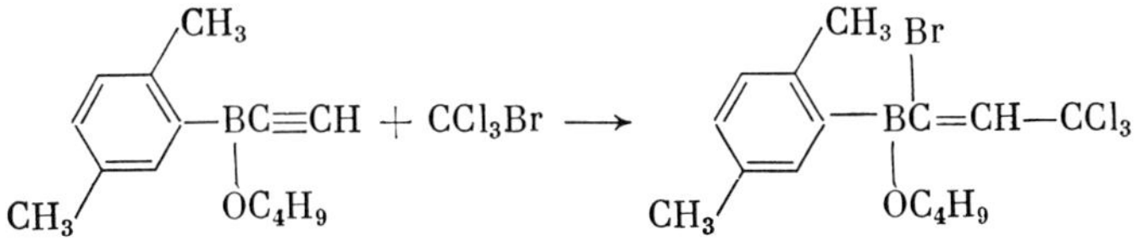

Under UV irradiation, hydrogen bromide adds to dibutyl ethynylboronate to form dibutyl 2-bromovinylboronate [25]:

$$HC{\equiv}CB(OC_4H_9)_2 \xrightarrow{HBr} \underset{58\%}{BrCH{=}CHB(OC_4H_9)_2}.$$

Bromine adds smoothly to ethynylboronates under UV irradiation to form 1,2-dibromovinylboronates [2].

Dibutyl ethynylboronate adds to cyclopentadiene, forming dibutyl bicyclo[2.2.1]hepta-2,5-dien-2-ylboronate [24, 35]:

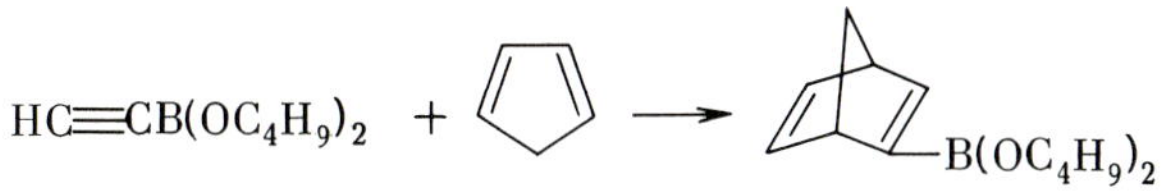

Dibutyl ethynylboronate reacts with ethyl diazoacetate to form 3-carbethoxy-5-pyrazolylboronic acid [36]:

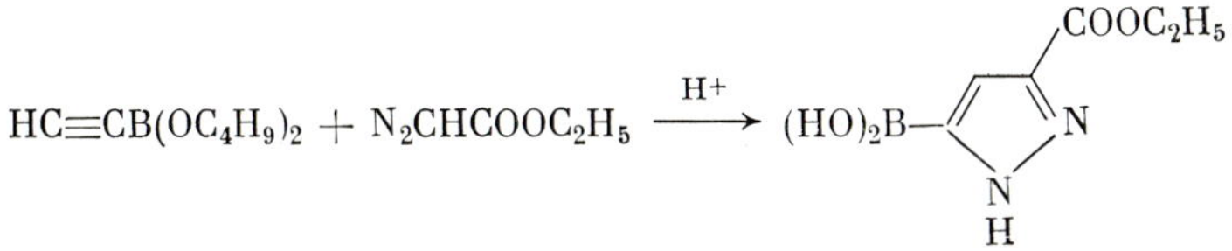

CHAPTER 4

Esters of propargyl- and allenylboronic acids

4.1 METHODS FOR THE PREPARATION OF ESTERS OF PROPARGYL- AND ALLENYLBORONIC ACIDS

The reaction of Grignard reagents obtained from propargyl bromides with methyl borate forms allenyl or propargyl compounds, depending on the nature of the starting bromides [37–40].

Table 1 shows the starting propargyl bromides, the organomagnesium compounds obtained from them, and the final organoboron compounds. As seen from this table, aliphatic propargyl bromides form allenyl boron compounds, while phenyl-substituted propargyl bromides lead to acetylenic boron compounds.

Table 1. Synthesis of organoboron compounds from propargyl bromides [38].

Propargyl bromide	RMgX	Organoboron compound	Yield %
$HC{\equiv}CCH_2Br$	$CH_2{=}C{=}CHMgBr$	$CH_2{=}C{=}CHB(OC_4H_9)_2$	36
$C_2H_5C{\equiv}CCH_2Br$	$C_2H_5C{\equiv}CCH_2MgBr$		
	$C_2H_5C(MgBr){=}C{=}CH_2$	$CH_2{=}C{=}C(C_2H_5)B(OC_4H_9)_2$	48
$C_3H_7C{\equiv}CCH_2Br$	$C_3H_7C{\equiv}CCH_2MgBr$		
	$C_3H_7C(MgBr){=}C{=}CH_2$	$CH_2{=}C{=}C(C_3H_7)B(OC_4H_9)_2$	33
$C_5H_{11}C{\equiv}CCH_2Br$	$C_5H_{11}C{\equiv}CCH_2MgBr$		
	$C_5H_{11}C(MgBr){=}C{=}CH_2$	$CH_2{=}C{=}C(C_5H_{11})B(OC_4H_9)_2$	22
$C_3H_7CHBrC{\equiv}CH$	$C_3H_7CH{=}C{=}CHMgBr$	$C_3H_7CH{=}C{=}CHB(OC_4H_9)_2$	40
$C_6H_5C{\equiv}CCH_2Br$	$C_6H_5C{\equiv}CCH_2MgBr$		
	$C_6H_5C(MgBr){=}C{=}CH_2$	$C_6H_5C{\equiv}CCH_2B(OC_4H_9)_2$	57
$C_6H_5C{\equiv}CCHBrCH_3$	$C_6H_5C(MgBr){=}C{=}CHCH_3$	$C_6H_5C{\equiv}CCH(CH_3)B(OC_4H_9)_2$	36

Cadiot, et al. [13, 41] used the cyclic borates *3* in the synthesis of esters of propargyl- and allenylboronic acids:

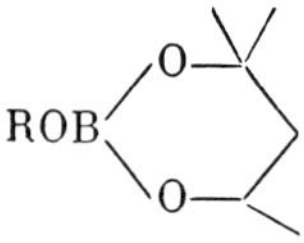

3

($R = C_2H_5$, C_4H_9).

The reaction is carried out in ether at −80°C, and the intermediately formed borates are hydrolyzed with hydrochloric acid:

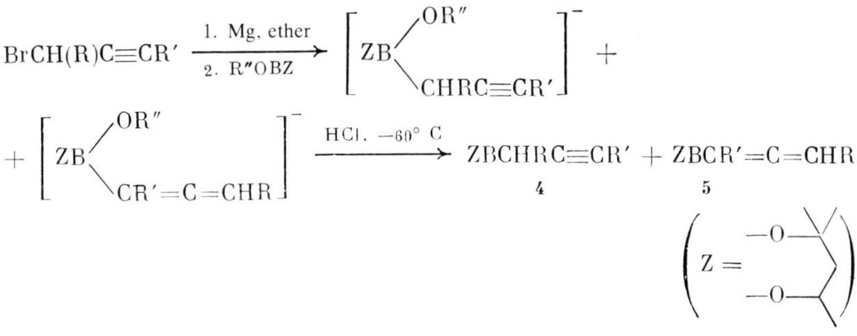

The yields and ratios of the esters of propargyl- (*4*) and allenylboronic (*5*) acids are shown in Table 2.

Table 2. Synthesis of the cyclic esters *4* and *5*.

Bromide	Radicals in the esters *4* and *5*	Ratio between *5* and *4*	Yield, %
$BrCH(CH_3)C{\equiv}CH$	$CH_3{-}CH{=}C{=}CH{-}$ $CH{\equiv}C{-}CH(CH_3){-}$	95 : 5	60
$BrCH_2C{\equiv}CSi(CH_3)_3$	$CH_2{=}C{=}C[Si(CH_3)_3]{-}$ $(CH_3)_3Si{-}C{\equiv}C{-}CH_2{-}$	24 : 76	76
$BrCH_2C{\equiv}CC(CH_3)_3$	$CH_2{=}C{=}C[C(CH_3)_3]{-}$ $(CH_3)_3C{-}C{\equiv}C{-}CH_2{-}$	15 : 85	65
$BrCH_2C{\equiv}CC_6H_5$	$CH_2{=}C{=}C(C_6H_5){-}$ $C_6H_5C{\equiv}C{-}CH_2{-}$	5 : 95	45
$BrCH_2C{\equiv}CH$	$CH_2{=}C{=}CH{-}$	100 : 0	70

4.2 CHEMICAL PROPERTIES OF ESTERS OF PROPARGYL- AND ALLENYLBORONIC ACIDS. SYNTHESIS OF ACETYLENIC AND ALLENIC ALCOHOLS

Favre and Gaudemar [39, 42, 43] investigated the reactions of dibutyl allenylboronate with aldehydes and ketones. In the case of the aldehydes, the reaction is accompanied by prototropic rearrangement, as a result of which the propargyl compounds *6* are formed:

$$\mathrm{RCHO} + \mathrm{CH_2{=}C{=}CHB(OC_4H_9)_2} \longrightarrow \underset{\underset{\mathbf{6}}{\mathrm{CH_2C{\equiv}CH}}}{\mathrm{RCH{-}O{-}B(OC_4H_9)_2}}$$

Ketones react with allenylboronates at 150–200°C. The course of the reaction depends on the nature of the ketone, the temperature, the reagent ratio, and the nature of the solvent. Only propargyl compounds (*7*) are obtained from acetone, acetophenone, and cyclohexanone, whereas, with other ketones (ethyl methyl ketone, isopropyl methyl ketone, diethyl ketone, etc.), a mixture of propargyl (*7*) and allenyl compounds (*8*) is formed:

$$\mathrm{RR'C{=}O} + \mathrm{CH_2{=}C{=}CH{-}B(OC_4H_9)_2} \longrightarrow$$

$$\underset{\underset{\mathbf{7}}{\mathrm{CH_2C{\equiv}CH}}}{\mathrm{RR'C{-}O{-}B(OC_4H_9)_2}} + \underset{\underset{\mathbf{8}}{\mathrm{CH{=}C{=}CH_2}}}{\mathrm{RR'C{-}O{-}B(OC_4H_9)_2}}$$

The results of the reaction of ketones with dibutyl allenylboronate are shown in Table 3 (see next page).

The ratio of the isomeric propargyl and allenyl borates changes slightly, depending on the ratio of the starting reagents used — with excess ketone, a larger amount of allenyl derivatives is formed.

The ratio of the isomeric compounds *7* and *8* is greatly affected by the nature of the solvent. Thus, when the reaction is carried out with diethyl ketone without a solvent or in benzene or ether, 59–63% of the allenyl compound *8* is obtained, whereas only the propargyl compound *7* is formed in dimethyl sulfoxide.

Table 3. Synthesis of unsaturated esters from ketones and dibutyl allenylboronate.

Ketone	Borate	Ratio of compounds *7* and *8*
CH_3COCH_3	$\begin{array}{l}(CH_3)_2C{-}OB(OC_4H_9)_2\\ \quad\mid\\ \quad CH_2C{\equiv}CH\end{array}$	100 : 0
$C_2H_5\overset{\overset{\displaystyle O}{\|}}{C}{-}CH_3$	$\begin{array}{l}CH_3(C_2H_5)C{-}OB(OC_4H_9)_2\\ \qquad\quad\mid\\ \qquad\quad CH_2C{\equiv}CH\end{array}$ $\begin{array}{l}CH_3(C_2H_5)C{-}OB(OC_4H_9)_2\\ \qquad\quad\mid\\ \qquad\quad CH{=}C{=}CH_2\end{array}$	78 : 22
$(CH_3)_2CH\overset{\overset{\displaystyle O}{\|}}{C}CH_3$	$\begin{array}{l}(CH_3)_2CH(CH_3)C{-}OB(OC_4H_9)_2\\ \qquad\qquad\quad\mid\\ \qquad\qquad\quad CH_2C{\equiv}CH\end{array}$ $\begin{array}{l}(CH_3)_2CH(CH_3)C{-}OB(OC_4H_9)_2\\ \qquad\qquad\quad\mid\\ \qquad\qquad\quad CH{=}C{=}CH_2\end{array}$	49 : 51
$(C_2H_5)_2C{=}O$	$\begin{array}{l}(C_2H_5)_2C{-}OB(OC_4H_9)_2\\ \qquad\mid\\ \qquad CH_2C{\equiv}CH\end{array}$ $\begin{array}{l}(C_2H_5)_2C{-}O{-}B(OC_4H_9)_2\\ \qquad\mid\\ \qquad CH{=}C{=}CH_2\end{array}$	41 : 59

Hydrolysis of the borates *6*, *7*, and *8* produces the corresponding acetylenic and allenic alcohols [42, 43]:

$$\begin{array}{lcl}
R{-}\underset{\displaystyle CH_2C{\equiv}CH}{\underset{\mid}{CH}}{-}OB(OC_4H_9)_2 & \longrightarrow & R{-}\underset{\displaystyle CH_2C{\equiv}CH}{\underset{\mid}{CH}}OH \\
RR'\underset{\displaystyle CH{=}C{=}CH_2}{\underset{\mid}{C}}{-}OB(OC_4H_9)_2 & \longrightarrow & RR'\underset{\displaystyle CH{=}C{=}CH_2}{\underset{\mid}{C}}OH
\end{array}$$

The reaction of aldehydes and ketones with esters of different organoboron acids of the propargyl and allene types and with their mixtures have been investigated [40, 42, 44]. The related data are presented in Table 4.

The reactions without rearrangement are usually considered as processes going through a four-center transition state, whereas, for reactions with rearrangement, a six-center transition state is assumed.

The cyclic ester of 3,3-dimethyl-1-allenylboronic acid *9*, under the action of a catalytic quantity of HCl, undergoes rearrangement to a

Table 4. Synthesis of alcohols from aldehydes and esters of propargyl- and allenylboronic acids.

Aldehyde	Borate	Alcohol	Yield, %
C_2H_5CHO	$C_3H_7CH=C=CHB\langle$	$CH\equiv CCHCH(OH)C_2H_5$ | C_3H_7	70
C_2H_5CHO	$C_6H_5C\equiv CCH_2B\langle$	$CH_2=C=CCH(OH)C_2H_5$ | C_6H_5	52
C_2H_5CHO	$C_6H_5C\equiv CCHB\langle$ | CH_3	$CH_3CH=C=CCH(OH)C_2H_5$ | C_6H_5	50
C_2H_5CHO	$C_2H_5C\equiv CCH_2B\langle$ $CH_2=C=CB\langle$ | C_2H_5	$C_2H_5C\equiv CCH_2CH(OH)C_2H_5$ $CH_2=C=CCH(OH)C_2H_5$ | C_2H_5	60
C_2H_5CHO	$C_3H_7C\equiv CCH_2B\langle$ $CH_2=C=C-B\langle$ | C_3H_7	$C_3H_7C\equiv CCH_2CH(OH)C_2H_5$ $CH_2=C=CCH(OH)C_2H_5$ | C_3H_7	78
C_3H_7CHO	$CH_2=C=C-B\langle$ | CH_3	$C_3H_7CHCH_2C\equiv CCH_3$ | OH $C_3H_7CHC=C=CH_2$ (93 : 7) | \\ OH CH_3	69

mixture of *cis* and *trans* isomers (2.4 : 1) of isoprenylboronic acid, *10* and *11*, each of which may be isolated in the pure state [45].

The action of HCl on the ester *9* or the esters *10* and *11* forms *cis*-(*12*) and *trans*-2-(3-chloro-3-methyl-1-butenyl)-4,4,6-trimethyl-1,3,2-dioxaborinanes (*13*):

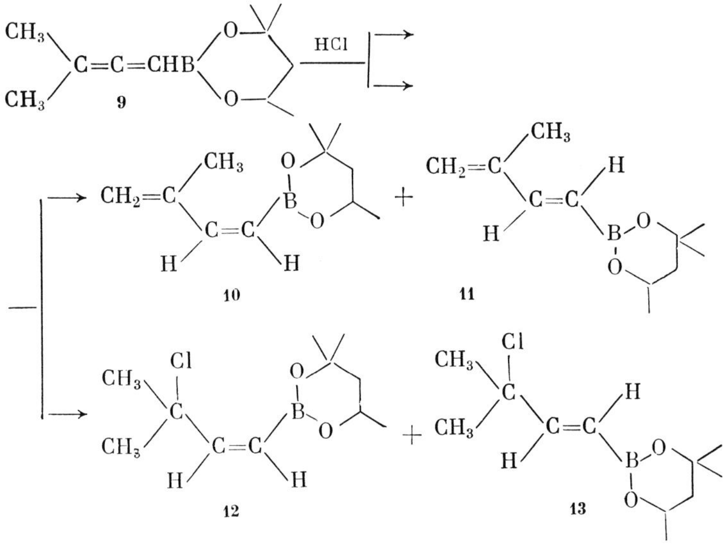

Cyclic esters of allenyl- and 3-methylallenylboronic acids react with hydrogen bromide to form a mixture of *Z*- and *E*-isomers of esters of 2-bromoalkenylboronic acids and of products of allene-acetylene rearrangement — esters of 2-alkynylboronic acids. The addition of a second mole of hydrogen bromide leads to esters of 2,2-dibromoalkylboronic acids. The cyclic ester of 3,3-dimethylallenylboronic acids reacts with one mole of hydrogen bromide to form a mixture of *Z*- and *E*-isomers of esters of 3-bromo-3-methylbuten-1-ylboronic acid; with a second mole of hydrogen bromide, an ester of 1,3-dibromo-3-methylbutylboronic acid is formed [57].

Esters of 3-methyl-1,2-butadienylboronic and 3-methyl-1,3-butadienylboronic acid react with mercury (II) acetate to form the difficultly accessible diene mercury compounds-3-methyl-1,2-butadienylmercury chloride and *trans*-3-methyl-1,3-butadienylmercury chloride [58].

Some other 1,3-dienylboranes are obtained by the addition of BCl_3 to vinylacetylenes. This method has been used for the synthesis of dichloro(2-chloro-1,3-butadien-1-yl)-borane, which is converted by the action of methanol to dimethyl 2-chloro-1,3-butadien-1-ylboronate [46, 59]

$$CH_2{=}CH{-}C{\equiv}CH + BCl_3 \longrightarrow CH_2{=}CH{-}CCl{=}CHBCl_2 \xrightarrow{CH_3OH}$$
$$\longrightarrow CH_2{=}CH{-}CHCl{=}CH{-}B(OCH_3)_2.$$

The photochemical cyclization of dialkyl-1,3-dienylboranes to substituted boracyclopent-3-enes, the protonolysis of which leads to homoallylic alcohols, has been described [60].

The *ate* complexes obtained from 9-alkyl-9-borabicyclo[3.3.1]nonanes and 1-lithio-1-methoxyallene form the corresponding 1-alkyl-1-methoxycyclopropanes under the action of acetic acid [61].

REFERENCES

1. E. C. Ashby and W. E. Foster, *J. Org. Chem.* **29,** 3225 (1964).
2. J. Hooz and R. B. Layton, *Can. J. Chem.* **50,** 1105 (1972).
3. J. Soulié and A. Willemart, *Compt. Rend.* **251,** 727 (1960).
4. D. Giraud, J. Soulié and P. Cadiot, *Compt. Rend.* **254,** 319 (1960).
5. J. Soulié and P. Cadiot, *Bull. Soc. Chim. France* **1966,** 1981.
6. J. Soulié and P. Cadiot, *Bull. Soc. Chim. France* **1966,** 3846.
7. M. F. Lappert and B. Prokai, *J. Organometal. Chem.* **1,** 384 (1964).
8. R. Köster, H.-J. Horstschäfer and P. Binger, *Ann.* **717,** 1 (1968).
9. A. A. Petrov, V. S. Zavgorodnii and V. A. Kormer, *Zh. Obshch. Khim.* **32,** 1349 (1962).
10. H. Hartmann and K.-H. Birr, *Z. Anorg. Chem.* **299,** 174 (1959).
11. J. Soulié and P. Cadiot, *Bull. Soc. Chim. France* **1966,** 3850.
12. J. Soulié, *Compt. Rend., C* **262,** 376 (1966).
13. A. L'Honore, J. Soulié and P. Cadiot, *Compt. Rend., C* **275,** 229 (1972).
14. T. Totani, K. Tori and H. Watanabe, *J. Chem. Soc. Japan* **68,** 1445 (1965); C. A., **63,** 17828 (1965).
15. B. Wrackmeyer and H. Nöth, *Z. Naturforsch.* **29b,** 564 (1974).
16. G. Wittig and P. Raff, *Ann.* **573,** 195 (1951).
17. P. Binger, G. Benedikt, G. W. Rotermund and R. Köster, *Ann.* **717,** 21 (1968).
18. R. Köster, G. Benedikt, W. Fenzl and K. Reinert, *Ann.* **702,** 197 (1967).
19. G. W. Rotermund and R. Köster, *Ann.* **686,** 153 (1965).
20. P. Binger and R. Köster, *Tetrahedron Letters* **1965,** 1901.
21. V. A. Sazonova and N. Ya. Kronrod, *Zh. Obshch. Khim.* **26,** 1876 (1956).
22. U. Krüerke, *Z. Naturforsch.* **11b,** 606 (1956).
23. U. Krüerke and Z. Naturforsch., **11b,** 676 (1956).
24. D. S. Matteson and K. Peacock, *J. Am. Chem. Soc.* **82,** 5759 (1960).

25. D. S. Matteson and K. Peacock, *J. Org. Chem.* **28,** 369 (1963).
26. V. S. Zavgorodnii and A. A. Petrov, *Zh. Obshch. Khim.* **31,** 2433 (1961).
27. L. G. Sharanina, V. S. Zavgorodnii and A. A. Petrov, *Zh. Obshch. Khim.* **36,** 1275 (1966).
28. British Patent 884,547 (1961); *C. A.* **57,** 7308f (1962).
29. D. S. Matteson and K. Peacock, *J. Organometal. Chem.* **2,** 190 (1964).
30. D. S. Matteson and K. Peacock, *J. Organometal. Chem.* **2,** 192 (1964).
31. B. M. Mikhailov and P. M. Aronovich, *Izv. Akad. Nauk SSSR, Otd. Khim. Nauk* **1956,** 322.
32. B. M. Mikhailov and V. A. Vaver, *Dokl. Akad. Nauk SSSR* **109,** 94 (1956).
33. B. M. Mikhailov and T. K. Kozminskaya, *Izv. Akad. Nauk SSSR, Otd. Khim. Nauk* **1959,** 80.
34. B. M. Mikhailov and T. A. Shchegoleva, *Dokl. Akad. Nauk SSSR* **108,** 481 (1956).
35. D. S. Matteson and J. O. Waldbillig, *J. Org. Chem.* **28,** 366 (1963).
36. D. S. Matteson, *J. Org. Chem.* **27,** 4293 (1962).
37. E. Favre and M. Gaudemar, *Compt. Rend., C* **262,** 1332 (1966).
38. E. Favre and M. Gaudemar, *Bull. Soc. Chim. France* **1968,** 3724.
39. E. Favre and M. Gaudemar, *Compt. Rend., C* **263,** 1543 (1966).
40. J. Blais, J. Soulie and P. Cadiot, *Compt. Rend., C* **271** 589 (1970).
41. J. Blais, A. L'Honore, J. Soulié and P. Cadiot, *J. Organometal. Chem.* **78,** 323 (1974).
42. E. Favre and M. Gaudemar, *J. Organometal. Chem.* **76,** 297 (1974).
43. E. Favre and M. Gaudemar, *J. Organometal. Chem.* **76,** 305 (1974).
44. E. Favre and M. Gaudemar, *Compt. Rend., C* **272,** 111 (1971).
45. B. M. Mikhailov, M. E. Gurskii and M. G. Gverdtsiteli, *Izv. Akad. Nauk SSSR, Ser. Khim.* **1976,** 2062.
46. B. M. Mikhailov, L. S. Vasil'ev, V. V. Veselovskii and V. G. Kiselev, *Izv. Akad. Nauk SSSR, Ser. Khim.* **1975,** 725.
47. W. Siebert, R. Full, T. Renk, and A. Ospici, *Z. Anorg. Allg. Chem.* **418,** 273 (1975).
48. N. M. D. Brown, F. Davidson and J. W. Wilson, *J. Organometal. Chem.* **209,** 1 (1981).
49. H. C. Brown and J. A. Sinclair, *J. Organometal. Chem.* **131,** 163 (1977).
50. R. Köster, H.-J. Horstschäfer, P. Binger and P. K. Mattschei, *Lieb. Ann. Chem.* **1975,** 1339.
51. J. A. Sinclair, G. A. Molander and H. C. Brown, *J. Am. Chem. Soc.* **99,** 954 (1977).
52. G. A. Molander and H. C. Brown, *J. Org. Chem.* **42,** 3106 (1977).
53. G. A. Molander and H. C. Brown, *Synthesis* **1979,** 104.
54. C. A. Brown and E. Negishi, *J. Chem. Soc., Chem. Comm.* **1977,** 318.
55. R. Köster, A. Bussmann and G. Schroth, *Lieb. Ann. Chem.* **1975,** 2130.
56. K. Uchimoto, Y. Yabuki, K. Okada and H. Nozaki, *Tetrahedron Lett.* **1976,** 3969.
57. B. M. Mikhailov, M. E. Gurskii and M. G. Gverdtsiteli, *Izv. Akad. Nauk SSSR, Ser. Khim.* **1978,** 1580.
58. B. M. Mikhailov, M. E. Gurskii and M. G. Gverdtsiteli, *Izv. Akad. Nauk SSSR, Ser. Khim.* **1976,** 2402.
59. B. M. Mikhailov, V. V. Veselovskii, L. S. Vasil'ev and V. G. Kiselev, *Izv. Akad. Nauk SSSR, Ser. Khim.* **1979,** 1580.
60. G. Zweifel, S. J. Backlund and T. Leung, *J. Am. Chem. Soc.* **99,** 5192 (1977).
61. N. Miyaura, T. Yoshinari, M. Itoh and A. Suzuki, *Tetrahedron Lett.* **21,** 537 (1980).

PART VI

ESTERS OF THIOBORIC AND ORGANOTHIOBORON ACIDS

The first mention of a sulfur-containing organoboron compound dates back to 1878, when Councler [1] obtained a compound that he assumed to be triisobutyl thioborate from triisobutyl borate and P_2S_5. Starting in 1908 [2], occasional investigations in this area were carried out over a period of fifty years, and did not become systematic in nature until 1959. At the present time, simple and convenient methods are known for the synthesis of the main types of compounds with B—S bonds, such as thioborates, esters of organothioboron acids, alkylmercaptoboranes, compounds with several different functions, sulfur-containing heterocyclic boron compounds of different types, etc.

Sulfur-containing boron compounds show a high reactivity and are used successfully for the preparation of different classes of boron derivatives. They are also used widely in the synthesis of sulfur-containing and a number of other classes of organic compounds.

Problems of thioborane chemistry have been discussed in review articles [3–5] and in a book by Gerrard [6]. Compounds with B—S bonds but without boron-carbon bonds have been considered in a monograph by Steinberg [7].

PART VI

CHAPTER 1

Thioborates

1.1. METHODS FOR THE PREPARATION OF THIOBORATES

Alkyl and aryl thioborates are obtained most conveniently by heating mercaptans or thiophenols with boron trichloride [8–11]. The thioborate yield increases significantly if the reaction is carried out in the presence of tertiary amines or pyridine [8]:

$$BCl_3 + 3RSH + 3R'_3N \longrightarrow \underset{70\text{—}90\%}{B(SR)_3} + 3R'_3NHCl.$$

Thioborates are also obtained from BCl_3 and silver mercaptides [11] or lead mercaptides [11–13], by heating boron tribromide with sodium mercaptide [14], by the action of mercaptans on trimers of alkylmercaptoboranes [15], dialkylmercaptoboranes [16, 17], trialkylamine-boranes [18, 19], penta- and decaboranes [20], or B_2S_3 [21]; and also by the sequential treatment of sodium borohydride with sulfur and a mercaptan [22].

1.2 CHEMICAL PROPERTIES OF THIOBORATES. USE IN ORGANIC SYNTHESIS

Thioborates are easily hydrolyzed with water to give boric acid and mercaptans [14, 18] and form alkyl borates and mercaptans with alcohols [11].

Trimethyl [8] and triethyl thioborates [23] react with pyridine to form 1:1 adducts, which decompose to the starting components upon heating.

In the reaction of thioborates with primary or secondary amines, a sequential replacement of the alkylmercapto groups by the corresponding alkylamino groups takes place; as a result, depending on the reagent ratio, N-substituted bis(alkylthio) aminoboranes, alkylthiodiaminoboranes, or tris(alkylamino)boranes are obtained [23].

The action of ethylenediamine on triethyl thioborate (1:1) forms a polymer [23]. Triketiminoboranes are obtained from trialkyl thioborates and ketimines [24].

Trialkyl thioborates enter into exchange reactions with esters of organoboron acids [25–27] and trialkylboranes [26, 28, 29], being converted to esters of organothioboron acids.

Complexes with metal carbonyls, $B(SCH_3)_3 \cdot Cr(CO)_3$ [30] and $(CH_3)_2BSCH_3 \cdot Cr(CO)_5$ [31], are obtained from thioboranes; complexes of cyclic thioboranes have been described, for example, in [32].

The spectral properties of thioboranes are discussed in [33, 34].

The high reactivity of thioborates shows up in their reactions with different classes of organic compounds. Thioborates add to some activated olefins and acetylenes, nitriles, and carbonyl compounds.

When thioborates are heated with vinyl ethers, vinyl alkyl sulfides and alkyl borates are obtained [35]:

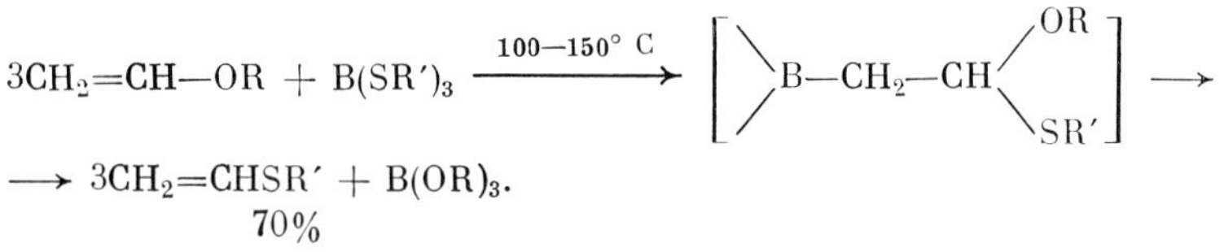

The reaction proceeds by addition of the thioborate to the double bond, followed by the elimination of an alkoxyborane.

Thioborates add exothermally to ethoxyacetylene to form the substituted vinylboranes *1*, which show a high reactivity [36–38]:

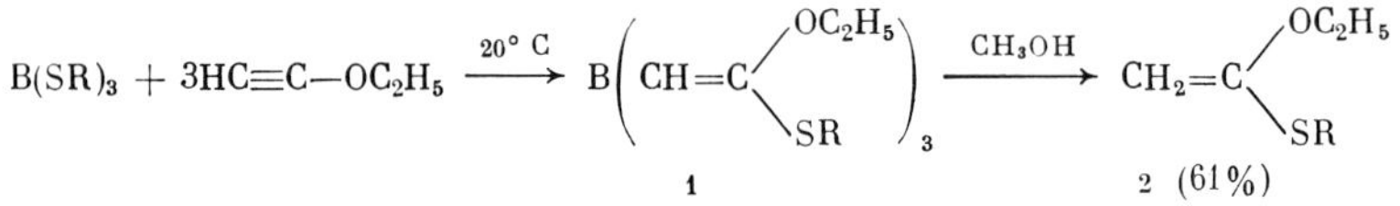

1 2 (61%)

In the compounds *1*, two B—C bonds are ruptured by methanol at room temperature, producing the 1-alkoxyvinyl alkyl sulfides *2* [37, 38]. The reaction of the vinylboranes *1* with aromatic aldehydes proceeds by replacement of the carbonyl oxygen by two organic radicals and leads to boric anhydride, B_2O_3, and 1,4-unsaturated compounds $ArCH[CH{=}C(OC_2H_5)SR]_2$, the acid hydrolysis of which produces 3-arylglutaric acids [36].

Alkyl thioborates react with aliphatic, α,β-unsaturated, aromatic aldehydes and ketones at room temperature to form thioacetals or thioketals and boric anhydride [39–42]:

$$3RCHO + 2B(SR')_3 \longrightarrow \underset{50\text{—}90\%}{3RCH(SR')_2} + B_2O_3,$$

$$3R_2CO + 2B(SC_2H_5)_3 \longrightarrow \underset{75\text{—}90\%}{3R_2C(SC_2H_5)_2} + B_2O_3,$$

$$3C_6H_5CH{=}CHCHO + 2B(SC_2H_5)_3 \longrightarrow \underset{59\%}{3C_6H_5CH{=}CHCH(SC_2H_5)_2}.$$

These reactions take place in neutral medium (in hydrocarbons or without a solvent), which is particularly important for the introduction of protecting groups into functional carbonyl compounds.

The treatment of α,β-unsaturated aldehydes and ketones with triphenylthioborane leads to substituted 1,3-diphenylthioalkenes. 1,3-Diphenylthio-1-hexene has been synthesized from 2-cyclohexenone in this manner [142].

Dimethylformamide reacts with butyl thioborate similarly to aldehydes and ketones, forming $(CH_3)_2NCH(SC_4H_9)_2$ [35]. At the same time, perfluoroacetone adds trimethyl thioborate to form the borate ester $B[OC(CF_3)_2SCH_3]_3$ [43].

Methyl diphenylthioacetate has been obtained by heating diphenyl ketene with trimethyl thioborate in ether, followed by hydrolysis [44]:

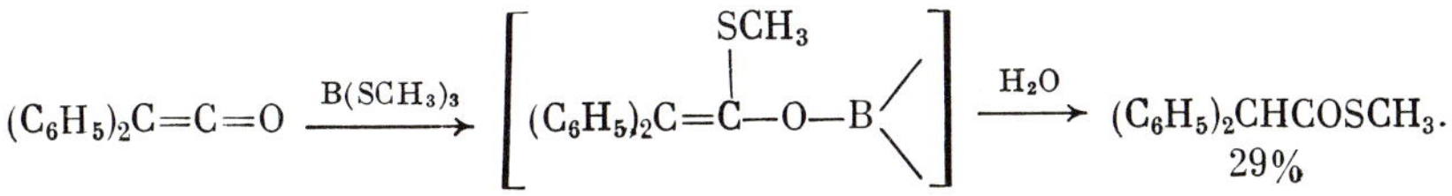

The reaction probably proceeds analogously to the reaction of the thioesters R_2BSR with ketene (discussed in Chapter 2, Section 2) and leads to vinyloxyborane, which is then easily hydrolyzed.

Triphenylselenoborane [143] and trimethylselenoborane [143, 144] react with aldehydes and ketones [143, 145] and also with acetals [146] to form selenoacetals and selenoketals, $RR'C(SeR'')_2$. $CH(SeC_6H_5)_3$ has been obtained from the ortho ester $CH(OC_2H_5)_3$ [146]. α-Oxiranes are decomposed by the thioborate $(C_2H_5S)_3B$ to form $C_2H_5SCH_2CH(OH)R$ [147]. The ring scission of oxiranes by the action of $(RSe)_3B$ has been described in [148]. The treatment of olefin oxides with $(RSe)_3B$ and aqueous sodium bicarbonate produces allylic alcohols [148].

Thioborates, like esters of dialkylthioborinic acids [45], add to some nitriles to form monomeric or dimeric iminoboranes. Thus, the reactions of trimethyl and triphenyl thioborate with fluoroacetonitrile lead to the dimeric compounds *3*, which are converted to the monomers when dissolved in CCl_4 [46]:

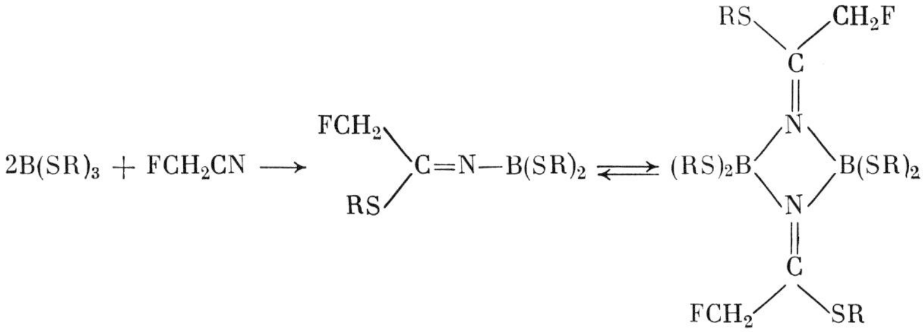

3

Trichloroacetonitrile adds thioborates to form the monomeric products $CCl_3(RS)\,C{=}NB(SR)_2$; the monomeric sulfur-containing iminoborane $(CH_3S)_2C{=}NB(SCH_3)_2$ is obtained by the reaction of trimethyl thioborate with methyl thiocyanate, CH_3SCN. The product of the addition of thioborates to acetonitrile is unstable and readily dissociates.

Diazomethane polymerizes under the action of thioborates, forming polymethylene. Phenyldiazomethane reacts with trimethyl thioborate, forming (after hydrolysis) benzyl methyl sulfide in 40% yield and a small amount of *trans*-stilbene [44]:

$$C_6H_5CHN_2 + B(SCH_3)_3 \xrightarrow[-N_2]{} \underset{\displaystyle B(SCH_3)_2}{C_6H_5\underset{|}{C}H{-}SCH_3} \xrightarrow[-CH_3SH]{H_2O} C_6H_5CH_2SCH_3 + B(OH)_3.$$

The action of trimethyl thioborate on diazoacetic ester forms ethyl *S*-methylmercaptoacetate, $CH_3SCH_2COOC_2H_5$, in 28% yield; methyl diphenylthioacetate (30%) is formed from benzoylphenyldiazomethane:

$$C_6H_5COCN_2C_6H_5 \longrightarrow [(C_6H_5)_2C{=}C{=}O] \xrightarrow[2.\ H_2O]{1.\ B(SCH_3)_3} (C_6H_5)_2CH{-}COSCH_3.$$

As suggested, the intermediate product of this reaction is diphenylketene.

Carboxylic acids form thioesters when heated with triethyl thioborate [47, 149]:

$$RCOOH + B(SC_2H_5)_3 \longrightarrow \underset{72\text{—}81\%}{RCOSC_2H_5}.$$

The same compounds are obtained by the treatment of anhydrides with esters of dialkylthioborinic acids [48]. The reaction of thioborates with sulfenic esters leads to unsymmetrical disulfides [49]:

$$(RS)_3B + 3C_6H_5SOCH_3 \longrightarrow \underset{75\text{—}82\%}{3RSSC_6H_5} + B(OCH_3)_3.$$

An analogous reaction with boron amines $B—NR_2$ has been used for the preparation of sulfenamides, $C_6H_5SNR_2$ [50].

In aromatic aldehydes, replacement of an oxygen atom by sulfur takes place under the action of B_2S_3, resulting in the formation of trithioaldehydes, $(ArCHS)_3$ [51]. Benzophenone does not react with B_2S_3, while acetophenone forms 2,4-diphenylthiophene:

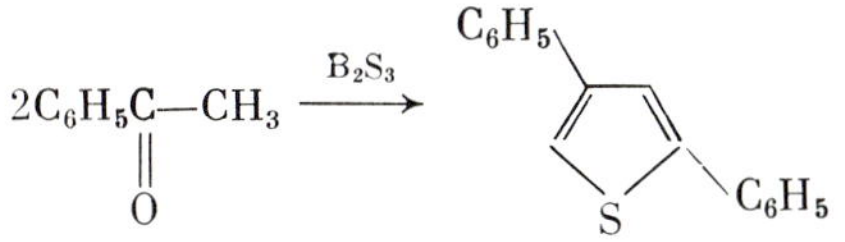

Methyl benzoate is converted to tetraphenyl-1,4-dioxa-2,5-cyclohexadiene by the action of boron sulfide.

Boron sulfide splits off two alkoxy groups from orthoesters, forming esters of thio acids and triethyl borate [52].

$$RC(OC_2H_5)_3 + B_2S_3 \longrightarrow RC(=S)—OC_2H_5 + B(OC_2H_5)_3.$$

CHAPTER 2

ESTERS OF ORGANOTHIOBORON ACIDS

Esters of organothioboron acids, R_2BSR' and $RB(SR')_2$, are readily available compounds. They are very reactive and find broad use in the synthesis of different types of organoboron compounds and also of some classes of sulfur- and oxygen-containing organic compounds.

2.1 METHODS FOR THE PREPARATION OF ESTERS OF ORGANOTHIOBORON ACIDS

A general method for the synthesis of esters of organothioboron acids consists of the action of mercaptans or metal mercaptides on organoboron halides.

When dichloroalkyl- or dibromoalkylboranes are boiled with excess butyl mercaptan, butyl alkylthioboronates are smoothly obtained [53]:

$$RBX_2 + 2n\text{-}C_4H_9SH \longrightarrow RB(SC_4H_9\text{-}n)_2 + 2HX.$$

Dibutyl phenylthioboronate has been synthesized from dichlorophenylborane by an analogous method. It should be noted that, in the reaction of an alkyldichloroborane, with even an excess of ethyl mercaptan, the formation of esters of alkylthioboronic acids is always accompanied by products of incomplete substitution — ethyl esters of chloro(alkyl)thioboronic acids [54, 55]. The latter are smoothly obtained by the reaction of alkyldichloroboranes with mercaptans in a 1:1 ratio or by an exchange reaction between $RBCl_2$ and $RB(SR')_2$ [54]. The esters of chloro- and bromoalkylthioboronic acids are thermally stable and can be distilled without disproportionation. Unsymmetrical compounds, such as the esters of dialkylamino(aryl)thioborinic acids, have been synthesized by the reaction of mercaptans with dialkylamino(aryl)boranes [56].

The thioesters $RB(SR')_2$ can also be prepared by other methods [4, 5], for example, by the action of mercaptans on trimethylamine-alkylboranes, $RBH_2 \cdot N(CH_3)_3$ [18, 57], or by the reaction of alkyldiiodoboranes with dialkyl disulfides (but not with di-*tert*-butyl disulfide) [58].

Esters of dialkyl- [28, 59–61] and diarylthioborinic [53] acids, R_2BSR', are obtained by heating the corresponding monohalides with mercaptans [28, 53, 59] or with sodium [60], magnesium [150], or lead [61] mercaptides, and also by the reaction of tetraalkyldiboranes, $(R_2BH)_2$, with mercaptans [14, 62]. The methyl esters R_2BSCH_3 have been synthesized by the action of dimethyl disulfide on dialkyliodoboranes [63]; the reaction proceeds with the elimination of iodine. The diboron thioester $(C_4H_9)_2BSCH_2SB(C_4H_9)_2$ has been obtained in 47% yield by the reaction of $(CH_2S)_n$ with $(C_4H_9)_2BI$ [63]. The properties of cyclic thioboranes have been discussed in a review article [64]. Compounds of the type $R_2BSC(S)N(CH_3)_2$, formed by the action of carbon disulfide on $R_2BN(CH_3)_2$ [5, 65], have an intramolecular chelate structure [63].

The reaction of dialkyliodoboranes with sulfur, H_2S_2, or Na_2S_2 leads to bis(dialkylboryl) disulfides, R_2BSSBR_2, which, when treated with R_2BX, form anhydrides of dialkylthioborinic acids, R_2BSBR_2 [66].

Bis(dimethylboryl) sulfide, $(CH_3)_2BSB(CH_3)_2$, has been synthesized from bromodimethylborane and $(CH_3)_3Sn—S—Sn(CH_3)_3$ in a ratio of 2:1; with a reagent ratio of 1:1, the thioborane $(CH_3)_2B—S—Sn(CH_3)_3$ is obtained [61]. The latter can be prepared by the action of chlorotrimethylstannane on dimethylthioborinic acid in the presence of triethylamine:

$$(CH_3)_2BSH + (CH_3)_3SnCl \xrightarrow{R_3N} (CH_3)_2B{-}S{-}Sn(CH_3)_3.$$

The stannyl group in the boron-tin sulfide is easily replaced by the action of phosphorus or arsenic halides, resulting in the formation of dimeric phosphorus or arsenic thioesters:

$$(CH_3)_2BSAsR_2 \xleftarrow{R_2AsCl} (CH_3)_2BSSn(CH_3)_3 \xrightarrow[-(CH_3)_3SnCl]{R_2PCl} [(CH_3)_2BSPR_2]_2$$

The exchange reaction of bromodimethylborane with bis[tetracarbonylmanganese(trimethylstannyl)sulfide] has been used for the synthesis of thermally stable dimeric tetracarbonylmanganese dimethylboryl sulfide [61]:

$$[(CO)_4Mn—S—Sn(CH_3)_3]_2 + 2(CH_3)_2BBr \xrightarrow[-(CH_3)_3SnBr]{} [(CO)_4Mn—S—B(CH_3)_2]).$$

The most convenient method for the synthesis of esters of dialkylthioborinic acids is the reaction between trialkylboranes and mercaptans [67–69]:

$$R_3B + R'SH \xrightarrow{O_2} R_2BSR' + RH.$$

The reaction is initiated by peroxide compounds and UV light [70]. Trialkylboranes freed of peroxides by heating to 100°C do not react with mercaptans below 150°C; however, the traces of air contained in the pieces of procelain or diatomaceous brick ("boiling chips") are sufficient to initiate the process. The reaction proceeds at an even greater rate, almost instantaneously, in the presence of traces of hydrogen peroxide or of alkyl hydroperoxide [70]. Thus, the reaction is a free-radical chain process, including a stage of chain propagation by bimolecular homolytic substitution (S_N2) of the alkyl radical on the boron atom by an alkylthio radical [71]:

$$R_3B \xrightarrow[\text{peroxide}]{O_2} R^\bullet,$$

$$R^\bullet + R'SH \longrightarrow R'S^\bullet + RH.$$

$$R'S^\bullet + R_3B \longrightarrow R_2BSR' + R^\bullet.$$

In unsymmetrical trialkylboranes, tertiary alkyl groups are split off most easily under the action of butyl mercaptan, with secondary and normal alkyl groups being split off with greater difficulty: *tert*-R > *sec*-R > *n*-R [72]. At the same time, the rate of the reaction of tributylborane with mercaptans at 50°C is approximately 70 times greater than that of tri-*sec*-butylborane, as a result of steric hindrance [73].

The above method for the synthesis of esters of dialkylthioborinic acids was used in [74–77, 125].

Unlike mercaptans, thiophenol reacts with trialkylboranes only upon heating; in this process, olefins and hydrogen are obtained along with the saturated hydrocarbons RH [67, 68].

Dialkylthioborinates can be obtained by heating trialkylboranes with sulfur [78, 79]:

$$R_3B + S \xrightarrow{140^\circ\ C} \underset{40\%}{R_2BSR} \xrightarrow{R'OH} R_2BOR' + RSH,$$

and also by exchange reactions between trialkylboranes and thioborates [25, 28, 29], in the presence of diborane [28] or tetraalkyldiboranes [25] as catalysts:

$$R_3B + B(SR')_3 \rightleftarrows R_2BSR' + RB(SR')_2,$$
$$R_2BSR' + B(SR')_3 \rightleftarrows 2RB(SR')_2.$$

By this method, 1,4-bis(1-borolyl)butane has been converted to 1-ethylthioborolane and the latter to an ester of 1,4-butanedithioboronic acid [80]:

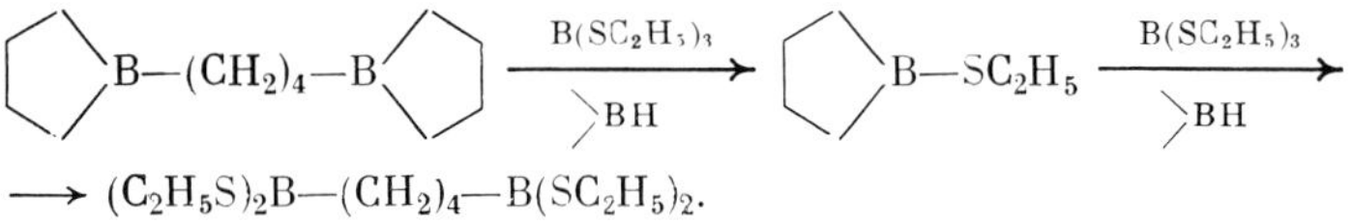

The most convenient method for the synthesis of esters of organothioboron acids involves exchange reactions of esters of organoboron acids with thioborates or esters of organothioboron acids [26, 27, 81, 82]:

$$3RB(OR')_2 + 2B(SR'')_3 \rightleftarrows 3RB(SR'')_2 + 2B(OR')_3,$$
$$3R_2BOR' + B(SR'')_3 \rightleftarrows 3R_2BSR'' + B(OR')_3.$$

These processes, which proceed without a catalyst, are reversible. The equilibrium may be displaced by removal of the most volatile component under vacuum. This method has been used for the synthesis of a number of alkanediboron and triboron compounds [80, 83, 84].

A large number of different heterocyclic sulfur-containing boron derivatives are known. For example, derivatives of 1,2,5-thiadiborolane of the type *4* have been obtained from 3,4-diiodothiophene and RBI_2, followed by treatment with sulfur [85].

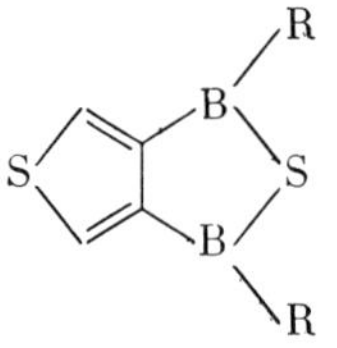

4

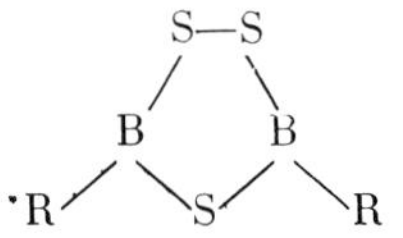

5

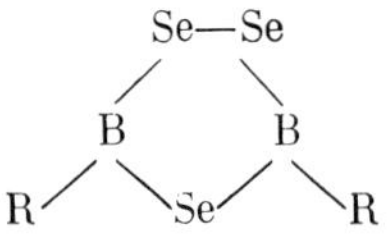

6

The trithiadiborolanes *5* [86] are formed by heating RBI_2 with di-*tert*-butyl disulfide [58], sulfur [87, 88], and Na_2S_2 [89]. Their selenium analogs (*6*) can be obtained by heating selenium with trialkylboranes [90], tetra- or dialkyldiboranes or RBI_2, and by other methods [91].

2.2 CHEMICAL PROPERTIES OF ESTERS OF ORGANOTHIOBORON ACIDS. USE IN ORGANIC SYNTHESIS

The esters of the thio acids R_2BSR' and $RB(SR')_2$ are highly reactive, since the energy of the B—S bond (89.2 kcal/mole) is lower than the energy of the B—O and B—N bonds (120–130 and 106 kcal/mole, respectively) [65]. The thioesters react readily with nucleophilic protonic reagents. These reactions, which proceed with the elimination of a mercaptan, are widely used for the preparation of different classes of boron derivatives containing oxygen, nitrogen, or sulfur.

The thioborinates are easily hydrolyzed by water and are converted by the action of alcohols to esters of borinic acids and mercaptans [68, 78, 79, 92–94]:

$$R_2BSR' + R''OH \longrightarrow R_2BOR'' + R'SH.$$

With ammonia [74, 78, 82, 93, 95] and amines [78, 79, 82, 95], the dialkylthioborinates react at room temperature to form aminodialkylboranes, R_2BNH_2 or R_2BNR_2; under the action of hydrazine, compounds of the type $R_2BNHNHBR_2$ are obtained [78, 96]. The reactions of diarylthioborinates [53, 97] and 1-alkylthioborolanes [92, 98] with amines proceed in an analogous manner.

The action of ammonia on alkylthioboronates forms *B*-trialkylborazenes, $(RBNH)_3$ [53, 99, 100], and diaminoalkylboranes, $RB(NHR')_2$, are obtained with primary amines [100, 101]. At the same time, the reaction with secondary amines proceeds with the replacement of only one SR group, as a result of which esters of dialkylamino(alkyl)thioborinic acids are formed; the latter react with primary amines, splitting off RS and R_2N groups, to produce *N*-substituted diamino(alkyl)boranes [100–102]. The thioesters $RB(SR')_2$ form polymers with hexamethylenediamine or hydrazine [103], while the reaction with ethanolamines leads to heterocyclic compounds with a system of O—B—N bonds [104]. Tetraethyl 1,3-propanedithioboronate is converted by the action of ammonia and primary amines to the corresponding 2,5-diamino-1-aza-2,5-diborinanes [105], while 2,7-bis(methylamino)-1-methyl-1-aza-2,7-diborepane has been synthesized from an ester of 1,4-butanedithioboronic acid and methylamine [83].

The reactions of ketimines with dialkyl- or diarylthioborinates or with alkyl- or arylthioboronates lead to monoiminoboranes ($R_2BN{=}CR_2$) [106, 151–153] or diiminoboranes [24, 154], while amidinoboranes have been synthesized from the thioesters R_2BSR' and amidines [107, 155].

Acetamidodialkylboranes (*7*), formed by the action of amides on $R_2BSC_4H_9$, add to nitriles with the formation of chelated dialkylborylacylamidinates (*8*) [108–110]:

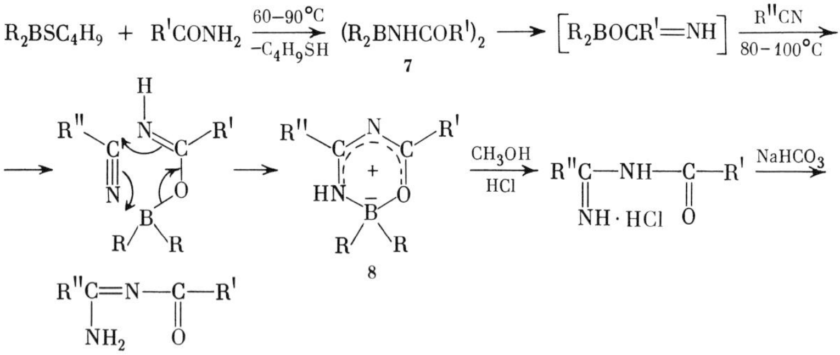

The chelates *8* react with hydrogen chloride in methanol to form *N*-acylamidine hydrochlorides [111, 112], which can be converted to the free bases by the action of a solution of sodium bicarbonate.

When the thioesters R_2BSR' are heated with cyclohexanone anil, the organoboron enamines *9* are formed [35]:

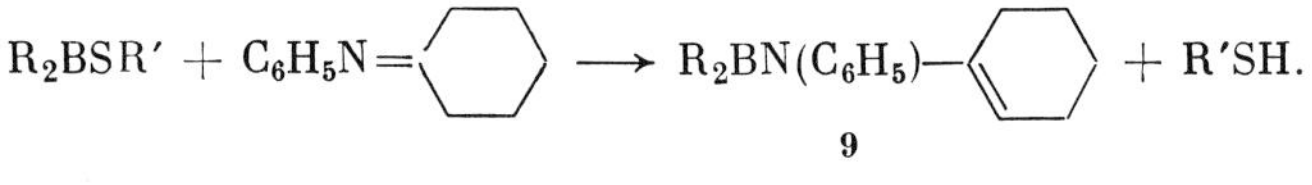

9

In this case, the anil probably reacts in the enamine form. In an analogous manner, *N*-allylcyclohexylideneamine has been converted to (*N*-allyl)cyclohexenylamino(dibutyl)borane which, at 150–180°C in the presence of tributylborane, is converted to 1-cyclohexenyl-2-

butyl-1,2-azaborolidine [113]; the latter compound has also been synthesized by heating 2-butyl-1,2-azaborolidine with cyclohexanone in benzene.

Boron enamines of the type *9* react on heating with nitriles to form the chelates *10* [114].

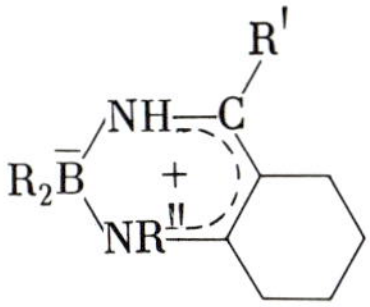

10

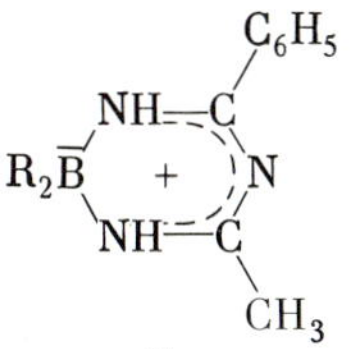

11

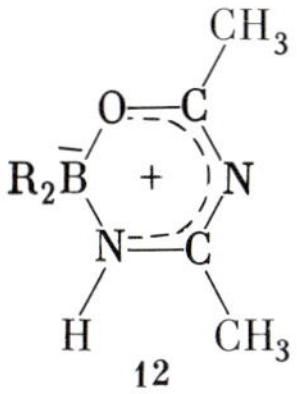

12

The reaction of the thioesters R_2BSR' with benzamidine and nitriles leads to the dialkylborylimidoylamidinates *11* [115]. By heating diacetamide ($CH_3CONHCOCH_3$) with esters of dialkylthioborinic acids, dialkylboryldiacetamidates can be synthesized [116] which, under the action of water, alcohols, or acetylacetone, decompose with the formation of the initial diacetamide. The methanolysis of *12* leads to $R_2BOCH_3{\cdot}NH{=}C(CH_3)NH_2$ [110].

The chelates *10a*, obtained by heating a mixture of cyclohexanone anil, $R_2'BSR''$, and excess nitrile in an autoclave at 170–180°C, are converted by boiling with HCl in butanol, followed by treatment with $NaHCO_3$ solution, to 9-alkyl-1,2,3,4-tetrahydroacridines in 40% yield [117, 156]:

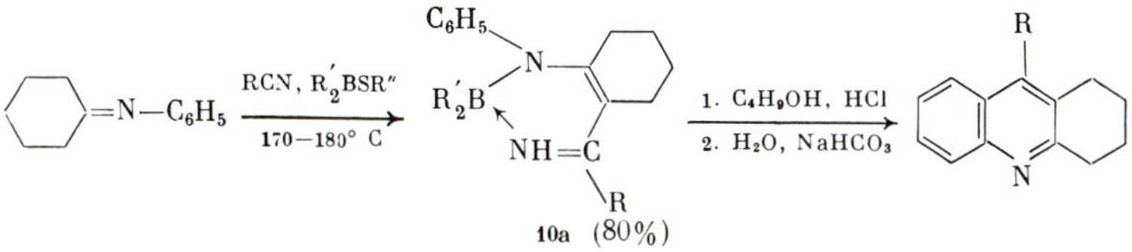

A method for the synthesis of anilides of 2-alkyliminocyclohexanecarboxylic acid from cyclohexanone anil, esters of dialkylthioborinic acids, and phenyl isocyanate has been proposed [157].

A method for the synthesis of (2-oxocyclohexyl)dialkylcarbinols from enamino dialkylboranes and isonitriles has been developed [158].

Dialkylthioborinates are easily transesterified by heating with higher mercaptans; they react with H_2S at 140–180°C to form dialkylthioborinic acids; with hydrocyanic acid, they form associated cyano(dialkyl)boranes [118]:

$$R_2BCN \xleftarrow[-R'SH]{HCN} R_2BSR' \xrightarrow[-R'SH]{H_2S} R_2BSH$$

Dialkylthioborinic acids and their selenium analogs, R_2BSeH, are also obtained by the action of H_2S or H_2Se on dialkyliodoboranes [119].

The action of bromine on esters of dialkylthioborinic acids forms dialkylbromoboranes and disulfides [119, 120, 160].

$$2R_2BSR' + Br_2 \xrightarrow{-63° C} \underset{78—98\%}{2R_2BBr} + R'S—SR'.$$

Dialkyl disulfides, RSSR, are also obtained by the action of thioboranes on *N*-sulfinylamines [159].

Methylthio(dimethyl)borane reacts slowly with methyl iodide at 20°C to form trimethylsulfonium iodide [121]:

$$(CH_3)_2BSCH_3 + CH_3I \xrightarrow[-(CH_3)_2BJ]{} (CH_3)_2S \xrightarrow{CH_3J} (CH_3)_3\overset{+}{S}\overset{-}{I}.$$

The alkylmercapto group in the thioesters $RB(SR')_2$ is easily replaced by halogen by the action of mercuric chloride [18]:

$$RB(SR')_2 + 2HgCl_2 \xrightarrow{80° C} \underset{60\%}{RBCl_2} + 2R'SHgCl.$$

This reaction has been used for the quantitative determination of B—S bonds in thioboranes.

The reaction of trialkoxystibines with thioboranes leads to tris(alkylthio)stibines, $(RS)_3Sb$, and boronate esters [122]:

$$C_6H_5B(SR)_2 + (R'O)_3Sb \longrightarrow C_6H_5B(OR')_2 + \underset{62—92\%}{(RS)_3Sb}.$$

Esters of dialkylthioborinic acids are capable of reacting with different types of unsaturated organic compounds. Similarly to thioborates, they react with alkyl vinyl ethers to form vinyl sulfides and esters of dialkylborinic acids [35]:

$$R_2BSC_4H_9 + CH_2{=}CHOC_4H_9 \longrightarrow \left[R_2BCH_2CH\begin{matrix} OC_4H_9 \\ SC_4H_9 \end{matrix} \right] \longrightarrow$$

$$\longrightarrow CH_2{=}CHSC_4H_9 + R_2BOC_4H_9.$$

Substituted vinylboranes have been obtained by the reaction of thioborinates with ethoxyacetylene [36–38]:

$$R_2BSR' + HC{\equiv}COC_2H_5 \longrightarrow R_2BCH{=}C(SR')OC_2H_5.$$

Esters of organothioboron acids have been used for the preparation of thioacetals and thioketals [39, 40]:

$$2(C_{10}H_7)_2BSC_4H_9 + (CH_3)_2CO \longrightarrow \underset{85\%}{(CH_3)_2C(SC_4H_9)_2} + (C_{10}H_7)_2BOB(C_{10}H_7)_2,$$

$$CH_3CH{=}CHCHO + C_6H_5B(SC_2H_5)_2 \longrightarrow \underset{67\%}{CH_3CH{=}CHCH(SC_2H_5)_2} + {}^1/_3(C_6H_5BO)_3.$$

These reactions proceed readily in hydrocarbons and represent a convenient method for the protection of carbonyl groups.

2-Alkyl-1,2-thiaborolanes add to benzaldehyde with the formation of 2-phenyl-4-alkyl-1,3,4-thiaoxaborepanes (*13*), which decompose to the starting substances under the action of piperidine [123]:

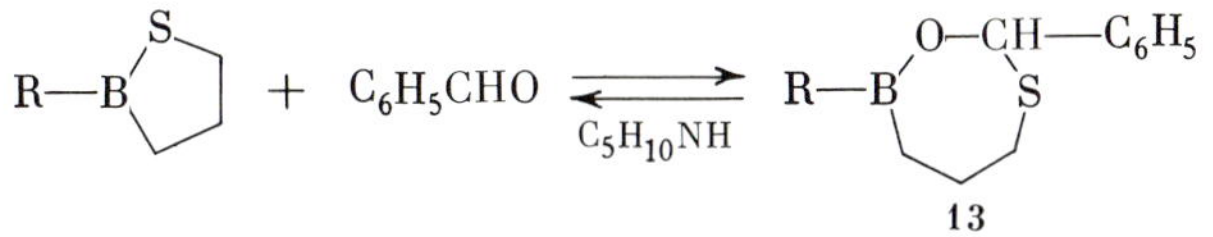

13

Ketene reacts with phenyl dibutylthioborinate at −25°C in carbon tetrachloride to form the vinyloxyborane *14*, the hydrolysis of which leads to thioacetate esters [124, 125]:

$$(C_4H_9)_2BSR + CH_2{=}C{=}O \longrightarrow \underset{\mathbf{14}}{CH_2{=}C\begin{matrix} SR \\ OB(C_4H_9)_2 \end{matrix}} \xrightarrow{H_2O} CH_3\underset{\underset{O}{\|}}{C}SR$$

$$(R = n\text{-}C_4H_9,\ C_6H_5).$$

The boron vinyl esters *14* add to the carbonyl group of aldehydes and ketones to form, after hydrolysis, *S*-esters of β-hydroxy acids [124–126] (See also Part I, Ch. 9.2.2):

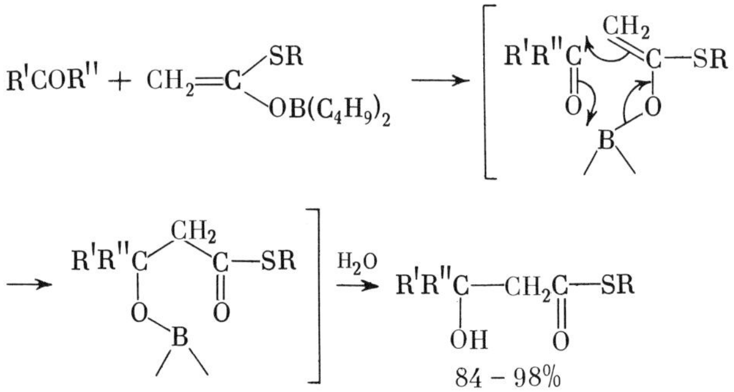

When acetoacetic ester is used as the carbonyl component, the *S*-butyl ester of β-hydroxy-β-methylthioglutaric acid, which is an analog of coenzyme A, is obtained [125]:

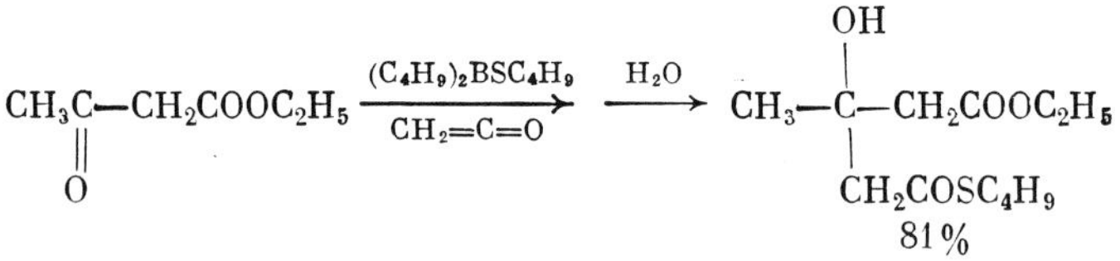

The vinyl esters *14* react analogously with organic nitriles, forming esters of β-iminothio acids in a yield of 35–40% [124]:

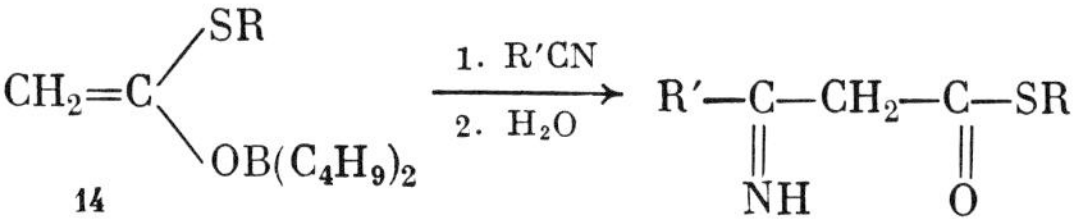

The *S*-butyl ester of 2,2,4-triphenyl-3-iminothiobutyric acid has been obtained in 49% yield from diphenylketene, butyl dibutylthioborinate, and benzyl cyanide [127].

The reaction of methyl vinyl ketone with phenyl dibutylthioborinate proceeds via the 1,4-addition of the B—SC_6H_5 fragment, resulting in the formation of the sulfur-containing vinyloxyborane *15* [124]:

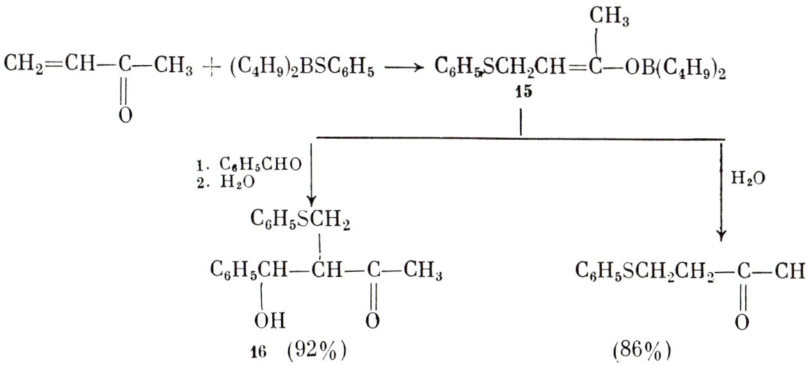

The mechanism of the 1,4-addition of thioboranes has not been cleared up. It is not impossible for the reaction to take place by a radical scheme, with the intermediate formation of RS radicals. The compound *15* is converted by the action of water to 4-phenyl-2-thiobutanone; its reaction with benzaldehyde leads to the 1,3-hydroxy ketone *16*.

Vinyloxyboranes of the type *17* are apparently intermediate products in the reaction of butyl dialkylthioborinates with organic isocyanates and benzyl cyanide, leading to the *N*-substituted *N*-butylthiocarbonylphenylacetamidines *18* [128]:

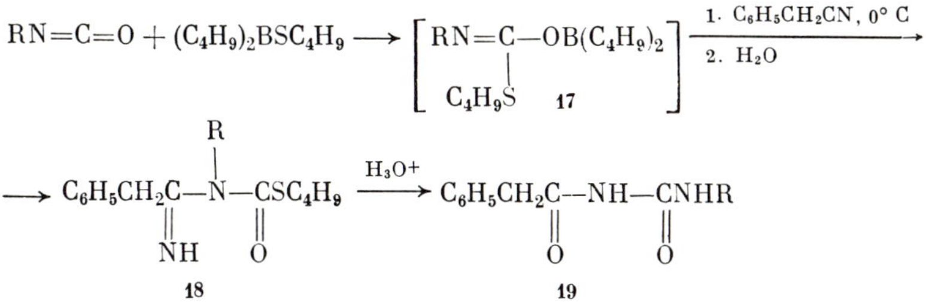

(R=alkyl, cycloalkyl, aryl)

To accelerate the addition process, the authors used catalytic quantities of BF_3 etherate and dry air. The yield of compounds *18* is 35–83%; by acid hydrolysis, they are quantitatively (with rearrangement) converted to the corresponding acylureas *19*. The reaction of isocyanates with thioborates proceeds analogously. According to the data of [129], cyclic esters of arylthioboronic acids add to isocyanates at the C=N bond.

The ester $(C_4H_9)_2BSC_4H_9$ adds to the C=N bond of phenyl isothiocyanate to form the compound *20*, which is converted by hydrolysis to butyl dithiophenylcarbamate [127, 128]:

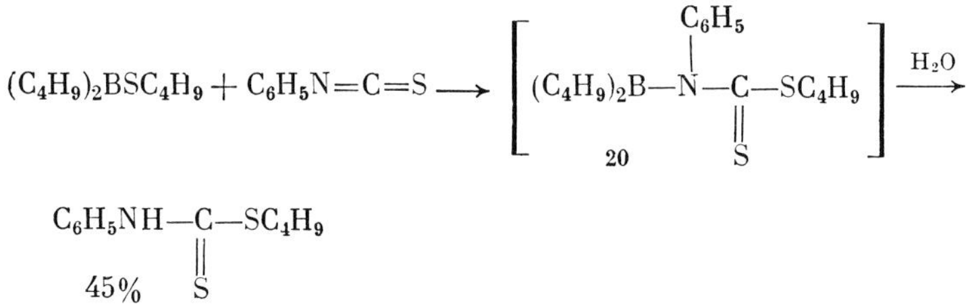

An analogous reaction with dicyclohexylcarbodiimide leads to *S*-butyl-*N*,N′-dicyclohexylisothiourea, $C_6H_{11}N{=}C(SC_4H_9)NHC_6H_{11}$.

Esters of dialkylthioborinic acids add to acetonitrile to form the sulfur-containing iminoboranes *21* in the form of a mixture of *cis* and *trans* isomers [45]:

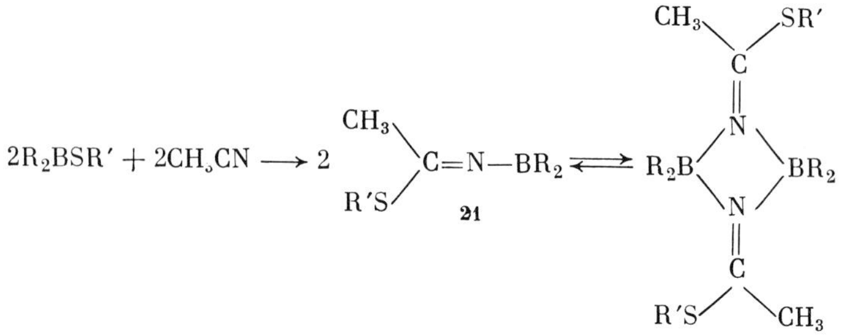

On heating in CCl_4, an equilibrium between the monomeric and dimeric forms sets in. The compounds *21* are also obtained by the reaction of trialkylboranes with organic thiocyanates, R′SCN [130].

Aliphatic and aromatic nitriles slowly add $RSBCl_2$ to form the dimeric iminoboranes $[CH_3S(R)C{=}N{-}BCl_2]_2$, many of which show a high thermal and hydrolytic stability [130]. Compounds of this type are also obtained by heating phenyl and chloromethyl thiocyanates, RSCN, with BCl_3 [130] or by the reaction of halodiiminoboranes with mercaptans [130, 131]. The exothermal reaction of bromo(bromotrichloromethylimino)methylborane with methyl thiocyanate leads to methyl bromide and compound *21a*, in which the NCS groups are bonded to carbon and boron via a nitrogen atom [131]:

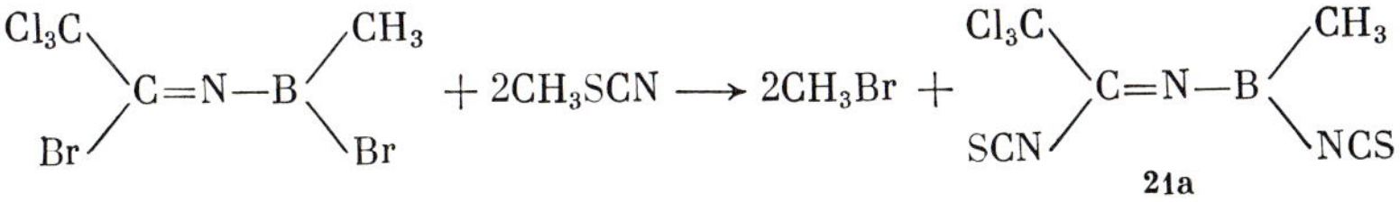

The α-iminoboranes *22* formed in the reaction of ethyl or phenyl diethylthioborinates with isonitriles are very active compounds and add a second molecule of thioester to form the sulfonium heterocyclic compounds *23*, which isomerize to compounds of the type *24* upon heating [132, 133]:

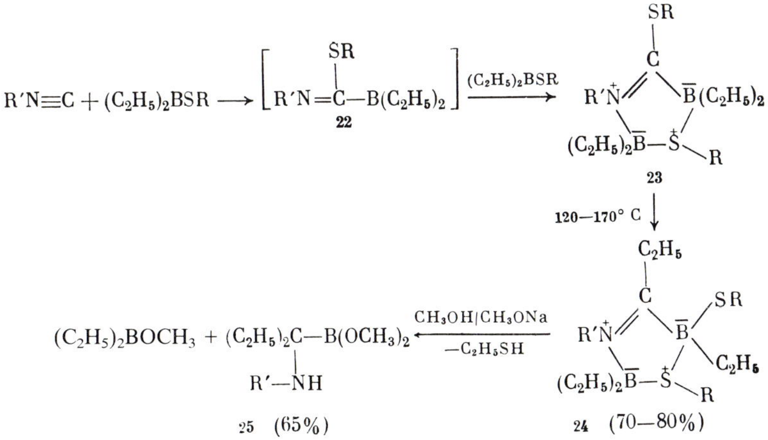

When heated with sodium methoxide in methanol, the heterocyclic compounds *24* are cleaved with rearrangement, being converted to the α-aminoboranes *25*; under these conditions, the compound *24* ($R{=}CH_3$) forms the oxygen analog *24* ($SR{=}OCH_3$).

Thioboranes split olefin oxides to form 2-hydroxy sulfides [134], with the direction of the reaction being determined by the structure of the oxide and, in part, by the nature of the solvent:

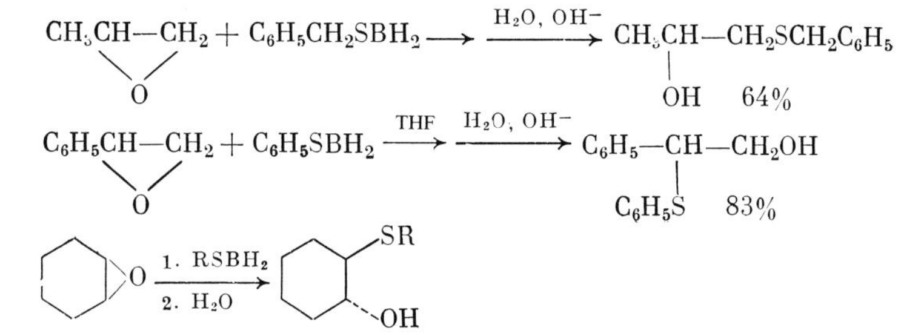

For the example of the oxides of *cis*- and *trans*-2-butene and stilbene and optically active styrene oxide, it has been shown that the opening of the oxirane ring takes place primarily or completely with inversion [135].

As previously noted, trialkylboranes,when heated with sulfur to 130–150°C, form polysulfides and thioborinate esters [78, 79]. This reaction, in combination with hydroboration, represents a method for the synthesis of mercaptans from olefins:

$$R_2C{=}CH_2 \xrightarrow{B_2H_6} (R_2CHCH_2)_3B \xrightarrow[\text{2. } H_2O \text{ or } R'OH]{\text{1. S, 130–150° C}} R_2CHCH_2SH.$$

Japanese chemists [136, 137] have used the reaction of triphenyl- and trialkylboranes with sulfur to prepare disulfides in 47–64% yield:

$$R_3B \xrightarrow[\text{2. } OH^-,\ O_2]{\text{1. S, 130° C}} RSSR.$$

Tributylborane, when irradiated with sunlight, reacts with 4,4′-bis(dimethylamino)benzothiophenone, forming tetra(4-dimethylaminophenyl)ethylene and butyl dibutylthioborinate in quantitative yield [138]:

$$2R_3B + 2Ar_2C{=}S \xrightarrow{h\nu} Ar_2C{=}CAr_2 + 2R_2BSR \xrightarrow{H_2O} RSH.$$

$$(Ar = 4\text{-}(CH_3)_2NC_6H_4).$$

When tributyl- [139] or trioctylborane [140] is heated with phenyl- or *p*-tolylsulfenyl chloride in dibutyl ether, unsymmetrical sulfides are obtained [139]:

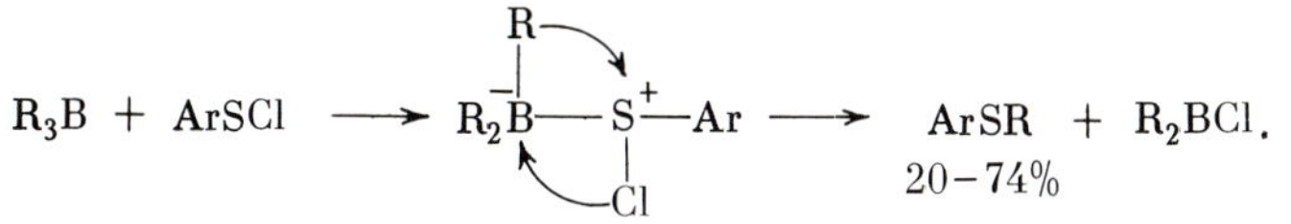

Sulfides are also obtained by the reaction of trialkylboranes with disulfides [141]:

$$R_3B + 2CH_3SSCH_3 \longrightarrow \underset{68-95\%}{2RSCH_3} + RB(SCH_3)_2.$$

The reaction is initiated by oxygen (in tetrahydrofuran) or by light (in hexane). 1-Methylthio-*cis*-bicyclo[3.3.0]octane has been obtained from *cis*-bicyclo[3.3.0]oct-1-yl(dimethyl)borane by this method [161].

Symmetrical and unsymmetrical sulfides can be obtained by the reaction of mercaptans with aldehydes or ketones in the presence of $C_5H_5N{\cdot}BH_3$ and trifluoroacetic acid [162].

The reaction of trialkylboranes with selenium dioxide in tetrahydrofuran leads to dialkyl selenides, R_2Se, while dialkyl diselenides, $(RSe)_2$, are obtained in aqueous tetrahydrofuran [163].

REFERENCES

1. C. Councler, *J. Prakt. Chem.* **18,** 371 (1878).
2. H. E. Cocksedge, *J. Chem. Soc.* **93,** 2177 (1908).
3. B. M. Mikhailov, *Usp. Khim.* **28,** 1450 (1959)
4. B. M. Mikhailov, *Usp. Khim.* **37,** 2121 (1968); B. M. Mikhailov, in: *Progress in Boron Chemistry,* Vol. 3, ed. R. J. Brotherton and H. Steinberg, Oxford, Pergamon Press (1970), 313.
5. R. H. Cragg and M. F. Lappert, *Organometal. Chem. Rev.* **1,** 43 (1966).
6. W. Gerrard: *The Organic Chemistry of Boron,* London-New York, Academic Press (1961)
7. H. Steinberg: *Organoboron Chemistry,* Vol. 1, New York, Interscience Publishers (1964), p. 817.
8. B. M. Mikhailov and Yu. N. Bubnov, *Izv. Akad. Nauk SSSR, Otd. Khim. Nauk* **1962,** 1378.
9. L. L. Petterson, B. J. Brotherton and J. L. Boone, *J. Org. Chem.* **26,** 3030 (1961).
10. D. M. Young and C. D. Anderson, *J. Org. Chem.* **26,** 5235 (1961).
11. J. Goubeau and H. W. Wittmeier, *Z. Anorg. Chem.* **270,** 16 (1952).
12. R. H. Cragg, J. P. N. Husband and A. F. Weston, *J. Inorg. Nucl. Chem.* **35,** 3685 (1973).
13. R. H. Cragg, J. P. N. Husband and A. F. Weston, *J. Chem. Soc. (Dalton)* **1973,** 568.

14. A. B. Burg and R. J. Wagner, *J. Am. Chem. Soc.* **76,** 3307 (1954).
15. B. M. Mikhailov, T. A. Shchegoleva, E. M. Shashkova and V .D. Sheludyakov, *Izv. Akad. Nauk SSSR, Otd. Khim. Nauk* **1962,** 1218.
16. B. M. Mikhailov and T. A. Shchegoleva, *Izv. Akad. Nauk SSSR, Otd. Khim. Nauk* **1959,** 1868.
17. B. M. Mikhailov and T. A. Shchegoleva, *Dokl. Akad. Nauk SSSR* **131,** 843 (1960).
18. M. F. Hawthorne, *J. Am. Chem. Soc.* **83,** 1345 (1961).
19. B. Egan, S. Shore and J. Bonnell, *Inorg. Chem.* **3,** 1024 (1964).
20. L. I. Zakharkin, V. I. Stanko and O. Yu. Okhlobystin, *Izv. Akad. Nauk SSSR, Otd. Khim. Nauk* **1961,** 2083.
21. J. Brault and J. M. Lalancette, *Can. J. Chem.* **42,** 2903 (1964).
22. J. M. Lalancette and A. Lachance, *Can. J. Chem.* **47,** 859 (1969).
23. T. A. Shchegoleva, E. M. Shashkova and B. M. Mikhailov, *Izv. Akad. Nauk SSSR, Otd. Khim. Nauk* **1961,** 918.
24. B. M. Mikhailov, G. S. Ter-Sarkisyan, N. N. Govorov and Z. Yanoushek, *Izv. Akad. Nauk SSSR, Ser. Khim.* **1975,** 484.
25. B. M. Mikhailov and L. S. Vasil'ev, *Dokl. Akad. Nauk SSSR* **139,** 385 (1961).
26. B. M. Mikhailov and L. S. Vasil'ev, *Izv. Akad. Nauk SSSR, Otd. Khim. Nauk* **1961,** 2101.
27. B. M. Mikhailov and L. S. Vasil'ev, *Zh. Obshch. Khim.* **35,** 1073 (1965).
28. A. Burg and F. M. Graber, *J. Am. Chem. Soc.* **78,** 1523 (1956).
29. B. M. Mikhailov, T. A. Shchegoleva and E. M. Shashkova, *Izv. Akad. Nauk SSSR, Otd. Khim. Nauk* **1961,** 916.
30. H. Nöth and U. Schuchardt, *J. Organometal. Chem.* **24,** 435 (1970).
31. W. Erl and H. Vahrenkamp, *Chem. Ber.* **103,** 3563 (1970).
32. W. Siebert, G. Augustin, R. Full, K. Krüger and Yi-Hung Tsay, *Angew. Chem* **87,** 286 (1975).
33. H. Vahrenkamp, *J. Organometal. Chem.* **28,** 181 (1971).
34. A. Cabana, J. Brault and J. M. Lalancette, *Spectrochim. Acta* **22,** 377 (1966).
35. V. A. Dorokhov and B. M. Mikhailov, *Izv. Akad. Nauk SSSR, Ser. Khim.* **1966,** 364.
36. B. M. Mikhailov and T. A. Shchegoleva, *Izv. Akad. Nauk SSSR, Ser. Khim.* **1971,** 661.
37. B. M. Mikhailov, T. A. Shchegoleva, E. M. Shashkova, L. I. Lavrinovich and V. S. Bogdanov, *Zh. Obshch. Khim.* **44,** 2185 (1974).
38. B. M. Mikhailov, T. A. Shchegoleva and E. M. Shashkova, *Zh. Obshch. Khim.* **44,** 2193 (1974).
39. B. M. Mikhailov and N. S. Fedotov, *Izv. Akad. Nauk SSSR, Otd. Khim. Nauk* **1962,** 999.
40. R. H. Cragg and J. P. N. Husband, *Inorg. Nucl. Chem. Letters* **6,** 773 (1970).
41. R. H. Cragg and J. P. N. Husband, *Inorg. Nucl. Chem. Letters* **7,** 221 (1971).
42. F. Bessette, J. Brault and J. M. Lalancette, *Can. J. Chem.* **43,** 307 (1965).
43. E. W. Abel, D. J. Walker and J. N. Wingfield, *Inorg. Nucl. Chem. Letters* **5,** 139 (1969).
44. C. D. Gutsche and K. Kinoshita, *J. Org. Chem.* **28,** 1762 (1963).
45. B. M. Mikhailov, V. A. Dorokhov and I. P. Yakovlev, *Izv. Akad. Nauk SSSR, Ser. Khim.* **1966,** 332.
46. A. Meller and A. Ossko, *Monatsh.* **101,** 1104 (1970).
47. A. Pelter, T. Levitt and K. Smith, *Chem. Comm.* **1969,** 435.
48. V. A. Dorokhov and B. M. Mikhailov, *Izv. Akad. Nauk SSSR, Ser. Khim.* **1970,** 1804.

49. R. H. Cragg, J. P. N. Husband and A. F. Weston, *Chem. Communs.* **1970,** 1701.
50. R. H. Cragg, J. P. N. Husband, G. C. Jones and A. F. Weston, *J. Organometal. Chem.* **44,** C37 (1972).
51. S. Jerumanis and J. M. Lalancette, *Can. J. Chem.* **42,** 1928 (1964).
52. J. M. Lalancette and J. Beauregard, *Tetrahedron Letters* **1967,** 5169.
53. B. M. Mikhailov, T. K. Kozminskaya, N. S. Fedotov and V. A. Dorokhov, *Dokl. Akad. Nauk SSSR* **127,** 1023 (1959).
54. B. M. Mikhailov and T. K. Kozminskaya, *Izv. Akad. Nauk SSSR, Otd. Khim. Nauk* **1962,** 256.
55. B. M. Mikhailov, E. A. Shagova and T. V. Potapova, *Izv. Akad. Nauk SSSR, Ser. Khim.* **1970,** 2048.
56. B. M. Mikhailov and V. A. Dorokhov, *Izv. Akad. Nauk SSSR, Otd. Khim. Nauk* **1962,** 1213.
57. M. F. Hawthorne, *J. Am. Chem. Soc.* **82,** 748 (1960).
58. M. Schmidt and F. R. Rittig, Angew. Chem, 82, 774 (1970).
59. B. M. Mikhailov and M. E. Nikolaeva, *Izv. Akad. Naur SSSR, Ser. Khim.* **1963,** 1368.
60. N. Ashikari, *Bull. Chem. Soc. Japan* **32,** 1157 (1959).
61. H. Vahrenkamp, *J. Organometal. Chem.* 28, 167 (1971)..
62. B. M. Mikhailov, A. A. Akhnazaryan and L. S. Vasil'ev, *Dokl. Akad. Nauk SSSR* **136,** 828 (1961).
63. M. Schmidt and E. Gast, *Z. Naturforsch.* **23b,** 1258 (1968).
64. H. Horn, *Chem. Ztg., Chem. App.* **94,** 467 (1970).
65. R. H. Cragg, M. F. Lappert, H. Nöth, P. Schweizer and B. P. Tulley, *Chem. Ber.* 100, 2377 (1967).
66. W. Siebert, E. Gast and M. Schmidt, *J. Organometal. Chem.* **23,** 329 (1970).
67. B. M. Mikhailov, V. A. Vaver and Yu. N. Bubnov, *Dokl. Akad. Nauk SSSR* **126,** 675 (1959).
68. B. M. Mikhailov and Yu. N. Bubnov, *Zh. Obshch. Khim.* **31,** 160 (1961).
69. K. Lang, West German Patent 1,079,634 (1960); *C. A.* **55,** 13316 (1961).
70. B. M. Mikhailov and Yu. N. Bubnov, *Izv. Akad. Nauk SSSR, Ser. Khim.* **1964,** 2248.
71. K. U. Ingold and B. P. Roberts: *Free Radical Substitution Reactions,* New York, Wiley-Interscience (1971).
72. A. G. Davies, T. Maki, B. P. Roberts, A. Pelter and D. N. Sharrocks, *J. Organometal. Chem.* **82,** 301 (1974).
73. A. G. Davies and B. P. Roberts, *J. Chem. Soc. B* **1971,** 1830.
74. L. F. Hohnstedt, J. P. Brennan and K. A. Reynard, *J. Chem. Soc., A* **1970,** 2456.
75. G. L. Gal'chenko, R. M. Varushchenko, Yu. N. Bubnov and B. M. Mikhailov, *Zh. Obshch. Khim.* **32,** 285 (1962).
76. G. L. Gal'chenko, M. M. Ammar, S. M. Skuratov, Yu. N. Bubnov and B. M. Mikhailov, *Vestnik MGU, Ser. Khim.* **1965,** No. 3, 10.
77. G. L. Gal'chenko, E. P. Brykina, V. K. Pavlovich, Yu. N. Bubnov and B. M. Mikhailov, *Zh. Fiz. Khim.* **43,** 2034 (1969).
78. B. M. Mikhailov and Yu. N. Bubnov, *Izv. Akad. Nauk SSSR, Otd. Khim. Nauk* **1959,** 172.
79. B. M. Mikhailov and Yu. N. Bubnov, *Zh. Obshch. Khim.* **29,** 1648 (1959).
80. B. M. Mikhailov, T. K. Kozminskaya and A. Ya. Bezmenov, *Izv. Akad. Nauk SSSR, Otd. Khim. Nauk* **1965,** 355.
81. B. M. Mikhailov, Yu. N. Bubnov and V. G. Kiselev, *Izv. Akad. Nauk SSSR, Ser. Khim.* **1965,** 68.

82. Yu. N. Bubnov, S. I. Frolov and B. M. Mikhailov, *Izv. Akad. Nauk SSSR, Ser. Khim.* **1968,** 824.
83. B. M. Mikhailov, L. S. Vasil'ev and A. Ya. Bezmenov, *Izv. Akad. Nauk SSSR, Ser. Khim.* **1965,** 712.
84. B. M. Mikhailov and V. F. Pozdnev, in the collection: Journal of General Chemistry. *Problems of Organic Synthesis,* Leningrad, Nauka (1965), p. 220.
85. B. A. Asgarouladi, R. Full, K. Schaper and W. Siebert, *Chem. Ber.* **107,** 34 (1974).
86. R. E. Sacher, W. Siebert, J. Nakovich and F. A. Miller, *J. Raman Spectrosc.* **1,** 285 (1973).
87. M. Schmidt, W. Siebert and F. R. Rittig, *Chem. Ber.* **101,** 281 (1968).
88. M. Schmidt, F. R. Rittig and W. Siebert, *Z. Naturforsch.* **25b,** 1344 (1970).
89. M. Schmidt and F. R. Rittig, *Z. Naturforsch.* **25b,** 1062 (1970).
90. B. M. Mikhailov and T. A. Shchegoleva, *Izv. Akad. Nauk SSSR, Otd. Khim. Nauk* **1959,** 356.
91. W. Siebert and F. Riegel, *Chem. Ber.* **106,** 1012 (1973).
92. B. M. Mikhailov and T. A. Shchegoleva, *Izv. Akad. Nauk SSSR, Otd. Khim. Nauk* **1961,** 1142.
93. B. M. Mikhailov and Yu. N. Bubnov, *Izv. Akad. Nauk SSSR, Otd. Khim. Nauk* **1960,** 1872.
94. B. M. Mikhailov and Yu. N. Bubnov, *Izv. Akad. Nauk SSSR, Otd. Khim. Nauk* **1960,** 1883.
95. B. M. Mikhailov and Yu. N. Bubnov, *Zh. Obshch. Khim.* **31,** 577 (1961).
96. B. M. Mikhailov and Yu. N. Bubnov, *Izv. Akad. Nauk SSSR, Otd. Khim. Nauk* **1960,** 368.
97. B. M. Mikhailov and N. S. Fedotov, *Zh. Obshch. Khim.* **32,** 93 (1962).
98. B. M. Mikhailov, T. K. Kozminskaya and L. V. Tarasova, *Dokl. Akad. Nauk SSSR* **160,** 615 (1965).
99. B. M. Mikhailov and T. K. Kozminskaya, *Izv. Akad. Nauk SSSR, Otd. Khim. Nauk* **1960,** 2247.
100. B. M. Mikhailov and F. B. Tutorskaya, *Zh. Obshch. Khim.* **32,** 833 (1962).
101. B. M. Mikhailov and T. K. Kozminskaya, *Zh. Obshch. Khim.* **30,** 3619 (1960).
102. B. M. Mikhailov and V. A. Dorokhov, *Izv. Akad. Nauk SSSR, Otd Khim. Nauk* **1962,** 1213.
103. B. M. Mikhailov and T. A. Kozminskaya, *Izv. Akad. Nauk SSSR, Ser. Khim.* **1965,** 439.
104. R. H. Cragg, *Chem. Communs.* **1969,** 832.
105. B. M. Mikhailov and V. F. Pozdnev, *Dokl. Akad. Nauk* **151,** 340 (1963).
106. B. M. Mikhailov, G. S. Ter-Sarkisyan, N. N. Govorov and N. A. Nikolaeva, *Dokl. Akad. Nauk SSSR* **1974,** 2389.
107. B. M. Mikhailov, V. A. Dorokhov, V. I. Seredenko and I. P. Yakovlev, *Izv. Akad. Nauk SSSR, Ser. Khim.* **1974,** 1665.
108. B. M. Mikhailov and V. A. Dorokhov, *Izv. Akad. Nauk SSSR, Ser. Khim.* **1970,** 1446.
109. B. M. Mikhailov, V. A. Dorokhov, V. S. Bogdanov, I. P. Yakovlev and A. D. Naumov, *Dokl. Akad. Nauk SSSR* **194,** 595 (1970).
110. V. A. Dorokhov, L. I. Lavrinovich, M. N. Bochkareva, V. S. Bogdanov and B. M. Mikhailov, *Zh. Obshch. Khim.* **43,** 1115 (1973).
111. B. M. Mikhailov and V. A. Dorokhov, *Izv. Akad. Nauk SSSR, Ser. Khim.* **1971,** 201.
112. B. M. Mikhailov, V. A. Dorokhov and L. I. Lavrinovich, *Zh. Obshch. Khim.* **42,** 1553 (1972).

113. V. A. Dorokhov, O. G. Boldyreva, V. S. Bogdanov and B. M. Mikhailov, *Zh. Obshch. Khim.* **42,** 1558 (1972).
114. V. A. Dorokhov and B. M. Mikhailov, *Dokl. Akad. Nauk SSSR* **187,** 1300 (1969).
115. B. M. Mikhailov, V. A. Dorokhov and V. I. Seredenko, *Dokl. Akad. Nauk SSSR* **199,** 1328 (1971).
116. V. A. Dorokhov, L. I. Lavrinovich and B. M. Mikhailov, *Dokl. Akad. Nauk SSSR* **195,** 1100 (1970).
117. B. M. Mikhailov, V. A. Dorokhov and O. G. Boldyreva, *Izv. Akad. Nauk SSSR, Ser. Khim.* **1973,** 2643.
118. B. M. Mikhailov and Yu. N. Bubnov, *Dokl. Akad. Nauk SSSR* **127,** 571 (1959).
119. W. Siebert, E. Gast, F. Riegel and M. Schmidt, *J. Organometal. Chem.* **90,** 13 (1975).
120. A. Pelter, K. Rowe, D. Sharrocks and K. Smith, *Chem. Comm.* **1975,** 531.
121. H. Vahrenkamp, *J. Organometal. Chem.* **28,** 167 (1971).
122. R. H. Cragg, M. Nazery and A. F. Weston, *Inorg. Nucl. Chem. Letters* **9,** 497 (1973).
123. V. A. Dorokhov, O. G. Boldyreva and B. M. Mikhailov, *Izv. Akad. Nauk SSSR, Ser. Khim.* **1970,** 191.
124. T. Mukaiyama, K. Inomata and M. Muraki, *J. Am. Chem. Soc.* **95,** 967 (1973).
125. K. Inomata, M. Muraki and T. Mukaiyama, *Bull. Chem. Soc. Japan* **46,** 1807 (1973).
126. T. Mukaiyama and K. Inomata, *Bull. Chem. Soc. Japan* **44,** 3215 (1971).
127. A. Mukaiyama, K. Inomata and S. Yamamoto, *Tetrahedron Letters* **1971,** 1097.
128. T. Mukaiyama, S. Yamamoto and K. Inomata, *Bull. Chem. Soc. Japan* **44,** 2807 (1971).
129. See [13].
130. A. Meller and A. Ossko, *Monatsh.* **102,** 131 (1971).
131. A. Meller and W. Maringgele, *Monatsh.* **102,** 121 (1971).
132. A. Haag and G. Hesse, *Intra-Science Chem. Rep.* **7,** 105 (1973).
133. A. Grote, A. Haag and G. Hesse, *Ann.* **755, 67** (1972).
134. D. J. Pasto, C. C. Cumbo and P. Balasubramaniyan, *J. Am. Chem. Soc.* **88,** 2187 (1966).
135. D. J. Pasto, C. C. Cumbo and J. Frazer, *J. Am. Chem. Soc.* **88,** 2194 (1966).
136. Z. Yoshida, T. Okushi and O. Manabe, *Tetrahedron Letters* **1970,** 1641.
137. T. Okushi, O. Manabe and Z. Yoshida, *J. Chem. Soc. Japan, Ind. Chem. Soc.* **73,** 2006 (1970).
138. M. Inatome and L. Kuhn, *Tetrahedron Letters* **1965,** 73.
139. P. M. Draper, T. H. Chan and D. N. Harpp, *Tetrahedron Letters* **1970,** 1687.
140. A. G. Davies and B. P. Roberts, *J. Chem. Soc., B* **1971,** 1835.
141. H. C. Brown and M. M. Midland, *J. Am. Chem. Soc.* **93, 3291** (1971).
142. T. Cohen, D. A. Bennet and A. J. Mura, *J. Org. Chem.* **41,** 2506 (1976).
143. D. L. J. Clive and S. M. Menchen, *J. Org. Chem.* **44,** 4279 (1979).
144. W. Siebert, W. Raf and R. Full, *Z. Naturforsch.* **30b,** 642 (1975).
145. D. L. J. Clive and S. M. Menchen, *Chem. Commun.* **1978,** 356.
146. D. L. J. Clive and S. M. Menchen, *J. Org. Chem.* **44,** 1883 (1979).
147. H. A. Klein, *Chem. Ber.* **112,** 3037 (1979).
148. A. Cravador and A. Krief, *Tetrahedron Letters* **1981,** 2491.
149. A. Pelter, T. E. Levitt, K. Smith and A. Jones, *J. Chem. Soc., Perkin I* **1977,** 1672.
150. F. Davidson and J. W. Wilson, *J. Organometal Chem.* **204,** 147 (1981).

151. B. M. Mikhailov, G. S. Ter-Sarkisyan, N. N. Govorov and N. A. Nikolaeva, *Izv. Akad. Nauk SSSR, Ser. Khim.* **1976,** 1820.
152. B. M. Mikhailov, G. S. Ter-Sarkisyan and N. N. Govorov, *Izv. Akad. Nauk SSSR, Ser. Khim.* **1976,** 1823.
153. B. M. Mikhailov, G. S. Ter-Sarkisyan, N. N. Govorov and N. A. Nikolaeva, *Zh. Obshch. Khim., Khim. Elementoorg. Soedin.* **1976,** 3.
154. B. M. Mikhailov, G. S. Ter-Sarkisyan and N. N. Govorov, *Izv. Akad. Nauk SSSR, Ser. Khim* **1976,** 1827.
155. V. A. Dorokhov, V. I. Seredenko and B. M. Mikhailov, *Izv. Akad. Nauk SSSR, Ser. Khim.* **1977,** 1593.
156. V. A. Dorokhov and O. G. Boldyreva, *Izv. Akad. Nauk SSSR, Ser. Khim.* **1978,** 2574.
157. V. A. Dorokhov, O. G. Boldyreva and B. M. Mikhailov, *Izv. Akad. Nauk SSSR, Ser. Khim.* **1980,** 1876.
158. V. A. Dorokhov, O. G. Boldyreva, A. S. Shashkov and B. M. Mikhailov, *Izv. Akad. Nauk SSSR, Ser. Khim.* **1976,** 1431.
159. A. Meller, W. Maringgele and H. Fetzer, *Chem. Ber.* **113,** 1950 (1980).
160. A. Pelter, K. Rowe, D. N. Sharrocks, K. Smith and C. Subrahmanyam, *J. Chem. Soc., Dalton* **1976,** 2087.
161. G. W. Kramer and H. C. Brown, *J. Chem. Soc.* **42,** 2832 (1977).
162. Y. Kikugawa, *Chem. Letters* **1981,** 1157.
163. A. Arase and Y. Masuda, *Chem. Letters* **1975,** 419.

SUPPLEMENTARY SOURCES OF INFORMATION

1. C. W. Allen and D. E. Palmer: "The Borabenzene Anion and Its Transition Metal Complexes", *J. Chem. Ed.* **55,** 497 (1978).
2. C. Aspisi, M. Bonato and M. Follet: "Diborane complexes and organoboranes: Their application in hydroboranization and reduction reactions," *Inf. Chim.* **212,** 217 (1981).
3. K. Avasthi, D. Devaprabhakara and A. Suzuki: "Non-catalytic hydrogenation via organoboranes," *J. Organomet. Chem. Libr.* **7,** 1 (1979).
4. H. C. Brown and J. B. Campbell: "Synthesis and Application of Vinylic Organoboranes," *Aldrichimica Acta* **14,** 3 (1981).
5. H. C. Brown, G. W. Kramer, A. B. Levy and M. M. Midland: "Organic Synthesis via Boranes," Wiley-Interscience, New York (1975).
6. H. C. Brown and E. Negishi: "Boraheterocyles via cyclic hydroboration," *Tetrahedron* **33,** 2331 (1977).
7. W. Buchner and H. Niederprüm: "Sodium borohydride and amine-boranes, commercially important reducing agents," *Pure Appl. Chem.* **49,** 733 (1977).
8. G. M. L. Cragg and K. R. Koch: "Organoborates in Organic Synthesis: The Use of Alkenyl-, Alkynyl-, and Cyanoborates as Synthetic Intermediates," *Chem. Soc. Rev.* **6,** 393 (1977).
9. G. Davidson: "Elements of Group 3," *Coord. Chem. Revs.* **34,** 74 (1981).
10. J. J. Eisch: "Rearrangements of Unsaturated Organoboron and Organoaluminum Compounds," *Adv. Organomet. Chem.* **16,** 67 (1977).
11. J. A. Gladysz: "Trialkylborohydrides in Organometallic Synthesis," *Aldrichimica Acta* **12,** No 1, 13 (1979).
12. R. O. Hutchins and F. Cistone: "Utility and Application of Borane Dimethylsulfide in Organic Synthesis," *Org. Prep. Proced. Int.* **13,** 225 (1981).
13. R. O. Hutchins and N. R. Natale: "Cyanoborohydride. Utility and Application in Organic Synthesis," *Org. Prep. Proced. Int.* **11,** 201 (1979).
14. G. W. Kabalka: "Catecholborane in Organic Synthesis," *Org. Prep. Proced. Int.* **9,** 131 (1977).
15. G. W. Kabalka: "Radionuclide Incorporation via Organoboranes," In: "Aspects of Mechanism and Organometallic Chemistry," Ed. J. H. Brewster, Plenum Press, New York (1978).
16. W. Kliegel: "Bor in Biologie, Medizin und Pharmazie", Springer, Berlin-Heidelberg, New York (1980).
17. R. Köster: "Organoboranes in Synthesis and Analysis," *Pure Appl. Chem.* **49,** 765 (1977).
18. C. F. Lane: "Reduction of Organic Compounds with Diborane," *Chem. Revs.* **76,** 773 (1976).
19. D. S. Matteson: "Methanetetraboronic and Methanetriboronic Esters as Synthetic Intermediates," *Synthesis* **1975,** 147.
20. D. S. Matteson: "Organometallic Reaction Mechanisms of the Non-transition Elements," Academic Press, New York-London (1974).
21. B. M. Mikhailov: "The Cyclic Coordination of Boron Compounds," *Pure Appl. Chem.* **49,** 749 (1977).
22. B. M. Mikhailov: "Boraheterocycles from Allylboranes," *Sov. Sci. Revs., Sect. B* **2,** 283 (1980).
23. B. M. Mikhailov: "The Chemistry of Boron-Cage Compounds," *Pure Appl. Chem.* **52,** 691 (1980).

24. E. I. Negishi: "Chemistry of Organoborates," *J. Organometal. Chem.* **108,** 281 (1976).
25. E. I. Negishi: "Organoborons and Organoaluminums (B, Al)," in: "Organometallics in Organic Synthesis, Vol. 1. General Discussion. Organometallics of Main Group Metals in Organic Synthesis. John Wiley & Sons, New York (1981), p. 286.
26. H. Nöth, B. Wrackmeyer: "NMR Spectroscopy of Boron Compounds," Springer, Berlin (1978), 461 pp.
27. T. Onak: "Organoborane Chemistry," Academic Press, New York-San Francisco-London (1975), 360 pp.
28. A. Pelter: "Rearrangements involving boron," Org. Chem. (N-Y), 42, "Rearrangements of ground and excited states," Pt. 2, 94, Swansea, Wales (1980).
29. A. Pelter and K. Smith: "Organic Boron Compounds," In: "Comprehensive Organic Chemistry. The Synthesis and Reactions of Organic Compounds." Ed. D. Barton and D. Ollis, Vol. 3: "Sulfur, Selenium, Silicon, Boron, Organometallic Compounds." Ed. D. N. Jones, Pergamon Press, Oxford, New York (1979), p. 687.
30. W. Siebert: "Boron heterocycles as ligands in transition metal chemistry," *Adv. Organometal. Chem.* **18,** 301 (1980).
31. V. I. Tel'noi and I. B. Rabinovich: "Thermochemistry of Organic Compounds of Non-Transition Elements," *Usp. Khim.* **49,** 1137 (1980).
32. J. Weil-Raynal: "Formation of Carbon-Carbon Bonds by Using Organoboranes," *Synthesis* **1976,** 633.
33. D. S. Matteson: "Boron: Boranes in Organic Synthesis. Annual Survey Covering the Year 1980," *J. Organometal. Chem.* **227,** 165 (1982).
34. K. Niedenzu: "Boron: Annual Survey Covering the Year 1980," *J. Organometal. Chem.* **227,** 199 (1982).

Author Index

Parentheses contain the reference number of the author's works, which is given in full on the page number in italics; the regular page numbers indicate the text pages on which the reference cited.

Subject Index